D0939911

EX LIBRIS
SOUTH ORANGE
PUBLIC LIBRARY

About Island Press

Island Press is the only nonprofit organization in the United States whose principal purpose is the publication of books on environmental issues and natural resource management. We provide solutions-oriented information to professionals, public officials, business and community leaders, and concerned citizens who are shaping responses to environmental problems.

In 2005, Island Press celebrates its twenty-first anniversary as the leading provider of timely and practical books that take a multidisciplinary approach to critical environmental concerns. Our growing list of titles reflects our commitment to bringing the best of an expanding body of literature to the environmental community throughout North America and the world.

Support for Island Press is provided by the Agua Fund, The Geraldine R. Dodge Foundation, Doris Duke Charitable Foundation, Ford Foundation, The George Gund Foundation, The William and Flora Hewlett Foundation, Kendeda Sustainability Fund of the Tides Foundation, The Henry Luce Foundation, The John D. and Catherine T. MacArthur Foundation, The Andrew W. Mellon Foundation, The Curtis and Edith Munson Foundation, The New-Land Foundation, The New York Community Trust, Oak Foundation, The Overbrook Foundation, The David and Lucile Packard Foundation, The Winslow Foundation, and other generous donors.

The opinions expressed in this book are those of the authors and do not necessarily reflect the views of these foundations.

Ecosystems and Human Well-being: Policy Responses, Volume 3

Ecosystems and Human Well-being: Policy Responses, Volume 3

Edited by:

Kanchan Chopra
Institute of Economic Growth Delhi, India

Rik Leemans
Wageningen University Netherlands

Pushpam Kumar
Institute of Economic Growth Delhi, India

Henk Simons
National Institute of Public Health and the Environment (RIVM) Netherlands

Findings of the Responses Working Group of the Millennium Ecosystem Assessment

Washington • Covelo • London

The Millennium Ecosystem Assessment Series

Ecosystems and Human Well-being: A Framework for Assessment
Ecosystems and Human Well-being: Current State and Trends, Volume 1
Ecosystems and Human Well-being: Scenarios, Volume 2
Ecosystems and Human Well-being: Policy Responses, Volume 3
Ecosystems and Human Well-being: Multiscale Assessments, Volume 4
Our Human Planet: Summary for Decision-makers

Synthesis Reports (available at MAweb.org)

Ecosystems and Human Well-being: Synthesis
Ecosystems and Human Well-being: Biodiversity Synthesis
Ecosystems and Human Well-being: Desertification Synthesis
Ecosystems and Human Well-being: Human Health Synthesis
Ecosystems and Human Well-being: Wetlands and Water Synthesis
Ecosystems and Human Well-being: Opportunities and Challenges for Business and Industry

Library of Congress Cataloging-in-Publication data.

Ecosystems and human well-being : policy responses : findings of the
Responses Working Group of the Millennium Ecosystem Assessment / edited by
Kanchan Chopra . . . [et al.].
p. cm.—(The Millennium Ecosystem Assessment series ; v. 3)
Includes bibliographical references and index.
ISBN 1-55963-269-0 (cloth : alk. paper)—ISBN 1-55963-270-4 (pbk. : alk. paper)
1. Human ecology. 2. Ecosystem management. 3. Ecological assessment
(Biology) 4. Environmental policy. 5. Environmental management.
I. Chopra, Kanchan Ratna. II. Millennium Ecosystem Assessment (Program).
Responses Working Group. III. Series.
GF50.E267 2005
333.95′16—dc22

2005017304

British Cataloguing-in-Publication data available.

Printed on recycled, acid-free paper ♻

Book design by Maggie Powell
Typesetting by Coghill Composition, Inc.

Manufactured in the United States of America
10 9 8 7 6 5 4 3 2 1

Millennium Ecosystem Assessment: Objectives, Focus, and Approach

The Millennium Ecosystem Assessment was carried out between 2001 and 2005 to assess the consequences of ecosystem change for human well-being and to establish the scientific basis for actions needed to enhance the conservation and sustainable use of ecosystems and their contributions to human well-being. The MA responds to government requests for information received through four international conventions—the Convention on Biological Diversity, the United Nations Convention to Combat Desertification, the Ramsar Convention on Wetlands, and the Convention on Migratory Species—and is designed also to meet needs of other stakeholders, including the business community, the health sector, nongovernmental organizations, and indigenous peoples. The sub-global assessments also aimed to meet the needs of users in the regions where they were undertaken.

The assessment focuses on the linkages between ecosystems and human well-being and, in particular, on "ecosystem services." An ecosystem is a dynamic complex of plant, animal, and microorganism communities and the nonliving environment interacting as a functional unit. The MA deals with the full range of ecosystems—from those relatively undisturbed, such as natural forests, to landscapes with mixed patterns of human use and to ecosystems intensively managed and modified by humans, such as agricultural land and urban areas. Ecosystem services are the benefits people obtain from ecosystems. These include *provisioning services* such as food, water, timber, and fiber; *regulating services* that affect climate, floods, disease, wastes, and water quality; *cultural services* that provide recreational, aesthetic, and spiritual benefits; and *supporting services* such as soil formation, photosynthesis, and nutrient cycling. The human species, while buffered against environmental changes by culture and technology, is fundamentally dependent on the flow of ecosystem services.

The MA examines how changes in ecosystem services influence human well-being. Human well-being is assumed to have multiple constituents, including the *basic material for a good life*, such as secure and adequate livelihoods, enough food at all times, shelter, clothing, and access to goods; *health,* including feeling well and having a healthy physical environment, such as clean air and access to clean water; *good social relations,* including social cohesion, mutual respect, and the ability to help others and provide for children; *security,* including secure access to natural and other resources, personal safety, and security from natural and human-made disasters; and *freedom of choice and action,* including the opportunity to achieve what an individual values doing and being. Freedom of choice and action is influenced by other constituents of well-being (as well as by other factors, notably education) and is also a precondition for achieving other components of well-being, particularly with respect to equity and fairness.

The conceptual framework for the MA posits that people are integral parts of ecosystems and that a dynamic interaction exists between them and other parts of ecosystems, with the changing human condition driving, both directly and indirectly, changes in ecosystems and thereby causing changes in human well-being. At the same time, social, economic, and cultural factors unrelated to ecosystems alter the human condition, and many natural forces influence ecosystems. Although the MA emphasizes the linkages between ecosystems and human well-being, it recognizes that the actions people take that influence ecosystems result not just from concern about human well-being but also from considerations of the intrinsic value of species and ecosystems. Intrinsic value is the value of something in and for itself, irrespective of its utility for someone else.

The Millennium Ecosystem Assessment synthesizes information from the scientific literature and relevant peer-reviewed datasets and models. It incorporates knowledge held by the private sector, practitioners, local communities, and indigenous peoples. The MA did not aim to generate new primary knowledge but instead sought to add value to existing information by collating, evaluating, summarizing, interpreting, and communicating it in a useful form. Assessments like this one apply the judgment of experts to existing knowledge to provide scientifically credible answers to policy-relevant questions. The focus on policy-relevant questions and the explicit use of expert judgment distinguish this type of assessment from a scientific review.

Five overarching questions, along with more detailed lists of user needs developed through discussions with stakeholders or provided by governments through international conventions, guided the issues that were assessed:

- What are the current condition and trends of ecosystems, ecosystem services, and human well-being?
- What are plausible future changes in ecosystems and their ecosystem services and the consequent changes in human well-being?
- What can be done to enhance well-being and conserve ecosystems? What are the strengths and weaknesses of response options that can be considered to realize or avoid specific futures?
- What are the key uncertainties that hinder effective decision-making concerning ecosystems?
- What tools and methodologies developed and used in the MA can strengthen capacity to assess ecosystems, the services they provide, their impacts on human well-being, and the strengths and weaknesses of response options?

The MA was conducted as a multiscale assessment, with interlinked assessments undertaken at local, watershed, national, regional, and global scales. A global ecosystem assessment cannot easily meet all the needs of decision-makers at national and sub-national scales because the management of any

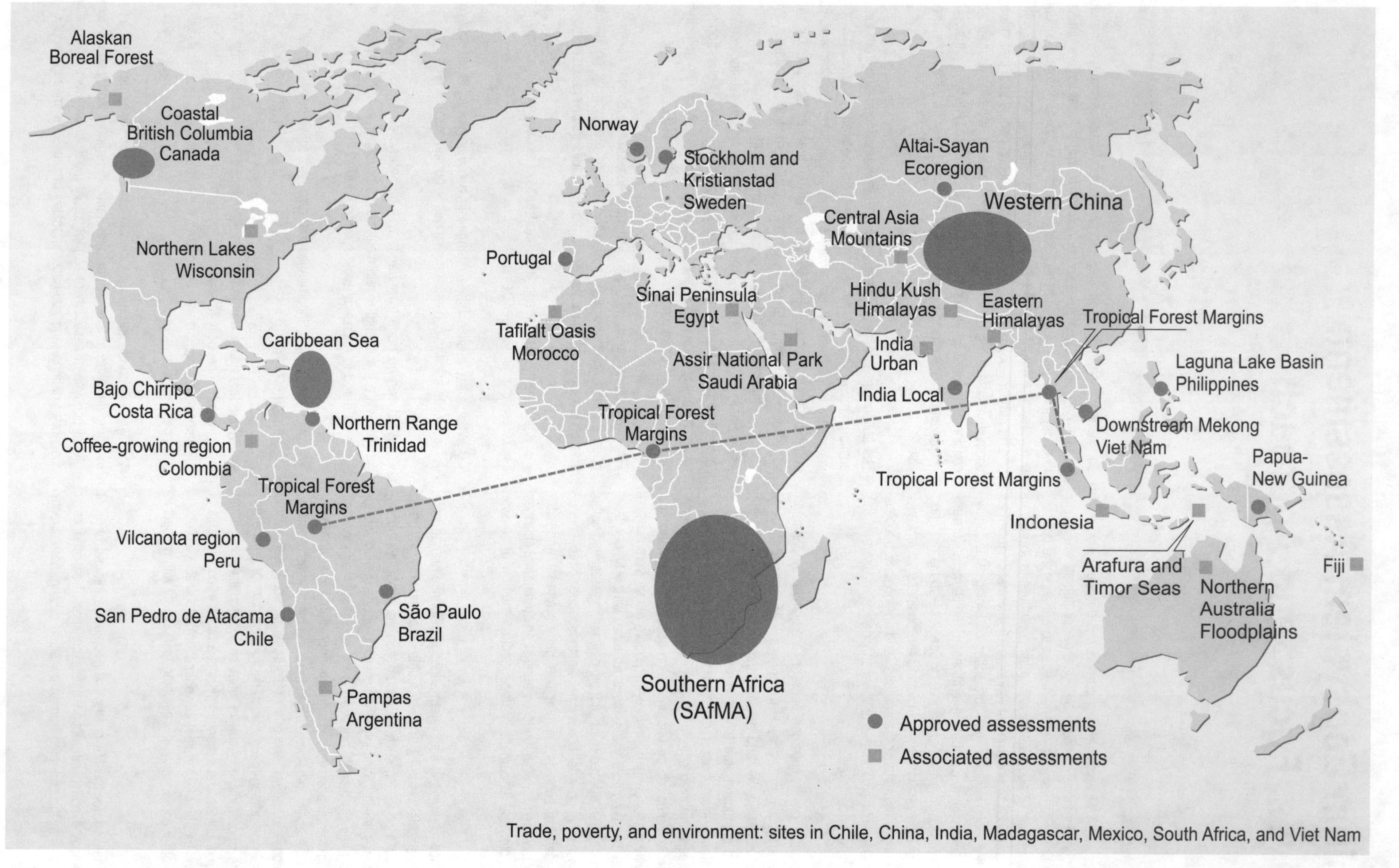

Eighteen assessments were approved as components of the MA. Any institution or country was able to undertake an assessment as part of the MA if it agreed to use the MA conceptual framework, to centrally involve the intended users as stakeholders and partners, and to meet a set of procedural requirements related to peer review, metadata, transparency, and intellectual property rights. The MA assessments were largely self-funded, although planning grants and some core grants were provided to support some assessments. The MA also drew on information from 16 other sub-global assessments affiliated with the MA that met a subset of these criteria or were at earlier stages in development.

SUB-GLOBAL ASSESSMENT	ECOSYSTEM TYPES										ECOSYSTEM SERVICES								
	COASTAL	CULTIVATED	DRYLAND	FOREST	INLAND WATER	ISLAND	MARINE	MOUNTAIN	POLAR	URBAN	FOOD	WATER	FUEL and ENERGY	BIODIVERSITY-RELATED	CARBON SEQUESTRATION	FIBER and TIMBER	RUNOFF REGULATION	CULTURAL, SPIRITUAL, AMENITY	OTHERS
Altai-Sayan Ecoregion			●	●	●			●			●		●	●		●		●	
San Pedro de Atacama, Chile			●		●						●	●		●			●	●	●
Caribbean Sea	●					●	●				●	●		●				●	
Coastal British Columbia, Canada	●			●	●			●			●			●		●	●	●	
Bajo Chirripo, Costa Rica		●		●	●						●	●		●		●		●	●
Tropical Forest Margins		●		●							●	●		●	●	●	●		●
India Local Villages		●		●	●						●	●	●	●		●	●	●	●
Glomma Basin, Norway		●		●	●			●			●		●			●		●	●
Papua New Guinea	●	●				●	●				●	●	●	●		●	●	●	●
Vilcanota, Peru		●	●					●			●	●		●			●	●	●
Laguna Lake Basin, Philippines		●		●	●						●	●		●	●			●	●
Portugal	●	●	●	●	●	●	●	●		●	●	●		●	●	●	●	●	●
São Paulo Green Belt, Brazil	●	●		●	●					●	●	●		●	●	●	●	●	●
Southern Africa	●	●	●	●	●					●	●	●	●	●		●		●	●
Stockholm and Kristianstad, Sweden		●			●					●	●	●		●	●	●	●	●	●
Northern Range, Trinidad	●			●	●			●			●	●		●		●	●	●	●
Downstream Mekong Wetlands, Viet Nam	●	●			●						●	●	●	●	●	●	●	●	●
Western China		●	●	●	●			●			●	●		●	●		●		●
Alaskan Boreal Forest				●	●						●					●		●	●
Arafura and Timor Seas	●					●	●				●			●	●				●
Argentine Pampas		●									●	●						●	●
Central Asia Mountains								●			●	●		●					●
Colombia coffee-growing regions		●						●			●	●		●				●	
Eastern Himalayas				●				●			●	●	●	●				●	
Sinai Peninsula, Egypt			●					●						●			●	●	●
Fiji	●					●					●	●	●						●
Hindu Kush-Himalayas					●			●				●		●			●	●	●
Indonesia	●					●	●				●			●					●
India Urban Resource										●	●	●	●	●	●			●	●
Tafilalt Oasis, Morocco		●	●								●	●						●	●
Northern Australia Floodplains					●						●	●		●			●	●	●
Assir National Park, Saudi Arabia		●		●				●			●						●	●	●
Northern Highlands Lake District, Wisconsin				●	●							●				●	●	●	●

particular ecosystem must be tailored to the particular characteristics of that ecosystem and to the demands placed on it. However, an assessment focused only on a particular ecosystem or particular nation is insufficient because some processes are global and because local goods, services, matter, and energy are often transferred across regions. Each of the component assessments was guided by the MA conceptual framework and benefited from the presence of assessments undertaken at larger and smaller scales. The sub-global assessments were not intended to serve as representative samples of all ecosystems; rather, they were to meet the needs of decision-makers at the scales at which they were undertaken. The sub-global assessments involved in the MA process are shown in the Figure and the ecosystems and ecosystem services examined in these assessments are shown in the Table.

The work of the MA was conducted through four working groups, each of which prepared a report of its findings. At the global scale, the Condition and Trends Working Group assessed the state of knowledge on ecosystems, drivers of ecosystem change, ecosystem services, and associated human well-being around the year 2000. The assessment aimed to be comprehensive with regard to ecosystem services, but its coverage is not exhaustive. The Scenarios Working Group considered the possible evolution of ecosystem services during the twenty-first century by developing four global scenarios exploring plausible future changes in drivers, ecosystems, ecosystem services, and human well-being. The Responses Working Group examined the strengths and weaknesses of various response options that have been used to manage ecosystem services and identified promising opportunities for improving human well-being while conserving ecosystems. The report of the Sub-global Assessments Working Group contains lessons learned from the MA sub-global assessments. The first product of the MA—*Ecosystems and Human Well-being: A Framework for Assessment,* published in 2003—outlined the focus, conceptual basis, and methods used in the MA. The executive summary of this publication appears as Chapter 1 of this volume.

Approximately 1,360 experts from 95 countries were involved as authors of the assessment reports, as participants in the sub-global assessments, or as members of the Board of Review Editors. The latter group, which involved 80 experts, oversaw the scientific review of the MA reports by governments and experts and ensured that all review comments were appropriately addressed by the authors. All MA findings underwent two rounds of expert and governmental review. Review comments were received from approximately 850 individuals (of which roughly 250 were submitted by authors of other chapters in the MA), although in a number of cases (particularly in the case of governments and MA-affiliated scientific organizations), people submitted collated comments that had been prepared by a number of reviewers in their governments or institutions.

The MA was guided by a Board that included representatives of five international conventions, five U.N. agencies, international scientific organizations, governments, and leaders from the private sector, nongovernmental organizations, and indigenous groups. A 15-member Assessment Panel of leading social and natural scientists oversaw the technical work of the assessment, supported by a secretariat with offices in Europe, North America, South America, Asia, and Africa and coordinated by the United Nations Environment Programme.

The MA is intended to be used:

- to identify priorities for action;
- as a benchmark for future assessments;
- as a framework and source of tools for assessment, planning, and management;
- to gain foresight concerning the consequences of decisions affecting ecosystems;
- to identify response options to achieve human development and sustainability goals;
- to help build individual and institutional capacity to undertake integrated ecosystem assessments and act on the findings; and
- to guide future research.

Because of the broad scope of the MA and the complexity of the interactions between social and natural systems, it proved to be difficult to provide definitive information for some of the issues addressed in the MA. Relatively few ecosystem services have been the focus of research and monitoring and, as a consequence, research findings and data are often inadequate for a detailed global assessment. Moreover, the data and information that are available are generally related to either the characteristics of the ecological system or the characteristics of the social system, not to the all-important interactions between these systems. Finally, the scientific and assessment tools and models available to undertake a cross-scale integrated assessment and to project future changes in ecosystem services are only now being developed. Despite these challenges, the MA was able to provide considerable information relevant to most of the focal questions. And by identifying gaps in data and information that prevent policy-relevant questions from being answered, the assessment can help to guide research and monitoring that may allow those questions to be answered in future assessments.

Contents

Foreword

The Millennium Ecosystem Assessment was called for by United Nations Secretary-General Kofi Annan in 2000 in his report to the UN General Assembly, *We the Peoples: The Role of the United Nations in the 21st Century.* Governments subsequently supported the establishment of the assessment through decisions taken by three international conventions, and the MA was initiated in 2001. The MA was conducted under the auspices of the United Nations, with the secretariat coordinated by the United Nations Environment Programme, and it was governed by a multistakeholder board that included representatives of international institutions, governments, business, NGOs, and indigenous peoples. The objective of the MA was to assess the consequences of ecosystem change for human well-being and to establish the scientific basis for actions needed to enhance the conservation and sustainable use of ecosystems and their contributions to human well-being.

This volume has been produced by the MA Responses Working Group and examines the strengths and weaknesses of various response options that have been used to manage ecosystem services, as well as identifying promising opportunities for improving human well-being while conserving ecosystems. The material in this report has undergone two extensive rounds of peer review by experts and governments, overseen by an independent Board of Review Editors.

This is one of four volumes (*Current State and Trends, Scenarios, Policy Responses,* and *Multiscale Assessments*) that present the technical findings of the Assessment. Six synthesis reports have also been published: one for a general audience and others focused on issues of biodiversity, wetlands and water, desertification, health, and business and ecosystems. These synthesis reports were prepared for decision-makers in these different sectors, and they synthesize and integrate findings from across all of the working groups for ease of use by those audiences.

This report and the other three technical volumes provide a unique foundation of knowledge concerning human dependence on ecosystems as we enter the twenty-first century. Never before has such a holistic assessment been conducted that addresses multiple environmental changes, multiple drivers, and multiple linkages to human well-being. Collectively, these reports reveal both the extraordinary success that humanity has achieved in shaping ecosystems to meet the need of growing populations and economies and the growing costs associated with many of these changes. They show us that these costs could grow substantially in the future, but also that there are actions within reach that could dramatically enhance both human well-being and the conservation of ecosystems.

A more exhaustive set of acknowledgements appears later in this volume but we want to express our gratitude to the members of the MA Board, Board Alternates, Exploratory Steering Committee, Assessment Panel, Coordinating Lead Authors, Lead Authors, Contributing Authors, Board of Review Editors, and Expert Reviewers for their extraordinary contributions to this process. (The list of reviewers is available at www.MAweb.org.) We also would like to thank the MA Secretariat and in particular the staff of the Responses Working Group Technical Support Unit for their dedication in coordinating the production of this volume, as well as the Institute of Economic Growth (India) and the National Institute of Public Health and the Environment (Netherlands), which housed this TSU.

We would particularly like to thank the Co-chairs of the Responses Working Group, Kanchan Chopra and Rik Leemans, and the TSU Coordinators, Pushpam Kumar and Henk Simons, for their skillful leadership of this working group and their contributions to the overall assessment.

Dr. Robert T. Watson
MA Board Co-chair
Chief Scientist, The World Bank

Dr. A.H. Zakri
MA Board Co-chair
Director, Institute for Advanced Studies,
United Nations University

Preface

The focus of the MA is on ecosystem services (the benefits people obtain from ecosystems), how changes in ecosystem services have affected human well-being in the past, and what role these changes could play in the present as well as in the future. The MA is an assessment of responses that are available to improve ecosystem management and can thereby contribute to the various constituents of human well-being. The specific issues addressed have been defined through consultation with the MA users. Broadly, the MA applies an integrated systems' approach to evaluate trade-offs involved in following alternate strategies and courses of action to use ecosystem services for enhancing human welfare.

The overall aims of the MA are to:

- identify priorities for action;
- provide tools for planning and management;
- provide foresight concerning the consequences of decisions affecting ecosystems;
- identify response options to achieve human development and sustainability goals; and
- help build individual and institutional capacity to undertake integrated ecosystem assessments and to act on their findings.

The MA synthesizes information from scientific literature, data sets, and scientific models, and utilizes knowledge held by the private sector, practitioners, local communities, and indigenous peoples. All of the MA findings have undergone two rounds of expert and governmental review.

This report of the MA Responses Working Group evaluates the current understanding of how human decisions and policies influence ecosystems, ecosystem services, and consequently, human well being. The assessment identifies and critically evaluates past, current, and possible future policy and management options for maintaining ecosystems (including biodiversity) and sustaining the flow of ecosystem services. The Responses Working Group is one of four MA working groups, each of which has contributed an assessment report. The Condition and Trends Working Group reviewed the state of knowledge on ecosystems, ecosystem services, and associated human well-being in the present, recent past, and near future. The Scenarios Working Group considered the evolution of ecosystem services during the first half of the twenty-first century under a range of plausible narratives. The Sub-global Working Group carried out assessments at different levels to directly meet needs of local and regional decision-makers and strengthen the global findings with finer-scale detail. Together, the working group reports provide local, national, regional, and global perspectives and information.

In the MA, responses are defined as the whole range of human actions, including policies, strategies, and interventions, to address specific issues, needs, opportunities, or problems. A response typically involves a "reaction to a perceived problem." It can be individual or collective; it may be designed to answer one or many needs; or it could be focused at different temporal, spatial, or organizational scales. In the context of managing ecosystems or ecosystem services, responses may be of legal, technical, institutional, economic, or behavioral nature and may operate at local/micro, regional, national, or international level at the time scale of days to hundred of years. The assessment focuses on responses that are intended to ensure that ecosystems and biodiversity are preserved, that desired ecosystem services accrue, and that human well-being is augmented. This is one of the major objectives of all conventions targeted by the MA, the Millennium Development Goals, and others.

Focus of the Responses Assessment Report

The Responses assessment report is rooted in the MA conceptual framework, which provides an understanding of the causes and consequences of changes in ecosystems across scales (local, regional, and global) and over time (MA 2003; see also Chapter 1 of this volume). *Ecosystems, ecosystem services, human well-being, and direct and indirect drivers initiating the links among them constitute the main elements of the MA conceptual framework.* (See Chapter 1 for definitions of these concepts.) Human responses are outcomes of human decisions and they influence and change the key connecting links between these elements. They determine how individuals, communities, nations, and international agencies intervene or strategize, ostensibly in their own interests, to use, manage, and conserve ecosystems. There are many ways to categorize responses, which are often determined by the problem at hand, the decision-maker/actor associated with, or the tradition of, the discipline.

The organizational scales of responses can be international (for instance, the U.N. conventions), multilateral and bilateral (important for transboundary problems), national, state/provincial, community (urban or rural), family, or individual. Decisions taken at each of these levels can affect ecosystems and ecosystem services. For example, national policies initiated to comply with international trade treaties can impact local ecosystems. The assessment methodology developed by the Responses Working Group is comprehensive enough to be used to assess responses at all scales, as and when they are relevant to the context of the particular ecosystem service being studied. The Responses assessment consists of a three-stage approach. The first stage focuses on factors that may either rule out a particular response or may define the critical preconditions for its success. Constraints that render a policy option infeasible are called the *binding constraints,* which are context specific. In the second stage, responses are compared across multiple dimensions, identifying compatibility or conflict between different policy objectives. Here the acceptable costs associated with the implementation of a response (the *acceptable trade-offs*) are identified. Finally, responses are evaluated from different perspectives in order to provide guidance that is the best balanced from the point of view of decision-making as shown in the illustration below:

As shown in the illustration, research, assessment, monitoring, and policy-making are all components of a continuing interactive

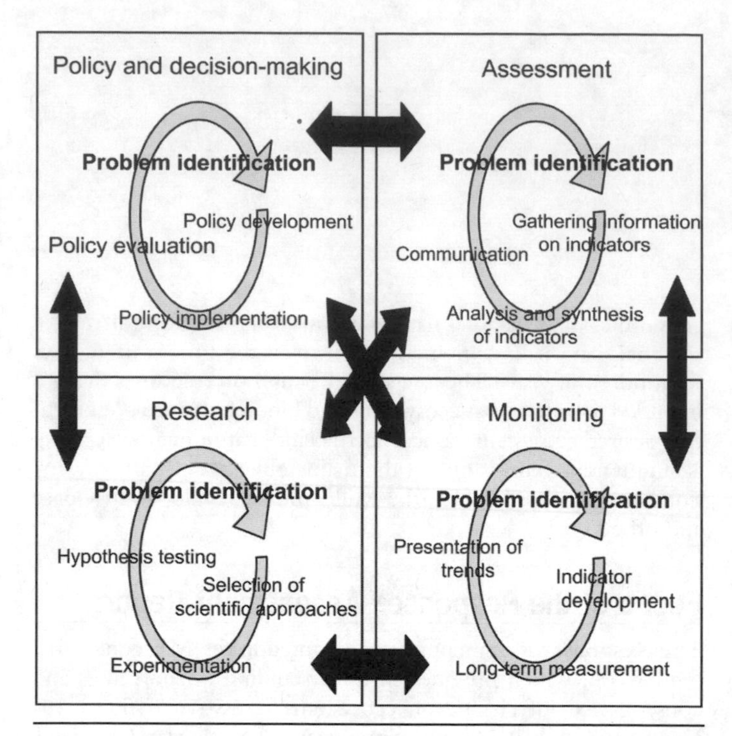

process to support development and implementation of responses. Decision-making starts by identifying a problem, followed by collating the research findings to help in defining and choosing policy options. (See Chapter 18 of this volume.) Policies are selected, implemented, and then evaluated for their effectiveness. The process is iterative and involves interaction with all kinds of information providers. Ideally, the decision-making cycle entails obtaining feedback from all categories of stakeholders. Similar loops exist for the research, monitoring, and assessment process, each with its characteristic objectives, approaches, and dynamics. Under the best circumstances, research insights should yield adequate monitoring networks and indicators of change, to be taken up for assessment toward an informed decision process. Understandably, the dynamics and timing of each of these cycles do not always evolve in perfect coordination with each other. The dynamic nature of information exchange and feedback to and from these processes and their stakeholders are integral to developing responses.

This implies that decision-making processes are liable to change over time to improve effectiveness. A number of mechanisms can facilitate this. Ecosystem dynamics will never be completely understood, socioeconomic systems will continue to change, and drivers can never be fully anticipated. It is important therefore that decision-making processes incorporate, wherever possible, procedures to evaluate outcomes of actions and assimilate lessons learned from experience. Debate on exactly how to go about doing this continues in discussions on adaptive management, social learning, safe minimum standards, and the precautionary principle. But the core message of all approaches is the same: acknowledge the limits of human understanding, give special consideration to irreversible changes, and evaluate the multiple impacts of decisions as they unfold.

Organization of this Volume

This assessment report has a large canvas to cover. Various response options are selected on the basis of the impact they have on a set of ecosystems and ecosystem services. The report examines these different societal responses and evaluates them by using diverse methodologies. The results are analyzed from diverse perspectives to draw key conclusions regarding their impact on human well-being.

To facilitate the analysis, this report is divided into three parts. Part I (that is, Chapters 1 through 4) introduces responses and focuses mainly on conceptual and methodological issues. Chapter 1 summarizes the MA conceptual framework and defines some important concepts. Chapter 2 discusses alternative typologies of possible responses. It differentiates responses by, actors, disciplines, drivers, and scales, and further characterizes them in terms of the instruments for intervention—such as economic, institutional, governance, and technological—thus highlighting the multidimensional nature of responses.

Chapter 3 elaborates on alternate methods of assessing responses. It sets up a framework that can be used to evaluate whether particular responses are effective and desirable from social, political, and economic perspectives. It indicates how social, political, and economic factors and their actors can act as constraints to the ability of responses or strategies to meet intended goals and avoid unintended consequences.

Chapter 4 highlights specific decision-making criteria in the above context. It also focuses on the role of uncertainty in assessing the effectiveness of responses. This uncertainty is partly a function of the methodology and tools applied but also an inherent characteristic of decision-making that is always a leap into the future.

Part II consists of ten chapters (5 through 14), each focusing on one or more ecosystem service. These chapters relate specific case studies from the literature and the sub-global assessments to the response typology and evaluation methodology outlined in Part I. Chapter 5 focuses on responses concerning biodiversity, which underlies all other ecosystem services. This chapter has a strong spotlight on ecosystem management and conservation.

Chapters 6, 7, and 8 dwell on the provisioning ecosystem services. Different responses at all major decision-making levels, which alter ecosystems providing these services, are presented and assessed. Special emphasis is laid upon the trade-offs and synergies between specific responses and their consequences. Responses that contribute to the sustainable use of these ecosystems are highlighted. In a similar vein, Chapters 9 through 13 focus on regulating services, and Chapter 14 assesses cultural ecosystem services. These chapters correspond to chapters pertaining to ecosystem services presented by the Condition and Trends Working Group. Together, the ecosystem services chapters in this volume and in MA *Current State and Trends* provide a complete overview of the current understanding of where, how, and why ecosystem services are changing; in what way the selected responses are having an impact on drivers, ecosystems, ecosystem services; and the different constituent parts of human well-being.

Taking an ecosystem service approach proved difficult for some of the chapters in Part II. For instance, few responses focus directly on managing ecosystems services toward climate regulation or waste management. Additionally, there has been no or little experience in treating the topics in some chapters (for example, waste management and climate regulations) as ecosystem services. Adhering too strongly to an ecosystem services approach could, in some cases, lead to too narrow a focus while the user audiences expect a broader treatment. This became apparent after the first review. We have therefore permitted a more user-oriented treatment of certain ecosystem services to allow for more comprehensive discussions of responses related to areas such as climate regulation, waste management, and disease control.

Chapter 15 deals with responses that address (provision of) ecosystem services across a number of systems simultaneously, explicitly including objectives to enhance human well being. Such integrated responses occurring across different scales could be oriented at different actors, generally employing a range of instruments for implementation. The assessment of sustainable management strategies and trade-offs between different responses is central here. The responses always integrate different aspects of ecosystems. Examples include integrated water, forest, or coastal management. Such responses may be at the international level in the form of framework conventions or at local levels in the form of concrete resource management projects. This chapter provides a comprehensive evaluation of such integrated responses.

Part III (Chapters 15 through 19) synthesizes the lessons learned from earlier chapters and provides an overarching evaluation of the interlinkages among drivers, ecosystems, ecosystem services, and ultimately, human well-being. Chapter 15 deals with responses that address (provision of) ecosystem services across a number of systems simultaneously, explicitly including objectives to enhance human well-being. Such integrated responses occurring across different scales could be oriented at different actors, generally employing a range of instruments for implementation. The assessment of sustainable management strategies and trade-offs between different responses is central here. The responses always integrate different aspects of ecosystems. Examples include integrated water, forest, or coastal management. Such responses may be at the international level in the form of framework conventions or at local levels in the form of concrete resource management projects. This chapter provides a comprehensive evaluation of such integrated responses.

The other chapters within Part III take on a specific aspect of human welfare for analysis such as material and social security, health, freedoms, and choice. Chapter 16 takes a strong human health perspective, while Chapter 17 emphasizes poverty reduction. The central questions in these chapters are:

- How have responses that were aimed at protecting ecosystems and their services, impacted the different constituents and determinants of human well-being?
- Did policies initiated at national levels for promoting well-being have negative impacts on ecosystems or on the accrual of ecosystem services?

These two chapters thus strongly emphasize the trade-offs and synergies between different responses.

Chapter 18 provides general "guidelines" for choosing responses, assessing the required information and decision-tools by discussing the relative strengths and weaknesses of alternate sources of information. Chapter 19 evaluates the Millennium Development Goals from a responses perspective. Sustainable use of ecosystems and thereby accrual of ecosystem services for human well-being is central to these chapters as in all others.

Kanchan Chopra
IEG, India

Rik Leemans
Wageningen University, Netherlands

Pushpam Kumar
IEG, India

Henk Simons
RIVM, Netherlands

Acknowledgments

We would like to express our sincere thanks to all authors of the Responses Working Group for their untiring enthusiasm during the entire process and for their efforts to ensure that all state-of-the-art knowledge is indeed assessed. It was a sheer pleasure to work with this extremely motivated group of social and natural scientists from across the globe.

This volume could not have been written without the very valuable guidance from the Board and Panel members of the Millennium Ecosystem Assessment. Dr. Walter V. Reid, Prof. Harold Mooney, and Dr. Angela Cropper were especially pivotal to the process and strongly contributed to the assessment as a whole through their many suggestions, penetrating comments, targeted styles, and constructive attitudes. We thank the MA Board and its chairs, Robert Watson and A.H. Zakri, and the members of the MA Review Board and its chairs, José Sarukhán and Anne Whyte, for their guidance and support for this working group. The wisdom and insights of all these individuals kept us focused, and their input was essential to harmonize the assessments of the different working groups.

Reviewers provided numerous constructive comments on, for example, structure, general content, specific statements, and inconsistencies between chapters. We greatly appreciate their efforts because these comments were instrumental in improving the overall quality of all chapters and the assessment as a whole. We further appreciate the daunting task of the review board to ensure that *all* comments, suggestions, and criticisms were addressed in a credible manner.

Finally, we are thankful to the Institute of Economic Growth (IEG), Delhi, and the Dutch Institute of Public Health and the Environment (RIVM) for supporting the chairs, hosting the Technical Support Unit (TSU) and facilitating the assessment in general. We are indebted to Meenakshi Rathore for her crucial support to the TSU at the IEG. In the last stage of the work, Shreemoyee Patra helped in editorial work for the volume and we are also thankful to her.

Special thanks are due to the other MA Secretariat staff who worked tirelessly on this project:

Administration
Nicole Khi—Program Coordinator
Chan Wai Leng—Program Coordinator
Belinda Lim—Administrative Officer
Tasha Merican—Program Coordinator

Sub-global
Marcus Lee—TSU Coordinator and MA Deputy Director
Ciara Raudsepp-Hearne—TSU Coordinator

Condition and Trends
Neville J. Ash—TSU Coordinator
Dalène du Plessis—Program Assistant
Mampiti Matete—TSU Coordinator

Scenarios
Elena M. Bennett—TSU Coordinator
Veronique Plocq-Fichelet—Program Administrator
Monika B. Zurek—TSU Coordinator

Engagement and Outreach
Christine Jalleh—Communications Officer
Nicolas Lucas—Engagement and Outreach Director
Valerie Thompson—Associate

Other Staff
John Ehrmann—Lead Facilitator
Keisha-Maria Garcia—Research Assistant
Lori Han—Publications Manager
Sara Suriani—Conference Manager
Jillian Thonell—Data Coordinator

Interns
Emily Cooper, Elizabeth Wilson

Kanchan Chopra
IEG, India

Rik Leemans
Wageningen University, Netherlands

Pushpam Kumar
IEG, India

Henk Simons
RIVM, Netherlands

Acknowledgment from the Co-chairs of the Assessment Board, Co-chairs of the Assessment Panel, and the Director

We would like to acknowledge the contributions of all the authors of this book and the support provided by their institutions that enabled their participation. We would like to thank the host organizations of the MA Technical Support Units—WorldFish Center (Malaysia); UNEP-World Conservation Monitoring Centre (United Kingdom); Institute of Economic Growth (India); National Institute of Public Health and the Environment (Netherlands); University of Pretoria (South Africa); Food and Agriculture Organization of the United Nations (Italy); World Resources Institute, Meridian Institute, and Center for Limnology of the University of Wisconsin–Madison (all in the United States); Scientific Committee on Problems of the Environment (France); and International Maize and Wheat Improvement Center (Mexico)—for the support they provided to the process.

We thank several individuals who played particularly critical roles: Rosemarie Philips for editing the report; Hyacinth Billings and Caroline Taylor for providing invaluable advice on the publication process; Maggie Powell for preparing the page design and all the figures; and Julie Feiner for helping to proof the figures and tables. And we thank the other MA volunteers, the administrative staff of the host organizations, and colleagues in other organizations who were instrumental in facilitating the process: Mariana Sanchez Abregu, Isabelle Alegre, Adlai Amor, Emmanuelle Bournay, Herbert Caudill, Habiba Gitay, Helen Gray, Sherry Heileman, Norbert Henninger, Toshi Honda, Francisco Ingouville, Humphrey Kagunda, Brygida Kubiak, Nicolas Lapham, Liz Leavitt, Christian Marx, Stephanie Moore, John Mukoza, Arivudai Nambi, Laurie Neville, Carolina Katz Reid, Liana Reilly, Philippe Rekacewicz, Carol Rosen, Anne Schram, Jeanne Sedgwick, Tang Siang Nee, Darrell Taylor, Tutti Tischler, Dan Tunstall, Woody Turner, Mark Valentine, Elsie Velez-Whited, and Mark Zimsky.

We also thank the current and previous Board Alternates: Ivar Baste, Jeroen Bordewijk, David Cooper, Carlos Corvalan, Nick Davidson, Lyle Glowka, Guo Risheng, Ju Hongbo, Ju Jin, Kagumaho (Bob) Kakuyo, Melinda Kimble, Kanta Kumari, Stephen Lonergan, Charles Ian McNeill, Joseph Kalemani Mulongoy, Ndegwa Ndiang'ui, and Mohamed Maged Younes. We thank the past members of the MA Board whose contributions were instrumental in shaping the MA focus and process, including Philbert Brown, Gisbert Glaser, He Changchui, Richard Helmer, Yolanda Kakabadse, Yoriko Kawaguchi, Ann Kern, Roberto Lenton, Corinne Lepage, Hubert Markl, Arnulf Müller-Helbrecht, Seema Paul, Susan Pineda Mercado, Jan Plesnik, Peter Raven, Cristián Samper, Ola Smith, Dennis Tirpak, Alvaro Umaña, and Meryl Williams. We wish to also thank the members of the Exploratory Steering Committee that designed the MA project in 1999–2000. This group included a number of the current and past Board members, as well as Edward Ayensu, Daniel Claasen, Mark Collins, Andrew Dearing, Louise Fresco, Madhav Gadgil, Habiba Gitay, Zuzana Guziova, Calestous Juma, John Krebs, Jane Lubchenco, Jeffrey McNeely, Ndegwa Ndiang'ui, Janos Pasztor, Prabhu L. Pingali, Per Pinstrup-Andersen, and José Sarukhán. We thank Ian Noble and Mingsarn Kaosa-ard for their contributions as members of the Assessment Panel during 2002.

We would particularly like to acknowledge the input of the hundreds of individuals, institutions, and governments who reviewed drafts of the MA technical and synthesis reports. We also thank the thousands of researchers whose work is synthesized in this report. And we would like to acknowledge the support and guidance provided by the secretariats and the scientific and technical bodies of the Convention on Biological Diversity, the Ramsar Convention on Wetlands, the Convention to Combat Desertification, and the Convention on Migratory Species, which have helped to define the focus of the MA and of this report.

We also want to acknowledge the support of a large number of nongovernmental organizations and networks around the world that have assisted in outreach efforts: Alexandria University, Argentine Business Council for Sustainable Development, Arab Media Forum for Environment and Development, Asociación Ixacavaa (Costa Rica), Brazilian Business Council on Sustainable Development, Charles University (Czech Republic), Chinese Academy of Sciences, European Environmental Agency, European Union of Science Journalists' Associations, EIS-Africa (Burkina Faso), Forest Institute of the State of São Paulo, Foro Ecológico (Peru), Fridtjof Nansen Institute (Norway), Fundación Natura (Ecuador), Global Development Learning Network, Indonesian Biodiversity Foundation, Institute for Biodiversity Conservation and Research–Academy of Sciences of Bolivia, International Alliance of Indigenous Peoples of the Tropical Forests, IUCN office in Uzbekistán, IUCN Regional Offices for West Africa and South America, Northern Temperate Lakes Long Term Ecological Research Site (USA), Permanent Inter-States Committee for Drought Control in the Sahel, Peruvian Society of Environmental Law, Probioandes (Peru), Professional Council of Environmental Analysts of Argentina, Regional Center AGRHYMET (Niger), Regional Environmental Centre for Central Asia, Resources and Research for Sustainable Development (Chile), Royal Society (United Kingdom), Stockholm University, Suez Canal University, Terra Nuova (Nicaragua), The Nature Conservancy (United States), United Nations University, University of Chile, University of the Philippines, Winslow Foundation (USA), World Assembly of Youth, World Business Council for Sustainable Development, WWF-Brazil, WWF-Italy, and WWF-US.

We are extremely grateful to the donors that provided major financial support for the MA and the MA Sub-global Assessments: Global Environment Facility, United Nations Foundation, David and Lucile Packard Foundation, World Bank, Consultative Group on International Agricultural Research, United Nations Environment Programme, Government of China, Ministry of Foreign Affairs of the Government of Norway, Kingdom of Saudi Arabia, and the Swedish International Biodiversity Programme. We also thank other organizations that provided financial support: Asia Pacific Network for Global Change Research; Association of Caribbean States; British High Commission, Trinidad and Tobago; Caixa Geral de Depósitos, Portugal; Canadian International Development Agency; Christensen Fund; Cropper Foundation, Environmental Management Authority of Trinidad and Tobago; Ford Foundation; Government of India; International Council for Science; International Development Research Centre; Island Resources Foundation; Japan Ministry of Environment; Laguna Lake Development Authority; Philippine Department of Environment and Natural Resources; Rockefeller Foundation; U.N. Educational, Scientific and Cultural Organization; UNEP Division of Early Warning and Assessment; United Kingdom Department for Environment, Food, and Rural Affairs; U.S. National Aeronautic and Space Administration; and Universidade de Coimbra, Portugal. Generous in-kind support has been provided by many other institutions (a full list is available at www.MAweb.org). The work to establish and design the MA was supported by grants from The Avina Group, The David and Lucile Packard Foundation, Global Environment Facility, Directorate for Nature Management of Norway, Swedish International Development Cooperation Authority, Summit Foundation, UNDP, UNEP, United Nations Foundation, U.S. Agency for International Development, Wallace Global Fund, and World Bank.

Reader's Guide

The four technical reports present the findings of each of the MA Working Groups: Condition and Trends, Scenarios, Responses, and Sub-global Assessments. A separate volume, *Our Human Planet*, presents the summaries of all four reports in order to offer a concise account of the technical reports for decision-makers. In addition, six synthesis reports were prepared for ease of use by specific audiences: Synthesis (general audience), CBD (biodiversity), UNCCD (desertification), Ramsar Convention (wetlands), business and industry, and the health sector. Each MA sub-global assessment will also produce additional reports to meet the needs of its own audiences.

All printed materials of the assessment, along with core data and a list of reviewers, are available at www.MAweb.org. In this volume, Appendix A contains color maps and figures. Appendix B lists all the authors who contributed to this volume. Appendix C lists the acronyms and abbreviations used in this report and Appendix D is a glossary of terminology used in the technical reports. Throughout this report, dollar signs indicate U.S. dollars and ton means tonne (metric ton). Bracketed references within the Summary are to chapters within this volume.

In this report, the following words have been used where appropriate to indicate judgmental estimates of certainty, based on the collective judgment of the authors, using the observational evidence, modeling results, and theory that they have examined: very certain (98% or greater probability), high certainty (85–98% probability), medium certainty (65%–58% probability), low certainty (52–65% probability), and very uncertain (50–52% probability). In other instances, a qualitative scale to gauge the level of scientific understanding is used: well established, established but incomplete, competing explanations, and speculative. Each time these terms are used they appear in italics.

Ecosystems and Human Well-being: Policy Responses, Volume 3

Summary: Response Options and Strategies

Core Writing Team: Kate Brown, Bradnee Chambers, Kanchan Chopra, Angela Cropper, Anantha K. Duraiappah, Dan Faith, Joyeeta Gupta, Pushpam Kumar, Rik Leemans, Jens Mackensen, Harold A. Mooney, Walter V. Reid, Janet Riley, Henk Simons, Marja Spierenburg, and Robert T. Watson.

Extended Writing Team: Nimbe O. Adedipe, Heidi Albers, Bruce Aylward, Joseph Baker, Jayanta Bandyopadhyay, Juan-Carlos Belausteguigotia, D.K. Bhattacharya, Eduardo S. Brondizio, Diarmid Campbell-Lendrum, Flavio Comim, Carlos Corvalan, Dana R. Fisher, Tomas Hak, Simon Hales, Robert Howarth, Laura Meadows, James Mayers, Jeffrey McNeely, Monirul Q. Mirza, Bedrich Moldan, Ian Noble, Steve Percy, Karen Polson, Frederik Schutyser, Sylvia Tognetti, Ferenc Toth, Rudy Rabbinge, Sergio Rosendo, M. K. C. Sridhar, Kilaparti Ramakrishna, Mahendra Shah, Nigel Sizer, Bhaskar Vira, Diana Wall, Alistair Woodward, and Gary Yohe.

CONTENTS

BOXES

TABLES

Introduction

The Millennium Ecosystem Assessment examines the consequences of changes to ecosystem services for human well-being. It assesses the conditions and trends in ecosystems and their services, explores plausible scenarios for the future, and assesses alternative response options. The assessment of the Condition and Trends Working Group affirms that, in the aggregate, changes to ecosystems have contributed to substantial gains in human well-being over the past centuries: people are better nourished and live longer and healthier lives than ever before, incomes have risen, and political institutions have become more participatory. However, these gains have been achieved at growing costs, including the degradation of many ecosystem services, increased risks of nonlinear changes, and the exacerbation of poverty for some groups of people. Persistent and significant local, national, and regional disparities in income, well-being, and access to ecosystem services continue to exist. The assessment of the Scenarios Working Group shows that the degradation of ecosystem services could grow significantly worse during the first half of this century and represents a barrier to achieving the Millennium Development Goals.

The question arises: What kind of action can we take? What policies can be developed and implemented by societies to enable them to move in chosen directions? In this report, we define "responses" to encompass the entire range of human actions, including policies, strategies, and interventions, to address specific issues, needs, opportunities, or problems related to ecosystems, ecosystem services, and human well-being. Responses may be institutional, economic, social and behavioral, technological, or cognitive in nature. Response strategies are designed and undertaken at local, regional, or international scales within diverse institutional settings. This report assesses how successful various response strategies have been and identifies the conditions that have contributed to their success or failure. Additionally, it derives lessons that can be applied to the design of future responses.

The MA conceptual framework (MA 2003) posits that people are integral parts of ecosystems and that a dynamic interaction exists between them and other parts of ecosystems, with the changing human condition driving, both directly and indirectly, changes in ecosystems and thereby causing changes in human well-being. (See Chapter 1, Box 1.2.) Direct and indirect drivers operate at different spatial, temporal, and organizational scales. Responses affect the direct and indirect drivers of change in ecosystems and thereby the services derived from ecosystems. In this framework, human–ecosystem interactions are dynamic processes and, as a result, drivers and responses co-evolve over time. Expansion of cultivated systems, for instance, was initially a response to the growing demand for food. Over time, this expansion of cultivation became a driver of change altering other ecosystem services, particularly as a result of habitat conversion, use of water for irrigation, and the excessive use of nutrients. A full assessment of the effectiveness of various responses must thus include the examination of the historical and contemporary contexts within which interactions between drivers and responses developed. The choice of the most effective set of response options needs to be informed not just by the impact of the response on a particular driver, but also by the interactions among different drivers themselves.

The effectiveness and impact of any response strategy depends furthermore on the interactions between the people who initiate the response and others who have a stake in the outcomes at local, regional, and global levels. Strategies initiated at the global level, such as through international conventions, for example, may have consequences on ecosystem services and human well-being at the local level.

The Responses Working Group assessed a wide range of responses and interventions undertaken by different decision-makers in many different economic, social, and institutional settings. In the sections that follow, this summary describes several key characteristics of successful responses, discusses methods for choosing responses, and reviews some of the more promising or effective responses. It also discusses some of the barriers to implementing promising responses; one barrier that deserves particular emphasis involves the limited number of trained people in many countries who are able to analyze response options and to develop and implement programs of action to address these problems. This assessment demonstrates the tremendous scope for actions that can help to enhance human well-being while conserving ecosystems; but without investment in the necessary human and institutional capacity, many countries will not be able to effectively pursue these options.

Characteristics of Successful Responses

Responses to environmental problems tend to be more successful when: a) there is effective coordination among the different levels of decision-making; b) transparent participatory approaches are used; c) the potential trade-offs and synergies among response strategies and their outcomes are factored into their design; and d) considerations of impacts on ecosystems and the potential contributions of ecosystem services are mainstreamed in economic policy and development planning.

Coordination across Sectors and across Scales

***Effective action to address problems related to ecosystem services requires improved coordination across sectors and scales.* [See especially 5, 17, 19]**

Almost any action affecting an ecosystem has consequences for many different services provided by that ecosystem. For example, a response designed to enhance the production of one ecosystem service, such as crop production, could harm other services such as water quality, fisheries production, or flood control. These trade-offs cannot be adequately addressed through traditional sectoral management approaches. Moreover, they cannot be adequately addressed through actions undertaken at a single scale, whether international, national, or local. Effective ecosystem management thus requires effective coordination, both among governmental institutions directly responsible for the environment and between those institutions and other sectors. [17]

Coordination among International Institutions

The cooperation among multilateral environmental agreements has improved in recent years, but considerable scope remains to increase the coordination and consistency among their objectives and actions. [17] To date, however, there has been relatively little effective coordination between MEAs and the politically stronger international economic and social institutions such as the World Bank (except in its role as an implementing agency of the Global Environment Facility), the International Monetary Fund, and the World Trade Organization. Despite their profound influence on the environment, economic and trade-related agreements have shown minimal commitment to environmental issues; neither have the poverty reduction strategies prepared by countries for the World Bank. Given the central importance of ecosystem services in achieving many Millennium Development Goals (in particular, the goals and targets related to poverty, hunger, disease, children's health, water, and environmental sustainability), the MDG process could in principle provide a means to better incorporate the environment into these other sectors, but little progress has yet been observed. [19]

Coordination across Decision-making Levels

International agreements are more likely to be translated into national policy if they include precise obligations, sanctions for violation, and monitoring provisions, and if they provide financial assistance for national implementation. While most MEAs meet some of these criteria, relatively few have sanctions for violation; in almost all cases, there is considerable scope for the agreements to be strengthened if the criteria were met more effectively. [17] For example, financial mechanisms such as the Global Environment Facility enable assistance to be provided through some ecosystem-related MEAs, but across the board these agreements would be more effective if greater assistance were available. Similarly, the Convention on Biological Diversity, the Convention to Combat Desertification, and the Ramsar Wetlands Convention could be strengthened if countries assumed additional outcome-focused obligations in addition to the more common planning and reporting obligations. The CBD, for example, has now established a specific outcome-focused target—the "2010 Target" to significantly slow the rate of biodiversity loss—but this target is not binding on individual countries.

Some steps have been taken by the ecosystem-focused MEAs to promote greater national implementation. For example, the national biodiversity strategies and action plans form a central implementation mechanism of the CBD and have resulted in some action at the national and local levels.[5] The CCD has encouraged the development of national action programs to combat desertification; 50 of these programs are now receiving international funding. While the CBD national biodiversity strategies and the CCD national action programs have stimulated and guided some actions and policy reforms, their primary impact has been within the environmental sector; they have been less effective in influencing action in other sectors. The overall effectiveness of the implementation of these and other MEAs could be strengthened if these planning processes were more effectively integrated into other processes such as decentralization and land reform, which generally have major effects on land use and desertification.

In general, international agreements dealing with ecological resources tend to be less successful than those concerning defense or trade because of the less obvious nature of reciprocal benefits to contracting parties, the major driving force in other agreements. Success of international legal instruments depends on the perception of the need for longer term cooperation. The design of the agreement and the manner in which the agreement was negotiated both play a role. Given the complexity of some negotiating processes and the lack of resources to enable the full participation of many developing countries in negotiations, some countries face serious challenges in ensuring adequate representation of their interests and perspectives in international agreements; this in turn undermines the effectiveness of the agreements. [17] Clearly, there exists an urgent need to augment developing-country capacity to participate in international negotiations.

Coordination at National and Sub-national Levels

At national and sub-national levels, effective responses to ecosystem degradation are constrained by the same weakness of cross-sectoral coordination and even coordination within the environmental sector. The implementation of many environmental conventions at a national level, for example, could be strengthened through more effective coordination among the national offices responsible for implementing different international agreements. More generally, at the national and sub-national levels successful response interventions often involve situation-driven integration across decision-making agencies. This type of integration tends to be found in situations where communities and lower level governments are given management and decision-making flexibility within broad enabling frameworks.

Participation and Transparency

***Insufficient participation and transparency in planning and decision-making have been major barriers to the design and implementation of effective responses.* [3, 4, 5, 7, 14, 15, 17]**

The importance of stakeholder participation is now widely recognized, although generally poorly implemented, at the international scale, as well as at the national and local scales. Although stakeholder participation can result in a slower and more costly process, it creates ownership in the policy being developed, commitment to successful implementation, and increased societal acceptance of the policy. Among international conventions, for example, the CBD states "management should be decentralized to the lowest appropriate level, and boundaries for management shall be defined by indigenous and local peoples, among others." The 1999 Ramsar Convention Conference of Parties adopted guidelines for the inclusion of local and indigenous people in the management of Ramsar wetlands. The problems associated with inadequate stakeholder participation are most apparent in the area of biodiversity conservation. Because local people are de facto the primary resource managers in most regions, working with local communities is essential to conserving biodiversity in the longer term. The establishment of protected areas, for example, is more effective when local communities have "bought in" to the protected area and have alternative livelihood opportunities or receive direct payments so that they are not harmed by creation of the proteced area. [5] This often requires the establishment of protected areas designed to support multiple uses of natural and cultural resources. Bottom-up decision-making processes rooted in a local and site-specific context have also enabled the negotiation of water agreements to become a catalyst for peace and cooperation. Note, for instance, that nation states belonging to very different political persuasions confirm water treaties such as the Nile treaty and the Indus Waters treaty. [7]

Important as stakeholder participation is, the financial costs and time needed for elaborate stakeholder processes can sometimes outweigh the benefits. Moreover, there is also the risk that "participation" can be co-opted into what are, at their core, centrally determined plans. This kind of "centralized decentralization" may well lead to the exclusion of disadvantaged groups even though they have been "consulted" in the decision process. Often this is the consequence of policies that do not take into account differences among stakeholders in preexisting situations. Examples are found in the watershed programs and the water user associations in India.

The introduction of participatory approaches in settings where people are not accustomed to such approaches must be accompanied by capacity-building among stakeholders if it is to succeed. The capacity created in this way must also be sustained. Key interventions include both public education and steps taken to strengthen social networks in order to facilitate the inclusion of all relevant forms of knowledge and information, including local and indigenous knowledge, in decision-making.

For participatory approaches to succeed, the stakeholders involved need access to information on both the resources being managed and the decision-making process. Effective monitoring, assessment, and reporting is therefore a key to success in allocating ecosystem services and implementing response options. Given the heterogeneity, constant change, and site-specific characteristics of ecosystem services and the human institutions through which they are managed, a fundamental but often overlooked need is for an independent and transparent process of assessment. Monitoring and assessment are critical components of pro-active adaptive management, as they can provide the feedback necessary to develop and continually improve implementation strategies as new information becomes available, constraints are identified, and enabling institutional structures put in place. Although considerable debate continues about the most effective mechanisms for stakeholder involvement in decision-making processes, all approaches agree on the same core elements: acknowledge the limits of human understanding, recognize knowledge gaps explicitly, give special consideration to irreversible changes, and evaluate the impacts of decisions as they unfold.

Trade-offs and Synergies

***Trade-offs and synergies among human well-being, ecosystems, and ecosystem services are the rule rather than the exception and this implies that informed choices must be made to achieve the best possible outcomes.* [5, 6, 7, 8, 11, 13, 15, 16, 17]**

The following categories of trade-offs are involved in managing ecosystem services:

- *Trade-offs between the present and future.* For example, some technologies developed to increase food production, such as the replacement of traditional cultivars with high yielding varieties or the excessive application of fertilizers and pesticides, have reduced the capacity of land and water systems to provide food in the future. [6] Similarly, some resource management practices yield economic benefits in the present, but defer costs to the future. Forest harvest, for example, provides immediate economic returns but may result in future costs in the form of degraded water quality or increased frequency of floods.
- *Trade-offs among ecosystem services.* The majority of response strategies have given priority to increasing the allocation of provisioning services, such as food production and water supply, often at the expense of regulating and cultural ecosystem services. For example, water has been impounded to enable increased irrigation and increased food production, but this reduces downstream water supplies, harms freshwater biodiversity, and degrades some cultural and recreational benefits provided by free-flowing rivers.
- *Trade-offs among constituents of human well being.* Responses are often directed at improving the material well-being constituent of human well-being to the neglect of other constituents of human well-being such as health and security. For example, increased use of pesticides can increase the production of food, but harm the health of farmworkers and consumers.
- *Trade-offs among stakeholders.* Ecosystems and their services are used differently by different groups of stakeholders: the needs of vulnerable groups are often marginalized in this process. For example, large scale commercial exploitation of forests for timber harvest often comes at the expense of the use of forests by local communities as a source of non-wood forest products. [8] Similarly, the conversion of mangrove forests to shrimp aquaculture benefits the farmers who have resources to invest in aquaculture operations, but harms the local fisherfolk who depend on capture fisheries associated with the mangroves.

Although negative trade-offs are common, positive synergies are also possible, and responses can be identified that create synergies and help in achieving multiple objectives. The long-term success of conservation strategies in areas where local people are dependent on the use of biological resources, for example, depends on meeting the needs of these communities. The exact nature of the synergy is more easily identified in specific ecological and societal contexts through an appropriate understanding of linkages between ecosystems and human well-being. Similarly, among the growing number of people who face health problems associated with obesity, reducing consumption of food would benefit both human health and reduce demand for ecosystem services.

Some potential and emerging synergies can only be realised if enabling institutions are created. For example, afforestation, reforestation, improved forest, cropland and rangeland management and agroforestry provide a range of opportunities to increase carbon sequestration. Similarly, slowing deforestation provides an opportunity to reduce carbon emissions. Such activities have the potential to sequester about 10 to 20% of projected fossil emissions up to 2050. [13] However, only a small part of this potential can be delivered with the institutions, technologies, and financial arrangements now in place. A large number of these issues remain undecided and prevent the use of forestry as a carbon management option.

Mainstreaming

***The quantity and quality of ecosystem services available are often determined to a greater extent by macroeconomic, trade, and other policies than by policies within the environmental sector itself.* [5, 6, 8, 17, 19]**

Some of the most significant drivers of change in ecosystem services and their use originate outside the sectors that have responsibility for the management of ecosystem services. For example, the availability of fish in coastal waters can be strongly influenced by government policies related to crop production or food price supports, since this will influence the amount of fertilizer and water used in crop production and hence the potential harmful impacts associated with nutrient pollution or changes in river flows. Similarly, trade policies can have significant impacts on forest product industries and thus on the management of forests. Indeed, this assessment finds that policies outside the forest sector are often more important than policies within the sector in determining the social and ecological sustainability of forest management. While inappropriate policies in other sectors can harm ecosystem services, changes in those policies can often also provide one of the most effective means for improving managment of ecosystem services. For example, reforms to the Common Agricultural Policy in Europe to incorporate environmental dimensions could significantly reduce pressures on some ecosystem services. [6]

In general, potential threats to ecosystem services and the potential contributions of ecosystem services to economic development and poverty reduction are not taken into account in development plans and trade policies. Very few macroeconomic responses to poverty reduction have considered the importance of sound management of ecosystem services as a mechanism to meet the basic needs of the poorest. The poverty reduction strategies that many developing countries are now preparing for the World Bank and other donors can be most effective if they include an emphasis on the links between ecosystems and human well-being, but few of the strategies incorporate these issues. [17] More generally, the failure to incorporate considerations of ecosystem management in the strategies being pursued to achieve many of the eight Millennium Development Goals will undermine the sustainability of any progress that is made toward the goals and targets associated with poverty, hunger, disease, child mortality, and access to water, in particular. [19]

Choosing Responses

Decisions affecting ecosystems and their services can be improved by changing the processes used to reach those decisions. [18]

The context of decision-making about ecosystems is changing rapidly. The new challenge to decision-making is to make effective use of information and tools in this changing context in order to improve the decisions. At the same time, some old challenges must still be addressed. The decision-making process and the actors involved influence the intervention chosen. Decision-making processes vary across jurisdictions, institutions, and cultures. Even so, this assessment has identified the following elements of decision-making processes related to ecosystems and their services that tend to improve the decisions reached and their outcomes for ecosystems and human well-being:

- use the best available information, including considerations of the value of both marketed and nonmarketed ecosystem services;
- ensure transparency and the effective and informed participation of important stakeholders;
- recognize that not all values at stake can be quantified, and thus quantification can provide a false objectivity in decision-making processes that have significant subjective elements;
- strive for efficiency, but not at the expense of effectiveness;
- consider equity and vulnerability in terms of the distribution of costs and benefits;
- ensure accountability and provide for regular monitoring and evaluation; and
- consider cumulative and cross-scale effects and, in particular, assess trade-offs across different ecosystem services.

A wide range of tools can assist decision-making concerning ecosystems and their services. [3, 4] The use of decision-making methods that adopt a pluralistic perspective is particularly pertinent, since these techniques do not give undue weight to any particular viewpoint. Examples of tools that can assist decision-making at a variety of scales, including global, sub-global, and local, include:

- *Deliberative tools (which facilitate transparency and stakeholder participation).* These include neighborhood forums, citizens' juries, community issues groups, consensus conferences, electronic democracy, focus groups, issue forums, and ecosystem service user forums.
- *Information-gathering tools (which are primarily focused on collecting data and opinions).* Examples of information-gathering tools include citizens' research panels, deliberative opinion polls, environmental impact assessments, participatory rural appraisal, and rapid rural appraisal.
- *Planning tools (which are typically used to evaluate potential policy options).* Some common planning tools are consensus participation, cost-benefit analysis, multicriteria analysis, participatory learning and action, stakeholder decision analysis, trade-off analysis, and visioning exercises.

Some of these methods are particularly well-suited for decision-making in the face of uncertainties in data, prediction, context, and scale. [4] Such methods include cost-benefit or multicriteria analyses, risk assessment, the precautionary principle, and vulnerability analysis. (See Table R1.) All these methods have been able to support optimization exercises, but few of them have much to say about equity. Cost-benefit analysis can, for example, be modified to weight the interests of some people more than others. The discount rate can be viewed, in long-term analyses, as a means of weighting the welfare of future generations; and the precautionary principle can be expressed in terms of reducing the exposure of certain populations or systems whose preferential status may be the result of equity considerations. Multicriteria analysis was designed primarily to accommodate optimization across multiple objectives with complex interactions, but this can also be adapted to consider equity and threshold issues at national and sub-national scales.

Scenario-building exercises provide one way to cope with many aspects of uncertainty, but our limited understanding of ecological and human response processes shrouds any individual scenario in its own characteristic uncertainty. [4] The development of a set of scenarios provides a useful means to highlight the implications of alternative assumptions about critical uncertainties related to the behavior of human and ecological systems. In this way, they provide one means to cope with many aspects of uncertainty in assessing responses. The relevance, significance, and influence of scenarios ultimately depend on the assumptions made in their development. At the same time, though, there are a number of reasons to be cautious in the use of scenarios. First, individual scenarios represent conditional projections based on specific assumptions. Thus to the extent that our understanding and representation of the ecological and human systems represented in the scenarios is limited, specific scenarios are characterized by their own uncertainty. Second, there is uncertainty in translating the lessons derived from scenarios developed at one scale—say, global—to the assessment of responses at other scales—say, sub-national. Third, scenarios often have hidden and hard-to-articulate assumptions. Fourth, environmental scenarios have tended to more effectively incorporate state-of-the-art natural science modeling than social science modeling.

Effective management of ecosystems requires coordinated responses at multiple scales. [15, 17] Responses that are successful at a small scale are often less successful at higher levels due to constraints in legal frameworks and government institutions that prevent their success. In addition, there appear to be limits to scaling up, not only because of these higher-level constraints, but also because interventions at a local level often address only direct drivers of change rather than indirect or underlying ones. For example, a local project to improve livelihoods of communities surrounding a protected area in order to reduce pressure on it, if successful, may increase migration into buffer zones, thereby adding to pressures. Cross-scale responses may be more effective at addressing the higher-level constraints and leakage problems and simultaneously tackling regional and national as well as local-level drivers of change. Examples of successful cross-scale responses include some co-management approaches to natural resource management in fisheries and forestry and multistakeholder policy processes.

Active adaptive management can be a particularly valuable tool for reducing uncertainty about ecosystem management decisions. [17] The term "active" adaptive management is used here to emphasize the key characteristic of the original concept (which is frequently and inappropriately used to

Table R1. Applicability of Decision Support Methods and Frameworks

Key: ++ = direct application of the method by design
+ = possible application with modification or (in the case of uncertainty) the method has already been modified to handle uncertainty
– = weak but not impossible applicability with significant effort

Method	Optimization	Equity	Thresholds	Uncertainty	Scale of Application		
					Micro	National	Regional and Global
Cost-benefit Analysis	+	+	–		√	√	√
Risk Assessment	+	+	++	++	√	√	√
Multicriteria Analysis	++	+	+	+	√	√	
Precautionary Principle*	+	+	++	++	√	√	√
Vulnerability Analysis	+	+	++	+	√	√	

*The precautionary principle is not strictly analogous to the other analytical and assessment methods but still can be considered a method for decision support. The precautionary principle prescribes how to bring scientific uncertainty into the decision-making process by explicitly formalizing precaution and bringing it to the forefront of the deliberations. It posits that significant actions (ranging from doing nothing to banning a potentially harmful substance or activity, for instance) may be justified when the degree of possible harm is large and irreversible.

mean "learning by doing"): the design of management programs to test hypotheses about how components of an ecosystem function and interact, in order to reduce uncertainty about the system more rapidly than would otherwise occur. Under an adaptive management approach, for example, a fisheries manager might intentionally set harvest levels either lower or higher than the "best estimate" in order to gain information more rapidly about the shape of the yield curve for the fishery. Given the high levels of uncertainty surrounding coupled socioecological systems, the use of active adaptive management is often warranted.

Promising Responses for Ecosystem Services and Human Well-being

Past actions to slow or reverse the degradation of ecosystems have yielded significant benefits, but these improvements have generally not kept pace with growing pressures and demands. Although most ecosystem services assessed in the MA are being degraded, the extent of that degradation would have been much greater without responses implemented in past decades. For example, more than 100,000 protected areas (including strictly protected areas such as national parks as well as areas managed for the sustainable use of natural ecosystems such as timber harvest or wildlife harvest) covering about 11.7% of the terrestrial surface have now been established. These protected areas play an important role in the conservation of biodiversity and ecosystem services, although important gaps remain in their distribution and management, particularly in marine and freshwater systems. Many protected areas lack adequate resources for management. Protected areas will not be completely effective until they are fully integrated into an ecosystem or landscape approach to management. [5]

An effective set of responses to ensure the sustainable management of ecosystems would address the indirect and direct drivers that lead to the degradation of ecosystem services and overcome a range of barriers. The barriers to be overcome include:

- inappropriate institutional and governance arrangements, including the presence of corruption and weak systems of regulation and accountability;
- market failures and the misalignment of economic incentives;
- social and behavioral factors, including the lack of political and economic power of some groups (such as poor people, women, and indigenous groups) who are particularly dependent on ecosystem services or harmed by their degradation;
- underinvestment in the development and diffusion of technologies that could increase the efficiency of use of ecosystem services and reduce the harmful impacts of various drivers of ecosystem change; and
- insufficient knowledge (as well as the poor use of existing knowledge) concerning ecosystem services and management, policy, technological, behavioral, and institutional responses that could enhance benefits from these services while conserving resources.

All these barriers are compounded by weak human and institutional capacity related to the assessment and management of ecosystem services, underinvestment in the regulation and management of their use, lack of public awareness, and lack of awareness among decision-makers of the threats posed by the degradation of ecosystem services and the opportunities that more sustainable management of ecosystems could provide.

The MA assessed 78 response options for ecosystem services, integrated ecosystem management, conservation and sustainable use of biodiversity, waste management, and climate change. Many of these options hold significant promise for conserving or sustainably enhancing the supply of ecosystem services; a selected number of promising responses that address the barriers just described are discussed here. (The full list of response options is presented in Appendix R1.) These responses in turn often require that the proper enabling conditions are in place. (See Box R1.) The stakeholder groups that would need to take decisions to implement each response are indicated as follows: G for government, B for business and industry, and N for nongovernmental organizations and other civil society organizations (including community-based and indigenous peoples' organizations and research institutions).

Institutions and Governance

Changes in institutional and environmental governance frameworks are sometimes required in order to create the

BOX R1

Enabling Conditions for Designing Effective Responses

Some examples of conditions that must be met in order to design and implement some of the response options identified in this assessment include:

- *supportive insurance and financial markets* are needed to ensure that economic value of ecosystem services is taken into account;
- *better information on who benefits and is harmed by changes in specific ecosystem services* is needed to enable the establishment of effective systems of payments for ecosystem services;
- *greater involvement of concerned stakeholders in decision-making* is required to ensure transparency and effective functioning of regulatory mechanisms;
- *appropriate forms of property rights* (mostly common property arrangements) need to be established to encourage private-public or community-state partnerships for resource conservation;
- *innovative partnerships among different knowledge-based institutions* need to be established to foster the integration of local and indigenous knowledge in decision-making processes; and
- *human and institutional capacity for assessing and acting on assessments* needs to be enhanced for decision-makers to have access to information they need concerning the management of ecosystem services.

enabling conditions for effective management of ecosystems; in other cases, existing institutions could meet these needs but face significant barriers. [2, 7, 11, 12, 15, 17] Many existing institutions at both the global and the national level have the mandate to address the degradation of ecosystem services but face a variety of challenges in doing so related to the need for greater cooperation across sectors and the need for coordinated responses at multiple scales (see the discussion above on Characteristics of Successful Responses). However, since a number of the issues identified in this assessment are recent concerns and were not specifically taken into account in the design of today's institutions, changes in existing institutions and the development of new ones may sometimes be needed, particularly at the national scale.

In particular, existing national and global institutions are not well designed to deal with the management of open access resources, a characteristic of many ecosystem services. Issues of ownership and access to resources, rights to participation in decision-making, and regulation of particular types of resource use or discharge of wastes can strongly influence the sustainability of ecosystem management and are fundamental determinants of who wins and who loses from changes in ecosystems. Corruption—a major obstacle to effective management of ecosystems—also stems from weak systems of regulation and accountability.

Promising interventions include:

- *Development of institutions that devolve (or centralize) decision-making to meet management needs while ensuring effective coordination across scales* (G, B, N). Problems of ecosystem management have been exacerbated by both overly centralized and overly decentralized decision-making. For example, highly centralized forest management has proved ineffective in many countries, and efforts are now being made to move responsibility to lower levels of decision-making either within the natural resources sector or as part of broader decentralization of governmental responsibilities. At the same time, one of the most intractable problems of ecosystem management has been the lack of alignment between political boundaries and units appropriate for the management of ecosystem goods and services. Downstream communities may not have access to the institutions through which upstream actions can be influenced; alternatively, downstream communities or countries may be stronger politically than upstream regions and may dominate control of upstream areas without addressing upstream needs.
- *Development of institutions to regulate interactions between markets and ecosystems* (G). The potential of policy and market reforms to improve ecosystem management is often constrained by weak or absent institutions. For example, the potential of the Clean Development Mechanism established under the Framework Convention on Climate Change to provide financial support to developing countries in return for greenhouse gas reductions, which would realize climate and biodiversity benefits through payments for carbon sequestration in forests, is constrained by unclear property rights, concerns over the permanence of reductions, and lack of mechanisms for resolving conflicts. Moreover, existing regulatory institutions often do not have ecosystem protection as a clear mandate. For example, independent regulators of privatized water systems and power systems do not necessarily promote resource use efficiency and renewable supply. [7] The role of the state in setting and enforcing rules continues to be important even in the context of privatization and market-led growth.
- *Development of institutional frameworks that promote a shift from highly sectoral resource management approaches to more integrated approaches* (G, B). In most countries, separate ministries are in charge of various aspects of ecosystems (such as ministries of environment, agriculture, water, and forests) and drivers of change (such as ministries of energy, transportation, development, and trade). Each of these ministries has control over different aspects of ecosystem management. As a result, there is seldom the political will to develop effective ecosystem management strategies, and competition among the ministries can often result in policy choices that are detrimental to ecosystems. Integrated responses intentionally and actively address ecosystem services and human well-being simultaneously, such as integrated coastal zone management, integrated river basin management, and national sustainable development strategies. Although the potential for integrated responses is high, numerous barriers have limited their effectiveness: they are resource-intensive, but the potential benefits can exceed the costs; they require multiple instruments for their implementation; and they require new institutional and governance structures, skills, knowledge, and capacity. Integrated responses at local levels have been successful in using the links between human well-being and ecosystems to design effective interventions, particularly where supportive higher level structures exist.

Economics and Incentives

Economic and financial interventions provide powerful instruments to regulate the use of ecosystem goods and services. [2] Because many ecosystem services are not traded in markets, markets fail to provide appropriate signals that might otherwise contribute to the efficient allocation and sustainable use of the services. Even if people are aware of the services provided by an ecosystem, they are neither compensated for providing these services nor penalized for reducing them. In addition, the people harmed by the degradation of ecosystem services are often not the ones who benefit from the actions leading to their degradation, and so those costs are not factored into management decisions. A wide range of opportunities exists to influence human behavior to address this challenge in the form of economic and financial instruments. Some of them establish markets; others work through the monetary and financial interests of the targeted social actors; still others affect relative prices.

Market mechanisms can only work if supporting institutions are in place, and thus there is a need to build institutional capacity to enable more widespread use of these mechanisms. [2, 6, 7, 8, 17] The adoption of economic instruments usually requires a legal framework, and in many cases the choice of a viable and effective economic intervention mechanism is determined by the socioeconomic context. For example, resource taxes can be a powerful instrument to guard against the overexploitation of an ecosystem service, but an effective tax scheme requires well-established and reliable monitoring and tax collection systems. Similarly, subsidies can be effective to introduce and implement certain technologies or management procedures, but they are inappropriate in settings that lack the transparency and accountability needed to prevent corruption. The establishment of market mechanisms also often involves explicit decisions about wealth distribution and resource allocation, when, for example, decisions are made to establish private property rights for resources that were formerly considered common pool resources. For that reason, the inappropriate use of market mechanisms can further exacerbate problems of poverty.

Promising interventions include:

- *Elimination of subsidies that promote excessive use of ecosystem services (and, where possible, transfer of these subsidies to payments for nonmarketed ecosystem services)* (G). Many countries provide significant agricultural production subsidies that lead to greater food production in countries with subsidies than global market conditions warrant; that promote the overuse of water, fertilizers, and pesticides; and that reduce the profitability of agriculture in developing countries. [7] Subsidies increase land values, adding to landowners' resistance to subsidy reductions. Similar problems are created by fishery subsidies. Although removal of production subsidies would produce net benefits, it would not occur without costs. The farmers and fishers benefiting directly from the subsidies would suffer the most immediate losses, but there would also be indirect effects on ecosystems both locally and globally. In some cases, it may be possible to transfer production subsides to other activities that promote ecosystem stewardship, such as payment for the provision or enhancement of regulatory or supporting services. Compensatory mechanisms may be needed for the poor who are adversely affected by the immediate removal of subsidies. Reduced subsidies within the OECD may lessen pressures on some ecosystems in those countries, but they could lead to more rapid conversion and intensification of land for agriculture in developing countries and would thus need to be accompanied by policies to minimize the adverse impacts on ecosystems there.
- *Greater use of economic instruments and market-based approaches in the management of ecosystem services* (G, B, N). Economic instruments and market mechanisms with the potential to enhance the management of ecosystem services include:
    - *Taxes or user fees for activities with "external" costs* (trade-offs not accounted for in the market). These instruments create incentives that lessen the external costs and provide revenues that can help protect the damaged ecosystem services. Examples include taxes on excessive application of nutrients or ecotourism user fees.
    - *Creation of markets, including through cap-and-trade systems.* Ecosystem services that have been treated as "free" resources, as is often the case for water, tend to be used wastefully. The establishment of markets for the services can both increase the incentives for their conservation and increase the economic efficiency of their allocation if supporting legal and economic institutions are in place. However, as noted earlier, while markets will increase the efficiency of the use of the resource, they can have harmful effects on particular groups of users who may be inequitably affected by the change. The combination of regulated emission caps, coupled with market mechanisms for trading pollution rights, often provides an efficient means of reducing emissions harmful to ecosystems. For example, one of the most rapidly growing markets related to ecosystem services is the carbon market [13]; in another example, nutrient trading systems may be a low-cost way to reduce water pollution in the United States [9].
    - *Payment for ecosystem services.* Mechanisms can be established to enable individuals, firms, or the public sector to pay resource owners to provide particular services. For example, in New South Wales, Australia, associations of farmers purchase salinity credits from the State Forests Agency, which in turn contracts with upstream landholders to plant trees, which reduce water tables and store carbon. Similarly, in 1996, Costa Rica established a nationwide system of conservation payments to induce landowners to provide ecosystem services. Under this program, the government brokers contracts between international and domestic "buyers" and local "sellers" of sequestered carbon, biodiversity, watershed services, and scenic beauty. These interventions are found to succeed, typically when a high degree of certainty exists with regard to the accrual of ecosystem services over time.
    - *Mechanisms to enable consumer preferences to be expressed through markets.* Consumer pressure may provide an alternative way to influence producers to adopt more sustainable production practices in the absence of effective government regulation. For example, certification schemes that exist for sustainable fisheries and forest practices provide people with the opportunity to promote sustainability through their consumer choices. Within the forest sector, forest certification has become widespread in many countries and forest conditions; thus far, however, most certified forests are in temperate regions, managed by large companies that export to northern retailers. [6] Certification and labeling is also being used at smaller scales. For example, the Salmon Safe initiative in Oregon, United States, certifies and promotes wines and other agricultural products from Oregon farms and vineyards that have adhered to management practices designed to protect water quality and salmon habitat. [7]

Social and Behavioral Responses

Social and behavioral responses—including population policy, public education, civil society actions, and empowerment of communities, women, and youth—can be instrumental in responding to ecosystem degradation. [2, 5, 6] These are generally interventions that stakeholders initiate and execute through exercising their procedural or democratic rights in efforts to improve ecosystems and human well-being.

Promising interventions include:

- *Measures to reduce aggregate consumption of unsustainably managed ecosystem services* (G, B, N). The choices about what individuals consume and how much they consume are influenced not just by considerations of price but also by behavioral factors related to culture, ethics, and values. Behavioral changes that could reduce demand for degraded ecosystem services can be encouraged through actions by governments (such as education and public awareness programs or the promotion of demand-side management), industry (such as improved product labeling or commitments to use raw materials from sources certified as sustainable), and civil society (such as public awareness campaigns). Efforts to reduce aggregate consumption, however, must sometimes incorporate measures to increase the access to and consumption of those same ecosystem services by specific groups such as poor people.
- *Communication and education* (G, B, N). Improved communication and education are essential to achieve the objectives of the environmental conventions, the Johannesburg Plan of Implementation, and the sustainable management of natural resources more generally. Both the public and decision-makers can benefit from education concerning ecosystems and human well-being, but education more generally provides tremendous social benefits that can help address many drivers of ecosystem degradation. For example, the Haribon Foundation in the Philippines has used communication, education, and mobilization of networks to motivate fishers and their communities to create marine sanctuaries to allow for fish populations to revive and restore declining catches; over 1,000 reserves have now been established. [5] Barriers to the effective use of communication and education include a failure to use research and apply modern theories of learning and change. While the importance of communication and education is well recognized, providing the human and financial resources to undertake effective work is a continuing barrier.
- *Empowerment of groups particularly dependent on ecosystem services or affected by their degradation, including women, indigenous people, and young people* (G, B, N). Women, indigenous people, and young people are all important "stakeholders" in the management of ecosystem services but, historically, each group has tended to be marginalized in decision-making processes. For example, despite women's knowledge about the environment and the potential they possess to improve resource management, their participation in decision-making has often been restricted by social and cultural structures. Similarly, the case for protecting young people's ability to take part in decision-making is strong as they will experience the longer-term consequences of decisions made today concerning ecosystem services. Greater involvement of indigenous peoples in decision-making can also enhance environmental management, although the primary justification for it continues to be based on human and cultural rights.

Technological Responses

Given the growing demands for ecosystem services and other increased pressures on ecosystems, the development and diffusion of technologies designed to increase the efficiency of resource use or reduce the impacts of drivers such as climate change and nutrient loading are essential. [2, 6, 7, 13, 17] Technological change has been essential for meeting growing demands for some ecosystem services, and technology holds considerable promise to help meet future growth in demand. Technologies already exist for reducing nutrient pollution at reasonable costs—including technologies to reduce point source emissions, changes in crop management practices, and precision farming techniques to help control the application of fertilizers to a field, for example—but new policies are needed for these tools to be applied on a sufficient scale to slow and ultimately reverse the increase in nutrient loading (recognizing that this global goal must be achieved even while increasing nutrient applications in relatively poor regions such as sub-Saharan Africa). Many negative impacts on ecosystems and human well-being have resulted from these technological changes, however. The cost of "retrofitting" technologies once their negative consequences become apparent can be extremely high, so careful assessment is needed prior to the introduction of new technologies.

Promising interventions include:

- *Promotion of technologies that increase crop yields without any harmful impacts related to water, nutrient, and pesticide use* (G, B, N). Agricultural expansion will continue to be one of the major drivers of biodiversity loss well into the twenty-first century. Development, assessment, and diffusion of technologies that could increase the production of food per unit area sustainably without harmful trade-offs related to excessive use of water, nutrients, or pesticides would significantly lessen pressure on other ecosystem services.
- *Restoration of ecosystem services* (G, B, N). Ecosystem restoration activities are now common in many countries and include actions to restore almost all types of ecosystems, including wetlands, forests, grasslands, estuaries, coral reefs, and mangroves. Ecosystems with some features of the ones that were present before conversion can often be established and can provide some of the original ecosystem services (such as pollution filtration in wetlands or timber production from forests). The restored systems seldom fully replace the original systems, but they still help meet needs for particular services. Yet the cost of restoration is generally extremely high in relation to the cost of preventing the degradation of the ecosystem. Not all services can be restored, and those that are heavily degraded may require considerable time for restoration.
- *Promotion of technologies to increase energy efficiency and reduce greenhouse gas emissions* (G, B). Significant reductions in net greenhouse gas emissions are technically feasible due to an extensive array of technologies in the energy supply, energy demand, and waste management sectors. Reducing projected emissions will require a portfolio of energy production technologies ranging from fuel switching (coal/oil to gas) and increased power plant efficiency to increased use of renewable energy technologies, complemented by more efficient use of energy in the transportation, buildings, and industry sectors. [13] It will also involve the development and implementation of supporting institutions and policies to overcome barriers to the diffusion of these technologies into the marketplace, increased public and private-sector funding for research and development, and effective technology transfer.

Knowledge and Cognitive Responses

Effective management of ecosystems is constrained both by a lack of knowledge and information concerning different aspects of ecosystems and by the failure to use adequately the information that does exist in support of management decisions. [2, 14] Although sufficient information exists to take many actions that could help conserve ecosystems and enhance human well-being, major information gaps exist. In most regions, for example, relatively little is known about the status and economic value of most ecosystem services, and their depletion is rarely tracked in national economic accounts.

At the same time, decision-makers do not use all of the relevant information that is available. This is due in part to institutional failures that prevent existing policy-relevant scientific information from being made available to decision-makers. But it is also due to the failure to incorporate other forms of knowledge and information, such as traditional knowledge and practitioners' knowledge, which are of considerable value for ecosystem management.

Promising interventions include:

- *Incorporate both the market and nonmarket values of ecosystems in resource management and investment decisions* (G, B). Most resource management and investment decisions are strongly influenced by considerations of the monetary costs and benefits of alternative policy choices. In the case of ecosystem management, however, this often leads to outcomes that are not in the interest of society, since the nonmarketed values of ecosystems may exceed the marketed values. As a result, many existing resource management policies favor sectors such as agriculture, forestry, and fisheries at the expense of the use of these same ecosystems for water supply, recreation, and cultural services that may be of greater economic value. Decisions can be improved if they include the total economic value of alternative management options and involve deliberative mechanisms that bring to bear noneconomic considerations as well.
- *Use of all relevant forms of knowledge and information in assessments and decision-making, including traditional and practitioners' knowledge* (G, B, N). Effective management of ecosystems typically requires "place-based" knowledge—information about the specific characteristics and history of an ecosystem. Formal scientific information is often one source of such information, but traditional knowledge or practitioners' knowledge held by local resource managers can be of equal or greater value. While that knowledge is used in the decisions taken by those who have it, it is too rarely incorporated into other decision-making processes and is often inappropriately dismissed.
- *Enhance and sustain human and institutional capacity for assessing the consequences of ecosystem change for human well-being and acting on such assessments* (G, B, N). Greater technical capacity is needed for agriculture, forest, and fisheries management. But the capacity that exists for these sectors, as limited as it is in many countries, is still vastly greater than the capacity for effective management of other ecosystem services. Because awareness of the importance of these other services has only recently grown, there is limited experience with assessing ecosystem services fully. Serious limits exist in all countries, but especially in developing countries, in terms of the expertise needed in such areas as monitoring changes in ecosystem services, economic valuation or health assessment of ecosystem changes, and policy analysis related to ecosystem services. Even when such assessment information is available, however, the traditional highly sectoral nature of decision-making and resource management makes the implementation of recommendations difficult. This constraint can also be overcome through increased training of individuals in existing institutions and through institutional reforms to build capacity for more integrated responses.

Appendix R1. Effectiveness of Assessed Responses

A response is considered to be *effective* when its assessment indicates that it has enhanced the particular ecosystem service (or, in the case of biodiversity, its conservation and sustainable use) and contributed to human well-being without significant harm to other ecosystem services or harmful impacts to other groups of people. A response is considered *promising* either if it does not have a long track record to assess but appears likely to succeed or if there are known means of modifying the response so that it can become effective. A response is considered *problematic* if its historical use indicates either that it has not met the goals related to service enhancement (or conservation and sustainable use of biodiversity) or that it has caused significant harm to other ecosystem services. Labeling a response as *effective* does not mean that the historical assessment has not identified problems or harmful trade-offs. Such trade-offs almost always exist, but they are not considered significant enough to negate the effectiveness of the response. Similarly, labeling a response as *problematic* does not mean that there are no promising opportunities to reform the response in a way that can meet its policy goals without undue harm to ecosystem services.

The typology of responses presented here is defined by the nature of intervention, classified as follows: institutional and legal (I), economic and incentives (E), social and behavioral (S), technological (T), and knowledge and cognitive (K). The actors who make decisions to implement a response are governments at different levels, such as international (GI) (mainly through multilateral agreements or international conventions), national (GN), and local (GL); the business/industry sector (B); and civil society, which includes nongovernmental organizations (NGO), community-based and indigenous peoples' organizations (C), and research institutions (R). The actors are not necessarily equally important.

The table includes responses assessed for a range of ecosystem services—food, fresh water, wood, nutrient management, flood and storm control, disease regulation, and cultural services. It also assesses responses for biodiversity conservation, integrated responses, and responses addressing one specific driver: climate change.

Response	Effectiveness			Notes	Type of Response	Required Actors
	Effective	Promising	Problematic			
Biodiversity Conservation and Sustainable Use						
Protected areas	■	■		PAs are extremely important in biodiversity and ecosystem conservation programs, especially in sensitive environments that contain valuable biodiversity components. At global and regional scales, existing PAs are essential but not sufficient to conserve the full range of biodiversity. PAs need to be better located, designed, and managed to ensure representativeness and to deal with the impacts of human settlement within PAs, illegal harvesting, unsustainable tourism, invasive species, and climate change. They also need a landscape approach that includes protection outside of PAs. [5]	I	GI GN GL NGO C R
Helping local people capture biodiversity benefits		■	■	Providing incentives for biodiversity conservation in the form of benefits for local people (e.g., through products from single species or from ecotourism) has proved to be very difficult. Programs have been more successful when local communities have been in a position to make management decisions consistent with overall biodiversity conservation. "Win-win" opportunities for biodiversity conservation and benefits for local communities exist, but local communities can often achieve greater benefits from actions that lead to biodiversity loss. [5]	E	GN GL B NGO C
Promoting better management of wild species as a conservation tool, including ex situ conservation	■			More effective management of individual species should enhance biodiversity conservation and sustainable use. "Habitat-based" approaches are critical, but they cannot replace "species-based" approaches. Zoos, botanical gardens, and other ex situ programs build support for conservation, support valuable research, and provide cultural benefits of biodiversity. [5]	T S	GN C NGO R
Integrating biodiversity into regional planning	■			Integrated regional planning can provide a balance among land uses that promotes effective trade-offs among biodiversity, ecosystem services, and other needs of society. Great uncertainty remains as to what components of biodiversity persist under different management regimes, limiting the current effectiveness of this approach. [5]	I	GN GL NGO
Encouraging private sector involvement in biodiversity conservation		■		Many companies are preparing their own biodiversity action plans, managing their landholdings in ways that are more compatible with biodiversity conservation, supporting certification schemes that promote more sustainable use, and accepting their responsibility for addressing biodiversity issues. The business case that has been made for larger companies needs to be extended to other companies as well. [5]	I	GN B NGO R
Including biodiversity issues in agriculture, forestry, and fisheries		■		More diverse production systems can be as effective as low-diversity systems, or even more effective. Strategies based on more intensive production rather than on the expansion of the area allow for better conservation. [5]	T	GN B

Response	Effectiveness			Notes	Type of Response	Required Actors
	Effective	Promising	Problematic			
Designing governance approaches to support biodiversity		■	■	Decentralization of biodiversity management in many parts of the world has had variable results. The key to success is strong institutions at all levels, with secure tenure and authority at local levels essential to providing incentives for sustainable management. [5]	I	GI GN GL R
Promoting international cooperation through multilateral environmental agreements		■		MEAs should serve as an effective means for international cooperation in the areas of biodiversity conservation and sustainable use. They cover the most pressing drivers and issues related to biodiversity loss. Better coordination between conventions would increase their usefulness. [5,15]	I	GI GN
Environmental education and communication	■	■		Environmental education and communication programs have both informed and changed preferences for biodiversity conservation and have improved implementation of biodiversity responses. Providing the human and financial resources to undertake effective work in this area is a continuing challenge. [5]	S	GN GL NGO C
Food						
Globalization, trade, and domestic and international policies on food		■	■	Government policies related to food production (price supports and various types of payments, or taxes) can have adverse economic, social, and environmental effects. [6]	E	GI GN B
Knowledge and education	■	■		Further research can make food production socially, economically, and environmentally sustainable. Public education should enable consumers to make informed choices about nutritious, safe, and affordable food. [6]	S K	GN GL NGO C
Technological responses, including biotechnology, precision agriculture, and organic farming		■		New agricultural sciences and effective natural resource management could support a new agricultural revolution to meet worldwide food needs. This would help environmental, economic, and social sustainability. [6]	T	GN B R
Water management		■		Emerging water pricing schemes and water markets indicate that water pricing can be a means for efficient allocation and responsible use. [6]	E	GN GL B NGO
Fisheries management		■	■	Strict regulation of marine fisheries is needed, both regarding the establishment and implementation of quotas and steps to address unreported and unregulated harvest. Individual transferable quotas also show promise for cold water, single-species fisheries, but they are unlikely to be useful in multispecies tropical fisheries. Given the potential detrimental environmental impacts of aquaculture, appropriate regulatory mechanisms need to supplement existing policies. [6]	I E	GN GL B NGO

Response	Effectiveness			Notes	Type of Response	Required Actors
	Effective	Promising	Problematic			
Livestock management		■	■	Livestock policies need to be reoriented in view of problems concerning overgrazing and dryland degradation, rangeland fragmentation and loss of wildlife habitat, dust formation, bush encroachment, deforestation, nutrient overload through disposal of manure, and greenhouse gas emissions. Policies also need to focus on human health issues related to diseases such as bird flu and BSE. [6]	T	GN B
Recognition of gender issues		■		Response policies need to be gender-sensitive and designed to empower women and ensure access to and control of resources necessary for food security. This needs to be based on a systematic analysis of gender dynamics and explicit consideration of relationships between gender and food and water security. [6]	S	GN NGO C
Fresh Water						
Determining ecosystem water requirements		■		In order to balance competing demands, it is critical that society explicitly agrees on ecosystem water requirements (environmental flows). [7]	I T	GN GL NGO R
Rights to freshwater services and responsibilities for their provision		■	■	Both public and private ownership systems of fresh water, and of the land resources associated with its provision, have largely failed to create incentives for provision of water services. As a result, upland communities have generally been excluded from access to benefits, particularly when they lack tenure security, and have resisted regulations regarded as unfair. Effective property rights systems with clear and transparent rules can increase stakeholders' confidence that they will have access to the benefits of freshwater services and, therefore, willingness to pay for them. [7]	I	GN B C
Increasing the effectiveness of public participation in decision-making		■		Degradation of freshwater and other ecosystem services has a disproportionate impact on those excluded from participation in decision-making. Key steps for improving participatory processes are to increase the transparency of information, improve the representation of marginalized stakeholders, engage them in the establishment of policy objectives and priorities for the allocation of freshwater services, and create space for deliberation and learning that accommodates multiple perspectives. [7]	I	GN GL NGO C R
River basin organizations		■		RBOs can play an important role in facilitating cooperation and reducing transaction costs of large-scale responses. RBOs are constrained or enabled primarily by the degree of stakeholder participation, their agreement on objectives and management plans, and their cooperation on implementation. [7]	I	GI GN NGO

Response	Effectiveness			Notes	Type of Response	Required Actors
	Effective	Promising	Problematic			
Regulatory responses	■			Regulatory approaches based on market-based incentives (e.g., damages for exceeding pollution standards) are suitable for point-source pollutants. Regulatory approaches that simply outlaw particular types of behavior can be unwieldy and burdensome, and may fail to provide incentives for protecting freshwater services. [7]	I	GN GL
Water markets		■		Economic incentives can potentially unlock significant supply- and demand-side efficiencies while providing cost-effective reallocation between old (largely irrigation) and new (largely municipal and instream) uses. [7]	E	GI GN B
Payments for watershed services		■		Payments for ecosystem services provided by watersheds have narrowly focused on the role of forests in the hydrological regime. They should be based on the entire flow regime, including consideration of the relative values of other land cover and land uses, such as wetlands, riparian areas, steep slopes, roads, and management practices. Key challenges for payment schemes are capacity-building for place-based monitoring and assessment, identifying services in the context of the entire flow regime, considering trade-offs and conflicts among multiple uses, and making uncertainty explicit. [7]	E	GN B C
Partnerships and financing		■	■	There is a clear mismatch between the high social value of freshwater services and the resources allocated to manage water. Insufficient funding for water infrastructure is one manifestation of this. Focusing only on large-scale privatization to improve efficiency and cost-recovery has proven a double-edged strategy—price hikes or control over resources have created controversy and, in some cases, failure and withdrawal. Development of water infrastructure and technologies must observe best practices to avoid problems and inequities. The re-examination and retrofitting/refurbishment of existing infrastructure is the best option in the short and medium term. [7]	I E	GI GN B NGO C
Large dams			■	The impact of large dams on freshwater ecosystems is widely recognized as being more negative than positive. In addition, the benefits of their construction have rarely been shared equitably—the poor and vulnerable and future generations often fail to receive the social and economic benefits from dams. Pre-construction studies are typically overly optimistic about the benefits of projects and underestimate costs. [7]	T	GN
Wetland restoration		■		Although wetland restoration is a promising management approach, there are significant challenges in determining what set of management interventions will produce a desired combination of wetland structure and function. It is unlikely that created wetlands can structurally and functionally replace natural wetlands. [7]	T	GN GL NGO B

Response	Effectiveness: Effective	Effectiveness: Promising	Effectiveness: Problematic	Notes	Type of Response	Required Actors
Wood, Fuelwood, and Non-wood Forest Products						
International forest policy processes and development assistance	■			International forest policy processes have made some gains within the forest sector. Attention should be paid to integration of agreed forest management practices in financial institutions, trade rules, global environment programs, and global security decision-making. [8]	I	GI GN B
Trade liberalization		■	■	Forest product trade tends to concentrate decision-making power on (and benefits from) forest management, rather than spreading it to include poorer and less powerful players. It "magnifies" the effect of governance, making good governance better and bad governance worse. Trade liberalization can stimulate a "virtuous cycle" if the regulatory framework is robust and externalities are addressed. [8]	E	GI GN
National forest governance initiatives and national forest programs		■		Forest governance initiatives and country-led national forest programs show promise for integrating ecosystem health and human well-being where they are negotiated by stakeholders and strategically focused. [8]	I	GN GL
Direct management of forests by indigenous peoples		■		Indigenous control of traditional homelands is often presented as having environmental benefits, although the main justification continues to be based on human and cultural rights. Little systematic data exist, but preliminary findings on vegetation cover and forest fragmentation from the Brazilian Amazon suggest that an indigenous-control area can be at least as effective as a strict-use protected area. [8]	I	GL C
Collaborative forest management and local movements for access and use of forest products		■	■	Government–community collaborative forest management can be highly beneficial but has had mixed results. Programs have generated improved resource management and access of the rural poor to forest resources, but have fallen short in their potential to benefit the poor. Local responses to problems of access and use of forest products have proliferated in recent years. They are collectively more significant than efforts led by governments or international processes but require their support to spread. [8]	I	GN GL B NGO C
Small-scale private and public-private ownership and management of forests		■		Small-scale private ownership of forests can deliver more local economic benefits and better forest management than ownership by larger corporate bodies where information, tenure, and capacity are strong. [8]	I	GL B C
Company–community forestry partnerships		■		Company–community partnerships can be better than solely corporate forestry, or solely community or small-scale farm forestry, in delivering benefits to the partners and the public at large. [8]	I	GL B C

Response	Effectiveness			Notes	Type of Response	Required Actors
	Effective	Promising	Problematic			
Public and consumer action	■			Public and consumer action has resulted in important forest and trade policy initiatives and improved practices in large forest corporations. This has had an impact in "timber-consuming countries" and in international institutions. The operating standards of some large corporations and institutions, as well as of those whose non-forest activities have an impact on forests, have been improved. [8]	S	NGO B C
Third-party voluntary forest certification		■		Forest certification has become widespread; however, most certified forests are in industrial countries, managed by large companies and exporting to Northern retailers. The early proponents of certification hoped it would be an effective response to tropical deforestation. [8]	I E	B
Wood technology and biotechnology		■		Wood technology responses have focused on industrial plantation species with properties suited for manufactured products. [8]	T	GN R B
Commercialization of non-wood forest products		■	■	Commercialization of NWFP has had modest impacts on local livelihoods and has not always created incentives for conservation. An increased value of NWFPs is not always an incentive for conservation and can have the opposite effect. Incentives for sustainable management of NWFPs should be reconsidered, including exploration of joint production of timber and NWFP. [8]	E	NGO B R
Natural forest management in the tropics		■		To be economic, sustainable natural forest management in the tropics must focus on a range of forest goods and services, not just timber. The "best practices" of global corporations should be assessed, exploring at the same time "what works" in traditional forest management and the work of local (small) enterprises. Considerable interest has developed in the application of reduced impact logging, especially in tropical forests, which lowers environmental impacts and can also be more efficient and cost effective. [8]	T	GI GN GL B NGO C
Forest plantation management		■	■	Farm woodlots and large-scale plantations are increasingly being established in response to growing wood demand and declining natural forest areas. Without adequate planning and management, forest plantations can be established in the wrong sites, with the wrong species and provenances. In degraded lands, afforestation may deliver economic, environmental, and social benefits to communities and help in reducing poverty and enhancing food security. [8]	T	GN GL B NGO R
Fuelwood management		■		Fuelwood remains one of the main products of the forest sector in developing countries. If technology development continues, industrial-scale forest product fuels could become a major sustainable energy source. [8]	T	GL B C

Response	Effectiveness			Notes	Type of Response	Required Actors
	Effective	Promising	Problematic			
Afforestation and reforestation for carbon management		■		Although many early initiatives were based on forest conservation or management, afforestation activities now predominate, perhaps reflecting the international decisions in 2001 to allow only afforestation and reforestation activities into the Clean Development Mechanism for the first commitment period. [8]	T E	GI GN B
Nutrient Cycling						
Regulations		■	■	Mandatory policies, including regulatory control and tax or fee systems, place the costs and burden of pollution control on the polluter. Technology-based standards are easy to implement but may discourage innovation and are generally not seen as cost-effective. [9]	I	GI GN
Market-based instruments		■		Market-based instruments, such as financial incentives, subsidies, and taxes, hold potential for better nutrient management, but may not be relevant in all countries and circumstances. Relatively little is known empirically about the impact of these instruments on technological change. [9]	E	GN B R
Hybrid approaches		■		Combinations of regulatory, incentive, and market-based mechanisms are possible for both national and watershed-based approaches and may be the most cost-effective and politically acceptable. [9]	I E	GI GN GL NGO C R
Flood and Storm Regulation						
Physical structures			■	Historically, emphasis was on physical structures/measures over natural environment and social institutions. This choice often creates a false sense of security, encouraging people to accept high risks. Evidence indicates that more emphasis needs to be given to the natural environment and nonstructural measures. [11]	T	GN B
Use of natural environment	■			Flood and storm impacts can be lessened through maintenance and management of vegetation and through natural or human-made geomorphological features (natural river channels, dune systems, terrace farming). [11]	T	GN GL NGO C
Information, institutions, and education	■			These approaches, which emphasize disaster preparedness, disaster management, flood and storm forecasting, early warning, and evacuation, are vital for reducing losses. [11]	S I	GN GL B C
Financial services		■	■	These responses emphasize insurance, disaster relief, and aid. Both social programs and private insurance are important coping mechanisms for flood disaster recovery. They can, however, inadvertently contribute to community vulnerability by encouraging development within floodplains or by creating cultures of entitlement. [11]	E	GN B

Response	Effectiveness: Effective	Effectiveness: Promising	Effectiveness: Problematic	Notes	Type of Response	Required Actors
Land use planning		■		Land use planning is a process of determining the most desirable type of land use. It can help to mitigate disasters and reduce risks by avoiding development in hazard prone areas. [11]	I	GN
Disease Regulation						
Integrated vector management		■		Reducing the transmission of infectious diseases often has effects on other ecosystem services. IVM enables a coordinated response to health and the environment. It uses targeted interventions to remove or control vector breeding sites, disrupt vector lifecycles, and minimize vector-human contact, while minimizing effects on other ecosystem services. IVM is most effective when integrated with socioeconomic development. [12]	I	GN NGO
Environmental management/ modification to reduce vector and reservoir host abundance		■		Environmental management interventions can be highly cost-effective and entail very low environmental impacts. [12]	I	GN B C R
Biological control/natural predators	■			Biological interventions can be highly cost-effective and entail very low environmental impacts. Biological control may be effective if breeding sites are well known and limited in number, but less feasible where these are numerous. [12]	T	GN B R
Chemical control			■	Insecticides remain an important tool and their selective use is likely to continue within IVM. However, there are concerns regarding the impacts of insecticides, especially persistent organic pollutants, on the environment and on human populations, particularly insecticide sprayers. [12]	T	GN B R
Human settlement patterns	■			The most basic management of human-vector contact is through improvements in the placement and construction of housing. [12]	T	GN NGO C
Health awareness and behavior	■			Social and behavioral responses can help control vector-borne disease while also improving other ecosystem services. [12]	S	C
Genetic modification of vector species to limit disease transmission			■	New "cutting-edge" interventions, such as transgenic techniques, could be available within the next 5–10 years. However, consensus is lacking in the scientific community on the technical feasibility and public acceptability of such an approach. [12]	T	GN B NGO R
Cultural Services						
Awareness of the global environment and linking local and global institutions	■			Awareness of the planet working as a system has led to an integrated approach to ecosystems. This process has emphasized the "human environment" concept and the discussion of environmental problems at a global scale. Local organizations also take advantage of emerging global institutions and conventions to bring their case to wider political arenas. [14]	S I	GI GN GL

Response	Effectiveness: Effective	Effectiveness: Promising	Effectiveness: Problematic	Notes	Type of Response	Required Actors
From restoring landscapes to valuing cultural landscapes		■		Landscapes are subject to and influenced by cultural perceptions and political and economic interests. This influences decisions on landscape conservation. [14]	S K	GL NGO C
Recognizing sacred areas	■	■		While linking sacred areas and conservation is not new, there has been an increase in translating "the sacred" into legislation or legal institutions granting land rights. This requires extensive knowledge about the link among the sacred, nature, and society in a specific locale. [14]	S	GL NGO C
International agreements and conservation of biological and agropastoral diversity		■		Increased exploitation and awareness concerning the disappearance of local resources and knowledge has highlighted the need to protect local and indigenous knowledge. Some countries have adopted specific laws, policies, and administrative arrangements emphasizing the concept of prior informed consent of knowledge-holders. [14]	I	GI GN
Integrating local and indigenous knowledge		■	■	Local and indigenous knowledge evolves in specific contexts and good care should be taken to not de-contextualize it. Conventional "best-practices" methods focusing on content may not be appropriate to deal with local/indigenous knowledge. [14]	K I	GN B NGO
Compensating for knowledge		■		Compensation for the use of local and indigenous knowledge by third parties is an important, yet complicated response. The popular idea that local and indigenous knowledge can be promoted by strengthening "traditional" authorities may not be valid in many cases. [14]	E K	GN B C
Property right changes		■	■	Communities benefit from control over natural resources but traditional leadership may not always be the solution. Local government institutions that are democratically elected and have real authority over resources in some cases may be a better option. There is a tendency to shift responsibilities back and forth between "traditional" authorities and local government bodies, without giving any of them real decision-making powers. [14]	I	GN GL C
Certification programs		■		Certification programs are a promising response, but many communities do not have access to these programs or are not aware of their existence. In addition, the financial costs involved reduce the chances for local communities to participate independently. [14]	I S	GI GN B
Fair trade		■		Fair trade is a movement initiated to help disadvantaged or politically marginalized communities by paying better prices and providing better trading conditions, along with raising consumers' awareness of their potential role as buyers. Fair trade overlaps in some cases with initiatives focusing on the environmental performance of trade. [14]	E S	GI GN GL NGO C

Response	Effectiveness: Effective	Effectiveness: Promising	Effectiveness: Problematic	Notes	Type of Response	Required Actors
Ecotourism and cultural tourism		■		Ecotourism can provide economic alternatives to converting ecosystems; however, it can generate conflicts in resource use and the aesthetics of certain ecosystems. Different ecosystems are subjected to different types and scales of impact from tourism infrastructure. Furthermore, some ecosystems are easier to market to tourists than others. The market value of ecosystems may vary according to public perceptions of nature. Freezing of landscapes, conversion of landscapes, dispossession, and removing of human influences may result, depending on views of what ecotourism should represent. Yet when conservation receives no budgetary subsidy, tourism can provide revenues for conservation. [14]	E	GL B C
Integrated Responses						
International environmental governance		■		Environmental policy integration at the international level is almost exclusively dependent on governments' commitment to binding compromises on given issues. Major challenges include reform of the international environmental governance structure and coherence among international trade and environment mechanisms. [15]	I E K T B	GI GN
National action plans and strategies aiming to integrate environmental issues into national policies		■		Examples include national conservation strategies, national environmental action plans, and national strategies for sustainable development. Success depends on enabling conditions such as ownership by governments and civil society, broad participation, both across sectors within the government and with the private sector, and at the sub-national and local scales. National integrated responses may be a good starting point for cross-departmental linkages in governments. [15]	I E K T	GN GL B NGO C
Sub-national and local integrated approaches	■	■		Many integrated responses are implemented at the sub-national level; examples include sustainable forest management, integrated coastal zone management, integrated conservation and development programs, and integrated river basin management. Results so far have been varied, and a major constraint experienced by sub-national and multiscale responses is the lack of implementation capacity. [15]	I E K T	GN GL NGO C
Waste Management						
Technologies for waste reduction, re-use, recovery, and disposal	■			These practices have enhanced ecosystem services, improved aesthetic conditions, restored habitats for human use and for biodiversity, increased public health and well-being, created jobs, and reduced poverty. [10]	T	GN GL B C
Compliance with waste management laws and regulations		■		Communities and industries are willing to comply with laws and regulations if there is clear understanding of the benefits and if all stakeholders are involved in the formulation of such laws. [10]	L	GN GL

Response	Effectiveness			Notes	Type of Response	Required Actors
	Effective	Promising	Problematic			
Environmental awareness and education		■		Environmental awareness and education have succeeded in allowing consumers and resource users to make informed choices for minimizing waste. Employers have introduced programs to encourage communities to reduce waste. [10]	S	GL C B
Indicators and monitoring		■	■	Industries and governments need to select indicators and standardize methods to monitor the sources, types, and amounts of all wastes produced. The practice of transparent, participatory, and accountable decision-making for ecosystem sustainability and human well-being is lacking in many countries. [10]	S	GN B NGO
Climate Change						
U.N. Framework Convention on Climate Change and Kyoto Protocol		■	■	The ultimate goal of the UNFCCC is stabilization of greenhouse gas concentrations in the atmosphere at a level that would prevent dangerous anthropogenic interference with the climate system. The Kyoto Protocol contains binding limits on greenhouse gas emissions on industrialized countries that agreed to reduce their emissions by an average of about 5% between 2008 and 2012 relative to the levels emitted in 1990. [13]	I	GI GN
Reductions in net greenhouse gas emissions		■		Significant reductions in net greenhouse gas emissions are technically feasible, in many cases at little or no cost to society. [13]	T	GN B C
Land use and land cover change		■		Afforestation; reforestation; improved forest, cropland, and rangeland management; and agroforestry provide opportunities to increase carbon uptake, and slowing deforestation reduces emissions. [13]	T	GN GL B NGO C
Market mechanisms and incentives		■		The Kyoto Protocol mechanisms, in combination with national and regional ones, can reduce the costs of mitigation for developed countries. In addition, countries can reduce net costs of emissions abatement by taxing emissions (or auctioning permits) and using the revenues to cut distortion taxes on labor and capital. In the near term, project-based trading can facilitate the transfer of climate-friendly technologies to developing countries. [13]	E	GI GN B
Adaptation		■		Some climate change is inevitable and ecosystems and human societies will need to adapt to new conditions. Human populations will face the risk of damage from climate change, some of which may be countered with current coping systems; others may need radically new behaviors. Climate change needs to be factored into current development plans. [13]	I	GN GL NGO C R

Part I

Framework for Evaluating Responses

Chapter 1

MA Conceptual Framework

BOXES

FIGURES

This chapter provides the summary of Millennium Ecosystem Assessment, *Ecosystems and Human Well-being: A Framework for Assessment* (Island Press, 2003), pp. 1–25, which was prepared by an extended conceptual framework writing team of 51 authors and 10 contributing authors.

Main Messages

Human well-being and progress toward sustainable development are vitally dependent upon improving the management of Earth's ecosystems to ensure their conservation and sustainable use. But while demands for ecosystem services such as food and clean water are growing, human actions are at the same time diminishing the capability of many ecosystems to meet these demands.

Sound policy and management interventions can often reverse ecosystem degradation and enhance the contributions of ecosystems to human well-being, but knowing when and how to intervene requires substantial understanding of both the ecological and the social systems involved. Better information cannot guarantee improved decisions, but it is a prerequisite for sound decision-making.

The Millennium Ecosystem Assessment was established to help provide the knowledge base for improved decisions and to build capacity for analyzing and supplying this information.

This chapter presents the conceptual and methodological approach that the MA used to assess options that can enhance the contribution of ecosystems to human well-being. This same approach should provide a suitable basis for governments, the private sector, and civil society to factor considerations of ecosystems and ecosystem services into their own planning and actions.

1.1 Introduction

Humanity has always depended on the services provided by the biosphere and its ecosystems. Further, the biosphere is itself the product of life on Earth. The composition of the atmosphere and soil, the cycling of elements through air and waterways, and many other ecological assets are all the result of living processes—and all are maintained and replenished by living ecosystems. The human species, while buffered against environmental immediacies by culture and technology, is ultimately fully dependent on the flow of ecosystem services.

In his April 2000 Millennium Report to the United Nations General Assembly, in recognition of the growing burden that degraded ecosystems are placing on human well-being and economic development and the opportunity that better managed ecosystems provide for meeting the goals of poverty eradication and sustainable development, United Nations Secretary-General Kofi Annan stated that:

> *It is impossible to devise effective environmental policy unless it is based on sound scientific information. While major advances in data collection have been made in many areas, large gaps in our knowledge remain. In particular, there has never been a comprehensive global assessment of the world's major ecosystems. The planned Millennium Ecosystem Assessment, a major international collaborative effort to map the health of our planet, is a response to this need.*

The Millennium Ecosystem Assessment was established with the involvement of governments, the private sector, nongovernmental organizations, and scientists to provide an integrated assessment of the consequences of ecosystem change for human well-being and to analyze options available to enhance the conservation of ecosystems and their contributions to meeting human needs. The Convention on Biological Diversity, the Convention to Combat Desertification, the Convention on Migratory Species, and the Ramsar Convention on Wetlands plan to use the findings of the MA, which will also help meet the needs of others in government, the private sector, and civil society. The MA should help to achieve the United Nations Millennium Development Goals and to carry out the Plan of Implementation of the 2002 World Summit on Sustainable Development. It has mobilized hundreds of scientists from countries around the world to provide information and clarify science concerning issues of greatest relevance to decision-makers. The MA has identified areas of broad scientific agreement and also pointed to areas of continuing scientific debate.

The assessment framework developed for the MA offers decision-makers a mechanism to:

- ***Identify options that can better achieve core human development and sustainability goals. All countries and communities are grappling with the challenge of meeting growing demands for food, clean water, health, and employment.*** And decision-makers in the private and public sectors must also balance economic growth and social development with the need for environmental conservation. All of these concerns are linked directly or indirectly to the world's ecosystems. The MA process, at all scales, was designed to bring the best science to bear on the needs of decision-makers concerning these links between ecosystems, human development, and sustainability.
- ***Better understand the trade-offs involved—across sectors and stakeholders—in decisions concerning the environment.*** Ecosystem-related problems have historically been approached issue by issue, but rarely by pursuing multisectoral objectives. This approach has not withstood the test of time. Progress toward one objective such as increasing food production has often been at the cost of progress toward other objectives such as conserving biological diversity or improving water quality. The MA framework complements sectoral assessments with information on the full impact of potential policy choices across sectors and stakeholders.
- ***Align response options with the level of governance where they can be most effective.*** Effective management of ecosystems will require actions at all scales, from the local to the global. Human actions now directly or inadvertently affect virtually all of the world's ecosystems; actions required for the management of ecosystems refer to the steps that humans can take to modify their direct or indirect influences on ecosystems. The management and policy options available and the concerns of stakeholders differ greatly across these scales. The priority areas for biodiversity conservation in a country as defined based on "global" value, for example, would be very different from those as defined based on the value to local communities. The multiscale assessment framework developed for the MA provides a new approach for analyzing policy options at all scales—from local communities to international conventions.

1.2 What Is the Problem?

Ecosystem services are the benefits people obtain from ecosystems, which the MA describes as provisioning, regulating, supporting, and cultural services. (See Box 1.1.) Ecosystem services include products such as food, fuel, and fiber; regulating services such as climate regulation and disease control; and nonmaterial benefits such as spiritual or aesthetic benefits. Changes in these services affect human well-being in many ways. (See Figure 1.1.)

The demand for ecosystem services is now so great that trade-offs among services have become the rule. A country can increase food supply by converting a forest to agriculture, for example, but

BOX 1.1

Key Definitions

Ecosystem. An ecosystem is a dynamic complex of plant, animal, and microorganism communities and the nonliving environment interacting as a functional unit. Humans are an integral part of ecosystems. Ecosystems vary enormously in size; a temporary pond in a tree hollow and an ocean basin can both be ecosystems.

Ecosystem services. Ecosystem services are the benefits people obtain from ecosystems. These include provisioning services such as food and water; regulating services such as regulation of floods, drought, land degradation, and disease; supporting services such as soil formation and nutrient cycling; and cultural services such as recreational, spiritual, religious and other nonmaterial benefits.

Well-being. Human well-being has multiple constituents, including basic material for a good life, freedom of choice and action, health, good social relations, and security. Well-being is at the opposite end of a continuum from poverty, which has been defined as a "pronounced deprivation in well-being." The constituents of well-being, as experienced and perceived by people, are situation-dependent, reflecting local geography, culture, and ecological circumstances.

in so doing it decreases the supply of services that may be of equal or greater importance, such as clean water, timber, ecotourism destinations, or flood regulation and drought control. There are many indications that human demands on ecosystems will grow still greater in the coming decades. Current estimates of 3 billion more people and a quadrupling of the world economy by 2050 imply a formidable increase in demand for and consumption of biological and physical resources, as well as escalating impacts on ecosystems and the services they provide.

The problem posed by the growing demand for ecosystem services is compounded by increasingly serious degradation in the capability of ecosystems to provide these services. World fisheries are now declining due to overfishing, for instance, and a significant amount of agricultural land has been degraded in the past half-century by erosion, salinization, compaction, nutrient depletion, pollution, and urbanization. Other human-induced impacts on ecosystems include alteration of the nitrogen, phosphorous, sulfur, and carbon cycles, causing acid rain, algal blooms, and fish kills in rivers and coastal waters, along with contributions to climate change. In many parts of the world, this degradation of ecosystem services is exacerbated by the associated loss of the knowledge and understanding held by local communities—knowledge that sometimes could help to ensure the sustainable use of the ecosystem.

This combination of ever-growing demands being placed on increasingly degraded ecosystems seriously diminishes the prospects for sustainable development. Human well-being is affected not just by gaps between ecosystem service supply and demand but also by the increased vulnerability of individuals, communities, and nations. Productive ecosystems, with their array of services, provide people and communities with resources and options they can use as insurance in the face of natural catastrophes or social upheaval. While well-managed ecosystems reduce risks and vulnerability, poorly managed systems can exacerbate them by increasing risks of flood, drought, crop failure, or disease.

Ecosystem degradation tends to harm rural populations more directly than urban populations and has its most direct and severe impact on poor people. The wealthy control access to a greater share of ecosystem services, consume those services at a higher per capita rate, and are buffered from changes in their availability (often at a substantial cost) through their ability to purchase scarce ecosystem services or substitutes. For example, even though a number of marine fisheries have been depleted in the past century, the supply of fish to wealthy consumers has not been disrupted since fishing fleets have been able to shift to previously underexploited stocks. In contrast, poor people often lack access to alternate services and are highly vulnerable to ecosystem changes that result in famine, drought, or floods. They frequently live in locations particularly sensitive to environmental threats, and they lack financial and institutional buffers against these dangers. Degradation of coastal fishery resources, for instance, results in a decline in protein consumed by the local community since fishers may not have access to alternate sources of fish and community members may not have enough income to purchase fish. Degradation affects their very survival.

Changes in ecosystems affect not just humans but countless other species as well. The management objectives that people set for ecosystems and the actions that they take are influenced not just by the consequences of ecosystem changes for humans but also by the importance people place on considerations of the intrinsic value of species and ecosystems. Intrinsic value is the value of something in and for itself, irrespective of its utility for someone else. For example, villages in India protect "spirit sanctuaries" in relatively natural states, even though a strict cost-benefit calculation might favor their conversion to agriculture. Similarly, many countries have passed laws protecting endangered species based on the view that these species have a right to exist, even if their protection results in net economic costs. Sound ecosystem management thus involves steps to address the utilitarian links of people to ecosystems as well as processes that allow considerations of the intrinsic value of ecosystems to be factored into decision-making.

The degradation of ecosystem services has many causes, including excessive demand for ecosystem services stemming from economic growth, demographic changes, and individual choices. Market mechanisms do not always ensure the conservation of ecosystem services either because markets do not exist for services such as cultural or regulatory services or, where they do exist, because policies and institutions do not enable people living within the ecosystem to benefit from services it may provide to others who are far away. For example, institutions are now only beginning to be developed to enable those benefiting from carbon sequestration to provide local managers with an economic incentive to leave a forest uncut, while strong economic incentives often exist for managers to harvest the forest. Also, even if a market exists for an ecosystem service, the results obtained through the market may be socially or ecologically undesirable. Properly managed, the creation of ecotourism opportunities in a country can create strong economic incentives for the maintenance of the cultural services provided by ecosystems, but poorly managed ecotourism activities can degrade the very resource on which they depend. Finally, markets are often unable to address important intra- and intergenerational equity issues associated with managing ecosystems for this and future generations, given that some changes in ecosystem services are irreversible.

The world has witnessed in recent decades not just dramatic changes to ecosystems but equally profound changes to social systems that shape both the pressures on ecosystems and the opportunities to respond. The relative influence of individual nation-states has diminished with the growth of power and influence of a far more complex array of institutions, including regional

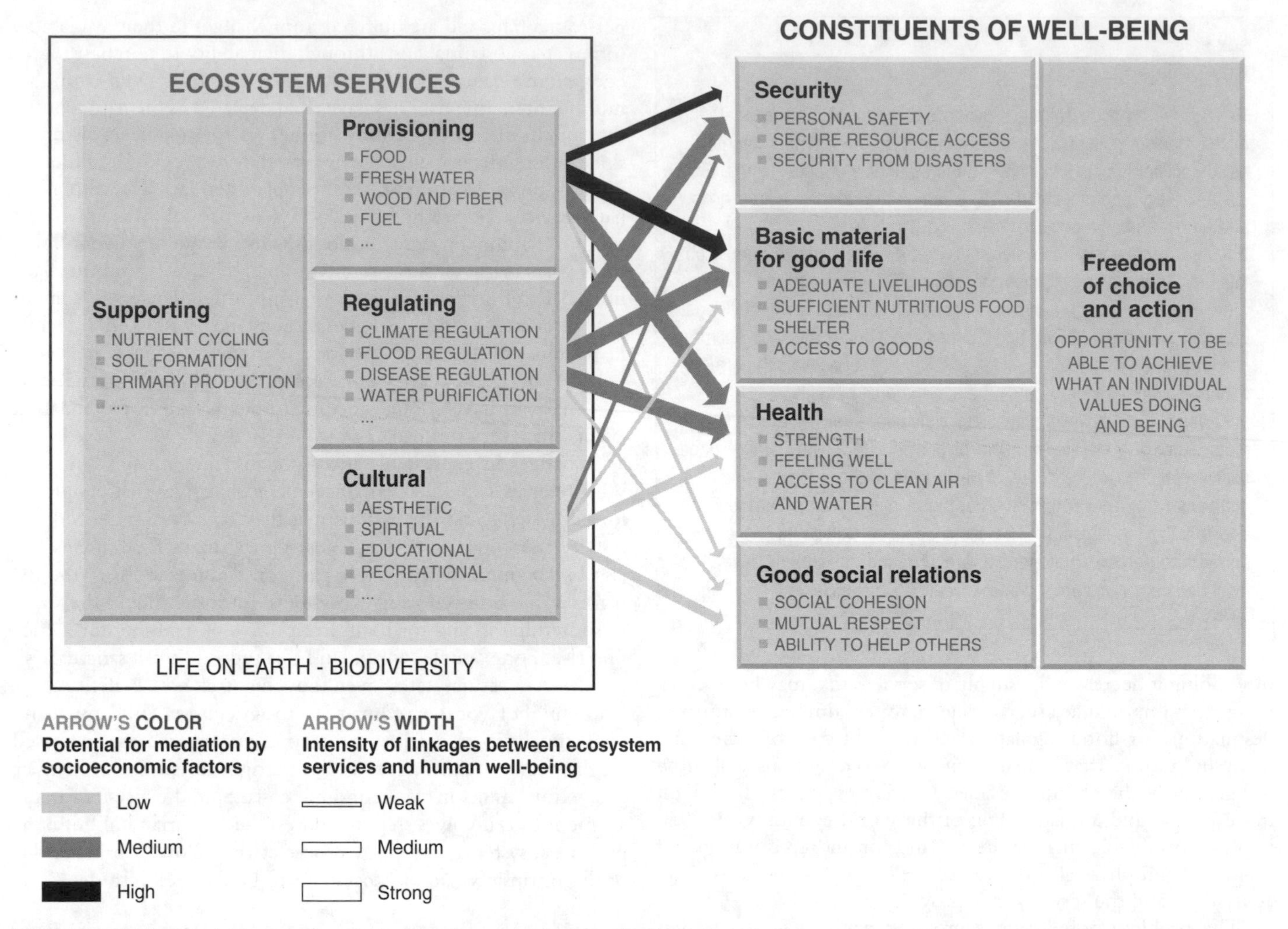

Figure 1.1. Linkages between Ecosystem Services and Human Well-being. This Figure depicts the strength of linkages between categories of ecosystem services and components of human well-being that are commonly encountered and includes indications of the extent to which it is possible for socioeconomic factors to mediate the linkage. (For example, if it is possible to purchase a substitute for a degraded ecosystem service, then there is a high potential for mediation.) The strength of the linkages and the potential for mediation differ in different ecosystems and regions. In addition to the influence of ecosystem services on human well-being depicted here, other factors—including other environmental factors as well as economic, social, technological, and cultural factors—influence human well-being, and ecosystems are in turn affected by changes in human well-being. (Millennium Ecosystem Assessment)

governments, multinational companies, the United Nations, and civil society organizations. Stakeholders have become more involved in decision-making. Given the multiple actors whose decisions now strongly influence ecosystems, the challenge of providing information to decision-makers has grown. At the same time, the new institutional landscape may provide an unprecedented opportunity for information concerning ecosystems to make a major difference. Improvements in ecosystem management to enhance human well-being will require new institutional and policy arrangements and changes in rights and access to resources that may be more possible today under these conditions of rapid social change than they have ever been before.

Like the benefits of increased education or improved governance, the protection, restoration, and enhancement of ecosystem services tends to have multiple and synergistic benefits. Already, many governments are beginning to recognize the need for more effective management of these basic life-support systems. Examples of significant progress toward sustainable management of biological resources can also be found in civil society, in indigenous and local communities, and in the private sector.

1.3 Conceptual Framework

The conceptual framework for the MA places human well-being as the central focus for assessment, while recognizing that biodiversity and ecosystems also have intrinsic value and that people take decisions concerning ecosystems based on considerations of well-being as well as intrinsic value. (See Box 1.2.) The MA conceptual framework assumes that a dynamic interaction exists between people and other parts of ecosystems, with the changing human condition serving to both directly and indirectly drive change in ecosystems and with changes in ecosystems causing changes in human well-being. At the same time, many other factors independent of the environment change the human condition, and many natural forces are influencing ecosystems.

The MA focuses particular attention on the linkages between ecosystem services and human well-being. The assessment deals with the full range of ecosystems—from those relatively undisturbed, such as natural forests, to landscapes with mixed patterns of human use and ecosystems intensively managed and modified by humans, such as agricultural land and urban areas.

A full assessment of the interactions between people and ecosystems requires a multiscale approach because it better reflects the multiscale nature of decision-making, allows the examination of driving forces that may be exogenous to particular regions, and provides a means of examining the differential impact of ecosystem changes and policy responses on different regions and groups within regions.

This section explains in greater detail the characteristics of each of the components of the MA conceptual framework, moving clockwise from the lower left corner of the Figure in Box 1.2.

1.3.1 Ecosystems and Their Services

An ecosystem is a dynamic complex of plant, animal, and microorganism communities and the nonliving environment interacting as a functional unit. Humans are an integral part of ecosystems. Ecosystems provide a variety of benefits to people, including provisioning, regulating, cultural, and supporting services. Provisioning services are the products people obtain from ecosystems, such as food, fuel, fiber, fresh water, and genetic resources. Regulating services are the benefits people obtain from the regulation of ecosystem processes, including air quality maintenance, climate regulation, erosion control, regulation of human diseases, and water purification. Cultural services are the nonmaterial benefits people obtain from ecosystems through spiritual enrichment, cognitive development, reflection, recreation, and aesthetic experiences. Supporting services are those that are necessary for the production of all other ecosystem services, such as primary production, production of oxygen, and soil formation.

Biodiversity and ecosystems are closely related concepts. Biodiversity is the variability among living organisms from all sources, including terrestrial, marine, and other aquatic ecosystems and the ecological complexes of which they are part. It includes diversity within and between species and diversity of ecosystems. Diversity is a structural feature of ecosystems, and the variability among ecosystems is an element of biodiversity. Products of biodiversity include many of the services produced by ecosystems (such as food and genetic resources), and changes in biodiversity can influence all the other services they provide. In addition to the important role of biodiversity in providing ecosystem services, the diversity of living species has intrinsic value independent of any human concern.

The concept of an ecosystem provides a valuable framework for analyzing and acting on the linkages between people and the environment. For that reason, the "ecosystem approach" has been endorsed by the Convention on Biological Diversity, and the MA conceptual framework is entirely consistent with this approach. The CBD states that the ecosystem approach is a strategy for the integrated management of land, water, and living resources that promotes conservation and sustainable use in an equitable way. This approach recognizes that humans, with their cultural diversity, are an integral component of many ecosystems.

In order to implement the ecosystem approach, decision-makers need to understand the multiple effects on an ecosystem of any management or policy change. By way of analogy, decision-makers would not make a decision about financial policy in a country without examining the condition of the economic system, since information on the economy of a single sector such as manufacturing would be insufficient. The same need to examine the consequences of changes for multiple sectors applies to ecosystems. For instance, subsidies for fertilizer use may increase food production, but sound decisions also require information on whether the potential reduction in the harvests of downstream fisheries as a result of water quality degradation from the fertilizer runoff might outweigh those benefits.

For the purpose of analysis and assessment, a pragmatic view of ecosystem boundaries must be adopted, depending on the questions being asked. A well-defined ecosystem has strong interactions among its components and weak interactions across its boundaries. A useful choice of ecosystem boundary is one where a number of discontinuities coincide, such as in the distribution of organisms, soil types, drainage basins, and depth in a waterbody. At a larger scale, regional and even globally distributed ecosystems can be evaluated based on a commonality of basic structural units. The global assessment being undertaken by the MA reports on marine, coastal, inland water, forest, dryland, island, mountain, polar, cultivated, and urban regions. These regions are not ecosystems themselves, but each contains a number of ecosystems. (See Box 1.3.)

People seek multiple services from ecosystems and thus perceive the condition of given ecosystems in relation to their ability to provide the services desired. Various methods can be used to assess the ability of ecosystems to deliver particular services. With those answers in hand, stakeholders have the information they need to decide on a mix of services best meeting their needs. The MA considers criteria and methods to provide an integrated view of the condition of ecosystems. The condition of each category of ecosystem services is evaluated in somewhat different ways, although in general a full assessment of any service requires considerations of stocks, flows, and resilience of the service.

1.3.2 Human Well-being and Poverty Reduction

Human well-being has multiple constituents, including the basic material for a good life, freedom of choice and action, health, good social relations, and security. Poverty is also multidimensional and has been defined as the pronounced deprivation of well-being. How well-being, ill-being, or poverty are experienced and expressed depends on context and situation, reflecting local physical, social, and personal factors such as geography, environment, age, gender, and culture. In all contexts, however, ecosystems are essential for human well-being through their provisioning, regulating, cultural, and supporting services.

Human intervention in ecosystems can amplify the benefits to human society. However, evidence in recent decades of escalating human impacts on ecological systems worldwide raises concerns about the spatial and temporal consequences of ecosystem changes detrimental to human well-being. Ecosystem changes affect human well-being in the following ways:

- ***Security*** is affected both by changes in provisioning services, which affect supplies of food and other goods and the likelihood of conflict over declining resources, and by changes in regulating services, which could influence the frequency and magnitude of floods, droughts, landslides, or other catastrophes. It can also be affected by changes in cultural services as, for example, when the loss of important ceremonial or spiritual attributes of ecosystems contributes to the weakening of social relations in a community. These changes in turn affect material well-being, health, freedom and choice, security, and good social relations.
- ***Access to basic material for a good life*** is strongly linked to both provisioning services such as food and fiber production and regulating services, including water purification.
- ***Health*** is strongly linked to both provisioning services such as food production and regulating services, including those that influence the distribution of disease-transmitting insects and of irritants and pathogens in water and air. Health can also be

BOX 1.2
Millennium Ecosystem Assessment Conceptual Framework

Changes in factors that indirectly affect ecosystems, such as population, technology, and lifestyle (upper right corner of figure), can lead to changes in factors directly affecting ecosystems, such as the catch of fisheries or the application of fertilizers to increase food production (lower right corner). The resulting changes in the ecosystem (lower left corner) cause the ecosystem services to change and thereby affect human well-being. These interactions can take place at more than one scale and can cross scales. For example, a global market may lead to regional loss of forest cover, which increases flood magnitude along a local stretch of a river. Similarly, the interactions can take place across different time scales. Actions can be taken either to respond to negative changes or to enhance positive changes at almost all points in this framework (black cross bars).

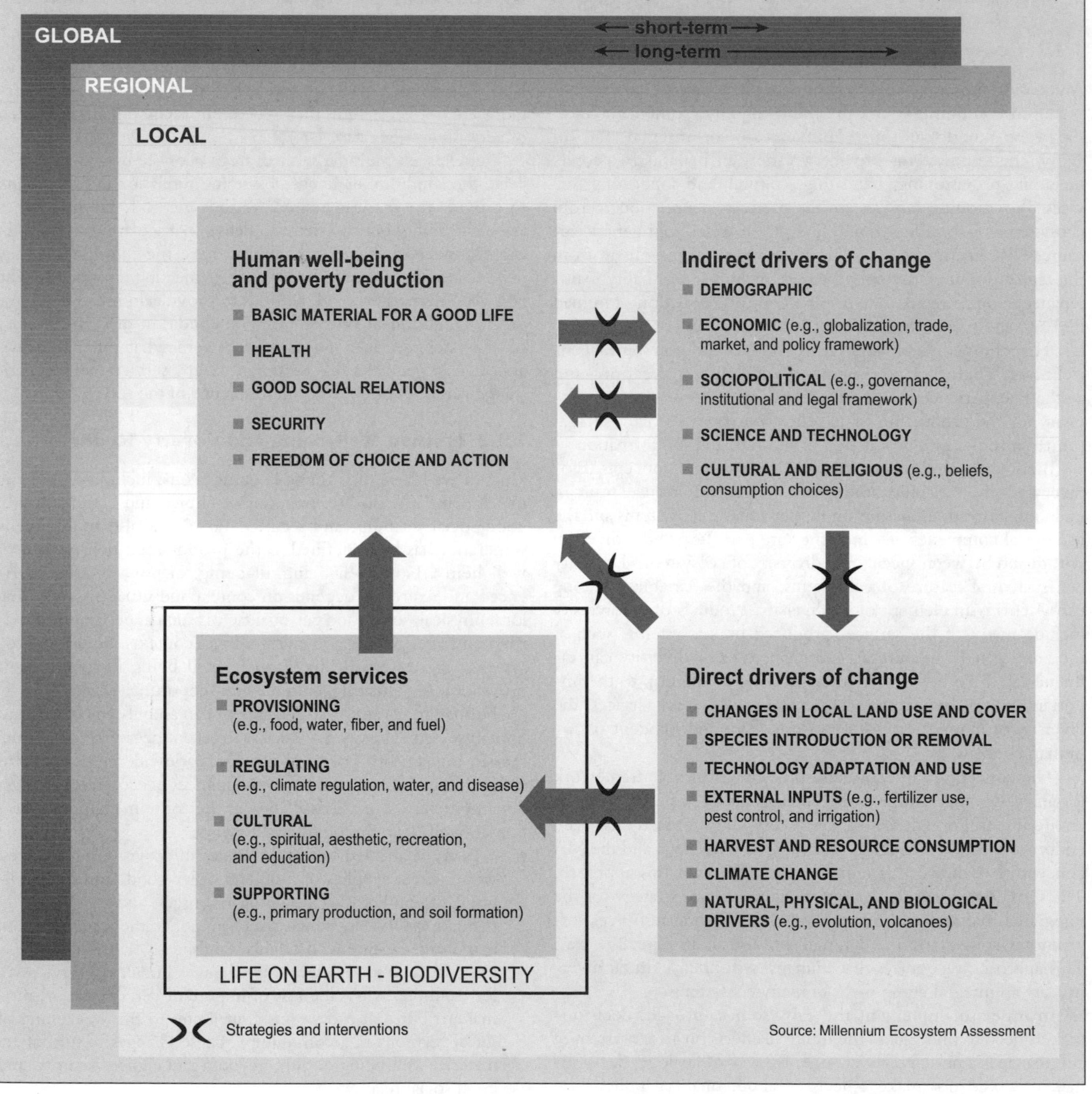

linked to cultural services through recreational and spiritual benefits.

- ***Social relations*** are affected by changes to cultural services, which affect the quality of human experience.
- ***Freedom of choice and action*** is largely predicated on the existence of the other components of well-being and are thus influenced by changes in provisioning, regulating, or cultural services from ecosystems.

Human well-being can be enhanced through sustainable human interactions with ecosystems supported by necessary instruments, institutions, organizations, and technology. Creation of these through participation and transparency may contribute to

BOX 1.3

Reporting Categories Used in the Millennium Ecosystem Assessment

The MA used 10 categories of systems to report its global findings. (See Table.) These categories are not ecosystems themselves; each contains a number of ecosystems. The MA reporting categories are not mutually exclusive: their areas can and do overlap. Ecosystems within each category share a suite of biological, climatic, and social factors that tend to differ across categories. Because these reporting categories overlap, any place on Earth may fall into more than one category. Thus, for example, a wetland ecosystem in a coastal region may be examined both in the MA analysis of "coastal systems" as well as in its analysis of "inland water systems."

Millennium Ecosystem Assessment Reporting Categories

Category	Central Concept	Boundary Limits for Mapping
Marine	Ocean, with fishing typically a major driver of change	Marine areas where the sea is deeper than 50 meters.
Coastal	Interface between ocean and land, extending seawards to about the middle of the continental shelf and inland to include all areas strongly influenced by the proximity to the ocean	Area between 50 meters below mean sea level and 50 meters above the high tide level or extending landward to a distance 100 kilometers from shore. Includes coral reefs, intertidal zones, estuaries, coastal aquaculture, and seagrass communities.
Inland water	Permanent water bodies inland from the coastal zone, and areas whose ecology and use are dominated by the permanent, seasonal, or intermittent occurrence of flooded conditions	Rivers, lakes, floodplains, reservoirs, and wetlands; includes inland saline systems. Note that the Ramsar Convention considers "wetlands" to include both inland water and coastal categories.
Forest	Lands dominated by trees; often used for timber, fuelwood, and non-timber forest products	A canopy cover of at least 40% by woody plants taller than 5 meters. The existence of many other definitions is acknowledged, and other limits (such as crown cover greater than 10%, as used by the Food and Agriculture Organization of the United Nations) are also reported. Includes temporarily cut-over forests and plantations; excludes orchards and agroforests where the main products are food crops.
Dryland	Lands where plant production is limited by water availability; the dominant uses are large mammal herbivory, including livestock grazing, and cultivation	Drylands as defined by the Convention to Combat Desertification, namely lands where annual precipitation is less than two thirds of potential evaporation, from dry subhumid areas (ratio ranges 0.50–0.65), through semiarid, arid, and hyper-arid (ratio $<$0.05), but excluding polar areas; drylands include cultivated lands, scrublands, shrublands, grasslands, semi-deserts, and true deserts.
Island	Lands isolated by surrounding water, with a high proportion of coast to hinterland	Islands of at least 1.5 hectares included in the ESRI ArcWorld Country Boundary dataset.
Mountain	Steep and high lands	As defined by Mountain Watch using criteria based on elevation alone, and at lower elevation, on a combination of elevation, slope, and local elevation range. Specifically, elevation $>$2,500 meters, elevation 1,500–2,500 meters and slope $>$2 degrees, elevation 1,000–1,500 meters and slope $>$5 degrees or local elevation range (7 kilometers radius) $>$300 meters, elevation 300–1,000 meters and local elevation range (7 kilometers radius) $>$300 meters, isolated inner basins and plateaus less than 25 square kilometers extent that are surrounded by mountains.
Polar	High-latitude systems frozen for most of the year	Includes ice caps, areas underlain by permafrost, tundra, polar deserts, and polar coastal areas. Excludes high-altitude cold systems in low latitudes.
Cultivated	Lands dominated by domesticated plant species, used for and substantially changed by crop, agroforestry, or aquaculture production	Areas in which at least 30% of the landscape comes under cultivation in any particular year. Includes orchards, agroforestry, and integrated agriculture-aquaculture systems.
Urban	Built environments with a high human density	Known human settlements with a population of 5,000 or more, with boundaries delineated by observing persistent night-time lights or by inferring areal extent in the cases where such observations are absent.

freedoms and choice as well as to increased economic, social, and ecological security. By ecological security, we mean the minimum level of ecological stock needed to ensure a sustainable flow of ecosystem services.

Yet the benefits conferred by institutions and technology are neither automatic nor equally shared. In particular, such opportunities are more readily grasped by richer than poorer countries and people; some institutions and technologies mask or exacerbate environmental problems; responsible governance, while essential, is not easily achieved; participation in decision-making, an essential element of responsible governance, is expensive in time and resources to maintain. Unequal access to ecosystem services has often elevated the well-being of small segments of the population at the expense of others.

Sometimes the consequences of the depletion and degradation of ecosystem services can be mitigated by the substitution of knowledge and of manufactured or human capital. For example, the addition of fertilizer in agricultural systems has been able to offset declining soil fertility in many regions of the world where people have sufficient economic resources to purchase these inputs, and water treatment facilities can sometimes substitute for the role of watersheds and wetlands in water purification. But ecosystems are complex and dynamic systems and there are limits to substitution possibilities, especially with regulating, cultural, and supporting services. No substitution is possible for the extinction of culturally important species such as tigers or whales, for instance, and substitutions may be economically impractical for the loss of services such as erosion control or climate regulation. Moreover, the scope for substitutions varies by social, economic, and cultural conditions. For some people, especially the poorest, substitutes and choices are very limited. For those who are better off, substitution may be possible through trade, investment, and technology.

Because of the inertia in both ecological and human systems, the consequences of ecosystem changes made today may not be felt for decades. Thus, sustaining ecosystem services, and thereby human well-being, requires a full understanding and wise management of the relationships between human activities, ecosystem change, and well-being over the short, medium, and long term. Excessive current use of ecosystem services compromises their future availability. This can be prevented by ensuring that the use is sustainable.

Achieving sustainable use requires effective and efficient institutions that can provide the mechanisms through which concepts of freedom, justice, fairness, basic capabilities, and equity govern the access to and use of ecosystem services. Such institutions may also need to mediate conflicts between individual and social interests that arise.

The best way to manage ecosystems to enhance human well-being will differ if the focus is on meeting needs of the poor and weak or the rich and powerful. For both groups, ensuring the long-term supply of ecosystem services is essential. But for the poor, an equally critical need is to provide more equitable and secure access to ecosystem services.

1.3.3 Drivers of Change

Understanding the factors that cause changes in ecosystems and ecosystem services is essential to designing interventions that capture positive impacts and minimize negative ones. In the MA, a "driver" is any factor that changes an aspect of an ecosystem. A direct driver unequivocally influences ecosystem processes and can therefore be identified and measured to differing degrees of accuracy. An indirect driver operates more diffusely, often by altering one or more direct drivers, and its influence is established by understanding its effect on a direct driver. Both indirect and direct drivers often operate synergistically. Changes in land cover, for example, can increase the likelihood of introduction of alien invasive species. Similarly, technological advances can increase rates of economic growth.

The MA explicitly recognizes the role of decision-makers who affect ecosystems, ecosystem services, and human well-being. Decisions are made at three organizational levels, although the distinction between those levels is often diffuse and difficult to define:

- by individuals and small groups at the local level (such as a field or forest stand) who directly alter some part of the ecosystem;
- by public and private decision-makers at the municipal, provincial, and national levels; and
- by public and private decision-makers at the international level, such as through international conventions and multilateral agreements.

The decision-making process is complex and multidimensional. We refer to a driver that can be influenced by a decision-maker as an endogenous driver and one over which the decision-maker does not have control as an exogenous driver. The amount of fertilizer applied on a farm is an endogenous driver from the standpoint of the farmer, for example, while the price of the fertilizer is an exogenous driver, since the farmer's decisions have little direct influence on price. The specific temporal, spatial, and organizational scale dependencies of endogenous and exogenous drivers and the specific linkages and interactions among drivers are assessed in the MA.

Whether a driver is exogenous or endogenous to a decision-maker is dependent upon the spatial and temporal scale. For example, a local decision-maker can directly influence the choice of technology, changes in land use, and external inputs (such as fertilizers or irrigation), but has little control over prices and markets, property rights, technology development, or the local climate. In contrast, a national or regional decision-maker has more control over many factors, such as macroeconomic policy, technology development, property rights, trade barriers, prices, and markets. But on the short time scale, that individual has little control over the climate or global population. On the longer time scale, drivers that are exogenous to a decision-maker in the short run, such as population, become endogenous since the decision-maker can influence them through, for instance, education, the advancement of women, and migration policies.

The indirect drivers of change are primarily:

- demographic (such as population size, age and gender structure, and spatial distribution);
- economic (such as national and per capita income, macroeconomic policies, international trade, and capital flows);
- sociopolitical (such as democratization, the roles of women, of civil society, and of the private sector, and international dispute mechanisms);
- scientific and technological (such as rates of investments in research and development and the rates of adoption of new technologies, including biotechnologies and information technologies); and
- cultural and religious (such as choices individuals make about what and how much to consume and what they value).

The interaction of several of these drivers, in turn, affects levels of resource consumption and differences in consumption both within and between countries. Clearly these drivers are changing—population and the world economy are growing, for instance, there are major advances in information technology and biotechnology, and the world is becoming more interconnected. Changes in these drivers are projected to increase the demand for and consumption of food, fiber, clean water, and energy, which will in turn affect the direct drivers. The direct drivers are primarily physical, chemical, and biological—such as land cover change, climate change, air and water pollution, irrigation, use of fertilizers, harvesting, and the introduction of alien invasive species. Change is apparent here too: the climate is changing, species ranges are shifting, alien species are spreading, and land degradation continues.

An important point is that any decision can have consequences external to the decision framework. These consequences are called externalities because they are not part of the decision-making calculus. Externalities can have positive or negative effects. For example, a decision to subsidize fertilizers to increase crop production might result in substantial degradation of water quality from the added nutrients and degradation of downstream fisheries. But it is also possible to have positive externalities. A beekeeper might be motivated by the profits to be made from selling honey, for instance, but neighboring orchards could produce more apples because of enhanced pollination arising from the presence of the bees.

Multiple interacting drivers cause changes in ecosystem services. There are functional interdependencies between and among the indirect and direct drivers of change, and, in turn, changes in ecological services lead to feedbacks on the drivers of changes in ecological services. Synergetic driver combinations are common. The many processes of globalization lead to new forms of interactions between drivers of changes in ecosystem services.

1.3.4 Cross-scale Interactions and Assessment

An effective assessment of ecosystems and human well-being cannot be conducted at a single temporal or spatial scale. Thus the MA conceptual framework includes both of these dimensions. Ecosystem changes that may have little impact on human well-being over days or weeks (soil erosion, for instance) may have pronounced impacts over years or decades (declining agricultural productivity). Similarly, changes at a local scale may have little impact on some services at that scale (as in the local impact of forest loss on water availability) but major impacts at large scales (forest loss in a river basin changing the timing and magnitude of downstream flooding).

Ecosystem processes and services are typically most strongly expressed, are most easily observed, or have their dominant controls or consequences at particular spatial and temporal scales. They often exhibit a characteristic scale—the typical extent or duration over which processes have their impact. Spatial and temporal scales are often closely related. For instance, food production is a localized service of an ecosystem and changes on a weekly basis, water regulation is regional and changes on a monthly or seasonal basis, and climate regulation may take place at a global scale over decades.

Assessments need to be conducted at spatial and temporal scales appropriate to the process or phenomenon being examined. Those done over large areas generally use data at coarse resolutions, which may not detect fine-resolution processes. Even if data are collected at a fine level of detail, the process of averaging in order to present findings at the larger scale causes local patterns or anomalies to disappear. This is particularly problematic for processes exhibiting thresholds and nonlinearities. For example, even though a number of fish stocks exploited in a particular area might have collapsed due to overfishing, average catches across all stocks (including healthier stocks) would not reveal the extent of the problem. Assessors, if they are aware of such thresholds and have access to high-resolution data, can incorporate such information even in a large-scale assessment. Yet an assessment done at smaller spatial scales can help identify important dynamics of the system that might otherwise be overlooked. Likewise, phenomena and processes that occur at much larger scales, although expressed locally, may go unnoticed in purely local-scale assessments. Increased carbon dioxide concentrations or decreased stratospheric ozone concentrations have local effects, for instance, but it would be difficult to trace the causality of the effects without an examination of the overall global process.

Time scale is also very important in conducting assessments. Humans tend not to think beyond one or two generations. If an assessment covers a shorter time period than the characteristic temporal scale, it may not adequately capture variability associated with long-term cycles, such as glaciation. Slow changes are often harder to measure, as is the case with the impact of climate change on the geographic distribution of species or populations. Moreover, both ecological and human systems have substantial inertia, and the impact of changes occurring today may not be seen for years or decades. For example, some fisheries' catches may increase for several years even after they have reached unsustainable levels because of the large number of juvenile fish produced before that level was reached.

Social, political, and economic processes also have characteristic scales, which may vary widely in duration and extent. Those of ecological and sociopolitical processes often do not match. Many environmental problems originate from this mismatch between the scale at which the ecological process occurs, the scale at which decisions are made, and the scale of institutions for decision-making. A purely local-scale assessment, for instance, may discover that the most effective societal response requires action that can occur only at a national scale (such as the removal of a subsidy or the establishment of a regulation). Moreover, it may lack the relevance and credibility necessary to stimulate and inform national or regional changes. On the other hand, a purely global assessment may lack both the relevance and the credibility necessary to lead to changes in ecosystem management at the local scale where action is needed. Outcomes at a given scale are often heavily influenced by interactions of ecological, socioeconomic, and political factors emanating from other scales. Thus focusing solely on a single scale is likely to miss interactions with other scales that are critically important in understanding ecosystem determinants and their implications for human well-being.

The choice of the spatial or temporal scale for an assessment is politically laden, since it may intentionally or unintentionally privilege certain groups. The selection of assessment scale with its associated level of detail implicitly favors particular systems of knowledge, types of information, and modes of expression over others. For example, non-codified information or knowledge systems of minority populations are often missed when assessments are undertaken at larger spatial scales or higher levels of aggregation. Reflecting on the political consequences of scale and boundary choices is an important prerequisite to exploring what multi- and cross-scale analysis in the MA might contribute to decision-making and public policy processes at various scales.

1.4 Values Associated with Ecosystems

Current decision-making processes often ignore or underestimate the value of ecosystem services. Decision-making concerning ecosystems and their services can be particularly challenging because different disciplines, philosophical views, and schools of thought assess the value of ecosystems differently. One paradigm of value, known as the utilitarian (anthropocentric) concept, is based on the principle of humans' preference satisfaction (welfare). In this case, ecosystems and the services they provide have value to human societies because people derive utility from their use, either directly or indirectly (use values). Within this utilitarian concept of value, people also give value to ecosystem services that they are not currently using (non-use values). Non-use values, usually known as existence values, involve the case where humans ascribe value to knowing that a resource exists even if they never use that resource directly. These often involve the deeply held historical, national, ethical, religious, and spiritual values people ascribe to ecosystems—the values that the MA recognizes as cultural services of ecosystems.

A different, non-utilitarian value paradigm holds that something can have intrinsic value—that is, it can be of value in and for itself—irrespective of its utility for someone else. From the perspective of many ethical, religious, and cultural points of view, ecosystems may have intrinsic value, independent of their contribution to human well-being.

The utilitarian and non-utilitarian value paradigms overlap and interact in many ways, but they use different metrics, with no common denominator, and cannot usually be aggregated, although both paradigms of value are used in decision-making processes.

Under the utilitarian approach, a wide range of methodologies has been developed to attempt to quantify the benefits of different ecosystem services. These methods are particularly well developed for provisioning services, but recent work has also improved the ability to value regulating and other services. The choice of valuation technique in any given instance is dictated by the characteristics of the case and by data availability. (See Box 1.4.)

Non-utilitarian value proceeds from a variety of ethical, cultural, religious, and philosophical bases. These differ in the specific entities that are deemed to have intrinsic value and in the interpretation of what having intrinsic value means. Intrinsic value may complement or counterbalance considerations of utilitarian value. For example, if the aggregate utility of the services provided by an ecosystem (as measured by its utilitarian value) outweighs the value of converting it to another use, its intrinsic value may then be complementary and provide an additional impetus for conserving the ecosystem. If, however, economic valuation indicates that the value of converting the ecosystem outweighs the aggregate value of its services, its ascribed intrinsic value may be deemed great enough to warrant a social decision to conserve it anyway. Such decisions are essentially political, not economic. In contemporary democracies these decisions are made by parliaments or legislatures or by regulatory agencies mandated to do so by law. The sanctions for violating laws recognizing an entity's intrinsic value may be regarded as a measure of the degree of intrinsic value ascribed to them. The decisions taken by businesses, local communities, and individuals also can involve considerations of both utilitarian and non-utilitarian values.

The mere act of quantifying the value of ecosystem services cannot by itself change the incentives affecting their use or misuse. Several changes in current practice may be required to take better account of these values. The MA assesses the use of information on ecosystem service values in decision-making. The goal is to improve decision-making processes and tools and to provide feedback regarding the kinds of information that can have the most influence.

BOX 1.4

Valuation of Ecosystem Services

Valuation can be used in many ways: to assess the total contribution that ecosystems make to human well-being, to understand the incentives that individual decision-makers face in managing ecosystems in different ways, and to evaluate the consequences of alternative courses of action. The MA uses valuation primarily in the latter sense: as a tool that enhances the ability of decision-makers to evaluate tradeoffs between alternative ecosystem management regimes and courses of social actions that alter the use of ecosystems and the multiple services they provide. This usually requires assessing the change in the mix (the value) of services provided by an ecosystem resulting from a given change in its management.

Most of the work involved in estimating the change in the value of the flow of benefits provided by an ecosystem involves estimating the change in the physical flow of benefits (quantifying biophysical relations) and tracing through and quantifying a chain of causality between changes in ecosystem condition and human welfare. A common problem in valuation is that information is only available on some of the links in the chain and often in incompatible units. The MA can make a major contribution by making various disciplines better aware of what is needed to ensure that their work can be combined with that of others to allow a full assessment of the consequences of altering ecosystem state and function.

The ecosystem values in this sense are only one of the bases on which decisions on ecosystem management are and should be made. Many other factors, including notions of intrinsic value and other objectives that society might have (such as equity among different groups or generations), will also feed into the decision framework. Even when decisions are made on other bases, however, estimates of changes in utilitarian value provide invaluable information.

1.5 Assessment Tools

The information base exists in any country to undertake an assessment within the framework of the MA. That said, although new data sets (for example, from remote sensing) providing globally consistent information make a global assessment like the MA more rigorous, there are still many challenges that must be dealt with in using these data at global or local scales. Among these challenges are biases in the geographic and temporal coverage of the data and in the types of data collected. Data availability for industrial countries is greater than that for developing ones, and data for certain resources such as crop production are more readily available than data for fisheries, fuelwood, or biodiversity. The MA makes extensive use of both biophysical and socioeconomic indicators, which combine data into policy-relevant measures that provide the basis for assessment and decision-making.

Models can be used to illuminate interactions among systems and drivers, as well as to make up for data deficiencies—for instance, by providing estimates where observations are lacking. The MA makes use of environmental system models that can be used, for example, to measure the consequences of land cover change for river flow or the consequences of climate change for the distribution of species. It also uses human system models that can examine, for instance, the impact of changes in ecosystems on production, consumption, and investment decisions by households or that allow the economy-wide impacts of a change in production in a particular sector like agriculture to be evaluated. Finally, integrated models, combining both the environmental and human systems linkages, can increasingly be used at both global and sub-global scales.

The MA incorporates both formal scientific information and traditional or local knowledge. Traditional societies have nurtured and refined systems of knowledge of direct value to those societies but also of considerable value to assessments undertaken at regional and global scales. This information often is unknown to science and can be an expression of other relationships between society and nature in general and of sustainable ways of managing natural resources in particular. To be credible and useful to decision-makers, all sources of information, whether scientific, traditional, or practitioner knowledge, must be critically assessed and validated as part of the assessment process through procedures relevant to the form of knowledge.

Since policies for dealing with the deterioration of ecosystem services are concerned with the future consequences of current actions, the development of scenarios of medium- to long-term changes in ecosystems, services, and drivers can be particularly helpful for decision-makers. Scenarios are typically developed through the joint involvement of decision-makers and scientific experts, and they represent a promising mechanism for linking scientific information to decision-making processes. They do not attempt to predict the future but instead are designed to indicate what science can and cannot say about the future consequences of alternative plausible choices that might be taken in the coming years.

The MA uses scenarios to summarize and communicate the diverse trajectories that the world's ecosystems may take in future decades. Scenarios are plausible alternative futures, each an example of what might happen under particular assumptions. They can be used as a systematic method for thinking creatively about complex, uncertain futures. In this way, they help us understand the upcoming choices that need to be made and highlight developments in the present. The MA developed scenarios that connect possible changes in drivers (which may be unpredictable or uncontrollable) with human demands for ecosystem services. The scenarios link these demands, in turn, to the futures of the services themselves and the aspects of human welfare that depend on them. The scenario building exercise breaks new ground in several areas:

- development of scenarios for global futures linked explicitly to ecosystem services and the human consequences of ecosystem change,
- consideration of trade-offs among individual ecosystem services within the "bundle" of benefits that any particular ecosystem potentially provides to society,
- assessment of modeling capabilities for linking socioeconomic drivers and ecosystem services, and
- consideration of ambiguous futures as well as quantifiable uncertainties.

The credibility of assessments is closely linked to how they address what is not known in addition to what is known. The consistent treatment of uncertainty is therefore essential for the clarity and utility of assessment reports. As part of any assessment process, it is crucial to estimate the uncertainty of findings even if a detailed quantitative appraisal of uncertainty is unavailable.

1.6 Strategies and Interventions

The MA assesses the use and effectiveness of a wide range of options for responding to the need to sustainably use, conserve, and restore ecosystems and the services they provide. These options include incorporating the value of ecosystems in decisions, channeling diffuse ecosystem benefits to decision-makers with focused local interests, creating markets and property rights, educating and dispersing knowledge, and investing to improve ecosystems and the services they provide. As seen in Box 1.2 on the MA conceptual framework, different types of response options can affect the relationships of indirect to direct drivers, the influence of direct drivers on ecosystems, the human demand for ecosystem services, or the impact of changes in human well-being on indirect drivers. An effective strategy for managing ecosystems will involve a mix of interventions at all points in this conceptual framework.

Mechanisms for accomplishing these interventions include laws, regulations, and enforcement schemes; partnerships and collaborations; the sharing of information and knowledge; and public and private action. The choice of options to be considered will be greatly influenced by both the temporal and the physical scale influenced by decisions, the uncertainty of outcomes, cultural context, and the implications for equity and trade-offs. Institutions at different levels have different response options available to them, and special care is required to ensure policy coherence.

Decision-making processes are value-based and combine political and technical elements to varying degrees. Where technical input can play a role, a range of tools is available to help decision-makers choose among strategies and interventions, including cost-benefit analysis, game theory, and policy exercises. The selection of analytical tools should be determined by the context of the decision, key characteristics of the decision problem, and the criteria considered to be important by the decision-makers. Information from these analytical frameworks is always combined with the intuition, experience, and interests of the decision-maker in shaping the final decisions.

Risk assessment, including ecological risk assessment, is an established discipline and has a significant potential for informing the decision process. Finding thresholds and identifying the potential for irreversible change are important for the decision-making process. Similarly, environmental impact assessments designed to evaluate the impact of particular projects and strategic environmental assessments designed to evaluate the impact of policies both represent important mechanisms for incorporating the findings of an ecosystem assessment into decision-making processes.

Changes also may be required in decision-making processes themselves. Experience to date suggests that a number of mechanisms can improve the process of making decisions about ecosystem services. Broadly accepted norms for decision-making process include the following characteristics. Did the process:

- bring the best available information to bear?
- function transparently, use locally grounded knowledge, and involve all those with an interest in a decision?
- pay special attention to equity and to the most vulnerable populations?
- use decision analytical frameworks that take account of the strengths and limits of individual, group, and organizational information processing and action?
- consider whether an intervention or its outcome is irreversible and incorporate procedures to evaluate the outcomes of actions and learn from them?
- ensure that those making the decisions are accountable?
- strive for efficiency in choosing among interventions?
- take account of thresholds, irreversibility, and cumulative, cross-scale, and marginal effects and of local, regional, and global costs, risk, and benefits?

The policy or management changes made to address problems and opportunities related to ecosystems and their services, whether at local scales or national or international scales, need to be adaptive and flexible in order to benefit from past experience, to hedge against risk, and to consider uncertainty. The understanding of ecosystem dynamics will always be limited, socioeconomic systems will continue to change, and outside determinants can never be fully anticipated. Decision-makers should consider whether a course of action is reversible and should incorporate, whenever possible, procedures to evaluate the outcomes of actions and learn from them. Debate about exactly how to do this continues in discussions of adaptive management, social learning, safe minimum standards, and the precautionary principle. But the core message of all approaches is the same: acknowledge the limits of human understanding, give special consideration to irreversible changes, and evaluate the impacts of decisions as they unfold.

Chapter 2

Typology of Responses

Coordinating Lead Authors: W. Bradnee Chambers, Ferenc L. Toth
Lead Authors: Indra de Soya, Jessica Green, Sofia Hirakuri, Hiroji Isozaki, Alphonse Kambu
Contributing Authors: Dagmar Lohan, Prisna Nuengsigkapian, Sergio Pena-Neira
Review Editors: Richard Norgaard, Ravi Prabhu

BOXES

TABLES

Main Messages

Responses are human actions to address specific issues, needs, opportunities, or problems in ecosystem governance and management. They encompass all policies, strategies, measures, and interventions that are established to change ecosystem status and processes directly, and those that modify direct or indirect drivers that shape ecosystem status and processes. Societies have been developing a wide range of responses to manage their interactions with ecosystems. They include legal, economic, social and behavioral, technological, and cognitive responses. An essential precondition for any response to work effectively is stable order based on social norms generally accepted by those at whom the response is targeted.

The main typology of responses is organized according to the dominant mechanism through which specific responses are intended to change human behavior or ecosystems characteristics. Most responses in practice involve an interaction of legal, social, economic, and technological elements to work effectively. Other dimensions along which responses can be classified are their effects on different direct and indirect drivers of ecosystem change, the actors typically using them, and the geographical scales and jurisdictional levels at which they are normally adopted.

The starting place for understanding responses is institutions. Institutions are not responses in their own right but create the framework and the medium by which responses can converge on direct and indirect drivers. Institutions, formal or informal, are found at multiples scales and are formed by various actors. In this sense, institutions are an important means for setting the rules of the game.

Legal responses have an overall function of providing the formal rules by which all other responses are framed and operationalized. Legal responses occur at a variety of levels internationally, nationally, and sub-nationally and are divided by well-recognized jurisdictional orders. International legal responses range across a variety of soft, customary, and codified rules, which in the last 30 years have materialized in a multitude of international agreements on environment and sustainable development. International responses rarely have direct effect at the national and sub-national levels without ratification or implementation through domestic legal responses. Domestic responses are also those made by decision-makers independent of international law making. Overall, nationally and sub-nationally, legal responses are typified by three broad categories of regulatory, administrative, and constitutional rules that either may be aimed directly at ecosystems change or could be rules outside the ecosystem sector but having direct bearing on ecosystems and human well-being (such as, economic sector responses). All of these rules remain static without implementation, compliance, and enforcement in their respective jurisdictions.

Economic responses work through the self-interest of people and their effort to improve their economic welfare, an important component of overall well-being. They either could be based on existing property rights or could create new ones. Economic responses interfere with the ways in which ecosystems services are traded in often-imperfect markets that also provide explicit valuation of traded items. Command-and-control instruments are straightforward and blunt when properly implemented. They are rarely cost-efficient, but in many cases they are the only feasible response option so cost-efficiency is irrelevant. Incentive-based instruments rely on the wisdom of the targeted individuals or groups (including private companies) to follow their self-interest and thereby find the cost-efficient way to reach the ecological target. Voluntarism-based instruments are based on self-control and they are often used either to prevent a stricter form of regulation or as a precursor to stricter regulation. Financial and monetary measures include diverse forms of transfer payments in exchange for implementing ecologically benign practices. International trade policies influence ecosystems management by regulating the flows of ecosystems goods and services across national borders.

Social, behavioral, and cognitive responses drive change by affecting the norms, values, attitudes, and knowledge of individuals and society. The provision of political rights and liberties empowers people, increasing transparency and awareness over matters of eco-system degradation. Education and public programs influence attitudes and norms that invariably drive change in relationships between society and nature; they also increase participation in public fora and debate. The empowerment of youth, women, and minority groups in society adds to knowledge through participation and inclusion. Participation leads to learning. The inclusion and legitimization of traditional knowledge has been widely recognized as valuable for addressing ecosystem protection issues.

Technological responses work through the products, devices, processes, and practices adopted in ecosystems management directly and in other human activities affecting ecosystems indirectly. They are applied in managing ecosystems, preventing degradation, as well as rehabilitating degradation that has already taken place. Providing incentives for innovation and technological research and development is a powerful response option that can sometimes have unexpected negative side effects.

Specific response options and their combinations can be used in different phases of ecosystem change for five main types of action: development, prevention, mitigation, adaptation, and rehabilitation. Ecosystem *development* is aimed at increasing the provision of selected ecosystems services, often at the expense of others and/or by transforming important features of the ecosystem. *Prevention* is an attempt to foreclose unwanted changes in the ecosystem before their commencement. *Mitigation* aims at slowing down and halting an already on-going transformation process. *Adaptation* recognizes that some kind and degree of change is inevitable and attempt to cope with the changing ecosystem conditions. *Rehabilitative* responses strive to improve degraded ecosystems in general or to restore them to a specific earlier status.

2.1 Introduction

The management of ecosystems, including the use of their services, and the regulation of human activities that impact ecosystems has been a major challenge for humanity through its long history. As long as human influences were limited to relatively small intrusions into ecosystems processes (below the maximum sustainable yield or the natural pollutant absorbing capacity of ecosystems), no intervention was required. However, as the scale of utilization of ecosystems services for human use and the magnitude of emissions of ecologically harmful materials have increased, the need for intervention and regulation of related activities has increased as well.

This chapter presents an overview of the wide range of responses societies have invented and use to regulate their interactions with ecosystems. The chapter introduces tools that can be used to respond to ecosystem-related problems and examines their links to human well-being and poverty reduction. It does not provide a formal assessment of the "state of knowledge" regarding the effectiveness of these various instruments (that is done in the chapters of Part II); rather, it defines and characterizes their enabling environment and interplay.

It is not possible to establish a typology of response options that can classify all interventions into strictly separated boxes. Most response options adopted in practice in contemporary societies combine elements from several clusters irrespective of whether the clusters are defined along disciplinary boundaries (law, economics, engineering, sociology); jurisdictional domains (financial, environmental, or public health regulation); or some other ordering principles. Accordingly, there are some overlaps in our typology as well. The basic principle underlying the classification system presented in this chapter is the primary intention and mechanism of the response. Is the intervention by public policy-makers and private stakeholders designed to change the behavior of the targeted community based on economic incentives, cognitive enlightenment and approval, or legal threat? In most cases, both the social execution mechanisms and the implementation processes entail components of several domains. For instance, most economic incentives need a legal framework to become effective, and many legal responses use monetary penalties as enforcement mechanisms even if their main effects work through legal threats.

The typology of response options draws on widely accepted typologies used in the disciplines we draw upon: law, economics, sociology, political science, engineering, psychology, social psychology, anthropology, and environmental ethics. Additional sources include interdisciplinary environmental studies that suggest typologies of responses in managing ecosystems, natural resources, and environmental problems, such as Kaufman-Hayoz et al. (2001) and Dietz and Stern (2002). All these typologies have their own merits and shortcomings. Since none of the typologies are sufficiently comprehensive to encompass the full range of response options relevant to the MA, we have combined and extended them, preserving adequate flexibility; chapters in Part II sort and appraise the response options in various sectors.

Depending on their main objectives, the response options can be adopted individually or combined in various ways to address problems in the five main types of ecosystem management: development, prevention, mitigation, adaptation, and rehabilitation. Ecosystem *development* aims to increase the provision of selected ecosystems services, often at the expense of others and/or by transforming important features of the ecosystem. *Prevention* attempts to foreclose unwanted changes in the ecosystem before they begin. *Mitigation* aims at slowing down and halting an already on-going transformation process. *Adaptation* recognizes that some kind and degree of change is inevitable and attempts to cope with the changing ecosystem conditions. *Rehabilitation* strives to improve degraded ecosystems in general or attempts to restore them to an earlier status. Any of these five types of actions may use different kinds of intervention (legal, economic, technological, etc.) at different scales (for example, international, national, local) by different actors (such as government, private sector, or community) to influence direct or indirect drivers of ecosystems change. Consequently, our typology and discussion of responses in selected contexts cuts across these five domains and contains relevant information for each.

In presenting a typology of responses, Chapter 2 first provides an overview of the range of intervention mechanisms. Subsequent sections look at the various response options by their impact on the various direct and indirect drivers of ecosystems change; by their availability to various actors to influence the ecosystems management activities of other actors; and by the scale of operation and jurisdictional context of the decision maker. Based on the relationships identified among response options on the one hand, and the drivers, actors, and scales on the other, the chapter's final section provides a synthesis of clusters of intervention opportunities by actors to influence specific drivers by using specific response options.

In each section, a matrix indicates the linkages between the response options and the components of disciplines, drivers, actors, and scales. These matrices demonstrate the multidimensional characteristics of most response options (for example, having roots in economics, law, and sociology or affecting institutions, individuals, and technologies simultaneously).

2.2 Typology of Reponses by Nature of the Intervention

This section describes response options by the nature of the intervention, including legal, economic, social, technological, and cognitive instruments and measures. In each subsection, reference is made to relevant scales, actors, and drivers. The temporal dimension of the response options is also discussed. Subsequent sections will examine response options in greater detail, with special reference to their interactions and the integrative approaches.

2.2.1 Institutional Framework as the Basis for Intervention

Our discussion of available responses and their effectiveness is framed within the larger context of institutions and their effects on human interaction shaping ecosystems change. Institutions moderate human behavior and thereby powerfully shape the nature of human interaction with nature. Institutions operate at various levels and scales, such as global, national, and sub-national levels and on the basis of both formal and informal rules. Ethics, values, and attitudes usually ascribed to larger cultural contexts also operate to moderate institutional behavior. Institutions are either absent or work badly when human impact on ecosystems are not being regulated in a desirable manner. The responses discussed below operate to build and strengthen larger institutional settings governing human interaction with nature.

According to the most widely used definition, institutions define the "rules of the game," which are humanly devised constraints for shaping human action (Hanna et al. 1996; Ostrom et al. 1994; Young 2002). Recent interest in institutions stems largely from transaction-cost economics that takes into account the costs associated with the multitude of transactions among individuals, and the ways in which economic actors seek to mitigate those costs (North 1990; Williamson 1985). Since the effects of environmental harm carry costs for humans, people have incentives to change the rules of the game in ways that reduce costs. In this vein, Garrett Hardin (1968) highlighted the "tragedy of the commons" and suggested changes in the rules of the game from "free access" to a system of specified private rights as solution to stem the degradation of the commons.

Problems associated with ecosystems and the implications for well-being make ecosystem health a truly global concern requiring a high degree of international cooperation. States have to cooperate in order to forge governance systems to address common problems, but many states may find it convenient to shirk obligations because, unlike in domestic society, there is no established body compelling states to act in a certain way. Many global environmental problems, such as the loss of biodiversity, depletion of fish stocks, or climate change are problems that require cooperation. When institutions function well, shirking and opportunism are minimized and the powerful are unable to appropriate rules for selfish ends that harm the collective interest and reduce overall well-being. Given that state sovereignty is a governing principle within the international system, states are reluctant to have others dictate the rules that govern the use of resources claimed as one's own, making cooperation difficult but not impossible. "Governance without government'' is possible, and selective regimes have proliferated in the international arena as a result of trying to do something about regional and global environmental problems (Levy et al. 1995; Young 2002).

International institutions moderate an anarchical state system and allow a high degree of cooperation by replacing power with legally binding rules (Slaughter et al. 1998). As many argue, international regimes remain an effective way of solving global environmental problems and the strengthening of these rules has many positive effects toward the evolution of a rule-based international system. In the past three decades since UNEP was created, the problems of ecosystems (natural as well as social) have generated a number of international institutions overseen by international organizations. The United Nations Environmental Program, for example, tackles global warming and the problem of biodiversity loss. Nongovernmental organizations lobbying for international regimes have contributed to the creation of a number of the new institutional arrangements.. The concept of ''sustainable development'' first put forth by the Brundtland Commission report ''Our Common Future'' (Brundtland 1987) led to agreements on biodiversity and climate change and are enshrined in the United Nations Conference on Environment and Development process, which has produced two international regimes: The Framework Convention on Biological Diversity and the Framework Convention on Climate Change

Subsequent international discussions have highlighted the need to integrate socioeconomic concerns with purely environmental ones. For example the Millennium Development Goal process is an effort to bring the socioeconomic dimension and notions of human well-being into questions of addressing environmental change. Moreover, institutions around environmental concerns, such as UNFCCC, have also influenced the actions and behavior of other agencies, such as the International Monetary Fund and World Bank, which are striving to bring their programs of action in line with these new rules. The Global Environmental Facility, which spawns collaboration among international actors at various levels, is a good example of broader institutions' influence on shaping the behavior of development actors in various arenas. Despite this rapid progress in global cooperative agreements, some demand an overarching institution, such as the World Trade Organization, to govern global ecosystem-problems.

Local institutions have the most direct bearing on many forms of ecosystem change. The degree to which national institutions affecting environmental problems exist and function effectively varies greatly from country to country. Much research on common pool resources addresses the question of rule-making among very small local communities that manage to ''govern without government'' (Ostrom et al. 1994; Powell 2000). There are hundreds of agreements on environmental issues that national governments implement effectively at the national level and many governments around the world now have ministries and other agencies devoted to safeguarding the environment.

Judging by the number of signatories to such conventions as well as participation in the GEF, there seems to be broad desire locally to accept the international rules of the game governing environmental issues. Global consensus and value change reflecting post-modern concerns seem to exert considerable pressure on governments to address environmental concerns and actively participate in regional and international environmental agreements. Local Agenda 21 is also a good example of how international agreements might be transformed into action at the sub-national level. National governments accede to international agreements like the Framework Conventions on Climate Change or Biological Diversity because of international and domestic pressure brought on by the proliferation of actors, fora, and regimes concerned about the health of the planet and local ecosystems.

The global change in the acceptance of the values of democratic government seems to be exerting pressure on national governments. Civil society awareness is one of the driving forces on the re-assessment of ecosystem protection, use, and utilization. NGOs in developing and industrial countries are in the forefront of the discussion of urban ecosystems, conservation, and social well-being. In this sense, since the 1970s, the movement on environmental justice in the United States and in Europe has created the necessity to address problems such as the interrelationship between people and their rural or urban ecosystems.

There is widespread belief that democracy as an organizing principle is a good safeguard for protecting ecosystems health and ensuring ecological justice. Many questions remain as to what types of national-level institutional arrangements matter in environmental questions, although there is much literature addressing the question of democratic design and policy outcomes in the field of political science (Lijphart 1999; Powell 2000). Much analysis also depends greatly on how questions of national institutions are addressed. For example, utilitarian models tend to see institutions as valuable tools by which one moderates established behavioral patters of individual (rational) actors. In such models, market mechanisms working through prices could be made to effect change and obtain the desired outcome. Thus, if a resource is being depleted, rising prices are expected to drive down demand and the market principle could be an effective tool.

However, an analysis considering social practice models might produce very different answers. Established tastes, cultural rites, and practices might be sticky and hard to change through price mechanisms, but perhaps more easily affected by education. Questions of effective transfer of information and education may in turn depend on questions of legitimacy. Are state schools or the temples and churches more effective purveyors of change? In such analyses, questions of social capital, networks, and informal economies tend to be more salient objects of analyses than simple market mechanisms around relative price change (DiMaggio 1994). For example, is the problem of addressing the destruction of rhinoceros horns in Africa one of price alone or one of broader institutional change involving education and cultural change? Clearly, interventions through institutional change apply at various levels, for various degrees of scale.

2.2.2 Legal Responses

Law plays an important role in environmental protection at both the international and the national levels. International law provides mechanisms, such as treaties, rules of customary law, judgments of international courts or tribunals, and general principles of international law, to protect the environment (Brownlie 1990). International legal agreements range from ''gentleman's agreements'' that go back to the nineteenth century (Klabbers 1996) to legally binding agreements. In general, there are two approaches: ''hard law'' and ''soft law.'' Commonly, treaties and custom create binding international law (that is, hard law), although custom has not been consolidated as a tradition to protect the environment (Birnie and Boyle 2002). Domestic environmental laws, likewise, use a set of regulatory techniques to achieve environmental objectives; these are also reviewed.

2.2.2.1 International Treaties

''Hard law'' refers to legally binding regulations that impose mandatory obligations on states, such as bilateral or multilateral treaties. In this case, participating states must implement and enforce

there is consensus against doing so. These mechanisms are not broad enough to address environmental issues directly, but they have handled disputes that involve trade and environment (Martin 2001). An amendment to GATT Article XX, or a quasi-judicial statement of understanding, is required to exempt any MEAs from trade rules (Jordan 2001).

Another promising regime for an amicable resolution is the new Permanent Court of Arbitration Optional Rules for Arbitration of Disputes Relating to Natural Resources and/or the Environment adopted in 2001. It will be of great value to conserving biodiversity (PCA 2004).

The CBD sets out a framework on liability and redress, including restoration and compensation, for damage to biodiversity (Article 14, 2.). The Cartagena Protocol on Biosafety to the CBD provides for a liability regime for damage resulting from the transboundary movements of living modified organisms (Article 27); In 2004, the Conference of the Parties decided to establish an open-ended ad hoc working group of legal and technical experts on liability and redress (UNEP 2004). Other international legal frameworks addressing redress issues in dangerous activities such as nuclear energy (the 1960 Convention on Third Party Liability in the Field of Nuclear Energy and the 1963 Vienna Convention on Civil Liability for Nuclear Damage); pollution (the 1977 Convention on Civil Liability for Oil Pollution Damage Resulting from Exploration for and Exploitation of Seabed Mineral Resources); and the transport of dangerous goods and substances (the 1989 Convention on Civil Liability for Damage caused during Carriage of Dangerous Goods by Road, Rail, and Inland Navigation Vessels and the 1999 Basel Protocol on Liability and Compensation for Damage resulting from Transboundary Movements of Hazardous Wastes and their Disposal).

These conventions each establish a strict liability regime, requiring demonstration of a causal link between the activity and damage. Few conventions on liability address the loss of biodiversity specifically, but the 1993 Convention on Civil Liability for Damage Resulting from Activities Dangerous to the Environment is a comprehensive convention dealing with liability and redress for environmental harm. The term "environment" encompasses natural resources both abiotic and biotic—air, water, soil, fauna, and flora—and the interaction between them (Article 2). Not yet in force, it is a promising convention to ensure adequate compensation for environmental damage (UNEP/CBD/ICCP/2/3).

A *monitoring system* is also valuable to implement treaties. This is vital to the regulatory control of emissions (United Nations 2003). Reporting under the Basel Convention (UN 1989) and the 1971 Ramsar Convention (UN 1972a) has been effective in Western and Central Europe. However, at the domestic level, the lack of monitoring has resulted in poor enforcement in Latin American, Caribbean, and Central Asian (UNEP 1999).

For better implementation of MEAs, an *environmental impact assessment system* has also been widely used at the international level. In addition to the general objective of the EIA, which is to ensure that development will not damage human health and the natural environment (Spellerberg 1991), the Convention on Biological Diversity calls for EIAs when activities are likely to have significant adverse impact on biodiversity (Article 14, 1). Several international agreements have EIA provisions, including the 1982 UNCLOS and the 1985 ASEAN Agreement on the Conservation of Nature and Natural Resources. The 1991 ECE Convention on Environmental Impact Assessment in a Transboundary Context ensures that an EIA is undertaken for activities likely to have a significant adverse transboundary environmental impact (Article 2(7)), and the 1991 Madrid Protocol on Environmental Protection to the Antarctic Treaty requires prior assessment of the environmental impacts for all activities listed in Annex I (Article 8).

Enforcement is the major concern of effective regulation. Most environmental conventions leave enforcement to the parties, which must enact the necessary national laws and enforce them in their territory. The obligation to enact such measures is a crucial part of the enforcement system. Along with formal regulatory enforcement is self-enforcement. In this case, individual states take the required measures to serve their own interests (Birnie 1992).

2.2.2.6 Domestic Environmental Regulations

Domestic legislation is critical since the implementation of treaties only occurs through the actions of the government agencies of each country. These agencies use a variety of tools to put the regulations into practice, both formally and informally. Formal methods emphasize coercive measures such as sanctions. Informal methods include certification systems or various voluntary measures (May et al. 1996). These measures rely on consumer preferences and corporate managers' aversion to shame (Campbell-Mohn et al. 1993).

"Command and control" is the most common regulatory means to achieve environmental objectives. Government simply imposes requirements on the conduct of individual actors; for instance, a government may set standards for the maximum level of a pollutant allowed at a facility. Command-and-control standards are clear and easy to enforce for those engaged in potentially polluting activities. Command-and-control regulations typically entail licenses and permits, and usually specify the pollutant, such as industrial pollution, discharges to sewers, and land contamination (Campbell-Mohn et al. 1993; Wolf et al. 2002). In addition, economic instruments and market mechanisms are commonly used for environmental protection. (Such economic instruments and detailed command-and-control interventions are discussed below in the section on economic responses.) The command-and-control approach remains the primary means of enforcement in the majority of countries in Africa, Asia, and the Pacific. However, in North America the trend is towards a policy mix with an emphasis on market-oriented mechanisms, public-private partnerships, and voluntary mechanisms (UNEP 1999).

National laws related to the protection of nature, including terrestrial and marine living resources, go back to 1597 (Birnie and Boyle 2002). Many laws were established in the late nineteenth to the early twentieth century. For instance, the law for protection of nature in national forests was established in 1915 in Japan (User Survey b, n.d.); the Reich Conservation Law was established in Germany in 1935 (SDUD n.d.). However, many countries developed their laws and regulations regarding protection of the environment and management of natural resources in the early 1970s. Legislation at that time was largely concerned with pollution control (water, air, and soil); later, it was expanded to other areas, such as nature conservation, the protection of public health, and the control of toxic substances and hazardous wastes. The development of domestic environmental legislation is partially a response to the obligations under MEAs (UNEP 1999). The bases of environmental law at the domestic level are often found in federal constitutions. The nature of the environmental issue at stake defines which tools will be used at the domestic level for implementing the laws.

2.2.2.7 Domestic Constitutional Law

In many countries, the constitution lays the basic principles for environmental regulation. Generally, the constitution prescribes the form of government, sets up political institutions, defines governmental functions, and establishes the rights and duties of citizens. Most legal interventions at the domestic level have some common features, that is, the constitution apportions power between the legislative, executive, and judicial branches of the government. The legislative branch enacts laws to regulate major environmental issues, such as air and water pollution, hazardous wastes, wetlands, endangered species, toxics and pesticides, energy reserves, and natural resources conservation (Jasper 1997). The executive branch converts the legal requirements and government policy into guidelines, memoranda, directives, and administrative orders and applies them. The judicial branch enforces the provisions of the environmental legislation (Bates 1995).

The constitutions of more than 100 countries guarantee a right to a clean and healthy environment (Kiss and Shelton 2004). They define adequate protection of the environment as essential to human well-being and to basic human rights. For instance, Article 225 of Brazil's constitution guarantees all citizens a healthy and stable environment (Brazil 1988). Peru also has environmental protection provisions in its constitution (Capítulo III, Peru 1993). Argentina's constitution states that "All inhabitants are entitled to the right to a healthy and balanced environment fit for human development" (Section 41, Argentina 1994). India's states that "the State shall endeavor to protect and improve the environment and to safeguard the forests and wild life of the country" (India 1949, Article 48A); in addition, Article 21 provides that no person shall be deprived of his life or personal liberty. The constitution of Bangladesh (1996) protects the right to life and personal liberty (Article 32), which implies the right to a safe and healthy environment (User Survey a n.d.).

Many constitutions establish procedures to assure the right to participation, right to information, transparency of process, and access to justice. Principle 10 of the Rio Declaration (1992) explicitly states the right of all concerned citizens to participate. It assures every individual access to information concerning the environment and the opportunity to participate in the decision-making process.

Public participation in the environmental decision-making process has been increasing in most countries. Nevertheless, levels of participation and procedures for involvement differ between industrial and developing countries. The participatory process has been stronger in industrial countries, but many countries still lack a minimal legislative framework for public participation. Industrial countries have adopted formal mechanisms for public participation, as with procedures for EIAs. These processes allow participation in the formulation, review, and evaluation of policies. Central and Eastern European countries with economies in transition in, and developing countries in Latin America, Asia, and the Pacific have improved in their public participation. But generally most regions need to improve their overall quality and breadth of participation in areas such as EIA and environmental decision-making.

In eastern and southern Africa, public participation comes through co-management of natural resources, as with the Communal Areas Management Programme for Indigenous Resources. Another practical example is that local people take part in reporting on the state of the environment in Lesotho, Malawi, South Africa, and Zimbabwe. Regarding the availability and access to environmental information, African countries are implementing information systems and networks at the national and regional levels. In Africa, however, the participation of women and youth in decision-making is still seriously lagging (Eckerberg 1997; Pantzare and Vredin 1993; UNEP 1999).

Furthermore, the participation of NGOs has been vital to environmental protection. Their roles have ranged from raising public awareness to shaping policies through extensive participation in the negotiation of treaties, particularly in the area of climate change. Despite their observer status, they have influenced the content of the text (Sands 1994). Some NGOs provide legal assistance to citizens and indigenous communities. They also promote compliance with MEAs. They have raised awareness nationally and internationally in several ways; for instance, in Sri Lanka, NGOs have prevented logging and stopped the construction of a thermal plant; in India, a social movement protested the construction of the Narmada dam; in the Philippines, a consortium of 17 environmental NGOs has implemented a seven-year Conservation of Priority Protected Areas Project (UNEP 1999).

2.2.2.8 Environmental Impact Assessment as Measure for Regulation

The implementation of environmental law is carried out through a variety of regulatory techniques, the most widely used of which is the environmental impact assessment. An EIA is often required for activities that are likely to have a significant adverse impact on the environment and are subject to the purview of a competent national authority (UN 1992c).

Basic EIA requirements include the alternatives to be considered, the dissemination of information on projects, and public participation in the decision-making process. As discussed earlier, there is a trend to incorporate biodiversity considerations into EIA procedures. (See Chapter 4 for more detailed discussion.) Although most countries have legal provisions on EIA for major projects, biodiversity considerations are often insufficient in the EIA process because they are given low priority compared with economic and development considerations (UNEP 2001). At the domestic level, for instance, the Brazilian constitution requires that states and counties carry out EIA as a tool of environmental monitoring (Brazil 1988). In short, many countries have relied on command-and-control instruments rather than economic incentives, which are becoming more widely used (UNEP 1999). The major challenge is to determine which instruments need to be combined for the optimal effect in each country. Countries need to find the right mix of social control, regulation, and economic instruments for their situations (Hirakuri 2003; UNEP 1999). There is no substitute for sound public policy, however.

2.2.2.9 Domestic Legislation outside the Environmental Sector

Laws and public policies outside the environmental sector should be considered that are critical to the protection of the environment and sustainable development. These laws are usually linked to the public policies of the countries that promote economic development. Many of them are associated with infrastructure-related areas, such as agriculture, forestry, settlement and mining. In many cases, the main causes of deforestation in Latin America have been policy choices by governments and subsequent laws to implement those policies. For example, governments have often favored the conversion of the forests into agriculture or shift cultivation, cattle ranching, and other land uses through subsidies (Repetto 1990; WRI 1985). The agricultural expansion now causing deforestation in Africa and Asia is related to population growth (FAO 1997).

Other proximate causes of forest loss relate to industrial development, such as palm tree plantations or shrimp farming, shift cultivation, particularly in Asia (Inoue and Isozaki 2003). In Asia and the Pacific, land use law allowing conversion of forest to agriculture and commercial logging has caused environmental destruction. Further causes of environmental destruction include laws related to mining, construction of roads, irrigation, construction of hydroelectric dams, and urban expansion (FAO 1997). The settlement and exploitation of the Amazon rain forest, for instance, has been facilitated by the construction of highways cutting through the forest. Mining activities can seriously affect the environment. The extraction of minerals creates imbalances in nature. If the mining occurs in forested areas, the environmental impacts are major, leading to a change in the water balance and pattern of rainfall, sedimentation, river pollution, disruption in wildlife and fishery habitats, variation in the microclimate, and general disruption of the ecosystem. Thus to guarantee sustainable development, it is necessary to develop legislation that considers the protection of the environment (FOE 1989).

The mentioned subsidies can be characterized as "perverse" in that they cause damage to the environment and the economy rather than help society achieve desired goals (Myers and Kent 1998). As Myers and Kent (1998) point out, perverse subsidies are mostly seen in five main sectors—agriculture, fossil fuels, road transportation, water, and fisheries. Ultimately, these subsidies destroy biodiversity. Aware of this, the Fourth Conference of the Parties to the CBD stressed taking appropriate action against those incentive measures that threaten biodiversity. The COP encouraged Parties and international organizations, to identify perverse incentives and to consider removal or mitigation of their negative effects on biodiversity (Decision IV/10).

The next step in removing perverse subsidies, or mitigating their negative effects, was a call for the Fifth COP to the CBD to set up a Programme of Work to engage on this issue (Decision V/15). This led the Sixth COP to request that the Executive Secretary specify how to remove or mitigate perverse incentives in collaboration with relevant international organizations (Decision VI/15). The Seventh COP accepted the proposals as providing a useful general framework to address the perverse incentives in various economic sectors and ecosystems. The COP also encouraged parties and governments to use, on a voluntary basis, these proposals in implementing the incentive measures of Principles 2 and 3 of the Addis Ababa Principles and Guidelines for the Sustainable Use of Biodiversity (Decision VII/18 and its annex).

2.2.2.10 Domestic Enforcement System

Judicial review is a commonly used tool at the domestic level to ensure accountability of the regulators under command-and-control measures. Judicial review is used by persons with direct interest in the subject of the complaint. The courts merely have supervisory jurisdiction over the decision-making activities of the regulators; the final decision on the merits is made by the regulatory agency (Wolf et al. 2002).

Liability is a tool to compensate entities for economic harm and natural resources damage and, in some cases, to restore them. Violations of environmental regulations may result in civil or criminal liability (Campbell-Mohn et al. 1993; Handler 1994; Wolf et al. 2002). The violation of an international obligation generally gives rise to a victim's right to compensation for damage. However, it is difficult to ensure state responsibility in the field of environmental law. Few treaties provide the specificity needed, as in defining the exact nature of the violation that would give rise to liability. In pollution-related cases, it is easier to establish state responsibility because the violation can be measured easily. The loss of biodiversity, however, is difficult to quantify, as is the exact degree to which it will adversely affect the ecosystem (Birnie 1992).

In addition to traditional legal enforcement, *public environmental awareness raising, information dissemination* through education courses or publication of reports (for instance, TRAFFIC Report on illegal trade in endangered species), and *the participation of stakeholders* are other enforcement measures widely taken at the domestic level. Participation of environmental NGOs at regime meetings would improve effective implementation, which would ultimately relate to enforcement. Some NGOs have exercised the right to petition for court judgment before national courts to stop or prevent environmental harms. Such judicial procedures can help victims and facilitate the development of more effective domestic environmental policies and laws. Furthermore, experts, academicians, and mass media also play important roles in enforcement and in increasing public awareness of environmental needs (Rosendal 2000; Wuori 1997; Somsen 1998; Wolf 2002).

The *ombudsman system* is another tool to enhance enforcement and aid dispute resolution (Bertran 2002). Many countries have established ombudsman laws, which include the protection of the environment, including Greece, the Netherlands, and New Zealand. Although many countries have not enshrined the ombudsman in laws, their governments have established ombudsman offices (Yannis 2001). At the supranational level, the European Union set up an ombudsman in 1995 to deal with complaints about mal-administration by European community institutions and bodies (Seneviratne 2000). This is a voluntary, nonbinding and non-adjudicatory set of dispute settlement procedures that can be implemented by the United Nations. The ombudsman can deal with disputes among states or between states and citizens, including multinationals, indigenous groups, and NGOs (Koh 2004).

2.2.2.11 Summary: Legal Responses

The international law tradition recognizes that legal responses take place at many different levels, such as international treaties, soft law, and international customary law. These various levels of legal instruments are interlinked. Indeed, in order to implement environmental policies, we must pursue the different levels concurrently, and parallel hard law (international treaty with concomitant protocol) with soft law (Resolutions or Guidelines). The key elements to be considered at all these different levels of legal response are implementation, compliance, and enforcement. *Implementation* refers to the actions that states adopt to comply with international treaties through domestic regulations. *Compliance* means the extent to which countries abide by the obligations, both procedural and substantive, set out in international treaties; an example of procedural obligation is the requirement to report; whereas, the obligation to cease or control an activity is a substantive obligation (Jacobson and Weiss 1997). *Enforcement* means the actions taken by competent authorities to ensure compliance with the laws. The basic question is how to put the regulations into practice. To this end, domestic legislation plays a vital role in the implementation of/and compliance with environmental laws.

As already noted, enforcement varies in intensity and quality among countries, according to the degree of strictness of enforcement policies. The foremost obstacle has been the ineffective implementation of legislation. The major reasons identified in many countries are the lack of trained staff, political will, monitoring, and enforcement; in some countries, appropriate and applicable standards are lacking as well. Also, the lack of coordination among government institutions and inadequate funding have been stumbling blocks (UNEP 1999).

On the other hand, the cooperation of environmental authorities with the public is now being required in several instruments.

In particular, access to information and justice, and public participation in decision-making processes, contribute to effective monitoring, compliance, and enforcement. International obligations need to be reflected in domestic policy, which will ensure the effectiveness of the regime. The compliance and enforcement approach will be successful only if it takes into account the interdependence of the economic, environmental, social, political, and cultural factors that bear on the management of natural resources.

2.2.3 Economic Responses

A wide range of opportunities exists to influence human behavior with detrimental effects for ecosystems and their services in the form of economic and financial instruments. Some of them establish markets; others work through the monetary and financial interests of the targeted social actors; and yet others affect relative prices. The *feasibility, effectiveness, and efficiency* of such interventions *depend on the biophysical characteristics of the problem* and *the socioeconomic circumstances* in which they are adopted. This section summarizes economic and financial response options ranging from hard regulatory forms mainly applied at small geographical scales to softer mechanisms used in the larger geographical and/or jurisdictional context. The classification of economic responses draws on standard categories in the environmental economics literature (Pearce and Turner 1990; Tietenberg 1992; Perman et al. 1996; Wills 1997; Common 1996; Hanley et al. 1997; Stavins 2000; OECD 2001), but it also includes novel groups to draw attention to new tools and approaches in ecosystems management that have a decisive economic component (Dietz and Stern 2002).

Many ecosystems problems are caused by what economists call "perverse subsidies"; they entail various forms of direct or indirect monetary transfers that are economically inefficient and environmentally harmful. Subsequent chapters in this volume list numerous examples ranging from deforestation to the depletion of fisheries. The precursor of any new intervention to protect ecosystems services should be to check existing regulations and eliminate or at least mitigate perverse subsidies. The environmental effectiveness and the economic efficiency of new response policies and measures depend crucially on the extent to which perverse subsidies are still part of the regulatory regime.

If one considers the diversity of the uses of ecosystem services, the pollutants affecting them, their impacts, social and political situations, economic and institutional conditions, and other factors, it becomes clear that no single instrument is the best for all types of ecosystems problems and socioeconomic situations. The criteria for assessing and choosing the intervention options include the following (partly based on Perman et al. 1996):

- *Cost-efficiency*: the extent to which the response option can achieve the desired environmental objective at the lowest possible cost;
- *Dependability*: the extent to which the intervener can rely upon the instrument to achieve the specified target (in this context, the relative slopes of the functions depicting the costs and benefits of interventions are an important consideration in choosing between quantity- or price-based instruments);
- *Information requirement*: the extent and nature of information required for formulating the intervention and the cost of gathering this information;
- *Enforceability*: the kind and level of monitoring needed to keep track of the implementation of the response chosen; type of measures available to enforce the intervention;
- *Long-term effects*: the long-range impacts of the chosen response, that is, whether its influence is constant, increasing, or decreasing over time;
- *Dynamic efficiency*: whether the response option is providing a continued incentive to improve performance with respect to the original ecosystem management objective;
- *Flexibility*: whether the chosen response option can be adapted quickly and cheaply when new information becomes available, the underlying conditions change, or targets need to be modified; and
- *Distributive impacts*: how the response option affects the welfare of different social groups and what the prospects are that winners can compensate losers.

The relative weights of these criteria depend on the ecosystem problem and its management context. These weights will influence considerably the choice of the response option. (See Chapter 18.)

2.2.3.1 Command-and-Control Interventions

These response options prescribe specific forms and/or quantities in restricting access to and regulating use of ecosystems services or emitting environmentally harmful substances. While legal command-and-control interventions are primarily enforced by threatening noncompliance with anti-criminal measures (imprisonment of individuals, temporary suspension, or complete disbanding of the legal entity), their economic counterparts are promoted by (often increasingly severe) monetary penalties. The classification of command-and-control interventions into the legal or economic categories is difficult because there is considerable overlap between the two. Most command-and-control interventions imply financial penalties in cases of initial or minor noncompliance and most apply increasing penalties for persistent noncompliance. Some regulations might even extend the penalty to criminal instruments but this does not justify their classification as legal instruments.

Prohibition is the strictest form of command-and-control response. This instrument bans all or certain clearly defined forms of ecosystem use. In acute cases, harsh forms of prohibition bar physical access or entry to the protected ecosystem.

Explicit controls are usually introduced to protect landscapes, terrestrial or water-based ecosystems in an ecologically valuable region. Typical forms of explicit controls prescribe certain types of land use.

Zoning and designation can also imply some form of prohibition, but their main concern is to direct various types of ecosystem uses to clearly demarcated geographical areas.

Direct provision of ecosystem services implies that the intervener takes full control of the resource, determines the amount to be appropriated, and distributes the resource to the entitled community. This option is often used by communities to maintain the productivity of their resource base or by government agencies under the circumstances of severe shortages when rationing of an ecosystem service becomes necessary. Since the use of most ecosystem services is difficult to control and supervise directly, the success of these arrangements requires either a high level of moral cohesion among community members or a strong policing and penalty threat operated by the regulator.

Fixed quota systems can be used both for controlling the use of ecosystems services by individuals, households, or other users, and for regulating the amount of harmful emission from individual sources of pollution. The former entails establishing the total amount of ecosystem service that can be taken and setting up a quota or license for each resource user. The latter involves apportioning the emitted quantity to the various sources in order to derive a quota or license for each source as a fixed quantity allowed to be emitted. The success of this type of response requires effective monitoring and harsh penalties for noncompliance. If monitoring and penalties fail, the effectiveness and therefore the dependability of the quota-based instrument will be reduced. The attainment of an economically efficient regulation with a fixed quota system is possible in principle, but rather unlikely in practice. In controlling pollutant emissions, economic efficiency would require the regulator to know the marginal abatement cost function for each polluter (that is, the cost of abating an incremental unit of pollutant). The costs of collecting such information are in most cases prohibitive. In practice, fixed quota systems often lead to arbitrary distribution of the emission quotas, resulting in inefficient allocation. Moreover, this response option provides very weak incentives to foster dynamic efficiency because once a polluter meets the allocated emission quota, there remains no incentive whatsoever to reduce pollution any further.

Technology regulation is another possibility to protect ecosystems from overexploitation or excessive amounts of harmful pollutants. To reach these ambient standard values, the regulator targets the production process or the equipment emitting the pollutants. This takes the form of specifying minimum technology requirements. Prescriptions for dust removal from flue gas, specifications of minimum stack-heights, requirements that cars have catalytic converters, and cooling or treatment technologies for wastewater are examples of technology regulations via command-and-control systems. This response option is easy to implement and cheap to administer because monitoring and administration costs are low relative to the enforcement costs of other options. It can be effective when "end of pipe" solutions are easily available but not used by the polluter. The instrument is also dependable. However, in most cases it is not cost efficient because it is not focusing on the least-cost abatement opportunities (although this may be less important in special cases when the exact location of the point of discharge matters for the impacts). Moreover, this intervention is inflexible because once the prescribed technology or equipment is in place, it is in most cases difficult to undertake additional modifications. Technology regulation also does not promote dynamic efficiency, because the changes required by the regulation are completed with the installation of the prescribed technology or equipment.

The need for interventions in the use of ecosystems services and for responses to changes in ecosystems emerges in extremely diverse social, economic, political, and biophysical contexts. Different ecosystems issues require different policy instruments. The relative merits and shortcomings of the command-and-control instruments compared to those of other economic response options become extraneous when they are the only feasible or environmentally effective interventions. The name "command and control" does not imply any negative connotation. It simply specifies that with these instruments the regulator prescribes or prohibits some actions and controls compliance.

2.2.3.2 Incentive-based Interventions

The second range of economic response options uses economic incentives to entice users of ecosystem services to limit their resource use to the socially optimal level. Defined here in a simple social cost–benefit context, the socially optimal level of control is where the marginal cost of abatement equals the resulting marginal benefit. In the presence of market failures, the market price does not reflect those social marginal benefits and the polluter has no incentive to invest to reach the optimal level of control. The regulator can impose an emission charge to provide the incentive for the producer to increase the level of control to the social optimum. Such optimal taxes are called Pigovian taxes. Using economic incentives in this way may involve increasing the delivery of selected ecosystems services (for example, community woodlots) or combining the solution of economic problems with measures to address environmental concerns (for example, ideas to reform the European Union's Common Agricultural Policy to reduce the pressure of excess production and to foster biodiversity).

Tax and subsidy schemes can be uniform or differential. They have been widely used to close the gap between the socially optimal level of using ecosystem services and the level of use based on more narrow private benefits. Taxes are charged for each unit of appropriated ecosystem service (per cubic meter timber cut or per cubic meter water diverted) or for each unit of pollutant discharged (kilogram of sulfur dioxide emitted, milligram of water pollutant released), whereas discharge taxes are sometimes based on an input that can be easier and more precisely measured (carbon content of the fuel instead of carbon dioxide emission). In a subsidy scheme, the regulator pays subsidies to the polluter for the abatement effort. The introduction of taxes and subsidies produces the same effects: they both modify the relative prices of the products with which the appropriation or the use of ecosystem services or the emission of pollutants are associated. Due to distributional implications, however, the long-term effects of taxes and subsidies differ. Taxes close the gap between the social and the private marginal benefits of using an ecosystem or emitting pollutants. The efficient level of the tax is equal to the difference between the marginal private benefit and marginal social benefit so that users of ecosystem services will consider what they have been ignoring before the tax and thereby internalize the formally external costs of their activities. Accordingly, profit-maximizing actors will adjust their use of ecosystem services so that the marginal social benefit will be equal to the marginal social damage, because their post-tax marginal private benefits of the resource use will be equal to the marginal social benefit.

The principles and the operational mechanisms of instruments under many other names are similar to those of taxes and subsidies. On the levy side, the list includes incentive, distributive, user, product, and administrative charges, as well as deposit-return systems. Explicit user charges include license fees (harvesting, hunting, fishing), entrance fees, severance, and resources taxes. The arrangements on the grant side incorporate compensations, tax incentives (reducing tax burden), relief, exemptions, and tax deductions.

Tradable resource use and tradable emission permits have become increasingly popular in recent decades. This response option has four elements. The first involves a decision about the total amount of resource use or pollutant emission to be allowed. In a socially efficient regulation, the total amount of resource use or pollutant emission permits should be equal to the efficient level of resource use or pollution. If the efficient level is not known, some other basis should be used to define the total amount of permits. The second element of a tradable permit scheme is regulation. Any resource user or polluter is allowed to appropriate the resource or emit the pollutant only up to the quantity covered by the permits available to him; above that level, a serious threat of an expensive fine or penalty must be installed and implemented. The third element is a decision about the initial allocation of the total amount of permits among the resource users or polluters.

The final element is the need to guarantee free trading of the resource use or emission permits. An efficient control of ecosystem service use or pollutant emission can be achieved either via a tax rate or by issuing a certain quantity of tradable permits. Taxes set the price of emission or ecosystem service use while the tradable emission permits set the quantity of the allowed resource use or the quantity of pollutants to be emitted. The profound difference compared to quotas, licenses, or standards is the transferability and marketability of the use or emission rights. If a permit market is free, then both the price and the quantity of the ecosystem service use or the pollution will be efficient. Moreover, if the amount of permits corresponds to the economically efficient level of resource use or pollution, then the equilibrium price of the permit will indicate the shadow price of the ecological service or pollution at the socially optimal level. In terms of cost-effectiveness, the effects of permits will be the same as the effects of the optimal tax for subsidy scheme. The main difference is in the distributional effects.

The establishment of a tradable permit system is essentially equivalent to creating a market for ecosystems services that were used more intensely than the socially optimal level. Tradable permits can take the form of ambient permits, emission permits, pollution offset systems, tradable harvest, or catch quotas.

A comparison between command-and-control versus market instruments indicates that command-and-control responses directly regulate the quantity of ecosystem service use or quantity of pollution emission or regulate the technology that is leading to pollution emissions. In contrast, market-based instruments alter the relative prices or generate price incentives to achieve socially desirable levels of ecosystem service use or pollution emission. A closer look at the market instruments reveals that a resource use tax or an emission tax scheme can achieve the efficient target at the lowest social cost; in fact, it can achieve any target at the least social cost. Moreover, a tax set at any level can achieve some reduction in ecosystem service use or some level of pollution abatement. In addition, market instruments generate dynamically efficient incentives for behavior. Since all users or polluters face the same tax, these outcomes emerge from the profit-maximizing behavior of the affected actors.

In contrast, command-and-control instruments are blind to cost-efficiency. They would achieve a cost-efficient solution only by coincidence. Commenting on environmental standards, one of the most widely used command-and-control instruments, Pearce and Turner (1990, p. 103) point out that "[T]he problem with standard-setting is that it is virtually only by accident that it will produce an economically efficient solution." The basic reason is that the regulator does not know the marginal abatement cost function of each polluter. Tietenberg (1992:403) reviews eight empirical studies; Perman et al. (1996, pp. 238–9) add two more analyses in which the costs of pollution abatement using alternative instruments are compared. The ratios of the actual command-and-control costs to those of theoretically expected least-cost market-based instruments found by these studies vary between 1.07 and 22, with a median ratio of 4.18. Even if one considers that the cited dozen-or-so case studies compare actual command-and-control costs to theory-based calculations of the costs of market-based instruments and it is unrealistic to expect the latter to operate at the theoretical minimum costs, these studies provide obvious evidence that market-based instruments are overwhelmingly superior to command-and-control instruments in terms of cost-efficiency.

An important consideration in comparing the response options is concerned with the transactions costs. The expenses associated with establishing and operating the necessary monitoring schemes, administrating the behaviour of the targeted actors, and enforcing the implementation of the chosen instrument can be substantial. These transaction costs often influence the choice of the least-cost instrument. Since transaction costs may be substantially lower for technology standards, regulators often prefer this response option irrespective of their actual social costs in the broader sense. Yet command-and-control instruments might be required in special cases where the pollutants involved are not uniformly mixing and their exact places of emissions matter. Moreover, in many areas of ecosystems management, these instruments are the only feasible environmentally effective response option, making the question of cost-efficiency irrelevant.

Another crucial consideration in the choice of instruments is dependability. Comparing the main market-based instruments shows that if the aggregated abatement cost function is known with certainty, then the tax rate can be determined to reach the desired level of abatement and the instrument will be completely dependable. Similarly, under these circumstances the amount of permits can be determined, the permit price will be predictable, and the tradable permit scheme will also be completely dependable. The situation is different if the marginal abatement cost function is not known with certainty. In this case, the tax rate can be set but the amount of ecosystem service reduction or pollution abatement will be uncertain; under these circumstances, the tax scheme has uncertain effectiveness and is not dependable. In the same situation, a permit scheme will be dependable for the quantity of ecosystem service use or pollution emission, but the associated costs will be uncertain. In this case the permit price cannot be predicted.

Different response options have different distributional consequences. In the case of tradable permits, the distributional effects depend on the initial permit allocation method. If the permits are sold by auctioning them out, then the equilibrium permit price will be equal to the aggregated marginal abatement cost associated with the total number of permits. In this case, the net transfer of funds flows from the resource users or polluters to the tax authority. In contrast, if the permits are distributed freely based on some arbitrary rules (grandfathering based on historical records or equal per capita allocation), then some resource users or polluters will sell part of their permits and gain from the transaction, while others will need to buy permits and thus lose compared to the unregulated situation. Free distribution of permits also results in no net transfer from resource users or polluters to the tax authority. In comparison, resource or pollution taxes represent clear transfers from the users/polluters to the tax authority; therefore, their distributional effect is the same as that of auctioning out the permits. Under a subsidy scheme, on the other hand, funds are transferred from the government to the polluters or ecosystem service users to change their respective behaviors.

There are serious competitiveness concerns associated with ecosystem or environmentally oriented interventions. A unilateral tax is perceived to harm international competitive positions irrespective of whether it targets ecosystem service use or pollutant emissions. This fear often leads to perverse regulation when activities associated with internationally traded commodities are either not regulated or regulated only lightly, whereas goods and services not traded internationally are subject to fierce regulation.

2.2.3.3 Voluntarism-based Instruments

Recent years have seen an upsurge in new approaches to responding to problems associated with uses of ecosystem services and pollution emissions. They all rely on implicit sources of behavioral change and thus tend to be specified as voluntarism-based options (OECD 2003; Dietz and Stern 2002).

Information provision and education intends to influence the behavior of targeted individuals or communities by providing solid and scientifically based information about the ecosystem implications of certain behavior, with the expectation that this will trigger behavioral change when resonating with broadly established and accepted ethical norms and principles. Many educational and information-based responses go beyond this and attempt explicitly to trigger a change in values and preferences that, in turn, will lead to behavioral change toward a more benign use of ecosystem services.

Ecolabels represent a specific and prominent form of information provision. By drawing the consumer's attention to the environmental implications of using or consuming a specific product, ecolabels have become an effective form to promote green consumer behavior among environmentally conscious people (Dietz and Stern 2002).

Voluntary measures take the form of explicit and formal agreements between the regulator and the targeted agent. They can also take the form of agreements among actors otherwise competing in the provision of the same range of goods and services, concerning their own management of ecosystems or pollution emissions or setting the same rules and regulations for their suppliers. Among the range of voluntary measures, government-promoted voluntary programs and industry-wide codes of practice represent the two main clusters.

The crucial feature of voluntarism-based instruments is that they are not put in place as formal or compulsory intervention. Yet it is often the case that users of ecosystem services negotiate voluntary agreements among each other or with their regulator to prevent a more stringent or more costly regulation being imposed upon them. These kinds of response options are particularly useful in previously unregulated areas of ecosystem management. Stakeholders can experiment with different technologies and management procedures to comply with the voluntary agreement without running the risk of having to pay high penalties if they fail. At the same time, regulators can monitor the process and collect information about the technological options and economic costs of reducing the pressure on ecosystems. These positive features also imply important limitations. If consumers find the price differences too large and do not choose ecolabeled products, or if the companies fail to achieve their voluntary targets because they find it too expensive, more effective instruments will be needed.

2.2.3.4 Financial and Monetary Measures

Financial and monetary response options include a broad array of measures ranging from small-scale, locally oriented actions to grand international schemes. In some cases, the small locally oriented instruments are needed to implement the large-scale arrangements.

Microcredits can support arrangements to directly reduce the pressure on ecosystem services or to start-up alternative forms of livelihood that will reduce the pressure on ecosystem services indirectly. Microcredits are particularly attractive instruments in those cases when they simultaneously contribute to ecosystem protection and poverty alleviation.

Loans are usually provided at a somewhat larger scale. They can help local ecosystem users or resource operators make the, often modest, investments required to change their technologies from a harmful to a more benign one.

Funds set up with private endowments and public resources, can be sources of microcredits and loans, but they often also provide the resource for changing management practices of a targeted ecosystem. Depending on the nature and internal regulation of the fund scheme, the requirements for commercial viability of the sponsored activities and the conditions for repayment are usually less stringent than those of bank loans or other commercial credit forms.

Public financing can take the form of direct and indirect intervention. Direct public financing is explicitly oriented toward the protection, replacement, or provision of an ecosystem service. Indirect public financing can take the form of state guarantees or government indemnity. This arrangement reduces the risk premium charged by the credit provider to resource operators and makes the acquisition of the required financial resource more affordable to them.

Debt swaps are a relatively new international financial response option. Many developing countries reached high level of indebtedness in the 1980s and 1990s. In order to service their debts, they were forced to overexploit and sell their environmental resources. Debt swap is an arrangement to help these countries out of the debt trap. Foreign debts are cancelled in exchange for commitments to set aside and preserve valuable ecosystems.

2.2.3.5 International Trade Policy

The economic incentives to overexploit local ecosystems often stem from the effective demand for their goods and services in remote geographical regions. Harvesting, processing, and exporting such ecosystem services in developing countries is an important way to alleviate poverty, improve quality of life, and start the accumulation of local capital resources to foster economic development. Yet it remains a challenge both for providers and recipients of such ecosystem services to avoid exploitive use and degradation of the resource base and to manage the use of ecosystem services to satisfy distant demand in a sustainable way.

International trade agreements are the legal form of controlling the economic incentives for exploitive use of ecosystem services. They involve both source and recipient countries. Such agreements can include qualitative characteristics (for example, species, size, or age of the natural resource that can enter international trade), quantitative limitations (for example, the amount of the ecosystem service that can be removed and allowed to enter international trade flows), or technological characteristics (for example, the equipment or process adopted, the management practices followed—ranging from the size and grid density of fishing nets to selective versus clear-cut harvesting of timber) of the ecosystem goods and services that are allowed to enter international trade.

Import restrictions imposed by recipient countries, typically industrial countries, either restrict or ban altogether the amount of ecosystem goods and services permitted to enter their domestic markets. These policies can target specific ecosystem goods in general (like products associated with endangered species from any country) or exports from certain countries for clearly defined ecological/environmental management reasons.

Export restrictions are put in place by source countries in the form of outright ban, export tariffs, or quotas in order to protect their own ecosystem.

2.2.3.6 Summary: Economic Responses

Economic and financial interventions provide powerful instruments to regulate the use and avoid the overuse of ecosystem goods and services. The adoption of economic instruments usually requires a legal framework and, in many cases, a social or institutional intervention as well. The various types of economic interventions are combined in many cases to achieve an effective regulatory regime. For example, import restrictions (as part of international trade policies) are typically complemented by information provision such as ecolabeling (a voluntarism-based instrument), debt swaps, and/or loans from the recipient country to the exporting country to entice sustainable management of the underlying ecosystem (financial and monetary measures).

The choice of a viable and effective economic intervention mechanism is determined by the socioeconomic context. Resource taxes can be a powerful instrument to guard against the overexploitation of an ecosystem service but an effective tax scheme requires well-established and reliable monitoring and tax-collection systems. Similarly, subsidies can be effective to introduce and implement certain technologies or management procedures but they are totally inappropriate if the prevailing pattern of using public funds is "take the money and run."

2.2.4 Social and Behavioral Responses

Social and behavioral responses including population policy, public education and awareness, empowerment of communities, empowerment of women, empowerment of youth, and civil society disobedience have been instrumental to a certain extent in shaping ecosystems and human well-being. These are interventions stakeholders initiate and execute through exercising their procedural or democratic rights (Douglas-Scott 1996; also see discussion above) in efforts to improve ecosystems and human well-being. Such measures are for by major global environmental policies such as Agenda 21 (United Nations 1992) and the Plan of Implementation of the World Summit on Sustainable Development (United Nations 2003).

These kinds of responses demonstrate the commitment and participation of a wide range of actors. Support structures have been used also to facilitate positive outcomes, as seen in the Amagasaki and Kawasaki pollution lawsuits in Japan, where through courts, good lawyers, medical experts, and scientists were mobilized to assist in successfully fighting for victims and preventing pollution. Furthermore, the lawyers' association and the pollution victims raised money for the legal battles. In these cases, commitment and support structures made victory possible, but there may be situations where resources are lacking and failure is possible. Moreover, social and behavioral interventions can be instrumental in conservation efforts when facilitated with the appropriate resources.

2.2.4.1 Population Policies, including Family Planning

Population growth can be a contributing factor to many social problems including environmental issues. For example, population pressure on arable land in the Asia Pacific region is partly responsible for land degradation (UNEP 1997). The same could be said for deforestation and the ever-decreasing biodiversity. Population growth since 1950 has been on a steady rise meaning that there will be more mouths to feed and resources will continue to deplete. The world's population in 2001 was 6.2 billion (Worldwatch Institute 2002) and it is predicted to stabilize at 7.8 billion in 2025, but only when appropriate policies and family planning measures exist (Worldwatch Institute 2001).

A range of stringent (fines and punishment) and lax or incentive-based measures (contraceptives, family planning, and educational programs) are already available to countries. China's one-child policy is one illustration of the stricter measures that has proven successful by reducing the number of children per woman from 6 to 2.5. Despite its success, there are criticisms on the restriction of individual rights and freedoms. India pursues softer measures, including birth control and education programs to curb population growth, and has been also successful (User Survey a, n.d.). Furthermore, it must be noted that while population policies can be beneficial in achieving their primary objective(s), they can also affect adversely social and economic aspects of countries, such as the human rights violations seen in China or the fear of less manpower to support the older generations in the future as is predicted for Japan.

2.2.4.2 Public Education and Awareness

Public education and awareness can play important roles in improving conservation efforts (Goodale 1995). A learned and informed society can make sound decisions about conservation (UNEP 1999). Both textbook learning and the media can facilitate this need. However, there are many who are denied formal education, access to the media, access to schools, teachers, and effective learning methods. In response, Chapter 36 of Agenda 21 (United Nations 1992) and paragraph 109 of the Plan of Implementation of WSSD, among others, stress the need to improve inadequacies in public education and awareness. Poverty of resources, ineffectiveness of the general education system (UNEP 1999; UNESCO and the International Association of Universities 1986), and the deficient utilization of full parameters of both formal and informal education (Goodale 1995) tend to be setbacks. Given the practical sociocultural, environmental, and economic difficulties in many societies, an emphasis in moving learning from classrooms and educational centers to more accessible settings can be more effective. Approaches and methods used tend to be impractical. Some educationists stress that effective learning involves proactive and participatory approaches (OECD 1993).

Despite challenges, countries and organizations alike have begun addressing environmental education and awareness (UNEP 1999). India's Lower House for instance, passed a bill in 2001 to make education a fundamental right. Other programs in India promote informal education and awareness about the environment including eco-clubs or enviro-clubs for first-hand experiences (TERI 2003). OECD reports that countries in Europe including Austria, Italy, Netherlands, Norway, and the United Kingdom have integrated environmental education into their systems. It also reports cases of both formal and informal environmental education and public awareness in developing countries, including Kenya, Senegal, Malaysia, the Philippines, Brazil, and Ecuador. IUCN, the World Wildlife Fund, and UNESCO's Man and the Biosphere Program all run environmental education and awareness programs (OECD 1993). Furthermore, the Ubuntu Declaration, which calls for greening school curriculums, is also a progressive step (UNU 2003).

Efforts are made but deficiencies still prevail, especially poverty of resources including infrastructure, teaching staff, and finances. Learning can be encouraged more at all levels across different age groups using all methods in a variety of settings. More importantly, constancy in some of the initiatives in public education and awareness is lacking. The need now is for stakeholders to address these issues in efforts to institutionalize environmental education and awareness to a greater extent wherever it is seen to be lacking (UNEP 1999).

2.2.4.3 Changing Values and Attitudes: Empowerment of Communities, Women, and Youth

As new concerns about environmental problems emerge, values and attitudes of people may transform simultaneously to meet the demands. Changes can be induced internally or imposed externally by factors including technological advancement, crime, gender issues, war, religion, education and awareness (UNEP 1999), and regulation. Also, changes depend on perceptions, experiences, and opportunities. For instance, environmental pollution and loss of biodiversity have caused many to change their values and attitudes toward the environment. The series of global and environmental policy initiatives that evolved during the 1970s, including the Stockholm Declaration on Human Environment, the minimum environmental regulation at the domestic levels, and stakeholders ranging from governments to industry designing and implementing solutions to environmental problems, signify the change in values and attitudes of society toward environmental concerns (Worldwatch Institute 2003).

A specific example is the "polluter pays" principle where certain environmental goods and services that were once considered as free are no longer free to use; polluting industries now are required to internalize costs for using environmental goods and services. A second example is the case of mechanisms for recycling of wastes in countries including Japan (Clean Japan Center 1999) Germany, and the United States (Council on Environmental Quality 1997) where some are required by law to recycle wastes. Some geographic regions, including the Arabian Peninsula have mechanisms to recycle wastewater, solid waste, and waste paper (UNEP 1999). But where values and attitudes are critically important to the improvement of environmental protection and sustainable development, their progress is by no means ever assured.

2.2.4.3.1 Empowerment of indigenous and local communities

Anthropologists and ethno-botanists claim that indigenous and local communities are experts in their environment and know well how it functions (Hugh-Jones 1999). Their practices may not be the best, but they have some beneficial attributes for biodiversity conservation. In fact, some practices of indigenous and local peoples are partly responsible for the remaining biodiversity. Despite these facts, indigenous and local communities have been struggling with oppression. Their participation in decision-making processes is limited. Their practices, beliefs, and ideologies have been disregarded. Their rights to land and property have been confiscated, disconnecting them from their land and environment.

Recently, however, indigenous and local communities—with their practices, knowledge, and innovations that contribute to conservation—have been given some recognition at the global level. Article 8 (j) of the CBD, for example, laid foundations for the use of practices of indigenous and local communities in biodiversity conservation (CBD 2003). Principle 22 of the Rio Declaration, Chapter 26 of Agenda 21, Principle 5 of the Forest Principles, Articles 16 and 17 of the CCD, and the World Heritage Convention (IUCN 1997) have all taken some measures to enhance the participation of such communities. Participation goes beyond the mere presence in processes and meetings to include the use and incorporation of some of practices, ideologies, values, and laws into the mainstream processes and systems. Possessing stronger land and property rights is one factor that can contribute to indigenous and local communities' participation in conservation efforts. These communities are among the least likely to receive some form of education. Occasionally, some of their may be unsustainable (Forrest 1999), but through education and awareness raising, unsustainable practices could be corrected. Finally, funding is needed to encourage conservation, but indigenous and local communities are one of the underprivileged groups that do not possess the funds to carry out activities. (See Box 2.1 for an example of sustainable resource management in Papua New Guinea.)

2.2.4.3.2 Empowerment of women

Women are most knowledgeable about their environments through tasks such as food gathering, gardening, washing, clothes making, and preservation. They are also the first educators of their children; what they learn through their interaction with their environment they pass on to their children. Women are also traditional healers who are knowledgeable of medicinal plants. They not only know about using medicinal plants, but are also in a better position to manage medicinal plants and the general environment in a sustainable way. (See Box 2.2.) Despite their knowledge about the environment and the potential they possess, they have been among the most suppressed groups. Participation of women in decision-making has been restricted by social and cultural structures. For example, in most societies, women are excluded from land tenure. However, recently there is a growing recognition of the role of women in conservation of biodiversity. They are beginning to organize themselves in numerous ways to contribute to development. Their empowerment and participation in shaping ecosystems and human well-being is crucial.

2.2.4.3.3 Empowerment of youth

Today's youth are tomorrow's leaders. For them to determine and lead society, they must be physically, spiritually, and mentally fit. Although there is the obvious need for empowering youth, a se-

BOX 2.1

Case Study: Land and Environmental Ethics in Melanesia

The Melanesians and their ancestors were never ecologists, but their land and environmental ethics are based on reciprocity and balance. This has contributed to conservation and sustainable use of biodiversity and their well-being to a certain extent. Consequently, Papua New Guinea now maintains approximately 7% of the world's biodiversity. The contributing factors for biodiversity conservation in Papua New Guinea are social norms, rules, values, and animistic beliefs. In Papua New Guinea, certain tribes believe that they have descended from certain animals or plants, which compels them to preserve the plants or animals. Furthermore, when cutting banana leaves, one is encouraged to use the mature leaves rather than the younger ones. Some of these practices contribute to conservation and sustainable use. As more than 70% of people in Melanesia live in villages and practice these lifestyles, encouraging the use of management practices people are familiar with might be an effective way forward rather than imposing externally designed management practices.

BOX 2.2

Case Study: WAINIMATE and Traditional Medicine

The WAINIMATE is an association of female traditional healers with the purpose of recognizing and valuing women's knowledge of conservation of medicinal plants. Studies conducted in Fiji and other parts of the world showed that women knew more plants than men. The WAINIMATE is one such group whose members are knowledgeable on medicinal plants and their uses. Using plants shows that one possesses the traditional knowledge associated with the plants and also the know-how to conserve and use them sustainably. One of the tasks of the WAINIMATE is ensuring that women know that traditional medicines are safe and effective for treating diseases. This initiative is also found to benefit local people in Fiji who cannot afford chemical drugs. Women could contribute tremendously to society if they had the opportunity. WAINIMATE is said to put out their first traditional medicine handbook.

(Communication with Wana Domokamica, Traditional Healer and Member of WAINIMATE, June 2001.)

ries of problems act counter to their empowerment. The youth of today face problems including unemployment, lack of support (Neumann 2000), lack of proper education, disease, and crime. In 2000, the International Labor Organization estimated that 70 million young people, mostly in industrial countries, are unemployed and the number is increasing. A number of sexually transmitted diseases, including HIV/AIDS, are common among youth. The Henry J. Kaiser Family Foundation reported in May 2000 that teens and young adults comprise one third of the 40 million people living with HIV/AIDS throughout the world. The Foundation predicts that this is only a tip of the iceberg and the figure will rise in the future (Henry J. Kaiser Family Foundation 2003; UNFPA 2002).

In many developing countries, deteriorating economic conditions have made education an expensive proposition, and youth do not attend school because parents cannot afford fees. A "user pays" policy in some countries (for example, Papua New Guinea) further exacerbates the problem. When youth are not in school or employed, they generally do not get much attention; when their energies are not adequately channeled, they may end up engaging in criminal activities. In the United States, for example, crime among youth is high; a report in 2000 by the Federal Bureau of Investigation indicated that 55.1% of the crimes committed throughout the country were by people below the age of 25 (FBI n.d.). Many issues threatening the youth today in both developing and industrial societies work counter to their empowerment and need special attention.

2.2.4.4 Civil Society, Disobedience, and Protest

Paragraph 150 of the WSSD Plan of Implementation calls for partnerships and the participation of all actors. History has witnessed major civil disobedience and protests associated with numerous issues ranging from the French Revolution to the Civil Rights Movement in the United States. Martin Luther King, Jr., Mahatma Gandhi, Nelson Mandela, Chico Mendes, and Rosa Parks have all participated in various forms of civil disobedience that have left milestone changes in human history.

While boycotts and bans initiated by NGOs may not always be effective, the role of civil disobedience and protests in guiding the world back to the right track when it is heading in a risky direction has historically been a significant one. Thus civil society disobedience is not about breaking the rule of law, but about alerting governments to the consequences of their inaction and bringing to light some of the hidden issues. There have been various successful as well as unsuccessful movements, both violent and nonviolent in nature, throughout the world. A few examples of such movements are The Ogoni people of Nigeria agitating against the oil company polluting their environment (Beauchemin 2001), the Bougainville people of Papua New Guinea opposing the government and an Australian mining company to stop environmental degradation and claim compensation for use of customary land (William 1998), Chico Mendes and his people in Brazil fighting to protect the rain forest of Brazil (see Box 2.3), and the Chipko Movement in India. Another recent example is Greenpeace protests on behalf of the environment that led to the bombing of its ship, the Rainbow Warrior, by the French Intelligence Agency. Although civil society disobedience and protests have been instrumental in driving change and maintaining balance, violent acts are not encouraged.

2.2.4.5 Summary: Social and Behavioral Responses

No one social and behavioral intervention alone can influence conservation of ecosystems effectively and enhance human well-being positively. A combination of interventions is necessary. To induce social and behavioral change, a step-by-step process is required. For instance, for women, civil society, local communities, and youth to be able to change their attitudes and mentality toward conservation and well-being, incentive-based initiatives can be useful. One such incentive that is lacking is balancing rights and responsibilities of stakeholders. Clear rights and responsibilities create an intimacy that can be the driving force for change. Education and awareness campaigns are also crucial and have been used to stress the positive and negative consequences of why one has to behave in a certain manner. Thus in education and awareness campaigns, up-to-date and accurate information can facilitate the tasks effectively. Furthermore, without support structures, social and behavioral interventions cannot succeed.

BOX 2.3

Case Study: The Chico Mendes Extractive Reserve
(My Hero 2003; Environmental Defense 2003)

A rubber tapper, environmentalist, and union leader, Chico Mendes lived with his people in the Brazilian Amazon tapping rubber and collecting Brazil nuts for the last 100 years. Around the same time and area, rich cattle farmers and industrialists began expanding their activities into the rain forest, causing major threats to the forests. In response, Chico Mendes and his people started a group to prevent the destruction of rain forest caused by the cattle ranchers and miners. His group fought for an extractive reserve. During peaceful protests they often encountered opposition and threats from their opponents and the government. The confrontation eventually led to Chico Mendes' assassination in 1988. After his death, pressure from both within and outside prompted Brazil to consider the work and concerns of Chico Mendes and his people leading to the establishment of the Chico Mendes Extractive Reserve, which now covers 97,057,000 hectares.

2.2.5 Technological Responses

Technological responses are intended to influence the tools (hardware) and procedures (software) people use in their direct interventions with ecosystems goods and services (for example, fishing and logging) and in all other activities that affect ecosystems indirectly (for example, emissions of pollutants). Technology can play a critical role in responding to ecosystem-related problems by providing a link between human activities and the natural resource base. When harnessed to its full potential and developed with ecosystem objectives in mind, technology can provide sustainable alternatives to polluting industrial processes and harmful commercial practices. With applications ranging from cleaner and more efficient production processes; to oil and chemical pollution control, containment, and recovery; to the potential for sustainable agricultural, forestry, and fisheries practices, technology can provide many environmental and economic benefits.

For the purposes of this chapter, technology is defined as the products, devices, processes, and practices associated with the management of ecosystems with special emphasis on harvesting and using their goods and services, or human activities emitting harmful substances into the ecosystems. In this sense, technology-related command-and-control responses comprise a subset of the more general class of technological responses discussed below.

Technology-related aspects are often included in other response options, for example, prescribing technological specifications as part of command-and-control interventions or under international trade policies. Nevertheless, most technological response options are concerned with local interventions in biochemical processes of the ecosystems or in harvesting their services. This section reviews technological response options along two ordering principles: the target and the timing of interventions. We draw on a diverse range of technology and technological development literature, including Rosenberg et al. (1992), Stoneman (1995), Rosenberg (1994), Grubler (1998), Grubler et al. (2002).

2.2.5.1 The Target of Responses

Products are targeted by technological responses in the form of restrictions concerning the quantities and/or quality of the ecosystem product allowed to be removed. They range from specifications of the age and size of living organisms that can be harvested to the complete ban on harvesting endangered species.

Devices can be an effective target of technological responses. Banning the use of harmful devices or prescribing the use of environmentally benign devices are convenient ways both to protect the targeted species of the ecosystem service (leaving the young generation of the targeted species behind for regeneration) and to prevent the removal of other species that comprise important components of the ecosystem.

Processes are another domain where technological responses can be effective. The sequence of certain operations in the field or the timing of harvest can make the difference between sustainability and collapse while removing the same amount of ecosystem goods and services for human use. ISO 14000, which is a series of voluntary standards in the environmental field, is a good example of such a response option.

Practices as technological responses represent a broader range of interventions often involving a combination of devices, processes, and practices. Purposeful or unintended introduction of new or alien species in an ecosystem or biological control of ecosystem processes as well as the clear cutting versus selective cutting of forests are examples of practice-related technological responses.

2.2.5.2 Timing of Responses

Different types of technological responses are appropriate, effective, and promising in different phases of ecosystem status.

Preventive technological interventions can be effective when the first signs of unfavorable ecosystem changes or deterioration of ecosystem quality are detected. Whether direct interventions in the biophysical processes or indirect regulation of the harvesting technologies and practices, preventive measures can help guide toward stewardship and sustainable management that satisfies human needs and at the same time preserves the integrity and productivity of the ecosystem.

Operative technological interventions incorporate a wide range of responses that have been or could be used as part of a meaningful adaptive ecosystem management strategy. They involve: monitoring the response of the ecosystem to human interventions; monitoring changes in the underlying biochemical processes; assessing the unfavorable or undesirable trends; and introducing appropriate technological measures to correct them.

Rehabilitative technological responses intend to correct the consequences of earlier mismanagement or misuse of ecosystem services. An explosion of technological measures has taken place over the past two decades to renew, restore, or rehabilitate degraded ecosystems. Literally hundreds of technological measures have been devised to redevelop soils, surface and subsurface water bodies, forests and other terrestrial ecosystems, mangroves, wetlands, fisheries, and animal populations.

2.2.5.3 Summary: Technological Responses

The targets of technological responses include products, devices, processes, and practices. Any or several of these are required in different stages of ecosystem management including preventive, operative, and rehabilitative phases. In order for technology to serve as an effective option for resolving ecosystem-related problems, an enabling environment needs to be nurtured that allows environmental technologies to be pursued, developed, disseminated, and integrated into society. The creation of such an enabling environment involves social, legal, and economic aspects and their interactions.

Technological responses represent powerful intervention mechanisms in ecosystems management. Yet in the past, they have often turned out to be a double-edged sword. Most technological interventions provided solutions to the targeted problem, but some have created undesirable side effects that may have been more severe than the original problem. As experience accumulates and technological assessment practices improve, such risks are expected to decline. Nevertheless, the ecosystems themselves are changing and it will never be possible to eliminate all uncertainties associated with technological interventions; therefore a reasonable degree of precaution is warranted when considering and adopting them.

2.2.6 Cognitive Responses

Arguably, the principal cognitive responses to ecosystem-related problems are either traditional in nature or scientific knowledge. While other cognitive responses, such as society's reaction to environmental change, and different actors' experiences and skills in addressing ecosystem-related problem must be noted, this section focuses on traditional wisdom and scientific knowledge. The section reviews the legitimization of traditional and scientific knowledge, as well as the acquisition of scientific knowledge, and considers how both types of knowledge can be used to respond to ecosystem-related problems.

Traditional knowledge refers to knowledge held by members of a distinct culture and to which numerous members of the culture contribute over time. It is acquired through past experiences and observations, and through means of inquiry specific to the culture, and generally concerns the culture itself or its local environment. *Scientific knowledge* stems from experimental and theoretical studies about the natural and social sciences. *Legitimization* is official acceptance and/or recognition that can lead, in the case of traditional and scientific knowledge, to the development of policies and measures based on the knowledge legitimized.

2.2.6.1 Legitimization of Traditional Knowledge

Traditional knowledge is relevant to responding to ecosystem-related problems as it encompasses extensive understanding of local flora, fauna, and ecological processes; the practice of selective breeding; utilization of plant and animal species for medicinal, agricultural, and other purposes, and consequently provides traditional peoples with the ability to contribute to the implementation of conservation policies (Mugabe 1999, p. 4; Roht-Arriaza 1996, p. 928). Extensive knowledge of local ecosystems has led to many instances in which traditional knowledge and practices have formed the basis for developing agricultural and other products, and in which traditional remedies have given rise to the pharmacopoeia of modern medicines. An example from agriculture is an insecticide based on active ingredients of the neem plant, whose particular characteristics were discovered thousands of years previously by indigenous Indian farmers. In the area of traditional medical knowledge, quinine, now commonly contained in medication to prevent malaria, has long been used by Andean indigenous peoples to cure fever (Roht-Arriaza 1996, pp. 921–22).

Traditional practices are also an important source of knowledge for sustainable development. Having gone through processes of trial and error, traditional practices have adapted to local needs and local ecosystems. Numerous examples of traditional practices contributing to sustainable development have been recorded in agriculture (Brookfield et al 2003), water management, and other areas. Growing recognition of the value of traditional knowledge and the interest in it in the biotechnology, pharmaceutical, and human health care industries over the last two decades has resulted in a correspondingly greater acknowledgment of traditional knowledge in international environmental law and policy, thereby contributing to its legitimization.

Traditional knowledge was first addressed at the international level at the 1992 United Nations Conference on Environment and Development, which stated that traditional peoples are central to environmental management and development due to their knowledge and practices, and further stipulated that they be empowered (Rio Declaration, Principle 22; Agenda 21, Chapter 26, Para. 3(a)(iii)). Since then, a number of legally binding instruments concerning or including provisions on traditional knowledge have been adopted, and programs of work developed. The International Labor Organization's 1991 Convention 169 on Indigenous and Tribal Peoples in Independent Countries, for instance, highlights the contribution made by indigenous and tribal peoples to the "ecological harmony of humankind," and notes that traditional knowledge shall be incorporated into educational programs and services for the peoples concerned (ILO Convention 169, Preamble and Article 27(1)). The Convention on Biological Diversity provides that contracting parties shall, as far as possible and appropriate, and subject to their national legislation, respect, preserve, and maintain traditional knowledge relevant to the conservation and sustainable use of biodiversity, as well as to promote its wider application (CBD, Article 8(j)). The Convention to Combat Desertification stipulates that contracting parties shall, subject to their national legislation, exchange information on traditional knowledge and ensure its adequate protection (CCD Article 16(g), UNCCD 1992). Further, parties are to support research activities that protect, enhance, and validate traditional knowledge (CCD, Article 17(1)(c)). Other instruments contributing to the legitimization of traditional knowledge at the international level include the Declaration on Science and the Use of Scientific Knowledge adopted by the 1999 World Conference on Science (Declaration on Science, Preambular Paragraph 26) and the Draft United Nations Declaration on the Rights of Indigenous Peoples (Draft UN Declaration, Preambular Paragraph 9).

In addition to acknowledging traditional knowledge in international environmental treaties and declarations, some multilateral development banks have adopted policies that address the importance of traditional knowledge. This has been done largely as a consequence of criticism of the detrimental impact of MDB-funded projects on traditional peoples. For example, the Bayano hydroelectric dam in Panama led to the forced relocation of 2,000 Kuna and 500 Embera indigenous people from their traditional territories (World Commission on Dams 1999, p. 15). The World Bank adopted Operational Directive 4.20 in 1991, which provides policy guidance to ensure that development projects benefit indigenous peoples and avoid or minimize adverse effects. The Directive emphasizes participation of indigenous peoples in development projects, stating that traditional knowledge be incorporated into the project approach of any project affecting indigenous peoples (World Bank 1991, Paragraph 8). The Inter-American Development Bank and the Asian Development Bank also refer to traditional knowledge (Inter-American Development Bank 1990, Guiding Principle C1(b); Asian Development Bank 1998, Paragraph 2(iii) Appendix).

An aspect central to legitimizing traditional knowledge is the recognition of its origins to traditional peoples as well as the recognition of its utility and relevance in an array of applications at broad levels. Possibly the main controversy surrounding the debate on granting intellectual property rights to traditional knowledge holders is the question whether the current international framework on intellectual property is an adequate forum for addressing the protection of traditional knowledge (Barsh 2001, p. 153). This question and the multitude of concerns arising out of it must be given close consideration in the future.

2.2.6.2 Knowledge Acquisition (Scientific Research) and Acceptance (Legitimization)

Scientific knowledge is pertinent to responding to ecosystem-related problems as it generates relevant information on the functioning of ecosystems, and identifies modes of application of this information, which can contribute to the protection of ecosystems and their components.

Scientific knowledge is commonly acquired through recorded observations of present events, through the analysis of information on past and future events, as well as through experimental studies. In order to respond to ecosystem-related problems based on scientific information, decision-makers both at the national and at the international level consult and are advised by a variety of bodies. A central role is played by scientific advisors working within governments, and by bodies specifically set up by governments to provide them with requested information and advice, such as the Center for Global Environmental Research, which conducts environmental research for the Japanese government, or TERI, which plays a similar role in India. National-level advisory bodies are complemented at the international level, advisory bodies established by intergovernmental processes, such as the Intergovernmental Panel on Climate Change, established in 1988 by UNEP and the World Meteorological Organization. Outside institutions providing scientific information include nongovernmental, inter-governmental, and industry organizations, research institutes, and universities.

Acquisition of scientific knowledge through government scientific advisors and advisory bodies is done by submitting information requests to the advisors and considering the information received. In addition to such mechanisms, outside bodies provide information during stakeholder meetings, through the dissemination of papers and by lobbying government representatives at conferences (Yamin 2001, p. 151; French 1996, p. 255–56). A key aspect in policy-makers' acquisition of scientific knowledge is the identification of the most relevant organizations and institutions. This is of particular importance in dealing with ecosystem degradation and protection due to the large variety of topics this encompasses, and consequently of organizations working on associated issues. Once identified, cooperation with the organizations and institutions must be ensured by, among other instruments, establishing effective and continuous communication. Communication between the IPCC and the decision-making body of the UNFCCC, for example, takes the form of both organizations attending and addressing each other's sessions, with the IPCC presenting its reports within a given time frame at meetings of the decision-making body. These reports are also sent to national governments, and meetings are held among senior officials, thus providing representatives of the decision-making body with an opportunity to submit requests for scientific information.

With the legitimization of scientific knowledge being achieved through its acceptance by policy-makers, it is of interest to consider those aspects that contribute to this acceptance. The two principal features are the credibility and policy relevance of the knowledge presented, which are advanced through a series of characteristics pertaining to the mandate, procedure, and membership of the body in question. *Policy relevance* is achieved, for example, through effective communication. Information is presented in nontechnical language in a manner understandable and relevant to policy-makers. Information requests submitted by policy-makers early in the scientific assessment process are responded to adequately (Levy 1993, p. 406; Meffe 1998, p. 742).

The *credibility* of the information presented is determined by its quality, the transparency of the scientific bodies' procedures, and policy-makers' buy-in into the information. The most widely used quality-assurance is to submit working papers to peer review prior to being published; the expertise of advisory body members is also key in contributing to the quality of the knowledge produced (Kimball 1996, pp. 100–01, 140, 144; Peterson 1998, pp. 429–30). Policy-makers' buy-in is in turn attained by including a government component in the body producing the scientific knowledge and, in the case of knowledge produced for international bodies, by ensuring geographic representation of the advisory body members (Agrawala 1998, p. 628). The Animal and Plants Committees established under CITES, for example, are each composed of ten regional representatives.

Finally, it must be noted that the legitimization of scientific knowledge is not a guarantee for being employed to address ecosystem-related problems, as governments may be unwilling to act on the basis of this knowledge if it stands in conflict with other concerns such as economic or political ones.

2.2.6.3 Summary: Cognitive Responses

This section described the role that the legitimization of traditional and scientific knowledge, as well as the acquisition of scientific knowledge, plays as a response on ecosystem-related problems. Traditional knowledge is relevant as it encompasses extensive understanding of local ecosystems and how they can be effectively managed and conserved. This knowledge is not always applied as widely as it might be as policy-makers are often unaware of its value. Scientific knowledge is also important as it responds to ecosystem-related problems by generating relevant information on the functioning of ecosystems, and it identifies modes of application of this information, which can contribute to decision-making and to the protection of ecosystems and its components.

2.2.7 Typology of Responses: Summary

Policy-makers have available an array of responses for sustainable management of ecosystems for ensuring human well-being. These responses are classified according to a typology of legal, economic, social and behavioral, technological and cognitive interventions. The chapter presents the typologies as one-dimensional and does not account for potentially complex interplay among many of the responses. Since the direct and indirect drivers of ecosystem change also interact in complex ways, choosing the most effective responses may depend on identifying interplay among the drivers. This is not the focus of this section. Instead, it offers a snapshot of the basic functional relationships between responses and how they work systemically together.

The responses are guided by an institutional framework that sets the rules of the game. The rules may be formal or informal. Legal responses serve a "command and control" function. Formal laws govern much of the operationalization of many of the other responses. At the international level, law tends to be weaker but is an area increasing in scope and function. Even when strong international legal responses do exist and are applied, effectiveness is highly dependent upon enforcement systems and the nature and degree of national-level acceptance. Conversely, domestic laws are usually backed up by strong enforcement systems. In general, ecosystem-related legislation, whether domestic or international, has tended to be weaker than economic and social legislation. With growing recognition of the dangers of environmental degradation and the need to protect ecosystems for intra- and intergenerational well-being, legal responses would gain strength. All legal responses, no matter what the scale, usually remain static without implementation, compliance, and enforcement in respective jurisdictions.

Economic and financial interventions are an effective policy tool to regulate the use and overuse of ecosystem goods and services. These response options are based on the premise that human beings are driven to maximize their economic welfare. Thus market mechanisms framed within the context of legal rules provide powerful incentives for people to moderate their behavior. Manipulating economic and financial factors can powerfully alter how ecosystems goods and services are valued and traded. The various types of economic interventions are combined in many cases to achieve an effective regulatory regime. The effectiveness of the economic intervention mechanism, however, is moderated by the fact that socioeconomic conditions vary from society to society.

Fundamentally the objective of legal and economic responses is to change human behavior by changing incentives. But human behavior can also change according to changing norms and values driven by cognitive factors. For example, by empowering people in the political realm, harm to ecosystems because of the corruption of a few can be mitigated. Women, civil society, local communities, and youth tend to demonstrate a strong aptitude for ecosystem stewardship because they are more directly dependent on ecosystem services for sustenance. Through the conferral of rights, liberties, and responsibilities, and through education and information dissemination, disempowered people gain advantages so as to protect their ecological patrimony. Participation and inclusiveness are important for instilling attitudes of stewardship.

Technological responses allow humans to mitigate their effects on ecosystems by allowing less dependence on them, by lowering anthropogenic impact, or by helping to restore degraded ecosystems. Technology, however, carries with it risks that cannot be fully accounted for in practice. Moreover, the right technology is often times unavailable in an equitable manner. The risk of side effects and unintended consequences of technological fixes make it imperative that proper evaluation and risk assessment be carried out before resorting to this response.

Knowledge underlies all types of responses. Institutional change is sometimes necessary in order to adapt to changes in the social and physical world. Such change is often instructed by new knowledge. Legal instruments must reflect new knowledge so that law is not illegitimate, leading to non-conformity and revolt. Knowledge and learning are also important factors in determining how market conditions change and thereby altering existing relationships of humans with nature. Knowledge is fundamental to belief systems, attitudes, values, and norms. Given the role that knowledge plays in forging cognitive processes, creating knowledge, applying it to concrete problems, and disseminating it are also important options for policy response.

2.3 Responses by Impact on Drivers

The MA conceptual framework (MA 2003) identifies the numerous and diverse events and processes affecting ecosystems as drivers. A driver is defined as any natural or human-induced factor that directly or indirectly causes a change in an ecosystem. These factors are structured into two broad categories: direct drivers and indirect drivers. *Direct drivers* are factors that unequivocally influence ecosystem processes and therefore can be identified and measured to differing degrees of accuracy. Direct drivers can be of anthropogenic or natural origin. In contrast, *indirect drivers* operate more diffusely, from a distance, and often by altering one or more direct drivers. Accordingly, an indirect driver can seldom be identified through direct observation of the ecosystem; its influence is established by understanding its effect on a direct driver.

This section assesses the main categories of responses defined above with respect to their impacts on the direct and indirect drivers of ecosystems change. The assessment is organized around two matrices in which response options and drivers represent the two axes. Each cell contains a pair of entries. The first entry indicates the effectiveness of the response in influencing the driver: to what extent can the given response be expected to modify the driver. On a scale of 1 to 5, higher marks indicate the higher expected effectiveness of the response. It is important to note that effectiveness in the following discussion is an assessment of the expectation that the given intervention is capable of bringing about the desired change in individual and/or social behavior. It is not to be confused with economic efficiency (cost-effectiveness) that measures the costs of an intervention. The second part of each entry shows the proximity of the response option to the targeted driver: how long is the chain of the cause-effect mechanisms from the response to the driver. The smaller the number of transmission steps in the response process is, the higher the mark assigned to the response option. The entries denote general assessments that represent the average performance of interventions across a diversity of ecosystems and social conditions. Ecosystem-specific assessments in chapters in Part II of this volume fluctuate around these values accordingly; the broad appraisal below provides a useful background to the more detailed discussions in Part II.

It is important to point out that entries in these tables, especially those concerning the effectiveness marks, represent estimated average values under "normal" socioeconomic conditions: rule of law and order, void of war and chronic corruption. Even under such conditions, the actual effectiveness may well vary somewhat depending on the prevailing sociocultural circumstances. Nevertheless, the tables broadly reflect current thinking about the possibilities of having an impact on different kinds of direct and indirect drivers of ecosystems change.

2.3.1 Direct Drivers

The list of direct drivers defined by the MA includes land use and land cover, species introduction and removal, technology adoption and use, external inputs, harvest and consumption, climate change, and natural physical and biological drivers. Table 2.1 presents the estimated effectiveness and the length of the causal chain of the various legal, economic, social, technological, and cognitive responses in influencing these direct drivers. Natural drivers are not included in this table because they by definition cannot be affected by response options.

Land use and *land cover* as direct drivers of ecosystems change can be most effectively controlled by domestic regulatory law and command-and-control interventions. These measures typically take the form of land zoning and of establishing natural parks or nature reserve areas. Empowering communities so that they can take the responsibility for their own lands can also be an effective option, especially if this is coupled with arrangements to legitimize traditional knowledge. In some cases, international treaties (like the CCD and the Ramsar Convention) provide the broader framework and additional motivation for domestic regulation but their effectiveness remains relatively low compared to domestic, national, and local interventions.

International agreements are also part of the response strategies in preventing *species introduction and removal* as well as external inputs, but domestic arrangements are more powerful in these cases as well. Domestic regulation and command-and-control type interventions at the national and local levels are most effective legal and economic responses. Empowerment of communities and the legitimization of traditional knowledge appear to be especially promising responses to control species introduction/removal and external inputs at the local level.

Similar to the cases of the previous two direct drivers, the *adoption and use of specific technologies* are most effectively controlled through domestic legal, economic, and social responses. Outright ban of ecologically harmful technologies and promotion of environmentally benign technologies by financial and monetary measures usually work. When empowered to manage their own resources, communities are also expected to make their technological choices by considering protection requirements. International trade policies, nevertheless, are likely to have increasing effects on technology adoption and use in appropriating ecosystems goods and services for international markets.

Harvest and consumption represent a mixed driver category. While harvest decisions are mostly made locally, part of the demand for ecosystems products can be triggered by remote consumption preferences. Domestic regulatory law and command-and-control interventions are conceived to be effective instruments to control harvesting technologies and the amount allowed to be harvested. Empowered communities are often capable of harmonizing harvest and protection concerns.

Climate change as a driver of ecosystems change represents a distinct case. An international legal framework is required to put the corresponding domestic regulatory mechanisms in place. Command-and-control as well as incentive-based instruments can be helpful in the implementation. International trade policies can usefully complement both international legal and domestic economic responses. Incentives for innovation and technological R&D are considered to be very effective response options, especially over the long term. In contrast, social responses, especially community-oriented measures, seem to be less important here.

2.3.2 Indirect Drivers

The MA identifies a broad set of indirect drivers that play a role in ecosystems changes. The list comprises demographic, economic, sociopolitical factors, science and technology, as well as values, culture, and religion. Table 2.2 shows the effectiveness of the response options in influencing these drivers and their proximity to the drivers.

Demographic drivers can mainly be influenced by domestic interventions, primarily by constitutional and regulatory law and explicit population policies and public education; economic command-and-control regimes, incentive-based, and financial measures can also play some role. The influence of international law on demographic drivers is limited to their implications for international migration flows (magnitude, direction).

Several response options in each main response category can effectively influence *economic drivers* of ecosystems change. Inter-

Table 2.1. The Relationship between Response Options and Direct Drivers of Ecosystem Change. The first number indicates the potential effectiveness of the response options to influence the driver. The second number shows the proximity of the response option to the driver.

	Direct Drivers					
Responses	Land Use, Land Cover	Species Introduction and Removal	Technology Adoption and Use	External Inputs	Harvest and Consumption	Climate Change
Legal						
International treaties	2/1	3/1	2/1	1/1	1/1	5/4
International soft law	1/1	1/1	1/1	1/1	1/1	3/4
International customary law	1/1	1/1	1/1	1/1	1/1	3/4
International agreements outside the environmental sector	2/1	2/1	2/1	2/1	2/1	2/3
Domestic environmental regulations	5/5	5/5	5/5	5/5	5/5	5/5
Domestic administrative law	4/4	3/3	3/3	3/3	3/3	3/3
Domestic constitutional law	5/2	1/2	1/2	1/2	1/2	2/1
Domestic legislation outside the environmental sector	3/4	3/4	3/4	3/4	3/4	3/4
Economic						
Command-and-control interventions	5/5	5/5	5/5	5/5	5/5	5/5
Incentive-based	4/5	3/4	5/5	5/5	5/5	4/3
Voluntarism-based	3/5	2/3	2/4	3/3	4/3	2/3
Financial/monetary measures	4/5	3/4	5/5	2/3	3/3	3/3
International trade policies	4/4	4/4	3/2	1/2	4/3	4/4
Social and Behavioral						
Population policies	3/3	3/3	2/2	3/2	3/3	3/3
Public education and awareness	3/5	3/5	3/5	3/5	3/5	3/5
Empowering communities	4/5	4/5	4/5	4/5	4/5	3/4
Empowering women	4/5	4/5	4/5	4/5	4/5	3/4
Empowering youth	3/4	3/4	3/4	3/4	3/4	3/4
Civil society protest and disobedience	2/5	1/5	1/5	1/5	1/5	1/3
Technological						
Incentives for innovation and R&D	3/3	2/2	3/5	2/2	4/5	5/3
Cognitive						
Legitimization of traditional knowledge	3/5	3/5	1/5	1/5	3/5	1/5
Knowledge acquisition and acceptances	3/5	3/5	3/5	3/5	3/5	3/5

national environmental treaties, customary law, and non-environmental agreements influence the ways in which ecosystems goods and services are harvested, used, and traded beyond national borders. Domestic environmental regulation and non-environmental legislation are the key domestic legal instruments to alter economic drivers. Obviously, the full arsenal of incentive-based, financial and monetary, and command-and-control interventions can be used rather effectively to induce changes in the economic drivers of ecosystem change. Some social responses are also available in the form of public education and awareness, and community empowerment.

The *sociopolitical drivers* are more closely tied to the local and national social and political conditions, but an increasing number of international arrangements, especially international customary

Table 2.2. The Relationship between Response Options and Indirect Drivers of Ecosystem Change. The first number indicates the potential effectiveness of the response options to influence the driver. The second number shows the proximity of the response option to the driver. Blank cells mean the response is not applicable to the driver.

	Indirect Drivers				
Response	**Demographic**	**Economic**	**Sociopolitical**	**Scientific and Technological**	**Cultural and Religious**
Legal					
International treaties		4/4	3/4	2/4	
International soft law	1/3	1/4	2/4	3/4	2/4
International customary law	1/1	5/5	5/5		2/4
International agreements outside the environmental sector	2/1	3/4	3/3	2/3	2/2
Domestic environmental regulations	5/5	5/5	5/5	4/4	5/5
Domestic administrative law	2/3	2/3	2/3	2/3	2/3
Domestic constitutional law	4/5	3/4	5/5	5/5	
Domestic legislation outside the environmental sector	5/5	5/5	5/5	4/4	5/5
Economic					
Command-and-control interventions	5/5	5/5	4/5	4/5	5/5
Incentive-based	3/5	5/5	4/4	4/4	3/4
Voluntarism-based	2/4	2/4	2/3	4/4	4/5
Financial/monetary measures	3/5	5/5	3/4	4/5	3/4
International trade policies		4/5	4/5	4/5	3/4
Social and Behavioral					
Population policies	5/5	1/3			
Public education and awareness	4/5	4/5	4/5	4/5	4/5
Empowering communities	3/4	4/4	5/5	3/4	4/5
Empowering women	3/4	4/4	5/5	3/4	4/5
Empowering youth	2/3	3/3	4/4	2/3	3/4
Civil society protest and disobedience	1/5	1/5	1/5	1/5	1/5
Technological					
Incentives for innovation and R&D		4/4	3/3	5/5	
Cognitive					
Legitimization of traditional knowledge	2/5	2/5	2/5		3/5
Knowledge acquisition and acceptances	2/5	2/5	2/5	2/5	2/5

law, might affect them. Nonetheless, domestic environmental and non-environmental regulations remain the main instruments to sway sociopolitical drivers in the legal realm. Not surprisingly, the social response options are likely to play the key role when such drivers of ecosystems change need to be addressed with education and empowerment of communities as the most promising ones.

Scientific and technological drivers appear to be more difficult to influence. The various economic responses and domestic legal regulation are the most promising avenues but public education and awareness raising about the ecological impacts of the technologies may also be effective responses in some cases.

Possibly the most controversial drivers and also the most difficult to control are the *cultural and religious drivers.* Two possible strategies are apparent from Table 2.2. The first one is blunt prohibition or prescription in the form of domestic constitutional or regulatory laws or by economic command-and-control interventions. The second avenue is to influence cultural and religious drivers through public education and awareness raising.

2.3.3 Responses and Drivers: Summary

It is important to recall that the effectiveness marks in Tables 2.1 and 2.2 indicate the prospects for success to achieve a given ecological, technological, or biophysical target. These values say nothing about the social costs of implementing the given response, let alone the economic efficiency (cost effectiveness) of the response option. These issues are discussed in Chapter 3. The economic efficiency of any response option crucially depends on the socioeconomic and institutional context in which, and on the resource/ecosystem problem to which, it is applied and thus it is impossible to assess at the general level of discussion in this chapter. Chapters in Part II of this volume provide more detailed discussion of the relative merits of applying different response options to different ecosystems goods and services.

Two clusters of response options emerge as potentially effective in altering direct drivers. Considering the effect of indirect drivers of ecosystems change on direct drivers, it seems to be difficult to override those effects and use responses other than prescriptive regulatory measures like domestic legal regulation or command-and-control instruments, although in some cases incentive-based economic responses may work as well. Land zoning, the prohibition of introducing or removing species, or the ban of certain technologies or inputs are hard and blunt tools to achieve clearly defined ecological objectives. The economic efficiency of such responses can nonetheless be improved by using incentive-based instruments (like tradable permits) in the implementation phase.

The second cluster of promising responses to affect direct drivers can be found in the social domain. Empowering communities and social groups close to and crucially depending on the ecosystem or resource base can be effective in mitigating ecosystems problems when communities establish generally accepted rules of access to, and harvest or use of, ecosystems goods and services.

Indirect drivers are the ultimate causes of ecosystems problems but they involve a broad range of demographic, economic, social, and technological factors. In a globalizing world, there is an increasing influence on indirect drivers from international legal and economic agreements. Nevertheless, the domestic regulatory responses and the domestic legislation outside the environment sector including them stand out as the most effective options. They are usefully complemented by economic responses, especially those based on incentives or involving command-and-control measures.

In the domestic realm, the pattern is rather obvious: economic responses, especially command-and-control, incentive-based, and financial and monetary responses dominate the options to affect economic driving forces. Social responses, primarily empowering communities, are most effective in influencing sociopolitical as well as cultural and religious drivers. Incentives for innovation and research and development are the most direct and most effective ways to sway scientific and technological drivers but economic responses, especially incentive-based ones, are also valuable.

2.4 Responses by Actors

2.4.1 Key Responses Available to the Government Sector

Governments at all levels, through laws, regulations, and other policy decisions, are key actors in the protection of ecosystems. Their actions can be direct or indirect.

Direct actions by governments to protect ecosystems by limiting or prohibiting commercial exploitation are the most easily understood and analyzed. Protection of land as parks, wilderness areas, etc., is the most obvious and visible of such actions. Such protection is often accompanied by efforts to counteract previous ecosystem degradation. There are also many other approaches, which, while stopping short of full protection, promote sustainable development and use of ecosystems, for example, community-based natural resource management (Viet et al. 2001).

At both the national and international levels, governments also act to protect species, habitats, and specific land types with policies that do not fully protect land. These policies either limit the types of activities that can occur on the land or promote activities that will limit or reverse ecological damage. CITES (www.cites.org) is an example of an international policy to protect species; the U.S. Endangered Species Act (www.endanagered.fws.gov) and India's National Policy and Action Plan on Biodiversity, 2000 (Indian Government 2001) are examples of policies at the national level. The Convention to Combat Desertification (www.unccd.int), with its requirement for national, sub-regional, and regional action plans to limit and reverse the spread of desertification, is an example of a policy at both the national and international levels, that is aimed at protecting both habitats and land types. U.S. regulations on the protection of wetlands (Clean Water Act, Section 404; www.epa.gov/owow/wetlands/regs/sec404.html) and the Central American Forest Convention (Aguilar and Gonzalez 1999) are examples of such policies at national and regional levels.

Many government policies implemented for other reasons affect ecosystems. Some of these policies have negative impacts on ecosystems, and their removal can be a response option. Examples of policies that can have a negative impact on ecosystems include:

- building of roads, dams, and other civic infrastructure that directly destroy habitats or open areas to more intensive settlement;
- agricultural policies, including subsides and unsustainable irrigation, that promote the cultivation of marginal land or overuse of existing farmland; and
- economic development policies that promote urbanization and strain water supply and other resources.

While these negative outcomes are often considered "unintended consequences," analyzing policies from a sustainable development perspective will often warn of potential negative outcomes. Environmental impact assessments are a useful tool in conducting the environmental portion of such an analysis.

2.4.2 Key Responses Available to the Private Sector

The private sector, that is, business and industry, is often portrayed simply as an exploiter of ecosystem goods and services. However, as a major user of ecosystem goods and services, the private sector can play an important role in the protection of ecosystems. The private sector acts at three levels: in partnership with governments, in partnership with other stakeholders, and on its own.

Partnerships between the private sector and government occur both formally and informally. An example of a formal arrangement was the partnership between TotalFinaElf and the Bolivian National Oil Company, YPFB, to minimize the ecological impacts of oil exploration in Bolivia's Madidi National Park (www.ipieca.org/downloads/biodiversity/sens_envir_case_studies/TotalFina Elf_bolivia.pdf).

Informal partnerships can develop in a number of ways. In 1970, S.C. Johnson began purchasing pyrethrum, a natural insecticide, for use in its products from the Pyrethrum Board of Kenya, the agency that controls and operates the pyrethrum business in Kenya. Over the years this relationship grew from a simple supplier–purchaser interaction into a collaborative effort with a strong degree of knowledge and technology exchange. Promotion of pyrethrum cultivation is beneficial to ecosystems in two ways: pyrethrum requires little fertilizer or pesticide input and it produces a natural product that can be used to reduce insecticide usage for other applications (www.wbcsd.org/web/publications/technology-cooperation2.pdf, pp. 39–46).

Partnerships between the private sector and other stakeholders can be very effective in encouraging more sustainable use of ecosystems. This was the case when Bayer CropScience conducted a pilot program of its integrated crop management program in Brazil. The goals of this program were to use the full range of weed and pest control techniques to reduce dependence on chemical agents and lessen potential impacts on ecosystems. Other stakeholders included local government authorities and farmers' associations (see www.wbcsd.org/web/publications/technology-cooperation 2.pdf, pp. 9–17). The pilot was successful and was used as a basis for a larger program in Guatemala.

Finally, private sector companies often act on their own in undertaking efforts to preserve and enhance the ecosystems in which they are working. Since environmental law and regulation is now comprehensive in most parts of the world, these efforts usually entail going beyond specific legal requirements. Rio Tinto has done this in Madagascar where it has put in place a team of Malagasy environmental professionals to carry out research and to monitor the progress on restoration of a biodiverse area in which the company is mining ilmenite. The goal is restoration of the forests and wetlands that are important not only as habitats but for the economic well-being of the local community (www.wbcsd .org, see case studies).

2.4.3 Key Responses Available to the Local Community

The importance of traditional and local managers in stewarding ecological resources is obvious. They are the actors who have to implement many government policies and their commitment, or lack thereof, to these policies can determine their success or failure. For example, many anti-desertification policies tried in the Sahel in 1970s and 1980s failed because they did not take local socioeconomic factors into account (OCEE 1996). Conversely, the Kikori Integrated Conservation and Development Project, a partnership between ChevronTexaco and the World Wildlife Fund in Papua New Guinea, which protected some of the world's rarest wildlife and promoted the sustainable development of local communities, was a success because it involved local communities in project planning and execution. The World Bank called this project "a model for other resource developers operating in ecologically sensitive areas" (www.ipieca.org/downloads/biodiversity/sens_envir_case_studies/ChevTex_PNG.pdf).

The knowledge that traditional and local managers bring as part of an informed public participation process can be invaluable in defining the ecological risks and ways of avoiding them.

2.4.4 Key Responses Available to NGOs

Advocacy groups, traditional environmental groups as well as social justice groups, play an important role in education and awareness-raising. They are often the first to call attention to the potential ecological impacts of proposed developments, and often play an important role in developing detailed information about the magnitude and extent of potential impacts.

Advocacy groups can play an important role in empowering local communities and other stakeholders. The benefits of public participation can be achieved only when the public has sufficient information about an issue to make an informed decision. For example, the economic benefits of development are usually well advertised, but their ecological costs may be hidden. Advocacy groups can provide information on those ecological costs, allowing local communities and other stakeholders to make informed choices.

Advocacy groups also play a critical role in mobilizing stakeholders at the national and international level. Again, education and public awareness are the key factors. Successful campaigns to save baby harp seals and raise the level of concern about endangered species could not have occurred without the international education and public awareness campaigns undertaken by environmental advocacy groups. Advocacy groups also develop and promote innovative approaches to ecosystem protection, for example debt-for-nature swaps which have been promoted by WWF and Conservation International (www.fao.org/docrep/w3247e/w3247e06.htm).

While advocacy groups often assume an adversarial posture, they also work in partnership with governments and the private sector to achieve mutual goals. An example is the partnership between the government of Bolivia, Fundacion Amigos de la Naturaleza, the Nature Conservancy, and three U.S. energy companies to protect over 60,000 hectares of the Noel Kempff Mercado National Park, one of the most biological diverse areas in the world (http://nature.org/initiatives/climatechange/work/art4253.html).

Finally, advocacy groups can act on their own to protect ecosystems. For example, the Nature Conservancy (www.nature.org) has bought or otherwise protected over 40 million hectares of threatened ecosystem.

2.4.5 Responses and Actors: Summary

Table 2.3 shows the relationship between responses and actors. Across the top of the table are the actors that range from governments to civil society groups. Down the rows are the various responses, from legal to cognitive responses. There are two numbers in each cell; the first number is the availability of the response to the actor. The numerical range is from one to five; a higher number indicates that the response is readily available to the actor while a lower number actor indicates that the response is either not available or seldom used. The second number in the cell shows the effectiveness that the actor has in using the corresponding response. A high number shows that the actor could effectively use the response and a lower number shows that the actor would have little effective result from using the response.

The tallies indicate some clear patterns of availability and effectiveness. Governments predominately have the widest range of responses available to them compared to other actors. Legal responses are only available to governments though other actors may be able to challenge legal responses through dispute settlement and judicial action or through influencing law-making negotiations through education and lobbying. The predominance that the government has over legal response options results from its control of the authority to make laws and the economic power to implement decisions. The effectiveness of these responses may vary and, though the response may indeed have the potential to change behavior, the implementation may be subject to socioeconomic factors that result in outcomes that are less than effective.

Social, economic, technological, and cognitive responses are generally only available to nonstate actors. The private sector tends to exercise control over financially based responses where it can create incentives for technological change, such as research and development, or where it has the financial power to implement the response. Incentive-based research and development is an important response for the private sector and one, in which it exercises considerable control; if used for the development of new products to protect and conserve ecosystems, it can be an effective response. Volunteer-based responses also tend to be an effective response, as business and industry prefer self-regulation and the

Table 2.3. The Relationship between the Responses and the Actors. The first number in a cell is the availability of the response to the actor. The second number shows the effectiveness the actor has in using the response. Blank cells mean the response is not applicable to the actor.

Response	Actors			
	Government	Private Sector	Local Communities	NGOs
Legal				
Treaties	5/5			
International soft law	2/5			
International customary law	3/5			
International agreement; legislation outside environment sector	3/5			
Domestic environmental regulations	5/5			
Domestic administrative law	3/5			
Domestic constitutional law	4/5			
Domestic legislation outside the environmental sector	4/5			
Economic				
Command-and-control interventions	5/5			
Incentive-based	5/5	5/5	2/3	2/4
Voluntarism-based	3/5	4/5	4/5	4/4
Financial/monetary measures	5/5	5/4	3/3	3/3
International trade policies	4/5			
Social and Behavioral				
Population policies	5/4	3/4	4/3	3/4
Public education and awareness	5/3	4/5	4/5	4/5
Empowering youth	3/5	4/5	4/5	4/5
Empowering communities	3/5	4/3	5/5	5/5
Empowering women	3/5	4/3	5/5	5/5
Civil society protest and disobedience			1/5	1/5
Technological				
Incentives for innovation R&D	5/4	5/5	5/4	5/4
Cognitive				
Legitimization of traditional knowledge	5/2		5/5	5/5
Knowledge acquisition and acceptances	5/3	4/3	3/2	4/4

flexibility to choose their own response instead of having these responses imposed by government. Education and awareness raising can also be effectively used by the private sector, though often these types of responses are not employed for the betterment of ecosystems but for marketing and sales.

Local communities and NGOs tend to have at their disposal social polices that educate, empower, or provide information and knowledge to change values, perceptions, and attitudes. Civil disobedience and protest may also be readily available to these actors, but the effectiveness of these responses is normally very low compared to other responses. Local communities and NGOs also play an important role in the legitimization and use of traditional knowledge, which is critical for understanding the complex systematic relationships between humankind and nature—a relationship that is not always understood by modern or scientific knowledge.

Innovation incentives for research and development are available to all actors, but in very different aspects. Whereas government and the private sector play important roles in providing the means for innovations, NGOs and local communities can set agendas either by defining the necessity for new technology and the need for practical applications of technology or by promoting greener ecosystem technologies through education, dissemination, and lobbying.

In discussing the various actors and the key responses available to them, this section has shown the limitations of responses available to nonstate actors compared with those available to governmental actors. Nevertheless each actor plays an important role in implementation and propagation of behavioral and ecosystem change.

2.5 Response Options by Scale of Operation of Decision-maker

Ecosystem related problems require policy responses that correlate to the scale of the problem. In the natural sciences, scale has traditionally been a prominent issue; but only recently has this issue come to the fore in discussion of policy responses to environmental problems. This section examines the different response options according to their scale and discusses appropriate pairing of the scale of the environmental problems and responses. Response options are determined by their jurisdictional reach, or by the decision-makers' authority to craft such a response. Thus the section examines the scale of various state-sanctioned response options, as well as the physical and political considerations that affect them.

2.5.1 Global/Universal Responses

Global responses to ecosystem problems are warranted when those problems are universal in nature—potentially affecting all people and ecosystems of the planet. Although there are numerous problems of this nature, there are few truly universal response options. Customary international law, defined above as, "a general practice accepted as law," is the main response option that is universal, for customary law is binding on all states, irrespective of their accession status to a particular treaty. The majority of global environmental problems, however, are addressed through multilateral solutions.

2.5.2 Multilateral Agreements

In contrast to customary international law, multilateral treaties are binding only on those parties that sign and ratify them. Much of the body of international environmental law has arisen through multilateral treaty-making. Examples range from the Basel Convention on the Control of Transboundary Movements of Hazardous Wastes and their Disposal to the Convention on International Trade and Endangered Species of Wild Fauna and Flora.

Climate change, often cited as one of the most complex environmental challenges facing society, demonstrates the need for multilateral response options. The Intergovernmental Panel on Climate Change noted in its Third Assessment Report that changes to atmospheric composition will have consequences for future levels of mean temperature, precipitation, sea level, and the occurrence of extreme events (IPCC 2001). Because climate change has the potential to affect all human beings, the policy response has been to draft multilateral legal instruments. The UNFCCC and the Kyoto Protocol rely on Hardin's model of "mutual coercion, mutually agreed upon," whereby parties mutually agree to limit their emissions of carbon dioxide. Other multilateral responses that ascribe to mutual coercion and preserve open access to resources include the Montreal Protocol, the CBD, and the Antarctic Treaty.

Most often, multilateral response options are appropriate for common pool resource problems, when the resource in question is both rival (one person's consumption will diminish another person's) and non-excludable (under current policy arrangements, no one can be barred from consuming said resource). Multilateral responses can include Hardin's model of mutual coercion, though the effectiveness may diminish with complex ecosystems or numerous actors. Restricting the transboundary movement of hazardous waste, for example, provides an incentive to reduce the production of this waste and to ensure its safe disposal, either within a party's borders or with the explicit prior informed consent of the recipient party.

However, this model of cooperation under anarchy presents a number of problems (Oye 1986). The greater the number of actors involved, the more difficult cooperation becomes. In addition, lengthy time horizons, as with the issue of climate change, often provide a disincentive for cooperation. However, actors who are forced to negotiate with each other over time, and across a number of issues, are more likely to cooperate, and less likely to free-ride.

2.5.3 Plurilateral Agreements

Plurilateral agreements address regional problems, often transboundary in nature. These issues require the participation of those parties affected by the problem, but need not involve states beyond the area of that ecosystem. For example, the United Nations Conference on Straddling Fish Stocks and Highly Migratory Fish Stocks, creates a framework and specific guidelines for regional and sub-regional management of migratory fish stocks. It builds upon an existing multilateral treaty, the U.N. Convention on the Law of the Sea, but adds much stronger incentives for compliance. Vessels found to be in violation may be banned from fishing the area covered by the regional agreement.

The Regional Seas Programme, run by UNEP, is another example of a set of plurilateral agreements. Of the seventeen separate bodies of water covered by the various programs, twelve have corresponding conventions and, in most cases, one or more protocols. These have been enacted to tailor policy responses to the environmental characteristics of and threats to each specific body of water. Each program includes the participation of all nations surrounding a given sea and takes measures to protect it against threats, including pollution, overexploitation, invasive species, and global change.

2.5.4 Bilateral Agreements

Bilateral agreements, like plurilateral agreements, are often established to respond to transboundary problems. One of the older examples is the International Joint Commission, established in 1909 by the United States and Canada to manage shared water bodies. Years later, in 1972, the two countries created the Great Lakes Water Quality Agreement, to manage the transboundary freshwater Great Lakes. This formal bilateral agreement, amended in 1978 and 1987 to set more stringent goals for ecosystem management, lay the foundation for an extensive network of actors to cooperate in joint management efforts (Karkkainen 2004). Similar bilateral agreements have been established in southern Africa to create joint water commissions for the management of shared watercourses; on the whole, these demonstrate a clear political commitment by the states involved, to create frameworks to facilitate joint management of watercourses water projects (Giordano and Wolf 2003).

2.5.5 National Policies

Though multilateral agreements involve the consensus of a group of nations, such instruments often fall to the national level for implementation. For example, the frequent amendments of the Montreal Protocol kept pressure on parties to phase out chlorofluorocarbons and other ozone-depleting substances, which in turn increased the effectiveness of the agreement. However, national policies need not always be in response to an international mandate. They are more often the product of public opinion or policy priorities of law-makers of that nation. For example, though the Kyoto Protocol has yet to enter into force, a number of nations have taken extensive measures to attempt to meet the negotiated reductions within the timetable of the first commitment period.

2.5.6 State/Province-Level Responses

Similarly, states or provinces can take actions on the basis of a national mandate or simply implement policies autonomously. In the United States, the state of California, for example, has adopted a number of its own policies to curb greenhouse gas emissions, despite the fact that the United States has stated that it will not accede to the Kyoto Protocol. Thus California is not legally required to take these measures, but rather chooses to do so of its own accord. Nonetheless, in the past two years, California has established a greenhouse gas registry, and will require car manufacturers to meet certain fleet standards from 2009 onward.

2.5.7 Local Responses

Like state and national response options, local responses to environmental problems can be in response to a mandate from a larger jurisdiction, or can be autonomous, taken to resolve issues specific to that area. An effort between communities and localities to jointly manage forest tracts in India is such an example (Sarin 1995); localities responded to a very specific need of residents in surrounding communities to reap greater benefits from their natural resource bases. On the local scale, voluntarism-based responses have emerged to address climate change. Many cities, for example, have conducted assessments of their greenhouse-gas emissions and implemented steps to try to reduce them—even in the absence of federally mandated regulation. To provide sufficient incentive, responsibility to manage these resources was partially devolved to residents themselves.

2.5.8 Challenges and Issues

Response options to ecosystem management problems have a considerable range—from the largest most complex ecosystems involving many actors to local or municipal ecosystems involving relatively few actors. The scale of the intervention varies with the ecosystem and with drivers of change, both direct and indirect. The challenge is not only to match the response option to the scale of the problem and drivers, but also to ensure that those problems with multiple responses do not conflict with each other.

Table 2.4 presents some of the challenges and issues discussed in this section; it differs slightly from the preceding tables. Since the objective of this table is to identify the scales at which response options are available, it does not assess proximity of the response option to its target. The absence of a number indicates that the response is not available at that scale.

The first section of the table indicates the appropriateness of specific types of legal responses at different scales. A "5" denotes an available response. The table shows that applicable responses cluster in two areas. Unsurprisingly, international legal responses are appropriate at the supra-national level—including multilateral, plurilateral, and bilateral responses. Domestic legal responses, including regulatory, administrative, and constitutional responses, are available at the national and sub-national level.

The section on legal responses, while indicating the available responses at different scales, does not distinguish among them in terms of effectiveness; for this reason, they are all assigned the same effectiveness value of 5. Without specifying the driver of change, it is difficult to identify the appropriate scale of response. For example, though a multilateral treaty may be most effective for a global commons problem such as protection of the world's oceans, it would not be an effective response to manage a river basin shared by two nations. Thus there are nuances to the effectiveness of legal response options not captured by the table.

The second section of the table indicates that economic responses are available at all scales; the numbers indicate the perceived effectiveness relative to the other economic responses. At the supra-national level, command-and-control interventions are likely to be used by the private sector, to standardize practices throughout large multinational corporations, and possibly, by international organizations such as the International Organization for Standardization.

On the national and sub-national levels, command and control will likely be a response option most used by governments. Incentive and voluntarism-based interventions are flexible response options available at all levels to a variety of actors. Financial and monetary measures are likely to be the purview of governments and the private sector, available to them at all levels of scale. Finally, international trade policies are available to governments, only at the supra-national scale.

The third portion of the table illustrates that social policies are likely to increase in effectiveness as the response approaches the local level. Since social response options are directed at individuals' beliefs and behaviors, targeted interventions are more likely to be effective; thus effectiveness of almost all social responses increases as the scale moves toward the local. Population policies are one exception to this pattern. Because population policies are politically sensitive, and often controversial, it follows that such decisions would be taken at the national level; political compromise in the context of a supra-national response is not likely. Thus there are no response options available at the supra-national level.

Both technological and cognitive response options are available at all scales, and may be equally effective at the supra-national, national, and sub-national levels.

In general, the scale of the response option is determined in part by the interaction between domestic and international interests. This two-level game, satisfying the political requirements on both the national and international levels, can become quite complex with many actors and competing national interests. The result is a smaller number of acceptable outcomes that satisfy all players involved (Putnam 1988). Thus the range of response options acceptable to all parties involved is smaller than it would be with fewer actors, or fewer pressures from the domestic level.

Another important consideration in scaling response options is the need for similar or complementary policies elsewhere in the hierarchy of response options. For example, multilateral agreements must be implemented by national, and sometimes sub-national policies. If the bureaucratic, political, legal, or economic infrastructures are insufficiently developed, that nation may not be capable of carrying out its obligations under the multilateral agreement. Sub-national responses, such as those on the local or municipal level, also need support from further up the hierarchy. For example, many failures of sustainable forestry management by communities have been attributed to the lack of property rights in the region or nation (Church 1996; Ruitenbeek 1998). It is also important to consider the use of multiple types of actors at different scales, as appropriate, to help surmount these difficulties. Conversely, sub-national responses may need to be harmonized with responses at national and supra-national levels. Thus there is an added challenge of "scaling up" these responses, so that they do not conflict with (and, at best, are in harmony with) larger-scale interventions (Ostrom et al. 1999).

2.5.9 Response Options by Scale: Summary

This section has discussed the different scales at which response options can be implemented, and the appropriateness of these scales for different types of responses. It has also outlined some of the challenges involved in determining and implementing response options at the appropriate scale. First, in some cases—such as when responding with legal instruments—knowledge of the drivers of change may be a necessary prerequisite for evaluating the relative effectiveness of different response options. Second,

Table 2.4. Challenges and Issues. Since scale is the focus, this table does not assess proximity of the response option to its target. Blank cells indicate that the response option is not available at that scale.

	Scale					
Response	**Multilateral**	**Plurilateral**	**Bilateral**	**National**	**State/Province**	**Local**
Legal						
International treaties	5	5	5			
International soft law	5	5	5			
International customary law	5	5	5			
International agreements; legislation outside environment sector	5	5	5			
Domestic environmental regulations				5	5	5
Domestic administrative law				5	5	5
Domestic constitutional law				5	5	5
Domestic legislation outside environmental sector				5	5	5
Economic						
Command-and-control interventions	5	5	5	5	5	5
Incentive-based	4	4	4	4	4	4
Voluntarism-based	3	3	3	3	3	3
Financial/monetary measures	4	4	4	4	4	4
International trade policies	4	4	5			
Social and Behavioral						
Population policies				4	4	4
Public education and awareness	2	2	3	3	3	4
Empowering communities	2	2	3	3	3	4
Empowering women	2	2	3	3	3	4
Empowering youth	1	1	2	2	2	3
Civil society protest and disobedience	2	1	1	3	3	3
Technological						
Incentives for innovation and R&D	4	4	4	4	4	4
Cognitive						
Legitimization of traditional knowledge	2	2	2	2	2	2
Knowledge acquisition and acceptances	2	2	2	2	2	2

the interaction between political interests at the domestic and international levels may help determine the array of responses available. Third, response options on different scales at a minimum must not conflict with each other and ideally should be complementary.

2.6 Synthesis

Considering the immense variety of ecosystems, the problems and challenges emerging in using their goods and services to improve human well-being, and the vast diversity of socioeconomic conditions under which they must be managed, it is an almost hopeless attempt to derive generally valid observations concerning the most promising responses. Running the double risk of being far too general yet still being wrong because under special circumstances counterexamples could be cited, this section presents some general patterns concerning the most promising responses available to the four main actor groups (government, private sector, NGOs, local communities) to induce changes in the direct and indirect drivers in response to feared, emerging, or prevailing problems with ecosystems.

Table 2.5 provides a synthesis of earlier tables in this chapter by compiling those responses that appear to be most effective in the hands of given actors to achieve desired changes in a driver of ecosystem change.

National governments play a central role in devising and implementing responses for several reasons. First, they control the domestic legal instruments ranging from constitutional to regulatory and administrative legislation. Second, they provide the context for other domestic responses. Third, they must utilize domestic legal instruments to implement most responses. Fourth, they provide the bridge from the international environmental and other agreements affecting the use of ecosystems goods and services to the national actors targeted by those agreements. National governments also control most economic responses, of which incentive-based and command-and-control measures are the most

Table 2.5. The Most Effective Response Options Available to Four Main Actor Groups to Influence Direct and Indirect Driving Forces

Actors	Government	Private Sector	NGOs	Local Community
Direct Drivers				
Land use, land cover	command-and-control regulatory incentive-based	voluntarism-based	education knowledge acquisition and acceptance	empowerment education legitimization of traditional knowledge
Species introduction and removal	command-and-control regulatory international treaty	voluntarism-base	deducation knowledge acquisition and acceptance	education empowerment legitimization of traditional knowledge
Technology adoption and use	command-and-control regulatory incentive-based financial/monetary international trade	financial/monetary incentive-based technology R&D	voluntarism-based education knowledge acquisition and acceptance	empowerment education legitimization of traditional knowledge
External inputs	command-and-control regulatory	incentive-based voluntarism-based	education knowledge acquisition and acceptance	education empowerment
Harvest	command-and-control regulatory	incentive-based technology R&D	education	legitimization of traditional knowledge empowerment
Climate change	international treaty command-and-control regulatory	voluntarism-based	education	voluntary-based empowerment
Indirect Drivers				
Demographic	domestic regulations domestic constitutional law		education empowerment	empowerment education
Economic	international trade policies incentive-based command-and-control financial/monetary	financial/monetary	voluntarism-based	voluntarism-based
Sociopolitical	international customary law domestic constitutional law domestic environmental regulations	voluntarism-based	voluntarism-based education	empowerment education
Scientific and technological	international soft law domestic environmental regulations incentives for innovation and R&D public education	incentives for innovation and R&D	education	traditional knowledge empowerment education
Cultural and religious	domestic constitutional law domestic regulatory law public education command-and-control	education empowerment	traditional knowledge education	empowerment traditional knowledge education

effective ones. Governments initiate national research and technological development programs and operate the basic education systems.

Therefore, national-level decision-making has a special role in several respects. First, even the best-designed local or regional actions are likely to be ineffective in the absence of proper coordination (for example, a stringent and enforced protective measure in one region may simply shift the harmful activity to another region). Second, the key legislative power is anchored at the national level (although the distribution between the federal and state levels varies in federal states). Finally, nation states are the recognized parties in the increasing number of international negotiations and agreements (from bilateral to global) concerned with ecosystem and biodiversity management.

At the other end of the spectrum, local communities are increasingly seen as the most appropriate guardians of their own ecosystems and resources. The empowerment of communities at large or special groups like women or youth emerges as a potentially effective response option from our assessment in the preceding sections. Their effectiveness can be further strengthened by education and information provision on the one hand and by the legitimization of traditional knowledge on the other.

NGOs can do a lot to help communities both at the production/harvest end and at the consumption end of ecosystems use. NGOs' contribution depends on an open and participatory process, the level of democratization in a country; and political comfort in engaging in an open dialogue and receptiveness to criticism. Their most effective response options are education, knowledge acquisition and acceptance, and encouraging voluntarism-based actions in the local communities and among consumers. On the production/harvest side, these activities help resource operators in making informed choices about land use, the introduction or removal of species, and the application of technologies. On the consumer side, it is mostly awareness raising about the implications of certain consumption patterns. NGOs can also target the private sector with these instruments.

When the national government gives the proper and clear signals and provides an operational framework, the private sector can rely on powerful response options to influence both direct and indirect drivers. Incentive-based instruments and voluntary measures can be used to have an effect on land use and land cover change or the application of external inputs. Financial and monetary instruments as well as incentives for innovation, research, and development can be used to shape harvesting practices and the adoption and use of technologies.

Another general pattern emerging from Table 2.5 is that the greatest potential for responses by NGOs and communities are related to the direct drivers. Except for climate change, these are decisions about local and regional resources and they are made locally. In contrast, the larger-scale responses concerning demographic, economic, political, and science/technology drivers are shaped by governments and, to some extent, by the private sector. The instruments available to NGOs to influence indirect drivers are fewer and relatively weaker.

Any typology involves some degree of (over)simplification. The typologies presented in this chapter are no exceptions. But it is important to point out that none of the response options presented and discussed here comes in a sterile or stand-alone form. The tools available to different actors complement each other and constitute a set of measures the final outcome of which will eventually guide the choices and decisions of consumers, the resource operators, and the intermediaries between them. The internal consistency of such packages is crucial. Therefore it is necessary to understand the effect mechanisms and the outcomes the various response measures may trigger, especially their potential unintended consequences, whether technological or social nature. Chapters in Part II explore the responses in detail but it is important to keep in mind their interactions.

References

Agrawala, S., 1998: Structural and process history of the intergovernmental panel on climate change, *Climatic Change,* **39,** pp. 621–642.

Aguilar, G. and M. Gonzalez, 1999: Regional legal arrangements for forest: the case of Central America. In: *Assessing the International Forest Regime,* Tarasofsky, R.G. (ed.), IUCN environmental policy and law paper no. 37, IUCN, Gland, Switzerland.

Argentina, 1994: Constitución Nacional del Argentina. Cited 3 December 2003. Available at http://www.uni-wuerzburg.de/law/ar00000_.html.

Asian Development Bank, 1998: *Policy on Indigenous Peoples.* Asian Development Bank Policy Paper, Manila, Philippines.

Bangladesh, 1996: *The Constitution of the People's Republic of Bangladesh.* Cited 3 December 2003. Available at http://www.bangladeshgov.org/pmo/constitution.

Barsh, R.L., 2001: Who steals indigenous knowledge? *American Society of International Law Proceedings,* **95,** pp. 153–161.

Bates, G.M., 1995: *Environmental Law in Australia,* 4th ed., Butterworths, Australia.

Beauchemin, E., 2001: Sabotage or negligence: Radio Netherlands. Available at http://www.rnw.nl/humanrights/html/ogoni011228.html.

Bertran, M., 2002: Judiciary ombudsman: solving problems in the court, *Fordham Urban Law Journal,* **29,** pp. 2099–2116.

Birnie, P., 1992: International environmental law: Its adequacy for present and future needs. In: *The International Politics of the Environment, Actors Interests, and Institutions,* A. Hurrel and B. Kingsbury (eds.), Clarendon Press, Oxford, UK, pp. 51–84.

Birnie, P. and A. Boyle, 2002: *International Law and the Environment,* 2nd ed., Oxford University Press, Oxford, UK.

Brazil, 1988: *Constitution of the Federative Republic of Brazil.* Adopted 5 Oct. 1988. Political database of the Americas. Available at http://www.georgetown.edu/pdba/Constitutions/Brazil/english98.html. Accessed, 5 March 2005.

Brookfield, H., H. Parsons, and M. Brookfield, 2003: *"Agrodiversity": Learning From Farmers Across the World,* UNU Press, Tokyo, Japan.

Brownlie, I., 1990: *Principles of Public International Law,* 4th ed., Clarendon Press, Oxford, UK.

Bruntland, G. (ed.), 1987: *Our Common Future: The World Commission on Environment and Development,* Oxford University Press, Oxford, UK.

Campbell-Mohn, C., B. Breen, W. Futrell, J.M. McElfish, and P. Grant (eds.), 1993: *Environmental Law: From Resources to Recovery,* Environmental Law Institute, West Publishing Co., St. Paul, MN.

Cartagena, 2000: Cartagena Protocol on Biosafety to the Convention on Biological Diversity, October 2000. Available at http://www.biodiv.org/doc/legal/cartagena-protocol-en.pdf.

Cassese, A., 2001: *International Law,* Oxford University Press, Oxford, UK.

Caubet, C. G., 1991: *As Grandes Manobras de Itaipu: Energia, Diplomacia e Direito na Bacia do Prata,* Editora Acadêmica, São Paulo, Brazil.

CBD (Convention on Biological Diversity), 2003: *Handbook of the Convention of Biological Diversity,* 2nd ed., Secretariat of the CBD, Montreal, Canada.

Chayes, A. and A.H. Chayes, 1991: Adjustments and compliance processes in international regulatory regimes. In: *Preserving the Global Environment: The Challenge of Shared Leadership,* J.T. Matthews (ed.), W.W. Norton & Company, The American Assembly, and World Resources Institute, New York, NY, pp. 279–308.

Church, P.E., 1996, Forestry and the environment: An assessment of USAID support for forest stewardship, USAID Evaluation Highlights, no. 59, USAID Bureau for Policy and Program Coordination, Center for Development Information and Evaluation, US Agency for International Development, Washington, DC.

Clean Japan Center, 1999, *Recycling Fact Book: Waste Management and Recycling in Japan,* Clean Center Japan, Tokyo.

Common, M., 1996: *Environmental and Resource Economics: An Introduction,* Longman, London, UK.

Council on Environmental Quality, 1997: *Environmental Quality: The 25th Anniversary Report of the Council on Environmental Quality,* Executive Office of the President, US Government Printing Office, Washington, DC.

Council of Europe, 1950: *European Convention for the Protection of Human Rights and Fundamental Freedoms,* 213 UNTS (United Nations Treaty Collection) 222.

Council of Europe, 1979: *Convention of the Conservation of European Wildlife and Natural Habitats,* 19 September 1979. Cited 7 May 2004. Available at http://www.ecnc.nl/doc/europe/legislat/bernconv.html.

Dietz, T. and P.C. Stern, 2002: Exploring new tools for environmental protection. In: *New Tools for Environmental Protection: Education, Information, and Voluntary Measures,* T. Dietz and P.C. Stern (eds.), National Academy Press, Washington, DC, pp. 3–15.

DiMaggio, P., 1994: Culture and economy. In: *The Handbook of Economic Sociology,* N. J. Smelser and R. Swedberg (eds.), Princeton University Press, Princeton, NJ, and Russell Sage Foundation, New York, NY.

Douglas-Scott, S., 1996: Environmental rights in the European Union: Participatory democracy or democratic deficit? In: *Human Rights Approach to Environmental Protection,* B. Alan and M. Anderson (eds.), Clarendon Press, Oxford, UK, pp. 109–128.

Eckerberg, K., 1997: National and local policy implementation as a participatory process. In: *International Governance on Environmental Issues,* M. Rolén, H. Sjoberg, and U. Svedin (eds.), *Environment and Policy, Vol. 9,* Kluwer Academic Publishers, Dordrecht, The Netherlands, pp. 119–137.

Environmental Defense, 2003: Chico Mendes sustainable rain forest campaign: Environmental defense, New York. Cited 18 March 2003. Available at http://www.environmentaldefense.org/article.cfm?contentid=1548.

FAO (Food and Agriculture Organization of the United Nations), 1997: *State of the World's Forests 1997,* FAO, Rome, Italy.

FAO, 2001: International treaty on plant genetic resources for food and agriculture [online]. Adopted by the thirty-first session of the FAO Conference, November 2001, FAO, Rome, Italy. Available at ftp://ext-ftp.fao.org/ag/cgrfa/it/ITPGRe.pdf.

FBI (Federal Bureau of Investigation), n.d.: Crime in the United States 2000, *Uniform Crime Reports.* Available at http://www.fbi.gov/ucr/ucr.htm#cius.

FOE (Friends of the Earth), 1989: Destruction of rain forest in the Brazilian Amazon and the role of Japan: Briefing paper presented at the International People's Forum on Japan and the global environment, 8–9 September 1989, FOE, Tokyo, Japan.

Forrest, S., 1999: *Global Tenure and Sustainable Use,* Sustainable Use Initiative-IUCN, Washington, DC.

French, H., 1996: The role of non-state actors. In: *Greening International Institutions,* J. Werksman (ed.), Earthscan, London, UK, pp. 251–258.

Goodale, G., 1995: Training in the context of poverty alleviation and sustainable development. In: *Empowerment Towards Sustainable Development,* N. Singh and V. Titi (eds.), Fernwood Publishing Ltd., Halifax, NS, Canada, and Zed Books Ltd., London, UK, and New Brunswick, NJ, pp. 82–91.

Grubler, A., 1998: *Technology and Global Change,* Cambridge University Press, Cambridge, UK.

Grubler, A., N. Nakicenovic, and W.D. Nordhaus (eds.), 2002: *Technological Change and the Environment,* Resources for the Future and International Institute for Applied Systems Analysis, Washington, DC.

Handler, T. (ed.), 1994: *Regulating the European Environment,* Chancery Law Publishing, London, UK.

Hanley, N., J.F. Shogren, and B. White, 1997: *Environmental Economics in Theory and Practice,* Macmillan, Houndmills, Basingstoke, UK.

Hanna, S.S., C. Folke, and K.G. Mäler, 1996: Property rights and the natural environment. In: *Rights to Nature: Ecological, Economic, Cultural, and Political Principles of Institutions for the Environment,* S.S. Hanna, C. Folke, and K.G. Mäler (eds.), Island Press, Washington, DC.

Hardin, G., 1968: The tragedy of the commons, *Science* **162,** pp. 1243–1248.

Henry J. Kaiser Family Foundation, 2003: HIV/AIDS policy fact sheet, The Henry J. Kaiser Family Foundation, Menlo Park, CA and Washington, DC. Cited 18 March 2003. Available at http://www.kff.org/content/2002/6039/6039.pdf

Hirakuri, S.R., 2003: *Can Law Save the Forest? Lessons from Finland and Brazil,* Center for International Forestry Research, Jakarta, Indonesia.

Hughes, E.L., 1996: Forests, forestry practices, and the living environment. In: *Global Forests and International Environmental Law,* Canadian Council of International Law, Kluwer Law International, pp. 79–125.

Hugh-Jones, S., 1999: "Food" and "drugs" in north-west Amazonia. In: *Cultural and Spiritual Values of Biodiversity,* D.A. Posey (ed.), United Nations Environment Programme, Nairobi, Kenya, and Intermediate Technology Publications, London, UK.

ILO (International Labour Organisation), Convention 169 concerning indigenous and tribal peoples in independent countries, 1989: 28 ILM 1077 (1990) [International Labour Migration].

IMO (International Maritime Organization), 2000–04: Ballast water convention. Adopted. Cited 30 April 2004. Available at http://globallast.imo.org/index.asp?page=mepc.htmandmenu=true.

India, 1949: The Constitution of India. Cited 3 December 2003. Available at http://www.constitution.org/cons/india/const.html.

Indian Government, 2001: *National Action Programme to Combat Desertification, Vol. 1,* Submitted to the United Nations Convention to Combat Desertification, pp. 72–73.

Inoue, M. and H. Isozaki (eds.), 2003: *People and Forest: Policy and Local Reality in Southeast Asia, the Russian Far East, and Japan,* Kluwer Academic Publishers, Dordrecht, The Netherlands.

Inter-American Development Bank, 1990: *Strategies and Procedures on Sociocultural Issues as Related to the Environment,* Inter-American Development Bank, Washington, DC.

IPCC (Intergovernmental Panel on Climate Change), 2001: *Summary for Policymakers,* IPCC Working Group I, Geneva, Switzerland.

IUCN (International Union for Conservation of Nature and Natural Resources – The World Conservation Union), 1997: *Indigenous Peoples and Sustainability: Cases and Actions,* IUCN Inter-Commission Task Force on Indigenous Peoples and International Books, pp. 41–45.

Jacobson, H.K. and E. B. Weiss, 1997: Compliance with international accords-achievements and strategies. In: *International Governance on Environmental Issues,* M. Rolén, H. Sjöberg, and U. Svedin (eds.), Kluwer Academic Publishers, Dordrecht, The Netherlands, pp. 78–110.

Jasper, M.C., 1997: *Environmental Law,* Oceana Publications, Inc., Dobbs Ferry, NY.

Jordan, B., 2001: Building a WTO that can contribute effectively to economic and social development worldwide. In: *The Role of the World Trade Organization in Global Governance,* G. Sampson G.P. (ed.), United Nations University Press, Tokyo, Japan.

Jurgielewicz, L.M., 1996: *Global Environmental Change and International Law,* University Press of America, Landam, MD.

Karkkainen, Bradley, 2004: Post-sovereign environmental governance, *Global Environmental Politics,* **4(1),** pp. 72–96.

Kaufman-Hayoz, R., C. Batting, S. Bruppacher, R. Difila, A. GiGuilio, et al., 2001: A typology of tools for building sustainable strategies. In: *Changing Things—Moving People: Strategies for Promoting Sustainable Development at the Local Level,* R. Kaufman-Hayoz and H. Gutscher (eds.), Birkhauser, Basel, Switzerland, pp. 33–107.

Kimball, L.A., 1996: *Treaty Implementation: Scientific and Technical Advice Enters a New Stage,* American Society of International Law, Washington, DC.

Kiss, A. and D. Shelton, 2004: *International Environmental Law,* 3rd ed., Transnational Publishers, Inc., Ardsley, NY.

Klabbers, J., 1996: *The Concept of Treaty in International Law.* Kluwer Law International, Dordrecht, The Netherlands.

Koh, Kheng-Lian, 2004: Personal communication, September 2004.

Levy, M.A., 1993: Improving the effectiveness of international environmental Institutions. In: *Institutions for the Earth: Sources of Effective International Environmental Protection,* R.O. Keohane, P.M. Haas, and M.A. Levy (eds.), MIT Press, Cambridge, MA, pp. 397–426.

Levy, M., O.R. Young, and M. Zürn, 1995: The study of international regimes, *European Journal of International Relations,* **1,** pp. 267–330.

Lijphart, A., 1999: *Patterns of Democracy,* Yale University Press, New Haven, CT.

MA (Millennium Ecosystem Assessment), 2003: *Ecosystems and Human Well-being: A Framework for Assessment,* Island Press, Washington, DC, 245 pp.

Martin, C., 2001: The relationship between trade and environment regimes: What needs to change? In: *The Role of the World Trade Organization in Global Governance,* G. Sampson G.P. (ed.), United Nations University Press, Tokyo, Japan.

May, P.J., R.J. Burby, N.J. Ericksen, J.W. Handmaer, J.E. Dixon, S. Michaels, and D.I. Smith, 1996: *Environmental Management and Governance, Intergovernmental Approaches to Hazards and Sustainability,* Routledge, London, UK.

Meffe, G.K., 1998: Conservation scientists and the policy process, *Conservation Biology,* **12(4),** pp. 741–742.

Mugabe, J., 1999: *Intellectual Property Protection and Traditional Knowledge,* ACTS Press, Nairobi, Kenya.

Myers, N. and J. Kent, 1998: *Perverse Subsidies: Taxes Undercutting Our Economies and Environments Alike,* International Institute for Sustainable Development, Winipeg, MB, Canada.

My Hero, 2003: Earth keeper hero: Chico Mendes. My Hero, California. Cited 18 March 2003. Available at http://myhero.com/hero.asp?hero=c_mendes.

NAFTA (North American Free Trade Agreement), 1992: *North American Free Trade Agreement,* NAFTA Parts 1–3, 32 ILM 289 (1993); Parts 4–8, 32 ILM 605 (1993), Winnipeg, MB, Canada.

Neumann, K., 2000: *Young People and the Environment in Europe: A Guide,* Canopus Foundation, Berlin, Germany.

North, D.C., 1990: *Institutions, Institutional Change, and Economic Performance,* Cambridge University Press, Cambridge, UK.

OAS (Organization of American States), 1969: The American convention on human rights, OAS Doc.36 Off.Rec.OEA/ser.L./V II.23 Doc. Rev.2, Washington, DC.

OCEE (Overseas Environmental Cooperation Centre), 1996: *Handbook on Desertification Control,* Tokyo, Japan, pp. 8.

OECD (Organisation for Economic Co-operation and Development), 1993: *Environmental Education: An Approach to Sustainable Development,* OECD, Paris, France.

OECD, 2001: *Sustainable Development: Critical Issues,* OECD, Paris, France.

OECD, 2003: *Voluntary Approaches for Environmental Policy Effectiveness, Efficiency and Usage in Policy Mixes,* OECD, Paris, France.

Ostrom, E., R. Gardner, and J. Walker, 1994: *Rules, Games, and Common-Pool Resources,* University of Michigan Press, Ann Arbor, MI.

Ostrom, E., J. Burger, C.B. Field, R.B. Norgaard, D. Policansky, 1999: Revisiting the commons: Local lessons, global challenges, *Science,* **284,** pp. 278–282.

Oye, K., 1986: *Cooperation under Anarchy,* Princeton University Press, Princeton, NJ.

Pantzare, M. and M. Vredin, 1993: *The CAMPFIRE Programme in Nyaminyami,* B.Sc. Thesis and MFS-report, September, UMEÅ Business School, Department of Economics, Umeå, Sweden.

Pearce, D.W. and R.K. Turner, 1990: *Economics of Natural Resources and the Environment,* Harvester Wheatsheaf, New York, NY.

PCA (Permanent Court of Arbitration), 2004: Optional rules for arbitration of disputes relating to natural resources and/or the environment._Cited 17 May 2004. Available at http://www.pca-cpa.org/PDF/ENRrules.pdf.

Perman, R., Y. Ma, J. McGilvray, 1996: *Natural Resources and Environmental Economics,* Longman, London, UK.

Peru, 1993: *Constitución Política del Peru,* actualizada hasta reformas introducidas por la Ley 27365, del 02/11/2000.

Peterson, M. J., 1998: Organizing for effective environmental cooperation, *Global Governance,* **4(4),** pp. 415–438.

Powell, B.J., 2000: *Elections as Instruments of Democracy: Majoritarian and Proportional Visions,* Yale University Press, New Haven, CT.

Putnam, R., 1988: Diplomacy and domestic politics: The logic of two-level games, *International Organization,* **42.**

Redgwell, C., 2001: Non-compliance procedures and the climate change convention. In *Interlinkages: The Kyoto Protocol and the International Trade and Investment Regimes,* W. Bradnee Chambers (ed.), United Nations University Press, Tokyo, Japan, **43.**

Repetto, R., 1990: Deforestation in the tropics, *Scientific American,* **262(4).**

Roht-Arriaza, N., 1996: Of seeds and shamans: The appropriation of the scientific and technical knowledge of indigenous and local communities, *Michigan Journal of International Law,* **17,** pp. 919–965.

Rosenberg, N., 1994: *Exploring the Black Box: Technology, Economics, History,* Cambridge University Press, Cambridge, UK.

Rosenberg, N., R. Landau, and D.C. Movery (eds.), 1992: *Technology and the Wealth of the Nations,* Stanford University Press, Stanford, CA.

Rosendal, K.G., 2000: *The Convention on Biological Diversity and Developing Countries,* Kluwer Academic Publishers, Dordrecht, The Netherlands.

Ruitenbeek, J., 1998: Rational exploitations: economic criteria and indicators for sustainable management of tropical forests, Occasional paper no. 17, CIFOR, Bogor, Indonesia.

Slaughter, A-M., A. S. Tulumello, and S. Wood, 1998: International law and international relations theory: A new generation of interdisciplinary scholarship, *The American Journal of International Law,* **92(3),** pp. 367–397.

Sand, P., 1991: International cooperation: The environmental experience. In: *Preserving the Global Environment: The Challenge of Shared Leadership,* J. T. Matthews (ed.), W.W. Norton and Company, The American Assembly and World Resources Institute, pp. 236–279.

Sands, P. (ed.), 1994: *Greening International Law.* The New Press, New York, pp. 413.

Sarin, M., 1995: Joint forest management in India: Achievements and unaddressed challenges, *Unasylva,* **46,** pp. 30–36.

SDUD (Senate Department of Urban Development), n.d.: Nature protection areas and landscape protection areas, SDUD, Berlin, Germany. Cited 10 September 2004. Available at www.stadtentwicklung.berlin.de/umwelt/umweltatlas/eda506_01.htm.

Seneviratne, M., 2000: Ombudsmen 2000: Paper presented in the author's inaugural lecture at Nottingham Law School on 17 April 2000, Nottingham, UK. Cited 12 September 2004. Available at http://www.bioa.org.uk/BIOA-New/Ombudsmen-2000-Mary%20Seneviratne.pdf.

Somsen, H., 1998: Dynamics, process, and instruments of environmental decisionmaking in the European Union. In: *Law in Environmental Decision-Making, National European, and International Perspectives,* T. Jewell and J. Steele (eds.), Clarendon Press, Oxford, UK, pp. 161–205.

Spellerberg, I.F., 1991: *Monitoring Ecological Change,* Cambridge University Press, Cambridge, UK.

Stavins, R.N., 2000: *Economics of the Environment: Selected Readings,* Norton, New York, NY.

Stoneman, P., (ed.) 1995: *Handbook of the Economics of Innovation and Technological Change,* Blackwell, Oxford, UK.

Susskind, L. and C. Ozawa, 1992: Negotiating more effective international environmental agreements. In: *The International Politics of the Environment, Actors, Interests, and Institutions,* A. Hurrel and B. Kingsbury (eds), Clarendon Press, Oxford, UK, pp. 141–165.

TCC (Trade Compliance Center), 2003: Bilateral investment treaties [online]. Cited 3 December 2003, Washington, DC. Available at http://www.tcc.mac.doc.gov/cgi-bin/doit.cgi?226:54:589141452:15.

TERI (The Energy and Resource Institute), 2003: Environmental education and awareness [online], TERI, New Delhi, India. Cited 12 March 2003. Available at http://www.terrin.org/division/padiv/eea/eea.htm.

Tietenberg, T., 1992: *Environmental and Natural Resource Economics,* HarperCollins, New York, NY.

UN (United Nations), 1945: *United Nations Charter.* Adopted 26 June 1945. Entered into force 24 October 1945. Amended by G.A. Res. 1991 (XVIII) 17 December 1963; Res. 2101 of December 1965; and Res. 2847 (CCVI) of December 1971, New York, NY.

UN, 1948: *Universal Declaration on Human Rights,* General Assembly Resolution 217 A (III) of 10 December 1948, New York, NY.

UN, 1966a: *International Covenant on Civil and Political Rights,* G.A.Res.2200A(XXI), UN doc A/6316 (1966), 999 UNTS 171. Reprinted in 6 ILM 368, New York, NY.

UN, 1966b: *International Covenant on Economic, Social and Cultural Rights,* G.A.Res.2200A (XXI), UN doc A/6316 (1966), 993 UNTS 3. Reprinted in 6 I.L.M. 360, New York, NY.

UN, 1972a: *Stockholm Declaration on Human Environment,* United Nations Conference on the Human Environment, 16 June 1972, U.N.Doc.A/CONF.48/14, 11 I.L.M. 1416, New York, NY.

UN, 1972b: *Convention on Wetlands of International Importance Especially as Wildlife Habitat,* 11 ILM 969, 2 February 1971, New York, NY.

UN, 1973: *Convention on International Trade in Endangered Species of Wild Fauna and Flora* (CITES), 993 U.N.T.S. 243 (1973), 3 March 1973, New York, NY.

UN, 1982: *The United Nations Convention on the Law of the Sea,* 21 I.L.M. 1261, 10 December 1982, New York, NY.

UN, 1987a: *Vienna Convention for the Protection of the Ozone Layer,* 26 I.L.M. 1516 (1987), New York, NY.

UN, 1987b: *Montreal Protocol on Substances that Deplete the Ozone Layer,* 26 I.L.M.1541 (1987), 16 September 1987, New York, NY.

UN, 1989: *Basel Convention on the Control of Transboundary Movements of Hazardous Wastes and Their Disposal,* 28 I.L.M. 649 (1989), 22 March 1989, New York, NY.

UN, 1992a: Agenda 21 [online], United Nations Division of Sustainable Development, New York, NY. Cited 12 March 2003. Available at http://www.un.org/esa/sustdev/agenda21text.htm.

UN, 1992b: *Rio Declaration on Environment and Development,* A/CONF/151/26 (Vol. I), New York, NY.

UN, 1992c: *Statement of Principles for a Global Consensus on the Management, Conservation and Sustainable Development of All Types of Forests,* UN DocA/CONF.151/6/Rev.1, 31 I.L.M. 881 (1992), 13 June 1992, New York, NY.

UN, 1995: *Agreement for the Implementation of the Provisions of the United Nations Convention on the Law of the Sea of 10 December 1982 relating to the Conservation and Management of Straddling Fish Stocks and Highly Migratory Fish Stocks,* A/RES/51/35, 77th plenary meeting, 9 December 1996, New York, NY.

UN, 1997: *Convention of the Non-navigational Uses of International Watercourses 1997,* UN doc.A/51/869, Adopted by the UN General Assembly, Resolution 51/229 of 21 May 1997, New York, NY.

UN, 2002: The Right to Water. The International Covenant on Economic, Social and Cultural Rights. Twenty-ninth session, Geneva, 11–29 November 2002, Agenda item 3, U.N.Doc.E/C.12/2002/11, 26 November 2002.

UN, 2003: Plan of implementation [online], United Nations Department of Economic and Social Affairs, Division of Sustainable Development, New York, NY. Cited 12 March 2003. Available at http://www.johannesburgsummit.org/html/documents/summit_docs/2309_planfinal.htm.

UNCCD (United Nations Convention to Combat Desertification), 1992: *Convention to Combat Desertification,* 33 ILM 1328 (1994), 17 June 1994, New York, NY.

UNCLOS, 1982: The United Nations Convention on the Law of the Sea. 10 December 1982, 21 I.L.M. 1261.

UNEP (United Nations Environment Programme), Division of Technology, Industry and Economics, International Environmental Technology Centre website. Available at http://www.unep.or.jp/Ietc/EST/Index.asp.

UNEP, 1995: *Report of the Seventh Meeting of the Parties to the Montreal Protocol on Substances that Deplete the Ozone Layer,* UNEP/OzL.Pro7/12, 5–7 December 1995, Vienna, Austria.

UNEP, 1997: *Global Environmental Outlook 1,* Oxford University Press, New York, NY, pp. 44.

UNEP, 1998: *Report of the Tenth Meeting of the Parties to the Montreal Protocol on Substances that Deplete the Ozone Layer,* UNEP/Oz.L.Pro10/9, Decision X/10, 23–24 November 1998, Cairo, Egypt.

UNEP, 1999: UNEP global environment outlook 2000 [online]. Cited 7 July 2003. Available at http://1www1.unep.org/geio-text/0141.htm.

UNEP, 2001: *Global Biodiversity Outlook 2000,* The Secretariat of the Convention on Biodiversity, UNEP/CBD/ICCP/2/3, New York, NY.

UNEP, 2003: *Report of the Fifteenth Meeting of the Parties to the Montreal Protocol on Substances that Deplete the Ozone Layer,* UNEP/OzL.Pro15/9, 10–14 November 2003, Nairobi, Kenya.

UNEP, 2004: *Liability and Redress,* UNEP/CBD/BS/COP-MOP/1/9, Conference of the parties to the convention on biological diversity serving as the meeting of the parties to the Cartagena protocol on biosafety. First meeting, 23–27 February 2004, Kuala Lumpur, Malaysia.

UNESCO and the International Association of Universities, 1986: *Universities and Environmental Education,* UNESCO and the International Association of Universities, Paris.

UNFCCC (United Nations Framework Convention on Climate Change), 1997: *Kyoto Protocol to the United Nations Framework Convention on Climate Change,* UNdocFCCC/CP/1997/7/Add.1, New York, NY.

UNFPA (United Nations Population Fund), 2002: *State of World Population 2002: People, Poverty and Possibilities,* UNFPA, New York, NY.

UNU (United Nations University), 2003: World's foremost education, scientific organizations call for greening of school curriculums [online]. UNU, Tokyo, Japan. Cited 1 September 2003. Available at http://www.unu.edu/hq/rector_office/press-archives/press2002/pre37.02.html.

User Survey a, n.d.: Six billion and counting: The world population crisis. Cited 14 March 2003. Available at http://www.geocities.com/Area51/Dimension/7689/six.html.

User Survey b, n.d.: Compendium of summaries of judicial decisions in environment related cases, SACEP/UNEP/NORAD publication series on environmental law and policy no. 3. Cited 5 December 2003. Available at http://www.unescap.org/drpad/vc/document/compendium/bg2.htm.

User Survey c, n.d.: The history of the protection of nature in national forests. Cited 11 September 2004. Available at http://homepage2.nifty.com/fujiwara_studyroom/english/policy/policy_storag e/nationl_forest_nc.pdf.

Viet, P.G., A. Mascarenhas, and O. Ampadu-Agyei, 2001: African development that works. In: *A Survey of Sustainable Development: Social and Economic Dimensions,* J.M. Harris, TA Wise, K.P. Gallager, and N.R. Goodwin (eds.), Island Press, Washington, DC, pp. 322–325.

Williams, C. (ed.), 1998: *Environmental Victims: New Risks, New Injustice,* Earthscan, London.

Williamson, O.E., 1985: *The Economic Institutions of Capitalism: Firms, Markets, Relational Contracting,* The Free Press, New York, NY.

Wills, I., 1997: *Economics and the Environment: A Signaling and Incentives Approach,* Allen & Unwin, St. Leonards, NSW, Australia.

Wolf, G. and A. T. Wolf, 2003: Transboundary freshwater treaties. In: *International Waters in Southern Africa,* Mikiyasu Nakayama (ed.), UNU Press, Tokyo, Japan.

Wolf, S., A. White, and N. Stanley, 2002: *Principles of Environmental Law,* 3rd ed., Cavendish Publishing Limited, London, UK.

World Bank, 1991: Operational directive 4.20: Indigenous peoples, World Bank, Washington, DC.

World Commission on Dams, 1999: Summary report of regional consultations: Large dams and their alternatives in Latin America, WCD, New York, NY. Cited 25 February 2003. Available at http://www.damsreport.org/docs/kbase/consultations/consult_latin.pdf.

Worldwatch Institute, 2001: *Vital Signs 2001,* Worldwatch Institute, Washington, DC.

Worldwatch Institute, 2002: *Vital Signs 2002,* Worldwatch Institute, Washington, DC.

Worldwatch Institute, 2003: *State of the World 2003,* Worldwatch Institute, Washington, DC.

WRI (World Resources Institute), 1985: *Tropical Forests: A Call for Action, Part I The Plan,* Report of an international task force convened by the World Resources Institute and The World Bank, Washington, DC, and the United Nations Development Programme, New York, NY.

Wuori, M., 1997: One of formative side of history: The role of non-governmental organizations. In: *International Governance on Environmental Issues,* M. Rolén, H. Sjoberg and U. Svedin (eds.), *Vol. 9, Environment and Policy,* Kluwer Academic Publishers, Dordrecht, The Netherlands, pp. 159–172.

Yamin, F., 2001: NGOs and international environmental law: A critical evaluation of their roles and responsibilities, *Review of European Community and International Environmental Law,* **10(2),** pp. 149–162.

Yannis, M., 2001: *The Role of the Ombudsman for the Protection of the Environment,* Proceedings of the Ombudsman workshop held on 18 May 2001, Athens, Greece. Cited 12 September 2004. Available at http://www.synigoros.gr/anpe/assets/ombudsman_workshop_en.doc.

Young, O.R., 2002, *The Institutional Dimensions of Environmental Change: Fit, Interplay, and Scale,* MIT Press, Cambridge, MA.

Chapter 3

Assessing Responses

Coordinating Lead Authors: Dana R. Fisher, R. David Simpson, Bhaskar Vira
Lead Authors: W. Bradnee Chambers, Debra J. Davidson
Review Editors: Richard Norgaard, Ravi Prabhu

BOXES

FIGURES

TABLES

Main Messages

Assessment of responses should distinguish between constraints that render a policy option infeasible and the acceptable consequences or side effects of a chosen strategy. We are proposing a multistage assessment process, which focuses first on those factors that may either rule out a particular response or be critical preconditions for its success (*binding constraints*). Responses are then compared across multiple dimensions in order to identify unintended impacts, focusing on identifying compatibility or conflict between different policy objectives. The pursuit of a specific objective may sometimes involve compromising another policy goal. Such considerations, while important, may be seen by decision-makers as acceptable costs associated with the implementation of an option (*acceptable trade-offs*).

Evaluating the relative success of responses requires an assessment of enabling conditions, binding constraints, and acceptable trade-offs across a number of domains. These include the political, which encompasses the legitimacy of and the political context for the response; institutional, which refers to the capacity for governance and implementation; economic, which looks at the availability of resources as well as the aggregate and distributional impacts of policy options; social, which refers to the broad social environment and preconditions for a response; and ecological, which defines systemic preconditions and constraints for a response. As many other chapters of the MA consider the ecological domain in detail, we will, while recognizing the central importance of ecological considerations, focus on the other four domains in this chapter.

The assessment of responses needs to recognize trade-offs between objectives. It is unlikely that all strategies will be able to satisfy diverse and often competing policy objectives. Resolving the trade-offs between these different objectives presents a significant challenge to determining appropriate responses. In some instances, it may be possible to make a "binary" decision: so long as some standard is satisfied, the choice among approaches can be made on other grounds. In other situations, a gain toward achieving one objective may need to be weighed against a negative outcome in some other domain.

Some responses may constitute "win-win" opportunities. While trade-offs between objectives are likely to occur, synergies are certainly possible. Some responses may constitute "win-win" opportunities. Policy-makers ought to remain alert for such opportunities and move aggressively to act upon them, but also remain guarded concerning the prospects of options that may "sound too good to be true."

Aggregating response impacts across different dimensions is a subjective process. Quantitative assessment techniques are not necessarily preferable to qualitative methods. Aggregating impacts across different dimensions (political, institutional, economic, social, and ecological) is difficult. Quantification may provide a "false" objectivity to what is essentially a subjective process. Decision-makers must, in the final analysis, make some assessment of the "weights" to be assigned to each factor and compare impacts along dimensions that are typically incommensurable.

Assessment methods must be sensitive to a plurality of perspectives. The assessment of responses needs to be multidimensional, involve inputs from multiple disciplines, and must attempt to integrate the perspectives of multiple decision-makers. Techniques that adopt a pluralistic disciplinary perspective are particularly pertinent, as they do not privilege any particular viewpoint.

A number of pluralistic decision-making tools and techniques are available. These tools can be employed at a variety of scales—global, sub-global, and local. This chapter presents a simple listing of tools that are available, as well as a preliminary analysis of their most appropriate scale(s) of application. In particular, a distinction can be made between *deliberative* tools, which facilitate the process of dialogue over responses; *information gathering* tools, which are primarily focused on collecting data and opinions; and *planning* tools, which are typically employed for the evaluation of potential policy options.

Assessment is a dynamic and adaptive process, which needs to be constantly updated in light of new information, as well as feedback from the social and ecological systems in which a response is implemented. Techniques such as adaptive management and adaptive co-management have been deployed usefully to create flexible and resilient systems of resource management. The advantage of such approaches is that they are able to deal with new empirical circumstances while ensuring that responses reflect the perspectives and interests of a wide variety of stakeholders.

The assessment process is only as good as the overall decision-making environment within which it is embedded. Trade-offs, choices, and synergies are often hidden or neglected in policy dialogue. Solutions to many intractable problems are likely to be context-specific, and it may not be easy to achieve consensus among stakeholders about the suitability of specific responses. A process in which choices and trade-offs are transparent is desirable, as it is most likely to allow decision-makers to choose locally appropriate responses that are congruent with their desired goals.

Because stakeholders will be affected differently, and may have differences of opinion about the relative desirability of different response strategies, consensus will be difficult. The potential for conflict is particularly high where there is disagreement among stakeholders over the objectives of intervention as well as the means to achieve these ends. For instance, while environmental ministries may prioritize ecosystem integrity, economic ministries may privilege economic growth. Bureaucracies may prefer centralized authority structures, while grassroots organizations may be more comfortable with inclusive and participatory approaches. Some of these differences may be reconcilable, but in other cases it may not be possible to achieve consensus among stakeholders.

3.1 Introduction

In the MA conceptual framework (MA 2003), responses are defined generically as *human actions, including policies, strategies, and interventions, designed to respond to specific issues, needs, opportunities, or problems*. Responses are seen in the context of perceived needs or problems. In the specific context of the MA, these needs or problems relate to the preservation of ecosystems and biodiversity, the accrual of desired ecosystem services, and the improvement of human well-being.

This chapter evaluates the human influences on responses that must be considered by decision-makers in response assessment. In this instance, we employ the term decision-makers broadly, to include all individuals who are in a position to promote an ecological response option, at the local, regional, national, or international level. There are at least two distinct, but interrelated, reasons why decision-makers need to evaluate responses. The first is to improve policy-making by learning from experience. Here, the decision-maker seeks to understand the reasons for perceived success and failure, and considers how such conditions can be replicated for future policy-making that is targeted at enhancing human well-being and ecosystems.

The second reason is to understand the impact of any particular response or set of responses. The need is to identify the linkages between the chosen responses and their effects on a wide range of proximate social, political, economic, and ecological variables, and ultimately on human well-being and ecosystems. We include a discussion of methods that may be employed to assess these variables, in order to maximize the potential success of responses and minimize the potential unintended effects that may arise in relation to response implementation. Evaluating responses can be a complex and costly endeavor, because of the need to understand the multidimensional impact of any chosen strategy, and the multiple actors and interests that may be involved in the process.

In the past, responses have fallen short of their intended goals due to, for example, an inadequate estimation of the skills and resources required for implementation, or a lack of understanding of the sources of cultural resistance to the behavioral changes required. Other responses, whether or not they meet intended ecological goals, can have extremely disruptive social consequences, such as when land tenure allocations are abruptly altered, causing conflict among pre-existing user groups. This chapter stresses the importance of trying to understand how social factors can hinder responses and how responses can lead to unintended social consequences. It proposes the use of evaluation methodologies that stress the employment of multiple criteria and a plurality of inputs into the decision-making process. Such methodologies are relevant to the assessment of a variety of responses, but are intended to be applied in the present context to responses that are targeted at the flows of services from ecosystems, as well as those that are implemented in other social sectors that may have indirect implications for ecosystem services and human well-being.

Understanding the relative success of responses requires an assessment of the enabling conditions and *binding constraints* that determine which specific objectives can be pursued, because they either may rule out a particular response or may represent critical preconditions for its success. Binding constraints are factors that render a policy option infeasible. These are distinguished from what we call *acceptable trade-offs:* unintended impacts associated with the implementation of a response that may be deemed acceptable because they are outweighed by benefits of the response. What is considered a binding constraint, and what is considered an acceptable trade-off, is in all instances context-specific; the proposed assessment method is not intended to elicit generalizations, but intended for use on a case-by-case basis. As a result, we purposefully avoid establishing a list of specific indicators, as these are expected to vary according to the specific contexts under consideration. More importantly, what is considered a binding constraint, and what is considered an acceptable trade-off, may also be seen differently by different stakeholders *within* cases. While transparent processes that utilize a deliberative democratic format have been shown to be tremendously successful in eliciting stakeholder support for a common course of action, deliberation is the key to decision-makers' understanding of different perspectives on any particular response. We recognize that, in some cases, the differences among perspectives may be so great that a resolution is not possible.

This chapter proposes a three-stage assessment process that focuses first on identifying the multiple human impacts associated with responses, along five domains described further below. In the first stage of assessment, those impacts that pose binding constraints are identified. These factors may explain the failure of a previous response, or may rule out its adoption in proposed planning processes, and will require either the selection of an alternative response or significant investments in creating more favorable conditions. If the impacts identified do not impose binding constraints on a particular response, they may be considered acceptable trade-offs, which may include both positive synergies and negative consequences within these domains. In the second stage of the assessment process, these potential trade-offs, and their acceptability in relation to the response, are identified. In this step, responses are compared across multiple dimensions, focusing on identifying compatibility or conflict between different policy objectives. Once these two steps are completed, decision-makers are ready for the third and final stage in the evaluation process, which entails the selection of preferred responses.

The assessment procedure is designed for use by diverse decision-makers at multiple spatial and temporal scales: from the local to the global, for the analysis of previous or current responses, as well as for the evaluation of the feasibility of proposed policies or responses to be implemented in the future. Further, the assessment procedure is intended to be a dynamic process, whereby new information and systemic feedback creates policy learning and the evolution of responses in an adaptive manner.

The assessment method is outlined in Figure 3.1. It is important to note that although this is a new process for assessment, the tools and methods that it draws on are based on the existing literature. Because it is an assessment of what exists, no new techniques/tools are being developed. In both stages of assessment, binding constraints and acceptable trade-offs should be evaluated in relation to five domains:

- the *political,* encompassing the legitimacy of the response and the political context in which the response would be implemented;
- the *institutional,* referring to the capacity for governance and implementation;
- the *economic,* referring to the aggregate and distributional consequences of the response for income and wealth, and economic conditions, including stability of property rights and the efficient use of available resources;
- the *social,* including the broad equity issues associated with a response; and
- the *ecological,* including the ecosystemic preconditions and context within which a response is being considered.

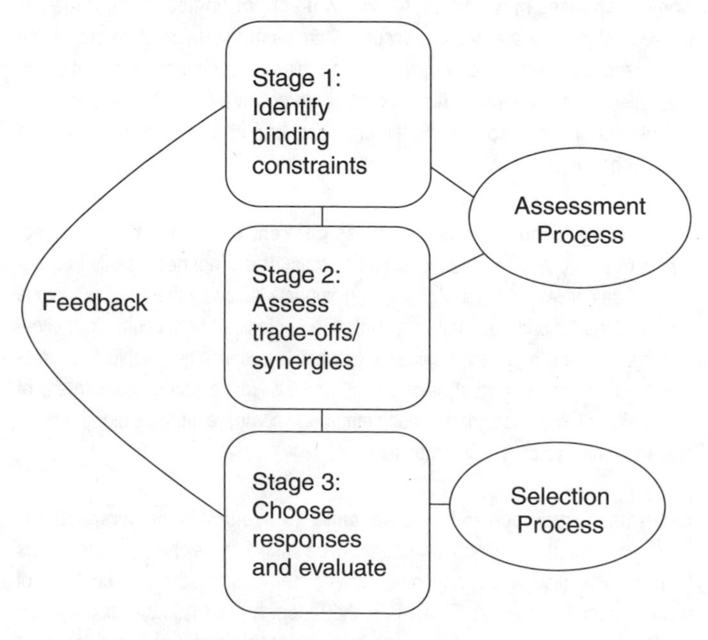

Figure 3.1. The Assessment Method

First and foremost, political context matters. Responses, whether they are limited to a local region, or are national or international in scope, have the potential to generate heated political debate and, in some cases, sufficient opposition to prevent further progress. On the other hand, the institutional sponsor of a given response may depend upon relations with other political stakeholders for resources and support. The following section of this chapter provides a social-scientific understanding of the political environment, paying particular attention to the political feasibility of responses and the potential sources of political opposition. Even those responses that are politically viable may not be effective because they may be beyond the capacity of the organizations that are assigned responsibility for their implementation. Subsequent sections discuss: the need to assess the institutional capacity for implementation at all potential levels of governance, including the local, provincial, national, and international; the need for economic analysis and how different options may perform relative to a range of economic criteria; and the social implications of responses, especially the "unintended consequences," or social externalities, that emerge as a result adopting particular policy choices.

Although these domains are treated as independent for the purposes of analysis, it is important to recognize their interrelationship in practice. The social context describes a set of broad parameters within which economic, political, and institutional activities function. We know that economic shifts, for example, inevitably influence the social, institutional, and political domains, just as each of these domains exerts influence on all others. Moreover, all human activities take place within an ecological context. Activities in any of the domains discussed here have direct implications for ecosystems, which in turn set ultimate boundaries on the range of human activities that can be sustained. A discussion of assessment of the ecological domain clearly warrants extensive treatment unto itself, and is considered in other chapters. Consequently, although we refer to this domain in several tables as a necessary feature of any response assessment, specific details are not considered further in this chapter.

The final section of Chapter 3 outlines in greater detail our evaluative method, which emphasizes that, in any *particular* policy environment, an effective assessment of responses should be multidimensional, involve inputs from multiple disciplines, and attempt to integrate the perspectives of multiple decision-makers. In such a pluralistic environment, it is possible that difficult trade-offs and choices between alternatives will dominate decision-making, although there may be opportunities for synergy. The suggested methods are intended to make these trade-offs, choices, and synergies explicit, since they are often hidden or neglected in policy dialogue. Difficult choices are often involved in decision-making, and it is usually not possible for strategies to achieve all desirable policy objectives. Solutions to these often-intractable problems are likely to be context-specific, and it may not be easy to achieve consensus among stakeholders about the suitability of specific responses. *However, it is desirable to follow a process in which these choices and trade-offs are made transparent, to enable decision-makers to choose responses that are appropriate to the context and congruent with their desired goals.*

3.2 Political Factors

Since responses are understood here as conscious efforts to change existing social structures or behavior, it is important to consider the political environment in which such changes are to be implemented. Responses may be difficult to introduce if the political conditions are unfavorable. Decision-makers need to assess the feasibility of responses based on the political environment in which these options are to be implemented. The political environment is defined by the actors and interests who have a stake in the response and by the political structures within which their strategies are pursued, including the process by which issues command attention and become part of a policy agenda. If the external political environment is assessed to be favorable, it may be possible to introduce a desired response; alternatively, it may be necessary to invest resources in political activities that would create a more conducive climate for implementation, or to alter the response so that it is more appropriate for a given political climate. Assessment of the political domain thus involves the identification of stakeholders, an evaluation of the relative power of each to influence ecological responses, and a characterization of the political structures involved.

3.2.1 Stakeholders

The first step in assessing the political feasibility of a given response involves the identification of those individuals and groups who are likely to be actively involved in, or may be affected by, either the formulation or implementation of a given response. These stakeholders include, but are not limited to, political actors, interest groups, social movements, implementing groups, political activists, power-brokers, and consumers. It is necessary to look at the specific roles that different stakeholders play in the political processes surrounding ecological response strategies. It is also important to recognize that the implementation of some responses involves multiple scales of decision-making. Building a dam to protect a flood plain in a rural area in a developing country, for example, might involve stakeholders at the local level of the village that is located in the flood plain, as well as the international financial organization that will fund the project.

Understanding the stakeholder community requires going beyond simply identifying potential stakeholders and their interests, to include an assessment of the relative power of each. When considering responses, it is necessary to identify the key stakeholders who are relevant to the strategies under consideration, and their potential for political mobilization. In general, stakeholders include those who already have the political influence to affect the ecological response and are motivated to employ it; those who wish to affect the ecological response and are actively seeking to acquire the political influence to do so; and those who are affected by the ecological response but are not actively seeking, or cannot reasonably be expected to acquire, policy influence.

Measuring power can be a complex task, due to the elusive nature of its exercise. In the past, many political scientists simply evaluated the relative ability of organizations to have their interests addressed in the political arena; clearly those whose agenda received the greatest level of support from policy-makers and other elected officials have the most power. Reliance on observations of such visible expressions of power, however, leaves many groups at the lower end of the power spectrum unidentified. Many individuals, despite the existence of grievances, simply do not participate, because they do not believe their efforts will pay off. Other groups have become so disempowered that they internalize the existing power structure and come to accept their lack of power as justified. This "third dimension of power" (Lukes 1974) often expresses the condition of oppressed groups. In many instances, the lack of power of these groups can itself pose a hindrance to ecological responses, as their participation may be necessary for implementation.

The most effective means to assess the relative power of stakeholders is through identification of their concerns and interests, and an evaluation of the respective success of each stakeholder in their efforts to pursue their concerns in the political arena, while paying particular attention to groups outside the political arena whose interests are clearly not being addressed. Further, it needs to be kept in mind that "stakes" are continually being renegotiated and redefined, so that this process is inherently dynamic.

In addition to stakeholders in civil society, the state is central to an understanding of the political environment within which responses are considered. Not only is the state itself a stakeholder, the structure of the nation-state (discussed further) defines relations between state and society by, for example, defining the level of tolerance and legal parameters of organized protest, and the specific steps involved in policy formulation. As such, the specific structure of a given nation-state has tremendous influence over which stakeholders are accorded influence, as well as the extent of, and the nature of, their influence. Nation-states are far from uniform. They not only vary in structure and function across the globe, but a given nation-state should also not be treated as a singular social actor. A state is a combination of actors and institutions, encompassing manifold activities that include everything from political fundraisers, legislative committee hearings, and consultative meetings, to policy implementation on the ground (Laumann and Knoke 1987, p. 381; Chubb 1983). Ecological responses include state involvement with civil society at multiple scales—local, sub-global, and global—which makes the relationship among organized interests and the state in its multiple forms all the more complex.

In other words, responses—which in many cases take the form of political decisions—are the product of the interrelations of multiple stakeholders and state institutions (Fisher 2004). These stakeholders, however, are not only working to affect the state, they are also influenced by the state themselves (Chubb 1983; see also Austen-Smith and Wright 1994).

3.2.2 Political Structures

Just as the list of stakeholders may vary in different geographical contexts, at different scales, and according to the specific political issues in question, the political structures that define stakeholder relations with the state, and establish the process within which policy-making occurs, will also vary. In some countries, it may be acceptable practice for citizens to hold rallies and demonstrations, or to litigate against the government, while in other countries it is not. Other forms of political activity used by stakeholders may include support of legislative candidates, distribution (and receipt) of informational material designed to sway public opinion, lobbying, testifying at public hearings, signing petitions, writing letters to legislators, or serving on citizen advisory panels. In short, these opportunity structures determine the distribution of power in a social system, and are defined by: cultural/traditional institutions; behavioral norms; legal/constitutional mandates; formal political structures that determine the "rules of political engagement"; and the influence of international regimes.

Political scientists and sociologists often distinguish among several stages in the policy-making process, but emphasize the importance of the first two (which can overlap in practice)—agenda setting and policy formulation. Of all the possible issues of concern among members of a social system, only a small number ever make it to the political agenda and become the focus of policy-making—a process heavily influenced by organizations capable of dominating the discourse and the selection and portrayals of political issues in the media. One of the more notable trends in environmental and ecological governance in recent years has been the tremendous growth in complexity, both of environmental concerns and of the political environment within which agenda-setting takes place. Many ecological concerns fail to receive adequate attention, simply because the immediate costs and latent benefits of many response options render them unattractive to elected officials in representative democratic systems, whose primary concern may be re-election within a short time horizon.

In past decades, governmental policy-makers have worked within a closed network of legislators, regulators, and, in some cases, relevant industry representatives, described as the "iron triangle" of regulation (Wilson 1980). Today, this iron triangle remains in place in regard to certain policy issues and in some regions. While industry organizations still tend to dominate in many contexts, agenda-setting in environmental and natural resource domains is coming under increasing scrutiny as groups in civil society, including environmental organizations, community-level justice organizations, the media, and scientific institutions, vie for influence. Not only have more social actors entered into political discussions, even the scale of environmental policy-making has expanded. Many environmental policies play out on the international stage at the same time that they are being negotiated internally within countries (Putnam 1988; Evans et al. 1993; De Sombre 2000).

Although each country is unique in its response, when policies are going through an international process, each country's response involves interaction with decision-makers within the international arena (for a full discussion, see Fisher 2004). At least in those democratic countries in which civil society is sufficiently strong, and the nation-state is sufficiently concerned about its own legitimacy, growing environmental awareness and activism can sometimes impinge upon this closed network of regulators and regulated, often eliciting defensive responses from both. (See Box 3.1.) In fact, there is increasing evidence that international environmental pressures can lead nation-states to build environmental capacity, regardless of the level of development in those countries (for example, Frank et al. 2000).

Interest groups in civil society that advocate for ecological responses include a variety of local, national, and international groups. The nature of many ecological concerns, however, often renders political support elusive. Such cases are frequently characterized by a small set of organized, concentrated economic interests opposed to particular ecological protective measures pitted against a very large, disorganized group of supporters (Olson 1965). This situation is especially true of ecological concerns that

BOX 3.1

Political Bargaining over Ecological Responses: "Job Blackmail"

One defensive response that has been employed by many resource-based companies has been termed "job blackmail" (Kazis and Grossman 1991): as resource-based industries face criticism from environmental groups, companies often emphasize the extent to which environmental protection measures have resulted in the loss of jobs. Although such tactics serve to forge an alliance between industry and the local communities from which their labor pool is drawn, others have claimed that job losses are more likely due to the rapid capital intensification of many resource-based industrial processes, and these tactics have only served to place blame on environmentalists, thereby shielding companies from criticism and labor unrest.

are cognitively ambiguous, are not perceived by politically salient actors as directly associated with livelihood, and are not captured readily by the more dominant conduits of such information.

Biodiversity and related "ecosystem services" are often very broad, and in some instances global, public goods. Seppanen and Valiverronen (2000) found that the destruction of biodiversity is an issue that has been difficult to popularize in industrial countries because it lacks a distinctive visual symbol that could encompass the concept. However, where destruction of biodiversity is linked with more immediate livelihood concerns (especially in developing countries), these issues enter the public agenda very rapidly, depending upon the political power of vulnerable communities and their supporters. A case in point is dam building in the Narmada valley in India, where the issues of displacement and ecological destruction came together to create a powerful movement against the dam (Roy 1999).

Perhaps the most important issue that determines the potential for ecological concerns to be placed on the political agenda and the subsequent formulation of policy is the power of the advocacy groups relative to other groups within government, industry, and civil society. As the highest national authority, the state ultimately must take action in the majority of ecological responses. Civil society organizations attempting to promote a response must in most circumstances convince the relevant state actors of the need for such a response. Ecological responses may also be introduced by the state itself. In both instances, given that ecological responses inevitably represent costs to other sectors of industry and/or civil society, the tendency for a state institution to promote this set of interests can be indexed by its *autonomy.* Autonomy is defined as the ability to determine a policy agenda despite external influence. The state may be completely autonomous if opponents of a proposed policy do not have sufficient strength (either in terms of numbers or political clout) to influence the regime. On the other hand, when groups opposed to a proposed response strategy have the power to threaten the regime, the state has no autonomy at all. In most instances, however, the autonomy of the state lies somewhere between these two extremes (Nordlinger 1981; Domhoff 1996). Where a nation-state is placed between such extremes is contingent upon its historical and cultural conditions, as well as the circumstances of a particular response. In the face of very strong opposition, a state institution may need to consider suitable compensation to "buy off" the opposition, or it may need to compromise on the proposed policy.

3.3 Institutional Factors: Capacity for Governance

While a large portion of research on policy effectiveness is focused on the relative power of stakeholders, *capacity* defines another essential element of governance that should not be taken for granted. History is replete with instances of powerful organizations falling short of their objectives due to a lack of capacity, most notably in the international arena. Governance is the sum of the many ways in which individuals and institutions, public and private, manage issues (Commission on Global Governance 1995). Implicit to this definition is the recognition that effective governance depends not on how any one institution performs or how any one set of actors interacts, but on how they perform and interact as a whole. With regard to ecological concerns, governance comprises a whole network of actors, involves a whole range of functions, and is underpinned by certain implicit or explicit principles, norms, rules, and decision-making procedures (Krasner 1983). It is important to note that local communities are important "institutions" to be included in any assessment, since they are implicated in the local implementation of ecological responses in most cases.

Capacity to govern can be defined as the ability of these institutions to execute responses effectively. If there is a high capacity for governance, a response has a better chance of being effective. The degree of effectiveness also depends on factors external to the institution. However, for an institution that lacks the skills, information, and resources necessary for the implementation of a response, outcomes are likely to be disappointing, regardless of the degree of support and enthusiasm expressed. Capacity for governance cannot be viewed as an artifact frozen in time and space, but as a process that changes over time. Institutional actors can learn, make compromises and change, and forge new relationships that can open the doors to additional skills and resources. For responses to be effective, they must be robust enough to adapt to these shifts. (See Box 3.2.)

In short, an assessment of the institutional domain entails an evaluation of the skills and resources possessed by the institutions that will bear responsibility for the implementation of a proposed response, relative to the skills and resources that would be required to implement that response. A gap between what is available and what is required may become a binding constraint and necessitate adjusting the proposed response in light of capacity limitations. Alternatively, the constraint may be overcome through a sustained effort in institutional capacity-building. As responses vary across manifold scales, the capacity to execute responses will depend on the institutions that operate at these scales: international (including regional and sub-regional); national (including provincial or state levels); and local (encompassing both urban and rural contexts) levels.

3.3.1 International Level

The spatial scale of several ecological concerns demands a response at the international level. Although efforts at international governance have multiplied exponentially in the past sixty years, international responses are enormously difficult to achieve, largely because the current international system of governance lacks the degree of stability and order that characterizes systems of governance at national and sub-national levels. The international system is characterized by the struggle for power between states, with a small number of states dominating this struggle (Strange 1983). The influence of certain non-state actors, such as financial

BOX 3.2

Institutional Resilience: The Ability to Adapt

Early international treaties, as well as many domestic policies, were not designed to take on new commitments, nor were they easily amended. As a result, many became stagnant and irrelevant to governments, and/or lost their effectiveness. Modern treaty-making, however, has incorporated a more adaptive approach, recognizing, for instance, that commitments by governments may strengthen when issues become better understood or when shifts in public opinion encourage governments to take action. Modern treaties contain various mechanisms that allow their parties to adapt or learn, or shift with societal norms and values. These include mechanisms such as framework and protocol approaches, learning systems such as education clauses, science and technology mechanisms that review progress in knowledge and advancement on the issue area (Chambers 2003a). Several recent domestic policy efforts have attempted to incorporate such an adaptive approach.

institutions and some nongovernmental organizations, has been increasing dramatically in recent years as well. Various institutions, both formal and informal, are designed to mitigate the influence of these dominant actors, including the formally recognized principle of sovereignty, numerous customary rules, and international treaties, as well as international governmental organizations such as the United Nations. These systems of international governance are often referred to as regimes, which serve as the frameworks through which international actors mediate their behaviors and play out their roles. These actors include states, sub-state actors, epistemic communities, business/industry, and civil society (for a full discussion of the limitations many actors face when trying to participate in international regimes, see Fisher and Green 2004).

While regimes can mediate actors' behaviors, the nature of the inter-state system renders these regimes only partially effective in determining outcomes and regulating the behavior of states and other actors. Several other factors come into play when determining the capacity of governance systems, including compliance; institutional legitimacy; implementation mechanisms; horizontal and vertical interlinkages between institutions; access to financial resources; and institutional adaptability.

Compliance describes the degree to which states follow formal rules and obligations dictated by international law. Though some international law is self-executing, requiring no ratifying legislation, most rules require implementation at the domestic level and thus national policy measures to ensure compliance. Such measures may include financial incentives, legislation, directives, procedures, or sanctions (Brown-Weiss and Jacobson 1998). In straightforward legal obligations, such as submitting progress reports, assessing compliance can be relatively easy. In other cases, however, evidence of compliance can be elusive, and the ability to apply sanctions at the international level can be problematic. *Assessing the potential for and/or evidence of compliance is an essential first step in evaluating effectiveness of responses at the international level.* (See Box 3.3.)

Another important aspect of governance is *legitimacy.* A number of attributes contribute to the perceived legitimacy of international governance regimes. These include the clarity of the rules, their "symbolic validation" (the states or entities responsible for creating the rules), their coherence (the interpretation of a rule according to some form of consistency) (Brown-Weiss and Jacobson 1998, p. 136), and their adherence to the existing hierarchy of rules. At the top of the hierarchy is the rule of recognition, which grants each country its sovereignty (Brown-Weiss and Jacobson 1998), and beneath these are "secondary rules" that guide making of constitutions, bills of rights, etc. Accordingly, if an international law is in adherence with these secondary rules then there is additional incentive for state compliance (Brown-Weiss and Jacobson 1998, p. 187). Actors are more likely to comply with international laws when they perceive those laws, and/or the institutions sponsoring them, to be legitimate (Franck 1990); hence legitimacy is an important component of any response assessment.

Although compliance is necessary, it is not sufficient to ensure response effectiveness. A particular international response strategy may fail regardless of the extent to which states are in compliance with international obligations. For example, experts agree that the 5.2% reduction of greenhouse gases called for in the Kyoto Protocol will not be enough to stave off climate change, and should only be viewed as a first step. It is also possible that a treaty may unintentionally create incentives to switch to other technologies that also have the potential to damage to the environment, such as increasing the use of nuclear power to reduce air pollutants associated with the burning of hydrocarbon fuels. The extent to which the proposed *implementation mechanism* is appropriate to the stated goals of the response is a key variable that must be assessed to determine whether adequate changes will occur in the behavior of the target group (Raustiala 2000). (See Box 3.4.)

Effective international governance also depends upon the nature of the *interlinkages between international institutions,* as several institutions must inevitably become involved in response formulation and implementation to ensure effectiveness. Unfortunately, international governance regimes are not conducive to the development of coordinated or synergistic approaches to collective environmental problem solving. The complexities of the issues involved, as well as the political nature of policy-making, mean that international responses are often negotiated in relative isolation. Negotiations are often carried out by specialized ministries or functional organizations in forums that are completely detached from the negotiating arena of other international agreements (Chambers 2003b). Even in this isolated context, the consensus building process that is necessary for effective multilateralism is difficult, but with the added burden of accounting for the multiple interrelations across policy domains—such as biodiversity protection and agriculture, for example—effective in-

BOX 3.3

Determinants of Compliance

Compliance may depend on an array of factors that vary from case to case, such as the intrusiveness of the activity; the characteristics of the accord; the negotiating environment; the actors involved; and the depth of the accord, which includes its obligations (binding or hortatory) as well as its precision (Brown-Weiss and Jacobson 1998). Consideration must also be given to the mechanisms for implementation, treatment of non-parties, the existence of free-riders, other countries' approaches to compliance, and the role of international organizations and the media. The "social, cultural, political and economic" conditions and how they influence compliance with the accord are also important considerations (Brown-Weiss and Jacobson 1998, p.7).

BOX 3.4

Assessing Implementation

To measure the effectiveness of implementation, most studies have looked at the implementation process both from the international and domestic levels and from the perspective of the state and civil society. These studies generally examine the use of international institutions to review implementation and the ways in which problems are resolved. The focus here is the "systems of implementation review" (Victor et al. 1998). This approach looks not only at the legal requirements set out in the agreements, but also at the participation of actors and the system-wide operating environment of the commitment—even in cases where formal procedures do not exist. Some scholars have, however, found that a focus on the establishment or diffusion of institutional forms of environmental protection may actually have little to say about the extent to which such measures or forms "have, or are likely to have, any definite connections with actual environmental protection outcomes" (Buttel 2000; see also Fisher and Freudenburg 2005). Several factors influence the effectiveness of implementation of international commitments at the national level; these may include the nature of the problem, configurations of power, institutions, nature of the commitment, linkages with other issues and objectives, exogenous factors, and public concern.

ternational governance is more often than not an elusive goal. Regardless of the difficulties associated with accounting for these interrelations, ignoring them has created global environmental institutions that are ineffective because they attempt to deal with extremely complex, interrelated systems—ecosystems—in piecemeal ways (Chambers 2003b). Therefore, instilling stronger mechanisms that facilitate interlinkages as well as intra-linkages between and across regimes, at the vertical level (ranging from global to local) and at the horizontal level (between regimes at the same level), is one means of improving response effectiveness (for example, Young 1999).

Adequate financing is, without question, one of the key factors for improving the capacity for governance. Financial resources are important for supporting an adequate implementation infrastructure, and for addressing the environmental problem itself. Financing is also important to ensure ratification and compliance on the part of developing countries, as in many instances these countries do not have the resources to meet the obligations. Not only the level of financing, but the institutional skills that are necessary to secure financing, as well as the efficiency of distribution of finances, are all important factors in determining the effectiveness of international responses.

3.3.2 Domestic (National) Level

At the national level, good governance is defined as the manner in which power is exercised in the management of a country's economic and social resources for development (World Bank 1992). Many determinants of the capacity for governance discussed above are equally relevant at the national level, including institutional legitimacy, implementation mechanisms, inter- and intra-linkages throughout the federal and state-provincial governance apparatus, and financial resources. Several additional issues have been highlighted as determinants of the domestic capacity for governance, including the largely administrative elements necessary to implement structural change, as well as political commitment (Leftwich 1994). In particular, effective ecological responses demand the following administrative features: a pluralist polity, in which multiple interests and ideologies can be represented (usually, but not always, through multi-party democratic systems); a clear separation between executive, legislative, and judicial functions that ensures the accountability and transparency of the decision process; and adherence to the rule of law. Furthermore, a committed and efficient public sector is needed that has the capacity to manage reform processes (World Bank 1992; see also Cardoso 2003). Countries lacking one or more of these features generally either are more resistant to the adoption of certain ecological policies and/or exhibit difficulties in their implementation.

In addition to administrative structure, a state's ability to deliver good governance is a function either of its political commitment or of the extent to which it has rationalized environmental concerns into its set of primary goals (Frickel and Davidson 2004). Although administrative structure can be assessed by focusing on the institutions of the state themselves, most decisions regarding ecological response options involve the interrelations among multiple social actors. Therefore, understanding political commitment is more complex, entailing an assessment of the state's relationship with society (for example, Habermas 1975). Particularly when addressing contemporary environmental concerns, a state institution may be dependent upon the expertise and resources available among groups in civil society for effective policy development, a condition known as "Embedded Autonomy" (Evans 1995). When addressing the reauthorization of the Clean Air Act in the United States, for example, the national government recognized that a cap-and-trade system that allowed companies to determine their own emissions policies would be most effective.

Organized nonstate actors that function at the peripheries of bureaucratic state systems can be particularly important sources of capacity (or hindrance) to the pursuit of ecological policies (for example, Adger et al. 2003). These may include, for example, the relationships among local economic interests and local tenure holders and regional regulatory offices (Davidson 2001). Assessment of nation-state capacity to implement ecological responses must include an evaluation of the ability of state institutions to forge facilitative relations with organizations in society that can bring additional resources to bear, while avoiding those relations that pose a hindrance (such as with interests opposed to ecological reform).

3.3.3 Local Level

While there has been extensive work that illustrates the success of community-based efforts in managing and implementing ecological responses, many scholars acknowledge that communities are not simply homogenous groups that work harmoniously to promote group objectives. Communities are more accurately seen as complex and dynamic institutions that are often characterized by internal differences and processes (Leach et al. 1999; Agrawal and Gibson 1999). In short, the debate over the effectiveness of community management has come full circle: from early pessimism about community action as exemplified in the work of Hardin (1968), to a relatively uncritical, and arguably idealistic, view of community-based conservation initiatives through the 1990s (Western and Wright 1994), to a contemporary recognition that community-based management regimes may be appropriate in some circumstances, but not in others (Agrawal 2001).

Scholars have highlighted several conditions upon which the ability of decentralized, locally embedded organizations to manage ecological responses is contingent, based on both field experiences and theoretical development (for example, Wade 1988; Ostrom 1990; Baland and Platteau 1996; Agrawal 2001). The conditions most frequently identified include: (1) perceived local benefits from cooperating; (2) clearly defined rights and boundaries for any natural resources implicated in the response; (3) knowledge about the state of those resources, including for example their extent, accessibility, and potential for regeneration; (4) small size of user groups; (5) low degree of heterogeneity of interests and values within user groups; (6) long-term, multilayered interaction across the communities and other governing institutions involved; (7) simple, unambiguous rules and adaptable management regimes; (8) graduated sanctions as punishment; (9) ease of monitoring and accountability; (10) conflict resolution mechanisms; (11) strong, effective local leadership; and (12) congruence with the wider political economy within which those communities function. These factors refer to characteristics of the resource itself (2 and 3), of the user group (4, 5, and 6) , and of the institutional arrangements for resource management (1, 7, 8, 9, 10, 11, 12). Interestingly, the literature has tended to neglect the role of technological factors while identifying conditions for successful local resource management; in part, this is a reflection of the institutional focus adopted by most authors in this tradition.

Assessments of community-level social capacity can be fairly intensive endeavors, and there are numerous variations within the social sciences regarding which community features should be used as indicators. For the purposes of evaluating the potential for local-level support for, and participation in, ecological responses, the conditions listed above can serve as a general guide, recognizing that some conditions will be more relevant than others in any given context. The characteristics of the natural resources themselves, and to a certain extent those of the user groups, are resistant to change, so it is particularly important to assess the appropriateness of a given response relative to these conditions as they might pose binding constraints. On the other hand, since institutional arrangements at the local level are conducive to change through the use of specific policy measures, intervention to enhance local capacity for governance is usually targeted at these institutional features.

3.4 Economic Factors

The economic effectiveness of responses needs to be considered for both pragmatic and philosophical reasons. From a pragmatic perspective, those who are promoting and financing ecological responses need to be concerned about the costs of the programs and the economic impact of those programs on affected groups. Money is the unit of account in which most governments and donors deal, and a limiting factor in what planners can achieve.

It is telling that one of the seminal treatises in formal economics is titled *The Theory of Value* (Debreu 1954). Unlike some other social scientists, economists are often not hesitant to offer value judgments concerning whether one outcome is "better" or "worse" than another

For these reasons, economic arguments and principles are often used to motivate responses to ecological degradation and to assess the effectiveness of responses. Individuals conducting assessments should, however, be aware of the strengths and weaknesses of economic tools. Among the strengths is a large body of formal theory demonstrating that social welfare—the "values" arising from production, consumption, and preservation—may, under certain conditions, be related to readily observable measures of economic performance: asset prices, incomes, and consumption. Even when these conditions are not met, economic tools may provide a useful, if perhaps not necessarily the only, metric for evaluating responses. Moreover, even if the economic paradigm does not always offer an overall objective with which all observers agree, it may still provide a useful means to an end by suggesting principles to reach any given goal in the most cost-effective manner.

Individuals engaged in assessments should also be aware of the weaknesses of received economic theory. The economist's standard measure of performance is rather narrow. One outcome is better than another, by this standard, if at least one person is made better off without making anyone else worse off. An outcome is *Pareto efficient* or *Pareto optimal* (after the Italian economist and sociologist Wilfredo Pareto [1848–1923]) if there is no better alternative under the no-one-is-made-worse-off criterion. Pareto optimal outcomes are, in general, not unique. Alternative outcomes with very different implications for different individuals may each be Pareto optimal. In short, Pareto optimality is about efficiency, not equity.

Attempts have been made to generalize from Pareto efficiency to a "Potential Pareto Improvement" criterion in which everyone could *in theory* be made better off, even if they may not necessarily be compensated in fact. Such approaches underlie benefit-cost analysis. Under a PPI criterion, the policy that maximizes the sum of monetary benefits net of costs across all individuals would be the "best," in as much as a redistribution of gains among individuals could assure that everyone would prefer it. While this prescription is implemented in many analyses, it remains controversial in some quarters, not least because the possibility of hypothetical improvements for everyone in society offers no assurance that everyone in society will actually realize an increase in his/her welfare.

Moreover, technical problems arise in applying the PPI criterion. Kaldor (1939) and Hicks (1939) proposed criteria by which policy changes would be socially preferred if "winners" could afford to compensate "losers" without the winners sacrificing all their gains (Kaldor) or if "losers" could not afford to bribe "winners" to forego their gains and still be better off without the policy change (Hicks). Following Scitovsky's (1941) observation that the criteria do not always favor the same policy, the suggestions have often been combined as the "Kaldor-Hicks criterion" that winners can compensate losers and losers cannot compensate winners. The PPI criterion has remained problematic, however, as other commentators uncovered other paradoxes and limitations (for example, Gorman 1955; Boadway 1974; Mishan 1981). Problems arise in considering the amount of *money* that would have to change hands in order for all parties to be indemnified fully against a policy that also changes *prices*. This calculation depends on whether it is made under the status quo or after a policy change, and may make PPI criteria difficult to apply in the consideration of major structural changes in economic organization. Of course, many environmental advocates believe that major structural changes will be required to achieve a sustainable future.

While the concept is problematic, justifications may still be offered to continue to base policy advice on PPI criteria. First, at least compared to alternatives that necessarily rely on subjective assessments of the moral worth of different individuals, BCA is relatively easy to apply and, because it respects the actual distribution of income rather than some subjective system of social weights, it is "objective." Second, it might be argued that since most societies have chosen relatively limited reallocations of wealth among their constituents as part of their general taxation and transfer policies, there is limited evidence that such societies really care enough about equity to make it an ancillary goal of their environmental policies. Third, some policy-makers may argue that that the "winners" and "losers" of policy action are sufficiently similar in either their economic circumstances or moral worth as to justify the working assumption that monetary gains and losses are equally socially valuable for each. Finally, with respect to the more technical problems of making logically consistent comparisons in a shifting landscape of relative prices, many analysts proceed under the often-implicit assumption that they are considering changes of small enough magnitude to obviate such concerns.

None of these arguments is wholly convincing and, indeed, some authors call for the abandonment of the benefit-cost paradigm (for example, Gowdy 2004). Perhaps a more balanced view is to regard BCA as one useful approach among others and to augment its prescriptions with additional criteria to address equity (Little 1950) and, perhaps, sustainability in the face of potentially irreversible ecological losses. Pezzey (1997), for example, argues that society should opt for the "optimal sustainable" path should a conflict arise between the objectives of maximizing net benefits and providing for future generations.

While the conceptual foundations underlying BCA remain somewhat unsettled, there are also practical impediments to conducting thorough and reliable BCAs. Economists have developed an elegant conceptual apparatus with which to interpret the prices of goods and services traded in existing markets. They have been less successful in the more difficult task of assigning prices to goods that are not yet transacted in markets.

We might digress to note that conducting a thorough BCA is not always required. In order to preserve an imperiled ecological system or resource, it ought to be sufficient to demonstrate that the benefits of its preservation exceed the costs. In some instances, this might be accomplished by establishing only one or a few of its many ecological values. More generally, however, rational policy requires that ecological assets be preserved when the sum of their values exceeds the costs of their preservation. Restricting analysis to only a subset of values is unlikely to motivate the preservation of all ecological assets that merit it. It should be noted that, to an economist, the "value" of something is not a measure of what it is worth to society to save it in total, but rather a measure of what society might be willing to give up to save a little more of it—the "marginal unit." Economists care about *marginal,* rather than *total,* values. Nevertheless, many environmental and resource economists have taken to the useful abbreviation of discussing the "total economic value" of ecological assets (Pearce and Turner 1990). In TEV, the totaling is done across attributes, rather than across quantities. These attributes may include the simultaneous and mutually compatible uses of, say a forest, for carbon sequestration, for recreation, and in appreciation of the biological diversity it harbors. The value of each attribute is calculated on the margin, but then these marginal values are summed across categories.

A result so fundamental as to be labeled the "First Welfare Theorem" in economics holds that a perfectly competitive economy is Pareto optimal (for example, Varian 1992). A perfectly competitive economy is one in which no actor can influence the prices at which he/she completes transactions, all agents are well-informed, and all commodities are privately owned. Under such circumstances, all goods and services trade at prices that reflect their worth to all members of society. Such circumstances essentially rule out inefficiency by construction: if one person valued something more than another, the two could enter into a mutually beneficial transaction to exchange it (although the formal demonstration of this result in a general setting involves the performance of a great deal of sophisticated mathematics; see, for example, Arrow and Hahn 1971).

The limitation inherent in this result is that public policy is enacted precisely because not all firms are competitive, not everyone is well informed, and not all commodities of interest are privately owned. In the context of ecological systems and services, economists are typically concerned with the second and third elements, particularly the third. The services provided by ecosystems are often "public goods," that is, they support people who bear little if any cost for their preservation yet share in their benefits. Public goods can lead to free riders—individuals who benefit without paying. The problem, in formal terms, is that those who provide ecosystem services by foregoing the destruction of natural ecosystems cannot assert ownership over the services such systems provide. A market economy allocates goods and services efficiently because it facilitates transactions by which one person can pay another to provide a desired good or service. Public goods give rise to a "market failure": there is typically no mechanism by which someone in North America can pay someone in Africa, for example, for the ecosystem services the latter provides the former.

Economists offer a prescription for such market failures by creating or simulating the "missing market" (for example, Kolstad 2000). If there *were* a market in which such transactions might occur, the efficiency properties of an ideal economy would be restored. Another problem arises, however. Just as the hypothetical North American might free-ride on the ecological contributions of the African, one North American might free-ride on the contributions made by another to compensate the African. Public goods are not adequately provided by private markets and must instead be allocated via political decisions and financed from public revenues. The classical argument is presented by Samuelson (1954), although others have argued that "the private provision of public goods" may not always be so inefficient due to altruism or other considerations (for example, Andreoni 1989; Bagnoli and Lipman 1992; and the numerous empirical accounts reported by Ostrom 1990).

Public goods lead to a problem of efficient allocation that does not arise in private transactions. The vaunted "invisible hand" of the market is said to result in the efficient allocation of private goods because each purchaser chooses to buy precisely the amount he wants at the market price, and each provider chooses to sell precisely the amount she wants at that price. There is no comparable mechanism for public goods. There is, however, some dissent on this point. The philosopher Sagoff (1994) suggests that a "missing market" is an economic oxymoron: transactions occur when the benefits of their consummation exceed the costs of their arrangement. But few economists are willing to take so extreme a view of the efficiency of economic arrangements, although some suggest a careful calculation of the administrative costs of allocating public goods and a corresponding constitutional reluctance to appropriate such functions (for example, Stroup 2003). Their point might best be summarized by noting that effort devoted to overseeing the allocation of public goods is itself a public good.

3.4.1 Valuing Benefits

If one accepts the benefit-cost approach to public decision-making—perhaps with some amendments or additions to ensure equity or sustainability—the next order of business is to determine the monetary consequences to be assigned to particular outcomes. Those falling under the rubric of "costs" are often relatively easy to infer. The costs of providing the services afforded by natural ecosystems are represented by the opportunity foregone to preserve them. These can often be inferred from, for example, the price of similar land that has been cleared and used for agriculture or other purposes. The benefits are much more difficult to measure. Some conservation organizations have adopted economic valuation as a response in itself, the idea being that demonstrating high economic values will motivate conservation (for example, Barbier et al. 1997). Yet the methods available for such studies remain imprecise, and, in many instances, spark fierce controversy.

While many economists would acknowledge the possibility that members of society can and do place large values on the preservation of biological diversity and natural habitats, they also despair of being able to measure such values accurately, at least in the near future. To quote a major study conducted by the U.S. National Academy of Sciences, "True public goods, such as biodiversity, species preservation, and national parks, present major conceptual difficulties" (Kokkelenberg and Nordhaus 1999, p. 8). Even when and if techniques can be established for conducting such valuations accurately, the same report states, "The overriding problem with all [valuation approaches] is that they require voluminous data and statistical analysis and can hardly be used routinely" (Kokkelenberg and Nordhaus 1999, p. 125). Perhaps the best hope, if such analyses cannot be "used routinely" is that representative studies can be identified and general results proposed.

Some approaches to ecological valuation have been widely reported. The most frequently cited, a study of global ecosystem service values by Costanza and numerous coauthors (1997) has been widely derided by economists (for example, Smith 1997; Pearce 1998; Toman 1998; Bockstael et al. 2000). Criticisms of the report by Costanza et al. have largely focused on its interpretations of economic concepts. Critics allege that the work confuses average and marginal values and reports an economic impossibility: a willingness to pay in excess of the ability to pay.

Other authors have pursued analyses on more modest geographical scales. Kremen et al. (2000) suggest that the public goods values arising from ecosystem preservation are larger at a global than a local scale, while Balmford et al. (2002) find that some local ecosystems are more valuable for the natural services they provide than they would be if converted to agricultural or residential use. While neither finding is implausible, it remains unclear how broadly representative each may prove to be.

Alternative valuation paradigms have also been suggested. Roughly a generation before the Costanza et al. work, ecologist Odum (1981) proposed that ecosystem services be valued at the cost of the energy embodied in their functions. More recently other researchers have suggested that the ecological costs of intensive urban development be proxied by the size of their "footprint" on the landscape, as determined by the area of land required to grow their food, provide their energy, dispose of their wastes, etc. (Rees and Wackernagel 1994). Whatever their other merits might be, such single-metric approaches are typically not accepted by mainstream economists. A celebrated result known as the "non-substitution theorem" (Samuelson 1951) establishes that economic values cannot in general be reduced to the contribution of a single primary commodity or input (see Dasgupta 2002 for an argument specifically in the context of environmental valuation). The nonsubstitution theorem is often presented as, and arose as, a refutation of Marx's labor theory of value.

There is a wide range of methods for BCA. Surveys of nonmarket valuation approaches can be found in a number of articles and books; among the most complete and authoritative are Freeman (2002) and the essays assembled by Mäler and Vincent (2004). Economists generally prefer to work with data arising from "revealed preferences"; that is, evidence arising from the decisions people have actually made when they have to pay for, and live with, their choices. A number of studies have been conducted in which the benefits of environmental improvement have been inferred from, inter alia, the increased harvests of timber, food, or fish afforded by environmental improvement (for example, Bell 1998; Barbier and Strand 1998); provision of goods, services, and amenities such as water (for example, Acharya 2000; Pattanayak and Kramer 2001); demand for recreational opportunities and services (for example, Hausman et al. 1995); the costs averted by the provision of ecosystem services (for example, Shogren and Crocker 1999); and the price premia commanded by land parcels situated in favorable positions vis a vis natural ecosystem services (for example, Irwin 2002; Thorsnes 2002). Sophisticated and difficult analyses are often required to tease out the value of ecosystem services from "revealed preference" data. Multiple imputations, each introducing its own statistical uncertainty, may be required to infer the values of goods and services that are not traded in markets from the prices of commodities that are. For an illustration of some of the data requirements and conceptual issues involved in such studies, see Irwin (2002).

While the data requirements and statistical sophistication required to conduct valuation through revealed preference are daunting, the alternative of attempting valuation more directly raises even more contentious issues. In recent decades, many environmental economists have turned to "stated preference" methods. This involves asking people what value they place on public goods such as biodiversity and ecosystem services. Current techniques involve variants such as asking how people would vote in a referendum to secure more biodiversity as well as pay higher taxes, or asking people to rank different tax-and-public-goods outcomes. Such an approach may be unavoidable in the calculation of *existence values* (values wholly independent of any present or future use or bequest). Yet stated preference approaches are anathema to many economists, as they do not require respondents to "put their money where their mouth is" by *actually* paying for what they profess to value. Regarding the general reliability of statements uttered without economic consequences, see Glaeser (2003). The evidence concerning whether people actually contribute what they claim they will is mixed; compare the contrasting findings of Strand and Seip (1992) with those of Vossler and Kerkvliet (2003). A fierce debate has also raged between those who allege that stated preferences reflect the "purchase of moral satisfaction" more than they do any carefully considered estimate of willingness to pay for the specific good in question (Kahneman and Knetsch 1992) and their opponents (for example, Smith 1993). For a perspective from outside economics on the elicitation of preferences, see Fischoff (2004).

Commenting on the use of stated preference methods in economics, V. Kerry Smith (1998) wrote:

> Indeed, there is a curious dichotomy in the research using [stated preference methods] for non-market valuation. Environmental economists actively engaged in non-market valuation continue to pursue very technical implementation or estimation issues, while the economics profession as a whole seems to regard the method as seriously flawed when compared with indirect methods. They would no doubt regard this further technical research as foolish in light of what they judge to be serious problems with the method.

Smith, who has contributed importantly to the literature on stated preference methods, probably overstates the case in characterizing the attitude of the "economics profession as a whole." A panel on which two Nobel Prize winners served provided a cautious but positive endorsement of stated preference methods (Arrow et al. 1993), and one of the journals published by the North American branch of the "profession as a whole" has devoted considerable space to a symposium on the subject (*Journal of Economic Perspectives* 1994). Still, stated preference methods, and, indeed, albeit to a lesser degree, nonmarket valuation methods in general, continue to spark controversy. An example of this tension is seen in a comment by Jerry Hausman, the 1985 winner of the American Economic Association's John Bates Clark Medal—awarded every two years to the individual regarded as the most accomplished American economist under 40 and often seen as a predictor of a future Nobel Prize; Hausman remarked that "Environmental economics is to economics what military music is to music" (*Business Week,* June 30, 1997). A benefit-cost assessment that relies on nonmarket valuation to establish a case for conservation is unlikely to be regarded as clearly dispositive by all commentators. Nor, it should also be noted, would a finding that conservation is *not* warranted, be based on the same methods.

Another very contentious issue in the calculation of benefits concerns the weighting of benefits received at different times. How should we weigh the interests of unborn generations in making current economic decisions? General practice has been simply to assume that future benefits should be discounted back to the present at a constant exponential rate. This gives rise to compound discounting and, with it, many often-cited anomalies; for example, discounted for 500 years at a rate of 5% per annum, the world's current total economic product would be worth considerably less than most used automobiles (the discounted present value computes to a little less than 500 dollars).

It is interesting to note the comments of some leading early economists on discounting. Ramsey, whose model of economic growth has dominated the analysis of that subject for almost a century, wrote that discounting "is ethically indefensible and arises merely from the weakness of the imagination" (1928). Koopmans (1960) provides a simple counter-argument. *Not* discounting, he wrote, might lead to the "tyranny of the future": current generations would be obliged to save for the unborn, accumulating capital to be enjoyed by presumably wealthier later generations. In addition to Koopmans' argument, exponential discounting has the desirable property of *time consistency*. One would not be tempted this year to revise savings or investment plans made last year simply because time has passed.

Whether or not later generations will, in fact, be wealthier is both pivotal and unknowable. One view of discounting is simply that a discount rate is a price like any other. Price ratios reflect ratios of marginal satisfactions achieved from consumption. If we choose to discount future relative to present consumption, that choice ought to reflect, at least in part, an expectation that we will not find ourselves in desperate straits in the future. Of course, if we did expect our future prospects to be worse, we would discount less. Moreover, recent research (Weitzman 1998; Gollier 2002; Newell and Pizer 2003) suggests that uncertainty about future production and consumption prospects ought also to motivate lower discounting.

While there are good reasons for discounting, there are also some problems. First, if we are interpreting discount rates as prices, the analysis of the previous paragraph applies only when a single decision-maker is deciding his/her own consumption plan over time. In as much as human lifetimes are limited, long-term decisions based on discounting necessarily involve one generation making a choice that affects others. While the literature is replete with references to the role of bequests from benevolent ancestors to their heirs (for example, Blanchard and Fischer 1989), it is an open question how best to protect the interests of one's progeny. Some researchers favor treating the present generation as agents empowered to save and invest for the future. Others feel the present generation should be required instead to preserve essentially the same configuration of natural assets as we now enjoy for future generations' use and enjoyment. These two views of the current generation's savings obligations represent in broad terms the schools of "weak" and "strong" sustainability, respectively (for example, Pezzey and Toman 2002). The issue of how best to preserve natural assets for future needs is also closely related to matters of option value, the precautionary principle, and the safe minimum standard (see below).

Some scholars would replace the discounted-present-value of utility formulation that has motivated most models of economic growth with a different criterion for social decision-making. Alternatives include a "Green Golden Rule" approach in which the objective would be to maximize sustainable long-term well-being and, implicitly, reduce the importance attached to the welfare of current generations (Beltratti et al. 1995). Graciela Chichilnisky (1996) proposed an objective that would combine traditional discounting with concern for the long term.

While such alternative approaches have merit, it seems reasonable to advise researchers and practitioners assessing response options to adopt a discounting approach, albeit one with conservatively low and, in the limit of the far-distant future, perhaps vanishingly small rates. This will be analytically more straightforward in making calculations. In as much as there are now compelling arguments in the literature for discounting distant-future gains at extremely low rates (Weitzman 1998; Gollier 2002; Newell and Pizer 2003), it seems reasonable to suppose that the apparatus of discounting can now be accepted in very broad terms without necessarily trivializing well-being in the distant future. The suggestion is sometimes made that analysts should—and people do—apply "hyperbolic" discount rates (see Heal 2000). With hyperbolic discounting, the net present value of a sum to be received in the future declines as a function of the *relative* time lag involved: the ratio of values assigned to events one and ten years in the future would be the same as that applied to events 10 and 100 years in the future. While survey and psychological evidence may support the approach, it may not matter much in practice if we adopt hyperbolic discounting or simply recognize the effects of manifest uncertainty on calculations concerning the far-distant future.

Finally, with respect to the estimation of benefits, it should be noted that uncertainty regarding the magnitude of benefits should not translate into a disregard for them. In fact, the opposite might be true for two reasons. First, just as the prudent household may purchase insurance against unavoidable risks, prudent resource managers may choose to indemnify themselves against the unknown ecological consequences of radical change by foregoing the associated short-term benefits. Second, even absent aversion to risk, it is generally wise to defer decisions with irreversible consequences until their expected benefits more than exceed their expected costs. The theory of option value, first developed in finance (for example, Dixit and Pindyck 1994), argues that there is a premium, or "option value," associated with retaining the option to wait for better information. This idea has been applied to environmental economics in general (Arrow and Fisher 1974) and the analysis of ecological assets in particular (Perrings and Walker 1997, Albers and Goldbach 2000). Some commentators put forward a stronger argument. The future is not simply unknown; it is unknowable. Hence, rather than assign option values to reflect known uncertainties, society should adopt a safe minimum standard or invoke a precautionary principle (for example, Ready and Bishop 1991). While the safe minimum standard and option value approaches are conceptually distinct, they may yield operationally equivalent policy prescriptions. (See Chapter 4 for a more detailed discussion of decision-making under uncertainty.)

3.4.2 Cost-effectiveness

If it is not possible to infer reasonable estimates of all values, it is not possible to perform a full BCA. In some instances, a partial analysis may suffice. There are circumstances in which measuring some of the benefits of ecosystem services and human well-being establishes that a response should be undertaken. The existence of other, as yet unquantified, values simply makes the case even stronger. Or society may simply have made the decision to allocate a certain sum to conservation or environmental improvement. The challenge then is to determine how to spend this sum most wisely.

In performing such cost-effectiveness analyses, the question is simply "how much do the relevant alternatives cost?" Economically relevant costs are *opportunity* costs, rather than financial, historical, or replacement costs. That is, the true measure of the cost of, for example, preserving a parcel of land as habitat for endangered species is the earnings (or more, generally, satisfaction) foregone in not exploiting it for other purposes. It does not matter if these are earnings that a landowner would "pay to herself" rather than to someone else if she were to convert the habitat to another use, nor would opportunity costs be obviated if a government were to condemn the land for public use rather than compensate the landowner for her lost earnings.

The economist's prescription for implementing environmental or conservation policy at least cost to society is typically to use "market based incentives" such as taxes on environmentally harmful products, subsidies to environmentally friendly ones, or a system of tradable permits or obligations in such products. The approach was first proposed by the British economist Pigou (1932)—hence the designation "Pigovian taxes" or "Pigovian subsidies"—and has been a staple of textbooks since, although the first significant policy applications did not materialize for many years. Subsequent research has noted the need for refinements to, for example, fine-tune taxes or permit prices to reflect spatial variation (Tietenberg 1978) or compensate for interaction with existing taxes on other commodities (Parry et al. 1997). However, the principal practical impediments to implementing cost-effective market-based incentives would appear to be largely political (Pearce 2004) and technological (Henriquez 2004). The latter barriers may be removed as capacity is enhanced to track environmental damage in real time and trace specific harms to their originators.

The question of cost-effectiveness has proved particularly contentious in international biodiversity conservation policy. Several authors have suggested that "direct payments" will prove more effective than "indirect approaches" such as those embodied in Integrated Conservation and Development Programs (Simpson and Sedjo 1996; Ferraro 2001; Ferraro and Kiss 2002). Direct payments are, in essence, Pigovian subsidies (Ferraro and Simpson 2002). Such approaches are being increasingly adopted by organizations such as Conservation International (CI 2000), and various nations and organizations are experimenting with payments for ecosystem services (Pagiola et al. 2002). These steps are being taken in large part in response to perceptions that "indirect approaches" have not proved cost-effective (Wells and Brandon 1992; World Bank 1997; Brandon 1998; Reid and Rice 1997; Southgate 1998; Oates 1999; Terborgh 1999; Terborgh et al. 2002). Others have pilloried such criticisms as attempts to "reinvent a square wheel" (Brechin et al. 2002) and might well regard economic injunctions to adopt direct incentives as similarly insensitive to cultural context.

In assessing the cost-effectiveness of conservation policies it seems reasonable to suggest that researchers be able to trace a clear and unambiguous trail from the response proposed to the conservation (and, as appropriate, development) outcome desired. Some arguments for direct incentives are stripped to their bare conceptual basics and, consequently, may ignore social and cultural factors. Yet conservation practitioners and project evaluators should be skeptical of broad interventions intended to solve multiple problems with single programs. Brandon (1998) comments trenchantly on the limitations of conservation interventions in addressing broader social ills.

3.4.3 Secure Property Rights

An important assumption underlying the claim that a competitive market economy is Pareto optimal is that property rights are well-defined. That is, someone owns each of the goods and services that people might trade with one another. The owner cannot be deprived of what he/she owns without agreeing to be compensated with something else. When it is impossible or infeasible to exclude other users, there is little incentive for one person to conserve resources for her future use, as it is likely that another will capture them in the interim. Hence, many commentators note that strengthening tenure—in essence, designating "owners" and empowering them to exclude others—will induce more conservation (for example, Scher et al. 2002; Terborgh et al. 2002).

There is, perhaps, more agreement among conservation practitioners and advisors on the importance of property rights than there is on the matter of how best to compensate owners for conservation. This could be troubling, as property rights alone may not suffice to motivate conservation. There are both costs and benefits to property rights. The benefits arise from security in investment: maintenance or improvement of property is repaid with future earnings. The costs arise from enforcement: property must be defended against those who would infringe upon it. The economic theory of property rights (Barzel 1997) suggests that property rights come to be defined when the benefits of their definition justify the costs of their enforcement.

Some commentators suggest that biodiversity might be protected better if property rights were more clearly established in its products. Establishing and strengthening property rights is, then, a potential response. This observation raises a question of cause and effect. If property rights come into being when the benefits of their establishment justify them, are the potential owners of biodiversity deprived of benefits because they cannot claim property rights, or do they not claim property rights because the potential benefits of doing so have not yet proved sufficient to compensate them? To give one example, some have argued that local people did not have sufficient incentives to conserve biodiversity because, at least absent participation in and protection from the Convention on Biological Diversity, they could not claim ownership of pharmaceutical preparations derived from natural products (for example, Vogel 1994). Others argue that property rights in such preparations have generally remained unspecified because the economic value of untested natural products is not high enough to justify their establishment (Simpson et al. 1996).

In general, practitioners and evaluators assessing the role of property rights in crafting effective responses should consider carefully the function to be served by stronger property rights. If it is to capitalize on as-yet unrealized values, it is reasonable to ask what impediments have prevented their realization to date. If it is to facilitate the receipt of payments for conservation, the source of payments must, of course, also be secured.

None of this discussion is intended to suggest that an adequate source of income in combination with secure ownership would *not* motivate effective conservation incentives. Geoffrey Heal (2000, 2002) has suggested that private and public goods might be, and in fact in many instances are being, "bundled" so as to make payments for the former and as an incentive for maintaining the latter. In southern Africa, for example, private farmers have found it more profitable to devote their lands to game animals than to raise cattle and other non-indigenous species (Bond 1993; Heal 2000; Muir-Leresche and Nelson 2000). By selling the right to enter, photograph, or hunt (in a regulated and sustainable fashion), such game ranchers are earning private returns while protecting a public good. Similar considerations may motivate private conservation of forest habitat in Costa Rica (Langholz et al. 2000) or the incorporation of onsite nature reserves at new housing developments (Heal 2000, 2002). It is certainly appropriate to consider roles for private markets in the economic assessment of responses.

Unfortunately, there are no simple policy prescriptions regarding the optimal or ideal system of property rights to protect and maintain ecological systems. A detailed review of the empirical literature on different property rights regimes concluded, "Success in the regulation of uses and users is not associated with any particular type of property-rights regime" (Feeny et al. 1990, p.12). An important implication of such work is to demonstrate that there are a variety of alternative institutional contexts under which ecosystems and their services may be appropriately managed. The case for secure property rights, therefore, should not be seen as a case that supports any *specific* system of property rights, such as privatization or nationalization.

As a final comment on property rights, we note again some themes already touched upon. Theoretical models of general competitive equilibrium often assume ownership of *everything* at every *time* and in every conceivable *state of nature* (more formally and compactly, "complete futures and contingency markets"). This is obviously impossible: future consumers cannot conduct transactions before they are born. Those alive at present might represent the interests of their descendents, and markets may reflect concern for future resource scarcity. However, different commentators differ as to whether such considerations constitute adequate safeguards for intergenerational equity.

Some similar issues may also arise in the consideration of intragenerational consequences of environmental and conservation policy. Instituting a widespread program of payments for the preservation of biodiversity might be expected to occasion a large transfer of wealth from richer to poorer nations as the relatively undisturbed land holdings of the latter increase in value. By the same token, a policy that obliged poor countries to conserve natural capital without corresponding compensation would exacerbate existing inequities. Researchers assessing the social consequences of responses should consider these factors, and note that changes in the distribution of wealth could have more profound impacts.

3.5 Social Factors

As discussed, stakeholders are those individuals or organizations who wish to affect a response, those who are able to affect it, and those who are affected by it. Groups characterized chiefly by the last criterion may not necessarily be actively engaged in the political process, but may nonetheless face direct or indirect impacts as a result of the implementation of a particular response. Any attempt to develop a strategic response to ecosystem maintenance will have implications for social systems beyond what can be considered strictly economic and political parameters. This section discusses a set of issues that are conventionally classified as relating to the equity of responses; it also introduces the idea of pluralistic belief systems as a potential source of inequity. A distinction is drawn between material inequity and cultural inequity. The latter relates to the ways in which different cultural groups, who perceive the environment and their relationship to nature in culturally distinctive ways, may also be impinged upon, with significant social impacts.

The distribution of gains and losses across stakeholders rarely occurs in a manner that is perceived as equitable by all. Devising and implementing ecological responses always entails gains and losses and, in many instances, these impacts have been distributed in markedly inequitable fashion, regardless of whether this outcome was intended by policy proponents. Equity issues emerge because of distributional impacts that affect stakeholders who can be distinguished along a number of dimensions (some of which may potentially be overlapping). These include gender, ethnicity, race, class, and generation. Some equity concerns may also arise from the pragmatic concern that power be given to, or retained by, those with the material wherewithal to "get the job done." In many contexts this is not solely an economic consideration, but may also involve ethnicity, gender, class, and other attributes. Equity concerns also emerge because groups may have beliefs and ideological opinions that are inadequately represented when choosing among responses and strategies. Although effectiveness is sometimes seen as a benefit to be traded off in terms of a reduction in equity, in some instances the inequitable social impacts of a response have the potential to impinge upon the effectiveness of the response itself.

3.5.1 Equity

Many stakeholders have divergent *material* interests, and these are differentially affected by policies that are targeted at meeting defined ecological objectives. While the economic paradigm treats such issues as distributional (and thus not relevant to the question of efficient resource allocation and management), there are usually profound implications for groups that suffer costs as a result of adoption of specific responses. Furthermore, if such groups are socially or politically marginalized, these costs are often not taken into account in determining the desirability of a response, and the groups are not compensated in any way.

Well-developed response strategies should ideally contribute to the reduction of material inequity between social groups. In many cases, however, a complete assessment of potential equity concerns is not carried out before implementing a policy. In some cases in the past, such policies have exacerbated existing social inequities, or have inadvertently caused new ones, by imposing changes in rights/access to resources or by providing (unintentionally or intentionally) compensation to certain groups. (See Box 3.5.) In many cases, there is a tension between the pursuit of ecological goals and social equity for one or more of three reasons. First, ecological goods are not uniformly distributed. Second, the pursuit of these goals more often than not imposes a greater degree of behavioral changes onto specific groups. Third, the ability to guarantee all individuals access to the necessary resources to support even a minimally sufficient standard of living, much less raise those standards to the quality of life enjoyed by those in industrial countries, simply not be physically possible at current global population levels, particularly in regions facing high population growth within ecosystems that are already strained beyond capacity.

3.5.2 Cognitive Differences and Ecological Beliefs

Social differences may emerge at the cognitive level, because of divergent perspectives, ideologies, or worldviews held by various

BOX 3.5

The Equity Implications of Forced Displacement

A particular issue in the discussion of equity relates to the forced movement of people and the displacement of existing livelihood strategies as a consequence of adopting a particular response. Any strategy that imposes constraints on the use of land and resources may impinge upon current lifestyles and livelihood strategies of a local population. This problem is particularly contentious in the conservation context, since the creation of areas that are reserved for the maintenance of habitats for biodiversity often impinges on existing land-uses. Such strategies may directly or indirectly encourage resettlement that, while relieving pressure on a particular conservation area, may only shift those pressures to other regions that may be equally stressed, or lead to decreased living standards when resettlement is not an option. In addition, population pressures (the need for living space, conversion of land to agriculture, and fuelwood collection) also present an imposing obstacle to the implementation of response strategies. This issue applies not only to developing countries, but has also become relevant in the industrial world, where absolute population pressures may not be an issue but the rapid growth of concentrated population centers in ecologically sensitive zones can place limits on the ability to exert control over ecosystem services.

stakeholders. Responses may be contested, not because of the perceived loss of material well-being, but because they imply a rejection of deeply held beliefs. In order to assess the social impacts of a particular response, policy-makers must recognize the existence of a plurality of worldviews and acknowledge that the adoption of a specific policy response may (inadvertently or deliberately) privilege specific worldviews. Thus, just as the material interests of stakeholders may be differentially affected by policy, it is also possible that the implementation of policy responses that privilege specific worldviews may elicit enormous resentment among cultural groups that are affected by the policy, which can hinder policy effectiveness, encourage social conflict, or reduce the potential for political viability of future responses. One of the most significant impacts of such oversight may be a failure to incorporate systems of knowledge and worldviews that may well be the most appropriate for the ecological goals sought.

Typically the worldviews that tend to be accorded privilege in ecological policy encompass an emphasis on certain environmental issues identified by predominantly North American and European environmental interests, and a reliance on Cartesian scientific methods and technological inputs. The issues identified by these interests may not be the priorities of peoples in other countries. (See Box 3.6.) The worldviews that become marginalized in the policy-making process include those of alternative environmental priorities, local knowledge, and ecological perspectives of peoples living in regions that are targeted, or at least implicated, by response strategies. This situation may not only lead to the undermining of important information and monitoring abilities, but can also lead to resentment among the affected groups. One such example can be seen in the case of whale hunting among the Makah people in Neah Bay, Washington, which is seen as a "cultural necessity" (Erikson 1999). This practice is bound up with traditions, rituals, and taboos integral to the tribe's cultural survival. In another example, some scholars have suggested that granting trade-related intellectual property rights to new plant and animal products is not only in conflict with the goals of the Convention of Biological Diversity, but may also undermine the lifestyles of indigenous peoples (Anuradha 2001).

BOX 3.6

The Post-materialism Hypothesis

Recent scholarship has debated the differences between ecological beliefs among the populations of advanced industrialized countries and those that are relatively poor. Many researchers use the notion of "post-materialism" to study the attitudes and role of civil society in industrialized countries (for example, Abramson 1997; Abramson and Inglehart 1995; Kidd and Lee 1997a, 1997b; Pierce 1997).

Abramson and Inglehart summarize the post-materialism thesis: it "assumes that the economic security created by advanced industrial societies gradually changes the goal orientations of mass publics" (p. 9). This is similar to the conventional wisdom among economists that environmental protection is a "luxury good" demanded in greater quantity by the wealthy (for example, Kolstad 2000; also Dasgupta 2003, who argues that the poor find themselves more dependant upon the services provided by natural environments). In general, scholars find that people with post-materialist values are more apt to prioritize environmental protection. Rather than focusing on social movement organizations and civic associations, the research on post-materialism looks at lifestyle issues and consumer behavior and how they are related to environmental protection.

However, the post-materialism thesis has come under criticism by scholars looking at global environmentalism (for example, Brechin 1999; Brechin and Kempton 1994; Dunlap and Mertig 1992; Guha and Martinez-Alier 1997; Adeola 1998). These scholars suggest that there are material and non-material attitudes that inform environmental beliefs in both rich and poor countries, and argue that it is inappropriate to assume that ecological issues will necessarily be a higher priority in advanced industrial nations.

3.6 Methods and Tools for Assessing Responses

For effective decision-making, it is necessary to examine the implications of chosen responses across the five domains: political, institutional, economic, social, and ecological. Failure to assess options across all these dimensions may cause several weaknesses, and strengths, to be overlooked. The material presented here has highlighted issues that emerge in assessing responses *within* any one of these dimensions, but an overall assessment must find ways of comparing outcomes *across* these different domains as well. The task of an integrated assessment is particularly difficult, and demands extensive resources, since it needs to recognize the multidimensional nature of impacts, but also requires methods that are sensitive to a plurality of perspectives from diverse intellectual disciplines. Further, decision-makers may differ on the relative desirability of response strategies, particularly when there are legitimate differences of opinion about both the objectives of intervention and the means to achieve these ends.

Along any one dimension, using any particular criterion for assessment, the evaluation process can distinguish between constraints that render a policy option infeasible, that is, the *binding* constraints, and those considerations that, although important, may be treated as costs associated with the implementation of an option that stakeholders might be willing to bear, that is, the *acceptable trade-offs*. The distinction is important, since classifying an impact as a binding constraint effectively rules out a particular response, while identifying acceptable trade-offs serves to alert a decision-maker to unintended consequences and potentially harmful side effects of an otherwise feasible strategy. (As noted earlier, the obvious caveat is that what is seen by a particular decision-maker as a binding constraint in a particular context may not be seen in the same light by other decision-makers.)

Recognizing that decisions are typically made in a pluralistic environment, this section presents some general principles that can guide decision-makers in assessing the effectiveness of responses. The principles that are outlined here simply follow a systematic thought process that makes explicit some of the trade-offs and choices that are inevitably involved in such decision-making. What this discussion emphasizes is that no single discipline, and indeed no particular perspective, can claim the greatest legitimacy in a debate. What is appropriate and desirable in one context may be completely unacceptable in another. To this extent, the effectiveness of responses is relevant to a specific scale of analysis, and to a particular spatial and temporal context.

Although we offer a comprehensive guide for response assessment, we recognize that in some instances decision-makers will not have sufficient time or resources to collect all the information that is necessary, particularly in instances where indicators are not readily available, and primary data must be gathered. Fieldwork in the social sciences can be costly as well as time-consuming, as it may involve a combination of historical research, participant observation, and personal interviews and/or surveys. In cases where resources fall short of what may be required, decision-makers have a number of options. They may explore cost-sharing options with other potential supporters, or garner local community participation in data gathering, for example. If decision-makers cannot raise enough resources to conduct a comprehensive assessment, we advise primary focus on potentially binding constraints, and the formulation of response strategies that are sufficiently flexible that they can be responsive to dynamic and uncertain conditions.

The framework involves three stages: identifying binding constraints, classifying acceptable trade-offs, and scoring potential responses against one another to choose the best option. The questions involved in the first stage of the assessment process—identifying binding constraints—are shown in Table 3.1. It provides examples of the types of issues that would allow decision-makers to identify whether processes in any one of the political, institutional, economic, or social domains could potentially be regarded as binding constraints in the context of specific responses. Clearly, if this is the case, the response itself is unlikely to be accepted, and the decision-making process must attempt to identify suitable and feasible alternatives. This procedure can be seen as the first filter that eliminates potential strategies that either fail to achieve important objectives, or are technically, morally, or politically unacceptable.

In the second stage of the assessment method, trade-offs can often be identified quickly. For example, autocratic responses may economize on public consultation that might dilute the impact of proposed responses targeted at specific, one-dimensional change.

Table 3.1. A Framework for Assessing Responses

Domain	Issue	Evaluation
Political	Can all potential stakeholders be identified?	Not likely to be a binding constraint, unless neglected stakeholders mobilize political opposition.
	Is the political context supportive?	If not, could be a binding constraint.
	Can the political context be changed?	If not, is a binding constraint.
Institutional	Is there adequate capacity for governance at an appropriate scale?	If not, could be a binding constraint.
	If not, can it be created?	If not, is a binding constraint.
Economic	Is the outcome cost-effective?	Could be a binding constraint if funds are limited.
	Are there secure and well-defined property rights?	Could be a binding constraint, if there are numerous competing demands on the resource.
Social	Is the outcome equitable, in a distributional/material sense?	Could be a binding constraint, if this is a high priority, or those disadvantaged by the response can effectively oppose it.
	Does the outcome violate the cultural norms of particular groups?	Not likely to be a binding constraint, unless consensus is an explicit objective.

These same trade-offs, however, are also likely to undercut the political feasibility and, in consequence, the equity of such approaches. Conversely, responses developed with the participation of numerous special interest groups may assemble the coalition of interests required for their implementation, but have limited impact due to the need to serve the disparate interests of the coalition.

This is not to say, of course, that synergies never arise. Some responses may constitute "win-win" opportunities in which, perhaps, native ecosystems might be maintained while enhancing local incomes, redressing inequities, or achieving other ends. Many examples have been proposed of such "win-win" opportunities (for example, Heal 2000; Daily and Ellison 2002). Changes in technology, consumer tastes, political liberty, and other factors may be expected to generate more such opportunities over time. While policy-makers ought to remain alert for such opportunities and move aggressively to act upon them, they must also retain a realistic outlook. One must always ask why an activity that affords multiple benefits has not yet been initiated. Closer inspection may reveal that there are, in fact, other stakeholders whose interests are imperiled by what may first appear to be a "win-win" opportunity. In other words, if it sounds too good to be true, it probably is (O'Brien and Leichenko 2003).

It seems unlikely, then, that instances could often be identified in which one response dominates all others in all categories (and, we might add, in the estimation of all stakeholders). Thus, even once unacceptable responses have been eliminated, the decision-maker will generally need to make further choices between a set of potentially feasible responses, each of which may have different specific implications for any one of these multiple domains. In other words, the third stage in the assessment process is to score potential responses against one another. Table 3.2 presents the Response Assessment Matrix, which provides a method for decision-makers to assess responses across the five domains identified in this chapter.

In order to make this matrix operational, sub-criteria must be developed within each of the five domains. These would then be associated with specific indicators that would help to assess the impact of the response. No general listing of such indicators is possible, however, since these are likely to reflect the specific resource and the particular local context, as well as the preferences of the decision-makers engaged in the evaluation process. In some contexts, such as the development of criteria and indicators for sustainable forest management, there is considerable progress toward developing a set of widely acceptable and measurable criteria and indicators (see Prabhu et al. 1999). In other areas, however, progress is much slower. Moreover, there are a number of issues for which readily accessible indicators either do not exist at present or their validity is disputed. Table 3.3 summarizes the main issues that relate to the assessment process and comments on the availability and acceptability of relevant indicators for each of these issues.

Table 3.2. Proposed Response Assessment Matrix

	Political	Institutional	Economic	Social	Ecological
Response 1					
Response 2					
Response 3					
. . . and so on					

Table 3.3. Indicators for the Impact of Responses

Domain	Issue	Availability of Indicators
Political	Identification of relevant stakeholders	Stakeholder analysis
	Balance of power between stakeholders	No indicators—subjective judgment
	Degree of policy support by stakeholders	No indicators—subjective judgment
	Relations between the state and key stakeholders—relative autonomy	No indicators—subjective judgment
Institutional	Legitimacy of international institutions	No indicators—subjective judgment
	Linkages between institutions	No indicators—subjective judgment
	Compliance with established rules and norms	Progress reports, voluntary submissions
	Implementation	Progress reports, voluntary submissions
	Institutional adaptability	No indicators—subjective judgment based on longevity and resilience of institutions
	Access to financial resources	Budget analysis
	Existence of pluralist democracy	Electoral participation rates, number of political parties
	Separation of executive, judicial, and legislative functions	Constitutional provisions
	Bureaucratic competence	No indicators—subjective judgment
	Capacity for local management	Presence of organized user groups
Economic	Cost-effectiveness	Benefit cost analysis, valuation studies
	Security of property rights	Legal framework
Social	Equity (material)	Distributional analysis of costs and benefits
	Conflicting worldviews and ideologies	Anthropological and cultural studies

Assessment of trade-offs can be qualitative, quantitative, or both, as shown in the examples in Tables 3.4 and 3.5. Resolving the trade-offs presents the greatest challenge to determining appropriate responses. In some instances, it may be possible to view indicators as binary: so long as some standard is satisfied, the decision between approaches can be made on other grounds. For example, a decision-maker might determine that, so long as local material inequities are not further exacerbated, the most cost-effective approach might be chosen. More generally, however, aggregating across these different dimensions is likely to be difficult. But it is not impossible, as is demonstrated by Brown et al. (2002), who use a matrix scoring method to reflect economic, social, and ecological considerations in the context of integrated coastal conservation and development. They involved a wide range of stakeholders in developing specific indicators across each of the domains; these were then converted into standardized scores that were aggregated using different weights in order to rank the alternatives. It is not clear that such methods are appropriate for all qualitative indicators, however, and their results are always highly sensitive to the weighting procedure that is adopted. Weights will also necessarily change over time, as circumstances change.

Table 3.4. Qualitative Assessment of Trade-offs and Synergies

	Political	Institutional	Economic	Social	Ecological
Response 1	++	+	−	+	+
Response 2	−	++	++	−	+
Response 3	+	−	+	−	−
. . . and so on					

This is not a counsel of despair by any means, however. We suggest only that decision-makers must, in the final analysis, make some subjective assessment of the weights to be assigned to each factor. Having done so, the construction of such matrices will help decision-makers focus on the important issues that emerge in choosing among alternatives. This process will also aid decision-makers in constructing alternative future scenarios, thereby making better-informed choices among the alternatives.

Although the assessment process outlined here suggests a rational, linear approach to evaluating options, it is important to emphasize that preferences, perceptions, information, and exogenous factors continually change, and an effective system must be responsive and flexible to such changes. The process of decision-making is an on-going one, not a one-time assessment of trade-offs and synergies between different objectives and the interests of different stakeholders. Thus the assessment of responses needs to accommodate feedback, as well as the possibility of learning and adaptation. There is a growing emphasis on the use of methods such as "adaptive management" and "adaptive co-management" to incorporate such dynamism into decision-making environments. (See Box 3.7.)

Such assessment processes can be undertaken by a number of diverse stakeholders at a variety of scales. In such circumstances, the use of decision-making methods that adopt a pluralistic perspective is particularly pertinent, since these techniques do not privilege any particular viewpoint. Box 3.8 presents an example of such a pluralistic decision making process, which has achieved significant progress towards sustainable adaptive forest management in Canada.

A number of pluralistic decision-making tools and techniques have been documented in the literature. These tools can be employed at a variety of scales, including global, sub-global, and local. Tables 3.6, 3.7, and 3.8 list some of the tools that are available, as well as a preliminary assessment of their most appropriate scale(s) of application. Although not necessarily complete, as new techniques and tools are constantly being developed, the *typology*

Table 3.5. Quantitative Assessment of Trade-offs and Synergies

	Political	Institutional	Economic	Social	Ecological
Response 1	4	2	−1	1	1
Response 2	−2	3	3	−2	2
Response 3	1	−1	2	−1	−2
. . . and so on					

BOX 3.7

Adaptive Management and Adaptive Co-management

Adaptive management (Holling 1978) draws upon a variety of techniques for the management of ecological resources that emphasize the wider system as an appropriate unit for analysis. These techniques are particularly sensitive to dynamism and feedback; they are holistic and process-oriented. Adaptive management recognizes the role of uncertainty in the decision-making environment, but uses adaptation (in the Darwinian sense) to allow responses to learn from these environmental variables, and thereby to produce more stable policy outcomes. Such systems are modeled across a variety of temporal and spatial scales, and usually involve the coupling of highly complex social and ecological systems (Berkes and Folke 1998; Berkes et al. 2003).

Adaptive co-management builds on these principles by emphasizing the importance of local ecological and social contexts within which decision-making operates. Ecological processes at the local scale display a particularity and variation that have an impact on the ways in which responses work in practice; incorporating such place-specific knowledge is an important element of an adaptive management strategy. Sensitivity to the social context demands the recognition of local communities as critical elements of any response strategy. Adaptive co-management systems have been described as "flexible, community-based systems of resource management tailored to specific places and situations and supported by, and working with, various organizations at different levels" (Olsson et al. 2004, p. 75). Responses developed through such processes emphasize the central role of learning, adaptation, and collaboration in the evolution of strategies, in an attempt to build more resilient social-ecological systems.

Assessments of empirical examples of such systems are increasingly seeking to understand why they are especially appropriate to the choice, implementation, and evaluation of responses in a dynamic world characterized by uncertainty and change (Olsson et al. 2004, for instance, evaluate two cases from Lake Racken in western Sweden and James Bay in Canada in order to identify the essential features of the adaptive co-management process).

is reasonably comprehensive. Distinctions are made between *deliberative* tools (Table 3.6), which facilitate transparency and stakeholder dialogue over responses; *information gathering* tools (Table 3.7), which are primarily focused on collecting data and opinions; and *planning* tools (Table 3.8), which are typically employed for the evaluation of potential policy options. Chapter 4 presents a more comprehensive overview of the specific issues that arise due to the presence of uncertainty in the decision-making environment, and reviews specific decision analytical frameworks that have been adopted in order to deal with such uncertainty.

3.7 Conclusion

Chapter 3 has attempted to outline a general guide to be used for assessing ecological responses through the identification of the impacts of responses across four domains: the political, institutional, economic, and social. Tools for assessing a fifth critical domain, the ecological, can be found in other chapters. These impacts may pose binding constraints to response efforts, or they may be considered acceptable trade-offs. We have consequently developed a three-stage assessment process, involving the identification of binding constraints, comparison of trade-offs, and the selection of a response that avoids binding constraints and minimizes negative trade-offs.

BOX 3.8

Canada's Model Forest Program: Integrating Multiple Dimensions into Responses

In an effort to develop innovative ideas and methods to promote more sustainable, adaptive forest management in Canada in a manner that ensures consideration of sociocultural and economic factors in any forest management response option, the federal government established Canada's Model Forest Program in 1992.

The Program now encompasses eleven model forests across Canada, selected to represent the diversity of ecosystems and social systems that characterize the Canadian forest milieu. Each model forest is designed to function as a living laboratory in which new, integrated forest management techniques are researched, developed, applied, and monitored in a transparent forum characterized by partnerships with stakeholders, including representatives from environmental organizations, native groups, industry, educational and research institutions, all levels of government, community-based associations, recreationists, and landowners, all of whom assist in the development of a research agenda, and participate in the research process.

Some accomplishments include: development of a voluntary wetland conservation program for private lands; establishment of protocols for reporting on socioeconomic indicators based on Statistics Canada census data; an ecosystem-based Integrated Resource Management Plan now being used by the province of Saskatchewan; production of a code of forestry practice booklet to help landowners understand and apply the principles of sustainable forest management; establishment of the Grand Lake Reserve to protect three eco-regions and habitat for the endangered Newfoundland pine marten. For more information visit the Model Forest Program Web site: http://www.nrcan.gc.ca/cfs-scf/national/what-quoi/modelforest_e.html.

Table 3.6. Deliberative Tools

	Scale of Application			
Tool	Global	Sub-global	Local	References
Area/neighborhood forums			√	Lowndes et al. 1998
Citizens' juries		√	√	Lowndes et al. 1998
Citizens' interactive panels		√	√	Richardson 1998; IPPR 1999
Community issues groups			√	Clarke 1998
Consensus conferences	√	√	√	IPPR 1999
Electronic democracy	√	√	√	IPPR 1999
Focus groups			√	Lowndes et al. 1998
Issue forums			√	Lowndes et al. 1998
Service user forums			√	Lowndes et al. 1998

Response assessments can be complex and costly endeavors, but their utility can be measured in the increased probability of response effectiveness. While in the past, assessments have tended to be undertaken from a disciplinary perspective, we emphasize the importance of maintaining an interdisciplinary approach to assessment to ensure that important impacts or assessment meth-

Table 3.7. Information Gathering Tools

Tool	Scale of Application			References
	Global	Sub-global	Local	
Citizens' research panels		√	√	IPPR 1999
Deliberative opinion poll		√	√	IPPR 1999
Environmental impact assessment		√	√	Taylor 1984
Participatory rural appraisal			√	Chambers 1983, 1997
Rapid rural appraisal			√	Chambers 1997

ods are not overlooked. Any evaluation of the human dimensions of ecological responses, however, will inevitably be characterized by subjectivity and difficulties associated with attempting to reconcile heterogeneous assessment methods.

No matter how comprehensive an assessment, decision-makers must be prepared for the likelihood that consensus is not always reached among all stakeholders involved in a particular response. Among the most important steps that should be taken to limit potential conflict are emphasizing an inclusive evaluation processes, so that assessment is not undertaken by elite decision-makers; maintaining transparency and accountability throughout the assessment process; and, ultimately, developing responses that are flexible enough to maintain effectiveness despite dynamic human conditions. In the end, however, consensus building may be difficult, but with work, it is likely to be possible.

Table 3.8. Planning Tools

Tool	Scale of Application			References
	Global	Sub-global	Local	
Consensus participation	√	√	√	Warner 1997
Cost-benefit analysis		√	√	Hanley and Spash 1993
Future search conferences		√	√	IPPR 1999
Innovative development			√	Del Valle 1999
Issue forums		√	√	Lowndes et al. 1998
Multicriteria analysis	√	√	√	Stirling and Maher 1999
Participatory learning and action			√	Guijt 1998; Holland 1998
Planning for real			√	IPPR 1999
Service user forums			√	Lowndes et al. 1998
Stakeholder decision analysis	√	√	√	Grimble et al. 1995; ESRC 1998
Trade-off analysis	√	√	√	Brown et al. 2002
Visioning exercises		√	√	Lowndes et al. 1998

References

Abramson, P.R., 1997: Postmaterialism and environmentalism: A comment on an analysis and a reappraisal, *Social Science Quarterly,* **78(1),** pp. 21–23.

Abramson, P.R. and R. Inglehart, 1995: *Value Change in Global Perspective,* University of Michigan Press, Ann Arbor, MI.

Acharya, G., 2000: Approaches to valuing the hidden ecological services of wetland ecosystems, *Ecological Economics,* **35,** pp. 63–74.

Adger, W.N., K. Brown, J. Fairbrass, A. Jordan, J. Paavola, et al., 2003: Governance for sustainability: towards a "thick" analysis of environmental decision-making, *Environment and Planning A.* In press.

Adeola, F., 1998: Cross-national environmentalism differentials: Evidence from core and noncore nations, *Society and Natural Resources,* **11,** pp. 339–364.

Agrawal, A. and C.C. Gibson, 1999: Enchantment and disenchantment: The role of community in natural resource conservation, *World Development,* **27(4),** pp. 629–649.

Agrawal, A., 2001: Common property institutions and sustainable governance of resources, *World Development,* 29, pp. 1649–1672.

Albers, H.J. and M.J. Goldbach, 2000: Irreversible ecosystem change, species competition, and shifting cultivation, *Resource and Energy Economics,* **22,** pp. 261–280.

Andreoni, J., 1989: Giving with impure altruism: Applications to charity and Ricardian equivalence, *Journal of Political Economy,* **97** (December), pp. 1447–1457.

Anuradha, R.V., 2001: IPRs: Implications for biodiversity and local and indigenous communities, *Review of European Community and International Environmental Law,* **10(1),** pp. 27–36.

Arrow, K. and A. Fisher, 1974: Environmental preservation, uncertainty, and irreversibility, *Quarterly Journal of Economics,* **88(1),** pp. 312–319.

Arrow, K., E.E. Leamer, P.R. Portney, R. Radner, R. Solow, et al., 1993: Report of the NOAA (National Oceanic and Atmospheric Administration) panel on contingent valuation, *Federal Register,* 15 January, **58(10),** pp. 4601–4614.

Arrow, K. J. and F.H. Hahn, 1971: *General Competitive Analysis,* Holden-Day, San Francisco, CA.

Austen-Smith, D. and J.R. Wright, 1994: Counteractive lobbying, *American Journal of Political Science,* **38,** pp. 25–44.

Bagnoli, M. and B.L. Lipman, 1992: Private provision of public goods can be efficient, *Public Choice,* **74** (July), pp. 59–77.

Baland, J.M. and J.P. Platteau, 1996: *Halting Degradation of Natural Resources: Is There a Role for Rural Communities?* Clarendon Press, Oxford, UK.

Balmford, A., A. Bruner, P. Cooper, R. Costanza, S. Farber, et al., 2002: Economic reasons for conserving wild nature, *Science,* **297(5583),** pp. 950–953.

Barbier, E.B., M.C. Acreman, and D. Knowler, 1997: *Economic Valuation of Wetlands: A Guide for Policy Makers and Planners,* RAMSAR Convention Bureau, Gland, Switzerland.

Barbier, E.B. and I. Strand, 1998: Valuing mangrove–fishery linkages: A case study of Campeche, Mexico, *Environmental and Resource Economics,* **12(2)** (September), pp. 151–166.

Barzel, Y, 1997: *Economic Analysis of Property Rights,* 2nd ed., Cambridge University Press, Cambridge, UK.

Bell, F.W., 1998: The state of nature as an input into the production function, *Journal of Economic Research,* **3,** pp. 1–20.

Beltratti, A., G. Chichilnisky, and G. Heal, 1995: The green golden rule, *Economics Letters,* **49(2)**: pp. 175–179.

Berkes, F. and C. Folke (eds.), 1998: *Linking Social and Ecological Systems: Management Practices and Social Mechanisms for Building Resilience,* Cambridge University Press, Cambridge, UK.

Berkes, F., J. Colding, and C. Folke (eds.), 2003: *Navigating Social-Ecological Systems: Building Resilience for Complexity and Change,* Cambridge University Press, Cambridge, UK.

Blanchard, O.J. and S. Fischer, 1989: *Lectures on Macroeconomics,* MIT, Cambridge, MA.

Boadway, R., 1974: The welfare foundations of benefit–cost analysis, *Economic Journal,* **84** (December), pp. 926–939.

Bockstael, N., R.J. Kopp, A.M. Freeman III, P.R. Portney, and V.K. Smith, 2000: On valuing nature, *Environmental Science and Technology,* **24(8),** pp. 1384–1389.

Bond, I., 1993: *The Economics of Wildlife and Landuse in Zimbabwe: An Examination of Current Knowledge and Issues,* World Wildlife Fund, Harare, Zimbabwe.

Brandon, K., 1998: Comparing cases: A review of findings. In: *Parks in Peril: People, Politics, and Protected Areas,* Chapter 13, K. Brandon, K.H. Redford, and S.E. Sanderson (eds.), The Nature Conservancy, Washington, DC, pp. 375–414.

Brechin, S.R. and W. Kempton, 1994: Global environmentalism: A challenge to the postmaterialism thesis? *Social Science Quarterly,* **75(2),** pp. 245–269.

Brechin, S.R. 1999: Objective problems, subjective values and global environmentalism: Evaluating the postmaterialist argument and challenging a new explanation, *Social Science Quarterly,* **80(4),** pp. 793–809.

Brechin, S.R., P.R. Wilshusen, C.L. Fortwangler, and P.C. West, 2002: Beyond the square wheel: Toward a more comprehensive understanding of biodiversity conservation as social and political process, *Society and Natural Resources,* **15(1),** pp. 41–64.

Brown, K., E.L. Tompkins, and W.N. Adger, 2002: *Making Waves: Integrating Coastal Conservation and Development,* Earthscan, London, UK.

Brown-Weiss, E. and H. Jacobson (eds.), 1998: *Engaging Countries: Strengthening Compliance with International Environmental Accords,* MIT Press, Cambridge, MA.

Buttel, F.H., 2000: World society, the nation-state, and environmental protection, *American Sociological Review,* **65(1),** pp. 117–121.

Cardoso, F., 2003: Civil society and global governance: contextual paper prepared by the panel's chairman. Available at www.un.org/reform/pdfs/cardosopaper13june.htm. Accessed 26 June 2003.

Chambers, R., 1983: *Rural Development: Putting the Last First,* Longman, London, UK.

Chambers, R., 1997: *Whose Reality Counts? Putting the Last First,* Intermediate Technology Publications, London, UK.

Chambers, W. Bradnee, 2003a: Towards an improved understanding of legal effectiveness of international environmental treaties, UNU/IAS working paper series, UN University/Institute of Advanced Studies International Environmental Governance, Yokahama, Japan.

Chambers, W.B., October, 2003b: How can interlinkages improve the legal effectiveness of international environmental agreements? UNU/IAS working paper series, UN University/Institute of Advanced Studies International Environmental Governance, Tokyo, Japan.

Chichilnisky, G., 1996: An axiomatic approach to sustainable development, *Social Choice and Welfare,* **13(2),** pp. 219–248.

Chubb, J.E., 1983: *Interest Groups and the Bureaucracy: The Politics of Energy,* Stanford University Press, Stanford, CA.

CI (Conservation International), 2000: Conservation international creates new market mechanism to conserve global forests, press release, 25 September, Conservation International, Washington, DC.

Clark, R., 1998: Community issues groups, UKCEED Bulletin, 55:19.

Commission on Global Governance, 1995: *Our Global Neighbourhood,* The Commission on Global Governance, Oxford University Press, Oxford, UK.

Costanza, R., R. d'Arge, R. de Groot, S. Farber, M. Grasso, et al., 1997: The value of the world's ecosystem services and natural capital, *Nature,* **387(6230),** pp. 253–260.

Daily, G. and K. Ellison, 2002: *The New Economy of Nature: The Quest to Make Conservation Profitable,* Island Press, Washington, DC.

Dasgupta, P., 2002: Modern economics and its critics. In: *Fact and Fiction in Economics: Models, Realism, and Social Construction,* U. Maki (ed.), Cambridge University Press, Cambridge, UK.

Dasgupta, P., 2003: *Human Well-Being and the Natural Environment,* Revised ed. with new appendix, Oxford University Press, Oxford, UK.

Davidson, D.J., 2001: Federal policy in local context: The influence of local state-societal relations on endangered species act implementation, *Policy Studies Review,* **18(1),** pp. 212–240.

Debreu, G., 1954: *The Theory of Value: An Axiomatic Analysis of Economic Equilibrium,* Cowles Foundation Monograph 17, Yale University Press, New Haven, CT.

De Sombre, E.R., 2000: *Domestic Sources of International Environmental Policy: Industry, Environmentalists and US Power,* MIT Press, Cambridge, MA.

Del Valle, A., 1999: Managing complexity through methodic participation: The case of air quality in Santiago, Chile, *Systemic Practice and Action Research,* **12(4),** pp. 367–380.

Diamond, P. and J. Hausman, 1994: Contingent valuation: Is some number better than no number? Symposium on Contingent Valuation, *Journal of Economic Persepctives,* **8(4)** (Fall), pp. 45–64.

Dixit, A. and R. Pindyck, 1994: *Investment Under Uncertainty,* Princeton University Press, Princeton, NJ.

Domhoff, G. William, 1996: *State Autonomy or Class Dominance? Case Studies on Policy Making in America,* Aldine de Gruyter, New York, NY.

Dunlap, R.E. and A.G. Mertig, 1992: *American Environmentalism: The U.S. Environmental Movement, 1970–1990,* Taylor and Francis, Philadelphia, PA.

Erikson, P.P., 1999: A whaling we will go: Encounters of knowledge and memory at the Makah Cultural and Research Center, *Cultural Anthropology,* **14(4),** pp. 556–583.

ESRC (Economic and Social Research Council), 1998: Strengthening decision-making for sustainable development, Report of a workshop held at Eynsham Hall, Oxford, 15–16 June, ESRC, Swindon, UK.

Evans, P.B., H.K. Jacobson, and R.D. Putnam, 1993: *Double-Edged Diplomacy: International Bargaining and Domestic Politics,* University of California Press, Berkeley, CA.

Evans, P., 1995: *Embedded Autonomy,* Princeton University Press, Princeton, NJ.

Feeny, D., F. Berkes, B.J. McCay, and J.M. Acheson, 1990: The tragedy of the commons: Twenty-two years later, *Human Ecology,* **18(1),** pp. 1–19.

Ferraro, P.J., 2001: Global habitat protection: Limitations of development interventions and a role for conservation performance payments, *Conservation Biology,* **15(4),** pp. 990–1000.

Ferraro, P.J. and R.D. Simpson, 2002: The cost-effectiveness of conservation payments, *Land Economics,* **78(3),** pp. 339–353.

Ferraro, P.J. and A. Kiss, 2002: Direct payments to conserve biodiversity, *Science,* **298(5599),** 29 November, pp. 1718–19.

Fischoff, B., 2004: Cognitive processes in stated preference methods. In: *Handbook of Environmental Economics, Volume II: Valuing Environmental Changes,* K-G. Mäler and J. Vincent (eds.), North Holland, Amsterdam, The Netherlands.

Fisher, D.R., 2004: *National Governance and the Global Climate Change Regime,* Rowman & Littlefield Publishers, Inc., Lanham, MD.

Fisher, D.R. and J.F. Green, 2004: Understanding disenfranchisement: Civil society and developing countries' influence and participation in global governance for sustainable development, *Global Environmental Politics,* **4(3).**

Fisher, D.R. and W.R. Freudenburg, 2005: Post industrialization and environmental quality: An empirical analysis of the environmental state, *Social Forces,* **83(1).**

Franck, T.M., 1990: *The Power of Legitimacy among Nations,* Oxford University Press, Oxford, UK.

Frank, D.J., A. Hironaka, and E. Schofer, 2000: The nation-state and the natural environment over the twentieth century, *American Sociological Review,* **65** (February), pp. 96–116.

Freeman III, A.M., 2002: *The Measurement of Environmental and Resource Values: Theory and Methods,* 2nd ed., Resources for the Future, Washington, DC.

Frickel, S., and D.J. Davidson, 2004: Building environmental states: Legitimacy and rationalization in sustainability governance, *International Sociology,* **19(1),** pp. 89–110.

Glaeser, E.L., 2003: Psychology and the market, Discussion paper 2023, December, Harvard Institute of Economic Research, Cambridge, MA.

Gollier, C., 2002: Discounting an uncertain future, *Journal of Public Economics,* **85(2),** pp. 149–166.

Gorman, W.M., 1955: The intransitivity of certain criteria used in welfare economics, *Oxford Economic Papers,* new series, **7,** pp. 25–35.

Gowdy, J.M., 2004: The revolution in welfare economics and its implications for environmental valuation and policy, Rensselaer Polytechnic Institute, Working paper. Available online at http://www.rpi.edu/dept/economics/www/faculty/gowdy/fff/Gowdy2004.pdf.

Grimble, R., M. K. Chan, J. Aglionby, and J. Quan, 1995: *Trees and Trade-offs: A Stakeholder Approach to Natural Resource Management,* Gatekeeper series no. 52, International Institute for Environment and Development, London, UK.

Guha, R. and J. Martinez-Alier, 1997: *Varieties of Environmentalism: Essays North and South,* Earthscan, London, UK.

Guijt, I., 1998: *Participatory Monitoring and Impact Assessment of Sustainable Agricultural Initiatives: An Introduction to Key Elements,* SARL discussion paper number 1, International Institute for Environment and Development, London, UK.

Habermas, J., 1975: *Legitimation Crisis,* Beacon Press, Boston, MA.

Haneman, W.M., 1994: Valuing the environment through contingent valuation, Symposium on Contingent Valuation, *Journal of Economic Persepctives,* **8(4)** (Fall), pp. 19–43.

Hanley, N. and C. Spash, 1993: *Cost-Benefit Analysis and the Environment,* Edward Elgar, Cheltenham, UK.

Hardin, G., 1968: The tragedy of the commons, *Science,* **162,** pp. 1243–1248.

Hausman, J.A., G.K. Leonard, and D. McFadden, 1995: A utility-consistent combined discrete choice and count data model assessing recreational use losses due to natural resource damage, *Journal of Public Economics,* **56,** pp. 1–30.

Heal, G., 2000: *Nature and the Marketplace: Capturing the Value of Ecosystem Services,* Island Press, Washington, DC.

Heal, G., 2002: Biodiversity and globalization, paper presented at 2002 Kiel Week Conference at Kiel University Institute for World Economics, 26 June 2002, Kiel, Germany.

Henriquez, B., 2004: Information technology: The unsung hero of market-based environmental policy, *Resources,* **152,** pp. 9–12.

Hicks, J.R., 1939: The foundations of welfare economics, *Economic Journal,* **49,** pp. 69.

Holland, J. (ed.), 1998: *Whose Voice? Participatory Research and Policy Change,* Intermediate Technology Publications, London, UK.

Holling, C.S. (ed.), 1978: *Adaptive Environmental Assessment and Management,* John Wiley, New York, NY.

Inglehart, R., 1995: Public support for environmental protection: Objective problems and subjective values in 43 societies, *PS: Political Science & Politics,* **28(1),** pp. 57–72.

IPPR (Institute of Public Policy Research), 1999: *Models of Public Involvement,* Institute of Public Policy Research, London, UK.

Irwin, E., 2002: The effects of open space on residential property values, *Land Economics,* **78(4),** pp. 465–81.

Journal of Economic Perspectives, 1994: Symposium on Contingent Valuation, with papers by P. Portney, The contingent valuation debate: Why economists should care, W.M. Hanemann, Valuing the environment through contingent valuation, and P. Diamond and J. Hausman, Contingent valuation: Is some number better than no number? *Journal of Economic Perspectives,* **8(4)** (Fall), pp. 3–64.

Kaldor, N., 1939: Welfare propositions. In: Economics and inter-personal comparisons of utility, *Economic Journal,* **49,** p. 549.

Kahneman, D., and J. Knetsch, 1992: Valuing public goods: The purchase of moral satisfaction, *Journal of Environmental Economics and Management,* **22(1),** pp. 57–70.

Kazis, R. and R.L. Grossman, 1991: *Fear at Work: Job Blackmail, Labor, and the Environment,* New Society Publishers, Philadelphia, PA.

Kidd, Q. and A-R. Lee, 1997a: Postmaterialist values and the environment: A critique and reappraisal, *Social Science Quarterly,* **78(1),** pp. 1–15.

Kidd, Q. and A-R. Lee, 1997b: More on postmaterialist values and the environment, *Social Science Quarterly,* **78(1),** pp. 36–43.

Kokkelenberg, E. and W. Nordhaus, 1999: *Nature's Numbers: Expanding the National Economic Accounts to Include the Environment,* National Academy Press, Washington, DC.

Kolstad, C.D., 2000: *Environmental Economics,* Oxford University Press, Oxford, UK.

Koopmans, T.J., 1960: Stationary ordinal utility and time preference, *Econometrica,* **28(2),** pp. 287–309.

Krasner, S. (ed.), 1983: *International Regimes,* Ithaca, NY.

Kremen, C., J. Niles, M. Dalton, G. Daily, P. Erlich, et al., 2000: Economic incentives for rainforest conservation across scales, *Science,* **288,** pp. 1828–1832.

Langholz, J.A., J.P. Lassoie, D. Lee, and D. Chapman, 2000: Economic considerations of privately owned parks, *Ecological Economics,* **33(2),** pp. 173–183.

Laumann, E.O. and D. Knoke, 1987: *The Organizational State: Social Change in National Policy Domains,* University of Wisconsin Press, Madison, WI.

Leach, M., R. Mearns, and I. Scoones, 1999: Environmental entitlements: Dynamics and institutions in community-based natural resource management, *World Development,* **27(2),** pp. 225–247.

Leftwich, A., 1994: Governance, the state and the politics of development, *Development and Change,* **25(2),** pp. 363–386.

Little, I.M.D., 1950: *A Critique of Welfare Economics,* Cambridge University Press, Cambridge, UK.

Lowndes, V., L. Pratchett, and G. Stoker, 1998: *Guidance on Enhancing Public Participation in Local Government,* Department of the Environment, Transport and the Regions, London, UK. Available at http://www.odpm.gov.uk/stellent/groups/odpm_localgov/documents/pdf/odpm_locgov_pdf_023830.pdf.

Lukes, S., 1974: *Power: A Radical View,* MacMillan, London, UK.

Mäler, K-G. and J. Vincent, 2004: *Handbook of Environmental Economics, Volume II: Valuing Environmental Changes,* North Holland, Amsterdam, The Netherlands.

MA (Millennium Ecosystem Assessment), 2003: *Ecosystems and Human Well-being: A Framework for Assessment,* Island Press, Washington, DC, 245 pp.

Mishan, E.J., 1981: A reappraisal of the principles of resource allocation. In: *Economic Efficiency and Social Welfare: Selected Essays on Fundamental Aspects of the Economic Theory of Social Welfare,* E.J. Mishan (ed.), Allen and Unwin, London, UK, pp. 3–13(a).

Muir-Leresche, K. and R.H. Nelson, 2000: *Private Property Rights to Wildlife: The Southern Africa Experiment,* Center for Private Conservation, Competitive Enterprise Institute, Washington, DC.

Newell, R.G. and W.A. Pizer, 2003: Discounting the distant future: How much do uncertain rates increase valuations? *Journal of Environmental Economics and Management,* **46,** pp. 52–71.

Nordlinger, E.A., 1981: *On the Autonomy of the Democratic State,* Harvard University Press, Cambridge, MA.

Oates, J.F., 1999: Myth and reality in the rain forest: How conservation strategies are failing in West Africa, University of California Press, Berkeley, CA.

O'Brien, K.L. and R.M. Leichenko, 2003: Winners and losers in the context of global change, *Annals of the Association of American Geographers,* **93,** pp. 89–103.

Odum, H.T., 1981: *Energy Basis for Man and Nature,* McGraw-Hill, New York, NY.

Olson, M., 1965: *The Logic of Collective Action; Public Goods and the Theory of Groups,* Harvard University Press, Cambridge, MA.

Olsson, P., C. Folke, and F. Berkes, 2004: Adaptive comanagement for building resilience in social-ecological systems, *Environmental Management,* **34(1),** pp. 75–90.

Ostrom, E., 1990: *Governing the Commons: The Evolution of Institutions for Collective Action,* Cambridge University Press, Cambridge, UK.

Pagiola, S., J. Bishop, and N. Llandell-Mills (eds.), 2002: *Selling Forest Environmental Services: Market-based Mechanisms for Conservation and Development,* Earthscan, London, UK.

Parry, I., W.H., Lawrence, H. Goulder, and D. Burtraw, 1997: Revenue-raising vs. other approaches to environmental protection: The critical significance of pre-existing tax distortions, *RAND Journal of Economics,* **28(4),** pp. 708–731.

Pattanayak, S.K. and R.A. Kramer, 2001: Worth of watersheds: A producer surplus approach for valuing drought mitigation in eastern Indonesia, *Environment and Development Economics,* **6,** pp. 23–146.

Pearce, D.W., 1998: Auditing the Earth, *Environment,* **40(2),** pp. 23–28.

Pearce, D.W., 2004: Environmental policy as a tool for sustainability. In: *Scarcity and Growth in the New Millennium,* R.D. Simpson, M.A. Toman, and R.U. Ayres (eds.), Resources for the Future, Washington, DC.

Pearce, D.W. and Turner, R.K., 1990: *Economics of Natural Resources and the Environment,* Johns Hopkins University Press, Baltimore, MD.

Perrings, C. and B. Walker, 1997: Biodiversity, resilience and the control of ecological-economic systems: The case of fire-driven rangelands, *Ecological Economics,* **22(1)** (July), pp. 73–83.

Pezzey, J.C.V., 1997: Sustainability constraints versus "optimality" versus intertemporal concern, and axioms versus data, *Land Economics,* **73(4),** pp. 448–466.

Pezzey, J.C.V and M.A. Toman, 2002: Progress and problems in the economics of sustainability. In: *Yearbook of Environmental and Resource Economics* 2002/3, H. Folmer and T. Tietenberg (eds.), Edward Elgar, Cheltenham, UK, pp. 165–232.

Pierce, J.C., 1997: The hidden layer of political culture: A comment on "post-materialist" values and the environment: a critique and reappraisal, *Social Science Quarterly,* **78(1),** pp. 30–35.

Pigou, A.C., 1932: *The Economics of Welfare,* 4th ed., MacMillan and Company, London, UK.

Portney, P., 1994: The contingent valuation debate: Why economists should care, Symposium on Contingent Valuation, *Journal of Economic Persepctives,* **8(4)** (Fall), pp. 3–17.

Prabhu, R., C.J.P. Colfer, and R.G. Dudley, 1999: *Guidelines for Developing, Testing and Selecting Criteria and Indicators for Sustainable Forest Management,* Centre for International Forestry Research, Jakarta, Indonesia.

Putnam, R.D., 1988: Diplomacy and domestic politics: The logic of two-level games, *International Organization,* **42,** pp. 427–460.

Raustiala, K., 2000: Compliance and effectiveness in international regulatory cooperation, *Case Western Reserve Journal of International Law,* **32,** pp. 387–440.

Ready, R.C. and R.C. Bishop, 1991: Endangered species and the safe minimum standard, *American Journal of Agricultural Economics,* **72(2),** pp. 309–312.

Rees, W., and M. Wackernagel, 1994: *Our Ecological Footprint: Reducing Human Impact on the Earth,* New Society, Philadelphia, PA.

Reid, J. and R. Rice, 1997: Assessing natural forest management as a tool for tropical forest conservation, *Ambio,* **26(6),** pp. 382–386.

Richardson, A., 1998: Health Panels. In: *Panels and Juries—New Government, New Agenda,* R. Sykes and A. Hedges (eds.), Social Research Association, London.

Roy, A., 1999: *The Greater Common Good,* India Book Distributors, Bombay, India.

Sagoff, M., 1994: Four dogmas of environmental economics, *Environmental Values,* **3(4)** (Winter), pp. 285–310.

Samuelson, P.A., 1951: Abstract of a theorem concerning substitutability in open Leontief models. In: *Activity Analysis of Production and Allocation,* T. Koopmans (ed.), Wiley, New York, NY, pp. 147–154.

Samuelson, P.A., 1954: The pure theory of public expenditure, *Review of Economics and Statistics,* **36,** pp. 387–389.

Scher, S.J., A. White, and D. Kaimowitz, 2002: *Making Markets Work for Forest Communities,* Policy brief, Forest Trends, Washington, DC.

Scitovsky, de, T., 1941: A note on welfare propositions in economics, *Review of Economics and Statistics,* **9,** pp. 77–88.

Seppanen, J. and E. Valiverronen, 2000: Nature in newspaper photographs: On the relations between text and image in environmental discourse, *Sosiologia,* **37(4),** pp. 330–348.

Shogren, J.F. and T.D. Crocker, 1999: Risk and its consequences, *Journal of Environmental Economics and Management,* **37(1),** pp. 44–51.

Simpson, R.D. and R.A. Sedjo, 1996: Paying for the conservation of endangered ecosystems: A comparison of direct and indirect approaches, *Environment and Development Economics,* **1,** pp. 241–257.

Simpson, R.D., R.A. Sedjo, and J.W. Reid, 1996: Valuing biodiversity for use in pharmaceutical research, *Journal of Political Economy,* **104(1)** (February), pp. 163–185.

Smith, V.K., 1993: Arbitrary values, good causes, and premature verdicts, *Journal of Environmental Economics and Management,* **22,** pp. 71–89.

Smith, V.K., 1997: Mispriced planet, *Regulation,* **20(3),** pp. 16–17.

Smith, V.K., 1998: Pricing what is priceless: A status report on non-market valuation of environmental resources. In: *International Yearbook of Environmental and Resource Economics 1997/8,* H. Folmer and T. Tietenberg (eds.), Edward Elgar, Cheltenham, UK: pp.156–208. Available at http://biodiversity economics.org/pdf/topics-602-00.pdf, p.42.

Southgate, D., 1998: *Tropical Forest Conservation: An Economic Analysis of the Alternatives for Latin America,* Oxford University Press, Oxford, UK.

Strand, J. and K. Seip, 1992: Willingness to pay for environmental goods in Norway: A contingent valuation study with real payment, *Environmental and Resource Economics,* **2,** pp. 91–106.

Strange, S., 1983: Cave! Hic dragones: A critique of regime analysis. In: *International Regimes,* S. Krasner (ed.), Brooks/Cole, Pacific Grove, CA, pp. 337–354.

Striling, A. and S. Maher, 1999: *Rethinking Risk: A Pilot Multi-Criteria Mapping of a Genetically Modified Crop in Agricultural Systems in the UK,* Science Policy Research Unit, Brighton, UK.

Stroup, R.L., 2003. *Economics: What Everyone Should Know About Economics and the Environment,* Cato Institute, Washington, DC.

Taylor, S., 1984: *Making Bureaucracies Think: The EIS Strategy of Environmental Reform,* Stanford University Press, Palo Alto, CA.

Terborgh, J., C. van Schaik, L. Davenport, and M. Rao (eds.), 2002: *Making Parks Work: Strategies for Preserving Tropical Nature,* Island Press, Washington, DC.

Terborgh, J., 1999: *Requiem for Nature,* Island Press, Washington, DC.

Thorsnes, P.G., 2002: The value of a suburban forest reserve: Estimates from sale of vacant residential building lots, *Land Economics,* **78(3),** pp. 426–441.

Tietenberg, T., 1978: Spatially differentiated air pollution emissions charges: An economic and legal analysis, *Land Economics,* **54,** pp. 265–277.

Toman, M.A., 1998: Why not to calculate the value of the world's ecosystem services and natural capital, *Ecological Economics,* **25(1)** (April), pp. 57–60.

Varian, H.R., 1992: *Microeconomic Analysis,* 3rd ed., Norton, New York, NY.

Victor, D. K. Raustiala, and E.B. Skolnikoff (eds.), 1998: *The Implementation and Effectiveness of International Environmental Commitments: Theory and Practice,* MIT Press, Cambridge, MA.

Vogel, J.H., 1994. *Genes for Sale: Privatization as a Conservation Policy,* Oxford University Press, Oxford, UK.

Vossler, C. and J. Kerkvliet, 2003: A criterion validity test of the contingent valuation method: Comparing actual and hypothetical voting behavior for a referendum, *Journal of Environmental Economics and Management,* **45(3),** pp. 631–649.

Wade, R., 1988: *Village Republics,* Cambridge University Press, Cambridge, UK.

Warner, M., 1997: Consensus participation: an example for protected areas planning, *Public Administration and Development,* **17,** pp. 413–432.

Weitzman, M.L., 1998: Why the distant future should be discounted at the lowest possible rate, *Journal of Environmental Economics and Management,* **36(3),** pp. 201–208.

Wells, M. and K. Brandon with L. Hannah, 1992: *People and Parks: Linking Protected Area Management with Local Communities,* World Bank, World Wildlife Fund, and US Agency for International Development, Washington, DC.

Western, D. and R.M. Wright, 1994: *Natural Connections: Perspectives in Community Based Conservation,* Island Press, Washington, DC.

Wilson, J.Q., 1980: *The Politics of Regulation,* Basic Books, New York, NY.

World Bank, 1992: *Governance and Development,* World Bank, Washington, DC.

World Bank, 1997: *Investing in Biodiversity: A Review of Indonesia's Integrated Conservation and Development Projects,* Indonesia and Pacific Islands Country Department (IPICD), Washington, DC.

Young, O., 1999: *Governance in World Affairs,* Cornell University Press, Ithaca, 224pp.

Chapter 4

Recognizing Uncertainties in Evaluating Responses

Coordinating Lead Author: Gary Yohe

Lead Authors: W. Neil Adger, Hadi Dowlatabadi, Kristie Ebi, Saleemul Huq, Dominic Moran, Dale Rothman, Kenneth Strzepek, Gina Ziervogel

Contributing Authors: Carmen Cheung, Daniel P. Faith, Robert Gilmore Pontius Jr., Mang Lung Cheuk

Review Editors: Neil Leary, Victor Ramos

BOXES

TABLES

Main Messages

Decisions about how to respond to external stresses are, of necessity, made under conditions of uncertainty. Incomplete information and imperfect knowledge about context and efficacy are facts of life, but well-established methods designed to help decision-makers cope with uncertainty exist. Applying them wisely can contribute directly to making decisions more effective.

Scenario analysis of the sort described in the *Scenarios* volume is one of many methods that can be employed to incorporate uncertainty into the evaluation of alternative responses to external stress. Sensitivity analysis, the construction of scenario trees, the augmentation of scenario trees with subjective probability distributions, and the estimation of response surfaces can all be applied in response evaluation. Each has its own strength, standing alone or used as part of a more integrated scenario analysis.

Cascades of uncertainty typically cloud our understanding of legal, market, institutional, and behavioral responses to change. Integrating across response strategies can mitigate and reduce elements of uncertainty, but it is unlikely that uncertainty can be eliminated in any important context. Decision-makers face pervasive uncertainty in choosing between responses. Each type of response has different sources of uncertainty. Regulatory responses have uncertain outcomes because of risk aversion in regulatory organizations, divergent stakeholder objectives, and diversity in human preferences for ecosystem services. Legal, institutional, and integrated responses exhibit uncertainty in the degree to which their implementation will be effective. All response strategies depend on stakeholders to establish their legitimacy, so governance structures introduce novel uncertainties. Combining and integrating response strategies often reduces implementation risks because integration can increase legitimacy and provide means for adaptive learning.

Uncertainty is manifest in surprise and unintended consequences. It is *well established* that our understanding of the complex systems within which response measures for ecosystem services are to be analyzed are clouded by uncertainty. As a result, responses can lead to unforeseen and often negative consequences. Ecosystems have intrinsic thresholds so that changes in their condition and feedback are often episodic and associated with changes in ecosystem function. Unintended consequences can arise even when many of the consequences of action are predictable because different decision-making bodies can cause negative spillovers into other areas even if they are successful in the pursuit of their own objectives.

Uncertainties expand when evaluations must be conducted beyond the bounds of historical experience. Projecting responses beyond the boundaries of historical experience brings the compounding effects of unknown contextual change to bear on their evaluation. Representations of uncertainty must then reflect the expanding implications of this uncertainty by conducting evaluations within hypothesized descriptions of the political, economic, social, and natural factors that will define future environments.

Uncertainty limits the ability of economic valuation methods to support collective decision-making for nontraded services, but not completely. Economic and institutional response strategies depend on decision tools that involve comparison of individual well-being across time and space. There are well-established methods for handling and quantifying uncertainties in these methods in contexts where markets exist. It is also well accepted that economic decision tools are limited in assessing responses for ecosystem services that are not traded in markets and where the values associated with them are not utilitarian in nature. The results of economic valuation techniques are not easily aggregated across scales; it follows that economic metrics cannot be applied in every circumstance where the relative merits of alternative responses are being contemplated. Other decision support techniques such as risk, multicriteria, and vulnerability analyses have also been designed explicitly to handle risk and uncertainty; they can sometimes more comfortably accommodate a diversity of decision-making contexts.

Uncertainties in the validation of vulnerability assessments arise because of scale and context specificity. Methods for determining vulnerability require quantitative and qualitative data that describe the drivers of change as well as the state of well-being for individuals and for social and ecological systems. They are important when the choice of a response tries to maximize the well-being of the most marginalized by identifying the most vulnerable people and places. There is debate as to whether vulnerability assessment methods allow for aggregation across scales because vulnerability depends on the position of the observer. It follows that uncertainties in vulnerability assessment can limit the degree to which findings from any particular context can be transported to another.

Scenarios provide one means of coping with many aspects of uncertainty, but our limited understanding of the ecological and human response process shrouds any individual scenario in its own characteristic uncertainty. Scenarios can be used to highlight the implications of alternative assumptions about critical uncertainties related to the behavior of human and ecological systems. At the same time, though, individual scenarios represent conditional projections based upon these specific assumptions. To the extent that our understanding of ecological and human systems represented in the scenarios is limited, specific scenarios are characterized by their own uncertainty. Furthermore, there is uncertainty in translating the lessons derived from scenarios developed at one scale (for instance, global) to the assessment of responses at other scales (for example, sub-national).

It is possible to integrate the political, economic, and social factors that impede or enhance the likelihood of success of any response with representations of the uncertainties that cloud our understanding of how they might work. The political, economic, and social factors identified in Chapter 3 map well into the determinants of adaptive (response) capacity, and applying a weakest link evaluation is appropriate. More specifically, the capacity to respond is fundamentally dependent on the factors that support the largest obstacle to response. Meanwhile, applying tools that explicitly recognize various sources of uncertainty can provide insight into the likelihood that the objectives of any response might actually be achieved without creating unintended consequences.

Scientific understanding of response mechanisms is frequently clouded by uncertainty, and this uncertainty affects the confidence with which descriptions of how these mechanisms can be expected to operate in a changing environment. Evaluating the relative strengths of the underlying determinants of response capacity can, however, provide a method by which this confidence can, itself, be assessed. When uncertainty can be quantified, standard thresholds can be applied to assign various degrees of confidence to specific conclusions. When only qualitative descriptors of uncertainty are available, confidence can still be conveyed in terms of the degree to which conclusions are or are not well established in theory and/or well supported by data and other evidence.

4.1 Introduction

Decision-makers face pervasive uncertainty in implementing response strategies as they try to manage ecosystem services. Uncertainty clouds their understanding of everything from how their response options might actually work to the methods that they use to assess their relative efficacy. This chapter takes this simple observation as a point of departure and tries to provide an insight into how valuation and decision-analytic frameworks can accommodate uncertainty. It also offers some guidance to those who want to assess how uncertainty combines with issues of political feasibility and governance (discussed in Chapter 3) to affect the confidence with which they can trust their conclusions about how best to respond. Both objectives recognize the fundamental truth that decision-makers have to make decisions even when uncertainty is extremely large; and both recognize that maintaining the status quo (that is, enacting no new response to one or more new sources of stress) is as much of a decision as moving robustly in many directions at the same time.

4.2 Cascading Uncertainties in Response Options and Assessment Methods

The different response strategies outlined in Chapter 2 and identified in subsequent chapters all operate in a landscape of uncertainty. Legal, economic, and institutional responses have fundamentally different types of risks and uncertainty associated with them. Uncertainty often reflects subjective views on the likelihood of various outcomes across a range of "states of nature." (See Box 4.1.) The corresponding levels of risk represent the product of these likelihoods and the consequences of their associated outcomes expressed in terms of ecosystem and/or human well-being and include the possibility of unintended consequences and other feedbacks. There are also other uncertainties associated with decision-making processes that include the possible divergence of opinion about approaches to risk across different stakeholders; these are concerns raised in Chapter 3. Taken together, all of these elements of uncertainty must be accommodated by methods and models used for the assessment of any response option including, of course, maintaining the status quo.

To illustrate the degree to which uncertainty is ubiquitous in the consideration of response strategies, Yohe and Strzepek (2004) have created a taxonomy of sources of uncertainty from a practical perspective anchored by methods and models that have been employed to describe and simulate critical connections between experience and expectation, drivers and state variables, and outcomes and consequences. To begin with, analytical methods and models are abstractions of the real world, and different approaches can produce wildly different answers to the very same questions. This simple phenomenon can be important in examining the relative merits of one particular model or another, but it introduces *model uncertainty* for analysts who are looking across model results for coherent views of the future. In addition, the ability of any particular model to offer a credible depiction of any connection is limited by the analyst's statistical ability to summarize perhaps vast but sometime paltry quantities of data that may or may not be particularly well defined. This can be called *calibration uncertainty.*

The limitations of calibrating a model are well understood by most practitioners, but they can be exacerbated when any one estimated model is used to produce *uncertain* predictions or projections of critical state variables. The difference between *prediction uncertainty* and *projection uncertainty* is, however, critical. To see the distinction, consider the simple case where a researcher has access to a set of historical data on a driving variable X and state variable Y whose causal relationship can be summarized by a linear relationship—a line whose points minimize the sum of the squared error of using the line rather than the data to represent the correlation between changes in X and resulting changes in Y. Now consider the question: what value of Y would be expected if variable X were to move along a particular trajectory over time? The best guess would be a series of points that lie along the line, but confidence about those guesses would fall as the trajectory took X farther away from the mean value of its historical range. Indeed, it would be possible to identify the boundaries of, for example, 95% confidence intervals for any possible value of X that lies within the extremes of the historical record with which the straight line was calibrated; these intervals would be credible representations of *prediction uncertainty.* If future drivers of change moved X outside the range of historical experience, then the value for Y read from the line would still be the best guess, but confidence would fall even more (the confidence intervals would expand even more) because the independent variable X would have moved beyond the realm within which the processes that sustained the relationship can be assumed to be valid. It is in this range that *projection uncertainty* presents itself.

The existence of projection uncertainty is the first recognition that underlying social and economic structures in many societies and contexts may change over time. Because these changes could occur even if the driving variable X did not exceed the boundaries of past experience, this sort of evolution of preferences and contexts is yet another way that the passage of time can undermine the credibility of using historically based modeling structures and methods as representations of future conditions. This is what might be called *contextual uncertainty,* and it can be enormous. It contains what most would understand as structural change, but it also includes value and preference uncertainties about which little is known at this point. Finally, *scale uncertainties* emerge when results for similar questions are compared across different geographic or temporal scales. The obvious point is that results generated at one scale cannot necessarily be scaled up or scaled down because emergent behaviors and baseline characteristics can vary dramatically. Box 4.2 illustrates a more subtle point: analyses conducted at different scales on the same data can produce different answers to the same question.

Turning finally to responses, it must be emphasized that the uncertainties that cloud our understanding of the connections be-

BOX 4.1

Subjective and Estimated Perceptions of Uncertainty and Risk

Computing the probabilities required to undertake a risk calculation is not always a simple matter. It may not, for example, be possible to conduct repeated trials. Nor is it always possible to produce theoretically based estimates of relative likelihood. Indeed, most interesting cases involve individuals' creating subjective views of probabilities from experience and/or careful reviews of scientific literature. Nonetheless, the output of a risk calculation will only be as good as its underlying data—estimates of outcomes in various states of nature, their consequences, and their associated likelihoods. How good are people at judging the critical outcomes and probabilities?

Slovic et al. (1979) authored an early investigation of this question that is still widely respected in the risk assessment literature. They found that experts systematically overestimate the chance of death associated with low-risk activities such as skiing and vaccination. Similarly, they underestimate the chance of death associated with high risk activities such as using handguns and smoking. Lay people showed the same systematic tendencies toward underestimation and overestimation, and their errors on the extremes of relative safety and extreme danger were actually more pronounced.

> **BOX 4.2**
>
> **Using Indices to Track Changes in Land Use**
>
> The concept of an index is simple and intuitive, especially for a single scale and level of aggregation. The mathematics of indices can be deceptively tricky, though, when researchers analyze the behavior of the indices at various scales and sundry levels of aggregation. For example, suppose a researcher wanted to create an index to indicate land cover change over three decades in a region of Massachusetts (USA) where the conversion from natural to built land has been a source of enormous concern. The first step would be to compare maps of land cover at two points in time, but what would be the best way to perform the calculation if each pixel contains more than one land cover type? Consider a pixel that is ¾ natural and ¼ built land cover at time *1* and then transitions to ¼ natural and ¾ built at time *2*. There are at least three reasonable ways to compute the transition from natural to built for this pixel.
>
> A common technique used by landscape scientists is to reclassify the pixel into the category that dominates. For this Boolean technique, the pixel would be classified as entirely natural in time *1* and as entirely built in time *2*. A second technique would be to assume that the proportions of natural and built land cover are distributed randomly within the pixel, so the probability that a patch of natural land within the pixel at time *1* transitions to built in time *2* is computed as ¾ times ¼, which is 3/16 of the pixel. This case reflects the multiplication rule for joint probabilities. A third technique would assume that the natural and built patches within the pixel persist from time *1* to time *2*. For example, this third technique would assume that the built land at time *1* is part of the built land at time *2*, and that the natural land at time *2* was part of the natural land at time *1*. Therefore, the only change within the pixel is a transition of one half of the pixel from natural to built land cover. This technique is based on a minimum rule of agreement, which is consistent with fuzzy set theory and is becoming accepted in landscape science.
>
> However, Pontius and Cheuk (2004) have shown that these three accepted methods to compute the transition from forest to built land cover for the given pixel could easily produce different trends for land use changes. The Boolean rule can behave chaotically at coarser resolution, the multiplication rule can detect larger transitions at coarser resolutions, but the minimum rule can be fairly stable across the same data. This is an example of how statistical results can be extremely sensitive to the selection of the units of analysis.

tween drivers and impacts are compounded by uncertainties that fog our understanding of how responses might work across a wide range of futures. The point here is simple: it is impossible to predict how any particular response might work in the future even along an assumed trajectory of how future gross impacts might evolve. Conversely, assessing the effectiveness of a response option within a single portrait of how the future might evolve (that is., conducting analysis as if the time trajectory of a gross impact on an ecosystem were known) would produce a dramatically overstated confidence in the time trajectory of net impacts.

4.3 Synthesis of Uncertainty in Identified Response Strategies

Subsequent chapters highlight a range of legal, institutional, and regulatory, behavioral, and market responses for fresh water, food, culture, and other services. The uncertainties and risks associated with these approaches are well documented. So, too, is the knowledge that uncertainty clouds our understanding of both the nature and characterization of the ecosystem services as well as the links between those services, human well-being, and the institutional context of any policy response. This section provides an overview of these uncertainties.

4.3.1 Uncertainty in Legal and Control Responses

The nature of uncertainty in legal, regulatory, and control responses to the management of ecological services is, first of all, associated with the characteristics of ecological systems and the degree to which they are amenable to human control. The risk factors associated with ecological management are typically interconnected. Holling et al. (1995) argue that ecological systems frequently have shared characteristics, perhaps most importantly the notion that changes in ecosystem function associated with human interference are not usually gradual. They are, instead, triggered by external perturbations and are therefore typically episodic. These authors also note that some of the functions that control ecosystems promote stability while others create destabilizing influences. It follows that regulating simultaneously for stability and resilience (that is, trying to achieve long-term stability while accommodating short-term variability) may be impossible. In addition, the spatial attributes of ecosystems are not uniform; they are skewed in their distribution and patchy at different scales. As a result, regulation cannot simply be aggregated across scales—what works for a single location will not *necessarily* work for a whole region. These characteristics of ecosystems often mean that management actions that are applied as blueprints across scales to maintain stability in ecosystem function can inadvertently lead to reduced resilience (Holling and Meffe 1996). Irrigation and homogenization of plant genetic stock in agriculture are good examples of the promotion of stability at the expense of resilience.

Second, there is uncertainty in outcome of control responses because regulatory environments can be quite diverse. Decisions are shaped by incentives, objectives, and attitudes toward risk, and these would be uncertain even if our understanding of natural systems were perfect. Nobody understands those systems perfectly, though, so a second layer of uncertainty must also be recognized. Taken together, these uncertainties can manifest themselves in three major ways. First, regulators are often averse to risk in their decision-making and concentrate on high-cost and extreme scenarios. Fear of making wrong decisions causes them to underemphasize the most likely outcomes or to take present day costs much more seriously than future costs. Either of these tendencies has implications for decision-making. In the literature about regulating invasive species, for example, Leung et al. (2002) argue for increased preventative action to reduce the threat to U.S. freshwater lakes. They estimate that preventing one species of invasive mussel moving into one lake would produce greater social and ecological benefit than the U.S. Fish and Wildlife Service spends on managing the invaders in all of the lakes in their jurisdiction. In this case, they argue an "ounce of prevention" is better than "a pound of cure."

Divergent utility functions and/or objectives of competing decision-makers also produce uncertainty in regulatory environments. In many cases, the need for regulation to encompass incommensurable objectives leads to what Ludwig et al. (2001) describe as "radical uncertainty"—even when consequences of action are predictable, objectives may be in conflict. Actions to reduce pollution loading in rivers in the Pacific Northwest region of the United States provide an example here—lowering levels of persistent pollutants has led to a population explosion in top predators from which disturbed river ecosystems may never recover.

Human demand for ecosystem function presents a third source of uncertainty. Human preferences for ecosystem functions change over time and space as economic and social circumstances change in unknown ways. The ultimate value to humanity of key functions may be infinite because of the interdependence of social and ecological integrity and continued existence, but some functions of nature are nonetheless more highly valued in particular societies, and these preferences change over time. Although it may be possible to observe historical changes in preferences for clean air and water, for example, Goulder and Kennedy (1997) argue that these changes provide no reliable guide either to future preference formation or to the cultural and social context within which the demand for ecosystem functions will arise. At the same time, however, even present day regulation of ecosystem functions is distorted by other interventions. Perverse subsidies for energy, water use, and agricultural commodities distort and create uncertainty in the regulation of environmental functions. Regulators act in social arenas where there may be preferences not only for the outcomes of regulation, but also for the social and political processes of decision-making. In institutional response, the legitimacy of their decision-making processes may be as important as the regulations that result.

4.3.2 Uncertainty in Institutional Responses to Ecosystem Protection

Institutional responses to threats to ecosystem services often involve a change in ownership or control of resources. In many instances, resources are privatized. In others, they are brought under government protection (or even ownership) because of perceived failures of private or collective management. Protected areas have effectively overturned previously held individual or common rights to resources, but this type of response can be limited by a number of uncertainties. There have been significant moves away from protected areas because of their perceived lack of legitimacy, particularly in the case of protected areas, which exclude previous users and even residents in these areas. As a result, the role of protected areas as a legitimate response is being reconsidered in some places, especially given the need for sustainable solutions that do not undermine social equity or impoverish and disadvantage poor rural communities in the developing world (see Brown 2002). Although they can be mitigated by clear and credible communication, the design and implementation of area protection face major uncertainties derived from three conditions: an incompatibility with the legitimacy of state appropriation; potential conflicts with other objectives of public policy and moral hazard; the lack of effective means of integrated implementation.

First, protected areas face uncertainty in conserving ecosystem services where there is a perceived lack of legitimacy of the de jure rights (rights in law). Regulations and rights to use resources may be incompatible with previous or de facto (rights in practice) management regimes. State appropriation of protected areas can therefore be a factor in the breakdown of traditional, usually communal, regimes of property rights and resource management (Bromley 1991). Indeed erosion of these systems of resource management has been shown to open the door to further environmental degradation, impoverishment, and demographic change. Moreover, expectations about what comes next after state appropriation can create additional uncertainty. For instance, the forestry sector in Nepal and the wetlands of Indonesia were once managed through local collective action institutions. State intervention led these institutions to unexpectedly allow open access because people expected further state appropriation (Bromley 1999; Adger and Luttrell 2000).

Second, uncertainty can be created by incompatibility with other policy areas. Perverse incentives in agriculture and other areas can, for example, undermine whatever responses are adopted for area protection. Sinclair et al (2002) find significant loss of bird diversity in agricultural land in the Serengeti compared to adjacent native savanna because of the reduction in insect abundance in the ground covering vegetation. Increases in agricultural land in the east African savanna habitats are not, however, being driven by agropastoral population growth, cattle numbers, or smallholder land use. Rather, the primary drivers are in the incentives to convert land by major landowners responding to agricultural policy. Homewood et al. (2001) demonstrate this result by differentiating between land cover change and its drivers in the Kenyan and Tanzanian parts of the Serengeti ecosystem.

Third, institutional responses involving changing property rights also lead to uncertainty if state resources are used to promote compliance with measures that may have been undertaken voluntarily. Agriculture and conservation policies in the United Kingdom have, for example, always been based on the primacy of private property. They are implemented through systems of incentives and compensation payments to private landowners. Areas such as Sites of Special Scientific Interest and others areas under the EU Habitats Directive are protected for a variety of functions and services. To conserve them on private land, owners are compensated for potential income rather than income actually foregone. Compensation for foregone income is estimated on the basis of prices for agricultural commodities that are inflated due to the workings of the Common Agricultural Policy in the EU member states. In other words, the perverse subsidies of the agricultural policy have further unintended consequences in making conservation payments more costly. Thus there is a significant moral hazard in such conservation responses based on the primary claims of private land (Bromley 1991).

4.3.3 Uncertainty in Institutional Responses that Engage Stakeholders

In recognition of the limitations in the traditional response options of controlling ecosystem services or changing the institutions of ownership, alternative approaches have emerged recently to integrate across response options. (See Chapter 15.) These alternative approaches often involve changing the basis of management to include wider sets of stakeholders. They aim to provide positive incentives for resource users and avoid the divergence between local and state objectives. There have been numerous attempts at sharing responsibility between regulators and resource users in fisheries, forestry, and other natural resource areas. In fisheries, for example, multistakeholder bodies like the U.S. Regional Fisheries Management Councils advise regulators on all aspects of ecosystem function and potential extraction rates (Brown et al. 2002; Berkes 2002). Alternatively, co-management institutional arrangements like the Australian Torres Strait Fisheries Management Committee involve sharing of power such that communities define their own management objectives. Either approach grants, in effect, user and ownership rights to local resource users and thereby allows them to develop sets of management rules. The hope is that these arrangements can resolve many of the resource conflicts that usually encumber the preservation of ecosystem functions (Bromley 1999).

There are uncertainties in these integrated multistakeholder responses. These include the potential for perverse incentive structures, incomplete or improper representation of stakeholders, and the inertia of governments in adopting co-management and, in effect, giving up their authority and power. Uncertainty in the sustainability of local collective action stems from the limitations of either state agency or local institutions to promote best practice. It is increasingly realized that simply allocating responsibility to local users is not necessarily a sufficient criterion for sustainability. Sustainable resource use requires a number of conditions. These include a favorable external environment such as appropriate technology and legal frameworks, a set of group and resource system characteristics such as defined boundaries and an identifiable set of stakeholders, as well as an agreed set of institutional arrangements that are easy to understand and enforce (Agrawal 2001). In the best cases, co-management overcomes the difficulties and limitations of both the government regulator and the local institution—the government agency provides the laws and the regulation of the external environment, while local institutions identify the legitimate stakeholders and enforce the rules (Berkes 2002). While effective monitoring and ex-post evaluation can help, significant risks and uncertainties involved in meeting the key criteria for sustainable management can remain.

Further uncertainty lies in the representation of stakeholders within any co-management arrangement. There are difficulties in "non-representation for contingent reasons" and "problems of the very possibility of representation" (Brown et al. 2002). Education programs can help, of course, but non-representation derives from inevitable bias in co-management in favor of the powerful and the articulate. Underlying ability and willingness and capacity to be heard and to articulate preferences are unevenly distributed across class, age, ethnicity, and gender. Other perspectives and interests, such as those of future generations and of the rights of ecosystems themselves, also suffer from the problems of the possibility of representation.

Much of the uncertainty in these response options derives from the unwillingness of regulators and government agents charged with conservation of ecosystem functions to embark on co-management and empowerment responses. Many environmental policy institutions fail to articulate a reason why all policy dialogue is presumed to be at the national level while the institutions of regional and local levels are assumed to be part of an implementation process (Bromley 1999). In many cases, national agencies are unwilling to share responsibility and power, even when they also perceive empowerment responses as means of reducing costs of enforcement and regulation. Further barriers to effective adoption of participatory and co-management arrangements in this area include a lack of legal and constitutional framework, a lack of trust in representation, and the resource cost of actually undertaking participatory management (for example, Tompkins et al. 2002). In terms of cost, despite perceptions to the contrary, co-management tends to be equally if not more resource intensive than traditional management, even when successful (Singleton 1998).

4.3.4 Comparing Effectiveness and the Case for Integration

Governance issues are the key to handling all of these uncertainties, both in reconciling values and in coping with surprises and unexpected consequences of interventions. Convergence and synergies in policy responses can be promoted through recognition of the legitimacy of processes alongside their efficiency, effectiveness, and equity. Legitimacy relates to the extent to which decisions are acceptable to participants on the basis of who makes and implements the decisions and how. Legitimacy can be gained as well as compromised through the process of making environmental decisions. There are no universal rules for procedures that guarantee the legitimacy of policy responses because cultural expectations and interpretations define what is or is not legitimate. When integration occurs, as argued in Chapter 15, the added legitimacy of horizontal and vertical response solutions increases the effectiveness of response for a number of reasons. First, including stakeholders gives greater sense of control over resources and hence reduces the overall cost of enforcement of protecting rights to the resource. Second, networks of stakeholders have value for other reasons, particularly in handling unforeseeable pressures and stresses on the ecosystems. Third, integrated responses lead to perceptions of greater value from the ecosystem services. (See Box 4.3.)

When the dynamics of ecosystem functions and the costs of amelioration are known, market-based policy responses have some advantage over regulatory or other forms of responses, primarily related to their efficiency or cost effectiveness in changing behavior among the agents causing loss of ecosystem functions. Regulation of the use of ecosystem function may inadvertently be

BOX 4.3

The Value of Stakeholder Perceptions in Decreasing Dissatisfaction with Response Impacts in Watershed Management

Stakeholder confidence in having access to benefits can determine their estimation of the value of watershed ecosystem services. This confidence can be hard to evaluate, but is necessary to increase the likelihood that the chosen response can be effective and gain stakeholder acceptance of the underlying processes. Willingness-to-pay is one method of decreasing uncertainty associated with stakeholder values. Some studies (for example, Koundouri, et al. 2003) have found a higher willingness-to-pay for responses about protection of wetlands along an international bird migration route under scenarios in which all of the relevant stakeholders participate (in this case, all countries along the migration route). Similarly, Porto et al. (1999) reported that domestic water users in Brazil (where nationwide river basin management policies had been adopted) were willing to pay more for water if the revenue from water fees were invested in the basin where the funds are generated and if users were able to participate in decisions about how the revenue was to be spent. Both of these examples speak to the sensitivity of contingent valuation results to the ownership ("property rights") assumptions of the participants.

O'Connor (2000) has argued that differences in willingness-to-pay often depend on the protection mechanism suggested, and whether it was regarded as fair and effective. This implies the need to develop effective institutional arrangements to control access, without which economic value cannot be captured. They are also a source of tremendous site-specific variation that needs to be considered to develop effective Payment Arrangements for Watershed Ecosystem Services (PWES) initiatives. Property rights, which define rights to particular streams of benefits as well as responsibilities for their provision, are critical because they determine whether those who pay the costs of management practices have access to any of the benefits, and therefore, see an incentive for conservation. Institutional arrangements also refer to relationships established among buyers, sellers, and intermediary organizations so as to reduce transaction costs. This evidence suggests that integrating stakeholder perceptions about response options can reduce uncertainty that might arise by ignoring key aspects of response options.

cost-effective when first designed, but market-based responses (of which taxes or tradable permits are the most common) generate dynamically efficient patterns of incentives on behavior over time. But these stylized arguments do not hold in the face of uncertainties described in earlier sections. Where the impacts of loss of ecosystem function are unknown, regulators and agents choosing between different market-based response options can underestimate or overestimate the necessary level of action and lead to both inefficient policy response and to the risk of nonlinear reorganization of ecosystem function (though the errors of over- or under-estimation need not be symmetric in magnitude or significance).

4.3.5 Unintended Consequences

History demonstrates the importance of attempting to evaluate, on as broad a scale as is feasible, the possible consequences of implementation of specific strategies, policies, and measures designed to protect and enhance ecosystem services. Unintended consequences associated with implementation of the original intervention can result in new significant problems that replace or compound the original issue. The capacity to respond effectively to threats to ecosystem services is dependent on the ability to foresee and anticipate surprise, as well as to deal with unexpected consequences of actions (Kates and Clark 1996). A systematic and thorough assessment of the dynamics of a particular issue (including the risks and benefits of doing nothing), the extent to which there are key uncertainties, and the magnitude of any potential adverse impacts can reduce the probability of unintended consequences (Ebi et al. 2005).

Without an assessment of the potential consequences of an intervention, beneficial steps may be inadvertently presented as cures. The determination of possible consequences of policies and measures should be across all relevant sectors and should consider current and potential future consequences. All implemented actions should include an on-going program for evaluation of both the program's effectiveness as well as any adverse consequences that could arise. Otherwise, recognition of a problem will be delayed, which can have adverse impacts on human well-being and the health of ecosystems. Two examples illustrate the importance of this issue.

In northern Ethiopia, micro-dams have been constructed to increase the availability of water for irrigation. After construction, Ghebreyesus et al. (1999) conducted a survey on the incidence of malaria (90% *Plasmodium falciparum*—the deadly strain of malaria) in at-risk communities close to dams and in control villages at similar altitudes but beyond the flight range of mosquitoes (primarily *Anopheles arabiensis*). The results showed that the microdams led to increased malaria transmission over a range of altitudes and seasons; the overall incidence of malaria for villages close to dams was 14.0 episodes per thousand child-months at risk, compared with 1.9 episodes in the control villages. These results could have been anticipated based on knowledge of the ecology of malaria, and appropriate measures to address the probable malaria problem could have been included in the development program. Including the health sector in the evaluation of the trade-offs of and responses to this irrigation development program could have prevented many children from suffering and dying from malaria.

A further example of unanticipated surprise is the on-going issue of the consequences of the installation of tubewells in Bangladesh and India to provide the population with access to clean drinking water. Beginning in the 1970s, tubewells were widely installed in an effort to provide a "safe" source of drinking water to populations experiencing high morbidity and mortality, especially among children, from water-related diarrheal diseases. At that time, standard water testing did not include tests for arsenic and it was not known that the groundwater accessed by these wells has naturally occurring high concentrations of arsenic. Unfortunately, it is still not a standard or routine test for rural water supplies. However, arsenic in drinking water was recognized as a problem prior to the installation of tubewells, so an evaluation program established at or soon after the installation of tubewells would have identified the problem much sooner. The U.S. Environmental Protection Agency set the current standard of 50 parts per billion in 1975, based on a Public Health Service standard originally established in 1942.

Possibly 30 million out of the 125 million inhabitants of Bangladesh drink arsenic-contaminated water (Hoque et al. 2000). Health consequences of exposure range from skin lesions to a variety of cancers. Because of the latency of arsenic-related cancers, it is expected that morbidity and mortality from historic and current exposures will continue for approximately 20 years after exposures are discontinued. Although a number of international initiatives are under way to help resolve this problem, solutions will likely take a decade or more. The installation of tubewells to reduce the burden of diarrheal diseases offers a variety of lessons for how to avoid unintended consequences of strategies, policies, and measures; more are likely to be learned as the resolution of this problem unfolds over time (Ebi et al. 2005). This situation clearly illustrates the risk of undertaking massive intervention programs without a determination of benefits and risks. Acute problems, such as access to safe drinking water, create pressure to find quick solutions. Programs should evaluate short-term responses while finding long-term solutions within the context of the underlying causes. Issues of scale and differences between absolute versus relative risks are imbedded in this lesson. A solution associated with a small risk implemented on a wide scale (such as the installation of tubewells) may have far more significant adverse health impacts than a solution with a larger risk implemented on a small scale.

A further lesson from the Bangladesh example relates to the process of implementation of interventions. Because tubewells were viewed as a technological fix, installation was implemented on a broad scale as rapidly as possible, not in an incremental or staged fashion that would incorporate regular evaluation of success (Ebi et al. 2005). Flexible and responsive approaches are needed in which new information and experience is properly evaluated and then used to appropriately modify interventions. Because arsenic contamination of drinking water is a classic second-generation problem, with the contamination discovered many years after the initial tubewells were installed, taking a staged approach to implementation could have had a much different result. The arsenic problem also reinforces the problem of reliance on a single or "silver bullet" technical solution to a problem instead of taking an integrated, multidisciplinary approach (Ebi et al. 2005).

4.4 Methods for Analyzing Uncertainty

Many techniques have been developed to include uncertainty in the evaluation of the relative efficacy of the various options that might be available to a system as it tries to respond to an external stress. Morgan and Henrion (1990) provide a concise overview of how to select a method for analyzing uncertainty. Presented here is a a quick summary of some of the more popular approaches. Since the MA has adopted a scenario-based approach, this section devotes most of its attention to scenarios, but other approaches are briefly summarized, with reference to their role in supporting scenario analysis.

4.4.1 Scenarios

The typology of uncertainties presented earlier in this chapter raises fundamental questions about our ability to foresee the impacts of particular response options, including both intended and unintended effects. This is closely related to our general uncertainty about what the future might hold. For many reasons, there has been an increasing use of scenarios to address complex issues involving socioecological systems and to explore how systems might respond to changes. The MA, for which both global and sub-global scenarios have been developed, is no exception. Thus it is important to understand how scenarios can and have been used to help us cope with the issue of uncertainty.

Berkhout and Hertin (2002, p. 39) argue that the future "needs to be thought of as being emergent and only partially knowable." Our uncertainty in knowing the future in general and, more specifically, the impacts of response options stems from three distinct types of indeterminacy: ignorance, surprise, and volition (Raskin et al. 2002). *Ignorance* refers to limits of scientific knowledge on current conditions and dynamics and is thus closely related to what this chapter has called model, calibration, prediction, and projection uncertainty. It implies that even if socioecological systems were deterministic in principle, our understanding of their future would still be uncertain. This is of particular concern for systems exhibiting chaotic behavior, where even slight changes in initial conditions can lead to dramatically different outcomes. Uncertainty due to ignorance is further compounded by *surprise,* the uncertainty due to the inherent indeterminism of complex systems that can exhibit emergent phenomena and structural shifts.

Finally, *volition* refers to the uncertainty that is introduced when human actors are internal to the system under study. Berkhout et al (2002) highlight the fact that because of conscious choice, the assumption of continuity made in the natural sciences is not applicable to social systems, implying that novelty and discontinuity are normal features of these systems. This compounds the types of uncertainty noted above, but is also a key aspect of what was referred to as contextual uncertainty. Moreover, the very process of ruminating on the future can influence these choices. Through this reflexivity, people work either to create the future they desire or to avoid that which they find objectionable.

As defined by Raskin et al. (MA *Scenarios,* Chapter 2), scenarios are "plausible, challenging, and relevant stories about how the future might unfold that can be told in both words and numbers. Scenarios are not forecasts, projections, or predictions." At best, scenarios might be considered conditional projections in that particular outcomes "reflect different assumptions about how current trends will unfold, how critical uncertainties will play out, and what new factors will come into play" (UNEP 2002; Robinson 2003). It is important to note that scenarios are not merely alternative runs of a model or a sensitivity analysis, although these can be important for looking at uncertainty within scenarios, as noted below.

If done properly, particularly when formal quantitative models are used, all of the underlying assumptions are made explicit in a scenario. Certain assumptions will take precedence, however. These represent the primary axes along which the scenarios will differ and are generally related to contextual issues. These fundamental differences provide the "logic" behind the individual scenarios and the scenario exercise as a whole (Schwartz 1996). The dimensions can be as simple as high versus low economic growth or high versus low population growth, but they can be much richer, reflecting amalgamations of more than one driving force, critical uncertainty, or new factor. A well-known example of this is the Special Report on Emissions Scenarios process of the IPCC, in which four scenario families were distinguished based upon two principle axes—degree of globalization versus regionalization and degree of emphasis on economic growth versus issues of environmental and equity (IPCC 2000). Less explicit framing of sets of scenarios, where multiple dimensions were considered but all combinations were not fully enumerated, can be found in the work of the Global Scenarios Group (Raskin et al 2002) and UNEP (2002). In these cases, the emphasis was on the most interesting and coherent combinations, recognizing that not all are plausible or worth exploring.

The establishment of a framework distinguishing different scenarios emphasizes the differences between scenarios and represents the principle way in which they have been used to address the issue of uncertainty. Looking deeper, though, it is clear that a scenario cannot be defined fully by one or two key assumptions. The other key driving forces and critical uncertainties must also be fleshed out. In doing so, it may be clear that other assumptions hold particular significance for the issues of concern. For example, in one of the IPCC SRES scenario families, a third axis related to energy technology was introduced in order to examine more explicitly what was considered a fundamental assumption related to the issue of greenhouse gas emissions (IPCC 2000). Furthermore, specific assumptions expressed in a scenario narrative can be consistent with a variety of quantitative representations. With respect to assumptions about particular driving forces, this relates to what was termed earlier as projection uncertainty; with respect to particular system relationships, it is more akin to model uncertainty. Finally, estimates of the results of particular assumptions, whether these are determined by qualitative reasoning or formal quantitative models, are subject to the other forms of uncertainty discussed above.

Although this has not necessarily been a key emphasis in scenario development, various strategies have been used to address this issue of within scenario uncertainty. In the IPCC SRES process, in addition to the particular case of different assumptions about energy technology in one scenario family, six modeling groups provided quantitative representations of the four primary scenarios. This resulted in the development of a total of 40 scenario realizations. Based on this, many of the key results, such as total greenhouse gas emissions and atmospheric concentrations, are presented as a range of estimates for each scenario family, rather than as a single trend line. In the case of the United Nations Environment Programme GEO3 scenarios, the original quantification of the four storylines was accomplished using a combination of modeling tools, with each tool taking responsibility for specific outcomes. Only afterwards has an analysis been undertaken to compare the results for consequences that were estimated by more than one tool (Potting and Bakkes, forthcoming). Finally, the Global Scenarios Group defines three classes of scenarios, each with two variants (Raskin et al 2002). These variants, however, differ to such a degree that they more truly reflect distinct scenarios, pointing out what can be a fuzzy boundary dividing what is called uncertainty across versus within scenarios.

The scenarios discussed above have all been undertaken at a global scale, with a limited amount of regional and local disaggregation. Within these scenarios, a wide range of variation at lower scales, in both driving forces and outcomes, is glossed over. At the same time, they are usually done over a long time period, with a large degree of smoothing of variability over shorter time periods. Finally, the elaboration of key actors is generally quite limited. As demonstrated in Strzepek et al. (2001) and Yohe et al. (2003), downscaling scenarios, both narrative and numerical, introduces additional uncertainties that must be recognized, but not necessarily probabilistically. In this situation, using a collection of scenarios to span a range of "not-implausible" futures can be useful in evaluating the relative robustness of alternative responses. (See Box 4.4 for a discussion of the development of the MA scenarios.)

4.4.2 Alternative Methods for Accommodating Uncertainty

A variety of other approaches can bring uncertainty to bear on response evaluations. Sensitivity analysis can, for example, be employed to compute the effects of changes in specific parameterizations and/or assumptions on important state variables and to construct measures of the relative importance of various sources of uncertainty. Sensitivity analysis can, therefore, be employed to explore the degree to which various alternatives might support scenarios that portray fundamentally different futures. Scenario trees like the SRES alternative story lines can thereby be created and differentiated not only by differences in their driving variables, but also by differences in their social, economic, political, and scientific contexts. More elaborate explorations of scenario trees sometimes attach probabilities to the various branches, but this can be dangerous; indeed, the authors of the SRES futures insist that no storyline is any more likely than another. When probability distributions for driving variables can be quantified, though, probabilistic portraits of wide ranges of possible outcomes for state variables can be produced in support of expected value

BOX 4.4

Uncertainty and Scenarios in the Millennium Ecosystem Assessment

The developers of the MA global scenarios and scenarios within the sub-global assessments have had to deal with the question of differences between and uncertainty within scenarios. These issues are discussed in much greater detail in other MA volumes, including *Scenarios* (Chapter 6), *Multiscale Assessments* (Chapter 9), and several sub-global reports. This box provides a brief summary of some of the key issues related to how uncertainty has been addressed in these exercises.

Three primary sources of information were used to help determine the primary axes along which the MA global scenarios differ: the specific needs expressed by the primary audiences for the MA; the insights drawn from interviews with leaders in nongovernmental organizations, governments, and businesses from around the globe; and explorations of ecological management dilemmas. In the end, the choice was made to focus on different possible strategies for achieving a sustainable and diverse future. One of the scenarios, called "Order from Strength," represents one plausible path for global breakdown wherein no clear strategy is pursued. In "Global Orchestration," by way of contrast, the focus is primarily on fair global policies. "Adapting Mosaic" focuses on local and regional flexibility, and "TechnoGarden" highlights technological innovation.

Within the sub-global assessments, a variety of uncertainties were articulated in the development of distinct scenarios, including uncertainties in both exogenous drivers and endogenous behavior. Two of the most commonly cited exogenous drivers were the nature of governance at higher scales and regional or international markets for products produced within the sub-global site. Among the most common endogenous uncertainties were the future of local institutional arrangements and the evolution of social attitudes toward the environment.

The development and presentation of the global scenarios has tried to address uncertainty in both the narrative storylines and their quantitative underpinning. There was an effort to be open and consistent in communicating issues of uncertainties in the global narratives. This was done, in part, by utilizing the scheme developed for handling uncertainty in the IPCC assessments, where particular expressions were associated with a level of, generally subjective, confidence. Also, the scenario developers explored how uncertain events could cause one scenario to branch into another.

Looking at the quantitative underpinning, as with the United Nations Environment Programme GEO3 scenarios, a number of different modeling tools were used to provide numerical estimates of key input and output indicators for the scenarios. This is somewhat different from the case for the IPCC emission scenarios, where each of the different modeling tools provided an independent and complete quantitative picture of the scenarios. In the MA, a process of harmonization was undertaken to ensure that the different tools used consistent sets of drivers. Specific tools were identified for calculating output indicators of ecosystem services and human well-being. The process included detailed assessments of the ability of different tools to forecast the indicators of interest. This usually resulted in a single tool for each indicator, but in particular cases different tools were used for different regions. Finally, the resulting estimates derived from the various tools were reviewed to assess their uncertainties and the potential influence these might have on the scenarios.

The scenarios in the sub-global assessments are much less clear about how they dealt with the question of uncertainties within individual scenarios. Most of the sub-global scenario exercises focused on the development of qualitative scenarios that rely solely on narratives, with the use of quantitative models being the exception. This limited the degree to which they were able to do the kind of sensitivity analysis seen in the global scenarios.

Finally, the issue of the development of multilevel scenarios, and the uncertainties inherent in doing so, is an issue that is more particular to the scenarios in the sub-global assessments. This development of multilevel scenarios goes beyond the incorporation of driving forces from higher scales, which has been present in all of the exercises; it also includes coping with the problem of actually linking scenarios at one scale to those at another, which has been much less common. Two approaches can be recognized here. The first is to embed the scenarios in the sub-global assessments in the MA (or other) global scenarios. This has been explored in the Portugal sub-global assessment, and may be more common in the later-starting sub-global assessments, as the MA global scenarios are now more developed. The other approach is to develop multiscale scenarios within a single sub-global assessment; this has been explored in the southern Africa sub-global assessment. The advantages and difficulties with both of these approaches, including issues of uncertainty in the linkages and feedbacks across scales, are just now being explored; the MA work is a productive first step in this process.

It must be noted, though, that the global scenarios are using elements from the sub-global narratives to add texture to their stories. Furthermore, particular issues of uncertainty related to scale arise in global scenarios in terms of modeling cross-scale effects and the presentation of quantitative results. Since the different models vary in their geographical breakdown, many of the quantitative results are presented at a highly aggregated scale. This poses a particular problem in the form of the loss of information about variability, which can undermine the confidence attributed to the conclusions drawn from the scenarios.

calculations of associated consequences and/or analyses of the robustness of various responses.

Finally, it is sometimes possible to provide estimates of response surfaces—empirically calibrated reduced-form relationships between driving variable and state variables—which can summarize the results of a large number of scenarios and/or alternative futures. These are particularly valuable when complex models are expensive (in time and money) to run; and they can also be especially useful in completing scenario interactions when the researcher is focusing attention on another part of the problem. A researcher interested in the detailed impacts of climate change on the likelihood of flooding might, for example, use a response surface representation of the energy sector and how it would respond to alternative population futures and different mitigation strategies in constructing the requisite connection between economic activity and flooding without building an elaborate energy model (Yohe and Strzepek 2004).

4.5 Decision Analytic Frameworks under Uncertainty

Every framework designed to support decisions about response options must be able to accommodate uncertainty in its application; their diversity is outlined in the MA framework volume (MA 2003, p. 196), with the advantages and disadvantages of each identified. For present purposes, it is sufficient to note that some frameworks have evolved to the point where uncertainty can be brought on board even if they were initially developed in deterministic environments. Others, though, were created with the explicit purpose of incorporating uncertainty into their structures. This section reviews how a few of the most important approaches to response decisions accomplish the proposition that uncertainty is ubiquitous.

4.5.1 Cost–Benefit Frameworks

Cast into a world of uncertainty, applications of the cost-benefit approach to selecting and designing response options require understanding of the range of possible outcomes in order to provide a full accounting of (potential) net benefits. The fundamental decision steps underpinning CBA are summarized in Hanley et al. (1997), among other places:

- derive estimates of costs and benefits,
- rank initiatives from high to low in terms of net benefits, and
- pursue as many initiatives (with positive net benefits) as possible within resource constraints.

If analysts can assess probabilities (even based on subjective judgments) across a range of outcomes that accommodate the full range of possibilities, then expected net benefits can be the basis of decision rankings. If analysts cannot assess probabilities, they can still assess robustness—the range of possible outcomes for which net benefits are positive—and perhaps identify critical thresholds for critical sources of uncertainty along which net benefits turn from positive to negative.

It is important to recognize from the outset that the cost-benefit approach to decision-making ignores the distribution of costs and benefits—an omission that can bring contextual uncertainty to the fore. Programs or projects are judged to be attractive as long as total (expected) benefits exceed total (expected) costs regardless of who bears the cost and who enjoys the benefit. Chapter 5 illustrates this point as it considers the range of costs and benefits that might be attributed to projects designed to restore or rehabilitate ecosystems. It highlights the potential need for compensating side-payments; especially if stakeholders are involved in restoration decisions, these compensation schemes add another layer of contextual uncertainty.

Choosing the correct discount rate is also an enormous issue when costs and benefits extend into the future because discounting can render long-term future effects almost irrelevant in the calculation of discounted net benefits. IPCC (1996) spent an entire chapter making a distinction between descriptive and prescriptive discounting for long time horizons—a distinction whose fundamental content is perhaps best exhibited by the Ramsey rule for inter-temporal optimization. According to this rule, intertemporal utility would be maximized if per capita consumption at the end of any year were discounted relative to consumption at the end of the previous year by the sum of a pure rate of time preference (a measure of impatience in consumption) and the product of the elasticity of the marginal utility of consumption (the rate at which utility changes as consumption grows) and the rate of growth of per capita consumption over the year in question.

Many analysts find the argument for including the second term convincing. Indeed, they commonly work with logarithmic utility functions for which the elasticity of marginal utility is equal to unity and the second term is simply the rate of growth of per capita consumption. The key to their conviction is that wealthier generations who will inhabit the future will attach smaller utility values to marginal changes in per capita consumption. Meanwhile, most attempts to measure the pure rate of time preference have focused attention on individuals' decisions over time, and those decisions do not shed much light on how society should weight the relative welfare of successive generations (Lowenstein 2002). The IPCC (1996) noted that many scholars think that the pure rate of time preference should be set at or close to zero when costs and benefits are extended well into the future. Weitzman (1998) reinforced their convictions by noting that low rates dominate expected discounted value calculations when they extend deep into the future; recently proposed hyperbolic approaches similarly guard against overly enthusiastic discounting.

Short-term calculations are less problematic. Markets are driven by private agents who discount the future at the return to private capital, that is, the opportunity cost of financial capital. If capital markets were perfect, then this discount rate would match the (short-term) pure rate of time preference. Capital markets are not perfect, of course, so the rate of return to private capital can exceed the pure rate of time preference for risky responses to market stresses (add a risk premium) or because the return to private capital is subject to (corporate) income taxation. In either case (and many others), the appropriate discount rate simply adds the effect of whatever distortion exists (such as risk or taxation) to the pure rate of time preference. Following Arrow and Lind (1970), Ogura and Yohe (1977) demonstrated that the marginal return to government investment (that is, the rate at which future costs and benefits of such an investment are discounted) could be allowed to fall below the pure rate of time preference if public investment would complement private investment and private capital markets were distorted by taxes. Their result simply recognizes that lower discount rates encourage investment (by making it more likely that discounted expected net benefits are positive) and thereby diminish the efficiency losses caused by existing economic distortions.

When the intensity of various initiatives can be modulated, the level that maximizes net benefits equates expected marginal cost with expected marginal benefit. Tol (2003) has observed, however, that marginal costs or marginal benefits may not be well defined under all plausible futures even if benefit and cost measures are, themselves, finite. In such cases, the paradigm breaks down. (See Box 4.5 for a discussion of issues of timing and uncertainty.)

Nonetheless, optimization techniques can inform not only how a policy intervention might be targeted, but also how it might be designed. Smit et al. (2000) emphasize the necessity, in the context of adaptation, of clearly understanding "Who is responding to what?" and "What do they know when they have to 'pull the trigger'?" Weitzman (1974) showed why they were right to do so. He envisioned a policy-maker who, on the basis of limited information, must choose between price- and quantity-based interventions in an effort to hold expected output at the optimal level. Economic agents could, however, respond to the price-based intervention by adjusting their outputs in response to changes in their environments that would materialize only after the policy intervention had been designed. Their outputs would vary under the price control, but they would be fixed under the quantity-based intervention. The price control would therefore increase expected private benefit to agents (otherwise, they would not adjust their outputs), but the associated variable output would also cause expected social cost to rise (if marginal social cost were rising). It turns out, therefore, that the policy design choice was critically dependent upon the relative size of these two increases.

4.5.2 Risk Assessment

The classic risk assessment approach to, for example, evaluating health or ecological risks, adopts a four-step risk paradigm:

- identify the hazard (could a particular agent or activity harm humans, animals, or plants?);
- assess the exposure-response relationship (to what degree can exposure to a hazard cause a response that could be harmful?);
- assess the level of exposure (to what degree are humans, animals, or plants exposed to the hazard); and
- characterize the associate risk (a reflection of the probability exposure times the associated consequences).

Each step involves a policy judgment; for example, the choice of one dose-response model over another is a "science-policy" choice (NRC 1994, 1996; Presidential Commission 1997).

The limitations of risk assessment should be recognized and understood. The underlying assumptions may, first of all, limit its applicability to complex environmental problems (Bernard and Ebi 2001). The assumption that a defined exposure to a specific agent causes a specific adverse outcome for identifiable exposed populations can, in particular, be questioned in many contexts. A health outcome may be distinctive and the association between immediate cause and its impact can be fairly clearly determined, but most outcomes associated with environmental exposures have many causal factors, which may be interrelated. These multiple, interrelated causal factors need to be addressed along with relevant feedback mechanisms in investigating complex disease/exposure associations, because they may limit the predictability of the health outcome and even the ability to estimate the degree of uncertainty in any risk estimation (Bernard and Ebi 2001).

While early risk assessments focused narrowly on determining the probability of harm, the general approach is evolving and becoming more relevant to complex environmental problems (Bernard and Ebi 2001). Recent assessments are considering social, economic, and political factors, and stakeholders are now expected to be involved throughout the risk assessment process to ensure that the characterization of risk addresses a broad range of concerns. Especially in light of increasing complexity, it is extremely difficult to make detailed and accurate assessments of risks

BOX 4.5

Cost–Benefit Analysis in the Presence of Uncertainty, Irreversibility, and Choice in Timing

Conventional theory and practice holds that a positive expected net present value (NPV) returned by a cost-benefit analysis tells the investor that it might be prudent to go ahead with an investment. In reviewing the applicability of CBA to natural systems, Aylward et al. (2001) recall a warning by Dixit and Pindyck (1994) in light of two hidden assumptions in the CBA. In the first case, the investment is reversible insofar as the investor can exit from the investment and recover the expenditure if the future (for example, future market conditions) turns out worse than expected. In the second case, the NPV rule assumes that there is no choice of timing if the investment is irreversible; that is, the investment is a "now or never" proposition. Most investment decisions do not fulfill either of these assumptions. Indeed, irreversibility and the possibility of postponing investment are very important characteristics of investments faced by firms and by society.

The value of delaying investment is equivalent to holding an "option" to invest the right, but not the obligation to invest, and thus can be called an option value. When an irreversible investment is made, the investor exercising the option effectively gives up the opportunity to wait for additional information (to reduce the uncertainty over the present worth or timing of the expenditure). This is the central point made by Dixit and Pindyck (1994): the opportunity cost of making a decision to go ahead with the investment is the loss of an option value. As a result, the NPV rule needs to be reworked so that the decision to invest is taken only when the benefits of the investment exceed the standard costs of investment *plus* the value of keeping the option alive. Dixit and Pindyck also show how the opportunity cost represented by the value of an option to invest can be very sensitive to uncertainties. Given that the growing literature on these options values shows that they can "profoundly affect" the decision to invest, they argue that these uncertainties may explain more of the variation in investment behavior than other variables such as discount rates.

The application of the theory of investment under uncertainty and irreversibility to natural systems (dams and water resources development, for example) is novel at this stage. Further investigation is needed to determine the applicability of these ideas to the project planning and evaluation process. Still, it seems likely that at least the insertion of a qualitative discussion and analysis of different alternatives in this regard may be useful at an early stage in the screening and ranking of projects. Indeed, it is possible to argue that stakeholder discussion of different scenarios for water and energy resources development should include these issues in an explicit fashion, given that they may have considerable bearing on the CBA outcomes.

In terms of specific areas for further investigation, it would be worth considering the extent to which, in practice, the passage of time is likely to reduce (or to increase) markedly the uncertainty about future values of the irreversible investments and divestitures associated with different options, particularly the environmental and social impacts. Attention should be paid to examining how the costs and benefits of investments may differ in terms of irreversibility, uncertainty, and timing. The objective here would be to see whether the different components of the alternatives under consideration are likely to have the same characteristics in this regard and thus can be bypassed, or whether important differences between alternatives are expected and should be accounted for in the decision process.

and hazards because of profound uncertainty in both the probability of an event occurring and the scale and nature of its consequences. These uncertainties may arise from a variety of factors (WHO EUR 1999), including:

- a lack of (credible) data in many situations;
- complexity in the interactions between humans and the environment, which typically means there are many possible causes for any adverse effect;
- complexity in space and time that makes it doubly difficult to establish causal connections;
- synergistic and/or cumulative effects that muddle our understanding of the combined effects of toxicants;
- the likelihood that hazards will appear from unpredicted sources; and
- diversity in the susceptibility to exposure across populations due to genetic, social, or environmental factors.

If a risk assessment fails to explicitly address these issues, it may give the illusion of an objectivity that is not justified.

4.5.3 Multicriteria Analysis

Issues derived from conflicting interests in the management of a given resource pose particular problems, especially when distributional implications are to be considered. Multicriteria analysis and its variants are often the formal framework of analysis used to decide among various response options under these circumstances. Formal MCA can trace its roots to Pareto at the end of the nineteenth century. In the 1970s, the development of multiobjective maximization methods permitted widespread application of quantitative MCA, especially in the management of water resources and (more recently given the advent of GIS) land-use planning. Multicriteria analysis depends on completing a number of concrete steps:

- identifying objectives,
- identifying options for reaching these objectives,
- identifying evaluation criteria,
- analyzing options against those criteria,
- making choices based on those analyses, and
- evaluation and feedback.

MCA's explicit recognition of a multiplicity of objectives and evaluation criteria gives it a potential advantage over economic paradigms based on cost, benefit, and efficiency in identifying sources of vulnerability to uncertainty and even to ignorance. Using the framework outlined above, it is possible to discover uncertainty in objectives, unexplored options, incomplete evaluation criteria, ignorance about system properties, and volatile rules of choice.

All of this complexity comes at a price, however. In allowing the sources of uncertainty to include ignorance and the incomplete representation of evaluation criteria and available options, practitioners of MCA usually assume, at least implicitly, that the objectives being considered, the means for reaching them, and the systems from which services are being derived are independent in time, space, and consequence from other questions and decisions regarding the interaction of natural and social systems. In making this assumption of independence, a large number of concerns about path dependency, cross-scale effects, and cumulative impacts can be missed entirely. (See Box 4.6.)

BOX 4.6

Scale Uncertainty in Multicriteria Analysis

Even if biodiversity patterns from place to place could be estimated well, uncertainty would remain in society's valuation of biodiversity relative to other needs. The application of MCA to explorations of biodiversity trade-offs is consistent with the notion that economic decision tools are limited in assessing responses for ecosystem services that are not traded in markets and where the values associated with them are not utilitarian in nature.

Faith (2002) illustrated how uncertainty might be reflected in multicriteria analyses. He considered, for example, a trade-off between devoting land to biodiversity conservation or other uses (for example, forestry production). The best outcome for a region could be found along a deterministic budget constraint by imposing a biodiversity target, but uncertainty may be addressed by sensitivity analysis. More specifically, analysis of trade-offs in Australia showed how some areas were always (or, alternatively, never) allocated to biodiversity conservation regardless of the relative weighting of various criteria within the biodiversity target.

Faith (2002) also calculated trade-offs under two scenarios for Papua New Guinea. In one case, he assumed no land-use constraints and concluded that that about 85% of the study's biodiversity target could be achievable at a very low cost. A second case represented a scenario in which areas already having some degree of high land-use intensity were assumed to be lost to biodiversity conservation. The capacity for cost-effective conservation was dramatically reduced. Indeed, the cost for the same conservation achievement level had more than doubled. Scenarios of this sort can be valuable tools with which to incorporate uncertainties into decisions about regional biodiversity trade-offs.

4.5.4 Precautionary Principle and "Safe Stopping Rules"

The precautionary principle is an approach used by policy-makers in which they consider taking action to protect a population from potential hazards with serious or irreversible threats to health or the environment before there is strong proof that harm will occur. In essence, the precautionary principle prescribes how to bring scientific uncertainty into the decision-making process by explicitly formalizing precaution and bringing it to the forefront of the deliberations (Marchant 2003). It posits that significant actions (ranging from doing nothing to banning a potentially harmful substance or activity, for instance) may be justified when the degree of possible harm is large and irreversible. Many factors influence these deliberations, including an assessment of the possible severity of the potential harm and the degree of scientific uncertainty associated with that assessment.

The application of the precautionary principle to environmental hazards and their uncertainties began with the Swedish Environmental Protection Act of 1969, with further elaboration in the German Clean Air Act of 1974 and the 1985 report on the Clean Air Act (Kheifets et al. 2001; Boehmer-Christiansen 1994; EEA 2001). Since the 1970s, the precautionary principle has been incorporated into over a dozen international environmental agreements, expressly incorporated into the legal framework of the European Union, and adopted into the domestic laws of numerous nations (Marchant 2003). This principle featured in the 1992 Rio Declaration on Environment and Development as Principle 15 (UN 1993): "In order to protect the environment, the precautionary approach shall be widely applied by States according to their capabilities. Where there are threats of serious or irreversible damage, lack of full scientific certainty shall not be used as a reason for postponing cost-effective measures to prevent environmental degradation."

Tamburlini and Ebi (2002) report that the European Commission's decision to ban beef from the United Kingdom represents a recent and dramatic application of the precautionary principle with a view to limiting the risk of transmission of bovine *spongiform encephalopathy.* The European Court of Justice ruled that this decision was justified by the seriousness of the risk and the urgency of the situation. The Commission did not, the Court ruled, act inappropriately in adopting a decision on a temporary basis pending improved scientific information. Its actions followed procedures approved in a communication on the precautionary principle authored by the Commission in February of 2000 (CEC 2000), which included the following guidelines for adopting measures on the basis of the precautionary principle:

- tailoring the measure to a chosen level of protection;
- applying the measure without discrimination (that is, treating comparable situations similarly);
- confirming that the measure was consistent with similar measures already taken;
- examining the potential benefits and costs of action or lack of action (including, where appropriate and feasible, an economic cost-benefit analysis);
- constructing review mechanisms by which new scientific data could be brought to bear on timely re-evaluation; and
- assigning the responsibility for producing the scientific evidence necessary for a more comprehensive risk assessment.

In this definition, the precautionary principle is "risk-oriented" in that it requires evaluations of risk that include cost and benefit considerations (Tamburlini and Ebi 2002). It is clearly intended for use in drafting provisional responses to potentially serious health threats until adequate data are available for more scientifically based responses. It also can be applied when there may be undue delay in the regulatory process.

The application of the precautionary principle does not mean that a scientific approach is not required. Nor does it mean that critical attributes of the risk can be ignored; these include irreversibility, magnitude of possible consequences, the probability of occurrence, the amount and type of uncertainty associated with the risk, societal benefits of the risk-creating activity, difficulty and costs of reducing risk, potential alternatives to the risk-creating activity, potential risk-risk trade-offs (that is, the degree to which proposed solutions create new risks), and public perceptions of the risk (Marchant 2003).

Proponents of the precautionary principle cite many examples of risks that were initially ignored or underestimated and later turned out to cause significant adverse human health impacts. There is the perception that environmental and health problems are growing more rapidly than society's ability to identify and mitigate them (Kriebel et al. 2001). In addition, increasing awareness of the potential for severe adverse effects due to global environmental change has weakened confidence in the abilities of decision-makers to identify and control risks in a timely and effective manner. Application of the precautionary principle is intended to prevent society from the costs of false negatives (that is, waiting to implement regulations when risks turned out to be real and significant); but increasing application of the precautionary principle raises the question of costs of false positives, that is, taking more regulatory action than turns out to have been required.

4.5.5 Vulnerability Analysis

Methods of vulnerability assessment have been developed over the past several decades in addressing natural hazards, food security, poverty analysis, sustainable livelihoods, and related fields. These assessments have helped to determine the baseline characteristics of those individuals, groups, or ecosystems that are sensitive to changes and shocks in the system. Vulnerability assessment can identify both general and specific vulnerabilities that enable targeted intervention and can guide future development projects. In identifying the most vulnerable systems or groups, measures to increase social resilience and ecosystem productivity can be prioritized; but the methods and tools used to undertake such assessments involve particular uncertainties. Model, calibration, and scale uncertainties are relevant to VAs. They are discussed in this chapter, MA *Current State and Trends* provides more detailed explanations of vulnerability and vulnerability assessments.

Vulnerability is a contested and ill-defined term. For example, vulnerability can be defined as the degree to which "an exposure unit is susceptible to harm due to exposure to a perturbation or stress and the ability (or lack thereof) of the exposure unit to cope, recover, or fundamentally adapt (become a new system or become extinct)" (Kasperson and Kasperson 2001, p. 21). By contrast, the climate change community uses the term in a significantly different manner. The Intergovernmental Panel on Climate Change defines vulnerability as, "The degree to which a system is susceptible to, or unable to cope with, adverse effects of *climate change,* including *climate variability* and extremes. Vulnerability is a function of the character, magnitude, and rate of climate variation to which a system is exposed, its *sensitivity,* and its *adaptive capacity*" (IPCC 2001a, p. 995). The implications of these varying definitions have been elaborated in attempts to provide synthesis between competing paradigms (Brooks 2003; Downing et al. 2003; Turner et al. 2003a). The competing definitions, when implemented to assess vulnerability, lead to uncertainty in exactly what is being measured and thus limit the scope of comparative studies.

Within the competing definitions, there is varying emphasis on common causal factors and outcomes. Some approaches emphasize the external conditions and impacts that lead to vulnerable states, such as the rate of climate change or the duration of a drought, for instance (IPCC 2001a). These approaches often assume that a change in the hazard would change the state of vulnerability of those impacted. An alternative approach focuses on the agents of interest and their underlying internal characteristic (Brooks 2003). This can be termed "social vulnerability." Studies in this vein explicitly consider livelihood elements (such as access to information and resources) as well as the broader socioeconomic environment (such as factors that determine the ability of the system and agents within the system) to cope with their existing situation and to respond to potential impacts (Adger 1999; Blaikie et al 1994; Turner et al. 2003b).

Calibration uncertainty is associated with diverse data used in VAs. Both quantitative and qualitative methods are used to evaluate vulnerability with their recognized limitations. Quantitative methods often entail the use of indicators. These indicators are frequently displayed in vulnerability and risk maps indicating the location of the most vulnerable areas (for example, Stephen and Downing 2001 and UNEP-GRID 2003). However, vulnerability indices should be treated with caution. Not only is the diversity and sensitivity of vulnerability hard to measure, but combining different variables of different scales and dimensions to produce one index can also mask many of the underlying processes.

Qualitative data can be equally difficult to validate. Oral histories, for example, are discussions with key local stakeholders that help to elicit information about local historical vulnerability. These histories rely on individual perceptions of the past state of the environment or society. These methods can be effective at gathering information on local vulnerabilities over past decades where there is limited data, but the nature of the information means that triangulation and validation is constrained within shared recall and memory. Other participatory methods that rely on involvement from multiple stakeholders can be subject to political motives and cultural perceptions that are often hard to tease out. It is important that these constraints are recognized and triangulation is undertaken to obtain multiple perspectives.

Scale uncertainty can occur when integrating data that assess different temporal and spatial scales and so the units of measurement are often inconsistent. For example, a village might be vulnerable to climatic variability if it does not have the means to cope with drought, but a household within the village may have planted drought-resistant maize and cope adequately with a dry season. Similarly, within that household the mother might ensure that her children eat the food first. The household may therefore appear to be resilient to drought even though many members within the household suffer. If, as a result, a family member fell ill the next year, then available labor would fall, resources would be spent on healthcare, and the household would become more vulnerable than other households in the village. Phenomena like these have different expressions at different scales that are important if vulnerability is to be captured adequately in a VA.

Although overlaying scales is difficult, VAs are beginning to move beyond static snapshots at a particular time and place to assessments that depict cumulative and long-term vulnerability at a variety of spatial scales. (See Box 4.7.) It has been recognized that vulnerability cannot be viewed as a static phenomena. Vulnerability is, rather, part of a dynamic process (Leichenko and O'Brien 2002). This recognition has implications for the level of uncertainty associated with vulnerability assessments because dynamic analyses require information about how processes and characteristics change over time.

BOX 4.7

RiskMap as a Vulnerability Analysis Tool: Different Lenses Lead to Different Outcomes

RiskMap is an interactive computer-aided tool that enables levels of vulnerability to be mapped within the dominant livelihood and household food economy zones (see Seaman 2000; Save the Children Fund 1997). It was developed in the early 1990s to predict, assess, and monitor famine. The input data to the program is qualitative and field-based.

Stephen (2003) reported that the program typically assesses the risk and dynamics of household livelihood security by describing income and reserves for three household categories (rich, modal, and poor) based on a variety of factors (including the normal pattern of employment, specific employment, livestock and other markets used, and the likely distribution of food and other goods among households). Surveys are conducted among experienced field-based and international staffs, in collaboration with local "informants," who generally live in the area and follow local market and trading patterns. Although this approach has its strengths, Stephen (2003) also noted that some users are uncertain about how the data can be validated.

The results describing who is vulnerable vary depending on the scale of analysis. In RiskMap, different indicators can be turned "on" and "off" and so determine which areas are mapped as vulnerable. This highlights the nature of vulnerability given a variety of specific definitions (see Downing et al. 2003). For example, if the indicator reflecting livestock-dependent households are used then a large area might be considered vulnerable. If access to aid is added, then a smaller area might be considered vulnerable. This highlights the importance of understanding the dynamics of vulnerability and the scope of the study, as these will determine the outcome of "vulnerable" groups.

4.5.6 Summary

Table 4.1 offers a summary of how the various decision-analytic frameworks reviewed in this section have been employed. Modeled after a similar IPCC table provided by Toth and Mwandosya (2001, Chapter 10), it shows that though all of the frameworks are able to support optimization exercises, few have much to say about equity. Cost-benefit analysis can, for example, be modified to weight the interests of some people more than others. The discount rate can be viewed, in long-term analyses, as a means of weighing the welfare of future generations; and the precautionary principal can be expressed in terms of reducing the exposure of certain populations or systems whose preferential status may be the result of equity considerations. Table 4.1 also suggests that only multicriteria analysis was designed primarily to accommodate optimization across multiple objectives with complex interactions, but MCA can also be adapted to consider equity and threshold issues at national and sub-national scales. Finally, the existence and significance of various thresholds for change can be explored by several tools, but only the precautionary principle was designed explicitly to address such issues.

All of these frameworks fall under what many view as decision analysis—a general rubric capturing a broad range of structures based on other representations of the losses or gains associated with external stress and the corresponding gains (diminished losses) or additional benefits of responding to those stresses. Several primary insights can be drawn from referring briefly to this wider perspective. One is that the expected outcome computed against any criteria across a wide range of futures is, given the enormous nonlinearities in most system responses, usually quite different from the outcome computed for the expected (best guess) future. Secondly, many decisions can be made iteratively in an adaptive management mode where new information is systematically incorporated into "mid-course" corrections through techniques that are as sophisticated as Bayesian updating of subjective probability distributions or as simple as excluding (or including) new possibilities.

In every case, the point is to make the best decision, where "best" is defined by the underlying evaluation criteria, given the available information and the information that is likely to become available as the future unfolds. In some cases, where decisions can be modified and adjusted easily, contingency-based rules designed to exploit future information can be the best choice. In others, where decisions made now involve investments with long lifetimes and/or reduce the set of feasible alternatives in the future, then the representations of uncertainty include representations of the distribution of future information. Lempert and Schlesinger (2000) offer an excellent example of how these considerations play out in the climate arena when profound uncertainties about the climate system and its interactions with the socioeconomic-political system (in the contexts of both drivers and impacts) are recognized.

4.6 Valuation Techniques under Uncertainty

Many of the decision-analytic approaches employed by analysts rely on the application of valuation techniques to market and nonmarket contexts. The idea is that relying exclusively on markets to provide estimates of value misses a wide range of other sources of human well-being that are not captured by markets per se. Bringing these sources to bear on the calculations that support various approaches to decision-making allows them to be informed by estimates that come closer to reflecting total economic value. In each case, uncertainty must be recognized and accom-

Table 4.1. Applicability of Decision Support Methods and Frameworks. Interpreting the *optimization, equity,* and *thresholds columns*: (*) designates direct applicability by design; (+) designates possible applicability with modification; and (−) designates weak but not impossible applicability with considerable effort. Interpreting the *uncertainty column*: (A) designates a method that has been modified to accommodate uncertainty and (E) designates a method that has been designed explicitly to handle uncertainty. Interpreting the *scale column*: (M) designates primary applicability at a micro scale; (N) designates applicability up to a national scale; (R) designates applicability to a regional or sector scale; (G) designates applicability up to a global scale; and (X) designates applicability at all scales. Interpreting the *domain column*: (M) designates primary application to mitigating the sources of stress; (A) designates primary application to adaptation; (B) indicates applicability to either mitigation or adaptation; (I) designates applicability to both mitigation and adaptation in an integrated way; and (X) designates applicability in all of the above.

Method	Optimization	Equity	Thresholds	Uncertainty	Scale	Domain
Cost–Benefit Analysis	+	+	−	A	X	B
Risk Assessment	+	+	*	E	X	X
Multicriteria Analysis	*	+	+	A	N and M	I
Precautionary Principle	+	+	*	E	X	X
Vulnerability Analysis	+	+	*	A	N and M	A

modated to a degree that is consistent with the needs of the decision-maker for precision. In some cases, uncertainty in the estimate of total economic value does not cloud the decision space because particular responses would be favored (or not) across the entire range (or at least most of it). In other cases, though, recognizing uncertainty in the estimates of economic value can lead to mixed and therefore contingent assessments. This section offers brief descriptions of how this is accomplished for some of the more popular methods.

4.6.1 Market-based Valuations

Techniques for estimating the values of goods and services that are derived from well-functioning markets are well established. Varian (2003) provides a concise description, and the intuitive underpinnings can be reviewed in Mansfield and Yohe (2004). Fundamentally, these techniques interpret demand curves (correlations between price and quantities willingly demanded) as marginal benefit schedules; that is, prices paid by consumers reflect the value that they place on the last unit demanded. They also interpret supply curves (correlations between price and quantities willingly supplied) as marginal cost schedules; that is, prices received by suppliers reflect the cost of producing the last unit delivered to the market. For any good, therefore, the area between these two curves from zero up to any specific quantity can be interpreted as a direct reflection of the net benefit achieved by society from the consumption of that quantity. It is the sum of "consumer surplus" (the amount that people would have been willing to pay for a given quantity if they had paid the marginal value of each unit rather than a single market clearing price) and "producer surplus" (the amount that firms receive in excess of the sum of the marginal cost of each successive unit).

Notwithstanding some technical details underlying this interpretation of net benefit (including the difference between using ordinary demand curves instead of compensated demand curves as the basis of benefit calculations), model uncertainty arises in these estimates because different market structures can produce different results. Estimates produced from a model based on the assumptions of perfect competition (for example, presuming that no actor in the market has power over the price actually charged by the market) can be dramatically different from estimates derived from a model that recognizes the game-theoretic strategic behavior of a limited number of suppliers who do have power over the price. So, too, can different assumptions about the degree to which market distortions and asymmetric information cause the specifications of the underlying determinants of demand and supply to deviate from efficiency (as opposed to adequacy and/or rights-based) norms. Even specifying the functional form of demand and supply schedules introduces model uncertainty.

Calibration uncertainty can also cloud market-based valuation estimates, since any empirical procedure will be able to explain only part of the variance in equilibrium prices and quantities even if it can handle pervasive identification problems and the aggregation of the preferences of a myriad of individual consumers and the marketing strategies of a collection of suppliers. In this case, though, paying attention to standard errors of parameter estimates and corresponding prediction errors can suggest an upper bound on uncertainty, but only given an underlying model specification. Prediction and projection uncertainty about the underlying determinants of demand and supply (prices of other goods, the distribution of income, the prices of inputs, and the pace of technological change, for example) and about how these drivers will evolve over time, can also cause trouble. Finally, contextual uncertainty can become particularly problematic. Issues about the persistence over time of existing distortions (wedges between actual marginal cost and benefit created by taxes, externalities, and other sources of omitted social cost, etc.) across an integrated economy must be raised, and valuation estimates will be critically sensitive to how these issues are resolved. Moreover, the degree to which valuation techniques and/or estimates are portable from one context to another depends on the degree to which underlying contextual structures are comparable.

4.6.2 Nonmarket Valuations

A variety of techniques have been developed to estimate the values of goods and services for which markets do not exist. Most are, nonetheless, firmly rooted in the market-based paradigm because they try to create "pseudo-demand curves" so that the calculus of consumer surplus just described can be applied directly. As a result, application of any of these techniques must begin by recognizing that all five sources of uncertainty can undermine confidence in the resulting value estimates—sometimes with a vengeance. In addition, each technique brings its own problems with consistency and potential bias to the table, so results in this area need careful interpretation if they are not to be misused. While directed specifically at contingent valuation techniques, this general concern was underscored by Diamond and Hausman

(1994) with the rhetorical question "Is some number better than no number?" Smith (1993) offers a solid appraisal of how to interpret nonmarket valuations of environmental goods; Section IV of Cropper and Oates (1992) similarly provides a presentation of the theoretical technicalities.

4.6.2.1 Hedonic Methods

Hedonic valuation methods are based on the notion that the value of certain properties for which there are no markets can be detected indirectly by calibrating their roles in supporting the prices of other goods and services. More specifically, this expectation holds that many of the underlying determinants of the demand for marketed goods like real estate or agricultural property might include variables that reflect things like environmental quality or climatic conditions. If the associations between the prices of marketed goods and these underlying characteristics can be quantified, then the value of these characteristics can be assessed indirectly by tracking changes in observed market prices.

Hedonic techniques have been employed in assessing the labor-market wage implications of negative characteristics of various locations (such as crime, pollution, congestion, extreme climate, and so on) as well as the benefits of positive attributes (educational opportunity, fine arts, mild climate, or sports facilities, for instance). Ridker and Henning (1967) offered a seminal analysis for sulfur particulates; Brookshire et al. (1982) and Bloomquist et al. (1988) provide more recent but well-respected estimates for nitrous oxide and particulate exposure. Mendelsohn et al. (1994) also applied hedonic techniques to assessments of the effects of long-term climate change in the context of maximally efficient adaptation by the agricultural sector across the United States. In so doing, they developed a controversial methodology that Mendelsohn and others have used in many other contexts.

The controversy over using hedonic techniques cannot be attributed entirely to the ravages of uncertainty, but prediction, projection, and contextual uncertainties certainly play a role. Application of hedonic techniques to climate change assumes, at least implicitly, that individuals at all locations have already adapted optimally to the current climatic conditions and that these adaptations can follow as long as climatic change pushes these conditions into new geographic areas. Recognition of this assumption is a source of concern when it was applied to climate change, and not simply because it contributes to model uncertainty or fails to report calibration difficulties. Results from hedonic applications to the climate arena depend critically upon apparently contradictory assumptions of how human systems will respond to change over time. More specifically, this application of hedonic techniques produces interpretable results only if the relative prices of market goods as well as the other determinants of demand do not change (a specific truncation of projection uncertainty). At the same time, however, the determinants of capacities that supported optimal (market-reflected) adaptations at the initial locations must change significantly; indeed, they must migrate completely to new locations (an equally rigid truncation of contextual uncertainty). Moreover, when the set of external stresses is expanded beyond the climate realm, second-best solutions create problems. It is impossible to understand precisely, in multi-stress contexts, what motivated the observed structure and so it is difficult to conclude with high confidence that an environmental stress was the cause.

Travel cost methods expand the hedonic approach, but their point is also to create a "pseudo demand curve"; see Brown and Mendelsohn (1984) or Kahn (1997) for a description of methodological details. In these exercises, relationships between how much individuals pay to travel to a particular location (like a lake or beach) and the number of times per year that they would be willing to make the trip are estimated. Some applications are based on the average number of trips (per capita) by residences of specific zones or regions; these require less data, but they produce only aggregate estimates. Other applications focus on individuals who actually visit, or could have visited, the study location. These provide more detailed information, but they are data intensive and subject to selection bias. Moreover, all applications must confront questions about what to include in their "willingness to pay" (for example, should they include the value of the time involved in traveling to the location, or just actual expenditures?). Once the data are collected, however, they can be used to produce "market" demand curves for specific populations for which travel costs represent the price of a nonmarket good. Environmental qualities anchor these demand curves, just as before, so changes in environmental parameters can be expected to shift demand. As in the hedonic construction, therefore, corresponding adjustments in price can be interpreted as estimates of the value of those changes.

4.6.2.2 Contingent Valuation

Contingent valuation techniques have been developed over time to produce valuations that are not tied to use values that can be observed from market interactions; but they, too, are designed to build demand curves where they do not exist. Mitchell and Carson (1989) were among the first to recognize the need to have some reflection of relative prices if natural resources were to be managed effectively. They invented CV, and thereby started the debate over whether or not survey results could be trusted. Hanemann (1994) offers a thorough description of how CV can be applied to a variety of circumstances; Portney (1994) as well as Diamond and Hausman (1994) provide coverage of the debate. For purposes of a cursory review, it is sufficient to emphasize that non-use values refer to the increases in individual utility generated by the satisfaction of knowing that something exists. The CV approach asks people to offer monetary estimates of those utility gains. A CV study must, therefore, describe the outcome to be valued, describe a (hypothetical) method of payment, and design a method of elicitation. Results across a large number of people are then summarized empirically as a demand curve and scaled-up so that it is representative of demand across a relevant population. At that point, all of the market-based valuation techniques described above can be applied directly.

The devil is in the details, of course, and so careful design is essential. Recent work suggests, for example, that practitioners do more than ask what something is worth; they create elicitation vehicles that carefully define the context of the valuation exercise and include questions whose answers allow some evaluation of internal bias; see Bateman and Willis (1995) and Bateman (2002) for examples and Arrow and Solow (1993) for some practical guidance. All these contributors to a growing literature confirm that the description of context must be constructed in a way that does not create biased reactions of respondents who might know nothing about the subject. The method of payment must be clearly understood to alleviate, at least to some degree, the concern that respondents never fully comprehend the method unless they see real money leaving their pockets.

One issue of particular importance is derived from the widely accepted result that the willingness to pay for an environmental improvement is generally smaller than the willingness to accept (compensation) to forego that improvement is perhaps the critical element of a long list of possible biases and design problems. Empirical support for this result can be found in Hammack and

Brown (1974), Rowe et al. (1980), and Knetsch and Sinden (1984). These authors derive theoretical support directly from diminishing marginal utility—a building block assumption of neoclassical economic theory. Other researchers, like Coursey et al. (1987), suggest that observed differences (for small payments in one direction or the other) are simple reflections of the fact that most people are more familiar with buying something than they are with selling it. In either case, these differences mean that the consumers of CV elicitations must do more than read the numbers; they must understand the entire elicitation process even before issues of uncertainty are raised. Moreover, prediction, projection, and contextual uncertainties can be particularly troublesome for contingent valuation, since individual responses to even a well-designed elicitation are extraordinarily context specific.

Despite all of these problems, Rothman (2000) has argued that valuation methods like contingent valuation can inform decision-makers in their consideration of various responses. The key lies in careful recognition of the level of precision required to support a particular decision. If, for example, a decision to implement a particular response were based on a cost-benefit calculation, then the issues raised here would be troublesome only if the estimates of net benefits derived from some analyses (say, from studies employing "willingness to accept" calculations) were positive while other estimates (for example, those derived from studies employing "willingness to pay" calculations) were negative.

4.6.3 Cross-cutting Issues

At least two cross-cutting concerns about all valuation techniques should be raised. First, all of the techniques noted above use the net-benefit interpretation of demand and supply structures to derive their fundamental measures of value; but this interpretation only makes sense in a utility (welfare) context when people pursue their own, well-defined best interest. As a result, every method is rooted firmly in the assumption of economic rationality (that people consume goods up to the point where the increase in their utility created by spending their last dollar on one good is the same as it would be if they spent that dollar on another good). But do people actually behave that way?

Second, each measure is also fundamentally derived from theories that describe individual decisions and produce individual demand curves for marketed goods or "pseudo-demand" curves for nonmarket goods. Scaling these representations of individual decisions up to structures that claim to represent collective behavior over entire markets, communities, or nations adds aggregation to the list of major sources of uncertainty; and it means that defining aggregation techniques adds another level of subjectivity to the results. Neither of these concerns is insurmountable, but both suggest that care needs to be taken in interpreting and applying valuation results in the decision-making process.

4.7 Synthesizing Political, Economic, and Social Factors in the Context of Uncertainty

The previous sections have offered reviews of valuation and decision-support tools that have been modified to accommodate uncertainty. A complete assessment of the ability to respond to external stresses cannot, however, stop with representations of the likelihood of success or the range of possible outcomes. If it is to be at all useful, processes that amplify our understanding of the cascade of uncertainty noted above need to feed into a structure where the factors that determine the feasibility of various response options (see Chapter 3) can be explored. In short, an integrated structure needs to be constructed so that synthetic analyses of response options can be conducted.

This section builds on the IPCC (2001a) notion of adaptive capacity and its determinants as identified by Yohe and Tol (2002) to suggest how this integration might be accomplished. While some have used this structure to produce mechanical indices of vulnerability based on the generic adaptive capacities of entire systems and to evaluate the feasibility that specific responses will accomplish their goal based on their specific adaptive capacities, it is perhaps best viewed as one way of organizing one's thoughts in an effort to try to understand why some responses work in some circumstances and not others.

4.7.1 Matching Political, Economic, and Social Factors to the Determinants of Responsive Capacity

Working from the IPCC (2001a) perspective that the vulnerability of any system to external stress is a function of exposure and sensitivity and that either or both of these manifestations can be influenced by its adaptive capacity, Yohe and Tol (2002) list seven determinants of (specific) adaptive capacity that are required to support any given response option:

(1) the availability of resources and their distribution across the population;
(2) the structure of critical institutions, the derivative allocation of decision-making authority, and the decision criteria that would be employed;
(3) the stock of human capital including education and personal security;
(4) the stock of social capital including the definition of property rights;
(5) the system's access to risk-spreading processes;
(6) the ability of decision-makers to manage information, the processes by which these decision-makers determine which information is credible, and the credibility of the decision-makers themselves; and
(7) the public's perceived attribution of the source of stress and the significance of exposure to its local manifestations.

Thinking of these determinants as the underlying components that support a system's ability to respond to a set of external stresses (that is, its *responsive capacity*) makes it clear that they simply add some detail to the critical factors for assessing responses identified in Chapter 3. Determinants 2, 6 and 7, for example, add some texture to the political factors described there. Determinants 1 and 5 reflect economic considerations, but also connect with determinants 3 and 4 to portray the significant role played by social factors.

4.7.2 A "Weakest Link" Approach to Evaluating Capacity

Taking the conceptual approach implied by the list of determinants to something applicable to systematic evaluation of various responses across site-specific and path-dependent contexts relies basically on the notion that a system's responsive capacity is fundamentally determined by the weakest link—the underlying determinant that provides the least support for the available responses in its ability to cope with variability and change in local environmental conditions. This hypothesis clearly requires some justification. Yohe and Tol (2002) reported some suggestive empirical results from international comparisons, but subsequent literature has been more persuasive. A growing body of literature has, for example, reached similar conclusions regarding income inequality and mortality (see, for example, Lynch et al. 2000;

Kaplan et al. 1996; Ross et al. 2000). Even more recently, McGuire (2002) looked for statistically significant explanations for variability in infant mortality across developing countries. Yohe and Ebi (2004) also noted a strong match between the prerequisites for prevention in the public health literature (where a weakest link hypothesis is well established) and the determinants of responsive capacity.

A review of economic literature also produces some supporting evidence. Rozelle and Swinnen (2004), for example, looked at transition countries across central Europe and the former Soviet Union to observe that countries which grew steadily a decade or more after their reforms have managed to (among other things) reform property rights *and* to create institutions that facilitate exchange and develop an environment within which contracts can be enforced and new firms can enter. Order and timing did not matter, but success depended on meeting all of these underlying objectives. Winters et al. (2004) similarly reviewed a long literature to conclude that the ability of trade liberalization to reduce poverty depends on the existence and stability of markets, the ability of actors to handle changes in risk, access to technology and resources, competent and honest government, *and* policies that promote conflict resolution and promote human capital accumulation.

4.8 The Challenge of Uncertainty: Creating, Communicating, and Reading Confidence Statements

Effective response options are the products of a process whose success is critically dependent upon an understanding of the key issue or problem of concern, the design of appropriate actions to address this issue, the effective implementation of the selected actions, and the honest monitoring and evaluation of outcomes to ensure that the actions achieve their goals without unintended consequences.

This chapter focused on the uncertainty inherent in implementation of responses, on the uncertainty inherent in our understanding of how ecosystems work, and on the interaction of these uncertainties with the tools that analysts employ to evaluate and choose between these response options. It began with a brief taxonomy of the sources of uncertainty, then looked back to Chapter 3 to see how the various critical factors of feasibility could be viewed as an anthropogenic source of uncertainty. From case to case, application of any of the decision-analytic frameworks and the evaluation methods reviewed earlier in this chapter will confront many if not all of these sources. Still, experience suggests that some sources of uncertainty can be expected to be more important for one framework than another. The example drawn from the climate change literature above was presented only to suggest how considerations of systems uncertainty might be integrated with the underlying factors that enable or constrain specific response options. As such, it is best viewed as a representation of the thought processes that MA authors conducted as they evaluated the confidence with which they could offer their conclusions.

The success or failure of present or past approaches to solving an issue is a function of the compatibility of the solution methods (including demand for services) and ecosystem dynamics. In a setting where demand is low relative to available system services (even at times of extreme natural stress), there will be "successful" management—but not necessarily because of a good management scheme. When the ecosystem is at the brink of state-change, even the wisest decision-making paradigms may be insufficient to prevent the inevitable. One challenge is recognizing the limits of our understanding of the processes underlying ecosystem dynamics and resilience. Another challenge is ensuring that management strategies for maximizing service extraction are not at the cost of system simplification and loss of effective resilience.

The most obvious opportunity is to learn from the mistakes and successes of the past. Each can be considered an experiment in how to frame and solve an ecosystem service management challenge. Systematic analyses of implemented response options are required to capitalize on this body of information and thereby to learn which factors enhanced the probability of success and which led to failures. Such learning holds the promise of teaching us how resources can be managed when baseline conditions are relatively stable and extra-scalar effects are relatively small before confronting the complication of significant shifts in baselines and simultaneous multiple stresses on ecosystems.

There are many possible response options to address a particular problem. Whether or not they can be effectively implemented depends on factors such as political feasibility; technological, economic, and social issues; and the capacity for governance. A better understanding of the process of the design and implementation of successful response options in the context of these factors is needed, including how barriers to implementation were overcome. Lessons learned could then be applied to other situations to reduce the negative consequences and take advantage of the opportunities that arise in the context of ecosystem management.

Processes and institutions need to be established to facilitate a learning-by-doing approach that includes monitoring of implemented response options and systematic evaluation of their results. These evaluations can be employed to investigate why some response options were effective while others were not. This is only possible if the original design included the establishment of necessary measures to collect the information required for a post hoc examination. It follows that the design of a response option should include actions to determine the effectiveness of the option in addressing the issue of concern and, accordingly, plans for collecting the required data—a process designed to keep track of the progress of implementation of a response option and its various components in relation to the goals established. It also follows that improved understanding of what works where will improve only if integrating analyses take careful account of spatial and temporal diversity.

Assessments of the sort presented in this volume are, of course, fundamentally the products of monitoring and evaluation exercises—nominally of our ability to manage ecosystem responses, but actually of our ability to understand exactly what is going on. This assessment is perhaps the most comprehensive and visible exercise of this sort, but its success is not guaranteed. The assessors who contributed to this work will only advance our long-term understanding of how ecosystem services support human well-being even as humans exert enormous stress on their potential longevity in an uncertain world if they are honest in identifying what is well established (*high certainty* conclusions), what is established but supported by incomplete analysis (*medium certainty*), what is subject to competing explanations (*low certainty*), what is entirely speculative (*very low certainty*), and what is entirely beyond the scope of our understanding (*severe gaps in knowledge*).

It is in the subsequent exploration of why the quality of our knowledge is so inconsistent that the thought-process described earlier in this chapter might be most valuable. Chapter 15, for example, will offer the conclusion that "integrated responses are gaining in importance in both developing and developed countries, albeit with mixed results." Since this point is based on fewer than 20 studies, though, it can be advanced as "established but incomplete" according to the guidelines for conveying confidence. Nevertheless, a systematic evaluation of these studies using a common thought template, can perhaps provide some insight into why the results of integrated responses have been so mixed—working in some site-specific and path-dependent contexts but not in others.

The burden for communicating these findings does not lie exclusively with the authors. Readers must also be honest in their assimilation of the assessment. They cannot, for example, seize on the negative studies of integrated responses and ignore the positive studies to support a general opposition to their implementation. Nor can they focus exclusively on the positive studies to advocate integrated responses ubiquitously. They must, instead, comprehend the implications of the full range of uncertainty described by the assessors of the full set of studies (as limited and as contradictory as they might be) and they must accept the various degrees of confidence reported by the authors as they make up their own minds. (See Box 4.8 for a discussion of one approach to determining how to deal with uncertain estimates.)

BOX 4.8

Defining Hotspots in an Uncertain World

"Biodiversity conservation" has uncertainty at its foundations—uncertainty not only about the identity and location of the many species (or other elements) that make up global biodiversity (*scale uncertainty*), but also about their values to humanity (*model and contextual uncertainty*). Sometimes estimates of possible future values simply are equated with our measures of variation (*calibration uncertainty*). But even here, surrogates or proxy information (for example, indicator species) are required because all components of biodiversity cannot be assessed in all places with any certainty. Response strategies (see Chapter 5) use surrogate information in ways that sometimes introduce new uncertainties about their adequacy (*prediction and projection uncertainty*).

For example, the advocacy of 25 global hotspots is one high-profile biodiversity response strategy; it highlights the inevitable uncertainties about surrogates. Myers (2003), in his recent review of this approach, notes differences of opinion about whether the identification of hotspots based on major taxa can be representative of patterns that would be found for other components of biodiversity. (See Chapter 4.) Myers addresses the concern that surrogacy is not "proven" but only "assumed" in this way: ". . . when will our research be able to 'prove' much about the 9.7 million invertebrates out of a putative planetary total of 10 million species, given that only around 1 million of them have been identified thus far?" An assessment of this literature might therefore suggest that the identification of "hotspots" is "established but incomplete."

References

Adger, W.N. and C. Luttrell, 2000: Property rights and the utilization of wetlands, *Ecological Economics,* **35,** pp. 78–91.

Adger, W.N., 1999: Social vulnerability to climate change and extremes in coastal Vietnam, *World Development,* **27,** pp. 249–69.

Agrawal, A., 2001: Common property institutions and sustainable governance of resources, *World Development,* **29,** pp. 1649–72.

Arrow, K. and R. Lind, 1970: Uncertainty and the evaluation of public investment decisions, *American Economic Review,* **40,** pp. 364–78.

Arrow, K. and R. Solow, 1993: Report of the NOAA panel on contingent valuation, *Federal Register,* **58,** pp. 4602–14.

Aylward, B., J. Berkhoff, C. Green, P. Gutman, A. Lagman, et al., 2001: *Financial, Economic and Distributional Analysis, Thematic Review III.1,* The World Commission on Dams, Cape Town, South Africa. Available at www.dams.org.

Bateman, I.J. and K.G. Willis (eds.), 1995: *Valuing Environmental Preferences,* Oxford University Press, Oxford, UK.

Bateman, I.J., 2002: *Economic Valuation with Stated Preference Techniques: A Manual,* Edward Elgar, Cheltenham, UK.

Berkes, F., 2002: Cross-scale institutional linkages for commons management: Perspectives from the bottom up. In: *The Drama of the Commons,* E. Ostrom, T. Dietz, N. Dolsak, P. C. Stern, S. Stonich, et al. (eds.), National Academy Press, Washington, DC, pp. 293–321.

Berkhout, F., and J. Hertin, 2002: Foresight future scenarios: developing and applying a participative strategic planning tool, *Greener Management International,* **37,** pp. 37–52.

Berkhout, F., J. Hertin, and A. Jordan, 2002: Socio-economic futures in climate change impact assessment: Using scenarios as "learning machines," *Global Environmental Change,* **12,** pp. 83–95.

Bernard, S.M. and K.L. Ebi, 2001: Comments on the process and product of the health impacts assessment component of the United States national assessment of the potential consequences of climate variability and change, *Environmental Health Perspectives,* **109(Suppl. 2),** pp. 177–84.

Blaikie, P., T. Cannon, I. Davis, and B. Wisner, 1994: *At Risk: People's Vulnerability and Disasters,* Routledge, London, UK.

Bloomquist, G.C., M.C. Berger, and J.P. Hoehn, 1988: New estimates of quality of life in urban areas, *American Economic Review,* **78,** pp. 89–107.

Boehmer-Christiansen, S., 1994: The precautionary principle in Germany: Enabling government. In: *Interpreting the Precautionary Principle,* T. O'Riordan and J. Cameron (eds.), Cameron and May, London, UK.

Bromley, D.W., 1991: *Environment and Economy: Property Rights and Public Policy.* Blackwell, Oxford, UK.

Bromley, D.W., 1999: *Sustaining Development: Environmental Resources in Developing Countries,* Elgar, Cheltenham, UK.

Brooks, N., 2003: Vulnerability, risk and adaptation: A conceptual framework, Tyndall Centre for Climate Change Research, Working paper 38.

Brookshire, D.S., M.A. Thayer, W.W. Schultze, and R.L. d'Arge, 1982: Valuing public goods: A comparison of survey and hedonic approaches, *American Economic Review,* **72,** pp. 165–77.

Brown, G. and R. Mendelsohn, 1984: The hedonic travel cost method, *Review of Economics and Statistics,* **66,** pp. 427–33.

Brown, K., 2002: Innovations for conservation and development, *Geographical Journal,* **168,** pp. 6–17.

Brown, K., E.L. Tompkins, and W. N. Adger, 2002: *Making Waves: Integrating Coastal Conservation and Development,* Earthscan, London, UK.

CEC (Commission of the European Community), 2000: *Communication from the Commission on the Precautionary Principle,* CEC, Brussels, Belgium.

Coursey, D., J.L. Hovis, and W.D. Schulze, 1987: The disparity between willingness to accept and willingness to pay measures of value, *Quarterly Journal of Economics,* **102,** pp. 679–90.

Cropper, M. and W. Oates, 1992: Environmental economics: A survey, *Journal of Economic Literature,* **30,** pp. 675–740.

Diamond, P. and J. Hausman, 1994: Contingent valuation: Is some number better than no number? *Journal of Economic Perspectives,* **8,** pp. 45–64.

Dixit, A.K., and R.S. Pindyck, 1994: *Investment under Uncertainty,* Princeton University Press, Princeton, NJ.

Downing, T.E., A. Patwardhan, R. Klein, E. Mukhala L., Stephen, et al., 2003: *Vulnerability Assessment for Climate Adaptation* (Adaptation Policy Framework Technical Paper 3), United Nations Development Programme, New York, NY.

Ebi, K.L., D. Mills, and J. Smith, 2005: A case study of unintended consequences: Arsenic in drinking water in Bangladesh. In: *Integration of Public Health with Adaptation to Climate Change: Lessons Learned and New Directions,* K.L. Ebi, Smith, J., Burton, I. (eds.). Taylor & Francis, London, 352 pp.

EEA (European Environment Agency), 2001: *Late lessons for early warnings: the precautionary principle, 1896–2000,* EEA, Copenhagen, Denmark.

Faith, D., 2002: Cost effective biodiversity planning, *Science,* **293,** pp. 193–94.

Ghebreyesus, T.A, M. Haile, K.H. Witten, A. Getachew, A.M. Yohannes, et al., 1999: Incidence of malaria among children living near dams in northern Ethiopia: Community based incidence survey, *British Medical Journal,* **319,** pp. 663–6.

Goulder, L.H. and D. Kennedy, 1997: Valuing ecosystem services: Philosophical bases and empirical methods. In: *Nature's Services: Societal Dependence on Natural Ecosystems,* G.C. Daily (ed.), Island Press, Washington, DC, pp. 23–47.

Hammack, J. and G. Brown, 1974: *Waterfowl and Wetlands: Toward Bioeconomic Analysis,* Johns Hopkins Press for Resources for the Future, Baltimore, MD.

Hanemann, M., 1994: Valuing the environment through contingent valuation, *Journal of Economic Perspectives,* **8,** pp. 19–43.

Hanley, N., J. Shogren, and B. White, 1997: *Environmental Economics in Theory and Practice,* Oxford University Press, New York, NY.

Holling, C.S. and G.K. Meffe 1996: Command and control and the pathology of natural resource management, *Conservation Biology,* **10,** pp. 328–37.

Holling, C.S., D.W. Schindler, B.W. Walker, and J. Roughgarden, 1995: Biodiversity in the functioning of ecosystems: An ecological synthesis. In: *Biodiversity Loss: Economic and Ecological Issues,* C. Perrings, K.G. Mäler, C. Folke, C.S. Holling, and B.O. Jansson (eds.), Cambridge University Press, Cambridge, UK, pp. 44–83.

Homewood, K., E.F. Lambin, E. Coast, A. Kariuk, I. Kikula, et al., 2001: Long-term changes in Serengeti-Mara wildebeest and land cover: Pastoralism, population or policies? *Proceedings of the National Academy of Sciences US,* **98,** pp. 12544–9.

Hoque, B.A., A.A. Mahmood, M. Quadiruzzaman, F. Khan, S.A. Ahmed, et al., 2000: Recommendations for water supply in arsenic mitigation: A case study from Bangladesh, *Public Health,* **114,** pp. 488–94.

IPCC (Intergovernmental Panel on Climate Change), 1996: *Climate Change 1995: Economic and Social Dimensions of Climate Change, Contribution of Working Group III to the Second Scientific Assessment of the Intergovernmental Panel on Climate Change,* Cambridge University Press, Cambridge, UK.

IPCC, 2000: *Emission Scenarios,* Cambridge University Press, Cambridge, UK.

IPCC, 2001: *Climate Change 2000: Impacts, Adaptation and Vulnerability,* Cambridge University Press, Cambridge, UK.

Kahn, J., 1997: *The Economic Approach to Environmental and Natural Resources,* Dryden Press, New York, NY, 515 pp.

Kaplan, G.A., E.R. Pamuk, J.W. Lynch, R.D. Cohen, and J.L. Balfour, 1996: Inequality in income and mortality in the United States: Analysis of mortality and potential pathways, *British Medical Journal,* **312,** pp. 999–1003.

Kasperson, J. and R. Kasperson (eds.), 2001: *Global Environmental Risk,* United Nations University Press/EarthScan, London, UK.

Kates, R.W. and W.C. Clark, 1996: Environmental surprise: Expecting the unexpected, *Environment,* **38,** pp. 6–34.

Kheifets, L. I., G.L. Hester, and G.L. Banerjee, 2001: The precautionary principle and EMF: Implementation and evaluation, *Journal of Risk Research,* **4,** pp. 113–25.

Knetsch, J.L. and J.A. Sinden, 1984: Willingness to pay and compensation demanded: Experimental evidence of an unexpected disparity in measures of value, *Quarterly Review of Economics,* **99,** pp. 507–21.

Koundouri, P., P. Pashardes, T.M. Swanson, A. Xepapadeas: 2003: *Economics of Water Management in Developing Countries: Problems, Principles and Policies,* Edward Elgar, Cheltenham, UK.

Kriebel D., J. Tickner, P. Epstein, J. Lemons, R. Levins, et al., 2001: The precautionary principle in environmental science, *Environmental Health Perspectives,* **109,** pp. 871–76.

Leichenko, R. and K. O'Brien, 2002: The dynamics of rural vulnerability to global change, *Mitigation and Adaptation Strategies for Global Change,* **7,** pp. 1–18.

Lempert, R. and M.E. Schlesinger, 2000: Robust strategies for abating climate change: An editorial essay, *Climatic Change,* **45,** pp. 387–401.

Leung, B., D.M. Lodge, D. Finnoff, J.F. Shogren, M.A. Lewis, and G. Lamberti, 2002: An ounce of prevention or a pound of cure: Bioeconomic risk analysis of invasive species, *Proceedings of the Royal Society London Series B,* **269,** pp. 2407–13.

Lowenstein, G. 2002: Time discounting and time preference: A critical review, *Journal of Economic Literature,* **40,** pp. 351–400.

Ludwig, D., M. Mangel, and B. Haddad, 2001: Ecology, conservation, and public policy, *Annual Review of Ecology and Systematics,* **32,** pp. 481–517.

Lynch, J.W., G.D. Smith, G.A. Kaplan, and J.S. House, 2000: Income inequality and mortality: Importance to health of individual income, psychosocial environment, or material conditions, *British Medical Journal,* **320,** pp. 1200–4.

MA (Millennium Ecosystem Assessment), 2003: *Ecosystems and Human Wellbeing: A Framework for Assessment,* Island Press, Washington, DC.

Mansfield, E. and G. Yohe, 2004: *Microeconomics* (11th ed.), W.W. Norton, New York, NY, 678 pp.

Marchant, G.E. 2003: From general policy to legal rule: Aspirations and imitations of the precautionary principle, *Environmental Health Perspectives,* **111,** pp. 1799–803.

McGuire, J., 2002: Democracy, social provisioning, and under-5 mortality: A cross-national analysis, Wesleyan University working paper, Department of Government, Middletown, CT.

Mendelsohn, R., W. Nordhaus, and D. Shaw, 1994: The impact of global warming on agriculture: A Ricardian approach, *American Economic Review,* **84,** pp. 753–71.

Mitchell, R.C. and R.T. Carson, 1989: *Using Surveys to Value Public Goods: The Contingent Valuation Method,* Resources for the Future, Washington, DC.

Morgan, M.G. and M. Henrion, 1990: *Uncertainty: A Guide to Dealing with Uncertainty in Quantitative Risk and Policy Analysis,* Cambridge University Press, Cambridge, UK.

Myers, N. 2003: Economic and environmental benefits of biodiversity, *Bioscience,* **53,** pp. 916–17.

National Research Council, CoRAoHAP, Board on Environmental Studies and Toxicology, Commission on Life Sciences, 1994: *Science and Judgment in Risk Assessment,* National Academy Press, Washington, DC.

National Research Council, CoRC, 1996: *Understanding Risk: Informing Decisions in a Democratic Society,* National Academy Press, Washington, DC.

O'Connor, M., 2000: Pathways for environmental evaluation: A walk in the (Hanging) Gardens of Babylon, *Ecological Economics,* **34,** pp. 175–94.

Ogura, S. and G. Yohe, 1977: The complementarity of public and private capital and the optimal rate of return to government investment, *Quarterly Review of Economics,* **46,** pp. 651–62.

Pontius, R.G., Jr. and M.L. Cheuk: A generalized confusion matrix for comparing soft-classified maps at multiple resolutions, *International Journal of Remote Sensing.* In press.

Portney, P., 1994: The contingent valuation debate: Why economists should care, *Journal of Economic Perspectives,* **8,** pp. 3–17.

Porto, M., R.L. Porto and Luiz Gabriel Azevedo: 1999: A participatory approach to watershed management: The Brazilian system, *Journal of the American Water Resources Association,* **35,** pp. 675–84.

Potting, J. and J. Bakkes (eds.): *The GEO-3 Scenarios 2002–2032: Quantification and Analysis of Environmental Impacts,* RIVM, Bilthoven, The Netherlands. In press.

Presidential/Congressional Commission on Risk Assessment and Risk Management, 1997: *Risk Assessment and Risk Management in Regulatory Decision-Making,* Washington, DC.

Raskin, P., T. Banuri, G. Gallopín, P. Gutman, A. Hammond, et al., 2002: *Great Transition: The Promise and Lure of the Times Ahead,* Tellus Institute, Boston, MA.

Ridker, R.G. and Henning, J.A., 1967: The determination of residential property values with special reference to air pollution, *Review of Economics and Statistics,* **49,** pp. 246–57.

Robinson, J., 2003: Future subjunctive: Backcasting as social learning, *Futures,* **35,** pp. 839–56.

Ross, N.A., M.C. Wolfson, J.R. Dunn, J-M. Berthelot, G.A. Kaplan, et al., 2000: Relation between income inequality and mortality in Canada and in the United States: Cross sectional assessment using census data and vital statistics, *British Medical Journal,* **320,** pp. 898–902.

Rothman, D., 2000: Measuring environmental values and environmental impacts: Going from the local to the global, *Climatic Change,* **44,** pp. 351–76.

Rowe, R.D., R.C. d'Arge, and D.S. Brookshire, 1980: An experiment on economic value of visibility, *Journal of Environmental Economics and Management,* **7,** pp. 1–9.

Rozelle, S. and J.F.M. Swinnen, 2004: Success and failure of reform: Insights from the transition of agriculture, *Journal of Economic Literature,* **42,** pp. 433–58.

Save the Children Fund, 1997: *RiskMap 2.1: A User's Guide,* London, UK.

Schneider, S.H., 1983: CO_2, climate and society: A brief overview. In: *Social Science Research and Climate Change: An Interdisciplinary Appraisal,* R.S. Chen, E. Boulding, and S.H. Schneider (eds.), D. Reidel, Boston, MA, pp. 9–15.

Schwartz, P., 1996: *The Art of the Long View: Planning for the Future in an Uncertain World,* Currency Press, New York, NY.

Seaman, J., 2000: Making exchange entitlements operational: The food economy approach to famine prediction and the RiskMap computer program, *Disasters,* **24,** pp. 133–52.

Sinclair, A.R.E., A.R. Mduma, Simon, and P. Arcese, 2002: Protected areas as biodiversity benchmarks for human impact: Agriculture and the Serengeti avifauna, *Proceedings of the Royal Society London Series B,* **269,** pp. 2407–13.

Singleton, S., 1998: *Constructing Cooperation: The Evolution of Institutions of Comanagement,* University of Michigan Press, Ann Arbor. MI.

Slovic, P., B. Fischoff, and S. Lichtenstein, 1979: Rating the Risks, *Environment,* **21,** pp. 14–39.

Smit, B., I. Burton, R.J.T. Klein, and J. Wandel, 2000: An anatomy of adaptation to climate change and variability, *Climatic Change,* **45,** pp. 223–51.

Smith, K., 1993: Nonmarket valuation of environmental resources: An interpretive appraisal, *Land Economics,* **69,** pp. 1–26.

Stephen, L. and T.E. Downing, 2001: Getting the scale right: A comparison of analytical methods for vulnerability assessment and household level targeting, *Disasters,* **25,** pp. 113–35.

Stephen, L., 2003: *Vulnerability and Food Insecurity in Ethiopia: Forging the Links Between Global Policies, National Strategies, and Local Socio-spatial Analyses,* PhD Dissertation, University of Oxford, Oxford, UK.

Strzepek, K., D. Yates, G. Yohe, R.J.S. Tol, and N. Mader, 2001: Constructing "not-implausible" climate and economic scenarios for Egypt, *Integrated Assessment,* **2,** pp. 139–57.

Tamburlini, G. and K. Ebi, 2002: Searching for evidence, dealing with uncertainties and promoting participatory risk management. In: *Children's Health and Environment: A Review of Evidence,* G. Tamburlini, O.S. von Ehrenstein, and Bertollini (eds.), European Environmental Agency, Copenhagen, Denmark, 2002.

Tompkins, E., W.N. Adger, and K. Brown, 2002: Institutional networks for inclusive coastal zone management in Trinidad and Tobago, *Environment and Planning A,* **34,** pp. 1095–111.

Tol, R.S.J., 2003: Is the uncertainty about climate change too large for expected cost-benefit analysis? *Climatic Change,* **56,** pp. 265–89.

Toth, F. and M. Mwandosya, (Convening lead authors), 2001: Decision-making frameworks, Chapter 10, *Climate Change 2001: Mitigation,* Cambridge University Press, Cambridge, UK.

Turner, II, B., R. Kasperson, P. Matson, J. McCarthy, R. Corell, et al., 2003a, A framework for vulnerability analysis in sustainability science, *Proceedings of the National Academy of Sciences US,* **100,** pp. 8074–79.

Turner, II, B., R. Kasperson, P. Matson, J. McCarthy, R. Corell, et al., 2003b: Illustrating the coupled human–environment system for vulnerability analysis: Three case studies, *Proceedings of the National Academy of Sciences US,* **100,** pp. 8080–85.

UN (United Nations), 1993: *Agenda 21: The UN Programme of Action from Rio,* United Nations, New York, NY.

UNEP and African Ministerial Conference on Environment, 2002: *African Environmental Outlook: Past Present and Future Perspectives,* Earthprint Limited, Hertfordshire, UK.

UNEP/GRID-Arendal, 2003: Project of risk evaluation, vulnerability, information and early warning (preview) [online]. Available at http://www.grid.unep.ch/activities/earlywarning/preview.

Varian, H., 2003: *Microeconomic Analysis,* W.W. Norton, New York, NY, 635 pp.

Weitzman, M.L., 1974: Prices versus quantities, *Review of Economic Studies,* **41,** pp. 50–65.

Weitzman, M.L., 1998: Why the far-distant future should be discounted at its lowest possible rate, *Journal of Environmental Economics and Management,* **36,** pp. 201–08.

WHO EUR (World Health Organization Regional Office for Europe), 1999: *Access to Information, Public Participation and Access to Justice in Environment and Health Matters,* document EUR/ICP/EHCO 02 02 05/12, WHO EUR, Copenhagen, Denmark.

Winters, L.A., N. McCulloch, and A. McKay, 2004. Trade liberalization and poverty: The evidence so far, *Journal of Economic Literature,* **42,** pp. 72–115.

Yohe, G and K. Strzepek, 2004: Climate change and water resource assessment in South Asia: Addressing uncertainties. In: *Climate Change and Water Resources in South Asia,* M.M.Q. Mirza (ed.), Taylor and Francis, Leiden, The Netherlands. In press.

Yohe, G. and K. Ebi, 2004: Approaching adaptation: Parallels and contrasts between the climate and health communities. In: *Integration of Public Health with Adaptation to Climate Change: Lessons Learned and New Directions,* K. Ebi, J. Smith and I. Burton (eds.), Taylor and Francis, Leiden, The Netherlands.

Yohe, G. and Tol, R., 2002: Indicators for social and economic coping capacity: Moving toward a working definition of adaptive capacity, *Global Environmental Change,* **12,** pp. 25–40.

Yohe, G., K. Strzepek, T. Pau, and C. Yohe, 2003: Assessing vulnerability in the context of changing socio-economic conditions: A study of Egypt. In: *Climate Change, Adaptive Capacity and Development,* J. Smith, R. Klein, and S. Huq (eds.), Imperial College Press, London, UK.

PART II

Assessment of Past and Current Responses

Chapter 5

Biodiversity

Coordinating Lead Authors: Jeffrey A. McNeely, Daniel P. Faith, Heidi J. Albers

Lead Authors: Ehsan Dulloo, Wendy Goldstein, Brian Groombridge, Hiroji Isozaki, Diana Elizabeth Marco, Steve Polasky, Kent Redford, Elizabeth Robinson, Frederik Schutyser

Contributing Authors: Robin Abell, Salvatore Arico, Robert Barrington, Florent Engelmann, Jan Engels, Pablo Eyzaguirre, Paul Ferraro, Sofia Hirakuri, Toby Hodgkin, Joy Hyvarinen, Pierre Ibisch, Devra Jarvis, Alphonse Kambu, Valerie Kapos, Izabella Koziell, Yumiko Kura, Sarah Laird, Julian Laird, Merab Machavariani, Susan Mainka, Thomas McShane, Vinod Mathur, K.S. Murali, Sergio Peña-Neira, Adrian Phillips, William Powers, Asha Rajvanshi, Ramanatha Rao, Carmen Revenga, Belinda Reyers, Claire Rhodes, Klaus Riede, John Robinson, Pedro Rosabal Gonzales, Marja J. Spierenburg, Kerry ten Kate

Review Editors: Gerardo Ceballos, Brian Huntley, Sandra Lavorel, Stephen Pacala, Jatna Supriatna

BOXES

FIGURES

TABLES

Main Messages

Biodiversity is the variety of all forms of life, including genes, species, and ecosystems. Biodiversity underpins ecosystem services: biological resources supply all of our food, much of our raw materials, and a wide range of goods and services, plus genetic materials for agriculture, medicine, and industry. **Biodiversity has value for current uses, possible future uses (option values), and intrinsic worth. Biodiversity conservation ensures future provision of un-named or "undiscovered" services, and so complements direct maintenance of recognized ecosystem services.**

Recent decades have witnessed significant loss of biodiversity, at a rate two to three times faster than has occurred in geological history. Responses to this crisis focus on the conservation of biodiversity and on the associated problems of sustainable use of biological resources and equitable sharing of benefits arising out of the use of genetic resources. **Effective biodiversity response strategies have a bearing on human well-being in two ways: (1) they conserve a source of current and future goods and services, and (2) they create synergies and trade-offs of biodiversity conservation with other needs of society, including sustainable use of biological resources.**

Assessments covering a wide range of responses highlight several overarching issues. One is that **difficulties in measuring biodiversity make response design difficult, and complicate assessments of the impact of responses.** The potential benefits of integrating biodiversity conservation with management and planning for environmental services are substantial, but few examples of successful implementation exist and measurement problems make assessment of gains uncertain. Few well-designed empirical analyses assess even the most common biodiversity conservation measures.

Measurement and valuation of biodiversity requires attention to local, regional, and global scales. **Biodiversity may be valued differently, and generate human well-being differently, at local versus global scales. Focusing exclusively on either global or local values often leads to a failure to adopt responses that could promote both values, or reconcile conflicts between the two.**

The success or failure of any response to conserve biodiversity will depend on the ecological and institutional setting in which it is applied. Even within the same ecosystem, heterogeneity in institutions, income opportunities, access to markets, and other characteristics of the socioeconomic setting can lead to very different reactions to a given response.

Establishing and managing protected area systems directly conserves biodiversity, but emphasis on establishing new protected areas rather than managing existing ones effectively, area-based targets rather than biodiversity itself, and lack of funding for enforcement and management limit their impact. Success of protected areas systems as responses requires better site selection, incorporating regional trade-offs, in order to avoid the ad hoc establishment that can leave some ecosystems poorly represented. It also requires adequate legislation and management, sufficient resources, greater integration with the wider region within which protected areas are found, and expanded stakeholder engagement. The "paper parks" problem remains: geographic areas may be labeled as some category of protected area but not achieve the promised form of management. Representation and management targets and performance indicators work best when they go beyond measuring total area apparently protected. Percent-area coverage for protected areas associated with the Millennium Development Goals provide a broad indicator, but regional/national-level planning requires targets that take into account trade-offs and synergies with other ecosystem services.

Response strategies based on capture of benefits by local people from one or more components of biodiversity (for example, products from single species or from ecotourism) have been most successful when they have simultaneously created incentives for the local communities to make management decisions consistent with (overall) biodiversity conservation. Response strategies designed to enhance the local benefits derived from a few biological resources also seek to promote management for broader biodiversity conservation (including protection of global values). But even when a product is potentially well-linked to overall biodiversity (as in benefits from biodiversity prospecting) the actual benefits may not flow to the community, which results in inadequate incentives for conservation management. Alternatively, conservation payments can create economic incentives for such management. Overall, long-term success for these response strategies depends on meeting the economic needs of communities whose livelihoods already depend to varying degrees on biological resources and the ecosystem services biodiversity supports.

Management and sustainable use of wild species, with direct links to livelihoods, will remain a key response. Targeted protection of particular species has had mixed success in protecting overall biodiversity. Reintroduction of species, though often very expensive, has been successful, but such success generally will require the consent and support of the people inhabiting the target area. Control or eradication of an invasive species once it is established has appeared extremely difficult and costly. Prevention and early intervention have been shown to be more successful and cost-effective. Successful prevention requires increased efforts in the context of international trade, and in raising awareness. Sustainable use programs must include consideration of social and economic issues as well as the intrinsic biological and ecological considerations related to the specific resource being used. Zoos, botanical gardens, aquaria, and other ex situ programs build support for conservation, support valuable research, and provide cultural benefits of biodiversity.

Incorporating biodiversity into integrated regional planning promotes effective trade-offs and synergies among biodiversity, ecosystem services, and other needs of society. The "ecosystem approach" points to bioregional planning approaches that can achieve trade-offs and synergies. However, developing a quantitative regional "calculus" of biodiversity can enable marginal gains/losses in biodiversity from different places, and from different response strategies, to be estimated as a basis for planning the use of land and water. Assessments highlight synergies and trade-offs when different responses are integrated into a coherent regional framework. Society may receive greater net benefits when setting of biodiversity targets takes all land and water use contributions into account. Within a regional planning structure, effective responses also ensure connectivity between protected areas, promote transboundary cooperation, and incorporate habitat restoration. Different land uses should be seen as part of a continuum of possibilities, linked in integrated regional strategies. The great uncertainty is about what components of biodiversity persist under different management regimes, limiting the current effectiveness of this approach.

Places managed by the private sector can be recognized as possibly contributing to regional biodiversity conservation. Some parts of the private sector are showing greater willingness to contribute to biodiversity conservation and sustainable use, due to the influence of shareholders, customers, and government regulation. Many companies are now preparing their own biodiversity action plans, managing their own landholdings in ways that are more compatible with biodiversity conservation, supporting certification schemes that promote more sustainable use, and accepting their responsibility for addressing biodiversity issues in their operations. Limitations include insufficient synthesis of lessons to date concerning best pathways to "encouragement" and ongoing distrust between conservationists and business. Influence of shareholders or customers is limited in cases where the company is not publicly listed or is government-owned.

Integrating biodiversity issues in agriculture, fishery, and forestry management encourages sustainable harvesting and minimizes negative impacts on biodiversity. Most effective are *in situ* approaches such as some examples of organic farming that have developed synergistic relationships between agriculture, domestic biodiversity, and wild biodiversity. However, assessments of biodiversity contributions from such management often look only at local species richness, and little is known about contributions to regional biodiversity conservation. Effective integration also has a regional focus as in strategies that intensify rather than expand total area for production, so allowing more area for biodiversity conservation.

Governance approaches to support biodiversity conservation and sustainable use are required at all levels, based on the idea that management should be decentralized to the lowest appropriate level. This has led to decentralization in many parts of the world, with variable results. Planning and priority setting at regional scales may require governance and financial inputs at these scales. The key to success is strong institutions at all levels, with security of tenure and authority at the lower levels essential to providing incentives for sustainable management.

International cooperation through multilateral environmental agreements requires increased commitment to implementation of activities that effectively conserve biodiversity and promote sustainable use of biological resources. Numerous multilateral environmental agreements have now been established that contribute to conserving biodiversity. The Convention on Biological Diversity is the most comprehensive, but numerous others are also relevant, including the World Heritage Convention, the Convention on International Trade in Endangered Species of Wild Fauna and Flora, the Ramsar Convention on Wetlands, the Convention to Combat Desertification, the United Nations Framework Convention on Climate Change, and numerous regional agreements. However, their effectiveness must be measured by their impacts at policy and practice levels. Attempts are being made (for example, through joint work plans) to create synergies between conventions. The link between biodiversity conventions and other international legal institutions that have a major impact on biodiversity (such as the World Trade Organization) remains weak.

Education and communication programs have both informed and changed preferences for biodiversity conservation and have improved implementation of biodiversity responses. Biodiversity communication, education, and public awareness have emerged as a self-standing discipline, though it requires further development. Where change in behavior requires significant personal effort or economic loss, communication and education needs to be accompanied by other measures that assure livelihood support. Strategic approaches to achieve management objectives need to reflect the benefits and perceptions of multiple stakeholders.

5.1 Introduction

This chapter assesses the trade-offs and synergies among global, national, and local interests in conserving biodiversity and using its components (biological resources) sustainably. Based on the assessment of key responses to biodiversity loss, the chapter also provides policy options for decision-makers in the relevant ministries and in the private sector.

5.1.1 Biodiversity Values and Relationship to Ecosystem Services

While this chapter focuses on responses specific to biodiversity, such responses inevitably have to consider trade-offs and synergies involving ecosystem services. In brief, this chapter views biodiversity responses as largely about considering option values at many scales, with strong links to ecosystem service values arising at each of these scales.

While any value generated by a single gene, species, or ecosystem can be seen as part of biodiversity value, this chapter treats the values of such individual components of biodiversity as opportunity costs or benefits to be considered as a key part of effective biodiversity responses, at many scales. Biodiversity responses capture options for future well-being, and these responses may also involve trade-offs and synergies with other more direct ecosystem services. Thus the chapter takes human well-being as its central focus for assessment while recognizing that biodiversity and ecosystems also have intrinsic value and that people take decisions concerning ecosystems based on considerations of both well-being and intrinsic value, with the latter including option values (Reid and Miller 1989). Determining biodiversity option values remains a challenge for policy-makers at all scales, from crop genetic diversity in agroecosystems to global existence values.

5.1.2 Local, National, Regional, and Global Biodiversity Values

The debate over the links of biodiversity to ecosystem services reveals conflict about the scale at which biodiversity generates value. Some argue that *local* biodiversity assessments are most useful and see *global* biodiversity values as ignoring the important local values of biodiversity, especially relating to ecosystem services. Vermeulen and Koziell (2002, p. 89), for example, see the focus on global values as a consequence of the fact that "the global consensus is that of wealthy countries," and recommend consideration of biodiversity in terms of services derived from it and not as an end in itself. But such an approach ignores valid non-local values. This chapter assesses values that derive from biodiversity at all scales, with full recognition of the trade-offs between different types and scales of value.

While trade-offs often seem daunting, this chapter's integrative perspective helps reconcile what can be conflicting requirements for uses at a given place. Consideration of global biodiversity implies value for what is *unique* at a place (or what is not yet protected elsewhere). Ecosystem services may well value exactly what makes that place *similar* to many others, even though this amounts to "redundancy" at the regional scale. But effective biodiversity responses can see both values as valid, with the within-place values seen as costs and benefits to be taken into account at the regional scale.

5.1.3 Goals, Main Points, and Structure of this Chapter

This chapter assesses responses that aim to conserve biodiversity and use biological resources sustainably, using case studies to determine what has or has not been successful under what conditions. Of many responses, this chapter assesses the nine responses that are most widely used and discussed: protected areas; local capture of benefits; wild species management; regional planning, agricultural/forestry and fishery policy; private sector activities; governance; multilateral environmental agreements; and education / communication. These responses may vary from the MA typology applied to ecosystem services—biodiversity is not considered an ecosystem service in the MA—and each response typically addresses more than one driver of biodiversity loss.

The chapter follows this list of responses in order, beginning with protected areas because virtually every country has a PA system as an important component of its efforts to conserve biodiversity. That assessment is followed by a section on responses that enable local people to capture biodiversity protection benefits because many of those responses are employed in and around protected areas to address conflict between local people and the aims of the protected areas. Responses under this heading include ecotourism, integrated conservation and development projects, and conservation payments. The section on the management of wild species deals with local people's reliance on particular species as well as addressing issues of managing invasive species and preventing extinctions. The next section assesses the role of regional planning in identifying synergies and trade-offs across the region that contains protected areas, intense land and water uses, managed ecosystems, and other potentially conflicting uses. Regional planning provides a mechanism for tying together responses that conserve biodiversity both within and beyond protected areas.

Agricultural and fisheries policies deal directly with biodiversity components, and their multiple impacts are assessed. The private sector's impacts on biodiversity are profound, and we assess efforts to minimize negative impacts, or to make them positive. Governance is the broadest response, and several multilateral environmental agreements contain provisions for the conservation of biodiversity by the signatories; although those countries then employ a range of responses to meet their obligations, this chapter assesses MEAs as a biodiversity conservation and sustainable use response in and of themselves. Finally, the chapter assesses the role of education and communication activities in generating support for conservation and sustainable use of biodiversity.

Several themes emerge from these assessments. First, the human well-being of local people dominates the assessment of many responses including PAs, governance, wild species management, and local capture of benefits. Although biodiversity at the global scale creates human well-being for people far removed from where valued biodiversity is found, the contribution of biodiversity to the well-being of local people is critical. Second, the divergence between local, national, regional, and global values of biodiversity (and the human well-being derived at these different scales) presents challenges to biodiversity conservation and sustainable use. A regional perspective can address conflict and create balance over geographic space. Finally, incorporating biodiversity benefits into management decisions permeates the assessments of agriculture, fisheries, MEAs, and private sector responses for conservation.

Taken as a whole, the following assessments lead to several conclusions:

- The current system of PAs is a valuable tool for conserving biodiversity, but these areas do not yet include all biodiversity components that require such protection. Better tools exist for selecting areas for inclusion in PA systems than are currently employed, and better management of individual PAs is required.
- For successful (global) biodiversity conservation, local people must be able to capture benefits from that conservation.
- Integrated conservation and development projects as currently designed rarely succeed in their objective of conserving biodiversity, yet their general concept remains valid. They need more realistic objectives and a stronger link to broader policy issues.
- Regional planning can achieve balance across areas to create a landscape that includes strict conservation areas, managed landscapes, intensively used areas, and other land and water uses.
- Direct incentives for biodiversity conservation usually work better than indirect incentives.
- More income must flow from the people and countries that value biodiversity from afar (at the global level) to the people and countries where much globally valued biodiversity is conserved, often at considerable opportunity cost.

These assessments suggest that in the coming years:

- Direct management of invasive species, particularly efforts to prevent invasions, will become a more important biodiversity conservation response.
- Biodiversity conservation in ecosystems managed for production will become increasingly important.
- Conservation payments appear promising even though they require more testing.
- Regional planning can be better utilized to integrate various biodiversity responses.
- Biodiversity conservation in light of climate change will pose many challenges, calling for improved management of habitat corridors and production ecosystems between protected areas, thereby enabling biodiversity to adapt to changing conditions.

5.1.4 Links to Multilateral Processes

5.1.4.1 Millennium Development Goals

The MDGs were endorsed by members of the United Nations in 2000 to address common pressing global concerns. The timeframe set to achieve the MDGs is 2015. In fact, the MDGs are already showing their potential of bringing together various actors to honor the commitment. For instance, in the African and Asian regions, finance ministers are using the MDGs to focus more resources on development. The New Partnership for Africa's Development is also using the MDGs and has begun reporting on progress toward their achievement. Environmental sustainability, including incorporation of biodiversity into development activities, is one of the eight goals. Like the WSSD Plan of Implementation, the MDGs cover broad and complex issues that will require better institutions and tools to achieve the goals. The Secretary General of the United Nations has initiated a Millennium Project to produce detailed advice for achieving the MDGs; he recognizes biodiversity as providing a foundation for achieving the goals.

5.1.4.2 World Summit on Sustainable Development

The 2002 World Summit on Sustainable Development produced a commitment by governments to address sustainable development. Paragraph 42 of the WSSD Plan of Implementation makes specific reference to the conservation of biodiversity, and many other parts of the Plan are relevant to biodiversity. Despite the commitment made to conservation of biodiversity and the idea of integrating it into development projects, the Plan is very broad and requires multiple institutions to collaborate for its successful implementation. The Plan's goal of reducing the rate of loss of biodiversity by 2010 is challenging because the current rate of loss is not known with any precision. Yet this goal had been adopted by the parties to the Convention on Biological Diversity at its sixth Conference of the Parties in April 2002, and the European Union has adopted an even stronger one (*halting* the loss by 2010). Further, the WSSD considers biodiversity as one of its five major issues; the others are water, energy, health, and agriculture.

5.1.4.3 World Trade Organization

Although the World Trade Organization has established a Committee on Trade and Environment to deal with environmental issues and has contributed to environmental policies such as the Rio Declaration, the WTO still insists that it was established primarily to deal with trade and not environmental issues. It has shown minimal commitment to environmental issues, including biodiversity. Despite its position, some progress can be assessed. One illustration is the tuna-dolphin case between the United States and Mexico (WTO 2003), which gave rise to some positive developments in favor of the environment and marine biodiversity in particular. The case has been brought under Article XX of GATT, which provides general exceptions, and the environment is one such exception (Lang 1995). Although the panel report on the case was not adopted, it nonetheless encouraged countries to take into account the impact trade could have on biodiversity and environment in general, and serves as a guide for countries to consider conservation of biodiversity. However, the reluctance of the WTO to integrate biodiversity and environmental concerns remains a challenge.

5.1.4.4 Other Processes

Numerous other international programs and events are dealing with or relevant to biodiversity. Prominent among them are convention reports (CBD, CCD, Ramsar, CITES, CMS); the Intergovernmental Panel on Climate Change; the *Global Environment Outlook* (UNEP/collaborating centers); the *World Resources Report* (UNEP, UNDP, World Bank, WRI); Earth Trends (WRI); IUCN Red Lists and Species Survival Commission Reports; the *Human Development Report* (UNDP); the *World Development Report* (World Bank); FAO Plant Genetic Resource Assessment; FAO reports on fisheries, forest, and agriculture; and the UNESCO-MAB program.

5.2 Assessing Protected Areas as a Response to the Loss of Biodiversity

5.2.1 Introduction

Protected areas are the most commonly used tool for in situ conservation of biodiversity, a role recognized under Article 8 of the CBD. A "protected area" is defined in Article 2 of the CBD as "a geographically defined area, which is designated or regulated and managed to achieve specific conservation objectives." Assessments of protected areas as a biodiversity conservation response strategy indicate that protected areas are an extremely important part of programs to conserve biodiversity and ecosystems, especially for sensitive environments that require active measures to ensure the survival of certain components of biodiversity. Most protected areas also contribute to a country's economic objectives in providing ecosystem services, supporting sustainable use of renewable resources, and generating tourism and recreation values. Ecosystem services generated and supported by protected areas include microclimate control, carbon sinks, soil erosion control, pollination, watershed protection and water supply, soil formation, nutrient recycling, inspiration, and a sense of place. Many protected areas are of great importance as tourist and recreation destinations—both nationally and internationally. As the world becomes increasingly urbanized, such values associated with protected areas seem likely to increase.

Protected areas also play an important role in ensuring the respect for, and recognition and maintenance of, important traditions, cultures, and sacred sites (Harmon and Putney 2003). Increasingly, protected areas are being used as mechanisms to promote peace-keeping efforts among nations, notably through transboundary protected areas and "peace parks" (Sandwith, Shine, Hamilton and Sheppard 2001).

Protecting specified areas from certain human uses has a long history including North Africa's age-old *hima* system (the reserves established by the Hafside dynasty in 1240 in Tunisia), and the marine protected areas established by local communities in the Pacific hundreds of years ago as a tool to preserve fishing areas (Kelleher 1999). All this indicates that traditional protected areas were an important resource management approach for the societies that established them.

The modern movement of protected areas is rightly associated with the establishment of the first national park in the United States at Yellowstone in 1872 (Phillips 2003). Through the evolution of the modern protected areas movement, some protected areas have been accurately criticized on the grounds that local communities have been excluded from resources that have traditionally supported their livelihoods. On the other hand, many protected areas have substantially contributed to the livelihoods of local people, offering options for alternative economic development (McNeely 1993; EC and IUCN 1999; EC/DFID/IUCN 2001).

Today more than 100,000 sites are recognized by IUCN as protected areas. (See Figure 5.1.) Together they cover about 11.7% of Earth's land area, equivalent to the whole of South America (IUCN/UNEP-WCMC 2003). However, the current global system of protected areas is not sufficient to conserve biodiversity, as a consequence of insufficient coverage, inappropriate location, insufficient management, and related economic and social factors. Some assessments of case studies (MacKinnon et al. 1986; Barzetti 1993) suggest improving the effectiveness of protected areas for biodiversity conservation, through a number of approaches, including: (1) strategies for effective protected areas design and management that are appropriate to their ecological, social, historical and political settings; (2) regional planning strategies that fully take into account the requirements of protected areas management within the context of the land/water uses surrounding them, and seeking trade-offs and synergies with ecosystem services; and (3) a better appreciation of the multiple economic values of protected areas, for local people, the nation in which they are located, and the world at large, and better ways of capturing these so that local people are not disadvantaged. Systems of protected areas that safeguard important biodiversity values at different levels (local, national, regional, and global) can contribute to human well-being when individual protected areas provide measurable benefits through services that complement "off-reserve" approaches.

5.2.1.1 Adequacy

Recent assessments have shown that, at the global and regional scales, existing protected areas, while essential, are not sufficient for conservation of the full range of biodiversity (UNEP/CBD/SBSTTA/9/5). Problems include lack of representativeness, impacts of human settlement within protected areas, illegal harvesting of plants and animals, unsustainable tourism, impacts of invasive alien species, and vulnerability to global change (IUCN 2003b).

Marine and freshwater ecosystems are even less well protected. Mulongoy and Chape (2004, p. 5) conclude that "recent assessments indicate that conservation of marine and coastal biodiversity is woefully inadequate, with less than 1% of Earth's marine ecosystems protected . . . major freshwater systems . . . are

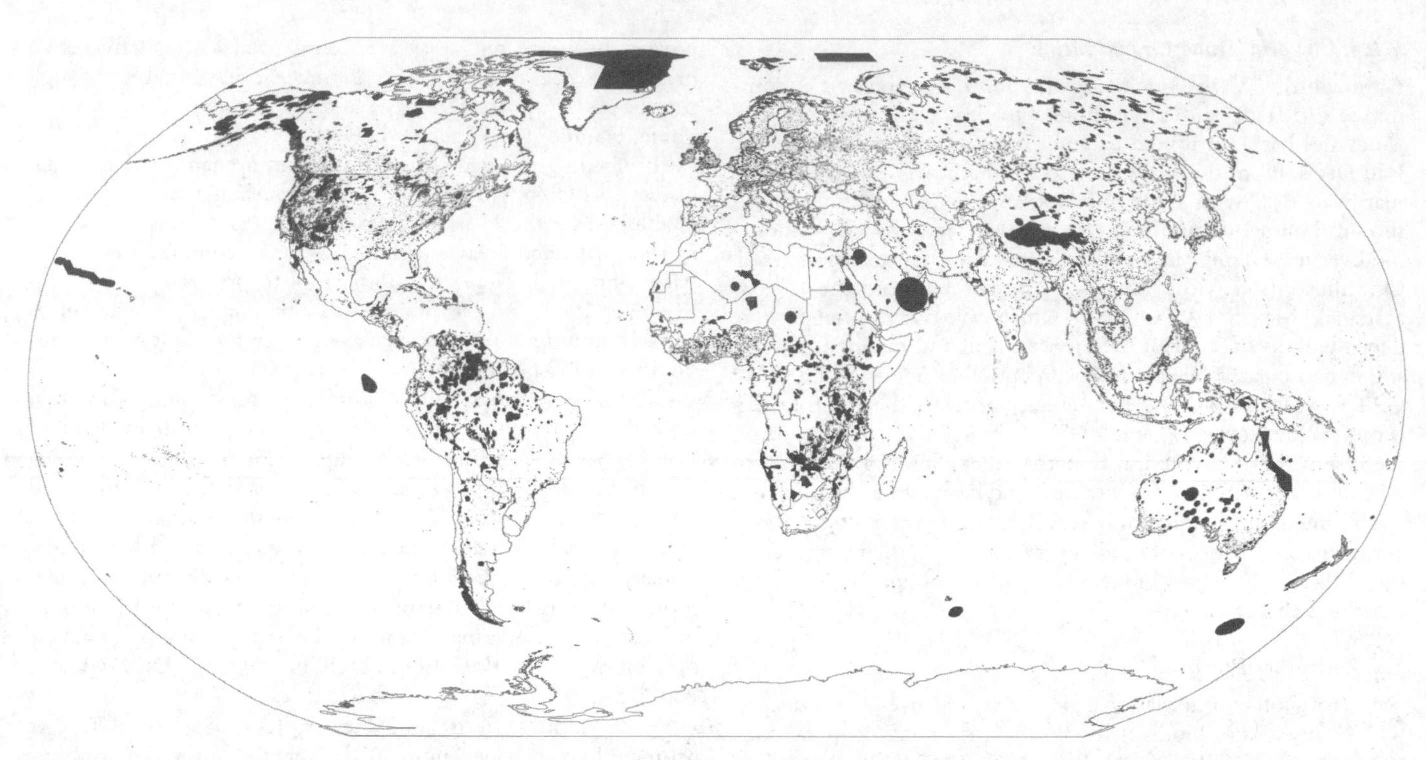

Figure 5.1. Global Network of Protected Areas. The global network of protected areas, according to 2004 World Database on Protected Areas. This map represents all protected areas recorded in the WDPA, except those for which no area information was provided. For some sites, only the central coordinates and total area were known (no polygon boundaries were available), and these are represented as circles. The WDPA includes sites with a variety of land management types and conservation effectiveness, including strictly managed reserves, areas subject to multiple uses, and indigenous reserves. (WDPA Consortium 2004)

also poorly represented.'' Further, global assessments based on area coverage are misleading in suggesting that biodiversity conservation is successful, because the coverage varies from country to country, some ecosystems are better protected than others, and certain kinds of biodiversity are best conserved outside of protected areas.

The country reports of the parties to the CBD provide numerous perspectives on the effectiveness of protected area systems. Stated needs include improved legislation, more effective management, more resources for protected area management, capacity building among protected area managers, effective integration between protected areas and the wider region, and effective involvement of all stakeholders in the establishment and management of protected areas (UNEP/CBD/AHTEG-PA/1/2). Furthermore, the 2003 fifth World Parks Congress identified the following critical factors of success: the sustainable financing of protected areas systems, adequate capacity of PA institutions and managers, and the application of scientific and traditional knowledge to protected areas planning and management (IUCN 2003a).

5.2.1.2 Basis for Assessments

Assessment of PA systems as a mechanism for biodiversity conservation requires attention to fundamental issues for the MA, including trade-offs and synergies at global, national, and local levels.

Because protected areas are a form of land/water use that explicitly rules out other resource uses, establishing them can involve conflict between the need for long-term biodiversity conservation (particularly for globally threatened and endemic species) and more immediate social and economic priorities. Access to products such as fuelwood and charcoal, medicinal plants, timber and game, are curtailed when local residents are prevented from entering a protected area. These restrictions could have cultural implications, depending on the degree to which local people are allowed access to, for example, sacred sites.

However, these conflicts could be minimized through adequate consultation and planning. One useful strategy is to promote the broader use of all IUCN Protected Areas Management Categories (IUCN 1994) that allows a gradation of the level of protection from areas strictly protected to those that support multiple uses of its natural and cultural resources. This could be a useful tool in establishing national systems of protected areas, as required under the CBD Article 8a (Davey 1998). Instead, most countries use only the more strictly protected categories at the national level, thereby foregoing benefits from additional areas established under categories allowing some forms of human use.

Most PAs provide multiple benefits, with different sites providing different mixes of benefits according to the objectives of their management (IUCN 1994). In addition to their conventional conservation objectives, protected areas are now expected to contribute more to social objectives. For example, the secretariat of the CBD (2004, p. 1) recognizes that ''a strong consensus has developed that protected areas need to make a solid contribution to poverty alleviation and sustainable development. The main challenge for using protected areas to alleviate poverty is how to find the right balance between the desire to live harmoniously with nature and the need to exploit resources to sustain life and develop economically.'' This approach demands the maintenance and enhancement of core conservation goals, equitably integrating them with the interests of all affected people, forging synergy between conservation, maintaining life support systems, and supporting sustainable development (IUCN 2003b).

Protected area planning and management with the participation of local communities is becoming more the rule than the exception, and it has been incorporated in the guidelines of development agencies (for example, the World Bank and the European Union) and key donors such as BMZ (the German Federal Ministry for Economic Cooperation and Development), GTZ (the German aid agency), and USAID. Innovative participatory and co-management arrangements for protected area management are being implemented in almost all regions of the world and considerable information and lessons learned from such initiatives are now available (Jaireth and Smyth 2003). Full stakeholder participation, including partnerships between civil society, government, and the private sector, has been also identified as a key guiding principle to integrate biodiversity conservation in development activities (EC/DFID/IUCN 2001).

Adequate attention to global biodiversity conservation will not follow easily from a focus on poverty alleviation. Lapham and Livermore (2003, p.13) state: "Biodiversity funding is now driven heavily by social and economic objectives, which are not necessarily synonymous with objectives such as avoiding extinctions or protecting unique and biologically diverse landscapes." Also (p. 20), "as poverty reduction becomes the driving force behind development assistance across all sectors, conservation appears to be falling by the wayside . . . Ramifications may include a reduced role for science in shaping biodiversity assistance priorities, decreased funding for crucial conservation activities, fewer projects with clear conservation outcomes, diminished biodiversity expertise within funding agencies, and less political attention to conservation." Similarly, Sanderson and Redford (2003, p. 390) are concerned that "in its new incarnation, poverty alleviation has largely subsumed or supplanted biodiversity conservation. This trend has gone largely unnoticed, but poses a significant threat to conservation objectives."

However, this conflict is not an irresolvable problem: an increasing number of successful cases show how protected areas can achieve both conservation and the *sustainable use* of biodiversity (MacKinnon 2001). Box 5.1 gives several examples. In other sections, this chapter argues for biodiversity to be more widely considered in other policies that may be aimed at poverty alleviation in much the same way that PAs, although largely focused on biodiversity conservation, also contribute to local and regional economies.

The degree to which global biodiversity conservation is "swept along" by the increased attention to poverty alleviation is largely unknown, with case studies supporting various conclusions. Vermeulen and Koziell (2002, p. 52) call for "indicators that are able to measure progress towards integrating different biodiversity values across the landscape." They acknowledge that local values may correspond to specific biological resources more than to biodiversity generally. Sanderson and Redford (2003, p. 390) argue that complementarity between global, national, and local values "can only be achieved if we respect the strengths and weaknesses of both conservation and poverty alleviation efforts and the trade-offs inherent in integrating them. Calls for 'pro-poor conservation' that ignore these trade-offs will end up in failure, with both the poor and biodiversity suffering."

5.2.2 Management of Protected Areas

While significant progress has been made in the establishment of a global network of protected areas, assessments suggest that many protected areas are not managed in ways that will enable them to achieve their objectives. The most appropriate way to manage protected areas depends very much on local conditions, and opinions vary as to the effectiveness of existing PA strategies (Brandon and Wells 1992; Terborgh et al. 2002). Some protected areas are "paper parks," with little or no investment in management. Yet an assessment of protected areas in Africa, the Caribbean, and the Pacific noted that even so-called "paper parks" play an important role in the development and further consolidation of national protected areas systems (EC and IUCN 1999). This role has been supported by recent research that suggests that simply establishing protected areas on a map does afford at least some protection to biodiversity (Bruner et al. 2001).

The 2003 fifth World Parks Congress identified increased PA management effectiveness as one of the key goals for the next decade (IUCN 2003a). A survey assessed that less than 25% of forest protected areas were well managed, with many forest protected areas having no management at all (IUCN 1999). The need to assess the effectiveness of protected area management and governance more systematically was also emphasized (Goal 4.2) in the CBD Programme of Work on Protected Areas, adopted at the seventh Conference of the Parties in Kuala Lumpur, Malaysia, in February 2004.

Recent assessments indicate several widespread weaknesses of PA management (WWF 2004). In a survey of management effectiveness of nearly 200 protected areas in 34 countries, only 12% were found to have implemented an approved management plan. The assessment concluded that PA design, legal establishment, boundary demarcation, resource inventory, and objective setting were relatively well addressed. But management planning, monitoring and evaluation, budget security, and law enforcement were generally weak among the surveyed protected areas. WWF judged poaching, encroachment by agriculture, ranching, urban development, logging, and non-sustainable collection of non-wood forest products to be the major threats. This situation is not confined to the developing world. A recent assessment of the ecological integrity of Canadian National Parks revealed many similar problems and identified long-term actions required to enhance the management of the system (Parks Canada Agency 2000), and in Europe, agricultural development policies have been identified as a major threat to protected areas management (Synghe 1998).

While today's PA management emphasis on community issues far exceeds what it was in the recent past, activities relating to people were often seen as poorly managed, especially with regard to tourism management. Thus the fifth World Parks Congress (Durban, South Africa, 2003) recommended promoting the role of tourism as a tool for biodiversity conservation and for support of protected areas through measures such as conservation education, encouraging stewardship among locals, and reinforcing local community development and poverty alleviation. But the WPC also noted that if tourism is not appropriately planned, developed and managed, it can contribute to the deterioration of cultural landscapes, threaten biodiversity, contribute to pollution and degradation of ecosystems, displace agricultural land and open spaces, diminish water and energy resources, and drive poverty deeper into local communities (IUCN 2003a).

WWF (2004) noted examples of protected areas working successfully with people, suggesting that success depends on a collaborative management approach between government and stakeholders, an adaptive approach that tests options in the field, comprehensive monitoring that provides information on management success or failure, and empowerment of local communities in a participatory system that provides direct access and ownership of resources. (See Box 5.2 for an example in Indonesia.) Others have given considerable attention to issues associated with indigenous and traditional peoples and protected areas. A long process

BOX 5.1

Benefits from Protected Areas: Marine Examples

Protected areas provide numerous benefits, including contributing to sustainable fisheries (see Bahamas and Samoa examples) and supporting recreation and tourism (Bonaire Marine Park and Merritt Island National Refuge).

Exuma Cays Land and Sea Park (45,620 hectares) in the *Bahamas* was established in 1958 covering both the terrestrial and marine environments associated with these islands. The Park became a no-take fisheries reserve in 1986. Research has shown that the concentration of conch in the park is 31 times greater than outside the park, providing several million conchs per year to areas outside the park available to be harvested by fishers. Additionally, tagged grouper from Exuma Park have been caught off of both north and south Long Island (Bahamas), indicating the Park is replenishing grouper stocks in areas as far as 250 kilometers away. Tagged spiny lobsters from the Exuma Park are found replenishing the marine environment of Cat Island, 100 kilometers away. The success of fisheries resource replenishment in the Exuma Park led the government to announce a policy decision in 2000 to protect 20% of the Bahamian marine ecosystem, doubling the size of the national protected areas system (WCPA News 2002).

In the Pacific Island of *Samoa,* like in many countries in the tropics, catches of seafood from coastal areas, lagoons, and inshore reefs have been decreasing over the past 10 years. Reasons for this decline include overexploitation, the use of destructive fishing methods (including explosives, chemicals, and traditional plant-derived poisons) and environmental disturbances. In order to address this problem, the Samoan Fisheries Division initiated in 1995 a community-based extension project in 65 villages which recognized the village *fono* (council) as the prime agency responsible for actions. A large number of villages (38) chose to establish small village fish reserves in part of their traditional fishing areas and decided to actively support and enforce government laws banning the use of explosives and chemicals for fishing. Some villages also set minimum size limits for capturing fishes. While many of the village reserves are small (ranging from 5,000 to 175,000 square meters), their number and the small distance among them forms a network of fish refuges. In just a few years, fisheries stocks have increased 30–40% and there are signs of recovery in reefs previously affected by destructive fishing methods. As the fish reserves are being managed by communities which have direct interest in their success, prospects for long-term sustainability of this initiative are high (King and Faasili 1998).

Bonaire Marine Park (2,700 hectares) was created in 1979 and cvers all reef areas around the island. While the resident population of the island is less than 15,000, 17,000 to 20,000 scuba divers visited the park every year, and diving tourism represented the main economic activity of the island. Total gross revenue from dive-based tourism was estimated to be $23.2 million in 1991. In addition to this, $340,000 was generated through taxes levied on divers. Visitor fees thus more than covered the cost of the establishment of the park ($518,000) and the recurrent management cost ($150,000) was. The park also generates employment for over 1,000 people. By 1994, the number of divers had increased to 24,081 and the total annual visits was about 70,000 (The Commonwealth of Australia 2003).

Located at Cape Canaveral, Florida, United States, *Merritt Island National Wildlife Refuge* contains two areas, with a total extension of 4,000 hectares, that have been closed to fishing since 1962. Before these areas were closed, commercial and recreational fishing in the areas was intensive, causing fish stocks to be heavily depleted. The value of this reserve for the recreational fishery outside its borders has been assessed by the number of record-size ("trophy") fish caught by recreational fishers. The area extending 100 kilometers to the north and south of the reserve provided 62% of record-size black drum, 54% of red drum, and 50% of spotted sea trout. Fish tagging studies showed that these species indeed moved out of the reserve into the surrounding waters (Roberts et al. 2001; Commonwealth of Australia 2003).

BOX 5.2

Komodo National Park, Indonesia

(http://www.komodonationalpark.org/downloads/CMP10.pdf)

The government of Indonesia has formed a team consisting of park rangers, navy, police, and fishery services that carries out a routine patrolling program. Since the team's inception in 1996, blast fishing has declined by over 80%, and marine biodiversity is reported to have increased dramatically. Law enforcement was emphasized because of the threat of outside fishers who used more destructive fishing practices. Even with the patrolling, it was found that a more intensive enforcement system had to be considered, as the improved condition of the marine park made it even more attractive to outside fishers. In combination with enforcement activities, exclusive fishing rights were proposed for traditional use zones to park inhabitants and for buffer zones to communities in the direct surroundings of the park.

of consultation among protected areas managers and indigenous peoples' groups assessed successes and failures in addressing the full integration of the concerns of indigenous peoples in PA planning and management. As a result, agreed principles, practical policy, and technical guidance on this issue have been developed and widely distributed (Beltrán 2000).

A strong consensus exists among protected area experts that the minimum critical ingredients for effective PA management are appropriate staffing, good public education and community outreach, and excellent enforcement capacity (Hockings et al. 2000). Because law enforcement is strongly correlated with management effectiveness, well-trained, well-equipped, and well-motivated teams of rangers and other field staff are fundamental. But to be effective, the local enforcement effort needs to be backed by a broader environment of good governance that ensures that laws and regulations are respected and enforced. Although protected areas management increasingly adopts an inclusive rather than exclusivc approach, law enforcement efforts will remain an essential element of effective management of protected areas, even where communities are involved in decision-making and are compensated or offered alternative livelihood opportunities.

An assessment of anthropogenic threats to 93 protected areas in 22 tropical countries reveals that most protected areas are under constant pressure from land clearance, grazing, fire, hunting, and logging (Bruner et al. 2001). The effectiveness of "protected area" was shown to be a function of the number of guards and the existence of "significant" sanctions for those caught breaking the law. Effectiveness did not appear to be affected by the number of people living in the park, the area's accessibility, local support for the protected area nor the involvement of local communities in park management. However, the authors used a narrow definition of "effectiveness" as the protection of biodiversity against human access, and ignored any adverse impact the protected areas may have had for the local communities' use of ecosystem services.

Despite conflicts between those responsible for the PA and those who historically have relied on resources within the PA, a growing consensus suggests that PAs will be more effective when:

- strong institutions define boundaries, access rights, and user participation,
- local communities have "bought in" to the protected area,
- local communities have alternative livelihood opportunities or receive direct payments and so are not made worse off by the PA, and
- sufficient resources are available for funding enforcement effort.

The funding of recurrent costs of protected areas management is a key problem. Trust funds to support protected areas are currently in place at least in Argentina, Bhutan, Bolivia, Brazil, Costa Rica, Ecuador, Indonesia, Jamaica, Mexico, Panama, and Peru. Payments for ecological services that are supporting protected areas are being implemented in Costa Rica and Brazil. However, the Global Monitoring Report (IBRD and World Bank 2004, p. 15) notes that "aid to developing countries to support improved environmental practices, both bilaterally and through multilateral vehicles, has declined after a short-lived increase following the 1992 Rio Convention." WWF (2004, p. 18) concludes that "environmental services provided by protected areas (such as provision of clean potable water) need to be recognized and paid for; national funds for protected areas must be strengthened; the budget of the Global Environment Facility (GEF) should be substantially increased in its replenishment, so as to meet the challenges of supporting the implementation of the [GEF's] Programme of Work."

The fifth World Parks Congress identified that $25 billion in additional annual support was required to establish and maintain an effective global system of protected areas. The CBD Ad hoc Technical Expert Group on Protected Areas (CBD 2003a) has argued that GEF funding is crucial for developing countries, because studies suggest that developing-country governments may only allocate only between $50 and $100 million annually towards direct costs of protected areas and conservation.

The multiple benefits provided by protected areas can help provide the basis for various innovative mechanisms for financing them. This highlights the need for effective valuation of these goods and services, which stimulate funding of protected areas. Such evaluations have been implemented, or are in the process of being undertaken, in many countries so as to counteract the general tendency to reduce budgets of protected areas.

5.2.3 Design of Protected Areas

Selection and design issues concerning individual terrestrial and marine protected areas include aspects relating to size, shape, connections, corridors, and edge effects. The application of all these aspects needs to be tailored to national and local environmental and socioeconomic circumstances. While "population viability analysis" clearly indicates protected area design considerations for some single species, PA design would benefit from an equivalent "biodiversity viability analysis" that takes into account the needs of *all* species in a region (Faith et al. 2003).

A recent extensive review of case studies relating to PA design (Environmental Law Institute 2003, p. 2) argues, "given the inherent complexity of ecological systems, scientists are understandably reticent about providing exact prescriptions for spatial (land and water use) planning and design because answers vary depending on the species, ecosystem, or scale in question." Nevertheless, even partial knowledge about species or ecosystem responses to human disturbance and fragmentation needs to be applied to land use decisions, ensuring that it is informed by the best available science. The review recommended certain "potential ecological threshold measures," which relate to habitat patch area, percent of suitable habitat, edge effects, and buffers. On the other hand, it would not give any guidance concerning a key design issue, corridors, given the current information. The assessment highlights the inadequacy of the information currently available to planners of protected areas.

An important consideration for PA design is the future impact of climate change. In this context, corridors and other design aspects to give flexibility to protected areas are regarded as good precautionary strategies.

5.2.4 Regional and Global Planning for PA Systems

In the past, many protected areas were selected simply because they were not suitable for agriculture or human settlement, or because they had scenic value. These factors help explain why existing PA systems are not representative of biodiversity. Recent developments in "systematic conservation planning" provide strategies to locate protected areas that maximize biodiversity representation and persistence, while minimizing conflict with competing land use needs.

One aspect of systematic conservation planning is "gap analysis," where "gaps" are habitat types that are under-represented in an existing network of protected areas (Noss 1996). Systematic PA selection seeks to fill such gaps (Pressey and Cowling 2001). Gap analysis now uses the principle of complementarity in setting priorities: the complementarity value of a place is indicated by those *additional* biodiversity elements it provides relative to an existing set of protected areas (Pressey et al. 1993).

The sophisticated framework for selecting sets of priority conservation areas on the basis of complementarities has been criticized as a scientific approach with limited practical use. For example, recent CBD documents describe it as too "data-hungry" for practical application, and such methods are seen by some to run counter to the need for more value-laden decision-making. Jepson and Canney (2003) argue against the need for experts, who are supposedly the only ones who can identify units of conservation. On the other hand, Noss (2003) and others see the involvement of "experts" and scientists as an essential element of the decision-making process. Overall, a systematic framework can make best possible use of all available data, however meager, and provide science-based decision support. The decisions themselves are matters of public choice, not science, so it should not be surprising if the ideal protected area system is hardly ever implemented (Noss 2003).

The nature of global values points to critical trade-offs and synergies with local values. Mulongoy and Chape (2004, p. 38) suggest, "within the network, individual protected areas are also designed to maximize their effectiveness . . . a protected area should usually be positioned to include the maximum biodiversity possible." But emphasis should be given to the complementarity value of the location, if global biodiversity values are the priority. Given that such marginal gains are not normally expressed in dollar terms, forms of multicriteria analysis have been used to explore regional trade-offs in the design of protected areas systems. This approach is in accord with overall response strategy options in the

MA focused on "integrated responses" (Brown et al. 2001). (See Chapter 15 of this volume.)

Case studies (Faith et al. 1996; Ando et al. 1998) have illustrated how better trade-offs (providing greater net benefits for society) can be achieved by explicitly taking variable opportunity costs of conservation into account in setting regional priorities for location of protected areas. In Uganda, a five-year $1 million program proposed expansion of the PA network in a way that minimized opportunity costs (Howard et al. 2000). The problem of "paper parks" might be addressed in part by having estimates of costs, including opportunity costs, presented "up front" when designing a protected area system. This provision is now applied in most countries, but often underestimates the difficulty in obtaining additional funding from external sources (grants, projects, etc.) to cover these costs.

Even in regions with very high potential conflict between biodiversity conservation and provision of other ecosystem services, planning based on trade-offs reveals a potential to achieve high conservation at remarkably small cost. In a Papua New Guinea planning study (Faith et al. 2001a, 2001b), high-value biodiversity areas often overlapped with high opportunity cost areas—areas presenting good opportunities for agriculture that would have to be forgone under conservation. Application of complementarity-based selection of priority areas for biodiversity conservation nevertheless identified a set of areas having low opportunity cost. The use of multicriteria analysis in the Papua New Guinea study revealed possible ways to minimize conflict among different land use needs, and found synergies in the location of priority protected areas. Areas that were given priority had high complementarities and also known distributions of rare or threatened species.

The few case studies available suggest that the use of multicriteria analysis to guide selection of such priority biodiversity conservation areas would (to some unknown degree) increase the regional net benefits provided by ecosystem services and biodiversity option values. To date such approaches appear to have seldom been applied.

In the Global Strategy for Plant Conservation (annex to decision VI/9), the sixth Conference of the Parties of the CBD specified (1) that by 2010 at least 10% of each of the world's ecological regions should be effectively conserved, implying increasing the ecological representation and effectiveness of protected areas; and (2) that protection of 50% of the most important areas for plant diversity should be assured through effective conservation measures, including protected areas.

Simple percent coverage targets at the global scale have several advantages (Hoft 2004) including easy compilation of the information base, applicability to various scales, and effective communicability. Such percent targets may be most applicable as a broad-scale indicator of performance, but with the caveat that achievement is expected to be via some country-scale spatial planning.

A complement to these approaches may be global priority setting that identifies regions where conservation planning is a priority. Such an approach may identify "narrowing" windows of opportunity for balanced planning (Faith 2001) as a way to incorporate well-being considerations into global biodiversity priorities. Although information on biodiversity at the species level in most freshwaters is poor (Revenga and Kura 2003), it has been possible to effectively identify "hotspots" or priority places for conservation action. At the global scale, "hotspots" point to the regions that may most urgently demand such systematic planning, because of high complementarity with other regions and high threat, the latter possibly indicating high opportunity costs of conservation.

Complementarity is important even at the global scale of planning and priority setting. The 25 global biodiversity hotspots identified by Conservation International are based on endemism, which is a special case of complementarity (Myers et al. 2000). Efficiency of resource allocation for this set of hotspots can be argued because high endemism values mean that each region contributes additional species (conserving such a "hotspot" also has been seen as a less cost-effective way to conserve biodiversity; see Ando et al. 1998). Similarly, the Global 200 priorities (WWF 2003) implicitly use complementarity in order to seek representation of all ecosystems within regional conservation and development strategies.

Any measure of the effectiveness of protected area systems that is based on area-coverage must consider that a system can be extensive in total area yet poorly represent the region's biodiversity (Pressey and Tully 1994). Allocating land for the protection of biodiversity has always faced, implicitly or explicitly, the opportunity costs of making the land unavailable for competing human uses. Because many protected areas were specifically chosen because they were not suitable for human use, the percentage of a country's surface that is a designated protected area says relatively little about the actual biological diversity protected (Pressey 1997; Barnard et al. 1998). This is further discussed later in this chapter.

Percent target strategies for protected area systems often call for percentage coverage of each forest or land cover type or biome within a region, in order to better address representativeness rather than just total amount of area. Percentage targets remain open to several criticisms (Faith et al. 2001b):

- They depend on nominated "types" that can be defined at different scales; the scale of types can determine the total area needed for any nominated percent target.
- Different land types vary in terms of internal heterogeneity or diversity; more diverse types arguably deserve a higher target.
- Different types vary in terms of likelihood of persistence in the absence of conservation action (for example, because of differences in geographic extent). Percent targets can run counter to types with greatest need for protection, because models of probabilities of persistence suggest that geographically extensive habitat types may have a reasonable probability of persistence of their components even in the absence of action, and so require a *smaller* not larger percentage area protection.

Rodrigues et al. (2004), in a species-based, global-scale, study of gaps in protected areas suggests that regions most in need of additional protected areas are not those indicated by application of percent targets. They analyzed 4,735 mammal species, 1,171 threatened bird species, 5,454 amphibian species, and 273 freshwater turtle and tortoise species. Of these, 149 mammals, 232 birds, 411 amphibians, and 12 tortoises are threatened with extinction and have habitats are not protected anywhere. This indicates that about 80% of birds and over 90% of the other taxa are contained within the protected area system, but significant numbers remain unprotected. A caveat is that the list of unprotected taxa, interpreted as broader biodiversity surrogates, may not predict where protected areas are most needed for biodiversity in general.

Biodiversity surrogates based upon best possible use of a combination of environmental and species (for example, museum collections) data may provide greater certainty in estimating biodiversity patterns. Such a "calculus" of global and regional biodiversity may allow biodiversity targets to be formulated in ways that integrate socioeconomic factors and avoid weaknesses of types and percent targets (discussed further below).

5.2.5 Assessment

The substantial progress in declaring land protected has successfully conserved much biodiversity, especially when protected areas are specifically designed to conserve particular species of concern. But many species and ecosystems are not included in the current system. There is a need to apply an integrated approach to land/water use management beyond protected areas, in addition to expanding PA systems to make them more representative. Even when the system is well designed, investments in managing the protected areas remain inadequate. The total cost of establishing and maintaining effective protected area systems has been estimated (Balmford et al. 2002), but the estimate is based on relatively old data, and the needs and costs of such systems urgently need updating. Further, pressures on protected areas are likely to increase as growing populations consuming more resources make more demands on the remaining natural habitats at a time of rapid climate change.

For protected areas to effectively address their conservation and development objectives several actions need to be taken urgently. With respect to policy, the following are needed:

- Build effective synergies between global conventions and agreements that are dealing with biodiversity conservation and sustainable development.
- Give priority to promoting the effective implementation of the CBD Programme of Work on Protected Areas, including increasing international funding.
- Promote the application of a full range of revenue generation and sustainable financing mechanisms for protected areas management, while removing the policy and institutional barriers to sustainable financing solutions.
- Adopt an ecosystem perspective and multisectoral approach to development cooperation programs that include support for protected areas and take into account the impact of activities in adjacent and upstream areas.
- Ensure that development cooperation supports the development of effective institutions for PA management, giving priority to build effective, transparent, accountable, inclusive, and responsive institutions.
- Identify and apply policy and institutional options that promote the fair and equitable sharing of costs and benefits of protected areas at local, national, regional, and international levels.
- Develop, through legal, policy, and other effective means, stronger societal support for PAs, based on the benefits and the value of the goods and services they provide.
- Ensure the availability, understanding, and use of accurate, appropriate, and multidisciplinary information by all key stakeholders dealing with PA planning and management.
- Give priority to promoting and applying the best scientific and traditional knowledge to PA planning and management.

A number of technical issues need consideration as well:

- Use better approaches, based on best practices, to design and plan protected areas, including the use of the whole range of IUCN protected areas management categories and their integration into land/water use planning.
- Promote application of tools and methods to assess PA management effectiveness in both terrestrial and marine protected areas as a tool to improve management.
- Complete global, regional, and national gap analyses using best-possible surrogate information for all of biodiversity, to improve representation and persistence of biodiversity conservation in protected area systems. Such analyses should also include the assessment of the social costs and benefits of establishing and managing such systems. Give priority to completing the global system in relation to marine protected areas, freshwater ecosystems, and desert and semi-desert ecosystems.
- Promote the effective application of the principles of good governance for effective protected areas management. These principles must contribute to the full participation of local communities and indigenous peoples in both management and decision-making processes on protected areas.
- Develop better tools to evaluate the impact/effectiveness of PAs on biodiversity conservation.

5.3 Helping Local People to Capture Biodiversity Benefits

One of the fundamental challenges for biodiversity conservation is that the benefits of that biodiversity protection often accrue to people who are far removed from the resources while the costs (especially in terms of lost access to resources) are primarily paid locally. Where people do receive benefits from biodiversity locally, which provides incentives for local management, those benefits may be different from the benefits that accrue to people living farther away. An example would be sustainable harvesting of a species by local people that would create the incentive to conserve the species, thus meeting the desire of people far away for the mere existence of the species. Even when the benefits are the same, economists contend that local people have little incentive to manage resources to provide benefits (here, protect biodiversity) beyond their own communities unless they have some means of capturing some of the value of those non-local benefits. "Capturing" means any method that allows local people to receive payment or compensation for undertaking biodiversity conservation or sustainable use that provides benefits to non-local people. The idea is that compensating local people for their biodiversity-friendly actions based on the value of those actions beyond the local area will improve their well-being and thereby maintain higher levels of biodiversity.

Mechanisms to promote local capture of national and global biodiversity benefits include economic incentives, integrated conservation and development projects, ecotourism, and benefit sharing. Establishing any of these mechanisms requires that a significant share of the cost to develop and maintain the institutional capacity to manage biodiversity be paid by international conservation organizations, donors, and nations. As Barrett et al. (2001, p. 501) say, "The global beneficiaries of biodiversity must not abdicate complete authority and responsibility to either tropical states or indigenous communities, but rather must work to improve the capacity of nested institutions to induce and enforce tropical conservation." But this is not in accord with current political reality, as donor funding for biodiversity continues to fall far short of the needs.

Because local people are de facto the primary resource managers in most regions, it can be concluded *with a high degree of certainty* that working with the local communities is essential to conserve biodiversity in the longer term. Local human populations are best placed to ensure effective husbandry of the resource, and because resources figure strongly in the livelihoods of rural populations, particularly the poor, these groups are particularly important stakeholders. Community-level benefits are central to sustainable management, particularly when the resource is large and distant from major administrative centers, and the relevant government departments are understaffed.

Yet community involvement in protecting biodiversity is most likely to be effective only with appropriate property rights systems in place. Land tenure and property rights are closely linked to the ability of local communities to capture the benefits of biodiversity conservation and hence to their incentives to protect biodiversity. Weak property rights undermine community involvement in the protection of biodiversity because the community is unable to restrict external access to local resources; and because the community has little incentive to adopt long-term strategies to manage these resources, decision-making tends to be short-term and opportunistic. For example, in the francophone territories in West Africa, forest residents have no authority and hence no ability to restrict the exploitation of game by "outside hunters" (Bowen-Jones et al. 2002). Hence any schemes to compensate the local community for biodiversity protection would be rendered ineffective. Not surprisingly, the most successful and well-documented cases of wildlife management in Africa come from the dry savanna zone in the south (the best known is CAMPFIRE, Communal Areas Management Programme for Indigenous Resources, in Zimbabwe), where, inter alia, the tenurial context is much more favorable.

In Ghana, encouraging local community management of wildlife resources has involved the proposal that the government Wildlife Division devolve property rights over wildlife to certain local communities, thereby providing an incentive for the community to conserve and manage the natural resource base (ULG Northumbrian Ltd 2000). Simply given this authority, the community would be expected to manage the resource to maximize its own benefits, not biodiversity benefits. However, giving the property rights to manage the resource to the local community provides a mechanism through which outside agencies concerned with biodiversity conservation can negotiate with the community, and through which the community can have the legal backing to protect the resource from "outsiders."

In practice, achieving improvements in both local people's well-being and biodiversity protection has proven elusive. This section assesses several important policy responses that have sought to bring in the two elements together.

5.3.1 Economic Incentives: Indirect versus Direct

"Incentives" broadly cover any mechanism for changing actions. Individuals make decisions based on preferences, opportunities, and constraints. Economic incentives can alter the outcome of a decision process by changing the constraints or the relative net benefits of the set of opportunities. The institutional and market setting in which the decision is made affects the relative values of opportunities or constraints. Incentives are recognized as a key issue for biodiversity conservation. For example, Article 11 of the CBD states that "each Contracting Party shall, as far as possible and as appropriate, adopt economically and socially sound measures that act as incentives for the conservation and sustainable use of components of biological diversity."

Incentives may be negative (disincentives), such as taxation or access and user fees, or positive, such as tax credits. Responses that create positive incentives work by altering an individual's behavior toward more conservation activities, generally by establishing a mechanism through which the individual captures, or is compensated for creating, some of the social benefits associated with conserving biodiversity. Negative incentives aim at reducing negative impacts on biodiversity by ensuring that full costs of resource exploitation are paid. Typically a combination of negative and positive incentives will be used to halt losses of biodiversity.

A combination of controls and positive incentives will be more cost effective than relying on one or the other, and hence will be more sustainable in the long run. Positive incentives can be used to compensate people for loss of access to resources within PAs. Wells et al. (1992) caution against making the unsupported assumption that people made better off by development projects will refrain automatically from illegal exploitation of a nearby PA; to increase conservation activities, the compensation must have some mechanism for creating a conservation incentive within the decision framework. In addition, incentives are unlikely to work without a monitoring and enforcement system.

Although many underlying principles for introducing incentives have been discussed, interventions must be case specific, and approaches typically will include a combination of incentive measures that may include economic and regulatory measures, as well as measures such as stakeholder involvement and public education, to build an enabling framework (OECD 1999). The classification of incentives as positive or negative relates to the actor's behavior. From the perspective of the implementation of incentives as a response, the distinction between indirect and direct incentives is more important. Both positive and negative incentives can be direct or indirect.

5.3.1.1 Indirect Incentives

Development interventions in or near endangered ecosystems *indirectly* seek to provide desirable ecosystem services through three mechanisms: (1) redirecting labor and capital away from activities that degrade ecosystems (for example, agricultural intensification); (2) encouraging commercial activities that supply ecosystem services as joint outputs (for example, ecotourism); or (3) raising incomes to reduce dependence on resource extraction that degrades the ecosystem. These mechanisms are not always successful. In the case of redirection of labor or "conservation by distraction," the response may not reduce the labor allocated to the degrading activity if other people are hired to take advantage of the opportunities provided (Muller and Albers 2004). Commercial activities that maintain ecosystem services may be successful on a limited basis but rarely is the demand for the outputs large enough to support more than a small fraction of the local population. Lastly, raising incomes leads to conservation only if the extracted products are "inferior" goods that are replaced by preferable and less degrading goods as incomes rise.

Community-based natural resource management initiatives, also called integrated conservation and development projects, are one type of indirect incentive intervention In order to truly integrate conservation-based and development-based projects, ICDPs must use the development project to create incentives for conservation, establishing a direct and on-going link between the two objectives. Given the issues with indirect incentives already described, it is not surprising that many assessments of ICDPs report that they have had limited success in achieving their joint conservation and development objectives (Wells and Brandon 1992; Ferraro et al. 1997; Wells et al. 1998; Oates 1999; Ferraro 2001; Terborgh et al. 2002). ICDPs have been assailed for several reasons: erroneous assumptions about the desires of local people to protect nature, ambiguous effects on conservation incentives, complex implementation needs, failure to recognize the role of the market setting, and lack of conformity with the temporal and spatial dimensions of ecosystem conservation objectives (Brandon 1998; Southgate 1998; Chomitz and Kumari 1998; Simpson 1999; Ferraro 2001; Terborgh and van Schaik 2002; Muller and Albers 2004).

ICDPs are intuitively appealing because they assume either that local people will forgo harvesting in the protected area if they are offered a development project (school, dispensary, road, etc.) as an incentive, or that local people harvest in the PA because they have no alternative, and they will stop if alternatives are provided. However, the assumption that development will automatically favor conservation is not supported by the evidence (Braken and Meredith 2000). An early assessment of ICDPs worldwide (Wells et al.1992) concluded that they were reasonably effective in meeting development objectives but very few had a significant positive impact on conservation.

An even more pessimistic view on ICDPs is exemplified in Terborgh et al (2002), who argue that a bitter lesson is that when rural communities derive substantial benefits from the sustainable use of natural resources, the improved local economy can set in chain a process that drives conservation and development apart.

ICDPs allow local populations to improve their well-being by capturing non-local people's willingness to pay for biodiversity conservation, but in practice ICDPs rarely turn that "capture" into on-going incentives for conservation. Integrating conservation and development objectives can help identify trade-offs (see discussion below) but the project's success must be used as an incentive for conservation.

In an example of how important the setting can be in determining success of responses, the WWF and the Royal Netherlands Development Agency's Tropical Forest Portfolio, composed of seven ICDPs in six countries, recognized that linking conservation and development is constrained by a variety of circumstances that collectively threaten projects such as these. Funding was provided to identify and better understand these constraints through rigorous monitoring, technical assistance, capacity development, and improved information exchange. Details of the portfolio and other ICDP experiences can be found in McShane and Wells (2004). The main portfolio lessons are:

- *Implementation must take place at different scales.* It is easier to integrate conservation and development at larger scales that have increased area for protection, buffer zone, and development activities. The challenge for practitioners is not to decide the best scale at which to operate, but rather the optimal combination of actions that are required at different scales.
- *The policy environment is as important as field-based approaches.* The success of conservation and development efforts depends on the policy and market setting. Without supportive laws, policies and regulations, and their enforcement, it is unlikely that these efforts will be either successful or sustainable.
- *Sound institutions are the foundation of effective resource management.* The institutional characteristics of conservation and development initiatives include several different aspects: legal and organizational frameworks, formal and informal property rights and rules that govern resource management, and the norms and traditions of the different stakeholders and actors. Such initiatives require institutional forms with the capacity to deal with ecological, social, economic, and even political change.
- *Acknowledge and negotiate trade-offs.* Rather than the "win-win" outcomes promoted (or assumed) by many practitioners, conflict is more often the norm, and trade-offs between conservation and development need to be acknowledged. Identifying and then negotiating trade-offs is complex, involving different policy options, different priorities for conservation and development, and different stakeholders. The challenge in negotiating these trade-offs is determining levels of acceptable biodiversity loss and stakeholder participation.

ICDPs are just one kind of indirect incentive. Another type is nature-based tourism. This is a popular enterprise whose goal is to allow local people to capture the value non-local people are willing to pay for the local environment by charging them for access and services (such as guided tours, meals, and housing). Nature-based tourism according to the World Tourism Organization includes all types of tourism where the primary motivation of tourists is the observation and appreciation of nature, as well as cultures. Nature-based tourism projects purport to create an incentive for conservation because the income generated is a function of the quality of the environment: fewer tourists will come to a degraded area than to a more attractive one.

Several market studies show that a preserved environment, well-managed protected areas, and high biodiversity are becoming important elements in the choice of a recreation destination. The World Tourism Organization has conducted market studies in Europe and North America which show that consumers are willing to pay for these characteristics particularly when they were guaranteed that money goes "towards preservation of the local environment and reversal of some of the negative environmental effects associated with tourism" (Goodwin 2003, p. 278).

As with other indirect methods of value-capture by local people, however, in most situations, tourism income is generated without creating incentives for biodiversity conservation for local people. In an assessment of ecotourism in the Ecuadorian Amazon, Wunder (2000) found that tourism as a local conservation incentive works only if it changes labor and land allocation decisions. Tourism can also have potential adverse impacts for the sustainable use of biological resources and their diversity. For example, the demand of tourists for water, fuel, food, or other needs can strain the local resource base. Similarly, noise, jeeps, and garbage from tour groups can degrade the natural environment. The CBD Guidelines on Biodiversity and Tourism Development (http://www.biodiv.org/programmes/socio-eco/tourism/guidelines.asp?page=6) provide impact assessment, management, and mitigation guidance. They are currently being tested on case studies all over the world.

In many areas, the people who gain the income from tourism are not in control of the natural resources and cannot protect the resources even though they have an incentive to do so. In Khao Yai National Park, Thailand, tourism provides conservation incentives in some villages but the tourist groups and their demand for services is simply too low to create such incentives in more than a small fraction of the villages surrounding that park. Also, most income from tourism accrues to tour companies and hotels staffed by non-locals, thereby creating no conservation incentive for locals. In other areas, such as Zabalo in Ecuador, the community came together to place and enforce limits on hunting in order to improve the ability of tourists to observe these species (Wunder 2000).

In southern Africa, tourism for viewing and hunting animals has spurred farmers to abandon farming and let the land regenerate to wildlife habitat. Heal (2002) reports that about 18% of land in southern Africa has been converted into "game ranches" to allow tour companies and local people to capture non-local values for biodiversity through tourism. Assessments suggest that many species have rebounded, particularly elephants, from low levels as a result of this form of tourism. Heal also notes that biodiversity protection does not occur only in these ranches but also on farmland because of the legal system concerning property rights for wild animals—a critically important component of success—in much of southern Africa, which encourages farmers to capture animals and then sell them to game ranches instead of killing them. Southern Africa appears to be one large example where the incentives for conservation created by tourism, combined with the institutional setting and value of alternative land and labor uses, are large and widespread enough to have had a positive impact on biodiversity conservation.

5.3.1.2 Direct Incentives

An alternative approach to encouraging the conservation of endangered natural ecosystems is to pay for conservation performance *directly*. In this approach, domestic and international actors make payments in cash or in kind to individuals or groups conditional on specific ecosystem conservation outcomes.

In many countries, tax incentives, easements, and tradable development permit programs are widespread. In fact, these financial incentives have been shown to be useful for conserving land voluntarily (Boyd et al. 1999).

Perhaps the most obvious payment scheme is the purchase of full property interests in which a landowner who may develop the land transfers the land to a party who wishes to conserve it. Although such a purchase may preclude other activities that are compatible with the conservation goal, full-interest acquisitions are the most institutionally straightforward of all the conservation payment mechanisms and the costs of monitoring and enforcing an agreement are relatively low.

Tax credits or other subsidies equal to the difference in value between developed and undeveloped uses can also remove land from development uses. Alternatively, instead of using tax credits as a reward, the government can use taxes to punish development. Many complications arise concerning the relationship between taxes and income flows from property, and with other tax issues. Like easements, and indeed most incentive-based responses, tax-based conservation incentives require monitoring in order to confirm that the taxpayers are maintaining the land as they claim to be maintaining it. A potential advantage of tax-based incentives is that most of the administrative resources and systems needed are largely already in place.

Another incentive response is the use of tradable development rights, which require a restriction on the amount of land that can be developed in a given area. A government can award landowners the right to develop some percentage of their land and then permit these development rights to be traded. Because property owners can, in effect, choose among themselves where development will ultimately be restricted, it leads to the least-cost development restrictions. Institutionally, TDRs are relatively complex because they require the establishment of a new market and will impose monitoring and enforcement requirements. Also, TDRs can be problematic if the ecosystem value of the land is highest on the properties where the cost of development is least. In such a case, the market will lead to development on the most ecologically valuable property. This problem can be corrected by introducing an additional level of complexity to the market—"trading differentials" that reflect property-specific ecological characteristics. But this change clearly implies an additional and formidable set of administrative challenges. Lastly, the costs of "thin markets" (few users of the market) add to the overall cost of implementation.

One potentially large drawback of both TDRs and tax incentives is the inability to target specific habitat types and even specific properties. Because effective conservation may depend not only on the total area preserved, but also on the configuration of conserved lands, the conservation efficacy of tax incentives and TDRs is difficult to predict. Still, these tools could be used in addition to regulations and easements that target particular parcels. Tax incentives, tradable rights, and rights purchases all rely on voluntary decisions made by property owners in response to incentives in order to promote conservation on properties where the value of alternative uses is lowest and thus conservation is attained at the lowest opportunity cost.

Another type of incentive is a conservation easement or "partial interest," which is a contractual agreement between a landowner and a conservation interest. In exchange for payment or a tax deduction, a landowner agrees to relinquish rights to future land development. Institutionally, easements involve complex contracting issues but are a well-established legal mechanism. From the perspective of biodiversity conservation where particular parcels are more important than the total amount of area, easements have the potentially important advantage of allowing sensitive parcels to be targeted. A chief complication arises from the need to monitor the terms of the easement contract, especially over long periods of time and as ownership changes.

Taken as a group, financial incentives have been useful methods for encouraging the private conservation of land, but few analyses exist that quantify the contribution of this land to biodiversity conservation.

In countries with less well-established property rights, legal institutions, and tax infrastructure, experimentation with direct payment initiatives has just begun. Examples include the use of forest protection payments in Costa Rica (Box 5.3), conservation leases for wildlife migration corridors in Kenya, conservation concessions on forest tracts in Guyana, and performance payments for endangered predators and their prey in Mongolia. South Africa and American Samoa have over a decade of experience with "contractual national parks," which are leased from communities. Other payment initiatives are being designed or are under way in Mexico, El Salvador, Colombia, Honduras, Guatemala, Panama, Russia, and Madagascar (Ferraro and Kiss 2003).

Proponents of the direct payment approach argue that such an approach is preferable to indirect approaches because it is likely to be more effective, efficient, and equitable, as well as more flexibly targeted across space and time (Simpson and Sedjo 1996; Ferraro 2001; Ferraro and Simpson 2002; Ferraro and Kiss 2002, 2003). Payments can be made for protecting entire ecosystems or specific species, with diverse institutional arrangements existing among governments, firms, multilateral donors, communities, and individuals (du Toit et al. 2004).

However, direct payments have also been criticized. Like indirect interventions, they require on-going financial commitments to maintain the link between the investment and the conservation objectives. They may also transfer property right enforcement responsibilities to local participants, which can lead to inter- and intra-community conflict. Others express concern that direct payments turn biodiversity into a commodity (Swart et al. 2003). To date, no rigorous analysis of direct payments in these settings assesses the amount of biodiversity that is protected or conserved by each program.

5.3.1.3 Combining Incentive Schemes

Although the above discussion of direct and indirect approaches suggests a dichotomous choice, in practice interventions cover a spectrum from those that are most to those that are least direct. In many settings, a combination of incentives and disincentives, and of indirect and direct mechanisms may prove best in order to alter decisions toward conservation and to compensate people for lost access to resources (Muller and Albers 2004). Whether incentives are direct or indirect, if they are not large enough they will not induce biodiversity conservation.

Although some combination of direct and indirect incentives may prove most useful in any given situation, the Pigouvian principle, which prescribes that damages to the environment be taxed according to the damage they do, suggests that direct incentives will generally be preferred because it is more efficient to provide incentives for the "appropriate" use of the factor in question, rather than to tangentially related things.

BOX 5.3

A Direct Approach: Costa Rica's *El Programa de Pago de Servicios Ambientales*

In response to substantial deforestation in the last fifty years, practitioners and policy analysts working in Costa Rica have developed a pioneering, nation-wide system of conservation payments to induce landowners to provide ecosystem services: El Programa de Pago de Servicios Ambientales, or PSA. With help from multilateral aid agencies, Costa Rica's natural resource managers broker contracts between international and domestic "buyers" and local "sellers" of sequestered carbon, biodiversity, watershed services, and scenic beauty.

Established in 1996, the PSA grew out of an existing institutional structure of payments for reforestation and forest management, but contained two notable changes: (1) payments were made for ecosystem services rather than for support to the timber industry per se, and (2) funds came from earmarked taxes and environmental service buyers via a newly created National Fund for Forest Financing (FONAFIFO) rather than from general revenue funds. Suppliers of services are primarily individual landowners, associations of landowners, or indigenous reserves. Buyers of services include the Global Environment Fund (biodiversity), Costa Rica's Office of Joint Implementation (carbon), domestic hydroelectricity and municipal water providers (watershed services), and Costa Rican citizens paying via a gasoline tax (for carbon, biodiversity, water, and scenic beauty). By 2001, over 280,000 hectares of forests had been incorporated into the PSA at a cost of about $30 million, with pending applications covering an additional 800,000 hectares. Typical payments have ranged from $35 to $45 per hectare per year for forest conservation (Castro et al. 2000; Chomitz et al. 1999; Ortiz et al. 2002).

The mere existence of direct payment initiatives, however, does not imply that practitioners who use them have been successful in achieving conservation and development objectives. Even in high-income nations, where direct payment programs are more established, empirical analyses about actual impacts are rare.

A recent study (Barton et al. 2003) explored prioritizing PSA environmental service payments to private landowners and suggested that gains could be made by integrating the program into a regional trade-offs framework that targets payments to landowners, taking both biodiversity contributions (for example, from national parks) and costs into account. Current payments are approximately based on national averages for opportunity costs of foregone cattle ranching and the direct financial costs of the different forestry activities which are promoted. The study compared existing PSA allocations to a cost-effective allocation based on biodiversity complementarity values—estimated biodiversity gains relative to the existing PSA areas and the national parks. Targeting of PSAs based on complementarity has provided a more cost-effective approach, and lends support to the idea of integrating various biodiversity responses (national parks, incentives schemes, etc.) into a regional trade-offs framework.

Similarly, some combination of positive and negative incentives may prove most useful in a given setting. In the case of biodiversity conservation, negative incentives include fines and penalties for misuse, such as hunting in parks or illegal conversion of habitat. Taxes and user fees also create negative incentives for biodiversity degrading activities and are widely used in situations where property rights are well-defined and markets are well-functioning. Such negative incentives fit with a "polluter pays" approach in that the activity that degrades biodiversity is discouraged directly. As in the case of pollution, taxes or fines (negative incentives) make the biodiversity degrading activity less attractive for all, and unprofitable for some.

One complication with negative incentives for biodiversity conservation, especially in poor, remote, and rural areas, is that the enforcement and implementation costs of such programs can be high and fall on the government. For example, it is not illegal to possess fuelwood, but it is illegal, and destructive of biodiversity, to extract fuelwood from some areas, and so the enforcer must employ expensive patrols to catch the extractor (nearly) in the act. Positive incentives are more state-based in that if the forest is still there, the payment is made. This structure has advantages for the government but does place the burden of enforcement on local people to keep everyone, including non-locals, from extracting.

Another problem with user fees, taxes on extraction, and fines is that they cause conflict between the local people, who may have traditionally used the resource or who may be heavily reliant on the resource, and biodiversity conservation actors (NGOs and government). Positive incentives have the advantage of generating goodwill and recognizing that local people often feel that they have rights to the resource, for which they are compensated through payments. Positive incentives for biodiversity conservation allow local people to capture some of the non-local benefits that biodiversity creates rather than putting them in the situation of bearing the costs of generating those non-local benefits.

Positive incentives may pose problems in implementation because some baseline must be established concerning who should receive what levels of payment, tax credit, or other incentive. The announcement of such a policy may induce migrants to enter the area and may encourage more destructive activities by residents seeking to demonstrate high levels of dependence on the resource and thus higher need for payment. In addition, payment policies increase the income of rural people and that income can increase demand for the resource, thereby decreasing the effectiveness of the incentive created. This "income effect" would, however, have to be quite large to completely offset the incentive created (Muller and Albers 2004). The characteristics of a given setting will determine whether these problems are more substantial than the costs of enforcement of tax/fine policies, the conflict between groups, and the socially wasteful avoidance activities people undertake to lower the fines that they pay.

The IUCN and other organizations believe that one of the most cost-effective approaches to biodiversity conservation is not the creation of new pro-biodiversity incentives but rather the removal of widespread and powerful anti-biodiversity incentives known as perverse incentives. Perverse incentives induce the reduction of biodiversity and are often an unintended side effect of a policy meant to address a different issue. "Perverse incentives can include subsidies, tax relief and below-cost resource pricing in the agricultural, energy, forestry, fisheries, mining and transport sectors, as well as marketing restrictions and seed distribution systems which encourage a narrower range of agricultural species and varieties" (IUCN n.d., p. 1). Such policies and the resulting perverse incentives abound across the world.

One oft-cited example is the perverse incentives created by the Brazilian government that caused rapid rates of deforestation in the Amazon. One policy lowered taxes on agriculture, which increased the profitability of converting land from forest to cropland. Another policy, also aimed at supporting agriculture, granted rights to unclaimed land to squatters who "use" the land for a length of time, which linked land ownership to clearing of the forest. These policies each created, probably unintentionally, perverse incentives to remove biodiversity (Binswanger 1989; Mahar 1989). Similarly, subsidies that encourage agriculture also discourage biodiversity conservation. Logging practices on government land in many countries also create perverse incentives that lead to too little conservation of forests and biodiversity.

Although the repeal or redirection of perverse incentives appears to be an important response option that could prove quite cost-effective, this important mechanism for biodiversity conservation is not widely used. In addition, such incentives continue to be created anew as side effects of new policies. Integrating biodiversity into regional planning and into agricultural and fisheries policy would limit the creation of new perverse incentives, and perhaps aid the removal of existing perverse incentives.

5.3.2 Importance of Community-based Responses and Implementation

5.3.2.1 Access and Benefit Sharing with Indigenous and Local Communities

Through benefit-sharing mechanisms, countries and sometimes, local communities can capture some of the non-local values of biodiversity. Such mechanisms may be implemented to address equity considerations, but unless the income captured provides on-going incentives for conservation to the effective resource managers, conservation will not occur. In addition, legal complications abound and may diminish the effective incentives for conservation.

Community-based management of natural resources can contribute significantly to human well-being, but conservation of biodiversity is rarely considered in these approaches. To the extent that community-based management, such as Joint Forest Management in India (Box 5.4), encourages long-run management as opposed to the degradation that often comes with open access forest, such management can both improve local human well-being and conserve biodiversity.

The CBD has developed a program of work on its Article 8(j), which concerns the knowledge, innovations, and practices of indigenous and local communities. Several of the 18 tasks outlined by the Working Group on 8(j) relate to access and benefit sharing.

The International Treaty on Plant Genetic Resources for Food and Agriculture has provisions on prior informed consent, benefit-sharing, and farmers' rights. While most benefits will be shared on a multilateral basis (rather than with the specific provider of genetic resources), benefits such as the exchange of information, access to and transfer of technology, capacity building, and even a commercial benefit-sharing package should be available to communities through the system. Communities may also benefit through involvement in conservation and sustainable use activities.

The sixth Conference of the Parties to the CBD adopted the so-called Bonn Guidelines on Access and Benefit Sharing. These guidelines were drafted to help parties develop and draft legislative, administrative, or policy measures on access and benefit sharing. They cover roles and responsibilities, suggested elements for transfer arrangements, possible other approaches to access and benefit sharing, capacity building, and the relation between the access and benefit sharing provisions and the agreement on Trade-related Aspects of Intellectual Property Rights and the WTO.

Considering access and benefit sharing merely in the context of the relationship between providers (local communities, national administrative bodies) and users (private companies) has not been successful. It is important to consider the broader institutional environment in which access and benefit sharing is nested (Tobin 2001; Rosenthal 2003). Important steps in the direction of the creation of a more appropriate institutional environment are

BOX 5.4

Community-managed Forests in India

Communities managing forests is no new phenomenon in India. Over 6,000 village committees have initiated efforts in the past century to conserve and use forest resources sustainably on their village common land, perhaps as a response to biomass scarcity (Murali et al. 2002). However, such efforts were isolated in nature and a government resolution on Joint Forest Management issued in 1990 opened new possibilities for these communities to strengthen their conservation of forest resources. The primary concern for communities to undertake forest conservation is the use of forest resources, in other words, to maintain the diversity of species that exist in forests.

JFM, envisaged that degraded lands would be protected through combined efforts of the community and the state would improve regeneration and enhance forest cover and meet community needs. The state would share the profits from timber sales as an incentive to the people. The motive was to enhance timber production through community effort despite the growing evidence that non-wood resources have more to contribute to the rural economy and state exchequer.

The effort toward participatory forestry management is more for livelihood security than for biodiversity conservation, though the idea was to do both. A comprehensive assessment of the ecological impacts of the community forestry program in India highlighted that more species were found in villages with self-initiated forest protection committees having a long history of protection (Murali et al. 2002; Ravindranath et al. 2000).

The species diversity in forest protected by committees formed prior to the 1990 resolution is high, indicating that the people maintained high species diversity for meeting their diverse biomass needs. The plantations raised by the forest department mainly constituted the fast growing non-native species such as *Eucalyptus, Acacia,* and *Casuarina,* to meet the immediate firewood demands. Furthermore, the area allocated for community protection was highly degraded, rendering the area less species rich. Thus JFM neglected to develop forest resources into biodiversity-rich resources. However, the species diversity was high in community forestry areas (Murali et. al. 2002).

The primary actors in the community forestry program are the forest department and the forest-dependent community. Influential actors at the policy level are donor agencies at the national and international levels, and nongovernmental organizations at the local level. The definition of success differs for all these four actors. The government may conclude community forestry to be successful if the forest is regenerated, while the community may feel success if some economic improvements are delivered. NGOs may see success if equity concerns are addressed, while the donor agencies may want a stable institution in place for sustainable use of forests. Thus concern for biodiversity conservation has not been given a priority. Overall, in the Indian context, it can be concluded that the participatory forestry program has returned to the country. However, more emphasis has been laid on improvement of livelihoods of the local people than on the conservation or enhancement of biodiversity.

the development of an international system for tracing the flow of genetic resources and the development of networks of codes of conducts for gene banks or botanical gardens. Finally, it is important to further develop the legal framework of intellectual property rights.

Two recent attempts successfully integrated the issue of IPR in the access and benefit sharing issue. First, the ITPGR regime proposes a system of farmers' rights, including research exemption and farmers' privilege in the development of new varieties. The treaty, however, does not address biodiversity conservation directly, as it is focused on food security, and concerns only a specified list of key species. Second, in India the 2001 Plant Variety Protection and Farmers' Rights Act (2001) grants plant breeders' rights to local communities (Lalitha 2004).

Acknowledging the Bonn Guidelines as an important first step, the seventh Conference of the Parties adopted Decision VII/19 on access and benefit sharing as related to genetic resources. It accepted the need for further work on the definition of certain terms in the Bonn Guidelines. It also discussed other approaches to assist with the implementation of the access and benefit sharing provisions of the Convention; measures related to prior informed consent, capacity building for access and benefit sharing, and the negotiation of an international regime on access to genetic resources and benefit sharing. The seventh Conference of the Parties also mandated the Ad Hoc Open-ended Working Group on Access and Benefit-sharing to negotiate a possible international regime on access and benefit sharing and adopted an Action Plan on capacity building for access and benefit sharing.

Overall, the trend is moving gradually toward more creative benefit sharing as experience and "best practice" in benefit sharing advance. Benefits shared through commercial partnerships today include the monetary (for example, fees, milestone payments, and royalties) and the non-monetary (for example, research collaborations, access to information and research results, training, technology transfer, and capacity building). They are spread across the short, medium, and long term, and vary by partnership and commercial sector, but a standard of "best practice" has emerged. The bulk of benefits for conservation and development resulting from commercial use of genetic resources have primarily resulted from the research process, and through an increasing use of partnerships between companies and source countries (usually represented by research institutions, only in a very few cases community groups). Direct benefits for conservation do not necessarily result from these partnerships, although some include payments to protected areas, and support national biodiversity inventories and biodiversity science necessary for national conservation plans. Indirect benefits include the promotion of sustainable economic activities based on the supply of genetic resources, and in some cases the supply of raw materials for manufacture (ten Kate and Laird 2002).

In the marine realm, as much as in the terrestrial one, the sharing of benefits arising from the utilization of biodiversity resources is intimately linked with the access to those resources.

5.3.2.2 Impact of the Setting on Effectiveness

The socioeconomic and institutional setting in which any of these responses are applied can significantly alter the outcome of the response. Understanding the potential interaction of a given response with the setting, local to national, can assist in determining which policies are likely to be effective in a particular setting.

From a theoretical perspective, the impact of improved market access on forest degradation and biodiversity is ambiguous (Omamo 1998; Key et al. 2000; Robinson et al. 2002). Without access to markets, most resource use will be for home consumption (Sierra 1999). As market access increases, the impact on the resource base, whether positive or negative, will depend on the relative strength of two effects. Some households will increasingly switch from purely subsistence extraction to commercial extraction, whereas other households, especially those with high opportunity costs of labor, may choose to purchase forest resources from the market rather than extract, using their labor for alternative activities (Robinson et al. 2002). In addition, policies or programs that improve market access to create economic incentives will typically interact with the distribution of labor opportunity costs (Robinson et al. 2002). The creation or improvement of roads allows a policy-maker to reduce market access costs directly (Bluffstone 1993; Cropper et al. 1999; Imbernon 1999). Resource use incentives change because roads reduce the cost both of accessing resources and of removing resources. Working in the opposite direction, the same roads also reduce the cost of accessing substitutes for forest resources (Robinson et al. 2002). The creation of roads also changes opportunities for labor, which may alter resource management decisions (Muller and Albers 2004).

5.3.3 Assessment

Positive incentives to induce local people to conserve biodiversity and use biological resources sustainably have the potential to improve local human well-being and protect biodiversity. These responses allow local people to capture non-local values of biodiversity and thereby place some of the cost of conserving biodiversity on those who value it outside the local area. The effectiveness of economic incentives for inducing biodiversity conservation, however, is strongly dependent on the setting in which the decision is made.

Economic incentives that use development activities cannot be all things to all people as the approach has so often been marketed to raise funds. Better management arises when trade-offs between biodiversity, income generation, and societal needs are realistically acknowledged. The promotion of "win-win" outcomes has been politically correct at best and naive at worst. Despite the importance of compensating people for the costs they bear, responses that only compensate people and do not create conservation incentives do not lead to biodiversity conservation.

A key constraint in identifying what works and what does not work to create economic incentives for ecosystem conservation is the lack of empirical data supporting or refuting the success of *any* approach. Project analyses focus on whether the project became self-sufficient or generated income, but almost never fully characterize the project's impact on biodiversity conservation. Few rigorous and systematic empirical evaluations assess whether an existing initiative to allow people to capture benefits from biodiversity is achieving the conservation and development objectives it purports to achieve. Empirical research on the use of economic incentives to achieve ecosystem conservation and economic development goals in low-income nations is a critical next step.

5.4 Promoting Better Management of Wild Species as a Conservation Tool

During the past 15 years, the 7,000 scientists affiliated with IUCN's Species Survival Commission have contributed to the development of more than 50 species action plans. These plans review the current situation for those taxa and suggest conservation actions needed to alleviate the threats to that species. A review of 42 of these action-plans reports some clear priorities among suggested conservation activities for the future (Schachter

1998). A clear majority of actions (54%) relate to the need for more research to fully understand the problems and potential solutions, while 15% relate to legislation and policy action. The majority of this policy action is related to gaze ting of protected areas, with a secondary emphasis on implementation of international multilateral agreements. Other recommendations are as follows: ecological management, such as control of invasive alien species, reintroduction of individuals and adaptation to climate change, and issues related to sustainable use each represent 7% of recommended actions and capacity building/public awareness activities account for 6%. Ex situ management recommendations represent 5% of suggested actions.

While the relative proportion of each of these management options may change with the taxa, generally speaking they are all employed in a broad-based conservation plan. Increased and improved knowledge is a critical tool for all the other management options. The following sections assess in more detail the experiences with these options.

5.4.1 Legislation and Policy Action

5.4.1.1 Protected Areas for Species Conservation

Protected areas have already been discussed and will not be discussed further here except to note a critical issue: whether a sample of species can act as adequate "surrogate" information for the general biodiversity patterns we need if we are to address "all" of biodiversity. One perspective on this problem is that if species conservation is a goal, providing protected areas based on a goal of representative habitat types is not an adequate strategy, as it can result in omission of species with restricted ranges and endemic species which often are most in need of conservation action (Brooks et al. 2004). An alternative view (Cowling et al., 2004) is that we will have to make best-possible use of all available data, and this will require combining species and habitat data in some appropriate ways (this view is discussed below).

5.4.1.2 Legislation

A few key international agreements are based on the species level of biodiversity, including CITES, the Convention on Migratory Species, and the International Convention for the Regulation of Whaling. The ICRW has come under considerable scrutiny, as the debate at the meetings of the parties has not been able to move beyond political agendas.

While CITES has had some notoriety in its role with respect to regulation of the ivory trade, and has succeeded in reducing international trade in some species, the "success" of this convention as a tool in conserving commercially important taxa such as fish and timber is yet to be proven. An IUCN report examining the effectiveness of CITES (IUCN 2000a) concluded that CITES had been effective in (1) providing a comprehensive database on international trade in wildlife, and (2) providing some incentives for conservation. However, the convention cannot be expected to have impact beyond its mandate of regulating international trade and domestic economic issues, and other pressures confound the effectiveness of CITES.

Despite the problems, these agreements provide an important opportunity for countries to debate issues relating to the sustainable use of their natural resources and to share ideas on the best ways to cooperate in this effort.

5.4.2 Ecological Management and Reintroduction

5.4.2.1 Reintroduction of Species

Reintroduction of species to their native habitats has become a major tool for species conservation. The principal aim of any reintroduction should be to establish a viable, free-ranging population in the wild (whether species, subspecies, or race) that has become globally or locally extinct (extirpated) in the wild. It should be reintroduced within the species' former natural habitat and range, the conditions that led to its previous demise should have been corrected, and should require minimal long-term management (IUCN 1998). IUCN guidelines consider the impact of reintroduction of species on human populations. Socioeconomic studies are recommended to assess impacts, costs and benefits of the reintroduction program to local human populations. In addition to the general guidelines, guidelines are available for specific taxa including elephants, non-human primates and galliforme birds (http://www.iucnsscrsg.org/pages/3/index.htm).

Reintroduction and restocking projects have been undertaken with more than 120 species. Typically such projects are carried out by a consortium of zoos and in cooperation with the coordinators of the relevant regional ex situ programs and taxon advisory groups. Examples of successful reintroduction or restocking projects, most of them involving several zoos, include Southern white rhino, Golden lion tamarin, Golden-headed lion tamarin, Mexican grey wolf, Black rhino, Przewalski's horse, European bison, Arabian oryx, Scimitar-horned oryx, Addax antelope, Sable antelope, Mhorr gazelle, Alpine ibex, Bearded vulture, California condor, Andean condor, Mauritius kestrel, White stork, Great eagle owl, Western Australian swamp turtle, Puerto Rican crested toad, Mallorcan midwife toad, Jersey agile frog, and many others. A compendium of reintroduction and reintroduction practitioners was completed in 1998 and highlighted projects that covered a broad spectrum of taxa and geographic regions. Plant reintroductions are reviewed by geographic region and 217 animal reintroductions are listed by taxa.

Lessons learned from some of these reintroduction projects have been compiled by Beck et al. (1994) and Reading et al. (2002). Beck noted that reintroductions of birds and mammals predominated, and that 48% of the projects reported involved species that were listed as threatened on the IUCN Red List. He reports a success rate of only 11% while noting that another study by Griffith et al. (1989) estimated a 38% success rate. Success depends mainly on a deep knowledge of the species biology and ecology, availability of suitable habitats (as remnants or restored habitats), an initial stock of individuals of high genetic and genetically-based phenotypic diversity, and a long-term monitoring of the reintroduced populations. Successful projects tend to be large in time scale and in numbers of species introduced. In addition, involvement of local people was found to be a key factor. When attempts fail, the reasons are probably related to the narrow focus on biological and technical aspects of the reintroduction (Reading et al. 2002).

5.4.2.2 Management of Invasive Alien Species: Prevention, Control, or Eradication

Invasive alien species are a growing threat to biodiversity and the 2003 IUCN Red List of Threatened Species documented several specific cases (www.iucnredlist.org). The CBD (Article 8h) calls on parties to "prevent the introduction of, control, or eradicate those alien species which threaten ecosystems, habitats, or species." Within that context a strong consensus agrees that preventing species invasions is the safest and most cost-effective approach to the problem of invasive species (Mooney and Hobbs 2000). However, the expansion of global trade is moving more species around the world more quickly and overwhelming efforts to prevent invasions.

Several models are available for ecological and economic assessment of controlling biological invasions (Higgins et al. 1997; Wadsworth et al. 2000; Perrings 2002) and cost-benefit analyses conclude that costs of eradication are always higher than costs of prevention.

Eradication and control of invasive species has taken the shape of many different strategies (Wittenburg and Cock 2001; Veitch and Clout 2002). Chemical control of invasive plant species, sometimes combined with mechanical removal like cutting or pruning, has been useful for controlling at least some invasive plants, but has not proven particularly successful in eradication. In addition to its low efficiency, chemical control can be expensive: in 1990 in the United Kingdom, the average cost of treating a hectare invaded by *Heracleum mantegazzianum* was $705 to $1,764 (Sampson 1994). Biological control of invasive species has also been attempted. The rationale behind this approach is to take advantage of ecological relationships like competition, predation, parasitism, and herbivory, between an invader and another non-native organism introduced as controlling agent. Results are mixed. For example, the introduction of a non-native predatory snail to control the giant African snail in Hawaii led to extinction of many native snails (Civeyrel and Simberloff 1996). Also, the prickly pear moth (*Cactoblastis cactorum*) used to fight the invading *Opuntia* species in Australia, has recently invaded the United States, posing a serious threat to the native *Opuntia* species (Stiling 2002). Some 160 species of biological agents, mainly insects and fungi, are registered for controlling invasive species in North America and many of them appear highly effective (Invasive.org, n.d.). At least some of the biological agents used are themselves potential invaders (Hoddle 2004).

Successful eradication cases have three key factors in common: particular biological features of the target species (for example, poor dispersal ability), sufficient economic resources devoted for a long time, and widespread support from the relevant agencies and the public (Mack et al. 2000).

When complete eradication is not possible, or it is not desired, as in the case of invading native species, some measures of "maintenance control" aimed at maintaining populations of the invading species at low, acceptable levels have been attempted. However, the chemical and mechanical controls used pose many problems, including the high cost and low public acceptance of some practices (Mack et al. 2000).

Although biological invasions are complex ecological, evolutionary and socioeconomic problems, a better understanding is being achieved, especially in ecology, both of invasiveness and habitat vulnerability to invasion. This knowledge is essential to determine how much effort to invest in controlling an invasive species that has already become established or to clarify the trade-offs managers and land planners will have to consider. Social and economic aspects have received less attention, perhaps because of difficulties in estimating the trade-offs involved in biological invasions (Perrings 2002). Developing models that include the different factors comprehensively and attempt to calculate cost-benefit ratios would be the best approach both for preventing and controlling biological invasions.

The Global Invasive Species Programme is an international response to address the problem, supported by the CBD Conference of the Parties. GISP has called for improved monitoring, better quarantine practices, an improved legal framework, greater attention to the problem of ballast water, better worldwide regulation of species trade, a more rational approach to biological invasions by the public and users of both native and non-native biodiversity, better mechanisms of control of established invasive species, and adequate monitoring and evaluation to test for success of eradication and control programs (McNeely et al. 2001). The CBD has adopted Guiding Principles on Invasive Alien Species (Decision VI/23) as a basic policy response, but it is too early to assess the effectiveness of implementation.

5.4.2.3 Adapting for Climate Change

A recent report has suggested that between 15% and 37% of species could be at risk of extinction due to the impacts of climate change (Thomas et al. 2004); an alternative view is that these estimates have a high degree of uncertainty, and many other reports have documented shifts in species distribution as a result of global change (Pounds et al. 1999; Parmesan and Yohe 2003; Root et al. 2003). Today's species conservation plans may effectively incorporate adaptation and mitigation aspects for this threat. Several potential tools are available to help assess species' vulnerability to climate change (IUCN 2003b). The first, which could be undertaken with or without detailed species information, is to produce a matrix of data available for species by geographic region, using point occurrence data or high-resolution grid data. The second method involves using global and/or regional climate models over different time scales, with impact models, to predict responses of species within particular habitats, and thereby appreciate expected habitat changes. A third method involves development of vulnerability criteria based on inherent biological characteristics of the species that would hinder adaptation to a changing climate. Some of these characteristics would include restricted ranges, poor dispersal, extreme habitat specialization, and susceptibility to climatic extremes. Further work on all these methodologies is needed before a full assessment can be prepared.

5.4.3 Sustainable Use Programs

Key multilateral environmental agreements such as the CBD include within their objectives sustainable use of natural resources. Sustainability can be defined as using resources "at rates within their capacity for renewal" (IUCN/UNEP/WWF 1991). At its simplest, the concept of "sustainable use" supposes with appropriate restraint and efficiency of harvesting, the wild species can be used without it becoming depleted (Mace and Hudson 1999). However, the term "sustainable use" also describes the approach of actively promoting use as a conservation strategy (Allen and Edwards 1995; Hutton and Dickson 2000). The argument is that promoting use, or allowing use to continue, encourages people to value wild resources. And when wild species and their habitats have value, this discourages the conversion of natural habitat to other competitive land uses.

Three management approaches parallel the three management goals for sustainable use of wild species: managing for the species, ecosystem-based management, and resource management. Conserving exploited species directly is the approach classically adopted in wildlife management (Caughley 1977; Beasom and Roberson 1985) and fisheries (Larkin 1977). Where the goal is species conservation, and where a specific population has a distinct identity and can be managed directly, the species management approach can be effective. However, managing for a single species is rarely a good substitute when the goal is ecosystem health, which is tied to the entire suite of species present in the area. Where human livelihoods depend on single species resources, species management can be effective, but where, as is frequently the case, people depend on a range of different wild resources, single species management is not the approach of choice.

Conserving exploited species when the management approach is ensuring resource availability to support human livelihoods is frequently unsuccessful. This is because optimal management for resources frequently requires overexploitation of particular wild species and an overall loss in biodiversity (Hulme and Murphree 1999). For example, the loss of large predators might be acceptable under a resource management approach in southern Africa grasslands if this allows private landowners to maintain high enough stocks of ungulates to be economically viable, and thus avoid conversion to other land uses. Maintaining resource availability can result in increasing the production from valued species at the expense of those species of less concern, even those, which have resource value. This increased specialization on certain species and homogenization of the resource base is akin to the conversion of a natural landscape into an agricultural landscape (Salwasser 1994; Freese 1998).

Regardless of the management goal and approach for sustainable use of wild species, the IUCN/SSC Sustainable Use Specialist Group has identified a set of considerations necessary to achieve successful sustainable use, based on global collective experience. To increase the likelihood that any use of a wild living resource will be sustainable requires consideration of the following principles (IUCN 2000b):

- The supply of biological products and ecological services available for use is limited by intrinsic biological characteristics of both species and ecosystems, including productivity, resilience and stability, which themselves are subject to extrinsic environmental change.
- Institutional structures of management and control require both positive incentives and negative sanctions, good governance, and implementation at an appropriate scale. Such structures should include participation of relevant stakeholders and take into account land tenure, access rights, regulatory systems, traditional knowledge, and customary law.
- Wild living species have many cultural, ethical, ecological, and economic values, which can provide incentives for conservation. Where an economic value can be attached to a wild living species, perverse incentives removed, and costs and benefits internalized, favorable conditions can be created for investment in conservation and sustainable use of the resources.
- Levels and fluctuations of demand for wild living resources are affected by a complex array of social, demographic, and economic factors, and are likely to increase in the coming years. Thus attention to both demand and supply is necessary to promote sustainable use.

Sustainable use of natural resources is an integral part of any sustainable development program, yet remains a highly controversial subject within the conservation community (Hutton and Leader-Williams 2003). Attention to all factors, beyond the biological and ecological characteristics of the resource involved is a key to success. In particular, care in establishing positive incentives for conservation and sustainable use is critical.

5.4.4 Communication/Awareness Raising

Education and communication for conservation is discussed at length later in this chapter. The principles outlined there are a useful basis for establishing communication strategy for species conservation.

5.4.5 Ex Situ Management

More than 1,000 zoos, aquaria and botanical gardens worldwide welcome in excess of 600 million visitors annually. The justification of these institutions is to complement in situ conservation in several ways including: (1) to provide increased knowledge about species that need conservation effort; (2) to raise awareness among the general public of the value of those species; (3) to raise funds for *in situ* action; and (4) to help build capacity both in country and abroad for global conservation. These objectives are enshrined in Article 9 (ex situ conservation) of the CBD. Successful captive management programs have also provided individuals for reintroduction programs. IUCN's Statement on the Management of Ex Situ Populations for Conservation (2002) provides specific direction to ensure that captive management of species contributes to in situ conservation.

As of September 2003, no less than 174 international studbooks for threatened species or subspecies, covering a wide range of taxa from Partula snails to large apes, were kept under the auspices of the World Association of Zoos and Aquariums. In addition, the regional zoo associations keep regional studbooks and have run, since 1981, cooperative ex situ population management programs for selected species. For example, the American Zoo and Aquarium Association, whose membership includes 218 accredited zoos and aquaria throughout North America, currently administers 106 Species Survival Plans® covering 171 species. However, the majority of these (64%) are for mammals, 13% for birds, and only 6% each for reptiles/amphibians and fish. No plants are included in these plans. The European Association of Zoos and Acquaria operates 138 European Endangered Species Programmes in which about 300 institutions from Europe and the Near East participate. Similar networks also exist in the Australasian Region (the Australasian Species Management Program of the Australasian Regional Association of Zoological Parks and Aquaria) and in Africa (the African Preservation Programme of the Pan African Association of Zoological Gardens, Aquaria, and Botanic Gardens). As extensive as some of the efforts to manage threatened species may be, they still represent a very small proportion of species diversity and primarily that of the charismatic megafauna.

That said, ex situ zoo populations can directly support the in situ survival of some species in a number of ways—through ongoing research to understand the biology and ecology of threatened species, training of specialists in conservation, public awareness raising and generation of resources for conservation on the ground. Finally, populations of captive specimens can provide the nuclei for reestablishment or reinforcement of wild populations in nature. The World Zoo Conservation Strategy emphasizes that such reintroductions and restocking projects, when properly applied (that is, in agreement with the IUCN/SSC Guidelines for Re-introductions), can bring great benefits to natural biological systems. However, while captive breeding programs are often touted as an important conservation contribution, especially when part of a reintroduction plan, they can also create uncontrolled demand for live specimens of endangered species. Clayton et al. (2000) presented a case study on trade in the endangered Indonesian Babirusa (*Babyrousa babyrussa*). International interest in the captive breeding of this species gave hunters and dealers the false impression of a potentially lucrative and officially sanctioned demand for any live Babirusas they might catch. Swift action by the Indonesian authorities halted this trade, but the study provides a warning about the damage that can be caused to the conservation of a species if management programs are instituted without a full understanding of the practicalities of its conservation, particularly interactions between the species and local people.

5.4.6 Assessment

This assessment of possible approaches to species conservation leads to the following conclusions:

- It is imperative to develop approaches for species conservation that take into account impacts on affected human populations. Management and sustainable use of wild species will remain a key response at the species and population level, with, in most cases, a direct link to livelihoods. It is therefore essential to design targeted approaches, with clear objectives, and measurable indicators for monitoring the outcome.
- Reintroduction of species, though often very expensive, has often had very good results. For many species, knowledge and technical expertise required for a successful reintroduction exist. However, reintroductions are unlikely to be successful without the consent and support of the people inhabiting the target area, so programs that consider and respond to local people's concerns are likely to be more successful and cost-effective. The success of some reintroduction efforts should not imply any weakening of conservation of species in their natural habitats.
- Control or eradication of an invasive species once it is established has appeared extremely difficult and costly. Prevention and early intervention have been shown to be always more successful and cost-effective than late responses. Successful prevention requires more efforts, especially in the context of international trade, and in raising awareness of the threat of invasive species.
- Sustainable use programs must include consideration of social and economic issues as well as the intrinsic biological and ecological considerations related to the specific resource being used.
- Zoos, botanical gardens, aquaria, and other ex situ programs are essential elements in building support for conservation, supporting valuable research, and providing cultural benefits of biodiversity to the visiting public.

It is noteworthy that the vast majority of these tools have been used on a very limited range of taxa—primarily the charismatic megafauna and some commercially important species such as fish. Moreover, we still know relatively little about the effectiveness of these tools for many plants, invertebrates, or species in the marine realm.

5.5 Integrating Biodiversity into Regional Planning

5.5.1 Introduction

The need to integrate conservation and sustainable use of biodiversity into relevant sectoral and cross-sectoral plans, programs, and policies is highlighted as a requirement in the CBD (Article 6b). This integration is commonly referred to as "the mainstreaming of biodiversity" and includes situations where biodiversity and economic gains can be simultaneously achieved, where biodiversity losses are exceeded by biodiversity gains, or a sectoral activity is dependent on sustainable use of biodiversity and the inclusion of biodiversity concerns into sectoral policies. The many examples of mainstreaming activities are showcased in several documents, including Pierce et al. (2002).

A major mechanism for mainstreaming is the incorporation of biodiversity into regional plans (discussed earlier). These plans are usually the outputs of a spatial systematic assessment of the region, identifying areas of conservation and development values, threats, constraints, and opportunities. Planning systems are an essential component of most sectors as they identify what happens where on the landscape, and ensure an effective land use system meeting the needs of the development sectors without compromising the needs of the environment. These regional plans have in the past been conducted by separate authorities with the conservation community usually identifying areas of biodiversity concern for conservation efforts and the development planning authorities assessing and identifying development opportunities in the area for farming, mining, tourism, etc.. In many instances these independent plans were based on different datasets and methodologies, and did not feed into one another. As is often the case with conservation and development, areas important to one are often important to the other (van Rensburg et al. 2004), resulting in conflicting demands of conservation and development sectors for the same land.

However, recent initiatives have shown that if one were to conduct these planning assessments concurrently for both development and conservation and have the planners talking to one another, then the options for trade-offs and win-win scenarios are increased (Faith et al. 1996; Ando et al. 1998; Cowling and Pressey 2003; Gelderblom et al. 2002). This realization has led to regional plans that cater to both development and conservation concerns. Many governments and planning authorities have identified this approach as an appropriate way of managing their natural resources.

At the heart of regional planning is the question of how best to accommodate development that meets the social and economic objectives of the region while ensuring that the condition of regional biodiversity is maintained. Sustainable development relies on biodiversity conservation as an integral part of regional policy and planning. Integrated regional planning is not new; these types of plans were already abundant in 1992 at a workshop on *The New Regional Planning* at the sixth IUCN World Congress on National Parks and Protected Areas in Caracas, Venezuela, where 50 case studies focused on integrated regional-scale planning in Africa, Asia, and North, Central, and South America.

Integrated regional plans focus on integrating sectors, scales, and responses and fall into the ecosystem approach described in detail by the CBD; this approach provides principles for integration across scales and across different responses. Its seventh Conference of the Parties, for example, has addressed "Principle 10," on achieving appropriate integration of biodiversity conservation and use of biological diversity. Central to its rationale is that "the full range of measures is applied in a continuum from strictly protected to human-made ecosystems" and that integration can be achieved through both spatial and temporal separation across the landscape, as well as through integration within a site.

The seventh Conference of the Parties made the following recommendations associated with Principle 10:

- develop integrated natural resource management systems and practices to ensure the appropriate balance between, and integration of, the conservation and use of biological diversity, taking into account long and short-term, direct and indirect, benefits of protection and sustainable use, as well as management scale;
- develop policy, legal, institutional, and economic measures that enable the appropriate balance and integration of conservation and use of ecosystems components;
- promote participatory integrated planning, ensuring that the full range of possible values and use options are considered and evaluated;
- seek innovative mechanisms and develop suitable instruments for achieving balance appropriate to the particular problem and local circumstances;
- manage areas and landscapes in a way that optimizes delivery of ecosystem goods and services to meet human requirements, conservation management, and environmental quality;

- determine and define sustainable use objectives that can be used to guide policy, management, and planning, with broad stakeholder participation.

Case studies evaluating implementation of the ecosystem approach are limited, and CBD has called for additional case studies. Some existing case studies (CBD 2003d, p. 8) have suggested a need to "dispel the myth that 'win-win' situations between development and conservation objectives were widely achievable, and concentrate instead on understanding how trade-offs and equitable compromises could be attained." Other lessons emerging from experiences so far suggest that mainstreaming the ecosystem approach would require increased take-up by parties to multilateral environmental, trade, and development agreements, and by financial institutions in their funding decisions (CBD 2003e).

Case studies point to the need for addressing trade-offs and synergies in regional planning. A recent review of experiences in 15 Asian countries (Carew-Reid 2002) found that biodiversity planning has greater influence if it is viewed more as a political and economic process in which hard decisions are made on resource allocation and use.

Regional plans that integrate biodiversity can be drawn up in a number of ways. Conservation priorities may be identified using standard procedures of systematic conservation planning mentioned earlier in this chapter. In this way, all development sectors are informed of development options and can direct development away from areas of high biodiversity conservation value. This is an improvement on the ad hoc planning that happened formerly in both the conservation and development sectors (Pierce et al. 2002).

The earlier section on protected areas assessed how a regional planning framework that incorporates complementarity can focus on the problem of integrating intrinsic and future values of biodiversity into a trade-off decision framework for regional planning. When global biodiversity values are integrated into multicriteria analysis, policy decisions about one "place" in a region are linked to overall regional net benefits/trade-offs. In this regional planning framework, the contribution of a place to global biodiversity conservation is necessarily estimated by its complementarity value. Other, local, values in a given place (which are sometimes related to biodiversity) enter the multicriteria analysis either as measurable additional benefits of conservation land/water use, or as opportunity costs of conservation in that place.

Resulting conservation priorities can form core conservation areas supported by buffer and transition zones under models like the biosphere reserve model. A key component of this partnership between development and conservation is often the formation of a cross-sectoral partnership between the conservation and planning authorities (Gelderblom et al. 2002; Cowling and Pressey 2003).

This focus on spatial priorities and trade-offs follows the ideas of Saunier and Meganck (1995), who highlight that integrated regional planning focuses on spatial units while cutting across sectors, instead of the older forms of planning which focus on sectoral units. Australia's use of Integrated Natural Resource Management planning at the catchment and subcatchment level follows a similar approach of conservation planning inputs into natural resource planning (Lowe et al. 2003). Both South Africa and Australia make use of the idea of "living landscapes" as the ideal end point of these integrated regional plans (Steiner 2000). "Living landscapes" as defined by Driver et al. (2003, p. 1) are "landscapes that support life of all forms, now and into the future." Lowe et al. (2003, p. 59) define a plan for a "living landscape" as one that "aims to protect a landscape's ecological health by integrating nature conservation into the farming landscape so that life on the land continues both for the flora and fauna and for farmers and their families."

5.5.2 Integration of Regional Response Strategies

Integrated regional planning relies not only on integrating different sectors, but also on the use and integration of a number of responses in the region. These responses are discussed in other sections and include protected area systems, promotion of local benefits, economic incentives, and mainstreaming biodiversity into development sectors like agriculture and the sustainable use of wildlife. Integration among these responses (or instruments) will promote effective trade-offs and synergies among regional values and sectors, and with global biodiversity values as well. Regional perspectives on payments to private landowners and accounting for biodiversity contributions from agricultural lands illustrate integration of strategies in a trade-offs/synergies framework. For example, local-global trade-offs benefit when location of protected areas seeks complementarity with the biodiversity contributions already provided by other land. Global biodiversity benefits from sustainable harvesting of native species, if quantified, can lower overall regional opportunity costs of biodiversity conservation. Subsidies for biodiversity-friendly agriculture can be targeted to those places where the consequent complementarity values, taking into account the region's other conservation efforts, are greatest.

Successful integration will be facilitated when global biodiversity gains and losses are quantified in a unified way over various response contexts, so that complementarity values can be calculated. Complementarity values resulting from alternative land uses then can be compared for effective decision-making. Biodiversity surrogates at present provide poor levels of confidence in estimating biodiversity gains and losses, particularly those resulting from management regimes such as ecoforestry and wildlife harvesting. Surrogates are often selected for their supposed ability to indicate rich sites or sets of sites with high overall biodiversity. Trade-offs-based planning requires surrogates that indicate both high and *low* complementarity values. Confidence in a low complementarity value means that a place might be assigned a land use that focuses on local rather than global benefits, promoting the ability to make trade-offs at the regional scale.

Regional-scale decision-making, over a range of responses, can focus on trying to retain the potential net benefits of the region, by looking at how scenarios of land use affect the capacity of the region to balance its competing objectives (Faith 2001). The useful concept of "irreplaceability" of places refers to the goal of retention of a region's capacity for biodiversity conservation. This term might be extended to sometimes refer to cases where a land use for a given place is "irreplaceable" for the region's capacity for balancing biodiversity conservation and other human well-being objectives.

Integration can play an important role in linking "reserve" and "off-reserve" conservation. A polarized debate rages about the relative value of formal protected areas versus lands that are more intensely used by people but that conserve (at least some) components of biodiversity. The two approaches are more properly seen as part of a continuum of possibilities, correcting weaknesses of both approaches by linking them in integrated regional strategies. For example, areas used to provide certain provisioning services can lead to destruction of habitat and biodiversity; but a regional perspective may help mitigate some lost biodiversity because a given area need only contribute certain "attributes" of biodiversity to overall regional biodiversity conservation. Formal protected areas are often vulnerable because they foreclose other opportunities for society, but a regional perspective in planning protected areas can minimize conflict through appropriate and balanced land-use allocations. It can also build on the biodiversity protection gains from the surrounding lands, thereby reducing some of the pressure for biodiversity protection in the face of other anticipated uses over the region. Rather than using the weaknesses of one approach to argue for the other, it is more effective to have an integrated regional strategy.

A recent South African case study (Cowling and Pressey 2003) illustrates some ways in which a regional assessment of biodiversity (including setting of targets for different habitat types) can take into account the expected status and contributions of lands outside of formal protection. However, the process of setting biodiversity targets suggests that biodiversity conservation may be served to the detriment of effective trade-offs that might have been achieved in the region. In that study, differential targets were set for different habitat types that served as surrogate biodiversity information. A type that was judged highly vulnerable to destructive use will not be expected to make as large an "off-reserve" contribution to overall regional biodiversity persistence, and so will be given a higher target (say, in percent area represented) for formal protection. A difficulty may be that trade-offs suffer: nonconservation opportunities will be unduly foreclosed because a habitat/vegetation type is given greater percentage protection simply because it can provide other benefits for society.

Setting higher protection targets on habitat types that are "threatened" because they are attractive for other uses can limit effective trade-offs between biodiversity and other ecosystem services (although such limitations may be appropriate in some settings). This loss occurs even when it can be assumed that some types are protected, to some extent, off-reserve. Box 5.5 looks at the place of biodiversity in environmental impact assessments.

5.5.3 Linking Protected Areas to the Landscape

Although the linking of protected areas to the landscape was discussed earlier, several additional points are relevant here. The land and water area required to satisfy the conservation of regional biodiversity are well known to be in excess of the land and water area made available for formal conservation (Rodrigues et al. 2004). For example, an assessment of protected areas in Indonesia (World Bank 2001) concluded that few of the protected areas in the country are large enough to maintain viable populations of their constituent species, and recommended stronger linkages of the protected areas system with surrounding areas. The advantages of conserving biodiversity on production or other off-reserve lands are therefore obvious (Pressey and Logan 1997). A land or water use may provide some "partial protection" of biodiversity and ensure the maintenance of biodiversity conditions of that region. Grazing and other light intensity forms of use (for example, wildflower harvesting) have been shown to be more amenable to biodiversity conservation (Pressey 1992; Scholes and Biggs 2004). This combination of conservation and development sectors in the same place naturally increases regional net benefits and forms part of the ideology of UNESCO biosphere reserves, which include combinations of core conservation areas surrounded and linked to zones of differing intensities of alternate land uses.

A recent South African study (Pence et al. 2003) found that 80% of the costs for acquiring protected areas might be saved by meeting biodiversity targets in part on private lands. Similarly, an integrated biodiversity trade-offs framework (Faith et al. 2001a, 2001b) suggests how such partial protection (for example, from private land) can contribute to the region's trade-offs and net benefits.

Arguments for such landscape-based synergies arise also from studies suggesting that clever arrangement of human-use habitats can promote biodiversity. This is embodied, for example, in "reconciliation ecology," which is based on the idea that habitat loss will imply species loss, but that "we can stop most of them by redesigning anthropogenic habitats so that their use is compatible with use by a broad array of other species" (Rosenzweig 2003, p. 194). Rosenzweig cites case studies demonstrating the potential of reconciliation ecology to increase compatibility of human use and biodiversity conservation.

BOX 5.5

Addressing Biodiversity Issues in Environmental Impact Assessment

Environmental impact assessment has been adopted by countries and financial and lending institutions, such as the World Bank and the Asian Development Bank, as a tool to assess development projects in many countries. EIA originally focused on pollution issues, but has expanded to include potentially adverse impacts on biodiversity.

The Convention on Biological Diversity has played a key role in encouraging governments to include biodiversity in national EIA frameworks. Article 14 of the CBD provides an explicit mandate for encouraging EIA as a planning tool for responsive environmental planning of development initiatives in a manner that ensures prevention or significant reduction in biodiversity resources and the enhancement of biological diversity wherever possible (UNEP 1998). However, the process often fails to incorporate biodiversity in full. Decision V/18 of fifth Conference of the Parties of the CBD called on its Subsidiary Body on Scientific, Technical, and Technological Advice (SBSTTA) to develop guidelines for incorporating biodiversity-related issues into legislation and/or processes on strategic environmental assessment and impact assessment.

Many specialized disciplines (for example, social impact assessment, technology impact assessment, health impact assessment) are gradually emerging within the EIA. It is now becoming increasingly common to also address impacts on biodiversity as a distinct category of assessments.

Checklists of biodiversity impacts are being developed (CEAA 1996; World Bank 1997; Rajvanshi 2003). Manuals are now available to guide data gathering and interpretation for decision-making (DEA 1992; UNEP 1996; EC 2001; UNEP 2002). Several good-practice guides (DOE 1993; CEAA 1996; World Bank 1997) and outline methods are available for assessing biodiversity impacts.

Several problems need to be addressed. A lack of formal requirements and inconsistent mechanisms of evaluating compliance constrain the role of EIA from a biodiversity perspective (IUCN 1999; Mathur and Rajvanshi 2001). Lack of information (regional biodiversity data and resource status reports), lack of clearly defined EIA terms of reference in relation to biodiversity, and weak enforcement of legislation are common barriers identified by most countries. In many situations, particularly for larger river basins, the appropriate region of analysis may extend across two or more countries, adding to the challenge of undertaking an EIA for a data-poor system.

Although techniques for eliciting biodiversity values are gradually being put into place, consideration of biodiversity is still not included as a "trigger" for EIA in most countries. If countries adapt their EIA legislation to address all threats to biodiversity, and if the lack of information can be addressed, EIA could play a greater role in biodiversity conservation.

Integrated coastal management plans and programs worldwide have shown that ICM plays an important role in maintaining natural resources and ecosystems. However, studies suggest that ICM practices have successfully provided a response to biodiversity conservation only in those cases having a significant degree of integration of sectoral policies in coastal areas (Cicin-Sain and Knecht 1998). In the context of ICM, policy integration is not an absolute, but arguably should be considered as a continuum, that is, from sectoral fragmentation to communication among sectors, coordination, harmonization, and eventually integration (Cicin-Sain 1993). A policy response in this direction has been agreed in the context of the CBD, whose Parties have advocated a better integration of the ecosystem approach into current integrated marine and coastal area management plans and programs (CBD 2000a).

Integrated river basin management, also known as integrated catchment management and integrated watershed management, is a landscape approach with the potential to address both biodiversity conservation and sustainable resource use considerations through implementation of a range of strategies and levels of protection. Although Gilman et al. (in press) found that IRBM has been variously defined by managers around the world, and that biodiversity conservation has only sometimes been the primary driver of IRBM efforts, IRBM is nonetheless grounded in the need to understand trade-offs, often of upstream versus downstream activities. With a river basin as the "landscape" of interest, it is possible to apply IRBM principles to identify complementary land and water uses: for example, a river identified as a priority for biodiversity protection might be designated "off-limits" for a new hydropower dam, but another river of lower priority might be designated a suitable alternative for new impoundments. Protected areas have been a relatively rare feature of IRBM efforts, but IRBM nonetheless has great potential for effective protected area design and management, both for aquatic and for terrestrial biodiversity conservation.

When assessing individual areas and their contribution to the off-reserve conservation of biodiversity, it is essential to assess the complementary contributions they make to conservation, that is, their contribution in terms of biodiversity not already represented in protected areas. For example, case studies that attempt to document biodiversity gains from organic agriculture, wildlife harvesting, or other land uses, often fail to address complementarity or how those gains fit with gains/losses elsewhere. The key to demonstrations of success in protecting global values of biodiversity would be increases in the marginal gains from those lands in the regional context. A review of case studies and approaches for "monitoring of biological diversity" (Yoccoz et al. 2001) documents the typical focus on species-richness, not complementarity. An increase in abundance of species, or even an increase in richness, is not as persuasive as an increase in the degree to which the land offers biodiversity gains *complementary* to those of other places (Faith and Margules 2002).

Failure to assess biodiversity contributions of managed lands using complementarity can misdirect conservation priority setting. For example, it has been argued that the European Union must prevent the decline of its "nature-rich" farmlands or it will fail to reach the MDG target of reducing species loss by 2010 (EEA 2004). Agricultural subsidies targeted at vulnerable farmland areas based on their "biodiversity" values (as estimated from the higher species' population sizes compared to abandonment of the land) would promote socially desirable levels of biodiversity from private land. However, the report at the same time acknowledges that, while land abandonment may lead to lower species diversity at field level, the natural habitats resulting from abandonment in fact may add to *overall* biodiversity at the regional scale. The call for priorities for these farmlands remains poorly justified in the absence of the contributions to decision-making that could be provided by estimates of complementarity contributions of different land uses, in different places.

In the absence of more exact information, sensitivity analysis of the contributions of off-reserve lands can influence the priorities for formal protection. The case study from Costa Rica shows how regional planning can integrate protected areas and payments for biodiversity conservation on private land. The study demonstrates that the effectiveness of conservation payments on private lands can be greatly increased through regional planning that targets payments using complementarity values, and can effectively complement efforts through formal protection in national parks in the region.

Approaches that use sensitivity analysis and simple surrogate information for biodiversity may help to bridge the gap between non-quantitative mainstreaming efforts and idealized complementarity-based approaches.

These assessments indicate that landscape links may increase the viability of protected areas, and so ensure their contributions to biodiversity in the broader region. More generally, land use / management decisions in different places in the landscape will imply incremental gains to biodiversity conservation in the region quite apart from any links to protected areas. Although these ideas of off-reserve management of biodiversity are theoretically appealing, the reality of implementation is far more complex. The establishment of biodiversity-friendly land use management on land (or water) of regional biodiversity concern requires the development of incentives for private and communally owned land (these incentives are discussed later in this chapter).

South Africa and Australia, as well as other countries making use of integrated regional planning, have learned many lessons on how to integrate biodiversity into regional planning (Lambert et al. 1995; Driver et al. 2003; Read and Bessen 2003; Lowe et al. 2003; Bennett and Wit 2001; Gelderblom et al. 2002). Conservation assessment involves identifying spatial priorities for conservation actions, which in turn forms a component of conservation planning; this planning should also involve the development of an implementation strategy and action plan (Knight and Cowling 2003).

There are several essential elements of successful regional plans. One of these elements is adequate scientific knowledge of the region, including defining boundaries that are sensible from a biological and administrative point of view and the collation of high quality data in the area. All successful plans highlight the need to involve all stakeholders from the beginning of the planning process. These stakeholders include communities in the region, as well as government and nongovernmental organizations responsible for implementation of these plans, As Driver et al. (2003, p. 11) highlight, planners need to "think implementation from the outset." All too often conservation plans end up as technical or academic exercises and do not lead to conservation action on the ground. In order to avoid this they suggest that an operational framework must be set up with the following key ingredients:

- ask "who wants this plan and what are the plan's aims?";
- pay attention to project design;
- involve implementing agencies in the conservation assessment team;
- involve stakeholders in a focused way that addresses their needs and interests;
- conduct the conservation assessment according to the best scientific principles; and

- interpret the conservation assessment results and mainstream the planning outcomes.

Monitoring and evaluation through the use of performance indicators is also of critical importance both in order to monitor the maintenance and recovery of biodiversity values and to encourage involvement (Lambert et al. 1995). Read and Bessen (2003) have identified several success and limiting factors in regional planning through 16 case studies based on an assessment of 154 projects as well as semi-structured interviews. These have been grouped into motivational, financial, and regulatory factors. Their recommendations for strategic action focus on: establishing clear values, priorities and cultures; understanding biodiversity values and threats; managing at the landscape scale; using science, information, and knowledge; building capacity; using a mix of mechanisms; and encouraging factors that drive integration.

5.5.4 Assessment

We can state with *high confidence,* based on 150 studies on large scale, regional planning for conservation linking networks of protected areas with other land uses (Bennett and Wit 2001), that a "landscape approach" that, for example, manages neighboring production forests as buffer zones and integrates protected areas with broader regional spatial planning, helps overcome stated limitations of protected areas on their own. Successful landscape approaches:

- focus on conserving biodiversity at the ecosystem, landscape. or regional scale, rather than in single protected areas;
- emphasize the idea of ecological coherence by encouraging connectivity;
- involve buffering of highly protected areas with eco-friendly land management areas;
- include programs for the restoration of eroded or destroyed ecosystems;. and
- seek to integrate economic land use and biodiversity conservation.

Overall, it is seen to be essential in these efforts to recognize the importance of regional context for implementing the ecosystem approach and monitoring progress. Our assessment suggests that the ecosystem approach implementation guidelines approved by CBD's seventh Conference of the Parties could add a requirement for a "calculus" of global and regional biodiversity. This would allow global biodiversity gains and losses to be quantified in a unified way over various response strategies, thereby clearly identifying the trade-offs involved at a regional level. Such a calculus of biodiversity depends on effective biodiversity surrogates. These will be based upon the best possible use of a combination of environmental and species (for example, museum collections) data, and will provide greater certainty in estimating such biodiversity gains and losses (Faith et al. 2003; Reyers 2004).

5.6 Encouraging Private Sector Involvement in Biodiversity Conservation

One of the most significant differences between the WSSD summits at Rio de Janeiro (1992) and Johannesburg (2002) was the greater presence of the business community as a major stakeholder at the 2002 WSSD. Although specified as an "actor" in the text of the CBD (Articles 10e, 16.4), business was perhaps slow to recognize its role in biodiversity conservation, and the CBD has been slow to recognize the link between industry and biodiversity. However, business has a wide impact on biodiversity, especially through the products and processes associated with mass consumption. How can business be given the inspiration, incentives, tools, and management systems to play an effective part in the biodiversity debate?

Companies exist to make a profit and thereby generate value for shareholders, so a company must have a business case for being involved in biodiversity conservation (Abbott et al 2004).

An embryonic business case is now emerging, with the following key elements:

- The activities of certain companies have significant impacts on biodiversity.
- As understanding increases about biodiversity and how ecosystems function, more evidence emerges of the potentially destructive impacts of some business activities.
- As with other environmental impacts, regulators and civil society increasingly require these negative impacts to be managed and, if possible, reduced or reversed, in order for a company to retain its license to operate from communities and regulators.
- Biodiversity impacts can be managed alongside a company's other environmental impacts, as part of an integrated environmental management system.
- Employees will prefer to work for a company that is a good corporate citizen.
- Companies with a better reputation will have easier access to investment capital.

This list of key elements, however, leaves out other positive aspects that depart from the conventional perspectives on "impacts." Land that is well-managed by a private sector company (for example, by a mining company) may make a measurable positive contribution to the conservation of regional diversity (see earlier discussion). This regional perspective has a second positive aspect. The private sector needs information about key biodiversity areas as part of its own regional planning, and so is encouraged to become part of partnerships that enhance the availability of such biodiversity data (for example, the "Proteus" scheme, see below). Table 5.1 highlights the particular interests of many sectors in biodiversity.

Some companies and sectors are dependent on biodiversity and healthy ecosystems in order to maintain their current operations. Obvious current examples include nature-based tourism and companies based on harvesting biological resources. Sustainability of supplies perhaps provides the most compelling case for business involvement in biodiversity conservation, although many companies remain uncertain as to precisely what they should do, once they have identified such a risk exposure.

A further source of uncertainty has been the perception that, since much biodiversity expertise lies especially with NGOs and academic institutions, simply funding biodiversity programs through such organizations should fulfill a company's requirements. Although the most enlightened companies have understood that biodiversity risks and impacts need to be managed alongside other such environmental issues within the company, corporate philanthropy remains a relatively common approach to dealing with biodiversity. While such funding is welcome, one perspective is that it remains relatively modest and is not a sufficient response to address fully the biodiversity impact of a company.

5.6.1 What Companies Are Doing

Once a company accepts that it has an important relationship with biodiversity, an increasingly standardized means of managing the issue, outlined in Bertrand (2002), becomes available. Essentially, the process is to align biodiversity management with a company's environmental management system. This approaches the problem

Table 5.1. Business Sectors with Direct Relevance to Biodiversity Conservation

Sector	Main Issue
Agriculture	increase food production while maintaining a healthy agroecosystem, integrate biodiversity and food production in more sustainable ways, include more efficient use of irrigation water
Aquaculture	minimize impacts on marine and freshwater biodiversity by, for example, culturing native species, minimizing risk of escape of individual animals, avoiding the conversion of sensitive or keystone habitats such as mangroves, and minimizing pollution
Engineering/ architecture/ planning	address the environmental impact of human settlements and the built environment (in the context of vulnerability to natural disasters and global change, but also related to biodiversity more generally)
Fisheries	minimize impacts on marine biodiversity and address overfishing to ensure the sustainability of the industry itself
Forestry	reduce impact of operations; important strategy: move toward more sustainable practices through market-based mechanisms such as certification
Hydropower	minimize impacts on aquatic and surrounding terrestrial systems through implementation of the World Commission on Dams recommendations and through engagement with regional planning efforts
Insurance/ financial sector	create incentives for the private sector to address biodiversity issues substantively; potential source of funds for restoration
Mining	minimize impacts and set industry standards on biodiversity generally and protected areas specifically
Oil and gas	minimize impacts and move toward "net benefit" concepts; mobilize them to encourage best practice in ancillary industries (e.g., shipping), and to address protected areas issues (e.g., "no go" commitments)
Shipping	address the spread of invasive alien species and the risk to biodiversity from shipping disasters (e.g., oil spills).
Tourism	encourage the tourism industry to be a force for good for biodiversity and to minimize impacts; standardization/certification of ecotourism practices
Water providers	integrate ecosystem management concepts

at the site level, where a biodiversity action plan (site-level BAP) enables the company to manage issues at a local level; at the company level, a company-wide biodiversity action plan enables the business to take a strategic approach to its relationship with biodiversity globally.

Certain sectors have an obvious and immediate relationship with biodiversity. For example, some companies have products, which are dependent on biological resources, such as fish and timber, or tourism. Other companies require access to mineral, oil or gas reserves that may be found in areas of high biodiversity, and often their extraction will have a significant impact on biodiversity. Many companies are major users of water, both as a component of their products and in the production cycle, and company interests in ensuring future sources of water may dovetail in part with the needs of aquatic ecosystems.

Such an obvious and immediate primary relationship with biodiversity has led to initiatives on both a company-by-company and a sectoral basis, the latter usually being linked to certification schemes or an industry-wide sustainable development program. These industry-led sectoral approaches are important, because they offer hope of reaching non-listed (that is, state and private) companies.

Examples of company initiatives include Unilever's objective to source all its fish from sustainable sources by 2005 and British Petroleum's biodiversity action plans. According to BP Australia (2000, p. 11), examples include, "Construction of wetlands at Bulwer Island refinery with different water depths and the planting of 17,000 seedlings since 1998. The sub-tropical wetland is now home to 96 species of birds. . . . The planting of the 1.4 million trees has provided protection to several hundred species of plants found only in the agricultural and woodland zones of southwestern Australia."

To give an idea of scale, in the United Kingdom, perhaps 40 of the FTSE-350 companies have a company-wide biodiversity action plan or take a strategic approach to biodiversity management. This number doubled between 2001 and 2003.

Some initiatives involve consortia of companies together with biodiversity conservation organizations. An example is the "Proteus" initiative, involving Anglo American, British Petroleum, and others as sponsors in collaboration with UNEP-WCMC (www.unep-wcmc.org). It plans to help make high-quality conservation information, such as that from museum collections, available to decision-makers via the Internet. The intention is to enable, for example, the overlay of company information on to biodiversity maps to assess potential environmental implications of business operations in different places.

Private sector efforts may help address perceived under-funding of the Global Environment Facility. Because the GEF focuses on financing the cost "increment" that will achieve global biodiversity benefits (the difference between national or local benefits that could be expected and the global biodiversity benefits arising from the project), there has been a perception that one limitation of GEF programs is that some national/local costs are left in need of funding (Horta and Round 2002).

5.6.2 What More Needs to Be Done

At the heart of company approaches lie three fundamental questions: What is the company's relationship with biodiversity? How can the impact of the company be measured? And what are the consequences for the value of the company? To date, as with many sustainability initiatives, the driving force has come from large companies in the private sector, driven by a combination of reputational and supply sustainability factors. This means that small and state-owned companies have been less involved in biodiversity initiatives, even though their collective impact may be greater. The business case made for and accepted by larger, private-sector, companies clearly might be extended to other companies as well. Efforts to involve other companies may be focused on voluntary and sectoral initiatives driven by the need to secure supply chains or maintain reputational value for an entire sector. This, for example, might give continued access to mining sites, or access to genetic resources.

Further developments are likely to focus on two main areas. First, the debate will move away from simply looking at the impact of companies on biodiversity, important though this is. Increasing emphasis will be given to ecosystem services, and how companies rely on them. This will require development of mechanisms for companies to understand their risk exposure and to manage those risks. Second, the biodiversity conservation community will accept that business has a role to play in the debate. This may be difficult to accept when controversial issues relating to genetically modified crops and to intellectual property cloud the debate. Nevertheless fully engaging the corporate sector is a necessary condition.

One tool which may encourage further companies to accept the challenge of managing biodiversity is an increasing ability to measure both the impact that biodiversity has on companies and the impact that companies have on biodiversity. A corollary of measurement would be a coherent cost-benefit analysis. Although individual companies will undoubtedly make the case to their own satisfaction, the development of such tools should help increase the involvement of companies. This may highlight a role for investors, who use such tools regularly in assessing companies' strengths and weaknesses in other mainstream financial areas.

Some risks to biodiversity may also arise from private sector involvement, for example, if the exploiting and the regulating party are the same. Exchange of personnel between governments and commercial enterprises can also lead to abuses that do not arise when the regulators are well distinct from the resource exploiters.

5.6.3 Assessment

Where decision-makers and the general public have accepted the role business can and must play, constructive dialogues have been established, leading to initiatives at the company, sectoral, or higher level. Much discussion has focused on biodiversity impacts, and attention to ecosystem services may be more likely to help business understand its impact and find methods to mitigate it.

Engaging business has been easiest where a business case exists for the company concerned. A key strategy is to engage businesses that do not have a direct link with biodiversity or ecosystem services, or do not face the consumer pressure that publicly listed companies face. Regulation is likely to remain a key tool for influencing business, but establishing a climate for companies to move beyond compliance is essential.

5.7 Including Biodiversity Issues in Agriculture, Forestry, and Fisheries

5.7.1 Introduction

Early farmers played an important role in creating and maintaining crop genetic diversity through the domestication and selection of crops suited to a wide range of environments (Harlan 1975). The livelihoods and well-being of millions of farmers still depends on this diversity (Richards 1986; Bellon 1996). The success of breeders in developing high yielding varieties has built on crop genetic diversity, identifying genes for improving adaptation, yield, and disease resistance (Plucknett et al. 1987). Moreover, substantial evidence is now accumulating on the way in which the continued maintenance of high levels of crop genetic diversity in agroecosystems, based largely on traditional cultivars, meets the needs of resource-poor farmers (Engels 1996). A similar dynamic has been followed with livestock (FAO 1999; Hall and Ruane 1993).

Pressures are growing on natural habitats that contain the wild relatives of crops or domesticated animals, and on farmers who maintain significant amounts of crop or animal genetic diversity in the form of local varieties. Increased population, poverty, land degradation, economic, and environmental change, combined with the introduction of modern varieties, have contributed to the erosion of genetic resources in both animals and crops (Prescott-Allen and Prescott-Allen 1982; Wilkes 1985; Pistorius 1997). The availability of large gene pools, including wild relatives, becomes even more important as farmers need to adapt over time to changing conditions that result from these pressures (Jarvis and Hodgkin 1999).

Agriculture is directly dependent on biodiversity, but agricultural practices in recent decades have focused on maximizing yields by focusing research and development on relatively few species, thus downplaying the importance of biodiversity. Subsequently, a large amount of genetic diversity has been lost. Conversion of natural habitats into domesticated ones has continued (Pagiola et al. 1997), and the use of chemical fertilizers and pesticides has continued to expand (Gunningham and Grabosky 1998). Both of these trends can be detrimental to biodiversity and harm agriculture in the long run rather than improve it. On the other hand, agricultural practices in some regions have developed landscapes that include considerable biodiversity; abandoning such lands can lead to the loss of at least some species.

The use of living modified organisms in agriculture is a topic of major international concern. As movements of LMOs pose potential risks to new environments they may enter, the Cartagena Protocol under the CBD (entered into force in 2003) sets out certain measures to be followed to avoid risks to biodiversity in general and agricultural biodiversity in particular. The Cartagena Protocol requires advanced informed agreement procedures and careful assessment of risks before allowing import of living modified organisms. Such risk assessment is based on the precautionary approach (Bail et al. 2002), using procedures and tools to help decision-makers make sound decisions that are compatible with agriculture, biodiversity, and trade.

This section focuses on biodiversity issues in agriculture (for example, maintaining genetic and crop diversity), forestry (for example, certification), and fisheries (for example, marine protected areas for biodiversity conservation), and defers to other chapters in this volume for a discussion of sustainable food production and fisheries (see Chapter 6) and sustainable forestry management (see Chapter 8).

5.7.2 Agriculture

5.7.2.1 In Situ Conservation Responses

In situ conservation of agricultural biodiversity has been defined as "the maintenance of the diversity present in and among populations of the many species used directly in agriculture, or used as sources of genes, in habitats where such diversity arose and continues to grow" (Brown 2000). It concerns entire agroecosystems, including the management of domesticated species (such as food crops or forage species) on fields or in home gardens, as well as their wild and weedy relatives that may be growing in nearby areas or natural ecosystems.

Supporting conservation and use of agricultural biodiversity requires an understanding of when, where, and how agricultural diversity will be maintained, who will maintain the material, and how those maintaining the material can benefit. This requires:

- measuring the amount and distribution of germplasm used by farmers within their agroecosystems;
- gaining an understanding of the processes used to maintain this germplasm;
- identifying the key persons or groups of people responsible for maintaining the germplasm;
- comprehending what factors influence these people to maintain diversity; and

- using the information and genetic materials for sustainable livelihoods and ecosystem health and services (Jarvis et al. 2000).

The International Plant Genetic Resources Institute's work on on-farm conservation of crop genetic diversity has involved over 20 countries and 30 crops. The UNU/PLEC program (United Nations University/People, Land management and Environmental Change), which focuses more at the landscape level, has also involved many countries in all parts of the developing world and in centers of agricultural and crop diversity (Brookfield et al. 2002). IPGRI (2001; see also Jarvis et al., in press) has recently prepared a state of the world review for the CBD on the current status and trends for management of crop diversity in agroecosystems by national programs and international initiatives, identifying some key issues.

First, assessments have shown that local cultivars and breeds on farm are complex and highly varied in their genetic structure (Achmady and Schneider 1995; Kshirsagar and Pandey 1996; Sebastian et al. 2000; CBDC-Bohol 2001). Different communities and cultures approach the naming, management, and distinguishing of local cultivars in different ways, and no simple relationship exists between cultivar identity and genetic diversity (Quiros et al. 1990; Zeven 1998; Cleveland et al. 2000). Considerable debate surrounds the use of farmer names as a basis for arriving at estimates of cultivar numbers (Jarvis et al. 2000).

Second, management practices are linked to the survival of certain cultivars. At high elevations in Nepal, farmers re-route cold water from the main valley rivers to raise the water temperature before irrigation so as to induce earlier flowering and timely maturation of their rice cultivars (Rana et al. 2000). The informal sector is an important provider of seeds needed for sustaining agricultural biodiversity (Almekinders et al. 1994). In Morocco, less than 13% of durum wheat seed and 2.5% of food legumes, in Nepal less than 3% of rice, and in Burkina Faso less than 5% of sorghum are bought as certified seeds each year from the formal sector, indicating that the majority of seeds used are from local crop diversity or from seed saved from earlier purchases (Mellas 2000; Ortega-Paczka et al. 2000; Kabore 2000). Improving on-farm seed storage was also shown to be important in the maintenance of traditional cultivars in the Philippines (Morin et al. 1998).

Third, many factors influence the choice of how many and which varieties to grow and on what proportions of crop area. In developing economies, crop cultivar diversity on farms has been attributed to risk avoidance or to management of such issues as, for example, climatic uncertainties or pest and disease problems (Bellon 1996; Pimentel et al. 1997); food security in relation to total food supplies and nutritional well-being (Johns 2002); income generation, providing products that can be sold in different markets or are of high value (Smale et al.1999; Gauchan et al. 2003); optimizing land use to ensure cultivars are available for difficult (stony, wet, cold) lands (Bellon and Taylor 1993; Rijal et al. 2000); and adaptation to changing conditions such as increasing drought (Sadiki et al. 2001). In advanced economies, diversity may be conserved through demand for specialized goods and services. Concerns for human and ecosystem health may influence societies to follow a policy goal of supporting local crop cultivars because of the social benefits they contain (Smale et al. 1999; Smale 2002).

Fourth, home gardens are important locations for agricultural biodiversity conservation, providing microenvironments that serve as refuges for crops and crop varieties that were once more widespread in the larger agroecosystem. Home gardens can serve as buffer zones around protected areas, as is the case with the Sierra del Rosario Biosphere Reserve in Cuba (Herrera and Garcia 1995). Farmers often use home gardens as a site for experimentation and introduction of new cultivars arising from exchange and interactions between cultures and communities, or as sites for domestication of wild species. These useful wild species are often moved into home gardens when their natural habitat is threatened, such as in the case of Loroco (*Fernaldia pandurata*) given the high rate of deforestation in Guatemala (Leiva et al. 2002). Studies of the genetic diversity of key home garden species in Cuba, Guatemala, Ghana, Indonesia, Sri Lanka, Venezuela, and Viet Nam have demonstrated that significant crop genetic diversity does exist in home gardens, and that home gardens can be a sustainable in situ conservation system (Watson and Eyzaguirre 2002).

Fifth, many options are available for increasing the benefits to farmers from local crop diversity (Jarvis et al., 2000). These options include (1) improving the material through participatory methods including participatory evaluation, improvement, and breeding (Soleri and Cleveland 2001; Joshi and Witcombe 1998; Castillo et al. 2000; Ceccarelli and Grando 2000; Bellon et al. 1999); (2) increasing consumer demand through public awareness, for example, through diversity fairs or nutritional awareness building (Gauchan et al. 2003; Johns 2002); (3) improving access to materials and information (Bellon 2001, Mazhar 2000); (4) adaptation to microniches and reduced agricultural inputs; (5) improving ecosystem health and services (CONSERVE 2001, Rijal et al. 2000); and (6) developing supportive policy recommendations (Gauchan et al. 2000, 2003; Correa 1999; Cromwell and van Oosterhout 2000).

5.7.2.2 Ex Situ Conservation Responses

One response to the loss of global crop diversity has been to conserve germplasm in ex situ conservation facilities. Over 6 million accessions of the world's major food plants are now conserved in over 1,300 gene banks worldwide, with about 90% of the accessions conserved in the form of seeds (FAO 1998). Various methodologies and approaches have been developed so that specific traits and alleles are conserved.

Seed banks have well advanced technologies for conserving and managing orthodox seeds (Engelmann and Engels 2002). However, for many developing countries, the maintenance of seed banks is difficult, as electricity supplies are unreliable and fuel is expensive. Various research projects have recently focused on the development of "low-input" alternatives to medium- and long-term cold storage (Engelmann and Engels 2002). One option is the development of the so-called "Ultra dry seed storage technology" (Zeng et al. 1998), which allows the storage of seed germplasm at room temperature, thereby obviating the need for refrigeration. Other research conducted on drying techniques such as sun and shade drying (Hay and Probert 2000) offers promising alternatives to improve the capabilities of resource-poor countries to conserve their seeds.

Field gene banks are the preferred method for species that produce short-lived seeds or are vegetatively propagated. Field gene banks have some drawbacks. Accessions are exposed to pests and diseases, natural hazards, and human error, all potential sources of erosion (Engelmann and Engels 2002). Field collection can pose a heavy burden on the national institutions, implying an urgent need for implementation of other measures to conserve plant genetic resources more effectively and cost-efficiently.

In-vitro techniques have been devised for the collection, multiplication, and short- and medium-term storage of plant germplasm (Engelmann 1997). In-vitro culture protocols have been published for well over 1,600 species (George 1996). Slow-growth storage is used routinely in a limited number of national, regional, and international germplasm conservation centers for a few species including bananas, some root and tuber crops, and temperate fruits (Engelmann 1999).

Cryopreservation, that is, storage at ultra-low temperature, usually that of liquid nitrogen (−196° Celsius), currently offers the only safe and cost-effective option for the long-term conservation of genetic resources of species. Techniques can now be considered operational on a routine and large-scale basis (Engelmann and Takagi 2000).

DNA storage is rapidly increasing in importance. DNA is now routinely extracted and immobilized into nitro-cellulose sheets. These advances have led to the formation of an international network of DNA repositories for the storage of genomic DNA (Adams 1997).

Pollen storage has also been considered as an emerging technology for genetic conservation (Towill and Walters 2000). In the past 10 years, cryopreservation techniques for pollen have been developed for a number of species (Hanna and Towill 1995) and cryobanks of pollen have been established for fruit tree species in a few countries (Ganeshan and Rajashekaran 2000).

Botanic gardens have long been the main center for the conservation of wild species (Heywood 1991). More recently, the contributions of botanic gardens in conserving different kinds of germplasm that are relevant to crop diversity have been recognized (Heywood 1999). Over 1,800 botanic gardens and arboreta are found in 148 countries worldwide and maintain more than 4 million living plants accessions (Wyse Jackson and Sutherland 2000). Many botanic gardens also have seed storage facilities, maintaining more than 250,000 accessions (Laliberte 1997).

Typically, ex situ conservation is carried out by universities or national institutions, which have developed facilities (storage, laboratories, information) and field production capacity necessary to undertake long-term commitments to store and make available the accessions. In addition, international gene banks are maintained by the research centers of the Consultative Group on International Agricultural Research. International collaboration on conservation and use of these plant genetic resources has been considerable, involving the work of the FAO Commission on Plant Genetic Resources and the development of the Global Plan of Action for the Conservation and Sustainable Utilization of Plant Genetic Resources for Food and Agriculture (FAO 1996) that was signed by about 150 countries. The International Treaty on Plant Genetic Resources was agreed in 2002 (FAO 2002) and is now in force.

Ex situ conserved materials are essential for the production of new improved cultivars and provide a basis for increasing productivity and the genetic diversity needed for production with, for example, reduced use of pesticides, fungicides, and herbicides, and improved water use efficiency. Genetic resources are sent from gene banks to users throughout the world, and developing-country users benefit considerably from these flows (Fowler et al. 2001). In areas affected by natural and man-made disasters, crop genetic diversity can help restore natural and agricultural ecosystems.

5.7.2.3 In Situ Conservation of Crop Wild Relatives in Natural Ecosystems

New crop cultivars are often obtained from wild or weedy materials. These processes continue to affect the genetic diversity of crops in centers of diversity as farmers adopt new genotypes into their farming systems (Jarvis and Hodgkin 1999; Altieri and Montecinos 1993; Quiros et al. 1992). In addition, farmers may bring wild varieties into their farming systems. Wild relatives of crop species (also called crop wild relatives) have already made substantial contributions to improving food production through the useful genes they contribute to new crop varieties (Hodgkin and Debouck 1992).

Genes that provide resistance to pests and diseases have been obtained from crop wild relatives and used in a wide range of crops, including rice, potato, wheat, and tomato. A classic example is the interspecific tomato hybrid between wild *Lycopersicon peruvianum* and cultivated *L. esculentum,* which led to scores of tomato varieties with resistance to root knot nematode (Rick 1963). Genes from crop relatives have been used to improve protein content in wheat and vitamin C content in tomato. Broccoli varieties producing high levels of anti-cancer compounds have been developed using genes obtained from wild Italian *Brassica oleracea.* Crop wild relatives have also been a source for genes for abiotic stress tolerance in many crops.

With the advances made in molecular genetics it is now possible to transfer genes between distantly related taxa or even taxa from different kingdoms, thereby broadening the value of crop wild relatives. Natural populations of many crop wild relatives are increasingly at risk. They are threatened by habitat loss, deforestation (for example, coffee in Ethiopia; Tadesse et al. 2002), and overgrazing and resulting desertification. Meilleur and Hodgkin (2004) have reviewed the current status and trends of conservation activities on crop wild relatives in about 40 countries throughout the world. Crop wild relatives are also traditionally found in agroecosystems in and around farms; the increasing industrialization of agriculture is reducing their occurrence.

While it is clear that the continuing development and deployment of more genetically uniform improved crop varieties has an effect on the amount and distribution of the diversity of traditional crop varieties in production systems, it is not clear what the effect of genetically modified varieties will be. Some are concerned that genetically modified crops will tend to reduce further the amount of crop diversity in production systems and that they might affect diversity of non-crop components (such as insect species), or that transgenes will move from crops to their close wild relatives. Others point to the dependence of conventional agriculture on agrochemicals as a major problem that genetically modified varieties could help overcome (Gepts and Papa 2003).

5.7.2.4 Eco-agriculture as a Response for Conserving "Wild Biodiversity"

Sustainable agriculture and sustainable management of crop diversity often depend on sustainable management of the surrounding natural ecosystem. One means of linking agriculture with other land uses is eco-agriculture, defined as a framework that seeks to achieve simultaneously improved livelihoods, conservation of biodiversity, and sustainable production at a landscape scale (McNeely and Scherr 2003).

Enhancing environmental responsibility as an aspect of on-farm management has driven the evolution of an array of sustainable agriculture and natural resource management models, including organic agriculture, agroecology, integrated crop management, and conservation farming. The relative economic, social, and environmental benefits of adopting any particular model are very situation specific, influenced by the needs, local use conditions, and resource capacities of individual practitioners, and also the nature of adjacent resource management strategies being implemented. Achieving meaningful benefits to biodiversity beyond farm level demands further coordination between strategies at the landscape scale. Eco-agriculture aims to build upon extant models and intentionally integrate the knowledge and activities of practitioners, policy-makers, researchers, educators, and extension services within the sectors of agriculture, conservation, and rural development. An integrated approach, encompassing a range of strategies, offers practitioners more choice to adopt the management system most appropriate to their needs.

Eco-agriculture can be supported through six overarching implementation strategies (McNeely and Scherr 2003):

- make space for biodiversity reserves within agricultural landscapes;
- develop simple, low-cost habitat niches and networks for wild biodiversity on and around farmlands;
- modify farming systems to mimic natural ecosystems;
- reduce pressure to convert further land to agriculture, enhancing the productivity of extant agricultural systems;
- reduce the use of external inputs within integrated pest, livestock and nutrient systems; and
- encourage soil, water, and vegetation resource management strategies with potential to benefit biodiversity.

These encompass activities that can be implemented by individual practitioners at a farm or ecosystem level and, at a landscape level, collaborative strategies that can enhance the adoption of strategically complementary approaches among neighboring land users.

Of the 36 cases reviewed by McNeely and Scherr (2003), 28 principally benefited poor, small-scale farmers. Enhanced ecosystem productivity and stability reduced production-associated risks, raised food and fiber production, and thus improved livelihood security. Net income increases were demonstrated in 15 cases, with other reviewed cases exhibiting significant economic potential. However, data on farm income impacts remain poor. The considerable overlap between regions where agricultural productivity increases are vital for food security and poverty reduction, and areas where wild diversity is richest, highlights eco-agriculture's significant potential to have positive impacts on rural poverty and biodiversity, provided that socioeconomic and political conditions are enabling.

5.7.3 Forestry

For a detailed discussion of sustainable forestry management, the reader is referred to Chapter 8 of this volume. Discussed here are two issues related to including biodiversity issues in the forestry sector.

5.7.3.1 Non-Wood Forest Products

Natural forest ecosystems are especially rich in biodiversity. However, unsustainable logging practices result from the lack of adequate planning and from biased policies focused on logs more than on the entire value of the natural forest. Despite these circumstances, various innovative measures that consider biodiversity in the natural forest ecosystems have emerged. One such measure is giving greater attention to non-wood forest products (IUCN 2000c). Unlike large-scale commercial logging, harvesting NWFPs can be less harmful environmentally, as it takes into account the entire value of the natural forest ecosystem, and can contribute to conservation of biodiversity. However, NWFPs require careful management and must take into account sustainable use practices in order to avoid overharvesting of certain species, as witnessed in the case of Brazil nuts in the Amazon, ginseng in North America, and rattan in Southeast Asia.

While policies geared toward promoting NWFP can be beneficial to biodiversity conservation and social and economic sectors, industries based on NWFP are only now emerging, and require more attention and full integration into forest policies of countries. This would balance the policies of countries that focus on large-scale commercial logging projects that are targeted at earning fast money, as can be witnessed in many temperate and tropical countries (Filer and Sekhran 1998).

5.7.3.2 Certification and Sustainable Forest Management

A measure that is voluntary in nature, but is gaining widespread recognition due to market pressures and conservation needs in the forestry sector, is certification for forest products harvested in a sustainable manner (Hirakuri 2003). Certification of forest products is market driven but at the same time contributes to conservation of biodiversity. Certification is now working in Europe, North America (Raunetsalo et al. 2002), and individual countries such as Brazil (Hirakuri 2003). The practice is expanding into other products such as coffee, where environmentally friendly cultivation practices that integrate biodiversity are encouraged. Coupled with certification is the forest tracing system used in sustainable forest management now found, at least in rhetoric, in countries such as Indonesia (Jakarta Post 2000), Russia (Forest.ru n.d.), Canada, and the United States to prevent illegal logging. The forest tracing system has contributed to SFM.

However, deforestation still continues in most parts of Africa, Latin America, Asia, and the Pacific, caused by poverty, population growth, economic growth, urbanization and the spread of agriculture (UNEP 1999). New and innovative initiatives including NWFP policies, certification, and SFM have to be designed to minimize deforestation. Furthermore, policies, legislation, and institutions need to be designed to support sustainable forest management.

5.7.4 Marine Reserves, Biodiversity, and Fisheries

This section focuses on marine reserves for biodiversity conservation, and the links to fisheries management. For a detailed discussion of fisheries for food, the reader is referred to Chapter 6 of this volume.

Fully protected marine reserves (a special kind of marine protected areas) can be defined as "areas of the ocean completely protected from all extractive and destructive activities." Marine protected areas are defined as "areas of the ocean designated to enhance conservation of marine resources." The level of protection within MPAs varies, and in many MPAs certain activities such as fishing may be allowed (Lubchenco et al. 2003, p. 53).

A meta-analysis of 89 studies (Halpern 2003) concluded that marine reserves, regardless of their size, in almost all cases lead to increases in density, biomass, individual size, and diversity (species richness) of species in the reserve (with the exception of invertebrates). The diversity of communities and the mean size of the organisms are between 20% and 30% higher compared to unprotected areas. The density of organisms is roughly double and biomass nearly triple. Proportional increases occur in reserves regardless of size. But for conservation and fisheries purposes, absolute increases in numbers and diversity are important (for example, to sustain viable populations, to ensure spill-over effects, and to protect against catastrophic events). It is therefore likely that at least some large reserves are required for biodiversity conservation.

The science of selection and design of marine reserves is now well developed (Roberts et al. 2003a; Hastings and Botsford 2003; Roberts et al. 2003b; Sala et al. 2002). Methods of MPA design and implementation have affirmed themselves as effective tools for biodiversity conservation (Crosby et al. 2000). These methods recognize the use of a whole set of tools—including no-take zones—that should be made available to all sectors of society that are concerned, directly or indirectly, with MPA design and implementation. In fact, the growing recognition of the importance of MPAs by sectors of society that are not, traditionally, conservation-driven, calls for an authoritative and at the same time adaptive approach for MPA implementation to achieve a balance between biodiversity conservation and economic development. Clear and pragmatic guidance with regard to MPA design and implementation is provided in a recent paper on the subject (Agardy et al. 2003). According to the authors, an appropriate mix of various management tools should be utilized in MPA setting and management, depending upon specific conditions and management goals.

Benefits to commercial fisheries outside the reserves are poorly documented so far and still the subject of debate (Ward et al. 2001), and benefits beyond the actual reserve limits (for example, benefits for fisheries) may be limited (Kura et al. 2004). Part of the problem is that few marine reserves have been strictly protected and monitored for a sufficiently long period that benefits in surrounding waters could show up. Even fewer reserves have been set up specifically to enhance a commercial fishery. In addition, monitoring and demonstrating the spillover effect is no easy matter, and documenting benefits to distant waters is even more difficult.

One understated strength of marine reserves is that they provide a clear example of an "ecosystem-based" approach to fisheries management, since they protect both fish and the ecosystems where they live. In marine reserves, all species—regardless of their commercial value, sex, or size—are protected. Reserves can also maintain the structure of marine communities intact, allowing important interactions among species to function unimpeded. This can provide a good complement to typical fishery management approaches that focus on maintaining a single species only (and only where there is a commercial incentive). This may be especially useful in the tropics, where many species may be commercially exploited in one fishery. A marine reserve approach, in this case, is probably easier to implement and enforce than trying to regulate the fishing effort or catch quota of each species separately (Ward et al. 2001; Roberts and Hawkins 2000).

Aquaculture is likely to grow further in importance, but is no panacea. It uses large amounts of wild fish, processed into fish food, and has other negative environmental impacts that need to be reduced (for example, the risk of invasive alien species through escape, diseases, impacts of genetically modified fish, conversion of natural ecosystems).

In 2000, the Conference of the Parties to the CBD charged its Ad Hoc Technical Expert Group on Mariculture with the task of assessing the consequences of mariculture for marine and coastal biodiversity and promoting techniques that minimize adverse impact. In 2003, the Group released its report, which, while stressing that all forms of mariculture affect biodiversity at all of its levels (mainly through habitat degradation, disruption of trophic systems, depletion of natural seedstock, transmission of diseases, and reduction of genetic variability, as well as the biodiversity-effects of pollutants and contaminants), also pointed out that, under certain circumstances, local mariculture activities can enhance biodiversity (CBD 2003b). A significant contribution by this group has been to agree on recommended methods and techniques for preventing the adverse effects of mariculture on biodiversity, the most important of which are proper site selection, optimal management practices (such as proper feeding) and technological enhancements, culturing different species together (polyculture), and the use of enclosed and, in particular, recirculating systems (CBD 2003b). The group also recommended the use of aquaculture-specific certification of the product, which highlights that the species in question has been produced according to guidelines, codes of practice (sometimes followed by eco-labeling), or quality standards such as organic mariculture (CBD 2003b).

5.7.5 Assessment

We can conclude with a *high degree of confidence* that the benefits from ex situ conserved genetic diversity are substantial, though much more work is required in this area to develop adequate appreciation and estimates of the full benefits. While the technology continues to improve, the major constraint is ensuring that an adequate range of genetic diversity is contained within the ex situ facilities, and that these remain in the public domain where they can serve the needs of poor farmers. In addition, ex situ facilities are unlikely to conserve the full range of genetic diversity of species. To achieve more effective conservation a complementary approach to conservation using in situ conservation should be adopted (Engelmann and Engels 2002).

The technologies for in situ conservation in domesticated landscapes are well developed. However, the economic incentives seem to favor a narrowing of genetic diversity and greater uniformity of crops. While attractive on the surface, this trend carries significant long-term dangers in terms of maintaining the capacity to adapt to changing conditions.

While the importance of wild relatives of domesticated plants and animals is well recognized, we conclude that very little is being done to carry out detailed inventories of their status and trends, and to ensure that protected areas are managed in ways that both conserve the wild relatives and make their genes available for use.

War, famine, and environmental disasters may limit the availability to poor farmers of many of the crop varieties that they have traditionally grown. Seed and other propagating materials may be lost or eaten, supply systems disrupted, and seed production systems destroyed. At the same time, aid organizations may distribute seed of new cultivars from a very narrow genetic base that often require rather different production practices than those practiced locally. The net effect of this can be substantial loss of traditional cultivars or changes in the numbers and types of varieties grown (Richards and RuivenKamp 1997).

Assessing the impact of eco-agriculture suffers from a lack of consistent, comprehensively documented research on eco-agricultural systems, particularly regarding agricultural production–ecosystem health interactions, but all 36 eco-agricultural initiatives reviewed by McNeely and Scherr (2003) demonstrate benefits to landscape and ecosystem biodiversity, while impacts on species biodiversity were very situation specific. The greatest benefits were realized when intentional ecosystem planning achieved coordinated adoption over large areas. However, even when adoption was limited to individual farm-level activities, significant benefits to "wild" biodiversity were recorded.

Conserving biodiversity in forest production systems has received considerable attention (Szaro and Johnson 1996; Lindenmayer and Franklin 2002), and the necessary policies and practices are well known. However, the incentives needed to put these into practice are still insufficient in most countries.

It can be stated with a *high level of confidence* that protecting marine areas is a good investment for the conservation of marine biodiversity. In some cases, MPAs may also help the recovery of fish stocks beyond reserve boundaries.

A scientific consensus on the science of marine reserves is emerging. Some of the key findings for biodiversity conservation are (Lubchenco et al. 2003):

- reserves conserve both fisheries and biodiversity;
- networks of reserves are necessary for long-term fishery and conservation benefits (a network of reserves provides significantly greater protection than a single reserve);
- reserves result in long-lasting and often rapid increases in the abundance, diversity, and productivity of marine organisms;
- reserves reduce the probability of extinction for marine species resident within them;
- increased reserve size results in increased benefits, but even small reserves have positive effects; and
- full protection is critical to achieve this full range of benefits.

It should be emphasized, however, that the importance of marine protected areas as a tool for in situ conservation depends also on factors outside the reserve boundaries, such as pollution, climate change, and overfishing.

5.8 Designing Governance Approaches to Support Biodiversity

5.8.1 Introduction

Designing governance approaches to support biodiversity is both a response in itself, and creates enabling conditions for other responses to succeed. For example, establishing laws for access to resources in PAs supports biodiversity conservation directly. Maintaining a well-functioning legal system enables other responses, such as PAs, to be effective because the enforcement of laws carries consequences without which the PA would become a paper park. This section assesses what kinds of governance work best for what aspects of biodiversity and under what conditions.

Governance as "the act or manner of governing" (*Oxford Concise Dictionary,* 8th edition, 1990) is used broadly here, and often relates to the exercise of governmental authority at various levels, but can also involve the exercise of some control or authority by other actors, for example indigenous peoples or the private sector. "Good governance" involves establishing and enforcing appropriate laws, developing management and other institutions, and maintaining a system that limits corrupt activities.

The CBD in its description of the ecosystem approach (Decision V/6) acknowledges that "the scale of analysis and action should be determined by the problem being addressed," and includes as two principles of the ecosystem approach that "management should be decentralized to the lowest appropriate level" and that "the ecosystem approach should be undertaken at the appropriate spatial and temporal scales." The similar approach in "subsidiarity" means that a higher level of authority should only act if the objectives of the intended action cannot be sufficiently achieved by a lower level of authority.

Many central governments have decentralized certain responsibilities, sometimes with insufficient attention to whether appropriate powers and responsibilities are being devolved from the center and whether the necessary local institutional infrastructure is in place to receive newly decentralized powers and obligations.

This decentralization can take different forms, varying in how much authority, accountability, and representation is assigned to the lower levels of governance. For example, since 2002, Mexico has decentralized the authority and responsibility for enforcement of federal environmental rules to states and local entities that demonstrate that they have the institutional capacity to take on those responsibilities. With respect to Mexico's megabiodiversity, such decentralization has required large investments in the management and institutional capacity at the local and state levels. The fact that biodiversity generates benefits beyond local and even regional boundaries implies that decentralization of biodiversity conservation management could shift management toward local benefit provision if proper national or international incentives and management structures are not nested with local management.

Devolution takes place when decision-making powers are devolved to local branches of the central state (prefects, administrators, or technical agents such as foresters). These upwardly accountable bodies are local administrative extensions of the central state. They may have some downward accountability built into their functions, but their primary responsibility is to central government. When authority and decision-making powers are devolved to local government authorities, issues of representation are at stake. One key question is how these local authorities are chosen. Are they elected by the communities they are supposed to represent, or are they appointed by central government?

Privatization is also often done in the name of decentralization and participation, and devolves public resources to private groups, such as individuals, corporations, management committees, NGOs, etc. These bodies may be accountable within certain legal and moral bounds, but their objectives are often determined by their members, not the public as a whole (Ribot 1999). Such privatization can lead to more exclusion than participation and to less public accountability.

The renewed focus on indigenous groups under the CBD's Article 8(j) has led to calls to reinforce "traditional" authorities in natural resource management. The role and legitimacy of these authorities may differ from community to community. Chiefs can be administrative auxiliaries of the state (hence upwardly accountable), dedicated to the local population (downwardly accountable), or autocratic local powers (Ribot 1999). Many decentralized and participatory environmental management policies and projects rely on NGOs, project or government-organized management committees, local project administrators, local government administrators, or technical service agents, to represent local communities in matters of natural resource decision-making (Ribot 1999, UNESCO 2000). When representative local government is in place, the empowering of alternative authorities (including "traditional authorities") undermines the function and ultimately the legitimacy of the local authorities (Ribot 1999). In some cases, this is aggravated by a constant shifting of power over resources from one set of authorities to another and back again (Spierenburg 2003).

Barrett et al. (2001) argue for stepping beyond the "false dichotomy" of community versus central government and recognize the value of diversity in approaches to governance. Community-based methods work best if social control at the local level is strong enough to restrict access to the resource. Government systems work best if they are run by a competent bureaucracy. Where authority should be placed depends on the resource to be managed and the relative strengths of the different levels of authority.

5.8.2 Examples of Governance Approaches in Biodiversity Conservation

Governance in practice requires institutions and a framework including rules on accountability, enforcement, reporting, and distribution of benefits. Sound institutions are essential for successful governance.

In 152 case studies of net loss of tropical forest cover, Geist and Lambin (2002) analyzed the underlying driving forces of tropical deforestation, including demographic, economic, technological, political, institutional, and cultural factors. Policy and institutional factors (including property rights, policy climate, and formal policies) were found to be an underlying cause of tropical deforestation in 78% of the case studies (second to economic factors, with 81%).

At the end of the 1990s, the Indonesian government devolved management responsibility for all forests outside protected areas to the district level within provinces (criteria and standards were still to be set by central government), in the context of new legislation that promoted regional autonomy. Many districts did not have the capacity to manage and enforce a sustainable forestry policy, resulting in logging concessions in biodiversity rich areas and illegal logging. The Indonesian Directorate of Nature Conservation acknowledged the problem and stated that the military may be needed to protect national parks instead of local police (Jepson et al. 2001).

Kellert et al. (2000) assessed the success of a number of Community Based Natural Resource Management approaches in Kenya, Nepal, and the United States using six variables, one of which was empowerment. They found that although all case studies intended to devolve authority from higher to more local levels, the actual extent of this devolution was uneven, "often questionably effective," or not equitable (with only small groups in local communities benefiting). A clear judicial and legislative mandate for devolution and well-developed institutions were identified as factors for success of the CBNRM scheme and, more specifically, its "empowerment" aspect.

Smith et al. (2003) investigated the correlation between quality of country governance and changes in three components of biodiversity (forests, African elephant, and black rhinoceros). The results (though less strongly for forests) confirm the link between corruption and conservation failure, emphasizing the need to strengthen institutions.

Local councils oversee the Masai Mara National Reserve in Kenya. Because few fees have been collected or effectively invested, local communities that were entitled to 19% of reserve revenues as compensation for human-wildlife conflict had received little or no money since the mid-1990s. The local council (Trans Mara County Council) has now contracted a private consortium to manage part of the reserve (ticketing, revenue collection, tourism management, security, and wildlife conservation), resulting in a significant increase in revenue collection and increased donor funding. If the consortium can ensure that the benefits do flow to neighboring communities, this becomes an example of a public-private partnership that successfully addresses governance problems (Walpole and Leader-Williams 2001; Caldecott and Lutz 1996).

5.8.3 Assessment

Where authority is devolved to a lower level because action at that level will be more effective (for example, better adapted to the resource that will be managed, or more capable to create incentives to conserve a resource), such devolution can allow local capture of benefits of biodiversity, combined with a sustainable management of the resource. Where a higher government level devolves authority for reasons not related to the achievement of the conservation goal (for example, to reduce the burden on a central government administration), or where devolution happens without institutional capacity at the lower level, it has not helped biodiversity conservation, and has even led to the loss of biodiversity.

Without good governance and strong institutions, the level at which authority is located may only marginally influence the success of responses to biodiversity loss. This statement, however, does not reduce the importance of the principles of decentralization or subsidiarity as a guiding principle for reasons of efficiency, democratic legitimacy, and ethics. Institutional diversity and nested institutional arrangements (Ostrom 1998) may be the way forward, as all levels of authority have their strengths and weaknesses (Ostrom et al. 1999).

The case studies reviewed identify the following issues as relevant when deciding on a governance approach:

- Not all functions can be decentralized usefully (Caldecott and Lutz 1996). For example, tangible benefits of biodiversity might most often be harvested locally, and local management may include the social control required to prevent overharvesting. At the same time, protection from, for example, armed poachers may require a higher-level authority.
- A necessary condition for successful decentralization is a national framework that supports it. Moreover, decentralization is a political process involving the redistribution of power, requiring a mediating body between the different levels (this can be central government, but also, for example, an NGO).
- Complications may be added where local people or authorities are unaware of some of the consequences of management options. Awareness and education are essential.
- Sound institutions and high quality of governance at all levels—including well-established tenure rules at the local level (Ostrom 1998)—are essential prerequisites for successful decentralization of environmental management.

5.9 Promoting International Cooperation through Multilateral Environmental Agreements

The most pressing global environmental issues—the loss of biological diversity, deforestation, invasive alien species, climate change, the loss of wetlands, overgrazing, the protection of international waterways, desertification, ozone depletion, and toxic waste—create major challenges for legislation-based responses. Various treaties have emerged in the past few decades to address these issues. These multilateral environmental agreements play a crucial role in the conservation and protection of the environment, and they are inextricably linked to the alleviation of poverty in developing countries. (See Box 5.6.)

But how effective have they been in protecting the environment? The effectiveness of multilateral environmental agreements has been widely discussed and well-documented (Jacobson and Brown Weiss 1997; Sand 1992; Werksman n.d.; May et al. 1996; Bilderbeek 1992; Cameron et al. 1996). Effectiveness varies according to the objective assessed, such as solving the problem, achieving the goals set out in the treaty, altering behavior patterns, and enhancing national compliance with the rules in international agreements (Birnie and Boyle 2002). Therefore, the effectiveness of different MEAs is influenced by the nature of the environmental problem and several other factors. Possible measures of success differ for the different MEAs. Boxes 5.7 and 5.8 discuss the relative success of two international agreements.

5.9.1 Key Factors Leading to Effective Implementation of Treaties

Several studies on implementation and compliance present similar findings on the effectiveness of environmental agreements. One empirical study indicates that although compliance has been low, the overall implementation of treaties is positive (Jacobson and

BOX 5.6

How Multilateral Environmental Agreements Affect Rural Poverty

Few multilateral environmental agreements address the poverty alleviation priority of developing countries, but the Convention on Biological Diversity, in its preamble, recognizes "that economic and social development and poverty eradication are the first and overriding priorities of developing countries." The CBD (Article 20.4) and UNFCCC (Article 4.7) also state that "eradication of poverty" is one of the commitments by the parties.

One of the key decisions of the seventh UNFCCC Conference of Parties (FCCC/CP/2001/L.24/Add.2) is the establishment of a Clean Development Mechanism, a mitigation measure that could assist developing countries achieve sustainable development, while recognizing economic growth as essential for alleviating poverty. Moreover, UNFCCC has recently adopted the COP 8 "Delhi Ministerial Declaration" (FCCC/CP/2002/L.6), which emphasizes the implementation of energy policies that support developing countries' efforts to eradicate poverty. The Convention to Combat Desertification also has several provisions toward alleviating poverty that create an enabling environment to achieve sustainability objectives. The CCD is considered one of the tools for poverty eradication, particularly in Africa (WSSD, Plan of Implementation para. 7(l)).

The Ramsar Convention does not specifically provide for the involvement of local people in wetland management. Nevertheless, the COP recommendation 6.3 calls for the inclusion of local and indigenous people in the management of Ramsar wetlands. Subsequently, in 1999 the COP adopted guidelines for establishing and strengthening local communities' and indigenous people's participation in the wetland management (Res. VII.8). This is the most systematic guideline on participatory management. As stated in the Ecosystem Approach Principles (Principle 2), management should be decentralized to the lowest appropriate level, and boundaries for management shall be defined by indigenous and local peoples, among others (Principle 7, CBD 2001–2004).

Brown Weiss 1997). The study shows that the governance features of a country determine the quality of its treaty implementation, and effective implementation and compliance involve numerous other factors, including the characteristics of the treaty, the political will to support it, the human resources committed to monitoring and reporting, the financial resources allotted, sanctions and enforcement, and country capacities. In addition, monitoring by civil society can encourage implementation.

Agreements that impose precise obligations are easier to assess, such as the Montreal Protocol on Substances that Deplete the Ozone Layer and CITES. On the other hand, agreements with vague obligations that do not establish clear standards make it difficult to judge the extent of compliance, such as the World Heritage Convention or the International Tropical Timber Agreement. A second characteristic involves the quantity of regulated objects. For instance, the Montreal Protocol deals with a limited number of substances, but CITES deals with thousands of species. This makes CITES difficult for customs officials to implement.

A second important factor for the implementation and compliance of MEAs is political will. Jacobson and Brown Weiss (1997) report that when the parties agreed to deepen their specific commitments, better implementation and compliance resulted.

Third, a reporting mechanism is essential (Chasek 2001). The parties to a convention need to report the measures used to comply with the obligations (Glasbergen and Blowers, 1995). National reports are critical because they provide the specific information to show that each country is meeting its obligation under a convention. Indeed, the ineffective monitoring of MEAs results in a lack of accurate, complete, and objective information on the performance of the parties (Werksman n.d.). A reporting system helps government officials understand their obligations under the treaties and the means that might be used to aid compliance. It has also been found that a standardized form improves the effectiveness of the reporting. For instance, the cooperation between CITES and the World Conservation Monitoring Center has shown that reporting improves effectiveness (Jacobson and Brown Weiss 1997).

Fourth, the availability of sufficient human resources to monitor compliance is essential. The secretariats of the conventions are expected to analyze the country reports submitted by the parties to the convention, but the secretariats are usually small in size—many less than 30 people—which makes it difficult to conduct a thorough and timely analysis (Jacobson and Brown Weiss 1997). CITES has been enhanced by the close monitoring for infractions through national and independent reporting by the secretariat. The infractions are reported in the Conference of the Parties and are widely publicized (Werksman n.d.).

A fifth factor for effective implementation and compliance is the availability of financial resources (Richardson 1992). A study in Cameroon shows that limitation of financial resources is the main reason for noncompliance with the procedural requirements of treaties (Jacobson and Brown Weiss 1997). To help assure their effectiveness, the Vienna Convention and the Montreal Protocol have an amendment to provide financial assistance in preparing inventories for the production and consumption of ozone-depleting substances, and in transforming production facilities (Jacobson and Brown Weiss 1997).

Some treaties have provisions for developed countries to assist developing countries in meeting their international obligations. For instance, the CBD (Article 20, 2) calls for "the developed country parties to provide new and additional resources to enable developing country parties to meet the agreed full incremental costs to them of implementing measures which fulfill the obligations of the convention."

A sixth factor for effectiveness is the establishment of sanctions. Mechanisms for dealing with noncompliance are important to enable punishment of violators (Chasek 2001). Typically, biodiversity-related treaties do not establish sanctions for noncompliance or for failure of parties to adhere to the procedural provisions (Chayes and Chayes 1991). The lack of monitoring and enforcement provisions in the treaties is a common shortfall of MEAs (Richardson 1992; Miles et al. 2002).

Although few MEAs impose any sanctions for noncompliance, most do include a dispute resolution mechanism. Formal dispute settlement mechanisms are rarely used, however, because countries have preferred to use negotiation to solve problems (Bilderbeek 1992). Most multilateral environmental regimes have no compulsory jurisdiction for dispute settlement. In the absence of a supranational regulatory institution, the implementation is carried out by national institutions (Sand 1991). However, the difficulty of monitoring and enforcement is compounded since implementation by the individual countries is so variable (Richardson 1992).

Table 5.2 lists 15 MEAs, which directly or indirectly relate to the protection and conservation of biological diversity. It also

BOX 5.7

The Convention on Biological Diversity

The 1992 Convention on Biological Diversity is arguably the most important multilateral environmental agreement dealing with biodiversity. An assessment of the effectiveness of the CBD needs to include consideration of progress towards the CBD's objectives, the external impact of the CBD, and the nature of the CBD as a mechanism (that is, what it may reasonably be expected to achieve).

A central strength of the CBD is the integrative nature of its three linked objectives of conservation, sustainable use, and benefit sharing. However, these are "process" rather than "outcome" objectives. Many of the CBD's provisions have vague formulations, whose content can only be tested in implementation (Rosendal 1995). The national biodiversity strategies and action plans, which form a central implementation mechanism, have resulted in positive actions at the national and local level, but experience varies. Weaknesses include the absence of a clear process for assessing, verifying, or discussing national reports on implementation (Global Forest Coalition 2002). The absence of consensual scientific knowledge in support of the CBD has been viewed as one of its greatest shortcomings (Le Prestre 2002).

Disaggregated, the CBD has had positive effects, notably at the national level because its main emphasis has been on national implementation. However, one could argue that a global response option is needed, one that can advance a coordinated and results-oriented international effort across the spectrum of complex, interacting forces that drive biodiversity loss. The CBD should be playing this role. In this respect, it has not yet succeeded.

In 2002, the target of achieving ". . . by 2010 a significant reduction of the current rate of biodiversity loss . . ." was included in the Strategic Plan for the Convention (Decision VI/26). This target date was reinforced by the Plan of Implementation adopted at the 2002 World Summit on Sustainable Development, which also confirmed the need for new and additional financial and technical resources for developing countries.

For the first decade of its existence, the CBD has been a process-oriented and relatively marginal convention, if viewed in a broad political context. Achievement of the 2010 target will require unprecedented political and financial commitments across a wide range of sectors. Other developments, such as the proposal for a new regime on benefit-sharing by the Group of Like-minded Megadiverse Countries, and the growing recognition of the importance of biodiversity to the Millennium Development Goals, raise the possibility of reinforcing the CBD to address the new agenda. This could involve the Conference of the Parties making use of Article 23.4(i), which states that it is to: "[c]onsider and undertake any additional action that may be required for the achievement of the purposes of this Convention in the light of experience gained in its operation." Currently, the CBD seems to stand at a crossroads, with its Biosafety Protocol recently entering into force, nearly 190 States Parties, and regular meetings of its Conference of Parties and various subsidiary bodies. On the other hand, government enthusiasm and funding seem to be waning, and relatively few nongovernmental organizations are stepping in to support decisive action to actually implement the CBD on the ground.

BOX 5.8

The Bolivian National Strategy for Biodiversity Conservation

Bolivia ratified the Convention on Biological Diversity in 1994, so the government needed to develop a corresponding framework for implementation. In a pluricultural and multiethnic country such as Bolivia, where many people depend directly on biodiversity services without being necessarily aware of the need for conservation, agreeing on biodiversity's role has been a major challenge.

The Bolivian process of developing its National Biodiversity Conservation Strategy was through committees led by the Ministry of Sustainable Development and Planning. Several committees were established involving specialists, experts, and representatives from the government and civil society. Hundreds of people participated in departmental, sectoral, and national workshops, raising the level of awareness of biodiversity of most of the decision-making levels of society and the government.

The Strategy was validated and approved by all participants in the process (civil society and the government) through an act signed at the concluding national workshop, and subsequently officially ratified through a Supreme Decree. Within the Strategy framework, Bolivia recognizes the strategic character of biodiversity to improve the quality of life of the population and promote national development, in addition to the need to promote its integration in development planning through plans and strategies at the national, departmental and sectoral levels. This includes linking strategies for the use of nonrenewable and renewable natural resources.

The objective of the Strategy is the conservation of biological resources, in particular those of ecological, economic, and cultural importance. It is recognized that the conservation of ecosystems, species, and genetic resources affected by destructive processes is fundamental to ensure the maintenance, functionality, productivity, and dynamism of the environment and to maintain the productive base of the country. At the same time the economic potential of biodiversity is recognized as a current and potential source of benefits at many levels in the medium and long term.

The Strategy has become a governmental policy and a principal challenge is to reach beyond traditional government and jurisdictional boundaries. Civil society has validated the Strategy, but does not feel sufficient ownership to promote its implementation. The Strategy was designed as a mechanism for multilateral and bilateral fundraising, and it will be possible to measure whether the future conservation funding was facilitated by the elements included in the strategy.

indicates where an MEA contains provisions related to implementation as discussed above.

Enforcement mechanisms differ among the treaties. In general, the expressions used in the treaties, which impose the obligations, are criticized as vague, thus lacking effective force (Boer 1998; Bilderbeek 1992). For instance, most of the treaties listed in Table 5.2 do not have clear provisions for implementation. Rather, they use expressions such as "to explore," "to encourage," or "where appropriate and feasible," which weaken the provisions.

Some treaties state explicitly that each country's domestic legislation should provide sanctions for noncompliance with the regulations of the agreements. For instance, the Basel Convention (Article 9.5) imposes strict trade sanctions. CITES (Article 8.1) requires the parties to take measures to penalize illegal trade, and to confiscate illegally traded specimens.

Table 5.2. Selected Provisions Related to Implementation and Enforcement of International Environmental Agreements (modified from Association on Scientific Uncertainty 2003)

	Convention	Enacting/ Strengthening of Domestic Laws and National Strategies	Identification of Policy Measures/ Performance standards	Notification/ Reporting System	Recommendations/ Accountability	Monitoring by Non-state Entity	Financial Resources	Eradication or Alleviation of Poverty	Non-compliance Measures	Sanctions	Dispute Settlement	Coordination with Other Conventions
1	Ramsar Convention (1971)	Art. 2; 3; 4; 5	Art. 4.1	Art. 2.5; 3.2	Art. 6.2(d)(e)	Art. 7.1	COP Resolution 4.3	COP Recommendation 6.3; Resolution VII.8				
2	WHC (1972)	Art. 5; 17; 18	Art. 11	Art. 11.1; 29	Art. 13.5	Art. 8.3	Art. 15–18		Operational Guidelines E. 46 (a), (b); 47.			
3	CITES (1973)	Art. 3–8; 10; 14	Art. 2; 3; 4.2; 5	Art. 8.7; 13.1; 13.2	Art. 8.8; 11.3(e); 12.2(h); 13.2; 13.3	Art. 11.6; 13.1		Resolution Conf. 10.9; 10.14; 10.15;		Art. 8.1	Art. 18.1, 2	Art. 14.2–14.6
4	CMS (1979)	Art. 3.4; 3.5; 5.5	Art. 5	Art. 3.7; 6.2; 6.3	Art. 3.6; 7.5(e)(g)(h)	Art. 8.2					Art. 5.4(e); 13.1, 13.2	Art. 12
5	Vienna Convention (1985)	Art. 2; 3.2		Art. 5	Art. 6.4(c)(f)	Art. 6.5					Art. 11.1; 11.2; 11.3(a) (b); 11.5	Art. 7.1(e)
6	Montreal Protocol (1987)	Art. 2; 4; 5	Art. 2; 4; 5	Art. 2.5; 2.7; 2.8(b); 4B.3 (Amendment); 5.4; 5.6; 7 (Amendment R); 9.34	B.4 (Amendment)	Art. 11.5	Art. 10; Article 1. T (Amendment); 13		Art. 8	Art. 8 and Art. 4	Art. 14	
7	Basel Convention (1989)	Art. 3; 4.1–4.10; 6–9		Art. 3; 4.1(a); 4.2(f); 5.1; 6.1; 6.4; 6.9; 6.10; 7; 11.2; 13		Art. 15.6	Art. 14			Art. 4.4; 9.5	Art. 20.1; 20.2; 20.3 (a) (b)	Art. 1.4; 4.12; 16.1(d)
8	UNFCCC (1992)	Art. 3; 4	Art. 4	Art. 4.1(a)(j); 4.2(b); 12	Art. 7.2(g)	Art. 7.6	Art. 4.2–10; 5; 11; 12.4; 12.5; 12.7; 21.3	Preamble; Art. 4.7			Art. 14.1; 14.2(a) (b); 14.5–14.7; 13	Art. 8.2(e)
9	CBD (1992)	Art. 6; 10;14.1; 15.7; 16.3; 16.4; 19.1; 19.2; 19.4	Art. 7(a); 8; 9	Art. 14.1(c)(d); 26		Art. 23.5	Art. 20; 21; 39	Art. 20.4			Art. 27.1; 27.2; 27.3(a) (b); 27.4	Art. 22; 23.4(h); 24.1(d); 32

10	ITTA (1994)	Art. 1(l); 25		Art. 29.2	Art. 12	Art. 3.1	Art. 1; 21; 25; 27; 28; 30.5; 34				Art. 31	Art. 14
11	Straddling Stocks Agreement (1995)	Art. 5; 6; 7.2; 10(c); 12.1; 14.1; 16; 18; 19; 21–23; 33.2	Art. 5; 6.3; 18.3	Art. 7.7; 7.8; 20.3; 21.4; 21.5–21.9; 21.12		Art. 10(h)	Art. 24–26				Art. 7.4; 10(k); 27; 29; 30.3; 30.4	Art. 4; 20.6; 44
12	UNCCD (1996)	Art. 4; 5; 9–11; 13–15;	Art. 10	Art. 26		Art. 22.7	Art. 4.2(h); 4.3; 6; 7; 9.2; 12; 13; 16–21; 26.7	Preamble; Art. 4.2.(c); 10.4.; 20.7.; Annex I, Art. 4.1.(a); 5.1.(a); 8.1.; 8.3.(a); Annex II, Art.4.2.	27		Art. 28.1; 28.2(a) (b); 28.3; 28.6	Art. 8; 22.2(i); 23.2(e)
13	Convention on Int'l Watercourses (1997)	Art. 5; 7; 20–23; 26.1; 27; 28		Art. 9; 12; 28.2							Art. 10.2; 33.1–33.8; 33.10 (a) (b)	Art. 3; 4; 8.2
14	Kyoto Protocol (1997)	Art. 2; 3; 5; 10	Art. 2; 3; 5; 10	Art. 3.2–3.5; 4.2; 7; 10(b ii)(f)	Art. 13.4(f)	Art. 8.5; 13.8	Art. 2.3; 3.5; 3.6; 3.10; 3.12–14; 6; 10; 11; 12; 13.4(g)		18		Art. 16; 19	
15	Cartagena Protocol (2000)	Art. 2.1; 2.2; 14.4;16; 17; 18; 25	Art. 11.6(a); 15.1; 18.2	Art. 8; 11.1; 11.5; 12.1; 13; 17; 20.3; 21.2; 25.3; 33	Art. 29.4(a)		Art. 28		34			Preamble para. 10; Art. 2.3; 14.3; 32

Notes:

1. Convention on Wetlands of International Importance as Waterfowl Habitat.
2. UNESCO Convention on World Heritage.
3. Convention on International Trade in Endangered Species of Wild Fauna and Flora.
4. Convention on Migratory Species.
5. Vienna Convention for the Protetion of the Ozone Layer.
6. Montreal Protocol on Substances that Deplete the Ozone Layer.
7. Basel Convention on the Transboundary Movements of Hazardous Waste and their Disposal.
8. United Nations Framework Convention on Climate Change.
9. Convention on Biological Diversity.
10. International Tropical Timber Agreement.
11. Agreement for the Implementation of the Provisions of UN Convention on the Law of the Sea (1982) Relating to the Conservation and Management of Straddling Fish Stocks and Highly Migratory Fish Stock.
12. United Nations Convention to Combat Desertification.
13. Convention on the Law of the Non-navigational Uses of International Watercourses.
14. Kyoto Protocol to the United Nations Framework Convention on Climate Change.
15. Cartagena Protocol on Biosafety to the Convention of Biological Diversity.

Most treaties have a reporting system and publish data on the relevant parties' follow-up of regime decisions. However, these data are often incomplete (Wettestad 1999). Some conventions do not have a specific requirement for reporting. In the case of the Ramsar Convention, this is compensated by a Conference of the Parties recommendation which stipulates that "all Parties should submit detailed national reports to the Bureau at least six months prior to each ordinary meeting of the Conference of the Parties" (Recommendation 2.1). In addition, Recommendation 5.7 on national committees mentions the opportunity for nongovernmental organizations to have input in the preparation of the report (Isozaki 2000). Nevertheless, many countries have not submitted their report because the recommendation does not specify content or guidance for preparing the report.

Together with a reporting system, notification from the concerned country is important for monitoring. In the case of CITES, it helps to detect cases such as illegal transboundary movements and illegal trade of specimens; it also helps to notify affected or potentially affected states that may suffer adverse effects on the conservation and sustainable use of biodiversity, and on human health (Biosafety Protocol, Article 17, 1).

Some treaties have concrete provisions on noncompliance, while others use other enforcement methods, such as imposing sanctions or making recommendations. The World Heritage Convention itself does not provide noncompliance measures, but the operational guidelines establish procedure for eventual deletion of properties from the World Heritage list when the property has deteriorated or when the necessary corrective measures have not been taken.

Some treaties provide for monitoring by non-state entities, such as nongovernmental organizations and experts in the concerned area. Most treaties have provisions to ensure coordination with other conventions.

5.9.2 Overcoming the Limitations

Implementation and compliance can be improved by taking into account the factors discussed above. Existing multilateral environmental agreements have been piecemeal, so the coordination of actions among various MEAs is essential for better implementation (Bilderbeek 1992). In this regard, the WSSD Plan of Implementation recognizes the need for cooperation between the relevant international organizations. As a priority to improve implementation, it advises actions "to encourage effective synergies among multilateral environmental agreements dealing with the protection and conservation of biodiversity, through the development of joint plans and programs with due regard to their respective mandates, regarding common responsibilities and concerns" (Paragraph 42 (c).

In fact, some conventions have been successfully implementing cooperative works in areas of the common interests among them. The Ramsar Convention has been promoting cooperation and coordination with other treaties to achieve the objectives of the convention. For instance, the Ramsar and World Heritage conventions have cooperated to identify and strengthen conservation of those sites of international importance, which are of mutual interest and benefit (Article II, Ramsar MOU with the World Heritage Convention, May 14, 1999). Furthermore, cooperation between the Ramsar Convention and the Convention on Migratory Species has been in effect since 1997, in terms of joint conservation action, data collection, storage and analysis, institutional cooperation, and new agreements on migratory species (Ramsar 1997). The Ramsar Convention has adopted its third joint work plan with the Convention on Biological Diversity, covering the period 2002–2006.

Although some treaties do provide financial resources, the terms usually avoid specifying concrete measures, such as how much assistance, to whom, and on what terms (Boer et al. 1998). For instance, to implement a convention like CITES requires significant financial resources for training, but that convention does not provide any means or financial mechanisms for doing so. Likewise, the Ramsar Convention does not have a provision for financial assistance. However, Conference of the Parties resolution 4.3 set up the Ramsar Small Grants Fund for Wetland Conservation and Wise Use in 1990 to provide assistance for wetland conservation initiatives in developing countries or countries with economies in transition. Since the level of funding has not been sufficient to fund many of the projects submitted to the fund, the Conference of the Parties has adopted measures to increase the wetland conservation fund (Kushiro Res. 5.8), to cooperate with the Global Environmental Facility and its implementing agencies (Brisbane Res. VI.10), and to consider receiving official development assistance and external funding to meet their obligations under the Convention (Brisbane Res. VI.6). Despite its modest funding mechanism, the Ramsar Convention has been exploring new ways to cope with and protect its listed wetlands.

In addition to formalized international cooperation among legally-binding instruments as discussed above, several initiatives among non-legally binding instruments help overcome the limitations. Expanding countries' participation in relevant bilateral, regional, and sub-regional agreements, initiatives, and networks is quite important for the effective implementation of MEAs. For instance, the Convention on Migratory Species has extended its work beyond its signatory parties. It has promoted regional actions and agreements among its parties, like the "Understanding Concerning Conservation Measures for Marine Turtles of the Atlantic Coast of Africa, Abijan, 1999."

One long-standing case is that in which Japan maintains bilateral agreements for the protection of migratory birds with the United States (1974), Australia (1981), China (1981), and Russia (1988). The countries exchange information on measures taken within each country and discuss the needs for further joint research. Another example of a successful regional agreement is the "Asia-Pacific Migratory Waterbird Conservation Strategy" started in 1996 with the objective to promote the conservation of migratory waterbirds and wetlands in the Asia-Pacific region (Wetlands International 2001), resulting in the establishment of major waterbird flyways in the Asia-Pacific Region. Another example is the International Coral Reef Initiative, a comprehensive framework of international cooperation for coral reef conservation and management, with a special focus on ecosystem and community-based management (UNEP–CAR/RCU 2000–2003). Box 5.9 discusses carbon sequestration as a policy response.

5.9.3 Assessment

Existing MEAs cover the most pressing drivers and issues related to the loss of biodiversity. Additional global agreements are therefore not required at this time, but better coordination between the existing conventions, especially at implementation level, would increase their success and avoid duplication or even contradictions that lead to inefficient use of the limited resources available. Regional instruments can be useful to address conservation issues, for example, at the scale of a river basin or a transboundary terrestrial conservation area, and have been shown to help implementation of global MEAs.

BOX 5.9

Assessing Carbon Sequestration as a Conservation Response in the Andes

Carbon sequestration is increasingly understood as an important global ecosystem service (Daily et al. 1997). Bolivia has gained experience with climate change mitigation through carbon sequestration through the Noel Kempff Climate Action Project. The project was co-designed and is executed, since 1997, by the Bolivian NGO Fundación Amigos de la Naturaleza (FAN), together with the government of Bolivia, the Nature Conservancy, and three energy companies (American Electric Power, PacifiCorp, and BP Amoco). The project is the largest forest-based carbon project in the world, protecting about 1.5 million hectares of tropical forests in the Bolivian Amazon for at least 30 years. The project was developed under the Activities Implemented Jointly pilot phase of the Kyoto Protocol and conserves natural forests that would otherwise have been subjected to continued logging and future agricultural conversion. It is expected to sequester seven million tons of carbon (Powers 2003; Brown et al. 2000).

For the first time in Bolivia, a market-based mechanism, rather than a donation, was to generate the funds needed to manage a large protected area. Carbon-sequestration-forest-conservation projects seemed to be an adequate response to the problem that nature conservation, in comparison to traditional land-use forms, does not provide sufficient benefits for local people in developing countries. Consequently, many local actors, such as indigenous communities and municipalities, developed a strong interest in carbon trading as an alternative and sustainable income. However, for diverse reasons conservation projects have yet to become eligible under the Clean Development Mechanism of the Kyoto Protocol, which for the time being allows only for forestry measures.

Nevertheless, the Noel Kempff Climate Action Project continues based on a voluntary commitment of the investors. The project has been and continues to be very important for the conservation of biodiversity, and is breaking ground to establish credible and verifiable methods to quantify greenhouse gas benefits of land-use change and forestry projects (Brown et al. 2000). Furthermore, household level economic analysis (Milne 2001) has demonstrated a net positive economic benefit to park-bordering communities, particularly through working to secure land tenure and facilitating "carbon tourism." Another study suggests that community benefits, while present, may have been overstated (May et al. 2003).

Currently, the CDM contains enormous potential for large restoration measures in degraded areas in the tropics. An array of native tree species could potentially restore the Andean montane ecosystems, assure the availability of water, prevent erosion and sedimentation, and support agricultural production (that is, through shade, nutrients, soil formation, etc.). One limiting factor has been the costs that could be covered neither by development projects nor by the local communities (Ibisch 2002). However, this might change completely should carbon credits yield a given income in the framework of the CDM.

The negotiation processes leading to the adoption of MEAs have succeeded in catalyzing political and scientific debate on environmental issues of international importance. The existence of MEAs has most likely also contributed to greater environmental awareness, though this does not mean that MEA provisions have been implemented on the ground.

The main issue, therefore, is implementation at the national level of existing MEAs (Bowker and Castellano 2002). It is worth reviewing the implementation of international treaties at the national level, especially the application of international environmental law by national courts. One study shows that national courts could play a supplemental role in implementation (Bodansky and Brunnée 1998). For example, in a Philippines timber court case, the Supreme Court ruled that the plaintiffs have standing to sue on behalf of their generation and subsequent generations (IC-SEA 1999); in a Tasmanian Dam case, the High Court of Australia held that the acceptance by Australia of an obligation under the World Heritage Convention as such sufficed to establish the power of the Commonwealth to make to fulfil the obligation (Bodansky and Brunnée 1998); and in Japan's Kogen Highway Plan case, the plaintiffs challenged the Hokkaido provincial government decision authorizing the construction of a road through Daisetsu National Park that threatened the "Naki rabbit" population and other wild flora and fauna (Isozaki 2000). These cases demonstrate that domestic courts play a vital role in the application of international environmental agreements.

National implementation of MEA provisions, compliance with reporting mechanisms, transparency of the reports, support to convention secretariats, and national capacity building are essential for success.

5.10 Education and Communication

5.10.1 The Case for Education and Communication

Policy-makers and biodiversity managers must deal with a vast array of external audiences and stakeholders, many of whom are not concerned with conservation. To at best reverse and at least mitigate detrimental human impact on ecosystems, policy-makers and natural resource managers must manage change in perceptions and actions. "Without communication, education and public awareness, biodiversity experts, policy makers and managers risk continuing conflicts over biodiversity management, ongoing degradation and loss of ecosystems, their functions and services. Communication, education and public awareness provide the link from science and ecology to people's social and economic reality" (Van Boven and Hesselink 2002, p. 3).

While much attention is usually given to bringing about change through individual-level learning, increasing attention is being given to change through organizational-level learning, whereby the institutions or governance structures are adapted to cope with the complexity and multilevel actions of sustainable development. This systems approach is often embodied in the term capacity development (Lusthaus et al. 2000), and makes use of the disciplines of communication and education to bring about innovation and transformation.

The benefits of investing in communication and education to manage change are widely recognized. At the international level, the environmental conventions include articles on public education and the Johannesburg Plan of Implementation cites education, awareness, and capacity building as means to achieve its objectives as well as to creating effective social institutions.

Regional case studies also document successful use of education and communication programs. For example, a study illustrated how a 12-year education program influenced the practices of eating seabirds in Quebec, Canada, documenting an increase in the populations of formerly threatened species of seabirds (Byers 2004). The Haribon Foundation in the Philippines has used communication, education, and mobilization of networks to motivate fishers and their communities to create marine sanctuaries to allow for fish populations to revive, since fishers were experiencing problems with declining catches. As a measure of success, over 1,000 reserves have been set up, resulting in economic benefits for fishers (Lavides et al. 2004).

Equally, when communication and education are not used, conservation efforts can be stymied and resources wasted. When caimans were reintroduced into a river in Uruguay, the local community, uninformed about the project, killed the animals, as they feared for their children's lives.

Information technology has facilitated cost-effective information sharing, enabling e-mail exchange among communities of practice, list servers among environmental journalists, on-line debates, e-learning and discussion forums. The Internet has been used to mobilize people quickly and in large numbers on specific issues.

A host of non-formal learning situations are provided in environment clubs, scouts, and adult and family education programs provided by museums, zoos, aquaria, botanical gardens, field studies centers, protected areas educational and interpretative programs, and ecotourism. These programs attract hundreds of millions of visitors annually, thereby contributing to developing a constituency for nature policy, though the extent to which these programs influence change in action for the environment has not been assessed here.

The success of communication and education efforts in terms of "awareness" are revealed by the fact that in most countries, nature conservation, environment, and sustainable development feature among the top ten—though not the top five—public concerns, (Hesselink 2003). However, some research suggests that this widely held concern is shallow, and support for biodiversity protection is easily eroded when countervailing considerations come into play, such as jobs, property rights, or human convenience (The Biodiversity Project 1999).

Biodiversity communication, education, and public awareness (CEPA, Article 13 of the CBD) is a powerful tool for mainstreaming biodiversity into sectoral practices, bringing local perceptions to the attention of the decision-making process, and potentially changing behavior (CBD 2000b, 2000c; CBD 2001, CBD 2003c). Biodiversity CEPA is more than environmental education, in that conventional educational approaches that have succeeded in raising environmental awareness are not adequate to reflect the complexity of the biodiversity concept (Hall-Rose and Bridgewater 2003). In the coastal marine area, many examples worldwide demonstrate that communication, education and public awareness activities do have a positive impact with regard to preventing the further erosion of ecosystems and reducing the main factors responsible for biodiversity loss, provided local communities are empowered with the capacity to take decisions on how to actually manage the ecosystems under consideration, on the basis of the information provided through CEPA programs (Mow et al. 2003). As with any other program, CEPA programs need regular evaluation, but they must also reflect the reality of the environmental, social, and economic context in which they are implemented.

5.10.2 Constraints Regarding the Use of Education and Communication

On the one hand, communication and education is a relatively weak instrument to bring about change if that change involves high barriers, such as great personal effort or economic loss. In these cases education and communication must be accompanied by other measures to ensure livelihood support. In organizational learning, education is often accompanied by incentives for promotion or assessing performance. On the other hand, it is evident that education and communication can be used more professionally, and the approaches used require some evaluation, reflection, and reconsideration.

Lessons have been drawn from the mistakes of imposing development or conservation solutions on populations that neglected the opinions and habits of the beneficiaries, leading to a lack of acceptance of the change or even to outright conflict (Mefalopulos and Grenna 2004). The result has been more willingness to involve stakeholders in formulating decisions, and a willingness to engage in partnerships and public-civil co-management of natural resources. This engagement is essential to developing trust between conservation organizations and the public (Stern 2004).

In most developed countries, with strong environment departments staffed by communication professionals, communication is used as a policy instrument to achieve policy and management objectives, as well as to mainstream environmental concerns in other sectors. Still difficulties can arise, as in the Netherlands, where the Nature Plan, largely conceived by ecologists, met with conflict from farmers who did not accept it, and had not been involved in its development. Despite informative and motivational communication, the plan did not create the desired acceptance because it neglected the "cultural factor" whereby people's rationality or perspectives take on those of the group to which they belong. The communication and policy formulating approach neglected the fact that people change as a result of discussion about issues that they think are important. (van Woerkoem et al. 2000).

Information on the state of the environment is available via the Internet, though information packaged as a support to decision-making is less well developed. In reviewing the impact of environmental information, Denisov and Christoffersen (2000) noted that it is not enough to tell people repeatedly that there are environmental problems; in the longer run, concrete information and ideas of what to do to resolve environmental problems are needed. Yet in Australia a study showed that having environment and wildlife information was not sufficient to drive interest within the community; rather, it is important to provide motivators and create relevance to encourage participants to actively seek information (NSW National Parks and Wildlife Service 2002). "Research in the field of environmental education and in commercial marketing has shown that there is no cause and effect progression from knowledge to attitude to behavior as educators have long believed" (Monroe et al. 2000, p. 3).

Since the 1970s, efforts have been made to integrate environmental education into the formal education systems with varying success. Schooling has focused on ecology, nature conservation, the impacts of pollution, and the need to recycle waste. More lately education is being challenged to deal with sustainable development, though the impact of education on long-term behavior for sustainable development is hard to assess. Palmer (1995) found that education, particularly at tertiary and upper secondary level, was the most important influence in developing commitment of only 9% of some 232 environmental educators, having much less impact than childhood contact with nature (29%) or the influence of parents, teachers, and other adults (26%).

5.10.3 Conditions for Success in Communication

Conduct research before implementation. Sometimes, communication means and media are decided on without assessing the critical target group that needs to be reached in order to effect change and how this should be done. Perceptions and social behaviors related to conservation issues and facts are not properly analyzed, and the communication systems used by the groups are not clearly defined. Lack of understanding of the relevant social factors is combined with poor practice in evaluation research, and a failure to use the latest in professional information on communication, media, and techniques (Encalada 2004).

Apply change models and appropriate communication. Most conservation practices that need to be promoted require a social response. This may be viewed as a revamped model of "diffusion of innovations" (Rogers 1983) in which communication contributes to: (1) creating consciousness about the existence of the innovation; (2) raising interest toward the innovation; (3) generating knowledge about the innovation; (4) motivating trying out the innovation; (5) helping achieve an appropriate evaluation of the tryout; (6) motivating decisions in favor of a solid adoption of the innovation; and (7) supporting with new and timely information for reevaluation of the adoption in order to consolidate it over time. Each of these steps requires different communication and it is important to apply the appropriate communication according to the stage of the process that people are going through.

Manage reputation and relationships. Stern (2004) suggests that as the global conservation community focuses much of its attention on attempting to provide alternative livelihoods to resource exploitation for residents living within the immediate vicinities of protected areas, careful attention must be paid to meaningful and appropriate engagement and communication with local populations. Results from his study (covering the United States, Ecuador, and the Virgin Islands) suggest that the ability to trust park managers is the most consistent factor associated with how local residents actually respond to national parks. Thus the ways in which parks and partner organizations engage local communities can make or break any projects designed to work with them. The most common explanations of distrust for park authorities included a lack of meaningful personal connection to these entities, a lack of genuine local involvement in park-related decisions or initiatives, complaints of broken promises made by park authorities and their partner organizations, and perceived inconsistency in park-related communication and in enforcement practices.

Manage stakeholder processes effectively. Social learning involves different actors with different interests being able to engage in dialogue. For this to occur, individuals need to be aware of, or be assisted to become aware of, the underlying assumptions and values that lead them to take a particular position. Conflict resolution and negotiation require individuals or groups to seek out common values, which requires being explicit about their assumptions. Reflection becomes a key tool in working through problem situations where values are in conflict and need to be reassessed.

To communicate effectively, deal with communication issues, not just with biodiversity issues. Each biodiversity conservation issue that management is addressing contains a specific communication issue. The communication issue is about how the people concerned relate to the biodiversity issue: what do they know, how do they feel, what do they perceive, what motivates their actions? Quite often a lot of technical information is communicated without giving any clue as to what the audience can do or contribute.

Communicate in understandable terms. One perspective on awareness programs is that they need to avoid jargon and technical terms such as "biodiversity" and "sustainability," which are abstract and remote from most people's lives. These "container" concepts arguably need both to be broken down into concrete issues that are closer to people's lives and to have actionable steps—a healthy river, a rich native bush land, sustained fish catch (Robinson and Glanznig 2003). However, there is little certainty that appreciation of biodiversity "option values" will follow from a focus on such current, concrete, issues.

Start with perceptions and motives of the people. Scientific facts often are communicated in the expectation of changing behavior. In Slovenia (Trampus 2003) conservationists wanted to stimulate people to conserve village ponds for biodiversity through various communication interventions. However, after asking a focus group to explore the ideas with the village people, the conservationists discovered that people were not motivated by biodiversity conservation but rather by cultural factors. These motives were used to promote the restoration of ponds. The engagement in action and the benefits felt as a community worked together to restore a pond had a strong motivating impact. Word spread from community to community with the result that many wanted to restore their ponds.

Create pride and involve in action. Based on work in some 35 tropical countries to stimulate conservation action, Manzanero (2004) has argued that the conditions for success include choosing a charismatic species, developing pride in that species, making a mascot, and sending a message to every segment of society, from religious leaders to children by way of music, stickers, and posters. Results from their approach include new protected areas, changes in legislation, change in behavior, and collective learning. Case studies are needed that examine whether biodiversity in general can benefit from this approach—either being "swept along" with efforts focused on the charismatic species, or itself being viewed as charismatic and a matter of pride.

5.10.4 Assessment

Communication and education are essential to achieve the objectives of the environmental conventions, the Johannesburg Plan of Implementation, and the sustainable management of natural resources more generally. Barriers to the effective use of communication and education include a failure to use research and apply modern theories of learning and change. While the importance of communication and education is well recognized, providing the human and financial resources to undertake effective work is a continuing barrier. Attention is often thrust on school education and providing information, yet evidence shows that more effective change strategies are required that address the individual, organizational, and institutional levels. More strategic approaches to achieve management objectives and policy need to consider the benefits and perceptions of the stakeholder, building relations with, and honoring input from, stakeholders.

5.11 Lessons Learned

5.11.1 Introduction

We have examined nine responses relating to biodiversity conservation, evaluating their strengths and weaknesses. These assessments are intended to contribute to decision-makers' understanding of the scientific basis and implications of decisions. Decision-making in practice almost always will involve more than consideration of biodiversity. While the responses were defined based on the broad goals of biodiversity conservation, sustainable use, and equitable distribution of benefits, discussion of strengths and weaknesses inevitably also addressed the degree of success in integrating these goals with demands of society for ecosystem services. Our assessments lead to several conclusions:

- The current system of PAs is a valuable tool for conserving biodiversity, but these areas do not yet include all biodiversity components that require such protection. Better tools exist for selecting areas for inclusion in PA systems than are currently employed, and better management of individual PAs is required.

- For successful (global) biodiversity conservation, local people must be able to capture benefits from that conservation.
- Integrated conservation and development projects as currently designed rarely succeed in their conserving biodiversity objectives, yet their general approach remains valid; they need more realistic objectives and a stronger link to broader policy issues.
- Direct incentives for biodiversity conservation usually work better than indirect incentives.
- Regional planning can achieve balance across areas to create a landscape that includes strict conservation areas, intensively used areas, and other land uses.
- More income must flow from the people and countries that value biodiversity from afar (at the global level) to the people and countries where much-valued biodiversity is conserved, often at considerable opportunity cost.

Problems affecting biodiversity often involve complex conflicts of interests, so solutions require approaches that synthesize contributions from numerous sectors (that may not always be used to cooperating). One solution is to establish localized and manageable points of intervention where practical solutions can be applied and tested. If successful, such a "small win" scenario can create a sense of control, lend credibility to conservation activities, and help build public confidence and enthusiasm (Heinen and Low 1992). A series of "small wins" can contribute to an overall strategy for conserving biodiversity or prompt political support for its wider application.

5.11.2 How "Biodiversity" Is Addressed in Responses

The MA conceptual framework (MA 2003, p. 7) "places human well-being as the central focus for assessment while recognizing that biodiversity and ecosystems also have *intrinsic* value and that people make decisions concerning ecosystems based on considerations of both well-being and intrinsic value." At the same time, "few decisions take account of indirect use value and very few take explicit account of existence values. As a result, many decisions about intervention into ecosystems are not based on the best possible information (p. 181)." This information problem is a critical one for biodiversity assessment, and for the success of trade-offs and synergies with other services. For example, one concern with the "hotspots" approach has been that any claimed efficiency is illusory if the indicator taxa are not broad indicators of more general endemicity patterns. We can never "prove" the value of any surrogate, but can make best-possible use of all available data in surrogacy strategies.

As part of strategies for addressing uncertainty, effective trade-offs (and synergies) of biodiversity and ecosystem services require more effective measurement or estimation of biodiversity at all scales. Common "mistakes" in designing biodiversity indicators have led to management strategies that have proven to be inconsistent and indefensible on the ground and have hidden trade-offs at the policy level (Failing and Gregory 2003). A general lesson is that poor measurement of biodiversity reduces the capacity to discover and implement good trade-offs and synergies between biodiversity and ecosystem services.

Sometimes responses to this information problem may overstate the "user needs" perspective and neglect the difficult problem of finding surrogates for global option values. For example, Failing and Gregory supposed that successful indicators for biodiversity vary depending upon the "end points" desired by the users in any particular context. The Royal Society (2003) similarly argues: "each biodiversity assessment would clearly identify: i) interested parties; ii) the attributes that those parties value and are seeking to measure. . . ." (p. 22). One example value includes commercial foresters' values placed on biodiversity "attributes" that are equated with "volume of timber that can be extracted" (p. 14). Such user-needs requirements may need to be balanced with efforts to measure intrinsic and option values that do not have immediate advocates.

The biodiversity of a place often is highly valued by the people living there, but these values may not be particularly relevant to global biodiversity values. For example, increased local diversity (genetic and species level) in agricultural systems often leads to better control of pests and diseases, but the consequent incentive for local protection of biodiversity may not link to any important global values. A lesson is that values at all scales are important in the design of response options, and decision-making can benefit from addressing trade-offs and synergies among them.

The pitfall of imagining diminishing returns from additional biodiversity is highlighted when considering the list of unanticipated services that may be important in the future. Biodiversity serves as a surrogate for a multitude of possible future services, and its conservation therefore can maintain options for the future.

We have seen that this "open-endedness" calls for trade-offs with other needs of society. The ecosystem approach has provided a framework for finding a balance among different needs, for example, through integrated natural resource management systems and through various policy, legal, institutional, and economic measures. Associated "mainstreaming" of biodiversity into other sectors has promoted balanced outcomes, even in cases where biodiversity gains are not measured.

Such trade-offs also may benefit from a "calculus" of biodiversity, so that gains and losses at the level of biodiversity option values can be quantified. A simple calculus must be based on surrogate information for general biodiversity patterns, so that these gains and losses ("complementarity values") are predictions of changes in amount of variation retained in a region. While it has been argued that biodiversity advocates have wrongly focused on "inventory" of species, genes, ecosystems (Norton 2001), a "calculus" of biodiversity that captures option values is appropriately based on "inventory." However, inventory and systematic efforts can be more strategic in filling knowledge gaps.

We have noted that arguments based on global biodiversity values ignore important local values of biodiversity relating to ecosystem services. An alternative to a preference for local values of "biodiversity" is to pursue balanced trade-offs and synergies among local, national, and global values. As long as local values and opportunities, whatever their source, are given appropriate weight, defining (or redefining) the "important" values of biodiversity as local not global is not an issue. Apparent conflict may be resolved also by realizing that often the local values and opportunities may have little to do with the biodiversity (biotic variation per se) of the place, instead linking to specific components of biodiversity (often valued species).

Clarifying local-versus-global values avoids misinterpretations about biodiversity's value. Examples can be put forward suggesting that *low* biodiversity, manipulated systems—such as wheat fields—provide most benefits to human well-being (Jenkins 2003). But such arguments, interpreted as casting doubt on links from biodiversity to human well-being, in fact highlight how biodiversity *does* matter. Individual low diversity places may be important in their complementary contributions to overall global biodiversity option values. Therefore, even the most dramatic successes in individual places at deriving extensive benefits from *low* biodiversity, manipulated systems provide no evidence that biodiversity is of less importance to human well-being.

We conclude this section by summarizing how a trade-offs/synergies perspective has suggested new perspectives on measuring biodiversity. First, good biodiversity surrogates must focus more on what matters in the context of trade-offs: do they predict general complementarity (marginal gain) values provided by a given place? This need is in accord with the general lesson that aggregate (global scale) estimates of ecosystem value are of limited use, given the fact that only marginal values are consistent with conventional decision-aiding tools (Turner et al. 2003). Second, more detailed information is required than that provided by conventional coarse-scale summaries such as species-area curves. The MA scenarios have made effective use of the idea of a species area curve: if some quantity of total area-extent of a given biome is not retained, then the curve implies that a certain proportion of species will be prone to extinction in the future. Biodiversity responses therefore might be seen as attempts to maintain a certain total area of each biome as sufficiently intact to support all the species found there. However, an amount of area lost could correspond to high or low biodiversity (and high or low opportunity costs); such curves do not distinguish well among these outcomes.

This chapter has focused more on trade-off curves, which substitute the "area" axis with the more informative "opportunity costs" axis. We have seen that a scenario for a region implying a total amount of, say, agricultural production does not necessarily imply (as a species-area relationship would suggest) that some given proportion of species is lost. Instead it implies a variable number that depends on the effectiveness of responses such as regional planning. A lesson in this chapter was that several aspects of responses can boost effectiveness: (1) regional planning may allocate forestry, agriculture, or other human uses in a way that least conflicts with biodiversity conservation; (2) human use may be carried out in a way that such places also make a contribution to regional biodiversity conservation and sustainable use; and (3) intensification of production may imply that a smaller total area conflicts with biodiversity conservation.

The contrast between an "area" focus and a "trade-offs" focus also has revealed lessons for how biodiversity targets are approached. Case studies have suggested how the same total amount of area protected can lead to large or small foregone opportunity costs, and large or small amounts of biodiversity protected. A percentage area target may be a good, rough, global-scale indicator, but at the scale of national/regional policies and planning, targets that are related to trade-offs will be more useful.

Trade-offs curves share a key property with species-area curves: the same incremental loss of intact area (say, to non-conservation uses) can imply a greater biodiversity loss the second time it occurs. An observed change in the rate of loss of intact area of a biome therefore can be a misleading indicator of actual rate of biodiversity loss. Accepting an observed reduced rate of area loss as indicating achievement of the 2010 biodiversity target could amount to acceptance of an increased rate of biodiversity loss (Faith, in press). Such curves also indicate a positive strategy for addressing the 2010 target. Even the same rate of area loss could correspond to a reduced rate of biodiversity loss, through response strategies that provide effective trade-offs and synergies where they do not yet occur. Further, effective trade-offs could mean that a greater gain in biodiversity results from a given level of increase in conservation area (Faith, in press).

5.12 Research Priorities

The biodiversity extinction rate is worrisome given that we do not have names for most of the 10 million plus species on the planet, and we do not know much even about the species that have been named. This section identifies some key questions that need to be answered if responses to the loss of biodiversity are to be more effective in the future.

5.12.1 How Does Biodiversity Underpin Ecosystem Services and Human Well-being?

Better quantification and integration of these benefits would provide greater impetus for biodiversity protection. Effective response strategies will help overcome the fact that ecosystem services currently are not fully captured in commercial markets (a caution highlighted in this chapter is that apparent links of "biodiversity" to some services, that in fact are not scientifically supported, may mean that pursuit of those services does not help biodiversity conservation). Better quantification and integration of benefits also would promote effective trade-offs and synergies in the regional integration of response strategies.

5.12.2 What Patterns of Biodiversity Represent Value for the Future?

A key dualism for understanding and designing responses is that biodiversity benefits are both global and local. Global loss is more a concern about long-term option values, and hence defines a critical knowledge gap that goes beyond current perceived services. Again, better quantification and integration of these non-use benefits provides greater impetus for biodiversity protection, because protection of an area may look more beneficial than some other land (or water) use when these values are taken into account. Research is critically needed to provide not only greater species distribution information (more species, more places), but also environmental data (to aid prediction of biodiversity patterns). An increase in systematic research is vital. Research can move beyond the conventional focus on "what is the total number of species?" to strategic filling of knowledge gaps, promoting a global calculus of biodiversity that allows statements about gains and losses in particular places.

In addition to better valuation of biodiversity benefits, better information about levels of uncertainty about biodiversity and its values could also greatly assist decision-making. For example, we may not know which species are most likely to go extinct but we may know something about the probabilities of losing certain functions or species overall. Incorporating that type of information into a description of our uncertainty about future biodiversity values into a decision framework for irreversible decisions under uncertainty could lead to better decisions about what irreversible actions to avoid and could identify what types of information would be most useful in refining decisions.

5.12.3 How Can Biodiversity Values Be Quantified?

Better quantification of biodiversity values (including option values) and of the services provided by existing or potential new protected areas will enable these values to be taken into account in land-use planning, policy-making, and other decisions about development. This research needs to recognize the dangers in directly estimating "dollar values" for all these benefits; option values arguably cannot be fully quantified this way, and even for local ecosystem services (say, provision of food) conversion to dollar values on "open markets" can easily underestimate true values to local people. Option values, at least, can be quantified in non-dollar ways, as gains in species representation or persistence from a given response option, and then fed into multicriteria analyses for trade-offs. Research is particularly needed on the gains/losses from human use lands, put into a regional context.

The importance of biodiversity and natural processes in producing ecosystem services upon which people depend remains largely invisible to decision-makers and the general public. Unlike goods bought and sold in markets, most ecosystem services do not yet have markets or readily observable prices. Assembling evidence on "non-market values" should be a high priority research topic. A substantial body of research in economics on non-market valuation is now available, though applying these methods to biodiversity is not fully developed. Existence value of species and other "non-use" values pose a difficult challenge to those who would try to measure the complete value of conserving biodiversity and natural processes. Despite the difficulty, it is worth gathering better evidence about benefits created by natural systems. One goal of such research could be to establish a system of biodiversity accounts to track changes in the status of biodiversity, in much the same way that national income accounts are used to track the status of national economies. For example, application of a calculus of biodiversity may provide one pathway for addressing the monitoring requirements of the 2010 biodiversity target.

A related point is that research is needed on how to establish and implement targets for biodiversity conservation and sustainable use. Our assessments have pointed to pitfalls in using "rates" of extinction, threatened species, area amounts, land-use threats, and other information. At the same time, our assessments point to the need for targets to somehow take into account the realities of trade-offs and synergies with other needs of society.

5.12.4 What Are the Social Impacts of Biodiversity Loss?

Greatly hindering any general assessment of the impact of responses on biodiversity loss, and resulting conservation, is the sheer lack of rigorous social science analysis of that impact. Empirical analysis of conservation responses lags significantly behind other social policy fields in its ability to draw inferences about the relationship between policy and biodiversity conservation. Conservation donors have only recently begun to even request evidence that their funds have the desired effect. With hundreds of millions of dollars spent annually, and with many species and rural livelihoods at stake, empirical analysis of the impact of the full range of policy options on biodiversity conservation, including the nine major ones assessed in this chapter, appears long overdue.

The available evidence allows us to say with *high certainty* that the rapid loss of biodiversity is a serious problem that threatens the functioning of natural systems and human well-being. Biodiversity is at risk largely because of human activity, but human well-being depends on the provision of ecosystem services from natural systems. Therefore, better management of human affairs, and better understanding and management of human interactions with the environment hold the key to finding solutions to conserving biodiversity, using biological resources sustainably, and ensuring equitable distribution of benefits derived. In order to successfully achieve these goals, more information is needed about how various human actions affect biodiversity and how biodiversity affects human well-being.

5.12.5 How Do Human Actions Affect Biodiversity and the Structure and Function of Ecosystems?

Understanding what human interventions in natural systems cause beneficial or detrimental changes is an important prerequisite for managing human interactions with the environment. Given the complexity of natural systems, our understanding of impacts and the ability to manage those impacts will be imperfect. Even so, increased understanding of the effect of human actions is of great value in trying to steer human actions towards less destructive practices while encouraging beneficial ones.

5.12.6 How Can Effective Incentives Be Designed for Conserving Biodiversity?

Market prices that do not incorporate the value of biodiversity or ecosystem services send the wrong set of signals to decision-makers. One solution to resolve the problem of incorrect incentives is to attempt "get the prices right" so that they truly reflect underlying values. Taxes on harmful activities and subsidies on beneficial activities are one means to shift prices to provide better signals of value. Often, powerful political forces will block tax measures, or will push for subsidies on activities that are harmful for the environment but beneficial to economic interests of a particular segment of society. Government regulatory approaches may face a similar set of political hurdles. In addition, if not carefully designed, regulations may have unintended consequences that are harmful to conservation. Research on the most effective ways to promote conservation and to coordinate actions through markets, government actions, and the supporting activities of non-governmental organizations is needed.

Understanding how to design, implement, and enforce conservation policy is particularly important in developing countries, which contain a large share of biodiversity, but often have weak institutions that may preclude effective enforcement of conservation laws. Developing countries also have great need for economic development to improve the well-being of their citizens. How economic development can occur while maintaining biodiversity and natural processes is one of the most important topics facing humanity at the start of the twenty-first century.

5.12.7 Who Gets to Make Decisions Affecting Biodiversity?

Decisions made by the current generation will shape the world that is handed down to future generations. Questions of sustainability and what constitutes responsible stewardship are important research topics. Many conservation benefits, such as carbon sequestration and providing habitat for the continued existence of species, provide global public goods. Yet local decision-makers often determine whether such benefits are provided, and may ignore important benefits that accrue outside their community. Allowing outside groups who may have a wider view to override local interests brings its own set of problems. How to put the slogan "think globally, but act locally" into practice is a recurring problem.

5.12.8 When Is It Better to Integrate or to Segregate Human and Conservation Activity?

A debate has flared in conservation circles in recent years between those who favor community-based conservation and integrated conservation and development projects versus those who favor emphasis on protected areas that seek to exclude people. Quantification of the value provided by existing or potential new protected areas, versus the value provided in landscapes that allow some economic activities, would provide guidance to land-use planning, policy-making, and other decisions about development. Response options typically must by their nature consider marginal gains in biodiversity conservation, posing a research challenge for quantification. The contribution of production or mixed-use lands to regional biodiversity protection is not well-indicated by the usual assessments of consequent species richness. Future assessments could instead examine how well these areas provide marginal gains and so be integrated with contributions from protected areas, conservation payments on private lands, and other policies aimed at reversing the loss of biodiversity.

References

Abbott, C., R. Barrington, R. Boyd, D. Hillyard, and K. Sargent, 2004: *Is Biodiversity a Material Risk for Companies?* ISIS Asset Management, London, UK, 59 pp. In press.

Achmady, L. and J. Schneider, 1995: Tuber crops in Irian Jaya: Diversity and need for conservation. In: *Indigenous Knowledge, Conservation of Crop Genetic Resources,* J. Schneider (ed.), Centro Internacional de la Papa/International Potato Center, Peru, pp. 71–78.

Adams, R.P., 1997: Conservation of DNA: DNA banking. In: *Biotechnology and Plant Genetic Resources Conservation and Use,* J.A. Callow, B.V. Ford-Lloyd, and H.J. Newbury (eds.), CAB International, Wallingford, Oxon, UK, pp. 163–174.

Agardy, T., P. Bridgewater, M.P. Crosby, J. Day, P.K. Dayton, et al., 2003: Dangerous targets? Unresolved issues and ideological clashes around marine protected areas, *Aquatic Conservation: Marine and Freshwater Ecosystems,* **13(4),** pp. 353–367.

Allen, C.M. and S.R. Edwards, 1995: The sustainable-use debate: Observations from IUCN, *Oryx,* **29,** pp. 92–98.

Almekinders, C.J.M., N.P. Louwaars, and G.H. de Bruijn, 1994: Local seed systems and their importance for an improved seed supply in developing countries, *Euphytica,* **78,** pp. 207–216.

Altieri, M.A. and C. Montecinos, 1993: Conserving crop genetic resources in Latin America through farmers' participation. In: *Perspectives on Biodiversity: Case Studies of Genetic Resource Conservation and Development,* S. Christopher, D.J. Potter, and J.I. Cohen (eds.), American Association for the Advancement of Science, Washington, DC, pp. 45–64.

Ando, A.J., J. Camm, S. Polasky, and A. Solow, 1998: Species distributions, land values, and efficient conservation, *Science,* **279,** pp. 2126–2128.

Association on Scientific Uncertainty, 2003: Report on measures for implementation of and compliance with MEAs with scientific uncertainty, Unpublished report.

Bail, C., R. Falkner and H. Marquard (eds.), 2002: *The Cartagena Protocol on Biosafety: Reconciling Trade in Biotechnology with Environment & Development?* Earthscan Publications Ltd., London, UK.

Balmford, A., A. Bruner, P. Cooper, R. Costanza, S. Farber, et al., 2002: Economic reasons for conserving wild nature, *Science,* **297,** pp. 950–953.

Barnard, P., C.J. Brown, A.M. Jarvis, A. Robertson, and L. Van Rooyen, 1998: Extending the Namibian protected area network to safeguard hotspots of endemism and diversity, *Biodiversity and Conservation,* **7,** pp. 531–547.

Barrett, C., C. Brandon, C. Gibson, and H. Gjertsin, 2001: Conserving tropical biodiversity amid weak institutions, *Bioscience,* **51(6),** pp. 497–502.

Barton, D.N., D.P. Faith, G. Rusch, J.O. Gjershaug, M. Castro, et al., 2003: Spatial prioritisation of environmental service payments for biodiversity protection, NIVA Report SNR 4746/2003, Norsk Institutt for Vannforskning/Norwegian Institute for Water Research [online]. Available at http://www.amonline.net.au/systematics/pdf/bioindicators2.pdf.

Barzetti, V. 1993: *Parks and Progress,* IUCN and the Inter-American Development Bank, Washington, DC.

Beasom, S.L. and S.F. Roberson (eds.), 1985: *Game Harvest Management,* Caesar Kleberg Wildlife Research Institute, Kingsville, TX.

Beck, B.B., L.G. Rapaport, M.R. Stanley Price, and A.C. Wilson, 1994: Reintroduction of captive born animals. In: *Creative Conservation,* P.J. Olney, G.M. Mace, and A.T.C. Feistner (eds.), Chapman & Hall, London, UK, pp. 265–286.

Bellon, M.R., 1996: The dynamics of crop infraspecific diversity: A conceptual framework at the farmer level, *Economic Botany,* **50,** pp. 26–39.

Bellon, M.R., 2001: *Participatory Research Methods for Technology Evaluation: A Manual for Scientists Working with Farmers,* Centro Internacional de Majoramiento de Maíz y Trigo/International Maize and Wheat Improvement Center Mexico, DF, Mexico.

Bellon, M.R., M. Smale, A. Aguirre, F. Aragón, S. Taba, et al., 1999: Farmer management of maize diversity in the central valleys of Oaxaca, Mexico: Methods proposed for impact assessment. In: *Assessing the Impact of Participatory Research and Gender Analysis,* N. Lilja, J.A. Ashby, and L. Sperling (eds.), CGIAR Programme on Participatory Research and Gender Analysis, Consultative Group on International Agricultural Research, Cali, Colombia.

Beltrán, J. (ed.), 2000: *Indigenous and Traditional Peoples and Protected Areas: Principles, Guidelines and Case Studies,* IUCN, Gland, Switzerland, Cambridge, UK, and WWF International, Gland, Switzerland, 133 pp.

Bennett, G. and P. Wit, 2001: *The Development and Application of Ecological Networks: A Review of Proposals, Plans and Programmes,* AID Environment and IUCN, Gland, Switzerland.

Bertrand, N. (ed.), 2002: *Business & Biodiversity: The Handbook for Corporate Action,* Earthwatch Institute (Europe), Oxford, UK/IUCN, Gland Switzerland, and World Business Council for Sustainable Development, Geneva, Switzerland.

Bilderbeek, S., 1992: *Biodiversity and International Law: The Effectiveness of International Environmental Law,* IOS Press, Amsterdam, The Netherlands.

Binswanger, A.P., 1989: *Brazilian Policies that Encourage Deforestation in the Amazon,* Working paper no. 16, World Bank, Washington, DC.

Birnie, P. and A. Boyle, 2002: *International Law & the Environment,* Oxford University Press, 2nd ed, New York, NY, 828 pp.

Bluffstone, R.A., 1993: *Reliance on Forests: Household Labor Supply Decisions, Agricultural Systems and Deforestation in Rural Nepal,* Ph.D. thesis, Boston University, Boston, MA.

Bodansky, D. and J. Brunnée, 1998: The role of national courts in the field of international environmental law, *RECIEL,* **7(1),** pp. 11–20.

Boer, B., R. Ramsay and D.R. Rothwell (eds.), 1998: *International Environmental Law in the Asia Pacific,* Kluwer Law International, The Hague, The Netherlands.

Bowen-Jones, E., D. Brown, and E.J.Z. Robinson, 2002: Assessment of the solution-orientated research needed to promote a more sustainable bushmeat trade in Central and West Africa, Report to the Wildlife & Countryside Directorate, Department for Environment Food and Rural Affairs, Department of the Environment, Transport and the Regions, London, UK.

Bowker, D.W. and M. Castellano, 2002: Enforcing international environmental treaties in domestic legal systems. In: *Transboundary Environmental Negotiation,* L. Susskind, W. Moomaw, and K. Gallagher (eds.), Jossey-Bass, San Francisco, CA, pp. 230–251.

Boyd, J.W., K. Caballero, and R.D. Simpson, 1999: *Law and Economics of Habitat Conservation: Lessons from an Analysis of Easement Acquisitions,* RFF discussion paper 99–32, Resources for the Future, Washington, DC.

BP Australia, 2000: Triple bottom line report. Available at http://www.bp.com.au/news_information/press_releases/triple_bottom_line_report.pdf.

Braken, T. and M. Meredith, 2000: Participation of local communities in management of totally protected areas, *Hornbill,* **4.** Available online at http://www.mered.org.uk/mike/papers/Comanagement_Hornbill_00.htm.

Brandon, K., 1998: Perils to parks: The social context of threats. In: *Parks in Peril: People, Politics, and Protected Areas,* K. Brandon, K.H. Redford, and S.E. Sanderson (eds.), The Nature Conservancy, Washington, DC.

Brandon, K.E. and M. Wells, 1992: Planning for people and parks: Design dilemmas, *World Development,* **20,** pp. 557–570.

Brookfield, H., C. Padoch, H. Parsons, and M. Stocking (eds.), 2002: *Cultivating Biodiversity: Understanding, Analyzing and Using Agricultural Diversity,* (Intermediate Technology Development Group) ITDG Publishing, London, UK.

Brooks, T.M., G.A.B. da Fonseca, and A.S.L. Rodrigues, 2004: Protected Areas and Species, *Conservation Biology,* **18,** pp. 616–618.

Brown, A.H.D., 2000: Population biology and social science. In: *Genes in the Field: On-Farm Conservation of Crop Diversity,* S.B. Brush (ed.), IPGRI/International Development Research Corporation/Lewis Publishers, Boca Raton, FL,, pp. 29–50.

Brown, K., W.N. Adger, E. Tompkins, P. Bacon, D. Shim, et al., 2001: Trade-off analysis for marine protected area management, *Ecological Economics,* **37(3),** pp. 417–434.

Brown, S., M. Burnham, M. Delaney, R. Vaca, M. Powell, et al., 2000: Issues and challenges for forest-based carbon-offset projects: A case study of the Noel Kempff Climate Action Project in Bolivia, *Mitigation and Adaptation Strategies for Global Change,* **5,** pp. 99–121.

Bruner, A., R. Gullison, R. Rice, and G. de Fonseca, 2001: Effectiveness of parks in protecting tropical biodiversity, *Science,* **291,** pp. 125–133.

Byers, B.A., 2004: Understanding and influencing behaviors in conservation and natural resources management, African Biodiversity Series No. 4, Biodiversity Support Program. Available at http://www.worldwildlife.org/bsp/publications/africa/understanding_eng/under standing1.html.

Caldecott, J. and E. Lutz, 1996: *Decentralization and Biodiversity Conservation: Issues and Experiences,* World Bank, Washington, DC.

Cameron, J., J. Werksman, and P. Roderick, 1996: *Improving Compliance with International Environmental Law,* Earthscan Publications Ltd., London, UK.

Carew-Reid, J. (ed.), 2002: Biodiversity planning in Asia: A review of national biodiversity strategies and action plans (NBSAPs), IUCN, Gland, Switzerland.

Castillo, G.F., L.M.R. Arias, R.P. Ortega, and F. Marquez, 2000: PPB, seed networks and grassroot strengthening in Mexico. In: *Conserving Agricultural Biodiversity In Situ: A Scientific Basis For Sustainable Agriculture,* D.I. Jarvis, B. Sthapit, and L. Sears (eds.), IPGRI, Rome, Italy, pp. 199–200.

Castro, R., F. Tattenbach, L. Gamez, and N. Olson, 2000: The Costa Rican experience with market instruments to mitigate climate change and conserve biodiversity, *Environmental Monitoring and Assessment,* **61,** pp. 75–92.

Caughley, G., 1977: *Analysis of Vertebrate Populations,* John Wiley & Sons, New York, NY.

CBD (Convention on Biological Diversity), 2000a: *Review of Existing Instruments Relevant to Integrated Marine and Coastal Area Management and their Implications for the Implementation of the Convention on Biological Diversity,* United Nations Environment Programme, Arendal, Norway.

CBD, 2000b: *Report of the CBD-UNESCO Consultative Working Group of Experts on Biological Diversity Education and Public Awareness on the Work of its First Meeting,* United Nations Environment Programme, Arendal, Norway.

CBD, 2000c: *Report of the CBD-UNESCO Consultative Working Group of Experts on Biological Diversity Education and Public Awareness on the Work of its Second Meeting,* United Nations Environment Programme, Arendal, Norway.

CBD, 2001: *Report of the CBD-UNESCO Consultative Working Group of Experts on Biological Diversity Education and Public Awareness on the Work of its Third Meeting,* United Nations Environment Programme, Arendal, Norway.

CBD, 2003a: *CBD Ad hoc Technical Expert Group on Protected Areas,* First meeting, 10–14 June 2003, Tjärnö, Sweden.

CBD, 2003b: *Report of the Ad Hoc Technical Expert Group on Mariculture,* United Nations Environment Programme, Arendal, Norway.

CBD, 2003c: *Report of the CBD-UNESCO Consultative Working Group of Experts on Biological Diversity Education and Public Awareness on the Work of its Fourth Meeting,* United Nations Environment Programme, Arendal, Norway.

CBD, 2003d: Expert meeting on the ecosystem approach: Review of the principles of the ecosystem approach and suggestions for refinement: A framework for discussion, UNEP/CBD/EM-EA/1/3. Available through www.biodiv.org.

CBD, 2003e: Expert meeting on the ecosystem approach: Proposals for development/refinement of the operational guidelines of the ecosystem approach, UNEP/CBD/EM-EA/1/4. Avialable through www.biodiv.org.

CBD, 2004: *Biodiversity Issues for Consideration in the Planning, Establishment and Management of Protected Area Sites and Networks,* Technical series no.15, Secretariat of the Convention on Biological Diversity, Montreal, Canada.

CBDC Programme–Bohol Project, 2001: *A Study on the Plant Genetic Resources Diversity and Seed Supply System of Bohol Island, Philippines,* Technical report no. 1., Southeast Asia Regional Institute for Community Education, Quezon City, Philippines.

CEAA, 1996: *A Guide on Biodiversity and Environmental Assessment,* Prepared jointly by Biodiversity Convention Office, Hull, Quebec, Canada, and the Ministry of Supply and Services, Ottawa, Canada.

Ceccarelli, S. and S. Grando, 2000: Barley landraces from the Fertile Crescent: A lesson for plant breeders. In: *Genes in the Field: On-Farm Conservation of Crop Diversity,* S.B. Brush (ed), IPGRI/International Development Research Corporation/Lewis Publishers, Boca Raton, FL, pp: 51–76.

Chasek, P., 2001: *Earth Negotiations: Analyzing Thirty Years of Environmental Diplomacy,* United Nations University Press, Tokyo, Japan.

Chayes, A. and A.H. Chayes, 1991: Adjustments and compliance processes in international regulatory regimes. In: *Preserving the Global Environment, The Challenge of Shared Leadership,* J. T. Matthews (ed.), The American Assembly & World Resources Institute, W.W. Norton & Company, New York, NY.

Chomitz, K., E. Brenes, and L. Constantino, 1999: Financing environmental services: The Costa Rican experience and its implications, *Science of the Total Environment,* **240,** pp. 157–69.

Chomitz, K.M. and K. Kumari, 1998: The domestic benefits of tropical forests: A critical review, *The World Bank Research Observer,* **13(1),** pp. 13–35.

Cicin-Sain, B. and R.W. Knecht, 1998: *Integrated Coastal and Ocean Management: Concepts and Practices,* Island Press, Washington, DC.

Cicin-Sain, B., 1993: Sustainable development and integrated coastal zone management, *Ocean and Coastal Management,* **21(1–3),** pp. 11–43.

Civeyrel, L., and D. Simberloff, 1996: A tale of two snails: Is the cure worse than the disease? *Biodiversity and Conservation,* **5,** pp. 1231–1252.

Clayton, L.M., E.J. Milner-Gulland, D.W. Sinaga, and A.H. Mustari, 2000: Effects of a proposed ex situ conservation program on in situ conservation of the Babirusa, an endangered suid, *Conservation Biology,* **14(2),** pp. 382–385.

Cleveland, A.D., D. Soleri, and S.E. Smith, 2000: A biological framework for understanding farmers' plant breeding, *Economic Botany,* **54(3),** pp. 377–394.

CONSERVE, 2001: *Impact of ecological pest management: Farmers' Field School (EPM-FFS) Training in the Three Municipalities of Arakan Valley Complex, Cotabato, Philippines,* Community-Based Native Seeds Research Centre, Inc., Cotabato, Philippines.

Correa, C.M., 1999: In situ conservation and intellectual property rights. In: *Genes in the Field,* S.B. Brush and A. Lewis (eds.), IPGRI/International Development Research Corporation/Lewis Publishers, Boca Raton, FL, pp. 239–260.

Cowling, R.M. and R.L. Pressey, 2003: Introduction to systematic conservation planning in the Cape Floristic Region, *Biological Conservation,* **122,** pp. 1–13.

Cowling, R.M., A.T. Knight, D.P. Faith, A.T. Lombard, P.G. Desmet, et al., 2004: Nature conservation requires more than a passion for species, *Conservation Biology,* **18(6),** pp. 1674–1676.

Cromwell, E. and S. van Oosterhout, 2000: On-farm conservation of crop diversity: Policy and institutional lessons from Zimbabwe. In: *Genes in the Field: On-Farm Conservation of Crop Diversity,* S.B. Brush (ed.), IPGRI/International Development Research Corporation/Lewis Publishers, Boca Raton, FL, pp. 217–239.

Cropper, M., C. Griffiths, and M. Mani, 1999: Roads, population pressures, and deforestation in Thailand, 1976–1989, *Land Economics,* **75(1),** pp. 58–73.

Crosby, M.P., R. Bohne, and K. Geenen, 2000: *Alternative Access Management Strategies for Marine and Coastal Protected Areas: A Reference Manual for their Development and Assessment,* US Man and the Biosphere Program, Washington, DC.

Daily, G.C., S. Alexander, P.R. Ehrlich, L. Goulder, J. Lubchenco, et al., 1997: Ecosystem services: Benefits supplied to human societies by natural ecosystems, *Issues in Ecology,* **2,** Ecological Society of America, Washington, DC.

Davey, A.G. (ed.), 1998: *National System Planning for Protected Areas,* IUCN World Commission on Protected Areas, University of Cardiff, Department of City and Regional Planning, Gland, Switzerland.

DEA (Department of Environmental Affairs), 1992: *The Integrated Environmental Management Procedure,* Guideline document no. 1, Integrated Environmental Management Guidelines Series, DEA, Pretoria, South Africa.

Denisov, N. and L. Christoffersen, 2000: *Impact of Environmental Information on Decision Making Processes and the Environment,* UNEP-GRID Arendal occasional paper 01 2001, UN Environmental Programme–Global Resources Information Database, Arendal, Norway.

Department of Environment, 1993: *Environmental Appraisal of Development Plans: A Good Practice Guide,* Department of Environment, London, UK.

Driver, A., R.M. Cowling, and K. Maze, 2003: *Planning for Living Landscapes: Perspectives and Lessons from South Africa,* Center for Applied Biodiversity Science at Conservation International, Washington, DC, and Botanical Society of South Africa, Cape Town, South Africa.

du Toit, J.T., B.H. Walker, and B.M. Campbell, 2004: Conserving tropical nature: Current challenges for ecologists, *Trends in Ecology & Evolution,* **19,** pp. 12–17.

EC (European Community), 2001: *Environmental Integration Manual, Volume 1: Procedures, Volume 2: Source Book,* Brussels, Belgium. Available through www.europa.eu.int.

EC and IUCN, 1999: *Parks for Biodiversity: Policy Guidance Based on Experience in ACP Countries,* IUCN, Gland, Switzerland, and Cambridge, UK.

EC, DFID, and IUCN, 2001: *Biodiversity in Development Project: Strategic Approach for Integrating Biodiversity in Development Cooperation,* EC, Brussels, Belgium/Department for Internatioal Development, Cambridge, UK/IUCN, Gland, Switzerland.

EEA (European Environment Agency), 2004: *High Nature Value Farmland: Characteristics, Trends and Policy Challenges,* EEA, Copenhagen, Denmark.

Encalada, M., 2004: Optimizing the use of research in order to consolidate communication planning for protected areas. In: *Communicating Protected Areas,* D.Hamú, E. Auchincloss, and W. Goldstein (eds.), Commission on Education and Communication, IUCN, Gland, Switzerland, and Cambridge, UK.

Engelmann, F. (ed.), 1999: *Management of Field and In Vitro Germplasm Collections,* Proceedings of a consultation meeting, 15–20 January 1996, International Center for Tropical Agriculture, Cali, Colombia, and IPGRI, Rome, Italy.

Engelmann, F. and H. Takagi (eds.), 2000: *Cryopreservation of Tropical Plant Germplasm: Current Research Progress and Applications,* Japan International Centre for Agricultural Sciences, Tsukuba, Japan, and IPGRI, Rome, Italy.

Engelmann, F. and J.M.M. Engels, 2002: Technologies and strategies for ex situ conservation. In: *Managing Plant Genetic Diversity,* J.M.M. Engels, V. Ramanatha Rao, A.H.D. Brown, and M.T. Jackson (eds.), IPGRI, Rome, Italy, and CAB International, Wallingford, Oxon, UK, pp. 89–103.

Engelmann, F., 1997: In vitro conservation methods. In: *Biotechnology and Plant Genetic Resources: Conservation and Use,* B.V. Ford-Lloyd, J.H. Newburry, and J.A. Callow (eds.), CAB International, Wallingford, Oxon, UK, pp. 119–162.

Engels, J.M., 1996: In situ conservation and sustainable use of plant genetic resources for food and agriculture in developing countries, IPGRI, Rome, Italy.

Environmental Law Institute, 2003: *Conservation Thresholds for Land Use Planners,* ELI, Washington, DC.

Failing, L. and Gregory, R. 2003: Ten common mistakes in designing biodiversity indicators for forest policy, *Journal of Environmental Management,* **68,** pp. 121–132.

Faith, D.P., 2001: Cost-effective biodiversity planning, *Science* 293 [online]. Available at http://www.sciencemag.org/cgi/eletters/293/5538/2207.

Faith, D.P.: Global biodiversity assessment: Integrating global and local values and human dimensions, *Global Environmental Change.* In press.

Faith, D.P. and C.R. Margules, 2002: Fine-scale complementarity analyses can reveal the extent of conservation conflict in Africa, *Science,* 21 **November 2001** [online]. Available at http://www.sciencemag.org/cgi/eletters/293/5535/1591#387.

Faith, D.P., P.A. Walker, and C.R. Margules, 2001a: Some future prospects for systematic biodiversity planning in Papua New Guinea—and for biodiversity planning in general, *Pacific Conservation Biology,* **6,** pp. 325–343.

Faith, D.P., P.A. Walker, J. Ive, and L. Belbin, 1996: Integrating conservation and forestry production: Exploring trade-offs between biodiversity and production in regional land-use assessment, *Forest Ecology and Management,* **85,** pp. 251–260.

Faith, D.P., C.R. Margules, P.A. Walker, J. Stein, and G. Natera, 2001b: Practical application of biodiversity surrogates and percentage targets for conservation in Papua New Guinea, *Pacific Conservation Biology,* **6,** pp. 289–303.

Faith, D.P., G. Carter, G. Cassis, S. Ferrier, and L. Wilkie, 2003: Complementarity, biodiversity viability analysis, and policy-based algorithms for conservation, *Environmental Science and Policy,* **6,** pp. 311–328.

FAO (Food and Agricultural Organization), 1996: *The Global Plan of Action for Conservation and Sustainable Utilisation of Plant Genetic Resources for Food and Agriculture,* FAO, Rome, Italy.

FAO, 1998: *The State of the World's Plant Genetic Resources for Food and Agriculture,* FAO, Rome, Italy.

FAO, 1999: *The Global Strategy for the Management of Farm Animal Genetic Resources,* FAO, Rome, Italy.

FAO, 2002: *The State of World Fisheries and Aquaculture,* FAO, Rome, Italy.

Ferraro, P.J., 2001: Global habitat protection: Limitations of development interventions and a role for conservation performance payments, *Conservation Biology,* **15(4),** pp. 990–1000.

Ferraro, P.J., and A. Kiss, 2002: Direct payments for biodiversity conservation, *Science,* **298,** pp. 1718–1719.

Ferraro, P.J. and A. Kiss, 2003: Will direct payments help biodiversity? Response, *Science,* **299,** pp. 1981–1982.

Ferraro, P.J. and R.D. Simpson, 2002: The cost-effectiveness of conservation performance payments, *Land Economics,* **78(3),** pp. 339–353.

Ferraro, P.J., R. Tshombe, R. Mwinyihali, and J.A. Hart, 1997: *Projets Intégrés de Conservation et de Développement: Un Cadre pour Promouvoir la Conservation et la Gestion des Ressources Naturelles,* Working paper no. 6, Wildlife Conservation Society, Bronx, New York.

Filer, C., and N. Sekhran, 1998: *Loggers, Donors and Resource Owners,* International Institute for Environment and Development, London, UK.

Forest.ru, n.d.: Principles of responsible timber trade of Russian wood. Cited 15 July 2003. Available at http://www.forest.ru/eng/sustainable_forestry/vision/guide.html.

Fowler, C., M. Smale, and S. Gaiji, 2001: Unequal exchange? Recent transfers of agricultural resources and their implication for developing countries, *Development Policy Review,* **19(2),** pp. 181–204.

Freese, C.H., 1998: *Wild Species as Commodities: Managing Markets and Ecosystems for Sustainability,* Island Press, Washington, DC.

Ganeshan, S. and R.K. Rajashekaran, 2000: Current status of pollen cryopreservation research: Relevance to tropical horticulture. In: *Cryopreservation of Tropical Plant Germplasm: Current Research Progress and Applications,* F. Engelmann and H. Takagi (eds.), Japan International Centre for Agricultural Sciences, Tsukuba, Japan, and IPGRI, Rome, Italy.

Gauchan, D., A. Subedi, and P. Shrestha, 2000: Identifying and analyzing policy issues in plant genetic resources management: Experiences using participatory approaches in Nepal. In: *Participatory Approaches to the Conservation and Use of Plant Genetic Resources,* E. Friis-Hansen and B. Sthapit (eds.), IPGRI, Rome, Italy, pp. 188–193.

Gauchan, D., M. Smale and P. Chaudhary, 2003: *Market Based Incentives for Conserving Diversity on Farms: The Case of Rice Landraces in Central Terai, Nepal,* Paper presented at 4th Biocon workshop, 28–29 August, Venice, Italy.

Geist, H.J., and E.F. Lambin, 2002: Proximate causes and underlying driving forces of tropical deforestation, *BioScience,* **52(2),** pp. 143–150.

Gelderblom, C.M., D. Kruger, L. Cedras, T. Sandwith and M. Audouin, 2002: Incorporating conservation priorities into planning guidelines for the Western Cape. In: *Mainstreaming Biodiversity in Development: Case Studies from South Africa,* S.M. Pierce, R.M Cowling, T. Sandwith, and K. MacKinnon (eds.), World Bank, Washington, DC, pp. 129–142.

George, E.F. (ed.), 1996: *Plant Propagation by Tissue Culture: Part 2. In practice,* 2nd ed., Exegetics Ltd., Edington, UK.

Gilman, R.T., R.A. Abell, and C.E. Williams: How can conservation biology inform the practice of integrated river basin management? *Journal of River Basin Management.* In press.

Gepts, P. and R. Papa, 2003: Possible effects of (trans)gene flow from crops on the genetic diversity from landraces and wild relatives, *Environmental Biosafety Resource,* **2,** pp. 89–103.

Glasbergen, P. and A. Blowers (eds.), 1995: *Environmental Policy in an International Context, Perspectives on Environmental Problems,* Arnold Publishing, London, UK.

Global Forest Coalition, 2002: Status of implementation of forest-related clauses in the CBD: An independent review and recommendations for action [online]. Cited 13 July 2003. Available at http://www.wrm.org.uy/.

Goodwin, H. and J. Francis, 2003: Ethical and responsible tourism: Consumer trends in the UK, *Journal of Vacation Marketing,* **9(3),** pp. 271–284.

Griffith, B., J.M. Scott, J.W. Carpenter, and C. Reed, 1989: Translocation as a conservation tool: Status and strategy, *Science,* **245,** pp. 477–480.

Gunningham, N. and P. Grabosky, 1998: *Smart Regulation Designing Environmental Policy,* Clarendon Press, Oxford, UK.

Hall, S.J.G. and J. Ruane, 1993: Livestock breeds and their conservation: A global overview, *Conservation Biology,* **(7),** pp. 815–8 25.

Hall-Rose, O. and P. Bridgewater, 2003: New approaches needed to education and public awareness, *Prospects,* **XXXIII(3),** pp. 263–272.

Halpern, B.S., 2003: The impact of marine reserves: Do reserves work and does reserve size matter? *Ecological Applications,* **13** (1 Supplement), pp. S117–137.

Hanna, W.W. and L.E. Towill, 1995: Long-term pollen storage. In: *Plant Breeding Reviews,* Volume 13, J. Janick (ed.), John Wiley, London, UK.

Harlan, J.R., 1975: *Crops and man,* American Society of Agronomy/Crop Science Society of America, Madison, WI.

Harmon, D. and A.D. Putney, 2003: *The Full Value of Parks: From Economics to the Intangible,* Rowman and Littlefield, New York, NY.

Hastings, A. and L. W. Botsford, 2003: Comparing designs of marine reserves for fisheries and for biodiversity, *Ecological Applications,* **13**(1 Supplement), pp. S65–70.

Hay, F. and R. Probert, 2000: Keeping seeds alive. In: *Seed Technology and Its Biological Basis,* M. Black and J.D. Bewley (eds.), Sheffield Academic Press, Sheffield, UK, pp. 375–404.

Heal, G., 2002: *Nature and the Marketplace,* Island Press, Washington, DC.

Heinen, J.T. and R.S. Low, 1992: Human behavioral ecology and environmental conservation, *Environmental Conservation,* **19(2),** pp. 105–116.

Herrera, M. and M. Garcia, 1995: *La Reserva de la Biosfera Sierra del Rosario, Cuba,* Doc. de trabajo, Programa Sur-sur, UNESCO, Paris, France.

Hesselink, F. (ed.), 2003: *Global Perceptions of Environment and Sustainable Development 2002–2003,* Corporate Communications Group, Commission on Education and Communication, IUCN, Gland, Switzerland.

Heywood, V.H., 1991: Developing a strategy for germplasm conservation. In: *Tropical Botanic Gardens: Their Role in Conservation and Development,* V.H. Heywood and P.S. Wyse Jackson (eds.), Academic Press, London, UK, pp. 11–23.

Heywood, V.H., 1999: The role of botanic gardens in ex situ conservation of agrobiodiversity. In: *Implementation of the Global Plant of Action in Europe: Conservation and Sustainable Utilization of Plant Genetic Resources for Food and Agriculture,* T. Gass, L. Frese, F. Begemann, and E. Lipman (eds.), Proceedings of the European Symposium, 30 June–3 July 1998, Braunschweig, Germany, IPGRI, Rome, Italy.

Higgins, S.I., J.K. Turpie, R. Costanza, R.M. Cowling, D.C. Le Maitre, et al., 1997: An ecological economic simulation model of mountain fynbos ecosystems: Dynamics, valuation and management, *Ecological Economics,* **22,** pp. 155–169.

Hirakuri, S., 2003: *Can Law Save the Forest? Lessons from Finland and Brazil,* Center for International Forestry Research, Jakarta, Indonesia, pp. 48–52 and 69–74.

Hockings, M., S. Stolton, and N. Dudley, 2000: *Evaluating Effectiveness: A Framework for Assessing the Management of Protected Areas,* IUCN, Gland, Switzerland, and Cambridge, UK, x + 121 pp.

Hoddle, M.S., 2004: Restoring balance: Using exotic natural enemies to control invasive exotic species, *Conservation Biology,* **18,** pp. 38–49.

Hodgkin, T. and D.G. Debouck, 1992: Some possible applications of molecular genetics in the conservation of wild species for crop improvement. In: *Conservation of Plant Genes: DNA Banking and in vitro Biotechnology,* R.P. Adams and J.E. Adams (eds.), Academic Press, San Diego, CA, pp. 153–181.

Hoft, R., 2004: Protected area coverage: A biodiversity indicator. In: *Biodiversity Issues for Consideration in the Planning, Establishment and Management of Protected Area Sites and Networks,* CBD Technical Series No.15, Secretariat of the Convention on Biological Diversity, Montreal, Canada.

Horta, K. and R. Round, 2002: The global environment facility: The first ten years: Growing pains or inherent flaws? Environmental Defense Fund (EDF) and Halifax Initiative [online]. Cited 24 February 2005. Available at http://www.newgreenorder.info/GEF_first_ten_years.doc.

Howard, P.C., T.R.B. Davenport, F.W. Kigenyi, P. Viskanic, M.C. Baltzer, et al., 2000: Protected area planning in the tropics: Uganda's national system of forest nature reserves, *Conservation Biology,* **14(3),** pp. 858–875.

Hulme, D. and M. Murphree, 1999: Communities, wildlife and the "new conservation" in Africa, *Journal of International Development,* **11,** pp. 277–285.

Hutton, J. and B. Dickson (eds.), 2000: *Endangered Species: Threatened Convention: The Past, Present, and Future of CITES,* Earthscan, London, UK.

Hutton, J.M., and N. Leader-Williams, 2003: Sustainable use and incentive-driven conservation: Realigning human and conservation interests, *Oryx,* **37,** pp. 215–226.

Ibisch, P.L., 2002: Evaluation of a rural development project in southwest Cochabamba, Bolivia, and its agroforestry activities involving Polylepis besseri and other native species: A decade of lessons learned, *Ecotropica,* **8,** pp. 205–218.

IBRD (International Bank for Reconstruction and Development) **and World Bank,** 2004: *Global Monitoring Report: Policies and Actions for Achieving the MDGs and Related Outcomes,* World Bank, Washington, DC.

IC-SEA, 1999: Standing to sue in the Philippines: A victory for future generations [online]. Cited 28 March 2003. Available at *www.icsea.org/sea-span/0399/RG0420LL.htm.*

Imbernon, J., 1999: A comparison of the driving forces behind deforestation in the Peruvian and the Brazilian Amazon, *Ambio,* **28(6),** pp. 509–513.

Invasive.org, n.d.: Invasive and exotic species of North America. Cited 25 October 2004. Available at http://www.invasive.org.

IPGRI (International Plant Genetic Resources Institute), 2001: On farm management of crop genetic diversity and the Convention on Biological Diversity's programme of work on agricultural biodiversity [online]. Available at http://www.biodiv.org/doc/meetings/sbstta/sbstta-07/information/sbstta-07-inf-07-en.pdf.

Isozaki, H., 2000: *International Environmental Law,* 1st ed., Shinzansha Publishing Co., Tokyo, Japan.

IUCN (International Union for Conservation of Nature and Natural Resources), 1994: *Guidelines for Protected Areas Management Categories: CNPPA with Assistance of WCMC,* Gland, Switzerland, and Cambridge, UK.

IUCN, 1998: *Guidelines for Re-introductions,* IUCN, Gland, Switzerland, and Cambridge UK.

IUCN, 1999: *Proceedings of the South Asian Regional Environmental Assessment Association (SAREAA),* IUCN Asia, Kathmandu, Nepal.

IUCN, 2000a: The effectiveness of trade measures contained in the Convention on International Trade in Endangered Species of Wild Fauna and Flora (CITES) [online]. Available at http://biodiversityeconomics.org/pdf/topics-405-00.pdf.

IUCN, 2000b: The IUCN policy statement on sustainable use of wild living resources (Resolution 2.29) adopted at the IUCN World Conservation Congress, Amman, October 2000 [online]. Available at http://www.iucn.org/themes/ssc/susg/policystat.html.

IUCN, 2000c: Non-timber forest products [online]. Available at http://www.iucn.org/themes/fcp/about/regional/ntfp.html.

IUCN, 2002: IUCN statement on the management of ex situ populations for conservation [online]. Available at http://www.iucn.org/themes/ssc/pubs/policy/exsituen.htm.

IUCN, 2003a: Recommendations of the Vth IUCN World Parks Congress, Durban, South Africa [online]. Available at http://www.iucn.org/themes/wcpa/wpc2003/index.htm.

IUCN, 2003b: Climate change and species, Unpublished meeting report.

IUCN, n.d.: Biodiversity brief 4: Incentive measures for the conservation and sustainable use of biodiversity. Available at http://www.wcmc.org.uk/biodev/briefs/BB4%20-%20web%20version.doc.

IUCN and UNEP-WCMC (World Conservation Monitoring Center), 2003: *United Nations List of Protected Areas,* IUCN and UNEP-WCMC, Cambridge, UK.

IUCN, UNEP, and WWF, 1991: *Caring for the Earth: A Strategy for Sustainable Living,* IUCN, UNEP, and WWF, Gland, Switzerland.

Jacobson, H.K. and E. Brown Weiss, 1997: Compliance with international accords—achievements and strategies. In: *International Governance on Environmental Issues,* M. Rolén, H. Sjoberg, and U. Svedin (eds.), Environment & Policy, Volume 9, Kluwer Academic Publishers, The Netherlands.

Jaireth, H. and D. Smyth (eds.), 2003: *Innovative Governance: Indigenous Peoples, Local Communities and Protected Areas,* Ane Books, New Delhi, India.

Jakarta Post, 2000: Tracing large-scale illegal logging business in Kalimantan, Jakarta [online]. Cited 22 February 2005. Available at http://www.dayakology.com/publications/articles_news/eng/tracing.htm.

Jarvis, D.I., L. Myer, H. Klemick, L. Guarino, M. Smale, et al., 2000: *A Training Guide for In Situ Conservation on-Farm: Version 1,* IPGRI, Rome, Italy.

Jarvis, D. and T. Hodgkin, 1999: Wild relatives and crop cultivars: Detecting natural introgression and farmer selection of new genetic combinations in agroecosystems, *Molecular Ecology,* **9(8),** pp. 59–173.

Jarvis, D.I., V. Zoes, D. Nares, and T. Hodgkin: On-farm management of crop genetic diversity and the Convention on Biological Diversity's programme of work on agricultural biodiversity, *Plant Genetic Resources Newsletter.* In press.

Jenkins, M., 2003: Prospects for biodiversity, *Science,* **302,** pp. 1175–1177.

Jepson, P. and S. Canney, 2003: Values-led conservation, *Global Biogeography and Ecology,* **12,** pp. 271–274.

Jepson, P., J.K. Jarvie, K. MacKinnon, and K.A. Monk, 2001: The end for Indonesia's lowland forests? *Science,* **292,** pp. 859–861.

Johns, T., 2002: Plant genetic diversity and malnutrition: Practical steps in the development and implementation of a global strategy linking plant genetic resource conservation and nutrition, *African Journal of Food and Nutritional Sciences,* **2(2),** pp. 98–100.

Joshi, A. and J.R. Witcombe, 1998: Farmer participatory approaches for varietal improvement. In: *Seeds of Choice: Making the Most of New Varieties for Small Farmers,* J.R. Witcombe, D.S. Virk, and J. Farrington (eds), Oxford and IBH Publishing Co. Pvt. Ltd., New Delhi, India.

Kabore, O., 2000: Burkina faso: PPB, seed networks and grassroot strengthening. In: *Conserving Agricultural Biodiversity In Situ: A Scientific Basis for Sustainable Agriculture,* D.I. Jarvis, B. Sthapit, and L. Sears (eds), IPGRI, Rome, Italy.

Kelleher, G., 1999: *Guidelines for Marine Protected Areas,* IUCN, Gland, Switzerland, and Cambridge, UK, xxiv + 107 pp.

Kellert, S.R., J.N. Mehta, S.A. Ebbin, and L.L. Lichtenfeld, 2000: Community natural resource management: Promise, rhetoric, and reality, *Society and Natural Resources,* **13,** pp. 705–715.

Key, N.D., E. Sadoulet, and A. de Janvry, 2000: Transactions costs and agricultural household supply response, *American Journal of Agricultural Economics,* **82(2),** pp. 245–259.

King, M. and U. Faasili, 1998: A network of small, community-owned village fish reserves in Samoa, *Parks,* **8(2),** pp. 11–16.

Knight, A. T., and R. M. Cowling, 2003: Conserving South Africa's "lost" biome: A framework for securing effective regional conservation planning in the subtropical thicket biome, Report 44, Terrestrial Ecology Research Unit, University of Port Elizabeth, Port Elizabeth, South Africa.

Kshirsagar, K.G. and S. Pandey, 1996: Diversity of rice cultivars in a rainfed village in the Orissa State of India. In: *Using Diversity* [online], L. Sperling, and M. Loevinsohn (eds.), International Development Research Centre. Available at http://www.idrc.ca/library/document/104582.

Kura, Y., C. Revenga, E. Hoshino, and G. Mock, 2004: *Fishing for Answers: Making Sense of the Global Fish Crisis,* World Resources Institute, Washington, DC.

Laliberte, B., 1997: Botanic garden seed banks: Genebanks world wide, their facilities, collections and networks, *Botanic Gardens Conservation News,* **2(9),** pp. 18–23.

Lalitha, N., 2004: Diffusion of agricultural biotechnology and intellectual property rights: Emerging issues in India, *Ecological Economics,* **49(2),** pp. 187–198.

Lambert, J.A., J.K. Elix, A. Chenowith, and S. Cole, 1995: *Bioregional Planning for Biodiversity Conservation: Approachs to Bioregional Planning,* Part 2, Background papers to the conference, 30 October–1 November 1995, Melbourne, Department of the Environment, Sport and Territories, Canberra, Australia, pp. 15–75.

Lang, W. (ed.), 1995: *Sustainable Development and International Law,* Graham & Trotman Ltd., London, UK.

Lapham, N.P. and R.J. Livermore, 2003: *Striking a Balance: Ensuring Conservation's Place on the International Biodiversity Assistance Agenda,* Conservation International, Washington, DC.

Larkin, P.A., 1977: An epitaph for the concept of maximum sustained yield, *Transactions of the American Fisheries Society,* **106,** pp. 1–11.

Lavides, M., A. Plantilla, N.A. Mallari, B. Tabaranza Jr., B. de la Paz, et al., 2004: Building support for and beyond protected areas in the Philippines: A Haribon's journey of transformations. In: *Communicating Protected Areas,* D.Hamú, E. Auchincloss, and W. Goldstein (eds.), Commission on Education and Communication, IUCN, Gland, Switzerland, and Cambridge, UK.

Le Prestre, P., 2002: The Convention on Biological Diversity: Negotiating the turn to effective implementation, *Isuma,* **3(2),** Les Presses del'Université de Montréal, Montreal, Canada. Cited 13 July 2003. Available at www.isuma .net.

Leiva, J.M., C. Azurdia, W. Ovando, E. Lopez, and H. Ayala, 2002: Contribution of home gardens to *in situ* conservation in traditional farming systems: Guatemalan component. In: *Home Gardens and In Situ Conservation of Plant Genetic Resources in Farming Systems,* J.W. Watson and P.B. Eyzaguirre (eds.), Proceedings of the Second International Home Gardens Workshop, 17–19 July 2001, Witzenhausen, Germany, pp. 56–72.

Lindenmayer, D.B. and J.F. Franklin, 2002: *Conserving Forest Biodiversity: A Comprehensive Multiscaled Approach,* Island Press, Washington, DC.

Lowe, K., J. Fitzsimons, A. Straker, and T. Gleeson, 2003: *Mechanisms for Improved Integration of Biodiversity Conservation in Regional Natural Resource Management Planning within Australia,* Report to Department of Environment and Heritage, Canabera, Australia.

Lubchenco, J., S.R. Palumbi, S.D. Gaines, and S. Andelman, 2003: Plugging a hole in the ocean: The emerging science of marine reserves, *Ecological Applications,* **13(1)** Supplement, pp. S3–7.

Lusthaus, C., M. Adrien, and P. Morgan, 2000: *Integrating Capacity Development into Project Design and Evaluation: Approach and Frameworks,* Monitoring and evaluation working paper 5, Global Environmental Facility, Washington, DC.

MA (Millennium Ecosystem Assessment), 2003: *Ecosystems and Human Well-being: A Framework for Assessment,* Island Press, Washington, DC, 245 pp.

Mace, G.M. and E.J. Hudson, 1999: Attitudes towards sustainability and extinction, *Conservation Biology,* **13,** pp. 242–246.

Mack, R.N., D. Simberloff, W.M. Lonsdale, H. Evans, M. Clout, and F.A Bazzaz, 2000: Biotic invasions: Causes, epidemiology, global consequences, and control, *Ecological Applications,* **10,** pp. 689–710.

MacKinnon, J., K. Mackinnon, G. Child, and J. Thorsell, 1986: *Managing Protected Areas in the Tropics,* IUCN, Gland, Switzerland, 295 pp.

MacKinnon, K. 2001: Editorial, *Parks,* 11(2), IUCN, Gland, Switzerland.

Mahar, D.J., 1989: *Government Policies and Deforestation on Brazil's Amazon Region,* World Bank, Washington, DC.

Manzanero, R. 2004: Promoting protection through pride. In: *Communicating Protected Areas,* D.Hamú, E. Auchincloss, and W. Goldstein (eds.), Commission on Education and Communication, IUCN, Gland, Switzerland, and Cambridge, UK.

Mathur, V.B and A. Rajvanshi, 2001: *Integrating Biodiversity in Impact Assessment: National Case Study for India,* United Nations Development Programme/ Biodiversity Planning Support Programme, Nairobi, Kenya.

May, P., E. Boyd, F. Veiga, and M. Chang, 2003: *Local Sustainable Development Effects of Forest Carbon Projects in Brazil and Bolivia: A View from the Field,* Instituto Pro–Nautra/ International Institute for Environmental Development, London, UK.

May, P.J., R.J. Burby, N.J. Ericksen, J.W. Handmer, J.E. Dixon, et al., 1996: *Environmental Management and Governance, Intergovernmental Approaches to Hazards and Sustainability,* Routledge, London, UK.

Mazhar, F., 2000: Seed conservation and management: Participatory approaches of Nayakrishi seed network in Bangladesh. In: *Participatory Approaches to The Conservation and Use of Plant Genetic Resources,* E. Friis-Hansen, and B. Sthapit (eds.), IPGRI, Rome Italy, pp. 149–153.

McNeely, J.A and S.J. Scherr, 2003: *Ecoagriculture: Strategies to Feed the World and Save Wild Biodiversity,* Island Press, Washington, DC.

McNeely, J.A. (ed.), 1993: *Parks for Life: Report of the IVth World Congress on National Parks and Protected Areas,* IUCN, Gland, Switzerland.

McNeely, J.A., H.A. Mooney, L.E. Neville, P. Schei, and J.K. Waage (eds.), 2001: *A Global Strategy on Invasive Alien Species,* IUCN in collaboration with the Global Invasive Species Programme, Gland, Switzerland, and Cambridge, UK.

McShane, T.O. and M.P. Wells (eds.), 2004: *Getting Biodiversity Projects To Work: Towards More Effective Conservation and Development,* Columbia University Press, New York, NY.

Mefalopulos, P. and L. Grenna, 2004: Promoting sustainable development through strategic communication. In: *Communicating Protected Areas,* D.Hamú, E. Auchincloss, and W. Goldstein (eds.), Commission on Education and Communication, IUCN, Gland, Switzerland, and Cambridge, UK.

Meilleur, B.A. and T. Hodgkin, 2004: In situ conservation of crop wild relatives, *Biodiversity and Conservation,* **13,** pp. 663–684.

Mellas, H., 2000: Morocco: Seed supply systems: Data collection and analysis. In: *Conserving Agricultural Diversity In Situ: A Scientific Basis for Sustainable Agriculture,* D. Jarvis, B. Sthapit, and L. Sears (eds.), IPGRI, Rome, Italy.

Miles, E.L., A. Underdal, S. Adresen, J. Wettestad, J.B. Skjærseth, et al., 2002: *Environmental Regime Effectiveness,* The MIT Press, Cambridge, MA.

Milne, M., 2001: *Forest Carbon Projects and Livelihoods: An Assessment Phase Project of Two AIJ,* Center for International Forestry Research, Jakarta Indonesia/ European Union, Brussels, Belgium/Japan International Cooperation Agency, Tokyo, Japan/ United States Agency for International Development, Washington, DC.

Monroe, M., B. Day, M. Grieser, 2000: *Environmental Education and Communication for a Sustainable World,* GreenCOM, Washington, DC.

Mooney, H.A., and R.J. Hobbs, (eds.), 2000: *Invasive Species in a Changing World,* Island Press, Washington, DC.

Morin, S., J.L. Pham, S. Sebastian, G. Abrigo, D. Erasga, et al., 1998: The role of indigenous technical knowledge in on-farm conservation of rice genetic resources in Cagayan Valley, Philippines. In: *People, Earth and Culture: Readings in Indigenous Knowledge Systems on Biodiversity Management and Utilization,* C. Apolinar, F. Baradas, R. Serran, and E. Belen (eds.), Philippine Council for Agricultural, Forestry and Natural Resources, Research and Development, People, Earth and Culture, Los Banos, Laguna, Philippines.

Mow, J.M., M. Howard, C.M. Delgado, and S. Talbet, 2003: Promoting sustainable development: A case study of the Seaflower Biosphere Reserve, *Prospects,* **XXXIII(3),** pp. 303–312.

Muller, J. and H.J. Albers, 2004: Enforcement, payments, and development projects near protected areas: What works where? *Resource and Energy Economics,* **26,** pp. 185–204.

Mulongoy, K.J. and S. Chape, 2004: *Protected Areas and Biodiversity: An overview of key issues,* Secretariat of the Convention on Biological Diversity, Montreal and UNEP-World Conservation Monitoring Centre, Cambridge, UK.

Murali, K.S., I.K. Murthy, and N.H. Ravindranath, 2002: Joint forest management in India and its ecological impacts, *Environmental Management and Health,* **13(5),** pp. 512–528.

Myers, N., R. Mittermeier, C. Mittermeier, G. Fonseca, and J. Kent, 2000: Biodiversity hotspots for conservation priorities, *Nature,* **403,** pp. 853–858.

Norton, B. G., 2001: Conservation biology and environmental values: Can there be a universal earth ethic? In: *Protecting Biological Diversity: Roles and Responsibilities,* C. Potvin, M. Kraenzel, and G. Seutin (eds.), McGill-Queen's University Press, Montreal, Canada.

Noss, R.F. 1996: Ecosystems as conservation targets, *Trends in Ecology and Evolution* **11,** 351 pp.

Noss, R.F., 2003: A checklist for wildlands network designs, *Conservation Biology,* **17,** pp. 1270–12575.

NSW National Parks and Wildlife Service, 2002: *Urban Wildlife Renewal Growing Conservation in Urban Communities Research Report,* Sydney, New South Wales, Australia.

Oates, J.F., 1999: *Myth and Reality in the Rain Forest: How Conservation Strategies Are Failing in West Africa,* University of California Press, Berkeley, CA.

OECD (Organization for Economic Co-operation and Development), 1999: *Handbook of Incentive Measures for Biodiversity: Design and Implementation,* OECD, Paris, France.

Omamo, S.W., 1998: Transport costs and smallholder cropping choices: An application to Siaya District, Kenya, *American Journal of Agricultural Economics,* **80,** pp. 116–123.

Ortega-Paczka, R., L. Dzib-Aguilar, L. Arias-Reyes, V. Cob-Vicab, J. Canul-Ku, et al.: Mexico: Seed supply systems: Data collection and analysis. In: *Conserving Agricultural Biodiversity In Situ: A Scientific Basis for Sustainable Agriculture,* D. Jarvis, B. Sthapit, and L. Sears (eds.), IPGRI, Rome, Italy, pp. 152–154.

Ortiz, M.E., L.S. Mora, and C.B. Carvajal, 2002: *Impacto del Programa de Pago de Servicios Ambientales en Costa Rica como Medio de la Reducción de la Pobreza* *en los Medios Rurales,* Escuela de Ingenieria Forestal, Instituto Tecnológico de Costa Rica, Cartago, Costa Rica.

Ostrom, E., 1998: Polycentricity, and incentives: Designing complexity to govern complexity. In: *Protection of Global Biodiversity: Converging Strategies,* L.D. Guruswamy, J.A. McNeely (eds.), Duke University Press, Durham, NC.

Ostrom, E., J. Burger, C.B. Field, R.B. Norgaard, and D. Policansky, 1999: Revisiting the commons: Local lessons, global challenges, *Science,* **284,** pp. 278–282.

Pagiola, S., J. Kellenberg, L. Vidaeus, and J. Srivastava, 1997: *Mainstreaming Biodiversity in Agricultural Development: Toward Good Practice,* World Bank, Washington, DC.

Palmer, J.A., 1995: *Influences on Pro-environmental Practices: Planning Education to Care for the Earth,* IUCN Commission on Education and Communication, Gland, Switzerland.

Parks Canada Agency, 2000: *Unimpaired for Future Generations? Protecting Ecological Integrity with Canada's National Parks, Vol. I: A Call to Action, Vol. II: Setting a New Direction for Canada's National Parks,* Report of the Panel on the Ecological Integrity of Canada's National Parks, Ottawa, ON, Canada.

Parmesan, C. and G. Yohe, 2003: A globally coherent fingerprint of climate change impacts across natural systems, *Nature,* **421,** pp. 37–42.

Pence, G., M. Botha, and J.K. Turpie, 2003: Evaluating combinations of on- and off-reserve conservation strategies for the Agulhas Plain, South Africa: A financial perspective, *Biological Conservation,* **112,** pp. 253–273.

Perrings, C. 2002: Biological invasions in aquatic systems: The economic problem, *Bulletin of Marine Science,* **70,** pp. 541–552.

Phillips, A., 2003: Turning ideas on their head: The new paradigm for protected areas, *The George Wright Forum,* **20(2),** pp. 8–32.

Pierce, S.M., R.M. Cowling, T. Sandwith, and K. MacKinnon (eds.), 2002: *Mainstreaming Biodiversity in Development: Case Studies from South Africa,* Biodiversity series, Impact studies, World Bank Environment Department, Washington, DC.

Pimentel, D., C. Wilson, C. McCullum, R. Huang, P. Dwen, et al., 1997: Economic and environmental benefits of biodiversity, *Bioscience,* **47(11),** pp. 747–757.

Pistorius, R., 1997: *Scientists, Plants and Politics: A History of the Plant Genetic Resources Movement,* IPGRI, Rome, Italy.

Plucknett, D.L., N.J.H. Smith, J.T. Williams, and N.M. Anishetty, 1987: *Genebanks and the World's Food,* Princeton University Press, Princeton, NJ.

Pounds, J.A., M.L.P. Fogden, and J.H. Campbell, 1999: Biological response to climate change on a tropical mountain, *Nature,* **398,** pp. 611–615.

Powers, W., 2003: Bolivia successfully innovates in carbon sequestration. In: *Biodiversity: The Richness of Bolivia: State of Knowledge and Conservation,* P.L. Ibisch and G. Mérida (eds.), Ministerio de Desarrollo Sostenible/Editorial FAN (Fundación Amigos de la Naturaleza), Santa Cruz, Bolivia.

Prescott-Allen, R. and C. Prescott-Allen, 1982: The case for in situ conservation of crop genetic resources, *Nature and Resources,* **18,** pp. 15–20.

Pressey, R.L. and V.S. Logan, 1997: Inside looking out: Findings of research on reserve selection relevant to "off-reserve" nature conservation. In: *Conservation Outside Reserves,* P. Hale and D. Lamb (eds.), University of Queensland, Brisbane, Australia.

Pressey, R.L., 1992: Nature conservation in rangelands: Lessons from research on reserve selection in New South Wales, *Rangelands Journal,* **14,** pp. 214–226.

Pressey, R.L. 1997: Priority conservation areas: Towards an operational definition for regional assessments. In: *National Parks and Protected Areas: Selection, Delimitation and Management,* J.J. Pigram and R.C. Sundell (eds.), University of New England, Centre for Water Policy Research, Armidale, Australia, pp. 337–357.

Pressey, R.L. and R.M. Cowling, 2001: Reserve selection algorithms and the real world, *Conservation Biology,* **15,** pp. 275–257.

Pressey, R.L. and S.L. Tully, 1994: The cost of ad hoc reservation: A case study in western New South Wales, *Australian Journal of Ecology,* **19,** pp. 375–384.

Pressey, R.L., C.J. Humphries, C.R. Margules, R.I. Vane-Wright, and P.H. Williams, 1993: Beyond opportunism: Key principles for systematic reserve selection, *Trends in Ecology and Evolution,* **8,** pp. 124–128.

Quiros, C.F., R. Ortega, L.W.D. Van Raamsdonk, M. Herrera-Montoya, P. Cisneros, et al., 1992: Amplification of potato genetic resources in their centre of diversity: The role of natural outcrossing and selection by the Andean farmer, *Genetic Resources and Crop Evolution,* **39,** pp. 107–113.

Quiros, C.F., S.B. Brush, D.S. Douches, K.S. Zimmerer, and G. Huestis, 1990: Biochemical and folk assessment of variability of Andean cultivated potatoes, *Economic Botany,* **44(2),** pp. 254–266.

Rajvanshi, A., 2003: *Proceedings of the Workshop on EIA Studies for Developing Projects,* CPCB Publication Series, Ecological Impact Assessment EIAs/03/2002/-2003, Central Pollution Control Board Publication, New Delhi, India.

Ramsar, 1997: Memorandum of understanding with the Bonn Convention [online]. Available at *http://www.ramsar.org/key_cms_mou.htm.*

Rana, R.B., D. Rijal, D. Gauchan, A. Subedi, B. Sthapit, et al., 2000: *Agroecology and Socioeconomic Baseline Study of Begnas Eco-site, Kaski, Nepal,* Nepal Agricultural Research Council/ Local Initiatives for Biodiversity, Research and Regional Development/IPGRI, Kathmandu, Pokara, and Lumle, Nepal.

Raunetsalo, J., H. Juslin, E. Hansen, and K. Forsyth, 2002: *Geneva Timber and Forest Discussion Papers: Forest Certification Update for the UNECE Region, Summer 2002,* United Nations publication, Geneva, Switzerland, pp.1–34.

Ravindranath, N.H., K.S. Murali, I.K. Murthy, P. Sudha, S. Palit, et al., 2000: Summary and conclusions. In: *Joint Forest Management and Community Forestry in India: An Ecological and Institutional Assessment,* N.H. Ravindranath, K.S. Murali and K.C. Malhotra (eds.), Oxford and IBH Publication, New Delhi, India, pp. 279–318.

Read, V., and B. Bessen, 2003: *Mechanisms for Improved Integration of Biodiversity Conservation in Regional NRM Planning,* Report prepared for Environment Australia, Canabara, Australia.

Reading, R.P., T.W. Clark, and S.R. Kellert, 2002: Towards an endangered species reintroduction paradigm, *Endangered Species Update,* **19(4),** pp. 142–146.

Reid, W. and K. Miller, 1989: *Keeping Options Alive: The Scientific Basis for Conserving Biodiversity,* World Resources Institute, Washington, DC.

Revenga, C. and Y. Kura, 2003: *Status and Trends of Biodiversity of Inland Water Ecosystems,* Secretariat of the Convention on Biological Diversity, Technical series no. 11, Montreal, Canada.

Reyers, B., 2004: Incorporating anthropogenic threats into evaluations of regional biodiversity and prioritization of conservation areas in the Limpopo Province, South Africa, *Biological Conservation,* **118,** pp. 521–531.

Ribot, J., 1999: Decentralisation, participation and accountability in Sahelian Forestry: Legal instruments of political–administrative control, *Africa,* **69,** pp. 23–65.

Richards, P. and G. RuivenKamp, 1997: *Seeds and Survival: Crop Genetic Resources in War and reconstruction in Africa,* IPGRI, Rome, Italy.

Richards, P., 1986: *Coping with Hunger, Hazard and Experiment in an African Rice-farming System,* Allen & Unwin Ltd., London, UK.

Richardson, E. L., 1992: Climate change: Problems of law-making. In: *The International Politics of the Environment, Actors Interests, and Institutions,* A. Hurrel and B. Kingsbury (eds.), Clarendon Press, Oxford, UK.

Rick, C.M., 1963: Barriers to interbreeding in *Lycopersicon peruvianum, Evolution,* **17,** pp. 216–232.

Rijal, D., R. Rana, A. Subedi, and B. Sthapit, 2000: Adding value to landraces: Community-based approaches for in situ conservation of plant genetic resources in Nepal. In: *Participatory Approaches to the Conservation and Use of Plant Genetic Resources,* E. Friis-Hansen and B. Sthapit (eds.), IPGRI, Rome Italy, pp. 166–72.

Roberts, C.M. and J.P. Hawkins, 2000: *Fully-protected Marine Reserves: A Guide,* WWF, Washington, DC, and University of York, Environment Department, York, UK.

Roberts, C.M., J.A. Bohnsack, F. Gell, J.P. Hawkins, and R. Goodridge, 2001: Effects of marine reserves on adjacent fisheries, *Science,* **294,** pp. 1920–1923.

Roberts, C.M., S. Andelman, G. Branch, R.H. Bustamante, J.C. Castilla, et al., 2003a: Ecological criteria for evaluating candidate sites for marine reserves, *Ecological Applications,* **13** (1 Supplement), pp. S199–215.

Roberts, C.M., G. Branch, R.H. Bustamante, J.C. Castilla, J. Dugan, et al., 2003b: Application of ecological criteria in selecting marine reserves and developing reserve networks, *Ecological Applications,* **13** (1 Supplement), pp. S215–228.

Robinson, E.J.Z., J.C. Williams, and H.J. Albers, 2002: The influence of markets and policy on spatial patterns of non-timber forest product extraction, *Land Economics,* **78(2),** pp. 260–71.

Robinson, L. and A. Glanznig, 2003: *Enabling EcoAction: A Handbook for Anyone Working with the Public on Conservation,* Humane Society International, WWF, Australia, and IUCN, Gland, Switzerland.

Rodrigues, A.S.L., S.J. Andelman, M.I. Bakarr, L. Boitani, T.M. Brooks, et al., 2004: Effectiveness of the global protected area network in representing species diversity, *Nature,* **428,** pp. 640–643.

Rogers, E.M., 1983: *Diffusion of Innovations,* The Free Press, New York, NY.

Root, T.L. et al., 2003: Fingerprints of global warming on wild animals and plants, *Nature,* **421,** pp. 57–60.

Rosendal, G.K., 1995: The Convention on Biological Diversity: A viable instrument for conservation and sustainable use? In: *Green Globe Yearbook of International Co-operation on Environment and Development 1995,* H.O. Bergesen, G.Parmann, and O.B. Thommessen (eds.), Oxford University Press, Oxford, UK, pp. 69–81.

Rosenthal, J., 2003: Politics, culture and governance in the development of prior informed consent and negotiated agreement with indigenous communities [online], Presented at the Conference on Biodiversity, Biotechnology and the Protection of Traditional Knowledge, Washington University School of Law, Saint Louis, 4–6 April 2003. Available at www.law.wustl.edu/centeris/confpapers/index.html.

Rosenzweig, M.L., 2003: Reconciliation ecology and the future of species diversity, *Oryx,* **37,** pp. 194–205.

Sadiki, M., L. Belqadi, M. Mahdi, and D.I. Jarvis, 2001: Identifying units of diversity management by comparing traits used by farmers to name and distinguish faba bean (*Vicia faba* L.) cultivars with measurements of genetic distinctiveness in Morocco, *Proceedings of the LEGUMED Symposium on Grain Legumes in the Mediterranean Agriculture,* 25–27 October 2001, Rabat, Morocco.

Sala, E., O. Aburto-Oropeza, G. Paredes, I. Parra, J.C. Barrera, et al., 2002: A general model for designing networks of marine reserves, *Science,* **298,** pp. 1991–1993.

Salwasser, H., 1994: Ecosystem management: Can it sustain diversity and productivity? *Journal of Forestry,* (**August**), pp. 6–10.

Sampson, C., 1994: Cost and impact of current control methods used against *Heracleum mantegazzianum* (giant hoogweed) and a case for investigating biological control. In: *Ecology and Management of Invasive Riverside Plants,* L.C. de Waal, L.E. Child, P.M. Wade, and J.H. Brock (eds.), John Wiley and Sons, Chichester, UK, pp. 55–56.

Sand, P., 1991: International cooperation: The environmental experience. In: *Preserving the Global Environment: The Challenge of Shared Leadership,* J.T. Matthews (ed.), The American Assembly & World Resources Institute, W.W. Norton & Company, New York, NY.

Sand, P., 1992: *The Effectiveness of International Environmental Agreements,* Grotius Publications Limited, Cambridge, UK.

Sanderson, S.E. and K.H. Redford, 2003: Contested relationships between biodiversity conservation and poverty alleviation, *Oryx,* **37,** pp. 389–390.

Sandwith, T, C. Shine, L. Hamilton, and D. Sheppard, 2001: Transboundary protected areas for peace and cooperation, *Best Practice Protected Area Guidelines Series,* No. 7, IUCN, Gland, Switzerland.

Saunier, R.E. and R.A. Meganck (eds.), 1995: *Conservation of Biodiversity and the New Regional Planning,* Organization of American States and IUCN, Washington, DC.

Schachter, J., 1998: *Review of Action Distribution in 42 Action Plans,* Unpublished IUCN Species Survival Commission intern project.

Scholes, R.J. and R. Biggs (eds.), 2004: *Ecosystem Services in Southern Africa: A Regional Assessment,* Council for Scientific and Industrial Research, Pretoria, South Africa.

Sebastian, R.L., E.C. Howell, G.J. King, D.F. Marshall, and M.J. Kearsey, 2000: An integrated AFLP and RFLP *Brassica oleracea* linkage map from two morphologically distinct doubled-haploid mapping populations, *Theoretical and Applied Genetics,* **100,** pp. 75–81.

Sierra, R., 1999: Traditional resource-use systems and tropical deforestation in a multi-ethnic region in north-west Ecuador, *Environmental Conservation,* **26(2),** pp. 136–145.

Simpson, R.D. and R.A. Sedjo, 1996: Paying for the conservation of endangered ecosystems: A comparison of direct and indirect approaches, *Environment and Development Economics,* **1,** pp. 241–257.

Simpson, R.D., 1999: The price of biodiversity, *Issues in Science and Technology,* **XV(3),** pp. 65–70.

Smale, M., 2002: The conceptual framework for economics research in IPGRI's global *in situ* conservation on-farm project. In: *The Economics of Conserving Agricultural Biodiversity on Farms: Research Methods Developed from IPGRI's Global Project, "Strengthening the Scientific Basis of In Situ Conservation of Agricultural Biodiversity,"* M. Smale, I. Mar., and D.I. Jarvis (eds.), Proceedings of a workshop, Gödöllo, Hungary, 13–16 May 2002, IPGRI, Rome, Italy.

Smale, M., M. Bellon and J.A. Aquirre, 1999: *The Private and Public Characteristics of Maize Landraces and the Area Allocation Decisions of Farmers in a Centre of Crop Diversity,* CIMMYT economics working paper 99–08, Centro Internacional de Majoramiento de Maíz y Trigo/International Maize and Wheat Improvement Center, Mexico, D.F., Mexico.

Smith, R.J., R.D.J. Muir, M.J. Walpole, A. Balmford, and N. Leader-Williams, 2003: Governance and the loss of biodiversity, *Nature,* **426,** pp. 67–70.

Soleri, D. and D.A. Cleveland, 2001: Farmer's genetic perceptions regarding their crop populations: An example with maize in the central valleys of Oaxaca, Mexico, *Economic Botany,* **55,** pp. 106–128.

Southgate, D. 1998: *Tropical Forest Conservation: An Economic Assessment of the Alternatives in Latin America,* Oxford University Press, New York and Oxford, UK.

Spierenburg, M., 2003: Natural resource management in the communal areas: from centralisation to de-centralisation and back again. In: *Zimbabwe, Twenty Years of Independence: The Politics of Indigenisation,* S. Darnolf and L. Laakso (eds.), Palgrave (MacMillan), London, UK, pp. 78–103.

Steiner, F., 2000: *The Living Landscape: An Ecological Approach to Landscape Planning,* McGraw-Hill, New York, NY.

Stern, M., 2004: Understanding local reactions to protected areas. In: *Communicating Protected Areas,* D.Hamú, E. Auchincloss, and W. Goldstein (eds.), Commission on Education and Communication, IUCN, Gland, Switzerland, and Cambridge, UK.

Stiling, P., 2002: Potential non–target effects of a biological control agent, prickly pear moth, Cactoblastis cactorum (Berg) (Lepidoptera: Pyralidae), in North America, and possible management actions, *Biological Invasions,* **4,** pp. 273–281.

Swart, J., 2003: Will direct payments help biodiversity? *Science,* **299,** 1981–1982.

Synghe, H. (ed.), 1998: *Parks for Life 1997: Proceedings of the IUCN/WCPA European Regional Working Session on Protecting Europe's Natural Heritage,* IUCN, Rome, Italy/ Federation of Nature and National Parks in Europe, Grafenau, Germany/ Bundesamt fur Naturschutz, Bonn, Germany.

Szaro, R. and D. Johnson (eds.), 1996: *Biodiversity in Managed Landscapes: Theory and Practice,* Oxford University Press, Oxford, UK.

Tadesse, W.G., M. Denich, D. Teketay, and P.L.G. Vlek, 2002: Diversity of traditional coffee production systems in Ethiopia and its need for in situ conservation. In: *Managing Plant Genetic Diversity,* J. Engels, V.R. Rao, A.H.D. Brown, and M. Jackson (eds.), CAB International, Oxon, UK, pp. 237–247.

ten Kate, K. and S.A. Laird, 2002: *The Commercial Use of Biodiversity: Access to Genetic Resources and Benefit Sharing.* Earthscan Publications, London.

Terborgh, J. and C. van Schaik C., 2002: Why the world needs parks. In: *Making Parks Work: Strategies for Preserving Tropical Nature,* J. Terborgh, C.P. van Schaik, L. Davenport, and M. Rao (eds.), Island Press, Washington, DC, pp. 3–14.

Terborgh, J., C. van Schaik, L. Davenport, and M. Rao (eds.), 2002: *Making Parks Work, Strategies for Preserving Tropical Nature,* Island Press, Washington, DC.

The Biodiversity Project, 1999: *Life, Nature, The Public, Making the Connection: A biodiversity Communications Handbook,* The Biodiversity Project, Madison, WI.

The Commonwealth of Australia, 2003: The benefits of marine protected areas [online]. Available at http://www.iucn.org/themes/wcpa/wpc2003/pdfs/programme/cct/marine/mpasfisherie saut.pdf.

The Royal Society, 2003: *Measuring Biodiversity for Conservation,* The Royal Society, London, UK.

Thomas, C.D., A. Cameron, R.E. Green, M. Bakkenes, L.J. Beaumont, et al., 2004: Extinction risk from climate change, *Nature,* **427,** pp. 145–148.

Tobin, B., 2001: Redefining perspectives in the search for protection of traditional knowledge: A case study from Peru, *RECIEL,* **10(1),** pp. 47–64.

Towill, L.E. and C. Walters, 2000: Cryopreservation of pollen. In: *Cryopreservation of Tropical Plant Germplasm: Current Research Progress and Applications,* F. Engelmann and H. Takagi (eds.), Japan International Centre for Agricultural Sciences, Owashi, Japan, and IPGRI, Rome, Italy, pp. 115–129.

Trampus, T., 2003: When planning behind the desk does not work, CEC case studies [online]. Available at www.iucn.org/cec.

Turner, R.K., J. Paavola, P. Cooper, S. Farber, V. Jessamy, et al., 2003: Valuing nature: Lessons learned and future research directions, *Ecological Economics,* **46,** pp. 493 – 510.

ULG Northumbrian Ltd., 2000: Considerations in the development of a community-based wildlife management programme, Unpublished report.

UNEP (United Nations Environment Programme), 1996: *EIA Training Resource Manual,* 1st ed., UNEP, Nairobi, Kenya.

UNEP, 1998: Impact assessment and minimizing adverse impacts: Implementation of Article 14, UNEP/CBD/COP/4/2. Available through www.biodiv .org

UNEP, 1999: *Global Environment Outlook,* Earthscan Publications Ltd, London, UK.

UNEP, 2002: *EIA Training Resource Manual,* 2nd ed., United Nations Environment Programme, Nairobi, Kenya.

UNEP, 2003: Status and trends of, and threats to, protected areas, UNEP/CBD/SBSTTA/9/5. Available through www.biodiv.org.

UNEP – CAR/RCU (Cartagena Convention/Regional Coordination Unit) 2000–2003: The International Coral Reef Initiative, UNEP – CAR/RCU.

Available at http://www.cep.unep.org/programmes/spaw/icri/ICRIinto .htm.

UNESCO (United Nations Educational, Scientific and Cultural Organization), 2000: *Survey on the Implementation of the Seville Strategy/Enquête sur la Mise en Oeuvre de la Stratégie de Séville,* SC.00/CONF.208/3, SC.2000/CONF.208/ CLD.5, UNESCO, Paris, France.

Van Boven, G. and F. Hesselink, 2002: Mainstreaming biological diversity: The role of communication, education and public awareness [online]. Available at http://www.iucn.org/webfiles/doc/CEC/Public/Electronic/CEC/Brochures/CECMainstreaming_anglais.pdf.

Van Rensburg, B.J., B.F.N. Erasmus, A.S. van Jaarsveld, K.J. Gaston, and S.L. Chown, 2004: Conservation during times of change: Interactions between birds, climate and people in South Africa, *South African Journal of Science,* **100,** pp. 266–72.

Van Woerkoem, C., F. Hesselink, A. Gomis, and W. Goldstein, 2000: Evolving role of communication as a policy tool. In: *Communicating the Environment,* M. Oepen and W. Hamacher (eds.), Deutsche Gesellschaft fur Technische Zusammenarbeit, Eschborn, Germany.

Veitch, C.R. and M.N. Clout, 2002: *Turning the Tide: The Eradication of Invasive Species,* IUCN SSC Invasive Species Specialist Group, IUCN, Gland, Switzerland, and Cambridge, UK, viii + 414 pp.

Vermeulen, S. and I. Koziell, 2002: *Integrating Global and Local Values: A Review of Biodiversity Assessment,* International Institute for Environment and Development, London, UK.

Wadsworth, R.A., Y.C. Collingham, S.G. Willis, B. Huntley, and P.E. Hulme, 2000: Simulating the spread and management of alien riparian weeds: Are they out of control? *Journal of Applied Ecology,* **37,** pp. 28–38.

Walpole, M.J. and N. Leader-Williams, 2001: Masai Mara tourism reveals partnership benefits, *Nature,* **413,** 771 pp.

Ward, T.J., D. Heinemann, and N. Evans, 2001: *The Role of Marine Reserves as Fisheries Management Tools: A Review of Concepts, Evidence and International Experience,* Department of Agriculture, Fisheries and Forestry, Canberra, Australia [online]. Available at http://www.affa.gov.au/content/publications .cfm?ObjectID=7391258C-618D-4964-AC079D2D3FABDE69.

Watson, J.W. and P.B. Eyzaguirre (eds.), 2002: *Home gardens and In Situ Conservation of Plant Genetic Resources in Farming Systems,* Proceedings of the Second International Home Gardens Workshop, 17–19 July 2001, Witzenhausen, Germany.

WCPA News, 2002. [online] Available at http://www.iucn.org/themes/wcpa/newsbulletins/news-may02.htm.

Wells, M., K. Brandon, and L. Hannah, 1992: *People and Parks: Linking Protected Area Management with Local Communities,* World Bank, WWF, and US Agency for International Development, Washington, DC.

Wells, M., S. Guggenheim, A. Khan, W. Wardojo, and P. Jepson, 1998: *Investing in Biodiversity: A Review of Indonesia's Integrated Conservation and Development Projects,* World Bank, East Asia Region, Washington, DC, pp. 1–119.

Werksman, J. n.d.: Five MEAS, five years since Rio: Recent lessons of the effectiveness of multilateral environmental agreements [online]. Accessed on 23 October 2004. Available at http://www.ecouncil.ac.cr/rio/focus/report/english/field.htm.

Wetlands International, 2001: Promoting the conservation of migratory waterbirds in the Asia Pacific region. Cited 12 July 2003. Available at http://www.wetlands.org/IWC/awc/waterbirdstrategy/default.htm.

Wettestad, J., 1999: *Designing Effective Environmental Regimes,* Edward Elgar, Cheltenham, UK, and Northhampton, MA.

Wilkes, H.G., 1985: Teosinte the closest relative of maize, *Maydica,* **30,** pp. 209–223.

Wittenburg, R., and M.J.W. Cock (eds.), 2001: *Invasive Alien Species: A Toolkit of Best Prevention and Management Practices,* CAB International, Wallingford, Oxon, UK, pp. xii–228.

World Bank, 1997: *The Environmental Assessment Source Book, Update no. 20,* The World Bank, Washington, DC.

World Bank, 2001: *Indonesia: Environment and Natural Resource Management in a Time of Transition,* World Bank, Washington, DC.

World Trade Organization, 2003: Mexico etc Versus US: "Tuna-dolphin," Geneva [online]. Available at http://www.wto.org/english/tratop_e/envir_e/edis04_e.htm.

Wunder, S., 2000: Ecotourism and economic incentives: An empirical approach, *Ecological Economics,* **32(3),** pp. 465–480.

WWF International (World Wildlife Fund), 2003: The global 200 [online]. Available at http://www.panda.org/about_wwf/where_we_work/ecoregions/global200/pages/endangered.htm.

WWF International, 2004: *How Effective are Protected Areas? A Preliminary Analysis of Forest Protected Areas,* Report prepared for the Seventh Conference of Parties of the Convention on Biological Diversity, WWF, Gland, Switzerland.

Wyse Jackson, P.S. and L.A. Sutherlands, 2000: *International Agenda for Botanic Gardens in Conservation,* Botanic Gardens Conservation International, Surrey, UK.

Yoccoz, N.G., J.D. Nichols, and T. Boulinier, 2001: Monitoring of biological diversity in space and time, *Trends in Ecology and Evolution,* **16,** pp. 446–53.

Zeng, G.H., X.M. Jing, and K.L. Tao, 1998: Ultradry seed storage cuts costs in gene bank, *Nature,* **393,** pp. 223–224.

Zeven, A.C., 1998: Landraces: A review of definitions and classifications, *Euphytica,* **104(2),** pp. 127–139.

Chapter 6

Food and Ecosystems

Coordinating Lead Authors: Mahendra Shah, Anastasios Xepapadeas

Lead Authors: Rose Emma Mamaa Entsua-Mensah, Günther Fisher, Alexander Haslberger, Frank Jensen, M. Monirul Qader Mirza, Eftichios Sartzetakis, Henk Simons

Contributing Authors: Christopher L. Delgado, Ernesto Gonzalez-Estrada, Simon Hales, Yumiko Kura, Helen Leitch, John McDermott, Jennifer Olson, Don Peden, Tipparat Pongthanapanich, Thomas F. Randolph, Robin Reid, D. Romney, Harrij T. van Velthuizen, David Wiberg

Review Editors: Arsenio M. Balisacan, Peter Gardiner

BOXES

FIGURES

TABLES

Main Messages

There exists a fundamental trade-off between the need to increase food production and the need to sustain, in the long run, the capacity of the ecosystems to support food production. Food production is by far the largest user of ecosystems and their provisioning services. It has the largest impact on ecosystems and biodiversity. Intensive exploitation of ecosystems to satisfy needs for food might erode the productive capacity of these ecosystems through soil degradation, water depletion or contamination, collapse of fisheries, and biodiversity loss. While the recent trends in the slowdown of demographic growth and potential advances in agricultural technology are encouraging, there are also increased pressures on the resource base (land, water, fisheries, and biodiversity), extensive habitat destruction, deforestation, and loss of biodiversity and agrobiodiversity, and potentially serious long-term effects from the regional impacts of climate change.

The impacts on ecosystems from attempts to increase food production have resulted largely from secondary effects and, as such, they represent negative externalities. Expansion of agricultural land in many regions is difficult. Farming land is disappearing because of land degradation and urbanization. Water resources are under pressure because of excessive use and contamination. Governments are faced with the challenge of ensuring that the agricultural water supply is sustainable and that ecosystems contributing to that sustainability are protected. The substantial increase in the use of agrochemicals, such as pesticides, in developing-country agriculture has resulted in environmental contamination, severe health hazards to farmers, and unprofitable crop production. There are emerging problems of overgrazing and dry land degradation, rangeland fragmentation and loss of wildlife habitat, dust formation, bush encroachment, deforestation, nutrient overload through disposal of manure, and greenhouse gas emissions. Critical issues relate to human health arising from the threat of diseases such as bird flu in poultry and BSE in cattle. Capture fisheries are facing overexploitation and stock depletion.

The emergence of water pricing schemes and the establishment of water markets in different parts of the world shows that water pricing is a response that promotes efficient allocation and responsible use. In the context of negative externalities from excessive use of agrochemicals, integrated pest management, utilizing biological rather than chemical agents, provides avenues for sustainable production without damaging side effects. Organic farming can contribute to enhancing sustainability of production systems and agrobiodiversity. Agroforestry is a contributory technology for increased food production, using nitrogen-fixing trees to increase soil fertility and nutrient cycling. On overfishing, strict regulatory mechanisms, enforcing fishing quota and fishing capacity reduction, are urgently needed. Aquaculture has an important role to play in meeting fish food demand. In that context, government support in combination with private sector investments is important. However, given the potential detrimental environmental impacts of aquaculture, appropriate regulatory mechanisms need to supplement polices.

The information and communication revolution has a significant role in mitigating impacts on ecosystems and sustaining their capacity for future generations through evolving national and international agricultural knowledge systems. The design of policies needs to cautiously take into account "worst-case" scenarios and abrupt changes in ecosystems, because many impacts of ecological trade-offs in increased food production are uncertain. Comprehensive knowledge of the natural resources and the environment, science, and technology, as well as the socioeconomics of sustainable agricultural development, is essential to design policy frameworks. It calls for comprehensive assessments, education, and knowledge dissemination at the local, national, and international level, in order to ensure that farmers can produce food in a manner that is environmentally, economically, and socially sustainable and that consumers have the opportunities to make choices regarding food that is nutritious and healthy, safe, and affordable.

Integrated agroecological approaches in scientific research, policy design, and appropriate regulatory frameworks at all levels from local to global are important for enhancing sustainability. New analytical approaches of decision-making, which integrate ecological and socioeconomic systems and environmental change, play an important role in generating science-based policy analysis to assess response options towards sustainable food and ecosystems. **New agricultural sciences combined with effective natural resource management could support a future agricultural revolution to meet worldwide food needs in the twenty-first century.**

Modern methods of biotechnology, such as market-assisted methods of breeding, as well as molecular methods for the preservation of germplasm diversity, are important scientific tools. Responses need to determine the level of risks to human health and the environment for their optimal deployment. Response polices need to be gender sensitive. Women play a substantive role in food production and food preparation. **The national and international public research needs to be substantially strengthened to meet the challenge of environmentally sound food production, especially in view of the growing role of the private sector and the privatization of agricultural research, which often lacks an environmental focus and whose profit orientation has thus far excluded the needs and crops of poor farmers.**

Agricultural research needs to give high priority to developing mitigation and adaptation options related to climate change and variability in tropical areas, particularly the developing countries with the least capacity to cope. **Changes in international agricultural trade, as well as the conditions of access to world markets offer both obstacles and opportunities to developing countries, which are particularly vulnerable.** Different approaches to trade and production have opened huge gaps within the developing world in terms of productive capacity and international marketing. The challenge is to make agricultural knowledge support efficient and policies environment friendly through the integration of biological, biochemical, agroecological and environmental, socioeconomic, and information sciences.

The Common Agricultural Policy constitutes a major response from the European Union aimed at securing food supply and enhancing the well-being of the rural communities. The introduction of an environmental dimension into the CAP shows direct recognition of the fact that agricultural policies need to correct undesirable environmental pressures. **Government policies developed around food production (price supports and various types of payments, or taxes) can have adverse economic, social, and environmental effects.** This issue is highlighted in the case of the international sugar market, one of the most heavily distorted international markets due to EU and U.S. support policies.

6.1 Introduction

The purpose of this chapter is to (1) present the important drivers associated with the provisioning of food (crops, livestock, fish) and the main issues related to the food provisioning services of ecosystems, (2) analyze and assess major responses, mainly in the form of economic, technological, and institutional interventions, undertaken in order to enhance and/or secure food provisioning services, (3) present through case studies the structure and the major impacts of selected responses on ecosystems and human well-being, and (4) provide some conclusions derived from the analysis and assessment.

The main issues associated with the food sector relate to the structure of consumption and production. Consumption patterns relate to both hunger and overconsumption, directly affecting human well-being. The structure of production systems directly affects food supply and also, through the production processes and the inputs used, has important impacts on resource availability and the state of the ecosystems. This section examines important direct and indirect drivers, as well as consumption patterns, production systems, and likely impacts on ecosystems associated with cropping, fisheries, and livestock husbandry.

A number of drivers affect the current and future capacity of ecosystems associated with food provisioning services. They also affect food needs and food provisions as determinants of human well-being. The focus here is on important drivers, which are interrelated and are within the wider system—drivers that are endogenous, depending on the spatial and temporal scale of analysis. Thus responses affect direct drivers such as changes in resource availability (land, water, fish biomass, biodiversity), intensification of production, and climate change, but also indirect drivers such as population growth or international trade regimes.

6.1.1 Population

An issue of great importance is the challenge of tripling the global food production in the poorest societies, some of which have been doubling their numbers in as little as thirty years. The generally accepted point (Arrow et al. 2004; Ehrlich and Ehrlich 2002) is that population and consumption jointly impose increasing strains on the ecosystems which are supporting them. (For more extensive information on population developments, see MA *Scenarios,* Chapter 7.)

Long-term predictions, on the other hand, regarding world population growth (UN 2003; IIASA 2004) indicate a slowdown in the annual population growth rate. This might imply a possible slowdown in the growth of demand for food. FAO (2003b) projects progress in raising food consumption levels, in improving nutrition, and in reducing the proportion of undernourished people. However, initially, this might not show as a decline in the number of undernourished people, due to population growth. Also, rapid population growth in some regions not only pushes up demand for food, but may also cause the agricultural resource base to diminish due to overexploitation and conversion of arable land for residential, infrastructure, and industrial uses. As a result, the developing world's capacity to expand food production may well be shrinking in some regions rather than expanding. In many developing countries, with limited prospects of bringing additional land into production, intensification will be required to enable higher productivity and at the same time ensure the sustainability of ecosystems and services.

Food security is also affected by patterns of trade and protection. The removal of trade barriers to developing countries, especially in the agricultural products that they have a comparative advantage in producing, along with reduced tariffs for processed agricultural commodities is expected to benefit them. Globalization of markets is also a stimulus to competitiveness and local forms of production. It will be important for developing countries to promote production systems for which their agroecosystems are suited and which can be sustained.

6.1.2 Natural Resources

Arable Land: About 4 billion hectares of the world's land is suitable for arable agriculture. (FAO 2003b), and nearly all land well suited for intensive agriculture is currently in cultivation (MA *Current State and Trends,* Chapter 26). In developing countries, population growth resulted in a substantial decline of arable land per capita. Furthermore, farming land is lost because of land degradation and urbanization. Land degradation is mainly due to soil erosion, loss of nutrients, damages from inappropriate farming practices, and misuse of agricultural chemicals (FAO 1995a). Urbanization affects food production by converting arable land, and by reducing labor input to the agricultural sector. China, for example, lost around one million hectares of arable land between 1987 and 1995, due to construction (Fischer et al. 1998).

Water: Approximately 70% of all fresh water withdrawals is currently used for agriculture. While rain-fed croplands might consume more or less water than the natural vegetation/soil conditions they replaced, irrigated areas consume significantly more. Irrigation systems divert 20–30% of the world's available water resources, but chronic inefficiencies in distribution mean that only some 40–50% of that water is actually used in crop growth (MA *Current State and Trends,* Chapters 7, 26). According to a study by the International Water Management Institute, irrigation water withdrawals will need to increase by 17% (Seckler 2000) as the area of irrigated land expands by a further 22 % in order to feed the population of 2025. Other projections based on different scenarios vary widely, with some even predicting a decrease (Cosgrove and Rijsberman 2000). Of greater concern than increased demands for irrigation water alone are the far higher increases in municipal and industrial water demands that compete for the limited resource and often draw water away from agriculture. The same IWMI projection predicts increases of 60% for industrial and 80% for municipal withdrawals.

In 1999, the irrigated area as a proportion of irrigation potential was 50% for the developing world, where most irrigation development is expected to occur. The figure is 13% for sub-Saharan Africa and more than 80% for South Asia, excluding India. By 2030, 60% of all land with irrigation potential in developing countries will be in use (FAO 2003b). Poor drainage and irrigation practices have led to water logging and salinization of approximately 30% of the world's irrigated lands (FAO 2001a). Groundwater from shallow aquifers is an important source of irrigation water, but its sustainable use is at risk due to over-pumping of aquifers, pollution from agrochemicals and the mining of fossil groundwater (FAO 2001a). Fertilizers and pesticides are a major cause of water pollution generally, and the nutrients from fertilizers are causing severe problems of eutrophication in surface waters worldwide.

About 10% of irrigation water in developing countries comes from reused wastewater. For irrigation use, wastewater should receive treatment, but in lower-income countries, raw sewage is often used directly, leading to a variety of health and environmental problems. Also, crops grown using untreated wastewater cannot be exported and access to local markets, at least partially, is restricted.

Fisheries: FAO estimates that currently 71–78% of fish stocks are fully exploited, overexploited, or recovering from depletion. The increase of the world's fish catches during 1950 to 1990 was followed by a decline in productivity due to overfishing (Jackson 2001). Overexploitation occurred due to rapid expansion of the world's fishing fleet, enormous advances in fishing technologies, poor understanding of fish population dynamics (or little concern for ensuring sustainable yields), and a failure to introduce effective management systems (see Alverson et al. 1994).

Biodiversity: Species extinction and biodiversity loss is caused by habitat destruction due to a variety of economic and social driving forces including poverty, human population growth, unsustainable human production and consumption patterns, as well as legal, institutional, and cultural aspects (Folke et al. 1992). In relation to food production, there are two aspects of biodiversity loss that can be regarded as constraining feedbacks to the attainment of food security. One is associated with habitat destruction due to expansion of cropland, pastures, and aquaculture; the other is loss of genetic diversity in agriculture due to specific crop choices and cultivation practices. Habitat destruction is also associated with tropical deforestation.

Loss of genetic diversity in agriculture is associated with changes due to factors such as domestication and development of genetically uniform varieties, the preference of farmers and consumers for certain breeds or varieties, global consolidation of the seed grain industry, and the adoption of high-yield varieties as part of the Green Revolution (Heal et al. 2002).

6.1.3 Agrobiodiversity

The analysis of the history of breeding of major crops such as wheat, maize, or rice shows that conventional breeding focused on objectives of increased productivity, increased resistance to diseases and pests, and enhanced quality with respect to nutrition and food processing. After the spread of crop ancestors from centers of origin and diversification of lines, breeding made use of natural variations and, later on, of the traits found in different lines or close relatives.

Modern methods of breeding have significantly increased yields of cereal crops in general. However, yield growth has slowed in the most intensively cultivated irrigated areas, and land quality has also declined.

To overcome these problems, scientists and farmers considered new ways to attain the overall objectives of improved yields and sustainable agricultural systems. They also looked for methods for improvements in health risks and environmental problems created by the need for high amounts of chemicals in many areas. Additional breeding objectives are stress tolerances, for example, drought or salt tolerance, or improved pest tolerance. Research programs have been implemented to arrive at an improved understanding of basic physiology and genomics.

International efforts to conserve agroecological diversity are required, along with wild gene pools, on which further genetic enhancement may depend. However, gene banks, without continuous propagation of races, do not suffice to conserve diversity. Presently, (for example, for rice) tens of thousands of land races are known in many areas worldwide. Additionally, some species (for example, some tree species) cannot yet be conserved ex situ.

6.1.4 Climate Change and Extreme Events

Climate change, climate variability, and extreme events will have significant impacts on global and regional food production, supply, and consumption. The impacts of climate change on natural and human systems have been classified according to the concepts of: *vulnerability*—the extent to which climate change may damage or harm a system; *sensitivity*—the degree to which a system will respond to a climate change; and *adaptability*—the degree to which practices, processes, or structures can be adapted to climate change.

A recent report by the International Institute for Applied Systems Analysis (Fischer et al. 2002b) commissioned by the United Nations for the World Summit on Sustainable Development, with a global coverage of all countries, integrates spatial agroecological potentials into a world economic and trade policy model framework. It evaluates the impact of climate change projections by some major climate models (General Circulation Models), as well as the socioeconomic development portrayed in the special report on emission scenarios of the IPCC Third Assessment. The results highlight that by the 2080s, the world's boreal and arctic ecosystems are likely to shrink by as much as 60% due to a northward shift of thermal regimes, and the unfavorable semiarid and arid land areas in developing countries may increase by up to 10%. Also, potential agricultural land will increase in North America (up to 30%) and the Russian Federation (up to 55%), but significant losses are projected for Africa, particularly in northern (up to 75%) and southern Africa (55%).

Analyses of the effects of climate change on food production potential show that there is actually a possibility of improvement in industrial countries. However, many developing countries, where food production is already insufficient, would suffer a further decline, resulting in aggravated malnutrition and famine problems.

While future food security depends mainly on political and socioeconomic conditions, climate change might affect the availability and distribution of food production and people's access to food. Mitigation measures are mainly reflected in the discussion about the Kyoto Protocol. However, it is crucial for adaptation measures to be elaborated, especially in developing countries, which are expected to suffer the most from climate change.

On the other hand, since land use change and agriculture are responsible for about a fifth of greenhouse gas emissions, agriculture could play an important role in mitigating climate change (FAO 2003c). Mitigation measures include mainly changes in cultivation patterns, reduction of fertilizer use, improvement of livestock diet and better manure management, alternatives to slash and burn land expansion, more efficient use of water resources, and promotion of carbon sequestration.

Extreme weather events—floods, droughts and cyclones—have emerged as the biggest threats to crop production, food availability and security in some developing countries. Here climate variability and weather fluctuations can create famine-like situations. Even in industrial countries farmers often go bankrupt because of crop losses due to extreme weather events.

6.2 Food Systems

The challenge of meeting safe and healthy food needs comprises the effective and sustainable functioning of the range of systems from production to consumption. The performance of agriculture over the last 100 years has been phenomenal. World population has increased almost four-fold. Today, global food production is sufficient to meet the world's food needs, and yet there is concern with regard to unhealthy food consumption, be it too little to too much. On the production side, the increasing degradation of land, water, and biological resources poses a major challenge of mobilizing agricultural science and technology. These need to manage natural resources and reduce social and economic vulnerability.

6.2.1 Food Consumption

Food is a basic human right and everyone should have access to nutritious, safe, and affordable food for a healthy and productive life. Although, at the global level, there has been significant progress in increasing average food consumption over the last 30 years, there are still some 840 million chronically undernourished people, mainly in the developing countries. At the same time, there is an emerging problem of overconsumption and changing lifestyles, resulting in obesity that is affecting more than 500 million people worldwide. The driving forces of changes in food consumption include demographic changes, urbanization, increasing levels of income, globalization and international trade, as well as rapidly changing consumer preferences.

The challenge in the developing world is to eradicate hunger, as some 15% of the total population is consuming 10–20% less food than the recommended minimum requirement. A priority focus on meeting food needs from domestic production is also important with regard to supporting domestic producers and reducing dependence on scarce foreign exchange. In the industrial and transition countries, the majority of the population already lives in urban environments, served by food supermarkets with a wide variety of domestic food products, as well as products from around the world. The number of people living alone has increased dramatically in developed countries and the traditional home family meal is declining. Health concerns are becoming predominant among rapidly aging populations. Food processors, distributors, and retailers are targeting and changing food consumption patterns, using the media and labeling, to market more and more processed and easy-to-cook foods. Campaigns that selectively highlight links between food and health, labor, and time saving often fail to give information about issues such as unhealthy levels of salt, sugar, and fat contents.

Cereals account for about half of total food energy consumption in the developing countries. Wheat consumption has increased the most, and many developing countries are meeting this demand through imports from the industrial countries, which continue to heavily subsidize producers. In the next 30 years, wheat imports by developing countries from temperate industrial countries are projected to increase some 2.5-fold to a total value of some $25 billion. Global food consumption of coarse grains continues to decline, but it is nutritionally of critical importance in many sub-Saharan countries, often accounting for over 70% of total calorie consumption. Demand for cereals in sub-Saharan Africa is expected to change with growing urbanization: demand is expected to increase for easy-to-prepare cereals like rice, and cereal products like bread, and expected to decrease for the local sorghum and millet, whose preparation is time-consuming (FAO 2003b).

Consumption of meat in developing countries more than doubled over the last two decades. Milk consumption in developing countries also increased considerably. It is more than triple the increase in the industrial countries. These trends, taken together, have been dubbed a "Livestock Revolution" (Delgado et al. 1999; Delgado et al. 2003). Although the massive increases in animal product consumption in the developing world are impressive, these countries still have a long way to go before they approach consumption levels of industrial countries. For instance, animal products comprised only 13% of calories consumed in the developing world in 2000, compared to 26% in developed countries (FAO 2000). Nonetheless, it is clear that diets are diversifying rapidly, with increasing consumption and production of animal products in developing regions. Poultry consumption will grow faster than other meats. Animal source foods have a positive impact on the quality and nutrient enhancement of the diet and can prevent or ameliorate many nutrient deficiencies. They are one of the few instruments for addressing the "hidden hunger" of nutrient deficiencies that exact a heavy toll on both societies and individuals in terms of mental ability late in life, energy levels, and susceptibility to diseases (Neumann et al. 2002). Alternatives to animal source foods for improving nutrition are few in the vast areas of poor developing countries, where distributing daily nutritional supplements is not feasible and green leafy vegetables are only available in local markets a few weeks a year.

Worldwide, more than 1 billion people rely on fish as an important source of animal protein. Fish provides at least 30% of their animal protein intake (FAO 2002a). Fish proteins are essential and critical in the diets of some densely populated countries where the total protein intake may be low. In countries such as Ghana, Indonesia, Sierra Leone, Bangladesh, Republic of Congo and Cambodia, it contributes to more than 50% of the total animal protein in the diet. In Japan, Iceland, and some small island states, fish is the major food source because of the lack of locally grown alternative protein foods. Besides, the people have developed and maintained a preference for fish. During the past decades, per capita fish consumption has expanded globally, along with economic growth and well-being. The consumption per capita of fish is larger in the United States and Europe than in Asia and Africa. In well-off industrial economies, the image of fish is changing. It is moving away from being a basic food and is becoming a culinary specialty. In industrial countries, economic growth has caused a growing proportion of fish to be consumed outside the home and in the form of ready-to-eat products. In volume terms, fish trade is still dominated by intermediate products, mostly in frozen form with a few standard categories of cured and canned products.

World food consumption has made substantial progress in terms of average daily per capita calorie consumption, rising from 2,400 calories to 2,800 in the last three decades. This has also been accompanied by major dietary changes comprising shifts toward increasing shares of meat, fat, and sugar, which in the developing countries rose from 20% of all food consumed to over 28% in the last four decades. The driving forces of these changes include transformation of lifestyles, traditions and culture, time pressures, demographic changes, economic growth, international trade and globalization of food markets—particularly through media targeting by the emerging transnational food companies. Major food safety incidents have also increased consumer concerns in recent years, leading to changes in consumer perceptions and food purchasing patterns.

Food consumption patterns vary from country to country and also within countries. Governments, civil society, particularly businesses, have to take responsibility to ensure that changes in food consumption patterns lead to good nutrition. Governments in industrial countries are already implementing standards for food quality, balanced nutritional content, labeling, etc., to reduce health risks. In contrast, many developing countries have few regulatory systems or public awareness campaigns to empower consumers to make the right food choices.

6.2.2 Food Production

Food production depends on natural resources (water, land, biodiversity), farm inputs (capital, power, chemical inputs, and seeds), and human inputs (labor, management skills, institutions).

Water: The greatest potential for increasing food production in developing countries lies in rain-fed agriculture. Low-cost technologies that allow judicious supplemental irrigation to bridge dry spells include treadle pumps. Trickle and seep-hose systems could substantially increase productivity of rain-fed agriculture. Irrigated agriculture has long been synonymous with high productivity. The 20% of the farmland that is irrigated produces 40% of the current food supply. In irrigated agriculture, the different types of irrigation used (surface or flood irrigation, sprinkler, drip, underground and sub-irrigation) have very important impacts on irrigation efficiency, the availability of water resources, and the state of ecosystems (for example, wetlands).

Land: The need to improve soil-water-plant nutrient management is addressed by Conservation Agriculture (Dixon et al. 2001). Conservation Agriculture contributes to environmental conservation with enhanced and sustained agricultural productivity. It ensures the recycling and restoration of soil nutrients and organic matter and optimal use of rainfall through retention and better use of biomass, moisture, and nutrients.

Biodiversity: The genetic resources for food and agriculture, which is the basis for world food security, have their base in biodiversity. Strong claims suggest that biodiversity promotes resilience and productivity of ecosystems. However, global biodiversity is changing at an alarming rate because of land conversion, inappropriate land use, climate change, pollution, unsustainable harvesting of natural resources, and introduction of exotic species (Pimm et al. 1995; Sala et al. 2000).

Inputs: The intensive use of *chemical inputs,* for example, fertilizers and pesticides, impacts ecosystems by changing the resource base and variety and level of services.

A number of trends and responses are evident in the case of *human inputs,* which include labor, management skills, and institutional arrangements. Production systems in many developing countries rely more on human labor than on mechanization. Hence, rural-to-urban migration of young adult males results in the burden of hard labor falling on women and the elderly (FAO 2001c). Investing in human capital in agriculture not only improves production efficiency, but also facilitates the adoption of new techniques, regulations, and practices that conserve and protect environmental resources. The recent trends of reducing agricultural extension services may not only slow agricultural productivity increases and provision of agricultural services, but may also negatively affect environmental protection and conservation of resources.

In the context of institutions and governance, a number of developing countries have undertaken reforms in support of agriculture including structural adjustment programs, poverty reduction strategies, fair commodity prices for products, increased levels of schooling, promoting gender equality, and reducing the scourge of HIV/AIDS. These institutional responses are important for increasing production, alleviating hunger and poverty, and achieving food security.

Detailed information and figures on status and trends is provided in MA *Current State and Trends,* Chapters 8 (Food), 26 (Cultivated Systems), and 18 (Marine Fisheries Systems).

6.2.2.1 Crops

Existing production systems include rain-fed, irrigated, wetland, and peri-urban farming systems. *Intensification* of existing production systems is a more realistic alternative for enhancing food production than undertaking further extensions (FAO 2003b, p.126). Intensification aims at increasing yields as a result of greater use of external inputs. Improved varieties and breeds, utilization of unused resources, improved labor productivity, irrigation, and better control of pests and diseases may also aid intensification. *Extensification* possibilities should be carefully considered, as much of the potentially available additional arable land is presently under tropical forests (in Africa and South America). The use of these lands for cultivation would be detrimental to biodiversity conservation, and increase greenhouse gas emissions causing regional climate and hydrological changes

The findings of MA *Current State and Trends,* Chapter 8, state that over the forty-year period, 1961–2001, the total output of crops expanded by some 235% globally. This indicates an average increase of just over 2% per year, always keeping ahead of global population growth rates. Output growth varied by region and over the period as a whole. Many middle-income and richer countries have seen a gradual slowing down in the growth of crop output in line with the deceleration of population growth and the attainment of generally satisfactory levels of food intake. Decelerating growth patterns in crop output have been most evident in developed countries and in Asia. Since food crop production has not grown as markedly, and population growth rates remain high, sub-Saharan Africa remains the only region in which per capita food production has not seen any sustained increase over the last three decades.

The cereal sector remains singularly important in several ways; in 2001, production of the principal cereal crops were: rice (381 million tons), maize (278 million), wheat (264 million), sorghum (44 million), millet (28 million), and barley (23 million). Cereals provide almost half of the calories consumed directly by humans globally (48% in 2001). Cereal production comprises about 58% of the world's harvested crop area, and an often disproportionately larger share of the usage of fertilizer, water, energy, and other agrochemical inputs. With regard to current trends, following a peak in foodstuff prices in 1996, a strong growth in crop output in 1999 was registered by both industrial and developing countries, but since then the general pattern of growth deceleration has resumed. In industrial countries, output actually declined in 2001 and 2002. In the case of cereals, global output levels have stagnated since 1996, while grain stocks have been on the decline.

During the last four decades, the best prospects of increasing food production were from raising yields on already cultivated lands, safeguarding against land degradation, and minimizing conversion of high-productive cultivated lands in the process of urbanization and economic transformation. Systematic research on productivity enhancement led to the Green Revolution, whose main components were increased use of high-yielding varieties of grain (primarily wheat and rice) and increased use of inputs such as fertilizers, energy, irrigation water, and pesticides. (See MA *Scenarios,* Chapter 7, for a more extensive discussion of the Green Revolution.)

Subsistence production is practiced by smallholders producing mainly for self-consumption with limited surplus production, and often constrained by lack of access to markets. Poverty is often severe among smallholder families. Their vulnerability is high since many cultivate poor soils and in areas prone to drought. Generally, low inputs (such as in finance, labor, seeds, and fertilizers) have led to low production, hunger, and thus poverty and low economic growth. Low productivity also negatively affects health and education, which in turn lowers productivity. Some traditional farming systems have improved yields and have been safeguarding the resource base by upgrading and diversifying cropping and adopting integrated pest management. For example, Indonesian rice farmers who adopted IPM, which reduces the need for pesticides, achieved higher yields than those who relied solely on pesticides (FAO 1996).

6.2.2.2 Livestock

Livestock and livestock products are estimated to make up over half of the total value of agricultural gross output in the industrialized countries, and about a third of the total in developing countries, but this latter share is rising rapidly (FAO 2003b). While growth rates in industrial countries have hovered at just over 1% for the past 30 years, growth rates in developing countries as a whole have been high and generally accelerating. As with many other global and developing-country trends, the situation in East Asia (and within that region, China) exerts a strong influence, where livestock product growth rates of over 7% per year have persisted for some 30 years, admittedly from a low base. As with crops, two regions draw attention: the transition economies and sub-Saharan Africa. The transition economies exhibit the same pattern of slow long-term shrinkage of output, followed by collapse in the early 1990s. Sub-Saharan Africa, faced with the world's highest stresses of poverty, malnutrition, and population growth, and continuing insecurity, particularly in pastoral areas within the sub-continent, has made slow progress and per capita output has hardly increased at all (Ehui et al. 2002). (See also MA *Current State and Trends,* Chapter 8.)

Technologies for sustainable animal agriculture are available for most of the world's livestock production systems. If applied, they will restore the balance between land and livestock and close nutrient cycles, thus reducing land degradation and nutrient loading of water resources. By restoring the balance between land and livestock, they will also address the social and health effects of the Livestock Revolution. However, these technologies will only be adopted if an appropriate policy framework is established.

Three broad types of production systems can be distinguished:

Industrial production systems: Industrial production of pork, poultry, and (feedlot) beef and mutton is the fastest growing form of animal production. In 1996, it provided more than half the global pork and poultry meat (broiler) production and 10% of the beef and mutton production. This represented 43% of total global meat production, up from 37% in 1991–93. Moreover, it provided more than two thirds of the global egg supply. Geographically, the industrial countries dominate intensive industrial pig and poultry production, accounting for 52% of global industrial pork production and 58% of poultry production.

Mixed farming systems: Mixed farming systems, the largest category of livestock systems in the world, cover about 2.5 billion hectares of land, of which 1.1 billion hectares is arable rain-fed cropland, 0.2 billion hectares is irrigated cropland, and 1.2 billion hectares are grassland. Mixed farming systems produce 92% of the world's milk supply, all buffalo meat and approximately 70% of the sheep and goat meat. About half of the meat and milk produced in this system is produced in the OECD countries, Eastern Europe, and the Commonwealth of Independent States, and half comes from the developing world. Over the last decade, meat production from mixed farming systems grew at a rate of about 2% per year and thus remained below global growth in demand.

Grazing systems: Grazing systems supply about 9% of the world's production of beef and about 30% of the world's production of sheep and goat meat. For an estimated 100 million people in arid areas, and probably a similar number in other zones, grazing livestock is the only possible source of livelihood. For the world's tropical rangelands, most attention has been on the arid lands, because of their perceived heavy degradation. However, recent findings stress the high prevailing level of productivity of meat and milk per unit area of land, the strong resilience of these arid rangelands, and the importance of traditional mobile grazing practices in maintaining the resource base. For the subhumid tropical savannas, human and livestock population pressure is lower. Finally, livestock-induced deforestation in the humid tropics has received much attention. Past driving forces behind the slashing and burning of tropical rain forest concerned export subsidies, subsidized interest rates for ranch establishment, and land tenure laws, which induced land speculation. More recently, the main driving force for deforestation has shifted toward smaller farmers, and food production for local consumption in forest margins as part of mixed farming systems.

6.2.2.3 Fisheries

Global fisheries landings peaked in the late 1980s and are now declining despite increasing effort and fishing power, with little evidence of this trend reversing under current practices. At the beginning of the twenty-first century, the biological capability of commercially exploited fish stocks was probably at an historical low. FAO (2003b) has reported that about half of the wild marine fish stocks, for which information is available, are fully exploited and offer no scope for increased catches. Of the rest, 25% are underexploited or moderately exploited. The remaining quarter is either overexploited or significantly depleted. Today, about 90% of wild fish come from the sea, the remainder from lakes and rivers. Of the fish caught at sea, probably about 10% (by volume) are caught in the high seas (that is, the areas outside the 200 nautical mile exclusive economic zone claimed by most countries bordering the sea). The vast majority of catches are obtained from waters on the continental shelf.

Although information on inland fisheries is less reliable than for marine capture fisheries, it appears that freshwater fish stocks are recovering somewhat from depletion in the Northern Hemisphere, while the large freshwater lakes in Africa are fully exploited, and in parts are overexploited. Some fish species exhibit more dramatic threshold effects, appearing less able to recover, than others.

Nine out of ten full-time fishers conduct low-intensive fishing (a few tons per fisher per year), often in species-rich tropical waters. Their counterparts in industrial countries probably number less than 1.5 million (FAO 1997) and generally produce several times that quantity per year, but they are not many and their numbers are falling as fishing is seen as a dangerous and uncomfortable way to earn an income. As a result, in some industrial countries, fishers from economies in transition or from developing countries are replacing local fishers.

During the past fifty years, aquaculture has become a globally significant source of food. By the end of the last century, it contributed roughly one third (by volume) of all fish consumed as food. The variety of supply from aquaculture is much below that of capture fisheries: only five different species of Asian carp comprise about 35% of world aquaculture production.

6.2.3 Impacts on Ecosystems

Crop production using the methods and inputs described above has major impacts on ecosystems, increasing their vulnerability. These impacts affect directly and indirectly, via ecological feedbacks, the resource base (land, water, biodiversity) through: direct use of resources as inputs, degradation due to agricultural pollution, effects on ecosystems' resilience (including processes and functions such as regeneration and self-cleaning capacities), or productivity. Crop production also has effects on human health and the health of other species.

For marine systems, the key factors that impact the ecosystem are salinity, ocean currents, and temperature changes. For inland waters, hydrological changes (for example, caused by dams, water abstraction), and water quality changes (including eutrophication, anoxia, water acidity, pollution, and toxic events) are the key factors.

Human impacts on the world's oceans, mainly through fisheries, have been substantial, leading to concerns about the extinction of marine taxa. For commercially exploited species, it is often argued that economic extinction of exploited populations will occur before biological extinction. However, this is not the case for non-target species caught in multispecies fisheries or for species with a high commercial value, especially, if this value increases as the species becomes rare. The perceived high potential for recovery, high variability, and low extinction vulnerability of fish populations have been invoked to avoid listing commercial species of fishes under international threat criteria. There is a need to learn more about recovery, which may be hampered by negative population growth at small population sizes or ecosystem shifts, as well as spatial dynamics and connectivity of subpopulations before the nature of responses to depletions is understood (Dulvy et al. 2003).

Livestock can have both positive and negative impacts on ecosystems around the globe. The positive impacts are mainly confined to smallholder farming systems where livestock provide a way to improve nutrient cycling and plant available nutrients. In pastoral systems, livestock may also provide unexpected benefits to wildlife where grazing pressure is light to moderate. Livestock production is also a main driver for massive transport of nutrients from developing to industrial countries, in the form of livestock feed. On balance, however, livestock impacts on ecosystem goods and services are largely negative, through impacts such as deforestation nutrient overloading, greenhouse gas emissions, nutrient depletion of grazing areas, dryland degradation from overgrazing, dust formation, and bush encroachment.

6.3 Responses: Selection and Analysis

The major responses associated with food provisioning services of ecosystems and their ecological feedbacks comprise a large variety of policy interventions and responses at the local, national, and international level. These address the complex and intertwined social, environmental, and economic issues. The responses are in a sense interventions induced by changes in drivers such as population and demography, economy and environment and natural resources, as well as science and technology.

In analyzing and assessing responses we seek to identify impacts on ecosystems, since all responses examined have impacts not only on food provisioning services, but also on supporting services through the ecological feedbacks. Impacts on ecosystems can result from unintended ecological feedbacks, from an intervention that was aimed at increasing food production (for example, the Green Revolution led to increased use of fertilizers, pesticides, and irrigation water); or they can be direct impacts aimed at correcting or preventing negative effects of existing responses (for example, the environmental component of the EU's Common Agricultural Policy).

In the context of food provisioning, we examine responses which can be associated with: (1) impacts on human well being; (2) the evolution of the economy and its institutions, including issues such as globalization, trade agreements, food related policies, and the design of agricultural policies; (3) knowledge and education related to food production and consumption; (4) technological change; and (5) impacts on the resource base (for example, water, fisheries).

The responses represent interrelated economic/financial, institutional, technological, social, and legal interventions covering the whole spectrum of the MA typology. By affecting food provision and food security, these responses have a direct impact on human well-being as well as the functionality and viability of ecosystems and ecosystem services. Actors initiating the response are mainly the state or international bodies, while the scale of operation ranges from local to global. Examples of the wide variety of responses impacting upon food supply and consumption are listed in Table 6.1.

6.3.1 Recognition of Gender Issues

Women play an essential role in achieving food and water security. (See Box 6.1.) While women play a critical role and have multiple responsibilities within the household and communities in securing healthy nutrition, their realties are often ignored at all levels of decision-making. Women farmers account for some 60–80% of food production in many developing countries. Women often spend more then nine hours a day fetching water and fuelwood and preparing food. They produce more than half the world's food and own 1% of the land. Response polices need to be gender sensitive and designed to empower the women by providing knowledge and ensuring access and control of resources toward achieving food security. This needs to be based on a systematic analysis of gender dynamics and explicit consideration of relationships between gender and food and water security.

6.3.2 Globalization, Trade, Domestic and International Policies on Food

Changes in world food production and international trade, as well as the conditions of access to world markets offer both obstacles and opportunities to developing countries. Different approaches to trade and production have opened huge gaps within the developing world in terms of productive capacity and international marketing (Stallings 1995). Some developing-country businesses have become foreign investors (for example, from the East Asian newly industrialized countries) while others cannot even sell in domestic markets without protection (Asia, South Africa, much of Latin America). As protection declines in the process of globalization, the situation of the latter countries becomes more precarious.

Regarding food and nutrition, in many cases, advocates of globalization favor export-oriented agriculture, often from large-scale operations, and modern food marketing methods including the use of packaged foods. On the other hand, excessive reliance on global markets entails dangers for poor countries (which are price takers and concentrate on a few exportable food commodities) when world markets become weak.

Another related issue is that increased global food production does not guarantee adequate access to food at either the household or the national level. The World Food Summit (in 1996) identified access to food—rather than the globally produced amount of food—as the key issue for food security.

Over the years, a complicated web of government policies has been developed around food production. Among the main goals of these policies are the support of domestic farmers' income, the support of domestic production, provision of research and development, security of food quality, and—more recently—protection of the environment. While some of these policies could have positive welfare effects, the traditional policies that subsidize agricultural production do have adverse effects. This section aims at tracing some of the effects of commonly used economic policies on food provisioning, human well-being, and the ecosystems. The most commonly used instruments of agricultural policy are listed in Table 6.2.

Figure 6.1 shows total support policies by type in the OECD countries for the period 2000–2002. The policies could have adverse economic, social, and environmental effects, that is, negative effects on sustainable development. On the economic side, they impose an extremely high cost. In 2002, this cost totaled $235 billion in OECD countries, of which $100 billion was accounted for by the European Union and $40 billion by the United States (OECD 2003b). They also distort market forces by diverting resources from their most productive utilization and lead to overproduction. Furthermore, they distort the terms of trade, reducing the profitability of agricultural production in developing countries. Finally, they promote overuse of certain inputs such as fertilizers and pesticides.

On the social side, they make farmers overly dependent on taxpayers for their livelihood, and they change wealth distribution and social composition by benefiting large corporate farms to the

Table 6.1. Examples of Policy Interventions and Responses Affecting Food Production, Distribution, Access, and Consumption. In *target* column, E = ecosystems; W = well-being. *Purpose:* A = access; D = distribution; P = production; R = resource base. *Type:* Ec = economic; F = financial; K = knowledge; L = legal; T = technological. *Level:* G = global; N = national; L = local.

Target	Objective	Purpose	Type	Intervention	Main Actor	Level	Impact on Well-being	Impact on Ecosystem
W	child nutrition	D	Ec	school feeding program		N	++	none
W	acute hunger	A	Ec	emergency food aid	WFP, bilateral	N	++	none
W	poverty, chronic hunger	A		"food stamps"	government	N, L	++	none
W	poverty	D	Ec	"food for work", etc.	government, NGOs	N	++	+/−
W	poverty	D	Ec	consumer price subsidy	government	N	+	none
W	health	P	Ec	food quality standards	WHO, government	G, N	+	none
W	health	P	Ec	labeling	government	N	+	+
E	resource management	D	Ec	integrated land use planning	government	N, L	+	+
E	biodiversity, ecosystem protection	P, R	Ec	protected areas	IUCN, government	G, N, L	+	++
E&W	poverty reduction and biodiversity conservation	D, P, A	Ec	direct payments for services provision by rich to poor	government, NGOs, private sector	G, N, L	+	+
E	land resources	D, R	T	land reclamation	government	N, L	+	−
E	land use	R	Ec, T	"grain for green"	government	N, L	+	+
W	land tenure	D	L	land ownership, access rights	government	N	+/−	+/−
W	land tenure	D	Ec	land, property tax	government	N, L	+/−	+/−
E	water supply	D	T	reservoir construction	government	N	+/−	−
W	water use	D	L	water rights	government	N, L	+/−	+/−
W	water use	A	Ec	water pricing	government, private sector	N, L	+/−	+
W	farm income	D	Ec	producer subsidy	government	N	+	+/−
W	revenue creation	P	F	producer tax	government	N, L	−	+/−

E	resource management	P	L	production quota	government	G, N	+	+
E	fishery management	P	L	total allowable catches	government	G, N	−/+	+
E	fishery management	P	Ec	tradable quota	government	G, N	−/+	+
E	fishery management	P	L	decommissioning schemes	government	G, N	−/+	+
W	agricultural production	P, D	F	credit	government, private sector, NGO	N, L	+	+/−
E	sustainable production	P, R	K	R&D, extension services	government, private sector, NGO	N	+	+/−
E	crop production	R	T	irrigation development	government, private sector	N	+	−
W	transport	A, D	T	infrastructure development	government, private sector	N	+	−
E	air, water, soil quality; health	P	Ec	agri-environmental regulation	government	N	+	+
E, W	water quality; health	P	Ec	input ceilings	government	N	+	+
E, W	water quality; health	P	Ec	environmental taxes, charges	government	N	+	+
W, E	farm income, supply, environment	A, P	Ec	set-aside programs	government	N	+	+
E	protection	P	L, Ec	trade quota	government	N	+/−	+
E	protection	P	Ec	tariffs	government	N	+/−	+
W	economic efficiency	D	Ec	trade agreements	government	G, N	+/−	+/−
E, W	watershed management	R	L	transboundary agreements	government	N	+	+

BOX 6.1

Case Study on Gender and Agriculture (www.fao.org/gender)

Often the most fundamental problem in policy and planning for the food and agriculture sector is to get those in decision-making positions to agree that there is a gender issue. Decision-makers either consider that "gender" is not a useful category for the purpose of economic policy and planning or refer to the lack of gender-disaggregated information and data as preventing the incorporation of gender in analytical work.

In developing countries, rural women are the main producers of staple crops like rice, wheat, and maize. These crops often provide up to 90% of the food intake of the rural poor. The contribution of the women in secondary crop production, such as legumes and vegetables, is even greater. Grown mainly in home gardens, these crops provide essential nutrients and are more often than not the only food available during the lean seasons or if the main harvest fails. Also, once the harvest is in, rural women provide most of the labor for post-harvest activities, taking responsibility for storage, handling, stocking, processing, and marketing.

In the livestock sector, women feed and milk the larger animals, while raising poultry and small animals such as sheep, goats, rabbits, and guinea pigs. In many countries, it is mostly the women who are engaged in inland fishing and aquaculture. They perform most of the work of feeding and harvesting fish, as well as processing and marketing the catch.

FAO studies demonstrate that while women in most developing countries are the mainstay of the agricultural sectors—the labor force for the farm and food systems—they have been the last to benefit from, or in some cases have even been negatively affected by, the prevailing economic growth and development processes and policies. Gender bias and gender blindness persist: farmers are still generally perceived as male by policy-makers, development planners, and agricultural service deliverers. As a result, women find it more difficult than men to gain access to valuable resources such as land, credit, agricultural inputs, technology, extension, training, and services. These are the very resources that could enhance their productive capacity.

Technology does not always benefit women. All too often, technology developed in response to the needs of commercial farmers—who are mostly men—actually works to the disadvantage of those who are already disadvantaged, especially women from poor or landless families. In Bangladesh, milling rice with a foot-operated mortar and pestle had traditionally provided the only source of income to many poor, landless women, particularly widows and divorcees. The introduction of mechanical hullers reduced the labor input from 270 hours per ton of rice to 5 hours per ton, thus freeing some 100,000 to 140,000 women (in relation to some 700 mills) for other lucrative work.

Agricultural extension programs ensure that information on new technologies, plant varieties, and cultural practices reach the farmers. However, in the developing world, extension and training services are primarily directed toward the men. Female farmers receive only 5% of all agricultural extension services worldwide, and only 15% of the world's extension agents are women. In Egypt, for example, women account for 53% of agricultural labor but only 1% of Egyptian extension officers are women. The resulting lack of information undermines women's productivity as well as their ability to safeguard the environment by using natural resources in a sustainable way.

Communication is a force for change. Information targeted at rural farmers can help them increase the quantity and improve the quality of the food they produce. Just as important is the information collected from them. Many development efforts fail women in particular, because planners have a poor understanding of the role women play in farming and household food security. They do not take the time to learn more about the activities and needs of the women from the women themselves.

Actions that can enfranchise and empower women in agriculture include:

- reform of inheritance and land tenure laws that limit ownership and use of land by women;
- mobilizing banks and credit institutions to lend to women even if they are constrained by lack of collateral of property and land;
- training of women agricultural extension agents and targeting extension services to women farmers;
- expanding and strengthening education programs directed toward girls and women;
- incorporating the needs and priorities of women in agricultural research and technology programs; and
- facilitating membership of women in agricultural cooperatives and farmer's organizations.

detriment of smaller family farms. While a primary aim of support polices is to support the income of farmers, in fact, only some 23% of the total expenses in price supports translates into additional income for farm households, as Figure 6.2 illustrates (OECD 2003c).

Table 6.2. Commonly Used Instruments of Agricultural Policy

Market price supports (minimum prices on selected products)	application of tariffs on imports
	purchasing predetermined quantities at minimum price
Payments to support agricultural income	based on farmers' output
	based on area planted/animal numbers
	based on historical entitlements
	based on input used
	based on input constraints
	based on overall farming income

In the context of policies affecting food provisioning and sustainable agriculture, research and development is of critical importance. Although in the past, public sector investment in agricultural research and development was significant, in recent years private sector investment is gradually increasing.

The problems of biodiversity loss and biosafety are related to the intensification of agriculture (including fishery, forestry, and animal husbandry), the increased role of the private sector in defining the research agenda, and the lack of regulatory mechanisms. Loss of biodiversity results in two ways. Directly, since industrial agriculture promotes the use of a selected small number of species on which all research is concentrated, and indirectly through the destruction of habitat and land conversion (FAO 2002b). It is well observed that, especially for the most commercial crops (such as rice, wheat, and peas), a small number of varieties account for a relatively large share of the total production, leading to a rapid decline in genetic diversity. The problem is intensified by the fact that the majority of subsidies and support systems are directed toward particular crops and livestock. Thus subsidization needs to be reduced and diversified, while investment in R&D need to be directed to support not only industrial agriculture, but also

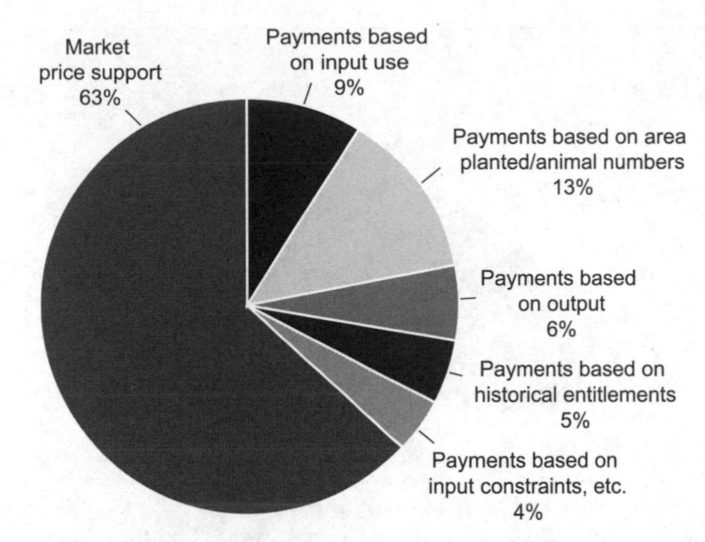

Figure 6.1. Composition of Agricultural Support Policies (OECD 2003)

alternative sustainable means of production that promote biodiversity.

The issue of biosafety in agriculture has become a very important one since it could affect human health and have long-term effects on sustainability of agriculture and food safety. Thus there is an urgent call for public investment in assessing and monitoring the possible effects of using genetically modified organisms in agriculture.

The WTO Agreement on Agriculture, which is under development, emphasizes the reduction of subsidy policies. Although trade liberalization could have positive effects, its overall impact on environment is ambiguous. Increased trade flows will affect the scale of agricultural activities and the structure of production in different countries, the mix of inputs and outputs, the production technology, and finally the regulatory framework. These adjustments, in turn, will impact on the international and domestic environment. International environmental effects include transboundary spillovers (such as greenhouse gas emissions), changes in international transport flows, and the potential introduction of nonnative species, pests, and diseases alongside agricultural products. Domestic environmental effects include ground and surface

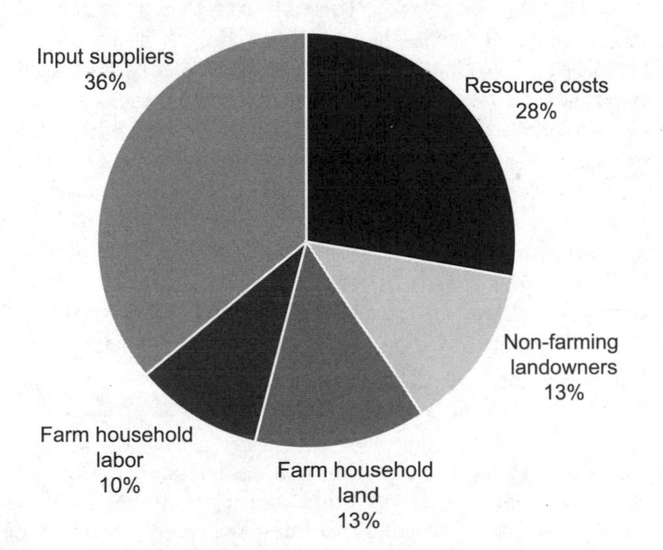

Figure 6.2. Allocation of Price Support (OECD 2003a)

water pollution from fertilizer and pesticide runoffs, and changes in land use that affect landscape appearance, flood protection, soil quality, and biodiversity (Walkenhorst 2000).

In general, there are traditional conflicts between free trade and environmental goals, and arguments are made supporting the view that the contribution of agriculture to environmental degradation could increase with trade liberalization. However, many studies find that the majority of benefits from trade liberalization in agricultural products will go to consumers in the industrial countries (FAO 2002b). Thus economic policies and institutions need to be developed in order to limit the adverse effects while enabling collection of the benefits from trade liberalization. Box 6.2 provides a detailed case study of distortions in the sugar market.

The Common Agricultural Policy is the most important and the most comprehensive sectoral policy ever developed in the European Union, and a forceful instrument of European integration. The CAP was developed with the aims of allowing free competition between farmers in member countries, eliminating as far as possible unequal treatment in different areas, and providing help in the modernization and development of European agriculture (European Commission 1997), although promotion of free competition among EU farmers, and especially between EU farmers and the rest of the world, has hardly been achieved. Box 6.3 discusses the CAP's evolution, including its incorporation of environmental goals.

6.3.3 Knowledge and Education

The food system of the world is going through a rapid and substantive transition. Knowledge and education are essential to achieve a sustainable food system, ensuring that farmers can efficiently produce food that is socially, economically, and environmentally sustainable and that consumers can make informed choices of food that is nutritious, safe, and affordable.

6.3.3.1 Sustainable Food Production Knowledge System

Historically, farmers produced food for their own needs and sold any surplus in the domestic market. This is still the norm among millions of poor farmers in the developing world. Local knowledge of resource conserving farming practices aimed at producing and harvesting different crop varieties, livestock, and fish to meet the needs of the farmers and the local markets have been at the core of traditional agricultural systems. These practices were in equilibrium with the environment.

The unprecedented increase in population and income growth during the last half-century led to increasing food demand, with changing consumption patterns. National and international agricultural research efforts responded by developing high-yielding crop varieties, intensifying livestock production systems and freshwater and marine fishing. The high-yielding varieties from the Green Revolution of the 1960s contributed to a doubling of world food production. However, over time environmental and social problems associated with high levels of inputs, monoculture systems, inefficient and polluting use of water, and the inability to reach many small farmers have come to the fore. Intensive livestock feeding systems have given rise to serious food safety and health concerns. Some marine fish stocks are already under threat of extinction due to overfishing.

The ongoing trade liberalization and globalization of food systems and lack of progress in WTO agricultural negotiations is contributing to widening disparities. Many producers in developing countries cannot compete against the large subsidy induced production and exports of many developed countries. At the na-

BOX 6.2

International Markets and Trade: The Case of Sugar Markets

The sugar market is one of the most heavily distorted agricultural markets. EU and U.S. support policies are primarily responsible for this distortion. The support policies not only fail to achieve their original intent of providing support to local small farmers, but also impose high costs on local consumers and taxpayers, and even higher costs on developing countries. The case of the sugar markets is a clear demonstration of the problems created by the agricultural support policies and illustrates in the most profound way the unfairness of the international trade system in its current state.

Description of the world market

Sugar is produced in more than 100 countries; global production in the year 2001 exceeded 130 million tons. More than 70% is produced from sugar cane, and the rest from sugar beet. The cost of producing sugar from beet is double that of producing it from cane. Brazil and India, both producing sugar from cane, are currently the leading producers followed, by the European Union of 15 countries. Figure A shows the main sugar producers in 2001; in that year, the top ten producing countries accounted for almost 70% of the total production. Production shares have little to do with differences in cost of production among countries, since they are strongly influenced by support policies.

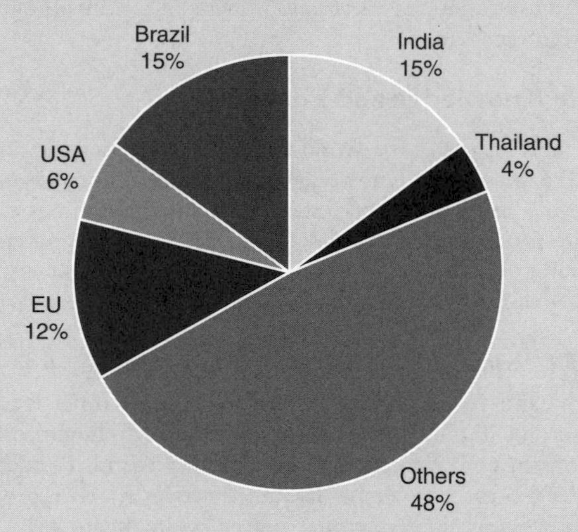

Figure A. World Production of Raw Sugar, 2001 (Statistical and Economic Information 2002, available at http://europa.eu.int/comm./agriculture/agrista/2002/table_en4321.pdf)

World market and support policies

Approximately 28% of the world's sugar is traded in world markets. The export market is very concentrated. The world's top five exporters (Brazil, the European Union, Australia, Thailand, and Cuba) supply approximately 72% of all world free market exports. The main exporter of raw sugar is Brazil, with 2% of world exports; followed by the EU-15 countries, with 15%, and Australia (10%), Thailand (9%), and Cuba (8%) (EU 2003c). While trade in raw sugar has been declining from the mid-1970s to the mid-1990s, the trade in refined sugar has steadily increased. The European Union is the main exporter of white sugar as Figure B illustrates.

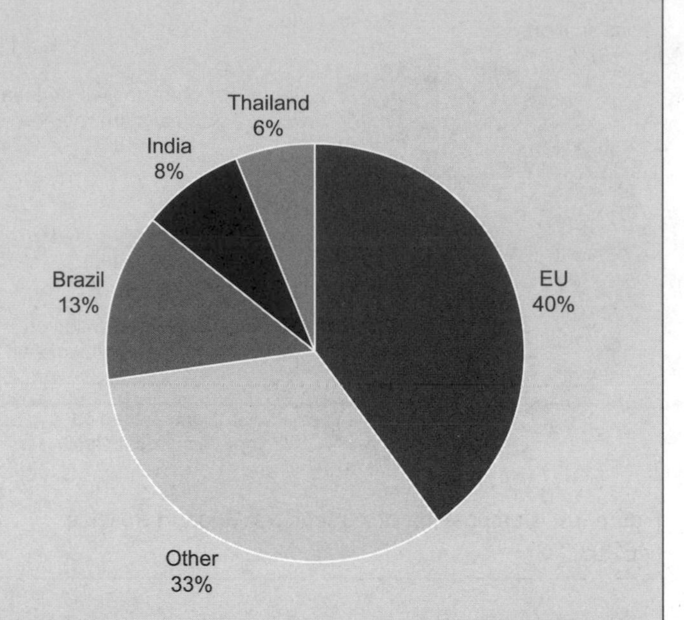

Figure B. World Exports of Refined Sugar, Share of Markets 2000–2001 (Oxfam Briefing Paper 27, available at http://www.oxfam.org.uk/what_we_do/issues/trade/bp27_sugar.htm)

Through the price support system (intervention prices, import duties, and export refunds) and its quota system, the European Union has managed to insulate its production from the world market, so that the prices received by EU producers for the quota production are two to three times higher than the world prices. Furthermore, through a system of production levies and export refunds, even production above the quota receives at least twice the world market price, adding to the pressure on the world market price. Tariffs for sugar imports are the highest in the European Union, reaching up to 140%, or 419 per ton (EU 2003a), effectively blocking all free market imports. EU imports raw sugar for processing at the high EU price on a preferential basis from a small number of developing countries. This web of policies has allowed the European Union to be a main player on the world sugar market, with a share of 12% of the production, 12% of the consumption, 15% of the exports, and 5% of the imports of the world (EU 2003a).

Although the United States remains a net importer of sugar, it has managed through a set of support policies to increase its production and as a result to restrict imports. The high support prices for internal production and the tariffs on sugar imports that reach up to 150% have restricted imports to just above 10% of the total demand in the United States. Currently, with world prices below 10 cents per pound on a raw basis, the EU support price is in excess of 30 cents per pound for raw sugar, while in the United States the minimum support price is 18 cents per pound for raw cane sugar and 22 cents per pound for refined beet sugar (Schmitz 2003). The support prices are expected to increase under the sugar program in the 2002 U.S. Farm Bill. This program contains price supports, payments in-kind, tariff rate quotas, and storage facility loan programs. The low U.S. imports add to the decrease in the world demand for sugar and thus to the decrease in the world price.

tional level, governments need to invest in facilitating participatory and transparent utility-oriented knowledge systems that empower farmers to adopt sustainable food production systems. Agricultural knowledge systems should give particular attention to integrating modern and traditional knowledge. The development in geographical information systems, including remote sensing, offers opportunities to build natural resources databases, critical for spatially relevant agricultural assessments. In many developing countries where agriculture is an important sector of the economy, there is an urgent need for investments in education to

Internal effects of support policies

The support policies in the United States, but most importantly in the European Union, receive strong criticism both domestically and internationally. Domestically, the main points of criticism concern distributional issues, as well as the extremely high cost to consumers and taxpayers. For example, the annual report of the European Court of Auditors mentions the ". . . high cost to consumers and overproduction in the EU . . . that continue to exist despite the . . . successive renewals of the common market for sugar" (European Court of Auditors, 2002, p. 71). While consumers and taxpayers bear the cost of the support policies, the benefits are reaped mainly by the highly concentrated sugar processing industry. In eight out of the fourteen sugar-producing countries, there is just one company controlling the quota. Despite the fact that sugar beet is one of the most profitable arable crops in the European Union, the quota system is such that it favors the larger sugar beet businesses. Small farmers receive a relatively small portion of the total benefits.

Environmental impact

Because of the support policies, sugar beet production is relatively intensive in Europe with negative effects on the environment. Sugar beet is commonly grown in rotation with other crops such as wheat, and it is generally found in the most productive arable regions of the European Union. Production is highly mechanized. It involves a particularly high level of herbicide use compared to other major temperate crop types, which reduces the presence of weeds, and probably other wild species (Baldock et al. 2002; DEFRA 2002). The high levels of nitrates potentially released from the leaves of the plant pose a risk for the pollution of groundwater and surface water (Baldock et al. 2002). Finally, the mechanized harvesting of the sugar beet has led to high levels of soil loss from the land and some areas where beet is grown are also vulnerable to erosion by wind (DEFRA 2002).

Impact on developing countries

The EU support policies have a positive effect on only the seventeen countries that enjoy preferential access to the EU market. Of those countries, four—Madagascar, Malawi, Tanzania, and Zambia—are least developed countries and their quota is only 4% of the total EU imports. In contrast, 80% of the benefits go to just five non-LDC countries—Mauritius, Fiji, Guyana, Swaziland and Jamaica. Other sugar producing LDC countries, such as Mozambique and Senegal, have no access to the EU market for raw sugar. At the same time, developing a sugar processing industry in these countries is not viable given the amount of EU exports. EU support policies also create problems for a number of low-income sugar producing countries such as Cuba and South Africa. The restricted access to the EU and U.S. markets and the reduction in the world price of sugar have had devastating effect on both the processing industries, as well as the farming business of these countries.

The support policies not only harm the economies of these countries as a whole, but they have a devastating effect at the community level since the farmers in most of these countries are small producers with no alternative source of income. Despite the fact that some of these countries, such as Mozambique and South Africa, have the lowest cost of production, sugar farming cannot guarantee a viable income to small farmers. These farmers cannot afford to harvest and transport sugar cane to the mills because of the extremely low world price and, as a result, are forced to give up farming and live in poverty. Sugar farming plays a less important role in the economies of developed countries than in those of developing countries. In the EU, employment through agriculture was just over 4% in 2000 (OECD 2001), while in the developing world an average of 50% of the people make their living from farming and agriculture (FAOSTAT database, August 2001). Allowing for a more open world market for sugar will be very important for the successful development of, and subsequently reducing poverty in, some LDC countries.

Potential changes

Apart from the cost to the sugar producing LDC countries, the support polices have negative effects on the economies of some of the low cost producing countries. A number of these countries, namely Australia, Brazil, and Thailand, have recently filed a request with the World Trade Organization to determine whether EU sugar production and export subsidies are legal under existing trade treaties. This move puts pressure on the European Union to change its support policies. Although, technically the European Union is within the bounds of the Uruguay round of agreements on agricultural products, it is quite obvious that its policies distort the international sugar market. This is unquestionable in the case of quota production, since the European Union exports the excess production at a price that is between one third and one half of the domestic guaranteed price. For example, in mid 2002, EU processors of white sugar were guaranteed a price of at least $620 while the world market price was just $180 (Oxfam 2002).

Similar arguments hold for the exports of white sugar, produced from the import of cane sugar from developing countries, as well as for the non-quota exports. Without the support policies, EU exports would most likely be eliminated because of the high costs relative to its main competitors. Not only would EU exports be eliminated, but the European Union would also cover most of its demand from imports. In such a case, although production would shift to more efficient producers, the world price would increase because the demand would be higher and the subsidized production would not exist. However, because the market would be thicker, the world price would be more stable (van der Linde et al. 2000).

Although a major relaxation in U.S. and EU support policies is needed, there is little evidence that it will be realized any time soon. The sugar sector is the only one that was not affected by the 1992 reform process of the EU Common Agricultural Policy. The CAP promoted competitiveness by compensating institutional price cuts with direct income payments (EU 2003b). The failure of the "Everything but Arms" initiative of the European Union and the new Farm Bill in the United States provide an indication that no major changes should be expected in the near future. Equally disappointing is the very slow movement in the current negotiations under the Doha Round for an Agreement on Agriculture. However, changes under the CAP reform proposals could have some positive effects, if price supports are replaced by direct income payments.

produce a cadre of agricultural researchers, as well as provide training of farmers and extension services personnel. Without this capacity, agricultural development to provide livelihoods for a substantial proportion of the populations in developing countries cannot succeed. And in turn, without a strong agricultural foundation, many developing countries cannot develop other sectors of the economy and services.

At the international level, the agricultural knowledge system must facilitate the exchange and sharing of national-level information and experiences. The international agricultural research sys-

BOX 6.3

The Common Agricultural Policy

The basic principles of the Common Agricultural Policy were set out in the Rome Treaty in 1957. The replacement of national agricultural policies with a common one was looked upon as a way of combining efforts to secure the supply of agricultural products to the consumer and provide a better standard of living to the agricultural community. In this sense, and following the historical evolution of the CAP in association with the enlargement and the integration processes in Europe, the CAP can be regarded as a response with a major impact on human well-being and poverty reduction in the EU countries.

Impacts on ecosystems

The CAP has contributed to a large degree to the modernization of agriculture in the European Union, but this modernization has been accompanied by damaging effects on the environment. In particular, the politically stimulated intensification of agricultural production has led to surpluses in certain products and to environmental degradation.

An example of the impact of agriculture on the environment is the change in the "Kempen" landscapes. These are high diversity enclosed areas found in Flanders (Belgium), southern and eastern Netherlands, North-Rhine-Westfalia (Germany) and Les Landes (France). They have a patchwork layout of woods, heath, swamps, mixed crops, scattered farmsteads and roads. Intensification of agriculture, use of fertilizers, manure disposal, and fragmentation of wild life habitats pressurized and increased the vulnerability of the ecosystems, by increasing the risks of soils dying out and also of groundwater pollution.

The increased attention regarding environmental conditions in the last decades resulted in attempts to introduce environmentally friendly policies in the CAP. These attempts marked the beginning of an ongoing process of integrating environmental concerns into agriculture and developing a unified agri-environmental policy framework.

Environment and the Common Agricultural Policy

The first attempts for environmental protection at the EU level started in 1972, since there was no mention of environmental policy in the Treaty of Rome. During the 1970s and the 1980s, the member states adopted more than 200 measures aimed mainly at reducing vehicle emissions, industrial and agricultural emissions, and effluents and noise. The environmental dimension was incorporated into the Single European Act, which came into effect in 1987, and provided the legal basis for environmental policy. It introduced three environmental policy objectives: (1) preserving, protecting and improving the quality of the environment, (2) protecting human health, and (3) prudent and rational utilization of natural resources. It also introduced four principles: precaution, prevention, rectification at the source, and polluter-pays.

It is important to contrast these objectives and principles with the corresponding objectives and principles of the CAP. The objectives of the CAP are: (1) increased agricultural productivity, (2) fair standard of living for the agricultural community, (3) stabilization of markets, (4) availability of supplies, and (5) reasonable consumer prices. The CAP principles are: market unity, financial solidarity, and community preference.

Since then, important statements issued by the Commission introduced the idea of controlling agriculture and protecting the environment. The Maastricht Treaty (1993) recognized environmental policy as a common policy, endorsed the sustainability principle, and set as an obligation the integration of environmental requirements in all EU policies. In the Fifth Environmental Action Program (1992–2000), agriculture is one of the five target sectors, and a fundamental objective is the achievement of sustainable agriculture, through the conservation of natural resources such as water, soil, and genetic resources. CAP reform in 1992, by encouraging farmers to use less intensive production methods, provided a way of reducing environmental pressures and unwanted surpluses. Furthermore, it included direct agri-environmental and afforestation methods. As for the structural aspects of CAP, environmental policy was recognized as a major component of the EU's rural development policy. In 1993, the assessment of environmental effects of activities was made compulsory.

The specific environmental measures associated with the CAP can be found in all three aspects of the CAP: the Common Market Organization, the accompanying measures, and the structural measures. Regarding the CMO, the set-aside program for cereals, oilseeds, and protein crops has direct environmental impacts. The set-aside program involves compensatory payments to farmers. It is beneficial to the environment since it reduces pressures from farm activities. On the other hand, to prevent negative environmental consequences, if land is left fallow, member states need to apply appropriate environmental measures. Under the set-aside and non-food production scheme, farmers are allowed to grow non-food

tem has a particular responsibility toward training and capacity building in developing countries. This is particularly important for new agricultural research and technology that not only requires a long time horizon, but is also highly capital and knowledge intensive.

6.3.3.2 Sustainable Food Consumption Knowledge System

The world's food consumption system is also going through a radical transformation. Consumers are increasingly separated from the food production systems. They need the knowledge and education to make informed food choices. This includes ethical, moral, and welfare, as well as economic and environmental, considerations.

Consumer concerns go well beyond basic human health. The quality of food and how it is produced; animal welfare; modern technology, and environmental, ethical, and cultural differences—all feature in the growing public debates about food quality and safety. Chemical and hygiene, as well as food security issues also cause concern.. We are faced with the problems of under- and overconsumption and a growing trend toward consumption of unhealthy processed foods. The emerging problems of overconsumption of the wrong kinds of foods is more and more driven by corporate food processing and marketing companies that use the media to change peoples' eating habits and taste. An example of this is the rapidly increasing incidence of obesity and diabetes due to consumption of high-sugar-content processed food combined with lifestyle changes with little exercise and physical activity. Government budgets increasingly have to deal with such health ailments and they have a responsibility to implement regulatory systems that ensure the availability of healthy and safe foods. The future food crisis may well be one of poor nutrition and related serious health issues.

The food consumption knowledge system needs to ensure that:

- scientific research findings on the implications of food technologies for human health and for the environment are presented clearly and underpin knowledge-based policy considerations, while recognizing that scientific evidence is often incomplete and equivocal;

products on set-aside land while still receiving the set-aside premium. These non-food products could have a positive impact on the environment, since they can be used as biomass or biofuel raw materials like fiber or ingredients for pharmaceutical products, thus reducing pressure on nonrenewable resources. Under the cropland set-aside and the long-term environmental set-aside, introduced by the accompanying agri-environmental measures, farmers could set aside land for twenty years in order to create biotopes or small natural parks.

The objectives of the agri-environmental measures, introduced with the accompanying measures, are to combine beneficial effects on the environment with a reduction in agricultural production, and to contribute to agricultural income diversification and rural development. In the context of these objectives, member states could provide aid for farmers who: (1) reduce the use of fertilizers, or introduce organic farming, (2) change to more extensive forms of crops, including forage production, (3) reduce the number of sheep and cattle per forage area, (4) follow environmentally friendly farming practices, (5) ensure the upkeep of abandoned farmlands, (6) set aside farmland for at least twenty years to establish biotope reserves and natural parks, or to protect hydrological systems, and (7) manage land for public access and leisure activities. Examples of such measures in action are the management of salt marshes in coastal lands in the United Kingdom, the program for the protection of flower species in Germany, the reduction in use of nitrogen fertilizers in Denmark, and the maintenance of grassland areas for extensive livestock farming in France.

Agricultural structural measures stress the environment as an essential part of sustainable rural development. These measures include horizontal measures, such as promotion of organic farming and better use of by-products and waste recycling. Specific regional measures promote objectives such as water management, soil conservation, combating erosion, biodiversity conservation, and landscape protection (see Leader initiative, European Commission 2003).

The new fundamental CAP reform adopted by the EU farm ministers on June 26, 2003, addresses these issues. Its key elements include:

- a single farm payment for EU farmers, independent of production; limited coupled elements may be maintained to avoid abandonment of production;
- the linkage of this payment to respect for the environment, food safety, animal and plant health and animal welfare standards, as well as the requirement to keep all farmland in good agricultural and environmental condition ("cross-compliance");
- a strengthened rural development policy with more EU money, new measures to promote environmental quality and animal welfare, and assistance to farmers in meeting EU production standards;
- a reduction in direct payments ("modulation") for bigger farms to finance the new rural development policy;
- a mechanism for financial discipline to ensure that the farm budget fixed until 2013 is not overshot; and
- revisions to the market policy of the CAP including asymmetric price cuts in the milk sector; reduction of the monthly increments in the cereals sector by half, with the current intervention price being maintained; and reforms in the rice, durum wheat, nuts, starch potatoes, and dried fodder sectors.

Conclusion

The CAP constitutes a major response by the European Union aimed at securing food supply on the one hand, and enhancing the well-being of the rural communities on the other. The intensification of agriculture that followed the introduction of the CAP undoubtedly created environmental pressures and degradation of European ecosystems. In response, the EU introduced policies that affect the environment both indirectly, through the land set-aside programs, and directly through agri-environmental measures, structural measures, environmental policy related to agriculture, and nature and resource conservation measures. It is expected that citizens could thus enjoy a higher provision of environmental services and a greater variety of products obtained through environmentally friendly practices.

Overall, the introduction of the environmental dimension into the CAP can be regarded as a direct recognition of the fact that the provision of food by ecosystems within a policy framework that attempts to protect both consumers and producers might create undesirable environmental pressures. This policy framework design needs, therefore, to include appropriate environmental policy objectives and measures for achieving them.

- food regulations including labeling are consistent with scientifically defined risks to health and the environment. The similarities and differences in regulation across countries need to be analyzed in relation to rigorously defined and agreed standards; and
- governments, the scientific community, the private sector, and civil society are transparent in presenting information on food risks and in putting in place measures to address these risks.

The information and communication revolution has a significant role to play in evolving national and international agricultural knowledge systems. While a third to a half of the population in the developed world has access to the Internet, in Asia and Africa this proportion is 0.5% of the population. The Internet provides a worldwide network for sharing of agricultural knowledge systems and it is essential that the wide digital divide be given priority attention.

It is not just a question of knowledge generation and transfer, but also an interaction of knowledge networks involving multiple stakeholders, namely the farmers, buyers, transporters, processors, distributors, retailers, and consumers. They all need to be involved in the development of agricultural knowledge systems.

In the knowledge economy, agricultural research and technology are as much social and economic activities as they are technical. Openly communicating with the broad public on an ongoing basis about the flows of new knowledge, its utility and potential socioeconomic implications, is essential. Given that future outcomes of new knowledge cannot be fully anticipated in advance, the agricultural knowledge system must be transparent and responsive and must foster trust.

6.3.3.3 Integrating Ecological and Socioeconomic Responses

The scientific and development policy community at the national and international level must work expeditiously toward the goal of achieving health-enhancing food systems that are socially, economically, and environmentally viable and sustainable. This will require multidisciplinary analytical capacity building, focusing on a systemic combination of relevant sciences, including biological and biochemical, agroecological and environmental, social and economic, as well as informatics. (See Box 6.4.)

BOX 6.4

Integrated Assessment: Agroecology, Economy, and Climate Change

The Food and Agricultural Organization of the United Nations and the International Institute for Applied Systems Analysis have over the last two decades developed integrated ecological and economic analytical tools and global databases. The focus has been on multidisciplinary scientific research, analyzing the current and future availability and use of regional and global land and water resources, in the face of local, national, and super-national demographic, socioeconomic, international trade and globalization, technological, and environmental changes, including climate change and climate variability.

AEZ/BLS (Agroecological Zones/Basic Linked System) combines a spatially explicit biophysical model of potential productivity of global land resources with a 34-region, 10-sector general equilibrium model. The spatial component allows a more detailed and realistic accounting for available land, its potential productivity, and the effect of future climate change on productivity.

The AEZ methodology for land productivity assessments follows an environmental approach; it provides a framework for establishing a spatial inventory and database of land resources and crop production potentials. This land-resources inventory is used to assess, for specified management conditions and levels of inputs, the suitability of crops/and utilization types in relation to both rain-fed and irrigated conditions. It also quantifies expected production of cropping activities relevant in the specific agroecological context. The characterization of land resources includes components of climate, soils, landform, and present land cover. Crop modeling and environmental matching procedures are used to identify crop-specific environmental limitations, under various levels of inputs and management.

Results of the AEZ/BLS integrated ecological–economic analysis of climate change on the world food system includes quantification of scale and location of hunger, international agricultural trade, prices, production, land use, etc. The analysis assesses trends in food production, trade, and consumption, and the impact on poverty and hunger of alternative development pathways and varying levels of climate change.

Following accession to the World Trade Organization, China is facing the challenge of defining transition strategies that maintain a socially sustainable level of rural incomes and employment, meet the needs of rapidly growing urban populations, are environmentally sustainable, and meet international commitments. A detailed case study of China (the CHINAGRO project) takes into account two prominent trends: China's increasing international trade relations as a result of its accession to the WTO and the change in dietary patterns due to rapid per capita income increases and fast urbanization. The project analyses the impacts of these trends on the agricultural sector and on the livelihoods of the rural population depending on agriculture.

Specifically, one of the issues under study is whether, in light of the fast-rising demand for animal proteins by Chinese consumers, and the sustained rural to urban migration, the country needs to aim at (1) self-sufficiency in cereals, protein feeds, and meat, including animal feed; or (2) importing feed; or (3) importing meat. A second issue under investigation is, not surprisingly, how the WTO accession, the Doha Round, and more generally China's opening to world trade will affect the agricultural economy of the country, and what feedbacks to the world market and hence consumers and producers in other regions can be expected. A third issue is to assess the implications of major ongoing infrastructural projects, in particular those aiming at redirecting water flows.

Agroecology has emerged as the integrated discipline that provides a holistic approach to manage agroecosystems and the sustainable use of natural resources. It provides guidelines to develop diversified agroecosystems—systems that take advantage of the effects of the integration of plant and animal biodiversity enhancing complex interactions and synergisms and optimizes ecosystem functions and processes, such as biotic regulation of harmful organisms, nutrient recycling, and biomass production and accumulation. Agroecology is of particular relevance to small farmers, emphasizing a development methodology that encourages participation, use of traditional knowledge, and adaptation of farm enterprises that fit local needs and socioeconomic and biophysical conditions (Altieri 1996).

6.3.4 Technological Responses

The agricultural science and research challenge is to combine the best of conventional breeding with safe and ethical molecular and cellular genetics research and biochemistry, to develop nutritionally enhanced and productive germplasm. (See Box 6.5.) The specific food crops of the poor, including coarse grains, roots and tubers, and plantains and bananas should be given the highest priority. Considerable scope exists for environmentally sound fish farming and intensive livestock production, with due consideration to health hazards and animal welfare consideration.

Technological responses in agriculture have had a massive, often regional, impact on the environment, human health, and development in the past. Recent technological responses will globally affect human health, food security, food safety, environment and environmental health, and socioeconomic and ethical issues. Food production systems have rapidly developed into globalized trading systems. Technological responses are considered inevitable for future food security and adaptation to local agroecological, socioeconomic, or ethical needs. Risk assessment, risk management, and risk communications are central elements in developments of the food production system. While risk assessment is based on science, scientific evidence and analysis cannot always provide immediate answers to questions posed.

6.3.4.1 Crop Breeding Strategies

Crop breeding strategies are highly dependent upon preservation of diversity of crops and wild relatives. There is growing scientific and public concern about a rapid decline of diversity, for example, of land races. There are two major alternatives for the conservation of genetic resources: in situ and ex situ.

In situ conservation refers to the conservation of important genetic resources in wild populations and land races, and it is often associated with traditional subsistence agriculture. It is concerned with maintaining the population of various species in the natural habitats where they occur, whether as uncultivated plant communities or in the fields of the farmers as part of existing agroecosystems. In situ conservation of crop plants involves the conservation on-farm of local crop cultivars (or landraces) with the active participation of farmers.

If the focus is only on agricultural varieties, the approach is only partially effective because traditional crop varieties, though much more diverse than elite varieties, are themselves much less diverse than wild populations and wild relatives. An attractive approach is to combine nature reserves focused on protection of

BOX 6.5

International Agricultural Research

The fundamental considerations include identification of the priority crops of relevance to the poor, what combination of public and private sector science can be most efficient and effective, and also where research effort can best be located.

Developed countries traditionally provided generous support to national agricultural research in that they created a wide and strong scientific capacity and produced products and innovations that were freely available. The intensified development of private sector research, accompanied by intellectual property protection and the trends in reduced public funding, calls for new research partnerships.

Developing countries such as China, India, Brazil, etc., have the scientific capacity and resources to research at the forefront of agricultural science. However, for most developing countries the scientific research capacity is severely limited and establishing national agricultural research systems (NARS) is generally a non-viable proposition. The poorest developing countries have less then one scientist and engineer per 10,000 people, in comparison to about 70 in the United States and Japan.

A decentralized global agricultural research partnership would be an effective and efficient way forward to produce the necessary research innovations of relevance to food-insecure countries. This could comprise a network of regional agricultural research centers linked to advanced NARS institutions in developing and developed countries, as well as, the international agricultural research system, namely the Consultative Group on International Agricultural Research. The multilateral institutional cosponsors of the CGIAR—the World Bank, the Food and Agriculture Organization of the United Nations, the United Nations Development Programme, and the International Fund for Agricultural Development—and bilateral donors must give serious consideration to creating regional agricultural research centers of excellence. This would also be of benefit to these aid partners in terms of complementing their own individual development efforts to reduce hunger and poverty.

The CGIAR has a particularly important role to play in this. It started out some 30 years ago with a strategic core research focus on productivity enhancement. However, over time there has been a shift toward programs of natural resource management, strengthening NARS, and policy research. At present, productivity-increasing research accounts for only about a third of CGIAR's annual expenditure of some $370 million. This level of funding is disparately miniscule in comparison, for example, to some $350 billion spent annually on agricultural subsidies by the developed countries.

In the developed countries, agricultural research funding amounts to some $14 billion, comprising half in the public sector and half in the private sector. For the developing countries, the total expenditure is about $8 billion. CGIAR funding is a minute share of these levels of expenditure and yet this international public good research system has demonstrated its major impact during the last 30 years and its beneficiaries have been farmers not only in developing countries, but also in the developed countries.

---

wild races and wild relatives with traditional agricultural practices. However, we should not expect traditional farmers to forgo the substantial economic benefits that may attend the switch to elite varieties. Hence, this may require direct economic subsidy or conservation of traditional varieties in some other way.

Ex situ conservation refers to the conservation of genetic resources off-site in gene banks, often in long-term storage as seed. A key international agreement, which comes up for renewal every four years, governing many of the world's most important crop diversity collections was recently renewed for an additional four years by 165 countries, ensuring that this diversity, which is critical for crop improvement, will remain in the public domain for the foreseeable future. Today, the Future Harvest Centers of the CGIAR conserve more than 500,000 samples of seeds and other plant parts in storage facilities called gene banks. The Centers do not own the material in the collections, but serve as trustees or custodians for them on behalf of the world community.

However, seeds of many important tropical species are recalcitrant, that is, difficult or impossible to store for long periods. Many crop plants are clonally propagated. Storing seed does no good, and tissue culture techniques for long-term storage are poorly developed. New technologies require to be explored to improve the possibilities for ex situ protection of diversity and in situ conservation policy and methods are more critical for such species.

6.3.4.2 Precision Agriculture

Precision agriculture or site-specific management refers to the differential application of inputs to cropping systems or tillage operations across a management unit (field). Input applications may vary either spatially or temporally within management units. The methods involved include application via predefined maps based on soil or crop condition or sensors that control application as machinery traverses the field. As monitoring systems, such as global positioning systems, allow monitoring by square meters instead of square kilometers, traditionally spatial changes across the field allows precocious control of chemical, fertilizer application, irrigation, or pest management.

Positive impacts on environmental quality will start to emerge as tools become available to apply chemicals, fertilizers, tillage, and seed differentially to a field, as well as tools to collect the yield or plant biomass by position across the field. Remote sensing technology will allow observation of the variation within a field throughout the growing season relative to the imposed management changes. Monitoring equipment to capture the surface water and groundwater samples needed to quantify the environmental impact through surface runoff or leaching is available. Technology to capture the volatilization of nitrogen or pesticides from the field into the atmosphere from modified practices exists (Hatfield, 1991).

Until a few years ago, precision agriculture was thought to have potential only in areas where technical facilities, as well as, field structure were compatible. Farmers in sub-Saharan Africa, however, have been practicing precision farming for centuries (Brouwer and Bouma 1997). A better knowledge of field level ecological variability, gained in part by making use of modern statistical techniques, can help farmers and researchers increase nutrient user efficiency through improved precision agriculture, also in low-input, low technology situations (Brouwer and Powell 1998; Voortman and Brouwer 2003; Voortman et al. 2004). It would also allow farmers to maintain, at least partially, the spatial variability that can contribute to risk-reduction in their production systems (Brouwer et al. 1993). Furthermore, adding science to traditional knowledge and practices is relevant and important to optimizing systems for production and to sustain them.

6.3.4.3 Genetically Modified Organisms in Agriculture

Modern methods of biotechnology include genetic modification to enable the development of crops, animals, or bacteria that exhibit traits which could not be introduced with classical breeding methods.

Only products derived from a limited number of genetically modified organisms (such as cotton, maize, oil seed rape, papaya, potato, rice, soybean, squash, sugar beet, tomato, wheat, and carnations) have been approved as yet in some countries. From these products, only a few products such as herbicide- and insect-resistant maize (BT maize), soybeans and oil seed rape—are on the international markets at present. During the six-year period 1996–2001, the major trait incorporated into genetically modified organisms was herbicide tolerance, with insect resistance being second. Major new developments include:

- in the near future, most market introductions of new transgenic crops will concern agronomic traits, especially herbicide resistance and insect resistance;
- altered nutrition and composition (for example, Vitamin A or iron deficiency);
- genetic modification of plants used to produce vaccines for human and animal illnesses;
- salt tolerant and drought resistant crops;
- transgenic crops in which the introduced trait is active in only one generation, so-called "Genetic Use Restriction Technologies";
- the first transgenic animal for food purposes that is likely to be licensed is fast- growing Atlantic salmon. Other fish in which genes for growth hormones have been introduced experimentally include carp, trout, tilapia, and wolf-fish. It should be emphasized that licensing of such transgenic species will require assessment of impacts and risks to the aquatic environment, because of their connectivity and the relative competitiveness of strains;
- most efforts in creating transgenic arthropods, such as insects for food-related uses, are in the area of pest control (for example, transgenic, sterile male plague insects have been produced experimentally); and
- soil bacteria-promoting crop development.

6.3.4.3.1 The scientific gene modification debate and concerns about biodiversity

While genetically modified organisms, generated following the purposeful introduction of exogenous DNA into plants or animals, should in theory present no more risks than plants or animals improved through selective breeding approaches, genetically modified organisms have a better predictability of gene expression than conventional breeding methods. Moreover, the transgenes are not conceptually different than the use of native genes or organisms modified by conventional technologies. Today, extensive areas have been planted with genetically modified crops in the United States and China, without untoward effects.

Biotechnology may help achieve the productivity gains needed to feed a growing global population; impart resistance to insect pests, diseases, and abiotic stress factors; and improve the nutritional value and enhance the durability of products during harvesting or shipping. New crop varieties and biocontrol agents may reduce reliance on pesticides, thereby reducing the crop protection costs of farmers and benefiting both the environment and public health. Research on genetic modification to achieve appropriate weed control can increase farm incomes and reduce the time women farmers spend on weeding and thus allow more time for childcare.

Biotechnology would also offer cost-effective solutions to micronutrient malnutrition, such as vitamin A and iron. Research in biotechnology on increasing the efficiency of utilizing farm input could lead to the development of crops that use water more efficiently and extract phosphate from the soil more effectively. The development of cereal plants capable of capturing nitrogen from the air could contribute greatly to plant nutrition, helping poor farmers, who often cannot afford fertilizers. By increasing crop productivity, agricultural biotechnology could help reduce the need to cultivate new lands and conserve biodiversity. If the appropriate policies are put into place, productivity gains could have the same poverty-reducing impact as the Green Revolution,.

The debate over genetic modification has highlighted the potential impacts on human health, environment, agrobiodiversity, and economic aspects, stemming mainly from impacts from random DNA integration into the genome of recipient organisms.

Human health: Impact on human health occurs through the formation of new products with allergenic or other effects. Although no direct risks for human health have been observed with genetically modified foods, the concerns have resulted in the establishment of risk assessment measures, recently established in CODEX guidelines on genetically modified foods (Haslberger 2003).

Environment: Impacts on the environment occur in several ways—by direct competitive effects through faster growing plants, animals or fish compared with wild species; by indirect effects (such as insect- resistant genes incorporated into plants reducing the activity or health of natural insect pollinators); by effects on wild relatives through the transfer of transgenes to wild species causing some change in function (such as inducing herbicide resistance in weedy species). There is however still controversy among scientists, as different outcomes have been reported on issues such as insecticide/pesticide use, yield increases, and environmental benefits (for example, Obrycki et al. 2001; Hilbeck 2001; Dewar et al. 2003). Different local agroecological conditions may contribute to different outcomes in the use of such crops; thus, their deployment may require careful case-by-case consideration. Risks to the environment of the transboundary movement of genetically modified organisms are being dealt with under the Cartagena Protocol of Biosafety.

Agrobiodiversity: For crops, the process of seed trading and transport can contribute to a potential spread of transgenes. Outcrossing of recombinant DNA could result in a significant transfer of recombinant DNA to wild or weedy plants, especially, in centers of origin of crops or in areas of high species diversity of plants related to the crop plant. Genetically engineered insects, shellfish, fish, and other animals that can easily escape, are highly mobile and form feral populations easily. They are of concern, especially, if they are more successful at reproduction than their natural counterparts. For example, it is possible that transgenic salmon with genes engineered to accelerate growth released into the natural environment could compete more successfully for food and mates than wild salmon, thus endangering wild populations. Thus particular guidelines, sterile release strategies, and other controls will be necessary to fully exploit the potential production advantages.

Economics: The mixing of genetically modified and unaltered crop products make the produce impossible to sell in markets unwilling to buy such products. This is currently the case as customers in some regions differentiate GM foods as unethical or unsafe, compared with agricultural products derived from plants and animals improved through conventional breeding programs.

The biggest risk of modern biotechnology for developing countries is that technological development may bypass poor farmers because of a lack of enlightened adaptation. It is not that biotechnology is irrelevant, but research needs to focus on the problems of small farmers in developing countries. Private sector research is unlikely to take on such a focus, given the lack of future profits. Without a stronger public sector role, a form of scientific apartheid may develop, in which cutting edge science becomes oriented exclusively toward industrial countries and large-scale farming.

The focus of biosafety regulations needs to be on safety, quality, and efficacy. The need and extent of safety evaluation may be based on the comparison of the new food and the analogous food, if any. In relation to environment, one has to look at the interaction of the transgenes with the environment. The potential of recombinant technologies allows a greater modification than is possible with conventional technologies. In most of the developing countries, there is no system in place to regulate the production and use of genetically modified organisms. The management, interpretation, and utilization of information will be an important component of risk assessment, and determine the effectiveness and reliability of this technology.

While modern biotechnology offers promise to increase productivity and protect natural resources and ecosystems, the risks of such events occurring need to be evaluated in a scientific manner. Strategies (such as the production of self-limiting populations of genetically modified organisms) need to be developed in the first instance, to limit the spread/escape of new materials. Among the measures needed:

- evaluations of risks need to be pre-planned and evidence-based;
- communication strategies for the results of such trials need to be developed for policy makers and for the general public and implemented; and
- future policy needs to be formulated on the basis of evidence and cost-benefit analyses, which include levels of estimated risk.

6.3.4.3.2 Analysis and assessment

Precision agriculture and integrated agricultural systems are generally believed to have the potential for supportive effects on sustainability, according to their use in specific agroeconomic conditions (for example, farm scales). Modern biotechnology is purported, from a technical perspective, to have a number of products for addressing certain food security problems of developing countries (Conway 1999; Skerrit 2000). The availability of such products could have not only an important role in reducing hunger and increasing food security, but also the potential to address developing world health problems. However, some governments believe the risks (safety, environmental, and/or economic) associated with modern biotechnology far outweigh the benefits.

Modern methods of biotechnology, as well as molecular methods for the preservation of germplasm diversity are generally accepted as important tools for improved sustainability in agriculture (Shah and Strong 2000).

The use of GM organisms in food production has developed into a significant part of agriculture in some countries. Scientific proof of advantages such as pest reduction can be shown for some crops in some areas, but many scientific uncertainties about advantages or risks (for example, out-crossing) are still evident. Present experiences suggest that it may not be possible to assess advantages or risks of genetically modified organisms in food production in general, but rather that they must be addressed case by case for specific agroecological or even socioeconomic conditions. Improved regulations, which allow a regional differentiated use of certain products in addition to a globalized trading system, may be desirable.

6.3.4.4 Sustainable Food Production Systems and Organic Farming

Improving the sustainability of complex food production systems requires a thorough understanding of the relationships between food consumption behaviors, processing and distribution activities, and agricultural production practices, as well as, a good understanding of the links between societal needs, the natural and economic processes involved in meeting these needs, and the associated environmental consequences. The ultimate goal is to guide the development of system-based solutions. Indicators covering the life cycle stages include origin of (genetic) resource; agricultural growing and production; food processing, packaging and distribution; preparation and consumption; and end of life.

Current trends in a number of indicators threaten the long-term economic, social, and environmental sustainability of the food system. Key trends include: (1) rates of agricultural land conversion, (2) income and profitability from farming, (3) degree of food industry consolidation, (4) fraction of edible food wasted, (5) diet-related health costs, (6) legal status of farm workers, (7) age distribution of farmers, (8) genetic diversity, (9) rate of soil loss and groundwater withdrawal, and (10) fossil fuel use intensity. Effective opportunities to enhance the sustainability of the food system exist in changing consumption behavior, which will have compounding benefits across agricultural production, distribution, and food disposition stages (Heller and Keoleian 2003), as well as alternative agricultural practices. One way of doing this is by enhanced breeding methods enabling improved traits for specific socioecological situations. Another is integrated organic farming.

6.3.4.4.1 Principles of organic farming and standardization

Organic farming management relies on developing biological diversity in the field to disrupt the habitat for pest organisms, and the purposeful maintenance and replenishment of soil fertility. Organic farmers are not allowed to use synthetic pesticides or fertilizers. Organic farming represents an alternative and more holistic view of agriculture and food production, and directly addresses the problems faced in many areas of conventional agricultural practice. Concerns about the environment and nature, livestock welfare, and food quality are thus essential elements of the philosophy behind organic farming.

The special values and principles of organic farming stem from the recognition that human society is an integrated part of nature, and that—due to the complexity of the socioecological systems—we have incomplete knowledge of the far reaching and future consequences of our actions. Based on these fundamental assumptions, three general principles—the cyclical principle, the precautionary principle, and the nearness principle of action and development—can be set out.

There is a special conception of sustainability in organic farming, which has been termed "functional integrity." Functional integrity corresponds to a systemic view, seeing agriculture as a complex system of production practices, social values, and ecological relations. The functional integrity of the system depends on the use of cyclical processes and the reproduction of crucial elements, such as soil fertility, crops, livestock, nature, and human institutions. As a principle of action, this is sometimes expressed in terms of the development of system's harmony with nature.

There is also a special conception of risk decisions and prevention in organic farming, which can be characterized in the form of the precautionary principle, which involves a self-reflective awareness of the limits of knowledge and control, and strategies for handling ignorance and uncertainty. The principle is implemented by acting before conclusive scientific understanding is available, and involves early detection of dangers through comprehensive research, and promotion of cleaner technologies.

6.3.4.4.2 Evidence for enhanced sustainability in organic farming

In a 21-year study of agronomic and ecological performance of biodynamic, bioorganic, and conventional farming systems in Central Europe, crop yields were found to be 20% lower in the organic systems, although input of fertilizer and energy was reduced by 34–53% and pesticide input by 97 %. However, at the global level, the scope for and adoption of organic agriculture is likely to be very limited since, for example, a 20% decline in crop yields would have serious consequences on food supplies. Enhanced soil fertility and higher biodiversity found in organic plots may render these systems less dependent on external inputs (Maeder et al. 2002). A long-term project (1992–1997) in the United Kingdom comparing conventional and integrated arable farming systems found that, in terms of total energy used, the integrated system appears to be the most efficient. However, in terms of energy efficiency, energy use per kilogram of output, the results were less conclusive (Bailey et al. 2003).

Results comparing data from organic or conventional farming are mostly a matter of intensive debate because of various specific or local aspects; while organic farming certainly shares many risks with conventional methods (for example, mycotoxin residues), the increase of organic farming has undoubtedly resulted in enhanced sustainability indicators as well as in an improved focus of the public perception in these problems. According to an FAO report (1999a), the unique aspect of organic farming is that almost all synthetic inputs are prohibited. Crop rotations are mandated and proper use and management of manure is essential. Organic farming helps conserve water and soil on the farm. Reduction in the use of toxic pesticides, which the World Health Organization estimates poison 3 million people each year, is important with regard to health risks of farm families.

In several developed countries, organic agriculture already represents a significant portion of the food system. According to a study the International Federation of Organic Agriculture Movements (IFOAM 2004), currently more than 24 million hectares of farmland are under organic management worldwide. FAO (1999a) reported that in developing countries, under the right circumstances, the market returns from organic agriculture can potentially contribute to local food security by increasing family incomes and some of the developing countries have begun to seize the lucrative export opportunities presented by organic agriculture.

The dramatic increase in the use of agrochemicals in developing countries in recent decades and concentration on cash crops has often resulted in environmental contamination, severe health problems, and unprofitable crop production. The need for changes has resulted in a need for alternatives: crop production systems which do not rely heavily on chemical inputs, but which nevertheless produce economically viable yields while minimizing environmental impacts. Integrated pest management is one such system that has been successfully implemented on a wide range of crops and agroclimatic zones. Many aid and development agencies have adopted IPM as the model for the agricultural development they support, and the OECD Development Assistance Committee encourages its member states to support IPM. (See Box 6.6.)

6.3.5 Water Management

Governments and water managers are faced with the need to increase water supply to meet a still expanding population's increasing demand for food and water, while at the same time insuring that the water supply is sustainable and that ecosystems contributing to that sustainability are protected. Although the total available fresh water of the world is considered, in the aggregate, sufficient to satisfy today's demand, the uneven distribution of the world's freshwater resources and the current mounting pollution of many waterways and aquifers result in a situation where at least 30 countries are considered water stressed (with freshwater resources less than 1,700 cubic meters per capita); 20 countries are water scarce (with less than 1,000 cubic meters per capita) (Rosegrant 1995). In regions where water is already stressed or scarce, meeting increasing demand for all water uses including ecosystem protection becomes increasingly difficult and expensive, particularly under the traditional approach of constructing new water supply projects. However, there are many options for providing the necessary water, many of which may be better and cheaper than constructing new projects. Some response options are listed below.

In terms of *supply-side management,* options include: constructing additional water storage and distribution systems; making better use of natural systems, such as wetlands and ground cover, to reduce erosion, store and filter water, and recharge aquifers; improving the efficiency of existing storage and distribution systems; improving water management techniques and institutions; importing bottled water; and desalinating seawater.

Demand side management options include: increasing water productivity; improving water pricing; importing more food rather than growing it; applying water quotas; using economic incentives to reduce withdrawals and pollution; improving water quality regulations; and initiating a pollution permits market.

6.3.5.1 Water Pricing in Irrigated Agriculture

Water pricing is one of the most important elements of recent water management frameworks, because it is the basis for achieving efficient allocation of water resources. Conversely, inappropriate water prices could encourage inefficient use of water and contribute to water shortages or depletion of water resources in the long run and degradation of the environment and the ecosystems (for example, Koundouri et al. 2003; Pashardes et al. 2002; Chakravorty and Swanson 2002).

Efficient pricing is also very important in the management of groundwater where, in addition to standard pumping and distribution costs, there are costs associated with externalities. These costs can be classified as (Howe 2002): (1) *contemporary pumping externalities* associated with the fact that individual pumping affects other groundwater users in the vicinity by lowering their water table and increasing their pumping costs; (2) *intertemporal externalities* stemming from the fact that pumping groundwater now affects its future availability; and (3) *groundwater quality externality* resulting when water has different quality characteristics in different parts of the aquifer and when pumping causes salt water intrusion.

Typically water prices in agriculture, when they exist, cover more or less the variable cost of water supply, while public authorities cover fixed costs. Sometimes prices are set according to some notion of farmers' "ability to pay." The structure of water pricing systems usually takes one of the following forms (Tsur and Dinar 1997): standard volumetric and fixed tariffs, area-pricing, tiered or block-rate pricing, land betterment levy pricing or passive trading, volumetric pricing with bonus, or water markets.

BOX 6.6

Integrated Pest Management

Definitions of IPM cover a range of approaches: from safe use of pesticides to elimination of virtually all pesticide use. The presence of pests does not automatically require control measures, as damage may be insignificant. A system of non-chemical pest methodologies needs to be considered before a decision is taken to use pesticides. Suitable pest control methods should be used in an integrated manner and pesticides need to be used on an as-needed basis only, and as a last-resort component of an IPM strategy. In such a strategy, the effects of pesticides on human health, the environment, and sustainability of the agricultural system and the economy need to be carefully considered. IPM programs are designed to generate independence and increased profits for farmers, and savings on foreign imports for governments.

IPM enables farmers to make informed decisions to manage their crops. Successful IPM programs replace reliance on most spraying, including calendar spraying of pesticides. It builds on the knowledge of women and men farmers of crop, pest, and predator ecology, to increase the use of pest-resistant varieties, beneficial insects, crop rotations, and improved soil management. Supportive agricultural research, training of extension workers and farmers, and farmer participation in pest management solutions, are key elements. IPM programs encourage access to information on non-chemical alternatives. Government adoption of IPM, as part of its agricultural policy, will move IPM from the level of individual projects to a more common approach, and will bring national benefits.

IPM programs involve farmers and field staff from national and local government units and nongovernmental institutions, enhancing ecological awareness, decision-making and other business skills, and farmer confidence. IPM thus has long-lasting socioeconomic benefits far beyond the field of plant protection.

Pesticide Problems Avoided with IPM

Hazards to Health	Hazards to Environment	Crop Production Problems
Acute poisoning: 3 million poisonings including 20,000 unintentional deaths occur annually (WHO)	Contamination of drinking water and ground water	Pesticide resistance: 520 species of insects and mites, 150 plant diseases, and 113 weeds are resistant to pesticides (FAO)
Symptoms of acute poisoning include severe headaches, nausea, depression, vomiting, diarrhea, eye irritation, severe fatigue, and skin rashes	Water contamination kills fish	Resistance can create a treadmill syndrome, as farmers use increasing inputs to little effect, while elimination of beneficial insects causes secondary pest outbreaks
Chronic ill-health problems can affect women and men, girls and boys exposed to pesticides whether because of their occupation or because they live near areas of use. Such problems can include neurological disorders, cancers, infertility, birth defects, and other reproductive disorders	Soil contamination	High costs of pesticides can lead to falling incomes for farmers: newer products are often safer, but more expensive
	Wildlife and domestic animals can be killed by spray drift or by drinking contaminated water	Farming communities lose knowledge of good horticultural practices and become dependent on expensive external inputs
	Exposure may also cause infertility and behavioral disruption	
	Persistence in the environment and accumulation in the food chain leads to diverse environmental impacts	
	Loss of biodiversity in natural and agricultural environments	

Water markets have become an increasingly important mechanism for efficient and flexible water allocation. (See Box 6.7 for selected examples.) Water markets and tradable water rights give water a value separate from land and provide incentives to use water more efficiently, since water saving can be sold for extra revenues or can be used to further increase production. Water markets are promoted by international organizations such as the World Bank and have been pursued within many developing countries (Thobani 1997).

In developing countries, the practicality and true ecological and livelihood impact of water pricing and markets is under scrutiny. Given this, a broader term of economic incentives will certainly be important. These could include positive incentives for farmers to save water, rather than penalizing the rural poor when it is often the urban wealthy who benefit from low food prices and could better afford the cost of dealing with negative externalities. (Box 6.8 shows how one system works.)

In addition to the issue of efficient water use, attention must be paid to the possible negative effects of irrigation (FAO 2002b). Irrigation of farmlands in areas with water scarcity could cause degradation of water-based ecosystems, such as wetlands and forests. Regional transfers of irrigation water could cause problems both in the withdrawal and the receiving regions. Intensive irrigation farming in arid and semiarid areas leads to water pollution through chemical runoff into surface water or percolation into groundwater. Overirrigation also often results in soil salinity problems, for example, the Indus Basin in Pakistan.

6.3.5.2 Responses to Water Pollution

6.3.5.2.1 The nature of agricultural non-point source pollution

Agriculture is the single largest user of water resources. Except for water lost through evapotranspiration, agricultural water is recycled back to surface water and/or groundwater. However, agriculture is both a cause and a victim of water pollution. It is a cause through its discharge of pollutants and sediment into surface and/or groundwater; through net loss of soil from poor agricultural practices; through salinization and water logging of irrigated land; and through salt water intrusion in coastal aquifers due to over pumping. It is a victim through use of wastewater and polluted surface and groundwater, which contaminate crops and transmit disease to consumers and farm workers. This section examines responses related to the regulation of agricultural water pollution, which is probably the most representative of the so-called non-point source pollution problems.

BOX 6.7

Experience with Water Markets (Bjornlund and McKay 2002)

Water markets have emerged in both developed and developing countries. Some examples are:

- In Mexico, permanent and temporary water trading took effect in 1994 with a new Water Act. Water transfers can be freely made and at low cost within water users' organizations, and on a communal basis. Inter-district transfers require government approval.
- Informal water markets have operated within irrigation districts in India and Pakistan, often illegally. Payment for water takes place through arrangements involving two-way share farming, where one party supplies water and the other supplies the land, and all other costs and profits are shared. Another possibility is the three-way share farming, where one party supplies land, one water, and one labor, and all other costs are shared. These informal markets have increased water availability and supply reliability, and promoted more socially equitable outcomes from irrigated farming within the community.
- In the United States, most water trade involves transfers from agriculture to urban users, or to industrial users, such as mining and power generation. Many water sources are organized within mutual companies providing water through networks of canals. Irrigators own shares in the company, giving them rights to a certain volume of water. Shares and the associated water are freely transferable within the area. Water banks, which prioritize buyers and facilitate sale of large volumes of water, have also developed in some western states. The U.S.US experience shows that the more well defined property rights are, the more water markets develop; and the larger the trading area and the more diverse the users groups, the more efficient they are.
- In Chile, the water code established in 1981 separated water rights from the land, made these rights freely transferable, and distinguished between consumptive and non-consumptive rights, the latter mainly for power generators.

The significance of non-point-source-type pollution is indicated by the fact that part of the degradation of many of the world's lakes and reservoirs can be traced to this type of pollution. (See Box 6.9 for a case study examining the Aral Sea.) Degradation is caused by a number of factors including nutrient loading due to intensive farming practices; toxic substances entering the water bodies as agricultural runoff along with forestry drainage, which includes a range of toxic pesticides and herbicides; accelerated sedimentation caused by farming on fragile soils and steep slopes, forestry activities, construction activities and urban drainage; acidification of aquatic systems from emissions of sulfur dioxide and nitrous oxides due to acid rain or through leaching from affected land. In a non-point source pollution problem, an environmental regulator can measure the ambient pollution at specific "receptor points," but cannot attribute any specific portion of the pollutant concentration to a specific discharger. Therefore, the problems that characterize a non-point-source pollution problem are mainly informational; Braden and Segerson (1993) have identified two broad classes of problems: those related to monitoring and measurement, and those related to natural variability.

6.3.5.2.2 Regulation

The inadequacy of the standard instruments of environmental policy to deal with NPS pollution has led, in recent years, to the development of policy schemes appropriate for NPS pollution problems (Xepapadeas 1997, 1999). These schemes can be divided into two broad categories: (1) ambient taxes where the scheme is based on the observed ambient pollution, and (2) input based schemes, where the policy scheme consists of taxes applied to observable polluting inputs.

Actual policies against water pollution that are common in many countries (OECD 1994) include user charges for sewerage and sewage treatment, water effluent charges, and charges in agriculture, along with a number of more specific policies. These are general policies that do not readily conform to the stylized characteristics of the NPS pollution instruments discussed above; nevertheless there are features that attempt to address the non-observability of individual emissions.

Charges in agriculture are a more profound case of input-based schemes. Charges on fertilizers as applied in many countries are based on the nitrogen and phosphorus content of fertilizers, which are the main contributors to NPS pollution in surface water. A number of off-farm management methods also exist for reducing phosphorus runoff such as vegetation buffer stripes, riparian zones, and dredging of the lake sediment.

More specific policies aimed at addressing NPS pollution problems, especially in relation to agriculture also exist. For example, in Austria there are groundwater protection zones in which, if the water quality is reduced, farmers have to comply with certain management practices or change land use. Spain has zonal programs for reducing fertilizers, the Netherlands has a manure and ammonia policy, England and Wales have codes, which give farmers guidance on maintaining good agricultural practices. Ireland has a voluntary scheme for farmers to follow a specific nutrient management plan.

6.3.6 Fisheries Management

As described in this chapter and MA *Current State and Trends*, Chapter 18, capture or wild fisheries have been overexploited (and habitats damaged) to the extent that current global catch levels are stable or are actually reducing. A conundrum that has helped mask effective global action so far is that apparent global stability in the catches does not highlight severe regional instances of overfishing, or the reduction of sizes and trophic levels of the fish being caught. Shortfalls in capture fisheries and price increases have led to ill-advised exploitation of "new" fisheries (sometimes of long-lived fish like orange roughy which reproduce slowly). These have gone through rapid boom and bust cycles and exacerbated the global decline.

In addition to effects on the structure of fisheries, further environmental effects are being noted, for example, on sea beds as a result of trawling, and damage to habitats such as tropical coral reefs through overfishing and destructive fishing practices. Capture fisheries are associated with large government revenues through taxes and exports, and employment (often for poorer communities and coastal areas). Individual national responses to overfishing have been frequently insufficient, or actively protectionist of the industry through subsidies and poor enforcement of existing regulations. (See Box 6.10.) As a result, global fishing capacity is far in excess of what is economically viable. Probably the most important causes of fishery collapse have been poor decision-making and lack of political will. Of late, there is a growing awareness that the traditional approach to managing fisheries, which considers the target species as independent, self-sustaining populations, is in need of revision. The need to implement ecosystem-based fisheries management is now being emphasized (FAO 2002a).

BOX 6.8

Water and Mixed Crop–Livestock Systems

Mixed crop–livestock production characterizes most irrigation and rain-fed agriculture in developing countries. Pure crop production is largely restricted to developed countries. Discussed here are rain-fed mixed crop–livestock systems, but livestock are a fundamental and overlooked component of most irrigations systems in developing countries.

Water accounting tools enable understanding of water use in mixed crop–livestock production systems (see Figure below). Although the geographic scale of analysis is largely arbitrary, water enters into and exits from farming systems, agroecosystems, and river basins. Within these systems, available water supports agricultural production, non-agricultural human needs and ecosystem services. Water used for food production competes with other uses. Water that does not leave farming systems is stored and available for future use. Water that has been used but does not leave these systems remains available for re-use provided that its quality has not been reduced to unacceptable levels. Losses of water also include evaporation, discharge, and contamination. Transpiration is the most essential form of agricultural water use that drives both agricultural production and maintenance of wild biodiversity.

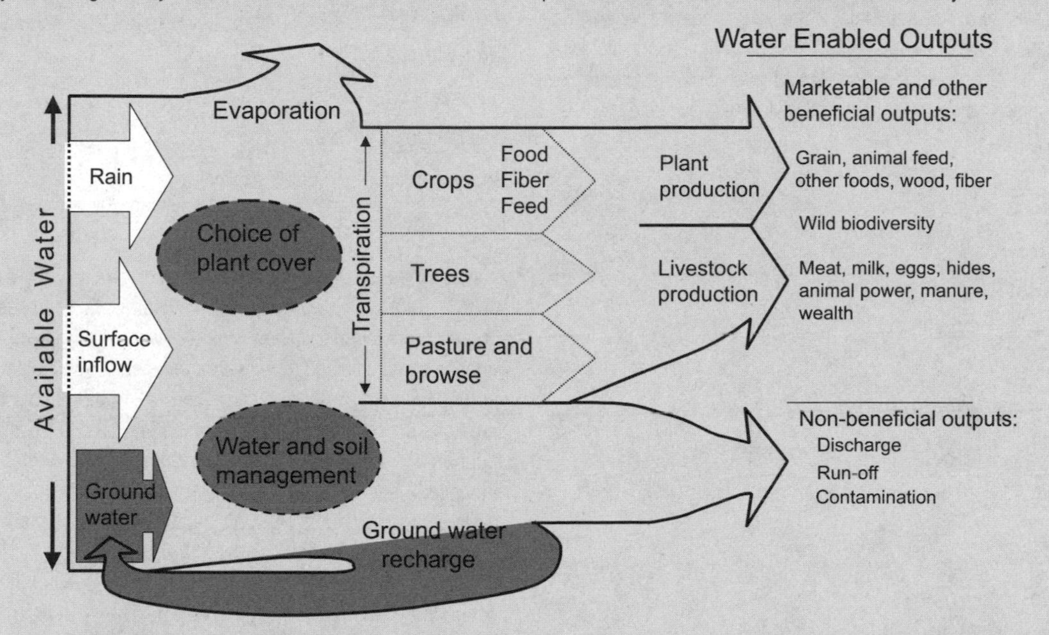

Water Accounting Framework Showing Relationships among Water Supply, Water Loss, Water Storage, and Livestock and Crop Production in Mixed Crop–Livestock Production Systems (modified from Molden et al. 2003)

Water discharged from upstream food producing systems affects downstream users. Excessive run-off causes downstream flooding while upstream food production can make water less available to downstream users. Yet upstream increases in infiltration can provide improved quality and a seasonally available downstream supply. Thus changing water use and productivity in one place may have both adverse and beneficial impacts elsewhere.

Farmers, planners, and policy-makers have many options for promoting a more efficient use of water. On the supply side, investments are possible in infrastructure to import water, and in development of water storage facilities such as ponds, dams and tanks. In addition, any land management activities or agricultural practices that encourage groundwater and soil-moisture recharge contribute toward storing or maintaining available water. On the demand side, water management is an important strategy to improve water use in mixed farming systems. This requires coherent policies, practices, and technologies that promote an optimal mix of plant species that are collectively responsible for enabling beneficial outputs, including animal and plant production and ecosystems services.

Choice of plant species serves to re-allocate water through benefit-producing transpiration pathways. Demand management also requires improving land management practices that promote groundwater and soil-moisture recharge through practices such as controlled grazing, maintaining vegetation cover, terracing, and conservation agriculture. This approach helps in retaining water for use in dry periods and reduces undesirable flooding downstream.

In developing countries, in contrast to industrial countries, livestock are not just productive commodities. They play a much larger role through provision of farm power and in many communities they represent wealth assets.

Overgrazing is often blamed for high rates of soil erosion, run-off, and flooding particularly in steep lands. Evidence from Ethiopia suggests strongly that the primary cause has been the replacement of grazing land with annual cropland requiring better integration of water management with crop and livestock production. Options for improvement include terracing, conservation agriculture, and de-stocking of livestock populations accompanied by action to increase the productivity of each animal. The framework provided in figure shows a framework that can help increase understanding of the interactions among people, water, crops, and livestock, and identify options to improve agricultural water productivity.

BOX 6.9

Agricultural Water Pollution Case Study: The Aral Sea Disaster

The Aral Sea lies in Central Asia; its basin includes Southern Russia, Uzbekistan, Tajikistan, Kazakhstan, Kyrgyzstan, Turkmenistan, Afghanistan, and Iran. The Aral Sea has no outlet, but equilibrium had been reached between the inflow and evaporation. In 1960, the Aral Sea was the fourth largest lake in the world, fed by the Amu Darya and Syr Darya rivers. The population of the area was 23.5 million in 1976 and has risen since then.

In the 1920s, the former Soviet Union started transforming the area into a major cotton producing area and used the river waters to irrigate the dry lands upstream from the Aral Sea. During the 1960s, the effects of the water diversion to massive irrigation schemes started to appear. The Aral Sea began to shrink, the shoreline retreated, and the salt concentration increased dramatically. The desiccation of the Aral Sea, derived from satellite remote sensing data, is shown here for the period 1960 to 2010 (projected).

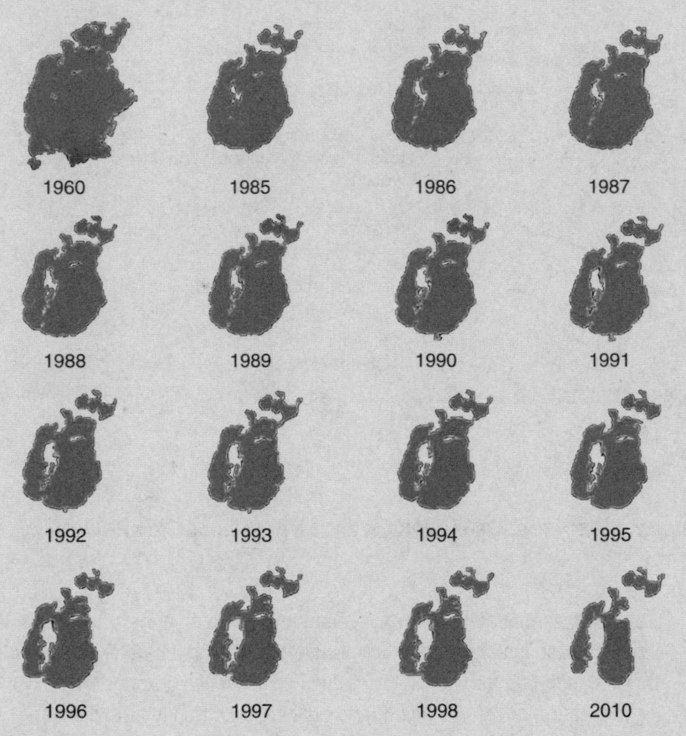

The Aral Sea has lost more than 60% of its area and approximately 80% of its volume (until 1998). The sea level has dropped about 18 meters in the same time period. Historical data indicate that the inflow to Aral Sea was 56 cubed kilometers per year prior to the intervention. During the period 1966–70, the inflow fell to 47 cubed kilometers per year; during 1981–85, it was 2. There was a huge increase in the salinity. In Uzbekistan, for example, the salinized area in 1982 was 36.3% of the total irrigated area; by 1985 it had risen to 42.8% of the total irrigated area.

The major ecological and water quality impacts include: salt content of major rivers exceeds standard by factors of two to three; contamination of agricultural products with agrochemicals; high levels of turbidity in major water sources; high levels of pesticides and phenols in surface waters; excessive pesticide concentrations in air, food products, and breast milk; loss of soil fertility; induced climatic changes; major decline and extinctions of animal, fish, and vegetation species; and destruction of commercial fisheries.

These developments have had a devastating impact on public health, as seen in this table showing public health impacts in the Aral Sea area since the 1980s (Ongley 1996).

Typhoid	29-fold increase
Viral hepatitis	7-fold increase
Paratyphoid	4-fold increase
Hypertonia, heart disease, gastric and duodenal ulcers	up 100%
Increase in premature births	up 31%
Morbidity and mortality (Karakalpakia, 1981–87)	
Liver cancers	up 200%
Gullet cancers	up 25%
Oesophageal cancers	up 100%
Cancer in young persons	up 100%
Infant mortality (1980–89)	up 20%

Agriculture is considered to be the root cause of the Aral Sea disaster (UNEP 1993). In particular, the agricultural practices resulted in effects such as: increase in irrigation area and water; withdrawals; use of unlined irrigation canals; rising groundwater; extensive monoculture and excessive use of persistent pesticides; increased salinization and salt runoff leading to salinization of major rivers; increased frequency of dust storms and salt deposition; discharge of highly mineralized, pesticide-rich return flows to main rivers; and excessive use of fertilizers.

UNEP responded to Russia's request to address the problem (World Bank 1997). Initial studies indicated that it was not possible to restore the Aral Sea. It is important to notice here that agricultural practices led to an *irreversible* change in the ecosystem. The Aral Sea Basin Program launched 19 projects in eight thematic areas for the purpose of attaining partial remedies. The main issues tackled include water and salt management, wetland restoration, and immediate project impact.

Water and salt management: Water management includes mainly water sharing as a transboundary issue, since there are five independent states involved, with the upstream states requiring water for electricity generation, and the downstream states requiring water for irrigation. Salt management is the most pressing problem, the land is losing productivity due to salinization and might be out of production, and salinity jeopardizes drinking water resources.

Wetland restoration: The purpose is to restore part of the Sea or to rehabilitate some ecosystems on the Sea's perimeter.

Immediate impact: The aim is to alleviate suffering in the disaster zone, by helping to provide clean water and fishing opportunities in the deltas, health care for people living near the Sea, and repair of the infrastructure (for example, schools, hospitals). Due to the size of the problem, additional joint action was taken by the World Bank, UNDP, UNEP and the European Union, all of which call for a synergy of the current efforts in the region and offer a wide opening for further initiatives.

Developments in the Aral Sea since the 1960s represent one of the greatest environmental disasters ever recorded with major social, economic, and ecological impacts. The Aral Sea provides a catastrophic example of how responses aimed at increasing production of the agricultural sector can generate feedbacks that devastate a once-productive region.

The *combined quantity-quality* water degradation proved to be devastating for the whole ecosystem. It is important to note here that agricultural pollution led to an *irreversible* change in the ecosystem; furthermore the change was *fast,* indicating the existence of *threshold* effects. Quantity-quality interactions, irreversibilities, and fast change after threshold points are issues discussed in theoretical models of pollution accumulation, and in this sense the Aral Sea disaster can be regarded as constituting a real life example of theoretical modeling.

In summary, the response of adopting massive agricultural development plans, without any precaution for detrimental side effects, created a major negative impact on human well-being and poverty in Central Asia. This in turn, seriously impeded the attainment of goals such as security, basic material for a good life, health, and good social relations.

BOX 6.10

The Collapse of the Newfoundland Cod Fishery

Combined stock and regulatory fluctuations, leading to eventual collapse, have been observed in fisheries. The Canadian cod fishery off the east coast of Newfoundland experienced its boom-bust phase in the mid-1950s. With the appearance of a new breed of factory-fishing, countries such as Germany (East and West), Great Britain, Spain, Portugal, Poland, the Soviet Union, Cuba, and countries in East Asia had legally fished to within 12 miles of the eastern Canadian and New England (U.S.) seaboards. Canada (and the United States), concerned that stocks were being reduced to almost nothing, passed legislation in 1976 to extend their national jurisdictions over marine living resources out to 200 nautical miles. Catches naturally declined in the late 1970s and stocks started recovering after the departure of the foreign fleets. However, national regulation did not set catch quotas at the late 1970s levels, and furthermore, new technology in the form of factory-trawlers, or draggers as they became known, became the mainstay of Canada's Atlantic offshore fishing fleet. As a result, the northern cod catch began a steady rise again, with a corresponding decline in stocks (MA *Current State and Trends,* Chapter 18).

By 1986, the stock decline was realized, and by 1988, there were scientific opinions recommending that the total allowable catch be cut in half. Possibly because of delayed regulatory response, by 1992, the biomass estimate for northern cod was the lowest ever measured. The Canadian Minister of Fisheries and Oceans had no choice but to declare a ban on fishing northern cod. For the first time in 400 years, the fishing of northern cod ceased in Newfoundland. The fisheries department issued a warning in 1995 that the entire northern cod population had declined to just 1,700 tons by the end of 1994, down from a 1990 biomass survey showing 400,000 tons (Greenpeace 2003).

This collapse illustrates the vulnerability of fish stock. It is a story that has been repeated in many other fisheries, including the California and Japanese sardine fisheries, and the Southwest African pilchard and North Sea herring fisheries.

Aquaculture has developed rapidly, particularly in Asia, but also in key countries in Europe, the United States, and Latin America to increase supplies. Over the past three decades, aquaculture has become the fastest growing food production sector in the world; it has increased at an average rate of 9.2% per year since 1970—an outstanding rate compared to the 1.4% rate for capture fisheries or the 2.8% rate of land-based farmed meat products (FAO 2002a; Kura et al. 2004).

In 2001, aquaculture produced 37.9 million tons of fishery products, nearly 40% of the world's total food fish supply and valued at $55.7 billion (FAO 2002a; Vannuccini 2003). Aquaculture production is expected to continue to grow in the future to meet the increasing demand for fish and fishery products. Aquaculture has become such a rapidly increasing sector by expanding, diversifying, and intensifying production, as well as by technological improvements in its operations.

However, the initially unregulated expansion of the fish farming industry has led to inappropriate land and water use in some cases, and breaks on potential levels of productivity through pollution, contamination, and disease losses. Although a more mature industry is developing, several issues must be addressed for aquaculture to start to balance losses in capture fisheries in a way that does not simultaneously damage the environment. Food safety and trade issues in aquaculture products mirror those for livestock products and have severe implications for developing countries that exploit fisheries for trade and also for food security purposes.

This section examines two types of responses, one related to the management of capture fisheries and the other related to aquaculture.

6.3.6.1 Capture Fisheries

6.3.6.1.1 The international framework for improving fisheries management

Much of the current depletion of marine fish stocks derives from the fact that oceans for hundreds of years have been managed as open access resources. These resources are highly vulnerable to overexploitation because there is no incentive for individual fishers to restrain their harvest (MA *Current State and Trends,* Chapter 18).

The United Nations Convention on the Law of the Sea provides coastal countries sovereignty over marine resources within 200 nautical miles of their coast so that the responsibility to manage coastal fisheries in a sustainable manner is squarely in the hands of coastal nations. Not all nations have adequate fisheries management plans and laws in place. Even when they do, implementation and enforcement often fall short, and fisheries are still subject to overfishing. Also, distant water fleets from industrial countries have been able, through payment of various fees, to access the exclusive economic zones of developing countries in many cases and rapidly deplete their resources

The FAO Code of Conduct for Responsible Fisheries provides voluntary guidelines according to the following principles:

- manage stocks using the best available science;
- apply the "precautionary principle," using conservative management approaches when the effects of fishing practices are uncertain;
- avoid overfishing; prevent or eliminate excess fishing capacity;
- minimize waste (discards) and bycatch;
- prohibit destructive fishing methods;
- restore depleted fish stocks;
- implement appropriate national laws, management plans, and means of enforcement;
- monitor the effects of fishing on all species in the ecosystem, not just the target fish stock;
- work cooperatively with other states to coordinate management policies and enforcement actions; and
- recognize the importance of artisanal and small-scale fisheries, and the value of traditional management practices.

More than 150 countries have formally embraced the Code since it was introduced in 1995 (FAO 1995b). To augment the general provisions of the Code, the FAO has issued a number of "technical guidelines for responsible fisheries" that look at certain important subjects in depth and interpret the Code with greater specificity. For example, the FAO has issued technical guidelines on applying the precautionary principle, integrating fishery management into coastal area management, developing aquaculture responsibly, and applying an "ecosystem approach" to fisheries, among other topics (FAO 2001a). In addition, it has overseen the development of four International Plans of Action, which consist of a set of recommendations on how nations should cooperate to track a given problem, assess its magnitude, and develop individual national plans of action to address the problem. So far, IPOAs on reducing seabird bycatch, conserving shark fisheries, reducing fishing capacity, and reducing illegal, unreported, and unregulated fishing have been approved by FAO member nations (FAO 2002a).

The elaboration of the principles in the Code and their general acceptance as norms by nations provides the important framework for more sustainable fishing. However, the Code of Conduct and the IPOAs are all voluntary agreements, free of legal mandates or enforcement mechanisms. Global action is undermined by nations that fail to fully implement or enforce them (Kura et al. 2004). Clearly, elaborating and implementing national plans in accordance with the Code of Conduct for Responsible Fisheries, including provision for traditional and small-scale fisheries in different countries, is a continuing requirement.

6.3.6.1.2 Tools currently exploited to manage fisheries

Fishery management generally aims at preventing stock depletion and securing the standards of living in the fishing sector. In the future, achieving the necessary reduction in fishing capacity will require fishers and vessels to leave fishing, and thus establishment of alternative livelihoods. As a consequence, economic support programs will become an integral part of future management strategies.

A number of fishery management methods and practices have been developed to reduce or restrict the capacity of fleets. Those, such as the imposition of total allowable catch for the fishery, vessel catch limits, mesh and size restrictions on gear, license limitation, individual effort quotas, and buy backs of vessels or licenses to reduce fleet numbers can all be considered as limitations of fishing capacity. They contrast with methods that seek to adjust the incentives for fishing. These seek to provide individual or group incentives and market mechanisms for meeting output targets, with greater flexibility of operation. Various responses are sketched in Table 6.3. In practice, a mixture of input and output controls appropriate to the individual fishery is the best way of managing it (World Bank 2004).

While designed to manage fishing, the above responses sometimes have inadvertent effects or incite perverse behaviors in relation to resource exploitation and sustainability. For example, the allocation of licenses to fishers or fishing vessels, entitling them to harvest from one or more stocks, is the most widely used system for controlling the fleet capacity (Cunningham and Gréboval 2001). However, licensing programs are insufficient on their own to control a fleet's overcapacity. A major limitation is that they do not prevent licensed fishers from expanding the capacity of their vessels or adding new technology to increase their catch.

The most common application of catch controls is the total allowable catch. If set at the right level (no higher than the fishery's maximum sustainable yield), TACs can effectively reduce the direct pressure on a fish stock. However, TAC systems give fishers the incentive to fish as quickly and intensively as possible to maximize their share of the allowable catch. This competition often leads to overfishing and high bycatch rates (for species not specified in the TAC). To combat this, individual fishing quotas have been introduced, where a specific proportion of the TAC may be allocated to individual fishers to harvest at their own pace. In many instances, fishers are allowed to treat these individual transferable quotas as personal assets, with the legal right to buy or sell them. The theory behind ITQs is that fishers are more

Table 6.3. Main Policies for the Management of Open-access Fisheries

Policy	Description
Fishing effort regulation	In this policy, one of the inputs in the index for fishing effort is restricted (for example, number of days at sea).
Decommissioning schemes	The purpose of this policy is to bring the capacity in line with catch potentials. This is done by reducing the fleet capacity through subsidized buy backs.
Marine protected areas	The aim of this response is to protect some fragile parts of a marine area by banning fishing within these areas. Some examples of use of this regulatory instrument are the Shetland box and the Norway pout box (Holden 1996).
Total quotas or total allowable catches	In this policy, a total quota is imposed on the fishery and when this quota has been filled, the fishery is closed (Clark 1990). The total quota is often recommended to be set at a level where maximum sustained yield is reached. Total quotas have in some cases been used in conjunction with individual quotas (for example, in the case of Iceland and New Zealand).
Rations	Under a rations policy, the total quota is distributed in short time intervals on vessels reflecting seasonal variations in catch possibilities. Rations are used for some species in Denmark. However, the system of rations creates huge information requirements.
License systems	A license system normally specifies how much can be caught and the weight of this catch. The purpose is to control the catch of each individual vessel.
Individual quotas	This policy sets a non-transferable individual annual quota that cannot be changed during the year and may, therefore, be thought of as a property right. Indeed, property rights regulation is very popular within fisheries; more than 55 fisheries in the world are regulated by property rights.
Individual transferable quotas	Under this policy, individual quotas are made transferable between fishermen; ITQs are used in, for example, Iceland, the Netherlands, and New Zealand.
Taxes or landing fees	In this policy, either fishing effort or catch is used to compile the tax. In practice taxes are not popular among fishermen and there are severe implementation problems.

likely to use sustainable practices if they hold a long-term interest in the fishery in the form of a guaranteed percentage of the harvest. The introduction of ITQs has indeed brought benefits in some fisheries. Iceland and New Zealand both have comprehensive ITQ programs that are generally considered successful in reducing overall fishing effort and improving the efficiency of the industry as a whole (Hannesson 2002).

But individual transferable quotes still rely on setting a total allowable catch,, and suffer from the same scientific difficulties in determining a reliable estimate of sustainable yield. ITQs give fishers an incentive to "high-grade," or substitute larger (and more valuable) fish caught later in the day for smaller fish caught earlier. The smaller fish are usually discarded overboard—dead or dying. If these management rules are imperfectly implemented it is difficult for fisheries managers to follow the precautionary approach to protect fish stocks. The stability of the quota system is built on setting the TAC in advance of the fishing season, and not altering it as the season progresses (Copes 2000), therefore, managers have little flexibility to change if they realize mid-season that the TAC is too high and overfishing can result. ITQs may be appropriate to cold water, single species fisheries but they are unlikely to be useful in multi-species tropical fisheries where multiple TACs and ITQs would make the system impractical. Management by areas and output monitoring may be more sensible in this case.

6.3.6.1.3 *Time and area closures*

Time and area closures can be effective management tools for fisheries, but are usually combined with other regulations because, on their own, neither will reduce the overall pressure. *Closed seasons* are used to protect stocks at critical times in their lifecycle—such as when they are spawning—or as a way of lowering the total catch. A major disadvantage of establishing a closed season is that fishers will have an incentive to race for fish during the open season. *Closed areas* are used to help depleted stocks recover, or to protect biologically critical areas such as spawning grounds or juvenile nurseries. However, if taken in isolation, this approach does not necessarily decrease the overall fishing pressure, as fishers simply move to an adjacent open space, increasing fishing pressure there. Establishing marine reserves—one type of "marine protected area," where fishing and other human activities are restricted—is one approach that limits fishing effort in certain areas. Given the proven ability of marine reserves to nurture stocks within their boundaries, there is a growing expectation that they will also enhance commercial stocks in surrounding waters and beyond.

The biological benefits of marine reserves for organisms and ecosystems within the reserves are well documented. But their benefits to commercial fisheries outside the reserves are still the subject of debate (Ward et al. 2001). Part of the problem is that few marine reserves have been strictly protected and monitored for long enough to determine the effect of potential benefits in surrounding waters. In addition, monitoring and demonstrating the spillover effect is no easy matter, and documenting benefits to distant waters is even more difficult. Nonetheless, there is some evidence to support the idea that reserves can benefit fish stocks outside their borders. Case studies and research in localized reef systems show that the recovery that comes from the establishment of a reserve can affect areas immediately adjacent to the reserves (Ward et al. 2001; Polunin 2003). Similarly, after a five-year closure of about 25% of the George's Bank, stocks of several species have increased, including scallop, haddock, and flounder (Murawski et al. 2000). These improvements are now beginning to spill over into waters outside the closed areas (Paul Howard, New England Fishery Management Council, personal communication, cited in Gell and Roberts 2003).

There is still much we do not know about marine reserves or how to maximize their benefit. The rate and nature of recovery of different fish species within reserves is likely to vary considerably. On the other hand, an understated strength of marine reserves is that they provide a clear example of one type of ecosystem-based approach to fisheries management, since they protect both fish and the ecosystem where they live. This may be especially useful in the tropics, where many species may be commercially exploited in one fishery. Recognizing the wide-ranging benefits of an ecosystem approach to managing fisheries, some countries have started testing the concept of a marine reserve with a commercial fisheries goal in mind.

6.3.6.1.4 *Future requirements for better governance*

An institutional framework for improved governance of the fisheries sector requires international collaboration in the following (see World Bank 2004):

- the fisheries management system;
- the monitoring, control and surveillance system;
- the fisheries judicial system;
- an institutional framework linking different types of stakeholders, including small-scale fisheries;
- a system of allocation of user rights (to counteract the unregulated nature of open access fisheries);
- control and development instruments (to ensure equitable development, as many aspects of fishing rights, etc., tend to be appropriated by large scale entrepreneurs);
- establishment of protected areas where appropriate and following multiyear research;
- managing exploitation patterns (through regulation of fishing operations by the means discussed above);
- fishing vessel and effort reduction programs;
- restocking (as and when feasible and appropriate);
- promoting aquaculture (discussed below);
- food safety and ecolabeling; and
- promotion of alternative livelihoods to fishing.

At the World Summit on Sustainable Development in 2002, many countries made a commitment to replenish overfished marine stocks by 2015 to sustainable levels, reflecting the increasing belief that fishery resources must be managed and used in sustainable ways taking the ecosystem that nurtures them into account. There is growing recognition that principles, policies, and mechanisms for prioritizing and allocating uses of aquatic areas must be put in place so that the impacts of fisheries on other sectors and vice versa are taken into account. Traditional approaches to managing fisheries, which tend to consider the target species as independent and self-sustaining populations, have proven to be insufficient. The need to implement ecosystem based fisheries management is currently being emphasized (FAO 2002a), although how to achieve this remains a continuing challenge to research and applied management.

6.3.6.2 *Aquaculture*

There are many different kinds of aquaculture and each system has its own strengths and weaknesses, which may positively or negatively affect overall productivity and the environment (Kura et al. 2004). Aquaculture as an integrated farming practice has the possibility of augmenting nutritional and income security for small farmers. However, many forms of aquaculture involve transforming land, coastal, and freshwater ecosystems, with serious ecological consequences for ecosystem integrity, and thus the delivery of other ecosystem services. For example, the replacement of mangroves to establish shrimp aquaculture facilities has been responsible in a major way for the loss of the mangrove habitat, particularly in Southeast Asia and Latin America (Boyd and Clay 1998). The destruction of hundreds of thousands of hectares of mangrove forests reduces crucial coastal protection and filtering functions. Box 6.11 discusses fish farming in Bangladesh.

Intensification of aquaculture leads to the emergence of issues that parallel the intensification of terrestrial livestock production. High stocking densities, poor water quality, and poor seed quality can lead to outbreaks of disease, which then spread to other ponds through water exchange. Increased movements of live aquatic animals and products as the industry grows have made the accidental spread of disease more likely. Effluent from aquaculture and pens is often released directly into surrounding waterways, causing pollution problems stemming from fertilizer, undigested feed, and biological waste in the water. This effluent can contribute to eutrophication of downstream waters, harm benthic communities, and cause damage to water and soil quality (Funge-Smith and Briggs 1998). Disease can then lead to pond abandonment and land degradation. Antibiotic drugs and other pro-biotics can significantly degrade the surrounding, local environment, and even have health effects on humans.

In addition, farmed fish that escape into the wild can threaten native species by acting as predators, competing for food and habitat or interbreeding and changing the genetic pools of wild organisms. Traits bred into farmed fish are often different from those that confer reproductive fitness in the wild, and interbreeding between escaped farmed fish between escaped farmed fish and wild fish may result in the loss of important local adaptations (such as home river returning capacity in wild salmon). The risk is greatest for small populations that are already threatened. The majority of these effects, including disease (ectoparasite) transmission, have been documented in large-scale salmon rearing operations. Escaped fish are intrinsically harder to monitor and control than vegetable crops or terrestrial animals.

Nearly one third of the world's fish caught in the wild, such as small pelagic fish like anchovies and menhaden are not consumed directly by humans but rather "reduced" to fish meal and fish oil and consumed by farm-raised animals, such as chickens, pigs, and carnivorous fish in some aquaculture systems. Aquaculture consumes more fishmeal so far than terrestrial livestock and poultry, as these have increasingly switched to vegetable-based meals. Wild-caught fish are also used as seed fish in some developing-country aquaculture operations, posing risks to wild fish stocks by removing juveniles from the population, although this is expected to diminish with research on closing life-cycles in culture.

BOX 6.11

Shrimp Farming in Bangladesh

Bangladesh is one of the least developed countries in the world, with a per capita income of $350 (World Bank 2002a). Agriculture including fisheries contributes about 30% to the gross domestic product of the country (BBS 2000). The contribution of the fishery sector to GDP was slightly over 5% in the year 1998–99. The fish industry, and particularly shrimp, plays a major role in nutrition, employment, and foreign exchange earnings of the country. In Bangladesh, about 51% of animal protein is supplied by fish (see Figure below).

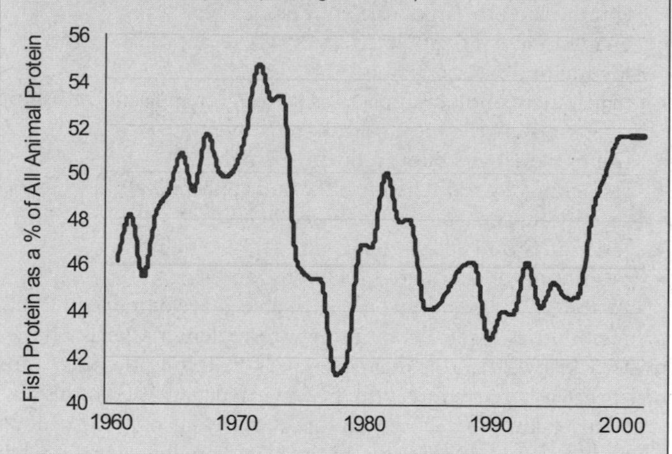

Contribution of Fish to Total Supply of Animal Protein in Bangladesh (Kura et al. 2004)

In recent years, there has been a marked increase in the contribution of fish to the total animal protein supply. About 1.2 million people are directly and indirectly employed in the fisheries sector, while another 11 million are indirectly employed (Vannucinni 1999). Bangladesh is a net exporter of fish and fishery products, which account for about 12% of the total of Bangladesh exports. Contribution of shrimp to total fish and fishery products is about 86%.

Loss of biodiversity is another important concern. According to Barkat and Roy (2001), the largest source of shrimp fry in commercial farming is wild fry. Using this process, thick nets are used to collect wild fry, following which the shrimp fry are sorted out and all the rest abandoned. In this crude process, millions of wild fry of other aquatic flora and fauna are lost, leading to reduced fish populations and ecological imbalances in the coastal region.

6.3.6.2.1 Future regulation of aquaculture

The major challenge regarding aquaculture will be to maintain the balance between support for further development of the sector and regulation to prevent potential adverse environmental and social impacts. Because the aquaculture industry has expanded so rapidly, the legal and political frameworks for maintaining it as a sustainable business have lagged behind. Article 9—Aquaculture Development—of the FAO Code of Conduct for Responsible Fisheries adopted in 1995, sets principles and guidelines for the sustainable development and management of aquaculture (Kura et al. 2004). Following these principles, many countries have started to implement national regulatory guidelines that address the environmental and social impacts from aquaculture in order to ensure its sustainability (FAO 2003d). Canada, for example, has developed a comprehensive Aquaculture Action Plan, which provides clear guidelines for applying regulatory responsibilities to aquaculture under the existing legislation (DFO 2001). The World Bank, Network of Aquaculture Centers, WWF, and FAO have initiated a process to provide guidelines for shrimp aquaculture and the environment (World Bank 2002b). These trends should be encouraged and unified.

Despite such progress, aquaculture-producing countries still face enormous challenges to support responsible practices. (See Box 6.12.) While there are examples of environmentally sound practices, one of the limiting factors is the lack of financial resources for some countries to take advantage of the advanced technology that lessens the impact of aquaculture on the surrounding environment (Emerson 1999). Thus national governments and development donors could assist through supporting formulation of comprehensive strategic development and zoning frameworks for aquaculture in coastal and inland settings. Integra-

BOX 6.12

Aquaculture in Africa

Aquaculture was introduced in much of the African continent around the 1950s as an innovation that would improve the economic and nutritional well-being of producers. In the former Belgian Congo, fishponds were built in mining areas to produce high protein food to feed miners (Moehl 1999). Fishponds were seen as an ideal component of integrated farming systems, as a fish crop was grown using by-products from the home and farm. Indeed, from Kenya to Sierra Leone, thousands of ponds were built, many to be abandoned after a few years of meager production.

In 1986, UNDP, FAO and the Norwegian Ministry of Development and Co-operation undertook a thematic evaluation of aquaculture. The evaluation recommended that future efforts should focus on a specific combination of species and aquaculture systems; identify geographical regions which correspond to specific species/systems combinations that should receive priority attention; pay close attention to recipient governments' effective commitment to aquaculture; and ensure systematic monitoring and evaluation of impact generated assistance provided.

In 1993, FAO, assisted by other collaborators, assembled a series of twelve national aquaculture reviews from countries responsible for 90% of the region's aquaculture production (Coche 1994). The major constraints identified by these reviews on the continental level were that (1) there were no reliable production statistics; (2) limited credit was available for small-scale farmers; (3) the technical level of the fish farmers was very low; (4) local feed ingredients were unavailable; (5) there was a lack of well-trained senior personnel; (6) transport costs were prohibitive; and (7) there was a lack of juvenile fish for pond restocking.

Today, Africa's fish and shellfish aquaculture production is slightly over 110,000 tons. It is only about 0.4% of the world's total production (Moehl 1999). In spite of the region's rich endowments, including untapped land, water, and human resources, African aquaculture remains undeveloped. The problems with regard to aquaculture in Africa are mainly institutional and technical. Institutional problems stem from frequently changing institutional homes for aquaculture and over-reliance on donor funds. Technical problems exist on two levels. Technologies were presented to farmers with little appreciation for what the farmers' needs are—a top down technology transfer. Secondly, when results were lower than anticipated, completely new technologies and/or culture species were sought, when in fact the initial technology had the capacity to produce more. In most cases, poor harvests were a result of poor management.

In spite of aquaculture's modest growth in Africa, the past three decades have not been without some tangible results. Aquaculture is now known throughout Africa, having evolved into a well-known, if not well-understood, production system. Fishponds are now an accepted component of farming systems on most of the continent (Moehl 1999).

Based on past experiences, the following lessons should be noted and incorporated into national development policies (Entsua-Mensah et al. 1999):

- Major government fish culture stations should be given financial autonomy and put under good management.
- Public infrastructure should be ultimately self-supporting.
- Farming inputs should not be distributed free to farmers, but should have at least a subsidized price.
- Technology should not be based on imported commodities (for example, hormones, feed, etc.)
- Selected culture species should be reproducible by the farmers themselves. On-station research to support small-scale aquaculture development should be based on inputs available to farmers and should be farmer-driven through joint activities.
- Sociocultural surveys should be conducted before introducing a new technology to a region.

There is, at present, a need for aquaculture policies (national development plans); national aquaculture information systems; demand-driven research that includes the socioeconomic aspects of research and development; reinforced linkages between research and development; adapted research on brood stock development; regional and sub-regional research and/or training centers involving NGOs in training and development; and training at all levels including practical training of farmers, technicians and extensionists (Entsua-Mensah et al. 1999).

tion of aquaculture and water management for complementary uses will help regulate environmental quality and resource sharing. Agencies can control and implement other environmental standards, guide species selection and effective hatchery operation, and help manage inputs and technical standards of safety in operation (for example, governing escapees) and the health of products (World Bank 2004).

Market incentives, such as certification for sustainably farmed products are leading to the development of guidelines. These guidelines are being promoted by producer organizations in industrial countries. This needs to be expanded, especially to developing countries, in order to promote the use of best practices to reduce the impacts of aquaculture.

6.3.6.2.2 Technological progress

Selective breeding approaches (as with crops and livestock) have been successful at creating improved breeds of fish for aquaculture (for example, salmon, tilapia, and carp), which will increase the yield and overall efficiency of aquaculture production (both for intensive as well as more extensive forms of aquaculture). Genetic transformation technologies may also be useful in the future for breed improvement. However, particular care must be taken in the use and release of genetically modified fish because of competition effects and ease of mixing with wild stocks in aquatic environments. Successful deployment of such strains may depend upon the successful extrapolation of sterile animal techniques that do not allow reproduction breed if they escape (Bartley 2000; Bartley et al. 2001). There is active research on protein feeds and biotechnologies that may increase the opportunities for developing non-fish-based aquaculture feed alternatives.

Advances in hatchery technology have raised the possibility of replenishing wild fisheries (restocking) from such sources (Munro and Bell 1997). Unfortunately, much of the research into stocking marine species is still at the experimental stage (Bartley and Casal 1999), and positive effects on augmenting fish populations are limited to a few specific examples. Restocking with *alien* species for the purposes of augmenting fisheries production can have disastrous effects on local biodiversity and need to be avoided from an environmental standpoint. The case of perch in Lake Victoria illustrates how the introduction of a non-native species for the purpose of increasing food production can lead to an ecological disaster. (See Box 6.13.) However, restocking of artificial reservoirs (with low indigenous biodiversity) can provide means for enhancing the production of fish for food.

BOX 6.13

Introduction of Non-native Species: The Case of Lake Victoria

Lake Victoria, with a surface area of 68,000 square kilometers and an adjoining catchment of 184,000 square kilometers, is the world's second largest body of fresh water (after Lake Superior), and the largest in the developing world. It is relatively shallow, with an average depth of 40 meters and a maximum depth of 80 meters. About 85% of the water entering the lake does so from precipitation directly on the lake surface, with the remainder coming from rivers draining the surrounding catchments. The most important of these rivers is the Kagera River, which contributes 7% of the total inflow. Tanzania, Uganda, and Kenya control 49%, 45%, and 6%, respectively, of the lake's surface.

Fishing pressure on Lake Victoria began to intensify in 1905 when the British introduced flax gill nets, which soon replaced papyrus nets and fish traps used by the local villagers. The most important fish species were the haplochromines (a type of cichlid) and the two endemic tilapias *Oreochromis esculentus* and *Oreochromis variabilis.*

By the 1960s, officials were actively stocking the Lake Victoria with Nile perch. Up until 1978, the Nile perch *Lates niloticus* accounted for less than 2% of the lake's fish biomass and the haplochromines accounted for 80% of the biomass. Between 1974 and 1978, there was hardly any stock assessment done in the Lake. In the 1980s, an abrupt change was noticed in Kenyan waters, and later on in Ugandan and Tanzanian waters also. The Nile perch suddenly jumped up to 80% of the lake's fish biomass, while the haplochromines dropped to 1%.

Ecological disaster

Before the 1970s, Lake Victoria had more than 350 species of fish from the cichlid family, of which 90% were endemic. However, the introduction of the Nile perch and tilapia caused a collapse in the lake's biodiversity. It also resulted in deforestation since wood was needed to dry the oily perch, while the haplochromines and other native cichlids could be sun dried. Forest clearing in turn increased siltation and eutrophication in the lake, putting in jeopardy the Nile and tilapia fishery (WRI 2001). A fishery that once drew on hundreds of species, mostly endemic, now relies on just three: a native pelagic cyprinid, *Rastrineobola argentea,* the introduced Nile perch *Lates niloticus,* and tilapia *Oreochromis niloticus* (Rabi 1996).

Socioeconomic impact

At first, official concern was for the problems the Nile perch posed on shore. The fish was big and could grow to nearly 2 meters, thus fishers needed bigger gear to deal with it. Also the villagers did not know how to cook the oily fish and could not sun-dry it. There were no markets for the fish, prices were low, and most of the fish was left on the beach to rot (Rabi 1996). Using U.N. funds, a Kenya Marine Fisheries Research Institute (team toured lakeside villages and Nairobi hotels, demonstrating how to fillet, freeze, smoke, and cook the fish. Foreign aid groups and investors moved in with processing plants and refrigerated trucks. The price for Nile perch soared and the local people could not afford the high prices. Shoes, belts, and purses were made from tanned perch hide and the dried swim bladder was sent to England for filtering beer and to Asia for making soup stock.

Before the introduction of the Nile perch, the native fish of Lake Victoria were harvested by small-scale fishers and processed and traded by women for local consumption. With the introduction of the Nile perch, large boats began to haul out the perch in tons on the open lake, where local fishers could not take their canoes. Many rich investors see the perch as being economically useful because it brings in dollars. The fish is sold to processing plants built along the Kenyan and Ugandan shore by investors from Asia, Europe and Australia. The fish is filleted, frozen, boxed, and loaded on trucks headed for the port of Mombassa, Kenya, where it is shipped to Europe and the Far East.

The traditional ways of the local people have been severely disrupted. They have moved to squatter camps near the fish-processing factories and are left with the scraps of *Lates* from the food-processing factories. The fleshy heads and tails are fried and sold to the local people as they are the only fish most local people can afford (Rabi 1996). Lake Victoria's Nile Perch Fishery generates as much as $400 million in export income, but few villagers around the lake benefit from it. While tons of perch find their way to restaurants in Europe, scientists have documented protein malnutrition around the lake (WRI 2001).

The revenues generated by the Nile perch fishery are much greater than those realized from the lake's native species. The distribution of wealth resulting from the Nile perch fishery is also different from the original artisanal fishery. Most of the local fishers are actually worse off, while large-scale operations that exploit the introduced species for foreign currency are doing well.

Conclusions

An estimated 30 million people depend on Lake Victoria, a lake whose natural resources are under increasing stress (Fuggle 2001). The population on the shore has grown fast over the past century, with corresponding increases in the demand for fish and agricultural products. Following the introduction of gill nets by European settlers at the beginning of the twentieth century, populations of indigenous fish species declined. Many were specially adapted to eat algae, decaying plant material, and snails that host the larvae of Schistosomes, which cause bilharzias in humans. As the lake started to eutrophicate, people became more vulnerable to disease.

As fish catches declined, non-native species were introduced, causing further stress to indigenous fish. The greatest impact resulted from the introduction of the Nile perch (*Lates niloticus*) in the 1960s as the basis of commercial freshwater fisheries. This had repercussions on the local fishing economy and distribution of wealth. Local people who had previously met most of their protein requirements from the lake began to suffer from malnutrition and protein deficiency. Although 20,000 tons of fish are exported annually to European and Asian markets, local people can only afford fish heads and bones from which the flesh has been removed.

Wetlands around the lake have been converted to grow rice, cotton, sugarcane, and their original function as natural filters for silt and nutrients has been lost. Run-off now carries soil and excess nutrients from the cultivated areas straight into the lake. The resulting algae growth clouds the surface water and reduces oxygen availability, seriously affecting the habitat of endemic fish species, which prefer clear waters, while their predator, the Nile perch, thrives in such murky waters. This further aggravates food insecurity in lakeside communities.

Increased nutrients, much of which are in the form of sewage, have stimulated the growth of the water hyacinth (*Eichornia crassipes*), one of the world's most invasive plants. This has seriously affected water transport and paralyzed many local fisheries. By the end of 1997, the 70% decline in economic activity reported at Kisumu port was attributable to water hyacinths choking the port and fish landings. The dense cover of water hyacinth also stimulated secondary weed growth and provided habitats for snails and mosquitoes.

Stock enhancement (to increase yields over normal levels) has been carried out in marine systems (notably for species of scallop), but is limited as a general approach by the economics of production, ability to harvest stocked individuals as adults, and returns on yields.

6.3.6.2.3 Marine reserves

Marine reserves have been proposed as a remedy for overfishing and declining marine biodiversity, but concern that such reserves would inherently reduce yields has hampered their implementation. However, some research has demonstrated that marine reserves deliver fishery benefits beyond their own boundaries because species inside the reserve rapidly increase in numbers, grow larger, and have more reproductive potential. Part of the population (as larvae or as adults) migrates outside the reserve, increasing yields for fishermen in the surroundings (Gell and Roberts 2003). The positive effects have been shown for resident and more mobile populations, but a wider application of the ecosystem and precautionary approaches is essential.

According to current research, the effectiveness of marine reserves in recovering fish stocks is influenced by a number of conditions, including the size, design, and location of the reserve, the life history and behavioral pattern of target fish species, how depleted the fish stock is when restoration begins, how much fishing has contributed to the decline of the fish stock, and how long the reserve remains closed to fishing (Ward et al. 2001).

Unfortunately, much remains unknown about marine reserves or how to maximize their benefit. We do not yet know if it is feasible to establish a network of reserves sufficient to recharge stocks and sustain the modern fishing industry at the same time, given that modeling studies indicate that as much as 20–50% of the range of a target fish population might have to be protected from all exploitation in order to sustain the fish stock over the long term (Ward et al. 2001). With this level of uncertainty, it will undoubtedly be very difficult for many politicians and fishers to support the kind of large and long-lasting closures in heavily fished waters, which fish recovery via a marine reserve system would call for. On the other hand, conventional fishery management approaches, such as quota systems and seasonal closures, also do not guarantee fish recovery and require concessions from fishers too. Moreover, the commitment nations made at the Johannesburg Summit to restore stocks is too ambitious to rely on these traditional approaches only, adding pressure to explore the marine reserve option further.

6.3.7 Livestock Management

The demand for livestock and livestock products has led to an intensification of livestock production systems, first in industrial countries and more recently in developing countries. Intensive (factory-farming) systems for dairy production and beef are most common in North America, while intensive poultry and swine production is common worldwide. The intensification of systems in developed countries has led to several concerns. These include nutrient pollution, rapid transmission of infectious disease agents, food safety, and animal welfare concerns. This section examines responses related to livestock production and their impacts on regional and global ecosystems.

6.3.7.1 Industrial or Intensive Livestock Production Systems

Environmental impacts of industrial or intensive livestock production are varied and important, mostly related to the disposal of manure. In the developing world, these systems are found mostly near urban areas in Asia and Latin America. Impacts include (de Haan 1997):

- Excretion of nitrogen and phosphate. Pigs and poultry excrete 65–70 % of their nitrogen and phosphate intake. Nitrogen can evaporate in the form of ammonia with toxic effects. Much of it is lost to the atmosphere as nitrous oxide, a greenhouse gas. Nitrates are leached into the groundwater posing human health hazards. Phosphorus saturation can also lead to eutrophication.
- Excretion and digestion of protein. Pigs excrete 70% of the protein in their feed, and chickens 55%. Ammonia from the digestion of protein acidifies the soil, causes acid rain, etc.
- The application of manure can cause N and P loss, N leaching into the water as nitrates, and contamination of surface waters leading to eutrophication.
- Anaerobic decomposition of manure releases large amounts of methane when stored in liquid form.
- Biodiversity may be reduced because of: (1) high demand for concentrate feed may create a need to clear more land for feed production, (2) the effect of waste on terrestrial and aquatic systems, and (3) requirement for uniform animals in large operations.
- In addition to these environmental problems, human health can be affected by zoonoses common in intensive production systems.

6.3.7.2 Mixed Crop–Livestock Systems: Increased Efficiency or Nutrient Mining?

The transfer of nutrients between soils, crops, and animals is an important environmental issue of modern agriculture. Animals harvest nutrients from the environment through the intake of feed, forages and crop residues. In time, some of those nutrients, whether metabolized or in their natural form, are recycled back to the ecosystem. The type of production system heavily influences the degree of interaction between crops and livestock, and the nature of nutrient cycling. Sere and Steinfeld (1996) classify livestock production as industrial, mixed farming or grazing systems. As systems transform from grazing to industrial, the extent of nutrient transfer between soil, crops/pastures and animals moves beyond the farm level to a national, regional and even global scale. In addition, the environmental impact of nutrient cycling by crop–livestock interactions is also determined by the state of nutrient balances in agricultural land. At a global scale, nutrient-deficient and nutrient-surplus regions are identified. Agricultural areas with serious nutrient depletion are widespread in Africa, Latin America, and marginal lands of Asia (Stoorvogel et al. 1993; Craswell et al. 2004). In contrast, nutrient-surplus regions are found in Western Europe, some areas of Eastern Europe, in the eastern and mid-western United States, in Southeast Asia and the large plains of China (Steinfeld et al. 1997; Craswell et al. 2004).

Thus in general terms, it is in industrial countries that nutrient-surplus areas are more common, and where more intensive industrial livestock production takes place, and in the developing world where drastic nutrient depletion of soil occurs and grazing and mixed farming systems are widely spread. It should be noted, however, that many of the environmental impacts of industrial-type livestock systems could also be found in large cities of developing countries with large concentrations of animals in peri-urban farms.

6.3.7.2.1 Industrial systems and use of manure in nutrient-surplus regions

A common feature of industrial systems is that the locations of feed production, livestock feeding, and manure and urine disposal are not geographically contained within the same farming system. Thus nutrients are exported from feed-producing regions and imported to the soils where manure is applied. Inappropriate manure application in nutrient-surplus areas can produce excess nitrogen and phosphorus, which leaches or runs off, polluting groundwater, aquatic, and wetland ecosystems, or leaves high levels of nitrates, phosphates. and potassium in the soil (Steinfeld et al. 1997; Zaccheo et al. 1997; Chamber et al. 2000; Craswell et al. 2004). Furthermore, deficient management of excessive amounts of manure/slurry can increase the production of volatile ammonia (Chamber et al. 2000) and methane (Yamaji et al. 2003).

6.3.7.2.2 Mixed farming/grazing systems and use of manure in nutrient-deficient regions

Due to the nature of mixed farming and grazing systems, nutrients voided in urine and manure have the potential to be recycled within the farming system. Several authors agree that manure application can be a valuable means of improving the characteristics of nutrient-poor soils; it has been reported that manure increases soil fertility, organic matter, pH, and water-holding capacity (Olsen et al. 1970; McIntosh and Varney 1973; Mugwira 1984; Fernández-Rivera et al. 1995; de Haan et al. 1997; Ehui et al. 1998). Urine can account for 40–60% of total N excretion, and although urinary-N is rapidly available for plant use, it is also easily lost through volatilization and leaching. Fecal-N on the contrary, is released in the soil at a slower rate, providing a sustained N supply that is better synchronized with crop demands. The amount of phosphorus voided in feces is relatively high and its excreted form is easily available for plant growth. Where soils are P-deficient, as is the case for most soils of sub-Saharan Africa, manure application on agricultural land improves P-availability for crops.

In grazing systems, most of the nutrients are recycled in situ, as urine and manure are spread in the areas where animals graze, although some of the nutrients are brought into the farm when animals are penned at night. While some volatilized N has the potential to return to the landscape after combining with rainfall, most P is not redistributed to grazing areas, but builds up where livestock is kept (Augustine 2003). Manure collection and management is easier in mixed farming than in grazing systems and enables farmers to make decisions on the use of manure—usually concentrating manure in the fields with valuable crops. With this practice, nutrients are exported from one field to the other and fertility gradients across fields within the same farm are generated (Prudencio 1993; Ramisch 2004).

Most of the plant material available for feeding cattle in mixed farming systems of the developing world is crop residue with poor nutrient content. In this case, feeding low-quality forage to livestock and subsequently using the manure as fertilizer helps to maintain the viability and sustainability of the system. In general, the lower the quality of the feed, the more beneficial the application of manure over the raw plant material (Delve et al. 2001). It is important to consider that, in the context of the sustainability of crop–livestock systems, the long-term benefits of manure application for soil characteristics are far more important than its short-term role as a nutrient provider (Murwira et al. 1995).

6.3.7.2.3 Nutrient cycling by crop–livestock interactions: policy and research for the future

The current state of agriculture and its global effects have focused increased attention on the way human-induced activities are altering the recycling of nutrients between soils, crops, and livestock. Crop–livestock interactions, which transfer nutrients through the collection, storage, and application of manure are causing an impact beyond the limits of the farming system, and their nature largely depends on regional economic characteristics.

Agriculture in developed countries, where nutrient-surplus regions prevail, is producing and adopting better guidelines for the storage and application of manure (Chambers et al. 2000; Salazar et al. in press). Some countries are enforcing the adoption of nutrient accounting schemes at the farm level (Breembroek et al. 1996; Craswell et. al. 2004), environmental costs are moving to the center of the debate of nutrient cycling, and legislation on water pollution and waste disposal is put in place. The international trade of feed and livestock products and their influence on national nutrient balances has also entered the global economic and environmental debate (for example, Lindland 1997; McCalla and de Haan 1998).

The panorama is quite different for the developing world where it has been forecast that the demand for livestock products will double by the year 2020 (Delgado et al. 1999), and that the intensification of crop–livestock systems in these countries is the path to meet this demand (McIntire et al. 1992; Steinfeld et al. 1997). As the intensification of mixed farming systems increases, the role of manure as the means to recycle nutrients will become essential. In some nutrient-deficient regions, however, manure itself cannot restore the fertility of soils that have been nutrient-mined continuously for many years. In these cases, calculations show that the amount of manure necessary to compensate for soil nutrient losses is, in practice, unattainable (Murwira et al. 1995). Research in crop–livestock systems in developing countries is focusing on better understanding of the use of manure and its interaction with other organic and inorganic fertilizers to enhance the viability of current production systems (for example, Thorne and Tanner 2002; Sanginga et al. 2003; Chikowo et al. 2004).

6.3.7.3 Pastoral Ecosystems of the Developing World: Causes, Change Processes, and Impacts

A suite of strong pressures within and outside pastoral systems are currently driving change in these systems in the developing world. Pastoral land use and the extent of rangelands around the world are contracting, principally through conversion to other land uses (croplands principally, but also into protected areas and urban land-use), but also because of a hypothesized loss in function of the remaining rangelands (Niamir-Fuller 1999). However, in Central Asia, the removal of the policy of collectivization has resuscitated traditional pastoralism in the last decade as pastoralists are no longer supplied with inputs like fencing and veterinary care, and thus the advantages of settled life are gone (Blench 2000). Globally, the tenure in pastoral systems is increasingly becoming privatized and customary political and management systems are becoming weaker (Galaty 1994; Niamir-Fuller 1999), limiting pastoral access to crucial key resources (swamps, deltas, riverine areas) in much of Africa and in Central Asia (de Haan 1999).

Some national governments invoke policies to settle pastoralists in villages (Galaty 1994). In Africa, several decades of drought (Nicholson et al. 1998) have coincided with high human population growth, adding further pressure on pastoral lands, especially those with higher potential for cultivation in the semiarid zone (Galaty 1994). In North Africa and West Asia, the expansion of irrigation has pushed pastoralists into very arid ecosystems, but this is not irreversible: the very opportunism inherent in pastoralism means that if the skills are not lost, pastoral systems can be revived (Blench 2000). Livestock development projects are also driving change in pastoral lands by opening up remote pastures with the spread of bore hole technology, and fragmentation of rangelands by veterinary cordon fences, particularly in southern Africa.

The impacts of the changes are numerous and include changes in land use, overgrazing, competition and synergies between livestock and grazing, carbon sequestration, and dust formation.

Changes in land use (expansion of cultivation and settlement): Expansion of cultivation fragments rangeland landscapes when farmers convert rangeland into cropland (Hiernaux 2000), leading to strong wildlife and vegetation losses (Serneels and Lambin 2001). Fragmentation can also occur when fence lines are built to prevent the spread of disease or to prevent wildlife from foraging in enclosed pastures. Often, fences exclude all domestic and wild grazers from key resources like swamps, riverine areas, and other productive areas.

Overgrazing: Overgrazing is the loss of ecosystem goods or services through heavy livestock grazing (over-trampling is implicitly included). Heavy livestock grazing occurs where livestock concentrate: around pastoral settlements, around water points, along animal tracks (Hiernaux 1996), and in open access pasture. If driven by cultivation, this overgrazing occurs in the wet season, when rangelands are most sensitive to grazing (Hiernaux 2000). In the Sudan, this concentration of livestock leads to loss of vegetative cover and accelerates erosion (Ayoub 1998). By contrast, heavy livestock grazing around pastoral settlements in arid areas (169 millimeters rainfall) of Namibia had minor impacts on woody vegetation and biodiversity, with impacts confined within the settlements themselves (Sullivan 1999). In the Kalahari, overgrazing around settlements often converts grassland into bush land within about two kilometers of the settlement (Dougill and Cox 1995). Across southern Africa, woody vegetation has replaced palatable grass species in heavily grazed areas (bush encroachment), caused by grazing pressure rather than climate (Perkins 1991; Skarpe 1990).

Competition and synergies between livestock and wildlife: By contrast, the impacts of heavy grazing on wildlife may be exactly the opposite of that on vegetation: greater impacts in drier than wetter ecosystems. Wildlife appears to avoid heavily grazed areas completely in arid northern Kenya (De Leeuw et al. 2001) but livestock mix more closely with wildlife in semiarid rangelands in southern Kenya. Around Samburu pastoral settlements in these arid lands, Grevy's zebra graze away from the settlements during the day, but move close to them during the night (Williams 1998). Disease transmission is a potential issue where livestock keeping and wildlife overlap)

Carbon sequestration: It is not clear whether current changes in rangelands (land-use change, overgrazing, fragmentation) are causing a net release or net accumulation of carbon, both above and below ground. Expansion of cultivation into rangelands probably strongly reduces carbon below ground, but may increase carbon above ground if farmers plant significant numbers of trees. If overgrazing converts grassland to bush land, then above-ground carbon will increase, but below ground carbon may decrease. In addition, rangelands are a significant carbon sink (IPCC 2000), but the potential of these areas for further sequestration may be difficult to realize (Reid et al. 2003).

Dust formation: Livestock grazing may be partly responsible for the dust plume that forms in the Sahel and moves west over the Atlantic Ocean each year (Nicholson et al. 1998), but this is unknown. The generation of dust in different land use types and in different geomorphologic positions needs to be measured to assess how livestock affects this source.

6.3.7.4 The Role of Livestock Production in Deforestation around the Globe

Pasture creation and cattle ranches have been blamed as major driving forces behind deforestation in Latin America, particularly in Brazil (Hecht 1985; Downing 1992; Kaimowitz 1996; Walker et al. 2000). The amount of land that has been converted from tropical rain forest to pasture, as well as the rate of conversion, is under reconsideration as the higher quality and resolution of new satellite imagery is aerially revealing more details of the extent of deforestation and clearance of secondary forests. Nevertheless, estimates are that approximately half of the deforestation in the Amazon is due to pasture generation.

Above and beyond the global environmental consequences, the conversion of land from humid tropical forest to pasture would appear to be inherently unsustainable given the low soil fertility and abundance of invasive weeds and woody species. The presence of the invasive, sometimes toxic species necessitates frequent burning. These factors behind pasture degradation reduce stocking densities and restrict the long-term use of the land for ranching (Walker et al. 2000). Heavily degraded land is difficult and uneconomic to recuperate, so the clearing for pasture often condemns the land to waste; it is estimated that half the area cleared in the Amazon for pasture has been abandoned (Hecht 1989; Faminow 1998). Despite the environmental implications and limitations, ranching continues to be profitable and conversion of forest to pasture continues (Arima 1997).

Intensifying production on the agricultural frontier may reduce the pressure to clear further forest. However, the assumption that intensifying productivity on the frontier will reduce deforestation might not hold (Hecht 1989; Angelson and Kaimowitz 1999). Increasing productivity on the forest edge may well prove to be an attractant, pulling new migrants to the area. Labor saving or income generating technologies may simply reduce production constraints allowing households to put more land under production, leading to additional land clearance. Perhaps most importantly, improved technologies do not address the underlying forces behind deforestation, such as migration push and pull factors, government programs, and the desire to clear land in order to claim ownership.

6.3.7.5 Livestock and Greenhouse Gas Emissions

The primary greenhouse gases emissions from agriculture are carbon dioxide (CO_2), methane (CH_4), and nitrous oxide (N_2O). Livestock contribute to emissions of these gases in several ways. At the animal level, rumen fermentation creates methane gas, which is 24.5 times more powerful as a greenhouse gas than CO_2 (IPCC 1995). Livestock emit 16% of all methane globally (IPCC 1995). Most of these emissions come from ruminants: cattle, sheep, and goats. Methane contributes 30% of the global warming potential of all agricultural emissions (Kulshreshtha et al. 2000). Besides livestock, the other main source of methane in agricultural systems is rice paddies (IPCC 1995). Livestock are responsible for about 110 million tons CH_4 per year, while rice paddies contribute about 60 million tons.

Methane production from livestock is controlled both by the level of productivity per animal and the quality of the feed. More productive animals emit less methane per unit product because a lower proportion of methane produced is used for maintenance (de Haan 1997). Methane production in livestock is also sensitive to diet quality and to the timing and amounts of forage fed (feed management strategies). Higher quality feed allows more efficient operation of the rumen and thus more efficient digestion; this results in more output per unit input (meat, milk) and reduced methane emissions per unit intake (Kurihara et al. 1999).

At the field level, improved forage for livestock has a role in reducing methane emissions, but may also function as a carbon sink. Fodder trees and shrubs sequester significant carbon above ground and store more carbon below ground than cereal-based cropping systems (IPCC 2000).

Another greenhouse gas, N_2O, with 320 times the warming strength of CO_2, is emitted during storage of liquid manure, produced either by ruminants or non-ruminants (IPCC 1995). Liquid manure is usually stored in highly intensified livestock systems for later use or sale. On the other hand, livestock can also reduce N_2O emissions by the substitution of manure for inorganic fertilizers. These fertilizers are a major source of N_2O emissions, and thus more substitution with organic manure is an important avenue for emission reduction.

At the landscape-level, the largest impact of livestock on greenhouse gas emission is through deforestation. Tropical forest systems are of great concern because they hold more than half of the world's above ground carbon (IPCC 2000). In the forests of Brazil, Peru, Cameroon, and Indonesia, pastures and grasslands hold 1% of the above ground carbon stocks of primary forest, less than short-term fallow cropping systems (Palm et al. 1999). These same pastures contain only 20% less carbon below ground compared with forests. Improved pastures do not improve carbon sequestration substantially over short fallow and degraded pasture systems (Palm et al. 1999).

Livestock also contribute to greenhouse gas emissions in rangelands, although conversion of rangeland to cropland is a large source. As rangelands contract and are converted to cropland, 95% of the above ground carbon can be lost and more than 50% of the below ground carbon (calculations based on IPCC 2000 figures). Within rangelands, overgrazing is the principle cause of loss of soil carbon (Ojima et al. 1993). Thus improved grazing management can have direct and substantial effects on soil carbon pools (IPCC 2000). In addition, heavy grazing and changes in fire regimes can convert grassland systems to bush land systems. This conversion may increase carbon above ground (Boutton et al. 1998), but it is not clear how carbon below ground is affected.

Lastly, the projected doubling of demand for livestock products over the next 20 years (Delgado et al. 1999) is likely to strongly increase the emissions of greenhouse gases into the atmosphere. Even though improvements in nutrition will likely lead to reduced emissions from livestock, the sheer increase in numbers necessary to meet demand will probably overwhelm any reductions in gas production through nutrition. The predicted rapid expansion of industrial livestock systems will create the potential for strongly increased production of nitrous oxide from excessive manure. Further, demand for livestock products is also likely to fuel more clearing of natural vegetation for ranching in the rain forest and to produce concentrate feed for burgeoning livestock populations.

6.4 Conclusion

The challenge of food and ecosystems in the twenty-first century will require comprehensive assessments and knowledge at the local, national, and international level of agroecological and socioeconomic conditions, in order to ensure that farmers can produce food in a manner that is environmentally, economically, and socially sustainable and that consumers have the opportunities to make choices regarding food that is nutritious and healthy, safe and affordable.

The scientific challenges of producing the food needs of the world will require targeted and prioritized agricultural research with the participation of farmers.

Specific research focusing on the differential vulnerability of farmers, as well as, ecosystems will be required. Additionally, the threats of global environmental change, for example, agricultural impacts of climate change, will need to be explicitly considered in mobilizing agricultural research efforts. This is especially true, since a long time horizon is required from initiation of research to local farm level implementation. The scientific challenge ahead is a formidable one and can only be met through national and international commitments to use the opportunities that science must provide in the coming decades.

In the second half of the twentieth century, the Green Revolution substantially reduced the risks of mass starvation and famines in the developing world. The international agricultural research efforts over two decades from the 1940s on focused on breeding high-yield dwarf wheat that increased yields two to threefold. In the following three decades (1960s to 1990s), this Green Revolution enabled world cereal production to increase threefold on about the same land acreage. Without this success, world farmers would have had to increase cereal harvested land area from 650 million hectares to over 1,500 million hectares, with all the environmental consequences of forest clearance and loss of biodiversity.

Beyond the main cereal crops, high yielding varieties of millet, sorghum, cassava, and beans among others were also developed in partnerships between national agricultural research systems in developing and developed countries together with CGIAR.

The scientific and technological experiences of the last half century, including the remarkable progress in science-based conventional breeding, will need to be combined with safe and ethical biological sciences—genomics and molecular genetics, physiology, and informatics research, as well as improved crop and land management systems, caring and environmentally sound livestock production, and fish farming. The developments in geographical information systems, including remote sensing and the increasing quality and coverage of sub-national, national, and global resource databases of soils, climate, land cover, etc., together with methodologies for crop, livestock, and fish productivity assessment and mathematical modeling tools need to be systemically integrated to ensure spatial sustainability. The consideration of a number of other issues, such as the increasing privatization of agricultural research and patenting, will also be critical to ensure that the millions of poor farmers are not bypassed by the new breakthroughs in agricultural science and technology.

The national and international agriculture research system faces a formidable challenge to harness the power of science, including:

- *Using science responsibly.* The emerging scientific tools of cellular and molecular biology can shorten the time and cut the costs required to develop innovative food varieties. Biotechnology tools can introduce genes that counter soil toxicity, resist insect pests, and increase nutrient content. Still, the questions of determining appropriate levels of risk and the ethics and societal acceptance of manipulating genetic material need to be resolved before the potential of biotechnology and genetic engineering can be fully realized.
- *Ensuring ecological sustainability.* New scientific tools will need to be combined with knowledge about natural resources in order to ensure sustainable and productive use and avoid inefficient water use, loss of arable lands and productivity declines,

deforestation, pollution and destruction of ecologically critical watersheds, loss of biodiversity, and health and environmental risks of intensive livestock production and fish farming.

- *Harnessing the Information Revolution.* The phenomenal potential of the information and communication revolution including the Internet, remote sensing, GIS, etc., can enable interactive global agricultural research systems combining the best of science with traditional knowledge.
- *Integrating ecology and socioeconomy.* The progress in understanding the functioning of ecological systems; the compilation of agricultural resources databases at sub-national, national, and global levels; and the development of analytical and mathematical modeling tools will be critical to enable spatially relevant application of the results of agricultural research to ensure that the best choice are made at the sub-national level in the context of national needs within a world food economy.

The poor need the deliverance of the promise of science, and without a global partnership and responsible commitment to productive and sustainable agriculture, there can be little progress toward reducing hunger, poverty, and human insecurity.

Governments, civil society, and the private sector around the world must provide the means for mobilizing science and research for food and agriculture. A participatory worldwide effort, building on the lessons and experiences of the last Green Revolution combined with the best of new agricultural sciences can enable the next agricultural revolution to meet world-wide food needs in the twenty-first century, with environmental, economic, and social sustainability.

One of the main lessons learned from the analysis of the responses is that the impacts on ecosystems from attempts to increase food production have been realized mostly as secondary effects, and as such they often represent negative externalities of agricultural production. These externalities have been ignored by small-scale agents, like individual farmers, in their decision-making processes, but also by governments in their effort to attain primary targets regarding food production. Externalities have also been ignored in non-cooperative situations emerging in international competition or in the presence of transboundary or global problems. Since these impacts have had a profound effect on the current state of well-being, but hold the potential for even more dramatic negative impacts on the capacity of the ecosystem to provide future services, it is essential that proper measures be undertaken in the present time.

Because the quantification of some of these impacts is uncertain, the design of policies could consider the precautionary principle, by taking into account "worst-case" scenarios and potential irreversibility in ecosystems, such as those experienced for example, in the Aral Sea disaster and the collapse of the Newfoundland Cod Fishery. New analytical approaches of decision-making under deep uncertainty, such as robust control methods, might prove helpful in designing policies following the precautionary principle.

The need to mitigate impacts on ecosystems and sustain their capacity for future generations makes necessary the introduction of appropriate regulatory frameworks at all levels from local to global, that will control for the externalities affecting the capacity of ecosystems to sustain their food provisioning services. Regulation is not without cost, but this cost basically represents the cost of using the services of the ecosystems for producing food. This service is currently largely unpaid, due to well-known reasons associated with missing markets and lack of well-defined property rights. Water pricing is an example of how governments are coming to grips with the valuation of scarce resources and essential environmental services. Other environmental services must be similarly valued and paid for to ensure their appropriate exploitation and the sustainability of production systems. If this cost is ignored, as has thus far generally been the case, then the capacity of ecosystems to maintain or even enhance their food provisioning services is at risk.

References

Altieri, M., 1996: *Agro Ecology: The Science of Sustainable Agriculture,* Westview Press, Boulder, CO.

Alverson, D.L., H.H. Freeberg, S.A. Murawski, and J.G. Pope, 1994: A global assessment of fisheries by-catch and discards, FAO Fisheries technical paper no. 399, FAO, Rome, Italy.

Angelson, A. and D. Kaimowitz, 1999: Rethinking the causes of deforestation: Lessons from economic models, *The World Bank Research Observer,* **14(1),** pp. 73–98.

Arima, E., 1997: Ranching in the Brazilian Amazon in a national context: Economics, policy and practice, *Society and Natural Resources,* **10(5),** pp. 433–51.

Arrow, K., P. Dasgupta, L. Goulder, G. Daily, P. Ehrlich, et al., 2004: Are we consuming too much? *Journal of Economic Perspectives,* **18(3),** pp. 147–72.

Augustine, D.J., 2003: Long-term, livestock-mediated redistribution of nitrogen and phosphorus in an East African savanna, *Journal of Applied Ecology,* **40,** pp. 137–49.

Ayoub, A.T., 1998: Extent, severity and causative factors of land degradation in the Sudan, *Journal of Arid Environments,* **38,** pp. 397–409.

Bailey, A.P., W.D. Basford, N. Penlington, J.R. Park, J.D.H. Keatinge, et al., 2003: A comparison of energy use in conventional and integrated arable farming systems in the UK, *Agriculture, Ecosystems & Environment,* 97(1–3), pp. 241–53. Available at http://www.sciencedirect.com/-aff1.

Baldock, D., J. Dwyer with J.M. Sumpsi Vinas, 2002: *Environmental Integration and the CAP,* A report to the European Commission, DG Agriculture, Institute for European Environmental Policy. Available at http://europa.eu.int/comm/agriculture/envir/report/ieep_en.pdf.

Barkat, A. and P.K. Roy, 2001: Marine and coastal tenure/community-based property rights in Bangladesh: An overview of resources, and legal and policy developments, Paper presented in the workshop on marine and coastal resources and community-based property rights, 12–15 June 2001, Anilo, Batangas, Philippines.

Bartley, D. and C.V. Casal, 1999: Impacts of introductions on the conservation and sustainable use of aquatic biodiversity, *FAO Aquaculture Newsletter,* **20,** pp. 15–7.

Bartley, D., 2000: *Responsible ornamental fisheries,* FAO Aquaculture Newsletter, Vol 24, pp. 10–14.

BBS (Bangladesh Bureau of Statistics), 2000: *Statistical Yearbook of Bangladesh,* BBS, Dhaka, Bangladesh.

Bjornlund, H. and J. McKay, 2002: Aspects of water markets for developing countries: Experiences from Australia, Chile and the US, *Environment and Development Economics,* **7,** pp. 769–95.

Blench, R. 2000: *You can't go home again, extensive pastoral livestock systems: Issues and options for the future,* Overseas Development Institute/FAO, London, UK.

Boutton, R.W., S.R. Archer, A.J. Midwood, S.F. Zitzer, and R. Bol, R., 1998: ^{13}C values of soil organic carbon and their use in documenting vegetation change in a subtropical savanna ecosystem, *Geoderma,* **82,** pp. 5–41.

Boyd, C., J. Clay, 1998: Shrimp aquaculture and the environment, *Scientific American,* **58,** pp. 59–65.

Braden, J., and K. Segerson, 1993: Information problems in the design of non-point-source pollution policy. In: *Theory, Modeling and Experience in the Management of Nonpoint-Source Pollution,* C. Russel and J. Shogren (eds.), Kluwer Academic Publishers, Dordrecht, The Netherlands.

Breembroek, J.A., B. Koole, K.J. Poppe, and G.A.A. Wossink, 1996: Environmental farm accounting: The case of the Dutch nutrients accounting system, *Agricultural Systems,* **51,** pp. 29–40.

Brouwer, J. and J. Bouma, 1997: *Soil and Crop Growth Variability in the Sahel,* Information Bulletin 49, International Crops Research Institute for the Semi-Arid Tropics, Sahelian Center, Naimey, Niger.

Brouwer, J., L.K. Fussell, and L. Herrmann, 1993: Soil and crop growth variability in the West African semi-arid tropics: A possible risk-reducing factor for subsistence farmers, *Agriculture, Ecosystems and Environment,* **45,** pp. 229–38.

Brouwer, J. and J.M. Powell, 1998: Micro-topography and leaching: Possibilities for making more efficient use of nutrients in African agriculture. In: Nutrient balances as indicators of productivity and sustainability in sub-

Saharan African agriculture, E.M.A. Smaling (ed.), *Agriculture, Ecosystems and Environment,* **71 (1/2/3),** pp. 229–39.

Chakravorty, U. and T. Swanson, 2002: Economics of water: Environment and development, Introduction to the special issue, *Environment and Development Economics,* **7,** pp. 733–50.

Chamber, B.J., K.A. Smith, and B.F. Pain, 2000: Strategies to encourage better use of nitrogen in animal manures, *Soil Use and Management,* **16,** pp. 157–61.

Chikowo, R., P Mapfumo, P. Nyamugafata, and K.E. Giller, 2004: Maize productivity and mineral N dynamics following different soil fertility management practices on a depleted sandy soil in Zimbabwe, *Agriculture, Ecosystems and Environment,* **102,** pp. 119–31.

Clark, C., 1990: *Mathematical Bioeconomics: The Optimal Management of Renewable Resources,* 2nd ed., Wiley Interscience, Hoboken, NJ.

Coche, A.G., B. Haight, and M. Vincke, 1994, *Aquaculture Development and Research in sub-Saharan Africa: Synthesis of National Reviews and Indicative Action Plan for Research,* Central Institute of Freshwater Aquaculture technical paper no. 23, FAO, Rome, Italy, 151 pp.

Conway, G., 1999: *The Doubly Green Revolution: Food for All in the 21st Century,* Penguin, London.

Copes, P., 2000: ITQs and fisheries management: With comments on the conservation experience in Canada and other countries, Address to the Unión del Comercio, la Industria y la Producción of Mar del Plata City, Argentina, 29 June 2000. Available as discussion paper 00–1, Simon Fraser University, Institute of Fisheries Analysis, Burnaby, BC, Canada, 21 pp.

Cosgrove, W. J., and F.R. Rijsberman, 2000: *World Water Vision: Making Water Everybody's Business,* Earthscan Publications, London, UK.

Craswell, E., U. Grote, J. Henao, and P. Vlek, 2004: Nutrient flows in agricultural production and international trade, Ecological and policy issues no. 78, Discussion paper on development, Center for Development Research, Bonn, Germany.

Cunningham, S., D. Greboval, 2001: Managing fishing capacity: A review of policy and technical issues, FAO Fisheries technical paper, FAO, Rome, Italy.

DEFRA (Department of the Environment, Food and Rural Affairs), 2002: Sugar Beet and the Environment in the UK, Report by the United Kingdom in accordance with Article 47(3) of Council Regulation 1260/2001, on the environmental situation of agricultural production in the sugar sector. Available at http://www.defra.gov.uk/corporate/consult/eisugar/report.pdf.

De Haan, C., 1999: Future challenges to international funding agencies in pastoral development: An overview. In: *International Rangelands Congress,* D. Freudenberger (ed.), Townsville, Australia, pp. 153–55.

De Haan, C., H. Steinfeld, and H. Blackburn, 1997: *Livestock and the Environment: Finding a Balance,* WRENmedia, Fressingfield, UK.

De Leeuw, J., M.N. Waweru, O.O. Okello, M. Maloba, P. Nguru, et al., 2001: Distribution and diversity of wildlife in Northern Kenya in relation to livestock and permanent water points, *Biological Conservation,* **100,** pp. 297–306.

Delgado, C., M. Rosegrant, H. Steinfeld, S. Ehui, and C. Courbois, 1999: *Livestock To 2020: The Next Food Revolution,* International Food Policy Research Institute, FAO, and International Livestock Research Institute, Washington, DC.

Delgado, C.L., 2003: Meating and milking global demand: Stakes for small-scale farmers in developing countries. In: *The Livestock Revolution: A Pathway from Poverty?,* A.G. Brown (ed.), Record of a conference conducted by the Academy of Technological Sciences and Engineering Crawford Fund, Parliament House, Canberra, Australia, 13 August 2003, A festschrift in honor of Derek E. Tribe, The ATSE Crawford Fund, Parkville, Victoria, Australia.

Delgado, C.L., N. Wada, M.W. Rosengrant, M. Siet, and M. Ahmed, 2003: *Outlook for Fish to 2020: Meeting Global Demand,* A 2020 Vision for Food, Agriculture and the Environment Initiative, October 2003, International Food Policy Research Institute, Washington, DC, and World Fish Center, Penang, Malaysia.

Delve, R.J., G. Cadisch, J.C. Tanner, W. Thorpe, P.J. Thorne, et al., 2001: Implications of livestock feeding management on soil fertility in the smallholder farming systems of sub-Saharan Africa, *Agriculture, Ecosystems and Environment,* **84(3),** pp. 227–43.

Dewar, A., M. May, I. Woiwood, L. Haylock, G. Champion, et al., 2003: A novel approach to the use of genetically modified herbicide tolerant crops, Proceedings of the Royal Society of London Series B, *Biological Sciences,* **270,** pp. 335–40.

DFO (Department of Fisheries and Oceans), 2001: *Aquaculture Action Plan: Enabling Aquaculture to Achieve its full Environmentally Sustainable Potential,* Department of Fisheries and Oceans, Ottawa, ON, Canada.

Dixon J., A. Gulliver, and D. Gibbon, 2001: *Farming Systems and Poverty: Improving Farmers' Livelihoods in a Changing World,* Food and Agriculture Organization of the United Nations, Rome, Italy, and World Bank, Washington, DC, p. 412.

Dougill, A. and Cox, J., 1995: Land degradation and grazing in the Kalahari: New analysis and alternative perspectives, Pastoral development network paper 38c, Overseas Development Institute, London, UK.

Downing, T., 1992: *Development or Destruction: The Conversion of Tropical Forest to Pasture in Latin America,* Westview Press, Boulder, CO.

Dulvy, N.K., Y. Sadovy, and J.D. Reynolds, 2003: Extinction vulnerability in marine populations, *Fish and Fisheries,* **4(1)** (March), p. 25.

Ehrlich, P. and A. Ehrilch, 2002: Population, development, and human natures, *Environment and Development Economics,* **7,** pp. 158–70.

Ehui, S., H. Li-Pun, V. Mares, and B. Shapiro, 1998: The role of livestock in food security and environmental protection, *Outlook on Agriculture,* **27,** pp. 81–7.

Ehui, S., S. Benin, T. Williams, and S. Meijer, 2002: Food Security in sub-Saharan Africa to 2020, Socioeconomics and policy research working paper no. 49, International Livestock Research Institute, Nairobi, Kenya.

Emerson, C., 1999: *Aquaculture Impacts on the Environment,* Cambridge Scientific Abstracts, Cambridge, UK.

Entsua-Mensah, M., A. Lomo, and K.A. Koranteng, 1999: *Review of Public Sector Support for Aquaculture in Africa,* Report prepared for the FAO Africa Regional Aquaculture Review, 22–24 September 1999, Shangri-La Hotel, Accra, Ghana.

European Commission, 1997: *Agriculture and Environment: CAP Working Notes,* EC, Brussels, Belgium.

European Commission, Agricultural statistics, statistical and economic information 2002. Available at http://europa.eu.int/comm/agriculture/agrista/2002/table_en/4321.pdf.

EU (European Union), 2003a: Reforming the European Union's sugar policy, Commission staff working paper, Commission of the European Communities, Brussels, Belgium. Available at http://europa.eu.int/comm/agriculture/publi/reports/sugar/fullrep_en.pdf.

EU, 2003b: Accomplishing a sustainable agricultural model for Europe through the reformed CAP: The tobacco, olive oil, cotton and sugar sectors, Commission staff working paper, Commission of the European Communities, Brussels, Belgium. Available at http://europa.eu.int/comm/agriculture/cap reform/com554/554_en.pdf.

EU, 2003c: Sugar: International analysis, production structures within the EU, Commission staff working paper, Commission of the European Communities, Brussels, Belgium. Available at http://europa.eu.int/comm/agriculture/markets/sugar/reports/rep_en.pdf.

European Commission, 2003: A Long Term Policy Perspective for Sustainable Agriculture: Environmental Impacts, Final report, GRP-P_158, EC, Brussels.

European Court of Auditors, 2002: Annual report concerning the financial year 2001, *The Official Journal of the European Communities,* **45** (November) C 295. Available at http://www.eca.eu.int/EN/RA/2001/ra01_1en.pdf.

Faminow, M., 1998: *Cattle, Deforestation and Development in the Brazilian Amazon: An Economic, Agronomic and Environmental Perspective,* CAB International, Oxford, UK.

FAO (Food and Agriculture Organization of the United Nations), n.d.: Gender, food and security. Available at www.fao.org/gender.

FAO, 1995a: *Dimensions of Need: An Atlas of Food and Agriculture,* FAO, Rome, Italy, pp. 16–98.

FAO, 1995b, *Review of the State of World Fishery Resources: Marine Fisheries,* FAO, Rome, Italy, pp. 1–56.

FAO, 1996: *Food for All,* FAO, Rome, Italy, p. 64.

FAO, 1997: *Review of the State of World Fishery Resources: Marine Fisheries,* FAO Fisheries Department, Rome, Italy.

FAO 1999: *Committee on Agriculture, 15th Session, Organic Agriculture,* FAO, Rome, Italy.

FAO, 2000: *Review of Public Sector Support for Aquaculture in Africa,* Report prepared for the FAO Africa Regional Aquaculture Review, 22–24 September 1999, Shangri-La Hotel, Accra, Ghana.

FAO, 2001a: *Crops and Drops: Making the Best of Water for Agriculture,* FAO, Rome, Italy, p. 22.

FAO, 2001b: *Food Supply Situation and Crop Prospects in Sub-Saharan Africa,* FAO, Rome, Italy.

FAO, 2002a: *The State of World Fisheries and Aquaculture,* FAO Fisheries Department, FAO, Rome, Italy.

FAO, 2002b: *World Agriculture: Towards 2015/2030,* Summary report, FAO, Rome, Italy.

FAO, 2003a: *Trade Reforms and Food Security: Conceptualizing the Linkages, Commodity Policy and Projections Service,* Commodities and Trade Division, FAO,

Rome, Italy. Available at http://www.fao.org/DOCREP/005/Y4671E/Y4671E00.HTM.

FAO, 2003b: *World Agriculture: Towards 2015/2030,* An FAO perspective, J. Bruinsma (ed.), FAO, Rome, Italy, and Earthscan, London, UK.

FAO, 2003c: *Impact of Climate Change on Food Security and Implications for Sustainable Food Production,* Committee on World Food Security, FAO, Rome, Italy.

FAO, 2003d: *Review of World Water Resources by Country,* FAO water report 23, FAO, Rome, Italy.

Fernández-Rivera, S., T.O. Williams, P. Hiernaux, and J.M. Poweel, 1995: Faecal excretion by ruminants and manure availability for crop production in semi-arid West Africa. In: *Livestock and Sustainable Nutrient Cycling in Mixed Farming Systems of sub-Saharan Africa, Volume ii.,* J.M. Powell, S. Fernández-Rivera, T.O. Williams, and C. Renard (eds.), ILCA (International Livestock Centre for Africa), Addis Ababa, Ethiopia, p.568.

Fischer, G., Y. Chen, and L. Sun, 1998: *The Balance of Cultivated Land in China During 1988–1995,* IR-98-047, IIASA, Laxenburg, Austria.

Fischer, G., M. Shah, and H. Velthuizen, 2002: *Climate Change and Agricultural Vulnerability,* World Summit on Sustainable Development, August, Johannesburg, South Africa.

Folke, C., K-G. Mäler, and C. Perrings, 1992: Biodiversity loss: An introduction, *Ambio,* **21(3),** p. 200.

Fuggle, R.F., 2001: *Lake Victoria: A Case Study of complex Interrelationships,* UNEP, Nairobi, Kenya.

Funge-Smith, J.S., and M.R.P. Briggs, 1998: Nutrient budgets in intensive shrimp ponds: Implications for sustainability, *Aquaculture,* **164(1–4),** pp. 117–33.

Galaty, J.G., 1994: Rangeland tenure and pastoralism in Africa. In: *African Pastoralist Systems: An Integrated Approach,* E.A. Roth (ed.), Lynne Reiner Publishers, Boulder, CO, pp. 185–204.

Gell, F. and C. Roberts, 2003: *Fishery Effects of Marine Reserves and Fishery Closures,* WWF, Washington, DC.

Greenpeace, 2003: Canadian Atlantic fisheries collapse. Available at http://archive.greenpeace.org/%7Ecomms/cbio/cancod.html.

Hannesson, R., 2004: The Privatization of the Oceans. In: D.R. Leal, ed., *Evolving Property Rights in Marine Fisheries,* Rowman and Littlefield, Lanham, MD, pp 25–48.

Haslberger, A., 2003: Codex guidelines for GM foods include the analysis of unintended effects, *Nature Biotechnology,* **21** (July), pp. 739–41.

Hatfield, J.L., 1991: Precision agriculture and environmental quality: Challenges for research and education. Available at www.arborday.org.

Heal G., P. Dasgupta, B. Walker, P. Ehrlich, S. Levin, et al., 2002: Genetic diversity and interdependent crop choices in agriculture, Beijer discussion paper, p. 170.

Hecht, S., 1985: Environment, development and politics: Capital accumulation and the livestock sector in eastern Amazonia, *World Development,* **13(6),** pp. 663–84.

Hecht, S., 1989: The sacred cow in the green hell: Livestock and forest conversion in the Brazilian Amazon, *The Ecologist,* **19(6),** pp. 229–34.

Heller, M., G. Keoleian, 2003: Assessing the sustainability of the US food system: A life cycle perspective, *Agricultural Systems,* **76,** pp. 1007–41.

Hiernaux, P., 1996: The crisis of Sahelian pastoralism: Ecological or economic? Pastoral development network paper 39a, Overseas Development Institute, London, UK.

Hiernaux, P., 2000: Implications of the "new rangeland paradigm" for natural resource management. In: *Proceedings of the 12th Danish Sahel Workshop,* Occasional paper 11, H. Adraansen, A. Reenberg, and I. Nielsen (eds.), Sahel-Sudan Environmental Research Initiative, Copenhagen, Denmark, pp. 113–42.

Hilbeck, A., 2001: Implication of transgenic insecticidal plants for insect and plant biodiversity, *Perspectives in Plant Ecology, Evolution and Systemics,* **4(1),** pp. 43–61.

Holden, 1996: *The Common Fisheries Policy,* Fishing New Books, Oxford, UK.

Howe, C., 2002: Policy issues and institutional impediments in the management of ground water: Lessons from case studies, *Environment and Development Economics,* **7,** pp. 625–42.

IFOAM (International Federation of Organic Agriculture Movements), 2004: *The World of Organic Agriculture: Statistics and Emerging Trends,* IFOAM, Bonn, Germany.

IIASA (International Institute for Applied Systems Analysis), 2004: *The End of World Population: Growth in the 21st Century,* W. Lutz, W.C. Sandersen, and S. Scherbov (eds.), Earthscan, London, UK.

IPCC (Intergovernmental Panel on Climate Change), 1995: *Second Assessment, Radiative Forcing of Climate, Summary for Policymakers,* IPCC, Cambridge University Press, Cambridge, UK.

IPCC, 2000: *Land Use, Land-use Change, and Forestry,* IPCC, Cambridge University Press, Cambridge, UK.

Jackson, J., M.X. Kirby, W.H. Berger, K.A. Bjorndal, et al, 2001: Historical overfishing and the recent collapse of coastal ecosystems, *Science,* **293(5530),** pp. 629–37.

Kaimowitz, D., 1996: *Livestock and Deforestation, Central America in the 1980s and 1990s: A Policy Perspective,* Center for International Forestry Research, Bogor, Indonesia.

Koundouri, P., P. Pashardes, T. Swanson, and A. Xepapadeas (eds.), 2003: *The Economics of Water Management in Developing Countries: Problems, Principles and Policy,* (co-ed. with) Edward Elgar Publishing, Cheltenham, UK.

Kulshreshtha, S.N., B. Junkins, and R. Desjardins, 2000: Prioritizing greenhouse gas emission mitigation measures for agriculture, *Agricultural Systems,* **66,** pp. 145–66.

Kura, Y., C. Revenga, E. Hoshino, G. Mock, 2004: *Fishing for Answers: Making Sense of the Global Fish Crisis,* WRI, Washington, DC.

Kurihara, M., T. Magner, R.A. Hunter, and G.J. McCrabb, 1999: Methane production and energy partitioning of cattle in the tropics, *British Journal of Nutrition,* **81,** pp. 263–72.

Lindland, J., 1997: The impact of the Uruguay round on tariff escalation in agricultural products, *Food Policy,* **22,** pp. 487–500.

Maeder, P., A. Fliebach, D. Dubois, L. Gunst, P. Fried, et al., 2002: Soil fertility and biodiversity in organic farming, *Science,* **296(5573),** pp. 1694–97.

MA (Millennium Ecosystem Assessment), 2005a: *Ecosystems and Human Well-being: Current State and Trends,* Island Press, Washington, DC.

McCalla, A. and C. de Haan, 1998: An international trade perspective on livestock and the environment. In: *Livestock and the Environment: International Conference,* A. J. Nell (ed.), Proceedings of the conference, 16–20 June 1997, Wageningen, The Netherlands.

McIntire, J., D. Bourzat, and P. Pingali, 1992: *CropLivestock Interaction in Sub-Saharan Africa,* World Bank, Washington, DC, p. 246.

McIntosh, J.L., and K.E. Varney, 1973: Accumulative effect of manure and nitrogen in continuous corn and clay soil: ii. Chemical changes in soil, *Agronomy Journal,* **65,** pp. 629–33.

Moehl, J., 1999: *Africa Regional Aquaculture Review,* Compendium.

Mugwira, L., 1984: Relative effectiveness of fertilizer and communal area manures as plant nutrient sources, *Zimbabwe Agricultural Journal,* **81,** pp. 81–9.

Munro, J. and J. Bell, 1997: Enhancement of marine fisheries resources, *Reviews in Fisheries Science,* **5,** pp. 185–222.

Murawski, S., R. Brown, H.-L. Lai, P.J. Rago and L. Hendrickson, 2000: Large-scale closed areas as a fishery-management tool in temperate marine systems: the Georges Bank experience. *Bulletin of Marine Science,* **66(3),** pp. 775–798.

Murwira, K.H., M.J. Swift, and P.G.H. Frost, 1995: Manure as a key resource in sustainable agriculture. In: *Livestock and Sustainable Nutrient Cycling in Mixed Farming Systems of sub-Saharan Africa, Volume II,* J.M. Powell, S. Fernández-Rivera, T.O. Williams, and C. Renard (eds.), International Livestock Centre for Africa, Addis Ababa, Ethiopia, p.568.

Neumann, C., D.M. Harris, and L.M. Rogers, 2002: Contribution of animal source foods in improving diet quality and function in children in the developing world, *Nutrition Research,* **22,** pp. 193–220.

Niamir-Fuller, M., 1999: International aid for rangeland development: Trends and challenges. In: D. Freudenberger (ed.), *International Rangelands Congress,* Townsville, Australia, pp. 147–52.

Nicholson, S.E., C.J. Tucker, and M.B. Ba, 1998: Desertification, drought and surface vegetation: An example from the West African Sahel, *Bulletin of the American Meteorological Society,* **79(5),** pp. 815–30.

Obrycki, J., J. Losey, O. Taylor, L. Jesse, 2001: Beyond insecticidal toxicity to ecological complexity, *BioScience,* **51(5),** pp. 353–61.

OECD (Organisation for Economic Co-operation and Development), 1994: *Managing the Environment: The Role of Economic Instruments,* OECD, Paris, France.

OECD, 1999: *The Price of Water: Trends in OECD Countries,* OECD, Paris, France.

OECD, 2001: *Agricultural Policies in OECD Countries, Monitoring and Evaluation 2001,* OECD, Paris, France. Available at http://www.blw.admin.ch/nuetzlich/publikat/e/monitoring.pdf.

OECD, 2003a: *OECD Agricultural Outlook 2003–2008,* OECD, Paris, France.

OECD, 2003b: *Farm Household Income, Issues and Policy Responses,* OECD, Paris, France.

Ojima, D.S., W.J. Parton, D.S. Schimel, J.M.O. Scurlock, and T.G.F. Kittel, 1993: Modeling the effects of climate and CO_2 changes on grassland storage of soil C, *Water, Air, and Soil Pollution,* **70,** pp. 643–57.

Olsen, P. J., R.J. Hensler, and O. J. Attoe, 1970: Effect of manure application, aeration and soil pH on soil nitrogen transformations and on certain soil test values, *Soil Science Society of America Proceedings,* **34,** 222–25.

Ongley, E., 1996: *Control of Water Pollution from Agriculture,* FAO, Rome, Italy.

Oxfam (Oxford Committee for Famine Relief), 2002: The great EU sugar scam: How Europe's sugar regime is devastating livelihoods in the developing world, Oxfam briefing paper no. 27. Available at http://www.oxfam.org/eng/pdfs/pr022508_eu_sugar_scam.pdf.

Palm, C., P. Woomer, J. Alegre, L. Arevalo, C. Castilla, et al., 1999: *Carbon Sequestration and Trace Gas Emissions in Slash-and-Burn and Alternative Land-Uses in the Humid Tropics,* Alternatives to Slash-and-Burn climate change working group final report, phase II, International Center for Rural Agriculture and Forestry, Nairobi, Kenya.

Pashardes, P., T. Swanson, and A. Xepapadeas (eds.), 2002: *Economics of Water Resources,* Kluwers Academic Publishers, Dordrecht, The Netherlands.

Perkins, J.S., 1991: *The Impact of Borehole Dependent Cattle Grazing on the Environment and Society of the Eastern Kalahari Sandveld, Central District, Botswana,* University of Sheffield, Sheffield, UK.

Pimm, S.I., G.J. Russel, J.L. Gittleham, and T.M. Brooks, 1995: The future of biodiversity, *Science,* **269,** pp. 347–50.

Polunin, N., N. Graham, 2003: Review of the Impacts of Fishing on Coral Reef Fish Populations, Western Pacific Fishery Management Council, Honolulu, HI.

Prudencio, C.Y., 1993: Ring management of soils and crops in the West African semi-arid tropics: The case of the Mossi farming systems in Burkina Faso, *Agriculture, Ecosystems and Environment,* **47,** pp. 237–64.

Rabi, M., 1996: TED case studies: Lake Victoria, Case number 388. Available at http://www.american.edu/projects/mandala/TED/victoria.htm.

Ramisch, J.J., 2004: Inequality, agro-pastoral exchanges, and soil fertility gradients in southern Mali, *Agriculture, Ecosystems & Environment,* Elsevier, London, UK.

Reid, R.S., P.K. Thornton, G.J. McCrabb, R.L. Kruska, F. Atieno, et al., 2003: Is it possible to mitigate greenhouse gas emissions in pastoral ecosystems of the tropics? *Environment, Development and Sustainability,* **6(1–2),** pp. 91–109.

Rosegrant, M.W., 1995: Dealing with Water Scarcity in the Next Century, 2020 Vision Brief 21, International Food Policy Research Institute, Washington, DC. Available at http://www.ifpri.org/2020/briefs/number21.htm.

Sala, O.E., F.S. Chapin III, J.J. Armesto, E. Berlow, J. Bloomfield, et al., 2000: Global biodiversity scenarios for the year 2100, *Science,* **387,** pp. 1770–74.

Salazar, F.J., D. Chadwick, B.F. Pain, D. Hatch, and E. Owen: Nitrogen budgets for three cropping systems fertilized with cattle manure, *Biosource Technology,* **96,** pp. 235–45. In press.

Sanginga, N., O. Lyasse, and J. Diels, 2003: Balanced nutrient management systems for cropping systems in the tropics: From concept to practice, *Agriculture, Ecosystems and Environment,* **100,** pp. 99–102.

Schmitz, A., 2003: *Commodity Outlook 2003: U.S. and World Sugar Markets,* Electronic Data Information Source document FE375, Institute of Food and Agricultural Sciences, Department of Food and Resource Economics, University of Florida, Gainsville, FL. Available at http://edis.ifas.ufl.edu/BODY_FE375.

Seckler, D., D. Molden, U. Amarasinghe, C. de Fraiture, 2000: *Water Issues for 2025: A Research Perspective,* International Water Management Institute Contribution to the 2nd World Water Forum, Colombo, Sri Lanka.

Sere, C. and H. Steinfeld, 1996: World livestock production systems: Current status, issues and trends, FAO animal production and health paper 127, FAO, Rome, Italy.

Serneels, S., and E.F. Lambin, 2001: Impact of land-use changes on the wildebeest migration in the northern part of the Serengeti-Mara ecosystem, *Journal of Biogeography,* **28,** pp. 391–407.

Shah, M., and M. Strong, 2000: *Food in the 21st Century: From Science to Sustainable Agriculture,* April, World Bank, Washington, DC.

Skarpe, C., 1990: Shrub layer dynamics under different herbivore densities in an arid savanna, Botswana, *Journal of Applied Ecology,* **27,** pp. 873–85.

Skerrit, J., 2000: Genetically modified plants: Developing countries and the public acceptance debate, *AgBiotechNet,* **2.**

Stallings, B., 1995: *Global Change, Regional Response: The New International Context of Development,* Cambridge University Press, Cambridge, UK.

Steinfeld, H., C. de Haan, and H. Blackburn, 1997: *Livestock and the Environment: Issues and Options,* Wrenmedia, Suffolk, UK.

Stoorvogel, J.J., E.M.A. Smaling, and B.H. Jansen, 1993: Calculating soil nutrient balances at different scales: I. Supra-national scale, *Fertilizer Research,* **35,** pp. 227–35.

Sullivan, S. 1999: The impacts of people and livestock on topographically diverse open wood- and shrublands in arid north-west Namibia, *Global Ecology and Biogeography,* **8,** pp. 257–77.

Thobani, M., 1997: Formal water markets: Why, when and how to introduce tradable water rights in developing countries, *The World Bank Research Observer,* **12(2).**

Thorne, P.J. and J.C. Tanner, 2002: Livestock and nutrient cycling in crop-animal systems in Asia, *Agricultural Systems,* **71,** pp. 111–26.

Tsur, J. and A. Dinar, 1997: The relative efficiency and implementation costs of alternative methods for pricing irrigation water, *The World Bank Economic Review,* **11,** pp. 243–62.

UN, 2003: *World Population Prospects, The 2002 Revision,* United Nations Population Division, New York.

UNEP (United Nations Environment Programme), 1993: *The Aral Sea: Diagnostic Study for the Development of an Action Plan for the Conservation of the Aral Sea,* UNEP, Nairobi, Kenya.

van der Linde, M., V. Minne, A. Wooning, and F. van der Zee, 2000: *Evaluation of the Common Organisation of the Markets in the Sugar Sector,* A report to the Commission of the European Communities, Netherlands Economic Institute, Agricultural Economics and Rural Development Division, Rotterdam, The Netherlands. Available at http://europa.eu.int/comm/agriculture/eval/reports/sugar/index_en.htm.

Vannuccini, S., 1999: *The Bangladesh Shrimp Industry,* FAO, Rome, Italy.

Vannuccini, S., 2003: *Overview of Fish Production, Utilization, Consumption and Trade,* Fishery Information, Data and Statistics Unit, FAO, Rome, Italy.

Voortman, R.L. and J. Brouwer, 2003: An empirical analysis of the simultaneous effects of nitrogen, phosphorus and potassium in millet production on spatially variable fields, *Nutrient Cycling in Agro-Ecosystems,* SW Niger, **66,** pp. 143–64.

Voortman, R.L., J. Brouwer and P.J. Albers, 2004: Characterization of spatial soil variability and its effect on millet yield on Sudano-Sahelian coversands in SW Niger, *Geoderma,* (**121**), pp. 65–82. Available online 31 December 2003.

Walkenhorst, P., 2000: *Domestic and International Environmental Impacts of Agricultural Trade Liberalisation,* OECD, Directorate for Food, Agriculture and Fisheries, COM/AGR/ENV(2000)75/FINAL, Paris, France. Available at http://econwpa.wustl.edu/eps/it/papers/0401/0401010.pdf.

Walker, R., E. Moran, and L. Anselin, 2000: Deforestation and cattle ranching in the Brazilian Amazon: External capital and household processes, *World Development,* **28(4),** pp. 683–99.

Ward, T., D. Heinemann, N. Evans, 2001: *The Role of Marine Resources as Fisheries Management Tool,* Department of Agriculture, Fisheries, and Forestry, Canberra, Australia.

Williams, S.D., 1998: *Grevy's Zebra: Ecology in a Heterogeneous Environment,* Ph.D. thesis, University College London, London, UK.

World Bank, 1997: *Environment Matters,* World Bank, Washington, DC.

World Bank, 2002a: World *Development Report 2002,* World Bank, Washington, DC.

World Bank, 2004: *Saving Fish and Fishers. Towards Sustainable and Equitable Governance of the Global Fishing Sector,* Report number 29090-GLB, Agriculture and Rural Development Department, World Bank, Washington, DC, p. 93.

World Bank/NACA/WWF /FAO (Network of Aquaculture Centres in Asia-Pacific/ World Wildlife Fund), 2002b: Shrimp farming and the environment, A World Bank/ NACA/ WWF/ FAO Consortium program. In: *To Analyze and Share Experiences on the Better Management of Shrimp Aquaculture in Coastal Area,* Synthesis report, (work in progress for public discussion), World Bank, Washington, DC.

WRI (World Resources Institute), 2001: Landmark report warns that degradation of Africa's ecosystems does not stop at national borders, WRI, Washington, DC. Available at www.wri.org/press/africa.

Xepapadeas, A., 1997: *Advanced Principles in Environmental Policy,* Edward Elgar Publishers, Cheltenham, UK.

Xepapadeas, A., 1999: Non-point source pollution control. In: *The Handbook of Environmental and Resource Economics,* J. van den Bergh (ed.), Edward Elgar Publishers, Cheltenham, UK.

Yamaji, K., T. Ohara, and H. Akimoto, 2003: A country-specific, high-resolution emission inventory for methane from livestock. In: Asia in 2000, *Atmospheric Environment,* **37,** pp. 4393–06.

Zaccheo, P., P. Genevini, and D. Ambrosini, 1997: The role of manure in the management of phosphorous resources at an Italian crop-livestock production farm, *Agriculture, Ecosystems and Environment,* **66,** pp. 231–39.

Chapter 7

Freshwater Ecosystem Services

Coordinating Lead Authors: Bruce Aylward, Jayanta Bandyopadhyay, Juan-Carlos Belausteguigotia

Lead Authors: Peter Börkey, Angela Cassar, Laura Meadors, Lilian Saade, Mark Siebentritt, Robyn Stein, Sylvia Tognetti, Cecilia Tortajada

Contributing Authors: Tony Allan, Carl Bauer, Carl Bruch, Angela Guimaraes-Pereira, Matt Kendall, Benjamin Kiersch, Clay Landry, Eduardo Mestre Rodriguez, Ruth Meinzen-Dick, Suzanne Moellendorf, Stefano Pagiola, Ina Porras, Blake Ratner, Andrew Shea, Brent Swallow, Thomas Thomich, Nikolay Voutchkov

Review Editors: Robert Constanza, Pedro Jacobi, Frank Rijsberman

---

BOXES

FIGURES

TABLES

Main Messages

Fresh water can make a greater contribution to human well-being if society improves the design and management of water resource infrastructure, establishes more inclusive governance and integrated approaches to water management, and adopts water conservation technologies, demand management, and market-based approaches to reallocation that increase water productivity. Rising human population and levels of socioeconomic development have led to a rapid rate of water resource development and the replacement of naturally occurring and functioning systems with highly modified and human-engineered systems. Meeting human needs for freshwater provisioning services of irrigation, domestic water, power, and transport has come at the expense of inland water ecosystems—rivers, lakes, and wetlands—that contribute to human well-being through recreation, scenic values, maintenance of fisheries and biodiversity, and ecosystem function.

The principal challenge is to balance these competing demands by acquiring the necessary institutional and financial resources and applying existing technologies, processes, and tools in order to increase the overall productivity of water for society. Agreement on rights and responsibilities with respect to the allocation and management of freshwater services is essential to reconcile diverging views on the degree of public and private participation. As agriculture comprises the major use of water resources globally, the choices between market reallocation and public/private investments in conservation will largely determine whether timely and cost-effective solutions will be found. The potential for climate change to alter availability and distribution of water supplies could be a further complicating factor.

In order to balance competing demands, it is critical that society explicitly agrees on ecosystem water requirements (environmental flows). Determination of ecosystem water requirements involves a societal decision of the desired condition of an ecosystem informed by data on the relationship between hydrology and ecosystem services, and followed by a cognitive or technical response to determine the water quantity and quality necessary to meet articulated objectives. This process is likely to be most successful when it is a collaborative one involving scientists, natural resource managers, and other stakeholders influenced by changes in the availability of the services provided by an ecosystem. Success in achieving outcomes is likely to take time to occur and measure. Any decision on ecosystem water requirements needs to be supported in national and regional water management policies and implemented through an adaptive management approach.

The shift from development of new supplies to emphasis on the reallocation of existing supplies, and integrated water resources management is fundamentally an issue of governance, which provides an entry point for broader policy reforms, as it implies the need to evaluate trade-offs among multiple and often conflicting uses across sectors, within the context of the entire flow regime. Key challenges in governance of freshwater services include democratic decentralization of decision-making processes and recognition or establishment of appropriate forms of property rights and responsibilities, which enables stakeholders to be adequately represented and to hold decision-makers accountable, which can make a difference in whether or not objectives are achieved. Appropriate forms of governance are also necessary to ensure equity, which is a fundamental enabling condition in the use of regulatory and/or market-based incentives for protecting freshwater services. River basin organizations can play an important role in facilitating cooperation and reducing transaction costs of large-scale responses.

Given the diversity of conditions in river basins, the more effective kinds of arrangements may be those that have evolved in response to site-specific conditions and extreme events, as these raise awareness of impacts and provide an opportunity to open policy debates. A key constraint will be negotiating with those who have vested interests in existing arrangements and who are resistant to change. Such reforms may be difficult and have high transaction costs, but may gain momentum from broader-scale social, economic, and political changes from which they may be inseparable, such as occurred after the fall of apartheid South Africa. They may also have added benefits—of strengthening democratic institutions and contributing to political stability, as occurred in the Danube Basin following the cold war period. Quality of information, obtained through independent and transparent assessment, is critical for increasing stakeholder confidence and, therefore, willingness to pay or otherwise cooperate in responses, and for identifying specific barriers to implementation.

Economic incentives have the potential to unlock significant supply- and demand-side efficiencies while providing cost-effective reallocation between old (largely irrigation) and new (largely municipal and instream) uses. Historic allocations of water have rarely taken account of its scarcity value or of the economic value of alternative uses. Payments and incentives for water conservation can increase water availability, just as pricing water at its full marginal cost can reduce demand. Functioning water markets can provide price signals for reallocation between different uses and also signals to guide conservation activities. Temporary trades have dominated experience with markets in developing and industrial countries alike. Experience with permanent transfers in Chile suggest that laissez-faire free markets without adequate consideration of third-party impacts will lead to adverse social and environmental consequences. Conversely, U.S. experience suggests that too much emphasis on water as a public and local resource is likely to greatly limit market activity, particularly permanent transfers. This will greatly constrain the achievement of economic efficiencies and ecosystem restoration. The Australian experience shows the potential of markets for reallocation to higher value uses. However, it demonstrates the need to be explicit about instream needs and to properly plan for the reintroduction of "unused" water that accompanies market development if environmental objectives are also to be met.

Key challenges in the development of payments for watershed service initiatives are to build capacity for place-based monitoring and assessment, to identify services in the context of the entire flow regime, to consider trade-offs and conflicts among multiple uses, and to make uncertainty explicit. Payment arrangements for ecosystem services provided by watersheds have been narrowly focused on the role of forests in the hydrological regime, they should instead be developed in the context of the entire flow regime, which would include consideration of the relative values of other kinds of land cover and land uses, such as wetlands, riparian areas, steep slopes, roads, and management practices. The value placed on watershed services will depend on stakeholder confidence in the effectiveness of proposed management actions for ensuring that the service continues to be delivered and that those who pay the costs will have access to the stream of future benefits. A precise determination of costs and benefits and their distribution presumes the ability to link actions and outcomes, so as to be able to demonstrate trade-offs. However, watershed processes are inherently variable and uncertain. Market mechanisms, on the other hand, tend to be more effective when uncertainty is low, because buyers like to know that they are getting what they pay for. Initiatives have often been based on myths and inappropriate generalizations about land and water relationships. While this may work in the short term, making uncertainty explicit is likely to be critical in managing buyer expectations and maintaining their cooperation in the long term.

A variety of public/private partnerships will have a role to play in financing water infrastructure for the provision of fresh water. There is a clear mismatch between the high social value of freshwater services and the resources that are being allocated to manage water. Insufficient funding to expand water infrastructure is one manifestation of this mismatch. Both inherent characteristics of the water sector (high fixed cost, low returns, long payback periods) as well as institutional problems (political interference, inadequate legal frameworks, poor management structures) explain the gap in funding infrastructure. A single-minded focus on large-scale privatization as a means of improving efficiency and cost-recovery has proven a double-edged political strategy—price hikes imposed by multinationals or acquisition of control over available resources have led to controversy and, in some cases, failure and withdrawal. To actually acquire "new" monies for investment, a fundamental change is required. At a national level, legal frameworks have to provide more certainty to the parties of long-term commitments. The water sector has to establish its priorities in a clear way and produce programs that include the definition of financing needs and sources. Finally, at the agency level, cost recovery must be improved and managerial and technical capacities enhanced.

New development of water infrastructure and technologies must observe best practices to avoid past problems and inequities; however, it is the reexamination and retrofitting/refurbishment of existing infrastructure that offers the most opportunity in the short and medium term. In regulated freshwater ecosystems, the optimal use of environmental flows will often require altered management of water infrastructure, supported by institutional arrangements across scales and actors. Once an environmental flow regime has been identified, management of freshwater ecosystems is likely to change. In highly modified and regulated systems, this may require decommissioning of dams or mitigating and altering dam operations and other water resource infrastructure, for example, managed flow releases. This predominantly requires an institutional response to facilitate these changes and may be accompanied by technological changes to retrofit infrastructure.

Table 7.1. Ecosystem Services Provided by Fresh Water and the Hydrologic Cycle. Many of the provisioning, regulatory, and cultural services can be enhanced through development of water resources (large-scale navigation can be increased by creating slackwater systems using dams); however, there are often off-setting losses or trade-offs between these service categories, such as loss of rapid transport downstream to locals or those seeking recreation.

Provisioning Services	Regulatory Services	Cultural Services
• Water (quantity and quality) for consumptive use (for drinking, domestic use, and agriculture and industrial use) • Water for nonconsumptive use (for generating power and transport/navigation) • Aquatic organisms for food and medicines	• Maintenance of water quality (natural filtration and water treatment) • Buffering of flood flows, erosion control through water/land interactions and flood control infrastructure	• Recreation (river rafting, kayaking, hiking, and fishing as a sport) • Tourism (river viewing) • Existence values (personal satisfaction from free-flowing rivers)

Supporting Services

- Role in nutrient cycling (role in maintenance of floodplain fertility), primary production
- Predator/prey relationships and ecosystem resilience

7.1 Human Well-being and Fresh Water

Ecosystem services are the benefits provided to people, both directly and indirectly, by ecosystems and biodiversity. In the Millennium Assessment, fresh water is a "provisioning" service as it refers to the human use of fresh water for domestic use, irrigation, power generation, and transportation. (See Table 7.1.) However, fresh water and the hydrological cycle also sustain inland water ecosystems, including rivers, lakes, and wetlands. These ecosystems provide cultural, regulating, and supporting services that contribute directly and indirectly to human well-being through recreation, scenic values, and maintenance of fisheries. Fresh water also plays a role in sustaining freshwater-dependent ecosystems such as mangroves, inter-tidal zones, and estuaries, which provide another set of services to local communities and tourists alike. This chapter explores how the trade-offs between these differing uses of fresh water and inland water systems can be balanced in the midst of increasing demand for all types of human benefit derived from fresh water.

7.1.1 Conditions, Trends, and Direct Drivers in Freshwater Services and Inland Water Ecosystems

In the past century, increasing human population and advancing levels of social and economic development have led to a rapid increase in the demand for freshwater provisioning services. In its natural state, fresh water varies considerably in terms of its availability in time and space. Water resources development—the construction of dams and irrigation channels, the construction of river embankments to improve navigation, drainage of wetlands for flood control, and the establishment of inter-basin connections and water transfers—has the aim of reregulating the natural hydrograph to meet human needs.

This has resulted in the replacement of naturally occurring and functioning systems with highly regulated and modified human-engineered systems. These "developed" systems have typically been designed solely for the satisfaction of the major human consumptive uses (irrigation or municipal and industrial use) or nonconsumptive use (hydropower and navigation).

These structural and capital-intensive responses—particularly large dams—have greatly augmented the natural availability of freshwater provisioning services. In the last 20 years alone, more than 2.4 billion people have gained access to water supply and more than 600 million have gained access to sanitation (World Water Commission 1999). At the same time, these supply responses have themselves become direct drivers of ecosystem degradation.

The impacts of water resource development are two-fold: less water remains in the ecosystem and the distribution and availability of the remaining water often has a different pattern from that present under natural conditions. It is estimated that the amount of water withdrawn from inland water systems has increased by at least 15 times over the past two centuries. (See MA *Current State and Trends,* Chapter 7, for a discussion of water withdrawals.) As a result, humans now control and use more than half of the continental runoff to which they have access. The impact of withdrawals, though, is not evenly spread and it is estimated that about 80% of the global population is living downstream of only 50% of Earth's renewable water supplies. (See MA *Current State and Trends,* Chapter 7.) Changes to the hydrograph and related physical, chemical, and biological processes have substantially degraded the condition of inland water ecosystems globally. (See MA *Current State and Trends,* Chapter 20.)

A related consequence of water resource development has been reduced water quality. Caused through the pollution of inland water ecosystems, this has occurred in parallel with the growth of urban, industrial, and agricultural systems. The major pollutants affecting water quality include nutrients, which drive eutrophication; heavy metals; nitrogen and sulphur based compounds, which cause acidification of freshwater ecosystems; organic compounds; suspended particles, both organic and inorganic; contaminants such as bacteria, protists, or amoebae; and salinity. According to the World Water Commission, more than half of the major rivers of the world are seriously polluted (WWC 1999). The presence of these pollutants depletes the capacity of rivers and associated inland and coastal ecosystems to provide clean water for social and economic uses.

Changes in the condition of freshwater and associated inland water ecosystems have also occurred at the hands of other direct drivers such as species introductions, land use change, and climate change. (See Table 7.2 and MA *Current State and Trends,* Chapters 7 and 20.)

7.1.2 Indirect Driving Forces

Most water-related problems, although caused by direct drivers such as water abstraction and pollution, are ultimately a product of indirect drivers. The development of water resources over the past century has been largely a result of the need to supply expanding populations with food, energy, and domestic and industrial water supplies and to facilitate opportunities for transport. Economic growth has further served to enhance the demand and consumption of freshwater services.

However, given the public as well as private good characteristics of fresh water, most water-related problems are ultimately a product of indirect drivers associated with the economic nature of fresh water in all its guises—and the manner in which this nature is accommodated or not by the institutional arrangements that govern the production, allocation, distribution, and consumption of freshwater services. The economic characteristics of fresh water, when combined with the dynamic nature of the hydrological cycle, present special challenges in the case of fresh water.

The potential for fresh water or ecosystems to have multiple uses, some of which will be private goods and others of which will either be perfect public goods or variations such as common pool or toll goods, creates this management challenge, as each type lends itself to a different management regime (Ostrom et al. 1993; Aylward and Fernández-González 1998). The market failure associated with public good characteristics suggests a need for mechanisms of social coordination in the form of institutional arrangements that can define, and adaptively manage, the level of provision and allocation of these goods and services that is desired by society.

Governance and the role of economic incentives are therefore critical indirect drivers with respect to balancing competing demands for freshwater. The inadequate governance associated with water resource development, particularly a single-minded, engineering-economic approach to the ecosystems services that inland water systems provide, has led to significant social and environmental impacts—impacts that have disproportionately affected the rural poor that rely on the natural functioning of inland water ecosystems (WCD 2000).

In the last two decades, increased attention has been paid to the importance of considering water as an economic commodity (Cosgrove and Rijsberman 2000). This has provoked considerable concern and controversy with respect to financing water infrastructure and water pricing, noticeably with regard to privati-

Table 7.2. Summary of Direct Drivers (Postel and Richter 2003)

Human Activity (Direct Driver)	Impact on Ecosystems	Services at Risk
Dam construction	alters timing and quantity of river flows. Water temperature, nutrient and sediment transport, delta replenishment, blocks fish migrations	provision of habitat for native species, recreational and commercial fisheries, maintenance of deltas and their economies, productivity of estuarine fisheries
Dike and levee construction	destroys hydrologic connection between river and floodplain habitat	habitat, sport and commercial fisheries, natural floodplain fertility, natural flood control
Diversions	depletes stream flow	habitat, sport and commercial fisheries, recreation, pollution dilution, hydropower, transportation
Draining of wetlands	eliminates key component of aquatic ecosystem	natural flood control, habitat for fish and waterfowl, recreation, natural water purification
Deforestation/land use	alters runoff patterns, inhibits natural recharge, fills water bodies with silt	water supply quality and quantity, fish and wildlife habitat, transportation, flood control
Release of polluted water effluents	diminishes water quality	water supply, habitat, commercial fisheries, recreation
Overharvesting	depletes species populations	sport and commercial fisheries, waterfowl, other biotic populations
Introduction of exotic species	eliminates native species, alters production and nutrient cycling	sport and commercial fisheries, waterfowl, water quality, fish and wildlife habitat, transportation
Release of metals and acid forming pollutants into the atmosphere	alters chemistry of rivers and lakes	habitat, fisheries, recreation, water quality
Emission of climate altering air pollutants	potential for changes in runoff patterns from increase in temperature and changes in rainfall	water supply, hydropower, transportation, fish and wildlife habitat, pollution dilution, recreation, fisheries, flood control

zation of municipal water supply, as well as with the application of market approaches, particularly with respect to the development of water markets and the use of payment systems for watershed services.

The discussion above highlights that water is in fact a resource that is often multifunctional and heterogeneous in nature. It is therefore not amenable to simple classification as either a public good or a private good. While water may be managed more successfully when its "economic" characteristics are recognized, due to its public good attributes, the solution will not be to treat it as a unidimensional commodity. Conversely, simply assuming fresh water is a public good in all contexts and uses is equally likely to lead to ruin. Rather, there is a need to respond to the inherent complexity of fresh water and work in an adaptable fashion toward site-specific solutions that accommodate the attributes and uses of fresh water in the local context.

7.1.3 Future Freshwater Challenges

While water resources development has increased the contribution of freshwater resources to human well-being, this is not to say that human needs for provisioning services are met today. Global figures indicate that 1.1 billion people do not have adequate access to good quality drinking water, 2.6 billion people lack access to sanitation, 800 million people do not have enough to eat, and 2 billion people do not have access to household electricity (UNESCO 2003; WHO and UNICEF 2004).

The International Food Policy Research Institute and the International Water Management Institute examined the impact of the world's growing human population on water and food (Rosegrant et al. 2002). With a global population of 8 billion people—a 2 billion increase—and a business-as-usual scenario, an overall increase in water withdrawals of 22% over 1995 levels is expected by 2025; this includes increases of 17% in the demand for water for irrigation, 20% in the demand for water for industry, and 70% in the demand for water by municipalities. Under a crisis scenario, a 37% increase in overall withdrawals is forecasted, while under a sustainability scenario, significantly less water—20%—is actually consumed in meeting future needs. Increased efficiency actually improves environmental flows and reliability of supply to agriculture.

The United Nations Millennium Development Goals set out ambitious targets with respect to human nutrition, education, and health, as well as for the environment. A target of halving the number of people who cannot access safe drinking water is the only explicit water-related MDG. Still the full range of freshwater ecosystem services will come into play—directly or indirectly—in achieving these goals (UN World Assessment Program 2003). The role of water in agriculture in meeting future demand and the MDGs is critical, and the ability to realize increased water productivity (getting more food for every drop of water) in irrigated and rain-fed agriculture will be of particular importance if water is to be freed up for ecosystems and other uses (Molden and Falkenmark 2003).

Meeting these provisioning needs will come at a significant cost. The report of the Camdessus Group to the Third World Water Forum in Kyoto, 2003, suggests that in developing countries, current spending on water services of $75 billion a year needs to be increased to about $180 billion if the water and sanitation MDGs are to be met (World Panel on Financing Water Infrastructure 2003). The food, water, and environment assessment currently being undertaken by the International Water Management Institute should yield additional information on the challenges and investments required with respect to agriculture and water (IWMI 2002).

Although needs and demands for provisioning services will continue to rise, it is increasingly accepted that any management response will need to deal with the potential for trade-off between the levels of different services provided by freshwater and associated inland water ecosystems. For example, the World Commission on Dams recently concluded that, although large dams have made an important contribution to development, this contribution has come at too high a price—referring to the high economic, social, and environmental costs (WCD 2000).

Efforts to "mitigate" the adverse impacts of traditional engineering responses to increasing supply are now the norm in most countries and, in many countries, efforts are now made to avoid these impacts in the first place by opting for other alternatives, including nonstructural approaches that dampen demand. At the same time, changing preferences in developed economies in favor of cultural and supporting services of inland water systems—such as tourism and recreation—imply that the balance of current effort is on ecosystem restoration rather than increasing provisioning services (for example, through new dams).

7.1.4 Optimizing Freshwater Ecosystem Services

Fresh water is a finite resource that cannot be distributed such that all the ecosystem services that it provides are maximized. This is the lesson of the environmental impacts observed across the world from water resource development. Ultimately, any development of water resources will involve a trade-off between provisioning, and the cultural, regulating and supporting services. In the past, the tendency was to sacrifice supporting, regulating, and cultural services in return for augmenting provisioning services. Increasing recognition of the consequences of such an approach has led to initiatives at all levels to address this issue and redress the balance.

However, current trends are likely to lead to a continued imbalance (see MA *Current State and Trends,* Chapters 7 and 20). Current trends to continue favoring provisioning services should reduce poverty. Nevertheless, due to the linkage between ecosystems and their cultural, regulatory, and supporting services, it is expected that poverty can only be reduced so far before feedback loops from ecosystem degradation will cascade back through these services, thereby reducing well-being, particularly for the poorest members of society. For example, in the Aral Sea, the benefits initially gained from increased agricultural productivity have been outweighed by the loss of fisheries and the impacts on human health, such as pulmonary diseases, from salt on the exposed seabed. Figure 7.1A shows the effect of these trends on freshwater ecosystem services that contribute to human well-being using a spider diagram.

An alternative is to strengthen the implementation of existing protective measures for the ecosystem. For example, full implementation of many conventions, laws, and rules would radically reverse ecosystem degradation and biodiversity loss. However, it is equally likely that full protection would greatly limit opportunities for meeting the continuing and growing needs for freshwater provisioning services with the probable consequence that poverty would increase. This scenario is shown in Figure 7.1B.

The alternative, of course, is to attempt to balance the services by optimizing across their full range. Figure 7.1C portrays such a generic scenario. Neither provisioning services nor the cultural, regulatory, and supporting services will themselves be maximized, but the overall effect would be to maximize welfare—subject to the constraint of targeting poverty reduction.

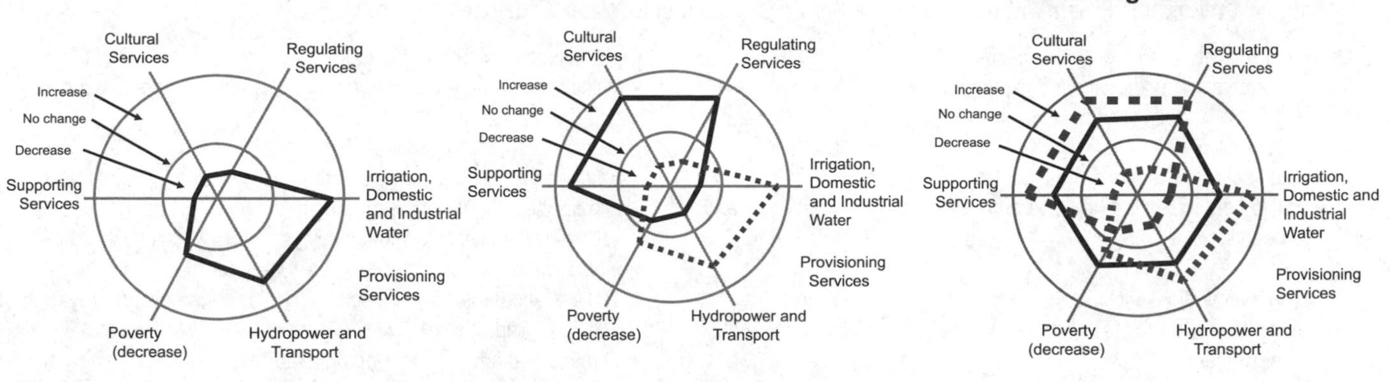

Figure 7.1. Balancing Trade-offs between Services and Impacts on Poverty

Clearly, with the challenges outlined above—both in terms of the demand and the need for appropriate institutional arrangements—the use of the term "optimizing" is very optimistic. In many contexts, efforts will be aimed at making the best of a difficult situation, striving for improvement and balance rather than optimization. Yet it is important to acknowledge, at least conceptually, that trade-offs among services present the possibility of optimization. While the responses selected to work in this direction will vary, it will be important to recall that the overall objective is to increase the overall productivity of all freshwater services and ecosystems.

One of the difficulties in assessing future needs is that there are many responses that are available. For example—as in the aforementioned IFPRI and IWMI forecasts—estimates of increases in the demand for water for agricultural purposes are highly variable due to differing assumptions about the productivity of water in agriculture (Molden and de Fraiture 2004). Raised to a higher level, the same concept applies. Inertia and difficulties in the governance of water is a significant problem, but if headway can be made, then there is at least the hope that existing technological and market approaches can reduce crisis estimates of the gap between supply and demand.

7.1.5 Selection of Responses for Assessment

With world population expected to stabilize in the middle of the next century, it is likely that the pressure on freshwater resources and inland water systems will never be greater than in the coming decades. In other words, it is critical that responses undertaken in the next 25–50 years accommodate increasing pressure for provisioning services, yet protect ecosystems, biodiversity, and their services. If provisioning services are consumed at the expense of ecosystems, the transition to a sustainable population will take place with a greatly reduced stock of natural capital with respect to freshwater services and inland water ecosystems.

A wide range of potential responses to the provision of fresh water and associated ecosystem services exist. Following on the typology presented in Chapter 2 of this volume, a rather exhaustive list of responses can be derived. (See Box 7.1.) This chapter distills responses from this broad list into those relating to governance and supply and demand management based on an identification of the key drivers that occur now and are expected in the future. A key message of this assessment is that underlying the selection of any responses must be recognition and consideration of the trade-offs implicit in selecting various response options.

The responses examined are outlined in Table 7.3. They include more traditional techniques, including supply-driven engineering responses and ways of applying these to yield more technically efficient, socially equitable, and environmentally responsible solutions. Alternatives are also examined that aim to increase the technical and economic efficiency of existing responses, rectifying inequities of past developments and restoring water quantity and quality of freshwater systems, as well as their diversity. The responses selected for assessment are intended to address a number of the critical future needs of policy- and decision-making as identified above and in the MA user needs analysis.

The responses were selected to illustrate different, but often complementary and mutually reinforcing, approaches. As a result, except for purposes of discussion, the responses are not always separate and discrete, but instead are often nested, one within the other or at different scales. Thus the choices made with respect to arrangements for property rights may determine the types of economic approaches that are used, which, in turn, may influence whether and what type of technology is ultimately applied in improving water resource management.

For example, improving the effectiveness of governance and developing appropriate institutional arrangements is an important response in itself, and is also an enabling condition for most, if not all, other responses. This is illustrated by market-based mechanisms, which would not work unless there are defined property rights and confidence that contractual agreements will be enforced. They may also rely on regulations that establish caps on water abstraction or pollutant emissions, and thereby create an incentive to use such approaches so as to reduce the costs of compliance.

While the limitation of responses that have proven inadequate in the past are covered, this is mostly to provide a context for the main focus, which is on what is being learned from new kinds of responses. These often require the development of new kinds of knowledge and other social capacities, and may present a whole different set of problems, not necessarily better or worse. For example, payments for watershed services have generated much enthusiasm, as they are seen as a more straightforward and direct approach to creating incentives for conservation that also helps to alleviate poverty, than integrated conservation and development projects, whose limitations are discussed extensively in Chapter 15; integrated responses often have numerous conditions that must be met for success. However, payments for watershed services, even if an improvement, present a different set of challenges and numerous unanswered questions as to their actual transaction costs.

We begin with issues of governance, which includes the challenges of determining ecosystem water requirements, rights, and

BOX 7.1

Response Options for Fresh Water and Related Services from Inland Water Ecosystems

Legal and regulatory interventions include:
- ownership and use rights at different administrative levels,
- regulation of pollution,
- regulation of environmental flows and artificial flood releases,
- legal agreements for river basin management, and
- regulations related to ecosystem and species conservation and preservation.

Economic interventions include:
- markets and trading systems for flow restoration and water quality improvements,
- payments for ecosystem rehabilitation,
- point source pollution standards and fines/fees, taxes,
- demand management through water pricing, and
- payments for watershed services.

Governance interventions include:
- participatory mechanisms (for example, watershed/catchment councils and farmer-based irrigation management systems),
- river basin organizations (international or regional scale),
- integrated water resource management and basin planning,
- private sector participation, and
- institutional capacity building (for example, for regulatory agencies).

Technological interventions include:
- water infrastructure projects (such as dams, dikes, water treatment and sanitation plants, desalinization),
- soil and water conservation technologies (such as physical and vegetative measures for soil and water conservation),
- end-use and transmission efficiency options (such as drip irrigation and canal lining/piping),
- demand management/technologies for higher end-use efficiency (such as low-flow showerheads, energy conservation programs/incentives), and
- research into water-saving technologies and breeding crops for drought tolerance.

Social, cultural, and educational interventions include:
- environmental education and awareness,
- making explicit the value of non-provisioning water ecosystem services, and
- research into land–water interactions in a watershed context.

responsibilities in the provision of fresh water, increasing the effectiveness of public participation, river basin organizations, and the role of regulation. Experience with the use of economic incentives for supply and demand management is then examined with particular reference to reallocating fresh water and conserving watershed hydrologic function, as well as mobilizing financing for the development of water supply infrastructure through public-private partnerships. Finally, supply infrastructure and technologies are covered briefly, with assessments of two in particular: a well-established response (large dams) and a new response (wetland restoration and mitigation). Other chapters address responses to issues of climate change (see Chapter 13) and issues of increasing the productivity of food production (Chapter 6).

7.2 Governance: Institutions for Managing Shared Waters

The objective of achieving a reasonable balance between the provisioning of fresh water and maintaining the underlying ecosystem processes that support fresh water and related ecosystem services implies a fundamental shift in water management approaches. This shift is from an emphasis on the continued development of new sources of water supply toward the reallocation of water among various uses and management of demand; recovery of costs of conservation management and research activities; and operations and maintenance of systems for delivery of water and sanitation services. This new emphasis is consistent with the integrated water resources management approach defined by the

Table 7.3. Response Options for Optimizing Human Well-being from Freshwater and Associated Inland Water Ecosystems

Governance	Supply Management	Demand Management
Defining ecosystem water requirements Property rights Participation in decision-making River basin organizations and transboundary management Regulatory	Economic incentives for reallocation and new supply • partnerships and financing • water markets • cap and trade systems • payments for watershed services Infrastructure • large dams • levees • locks and canals Technologies • wetland restoration • agricultural water conservation • desalinization • rainwater harvesting	Economic incentives for consumers • water pricing • payments and subsidies for on-farm and household water conservation Water conservation technologies • on-farm water efficiency and management improvements • municipal and industrial water measurement and savings devices

Global Water Partnership (2000), which talks of managing water so as to advance a country's social and economic development goals in ways that do not compromise the sustainability of vital ecosystems, by taking into account the broader ramifications of sectoral actions.

Integrated water resources management is a challenge of governance as well as of economics and of obtaining adequate scientific knowledge, as it requires making decisions regarding the trade-offs among the multiple and often conflicting uses and interests that often accompany changes in policy objectives and management practices. Implementation also requires the development of appropriate and effective institutional arrangements that provide stakeholders with the incentive to cooperate and comply with management plans.

Because of the vital importance of water to human society, problems in water management also reflect more general weaknesses in governance, and provide a point of departure for broader policy reforms, and vice versa. This is explicitly recognized, for example, in the European Water Framework Directive, which requires harmonization with policies in other sectors. The Water Framework Directive Common Implementation Strategy gives priority to achieving consistency of agriculture, transport, energy, internal markets, development, fisheries, and marine policies, with WFD water policy objectives. Conversely, events such as the fall of apartheid in South Africa made possible major reforms in water policy. In the Danube Basin, cooperation on basin-wide management issues only became possible with the end of the Cold War, where efforts to achieve consistency with the Water Framework Directory are intertwined with the development of democratic institutions.

As is pointed out in the World Water Development Report of the United Nations World Water Assessment Program (UNESCO 2003), there has been a significant expansion, over the past 20 years, of international programs and institutions that pertain to fresh water, beginning with Agenda 21, and these include:

- new national water laws,
- international agreements,
- the formation of river basin organizations,
- the establishment of the World Water Council, which sponsored the two World Water Forums and an independent World Water Commission to produce a World Water Vision, and
- establishment of the Global Water Partnership to form local and regional partnerships.

While these bodies have contributed to broader awareness of water problems and solutions, there is also a general acknowledgement that there is a wide gap between formal policies and actual practices. For example, as stated in a report regarding watershed management in Peru, while there has been an evolution of the concept of watershed management, toward more integrated and participatory approaches, which encompass social as well as environmental aspects over larger spatial areas, these are taken up more easily in international discourse than in actual watershed management practices (Bellido 2003). Continued effort to support more effective on-the-ground implementation is always useful, particularly if placed within a structured framework for learning and adaptive management.

Perhaps the most significant methodological challenge is to develop the capacity for a place-based approach to assessment, which is necessary to identify ecosystem functions that support the provision of valued ecosystem services in a specific context, and to select feasible and appropriate institutional arrangements. This requires consideration of the entire flow regime, which provides a framework for considering the full range of interests in the context of the specific ecosystem functions that support them, and for identifying trade-offs, including uncertainties. It can be regarded also as an institutional challenge in that impacts on livelihoods and other aspects of well-being are easily overlooked in the absence of effective participation and recognition of the rights of the various stakeholders.

The responses to the challenge of developing effective institutions addressed in this section include (1) determining ecosystem water requirements, (2) determining rights to freshwater services and responsibility for their provision, (3) increasing the effectiveness of public participation in decision-making, (4) developing river basin organizations, and (5) regulatory responses.

7.2.1 Determining Ecosystem Water Requirements

In middle-income developing and industrial countries, where basic needs for food, water, power, and transport have largely been met by past water resource development, attention is refocusing on the regulating, supporting, and cultural services that well-functioning inland water ecosystems can provide. Given the history of development of land and water resources, this often requires efforts to restore the natural function of watersheds, rivers, and wetlands. There has also been a push in some developing countries to consider ecosystem water requirements before installing water resource infrastructure and associated allocation systems. (See Box 7.2.)

The desire to determine ecosystem water requirements can be driven by anecdotal observations of a decline in certain ecosystem

BOX 7.2

The Lesotho Highlands Water Project (Brown and King 2003)

The Lesotho Highlands Water Project is aimed at regulating water resources across the headwaters of the Senqu River system in Lesotho through the establishment of large dams and weirs. River regulation is considered economically valuable for Lesotho and downstream in South Africa's Vaal region. The Lesotho Highlands Water Commission, a bi-national body, oversees the project.

In 1986, a treaty was signed to initiate the project, which included provision for renegotiation after 12 years. Renegotiation of the terms of the project were delayed in 1998 so that an assessment could be undertaken to predict the biophysical, sociological, and economic responses of major infrastructure works, as well as mitigation and compensation costs for affected subsistence groups that use the river. This was done using the Downstream Response to Imposed Flow Transformations (DRIFT) approach by examining responses under different flow regulation scenarios.

The major potential water conflict issues were the impact of regulation on the 39,000 subsistence users downstream of the dam sites and the potential impact on ecosystem services such as ecotourism. In general, it was predicted that the more water released downstream of storage, the less the change in river condition and associated socioeconomic impacts to riparian people, but the greater the impact on system yield, and hence, potential for revenue through development opportunities.

The result of the application of DRIFT was that flow releases from some of the major dams and weirs have been altered and the design and operation of news dams will be modified (as at the Mohale Dam), particularly to facilitate environmental release in the form of seasonal releases or small floods.

features and functions or by the results of applied analysis spawned by observation. A change in condition may be detected by local communities (for example, the decline in catch size of native fish and the change in vegetation composition on floodplains) or through government-sponsored surveillance monitoring programs conducted as a legislative requirement. Determining ecosystem water requirements may also be undertaken for undeveloped water resources, although this is uncommon given that most available water resources across the globe are already developed to some degree.

Many of the services provided by inland water ecosystems are supported by natural variability in spatial and temporal patterns in the distribution, abundance, and quality of water and the interaction between the basin, climate, geology, topography, and vegetation of which they are a product (Poff et al. 1997; King and Brown 2003). Other processes central to ecosystem functioning are also sustained by flow such as sediment and nutrient transport (Whiting 2002). The development of water resources, while often benefiting provisioning services, has modified these patterns and deleteriously impacted regulating, supporting, and cultural services. (See MA *Current State and Trends,* Chapters 8 and 23.)

Components of the flow regime are the natural patterns of variation in the quantity and timing of the flow of a river, including the natural disturbances associated with these flow patterns (Poff et al. 1997). The need to maintain adequate quantity and quality of water in inland water ecosystems is captured by the notion of ecosystem water requirements and management, which are referred to in much of the literature on river ecology, where this concept has the greatest currency as "environmental flows."

Put simply and based on concepts developed for rivers, environmental flows are the water that is left in a river ecosystem, or released into it, for the specific purpose of managing the condition of that ecosystem (Brown and King 2003). Applied more liberally, ecosystem water requirements can be taken to mean the water left in an inland water ecosystem, or released into it, for the specific purpose of meeting objectives with regard to the balance of services provided by that ecosystem. This could apply equally to coastal systems dependent on freshwater supply. While the definition of ecosystem water requirements provided here is broad, their determination has been developed almost entirely for riverine ecosystems.

A failure to determine and allocate water for ecosystems may not only result in the loss of the natural functions of the parent system, as has been observed following water resource development on the Colorado, Ganges, Nile, and Yellow rivers (Postel 2000), but also cause degradation of dependent ecosystems, as for example in the loss of biotic production in the Sea of Cortez at the mouth of the Colorado River (Baron et al. 2002). These and other examples from across the world have led to the recognition that the ecological sustainability of inland water ecosystems is threatened by the hydrologic alterations carried out by humans (Poff et al. 2003).

Defining ecosystem water requirements targets the direct drivers of resource consumption in the form of water abstraction and the reregulation of the timing of flows, the objective being to determine the quantity, quality, and timing of water that should remain "instream" or within an ecosystem for the variety of regulatory, supporting, and cultural services it sustains. Given the site-specific nature of what might constitute ecosystem water requirements, a wide range of approaches exist for implementing these flows. (See Box 7.3.)

BOX 7.3

Approaches to Implementing Environmental Flow Regimes

In rivers and streams, restrictive flow management—where abstractions and diversions are controlled—can be used to achieve an environmental flow, for example, the Murray-Darling Basin cap (Acreman and King 2003, see Box 7.5). Depending on abstraction levels and the identified needs of the system, water recovery may also be required, which can be achieved through reducing allocations for consumptive uses such as irrigation (Postel 2000). In this instance, active flow management may be necessary and involve management of "environmental water" through practices such as managed floods. This method is gaining support among river managers globally and has been implemented in rivers such as the Senegal, Phongolo, and Kafue in Africa (Acreman et al. 2000), Colorado in the United States (Webb et al. 1999), and the Murray in Australia (Siebentritt et al. 2004).

The institutions involved in undertaking ecosystem water requirement determinations are usually those that have the required governance and technical capacity. These are principally governments looking to inform natural resource management policy-making, or not-for-profit groups wanting to inform advocacy positions. The process will typically occur at a catchment scale, but will require the involvement of actors from a local through to a regional and national scale.

There are essentially two steps in most techniques used to define ecosystem water requirements: the setting of objectives and the definition of the water (quantity and quality) required to meet those objectives supported by technical advice on the condition of the river system under various flow regimes (Acreman and King 2003).

7.2.1.1 Defining Ecosystem Condition: A Balance between Ecosystem Services

Maximizing human well-being in freshwater dependent ecosystems requires a decision involving all the stakeholders on the desired condition of an ecosystem (Poff et al. 2003). Such a decision needs to balance human aspirations for water resource development for provisioning services, with the indirect water requirements needed for other ecosystem services (Acreman 2001).

Historically, an indirect and unconscious decision was made on the amount of water required to sustain an ecosystem. For example, damming a river and diverting water for irrigation prior to the development of our current understanding of flow ecology relationships effectively represented a decision to not supply water to other ecosystem services. The process of determining ecosystem water requirements aims to make a conscious and intentional decision on how water is distributed among different services.

Invariably a variety of views exist among stakeholders on what the desired level of ecosystem services should be and the trade-offs that are acceptable in improving human well-being. For example, reduced water for abstraction may mean less irrigation, but this may provide more water to achieve desired levels of fishing, boating, ecosystem resilience, and scenic values. Setting this balance has sparked debates across the world. Decision-makers responsible for water allocation may seek the minimum flow that must remain in a river to maintain the environment. This approach encompasses the notion of resilience, which refers to the thresholds within which changes to natural flow regimes can be adjusted. However, such thresholds below which the ecosystem abruptly degrades are very illusive and may not exist in reality (Acreman in press).

7.2.1.2 Methodologies for Defining Ecosystem Water Requirements

Recognition of the need for adequate quantity, quality, and timing of flows in river systems has led to the development of over

200 methodologies to determine ecosystem water requirements (Tharme 2001). Many of these originated in the United States, South Africa, and Australia (Postel and Richter 2003). All make the assumption that aspects of flow are critical for maintaining certain ecosystem features.

Methods vary from those with a focus on defining minimum instream flow requirements for a few high-profile species to more recent efforts to describe the full spectrum of flow characteristics needed to maintain the ecosystem condition (that is, the flow regime), including low flows, higher flows, large floods, specific rates of rise and fall in water levels, and annual variability in discharge (Poff et al. 1997; Postel and Richter 2003). In this context, the flow regime provides a framework for considering the full range of interests in the context of the specific ecosystem functions that support them, and for identifying and quantifying the trade-offs associated with various scenarios, both in biophysical and socioeconomic terms.

Despite the large number of methods, each of which is unique in some regard, many methods are broadly similar (Dunbar et al. 1997) and aim to draw links between hydrological and hydraulic features with ecological responses, whether of individual species and populations, or their habitats. (See Table 7.4.) Further differentiation can be made between bottom-up approaches that build a flow from a base of mandatory low flows and then add freshets and floods (Arthington et al. 1998) and top-down down methodologies that determine the maximum acceptable departure from natural flow conditions (Brizga 1998).

The current trend is toward more holistic approaches that incorporate a number of flow assessment methods and examine all aspects of the ecosystem. Application of these methods requires multiskilled teams, including hydraulic engineers, hydrologists, ecologists, and geomorphologists, as well as those with skills from the social sciences (Postel and Richter 2003). Where expertise is minimal, more interactive approaches may be considered that draw on local and traditional knowledge.

Table 7.4. Advantages and Disadvantages of Different Methods and Characteristics of Setting Environmental Flows

Method Type	Advantages	Disadvantages
Look-up table	inexpensive, rapid to use once calculated	not site-specific
		hydrological indices are not valid ecologically
		ecological indices need region-specific data to be calculated
Desk top	site-specific	long time series required
	limited new data collection	no explicit use of ecological data
		ecological data too time-consuming to collect
Functional analysis	flexible, robust, more focused on whole ecosystem	expensive to collect all relevant data and to employ wide range of experts
		consensus of experts may not be achieved
Habitat modeling	replicable, predictive	expensive to collect hydraulic and ecological data

One example of such an approach is the Downstream Response to Imposed Flow Transformation framework (see Box 7.4), which can inform a decision on the desired condition of an ecosystem through assessment of scenarios. The DRIFT framework differs from other approaches such as the Instream Flow Incremental Methodology and the Catchment Abstraction Management Strategy in that it explicitly considers the socioeconomic implications of different scenarios (Acreman and King 2003).

Public participation to allow consideration of stakeholder views is central to the determination of ecosystem water requirements and was incorporated into the Murray-Darling Basin Commission's "The Living Murray" initiative (Murray Darling Basin Commission 2003). This involved a coordinated process of community engagement combined with socioeconomic and biophysical assessments to determine the desired condition of the River Murray, expressed as ecological objectives and outcomes for "significant ecological assets" from across the river system. (See Box 7.5.) In November 2003, this led to a decision to take a "first step" in recovering up to 500 gigaliters of water to support meeting these objectives.

Definition of the stakeholders to be included in assessments of water requirements will influence greatly the effectiveness of public participation efforts, and satisfaction of stakeholders with the outcomes of this process. There are arguments to suggest that in general (inter)national non-users should have a role in regional/local decision making, especially when important, non-replaceable features of an ecosystem are involved. Ultimately the solution to this issue is a definition of the boundary conditions and the degrees of freedom for debate among stakeholders.

7.2.1.3 Effectiveness

A common conclusion of practitioners in this field is that there is no single best method for determining ecosystem water requirements. Instead, application of flow assessment methodologies should be tailored to the unique set of ecological, water develop-

BOX 7.4

DRIFT Flow Assessment Methodology

The DRIFT (Downstream Response to Imposed Flow Transformations) methodology is an interactive and holistic approach for advising on environmental flows for rivers targeted for water-management activities. The methodology employs experienced scientists from a range of biophysical disciplines and, where there are subsistence users of the river, engages a number of socioeconomic disciplines. It produces a set of flow-related scenarios for water managers to consider, which describe a modified flow regime; the resulting condition of the river or species; the effect on water resource availability for off-stream users; and the social and economic costs and benefits. The process involves one or more multidisciplinary workshops, attended by a range of affected stakeholders to develop agreed biophysical and socioeconomic scenarios.

The development of scenarios requires an assessment of biophysical, social, and economic data and may draw on results from other predictive models that assess the responses of specific biota to flow (as in the physical habitat simulation model or PHABSIM). DRIFT should run parallel with two other exercises that are external to it: a macroeconomic assessment of the wider implications of each scenario, and a public participation process, whereby people other than subsistence users can indicate the level of acceptability of each scenario.

BOX 7.5

The Murray-Darling Basin Cap
(Murray-Darling Basin Commission 2003)

In June 1993, the Murray-Darling Basin Ministerial Council directed that a study be undertaken into the issue of altered flows and their consequences for the rivers of the Murray-Darling Basin. This led to an audit of water use, which confirmed increasing levels of diversions and associated declines in river health. In response, the Council introduced an interim cap on water diversions in the Basin in 1995. This was established as a permanent cap on July 1, 1997. In agreeing to the implementation of the cap, the Council essentially made a decision about the balance between the social and economic benefits derived from development of the Basin's water resource and the water needs of the riverine ecosystem. Implementation of the cap is the responsibility of the member states of the Murray-Darling Basin Initiative. Water audit monitoring reports are prepared annually to insure state water use is consistent with the cap and an independent audit group reviews progress on its implementation.

ment, and socioeconomic conditions in a river basin (Postel and Richter 2003). This is because of the variation in enabling conditions and binding constraints between each method and the prevailing conditions in a river basin.

The major issues when applying a methodology are the data required, technical capacity, funding, and time available to complete assessments (Arthington et al. 1998; Tharme 2001). The current shift toward more holistic methods brings with it a greater reliance on detailed ecological, hydrological, and hydraulic data. In the absence of hydrological data, computerized simulation models may be used to synthesize estimates of natural flow (Postel and Richter 2003). However, where the link between ecology and flow is unknown, predictions are more likely to call on expert opinion. Reliance on expert opinion is encapsulated in the Expert Panel Approach, which has been used widely across the eastern states of Australia with considerable success (Brown and King 2003).

The need for expert opinion highlights the often imperfect knowledge of the relationships between the ecology and hydrology of an ecosystem, perhaps the greatest source of uncertainty in determining ecosystem water requirements. This uncertainty will be less where the focus is on a single species or ecosystem component; but where a range of ecological and flow relationships are integrated, uncertainty will rise.

Various assessment techniques embrace uncertainty within their design. In the Lesotho Highlands Water Project, uncertainty was managed through the use of severity rankings, enabling scientists to indicate within a range how great each described change would be; where uncertainty surrounded predictions, the range of severity ranking was expanded (King and Brown 2003). Another method of expressing certainty in data quality was used in the Murray Flow Assessment Tool (based on an Environmental Flows Decision Support System; see Jones et al. 2003), whereby the quality and source of ecological data used for the assessment were recorded and open to interrogation by all users.

Time and money also influence the choice of method and approach. Expert panel methods are often rapid to carry out (Brown and King 2003), and because little to no additional data collection is required, cost is largely a function of the time of the experts. In contrast, more interactive approaches that require data collection across ecological, hydrological, and socioeconomic disciplines require more time and money. For example, an application of DRIFT could take up to three years (as in the Lesotho Highlands Water Project; see King and Brown 2003) and cost at least $1 million.

More recent environmental flow assessments have involved a broader array of stakeholders, including greater representation of community representatives and those who possess local or traditional knowledge—often these are the people who may be most directly impacted upon by upstream developments such as dam construction and water abstraction. Therefore, another binding constraint on the success of determining ecosystem water requirements is the extent to which public participation is embraced by local, regional, and national governing bodies.

In many countries where ecosystem water requirements are determined, a strong and effective regulatory framework and capacity for environmental protection exists. This framework serves to insure that additional "development" takes account of social and environmental issues (that is, through environmental impact assessment), can be used by activists to halt projects that threaten scarce environmental resources (for example, endangered species legislation), and provides for environmental standards and redress for specific environmental pollutants (for example, clean water legislation). Such a regulatory framework may be effective for restraining the environmental excesses of development; however it is often not sufficient to provide for a proactive response to restore degraded systems.

The effectiveness of the process for determining environmental flow regimes is influenced by the ability of practitioners to choose the right method following consideration of the enabling conditions and binding constraints for each application. Hence, this part of the process is effective provided an appropriate approach and mix of methods is adopted. The real test of the effectiveness of determining ecosystem water requirements lies in their allocation or implementation and the ability of the resulting flow or water regime to achieve its predefined objectives.

Where objectives are simple, such as minimum stream flows, they may be easy to measure and achieve. However, where the objective is to maintain a desired condition for a number of ecosystem components, the task is more difficult. In many instances, ecological response will be slow. Mitsch and Wilson (1996) suggest that it may take 15–20 years before monitoring can reveal the success (or lack of it) of large-scale wetland restoration projects. Determining the success of environmental flow regimes against economic, social, and environmental goals at the scale of the ecosystem, river basin, or even catchment may take considerably longer.

The success of changed river management to meet aspects of ecosystem objectives can, however, be judged somewhat by responses measured over the short term, as for example in reforming the sand bars in the Colorado River in the United States (Webb et al. 1999) and the expansion of the native perennial grasses on the Murray floodplain in Australia (Siebentritt et al. 2004) and on the Logone floodplain in Africa (Scholte et al. 2000). These responses, although not evidence of meeting objectives for the entire ecosystem, do indicate that responses are possible and consistent with broader objectives.

In the absence of the desired response, care needs to be taken in differentiating between a failed flow regime (the expression of a certain combination of ecosystem water requirements), a delayed response, and the presence of a system that is not capable of responding. The latter may occur if the system has been too severely degraded or if other factors are having a mitigating effect. With regard to the latter, patterns of water quantity and quality are rarely the sole drivers of degradation. Others include habitat modification, reduced longitudinal and lateral connectivity, salinization, and eutrophication. As such, the allocation of ecosystem water requirements should form part of wider river basin management strategies, of which they may be a cornerstone.

7.2.1.4 Findings and Conclusions

There is growing support for the conclusion that there is no single best method for determining ecosystem water requirements (Acreman and King 2003). Instead, selection needs to be informed by site-specific factors that match a flow assessment technique with the enabling conditions and binding constraints relevant to a particular location/site and application.

With respect to success thus far with determining ecosystem water requirements, where the objectives set are purely hydrological and relate to setting minimum flows, there are numerous examples of success. However, there would seem to be few examples of the effectiveness of the more recent trend toward holistic approaches that set objectives for a range of features of the flow regime and the ecosystem, and social responses to ecosystem change. This is largely a function of the long lead times for ecosystem-scale response.

Through incorporation of an adaptive management approach, the effectiveness of a flow regime designed to meet certain ecosystem objectives may be improved through time. Adaptive management is an integrated response option that provides a way to build on a base of imperfect knowledge. Active flow management through the use of environmental water "allocations" provides an ideal opportunity to learn more about the functioning of freshwater ecosystems and their water requirements. Such events can be treated as experiments to test management hypotheses.

7.2.2 Rights to Freshwater Services and Responsibilities for their Provision

As discussed above, freshwater services may span the range from public to private good. Due to their public good attributes, the value placed on freshwater services by actual or potential beneficiaries depends not only on demand, but also on whether they have confidence in the effectiveness of proposed management actions and institutional arrangements needed to insure that the service is actually delivered, and that they will have access to the stream of benefits. Access to benefits is determined by property rights, which refer not only to private individual rights, but are defined more broadly as "the capacity to call upon the collective to stand behind one's claim to a benefit stream" (Bromley 1989), and may include common property rights of communities and public as well as private forms (Ostrom 1990). Absence of enforceable rights is defined as a condition of "open access."

Rather than a pure concept of "ownership," this broader concept of property refers to overlapping bundles of rights held by individuals, groups, or the state, which will depend on the characteristics of the resource and on valued uses. For example, some individuals may have use rights to bathe in a river or water their animals; a water users' association may have control or decision-making rights to divert some of the water (management) or exclude others from using it; and the state may claim the alienation rights to transfer the water.

Although the state may define property rights over these resources, in fact, there are multiple sources of land and water rights that need to be considered, including those that derive from:

- international law or treaties (the Ramsar Convention, for example, which affects rights and uses of wetlands);
- state (or statutory) law, which may in itself have many different rights defined for various uses and users;
- religious law and principles, such as Islamic principles of the "right to thirst" for humans and animals;
- regulations of particular development projects; and
- customary law and local norms.

Each type of property right is only as strong as the institution that stands behind it. While in some cases, the government has a very strong influence, it does not operate alone. In many cases, customary or religious law may be more influential. Because of this complexity, changes in state law alone do not automatically translate into changes in property rights on the ground (Meinzen-Dick and Pradhan 2002).

Instead, as the above suggests, changes in property rights generally occur as a result of contested processes between different kinds of overlapping and competing claims. Outcomes will inevitably depend on the political and economic power of those who stand to gain or lose, on scientific and technological advances that reduce transaction costs of controlling access to the service, the ability to transfer rights, and changes in social values as new kinds of problems emerge that threaten future provision of the service. For example, in the United States, riparian rights, which limited uses of water to levels that did not impair the natural flow (which was valued for supporting the operation of power mills), was replaced by rights to reasonable use needed to enable the higher consumptive uses required for irrigation, municipal, and industrial uses, thereby curtailing rights to natural flows and often to aquatic resources that rely on them (Sax 1993; Tarlock 2000). Because of new social goals of protecting the ecosystem, rights are again being contested and redefined (Fahrenthold 2004; Santopietro and Shabman 1990).

Depending on the outcome, this process of making and contesting claims on freshwater services can lead to the renegotiation of new kinds of rights and responsibilities in which the uses of land, water, and other natural resources are limited to those uses that do not impair service provision. They may also help to insure that those who pay the costs of management practices have access to their benefits through various forms of compensation. Without some form of entitlement, land users will not be in a position to enter into contractual agreements regarding land uses and management practices, nor will they have access to benefits from investments made in such practices, or be in a position to trade rights to fresh water for other goods and services (Tognetti 2001; Swallow et al. 2001).

The rise in values of these services may further disadvantage those who lack rights by leading to regulatory restrictions on their uses of water and land, and sometimes to their displacement, without any corresponding access to benefits. For example, a case study in Thailand suggests that dry season flows have diminished primarily because of a dramatic increase, both downstream and upstream, in dry season cultivation and irrigation of soybeans by those who own paddy fields. However, the focus of regulation intended to address the problem has been on the more vulnerable farmers, who are dependent on rain-fed slopes in areas where significant forest cover remains, who have the least significant impacts on hydrology, and who are regarded as guardians of resources rather than as legitimate users (Walker 2003). In many cases, protection of watersheds has been used by governments as justification for centralized control of land and water resources and regulatory conservation policies that exclude upland communities. This approach tends to be resisted by local populations using various means such as fire, which further reduce the effectiveness of regulatory approaches (Swallow et al. 2001; Blaikie and Muldavin 2004).

A key issue in defining rights and responsibilities is the extent to which land users should be compensated for the costs of management practices by those who directly benefit from them, and the extent to which such practices should be required through regulations, based simply on a responsibility not to pollute and not to harm others. In the former case, land users are paid for the costs of conservation management practices needed to insure continued provision of valued freshwater services, in addition to marketable agricultural and timber products. In the latter, to the extent there is compliance, the ability of farmers to stay in business will depend on the extent to which they are able to pass their costs on to consumers in the form of prices.

7.2.2.1 *Effectiveness*

The process of changing or redefining rights and responsibilities, so as to reflect new values placed on freshwater services, is primarily constrained by transaction costs. These are costs associated with the development and enforcement of rules or institutional arrangements that can effectively control access to specific services, monitoring the conditions of watershed processes that support the production of services, and the resolution of conflicts between new and existing multiple uses and values, as services become scarce.

The costs to develop and enforce rules, and monitor watershed processes, are largely related to the site-specific biophysical characteristics of the hydrological cycle as well as to those of the users and their social and economic context. These have implications for the kinds of arrangements that will be feasible and effective, the information needed to support management decisions, and the technical and institutional capacities that may need to be developed. For example, because of the relatively large size of upper watershed areas, conservation management practices may be necessary over large areas, well beyond the level of individual plots, before a significant change can be detected in the provision of services. It will also be necessary to target marginal and unproductive areas that contribute disproportionately to off-site impacts—steep slopes, gullies at the base of escarpments, river banks, forest margins, and paths and roads, for example. Given that these areas provide little if any return to individual landowners on investments in conservation practices, individual private property rights generally do not provide a sufficient incentive for landowners to make such investments. In the absence of a set of enforceable rules, such areas are, in effect, "open access" areas (Swallow et al. 2001). Emphasis on the registration of formal private rights also carries risks of overlooking informal rights found in existing practices and social norms for managing the water and land uses by which it is affected, of placing at a disadvantage those with less education and who have fewer social connections, and of increasing uncertainty by creating new rules without developing the capacity and willingness to enforce them (Meinzen-Dick and Bruns 2003).

A key factor in overcoming the above constraints and generally reducing transaction costs is to strengthen forums for negotiation of rights and responsibilities, which, particularly at smaller scales, are better suited than rigid legal frameworks for creating access rules that are responsive to site-specific and constantly changing conditions (Meinzen-Dick and Bruns 2003). Given that there are normally large differences in the relative power of various stakeholders, collective action may also be necessary just to enable marginalized stakeholders to obtain rights or even recognition of existing informal rights. By providing a space for regular patterns of interaction, such forums can also facilitate the development of trust—the *social capital* that enables stakeholders to collaborate and engage in the collective action needed for monitoring and stewarding large and marginal upper watershed areas. This may take the form of user associations and watershed councils that can negotiate on behalf of numerous users of water and land, as well as provide technical assistance in the preparation and implementation of joint management plans. This process may gain momentum from broader social and technological changes and fom effective decentralization of resource management; these are further discussed in subsequent sections.

A promising approach that has been identified in Indonesia and the Philippines is where upland populations are granted tenure that is conditional upon compliance with land management plans and co-management agreements (Swallow et al. 2001; Rosales 2003). In Indonesia, this approach only became possible with the fall of the Suharto regime, which enabled local populations to voice demands for change in the property rights regime and to resist coercion by the forestry department and commercial interests. The result is a current trend toward decentralization in which management rights are being granted in exchange for adhering to agreed upon management plans (Swallow et al. 2001).

A general area of conflict that will undoubtedly have implications for the outcomes of future initiatives is between efforts to recover costs of both delivery and conservation of fresh water through pricing and privatization, and the recognition of water as a more universal kind of human right, consistent with objectives of poverty alleviation. Both objectives, for example, are found in the fourth Dublin Principle, which recognizes water as an economic good, but states also that "within this principle, it is vital to recognize first the basic right of all human beings to have access to clean water and sanitation at an affordable price." This dilemma is embodied also in the South African constitution and water law. Whether or not water pricing and privatization have enabled governments to expand services to poor and previously unserved populations addressed below. However, the approach is severely constrained by the ability to pay, and has been actively resisted through protests in a number of countries, including South Africa and several Latin American countries.

Among the better known was a protest in Cochabamba, Bolivia, where water privatization was brought to a halt after it led to monthly water bills that were double or triple what they had been previously and that, for many, amounted to 20% of their salaries (Rothfeder 2003). Full-cost pricing was also the subject of controversy during the development of the EU Water Framework Directive, which, in its final form, only required that environmental costs be taken into account in determining water prices (Kaika and Page 2002). In South Africa, this issue was addressed by designing price structures that distinguish between different uses so as to reflect policy priorities of providing for basic human and environmental needs. However, as the case of South Africa shows, the major challenge lies in implementation. (See Box 7.6.)

7.2.2.2 *Findings and Conclusions*

The legitimacy of various conflicting claims to freshwater services will depend on the outcome of negotiation and conflict resolution. Establishing such a process is central to sustainable management of freshwater services because it allows for the more flexible and adaptive responses needed to cope with the complexity and uncertainty of watershed processes than formal and rigid legal frameworks.

Both public and private ownership of fresh water, and also of the land resources associated with its provision, have largely failed to create adequate incentives for provision of services associated with it because upland communities have generally been excluded from access to benefits, particularly when they lack tenure security and have resisted regulations regarded as unfair. Public and

BOX 7.6

Water as a Human Right and an Economic Good in South Africa

The constitution of South Africa guarantees access to sufficient water and a safe environment as fundamental human rights, contingent on the availability of resources on the part of the state. The state has a duty to take reasonable legislative and other measures to achieve the progressive realization of the right. Under the South African water law, a minimum quantity of water is reserved for both human and ecological needs prior to allocation to other uses, and is provided free of charge to municipal water authorities (DWAF 1999). However, the municipalities still charged users for distribution costs, by requiring the use of pre-paid cards to obtain water from meters. This practice led to an outbreak of cholera when those who could not pay obtained water from polluted puddles and canals (Thompson 2003). The government now provides a minimal "lifeline" of free water, equivalent to half of the minimum standard established by the World Health Organization.

Uses of water beyond the amount reserved for human and ecosystem needs must be registered and, with some exceptions for some existing uses, licensing and fees are required for stream flow reduction activities; irrigation, industrial, mining, and energy uses; and water services authorities. Exceptions for existing uses, and tensions with users not exempted, suggest that whether licensing can truly serve as a reallocation mechanism may depend less on formal changes in the law than on the strength of public participation. This implies the need to provide users with information about the distribution of costs and benefits and to build institutional capacity so as to strengthen the bargaining power and negotiation capacity of the poor (Schreiner and van Koppen 2000).

private ownership have also failed to create incentives for control of off-site impacts of land use practices. With the absence of the recognition of common as well as private and public forms of property rights, there is likely to be insufficient incentive for practices needed to insure continued provision of freshwater services.

Given that property rights are intended to provide security, and cannot arbitrarily be taken away, they do not change easily. Conversely, attempts to recognize or redefine rights will generally be a source of conflict. Such changes are often linked to political momentum generated by extreme events such as droughts, floods, chemical spills, and broader social and political changes, which create opportunities to open up policy debates, and to redefine rights and responsibilities.

Effective property rights institutions with clear and transparent rules can increase stakeholder confidence that they will have access to the benefits of freshwater services and, therefore, willingness to pay for them. Their development can be facilitated by an analysis and recognition of existing rights, both formal and informal, which is indicated by the assets to which various stakeholders have access. This will be more evident in the response to extreme conditions. The development of a feasible strategy for protecting freshwater services will also require an identification of anticipated transaction costs.

Special attention should be given to broad social changes and structural reforms, as these may provide special opportunities for changes in property regimes that are no longer relevant in the face of rapid global changes and the new kinds of problems that are associated with environmental degradation. Post-armed-conflict situations may present special opportunities for changes in policies and rights because such situations reflect a complete breakdown of legal and political systems that were inadequate to begin with. Increased dependence on natural resources in the face of such collapse can also heighten awareness of the role of freshwater and other ecosystem services in human well-being. As was suggested by Professor Kader Asmal, who chaired the World Commission on Dams (2000), in spite of conflicts that have been inherent in the management of increasingly scarce water resources, and the rhetoric of "water wars," the negotiation of water agreements have ultimately been a catalyst for peace and cooperation (Asmal 2000). One reason is that negotiations regarding management of water are rooted in specific water projects, found in a local and site-specific context, which gives them more weight than vague and undefined agreements that have broad scope, and provide a point of departure for a bottom-up decision-making process.

In the face of conflicting claims, allocation of water and development of water-related infrastructure tends to give disproportionate influence to those with the more tangible and dominant economic interests compared with livelihood and environmental concerns. Such differences in power are often difficult to overcome in the short term, even through collective action. Therefore, the process of defining rights to the benefits of freshwater services, and responsibilities for actions needed to insure continued provision, should be regarded as long-term and on-going.

7.2.3 Increasing the Effectiveness of Public Participation in Decision-making

Stakeholder participation may improve the quality of decisions because it allows for a better understanding of impacts and vulnerability, the distribution of costs and benefits associated with trade-offs, and the identification of a broader range of response options that are available in a specific context. If poverty alleviation and sustaining livelihoods are the primary objectives, it is also essential that stakeholder concerns be a starting point for determining what the specific objectives are, and in the development of responses to freshwater degradation.

An important distinction in participatory processes is between those who allow for effective participation and those who only give an appearance of participation in decisions that have, in effect, already been made. Ultimately, the effectiveness of stakeholder participation depends on whether it makes any difference in decision-making, whether it contributes to the establishment as well as the achievement of objectives, and whether it provides an opportunity to work through difficult issues rather than avoid them. It should also provide stakeholders with an opportunity to learn, and to reconsider the values they place on freshwater services. As the Great Lakes water quality agreement illustrates (see Box 7.7), whether or not stakeholders are able to have an active role in the process can also have implications for whether or not goals are achieved (Sproule-Jones 1999).

7.2.3.1 Effectiveness

Principle 10 of Agenda 21 defines participation to include access to information, participation, and justice. It may be limited by factors such as: geographic isolation, common in upper watershed areas; language and educational barriers; access to information that is timely and relevant; whether participation is made possible in the early phases of a process (planning and defining problems); whether the decision process provides an opportunity for deliberation and learning; and legal frameworks that define rights (land tenure, for example) and provide measures of recourse, all of which determine the relative bargaining power of various stakeholders.

BOX 7.7

Implementation of the Great Lakes Water Quality Agreement

The Great Lakes Water Quality Agreement was implemented by developing separate remedial action plans for each of 43 areas of concern. A comparison of the process of developing the plans in each area suggests a strong relationship between success in the restoration effort and the active involvement of stakeholders both in development and in oversight of implementation. Conversely, failure was associated with agency indifference or hostility toward the participatory process. In most cases, agencies regarded the planning process as a source of new resources and public support for traditional concerns, rather than as an incentive for new approaches. In some cases, stakeholder forums were held after the agencies had written the initial reports. In one exception, Hamilton Harbour, more active participation by stakeholders in the development of the plan, and attempts to make the document legally binding, became a source of conflict with the Ontario government (Sproule-Jones 1999).

Conflict played a positive role in improving the plan because it provided an opportunity to work through the more difficult issues and avoid lasting polarization. A low level of conflict in another area, the Saginaw Bay, is attributed to the limited focus of the steering committee, which limited its task to data review and avoidance of difficult issues. Stakeholder input was also done in a way that limited interaction among stakeholders and therefore the opportunity for an iterative learning process (MacKenzie 1996).

In sum, it was the decentralized nature of the overall program that allowed for different strategies to be pursued within the common domain of the planning framework, and for many lessons to be learned. The process of prioritizing plan recommendations also demonstrated their role in the overall basin-wide strategy, and led to greater interaction among governmental agencies.

The nature and quality of participation, and who participates, will be closely related to the scale and institutional context in which it takes place. This may range from informal community-based initiatives to more formal watershed councils and inter-organizational partnerships typically found at smaller scales, to river basin organizations found at larger scales, which often cross national boundaries. Smaller scales provide opportunities for more direct participation and face-to-face interactions among those who share relationships to a specific place, and who often have greater understanding rooted in local knowledge. Larger scales present the greater challenges of achieving adequate representation of sub-basin interests as well as of providing information. Both present challenges of accountability and of providing adequate measures of recourse.

A key enabling condition, without which participation is unlikely to be sustained, is effective decentralization of authority for management of freshwater services. This implies the transfer of authority in the form of constitutionally protected rights, downward accountability to those who are represented, and access to the benefits of natural resources (Ribot 2002). This can increase the capacity of local authorities to respond to the variability of site-specific conditions because it enables them to make decisions regarding the allocation of resources, which can also be a source of the financial autonomy that is needed to exercise authority (Kaimowitz and Ribot 2002).

One barrier to democratic decentralization is a widespread perception that local authorities do not have the capacity to manage freshwater resources. However, in the absence of the authority to actually do so, and access to the benefits of natural resources, they have not had the opportunity or the incentive to either gain experience or to demonstrate capacity (Ribot 2002). Learning will therefore be an important part of the process. In many cases, however, farmers and users have built and have been managing their own water sources for centuries, displaying considerable technical skill. The trick is getting this recognized by government agencies, and linking these local understandings to basin-scale management. Access to benefits is discussed more specifically in subsequent sections that pertain to financing and the development of economic incentives.

Broader participation in public policy decisions is mandated in a number of relatively recent international water agreements and in legislative initiatives. Recent years have also seen a proliferation of various kinds of river basin organizations, which are discussed further in the next section, as a more specific institutional form for participation. Among the more far-reaching initiatives is the EU Water Framework Directive, which requires public participation in the development of management plans for all river basin districts in EU countries. This trend toward a shift in water management policies, from command-and-control regulations under the authority of states, toward more integrated approaches that transcend their boundaries, is also leading to the creation of new institutional entities or rules of the game that engage different sets of actors.

This shift raises a number of unanswered questions as to the degree of representation reflected by those who participate, to whom they are accountable, and, ultimately, of the legitimacy of the process (Kaika and Page 2002). Governments at national levels tend to have geopolitically driven interests that often conflict with the well-being and livelihood interests of many of their citizens, to whom they are not always accountable. Local governments are not necessarily any more representative of or more accountable to local stakeholders than are more centralized ones. Independent voices such as those of NGOs may play an important role in supporting community-based efforts and in developing innovative policy approaches, but usually represent narrower interests and are not designed or intended to represent all stakeholders or to create substitutes for democratic processes. Given the scope of the topic, it is not possible to draw general conclusions. However, there are some lessons that can be drawn from selected case studies.

Although decentralization is an inherent goal of many basin-level initiatives, these are often implemented such that pre-existing top-down centralized structures remain in place. For example, in a case study of the state of Madhya Pradesh in India, plans supposedly developed by local watershed committees established for purposes of local participation tend to follow blueprints created by upwardly accountable technicians. These blueprints were driven by the need to meet tangible physical targets and deadlines that had already been decided upon (such as the number of trees planted, number of compost pits dug, hours of volunteer labor) and to obtain the approval of technical committees. Selection of sites for implementation of watershed conservation projects tended to be based on administrative expedience. As a result, areas of tenure dispute on the steeper slopes, where conservation efforts were most needed, tended to be avoided altogether (Baviskar 2002).

Recognition of the limitations of managing irrigation by way of centralized government, and the inadequacy of building irrigation infrastructure without corresponding local management capacity, has led to a trend toward institutional reform in irrigation, in which responsibilities for operations and maintenance of irrigation systems are devolved to user associations (Vermillion 1999). Although, in many cases, these transfers have been incomplete, or responsibilities have been transferred without corresponding rights, they provide a considerable base of experience and a basis for identifying elements that are critical for the success of such initiatives. In general, comparative case studies of the more comprehensive devolution programs have shown a reduction of costs to government, an increase of costs to farmers, an increase in fees collected, increased cost recovery, and variable impacts on productivity. However, some benefits were observed even in less comprehensive devolution programs. For example, in one case, acquiring water through a user association resulted in the reduction of costs to users even though fees were doubled because it eliminated other costs associated with numerous informal transactions, such as the need to pay bribes in return for receiving water allocations (Vermillion 1999).

7.2.3.2 Findings and Conclusions

Degradation of freshwater and other ecosystem services generally have a disproportionate impact on those who are, in various ways, excluded from participation in the decision-making process. Effective participation may therefore require concerted political pressure to overcome resistance by those in positions of power and authority, which may be brought about through various forms of collective action by stakeholders.

Whether or not initiatives to increase broad-based participation have a lasting impact will depend on the extent to which participation is institutionalized in democratic local government as a constitutional right (Ribot 2004). As discussed in the earlier section on rights and responsibilities, in the absence of momentum that may be generated by periods of crisis and broader social change, or other special opportunities for reform, this is generally a process of institutional development that may take time. However, even partial transfers of authority can have benefits. A focus on implementation provides an opportunity for adjustment as lessons are learned.

In the meantime, a key focus for improving participatory processes is to help level the playing field through measures to increase the transparency of information; improve the representation of marginalized stakeholders; engage them upfront in the establishment of policy objectives and priorities for the allocation of freshwater services, with which specific projects should be consistent; and create a space for deliberation and learning that accommodates multiple perspectives.

7.2.4 River Basin Organizations

Recognition that there are trade-offs among multiple interests and uses of fresh water provided by river basins, has led to a trend toward the formation of river basin organizations as a vehicle for basin-wide assessment, planning, management, and conflict resolution. RBOs may be formed to manage basins within individual countries as well as in an international transboundary context, which presents an added layer of management challenges and gives rise to sovereignty considerations.

Pivotal concerns for basin-wide organizations are typically issues of water allocation among multiple and often conflicting uses. Decisions about infrastructure development such as dams, reservoirs, and navigation canals are critical because these modify and divert flows of water that make possible agricultural irrigation and urban development, and are therefore a major driver of land use change. They also block flows of sediment that maintain downstream river channels and coastal areas. Other key concerns at the basin level are with cumulative impacts of land use changes and inputs of pollutants that are detectable over long distances.

A river basin focus was formally recommended at the International Conference on Water and the Environment held in Dublin in 1992, and also endorsed in the 1992 Rio Declaration and Agenda 21 (particularly Chapter 18). However, such organizations are not entirely new. Among the better known early examples are the Tennessee Valley Authority, a regional river basin authority formed in 1933, and the French system of water management, organized around river basins, which was established in 1964. The Mekong Committee, predecessor of the Mekong River Commission, was created in 1957, one of the earliest formal RBOs in the developing world (Ratner 2003). The TVA served as a model for several RBOs in other parts of the world, including regional development corporations in Latin America, river basin commissions in Mexico established toward the end of the 1940s and the beginning of the 1950s (Barkin and King 1986; Garcia 1999), and the Damodar Valley Authority in India. Globally, the International Network of River Basin Organizations lists 133 members in 50 countries at present. The trend is driven in part by donors such as the Inter-American Development Bank, which has already financed, or is likely to finance, more than 20 projects at the basin management level in Latin America and the Caribbean.

RBOs range in type from those that have the authority to plan, promote, and enforce their plans, to those that lack authority but play important roles in an advisory capacity in the assessment needed to inform planning, priority setting, stakeholder negotiations, and decision-making. Examples of the former include the TVA (see Box 7.8), the Murray-Darling Basin Commission, and the French system of water management, which is carried out by river basin committees, and, at sub-basin tributary levels, by local water committees, consistent with national water policies. The recently adopted EU Water Framework Directive also requires a comprehensive basin approach in which the territories of all European countries are assigned to river basin districts that have the authority for implementing the directive. Other countries that currently provide legislative mandates for a nation-wide basin approach are South Africa, Brazil, and Mexico.

In an alternate model, which can be considered an adaptive approach, many RBOs have evolved from a narrow and sectoral

BOX 7.8

The Tennessee Valley Authority

The Tennessee Valley Authority was formed during the Great Depression, and became a model of comprehensive river basin development to meet multiple objectives in support of economic development. The TVA was granted broad powers under which it developed a system of multipurpose dams for purposes of flood control, power generation, navigation, recreation, and maintenance of flows necessary for maintenance of water quality and aquatic habitat. An equally important mission in the earlier period was a program for multi-resource conservation and development, to protect the natural resource base that was recognized as essential for regional economic development (as, for example, by planting trees on eroded lands) and generally to promote human welfare in a poverty-stricken region. However, in the post-World War II period, power generation became the dominant mission, as it was largely self-financing. Although the TVA had had a number of successes and much popular support for its natural resource programs, these have been more vulnerable because they depend on Congressional appropriations for their budget (Miller and Reidinger 1998).

focus on water bodies and point sources of pollution, to a focus on entire watersheds based on inter-sectoral approaches, to the extent that interested parties are able to reach agreement and cooperate. Capacity building thus tends to occur as needed, in response to recognition of new types of problems that are beyond existing response capacities, or in response to extreme events that expose structural weaknesses and bring problems to broad public attention more rapidly, thereby enabling the development of new agreements and policies needed to resolve conflicts among multiple and diverse uses. Regulatory authority tends to rest with existing agencies and political jurisdictions, which may need to be politically compelled to effectively participate, or offered some form of incentive.

This kind of RBO tends to play an intermediary role by activities such as convening interested parties for purposes of strategic planning, management, and conflict resolution, preparing master plans, reviewing project proposals, proposing policies, creating an information system, and coordinating monitoring efforts. One well-known example of this is the Chesapeake Bay Program. (See Box 7.9.) A basin-wide entity that evolved in a similar fashion after the failure of attempts to create a basin-wide planning agency is the Laguna Lake Development Authority in the Philippines, which was given expanded regulatory powers step by step, as its focus shifted from fisheries promotion to watershed-level pollution control (FAO 2002).

Given the absence of an overarching legal authority and the need to rely on voluntary agreements among countries that occupy a basin, this alternate model also tends to be found in transboundary initiatives. A well-known example is the International Commission for the Protection of the Rhine, which was established in 1950 to address concerns about pollution. Over time, however, and in response to specific events, the ICPR began to also address ecosystem concerns, and to develop an integrated approach to river management that considers land use and spatial planning so as to begin to restore floodplains and wetlands, and thereby mitigate flooding by "making space for the river" (Wieriks and Schulte-Wülwer-Leidig 1997). In what may be the beginning of a more proactive approach, studies are also investigating the management implications of climate change, which is expected to bring about significant changes in the availability of water (Middelkoop et al. 2000).

A challenge that is particularly pronounced in transboundary water management, to which this section offers special attention, is the strengthening of provisions for various aspects of public involvement, which includes access to information, public participation, and access to justice or legal recourse. An important tool for public involvement is the development of a process for transboundary environmental impact assessment (Cassar and Bruch in press).

BOX 7.9

The Chesapeake Bay Program (Hennessey 1994)

The Chesapeake Bay Program has origins in an agreement made between two key states (Maryland and Virginia) in response to conflicts between oyster fisheries and discharges from urban areas. This was the beginning of a trend toward more inclusive inter-jurisdictional agreements that now cover most of the basin. Given the resistance of existing authorities to the creation of a new regional authority, the program adopted a multistate and federal cooperative governance structure. As concerns extended from the main water body to reduction of nutrient inputs from the upper basin areas, the agreement was expanded to include local governments, who retain responsibility for land use decisions and whose cooperation is needed to meet nutrient reduction targets.

7.2.4.1 Effectiveness

RBOs are constrained or enabled primarily by the extent to which all of the relevant stakeholders participate, are able to agree on objectives and management plans, and cooperate in their implementation. In a transboundary situation, the capacity to implement agreements and plans will depend on the level of commitment of individual countries to integrated river basin management and to ecosystem management, whether they have mutual or complementary interests, and relative bargaining power. Top-down governmental and institutional mechanisms need to be balanced with bottom-up considerations of transparency, public participation, and accountability. It also helps to have specific and measurable goals, such as the cap on water diversion in the Murray Darling Basin in Australia.

An illustrative case is the Mekong River Commission, whose member countries do not include China, which holds 22% of the basin area and is actively building dams. Using its superior bargaining power, it has been negotiating agreements to improve navigation with neighboring states independently, and has avoided joint review of the impacts of its dam-building program on the livelihoods of those in downstream areas, particularly in Viet Nam and Cambodia, which rely on the flooding patterns that sustain fisheries and production of rice, and also control seawater intrusion. China's perspective is that the regulation of the upper Mekong will benefit people by supporting navigation and irrigation activities, flood control, and dry season power generation, and also by containing erosion. Downstream countries have not been given the opportunity for an independent assessment. However, it is important to keep in mind that Thailand and Viet Nam have also been constructing dams and have found ways to avoid the joint review of water diversions that have already created problems downstream. Cambodia, which is the most dependent on freshwater services to support livelihoods, is also the poorest and has the least bargaining power (Ratner 2003).

In the case of the Danube, where the basin states have greater mutual interests and incentive to participate, transboundary management practice has even contributed to political stability—as demonstrated by efforts to establish a river basin agreement that were sustained even during the period of war between 1991 and 1995 (Murphy 1997). Implementation of that agreement, a jointly developed strategic action plan, and efforts to achieve consistency with the EU's Water Framework Directive are also playing an important role in the economic development and strengthening of democratic institutions in Eastern Europe (World Wildlife Fund 2002).

A key enabling condition is access to information and cooperation in the assessment process itself. An important planning tool, developed in recent years, is the transboundary environmental impact assessment (TEIA), which can be utilized to enhance the cooperation and management of shared waters in a transboundary context. A key difference between the TEIA and the EIA is that the TEIA can facilitate cooperation and dialogue across borders. If utilized effectively, it has the ability to provide local people and under-represented interests an opportunity to be heard and to participate in decision-making that affects their environment and livelihoods across borders where otherwise they would be excluded. TEIA is more challenging to implement than domestic EIA because it increases the need for institutional coordination, sensitivity to sovereignty, public participation across borders, varying domestic EIA standards, and cultural differences between involved parties.

Europe has developed the most authoritative, binding commitments to TEIA and several other countries are in the process of developing agreements. In 1991, the UNECE convened the Convention on Environmental Impact Assessment in a Transboundary Context, also known as the Espoo Convention (UNECE 1991), which is arguably the most authoritative and specific international legal codification of TEIA. The Convention's strength lies in the specific codification on issues such as harmonization of national laws, and on the negotiation of more specific bilateral agreements between states, that leave little doubt as to what is required to implement a TEIA. The overall objective is to enhance cooperation between states over shared water courses, harmonize individual EIA procedures between states, and advance nondiscrimination to insure that all affected people have the opportunity to participate equally (Knox 2002). Optimally, TEIA should accord the same protections and access to information to the public of neighboring states as to individuals within a country's own borders. The future development of TEIA will most likely be driven by example, and unfortunately, examples are scant at present. As more TEIAs are undertaken, experience in implementing these will increase (Cassar and Bruch in press)

A pattern often observed is the tendency for basin-level management to be dominated by the more tangible and economically dominant interests. For example, in the case of the Tennessee Valley Authority, which was intended to meet a broad range of multiple objectives related to both conservation and development, emphasis eventually shifted toward provision of hydropower and flood control. This was criticized for providing disproportionate benefits to populations concentrated in large downstream urban areas (Barrow 1998). A similar pattern is evident in a review of basin-level initiatives in five African countries (Barrow 1998). In Central America, watershed management concerns go back to the early part of the last century, but did not get placed at the top of political agendas until they were seen as threats to higher priority interests downstream—such as the sedimentation of large hydroelectric dams or the Panama Canal (Kaimowitz 2004).

7.2.4.2 Findings and Conclusions

Effectiveness of basin-level organizations will depend on the kinds of development paths made possible by water allocation decisions, acceptability of the resulting distribution of costs and benefits among stakeholders, whether these help to achieve objectives of poverty alleviation, and maintenance or restoration of at least those ecosystem processes that support the provision of desired services.

Given that conflicts often exist between basin-wide interests and those at local scales, at which impacts of management activities are experienced and tend to have direct livelihood implications, effectiveness of RBOs will also depend on whether sub-basin and community-level interests are adequately represented and are able to effectively participate in basin-scale decision-making. Sub-basin level organizations such as watershed councils, land-care groups, village level catchment committees, and associations of users and farmers play important roles in addressing problems associated with land and water relationships that are difficult to detect at larger scales, in the promotion of new land use practices, and in the direct involvement of stakeholders in face-to-face settings. In contrast, basin-scale actors tend to be representatives of interested parties, which may include government agencies, NGOs, and associations of resource users. So far, there is little evidence of successful scaling up from village to basin levels (Swallow et al. 2001).

Given that many problems in river basin management are the result of unanticipated consequences, and may not be obvious because they disproportionately affect marginalized stakeholders with little voice in decision-making, a key to RBO effectiveness will be in whether there is an independent and transparent process of assessment in the implementation phase, the relevance of such assessments to actual stakeholder concerns, and the ability to learn from them and to improve on past practices.

Ultimately, whether or not RBOs achieve the multiple objectives of integrated water resources management will depend on how particular initiatives are implemented in practice, an assessment of which will also require stakeholder insights. Other aspects of effectiveness, discussed in other subsections, include decentralization of decision-making processes and use of appropriate regulatory and financial instruments to create appropriate incentives and provide some measure of financial autonomy to RBOs.

Given environmental heterogeneity; embeddedness in social, economic, and political frameworks; different stages of socioeconomic development; and different management capacities and technical expertise, no single institutional model is likely to be equally applicable in all basins. RBOs are actually a mosaic of overlapping institutions.

Regional authorities, based on blueprints, tend to have many of the same weaknesses they were intended to address, such as centralized authority, domination by special interests, and application of sectoral approaches to multiple sectors. They may also be a response to trends among donor organizations toward financing basin-level projects rather than the most pressing problems. The more successful approaches have no blueprints, but tend to evolve in response to site-specific conditions, trends, and extreme events. A key factor, addressed in a previous section, is the democratic decentralization of authority, as it can enable responses more appropriate to their context.

An important enabling condition is that river basin organizations are inclusive of sub-basin level interests. This can at least potentially be achieved by supporting and reinforcing successful community-level initiatives—allowing them to "scale up," and further building a community-level response capacity through provision of information that enables them to effectively participate in responding to larger-scale threats.

Access to information is an important aspect of integrated river basin management that has been increasingly incorporated into water resources management policy and regulation, and is a mainstay of many IRBM policies. Likewise, public participation is being increasingly incorporated into IRBM policies. Access to justice is less commonly incorporated. Many institutions, including the Nile River Basin Initiative and the Mekong River Commission, have embraced principles of transparency and public participation in environmental decision-making. However, while these principles are increasingly common in institutional mandates governing transboundary waters, specific measures to achieve these goals often require more elaboration for implementation to be fully effective. Strengthening the institutional mechanisms to do this is increasingly recognized as the way forward to improve the management of shared water in a transboundary context. One of the ways this is being done is through the development of a process for transboundary environmental impact assessment.

7.2.5 Regulatory Responses

Command-and-control regulatory responses applied to freshwater services include technological, end-of-pipe controls and discharge permits that have been applied to point sources of pollution, regulation of non-point sources through instruments such as the establishment of total maximum daily loads under the U.S. Clean Water Act, and restrictions on land use for purposes of watershed protection. (Non-point sources are more thoroughly discussed in Chapter 6 of this volume.) Regulations also play important roles in creating cap-and-trade systems that serve to limit pollution or resource uses (discussed below). Regulatory approaches also generally support market-based approaches and other instruments by defining the "rules of the game."

7.2.5.1 Effectiveness

Regulatory approaches have generally been considered effective for reducing pollution loads from the more significant point sources. However, by itself, this approach is generally regarded as inadequate for addressing numerous small-point sources and non-point sources because it would require more extensive enforcement capacity, as well as site-specific information and authority for controlling land use (NAPA 2000). This is a problem even in countries with well-developed regulatory infrastructure. However, it is perhaps most dramatically illustrated in the Danube Basin, where the privatization in Eastern European countries that followed the end of the Cold War increased the number of economic actors and pollution sources, while the capacity for inspection and enforcement remained low, given the lack of local-level institutions (Koulov 1997).

Key limitations on regulatory instruments with respect to diffuse non-point sources are the lack of flexibility, the information base, and the capacity needed to address the site-specific nature of watershed problems. Regulatory bodies also often lack specific kinds of legal authority as well as the capacity needed to control the diverse kinds of land use activities associated with non-point sources, as this tends to be an authority exercised through local government planning processes, and in which individual landowners retain significant levels of discretion. For example, the U.S. Clean Water Act requires states to establish total maximum daily loads as a basis for allocating the burden of reductions among non-point source emitters. Given that federal agencies have no direct authority for regulating uses of land that result in non-point source emissions, or for water allocation, this provision may be the only source of authority through which such reductions can be legally compelled, and may be useful as an incentive for emission reductions. However, it is also regarded as unwieldy from a technical and administrative perspective, and as having the potential to paralyze efforts by citizens groups with paperwork (NAPA 2000).

TMDLs have been criticized both for the lack of criteria for determining whether objectives have been achieved, and for the lack of independent assessment. A key limitation is that uncertainty in watershed models used to link pollutant loads with water quality so as to demonstrate the effectiveness of required actions for meeting standards makes TMDLs vulnerable to court challenges. It is also difficult to determine whether water quality standards themselves have been achieved, given the variation in standards across states; the lack of a consistent, nationwide set of data on water quality; and the lack of consistent protocols for gathering such data (NAPA 2000). However, it should be kept in mind that, given the heterogeneity of environmental conditions, sources of pollution, and end uses, no single standard approach would be possible or desirable.

Establishment of protected areas in upper watersheds is also a form of regulatory control over land use, as is illustrated in the case of the Hindu Kush Himalaya region. (See Box 7.10.) Such regulatory abrogation of formal or informal property rights is generally regarded as ineffective because it fails to recognize the rights of local populations who have depended on such areas to support their livelihoods, and excludes them from access to benefits (as discussed above).

7.2.5.2 Findings and Conclusions

Regulatory approaches that involve market-based incentives such as damages for exceeding pollution standards are particularly suited to the point discharge of pollutants into water bodies. Regulatory approaches that simply outlaw particular types of behavior can be more unwieldy and ultimately burdensome, as they may fail to provide an incentive for finding more effective ways of achieving protection of freshwater services, as a way to avoid them. For example, if not paralyzed by procedures and technical requirements, efforts of citizens groups can complement TMDLs, when they are able to foster stakeholder agreements on actions to be taken and on funding priorities. An alternative to absolute regulatory land use controls is to provide some form of compensation to cover the cost of conservation practices, an approach discussed in the next section.

Non-point sources and small, scattered point sources are difficult to respond to adequately under both regulatory and economic approaches to water management because they require extensive monitoring. It is difficult, for example, to assess quantities of nitrates leaking from a given field, as it depends on rainfall, management practices, and other site-specific conditions. Because of the uncertainty with regard to non-point source emissions, regulatory measures may be more effective than economic instruments in controlling them. Where appropriate, they may also become more effective as scientific and technological advances make it cheaper to gather and disseminate information.

7.3 Economic Incentives for Supply and Demand Management

Typically, water as a resource has been undervalued and underpriced, while infrastructure projects for water resources development have been heavily subsidized. This disconnect has led to water being managed ineffectively for people and ecosystem services (Johnson et al. 2001). Economic incentives generally refer to the use of market-based instruments, incentive payments, and pricing strategies to alter the economic return from the use of scarce resources to better reflect the environmental and social impacts.

These strategies are being applied to manage fresh water in at least four different ways:

- using markets to reallocate water from existing, low-value uses to new, higher-value uses, such as from agricultural to urban or instream uses (water transfers and water banking, for example);
- developing cap-and-trade systems to avoid overexploitation of water resources, improve water quality and mitigate for ecosystem degradation (nutrient trading, groundwater mitigation banking, wetland mitigation banking);
- using incentive payments and water pricing to provide water and watershed managers with incentives to conserve water quantity and improve water quality as it is conveyed to the point of use, thereby providing a way to meet additional uses with the same amount of water (such as incentives for agricultural water conservation); and
- developing public/private partnerships for the financing of new supply infrastructure and technologies, particularly for municipal and industrial purposes.

BOX 7.10

Causes of Environmental Degradation in the Hindu Kush–Himalaya Region

A widely accepted theory in the 1970s was that accelerated erosion, sedimentation of river beds, and increasingly severe downstream flooding in the Hindu Kush–Himalaya region was driven by population growth, ineffective agricultural technologies, cultivation of steep slopes, forest clearance, overgrazed pastures, and unsustainable use of forests for fuelwood and fodder. This theory of Himalayan environmental degradation was found to be largely unacceptable by international experts on the grounds that such impacts were not significant when compared to the high rate of mass wasting and natural erosion, which delivers large amounts of sediment to river systems in episodic events (Bandyopadhyay and Gyawali 1994); the experts said there were more complex root causes (Jodha 1995; Kasperson 1995).

Nevertheless, the concept that accelerated erosion was largely due to unsustainable practices and population growth retain significant influence on policy-making in Asia. For example, in the mountains of India and China, the concept provides justification for government efforts to increase access and control over watershed forests, and to manage them in a way that supports national-level interests such as revenue generation and prevents sedimentation of dams. In India, watershed management is for the stated purpose of preventing accelerated erosion, flooding, and desiccation of water supplies, and has been carried out through conservation policies that had excluded local populations from access to benefits that support livelihoods, although there is now greater emphasis on more participatory modes of forest management (Vira 1999). In China, logging bans and grassland enclosures were adopted in response to large floods, and to prevent siltation of the Three Gorges Dam, as well as to restrict indigenous land use practices of shifting cultivation and nomadic pastoralism, which are seen as the culprits (Blaikie and Muldavin 2004).

Given the differences of scale, it is not at all clear that changes in upstream management practices would detectably or significantly reduce sedimentation of dams or distant downstream flooding, though it may significantly reduce more localized flooding. Although it is difficult to generalize the reasons for the extent to which land use practices and deforestation contribute to the flooding in the foothills, there is not much doubt regarding the absence of convincing ecological links of such upland degradation with the regular monsoon floods in the distant deltaic plains, such as in Bangladesh. According to a recent case study, neither the frequency nor the volume of flooding has increased in Bangladesh over the last 120 years. The study also found that following a period of heavy rainfall and catastrophic flooding along a tributary of the Ganges in Nepal, there was only an insignificant fluctuation of water levels in the Ganges itself, which could have been associated with local rainfall (Hofer 1997).

An alternate explanation for the flooding in Bangladesh is rainfall within Bangladesh itself, and in the Meghalaya Hills in the Brahmaputra Basin, which are located in India, north of Bangladesh, a place known for some of the highest rainfall in the world. It also has shallow soils and rocky surfaces, which leads to immediate runoff (Hofer 1997). Regulatory restrictions on land uses that support local livelihoods also do not address root causes of degradation that may be more significant, such as land use intensification to support a shift to production for markets rather than for local consumption, and conflicts associated with the nationalization of forests, privatization of common property, and development of roads and industries, which are in conflict with livelihood interests (Jodha 1995; Kasperson 1995).

These strategies may be used to manipulate the economic incentives affecting the production, allocation, or consumption of freshwater services. Direct payments for watershed protection change the incentive structure for management of upland areas with resulting changes on downstream availability and quality of water. The existence of functioning water markets places a financial opportunity cost on the holding of water rights and, therefore, makes the allocation of water rights more responsive to the economic values associated with different uses of water. Direct incentives for household or on-farm water conservation cause consumers to adopt water-saving technologies that reduce the overall demand on water delivery systems and their natural sources.

Financing for these strategies may involve a mix of public and private monies. When the benefits can be limited to those who pay for them, consumers of freshwater services or entities that generate power or distribute water often pay directly for these services. Acquiring ownership rights to use water or manage lands for watershed services through land and water markets, paying others to conserve water or improve land management, or acquiring mitigation credits through a cap and trade system (effectively trading one service for another) are examples of permanent and temporary ways to acquire access to freshwater services. Public financing can be used to provide freshwater services through these pathways, but often, its primary route is through the more direct route of funding water resource development that increases water supply to consumers or water and watershed conservation activities that improve water supply and quality. Public "funding" may also be obtained for actions to restore freshwater services through tax incentives to companies and individuals. Clearly, another important public role is to provide the enabling environment (property rights and regulations) for markets and cap and trade systems.

Nongovernmental entities and other private/public intermediaries also often play a role in gathering funds from service consumers, public sources, and philanthropists, and then disbursing funds to farmers and households. These intermediaries can help to reduce transaction costs inherent in establishing arrangements among numerous stakeholders spread out over large and remote areas, who may lack clear title to land or access to water by facilitating agreements among them, negotiating on their behalf and providing various kinds of legal and technical assistance, and assessing and disseminating appropriate information. Financing arrangements in practice often consist of a combination of sources and may finance not just freshwater services, but also a number of other ecosystem services.

As the scarcity of fresh water has increased relative to demand, attention has intensified across all the approaches and methods listed above. Considerable innovation has occurred in the last decade in the area of cap and trade systems (in water and other environmental services). Although these tools hold promise, it is still too early to assess their effectiveness. Box 7.11 reviews a number of the innovative ways that these created markets can affect resource use and environmental degradation or restoration. As cap and trade systems lead to the creation of markets, a proxy for their effectiveness is the extent to which water markets—which have a long history—have functioned in the reallocation of water rights. A thorough examination of water markets is, therefore, provided below.

BOX 7.11

Cap-and-Trade Systems for Water and Watersheds

Cap-and-trade systems have been applied effectively in controlling point source air pollution, but are relatively new as a tool for provisioning water and watershed services. Where ecosystem maintenance or restoration goals are well-defined, the employment of cap-and-trade systems may be an appropriate response. The cap-and-trade approach involves three steps: (1) determination of the cap, or the level of resource use or pollution that is allowable, (2) the allocation of use permits or pollution credits, and (3) the development of a market for the exchange of permits or credits between willing buyers and sellers. Key issues in designing these systems include the initial method for allocating rights and rules for transferability. Limits on transferability may be used to prevent concentration of rights in the hands of a few or to maintain rights within a particular community. However, such limits may also reduce efficiency by reducing the pool of buyers and sellers (Rose 2002). Cap-and-trade systems also require a strong regulatory infrastructure to insure that the caps are met.

These systems are being applied increasingly to the management of groundwater, surface water, wetlands, and water quality.

Groundwater Credit Trading: In basins where streams are fed largely by groundwater, once surface waters are fully allocated, additional groundwater withdrawals can have adverse affects on stream flow. Once a limit is placed on total groundwater withdrawals, groundwater pumping credits are created and traded so as not to further impair surface water flows. Such a system is in use in the Edwards Aquifer of Texas, where it has led to an active market in credits (Howe 2002).

Groundwater Mitigation, Credit Trading, and Banking: Another approach is to use a cap and trade system to achieve conjunctive management, which is the integrated management of surface and groundwater. The further development of groundwater sources can then be off-set, not just by reducing other groundwater withdrawals, but also by restoring stream flow or recharging aquifers. In 2002, the state of Oregon developed a mitigation cap-and-trade system for the Deschutes Basin, which has led to the development of markets for both temporary and permanent credits (see www.wrd.state.or.us).

Wetland Mitigation Banking: Wetland mitigation banking was developed in the United States by the U.S. Environmental Protection Agency in order to provide a more cost-effective and efficient option to meet regulations under Section 404 of the Clean Water Act. Mitigation banks may be established where a wetland has been "restored, created, enhanced, or (in exceptional circumstances) preserved" and the credits created can later be applied to areas in which wetlands are removed by development (USEPA 2003). Advantages include eliminating the temporal gap between when a wetland is created and one is eliminated by development, reducing costs of regulatory compliance, and reducing delays for development. As documented later in this chapter, criticism exists in the form of uncertainty about engineers' and biologists' expertise to recreate the intricate functions of a wetland and that the new wetlands may not be anywhere near the original wetlands. For example, in 2002, the New Jersey Department of Environmental Protection released a study that concluded its mitigation banking program, which had targeted two acres of restoration to every one acre lost, had actually resulted in a 22% net loss of wetland acreage and created only 45% of expected acreage (NJDEP 2002). While most wetland mitigation banks are federally supported, there is growing entrepreneurial interest and the first private bank was chartered in December 2002. The major markets that have been identified as targets for wetland mitigation banking include commercial land developers, airports, departments of transportation, oil and gas transmission line companies, and electric utilities (Zinn 1997). According to the U.S. Environmental Protection Agency, roughly 100 mitigation banks are in operation in the United States.

Nutrient Trading: Water quality trading is another response emerging in the United States to meet total maximum daily load regulations under the Clean Water Act. Under the Act, waterways must not exceed certain nutrient levels and states must develop plans for remediating the waterways back to established TMDL levels. Trading is limited to the immediate watershed in which the TMDLs are specified, though there are some exceptions. The Connecticut Nitrogen Exchange Program is one example of a nutrient trading program that emerged to reduce hypoxia in the Long Island Sound. In 1990, Connecticut, New York, and the EPA agreed upon a Comprehensive Conservation and Management Plan to reduce the level of dissolved oxygen in the Sound by 58.5% between 2000 and 2015 (USEPA 2003). To meet its commitment, Connecticut chose to implement a trading system among point and non-point sources, requiring 79 municipal, publicly owned treatment works to reduce nitrogen discharge by 64% from 2000 levels. The exchange is expected to save $200 million over 14 years; in its first year of operation, the program reduced nitrogen discharges by 15,000 pounds, or 50% of the target reduction (Rell 2003; Johnson 2003).

Nutrient trading has also developed in effluent, stormwater, and agricultural runoff. For example, in Australia's Murray-Darling Basin, efforts are under way to develop markets for salinity trading. The Basin's Salinity Debits and Credits Management Framework allows states that have contributed to the cost of projects to reduce salinity, thereby creating salinity credits, to implement measures that might increase salinity within agreed limits, employing salinity debits. Credits may be traded and are tracked through a Salinity Register (Murray-Darling Basin Commission 2001).

An important aspect of freshwater services provisioning is the conservation of existing supplies. There are substantial savings to be gained from improvements in municipal, industrial, and agricultural systems around the world (Gleick 2000). For example, in California, the potential for water conservation and efficiency improvements in just the residential, commercial, and industrial sectors is 33% (Gleick 2003). However, as the single largest use of water in the world, irrigated agriculture has been estimated to be only about 40% efficient on average and, therefore, a prime target for conservation measures (Postel 1997). The remaining 60%, lost through leaky or unlined canals, overwatering of crops, and inefficient technology, is often considered wasted (Molden and de Fraiture 2004). Gleick notes that as of 2000, only 1% of the world's irrigated land was under micro-irrigation; 95% had efficient drip irrigation or micro-sprinklers (Gleick 2003).

While significant savings exist through irrigation efficiency improvements, many studies suggest there may not be as much as often thought (Molden and de Fraiture 2004, p. 9). For example, the lost water typically returns to a waterway or recharges groundwater and is subsequently used downstream, either by other irrigators or ecological uses instream. Therefore, water productivity at a basin-wide level is likely much higher than when estimated at the irrigated agriculture level (Gleick 2000). In fact, improvements in irrigation efficiency may even harm downstream users as savings are used upstream of where they used to be available (Molden and de Fraiture 2004). While improvements in irrigation efficiency will be necessary to improve water productivity, they will not be sufficient to meet environmental, municipal, or other needs since savings, particularly in water-tight areas, will likely provide incentive for irrigators to increase intensity or production. Therefore, regulations or other means will be needed to ensure that savings are allocated to other uses as well as agriculture (Molden and de Fraiture 2004).

An international research program known as the Comprehensive Assessment of Water Management in Agriculture is currently under way, to be completed in 2006, and will provide a thorough assessment of a number of topics including that of conservation and management of water in irrigation. Pending the results of that Assessment, this section on payment approaches takes up an area of increasing innovation internationally, that of making payments to landowners for the watershed (hydrological) services that well-managed lands can provide to downstream users and communities.

The focus of this economic incentives section shifts, then, from rural land, water, and agricultural water management issues to economic incentive issues surrounding the need to provide water supplies to communities for domestic and industrial purposes. Even with greatly improved management of ecosystems upstream, there will still be an enormous unmet and increasing need in coming decades to provide the physical infrastructure to bring water to communities. A critical issue here is the financing that will be required to implement these upgrades and new systems and what will be the roles of the public and private sectors in carrying out this task.

A general recognition that regulatory approaches are, by themselves, inadequate for ensuring the continued delivery of fresh water has led to the recent interest in applying these market-based institutional arrangements. In part, this also reflects societal changes in attitudes and an increasing willingness of beneficiaries to pay for these services (Landell-Mills and Porras 2002). Efforts to develop such arrangements can also be considered part of a global trend of institutional change in water resources management aimed at improving the recovery of costs—both the operational costs of delivering basic water supplies and sanitation, as well as the costs of conservation and research activities (Saleth and Dinar 1999).

7.3.1 Water Markets

Bjornlund and McKay offer three compelling reasons to consider the use of markets for the provision of fresh water (Bjornlund and McKay 2002). First, tradable water rights create a value for water that is distinct from land and, therefore, able to be preserved in its own right. Second, full cost recovery pricing incorporates externalities associated with inefficient use and encourages inefficient users to leave the market. Third, the use of market forces rather than government intervention to facilitate reallocation reduces transaction costs and delays. While markets clearly respond to efficiency objectives, they have their limitations, particularly with regard to the importance of equitable solutions (Johnson et al. 2001). Water is often viewed as at least partially a public good (Thompson 1997). As a result, purely unfettered markets are not only unlikely to evolve, but probably undesirable. The support for market approaches is typically circumscribed by the requirement that these markets take place within a carefully constructed policy and regulatory framework (Howe 2002).

As countries develop, water transfers, and leasing and trading programs have emerged to address the need to reallocate water from traditional uses (primarily agriculture) to new and growing uses (municipal and environmental). The establishment of markets for the transfer of water depends on a number of enabling conditions that largely have to do with creating private property rights that are transferable. Institutional approaches to enhance transferability include water exchanges, water banks, and instream acquisitions programs. Cap-and-trade systems are more novel, but are rapidly being applied to resource and environmental issues related to water management. A description and context for each of these aspects of water markets is provided below, followed by an assessment of the effectiveness of markets as a reallocation and supply response.

While the attempt here to distinguish between various forms of market-based responses is useful for the purpose of analysis, in practice, one response may borrow from or encompass another, and some operational issues such as price discovery may apply across the board.

7.3.1.1 Water Transfers: Property Rights and Enabling Conditions

Globally, the range of systems that regulate the allocation and manage the use of water is broad, with the primary distinction being the degree to which the user has a private right to the use and ownership of water. In systems where water is owned as a public resource and use occurs only upon the issuance and renewal of a temporary permit, the use of markets to reallocate water is an unlikely response. In these systems, water allocation is achieved through regulatory control, government policy, and administrative process.

On the other end of the spectrum are systems where rights to use or ownership are extended to users and are tradable. The ability to trade in these rights will typically depend on their transferability and validity. In cases where water rights are not clearly defined and allocated, trading will be limited if it occurs at all. For example, in some basins in the western United States, water rights are not fully adjudicated or prior Indian reserve rights exist that have not been adjudicated or settled. Lack of clear property rights limits the transferability of these rights and can increase the risk to those engaging in water rights transfers.

The validity of the intended use on a transfer is another important consideration. Many countries and states manage water under a "beneficial use" doctrine, whereby water not beneficially used is lost to the user or right-holder. A notable exception is Chile, whose free-market Water Code no longer contains a requirement for beneficial use. The result of this policy was a prolonged and unsuccessful effort in the mid-1990s by the state governmental water agency to pass a tax on unused water rights (Bauer 2004). However, the more unfortunate result was that the large hydropower companies were able to file for large water rights on some rivers, thereby establishing monopoly rights on those rivers.

Where the beneficial use test applies, a key enabling condition for the use of markets in reallocation is the statutory provision that the "new" uses are beneficial. This is particularly the case where markets are used to reallocate surface water from out-of-stream uses to instream uses. For example, since the 1980s, the Pacific Northwest states of Oregon, Idaho, Montana, Washington, and California in the United States have adopted laws and regulations that allow water to be transferred instream as a beneficial use, in some cases even adopting instream water rights that reflect minimal needs for fish and wildlife (Landry 1998). A number of U.S. states also recognize recreation, aesthetics, and pollution mitigation as beneficial uses. Prior to these statutory changes, it would not have been possible to transfer the character of use to an instream use.

Ownership of instream rights is a further complicating issue. When water is reallocated to an agricultural or municipal user, the owner is clear. This may not be the case when water is reallocated to an instream use. Despite much interest in the creation of private "trusts" to hold these water rights, western U.S. states allowing instream beneficial uses have preferred to adopt a public trust doctrine, whereby these rights are held exclusively by the relevant state agency upon transfer. The buyer interested in restoring instream flow must therefore purchase the water right and transfer it instream by, in effect, turning it back to the state. Difficulties with this approach exist as conflict may develop between the roles of the state as administrator and as property right holder, and constraints on state budgets may impair efforts to insure that the instream flow rights are monitored and enforced as against out-of-stream rights (Aylward 2003).

Depending on the context, permits and rights may also be transferable on a temporary basis. Where water is leased in this manner, the ownership or use returns to the original user upon termination of the lease. With water that is leased instream, the buyer may become the lessee, but the leased right is often still held by the state for the duration of the lease.

7.3.1.2 *Water Exchanges*

Water exchanges vary in size and activity from full service operations offering brokerage, water rights information, and consulting services to small, nearly virtual bulletin board systems providing a place for buyers and sellers to connect. Bulletin board systems are pervasive wherever there is an agency (such as an irrigation district or company) that provides centralized services (in particular water delivery) to water users. However, it is in the Murray-Darling Basin of Australia where water exchanges have seen the most rapid pace of institutional development for trades between water users. Two distinguishing features of the exchanges in the Murray-Darling Basin are that they serve to transfer water outside of the traditional confines of a specific administrative or geographic area and they have pioneered the use of electronic auction techniques. These exchanges operate for the purpose of easing the transfer process and facilitating short-term trades that might not take place otherwise. While most trades that take place are temporary—and in some cases nearly instantaneous—the exchanges are beginning to place a few permanent trades (Bjornlund 2002).

Water exchanges emerged in Australia in response to a cap placed on water use within each state of the Murray Darling Basin. The cap was established in 1997 and limited surface water usage to 1994 levels, but left it up to the states to decide how to achieve the reduction. Trading has been active since 1997, particularly in drought years when all allocations are cut back (Bjornlund 2002). While water exchanges do not require a cap-and-trade system per se, it is worth emphasizing that, ultimately, markets evolve only in the presence of scarcity. With respect to water resources, scarcity may evolve in response to a cap on further appropriation or it may evolve due to a physical scarcity of water. In the western United States, water rights are allocated according to the prior appropriation doctrine of "first come, first served." No cap on appropriation of surface water rights is necessary to drive a water market in this case. "New" appropriations will be of little value when stream flow is already fully appropriated—in this case, there will be a natural tendency for new needs to be met by the reallocation of existing prior uses (provided of course that priority is conveyed along with the property right).

7.3.1.3 *Water Banks*

The term "water bank" has many interpretations but, in general, refers to an institutional arrangement for temporarily moving water from one use (or user) to another that involves the participation of an intermediary. Water banks are a feature of the American West most notably in Idaho, Texas, California, Oregon, and Washington (Clifford et al. 2004). The first formal rental pool in Idaho was set up in 1937, following a decade of informal water leasing between agricultural users (Howe 1997). The term "bank" may well reflect the fact that most large water banks are based on water stored in a reservoir. The water is therefore "banked" or stored until such time as it is purchased and used.

Water banks have become a preferred option in the western United States, as they operate within a confined area—often an irrigation district—and the water is unlikely to travel very far. Since irrigation districts have a vested interest in retaining the right to delivery water and, therefore, ensuring their customer base, they are more likely to find the temporary and limited nature of water banking an amenable option for storage management. The distinguishing feature between water banks and exchanges is that an exchange simply brokers water rights, whereas a bank will either hold rights or retain a role as a lessee of the rights.

7.3.1.4 *Instream Water Acquisition Programs*

The acquisition of water for instream flow restoration on the part of state agencies and local water trusts and conservancies has become a popular tool in the western United States, where rivers are affected drastically by summer withdrawals for irrigation, which often leads to the listing of species as threatened and endangered under federal law. (See Box 7.12.) Water leasing is particularly popular with water right holders because it avoids the permanent dedication of the water right instream. It may also be useful in extending the lifetime of a water right given that if a water right is not being put to a "beneficial" use, it can be subject to forfeiture (typically after five years). Leases also provide considerable flexibility as they may take many forms, including fixed terms, dry year options, forbearance agreements, conservation off-sets, and exchange or barter agreements. As in general with efforts to "purchase" water for instream flow restoration, leasing is typically found off main stems and may be particularly useful in small tributaries where a small quantity of water may make the difference between a stream that goes dry and a minimum flow level to support fish and recreational uses.

Instream flow acquisition is an attractive alternative in prior appropriation systems because many streams and upper tributaries are over-appropriated. This means that even if a "senior" water user does not choose to use water in a given year, there will be a "junior" user who will divert the water. Only if an "instream" water right is created that is of sufficient seniority to ensure that the state can protect the water from junior users can instream flows be assured. Ultimately, in order to ensure environmental flows sufficient to maintain habitat and species, permanent instream transfers of water rights will be necessary.

7.3.2 Effectiveness of Market Approaches

Water trading has a long but narrow history (Howe 2002). Water auctions were held in Spain as far back as the sixteenth century and water trading has occurred in Chile and the United States for over a century (Howe 2002). In the United States, the Carey Act in the early twentieth century provided incentives for the formation of private companies to appropriate and develop water resources in the American West. Transfer and leasing of water between agricultural users began shortly thereafter.

Reallocation of agricultural rights to other uses—particularly municipal uses—also has a long history, but a contentious one. Water trading in California in particular has an infamous history

BOX 7.12

The Deschutes River Conservancy (http://www.deschutesrc.org)

The Deschutes River Conservancy is a partnership initiated by Environmental Defense, the Confederated Tribes of the Warm Springs Reservation, and irrigation districts in central Oregon in the United States. DRC founders recognized the need for a private organization with ecosystem-determined goals and methods based on positive incentives, consensus, and local governance. Since approximately half of the Basin's land area is managed by federal agencies, it was clear that such a private organization would need the capacity to partner on projects with the federal agencies to be truly ecosystem- and basin-wide in scope. In March 1996, Senator Hatfield introduced S. 1662, authorizing federal agencies to work with this private nonprofit organization on ecological restoration projects using Bureau of Reclamation funds on a 50–50 cost-share basis.

The DRC mission is to restore streamflow and improve water quality in the basin through the use of voluntary market-based economic incentives with the aim of enhancing the quality of the region's natural resources and adding value to its economy. The DRC's Water Acquisitions Program uses instream provisions under Oregon water law to acquire and protect water rights instream. In nearly six years of operation, the DRC has increased local awareness and appreciation for the value of water rights through payments to irrigation districts for water conservation projects and providing a market for temporary and permanent instream transactions for water right holders. These programs have trebled flows in the main stem Deschutes and restored 50% of target flows for fish and wildlife in critical dewatered tributaries. The DRC also operates a full range of brokerage and administrative services, including fee-for-service assistance to private clients, a groundwater mitigation bank, and water banking services to irrigation district and municipal partners to more effectively and equitably manage water resources in the Basin.

DRC efforts to improve water quality are directed through its Riparian Habitat Restoration Program. Since 2002, the DRC has partnered with the Climate Trust of Oregon on an innovative water quality and carbon sequestration program. The Climate Trust is providing the DRC with $780,000 to help landowners in the Deschutes Basin restore riparian areas while reducing greenhouse gases in the atmosphere. The DRC is matching the Climate Trust's funding with federal and foundation funds. Additional support comes from leveraging U.S. Department of Agriculture Conservation Reserve Program funds. In the first year of the program, a large-scale restoration project in the headwaters of the Trout Creek Watershed replanted 100 hectares of contiguous riparian buffer with 71,000 native plants (willow, red osier dogwood, mountain alder, and ponderosa pine, among them). A 52-year conservation easement and monitoring plan now protects 10 kilometers of one of the principal remaining steelhead runs in the Deschutes Basin.

(Reisner 1986). The Owens Valley water "grab" by the Los Angeles Metropolitan Water District in 1905 presages the worst fears of those opposed to a free market in water. By quietly purchasing most of the water rights in the Owens Valley, the District severely curtailed agricultural activity in the valley. The action had immediately devastating effects on the local economy and way of life, but even longer-lasting negative impacts on the Mono Lake ecosystem and its migratory waterfowl. The assessment below begins by checking the extent to which water markets have led to significant reallocation and how these transfers have affected third parties.

7.3.2.1 Efficacy of Markets in Reallocation

Several countries have experimented with creating water markets for irrigation water, most notably the western United States, southeastern Australia, Mexico, and Chile (Bjornlund and McKay 2002). With the most laissez-faire system, Chile has relied largely on the natural evolution of markets. Australia and the U.S. states have tried to stimulate markets through the creation of water exchanges, water banks, and instream flow acquisition programs. Evidence of market activity, the balance of trade between permanent and temporary reallocation, pricing, and institutional innovations to facilitate trading are covered below.

7.3.2.1.1 Chile: The free market

In Chile, a formal water policy for markets (declared in 1976) and laissez-faire Water Code (1981) has promoted a free market in water rights largely unfettered by public interest concern (Bauer 2004). Despite the extensive nature of the system, temporary trades between farmers on the same canal are still the most frequent trades that take place, although these are often informal trades that do not depend on the Water Code (Bauer 2004, Bjornlund and McKay 2002).

A study of four areas in central and northern Chile, selected because they were expected to have active water markets, showed that in fact there was very little trading of water rights in three of the four study areas, with the exception being the Limarí area (Hearne 1995). The principal explanation for the lack of activity was that the rigid canal infrastructure made it costly to change water distribution, particularly among farmers. Further work in the Santiago, Chillán, and Bulnes water registries found annual trading to vary from 0.6% to 3% of total allocations (Rosegrant and Gazmuri 1994). In the Santiago registry, which had the highest level of activity, only a small amount of the water (3%) moved from agricultural to municipal use.

A later study of the Limarí water market found that the market operated efficiently and provided important benefits for both buyers and sellers (Hadjigeorgalis 1999). In Limarí, there is abundant evidence that water has been frequently reallocated to higher-value uses within the reservoir system. In addition, the market has provided farmers with the flexibility to manage some of the risks caused by uncertainties in water supplies and in agricultural markets. Poor farmers, for example, have been able to lease their water rights to other farmers during drought years, when water prices are high and income from irrigation is uncertain. Much of the temporary trading occurs informally between farmers on the same canal system and numbers are not available on the size of this market (Brehm and Quiroz 1995). However, Romano and Leporati note that the market has had negative distributive impacts, especially with regard to peasant farmers who have little bargaining power (Romano et al. 2002).

The Chilean experience thus suggests that the freedom to buy and sell water rights has led to the reallocation of water resources to higher-value uses only in certain areas and under certain circumstances (Bauer 2004). Brehm and Quiroz (1995) argue that lack of activity may simply suggest a close to optimal initial allocation of rights. Still, studies have identified a number of factors limiting market activity in Chile, including:

- constraints imposed by physical geography and rigid or inadequate infrastructure (limiting cross-canal transfer);
- legal and administrative uncertainty over the validity of water rights;
- cultural and social reluctance to conceptualizing water as an economic good;
- inconsistent and variable price signals; and

- concurrent subsidies for efficiency improvements (Bauer 2004; Bjornlund and McKay 2002; Brehm and Quiroz 1995).

Despite the lack of market activity, the principal benefit of Chile's Water Code is that secure private property rights have led to significant investment in the water sector for both agricultural and infrastructure development (Bauer 2004; Hearne and Easter 1995; 1997).

7.3.2.1.2 *The United States: Water banks and instream flow acquisition*

Permanent transfers of water have long occurred in western part of the United States. Allocation and overallocation of stream flows in the late nineteenth century in many areas of the West, combined with the application of the prior appropriation doctrine, enabled a market for transfers. Few historic measures of market activity exist; however. One investigation of transfers in the period 1975–84 in several states suggests widely varying activity. Only three transfers were found in California, where federal and state projects do not allow transfers outside of such projects. In Colorado and New Mexico, where markets are most developed, approximately 1,000 transfers were found, largely from agricultural to non-agricultural users.

Temporary transfers are far more prevalent and often facilitated by water banks. These banks, known as "officially sanctioned arrangements for short-term transfer," were first operated in Idaho in the 1930s (Howe 1997). Over the years, water banking on the Snake River, based on storage, has seen continued development. In the 1980s, much of the water was transferred from agriculture to power generation, with annual traded volumes in the range of 185 to 370 million cubic meters. As the price was set administratively ($2 per cubic meter during the late 1980s) and did not account for water availability, large supplies (with little demand) were obtained in good water years and minimal supplies in dry years (with high demand). This lack of a responsive pricing mechanism has been a major problem with this bank, as well as the Boise and Payette River banks, also operating in Idaho (Howe 1997).

California operated a "drought" water bank in 1991–92 and 1994. The principal purpose of the bank was to transfer water from irrigation to municipalities. In 1991, the bank bought over 1 billion cubic meters of water (largely stored water) without having a ready buyer, ending up with 505 million cubic meters of unsold water that year (MacDonnell 1994). Half the water came from fallow land, and the remainder from substituting groundwater for surface water and from storage. The bank paid $0.10 per cubic meter, and sold the water for $0.14 per cubic meter (Howe 1997). In effect, the bank acted as a speculator by holding water across seasons and may have actually competed with its customers. In subsequent years, the price was reduced significantly, with buying occurring at $0.04 and selling at $0.58 per cubic meter.

Despite the obstacles encountered with the drought banks, water marketing and trading has become an institution in California. It accounts for approximately 3% of all water use in the state, with agricultural districts from the Central Valley, Imperial, and Riverside counties supplying 75% of the water. Market drivers have not been municipal users, as is the case elsewhere, but rather changing environmental regulations (Hanak 2003). Trade is balanced between agricultural, environmental, and municipal demands, with direct environmental purchases for instream and habitat restoration (resulting mostly from federal and state programs) growing from 12% to 30% of demand between 1994 and 2001. As much as 50% of demand now comes from agricultural demand growth in the San Joaquin Valley and the remainder from municipal demand, which has actually declined in importance since 1994. However, municipal demand accounts for the longer term and permanent purchases—approximately 20% of the total (Hanak 2003).

There is some consensus on the conditions required for successful water banking efforts. Establishing a spot market (perhaps through an auction) is important in conveying necessary information on trades occurring in known quantities of available water (MacDonnell 1994; Landry 1998). Second, expected future flows should be sold in volumetric terms (something that is rarely done). Third, water rights may be leased, and in doing so, the delivery risk falls on the lessee, something that is commonly practiced in irrigation districts (MacDonnell 1994). Other factors that weigh heavily on the design and operations of a water bank are the homogeneity of the water rights, the ability to hold water over time (in storage, for example), and thinness of the market (MacDonnell 1994, pp. 4–15).

Although water banks in the United States have typically been devised as a means of shifting water to urban and power users, they have also been used to meet ecosystem needs. In 1993, the Bureau of Reclamation made the first purchase from the bank for the purposes of salmon restoration, acquiring 250 million cubic meters for this purpose (MacDonnell 1994). In the 1990s, explicit programs of water acquisitions for instream purposes and ecosystem needs were initiated, and programs to develop the institutional capacity for such efforts are now under way in a number of states, including one multistate program in the Pacific Northwest. (See Box 7.13.) In Australia, the realization that markets have not spurred instream flow restoration led to proposals for the creation of a large state-sponsored fund to acquire water rights for this purpose.

Instream leases are found to be very effective in situations where timeliness, low transaction costs, or temporary intermittent restoration is needed, or when water right holders choose not to exercise their water rights. In addition, leases serve to introduce water right holders, state agencies, and interested participants to the process of instream transfers. Between 1990 and 1997, over 2.5 million cubic meters of water was leased instream in the western states, constituting 84% of total water placed instream, including purchases and donations (Landry 1998). However, leasing constituted only 61% of total expenditures, confirming that leasing is a less costly method of placing water instream.

7.3.2.1.3 *Australia: Murray-Darling markets and water exchanges*

Permanent and temporary trading was introduced in the state of Victoria in the 1989 Water Act. Following finalization of regulations in 1991, the first transfers began in 1992. In South Australia, a moratorium on new water licenses in 1969 was followed by a period of readjustments to water rights to reflect actual or committed use such that all resources were allocated by 1976. Trading was subsequently introduced in 1983, first outside of large irrigation districts and then in 1989 to these districts. The Irrigation Act of 1994 permitted trade between districts and non-district areas. A cap of 2% of total entitlement limits movement of water out of districts (Bjornlund and McKay 2002).

The Australian experience reveals the predominance of water banking to facilitate temporary trades rather than permanent reallocation. Studies of permanent transfers in the Goulburn-Murray Irrigation District in the state of Victoria and River Murray show annual trades as a percentage of total allocated volumes ranging from 0.35% to 1.7% (Bjornlund and McKay 2002). The annual number of transfers in 1995–96 (the last two years of the study

BOX 7.13

Financing Flow Restoration in the Pacific Northwest of the United States (www.cbwtp.org)

In the Pacific Northwest, the public trust model as developed first by land trusts was extended to water rights through the Oregon Water Trust in 1993 and further extended in subsequent years to the Washington and Montana Water Trusts. These organizations have raised millions of dollars from members and interested foundations for the purposes of engaging in projects and transactions that return water rights to the public trust, that is, the water rights are dedicated instream either permanently or for a period as a lease. The Oregon Watershed Enhancement Board created by the state governor in 1996 as a means of heading off federal Endangered Species Act actions has funneled state funds (including revenues from vehicle license tags bearing a salmon insignia) to restoration efforts. Similar concern in the state of Washington over Endangered Species Act listings and other regulatory action prompted the state to begin providing its Department of Ecology with millions of dollars in public funds to undertake similar actions in 2000.

More recently, in 2003, the Bonneville Power Administration initiated a Columbia Basin Water Transactions Program to explore innovative strategies, including water rights transactions for environmental flows, as part of its obligations under the National Marine Fisheries Service Biological Opinion on the Columbia River System. In 2003 and 2004, the first two years of what was initially proscribed to be a five-year program, funds of up to $2.2 million and $4.2 million, respectively, were allocated. The Bio-Op calls for annual funding to reach at least $5 million, which would be a significant portion of the larger BPA Fish and Wildlife Program responsible for expenditure of $140 million annually. As administered by the National Fish and Wildlife Foundation, 11 local entities from Oregon, Washington, Montana, and Idaho have qualified to participate in the program. Although the funds are technically federal (BPA is effectively a federal parastatal agency), they are sourced from ratepayers, as BPA earns its revenues by producing and selling electricity in the Pacific Northwest states.

period) averaged 120–150 in each of the areas on annual traded volumes of 4,000 to 7,500 cubic meters. Subsequent increases in trading activity in the period 1997–98 in the GMID show 350 transfers and 25,000 cubic meters of permanent water trades, while there where 4,500 trades of 250,000 cubic meters of temporary water (Bjornlund and McKay 2002). Analysis confirms that the market for temporary water increases when allocations are tight (100%, for example)—up to 16% of the allocation was traded in 2001. In years with surplus water and large allocations (200%, for example), the percentage of water transferred temporarily is closer to 3–5% of allocations (Bjornlund 2002).

Price dispersion is noted in both markets, though it has decreased in the Murray over a ten-year period, suggesting a degree of market maturity (Bjornlund and McKay 2002). Prices for permanent transfers are consistently in the range of US$ 0.70 per cubic meter (AUS$ 0.99 per cubic meter) in that market. Temporary trades in 2000 and 2001 ranged from US$ 0.03 per cubic meter (AUS$ 0.04 per cubic meter) to US$ 0.14 per cubic meter (AUS$ 0.20 per cubic meter), depending on the balance of supply and demand (Bjornlund 2002).

Water exchanges now facilitate 10–40% of temporary trading levels in the three Murray-Darling Basin states (Bjornlund and McKay 2002). Further analysis of the GMID suggests that exchanges tend to be used for small and immediate transfers as well as transfers between distant parties, whereas private parties engaging in large, perhaps permanent and more local, transfers tend to prefer the business be done privately (Bjornlund and McKay 2002).

The temporary market can range from annual transfers to weekly transfers, allowing water users to access water when they need it and in response to timely information such as weather forecasts. In the temporary market, the market driver is water availability during the season, and the unit transferred is measured volumetrically; its delivery virtually assured (Bjornlund and McKay 2002). However, with permanent transfers, it is the water right, rather than the actual water, that is being purchased. Consequently, delivery is more uncertain and varies with the annual allocation determination. There are other limiting factors to the exchange as well, including alternatives and information requirements. The maximum price at which trades occur will be limited by commodity prices and alternative inputs; at some point, it is cheaper for farmers to buy grain than to buy water (Bjornlund 2002). Furthermore, the larger and more efficient an exchange is expected to be, the more that effective participation requires additional information inputs, which are not costless. (See Box 7.14 for a discussion of various pricing methods for water exchanges.)

There are specific conditions faced in the Australian case that may contribute to the inability of the exchanges to facilitate permanent transfers. An overall trend toward reducing industrial capitalization, particular tax benefits to leasing as opposed to purchasing or "investing" in water, the uncertainty over annual allocations and thus fluctuations in the value of water rights traded, and speculative interests create an environment in which the risk and volatility of permanent trades is too high to justify the investment (Bjornlund and McKay 2002).

7.3.2.2 Third-Party Impacts of Water Transfers

Third-party impacts include the full range of effects that a trade between a buyer and a seller have on other ("third") parties and include impacts on other diverters of water on the canal or river, as well as environmental, social, and regional economic impacts (Gould 1988; Howe et al. 2003; Thompson 1997).

In the United States, impacts on other water users typically are accounted for through transfer regulations (Howe 1997). While the assessment of physical impacts is difficult, it is the degree to which more diffuse impacts are dealt with that can affect the workings of water markets (Gould 1988; Thompson 1997). Efforts to facilitate the regulatory process have led to only marginal increases in water transfers. More significant obstacles include inconsistent legal rulings, opposition by government entities that control water and conveyance systems, and concerns of communities that export water (Thompson 1997). The conventional view of water as strictly a public good that ought to be controlled only by the public (like the government) is at odds with a market approach and has hamstrung the development of permanent transfers (Thompson 1997).

Viewed from another perspective, the result of a study by the U.S. National Academy of Sciences concluded that third-party impacts were inadequately addressed by state rules (particularly for interest historically underrepresented in western water allocation) and remain the primary impediment to water transfers in the West (NRC 1992).

7.3.2.2.1 Environmental impacts

One perverse outcome of the implementation of water markets in Australia has been the activation of volumes of water that were

BOX 7.14

Pricing Methods for Water Exchanges and Banks

A key enabling factor in the success of water exchanges or banks is the design of the pricing mechanism. The pricing methods that have been used most predominantly are the sealed double bid and the bulletin board approaches (Bjornlund 2002). In the sealed double bid approach, buyers and sellers submit price per unit and quantity bids. The auction operator arrays buyers' bids in descending price order and sellers' bids in ascending price order. One clearing price is determined for the market, and this price is paid to all bidders such that no buyer pays more than she is willing and no seller accepts less than she is willing (Bjornlund 2002).

While the sealed double bid approach is expected to offer greater efficiency benefits than the bulletin board approach, in practice, it has not been as successful as expected. In its first year, the Southern Riverina (Australia) Irrigation District Council's Water Exchange tried the sealed double bid approach only to have participants request a bulletin board approach instead the next year. Since the exchanges have been preferred for temporary trades, the bulletin board approach offered more accurate market information and fluid, timely transactions (Bjornlund 2002). Since 1950, the Northern Colorado Water Conservancy District has maintained a bulletin board system in which short-term rentals are mostly traded (MacDonnell 1994). However, a downfall of the bulletin board approach is path-dependence—it matters who meets whom first, and therefore the market is not as efficient as it could be.

Other price discovery mechanisms that may be used in exchanges or banks include the following (Howe 1997):

- standing offers to buy and sell at fixed prices,
- sealed bid double auctions (and repeated sealed bid double auctions),
- live or sealed bid auctions for unique water rights,
- contingent water markets, including agricultural lease-outs. Participants enter contracts where the execution is contingent upon conditions at a certain time in the future. Two scenarios are negotiated and execution of the agreement always occurs because one of the scenarios (it rains, it does not rain) occurs. The bank would facilitate bringing together buyers and sellers, and the attraction to this would be that the transfers would not be permanent, and
- futures water markets.

previously unused (Isaac 2001). Seeking to dispel the "markets are a panacea" myth, Isaac points out that the doubling of water prices within four years denotes an improved efficiency of the market, but only in the purely economic sense that more water is now going to the highest and best use, including water that was not previously used (Isaac 2001). However, this would appear to be less a failure of the market than a failure of regulation. In setting the cap, regulators apparently did not apply a look-back period for beneficial use, as is done in the United States, to insure that the water being traded had actually been used for a beneficial purpose.

Implementation of an approach within a system where rights are privately held may facilitate the transfer of water to instream purposes according to the relative economic merits of water instream and out-of-stream. However, a "free" market in water is unlikely to be sufficient to reach environmental flow objectives, given the social and economic incentives that favor out-of-stream uses (Aylward 2003). In fact, the free-market Chilean experience bolsters this conclusion. In 1985, the Chilean government passed a separate law to subsidize the development of small- to medium-scale private irrigation projects, because the Water Code's market incentives had failed to stimulate private investment or water conservation initiatives (Bauer 2004). Instead, it is important to provide a regulatory framework that guides the reallocation of water between in- and out-of-stream uses in the direction desired by society (Bjornlund and McKay 2002).

The effectiveness of water markets in meeting water quality concerns is also tenuous. One study found that in the GMID, water was being sold from higher salinity soils to lower ones, though this did not hold true in South Australia along the River Murray. There, they found that due to changes in flow and dilution effects, salinity levels in the River Murray actually rose (Bjornlund and McKay 2002).

Much of the recent activity, at least in the United States, for acquisition of water rights instream has been for the protection of endangered fish species (Landry 1998, p. 6). Therefore, it is important to determine whether there is evidence that water quantity and water quality have improved as a result of market forces. In the 1991 California Water Bank experience, transfers were found to have both positive and negative environmental impacts, though they are not quantified in a cost-benefit analysis (MacDonnell 1994). Possible negative effects included damage to bird and wildlife forage and habitat as a result of removing grain crops, reductions in groundwater recharge from lack of seepage, reduction in groundwater quality, increased subsidence, and negative impacts on fishery conditions in the Delta caused by increased pumping. Potential positive impacts included improved surface water temperature, quantity, and quality, and reduced fish entrapment, though no conclusive evaluation was conducted on these effects (MacDonnell 1994). Even more surprising is the fact that measurement of the amount of water added to the system from market activities was imprecise.

7.3.2.2.2 Socioeconomic impacts

The underlying premise of using markets to allocate resources is that they result in efficient resource use. A study of the 1991 California Water Bank found that the Bank created net benefits of $91 million, including $32 million in net benefits to the agricultural sector (Howitt et al. 1992, in MacDonnell 1994). Studies in two watersheds in Chile also suggest significant net gains from trade as water moves from low to high value agricultural uses (Hearne and Easter 1995). The distribution of these economic impacts can be of concern, however, when they are concentrated in specific localities. Analysis of the California Water Bank suggests that total adverse impacts can be reduced by spreading acquisitions over a large geographic area (Howitt et al. 1992, in Howe 1997).

For permanent transfers, another concern is the extent to which a capital asset (water) is sold to finance recurring costs. In an Australian survey regarding the motivation of buyers and sellers to enter into transactions, it was found that in the Goulburn-Murray Irrigation District, only 26% of sellers used the revenue from sale of water to improve irrigation practices, while the rest used the funds for general revenue or for debt reduction (Bjornlund and McKay 2002). Similar results from the other districts echo the fact that the money was used mostly for income generation rather than reinvestment in the land.

In Chile's Limarí River Basin, the result of markets is that many small farmers have sold their water rights to larger farmers or to agribusiness corporations. However, little research has been carried out to assess whether this is an equitable result (Bauer 2004). In general, however, the Chilean experience suggests that the imposition of an unfettered free market in water into a developing context with significant existing socioeconomic inequality will lead to further inequities. (See Box 7.15.)

Thompson (1997) provides a response to these impacts by questioning why the transitional impacts illustrated above seem to be regarded as "special" in the water context. In general, in the U.S. context, resources are not constrained from moving to meet changing market demand. Federal or state assistance during such a transitional period is typically available on an economy-wide basis, and if such assistance programs are inadequate, it can be argued that they should be increased across the board, regardless of what causes the economic dislocation (Thompson 1997).

Other third-party impacts that have been considered with respect to water markets include the land use and associated impacts from transfers that lay land fallow. The California legislature has rejected legislation three times to require compensation for communities negatively affected by water transfers. Rather, the trend has been to voluntarily incorporate funds for local communities as part of transfer and land fallowing programs (Hanak 2003).

Effectiveness of market-based payment arrangements for delivery of water and watershed services will largely depend on stakeholder willingness to pay for them. This, in turn, depends on the level of confidence in the effectiveness of management actions to provide the ecosystem services and institutional arrangements needed to ensure access to benefits by those who pay the cost. These governance issues are covered in greater depth earlier in this chapter and important enabling conditions for markets of any kind.

7.3.2.3 Findings and Conclusions

A survey of global experiences demonstrates a range of water market tools used to reallocate water to new and higher value uses. Historically, the explicit purpose of water markets has been to improve resource efficiency, either within agriculture or among agricultural, municipal, and hydroelectric power needs. However, markets are increasingly coming to incorporate ecosystem needs as well. Since the mid-1990s, efforts have been made in the United States to explore the potential of water markets to meet ecosystem needs. However, markets have developed slowly and experienced limited activity, particularly of permanent transactions, due to public concern over the importance of local control of water resources (Thompson 1997).

The Australian experience shows the potential of markets to reallocate water to higher value uses, but cautions that instream needs are still mostly a public good, and therefore targets must be explicit and properly planned for in order to be achieved. Similarly, the Chilean experience demonstrates that a purely laissez-faire approach to creating markets for water may fail to protect the public good characteristics and have negative social and environmental consequences (Bauer 2004). Furthermore, the Australian experience demonstrates the need to prevent an increase in gross water use through the activation of previously unused water as markets develop.

In order for ecosystem needs to be achieved through markets, what is needed is either an explicit purchasing program for instream purposes or a system to reduce water allocations (a cap-and-trade mechanism whereby allocations can be ratcheted downward, for example). Experiences in the United States, including Oregon's Klamath Basin and Colorado's Arkansas River Valley, with allocation reductions imposed by regulatory action, demonstrate the conflict that can emerge when regulatory reductions are imposed, even if market mechanisms are used as a method for the redistribution of allocations (Howe et al. 1990). A more promising approach may be to provide direct governmental funding for buying back water rights to be retired instream, as implemented through the Bonneville Power Administration in the U.S. Pacific Northwest and considered in the Australian case.

Because the benefits to improved stream flow and freshwater ecosystems are still inherently public goods, the role of good governance and complete property rights for water remain fundamental enabling conditions for well-functioning markets. While there is a role for the use of markets to develop efficient water allocations, there is also a role for governments to regulate in providing stable and appropriate institutions for these markets to operate (Johansson et al. 2002).

BOX 7.15

Third-party Impacts of Water Markets on the Chilean Poor
(Bauer 2004)

There are a number of reasons why the Water Code and water markets in Chile have harmed poorer farmers. First, the military government did not provide the public with information, advice, or help in adjusting to the new law. Peasants and small farmers often learned about the new rules and procedures for acquiring or regularizing water rights too late to take advantage of them or to adequately protect themselves. Even in cases in which poor farmers got to know about the procedures, they were unable to use them without legal, financial, and organizational assistance.

Second, poor farmers are generally unable to participate in the water market except as sellers (if they are fortunate enough to have legal title to water rights, which is uncommon). They lack the money or credit needed to buy water rights. Their main hope for access to additional water is to benefit from the increased return flows that could result from improved irrigation efficiency on the part of more prosperous irrigators upstream. However, downstream users have no legal claim to such unused surplus flows, which are therefore an unreliable and insecure source of water.

Third, peasants and small farmers lack the economic resources and social and political influence needed to defend their interests effectively in the current laissez-faire regulatory context, in which private bargaining power is crucial. This is a disadvantage in two areas: conflicts over water use and conflicts over regularizing water rights titles.

7.3.3 Payments for Watershed Services

Economic incentives used to insure the delivery of watershed services essentially consist of payments to landowners to alter land management practices in the expectation of downstream benefits. A review completed in 2002 identified 63 examples from around the world of the application of market-based approaches to the provision of watershed services (Landell-Mills and Porras 2002). Key services paid for have included ensuring regular flows of water, protection of water quality, and control of sedimentation.

The types of arrangement vary depending on the characteristics of the service, the scale of relevant ecosystem processes, and the socioeconomic and institutional context. These range from informal, community based initiatives to more formal contracts between individual parties, and to complex arrangements among multiple parties facilitated by intermediary organizations. They may also include a mix of market instruments, and regulatory and policy incentives that are more likely to become necessary at larger scales, when threats are beyond the response capacity of individual communities, and a common set of rules may be required for purposes of trade among a larger and more diverse group of actors (Rose 2002). The instruments typically used to pay landowners for improving land management and protecting watershed services include transfer payments between governments and landowners, tradable development rights, voluntary contractual arrangements, and product certification and labeling. (See Box 7.16 for examples of each of these payment mechanism.)

The use of cap-and-trade mechanisms in managing upstream-downstream pollutant and water quality issues, as well as in wetland mitigation is also increasing. The assessment below focuses on transfer payment schemes. These schemes are perhaps the most straightforward approach for a government to take in providing economic incentives to landowners consistent with achieving watershed management objectives. They also offer the advantage of being susceptible to implementation in a comprehensive fashion covering large areas (or countrywide) and, therefore, have seen the widest application to the provision of watershed services and land management more generally.

7.3.3.1 Effectiveness

A key issue with respect to these upstream efforts at providing freshwater and other ecosystem services is whether the payment promotes an activity that actually produces the intended end result downstream. The ability to demonstrate this is key to building stakeholder confidence and willingness to pay for services. However, given the complexity of watershed processes, it is also a key challenge.

In the tropics, the general presumption that maintaining forests or reforestation is a superior land use, not just for watershed services, but for biodiversity, carbon sequestration, and other ecosystem services, often has been sufficient to launch payments for services programs. Thus a main concern has been to find buyers able and willing to pay, rather than what is needed, to insure they are provided, in a specific context (Pagiola 2002; Landell-Mills and Porras 2002). Unfortunately, when it comes to the tropics, the linkage between forest cover and downstream hydrological function is not a simple matter (Bruijzneel 2004, Bonell and Bruijnzeel 2004). Adding in the complexity of the linkage between hydrological function and the ecosystem services that contribute to the economic well-being of society makes it even more challenging to design effective interventions (Aylward 2004).

Recent reviews of the scientific knowledge base that has supported many watershed payment initiatives in Costa Rica suggests that management is limited by the lack of reliable and precise

BOX 7.16

Payment Mechanisms for Watershed Services

Transfer Payments: Transfer payments tend to be used at the national level, or over large areas and heterogeneous conditions. The best-known examples are the US Conservation Reserve Program and similar initiatives in some European Union countries. Under these programs, farmers are compensated for conservation measures based on a number of criteria, which include water quality and soil conservation. It is also the approach used in the Costa Rica FONAFIFO program, which provides payments to forest owners for multiple environmental services. Because of the broad social benefits associated with multiple services, these are usually supported by taxes rather than user fees, but may still pool funds from a variety of sources. In another example, in New South Wales, Australia, associations of farmers purchase salinity credits from the State Forests Agency, which in turn contracts with upstream landholders to plant trees, which reduce water tables and store carbon (Perrot-Maître and Davis 2001; State Forests of New South Wales 2001; Sundstrom 2001).

Tradable Development Rights: Tradable development rights require a strong planning and regulatory capacity to identify zones where development is to be restricted or permitted, and to enforce those restrictions for the period of the agreement. A well-known example in the provision of freshwater services in the United States is in the New Jersey Pinelands. The Pinelands Development Credit Program created credits for landowners whose land uses were restricted as a result of zoning. The credits may then be sold to developers in areas designated for development. The purpose of the program was to protect economic interests in the region as well as a very large drinking water aquifer (Collins and Russell 1988).

Tradable development rights are also used in wetland mitigation banks and conservation easements. Conservation easements involve the acquisition of water and land rights and can be implemented quickly and for any time period. It is one of many instruments used to implement the New York City Watershed Agreement, in which the city invests in upstream watershed protection measures to protect its water source, rather than build a new filtration plant (Perrot-Maître and Davis 2001).

Voluntary Contractual Arrangements: Voluntary contractual arrangements are often straightforward and may stand alone when negotiated among individual parties, such as the case of an agreement between the La Esperanza Hydropower Company and the Monteverde Conservation League in Costa Rica, which is the sole owner of the forested area upstream from the plant (Rojas and Aylward 2002). Agreements among numerous parties require that more consideration be given to the establishment of decision-making entities for purposes of allocating funds to priority conservation measures. An example is a trust fund, such as the Fondo del Agua (FONAG), established in Quito, Ecuador, to protect two upstream ecological reserves. This fund is overseen by a stakeholder board. It allocates pooled funds and "in kind" support received from municipal entities, NGOs, and private sources (Echavarria 2002).

Payments for watershed services are often referred to as market-based approaches, and are sometimes confused with privatization, but in fact usually consist of a hybrid package of instruments, both public and private, which accommodates the variability and uncertainty of services provisioning. For example, transfer payments may be made through voluntary contractual arrangements between government and landowners who provide services, using funds derived from user fees voluntarily paid by downstream water users. This set-up was used in Valle del Cauca in Colombia, where such an arrangement was necessary because the Colombian government had developed plans for watershed management, but did not have the ability to fund them. Funds were provided by associations of irrigation farmers who voluntarily paid additional fees to support the provision of a reliable water supply during the dry season (Echavarría 2002).

Product Certification and Labeling: Another mechanism is product certification and labeling. In this case, landowners are rewarded for specified management practices only indirectly—through the potential to use certification to increase their share of the market and possibly to obtain a price premium. Certification and labeling requires intermediary organizations to establish standards for labeling and to certify practices. An example is the Salmon Safe initiative, which certifies and promotes wines and other agricultural products from Oregon farms and vineyards that have adhered to management practices designed to protect water quality and salmon habitat. This tool has seen only limited application to specific watershed services, but has general applicability to improvements in land management, one typical output of which is improved watershed services.

information on forest water linkages (Pagiola 2002; Rojas and Aylward 2003). Instead, most are based on conventional wisdom, secondary sources of information, and selective references to literature reviews on forest hydrology. Regardless of the source material, they tend to invariably support statements that protection of forests will increase water yields (Rojas and Aylward 2003). In some cases, such as in the Arenal Basin (Castro and Barrantes 1998) and in Heredia (Castro and Salazar 2000), the values of watershed protection are calculated based on the opportunity cost of returning cleared land to forest cover, with no attempt to model and assess links between land use and hydrology, and to estimate the marginal values of water in specific consumption and production activities. Case studies from other regions also report the general absence of the scientific data needed to support valuation (Munawir et al. 2003; Geoghegan 2003; Rosales 2003; Echavarria 2004; The et al. 2004; Johnson and Baltodano 2004).

Given the diversity of the contexts in which payment arrangements are being developed and applied, and the impossibility of obtaining complete information that links causes with effects, and management actions with outcomes, a key consideration is whether a program is able to maintain the flexibility needed to make adjustments as barriers to implementation are encountered, and as lessons are learned. An example of an initiative that has been allowed to evolve over time, and continues to make adjustments, is the development of the Costa Rican FONAFIFO program, which is designed to support the multiple services provided by forests, of which fresh water is one. This program, which has been in place for five years, was built on an earlier ten-year program of payments for reforestation and the lessons already learned regarding barriers to implementation, as well as institutional arrangements already in place. Originally motivated by reduced timber supplies, the program had already made several adjustments over time, both to reflect the broader objectives of protecting natural forests and to allow greater farmer participation (Pagiola 2002). Additional adjustments were made in 2002 to include agroforestry and indigenous reserves in the program (Rosa et al. 2003).

A recent assessment of the social impacts of the program in the Virilla watershed found that this continuous institutional innovation has had significant benefits in terms of strengthened capacity for integrated management of farm and forest resources, and has contributed to the protection of 16,500 hectares of primary forest, sustainable management of 2,000 hectares, and reforestation of 1,300,000 hectares, which have had spin-off benefits for the protection of biodiversity and prevention of soil erosion. However, there are also high opportunity costs, particularly for smaller landowners, who tend to rely more heavily on small areas of cleared forest, and to combine forestry with other activities such as shelter for cattle and shade coffee. Those with larger tracts received a greater advantage from the program because they were able to maintain a higher proportion of the land in forest (Miranda et al. 2003).

Most initiatives have focused on links between upper watershed land use practices and downstream urban water supplies, sedimentation of hydropower dams, and irrigation canals. Although it has generally been found difficult to provide economic justification for interventions at this scale, little attention has been given to the more local level impacts within micro-watersheds, where land and water relationships can be better understood and stakeholders can be more directly engaged. Although the values placed on improvement of water quality are modest, it has been suggested that land use interventions for this purpose may be justifiable as part of an integrated community resource management strategy (Johnson and Baltodano 2004).

Because of the difficulties discussed, current trends are toward just such small-scale pilot initiatives that may have the potential to be scaled up to address problems at larger scales as capacity is developed. The PASOLAC program, which is engaged in 10 pilot initiatives, works to improve land and water management by small producers in the hillsides of Nicaragua, El Salvador, and Honduras, and is helping to develop markets for watershed services through the local municipalities. As of 2003, there were agreements in place between upland producers and downstream organizations in San Pedro del Norte in Nicaragua; Tacuba, El Salvador; and Campamento and Jesús de Otoro in Honduras (Pérez 2003).

This kind of bottom-up approach is generally expected to help insure that regional scale organizations are more representative of and accountable to local livelihood interests. Many current initiatives are also developing action research and learning approaches that can support capacity building as well as the exchange of knowledge (IIED 2004; RUPES 2004).

7.3.3.2 Findings and Conclusions

Given the heterogeneity and constant change in ecosystems and in human institutions, the site-specific nature of watershed processes, which are dominated by randomly timed and extreme events, and the difficulty of linking multiple causes and effects, or predicting outcomes, an adaptive approach to management is required, which implies the need for on-going assessment to support decision-making. However, preliminary assessments based on generalizations can provide some thumb rules and working hypotheses from which to begin. Perhaps the most significant challenge in such initiatives is to develop the capacity for a place-based approach to assessment, which is necessary to identify ecosystem functions that support provision of valued ecosystem services in their landscape context, and to select payment and institutional arrangements that are feasible and appropriate to that context. To be effective, market-based initiatives also need to be viewed as part of a long-term process of building appropriate institutions, and in the context of broader issues of structural reform and social change.

Effectiveness of market-based payment arrangements for delivery of watershed services will ultimately depend on stakeholder willingness to pay for them. This in turn depends on their confidence in the effectiveness of management actions for providing valued services, and that of institutional arrangements needed to insure access to benefits by those who pay the cost, both of which require greater attention.

The use of market arrangements does not automatically protect ecosystems, insure provision of their services, achieve an equitable distribution of costs and benefits, or reduce poverty. These objectives need to be made explicit and addressed in the design of economic interventions, which are as much an issue of governance as of economics. Intermediary organizations, governmental and otherwise, often play important roles in leveling the playing field and reducing barriers to market access by the poor. They can also help to reduce the transaction costs inherent in establishing arrangements among numerous stakeholders spread out over large and remote areas, and who may lack clear title to land, by facilitating agreements among them, negotiating on their behalf, and providing various kinds of legal and technical assistance, as well as through assessment and dissemination of appropriate information.

Given that land and water relationships are more detectable at the scale of micro-watersheds, it has been difficult to provide economic justification for interventions at the larger scales needed to insure the delivery of freshwater services to hydropower facilities and large municipalities. Emphasis is shifting toward much smaller-scale initiatives where causes and effects can be better understood and stakeholders may become more directly engaged. This provides a better point of departure for development of the capacity to respond to larger-scale problems in a way that is more representative of and accountable to livelihood interests.

7.3.4 Partnerships and Financing

In order to meet growing and as yet unmet demand for freshwater provisioning services in a sustainable fashion, a number of tasks related to physical infrastructure and the supplying ecosystems need to be performed. These tasks include building, operating, and maintaining infrastructure; information gathering; weather forecasting; managing ecosystems; controlling pollution; and preventing erosion. Unfortunately, most of these tasks, most of the time, are underfunded (especially in the developing world).

The use of economic incentives in managing terrestrial and aquatic ecosystems that provide fresh water are covered above. This section examines critical issues in the financing of water infrastructure, largely for domestic and industrial use. Investing in water resource development may produce enormous social returns (if it is well designed), but the financial returns are slow and low compared to the financial returns of investment in such other sectors as energy and telecommunications.

This section provides answers to the following questions: How has water infrastructure been financed? Why is water infrastructure financing prone to so many problems? What have been the trends in public-private partnerships? What are the binding constraints and enabling conditions that affect long-term financing of infrastructure needed to manage and use freshwater resources in a sustainable way?

7.3.4.1 Water Infrastructure Financing

Financing infrastructure means using funds to acquire long-term physical assets. The costs of developing and using water resources are eventually paid by either water users or taxpayers, or aid donors. However, financing can come from several sources. Table 7.5 shows the main sources for financing, the instruments or means they use, and their performance in the recent past. Summary information on financing sources for water infrastructure in 2002 suggests that 69% of financing is from the domestic public sector, 17% from external aid, and the remaining 14% from the private sector.

There is not much ground for optimism that these sources will increase their funding of water infrastructure in the future, considering that reliable long-term financing requires sources to be reimbursed. This reimbursement can only come from one of three groups: donors, taxpayers, or water users. The good news has come from the aid sector. Aid money is likely to increase in view of all the recent international agreements in development financing motivated by the setting of global social, economic, and environmental goals (the Millennium Development Goals, for example). At the 2002 United Nations Conference on Financing for Development in Monterrey, Mexico, donor governments and international agencies committed themselves to increasing their aid by 25%. However, the other two reimbursement sources (taxpayers and water users) will have more difficulties in rising to the challenge. Only a small proportion of local water systems recover their full operating and maintenance costs, let alone investment costs. Money from taxpayers and water users is not likely to increase unless the institutional problems that beset the fiscal and water sectors of most developing countries are solved.

Three features of the water sector that make cost recovery from water users very difficult. First, fixed costs constitute a very large proportion (around 90%) of total costs. This means that water agencies can operate (once the infrastructure is there) even with very low budgets. Second, water distribution is locally a natural monopoly (an industry in which technical factors—like the requirement of a network of pipes—preclude the efficient existence of more than one producer). Third, decisions on water tariffs are politically sensitive in most of the developing world. When taken together, these conditions imply that local water systems can operate (in the short run) with very low budgets, they do not face competition, and are politically constrained to implement changes that would allow them to recover costs.

The problem gets more complicated when funding is provided in foreign currency and the revenues are in local currency. Most of the loans that have to be repaid in foreign currency have suffered from this problem. Existing cost recovery contracts and financial instruments do not provide water systems with enough coverage to deal with this risk, especially in the case of massive devaluations, such as the peso devaluation in Argentina. These problems may beset water agencies whether they are operated by private companies or by public agencies, or by a mix of public and private organizations.

Furthermore, many national governments have devolved the responsibility for providing water services to regional or local agencies. However, these agencies usually have a very limited ability to raise finance on their own. Most of them need the support of the national government, as, for example, in guaranteeing loans. In some countries, sub-national levels of government are not allowed to raise money themselves. Even when they are allowed to do so, they tend to be short of expertise in financial management.

There are important constraints in fiscal resources, too. The fiscal weakness of many developing countries is the most obvious constraint, but not the only one. The fiscal relationship between the national and the sub-national levels of government is unclear and unpredictable in several countries. Under these conditions, long-term commitments are not likely to arise or be successful.

Problems have affected projects from the private and public sectors. Public scandals, accusations of corruption, tariff increases, allegations of failure to deliver the promised capital investments, and claims of failure to increase services to poor communities have plagued private sector participation in the water sector. However, the public water sector is not immune to these problems either. Allegations of failure to reform, improve efficiency and financial sustainability, limit political patronage, or expand access to as well as quality of services are common in many parts of the world. Unfortunately, the debate has been polarized between those who see private sector participation as a panacea and those who want to ban it completely because they think that the private sector cannot play a positive role in the water sector. These rapidly diverging views have increased the political risk and uncertainty associated with private sector participation and have led to an impasse among stakeholders on how to improve access to and the quality of water and sanitation services. The remainder of this section assesses the need, scope, constraints, and enabling conditions in order to better understand the potential role of private sector participation in the water sector, based on work undertaken by OECD (2003b).

7.3.4.2 The Need for Public-Private Partnerships

Governments around the world face difficult economic and political choices posed by the urban water sector. Securing safe, reli-

Table 7.5. Financing Sources and Means

Sources	Means	Observations
Water users	through tariffs or by developing their own infrastructure	lack of access to large volumes of finance constrain this source
Public water authorities and utilities	from user charges, loans, and subsidies	resistance to cost-recovering tariffs constrain this source
Private water companies	from user charges, loans, subsidies, and equity	resistance to cost-recovering tariffs constrain this source; the pool of companies has shrunk
National and sub-national governments	through subsidies, loan guarantees, and proceeds of bond issues	by far the largest source, but lack of coherent water strategies and weak fiscal conditions of national and sub-national governments in developing countries constrain it
Financial institutions (domestic, international, and multilateral)	offering loans	low private yields and long return periods discourage commercial lending; additionally, bank lending has declined because banks are more averse to lending to emerging markets; in some sectors (such as irrigation and hydropower), hostility to large storage projects has constrained this source
Bilateral and multinational aid agencies	through grants or soft loans	international aid for water and sanitation fell in the last few years (from $ 3.5 billion in 1996–98 to $ 3.1 billion in 1999–2001)
Local communities, self-help and nongovernmental organizations	through community participation, micro-credits, and application of low-cost technologies	knowledge gap (project design and financial aspects)

able, reasonably priced water and sanitation services for all is one of the leading challenges facing sustainable development. Many governments—local or national—have failed to recognize that, once it is piped, water for domestic and industrial use is a finite natural resource and an economic good. Instead, they have subsidized its use through a long history of underpricing and opposition to full cost recovery. Recently, the South African government has chosen to see domestic water as a basic right and to treat it as a public good, and is providing a base amount of domestic water free to the population at large.

Unfortunately, the long-term consequence of not setting a market price reflecting the cost of water provision and its true value to society has been a failure to recover costs, which, when combined with insufficient general funds, has led to water systems that are often operated inefficiently and, where services are unreliable, lack coverage, regular maintenance, and good design. With 88% of the 1.1 billion increment in global population through to 2015 likely to live in urban areas, there is a serious need to not only repair ailing systems, but build and operate new ones (WHO and UNICEF 2000).

Many towns and cities in developing countries have unreliable piped water systems and experience regular supply interruptions. Furthermore, the quality of services provided by existing systems is deteriorating, mainly because of the high capital costs of infrastructure, low user charges, and diminishing government resources for addressing urban water issues. The lack of investment in water supply and wastewater treatment threatens the quality of the services provided to citizens mostly in developing countries, provokes the decline in environmental and health standards, and contributes to poor demand management. Revenues and income for water companies are generally insufficient and unpredictable (OECD and World Bank 2003).

If the Millennium Development Goals on water and sanitation are to be met, current spending on water services of $75 billion a year needs to be increased to $180 billion (World Panel on Financing Water Infrastructure 2003). However, this target will be difficult to meet with public funds alone, as both government budgets and overseas development assistance have shown decreasing trends recently. Some governments are therefore increasingly looking to a range of private sector partners to provide access to two key resources: (1) improved management systems and technical options, and (2) private investment funds.

In OECD countries, investment needs also will increase substantially over the next few years, requiring greater efficiency through better management and the use of new sources for investments. For instance, in the European Union, about $5 billion per year are currently spent on water and wastewater services, and capital investment is predicted to increase by 7% a year for the foreseeable future (Owen 2002).

7.3.4.3 Public–Private Partnerships

Since the mid-1990s, an important approach that has been gradually introduced in the water sector is the notion of partnerships between public and private agents. Public-private partnerships correspond to any form of agreement (partnership) between public and private parties. These should not be confused with privatization, where the management and ownership of water infrastructure are transferred to the private sector. There is a wide range of approaches for involving the private sector as a partner in improving the performance of water and sanitation systems. (See Table 7.6.) Some options keep the operation (and ownership) in public hands, but involve the private sector actors in the management, operation, and/or financing of assets. A common point of all these options is that the government always retains responsibility for setting and enforcing performance standards—regardless of the form of private involvement chosen. The fact that the water sector is one of the natural local monopolies means that a strong regulatory role is required to insure that performance standards are met and the interests of consumers protected.

Table 7.6. Allocation of Public/Private Responsibilities across Different Forms of Private Involvement in Water Services. Cell shading in the table is to be interpreted as follows: dark grey = public responsibility; light grey = shared public/private responsibility; white = private responsibility. (Analysis of the authors based on OECD 2000)

Form of Involvement	Asset Ownership	Capital Investment	Design and Build	Operation	User Fee Collection	Oversight of Performance and Fees	Typical Duration
Design and construct contracts							—
Service contracts							1 to 3 years
Management contracts							3 to 5 years
Lease contracts							8 to 15 years
Joint ventures							
Build, operate, transfer							15 to 25 years
Concession Contracts							20 to 30 years
Passive public investment							—
Fully private provision							—

There is no universal "right answer" on how to use private investment to help improve water services. Ultimately, governments need to devise arrangements that fit the local context, and some may decide that public-only is best. Where the private sector is hesitant to engage itself, it might be suitable to start with methods that involve low risk for the private operator (as in service contracts), moving only later toward more ambitious forms of involvement, if considered appropriate (OECD 2003a). The degree of risk-sharing will be an important determinant of private sector participation.

The most commonly cited advantages of private sector participation are that it brings technical and managerial expertise to the sector, improves operating efficiency, entails injections of capital and greater efficiency in its use, and increases responsiveness to consumer needs and preferences. It is often assumed that the private sector has significantly better access than governments to capital flows and to the technical know-how that is essential in the provision of critical water services.

In some OECD countries, public-private partnerships have existed in the water sector for more than a century, as in France; and in most other OECD countries, they have existed for more than a decade. The options range from limited private investment to full divestiture, predominantly in England and Wales. (See Table 7.7.) Water supply in France is in public ownership, but management is a mix of public and private systems. The French municipal authorities act as an economic regulator. In the United Kingdom, the ownership and management are private, but the economic regulator (Office of Water Services) is an independent body. The United States has a part public and part private ownership structure, but is dominated nevertheless by the public sector. There is a growing tendency in OECD countries for water systems to be managed by groupings of municipalities so as to organize supply at a larger scale. Other forms of consolidation have also been occurring. An example of this is the case of the Netherlands, which reduced the number of water boards from 210 in 1950 to 15 in 2002 (van Dijk and Schwartz 2002). These developments are likely to attract additional private sector interest for water due to the increased project size and potentially associated economies of scale.

Table 7.7. Share of Public–Private Partnerships in Key OECD Urban Markets (OECD 2003b, based on BIPE 2001)

Country	**Public Sector Management** *(percent of population served)*	**Private Sector Management** *(percent of population served)*
Germany	96	4
France	20	80
United Kingdom	12	88 (100 in England)
The Netherlands	100	—
United States	85	15

In the developing world, while the 1990s saw a significant increase in private sector participation in the water sector it is estimated that only 3% of the population in poor or emerging countries is provided with drinking water through private operators (Owen 2002). As shown in Figure 7.2, since 2000, the number of projects has decreased substantially. In effect, the important peaks in certain years correspond to the large concessions that took place in Buenos Aires in 1993, in Manila in 1997, and the

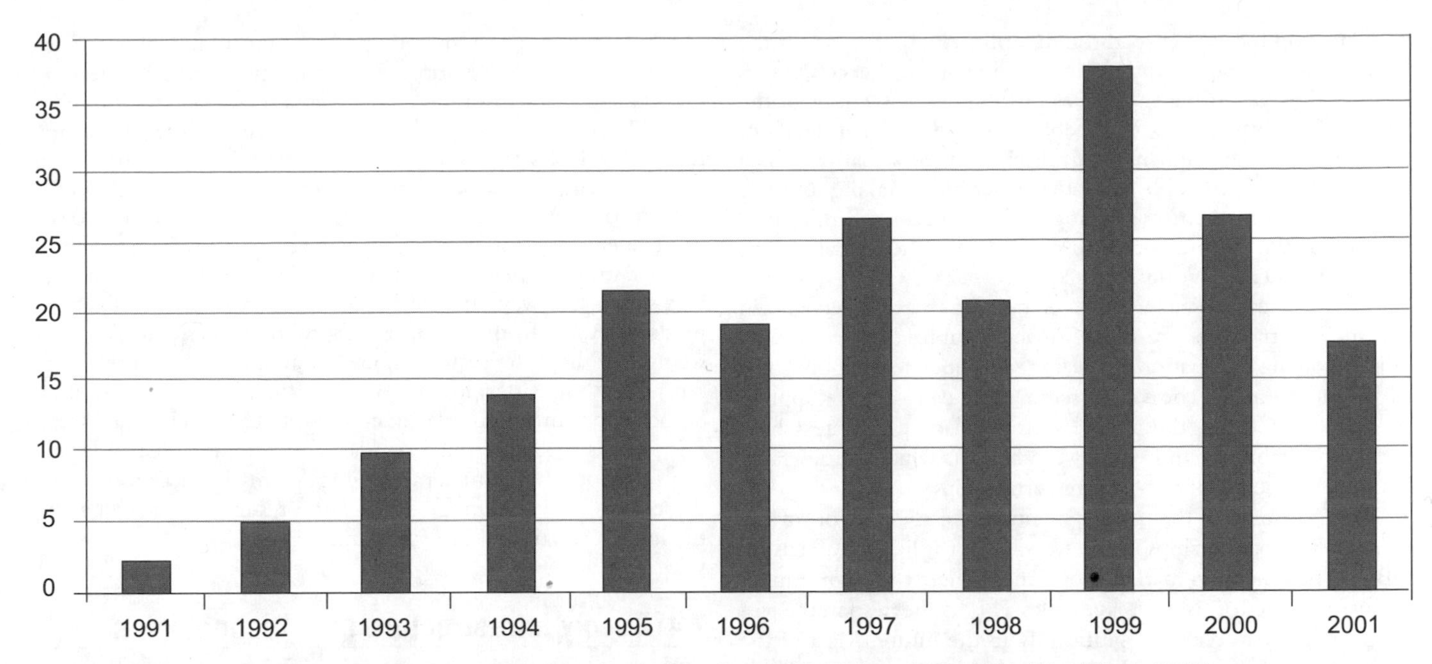

Figure 7.2. Number of Projects with Private Participation in Developing Countries, 1990–2001 (World Bank 2003)

privatization of the Chilean water utilities in 1999 (World Bank 2003b).

According to the World Bank, public-private partnerships are most common in Latin America, followed by East Asia and the Pacific, and Europe and Central Asia. While the nature of private sector participation may range from partial financing of investments to an increasing role in the operation of services, most countries have opted for the concession approach, in which the private sector participates in managing some services, but the public sector retains ownership of the system.

It is important to bear in mind that many examples of efficiently managed public water and sanitation utilities exist, and that the characteristics of the public sector differ among countries. Thus in many countries, it is not necessarily the public sector per se, but factors such as faulty incentive structures, politicization of appointments and management, and other bureaucratic weaknesses that contribute to poor performance. Despite widespread belief in the potential for efficient use of the private sector in some areas of service provision, empirical evidence of the relative merits of private and public management in the water sector is relatively limited (OECD 2003a).

7.3.4.4 *Effectiveness*

Recent experience with the involvement of the private sector mainly in non-OECD countries suggests that there are major obstacles that significantly hinder greater private sector participation in urban water services. Despite high hopes that private sector participation might help overcome the financing gap for achieving international goals for access to water and sanitation, an increasing number of water sector projects with private sector participation appear to be in crisis, often due to the difficult economic situation in the host country. The number of such projects has been decreasing and investment flows have been slowing over the last four years. This has triggered recognition by both public and private actors of a number of systemic problems in the design of projects, for which solutions need to be found. These include weak regulatory set-ups, lack of political support for private sector participation, need for long-term debt finance, low returns on investment, fragmented deal size, poor creditworthiness of local governments, poor contract and project design, and a frequently inappropriate allocation of risks between involved parties. Some of these key issues are highlighted below.

Regulatory frameworks in host countries are often insufficient and unstable. This generates significant uncertainty about future cash flows for the private operator, since essential cost elements (such as waste water treatment requirements) as well as revenues (such as tariffs) cannot be anticipated. This situation, together with the often weak levels of contract enforcement, is among the key reasons for the low use of public-private partnerships in many emerging market economies and developing countries. Technical assistance from donors can help to remove many of these obstacles by providing support for capacity building and institutional reform, but ultimately, political commitment is also needed (OECD and World Bank 2002).

Political commitment to public-private partnerships at all relevant government levels is essential, since water is perceived to be more than a simple good by both consumers and many politicians. This has sometimes been overlooked, leading to the rapid loss of political backing as soon as the projects encountered initial difficulties (OECD and World Bank 2002).

Networked water systems have extremely high capital costs, well in excess of other infrastructure services. They are mostly financed with debt, for as long a term as is commercially available. Given the high initial costs, extremely long pay-back periods are necessary, and it is essential that revenue streams be as secure as possible. Urban water services are also a business with relatively low rates of return on investment. Due to these sectoral specificities, private operators are particularly sensitive to the quality of the investment climate and the level of risk, which is an important obstacle to public-private partnerships in many regions of the world. Furthermore, in the last couple of years, the risk aversion in the infrastructure market in general has increased because of several events, including the September 11, 2001, attacks, the recent corporate bankruptcies, the reduced number of strategic investors, and the rating downgrades (OECD and World Bank 2002).

Finally, many public-private partnerships have encountered difficulties due to insufficient attention being paid to the social consequences of involving the private sector as they often implied tariff increases due to a move towards the full recovery of operation and maintenance costs through tariffs. Another reason is the popular mistrust of institutions involved in such projects. Unless continued access to water services of the poorest sections of the population is insured at a reasonable cost, and sufficient levels of transparency in decision-making insured, major social resistance must be expected to public-private partnerships. Making sure that social protection schemes are being developed prior to or in parallel with public-private partnerships is therefore a crucial success factor (OECD and World Bank 2002).

Even if these obstacles are overcome, it must be recognized that such partnerships are not a panacea. Public-private partnerships involving international private sector operators cannot solve all the problems in the water sector, nor can they be applied everywhere. Clearly, the private sector will only operate where certain profitability requirements can be met, which considerably limits the scope for public-private partnerships.

First, for some of the reasons mentioned earlier, major investment in such partnerships in the water sector is likely to focus on OECD and emerging market economies, where the environment for foreign investors is most favorable. This has been so in the past and is unlikely to change significantly in the future. Most of the applications have been in high- and middle-income countries, leaving least developed countries uncovered. For instance, less than 0.2% of all private sector investments in the water and sanitation sector of developing countries went to sub-Saharan Africa (United Nations Millennium Project 2003).

Second, there are only a limited number of international water operators, and their human and financial capacities allow for the management of only a limited number of projects. The three largest private operators account for more than 50% of the global market. Public-private partnerships in non-OECD countries, therefore, focus on urban areas that are likely to yield the most substantial revenue flows and offer the best opportunities to achieve significant economies of scale—typically large cities with populations of 500,000 or more.

While the potential arena of operation of international private operators is limited, opportunities for the involvement of new entrants may exist. This is particularly the case of domestic private sector companies in developing countries. The mobilization of these actors may help enlarge the scope of public-private partnerships.

7.3.4.5 Findings and Conclusions

There is a clear mismatch between the high social value of freshwater services for domestic use and the resources that are being allocated to manage water. Insufficient funding to expand water infrastructure is one manifestation of this mismatch. Both inherent characteristics of the water sector (high fixed cost, low returns, long pay-back periods) as well as institutional problems (political interference, inadequate legal frameworks, poor management structures) explain the gap in funding infrastructure. No single source will be able to bridge this gap on its own. There are several sources of funding water infrastructure and all have a role to play. In addition to a more creative use of existing financial instruments and the development of new ones, changes at different levels are needed in order to unleash financing sources. At a national level, legal frameworks have to provide more certainty to the parties of long-term commitments. The water sector has to establish its priorities in a clear way and produce programs that include the definition of financing needs and sources. Finally, at the agency level, cost recovery must be improved and managerial and technical capacities, enhanced.

In the future, public-private partnerships should take into account the following priorities. Governments should be clear on their strategies and priorities for the water sector, and plan accordingly. There must be an effort to optimize the use of existing financial vehicles and introduce new ones. Finally, a long-term sector strategy should be adopted in order to achieve more efficient urban water management (OECD and World Bank 2003).

The enabling conditions for adequate long-term financing of the infrastructure needed to manage and use freshwater resources in a sustainable way go well beyond the conditions of financial markets. Apart from having access to resources, the financial strength of the water sector requires clear and transparent priority setting as well as developing programs to meet these priorities. Some actions that would facilitate these tasks include establishing priorities as well as service standards in a transparent and clear way, developing and implementing programs and actions to meet those goals, and obtaining access to the resources that will allow the implementation of programs and actions.

7.4 Supply Infrastructure and Technologies

As discussed earlier, the demands for fresh water have grown drastically over the last few centuries and provided the stimulus for the emergence of physical infrastructure that regulates the natural flow characteristics of free-flowing rivers. In terms of infrastructure for the enhancement of storage, a large number of dams have been constructed all over the world. "In North America, Europe, and the former Soviet Union, for example, three-quarters of the 139 largest river systems are strongly or moderately affected by water regulation resulting from dams, inter-basin transfers, or irrigation withdrawals" (Gleick et al. 2001, p. 22). In addition, hundreds of thousands of kilometers of dikes and levees have been constructed with the purpose of river training and flood protection. While these structures have clearly provided increased supply of fresh water for many uses, as well as flood control, all too often, they have had debilitating effects on the surrounding ecosystems, their naturally occurring services, and their biodiversity.

A number of other well-developed and documented technologies are available for improving efficiency and water resource management (World Commission on Dams 2000):

- micro-watershed level conservation of rainwater through physical or vegetative land management;
- rooftop rainwater harvesting;
- water recycling and reuse;
- desalinization for domestic water supplies (in coastal areas);
- on-farm agricultural water conservation, such as sprinklers and drip irrigation;
- crop selection and irrigation management based on meteorological conditions;
- improving crop productivity (with same or lower water use) through technological inputs;
- improvements to irrigation and municipal system management and conveyance;
- household water-saving devices, such as low-flow showerheads and water efficient toilets;
- managed flood releases from reservoirs to simulate historic flooding and impacts on downstream landscapes and ecosystems; and
- improved reservoir management and technologies for reducing evaporation loss.

Another technology that may come into play in coming decades is that of inter-basin transfers, particularly large, mega transfers between major river systems. For example, in India and

China, transfer projects costing hundreds of billions of dollars are proposed. These projects do not recognize ecosystem water needs and downstream consequences (Bandyopadhyay and Perveen 2004).

In this section, brief assessments are presented of three very different response options from the infrastructure and technology field: large dams, wetland restoration and mitigation projects, and desalination. To some degree, these represent the predominant past approach to water development, a current ecosystem approach to ecosystem restoration (or at least maintenance), and a promising new supply technology

7.4.1 Large Dams

A common response to water supply augmentation is the construction of large dams, which are defined by the International Commission on Large Dams as those with a height greater than 15 meters from the foundation or those that are 5–15 meters high with a volume of more than 3 million cubic meters. More than 45,000 large dams exist worldwide (WCD 2000).

Large dams can be used to regulate, store, and divert water for agricultural production and consumptive use in urban and rural areas. They were seen as integral components of the Green Revolution and were promoted widely in this period. Over half of the world's large dams have been built for irrigation and water supply purposes. Beyond water supply augmentation, large dam use also includes hydroelectric power generation and flood control (WCD 2000).

Large dams have proved especially useful for providing greater security in the face of water scarcity and variable supplies of water, a feature of countries with semi-arid catchments, where flow is highly variable and characterized by periodic drought such as South Africa, Spain, and Australia (WCD 2000).

7.4.1.1 Effectiveness

A full assessment of a few trillion dollars worth of infrastructure is beyond the scope of the current assessment. Instead, the experience with large dams is briefly summarized, based on the results of the World Commission on Dams, a multistakeholder international assessment, which recently spent three years and over $10 million dollars to carry out just such a task.

The benefits attributed to large dams include water supply to growing populations; increased food production; electric power for domestic, industrial, and other uses, as well as navigation and flood control. However, the environmental and social impacts of large dams are also well-known and have led to the very controversy and stalemate that resulted in the call by different parties to the debate over dams for an independent commission.

The WCD report identified a number of central issues in the dam debate, including: performance (costs and benefits), environmental impacts and sustainability, social impacts and equity, economics and finance, and governance and participation.

Of particular relevance to ecosystem health and human well-being are the environmental, social, and economic issues raised in the report. Environmentally, the impact of large dams on freshwater ecosystems is widely recognized as being more negative than positive. The impacts include, amongst others: changes in flow and sedimentation patterns; irreversible loss of species and populations, such as upstream and downstream fisheries; loss of habitat and associated biodiversity and ecosystem services from floodplains, wetlands, and estuarine and marine ecosystems; and greenhouse gas emissions from decaying organic material in the flooded basin.

The direct social impact of large dams is striking—they have led to the displacement of 40–80 million people worldwide and terminated access by local people to the natural resources and cultural heritage in the valley submerged by the dam (WCD 2000). The perennial freshwater systems established by large dams also contribute to health problems. For example, epidemics of Rift Valley fever and bilharzias coincided with the construction of the Diama and Manantali dams on the Senegal River (World Bank 2003a). Aside from the direct impacts of large dams, the benefits of their construction have rarely been shared equitably—the poor, vulnerable, and future generations are often not the same groups that receive the water and electricity services and the social and economic benefits from dams (WCD 2000).

Large dams were also found wanting from an economic and financial perspective. Pre-construction studies are typically overly optimistic about the benefits of projects and underestimate the costs. In a sample of 248 dams compiled by the WCD, the average cost overrun was a full 50% of the originally estimated costs. Further, the simplistic economic cost-benefit analyses applied often fail to adequately integrate the social and environmental impacts into the planning cycle.

As a result of these findings, the WCD concluded that, "The positive contribution of large dams to development has, in many cases, been marred by significant environmental and social impacts which are unacceptable when viewed from today's values" (WCD 2000, p. 198). The construction of large dams remains a viable option for augmenting water supply, but the conclusions of the WCD report suggest that the large dam option is one that needs to be carefully examined given past experience.

An underlying principle in the WCD approach is the recognition of stakeholder rights and a negotiated decision-making process. The publication of the *Report of the World Commission on Dams* served to bring to light many of the social and ecological costs of major water infrastructure. The report implicitly suggests the need to replace the traditional engineering view of dams and development with a new and more widely acceptable approach (Bandyopadhyay et al. 2002). The core of this approach is based on the newfound strategic importance and economic significance of the ecosystem services provided by rivers and watersheds.

7.4.1.2 Findings and Conclusions

The construction of large dams started in the early twentieth century and peaked during 1960–70. For various reasons, including resistance by people's movements, changing trends in project finance, and growing concern over the environmental and social impacts, the level of dam construction during 1990–2000 fell to almost the level of all the dams constructed during 1901–50. Many proposed dams have been postponed or canceled in the last two decades (Gleick et al. 2001, p. 22). Since the WCD process concluded in 2000, a number of new large dams have moved forward or been proposed (Three Gorges in China, and Kárahnjúkar in Iceland, to name two), and a number of large projects have run into difficulties (such as Bujugali in Uganda and Nam Theun II in Laos). Given the long lead time necessary to plan, finance, and build large dams and the continued controversy over the WCD report, it is too early to say what impact the report will ultimately have on the future of large dams. All that can be said at this point is that the report has been widely circulated and discussed, and that it has enhanced the legitimacy of the position that continued reliance on a supply-driven engineering approach is not sufficient to overcome the challenges ahead (Bandyopadhyay et al. 2002).

7.4.2 Wetland Restoration and Mitigation

Wetland restoration is a broad response category. In a strict sense, wetland restoration refers to the process of inducing and assisting the abiotic and biotic components of an ecosystem to return to their original state (Bradshaw 1997). However, definitions of restoration also encompass (1) actions to improve the condition of a site that may not necessarily be in the direction of pre-existing conditions (Bradshaw 1997), and (2) mitigation projects that seek to create wetlands to replace those that may have been lost from human interventions (Zedler 2000). This assessment considers wetland restoration in its broadest sense and is thus consistent with the Ramsar Convention definitions of this term.

Wetland restoration approaches are numerous and include engineering solutions such as backfilling canals and the removal of contaminated groundwater, biological interventions including controlling the impact of feral fish and reestablishing wetland plants, through to hydrological management to increase the effective inundation across floodplains and reintroduction of drying cycles.

7.4.2.1 Effectiveness

Mitsch et al. (1998) suggest that wetland restoration has become controversial in part because of the uncertainty about what is necessary to create and restore wetlands, that is, what combination of processes leads to the establishment of a desired combination of wetland structure and function. Such understanding would include the germination requirements, seed viability, and seedling growth characteristics of target wetland plants (van der Valk et al. 1999). However, imperfect knowledge regarding ecosystems can produce unexpected outcomes, as demonstrated in the uncertainty shown in models used to predict changes in wetland structure and function (Klotzli and Grootjans 2001).

With respect to the restoration community's understanding of the influential factors, it can be confidently said that the outcome of restoration projects is influenced by variables such as landscape context and site selection, hydrological regime, the rate of development of ecosystem attributes, nutrient supply rates, disturbance regimes, seed bank condition, invasive species, and life-history traits (Zedler 2000).

One of the major disagreements among wetland scientists is in relation to the role of abiotic conditions, especially hydrology, versus life-history traits in determining wetland structure and function. Mitsch et al. (1998) suggest that restoring the hydrological regime, or more generally abiotic conditions, is sufficient to reestablish structural features, particularly vegetation (the self-design approach). This is in contrast to Galatowitsch and van der Valk (1996), who suggest that at least in prairie potholes, dispersal is likely to be more limiting.

These approaches have implications for the restoration techniques applied—self-design is likely to focus on recreating hydrological features where the design approach will see engineering and replanting strategies as important. While this represents a major area of debate, it does not undermine wetland restoration as a response to wetland degradation, rather it places greater emphasis on understanding the factors that limit the rehabilitation of a site.

Perhaps of greater importance than the process of restoration is the actual restorability of a site. This has been highlighted in recent studies with the finding that it simply may not be practical to restore some wetlands because of the extent of degradation. This is an issue predominantly relating to abiotic factors such as wetland soils, the composition of which may have been irreversibly altered through changes in pH and nutrient status. Even if the changes are reversible, the time taken might be decades to centuries (Zedler 2000).

Cost is also an issue for wetland restoration and will reflect the extent of degradation and the objectives for restoration. Where the drivers of degradation operate at a local scale and are easy to identify and rectify, the cost may only be in terms of the voluntary time and effort provided by community groups. In contrast, where degradation is due to a multitude of factors operating at a regional or catchment scale, the cost may be in hundreds of millions of dollars (as in the $685 million Florida Everglades restoration project; Young 1996). This is certainly the case for large-scale environmental flow projects that aim to restore large wetland areas.

The success of wetland restoration is ultimately determined by the ability of a project to meet its original goals. In this regard, it has been noted on a number of occasions that wetland restoration projects suffer from poorly stated or unstated goals and objectives (Zedler 2000). An example of a poorly stated goal is one that is too generic to meet the original intent of the project, for example, the increased diversity of wetland plants within five years. This may be achieved with little movement toward reference conditions or more detailed aspects of structure and function.

The setting of well-stated goals must form part of a broader, comprehensive, and rigorous process for planning, developing, implementing, and evaluating restoration projects. This is widely accepted among wetland management practitioners as a key to success. In this regard, numerous frameworks for designing wetland restoration projects have been articulated, much of which has been synthesized in Ramsar's *Principles and guidelines for wetland restoration* (www.ramsar.org). Consistent with these frameworks is an adaptive management approach discussed in more detail in Chapter 15 of this volume, which allows for iterative learning and a chance to build on imperfect knowledge.

Although there is no definitive answer as to the ingredients of a successful wetland restoration project, it can be said with high confidence that there is a positive correlation between successful restoration and (1) a clear identification of the drivers of degradation, (2) where a small number of drivers are active, (3) when drivers operate at a local scale, (4) where drivers are inexpensive and easy to mitigate, (5) where the trade-offs required to mitigate the drivers are minimal, and (6) where the degradation of the wetland is reversible.

Despite meeting some of these criteria, where functional equivalence to an original state is the goal of restoration, success may be difficult to measure. For example, Mitsch and Wilson (1996) reported that where goals are specific and relate to aspects of wetland functioning, the time required to measure success could be 15–20 years. This is consistent with van der Valk's (1981) work on prairie pothole wetlands indicating that succession in wetlands may occur in 25-year cycles. Mitsch and Wilson (1996) suggest that in the case of restoring or creating forested wetlands, coastal wetlands, or peat-lands, it may require even more time.

7.4.2.2 Findings and Conclusions

Achieving functional equivalence is important when creating new wetlands to replace those that are destroyed. The question asked often is: "Does the structure and function of the new wetlands replace that of the old?" As for restoring degraded natural wetlands, this is hampered by uncertainty in the role of different abiotic and biotic factors and the observation that each wetland is a product of the unique contributions of these factors. The conclusion of numerous studies is that created wetlands rarely perform the same functions or house the same biodiversity as the original site.

For this reason, it is unlikely that created wetlands are going to structurally and functionally completely replace destroyed wetlands. This may be equally the case for degraded natural wetlands and is reflected by the Ramsar Convention, which notes that "the maintenance and conservation of existing wetlands is always preferable and more economical than their subsequent restoration" and that "restoration schemes must not weaken efforts to conserve existing natural systems."

7.4.3 Desalination

Desalination is the production of fresh, low-salinity potable water from saline water source (seawater or brackish water) via membrane separation or evaporation. The mineral/salt content of the water is usually measured by the water quality parameter total dissolved solids in milligrams per liter or parts per thousand. The World Health Organization and the U.S. Environmental Protection Agency, under the Safe Drinking Water Act, have established a maximum TDS concentration of 500 mg/L as a potable water standard. This TDS level can be used as a classification limit to define potable (fresh) water. Typically, water of TDS concentration higher than 500 mg/L and lower or equal to 15,000 mg/L is classified as brackish. Natural water sources such as sea, bay, and ocean waters that usually have TDS concentration higher than 15,000 mg/L are generally classified as seawater. For example, Pacific Ocean seawater along the U.S. west coast has a TDS concentration of 33,500 mg/L, of which approximately 75 % is sodium chloride.

Approximately 97.5% of the water on our planet is located in the oceans and, therefore, is classified as seawater. Of the 2.5% of the planet's fresh water, approximately 70% is in the form of polar ice and snow, and 30 % is groundwater, river and lake water, and air moisture. Even though the volume of Earth's water is vast, less than 10 million of the 1,400 million cubic meters of water on the planet are of low salinity and are suitable for use after applying conventional water treatment only. Desalination provides a means for tapping the world's main water resource—the ocean.

7.4.3.1 Effectiveness

By 2004 over 17,000 desalting units with a total production capacity of 37.75 million cubic meters per day (10 billion gallons per day) have been installed in approximately 120 countries (Wangnick 2004). Desalination techniques predominantly use either thermal or membrane processes (Buros 2000). Thermal desalination technologies use a variety of forms of distillation, including multiple effect distillation, multistage flash distillation, or vapour compression distillation. Membrane separation is typically accomplished by reverse osmosis or electrodialysis technologies. (See Box 7.17.)

Most of the large seawater desalination facilities built in the past 10 years or currently undergoing construction are delivered under public-private partnership arrangement using build-own-operate-transfer method of project implementation. The BOOT project delivery method is preferred by municipalities and public utilities worldwide because it allows cost-effective transfer to the private sector of the risks associated with the number of variables affecting the cost of desalinated water, such as: intake water quality and the often difficult to predict effects on plant performance; permitting challenges; start-up and commissioning; fast-changing membrane technology and equipment market; and limited public sector experience with the operation of large seawater desalination facilities (Voutchkov 2004a).

BOX 7.17

Reverse Osmosis: Removing Salt using Semi-permeable Membranes

Reverse osmosis separates solutes from saline water by forcing it through a semi-permeable membrane under pressure. Unlike distillation no heating is required with most of the energy used to pressurize the saline feed water. The basic system components are pre-treatment, high-pressure pumps, membrane assembly, and post-treatment. Beyond its use for desalting, reverse osmosis can also be used for removal of other impurities and contaminants such as iron, lead, nitrate, endocrine disruptors, arsenic, disinfection by-products, bacteria, viruses, and other pathogens and emerging contaminants. The first RO membrane was first developed at the University of California at Los Angeles in the early 1960s by Loeb and was used to produce drinking water from seawater. This relatively thick membrane was made from cellulose acetate and required feed pressures in excess of 2,000 pounds per square inch (psi). Today RO membranes used to desalinate seawater are made of thin-film composite plastic materials and require about 800 to 1,200 psi, while brackish water applications may necessitate feed pressure ranging from 100 to 600 psi. The feed pressure required depends primarily on the total dissolved solids (TDS) concentration and the temperature of the water—lower TDS levels and warmer waters requiring lower feed pressures.

The largest RO membrane desalination plant worldwide in continuous operation today is located in the United Arab Emirates (Taweelah) and has capacity of 227,000 cubic meters per day. The largest plant in construction is the Ashkelon seawater desalination facility in Israel; this plant will be operational in the spring of 2005 and will have capacity of 395,000 cubic meters per day. The Yuma desalting plant in Arizona is the largest U.S. reverse osmosis brackish water desalination plant and can produce about 275,000 cubic meters per day. The largest seawater desalination plant in the United States is the Tampa Bay, Florida, facility, which has potable water production capacity of 95,000 cubic meters per day.

Up until the 1970s, desalination was predominately performed using distillation techniques, with some commercial units capable of producing up to 8,000 cubic meters per day. Subsequent technological improvements have seen an expansion in the use of membrane processes, especially reverses osmosis. Currently, multistage flash distillation only accounts for 36.5% of total installed desalination capacity worldwide, down from 51.3% a decade ago. By contrast, plants using RO have risen from 32.7% in 1993 to 47.2% today (Wangnick 2004). The increasing popularity of RO membrane desalination is driven by remarkable advances in the membrane separation and energy recovery technologies, and associated reduction of the overall water production costs.

Other less commonly used techniques for desalting water include ion-exchange methods, freezing, membrane distillation, and solar and wind driven systems (Buros 2000). Solar systems include solar stills, which heat and vaporize water from a ground level basin and then collect vapor from a sloping glass roof. This technique faces drawbacks such as high capital costs, vulnerability of glass to weather damage, and the large collection areas required (Gleick 2000). An alternative to these systems is to use wind and solar generated electricity to drive more traditional desalting processes.

The developments in seawater desalination technology during the past two decades, combined with transition to construction of large capacity plants, co-location with power plant generation facilities and enhanced competition by using the BOOT method of project delivery have resulted in a dramatic decrease of the cost of desalinated water. Recent large reverse osmosis desalination projects in the United States, Israel, Cyprus, Singapore, and the Middle East have installed costs of approximately $0.50 per cubic meter, down from $1.50 per cubic meter in the early 1990s (Voutchkov 2004b).

7.4.3.2 Findings and Conclusions

There is no single best technique for desalination. The selection of a desalination process depends on site-specific conditions, economics, the quality of water to be desalinated, the purpose of use, and local engineering experience and skill (Gleick 2000). Operating and capital costs in particular are influenced by the capacity and type of desalination plant, the quality of feed water and the energy required to drive the process (Buros 2000). Water from desalination is generally expensive, although costs have decreased in recent years to such an extent that in some areas of the United States desalting brackish water is cheaper than alternative measures such as piping conventionally treated water (Buros 2000).

Although, no major technology breakthroughs are expected to bring the cost of seawater desalination further down dramatically in the next several years, the steady reduction of desalinated water production costs coupled with increasing costs of water treatment driven by more stringent regulatory requirements, are expected to accelerate the current trend of increased reliance on the ocean as an environmentally friendly and competitive water source. This trend is forecasted to continue in the future and to further establish ocean water desalination as a reliable drought-proof alternative for many communities in the United States and worldwide.

References

Acreman, M.C., 2001: Ethical aspects of water and ecosystems, *Water Policy Journal,* **3,** pp. 257–265.

Acreman, M.C., and J.M. King, 2003: Defining water requirements. In: *Flow: The Essentials of Environmental Flows,* M. Dyson, G. Bergkamp, and J. Scanlon (eds.), International Union for Conservation of Nature and Natural Resources, Gland, Switzerland, and Cambridge, UK.

Acreman, M.C., F.A.K. Farquharson, M.P. McCartney, C. Sullivan, K. Campbell, et al., 2000: Managed flood releases from reservoirs: Issues and guidance, Report to DFID and the World Commission on Dams, Centre for Ecology and Hydrology, Wallingford, UK.

Arthington, A.H., S.O. Brizga, and M.J. Kennard, 1998: Comparative evaluation of environmental flow assessment techniques: Best practice framework, LWRRDC occasional paper 25/98, Land and Water Resources Research and Development Corporation, Canberra, Australia.

Asmal, K., 2000: Water is a catalyst for peace, Remarks of Prof. Kader Asmal, South Africa Minister of Education, Opening session, Stockholm Water Symposium Laureate Lecture, Convention Centre, 14 August 2000, Stockholm, Sweden.

Aylward, B., 2003: Financing environmental flows. In: *Flow: The Essentials of Environmental Flows,* M. Dyson, G. Bergkamp, and J. Scanlon (eds.), International Union for Conservation of Nature and Natural Resources, Gland, Switzerland, and Cambridge, UK.

Aylward, B., and A. Fernández González, 1998: Institutional arrangements for watershed management: A case study of Arenal, Costa Rica, Collaborative Research in the Economics of Environment and Development working paper series no. 21, International Institute for Environment and Development, London, UK.

Aylward, B., 2004: Land use, hydrological function and economic valuation. In: *Forest-Water-People in the Humid Tropics,* M. Bonnell and L.A. Bruijnzeel (eds.), Cambridge University Press, Cambridge, UK.

Bandyopadhyay, J., and D. Gyawali, 1994: Himalayan water resources: Ecological and political aspects of management, *Mountain Research and Development,* **14(1),** pp. 1–24.

Bandyopadhyay, J., and S. Perveen, 2004: Interlinking of rivers in India: Assessing the justifications. In: *Interlinking of Rivers in India: Myth and Reality,* S.K. Bhattacharya, A.K. Biswas, and K. Rudra (eds.), The Institution of Engineers, Kolkata, India.

Bandyopadhyay, J., B. Mallik, M. Mandal, and S. Perveen, 2002: Dams and development, *Economic and Political Weekly,* **37(4),** pp. 4108–4112.

Barkin, D., and T. King, 1986: *Desarrollo Economico Regional (Enfoque por Cuencas Hidrologicas de Mexico),* 5 edición, Siglo XXI Editores, Ciudad de México, México.

Baron, J.S., N.L. Poff, P.L. Angermeier, C.N. Dahm, P.H. Gleick, et al., 2002: Meeting ecological and societal needs for fresh water, *Ecological Applications,* **12(5),** pp. 1247–1260.

Barrow, C., 1998: River basin development planning and management: A critical review, *World Development,* **26(1),** pp. 171–186.

Bauer, C., 2004: *Siren Song: Chilean Water Law as a Model for International Reform,* Resources for the Future, Washington, DC.

Baviskar, A., 2002: Between micro-politics and administrative imperatives: Decentralization and the watershed mission in Madhya Pradesh, India, Paper presented at conference on decentralization and the environment, World Resources Institute, Bellagio, Italy.

Bellido, G.S., 2003: *Informe Nacional Peru,* Red Nacional de Manejo de Cuencas Hidrográficas-REDNAMAC (Red Nacional de Manejo de Cuencas/National River Basin Management Network), Lima, Peru.

BIPE (Bureau d'Information et de Prévision Économique)**,** 2001: Prix de l'eau: Eléments de comparaison entre modes de gestion en France et en Europe, BIPE, Paris, France.

Bjornlund, H., 2002: *Water Exchanges: Australian Experiences,* School of International Business, University of South Australia, Adelaide, Australia.

Bjornlund, H., and J. McKay, 2002: Aspects of water markets for developing countries: Experiences from Australia, Chile, and the US, *Environment and Development Economics,* **7,** pp. 769–795.

Blaikie, P.M., and J.S.S. Muldavin, 2004: Upstream, downstream, China, India: The politics of environment in the Himalayan region, *Annals of the Association of American Geographers,* **94,** pp. 520–548.

Bonell, M. and L.A. Bruijnzeel (eds.), 2004: *Forests, Water and People in the Humid Tropics.* Cambridge University Press, Cambridge, UK.

Bradshaw, A., 1997: What do we mean by restoration? In: *Restoration Ecology and Sustainable Development,* K.M. Urbanska, N.R. Webb, and P.J. Edwards, (eds.), Cambridge University Press, New York, NY, pp. 8–16.

Brehm, M. and J. Quiroz, 1995: *The Market for Water Rights in Chile: Major Issues.* World Bank Technical Paper Number 285. World Bank, Washington, DC.

Brizga, S.O., 1998: Methods addressing flow requirements for geomorphological purposes. In: *Comparative Evaluation of Environmental Flow Assessment Techniques: Review of Methods,* A.H. Arthington and J.M. Zalucki (eds.), LWRRDC occasional paper no. 27–98, Land and Water Resources Research and Development Corporation, Canberra, Australia.

Bromley, D.W., 1989: *Economic Interests and Institutions,* Basil Blackwell, Oxford, UK.

Brown, C., and J.M. King, 2003: Environmental flows: Concepts and methods. In: *Water Resources and Environment: Technical Note C.1,* R. Davis and R. Hirji (eds.), World Bank, Washington, DC.

Bruijnzeel, L.A., 2004: Hydrological functions of tropical forests: Not seeing the soil for the trees? *Agriculture, Ecosystems and Environment,* **104,** pp. 185–228.

Buros, O.K., 2000: *The Desalting ABC's,* Booklet for the IDA, International Desalination Association, Topsfield, MA.

Cassar, A.Z., and C. Bruch: Transboundary environmental impact assessment (TEIA) in international watercourse management, *New York University Environmental Law Journal.* In press.

Castro, E., and G. Barrantes, 1998: *Valoración Económico Ecológico del Recurso Hídrico en la Cuenca Arenal: El Agua un Flujo Permanente de Ingreso,* Área de Conservación Arenal (ACA), MINAE (Ministerio de Ambiente y Energía), Heredia, Costa Rica.

Castro, E., and S. Salazar, 2000: *Valor Económico del Servicio Ambiental Hídrico a la Salida del Bosque: Análisis de Oferta,* Empresa de Servicios Publicos de Heredia S.A., Guapiles, Costa Rica.

Clifford, P., C. Landry, and A. Larsen-Hayden, 2004: *Analysis of Water Banks in the Western United States,* Department of Ecology, Water Resources Program, Olympia, WA.

Collins, B.R., and E.W.B. Russell (eds.), 1988: *Protecting the New Jersey Pinelands: A New Direction in Land Use Management,* Rutgers University Press, New Brunswick, NJ, and London, UK.

Cosgrove, W.J., and F. Rijsberman, 2000: *World Water Vision: Making Water Everybody's Business,* World Water Council, World Water Vision and Earthscan, London, UK.

Dunbar, M.J., A. Gustard, M.C. Acreman, and C.R.N. Elliot, 1997: Overseas approaches to setting river flow objectives, Report to Environment Agency W6B **(96)4,** Institute of Hydrology, Wallingford, UK.

DWAF (Department of Water Affairs and Forestry), 1999: Establishment of a pricing strategy for water use charges in terms of Section 56(1) of the National Water Act, 1998, Government notice 1353, DWAF, Pretoria, South Africa.

Echavarria, M., 2002: Financing watershed conservation: The FONAG water fund in Quito, Ecuador. In: *Selling Forest Environmental Services: Market-based Mechanisms for Conservation and Development,* S. Pagiola, J. Bishop, and N. Landell-Mills (eds.), Earthscan, London, UK.

Echavarria, M., J. Vogel, M. Alban, and F. Meneses, 2004: The impacts of payments for watershed services in Ecuador: Emerging lessons from Pimampira and Cuenca, Markets for Environmental Services series no. 4, International Institute for Environment and Development, London, UK.

Fahrenthold, D.A., 2004: Maryland watermen mull suing over bay: Class action may focus on farming firms, municipal plants, *The Washington Post,* Washington, DC, 28 July 2004, Section A, p. 6.

FAO (Food and Agriculture Organization of the United Nations), 2002: Land-water linkages in rural watersheds, Electronic workshop, 18 September–27 October 2000, FAO, Rome, Italy.

Galatowitsch, S.M. and A.G. van der Valk, 1996: The vegetation of restored and natural prairie wetlands, *Ecological Applications,* **6(1),** pp. 102–112.

Garcia, L., 1999: *Review of the Role of River Basin Organizations (RBOs) in Latin America,* Inter-American Development Bank, Washington, DC.

Geoghegan, T., V. Krishnarayan, D. Pantin, and S. Bass, 2003: *Incentives for Watershed Management in the Caribbean: Diagnostic Studies in Grenada, Jamaica, St. Lucia and Trinidad,* The Caribbean Natural Resources Institute and the International Institute for Environment and Development, Laventille, Trinidad, and London, UK.

Gleick, P.H., 2000a: The changing water paradigm: A look at twenty-first century water resources development, *Water International,* **25(1),** pp. 127–138.

Gleick, P.H., 2000b: *The World's Water 2000–2001: The Biennial Report on Freshwater Resources,* Island Press, Washington, DC.

Gleick, P.H., A. Singh, and H. Shi, 2001: *Threats to the World's Freshwater Resources,* Pacific Institute, Oakland, CA.

Gleick, P.H., 2003: Global freshwater resources: Soft-path solutions for the 21st century, *Science,* **302,** pp. 1524–1528.

Global Water Partnership, 2000: Integrated water resources management, Technical Advisory Committee background paper no. 4, Global Water Partnership Technical Advisory Committee, Stockholm, Sweden.

Gould, G., 1988: Water rights transfers and third-party effects, *Land and Water Law Review,* **23,** pp. 2–41.

Hadjigeorgalis, E., 1999: Private water markets in agriculture and the effects of risk, uncertainty and transaction costs. Ph.D dissertation, University of California at Davis.

Hanak, E., 2003: *Who Should be Allowed to Sell Water in California? Third Party Issues and the Water Market,* Public Policy Institute of California, San Francisco, CA.

Hearne, R., 1995: The market allocation of natural resources: Transactions of water use rights in Chile, Ph.D. dissertation, Department of Agricultural Economics, University of Minnesota, St. Paul, MN.

Hearne, R., and K.W. Easter, 1995: Water allocation and water markets: An analysis of gains-from-trade in Chile, World Bank technical paper no. 315, World Bank, Washington, DC.

Hearne, R. and K.W. Easter, 1997: The economics and financial gains from water markets in Chile, *Agricultural Economics,* **15,** pp. 187–199.

Hennessey, T.M., 1994: Governance and adaptive management for estuarine ecosystems: The case of the Chesapeake Bay, *Coastal Management,* **22,** pp. 119–145.

Hofer, T., 1997: Meghalaya, not Himalaya, *Himal South Asia,* **10(5)** (September–October), pp. 52–56.

Howe, C.W., 1997: Increasing efficiency in water markets: Examples from the western United States. In: *Water Marketing: The Next Generation,* T. Anderson and P. Hill (eds.), Rowman and Littlefield Publishers Inc., Lanham, MD, pp. 79–100.

Howe, C.W., 2002: Policy issues and institutional impediments in the management of groundwater: Lessons from case studies, *Environment and Development Economics,* **7,** pp. 769–795.

Howe, C.W., and C. Goemans, 2003: Water transfers and their impacts: Lessons from three Colorado water markets, *Journal of the American Water Resources Association,* **39,** pp. 1055–1065.

Howe, C.W., J.K. Lazo, and K.R. Weber, 1990: The economic impacts of agriculture-to-urban water transfers on the area of origin: A case study of the Arkansas River Valley in Colorado, *American Journal of Agricultural Economics,* **72,** pp. 1200–1204.

Howitt, R., N. Moore and R. Smith: 1992: A retrospective on California's 1991 emergency drought water bank, Report prepared for the California Department of Water Resources, Sacramento, CA.

IIED (International Institute for Environment and Development), 2004: Developing markets for watershed protection services and improved livelihoods. Implementation Phase Planning Workshop Report. IIED, London. Available at: http://www.iied.org/docs/flu/waterworkshop.pdf.

Isaac, M., 2001: *Embedding Social Capital in the Construction of Water Markets,* American Water Resources Association/ International Water Law Research Institute, University of Dundee, Dundee, UK.

IWMI (International Water Management Institute), 2002: *Framework for the Comprehensive Assessment of Water Management in Agriculture,* IWMI, Colombo, Sri Lanka.

Jodha, N.S., 1995: The Nepal middle mountains. In: *Regions at Risk: Comparisons of Threatened Environments,* J.X. Kasperson, R.E. Kasperson, and B.L. Turner (eds.), United Nations University Press, Tokyo, Japan, New York, NY, and Paris, France.

Johnson, G., 2003: Connecticut's nitrogen trading program, Paper presented at USEPA's National Forum on Water Quality Trading, 22–23 July 2003, Chicago, IL.

Johansson, R.C., Y. Tsur, T.L. Roe, R. Doukkali, and A. Dinar, 2002. Pricing irrigation water: A review of theory and practice, *Water Policy,* **4(2),** pp. 173–99.

Johnson, N., C. Revenga, and J. Echeverría, 2001: Managing water for people and nature, *Science,* **292** (11 May), pp. 1071–1072.

Johnson, N.L., and M.E. Baltodano, 2004: The economics of community watershed management: Some evidence from Nicaragua, *Ecological Economics,* **49,** pp. 57–71.

Jones, G., A. Arthington, B. Gawne, T. Hillman, R. Kingsford, et al., 2003: *Ecological Assessment of Environmental Flow Reference Points for the River Murray System,* Interim report prepared by the Scientific Reference Panel for MDBC Living Murray Initiative, Murray-Darling Basin Commission, Canberra, Australia.

Kaika, M., and B. Page, 2002: The making of the EU water framework directive: Shifting choreographies of governance and the effectiveness of environmental lobbying, Working paper WPG 02–12, Oxford University, Oxford, UK.

Kaimowitz, D. and J.C. Ribot, 2002: Services and infrastructure versus natural resource management: Building a base for democratic decentralization. Conference on decentralization and the environment, Bellagio, Italy, 18–22 February 2002. Sponsored by the World Resources Institute.

Kaimowitz, D., 2004: Useful myths and intractable truths: The politics of the link between forests and water in Central America. In: *Forest-Water-People in the Humid Tropics,* M. Bonnell and L.A. Bruijnzeel (eds.), Cambridge University Press, Cambridge, UK.

Kasperson, J.X., R.E. Kasperson, and B. L. Turner, II (eds.)., 1995: *Regions at Risk: Comparisons of Threatened Environments,* United Nations University Press, Tokyo.

King, J.M., and C. Brown, 2003: Environmental flows: Case studies. In: *Water Resources and Environment: Technical Note C.2,* R. Davis and R. Hirji (eds.), World Bank, Washington, DC.

Klotzli, F., and A.P. Grootjans, 2001: Restoration of natural and semi-natural wetland systems in Central Europe: Progress and predictability of developments, *Restoration Ecology,* **9,** pp. 209–219.

Knox, J.H., 2002: The myth and reality of transboundary environmental impact assessment, *American Journal of International Law,* **96,** pp. 291–319.

Koulov, B., 1997: Environmental issues in the post-socialist period: Political change and environmental policy. In: *Bulgaria in Transition,* J. Bell (ed.), Westview Press, Boulder, CO.

Landell-Mills, N., and I. Porras, 2002: *Silver Bullet or Fools' Gold: Developing Markets for Forest Environmental Services and the Poor,* International Institute for Environment and Development, London, UK.

Landry, C.J., 1998: *Saving Our Streams through Water Markets: A Practical Guide,* Political Economy Research Center, Bozeman, MT.

Levy, S., 2003: Turbulence in the Klamath River Basin. *BioScience,* **53,** pp. 315–320.

MacDonnell, D.L.J., 1994: *Using Water Banks to Promote More Flexible Water Use,* Natural Resources Law Center, University of Colorado, Boulder, CO.

MacKenzie, S.H., 1996: *Integrated Resource Planning and Management: The Ecosystem Approach in the Great Lakes Basin,* Island Press, Washington, DC.

MDBC (Murray-Darling Basin Commission), 2001: *Basin Salinity Management Strategy 2001–2015,* Murray-Darling Basin Ministerial Council, Canberra, Australia.

MDBC, 2003: *Murray Darling Basin Initiative,* Murray-Darling Basin Ministerial Council, Canberra, Australia.

Meinzen-Dick, R.S., and R. Pradhan, 2002: *Legal Pluralism and Dynamic Property Rights,* Consultative Group on International Agricultural Research system-wide program on collective action and property rights no. 22, International Food Policy Research Institute, Washington, DC.

Meinzen-Dick, R.S., and B.R. Bruns, 2003: Negotiating transitions in water rights, *Water Resources Impact,* **5,** pp. 22–24.

Middelkoop, H., J. Rotmans, M.B.A. van Asselt, J.C.J. Kwadijk, and W.P.A. van Deursen, 2000: Development of perspective-based scenarios for global change assessment for water management in the lower Rhine Delta. UNESCO-WOTRO International Conference "Water for Society" organized by IHE, Delft, The Netherlands. Available at http://www. Icis.unimaas.nl/publ/.

Miller, B.A., and R.B. Reidinger, 1998: *Comprehensive River Basin Development: The Tennessee Valley Authority,* World Bank, Washington, DC.

Miranda, M., I.T. Porras, and M.L. Moreno, 2003: *The Social Impacts of Payments for Environmental Services in Costa Rica: A Quantitative Field Survey and Analysis of the Virilla Watershed,* Markets for Environmental Services Series No. 1, International Institute for Environment and Development, London, UK.

Mitsch, W.J., and R.F. Wilson, 1996: Improving the success of wetland creation and restoration with know-how, time, and self design, *Ecological Applications,* **6(1),** pp. 77–83.

Mitsch, W. J., X. Wu, R.W. Nairn, P.E. Weihe, N. Wang, et al., 1998: Creating and restoring wetlands: A whole ecosystem experiment in self-design, *BioScience,* **48,** pp. 1019–1030.

Molden, D., and M. Falkenmark, 2003: Water and the millennium development goals: Meeting the needs of people and ecosystems, *Stockholm Water Front: A Forum for Global Water Issues,* **4** (December), pp. 2–3.

Molden, D., and C. de Fraiture, 2004: Investing in water for food, ecosystems and livelihoods, Discussion draft blue paper, a comprehensive assessment of water management in agriculture, Stockholm International Water Institute, Stockholm, Sweden.

Munawir, S., S. Salim, A. Suyanto, and S. Vermeulen, 2003: *Action-Learning to Develop and Test Upstream-Downstream Transactions for Watershed Protection Services: A Diagnostic Report From Segara River Basin, Indonesia,* PSDAL-LP3ES (Pengembangan Sumberdaya Air dan Lahan-Lembaga Penelitian, Pendidikan dan Penerangan Ekonomi dan Sosial/Center for Water and Land Resources Development and Studies and International Institute for Environment Development, Jakarta, Indonesia, and London, UK.

Murphy, I.L., 1997: *The Danube: A River Basin in Transition,* Kluwer Academic Publishers, Boston, MA.

NAPA (National Academy of Public Administration), 2000: *Environment.gov: Transforming Environmental Protection for the 21st Century,* NAPA, Washington, DC.

NJDEP (New Jersey Department of Environmental Protection), 2002: Study shows New Jersey's freshwater wetlands mitigation program missed opportunity for net increase in acreage, Press release, NJDEP, Trenton, NJ.

NRC (Natural Resources Council), 1992: *Water Transfers in the West: Efficiency, Equity, and the Environment,* National Academy Press, Washington, DC.

OECD (Organisation for Economic Cooperation and Development), 2000: *Global Trends in Urban Water Supply and Waste Water Financing and Management: Changing Roles for the Public and Private Sector,* OECD, Paris, France.

OECD, 2003a: *Improving Water Management: Recent OECD Experience,* OECD, Paris, France.

OECD, 2003b: Public-private partnerships in the urban water sector, Policy brief, OECD, Paris, France.

OECD and World Bank, 2002: *Private Sector Participation in Municipal Water Services in Central and Eastern Europe and Central Asia,* OECD, Paris, France, and World Bank, Washington, DC.

OECD and World Bank, 2003: *Private Sector Participation in Municipal Water Services in Central and Eastern Europe and Central Asia: Facing a Crisis of Confidence in Private Sector Participation in the Water Sector: Measures to Overcome Obstacles to more Effective PSP,* OECD, Paris, France, and World Bank, Washington, DC.

Ostrom, E., 1990: *Governing the Commons: The Evolution of Institutions for Collective Action,* Cambridge University Press, Cambridge, UK.

Ostrom, E., L. Schroeder, and S. Wynne, 1993: *Institutional Incentives and Sustainable Development: Infrastructure Policies in Perspective,* Westview Press, Boulder, CO.

Owen, D.L., 2002: *The European Water Industry: Market Drivers and Responses,* CWC Publishing, London, UK.

Pagiola, S., 2002: Paying for water services in Central America: Learning from Costa Rica. In: *Selling Forest Environmental Services: Market-based Mechanisms for Conservation,* S. Pagiola, J. Bishop, and N. Landell-Mills (eds.), Earthscan, London, UK.

Pérez, C.J., 2003: Payment for hydrological services at a municipal level and its impact on rural development: The PASOLAC experience. Proceedings of FAO workshop on payment schemes for environmental services in watersheds, Arequipa, Peru, 9–12 June 2003.

Perrot-Maître, D., and P. Davis, 2001: *Case Studies: Developing Markets for Water Services from Forests,* Forest Trends, Washington, DC.

Poff, N.L., J.D. Allan, M.B. Bain, J.R. Karr, K.L. Prestegaard, et al., 1997: The natural flow regime, *BioScience,* **47,** pp. 769–784.

Poff, N.L., J.D. Allan, M.A. Palmer, D.D. Hart, B.D. Richter, et al., 2003: River flows and water wars: Emerging science for environmental decision-making, *Frontiers in Ecology and the Environment,* **1(6),** pp. 298–306.

Postel, S.L., 1997: *Last Oasis: Facing Water Scarcity,* W.W. Norton, New York, NY.

Postel, S.L., 2000: Entering an era of water scarcity: The challenges ahead, *Ecological Applications,* **10(4),** pp. 941–948.

Postel, S.L., and B. Richter, 2003: *Rivers for Life: Managing Water for People and Nature,* Island Press, Washington, DC.

Ratner, B.D., 2003: The politics of regional governance in the Mekong River Basin, *Global Change,* **15,** pp. 59–76.

Reisner, M., 1986: *Cadillac Desert,* Viking Penguin, New York, NY.

Rell, L.G.M.J., 2003: Lieutenant Governor Rell marks the success of Connecticut's innovative nitrogen trading program, Press release, Long Island Sound Study, Stamford, CT.

Ribot, J.C., 2002: *Democratic Decentralization of Natural Resources: Institutionalizing Popular Participation,* World Resources Institute, Washington, DC.

Ribot, J.C., 2004: *Waiting for Democracy: The Politics of Choice in Natural Resource Decentralization,* World Resources Institute, Washington, DC.

Rojas, M., and B. Aylward, 2002: *The Case of La Esperanza: A Small Private, Hydropower Producer and a Conservation NGO in Costa Rica,* Land and Water Linkages in Rural Watersheds Case Study Series, FAO, Rome, Italy.

Rojas, M., and B. Aylward, 2003: *What are We Learning from Experiences with Markets for Environmental Services in Costa Rica? A Review and Critique of the Literature,* Markets for Environmental Services Series No. 2, International Institute for Environment and Development, London, UK.

Romano, D., and M. Leporati, 2002: The distributive impact of the water market in Chile: A case study in Limari Province, 1981–1997, *Quarterly Journal of International Agriculture,* **41,** pp. 41–58.

Rosales, R.M.P., 2003: *Developing Pro-Poor Markets for Environmental Services in the Philippines,* Markets for Environmental Services Series No. 3, International Institute for Environment and Development, London, UK.

Rosa, H., S. Kandel, L. Dimas, and W.C. Méndez, 2003: *Compensation for Environmental Services and Rural Communities: Lessons from the Americas and Key Issues for Strengthening Community Strategies,* PRISMA (Programa Salvadoreno de Investigacion sobre Desarrollo y Medio Ambiente), San Salvador, El Salvador.

Rose, C.M., 2002: Common property, regulatory property, and environmental protection: Comparing community-based management and tradable environmental allowances. In: *The Drama of the Commons,* E. Ostrom, T. Dietz, N. Dolak, P. Stern, S. Stonich, et al. (eds.), National Academy Press, Washington, DC.

Rosegrant, M.W. and R. Gazmuri, 1994: *Reforming Water Allocation Policy through Markets in Tradable Water Rights: Lessons from Chile, Mexico and California.* EPTD Discussion Paper No. 6, IFPRI, Washington, DC.

Rosegrant, M.W., X. Cai, and S.A. Cline, 2002: *World Water and Food to 2025: Dealing with Scarcity,* International Food Policy Research Institute and International Water Management Institute, Washington, DC.

Rothfeder, J., 2003: Water rights, conflict, and culture, *Water Resources Impact,* **5,** pp. 19–21.

RUPES (Rewarding the Poor for Upland Environmental Services), 2004. The World Agroforestry Center, Bogor, Indonesia. Available at: http://www .worldagroforestry.org/sea/Networks/RUPES/index.asp.

Saleth, R.M., and A. Dinar, 1999: *Water Challenge and Institutional Response: A Cross-Country Perspective,* The World Bank Development Research Group, Rural Development and Rural Development Department, Washington, DC.

Santopietro, G.D., and L.A. Shabman, 1990: Common property rights in fish and water quality: The oyster fishery of the Chesapeake Bay, Designing Sustainability on the Commons: The First Annual Conference of the International Association for the Study of Common Property, 27–30 September 1990, Duke University, Durham, NC.

Sax, J.L., 1993: Property rights and the economy of nature: Understanding Lucas v. South Carolina Coastal Council, *Stanford Law Review,* **45,** pp. 1433–1455.

Scholte, P., K. Philippe, A. Saleh, and B. Kadiri, 2000: Floodplain rehabilitation in North Cameroon: Impact on vegetation dynamics, *Applied Vegetation Science,* **3,** pp. 33–42.

Schreiner, B., and B. van Koppen, 2000: From bucket to basin: Poverty, gender, and integrated water management in South Africa, Paper presented at Integrated Water Management in Water-Stressed River Basins in Developing Countries: Strategies for Poverty Alleviation and Agricultural Growth, International Water Management Institute, 16–21 October 2000, at Loskop Dam, South Africa.

Siebentritt, M.A., G.G. Ganf, and K.F. Walker, 2004: Effects of an enhanced flood on riparian plants of the River Murray, South Australia, *River Research and Applications,* **20,** pp. 765–774.

Sproule-Jones, M., 1999: Restoring the Great Lakes: Institutional analysis and design, *Coastal Management,* **27,** pp. 291–316.

State Forests of New South Wales, 2001: The war against salinity: Could forests be the answer? [online]. Accessed 25 October 2001. Available at http://www.forest.nsw.gov.au/_navigation/active_frame.asp?bodypath=/bush/feb00/feature/page14.asp.

Sundstrom, A., 2001: Salinity control credits: A comment [online], Nature Conservation Council of New South Wales. Accessed 3 October 2001. Available at http://www.nccnsw.org.au/veg/reference/salt.credits.html.

Swallow, B.M., D.P. Garrity, and M. van Noordwijk, 2001: *The Effects of Scales, Flows and Filters on Property Rights and Collective Action in Watershed Management,* International Food Policy Research Institute, Washington, DC.

Tarlock, A.D., 2000: Reconnecting property rights to watersheds. *William and Mary Environmental Law and Policy Review,* **(25)69.**

The, B.D., D.T. Ha, and N.Q. Chinh, 2004: *Rewarding Upland Farmers for Environmental Services: Experience, Constraints and Potential in Vietnam,* World Agroforestry Centre, Southeast Asia Regional Office, Bogor, Indonesia.

Tharme, R., 2001: A global perspective on environmental flow assessment: Emerging trends in the development and application of environmental flow methodologies for rivers, Paper presented at Environmental Flows for River Systems, at Cape Town, South Africa.

Thompson, B., 1997: Water markets and the problem of shifting paradigms. In: *Water Marketing: The Next Generation,* T. Anderson, and P. Hill (eds.), Rowman and Littlefield Publishers Inc., Lanham, MD, pp. 1–30.

Thompson, G., 2003: Water tap often shut to South Africa's poor, *The New York Times,* New York, NY.

Tognetti, S., 2001: *Creating Incentives for River Basin Management as a Conservation Strategy: A Survey of the Literature and Existing Initiatives,* World Wildlife Fund, Washington, DC.

UNECE (United Nations Economic Commission for Europe), 1991: Paper presented at Convention on Environmental Impact Assessment in a Transboundary Context, 25 February 1991, 1988 U.N.T.S. 310 (United Nations Treaty Collection).

UNESCO (United Nations Educational, Scientific and Cultural Organization), 2003: *The UN World Water Development Report: Water for People/Water for Life,* World Water Assessment Program, UNESCO, Paris, France.

United Nations Millennium Project, 2003: *Background Paper of the Millennium Project Task Force on Water and Sanitation,* United Nations Development Programme, New York, NY.

USEPA (United States Environmental Protection Agency), 2003: *Water Quality Trading Assessment Handbook: EPA Region 10's Guide to Analyzing Your Watershed,* EPA 910-B-03-003, EPA, Washington, DC.

Van der Valk, A.G., 1981: Succession in wetlands: A Gleasonian approach, *Ecology,* **62(3),** pp. 688–696.

Van der Valk, A.G., T.L. Bremholm, and E. Gordon, 1999: The restoration of sedge meadows: Seed viability, seed germination requirements, and seedling growth of *Carex* species, *Wetlands,* **19,** pp. 756–764.

Van Dijk, M.P. and K. Schwartz, 2002: Financing the water sector in the Netherlands: A first analysis, Paper presented at Financing Water Infrastructure Panel in The Hague, 8–10 October 2002, Netherlands Water Partnership, Delft, The Netherlands.

Vermillion, D.L., 1999: Property Rights and Collective Action in the Devolution of Irrigation System Management, In: *Collective Action, Property Rights, and the Devolution of Natural Resource Management: Exchange of Knowledge and Implications for Policy,* Deutsche Stiftung für Internationale Entwicklung, Puerto Azul, The Philippines.

Vira, B., 1999: Implementing Joint Forest Management in the Field: Towards an Understanding of the Community-Bureaucracy Interface, In: *A New Moral Economy for India's Forests?,* R. Jeffery and N. Sundar (eds.), New Delhi, Sage, pp.254–274

Voutchkov, N.V., 2004a: The ocean: A new resource for drinking water, *Public Works,* (**June**), pp. 30–33.

Voutchkov, N.V., 2004b: Tapping the ocean for reliable local water supply, *Asian Water,* (**December**), pp. 20–24.

Walker, A., 2003: Agricultural transformation and the politics of hydrology in Northern Thailand, *Development and Change,* **34,** pp. 941–964.

Wangnick, K., 2004: International Desalination Association worldwide desalting plants inventory report no. 18. Produced by Wangnick Consulting for International Desalination Association, Topsfield, MA.

WCD (World Commission on Dams), 2000: *Dams and Development: A New Framework for Decision-Making,* Earthscan, London, UK.

Webb, R.H., J.C. Schmidt, G.R. Marzolf, and R.A. Valdez (eds.), 1999: *The Controlled Flood in Grand Canyon,* American Geophysical Union, Washington, DC.

Whiting, P.J., 2002: Streamflow necessary for environmental maintenance, *Annual Reviews of Earth and Planetary Sciences,* **30,** pp. 181–206.

WHO and UNICEF (World Health Organization and United Nations Children's Fund), 2000: *Global Water Supply and Sanitation Assessment 2000 Report,* WHO and UNICEF, New York, NY.

WHO and UNICEF, 2004: *Meeting the MDG Water and Sanitation Target: A Mid-term Assessment of Progress,* WHO and UNICEF, Geneva, Switzerland, and New York, NY.

Wieriks, K., and Schulte-wulwer-Leidig, 1997: Integrated water management for the Rhine river basin: From pollution prevention to ecosystem improvement, *Natural Resources Forum,* **21,** pp. 147–156.

World Bank, 2003a: *Managed Flood Releases From Reservoirs: Summary Report for the Senegal River Basin,* Department for International Development Environment Research Programme, Project R7344, World Bank, Washington, DC.

World Bank, 2003b: *Private Participation in Infrastructure: Trends in Developing Countries in 1990–2001,* World Bank, Washington, DC.

World Water Commission, 1999: *A Water Secure World: Vision for Water, Life, and the Environment,* World Water Vision and World Water Council, Marseilles, France.

World Wildlife Fund, 2002: *Waterway Transport on Europe's Lifeline, the Danube: Impacts, Threats and Opportunities,* World Wide Fund for Nature, Gland, Switzerland.

Young, P., 1996: The "new science" of wetland restoration, *Environmental Science and Technology,* **30,** pp. 292–296.

Zedler, J.B., 2000: Progress in wetland restoration ecology, *Trends in Ecology and Evolution,* **15,** pp. 402–407.

Zinn, J., 1997: *Wetland Mitigation Banking: Status and Prospects,* CRS reports for Congress, 12 September 1997, Congressional Research Service, Washington, DC.

Chapter 8

Wood, Fuelwood, and Non-wood Forest Products

Coordinating Lead Authors: Nigel Sizer, Stephen Bass, James Mayers

Lead Authors: Mike Arnold, Louise Auckland, Brian Belcher, Neil Bird, Bruce Campbell, Jim Carle, David Cleary, Simon Counsell, Thomas Enters, Karin Fernando, Ted Gullison, John Hudson, Bob Kellison, Tage Klingberg, Carlton N. Owen, Neil Sampson, Sonja Vermeulen, Eva Wollenberg, Sheona Shackleton, David Edmunds

Contributing Authors: Patrick Durst, D.P. Dykstra, Thomas Holmes, Ian Hunter, Wulf Killmann, Ben S. Malayang III, Francis E. Putz, Patricia Shanley

Review Editors: Cherla Sastry, Marian de los Angeles

BOXES

TABLES

Main Messages

Strategies to address the impacts of forest product use on ecosystem health and human well-being are strongly affected by actions outside the forest sector. Some responses to problems related to forest products are achieving far more impact than others. Outcomes tend to be shaped as much or more by policies and institutions related to trade, macroeconomics, agriculture, infrastructure, energy, mining, and a range of other "sectors" than by processes and instruments within the forest sector itself. The objectives of some sectoral responses might be better achieved by non-forest measures; for example, land reform might benefit poor communities more than collaborative forest management. When considering responses, it is important to understand the degree to which each may be undermined or overridden by driving forces beyond the forest sector and the degree to which each can engage with and influence such forces.

Forest product trade tends to concentrate decision-making power over (and benefits from) forest management in the hands of powerful interest groups, rather than spreading it to include poorer and less powerful players. It "magnifies" the effect of governance, making good forest governance better and making bad forest governance worse. This threatens prospects for long-term sustainability. Both increased trade and trade restrictions can make impacts worse if underlying policy and institutional failures are not tackled. Trade liberalization can stimulate a "virtuous cycle" if the regulatory framework is robust and externalities are addressed.

International forest policy processes have made some gains within the forest sector. Attention now needs to turn to integration of agreed forest management practices in financial institutions, trade rules, global environment programs, and global security decision-making. The last decade saw many intergovernmental and civil society 'soft' policy responses to define sustainable forest management and to produce guidelines that could be interpreted locally. These responses included the United Nations Conference on Environment and Development, the International Tropical Timber Organization, and the Convention on Biological Diversity; they have both enabled much local progress and linked forest debates between local and global levels. Much critical intergovernmental policy work within the sector has been done. National policy and the interpretation and implementation of international policy at the national level are increasingly influenced by extra-sectoral policy and planning frameworks. Forest sector frameworks will have to adjust to more directly serve these wider goals or their influence will diminish.

Forest governance initiatives and country-led national forest programs are showing promise for integrating ecosystem health and human well-being where they are negotiated by stakeholders and strategically focused. Multilateral and bilateral accords to combat illegal logging, its associated trade, and the governance frameworks that might prevent it are becoming important venues for developing action plans and agreements. National forest programs are now being strongly promoted on the understanding that they follow a country-led approach. To be most effective, these programs should have multistakeholder involvement in forest decision making; be a means for cooperation, coordination, and partnership; promote secure forest resource access and use rights; involve research and traditional knowledge; and be built upon the study and policies on underlying causes of deforestation and degradation. In addition, they should include codes of conduct for business. They should have built in monitoring, evaluation, and reporting on their progress and effectiveness. To date, the new breed of national forest programs, although quite widespread, shows more promise than tangible results.

Local responses to problems of access and use of forest products have proliferated in recent years. They are collectively more significant than efforts led by governments or international processes but require their support to spread. A wide range of local responses have emerged "spontaneously" over the last decade, each with locally appropriate organizational forms and proven or potential impact in improving the contribution of ecosystems to human well-being and poverty alleviation. They often have a strong emphasis on gender equity. These include *campesino* forestry organizations in Central America, forest user groups in Nepal, the National Council of Rubber Tappers in Brazil, people's natural resource management organizations in the Philippines, and the Landcare movement in Australia. Policy frameworks could better assist such groups to build on what they are already doing and to enable new partnerships. Multistakeholder poverty-forests learning processes could be fostered with codes of conduct for supporting local initiatives. These could be integrated into national forest programs and poverty reduction strategies.

Government-community collaborative forest management can be highly beneficial but has had mixed results. Most collaborative management has promoted arrangements that maintain and even extend central government control. Local people generally have better legal access to forests and some have higher incomes but many have lost access and benefits. As a result the "co-management" response is shifting. Management increasingly involves not just a local group and the government but a range of stakeholders, and acknowledges overlapping systems of management and diverse interests. Local people are able to win more benefits for themselves where they have strong local organizational capacity and political capital to mobilize resources and negotiate for better benefits. NGOs, donors, federations, and other external actors also have a key role in supporting local interests. Where local groups manage their own forests without state intervention, however, they are not necessarily better off. Without government support, they often have difficulty implementing or enforcing their decisions. Improved formal access to forests has helped in many cases to protect a vital role of forests as safety nets for rural people to meet their basic subsistence needs. The benefits to be gained beyond the subsistence-level, however, are limited.

There is a widespread need for support to enable people in forest areas to secure their rights and strengthen their powers to negotiate fair division of control, responsibility, and benefits with other actors. Many governments have realized that they cannot secure a balance of public and private benefits from forests. Some have transferred control to private entities under lease agreements requiring public benefits to be guaranteed. Others have recognized, returned, or created rights for local communities to own forests, manage them, benefit from them, and bear certain costs and risks. Such communities often lack adequate recognition, powers, organization, capacity, and information to make use of these rights. Ways to cover the transaction costs of collective action are still sought. Checks and balances need to be in place to ensure that no group, including the local elite, controls benefits and decision-making. Processes are needed that acknowledge plural interests among the different groups and give special attention to livelihood needs of the poor. Culturally appropriate and technically sound cooperation between indigenous and non-indigenous organizations to reinforce natural resource management on indigenous lands is rare. This is much needed given the rapid growth in areas over which indigenous peoples have control.

Where information, tenure, and capacity are strong, small private owners of forests may deliver more local economic benefits and better forest management than larger corporate owners. Individuals and families have proven their potential to practice good forestry over the long term. However, many conditions are required for this to be effective. These include good knowledge, capacity to manage, market information, organization among smallholders to ensure economies of scale, long-term tenure, and transfer rights. Private ownership (or "family forestry") is common in Western Europe and in the southern United States, and is increasingly common in Latin

America and Asia. It may lead owners to assume a greater sense of responsibility and foster long-term thinking, prompting them to pursue sustainability, partly for risk reduction. Experience in Nordic countries and in many continental European countries shows the positive effects generated by information flow, education, and training and that it can be in the self-interest of owners to "do right."

Company–community partnerships can be better than solely corporate forestry, or solely community or small-scale farm forestry, in delivering benefits to the partners and the public at large. Companies may seek to improve long-term survival and competitiveness. Communities may prioritize gains such as secured land tenure or improved local infrastructure. Effects on equity and rural development are mixed. Financial returns often have proven insufficient to lift community partners out of poverty. Making the most of partnerships requires iterative approaches to developing equitable, efficient, and accountable governance frameworks (at the contract level and more broadly), raising the bargaining power of communities, particularly through association at appropriate scales, fostering the roles of brokers and other third parties (especially independent community development organizations), sharing the benefits of wood processing as well as production, and working toward standards that give equal opportunities to small-scale enterprises.

Public and consumer action has resulted in some important forest and trade policy initiatives and improved practices in some large forest corporations. Public and consumer action has been key in the development of forest and trade policy initiatives in "timber consuming countries" and in international institutions. The operating standards of some large corporations and institutions, as well as of those whose non-forest activities have an impact on forests, have been improved. Consumer campaigns have provided the underpinning for forest certification and served as a useful mechanism for bringing public attention to, and engagement with, issues that are often geographically remote. Such campaigns can potentially continue to play an important role both in maintaining public awareness of forestry issues and in encouraging improved forest management.

Forest certification has become widespread; however, most certified forests are in the "North," managed by large companies and exporting to Northern retailers. The early drivers of certification hoped it would be an effective response to tropical deforestation. There has been a proliferation of certification programs to meet different stakeholders' needs with the result that no single program has emerged as the only credible or dominant approach internationally. Many certification programs have developed group certification of small growers, or certification of regions with a single management regime. Stepwise approaches to certification, starting with legality verification, are now emerging and hold promise for wider applicability and adoption in tropical regions and Russia. National certification programs in Brazil, Malaysia, Indonesia, and elsewhere have increased adoption of this response in the "South."

Commercialization of non-wood forest products has achieved modest successes for local livelihoods but has not always created incentives for conservation. There has been significant growth in some NWFP markets. This has followed extension of the market system to more remote areas; increased interest in natural products such as herbal medicines, wild foods, handcrafted utensils, and decorative items; and development projects focused on production, processing, and trade of NWFPs. Few NWFPs have large and reliable markets. Those that do have tend to be supplied by specialized producers using more intensive production systems. Many other NWFPs are vital to the livelihoods of the poor but have little scope for commercialization. Such commercialization has achieved modest impacts for livelihoods through combinations of technical and capacity-building interventions to improve raw material production, processing, trade, and marketing, and through development of co-operatives, improved policy, and institutional frameworks. There are often problems, however, with stronger groups gaining control at the expense of weaker groups and with overexploitation of resources. Increased value does not automatically translate into effective incentives for conservation and can have the opposite effect.

Sustainable natural forest management in the tropics should be focused on a range of forest goods and services, not just timber, to be more economically attractive. Low-cost new technology has made a difference to some forest management functions. Diverse cultures can be expected to arrive at local solutions to securing both wood supplies and forest environmental services. While the "best practices" of global corporations are worthy of scrutiny, there is also much to be gained by exploring "what works" in traditional forest management and the work of local (small) enterprises. Since the early 1990s, considerable interest has developed in the application of reduced impact logging, especially in tropical forests, which lowers environmental impacts and can also be more efficient and cost-effective.

Development of farm woodlots and large-scale plantations is an increasingly widespread response to growing wood demand and as natural forest areas decline. Without adequate planning and management, the wrong growers, for the wrong reasons, may grow forest plantations in the wrong sites, with the wrong species and provenances. In areas where land degradation has occurred, afforestation may play an important role in delivering economic, environmental, and social benefits to communities and help in reducing poverty and enhancing food security. In these instances, forests and trees must be planted in ways that will support livelihoods, agriculture, landscape restoration, and local development. There is increasing recognition that semi-natural and mixed-species, mixed-age plantings can provide a larger range of products, provide "insurance" against unfavorable market conditions, reduce the effects and economic consequences of insect and disease attacks, harbor greater diversity of flora and fauna, contain the spread of wildfires, and provide greater variety and aesthetic value.

Fuelwood remains one of the larger outputs of the forest sector in the South. If technology development continues, then industrial-scale forest product fuels could become a major contributor to sustainable energy sources. Consumption of fuelwood has recently been shown to be growing less rapidly than earlier thought. This follows increasing urbanization and rising incomes as users switch to more efficient and convenient sources of energy. In some regions, including much of developing Asia, total fuelwood consumption is declining. Efforts to encourage adoption of improved wood burning stoves have had some impact in urban areas of some countries but little success in rural areas due to cultural and economic obstacles to their adoption. Recent attention to improved stoves has shifted from increasing efficiency of fuelwood use to reducing damage to health from airborne particulates and noxious fumes associated with the burning of wood and charcoal. In Northern regions, as renewable options gather more momentum and the technology becomes more fine tuned, it can be expected that dendro power options, using wood to fuel electricity generation, will become more competitive and investor friendly.

8.1 Introduction

This chapter assesses the impact on ecosystem health and human well-being of actions taken to influence the production and use of wood, fuelwood, and non-wood forest products (also known as non-timber forest products). These actions are responses to the ecosystem and human well-being conditions and trends associated with forest products that are assessed in MA *Current State and*

Trends (Chapters 9 and 21). The effectiveness of these responses is also assessed in relation to the possible scenarios in MA *Scenarios*.

The chapter discusses (1) driving forces of change in ecosystems that produce wood, fuelwood, and non-wood forest products, and the problems and opportunities they create; (2) interventions and actions to tackle the problems; (3) an assessment of selected responses; and (4) lessons learned. Other chapters in this and other MA volumes assess ecosystems and services closely linked to the provision of wood, fuelwood and non-wood forest products. Gaining a full picture of the state of forests and woodlands, the provisioning services of wood and NWFPs, and the human actions taken to address problems linked to wood and NWFPs requires looking at them as well. (See Chapters 5, 7, 15, 16, in this volume; MA *Current State and Trends,* Chapters 10, 13, 14, 17, and 24; and MA *Scenarios,* Chapter 10.)

8.1.1 Driving Forces of Change in the Ecosystems that Provide Forest Products

There is a range of strong proximate (or direct) drivers of change in the ecosystems that produce wood, fuelwood, and non-wood forest products. Some of these drivers are natural phenomena. Almost all interact in complex and unpredictable ways with human activities to influence the ability of wildlands, forests, plantations, and agricultural systems to produce wood, fuelwood and non-wood forest products.

Fire is the most immediate and dramatic agent of ecosystem change, and is an important process in many forest systems. Fire-affected forests have developed under characteristic fire regimes, ranging from frequent, non-lethal ground fires to infrequent, lethal, stand-replacing events (Pyne et al. 1996). Traditional societies used fire extensively to encourage the growth of food plants, to encourage new growth and attract animals for easier hunting, to control insects and disease, and to develop defensible space around villages (Pyne et al. 1996). Traditional forest management techniques stemming from Europe, combined with the fear of fire damage to wooden houses, fences, and settlements, and the desire to prevent the loss of valuable trees, led to increasingly effective fire prevention efforts in many forested areas, including North America, Europe, and Australia. These efforts, which often had the effect of removing fire as an ecosystem process, created significant ecological changes in many fire-adapted forests (Covington and Moore 1994). One result has been increasing concerns with forest health and the changing nature of wildfire, with greatly increased incidence of uncharacteristically large, intense, and severe fire events. These events, which may consume 5–20 times as much fuel as historical fires in these systems, can permanently damage soils (Giovannini 1994), alter ecosystem recovery rates (Cromack et al. 2000), create significant air pollution and human health impacts (Neuenschwander and Sampson 2000), and threaten significant population centers (NCWD 1994).

Both native and introduced *diseases, fungal infections, and insects* are important disturbance agents in forest ecosystems as well, and often these vectors interact with fire (Harvey 1994). While epidemics can occur in healthy forest ecosystems, most often in connection with periods of climate stress, they occur more frequently in forests where the vegetation is stressed and unhealthy due to overcrowding, lack of moisture or nutrients, or the invasion of ill-adapted species (NCWD 1994; Pyne et al. 1996). Large areas of uniform, mature forests in the boreal zone are similarly susceptible. Where trees have been killed by insect or disease epidemics, they are much more susceptible to large, uncharacteristic wildfires. Conversely, large areas of fire-killed timber are open invitations to insect epidemics that can then advance into adjoining unburned forests (Harvey 1994). These interrelated forest health problems are made worse in areas where forest management (or the lack of it) has created large, unbroken tracts of forest that lack age, structural, or species diversity (Sampson and Adams 1994).

Extreme weather, such as strong winds and floods can also be dramatic. Anthropogenic climate change is likely to exacerbate such weather events and to bring about more widespread shifts in the ecosystems that provide forest products (see MA *Current State and Trends,* Chapter 14). Unnatural changes such as simplification of ecosystems, dam building, and heavy pesticide use can exacerbate the natural forces described above.

Movements and migration of people, rising consumption of natural resources and land, changing human values, urbanization, and many other shifts in *human behavior* are having a huge impact on forests, farming, and use of wood. In many parts of the world, such as Southeast Asia and Africa, demographic change puts increasing pressure on land where wood is available or being produced. In wealthy countries, such as the United States and Japan, per capita demand for wood products continues to grow and already is many times greater than in poor countries.

Land and resource management practices are shifting as wood and related products are derived more intensively, such as through large-scale plantations of genetically cloned trees that grow faster than their natural ancestors. Ownership is shifting as large forestry enterprises continue to consolidate globally to achieve greater competitiveness through economies of scale, and as governments recognize traditional forest managers such as native peoples in South America. Protest is common from farmers groups, environmentalists, communities, and others over who owns and controls forest resources. Where violent conflict between political or ethnic groups occurs in rural areas, it often plays out in remote forests and woodlands. In several countries, governments and insurgent forces have used revenue from timber to finance military activities.

All of these proximate drivers of change are influenced by a range of underlying, interconnected processes; some of these and their possible impacts are examined here.

Globalization has impacts through trade liberalization, which changes the key centers of demand and production and enhances competition. This tends to concentrate wood and fiber production on intensive, controllable, and accessible land (though ownership of the land may be disputed by local communities) where costs of production are lower. Fewer, larger companies increasingly control a larger portion of wood and fiber production, processing, and trade. Products are increasingly standardized in form and quality. Meanwhile there is globalization of knowledge and advocacy about what is "good" or "responsible" production and awareness of issues associated with wood.

Governments still own much forest land, but *privatization* of forest resource ownership, fiber production, and forest management services, such as third-party certification, are dominant trends. This may improve the efficiency of production and the quality of products, but it also can result in declining access to resources for some of the world's poorest people.

Decentralization of authority and responsibility to local government, communities, and the private sector is common in many parts of the world, including in large forest-rich countries such as Indonesia and Brazil. This shifts power closer to the people most affected by local resource use and might improve management where local institutions are adequate and accountable.

Changing patterns of wood consumption are emerging along with new technologies, fashions, and substitutes. Engineered and more highly designed wood products are replacing simple solid wood, resulting in lower resource intensity for some uses such as home construction. Nonetheless, fuelwood continues to be the major source of energy for many poorer and rural families. The geography of consumption is also shifting as huge new import markets emerge in China and India, set to rival Europe and the United States as sources of growing demand.

Technology is changing the way wood is produced, processed, and used. Biotechnology is given increasing emphasis in commercial plantations with cloned trees to standardize production and quality and to increase growth rates. Much experimentation is done to develop new generations of "super trees" using genetic modification. These modified trees are being criticized by interest groups concerned about possible environmental impacts. Wood engineering is allowing the use of more species and smaller pieces of wood in processing. Wood-fiber-gasifying energy generators are also being developed and could one day produce large amounts of renewable electricity using trees harvested from fast-growing plantations.

Food production and processing have a large impact on forests and wood production. The dynamics that affect food production in turn affect the forest–farm interface geographically, economically, and socially.

Stakeholder values and opinions are changing. Environmental and social responsibility is increasingly mainstream and calls for pro-people and pro-environment approaches are ever stronger. There is also pressure for greater transparency of how forest and forestland are administered and managed. Increasingly there are expectations of multistakeholder approaches to decision-making by governments and increased partnership with civil society by business.

Yet governance systems that can manage forest stakeholder values effectively and equitably are often weak where their need is great. Where there is limited provision of social services, weak justice systems, and slow economic growth, the interests of the few come to dominate the many and there is little incentive for the local population to be loyal to national government. In some such contexts, violent conflicts have emerged.

8.1.2 Problems and Opportunities Created by the Driving Forces of Change

Ecosystems and human well-being face a range of problems as a result of the driving forces described above. The area of provisioning ecosystems is declining due to deforestation, desertification, and forest degradation. There is also declining quality of ecosystems (productivity, diversity, standing stock quality, and health support services), and increasing vulnerability of ecosystems (increase in fires, climate change, and pathogens). Resource extraction and management technologies for wood, fuelwood and non-wood forest products can have impacts on biodiversity, water quality, carbon storage, and cultural values.

Stakeholder equity problems are widespread. There is often inequitable access to wood, fuelwood and non-wood forest products; poor sharing of costs and risks of production; and conflicts and mistrust between stakeholders. Conservation efforts in some places creates burdens for others; for example, China is currently protecting its own natural forest and importing much wood from Russia and Indonesia, which, given forest governance weaknesses in those countries, leads to excessive and illegal harvesting.

Since many of the driving forces of change originate in processes beyond the forest sector (extra-sectoral), many of the problems in the use of forest products stem as much or more from extra-sectoral policies and institutions—trade, structural adjustment, poverty reduction, debt, agriculture, infrastructure, energy, mining—than from processes and institutions within the forest sector itself. Such extra-sectoral policies and institutions often override or undermine priorities negotiated by forest stakeholders.

Further problems with the current policies and institutions that constitute forest governance are abundant (WCFSD 1999; IPF 1996). These include the following:

- Forest rights are often insufficiently well negotiated, established, and legally and institutionally backed-up for effective and equitable forest management.
- Policies and investment conditions sometimes create perverse effects and make it impossible to tackle problems and realize opportunities associated with changing driving forces. Elsewhere policy "inflation" has occurred—with an excess of international precepts and lack of real capacity and mechanisms to deliver local benefits.
- Decentralization is often incomplete and coordination of institutional roles insufficient to support effective and equitable forest management.
- Smaller forest enterprises, fuelwood-dependent stakeholders, and users and managers of non-wood forest products, many very poor, are often "invisible" to policy processes (their values and forest management practices are ignored or misunderstood).
- Information about specific wood-producing ecosystems—including their location, extent, capability, and vulnerability—is inadequate, and forest research capabilities are weak.
- Corruption and weak regulation or enforcement lead to poor forest management in some places.

In addition, there are problems linked with the markets. Many pro-sustainability approaches are unviable financially. Viable approaches are not always socially and environmentally responsible and market prices often do not reflect social and environmental values, a situation worsened by competition between producers.

Despite these potential problems, there are also opportunities arising from anthropogenic driving forces. Technology allowing concentration of fiber and fuel production on small areas of land has the potential to release other areas for environmental and livelihood purposes, though this depends heavily on other factors. There is potential for cash-poor producers to access high-value markets as market information improves. There is greater transparency to forest resource information and strengthening of government-led reporting such as through the various criteria and indicators processes. Knowledge of sustainable practices is now being shared more easily among groups and nations. Decentralization offers opportunities to match wood production with local livelihood needs and constraints.

8.2 Overview and Selection of Responses

In the past, governments made the majority of responses to the issues summarized above through laws and regulations covering the ownership, management, and use of forests; the harvesting, transport, and trade of forest products; and the extraction and use of income from public lands. These responses were designed to shift the balance between public and private benefits toward the public end of the spectrum (for example, environmental services for public benefits, rather than wood production for private ends).

In the last three decades, a richer range of responses has emerged that spans a spectrum from "pure" public regulation to "pure" private, voluntary approaches. Across this spectrum, market-based approaches have emerged to allocate costs and benefits. Some nongovernmental responses, such as voluntary forest certification, are proving to be just as effective as state regulations.

Some approaches described here as "responses" are explicit policy instruments and intervention programs; others can be better seen as "spontaneous" local reactions and social movements.

Not all responses to change in the ecosystems that produce wood, fuelwood, and non-wood forest products are assessed here. Rather, fifteen responses have been selected for investigation on the basis of the following criteria: whether the response attempts to address a major problem or opportunity; whether it evokes political interest or contention; whether a major investment has been made in it; and whether there are strong indications of positive impact. The response options fall into the following four main types:

- *Multistakeholder and extra-sectoral policy processes.* These include international forest policy processes and development assistance; trade liberalization; and national forest governance initiatives and national forest programs.
- *Rights to land and resource management.* These include direct management of forests by indigenous peoples; collaborative forest management and local movements for access and use of forest products; small-scale private and public-private ownership and management of forests; and company–community forestry partnerships.
- *Demand-side, market-driven, and/or technological responses.* These include public and consumer action; third-party voluntary forest certification; wood technology; and commercialization of non-wood forest products.
- *Land management institutions, investment, and incentives.* These include natural forest management in the tropics; forest plantation management; fuelwood management; and carbon management.

The following sections assess the various response options in terms of their impact on ecosystem health and human well-being; the final section summarizes lessons learned.

Truly extra-sectoral responses, which have clearly improved impacts of forest product use on ecosystem health and human well-being in mind, are rare. Trade is one arena in which such responses are visible and these are discussed below, with some additional examples given in Chapter 15. Most of the responses discussed have an extra-sectoral dimension—relying on engagement with driving forces beyond the forest sector—and should be judged in part by their effectiveness in this.

A number of other important options are not addressed here. For example, *importing* wood is an option for an individual country that cannot produce wood cost-effectively. This shifts any ecosystem problems to another country, but is positive if comparative advantage can be realized. *Producing substitutes* for wood products (such as metals, plastics, concrete, and non-wood fibers) results in a different set of ecosystem issues (often agricultural, as in the case of non-wood fiber); the major drawback is that many substitutes may neither invest in renewable resources (the bulk of plastics manufacture is petroleum-dependent) nor exhibit the same degree of concern for ecosystem services that the various wood-producing sectors are increasingly doing. These alternatives are also often more energy and water intensive than wood (Hair et al. 1996; Koch 1991; Meil 1994).

Some key responses are omitted here because they are covered in other chapters (for example, protected areas, which are covered in Chapters 5 and 15). Some new "paradigms" gaining significant currency, such as ecosystem approaches and landscape restoration, are not included because their impacts have yet to become clear. Single powerful institutional frameworks, such as the World Bank's forest strategy and policy, are not covered directly but are treated indirectly where their influence is strong. Other key arenas of problem and opportunity in forest product impacts on ecosystem health and human well-being seem to lack major responses. For example, concerted initiatives to address these links from the standpoint of forestry labor are difficult to identify.

Implementation of the full set of responses assessed here is not the norm in the forest sector. Indeed some places demonstrate hardly any of these responses. Nevertheless, each of the selected responses has substantial and generally growing significance globally for the way wood, fuelwood, and non-wood forest products are developed and used.

8.3 Multistakeholder and Extra-sectoral Policy Processes

8.3.1 International Forest Policy Processes and Development Assistance

A host of international processes and initiatives engage with forest issues. Many are intergovernmental, some are civil society approaches, and others are driven by the private sector. They can be clustered in four groups: forest, environment, trade, and development policy.

8.3.1.1 Forest Processes

The core international policy process on forests includes the debate, negotiations, and decisions stretching from the 1992 Rio Earth Summit to the current United Nations Forum on Forests. UNFF's objective is to promote the management, conservation, and sustainable development of all types of forests and to strengthen the long-term political commitment to this end. It has been catalytic in developing a number of distinct forestry response options, which are considered elsewhere in this chapter. It has achieved the following (Bass 2003; Sizer 1994):

- kept forests on the international agenda, especially in the context of sustainable development;
- provided opportunities for collaboration and lesson learning at inter-sessional meetings on a wide range of technical and some cross-cutting issues;
- promoted consensus around a set of U.N. Forest Principles and identified 20 main voluntary "Proposals for Action" (incorporating a total of 270 detailed proposals that some countries find hard to interpret and thus implement);
- helped define and give legitimacy to country-led national forest programs as the main means to implement the Proposals for Action;
- developed sets of criteria and indicators for sustainable forest management that have provided a common language that has brought stakeholders closer together, but allowed national and local differences in interpretation. These have influenced the development of voluntary forest certification;
- sought to improve collaboration and coordination with other policy processes and international organizations under the Collaborative Partnership on Forests; and
- promoted NGO involvement in U.N. processes.

However, UNFF also has weaknesses. To date, it has:

- failed to reach agreement on the voluntary monitoring of implementation in ways that could provide evidence of direct impact;
- remained very sectoral, and has struggled to make any significant progress on key cross-cutting issues (finance, trade and environment, technology transfer).
- failed to achieve a consensus on the nature and justification for a legally binding instrument but will continue to absorb time and energy in an attempt to do so; and

- remained excessively dominated by governments, despite pioneering NGO involvement within U.N. processes establishing a multistakeholder dialogue and the Collaborative Partnership on Forests.

8.3.1.2 Environment Process

Of the key environment processes and initiatives, the Convention on Biological Diversity, the United Nations Framework Convention on Climate Change, and the Global Environment Facility have been most influential to date. The United Nations Convention to Combat Desertification is starting to have an impact through national action programs. The main impact of the CBD has been the development of national biodiversity strategies and action plans; its revised work program on forest biodiversity has potential, but its ambition far exceeds the resources available for its implementation. CBD's benefit sharing objective has been of great interest to many developing countries, but it has generated difficult debates about intellectual property rights and trade that go well beyond biodiversity. UNFCCC introduces the subject of markets for environmental services. The wide array of experiments to test market approaches for provision of watershed services, biodiversity, and carbon are creating a body of understanding that is reaching an ever-wider audience.

8.3.1.3 Trade

The International Tropical Timber Organization is a unique commodity agreement that balances concern for improving trade with conserving the resource base on which trade depends. It has been effective in its purpose of facilitating discussion and international cooperation on the international trade and utilization of tropical timber and the sustainable management of tropical forests (Poore 2003). ITTO has achieved the following:

- It was influential in the 1980s and early 1990s when it was effectively the only intergovernmental forum on forest issues.
- It captured public and political attention with its assessment of the sustainability of tropical forest management.
- It made a significant contribution to the concept of criteria and indicators.
- It developed a series of guidelines on management practices that has been well used.
- It has the potential to contribute to the development of trade in marketable environmental services of tropical forests.

Concern with forest law enforcement, governance and trade gathered pace in the late 1990s, when the scale and impacts of illegal logging, and the power of some forest industries to run amok, became better understood. The Group of 8 and other international forums took up the issue. The forest law enforcement, governance, and trade initiatives now under way address the governance, policy, and market failures that cause and sustain illegal logging and associated trade. The FLEGT processes took advantage of the political space created by an East Asia Ministerial Conference and the African ministerial process (where exporting countries spoke with a frankness not heard before, and importing countries acknowledged their role in sustaining demand for illegally logged timber). In addition to East Asia, FLEGT processes are also under way to varying degrees in Europe and Africa.

New multistakeholder regional initiatives are also emerging that hold promise to better address governance and enforcement issues. These include the Asia Forests Partnership (Sizer 2004). It is too early to assess the utility of these approaches.

As these processes evolve, they are more likely to need to grapple with more aspects of governance (Colchester et al, 2004). National forest programs are potentially the ideal integrating framework at national level. Internationally, interventions are likely to be needed from agencies previously little linked to forest issues—for example, the United Nations Security Council being called upon to take action on conflict timber.

8.3.1.4 Development

International development assistance for forestry has passed through four different phases, with considerable overlap, over the last 40 years: industrial forestry, social forestry, environmental forestry, and sustainable management of natural resources. Recently forestry assistance entered a fifth phase, framed by the new poverty agenda that emerged from ideas about how to reduce poverty based on providing opportunity (growth), empowerment, and security. Forestry assistance now links the United Nations Millennium Development Goals, with poverty reduction foremost among these, with a set of mechanisms and instruments for delivering aid that includes poverty reduction strategy papers, medium-term expenditure frameworks, sector-wide approaches, and direct budgetary support. The development community is still adjusting to these new changes. There has been a distinct move away from discrete sectoral projects and a sharp decline in related funding from the peaks reached during the early 1990s.

This decline has been particularly marked in rural development and within forestry. Poverty reduction strategies involve political choices. Where a national consensus is hard to reach and where urban biases exist, the voices of the rural poor are heard less distinctly. SWAPs favor social sectors where it is public expenditure that largely determines outcomes and where institutional relationships are manageable. Productive sectors and crosscutting themes like forestry do not sit comfortably with the SWAP model. Direct budgetary support places responsibility for choice of development strategy and sectoral allocation of resources in the hands of developing countries themselves.

Response options within the new poverty agenda must demonstrate that they contribute to growth (including reduced vulnerability), empowerment, and security. This will take many forms, including:

- helping to understand and express how forest-related interventions can be supportive of wider policy objectives;
- supporting institutional change in public sector organizations in ways that contribute to wider social and economic goals;
- scaling up community forestry as part of wider livelihood strategies, in ways that stress political and legal change as much as local forest management arrangements;
- helping community–company partnerships respond to market opportunities; and
- working with a range of partners to tackle illegal logging and associated trade.

8.3.1.5 Policy Challenges

Much critical intergovernmental policy work within the sector has been done. Short-term priorities are reaching agreement on how countries should monitor, assess, and report on forests and reaching a conclusion on a legally binding instrument. More attention should now be focused on policy implementation at the regional, eco-regional, and national levels. It is easier for countries to identify issues of common interest at the regional and eco-regional levels; in many cases, institutions or processes are available that can be used.

More attention is needed in the integration of agreed forest management principles and practices in multilateral financial institutions, trade rules, and the Global Environment Facility. The U.N. Security Council should play its part in curbing trade in conflict timber. National policy (and the interpretation and implementation of international policy at the national level) will be increasingly influenced by these and other extra-sectoral policy and planning frameworks. Forest sector frameworks will have to adjust, to more directly serve these wider goals, or their influence will diminish.

8.3.2 Trade Liberalization

Trade in forest products is growing rapidly, involves every country in the world, and is worth about US$330 billion annually. Conventional trade theory predicts economic benefits to both trading partners, which is broadly observed in forest product trade (Sedjo and Simpson 1999; USTR 1999). Three problems complicate matters: unanticipated levels of benefits and costs due to market imperfections; inequitable distribution of those benefits and costs; and disputed values ascribed to different types of benefits and costs, especially between market and non-market values (World Bank 2002; IIED 2003). Different interest groups perceive the relative importance of these problems differently, and consequently promote different initiatives to solve them.

8.3.2.1 Initiatives to Influence Forest Products Trade

Trade liberalization is the dominant economic paradigm; however, when non-tariff measures and effects of subsidies are taken into account, the net trend internationally is probably slightly toward increased protection rather than liberalization (Rice et al. 2000; Bourke 2003). In addition to forest products trade policy, and macroeconomic policies affecting interest rates, stability, and risk, significant effects are created by other policies. Logging bans displace logging problems to other locations and countries rather than solving problems (Brown et al. 2002). Forest tenure is affected by privatization, and decentralization measures are creating new trade players (White and Martin 2002). Sectors competing for inputs or land dictate whether there are any forest products to trade. Policies that support large-scale agriculture have had particularly significant effects (Hyde forthcoming).

There are more than a hundred regional agreements that affect forest trade in some way (IIED 2003). Regional trade agreements are the most prominent of these, including Asia-Pacific Economic Cooperation, the North American Free Trade Agreement, and the European Union. Regional mechanisms to control illegality in forest trade have also begun to receive support and provide platforms upon which to develop new ideas (see earlier discussion). Internationally, influence over trade is dominated by the World Trade Organization negotiations, which have not installed pro-forest principles and clarified forest trade uncertainties. Other international agreements influencing forest trade include those on forestry, climate change, trade in endangered species, biodiversity conservation, core labor standards, guidelines for multinational enterprises, and combating bribery.

Voluntary initiatives (demand-side processes such as certification and labeling, supply chain management and product campaigns; and supply-side initiatives such as environmental management systems, investment guidelines, and corporate citizenship) have made significant headway in recent years but their influence on trade is still relatively small.

8.3.2.2 Impacts

Trade liberalization and initiatives to influence its course in the forest sector have produced several strong trends:

- increasing consumption and production, and increasing trade as a percentage of production. These trends are particularly pronounced in developed countries and for highly processed products;
- a continuing strong segregation of trade into regional trade flows (Wardle and Michie 2001; Rytkonen 2003); and
- a transition of tropical countries from net exporters to net importers of wood (IIED 2003).

In terms of the maturity of markets, trade with regions in the early stages of market development increases unsustainable harvesting from open access and mature natural forests. It is only at the mature stage of market development that good forestry practice becomes economically attractive in comparison with agricultural land values and the cost of protecting property rights (Hyde forthcoming).

For most developing markets, existing regulatory capacity is too weak to control external demands on the resource, and trade liberalization is likely to result in an increase in unregulated logging (Sizer et al 1999). Where windfall resource rents occur, public sector corruption is often rife (Ross 2001; Wunder 2003). However, there is strong evidence that, where there is strong regulatory and institutional backup, reducing trade restrictions reduces public sector corruption (Richards et al. 2003). In some situations, trade liberalization may not bring about a real reduction in corruption, merely a change in the pattern of winners and losers.

Trade liberalization is usually promoted within a package of measures, and its impact depends on what else is in the package, such as state downsizing, decentralization, deregulation, privatization, concession bidding and forest taxation, and the capacity and will of the government to implement it. The way in which trade policies interact with these changes determines whether they improve or reduce policy and institutional capability for sound forest management (Seymour and Dubash 2000; Tockman 2001).

Recent analysis has concluded that the impacts on policies and institutions of trade liberalization are positive where there are robust policies and institutions (a virtuous cycle) and negative where they are weak (a vicious cycle). Trade appears to be a magnifier of *existing* policy and institutional strengths and weaknesses rather than a major driver of change (Anderson and Blackhurst 1992; Ross 2001; IIED 2003).

The forest products sector is less concentrated than many other industrial sectors, although in developing countries concentration is much more marked. There is a clear trend toward greater involvement of transnational companies in the sector, particularly for pulp and paper products, but their importance varies. Transnational companies have played a major role in the exports of tropical timber in West and Central Africa, and Southeast Asia, but in countries such as Brazil and the Philippines, they have not been a major factor driving development of the sector (ITTO 2002). Transnational companies may generate wealth through trade, which may provide the basis for improved policy and institutional frameworks in the forest sector (Young and Prochnik 2004). On the other hand, there is a tendency for more exploitative transnational companies to target weaker governance structures (Sizer and Plouvier 2000).

8.3.2.3 Policy Challenges

A range of policy and practice measures have been identified as priorities for improving the impact of trade on forest management (IIED 2003), including:

- revise distorted agricultural trade policies and improve regional development policies (this will have greater beneficial impacts on forestry practice than changes in forest or forest trade policies);

- improve engagement of "underpowered" groups in trade policy decision-making;
- ensure that institutional strengthening occurs before trade liberalization;
- require cost internalization as well as liberalization, and consider the case for protection to achieve the social component of sustainability;
- link trade to improved property rights;
- install policies for equitable and efficient allocation of forest land;
- develop graded incentives for value-added processing that are more closely linked to sustainable forest management;
- prevent tariff escalation on processed products; and
- promote foreign direct investment in responsible forest business.

The most effective way to improve the beneficial impacts of trade is to link trade liberalization to improved, impartially administered property rights—either nationally through decentralization or locally through the empowerment of local and community institutions (IIED 2003).

8.3.3 National Forest Governance Initiatives and National Forest Programs

The United Nations Food and Agriculture Organization estimates that about 190 countries are currently involved in national forest planning of various kinds. There have been two main sources of multistakeholder policy reform processes in recent times: responses to pressure from local levels and responses to international opportunity or to international soft law.

8.3.3.1 National Governance Reform Initiatives Affecting Forests

Significant policy change with many stakeholders involved has emerged from initiatives to support participatory forestry at the local level. Since the early 1970s, many projects have been based, often with donor support, on the notion that local people should be able to participate more in forestry development. The best of these projects subsequently resulted in increased local responsibility for forest resources, improved local rights, increased bargaining power of local actors at the national level, and multistakeholder policy reform as other actors recognized the imperative for it and came to the negotiating table. The greatest positive effects were probably felt in countries of low forest cover such as Nepal and Tanzania, where, as the capacity of local people to manage forests was given greater policy support, the condition of the resource also improved (Brown et al. 2002).

In Europe and North America, experience has been different. Reform has also been generally stimulated by business and environmental agendas. Differences in national government styles and cultures, and in the strength of business and civil society networks, have produced a wide range of national forest planning processes.

Translation to the national level of opportunities and agreements stemming from international policy dialogue has stimulated various approaches to forestry reform (Mayers and Bass 2004). These include the following:

- *National Forestry Action Plans.* National forestry action plans called for by the international Tropical Forests Action Plan were launched by FAO, UNEP, the World Bank, and the World Resources Institute in 1985. Never before had there been such multi-country attention aimed at benefiting tropical forests. Many donors and larger NGOs supported the initiative and at one point more than one hundred countries were implementing or developing national forestry action plans within the framework of TFAP. The TFAP could be characterized as a top-down, quick but comprehensive fix to the perceived tropical forest crisis, the perception being promoted by NGO and media concern about "deforestation." TFAP set a "standard" for a balanced forest sector for the next decade and defined a new liturgy for forestry aid planning. But in practice it resulted in fewer improvements than had been hoped. TFAP was not able to challenge the inequities and perverse policies that underlay deforestation, and then to build the necessary trust between governments, NGOs, local people, and the private sector. Its standardization within a global framework and the exigencies of the aid system that supported it meant that the TFAP did not adequately recognize diverse local perceptions, values, capacities, and needs. Because of such weak links between causes of problems and identified desired impacts (a persistent problem in the forestry context), TFAP in effect contained few measures that could be reasonably expected to achieve its objective of reducing deforestation (Shiva 1987; Sizer 1994).
- *Forestry Master Plans.* Forestry master plans were led mainly by the Asian Development Bank (with Finland as a frequent co-donor) and consisted of extensive studies of all parts of the sector. The studies were not very participatory nature, and they constituted the basis for a forest policy and investment plan principally directed at commercial functions. Agreement was reached with TFAP that a country could be involved with TFAP or forestry master plans but not both. The countries that used forestry master plans included Sri Lanka, Nepal, Bangladesh, the Philippines, Thailand, Pakistan, and Bhutan.
- *Forestry Sector Reviews.* Forestry sector reviews were required by the World Bank in a range of countries to qualify for sectoral support. Their format was similar to that of the forestry master plans. Countries that developed forestry sector reviews included Kenya, Malawi, and Zimbabwe. The long lists of policy prescriptions contained in forestry sector reviews were largely ignored once support had come and gone.
- *National Environmental Action Plans.* National environmental action plans were undertaken from the mid 1970s to the early 1990s at the behest of the World Bank; in some countries, they overlapped with forestry sector reviews. They were effectively a form of conditionality and today have been eclipsed by comprehensive development frameworks and poverty reduction strategy papers.
- *National Conservation Strategies.* National conservation strategies were popular in the 1970s and early 1990s when about 100 countries prepared them, many with technical support from IUCN and some showing creativity in both multistakeholder processes and practical linkage of environment and development. While many fell by the wayside, a few (such as the Pakistan National Conservation Strategy) are now providing a valuable platform for addressing economic growth and poverty alleviation.

Several initiatives stem from the UNCED 1992 multilateral environmental agreements and have a mixed record in influencing national forestry planning, including the following (OECD and UNDP 2002):

- *National Biodiversity Strategies and Action Plans.* National biodiversity strategies and action plans were stimulated by the requirements of the 1992 Convention on Biological Diversity. About 70 countries have completed them, some supported by the GEF. They often lack analysis of forestry's use of biodiversity as well as integration with other plans and strategies. A few highly participatory NBSAPs have considerable momentum and potential impact on forestry decision-making, for example in India and Guyana.

- *National Action Programs.* National action programs to combat desertification were a response to the 1994 Convention to Combat Desertification. Many dryland countries have developed NAPs, with 50 of them receiving funding from UNDP's Office to Combat Desertification and Drought. A few national action plans have analyzed and stimulated actions in forestry. They vary greatly but have tended to be developed by ministries of environment with only weak links to key processes such as decentralization and land reform that may have major effects on land use and desertification.
- *National Communications.* Annex 1 parties to the UNFCCC must submit periodic national communications to the UNFCCC Secretariat reporting on their actions to address climate change. By April 2003, some 100 developing countries had submitted such reports, with only a few covering carbon source and sink dimensions of forests.

Despite their best endeavors, the net effect of the multilateral environmental agreements is at best to provide a source of ideas to national-level debate about forests. They do not provide an integrated legal regime that views forests, and those that depend on them, in a holistic way. Countries both poor and wealthy are thus generally able to escape from their commitments. Two integrating frameworks currently holding sway in international debates aim to have more power at the national level:

- *National Sustainable Development Strategies.* National sustainable development strategies are to be adopted by all governments following the 1992 Earth Summit. The 2000 Millennium Development Goals were signed by 147 heads of state, accompanied by targets, including to "integrate the principles of sustainable development into country policies and programs and reverse the loss of environmental resources." There are few national sustainable development strategies, although the recent development of guidance and lessons for practitioners (OECD and UNDP 2002) may stimulate more.
- *Poverty reduction strategies.* Poverty reduction strategies were initially required by the IMF and World Bank as a basis for access to debt relief in highly-indebted poor countries. Poverty reduction strategy papers have been required by all countries supported by the International Development Association since July 2002. Interim poverty reduction strategy papers (I-PRSPs) are road maps to full PRSPs. As of April 2003, 26 full PRSPs and 45 I-PRSPs had been prepared. Bilateral donors are also increasingly subscribing to poverty reduction strategies and they have thus emerged as a central determinant of the development agenda in many countries. The recognition of forests as a development asset has so far been limited in many poverty reduction strategies. Of the 11 PRSPs and 25 I-PRSPs in sub-Saharan Africa, 74% touched on forestry issues but almost none were convincing about forests–poverty links and forests' future potential (Oksanen and Mersmann 2002).

8.3.3.2 National Forest Programs

National forest programs are now being strongly promoted on the understanding that they follow a country-led approach, rather than an international program or precept in the style of the TFAP (UNFF 2002; FAO 2004). The notion of NFPs was developed by the international Forestry Advisers Group (an informal group of aid agency forestry advisers), adopted by FAO (FAO 1996), then endorsed by the Intergovernmental Panel on Forests (Six-Country Initiative 1999).

All countries that have taken part in U.N. forest policy dialogues have adopted the requirement for a national forest program. It is consensus-based soft international law. Agenda 21, the UNCED action plan (UNCED 1992), invited all countries to prepare and implement national forest programs and stressed the need to integrate these activities within a global, inter-sectoral, and participatory framework.

The post-UNCED intergovernmental negotiations on forests stress the role of NFPs, and the current United Nations Forum on Forests action plan commits countries to pursuing NFPs (UNFF 2002). Regional approaches to pushing NFPs are also beginning in Europe (MCPFE 2002). Meanwhile the European Union requires countries to have NFPs or their equivalent in order to receive forest subsidies (Glück et al. 2003). The NFP concept currently promoted at the international level (FAO 2004; World Bank 2002) puts particular emphasis on the following:

- multistakeholder involvement in forest decision making;
- means for cooperation, coordination, and partnership;
- secure access and use rights;
- research and traditional knowledge;
- forest information systems;
- study and policies on underlying causes of deforestation/degradation;
- integrating conservation and sustainable use, with provisions for environmentally sensitive forests, and for addressing low forest cover;
- codes of conduct for the private sector; and
- monitoring, evaluating, and reporting on NFPs.

Although there is probably no example of a contemporary NFP that has achieved optimal systems for all of the above, Malawi, Uganda, Brazil, Costa Rica, Vietnam, India, Finland, Germany, and Australia are leading the way (Bird 2002; Humphreys 2004; Mayers et al. 2001; Savenije 2000; Thornber et al. 2001). However, it is too early to see significant results. Many NFPs were judged to be "stalled," due to lack of institutional, human, and financial capacity, as well as lack of adequate policies, poor institutional co-ordination, and deficient mechanisms for public participation (FAO 2004). Widespread agreement on the need for "country-driven, holistic" processes is not matched with implementation.

If NFPs are to succeed, they need to avoid the mistakes of many NFAPs, FMPs, FSRs and the like that remained exercises on paper only. They failed to catalyze the detailed actions expected of them, in general because they failed to engage with political and economic reality to show not only what needs to change, but also how it can change, and how such change can be sustained.

8.3.3.3 Policy Challenges

Experience suggests that the best hope lies in developing local *processes* and *systems* that bring together the best that exists locally, and filling gaps where needed with the help of international thinking (Mayers et al. 2001; OECD and UNDP 2002). These processes include the following:

- *political processes* that install and maintain forestry's potential and NFP priorities at a high level, and provide the means to revise policies;
- *participation systems* that enable equitable identification and involvement of stakeholders, including previously marginalized groups, and create space and responsiveness for negotiating, vision, roles, objectives, and partnerships;
- *local-benefit "screening" processes* that ensure that the forest sector keeps working to optimize its contributions to poverty-reduction and local livelihoods;

- *information and communication systems* that generate, make accessible, and use interdisciplinary research and analysis; form clear baselines; and get plans well communicated with strong "stories";
- *monitoring systems* that can pick up and communicate the key changes in forests and human well-being;
- *financial systems* that generate and manage adequate resources and ensure investment conditions, internalize externalities, and promote cost-efficiency;
- *human resource development systems* that promote equity and efficiency in building social and human capital, with an emphasis on holding on to tacit knowledge and promoting innovation;
- *extra-sectoral engagement processes* that put synergies and potential conflicts with other sectors and macro-plans at the heart of thinking and action; and
- *planning and process management systems* that demonstrate efficiency (strategic, not overly comprehensive actions with realistic timeframes), transparency, accountability, and therefore legitimacy in decision-making.

8.4 Rights to Land and Resource Management

8.4.1 Direct Management of Forests by Indigenous Peoples

Direct management of forests by indigenous and traditional peoples occurs in its purest form in two utterly different institutional contexts: where states exercise little or no effective control over territory, creating space for autonomous management of forest resources, or where a highly sophisticated state with an indigenous population acknowledges significant sovereignty to native polities. Canada and New Zealand, and to a lesser extent Australia and the United States, are examples of the latter. The former scenario is almost exclusively restricted to the tropical world. Most indigenous peoples inhabit a more ambiguous political and institutional landscape, where land tenure can be restricted to usufruct, conceded but heavily regulated, or denied altogether. Even where sovereignty is formally conceded to indigenous peoples, such as in Canada, its recognition in practice may be weak (Colchester 2004). In these contexts, complex interactions with governments and a surrounding non-indigenous civil society determine natural resource management, including management of forests (Redford and Mansour 1997).

8.4.1.1 Impacts of Forest Management by Indigenous Peoples

The processes of colonization and globalization have affected indigenous peoples for centuries, and provoked major changes for most of them, with transitions from permanent to shifting agriculture and back again, geographical displacement, rapid modifications in trading patterns, and economic articulation with the outside world. This universal historical experience is contrary to the many mythic representations of indigenous expertise in natural resource management as linked to very longstanding occupation of a particular natural environment.

The defining characteristics of natural resource management by indigenous peoples across cultures are flexibility, versatility, adaptability to change, and heavy investment in the training of resource management specialists with broad expertise. Indigenous natural resource management tends to be geared toward broadly based livelihoods composed of the simultaneous exploitation of multiple ecological niches and processes. Its defining characteristic is the ability to adapt effectively to the many externally forced changes of habitat and economy that history has imposed upon indigenous peoples.

For forests, this has usually involved a paradoxical combination of intensive but diffuse management—intensive in the sense that a wide variety of ecological processes in forests (succession, species composition, forest structure) are heavily manipulated by indigenous peoples, but diffuse in the sense that this manipulation is so geographically widespread that it often becomes difficult to draw the boundary between anthropogenic and natural forests. This has two common consequences: a mimicking of natural processes through cultural means, which underlies the greater integrity and functionality of forests in indigenous areas, and difficulty in handling specialization and intensification. This has become a perennial problem in sustainable development projects involving forest management or natural resource management in general in indigenous areas.

Indigenous control of traditional homelands is often presented as having environmental benefits by indigenous peoples and their supporters, although the dominant justification continues, rightly, to be based on human and cultural rights. While little systematic data yet exists, preliminary findings on vegetation cover and forest fragmentation from the Brazilian Amazon, where this work is most advanced, suggests that the creation of an indigenous area is at least as effective a protection strategy as the creation of a strict-use protected area.

However, many well-documented examples exist of local exhaustion of a particular natural resource in indigenous areas, for a variety of reasons (Robinson and Bennett 2000). The conquest of land and usufruct rights and expansion of indigenous areas systems is often followed by population increases and greater pressure on natural resources, at least in the short term. The very consolidation of cultural autonomy and a legal and property regime inherent in a successful indigenous claim to land opens up the possibility of new arrangements, such as leases, concessions, and compensation payments, whose net effect is to reduce direct natural resource management by indigenous peoples, or render it controversial.

8.4.1.2 Policy Challenges

There are many documented examples of successful environmental management in individual indigenous areas, either directly or in some form of shared management in which indigenous representatives have a significant say. Nevertheless, the non-indigenous institutions with technical expertise in natural resource management, both governmental and non-governmental, have generally failed to devote the same attention to the development of applied knowledge and methodologies for indigenous areas as they have to national parks.

Indigenous organizations across the world are often poorly informed about technologies and techniques that are routine for other resource management agencies—remote sensing, satellite imagery, zoning, monitoring, and formal management plans—that may have potential for reinforcing natural resource management in indigenous areas. In their absence, there is a shortage of quality field data to inform policy, a demand increasingly heard from indigenous organizations themselves.

Culturally appropriate and technically sound cooperation between indigenous and non-indigenous organizations to reinforce natural resource management on indigenous lands is rare; achieving it should be a concern for governments and civil society alike.

8.4.2 Devolution and Local Forest Management and Local Movements for Access and Use of Forest Products

Governments and donor projects have developed diverse institutional arrangements to provide rural people more formal rights to forests and their management. Millions of the rural poor have participated in local forest management policies and programs during the last two decades. The results have been mixed. Most arrangements have maintained and even extended central government control (Sundar 2001; Fisher 1999; Malla 2000; Balad and Platteau 1996; Edmunds and Wollenberg 2003; Shackleton et al. 2002). While local people generally have better legal access to forests and some have higher incomes, many have also suffered negative trade-offs (Sarin 2003). Forestry has not often been the best entry point for integrated resource management and rural development. Local people have usually not shown a consistent interest in forest conservation (Shackleton and Campbell 2001).

Triggered by these experiences and the increasing complexity of demands from different interest groups, local forest management policies are shifting. They increasingly involve not only collaborative management arrangements between a local group and the government, but a range of stakeholders and acknowledgement of overlapping systems of management and diverse interests. There is more emphasis on facilitating decisions through negotiation. There is also increasing recognition of the need for frameworks that better emphasize local peoples' rights to self-determination and enable more effective representation of the rural poor in negotiations. The rural poor and their federations and advocates are bringing a new sophistication to negotiations and increased demands for their voices to be heard (Singh 2002; Britt 1998; Colchester et al. 2003).

8.4.2.1 Scope and Scale of Local Forest Management Policies

Local forest management programs now occur around the globe. In India, more than 63,000 groups have enrolled in joint forest management programs to regenerate 14 million hectares. In Nepal, 9,000 forest user groups are trying to regenerate 700,000 hectares of forest. In Brazil, farmers participate in managing 2.2 million hectares as extractive reserves. Half the districts in Zimbabwe have CAMPFIRE (Communal Area Management Programme for Indigenous Resources) schemes. More than half of the natural forest in the Gambia (17,000 hectares) is under community forest management. The programs generally have resulted in significant levels of improved resource management and have improved access of the rural poor to forest resources, but have fallen short in their potential to benefit the poor (Upreti 2001).

The institutional arrangements of the different approaches to local management have strongly influenced how policies affect local people. Formal arrangements include corporate, legal organizations composed of rights holders (such as rubber tappers' organizations in Brazil, *ejidos* in Mexico, trusts in Botswana, conservancies in Namibia, and communal property associations in Makuleke, South Africa). There are also village committees facilitated by government departments, such as the village natural resource management committees in Malawi, and forest protection committees in India. The Gambia's "Community-controlled State Forests" program encourages communities that have designated community forests to help protect the surrounding state forest area in exchange for a share of the resulting income. In the Philippines and China, contractual agreements between the government and households or individuals have been developed where individuals exercise varying degrees of authority over species selection, harvesting practices, sale and consumption, and the distribution of benefits. In addition, there are local government organizations such as rural district councils in Zimbabwe and *panchayats* in India, and multistakeholder district structures aligned to line departments such as *Tambon* councils in Thailand and wildlife management authorities in Zambia. Arrangements allocate varying degrees of rights to forest and land. Many impose forest management requirements.

Self-initiated local responses to problems in access and use of forest products have also proliferated in recent years; they are collectively more significant than efforts led by governments or international processes, but they require the latter's support to spread. Such local organizations include *campesino* forestry organizations in Central America, forest user groups in Nepal, the National Council of Rubber Tappers in Brazil, people's natural resource management organizations in the Philippines, and the Landcare movement in Australia and elsewhere.

8.4.2.2 Effectiveness of Devolved Control

The degree of control transferred by the state under these different institutional arrangements has affected the outcomes for local people. Bureaucratic control was higher and the responsiveness of programs to local needs lower where arrangements allocated control to higher levels of social organization, local government, or district structures. In such cases, state interests in resource production, revenues, and environmental conservation more strongly overrode villagers' interests in livelihood needs. Existing capacities for management were weakened (Edmunds and Wollenberg 2003).

Local people were able to win more benefits for themselves where they had strong local organizational capacity and political capital to mobilize resources and negotiate for better benefits. NGOs, donors, federations, and other external actors had a key role in supporting local interests. Where local groups managed their own forests without state intervention, however, they were not necessarily better off, since without government support, they often had difficulty implementing or enforcing their decisions (Shackleton and Campbell 2001).

Although access to some important subsistence products improved, access to other important local resources such as timber or game remained restricted. Where financial benefits occurred, governments often failed to deliver on their promised share of incomes. Benefits from timber and valuable NWFPs were often reserved for, or at least shared with, the state or local elite (Shackleton and Campbell 2001). Only in a few exceptional cases did poor communities receive substantial financial benefits.

The improved formal access to forests has helped in most cases to protect a vital role of forests as safety-nets for rural people to meet their basic subsistence needs. However, the benefits to be gained beyond the subsistence-level were limited. Property rights would need to extend to more secure rights over valuable resources, for the poor to benefit substantially. Programs focused on organizing collective action around the management of a single resource such as forests may also divert effort from other sources of livelihood. Forests are not always the most important resource for poor people; the economic and social environment can create pressures to convert forests. Many of the poor might be better off with land reform measures that are not linked to forest management, but these programs are not in the interest of forest departments.

Co-management has demonstrated the difficulty of dividing roles and responsibilities, especially where the interests of the groups involved are highly divergent. Forest agencies have had varying experiences in organizing collective action. Romantic ideals about harmonious communities and the local knowledge and capacities of "traditional peoples" have been counterbalanced by the internal conflict and lack of leadership in many communities and the difficulty of organizing collective action where local social capital is weak (Stanley 1991; Gibson et al. 2000). Many co-management efforts rely on the role of outside agents to facilitate group action and sustaining group action has proven difficult. Other stakeholders such as local governments or NGOs often create their own sets of incentives or pressures for local people that work against co-management initiatives (Edmunds and Wollenberg 2003).

8.4.2.3 Policy Challenges

State officials and local people have had different expectations of what devolved management was supposed to achieve and how. Forest departments have mostly controlled the terms of devolution and co-management schemes. There is now a need to develop the institutional arrangements and capacities that enable people in forest areas to have the rights and power to bring about a fair division of control, responsibility, and benefits between government and local people. Checks and balances need to be in place to ensure that no one group, including the local elite, controls benefits and decision-making.

Frameworks for natural resource management that are developed more locally and then linked to national objectives have been shown to be more flexible and responsive to local interests. Relevant local stakeholders can develop these frameworks, with special support given to the disadvantaged poor to negotiate for their interests. Experience suggests that local responsiveness will be higher to the extent that effort is made to monitor and evaluate impacts and that institutional arrangements facilitate good communication and learning about these impacts among stakeholders. The learning process should include both local interest groups and national policy-makers to best manage different interests.

Policy frameworks could better assist self-initiated local responses to problems in access and use of forest products to build on what they are already doing, and to enable new partnerships. Multistakeholder poverty–forests learning processes could be fostered with codes of conduct for supporting local initiatives and integrating them in national forest programs and poverty reduction strategies.

8.4.3 Small-scale Private and Public–Private Ownership and Management of Forests

Small-scale private (non-industrial, non-community) ownership (or "family forestry") is very common in Western Europe and in the southern part of the United States. In Sweden, half of the forest area (with 60% of the production of wood) is owned by over a quarter of a million people. In Finland, over 75% is privately owned. An average holding in Sweden is around 50 hectares; in Finland, 30 hectares; in Germany, 7 hectares; and in France and Spain, below 5 hectares. Experiences from Scandinavia and from continental Europe indicate that privately operated forestry has strong sustainability credentials (National Board of Forestry 2001).

Since the discussion below is based on experiences mainly from Western Europe, some lessons may be possible to apply to forestry in other parts of the world, while some may not. Fundamental differences in the institutional framework and in culture will affect the outcomes. Private ownership is not merely a judicial matter—it is a matter of culture and tradition. More positively, some factors mentioned may be of importance also in countries with quite different institutions, such as local community or village control (or ownership) over the forest.

Small-scale private ownership may lead to closer management and more efficient economic use in the self-interest of the owner. Planting, pre-commercial thinning, and collection of firewood are well suited for do-it-yourself work. Gathering of berries and mushrooms, hunting, and recreational activities can often be conveniently combined with planning or supervision of forest production activities. Private ownership may lead to a greater sense of responsibility assumed by the owner, which may foster long term thinking such that sustainability is naturally sought, partly as a risk reduction strategy.

When the imperative of biodiversity conservation was brought to the fore in Scandinavia in the 1970s, a difference was observed between privately owned forests and large-scale corporate or public forests. In general, the private forests were more biologically varied (especially at a landscape level). This led, in Sweden, to private forest ownership being fully recognized in policy whereas previously large-scale forestry had been seen as the priority model (Klingberg 2004).

Constraints that have arisen, and ways in which they have been overcome primarily in the Scandinavian context, are assessed below (Klingberg 2004):

- *Efficiency.* Small holdings can be technically inefficient, leading some owners to cooperate with neighbors. Originally, these associations worked as wholesalers, assembling round wood and negotiating prices with large pulp mills and sawmills. Today they are large economic enterprises, organizing harvesting operations with modern machinery and professional staff, which single owners cannot afford. The associations have also invested in sawmills, pulp mills, and bioenergy production, thereby securing demand for wood harvested.
- *Knowledge and competence.* Lack of knowledge can result in mismanagement or even destruction of the holding. Do-it-yourself activities also tend to have higher accident rates than professional lumbering. Both the associations and the government work to solve these problems through training and information provision.
- *Raising standards.* The associations in Sweden are active in raising both forest production and the level of environmental protection. Certification is being pushed, with higher standards than those found in the legislation.
- *Long-term perspective.* A fundamental factor is the long-term thinking by many private owners, who plan to pass on their holdings to younger generations. Regenerating harvested forests is an established norm.
- *Combined activities.* Many small owners combine other jobs with the income from the forest, thus forming viable rural livelihoods.

Property rights are fundamental to the prospects for family forestry. Laws and regulations must back up smallholders' ownership and property rights. In many countries, ownership legislation and the system of land registry may not be conducive for private forestry holdings.

In the Nordic countries and many continental European countries, training and dissemination of knowledge has been used systematically to improve small-scale private forestry. For example, the Swedish Regional Forestry Boards have for over 50 years both been responsible for enforcing the Forest Act and for disseminating extension material and running study circles and courses with forest owners.

Boxes 8.1 and 8.2 provide assessment of two important examples of larger-scale private involvement in forest management—public-private partnerships and conservation concessions.

8.4.4 Company–Community Forestry Partnerships

8.4.4.1 Spread and Effectiveness of Company–Community Forestry Partnerships

In recent years, a range of partnerships has emerged between forestry companies and communities or groups of smallholders, and

BOX 8.1

Public–Private Partnerships in Forest Management: Some U.S. Case Studies

In the United States, 70% of commercial forests are privately owned. Hundreds of initiatives of public–private collaboration represent a new wave in forest management. Some examples:

- *The Conservation Fund* developed a broad public–private coalition to seize the opportunity represented by Champion International Corporation, which put 120,000 hectares of forestland on the market, much of it within the Adirondack Park of upstate New York. With funds from government and private forest management organizations, TCF purchased the entire 120,000 hectares. Today, some of the lands are protected while others operate as working forests with private forest managers where the future forested state is ensured by conservation easements.
- *Seven Islands Land Company* is among the oldest and largest owners of forests in Maine. With dozens of heirs and nearly 400,000 hectares, there are few similar ownerships anywhere in the world. The owners worked with the New England Forest Foundation to negotiate a conservation easement covering more than 280,000 hectares in northern Maine. The multiyear campaign to raise nearly $30 million dollars was successfully completed in 2002.
- *Safe harbor* is an approach developed by Environmental Defense to engage private owners in endangered species conservation. Private landowners enter a voluntary agreement with the U.S. Fish and Wildlife Service whereby the owner limits his exposure and potential impacts from threatened and endangered species. The parties identify a baseline and if future management results in an expansion of the population, the landowner's regulatory obligation is limited to the baseline population.

many are widespread globally. They vary widely in terms of types of forest products, types of partners, and the degree of development and equity between the partners. (See Table 8.1.) Outgrower schemes and joint ventures predominate, but several other kinds of arrangements, many informal, have arisen in response to local circumstances. Company–community partnerships are globally widespread, occurring in at least 23 countries in North America, South America, Europe, Africa, Asia, and the Pacific (Mayers and Vermeulen 2002).

Behind the range of partnership types lie a range of motives for entering into partnership. Globalization of investment, trade, and technology, coupled with increasing decentralization and grassroots demands for autonomy, provides strong impetus to both companies and communities. Neither party on its own can access and secure all the means for producing the goods and services it needs. Third parties are also pivotal participants in company–community deals: local and central government; development agencies and NGOs; providers of credit and insurance; certification bodies; and cooperatives, federations, and trade unions.

Evidence to date shows that partnerships can be better than solely corporate forestry, or solely community or small-scale farm forestry, in delivering the wide range of benefits now expected by the partners and by the public at large. Importantly, partnerships are able to provide superior economic returns to both partners in addition to public benefits. But direct economic returns are not always the most highly valued output to either partner. Partnerships are foremost a means to share the risks of production and marketing (Mayers and Vermeulen 2002).

Partnerships entail costs that can outweigh benefits under certain conditions, such as inappropriate government policy. (See Table 8.2.) Some impacts of company–community forestry partnerships remain debatable. Effects on local equity and rural development are mixed, and financial returns have often proven insufficient to lift community partners out of poverty, either through direct membership or through knock-on effects such as new employment and upstream/downstream small-scale business opportunities. Furthermore, equity in power between company and community partners is seldom achieved, and often actively avoided by the company partner in spite of the obvious reductions in risk of interacting with a more equal, legitimate partner (Mayers and Vermeulen 2002).

8.4.4.2 Policy Challenges

Making the most of partnerships centers on five key themes:

- iterative approaches to developing equitable, efficient, and accountable governance, both at the contract level and more broadly;
- raising the bargaining power of communities, particularly through association at appropriate scales;
- fostering the roles of brokers and other third parties, especially independent community development organizations;
- sharing the benefits of wood processing as well as production; and
- working toward standards (for example, in licensing requirements or certification) that give equal opportunities to small-scale enterprises.

8.5 Demand-side, Market-driven, and Technological Responses

8.5.1 Public and Consumer Action

8.5.1.1 Evolution of Public and Consumer Action

Consumer action emerged in the early 1970s as a means of addressing the global loss of forests. Initially, campaigns focused on tropical forests. As well as aiming to bring about actual changes in flows in the trade of commodities deriving from tropical forest areas, they were also used as a means of informing the public in countries such as the United Kingdom and the Netherlands about a distant environmental issue by identifying international trade linkages.

Wider public and political action also developed at this time, and for similar reasons. Various interest groups, such as Friends of the Earth in Europe and the Environmental Defense Fund in the United States, were identifying linkages between multilateral institutions such as the World Bank and bilateral donor agencies and forest-destructive programs in the tropics. As these programs were at least partly designed and managed by agencies accountable to industrial-country democratic governments and, ultimately, funded through taxpayers contributions, the public was encouraged to express concern and demand cessation of the funding of such damaging activities. Mass letter-writing campaigns urged governments to take the appropriate action within the relevant global institutions and to adopt suitable domestic policies and safeguards concerning the use of development cooperation funding.

Such actions continued to grow during the late 1970s and 1980s and, following the lead of interest groups in the United

BOX 8.2

Conservation Concessions

A "conservation concession" is a voluntary agreement whereby governments and other affected stakeholders are compensated for foregoing economic development on public lands. Conservation concessions are modeled after typical resource extraction contracts, such as logging concessions; however, rather than paying for the right to extract natural resources from public lands, the investor pays for the right to preserve the forest.

The conservation concession is a relatively new mechanism, and only two applications have been completed to date. The first is a 100,000 hectare area in Guyana. The second is a 135,000 hectare concession along the Madre De Dios River in Peru. Other conservation concession deals are at various stages of development in countries such as Indonesia, Mexico, and Bolivia. The rate of implementation is significantly impeded by two factors. First, the in-country capacity and organization of many developing-country NGOs to implement conservation concessions is poor; for example, capacity in resource valuation, contract law, and stakeholder analysis is often weak. Second, because of the general unavailability of financing for recurrent management costs or compensation payments, financing for conservation concessions tends to be available only on a project basis.

The components of a conservation concession contract that the conservation investor and the government must negotiate include the following:

- *Payments.* The cost of the conservation concession is calculated to reflect the "opportunity cost" of conservation. This includes the value of foregone employment and taxes incurred as a result of creating the concession.
- *Duration.* The duration of a conservation concession is flexible, but typically is the same as the duration for land use contracts that it is replacing.
- *Management plan and objectives.* The final component of the negotiated agreement is to develop a management plan for the concession area. The management plan includes a clear statement of the conservation objectives for the concession and performance indicators that demonstrate whether these objectives are being met.

The conservation concession approach is novel. Nevertheless, there is enough experience to identify some of the strengths and weaknesses of the concession relative to other conservation mechanisms. The conservation concession transfers the cost of conservation to stakeholders who are better able and willing to bear it. This apparent strength of the concession can also be a great limitation if resource rights are very valuable, for example, in Southeast Asian forests or in temperate forests with high commercial stocking. Conservation funding may simply be unable to compete with other land uses in these areas.

Because they are not permanent, conservation concessions may encounter less political resistance to implement. Concessions may also be useful to obtain an interim conservation status after which a more permanent mechanism may be sought. The temporary nature of a concession can also be a weakness in some contexts, as it cannot guarantee the permanent protection of any particular forest.

Accountability is one of the greatest strengths of the conservation concession. Annual payments are made only if periodic monitoring and evaluation indicate that the conservation objectives for the concession are being met. Increased accountability also brings with it a greater risk of detecting failure.

One of the strongest criticisms of the conservation concession approach is that it may inadvertently create perverse impacts. For example, countries may be unwilling to create new protected areas if they think that they can attract investors to finance conservation concessions. However, it should be possible to develop policies to mitigate this risk. For example, conservation concessions could be restricted to being a "phase two" conservation mechanism, used only after a country has established a representative network of protected areas.

Table 8.1. Typology of Company–Community Forestry Arrangements, by Partner

Company Type	Individual Tree Growers	Individual Tree Users	Group of Tree Growers	Group of Tree Users
Forest product buyer, processor (large-scale)	outgrower schemes for timber, pulp, commodity wood, or NWFPs farm forestry support and crop share arrangements	product supply contracts farmer outprocessing	outgrower schemes joint venture for timber or pulp corporate social responsibility project contracts by communities group certification with company support	product supply contracts community processing or farmer outprocessing
Forestry concession or plantation owner (large-scale)	land leased from farmers	co-management for NWFPs	concessions leased from communities corporate social responsibility project	co-management for NWFPs
Small local production or processing enterprise	credit/product supply agreements	product supply agreements	credit/product supply agreements joint ventures	product supply agreements
Environmental service company	forest environmental service agreements			

Table 8.2. Conditions under which Companies, Communities, and Landscapes Win or Lose in Partnership Arrangements

Outcome	Without Partnership Arrangements	With Partnership Arrangements
Companies lose	inadequate supplies from restricted land and resource access	transaction costs of developing deals too high
	high risk of non-cooperation or resistance from communities	process too complicated
Companies win	absence of pressure from communities, law, or market	secure supplies of raw materials and/or workforce
	profitable to buy community land, pay off local elites, and massage opinion with public relations	"social licence to operate" granted by communities and wider society
Communities lose	lack of livelihood-improving opportunities in rural areas	become locked into dependency, or ripped off by companies
	lack of legal or bureaucratic permissions to develop land or trees without companies	pushed into unwise or sub-optimal land uses
Communities win	livelihoods not skewed by single strategies, commodities, or markets	income generated or services provided where few other rural alternatives available
	self-determination unaffected by company agendas	capacity for community-run development options enhanced
Environmental deterioration	forest asset stripping by companies seeking out weak local governance	inappropriate trees used or natural forest felled
	non-forestry land uses may be less optimal or landscape-degrading	other land uses like grazing squeezed or displaced causing degradation
Environmental benefits	land use systems and product diversity more optimal without forestry	reduced micro-level erosion from forest land uses
	land and resource control pattern more sustainable without deals	more forest goods and services in the landscape

Kingdom and United States, were also taken up in other countries, including Australia, Germany, France, Japan, Denmark, Austria, and the Netherlands. Consumer action continued to focus on the tropical timber trade. Significantly, almost all campaigns took a nuanced approach, calling not for a total boycott of tropical rainforest timber but for a "selective boycott" of products that had not been derived from "sustainable sources." This provided the basis for the later development of forest certification schemes. Consumer action campaigns worked through networks of locally affiliated activists, reinforced through media campaigns that served to highlight the connections between high-profile retail and manufacturing companies and forest management problems in identified areas (such as Brazil and Sarawak). Specific actions included picketing of, and dramatic protests outside, retail outlets, and the application of stickers and posters to shops and wood products. One particular target of such activities was the trade in Brazilian mahogany, which research had shown was largely derived from illegal exploitation of areas supposedly protected for indigenous communities. By the early 1990s, such actions had substantially reduced levels of imports of Brazilian mahogany into some countries.

Related to consumer actions were efforts to ensure that local and national governments in tropical timber "consuming countries" adopted purchasing policies that encouraged the use of timber from "sustainable sources." As a result of these efforts (also by local and national activist networks), by 1992, several hundred local authorities, including major metropolitan authorities, throughout Europe, North America, and Japan, had adopted such policies. Several national governments (including those of Austria and the Netherlands) also moved toward such policies, though these evidently ran foul of both EU and GATT trading rules and were never fully implemented. They did, however, send a strong political message to tropical timber producing nations.

Campaigns aimed at the international financial institutions succeeded in drawing public attention to the role of agencies such as the World Bank, and by implication the governments that supported them, in specific projects with major impacts on tropical forests. In some cases, such projects were either halted or significantly altered to reduce environmental and social impacts. Public pressure also succeeded in bringing about multilateral policy change, most notably the adoption by the World Bank of a new forest policy in 1993 that prohibited the use of Bank funds for commercial logging operations in tropical moist forests (World Bank 1991).

Possibly one of the most important results of consumer action between the mid-late 1970s and the early 1990s was the development of forest certification and labeling schemes. In 1987, Friends of the Earth established the "Good Wood Seal of Approval" scheme, which aimed to help consumers distinguish between products derived from "environmentally and socially acceptable sources" and products derived from "destructive" sources. It was underpinned by the belief that, by developing guaranteed markets for "acceptable" products, possibly with a price premium, an incentive would be provided for forest managers to adopt sustainable forest management practices.

The establishment of the Forest Stewardship Council in 1993 coincided with a decline in mass public action concerning forests in a number of "consumer countries," and the onset of "media fatigue" on these issues. The source of "pressure" on the timber trade thus partly shifted from consumers and the wider public to trade groupings such as the "1995 Groups" organized by the Worldwide Fund for Nature. Through high-profile marketing campaigns, WWF has encouraged the public to selectively purchase FSC-certified wood products, while simultaneously working collaboratively with timber producing, manufacturing and retailing companies to assist them in gaining FSC certification.

However, during the later 1990s and continuing today, a more radical form of protest has emerged. Focusing on the trade in illegal and "conflict" timber (and capitalizing on various international research initiatives which have documented the wide extent of this problem), environmental pressure groups such as Greenpeace, Environmental Investigation Agency, and Global Witness have conducted high profile "naming and shaming" campaigns against specific forest sector corporations and government agencies. Such campaigns have been waged in many European countries, as well as in North America. These campaigns have contributed to the signing of bilateral agreements concerning illegal wood products between the government of Indonesia and various timber importing countries and the development of a draft policy by the European Union (EU FLEGT) concerning the use of voluntary licensing as a means of distinguishing legal from illegal wood.

Public attention has now shifted to other environmental issues, especially global climate change. The deliberations on global forest issues within the UN framework (such as the Intergovernmental Forum on Forests and UNFF) have not been seen by many civil society organizations as likely to result in significant improvements "on the ground," and therefore have not provided a useful focus for mass public action and political lobbying (UK TFF 1998). Much of the debate between the various stakeholders on issues such as forest management standards, conservation, and human rights takes place within the context of certification, in which the wider public is little involved.

8.5.1.2 Effectiveness of Public and Consumer Action

Public action has undoubtedly had a number of important and positive consequences (Elliot 2000). However, it is also evident that such actions are very difficult and costly to sustain, particularly as they are dependent on the use of mass media, which suffers from "issue fatigue." Because of the need for media attention, such campaigns have tended to focus on targets with a high public profile in the countries in which the campaigns take place, especially large companies with operations in tropical forest areas. The response from the target corporation is likely to be to be withdraw from the operation altogether (rather than improve its standards) and there have been instances where such operations have been taken over by other companies with lower operating standards (Amazon Financial Information Service 2001–2).

Consumer campaigns and media exposes—such as those concerning illegal logging—do not always fully address the underlying causes of forest loss and degradation, especially the problems of inequitable land tenure and forest community poverty in developing countries.

Where public and consumer actions concern tropical forest issues, they are dependent on a strong understanding of local conditions, which can usually only be derived through close working relationships between civil society organizations in northern and tropical countries. However, few NGOs in developing countries actually have the resources or capacity to sustain such work over long periods of time, and therefore there is a danger that public actions in Europe or North America strongly reflect the views and priorities of "northern" NGOs rather than groups in the countries concerned.

8.5.1.3 Policy Challenges

Consumer campaigns can potentially continue to play an important role both in maintaining public awareness of forestry issues and in encouraging improved forest management. However, it is likely that the only institutions likely to be capable of sustaining the information and media exercises necessary for such campaigns are governments. While these are pressing policy challenges, it would be unrealistic to assume that governments will prioritize them without further concerted work by NGOs to mobilize consumers.

8.5.2 Third-party Voluntary Forest Certification

Certification is the procedure by which a third party provides written assurance that a product, process, or service conforms to specified standards, on the basis of an audit conducted according to agreed procedures. It may be linked with product labeling for market communication purposes. Certification offers independent assessment of the quality of forest management in relation to prescribed standards. It is voluntary, the forest manager being driven primarily by the prospect of access to markets that demand forest products produced in a responsible way, but also by improvements to company reputation and capacity.

8.5.2.1 Current Status of Certification

Forest certification has evolved rapidly. In only a decade, it has become routine practice in an increasingly large range of countries and forest conditions, and several schemes have sprung up. Three concerns are uppermost in assessing its effectiveness as a response to forest problems. First, the early drivers of certification hoped it would be an effective response to tropical deforestation. Now, however, most certified forests are in the north, managed by larger companies and exporting to northern retailers. Second, there has been a proliferation of certification programs to meet different stakeholders' needs, with the result that no single program has emerged as the only credible or dominant approach internationally. Third, the competitiveness of small and medium-sized enterprises, which may have advantages for sustainability and local livelihoods, is called into question where certification becomes the preserve of larger companies only.

Forest management certification assesses the performance of on-the-ground forestry operations against a predetermined set of standards. If the forestry operations are found to be in conformance with these standards, a certificate is issued, offering the owner or manager the potential to bring products from the certified forest to the market as "certified" products. This market potential is realized by a supplementary certification, which assesses the chain of custody of wood from the forest, through timber processor to manufacturer, to importer, to distributor, to retailer. In this sense, forest certification is market driven.

Accreditation is the process of recognition against published criteria of capability, competence, and impartiality of a body involved in certification, and results in licenses to operate a particular certification scheme. It "certifies the certifiers."

Certification schemes often make provision for the following:

- *multiple source chain of custody* to enable certification for paper, composite wood, and other products. This may allow processors a mix of certified and uncertified material where this reflects local supplies, and so reduces cost. It may also favor mixture with recycled materials;
- *group certification of smallholders,* to allow several small enterprises to be covered by one certificate, held by the group manager. This can reduce costs, provided group members are sufficiently similar to create scale economies;

- *forest manager certification,* where a professional manager is responsible for several small areas; this, too, reduces costs and creates economies of scale;
- *recycled wood certification,* which accords certified status to reclaimed or recycled wood where chain of custody is known; and
- *ecological zone harmonization of national standards,* to ensure that standards covering similar ecological zones can be rationalized.

Since its emergence in 1993, the FSC has certified forests in all continents, with a rapid increase in the area covered. Numerous other international and national forest certification schemes have more recently been launched, including in the United States, Malaysia, and Brazil. (See Box 8.3.) Many local stakeholders wanted to take charge of the process of developing certification schemes, to ensure they were appropriate to their forest types, enterprise types, and governance systems (Confederation of European Paper Industries 2002).

Where there is contention over a certification scheme, it tends to concern one or some of the following:

- perceived dominance or exclusion of certain parties,
- perceived lack of comparability between standards in a given region, and
- the degree of challenge or "stretch" represented by the gap between normally applied legal standards and the particular certification standards.

Where once there was some hostility between schemes and their supporters, there has been increasing collaboration and mutual support. There is a genuine desire to see certification play a key role among the responses to forest problems. To the extent possible, the individual schemes are beginning to put their differences aside to find an enduring role for certification.

8.5.2.2. *Effectiveness and Policy Challenges in Certification*

Observations on the effectiveness of certification as a response option include the following (Bass et al. 2001; Eba'a and Simula 2002). First, overall effectiveness in reducing poor forest management and deforestation depends critically on the incentive effects of market-based certification. In practice, the high threshold levels of certification standards (and FSC's in particular) have provided incentives only to already "good" producers rather than to improving bad practice. However, these "good" producers also now meet all current legal requirements, including those that they might normally not bother to meet. Most of them have also tightened management systems, especially for managing environmental impact. Thus certification has encouraged competition between producers at the high end of competence (just above and below the "certified" threshold). However, there are few incentives to cause the really bad producers to change behavior and be certified. The need for several thresholds (step-wise or phased approaches) is now being discussed, along with ways to complement certification with instruments for illegal logging.

Second, at the level of their standards, most schemes are applicable to many types of forest. Most certification schemes have been able to develop and apply one overarching standard agreed by many stakeholders and there are considerable similarities between the standards. Certification has coped effectively with *complexity* (in standards and their interpretation) and yet also delivers a *simple* message to consumers and producers.

Third, in practice, larger producers find it easier to benefit from certification, as they have better access to information and markets, scale economies, formal management systems on similar forest types, and an ability to bear risks and costs. The area of certified forest under community or small enterprise management is correspondingly much smaller. Many certification schemes have responded with special schemes for group certification of small growers or for certifying entire regions with one management regime. But there are those that question why a small community group occasionally harvesting timber on its own land should be held as accountable as a major corporation harvesting each day on leased public land.

BOX 8.3

Selected Forest Certification Programs

The *Forest Stewardship Council's* objectives are to promote global standards of forest management, to accredit certifiers that certify forest operations according to such standards, and to encourage buyers to purchase certified products. FSC is one of the first institutions to have been deliberately designed to sustainable development principles. It is a membership organization, with decisions made through meetings of a General Assembly, which is divided into three equal chambers: social, environmental, and economic. All three chambers have Northern and Southern sub-chambers, each with half of the total chamber votes. Governments are not entitled to participate in FSC's governance, even as observers, although government employees have been very active participants in some FSC national initiatives. FSC has a set of ten principles and related criteria (P&C) covering environmental, social, economic, and institutional aspects of forest stewardship, which apply to all forests, both natural and plantations. These P&C serve as a basis for the development of national and regional forest management standards. Certification standards that are consistent with both the P&C and with FSC's process guidelines for standards development are eligible for FSC endorsement. Such standards have been developed by both FSC-organized national working groups and by independent processes. FSC owns a trademark, which may be used to label products from certified forests. It has so far certified 37 million hectares in 55 countries (as of April 2003).

The *Programme for Endorsement of Forest Certification Schemes* (previously called the Pan-European Forest Certification Framework) is a voluntary private-sector initiative, designed to promote an internationally credible framework for forest certification schemes and initiatives. Its criteria are consistent with the intergovernmental Pan-European Criteria and Indicators for Sustainable Forest Management, thereby attracting considerable support from both European and national governments. National certification schemes that meet PEFC requirements can apply for endorsement and the right to use the PEFC trademark for product labeling. National PEFC governing bodies set standards and operate national schemes, and are represented on the PEFC Council Board. The initiative was given strong impetus by Finnish, German, French, Norwegian, Austrian, and Swedish forest owners, who wished to ensure that small woodland owners are not disadvantaged by certification and that local conditions are accounted for. It was supported by the national forest certification schemes that had been emerging in some of these countries, which felt themselves to be individually too small to develop an adequate presence. PEFC started in 1999; as of June 2003, it had certified 47 million hectares in 14 countries.

At the level of individual countries, the number of *national certification programs* under development is increasing rapidly. These include the Sustainable Forestry Initiative in the United States, and systems in Canada, Brazil, Malaysia and Indonesia.

Fourth, certification is largely document-based, and is predicated on formal, structured means of planning and monitoring. In practice, this assumption is biased against traditional societies and "part-time" foresters. Some current certification standards and procedures cannot recognize good management in some of the complex land use systems of indigenous and community groups.

Fifth, some environmental and social services are produced at levels other than the forest management unit (such as the landscape or the nation), which may not be under the control of the certified enterprise but which require its active engagement. Further developments are needed to ensure that certification encourages and recognizes improved relations between the forest management area and surrounding land uses.

Sixth, certification is a cost-effective complement to traditional administrative regulation. In all countries, certification is, at a minimum, encouraging some companies to meet legal requirements. In some countries, state forest authorities support certification as a "privatized" form of forest monitoring, and are making incentives available. In countries where regulation and enforcement is weak, certification has ensured that at least some producers are meeting not only legal requirements but also higher standards, and that this is monitored.

Seventh, certification depends for success on its credibility. The key ingredients are participation in defining standards to ensure they reflect many stakeholders' needs, consultation of local stakeholders when certifying forest management, and verification by third parties using tried-and-tested mechanisms with precedents in other sectors. Proliferation of certification schemes, which is leading to consumer confusion and a reluctance of some firms to be certified at all, has undermined the credibility of the approach and prompted considerable efforts by the wood products industry to investigate the potential of mutual recognition among schemes.

Eighth, in practice, certified products command only a minority of the forest products market (about 4% globally in 2003), with highest market penetration in Western Europe. Certified producers tend to gain market access, rather than a price premium (although a premium is available in some segments). More needs to be done to educate consumers about sustainable forestry and certification if the demand is to rise significantly. However, if market benefits have proven elusive, other incentives for certification are becoming apparent, such as certification to secure access to resources such as land, finance, and insurance.

8.5.3 Wood Technology and Biotechnology

Wood technology responses to date have been focused primarily on species used in industrial plantations, which must have wood properties suited for the products to be manufactured (Zobel and van Buijtenen 1989). There is considerable variation within a species, from pith to bark at a given height in the tree, and from base to top of tree, among trees within a stand, among stands within a region, and among regions (Kellison 1967). The phenomenon holds true regardless of the property, whether it be basic density, fiber dimensions, cellulose content, lignin content, moisture content, resin content, or any other trait of interest. This variation allows for genetic selection for any trait of economic importance (Zobel and Talbert 1984).

The wood properties of greatest economic importance for industrial manufacturing are basic density, fiber dimensions (length, width, lumen diameter, cell wall thickness, microfibril angle), number of fibers per unit area, and cellulose content. Conventional breeding programs have been effective in changing commercially important wood properties (Zobel and Talbert 1984). The property that has received most attention is wood-specific gravity or wood density. The reason for concentrating on wood density is its correlation with chemical pulp yield, strength properties of paper and paperboard, and strength properties of solid wood products, especially lumber.

The trend is for wood production to be shifting from the temperate and boreal regions of the world to certain parts of the tropics and subtropics (Kellison 2001). The major reason for that trend is the high growth rates of the trees, almost all of which are exotics, at the lower latitudes. While it will be many years before pulp production from northern plantations is greatly reduced, these plantations will represent a declining share of the global market. The reduction may be quickened if depreciation and amortization continue to exceed capitalization, which in the North American industry for example, has been the case every year since 1996 with the exception of 1999 (Connelly et al, 2004).

From a biological standpoint, the major species groups that are being intensively managed in plantations are *Pinus, Eucalyptus,* and *Acacia.* The pines receiving greatest attention are *P. taeda* and *P. elliottii* from the southern United States, *P. radiata* from California, *P. caribaea* var. *hondurensis* from Central America, and *P. patula* from Mexico. The eucalypts species of greatest importance are *E. grandis, E. urophylla,* and *E. globulus, E. teriticornus,* and *E. camaldulensis,* all of which have their origin in Australia and the islands of Indonesia. The acacias, too, have Australia as their origin; they include *A. mangium, A. mearnsii, A. aulicoliformis,* and *A. crassifolia.*

Using the same silvicultural practices, forest productivity of the *Pinus* species is at least twice as great in the exotic environments as in their indigenous habitats, and the rotation ages are typically 20% shorter. Similarly, the species of eucalypts and acacias produce 20–to 60 cubic meters per hectare per year, with harvest ages ranging from 5 to 12 years. Only with the most intensive silviculture, including fertigation (the application of fertilizers and water in metered amount through a drip-irrigation system), can angiosperm plantations in the temperate zones approach these growth rates. Even where plantation forestry is practiced in the temperate zone, a cost disadvantage exists in the economics of producing the wood.

The advantages of plantation forestry in the tropics and subtropics for fiber production so far outweigh the opportunities in the temperate and boreal zones that the developing countries to either side of the equator will benefit at the expense of their northern neighbors. The prognosis is that the plantation forests in the temperate and boreal zones will increasingly be managed for solid wood products. Fiber processing will be only a by-product of saw log forestry. (See Box 8.4.)

8.5.4 Commercialization of Non-wood Forest Products

Commercialization of non-wood forest products has become a means, promoted by researchers, conservation and development organizations, and, more recently, governments, to achieve rural livelihood improvement in an environmentally sound way. The category NWFP includes all products that are derived from forests with the exception of timber. In practice, the definition of the term has been ambiguous and inconsistent (Belcher 2003). Some authors restrict the category to products of natural reproduction, while others include managed or cultivated products. Generally speaking, the category includes plant, animal, and fungus species used for fuel, food, medicine, forage, and fiber, that have valuable chemical components or that are used for ritual purposes.

BOX 8.4

Wood Products Manufacturing Technology: A U.S. Case Study

Over the past 20 years, most operating North American softwood sawmills have been re-equipped with a wide assortment of highly automated equipment optimized for processing small logs. Lumber recovery factors have increased by nearly 50%, and productivity has nearly doubled. As sawmill recoveries have improved and plywood production has declined, chip production from these wood products manufacturing facilities has also been reduced. During this same period, raw material demand from the pulp and paper industry has increased nearly 20 percent. The industry has satisfied nearly all of this demand with recycled fiber, and that trend is expected to continue.

Total demand for roundwood has almost doubled in the United States over the past 25 years, only part of which is due to increases in the demand for pulpwood. Most of the increase has come from rapid growth in the strand products industry, and most of that added demand has been for hardwood, which has helped create a market for this low-cost wood. Strand-based products use softwoods as well, and that industry will likely expand into softwood growing areas where the price is competitive with hardwood sources. In the 1990s, the trend away from large diameter logs accelerated with the virtual elimination of timber sales from federal lands in the Pacific Northwest. With the reduced availability of large diameter logs and the growth of oriented strand board and engineered products, the demand for (and relative value of) these large logs has also declined.

Shorter rotations are more economically competitive than the long rotations needed for large logs, and improved efficiency in processing smaller sawlogs plus rising prices for fiber grade logs combine to support intensively managed forests in the United States. These forests can produce nearly 100% more annual growth than forests managed to produce large sawlogs.

In 1980, plywood manufacturing was concentrated mainly in the southern and western part of the United States. The net effect of substituting oriented strand board for plywood has been a net migration of panel industry jobs from the west to the north.

Interest in NWFPs began in earnest in the late 1980s and early 1990s, in conjunction with increasing global concern about deforestation and rural poverty. Forests gained heightened appreciation as sources of multiple products and services, and as important sources of livelihood for forest-based people (de Beer and McDermott 1989; Falconer 1990; Plotkin and Fomolare 1992). Researchers began to document the tremendous range of products used by forest people. Optimistic comparisons suggested that total NWFP values approached or exceeded timber values from the same forests (e.g., Peters et al. 1989). More realistic assessments followed, giving lower estimates (Godoy and Bawa 1993), but a movement had started. Environmentalists and social activists championed the idea that NWFPs extracted from the forest could provide an environmentally sustainable basis for livelihoods, leading to the establishment of "extractive reserves" for rubber, Brazil nuts, and other NWFPs in the Brazilian Amazon beginning in 1990, and exploration of the potential for similar approaches throughout the tropics (Ruiz-Pérez and Arnold 1996; Neumann and Hirsch 2000).

The underlying assumptions, often implicit, were that NWFP harvesting is more benign and valuable than timber harvesting, that it benefits poor people, and that it provides incentives for local people to conserve forests. In fact, none of these premises is necessarily true, and positive outcomes are only likely under restricted conditions (Ruiz-Perez et al. 2004).

8.5.4.1 Constraints on Implementation

The vast majority of NWFPs are consumed directly by the people that collect them and their families. Some are important mainstays in the household economy. The ubiquitous use of bamboo in the construction of buildings and utility items in rural areas, or the regular consumption of wild meat and vegetables, are examples. Other NWFPs are used infrequently, but can be critically important as sources of food when other sources are unavailable. Such emergency foods can make the difference between life and death.

A smaller, but still considerable, number of NWFPs are produced for sale or barter. These include various fruits, nuts, and vegetables that are primarily traded in local and regional markets and "bush meats" that are traded in large quantities in urban markets (Brown and Williams 2003). Other products find demand in more distant markets. High value mushrooms are collected in remote forests in China and sold the next day in supermarkets in Tokyo, and various herbal medicines and essential oils are sold in the growing western health and beauty markets.

A combination of factors has led to growth in some NWFP markets. The extension of the market system to more remote areas has created both the demand and the opportunity for increased cash incomes by NWFP producers. Globalization and growing interest in various kinds of natural products such as herbal medicines, wild foods, handcrafted utensils, and decorative items have increased demand and trade in these products. And development projects have increasingly sought to increase income opportunities, including through the production, processing, and trade of NWFPs. Still, the majority of traded NWFPs are sold in relatively small quantities (per producer; collective quantities can be very large), and for relatively low prices by the raw material producers. They are important in helping households to meet current consumption needs, and some are relied on as the main or the only regular source of cash income. Few NWFPs have large and reliable markets, and these tend to be supplied by specialized producers using more intensive production systems (Belcher et al 2003).

There is strong evidence that the poorest of the rural poor are most dependent on NWFPs (Neumann and Hirsch 2000), that the poor frequently use NWFPs as an "employment of last resort" (Angelsen and Wunder 2003), and that NWFPs serve an important safety net function (McSweeney 2004). Cavendish (1998) explains this in terms of the economic characteristics of forest-dependent people and of the products themselves. Many forest products are available as common-property resources in traditional systems or as de facto open-access resources, in state forest lands for example. They can be harvested and used with little processing, using low cost (often traditional) technologies. Some NWFPs are likely to be available for direct consumption or sale when crops fail due to drought or disease, or when shocks hit the household such as unemployment, death, or disease (Cavendish 1998).

The same factors that tend to make them important in the livelihoods of the poor also limit the scope for NWFPs to lift people out of poverty. Markets for many of these products are small and many are "inferior products." Naturally reproducing products tend to be dispersed, with seasonal and annual fluctuations in quantity and quality of production. Individual harvesters are limited in the amount that they are able to harvest. Open access resources are highly susceptible to overexploitation. The remote locations where wild NWFPs tend to be produced often have poor market access. All of these factors put producers in a weak bargaining position relative to traders who typically provide transport, market connections, and credit to NWFP collectors in classic patron-client relationships. In some respects, such products can be viewed as "poverty traps" in that people rely on NWFPs because they are poor and do not have better alternatives, but they are unable to use these resources to break out of poverty (Neumann and Hirsch 2000).

As Dove (1993) noted, in those cases where NWFPs have high value, they tend to be appropriated by people with more power, more assets, and better connections. This might happen through coercion and physical control of the trade, but more often control is achieved through domestication, when market forces lead to intensified and specialized production.

Homma (1992) developed a simple economic model that shows how high demand for NWFPs can over time lead to overexploitation of the naturally regenerating resource base, production on plantations outside of forests, and increased competition from synthetic substitutes. Empirical studies such as that by Belcher et al (2003) found strong evidence for this trend in a comparison of commercially traded NWFPs.

8.5.4.2 Effectiveness of Commercializing Non-wood Forest Products

There have been successful efforts to promote NWFP commercialization through combinations of technical and capacity-building interventions to improve raw material production, processing, trade, and marketing, and through improved policy and institutional frameworks. Resource tenure is a key factor, and considerable effort has been invested to help communities gain recognized rights and responsibilities to manage and use forest resources (as discussed in the section on collaborative forest management).

Simple interventions can be very effective. Providing a weigh scale and information on commodity prices and quality requirements of wholesale buyers in a trading center can help remote producers gain a better bargaining position. Collective investment in a building for storage or in a drying machine gives producers of perishable commodities more flexibility in their marketing. Improvements in processing and marketing, to improve product quality and reach more valuable markets, add value, creates more income downstream in the market chain, and increases demand and earnings for raw material producers.

The empirical evidence is mixed. There are success stories where production has been improved, markets have increased, and income generation has improved. Problems may also arise however, with inequitable distribution of benefits, stronger groups gaining control at the expense of weaker groups, and overexploitation of resources.

On the conservation side, success has been limited. The idea that NWFP harvesting has a lower impact than timber harvesting may be true in extensive, subsistence-oriented systems. But as products enter commercial markets, pressure on the resource base increases. Open-access resources are notoriously susceptible to overexploitation, and species-level impacts can be severe. All cases based on naturally regenerating resources in one major study of commercial NWFP cases (Belcher et al. 2003) reported declining resource bases. Harvesting that reduces stocks (for example, agarwood, palm-heart, wood for carving), especially of slow growing, slow-reproducing species, typically has faster and more severe impacts than harvesting of flows (for example, fruit, nuts). But harvesting pressure can also reduce reproductive success (by removing flowers or fruit), threatening longer-term sustainability.

At the ecosystem level, the hypothesis that increasing NWFP value could provide incentives for forest conservation has not been confirmed. To be true, it would require that the people who benefit from NWFP production are major agents of deforestation or that they have influence over those agents, and that low-intensity NWFP production is the most economically rewarding use of the forest. In practice, this linkage is often missing. The intended beneficiaries of NWFP development activities often are not the main agents of deforestation and do not have control or even influence over decisions to log or convert forest. Increased value does not automatically translate into effective incentives for conservation (Salafsky and Wollenberg 2000). Moreover, successful commercialization may create incentives to intensify NWFP production through enrichment planting or cultivation. To the extent that this is done in natural forest areas, it will result directly in reduced biodiversity or outright conversion of the management unit to an NWFP plantation.

8.5.4.3 Policy Challenges

Many NWFPs do not have scope for commercial development but are extremely important in millions of households. This has not been recognized adequately; for example, the contribution of forests to livelihoods has been chronically overlooked in poverty reduction strategy papers (Oksanen and Mersmann 2002). These values alone may be enough to justify forest conservation and enhancement.

A smaller but still substantial subset of the NWFP category has important local, regional, or international markets. Some of these markets are growing, and there are opportunities to increase incomes and employment-generation through targeted policy and project-level interventions. Typically, NWFPs have been ignored by policy. They are often covered by forest regulations designed for timber management, for example, or are not considered at all. Management of naturally regenerating resources can be improved with policy that more effectively gives incentives for sustainable management. Rattan harvesting concessions in Asia, for example, are frequently allocated over large areas to non-local concessionaires for very short periods. The concessionaires thus have no incentive to harvest sustainably or to invest in regeneration and local people benefit only from low- paying jobs as harvesters. Basic biophysical research is lacking for many valuable NWFPs, constraining efforts to improve management. More investment is needed in this area. One promising area is joint-production of timber and non-wood products. Improvements will require appropriate sivicultural research and new kinds of company–community partnerships.

For livelihoods improvement, the key interventions may be in resource control and in market development and capacity building for small-scale producers to enable them to compete in tough markets. In this vein, it is necessary to keep in mind that the rural poor typically have diverse economic activities and are risk averse. NWFP-oriented interventions should try to keep other options open and not focus exclusively on one activity.

There are inherent contradictions between commercial development and biodiversity conservation, at least at the level of the management unit. Increased demand leads to overexploitation of naturally regenerating resources, especially under the open-access conditions that prevail in many natural forest areas. Where conditions allow, producers tend to increase their management intensity, moving toward cultivation in horticultural or plantation systems. At the management unit this means converting forest to domesticated systems, with associated biodiversity loss. Conservation objectives might be achieved if such systems successfully reduce pressure on remaining natural forest.

8.6 Land Management Institutions, Investment, and Incentives

8.6.1 Natural Forest Management in the Tropics

Whenever management was attempted with the intent to conserve and utilize natural forests, one model became dominant (Troup 1940). Based on the earlier concept of sustained yield, wood supply was designed to be continuous over generations, with harvests planned not to exceed growth. Maintaining environmental quality and safeguarding rural employment were other key objectives of this response. Knuchel (1953) provided an early description of the technical approach. However, the practice of natural forest management in the tropics, and in particular the wet tropics where stocks of high value timber species are found, has proved problematic (Bruenig 1996; Dawkins 1957; Putz et al. 2000). Controversy has long raged over the potential for "sustainable" forest management as a viable economic activity in the tropics (Leslie 1977; Poore 1989; Dawkins and Philip 1997), in part as a result of the restrictions it places on timber harvest levels. This dispute continues (Rice et al. 2000; Pearce et al. 2003). Land allocations or appropriations for other purposes, and overexploitation of other forest resources for subsistence use or commercial gain, have also undermined the prospects for long-term natural forest management.

Nevertheless, since the early 1990s, huge investments have been made to promote improved management of natural forests and see it put onto an operational footing in a large number of countries (ITTO 1998). Over the last decade, an increasing (although ill-determined) amount of tropical forest has come under some form of management, which aims to achieve product utilization while conserving the natural resource. Reduced impact logging techniques have been especially popular. (See Box 8.5.)

8.6.1.1. Constraints on Implementation

In a large number of tropical timber-producing countries, poor governance undermines the management system (Brown et al. 2002). Timber licensing systems are frequently opaque, subject to considerable political patronage, and the beneficiaries are not publicly known (Gray 2002; Sizer 1995 and 1996). As a result, forest managers have limited influence over those given the rights to harvest timber and find it difficult to exert sufficient control to safeguard ecosystem health. However, interest in, and support for, forest law enforcement has recently become a major policy concern, as the extent of illegal logging has become more widely known and recognized as a significant constraint on new forest management initiatives (World Bank 2002).

Forest management has tended to be more successful where no viable land-use alternative exists. However, even then, low yields together with heterogeneous species distribution patterns have limited the viability of natural forest management. In contrast to temperate regions, valuable tree species occur at very low stocking levels over much of the tropics, and their spatial distribution is poorly understood (and therefore difficult to predict). In addition, many tree species suffer from a high incidence of natural defect in the wood that precludes otherwise desirable trees from being felled. High levels of previous timber exploitation are a further limiting factor that is becoming increasingly important in forest areas where access is good. The considerable cost of specialized machinery for logging heavy tropical hardwoods also poses a constraint, particularly for small-scale operators. Sustainable forest management in the tropics is frequently uneconomic if viewed in timber production terms alone.

Natural forest management has proved difficult to implement on a large scale, especially where access is limited. An annual felling coupe of 500 hectares in mixed tropical forest seems about the maximum that can be managed within one planning unit, without exceptional levels of management inputs. Many timber concessions in the tropics exceed this limit, despite lacking staff with the necessary management skills and associated resources.

A history of forest management in the region is helpful. Natural forest management is an information demanding process, which relies heavily on written records due to the long-term nature of many of its constituent activities. Where management records have been lost, this has proved to be a serious constraint to reviving forest management after periods of neglect.

Staff continuity within many forest authorities has suffered during the diverse changes in their structure and function in recent years. Despite much attention to institutional reform, roles and responsibilities have not always been clarified or backed-up

BOX 8.5

Reduced Impact Logging

Reduced impact logging comprises a set of harvesting practices that reduce impacts to residual vegetation, soils, and other environmental attributes compared with unplanned harvesting practices. RIL can reduce damage by as much as 50% compared with conventional logging (Pinard and Putz 1996; Holmes et al. 2002; Killmann et al. 2002).

Typically, RIL requires thorough resource inventories and careful harvest planning. Roads, skid trails, and log landings are planned and constructed so that they adhere to engineering guidelines designed to minimize soil disturbance. Directional felling techniques are applied to minimize damage to the residual stand and stumps are cut low to reduce waste. Heavy machinery is required to remain on skid trails and roads in order to limit soil disturbance and damage to vegetation. A post-harvest assessment is essential in order to provide feedback to loggers, concession holders, and forest department personnel (Dykstra 2002, 2003).

RIL can also be more efficient and cost-effective than unplanned harvesting. In the Brazilian Amazon under RIL, the overall cost per cubic meter of wood produced was 12% less than under conventional logging (Holmes et al. 2002). However, under different conditions, applying RIL can be costly. In the Malaysian State of Sabah, profits reportedly fell substantially when a switch was made from conventional logging to RIL (Tay et al. 2002). Other studies confirm that log production under RIL is often 20% lower than under conventional logging, due mostly to restrictions on logging in environmentally sensitive areas (Killmann et al. 2002). Financial benefits associated with the application of RIL are largely due to better planning and improved supervisory control. To obtain these savings, technically competent planners, loggers, and supervisors are essential. Personnel with the skills needed to apply these practices are rare in many parts of the tropics, so human resource development is a critical requirement for the adoption of RIL.

with development of capabilities. The desirable separation of the functions associated with forest management, forest regulation, and revenue collection have often not been made. The decline in forest management expertise has diminished the capacity of institutions to adopt flexible responses and has led to standardization of forest management prescriptions. Without increased funding to strengthen forestry institutions, this situation will remain a significant constraint to the successful application of this response.

Forest management is a field activity where the sequencing of a number of operations is critical to success. However, difficult working conditions are frequent, and matters are made worse by the lack of attention given to the health and safety of field staff in many countries. Education and training requirements remain poorly addressed, resulting in a lack of appropriately trained staff and the non-functioning of local professional associations.

Finally, in the species-rich tropics botanical identification is a constraint, particularly where emphasis is now given to the management of rare and non-tree species. Another shortcoming is the lack of attention given to the regeneration of the forest, despite considerable research investment. Studies continue to have limited impact on the implementation of forest management in many countries. More could be done to design effective dissemination strategies of research results that target forest managers.

8.6.1.2 Effectiveness of Response

Natural forest management has been successful in maintaining ecosystem health when it has also provided direct benefits to local communities. State authorities, without the involvement of local communities, carry out much forest planning with forest revenue appropriated by central government. This approach became common in tropical forests with disappointing results, in that it was unable either to safeguard the forest resource or to support local human well-being. The situation is now slowly changing. Not only is this helping to conserve the forest ecosystem, it is making a wider contribution to human well-being by offering an example for application in other public sectors.

8.6.1.3 Policy Challenges

Diverse, locally tailored solutions are needed for securing both wood supplies and forest environmental services. Wherever such solutions are developed, governance frameworks should become sufficiently flexible to support them. There is a compelling case for governments to give greater weight to locally determined approaches that provide solutions to the trade-offs associated with the management of natural forests.

8.6.2 Tree Plantation Management

The global area of tree plantations was 187 million hectares in 2000, a significant increase over the 1990 estimate of 43.6 million hectares (FAO 2001a). Although plantations are equivalent to only 5% of global forest cover, they were estimated to supply about 35% of global roundwood in the year 2000, and it is predicted that this figure will increase to 44% by 2030. (See MA *Current State and Trends,* Chapters 9 and 21.) Plantations will play an increasing role as natural forest areas decrease (largely in developing countries), are designated for conservation or other purposes (largely in developed countries), or are economically inaccessible (CIFOR 2003).

In addition to wood, it is possible for forest plantations to provide other environmental, social, and economic benefits, including NWFPs such as honey, resin, and medicinal plants; combating desertification; absorbing carbon to offset carbon emissions; protecting soil and water; rehabilitating lands exhausted from other land uses; providing rural employment; and, if planned effectively, diversifying the rural landscape and maintaining biodiversity. These contributions have been recognized by a number of the U.N. conventions. Afforestation and reforestation qualify for support under the Clean Development Mechanism of the UNFCCC for development of carbon sinks, the Global Mechanism of the Convention to Combat Desertification, and the Global Environment Facility for rehabilitation of degraded lands under the CBD.

Trees are increasingly being planted to support agricultural production systems, community livelihoods, poverty alleviation, and food security. Communities and smallholder investors, including individual farmers, grow trees in shelterbelts, home gardens, and woodlots and in a diverse range of agroforestry systems to provide wood, non-wood forest products, fuelwood, fodder, and shelter.

There is a strong trend toward commercialization and privatization of state forest plantation resources in an endeavor to manage these resources more effectively and efficiently in response to free market forces. However, about half of the global forest plantation estate is grown primarily for environmental and ecological rehabilitation and protection, and so is not suitable for management for industrial purposes.

Plantation managers in many countries are under pressure to ensure that their forest plantations form an environmentally and socially friendly source of world roundwood, fiber, fuelwood, and non-wood forest products. Certification, government procurement policies, and public pressure in relation to forest plantation siting and management are behind this.

8.6.2.1 Constraints

Not all afforestation has positive economic, environmental, social, or cultural impacts. Without adequate planning and appropriate management, forest plantations may be grown in the wrong sites, with the wrong species or provenances, by the wrong growers, for the wrong reasons. Examples exist where natural forests have been cleared to establish forest plantations or where customary owners of traditional lands may have been alienated from their sources of food, medicine, and livelihoods. In some instances poor site and species matching and inadequate silviculture have resulted in poor growth, hygiene, volume yields, and economic returns. In other instances, changes in soil and water status have caused problems for local communities. Land-use conflicts can occur between forest plantation development and other sectors, particularly the agricultural sector and with communities who may be alienated from their traditional land resources.

8.6.2.2 Lessons Learned

Incentives (direct and indirect) have often been used by governments to encourage investment by the private sector to stimulate accelerated rates of afforestation. However, these have sometimes stimulated inappropriate activities (CIFOR 2003).

Surplus or marginal agricultural and degraded lands are increasingly targeted for afforestation. However, land-use conflicts can arise when the land perceived as available and accessible is actually used for grazing and provision of non-wood goods and services, often according to customary or traditional land-use rights (Anon. 2003).

Price pressures may threaten the range of forest plantation benefits as approximately half of all forest plantations are driven by wood profitability. There are early warning signs that leading countries in forest plantation development (New Zealand, Chile,

Australia, Finland, and Sweden) are feeling the pressure of depressed prices for a range of forest products.

There are strong pressures toward short rotation, fast growing, lower-valued forest plantation products, which provide fiber for breakdown and reconstitution into a wide range of products in the form required by the consumer. Productivity can be sustained through reduced impact harvesting and practices that reduce soil erosion, conserve water, and maintain soil fertility through subsequent rotations. Appropriate management techniques for planted forests can also help conserve or even enhance biological diversity. Protection from fires, insects, and disease is critical (FAO 2001b and 2001c; Evans and Turnbull 2004).

There have been serious concerns regarding large-scale monocultures. There is increasing recognition that semi-natural and mixed-species, mixed-age plantings can provide a larger range of products, provide "insurance" against unfavorable market conditions, reduce the effects and economic consequences of insect and disease attacks, harbor greater diversity of flora and fauna, contain the spread of wildfires, and provide greater variety and aesthetic value in the landscape (Evans 1999; CIFOR 2003).

8.6.2.3 Policy Challenges

In areas where land degradation has occurred, afforestation can play an important role in delivering economic, environmental, and social benefits to communities. In these instances, forests and trees must be planted in ways that will support livelihoods, agriculture, landscape restoration, and local development aspirations (Anon 2003).

Caution is widely urged on the complex issues of bio-security (particularly relating to invasive insects, diseases, and forest plant species and the adoption of sound phyto-sanitary procedures) and the application of biotechnology (genetic modification, cloned germplasm, hybrid stock). Both these issues have potential positive and negative impacts on forest plantation health, vitality, productivity, and sustainability. In unregulated situations, there is increasing evidence of insufficiently proven germplasm (insufficient laboratory, field and demonstration trials) being used and incidences of bio-prospecting, which increase the potential for genetic pollution.

8.6.3 Fuelwood Management

Woodfuel remains one of the larger outputs of the forest sector, in some situations the largest. However, consumption of fuelwood has recently been shown to be growing less rapidly than had been estimated earlier. Increasing urbanization and rising incomes are reflected in a slowing down in the rate of increase in use of fuelwood as users switch to more efficient and convenient sources of energy. In some regions, including much of developing Asia, total consumption is now declining. In others, it appears to be approaching a peak (FAO in press). Charcoal use, on the other hand, is still growing, forming a much larger proportion of the woodfuels total in Africa and South America (and some countries in Asia). Charcoal is the main transition fuel to which fuelwood users shift as they move up the "energy ladder," and it is often a major urban fuel. It is also an important industrial fuel in some situations.

8.6.3.1 Impacts on Ecosystems

Supplies of fuelwood and wood for charcoal are drawn from a much wider base than just forests. Information from 13 countries in Asia showed that, in five countries, more than 75% of fuelwood came from outside forests (RWEDP 1997). Much fuelwood production for sale is a by-product of land clearance for agriculture. Significant pressures on forest and woodland from woodfuel harvesting are mainly associated with areas supplying urban demand for charcoal (SEI 2002; Ninnin 1994). In dryland forests in parts of Africa, production of charcoal as the main wood output can materially alter the structure and productivity of the forests.

Overall, demand for fuel is seldom likely to deplete or remove forest cover on a large scale. There is not a "fuelwood crisis" of magnitude, and with such potentially dire consequences in terms of forest depletion, as to require major interventions to maintain or augment supplies (Dewees 1989). Areas of concern are generally limited to situations where there is concentrated and growing urban demand for charcoal.

8.6.3.2 Impacts on Users

Use of wood as a fuel may be less of a concern to the security of the forest estate than has in the past been feared, but it constitutes a large part of the contribution that forestry can make to livelihood security and poverty alleviation. Most use is still of a rural subsistence nature. Gathered supplies of fuelwood still constitute rural households' main source of domestic energy.

The poorest tend to be disadvantaged by shifts to bring remaining common pool resources under sustainable management. Fuelwood harvesting tends to be restricted in this process, and women's needs for fuelwood commonly have lower priority than those of men for forest products for sale. Women practice a range of measures to respond to reduced access to fuelwood supplies, and seldom list this high among their concerns, but it is still likely to involve a cost to them, if only in terms of increasing collection time or having to shift to less favored fuels.

8.6.3.3 Fuelwood Opportunities and Response Options

Though wood is the principal source of energy for cooking and heating for so many of the poor, it is the least efficient. Unless they have access to technology to convert wood and charcoal into modern forms of energy, real costs of energy from woodfuels can be high even for the poor. In contrast, industrial scale dendro power is gaining in interest in some parts of the world. (See Box 8.6.)

Considerable efforts have been devoted to encouraging adoption of improved wood-burning stoves. These have had some impact in urban areas of some countries, but little success in rural areas. Assessments indicate that lack of success was often due to failure to understand that users valued stoves for reasons other than fuel economy and that "improved" stove designs had not addressed these needs, or due to the constraints posed by the cost of purchasing stoves. Some evidence suggests that where stoves are seen as saving money (in towns) they are popular, but where they are merely saving time or biomass (in rural areas) men are not prepared to spend money purchasing them.

Recent attention to improved stoves has shifted from increasing efficiency of woodfuel use to reducing damage to health from airborne particulates and noxious fumes associated with the burning of wood and charcoal (IEA 2002).

The effective transfer and enforcement of local rights are important considerations. Issues that often remain to be resolved include the continuing role of forest departments, community leaderships with interests at variance with those of their members, and difficulties in devising and putting in place control and management mechanisms with transaction costs less than the value of the woodfuel.

The potential and constraints of woodfuel selling as a source of income for the poor are poorly recognized in forestry or poverty

BOX 8.6

Dendro Power

Dendro power involves the use of wood-based materials for power generation (RWEDP 2000). One useful feature of dendro power is its potential to use sustainably grown fuelwood . Interest in dendro power is gathering momentum due to its multiple benefits of renewable power, reforestation, and income generation (especially in rural areas). On a global scale, it has potential to reduce air pollution, and increase carbon sinks. It is considered to be an environmentally benign power source, with zero carbon emissions if properly managed. Dendro power is used on a limited scale in countries such as Sweden, Finland, Netherlands, United Kingdom, Brazil, United States, as well as in many Asian countries, including Thailand, China, India, Malaysia, Indonesia, and the Philippines.

At present, most of the biomass-fuelled electricity generation is through steam turbines with net efficiencies of about 20–25%. In thermo-chemical processes, the biomass product is heated to break it down into gases, liquids, and solids. These are considered to be higher value and more convenient products. Further processing produces gases and liquid fuels like methane and alcohol. Methane can be used in gasification processes to produce electricity and liquids that are used as transportation fuels. Gasification technologies have the potential for higher conversion efficiencies of up to 45%. Integration gasification combined cycles are the latest development that combines gas and steam turbines to produce even higher efficiencies.

The success of dendro power generation depends on its ability to supply adequate fuel at low costs on a regular basis without over-exploiting the source. The generation of power requires a huge quantity of wood. A project in the range of 20–40 megawatt requires some 12,000 hectares of fuelwood plantation, or a $50–100 million investment (Hulcher 1995, as quoted in Bhattacharya 2001). Fuel sources can be grown on degraded land, thereby utilizing land not suitable for other activities. The energy source can be grown and managed as dedicated plantations, or as agroforestry systems or in woodlots (Fernando 2003).

reduction initiatives. Market demand for woodfuels can provide an important source of income for the poor. But reliance on it can also impede progress out of poverty, especially with large and rapid structural changes in urban market demand for woodfuels. There is a need for better understanding of such changes, and how best to support producers.

There has been a general failure of control measures to put commercial woodfuel production on a more sustainable basis. Initiatives to raise prices closer to replacement values, and to capture some of this in ways that would contribute to meeting the costs of management and regeneration, have not had much success. Transaction costs of trying to control collection from natural forests, and to differentiate in the marketplace between fuelwood from natural and planted sources, are often too high compared to the value of the wood being traded. This might be overcome by implementing such controls more effectively. However, this would raise costs for producers and lead to higher prices for urban users, resulting in considerable hardship for the latter, and aggravating problems of underinvestment and poor productivity by the former (SEI 2002).

8.6.3.4 Policy Challenges

The need to incorporate woodfuels more fully into the forestry mainstream has not been adequately addressed, despite the growing focus on giving forestry a stronger livelihood orientation. At the policy level, more effective recognition of the needs of the landless and very poor is needed in the process of making decisions about changes in land tenure and use. These considerations can also reinforce the case for conversion of open access use into common property rights. While privatization can create a more favorable environment for those with rights to land to invest in woody production, it can severely disadvantage those without land, unless their needs are recognized and taken into account.

Significant constraints are too often imposed on those who can participate in production, and can create distortions to trade and markets: competition from subsidized woodfuel supplies from government forests; taxes and other charges to generate government revenue from fuelwood trade; restrictions imposed in the name of conservation and prevention of "excessive" forest harvesting; and other regulations governing private sale of and trading in woodfuels. Such interventions are often unnecessary, counterproductive, or poorly designed and implemented, and need to be critically examined.

8.6.4 Carbon Management

Though there is not yet agreement on the modalities for implementing carbon forestry projects under the Kyoto Protocol, a wealth of experience has been developed as a result of more than a decade of pilot programs. Although many of the early initiatives were based on forest conservation or management, afforestation activities now predominate, perhaps reflecting the international decisions to allow only afforestation and reforestation activities into the CDM for the first commitment period. Afforestation and deforestation activities are attractive from a development point of view, and their carbon benefits are real, measurable, and marketable. Countries are increasingly recognizing the importance of forest cover for their water and soil management and for reduced vulnerability to extreme climatic events.

There are a number of issues that remain undecided in relation to the implementation of carbon forestry activities. These can be broadly grouped into technical, policy, and market uncertainties.

- *Technical uncertainties*. Issues relating to the validity of land use activities as a carbon sink and the quantification of net greenhouse gas benefits remain controversial among the scientific and policy making community.
- *Policy uncertainties*. The lack of agreement, at the international level, on the eligibility of forestry activities in mitigating climate change has to date been a major factor in restraining the extent of project development on the ground.
- *Market uncertainties*. The market for purchasing forestry based carbon offsets or investing in projects has reflected the ongoing technical and policy uncertainties and controversies of the land use sector. In particular, the withdrawal of the United States from the Kyoto Protocol process has reduced the market for forestry-based Joint Implementation and CDM projects substantially.

The likely impact of JI and CDM is largely dependent on the specific rules still being developed and the response of the carbon market to increased supply of forestry-based carbon offsets. Despite the early stages of implementation of climate change initiatives, experience to date has identified some important lessons that could inform the future debate on these issues. (See Table 8.3.)

8.6.5 Fire Management

There is a major effort underway to re-introduce fire as an effective ecosystem process in those forest areas where the lack of fire

Table 8.3. Lessons Learned from Forest-based Carbon Sequestration Projects

Experiences and Lessons	Possible Action and Future Opportunities
Fragmentation: the carbon benefits of land-based activities tend to be dealt with in isolation, rather than with other benefits or objectives	***Integration:*** the integration of carbon benefits with other objectives, services, products, and benefits at the landscape level is essential.
Costs: the project development cycle has high transaction costs that act as a barrier to many projects, specifically small or development oriented projects	***Cost reduction:*** approaches are needed to reduce the costs of project development to individual initiatives (e.g., provision of seed capital, simplified procedures for technical analyses, bundling, etc.)
Scale: small projects often result in multiple local benefits but are often not feasible due to high costs and limited carbon products	***Bundling:*** the gathering together of small-scale projects under an umbrella scheme will result in the economy of scale, ensure local benefits are secured, and add robustness to smaller projects.
Limited funding: the income generated through the sale of carbon offsets is rarely enough to fund the development and implementation of projects	***Innovative financing:*** measures that attract additional financing are needed, for example, through integration with other objectives and conventions or higher pricing for additional benefits.

has contributed to forest health problems and the increasing occurrence of uncharacteristically severe wildfires (USDA Forest Service 2000). The objectives have been severalfold: protect human life and property in fire-adapted ecosystems, reduce ecological damage to forests, avoid excessive suppression costs, restore ecosystem integrity and health, protect wildlife habitats and biodiversity, and lower air pollution problems (Mutch et al. 1993; Neuenschwander and Sampson 2000; USDA Forest Service 2000). Significant technical and political obstacles must be addressed if the effort is to be successful. The technical obstacles generally revolve around the current fuel conditions in these forests, or the existence of large, uniform areas of unhealthy or mature stands. These require careful management interventions that either reduce fuels to levels that allow fire to burn in historically characteristic ways or break up large areas of uniform conditions so that landscape patchiness is restored (Covington and Moore 1994; Mutch et al. 1993; USDA Forest Service 2000).

While most of the techniques have been well tested at research plot levels, there is limited experience at the large landscape levels needing treatment in areas like the western United States, northern Canada, or Russia. These problems are made more complex in those areas where significant human populations exist. Even with fuel reductions and carefully prescribed burning to restore fire to its ecologically required levels, the amount of air pollution created may exceed what people will tolerate (Neuenschwander and Sampson 2000). Political opposition to the inevitable risks of using fire as a forest management tool, the considerable costs involved in effectively managing an active fire program, and the pollution and human health impact that will be intentionally generated are significant and will require carefully crafted strategic approaches that generate widespread public support if they are to be overcome (USDA Forest Service 2000).

8.7 Summary Lessons

Civil society and private sector players are becoming as important as government in developing responses; furthermore, their involvement helps ensure that policy outcomes are more durable. Urban and market players are increasingly significant. This reflects growing public concern to secure a range of ecosystem services from forests and other wood-producing ecosystems. Innovative responses, such as many forms of partnership to create balanced land use for wood and other benefits, and certification to assure such a balance, are offering new forms of "soft policy" that influence government strongly.

Consequently, *multistakeholder policy processes,* from local to international levels, are becoming significant in developing, debating, and reviewing response options. They are important in deciding on the balance between the public and private benefits to be obtained from wood-producing ecosystems. However, they are still often poor at identifying and involving marginal groups, for which brokers can be helpful. Many are also one-off, rather than installing continuous improvement systems that keep up with the dynamics of wood supply and demand and deal with change.

Ultimately, *public perceptions and beliefs* are key. For example, progress needs to be made in improving public understanding of the wide land use spectrum that potentially provides wood, and therefore of the legitimacy of plantations as wood-producing ecosystems, potentially freeing up other land for other ecosystem benefits.

There has been a strong trend toward *privatization or decentralization* of control over forests, forest management services, and enterprise. This, together with other forms of liberalization and structural adjustment, has helped to remove perversities that acted against sustainable wood supply. It has helped to create a wider range of "willing stewards" of forests and wood-producing lands but has not always conferred adequate rights and powers on them to enable them to exercise stewardship.

Market-based responses are redistributing rights to stakeholders, making them more effective in securing both wood supplies and other ecosystem services. Market approaches to allocating use rights to public lands, and voluntary certification, are helping to change the structure of wood industries. However, it is usually existing "good practice" companies that are benefiting. Step-wise incentives are needed to encourage the bulk of wood producers to gradually develop existing capacity from a low base, to cover transaction costs, and hence to improve forest management practice. Other responses are needed to "close doors" to bad practice; these are unlikely to be market-based, but will need legal action and enforcement.

To shift wood production toward sustainability is a challenge that goes beyond selecting individual "responses" toward restructuring governance of the sector. Progress is made by *coherent sets of interacting responses* that suit a particular case, country, wood market, or governance structure. A coherent, effective "set" of response options might differ depending on the prevailing context. (See Table 8.4.) Developing an effective set of responses is, therefore, largely a *governance and institutional development* question. Urgent requirements for institutional strengthening tend to be at the local level, for it is only through local institutions that sustainable forest management can be precisely defined and pursued, and decisions made on the balance with other activities. A clear institutional separation of forest regulation, management, enterprise, and revenue collection tends to be needed among government authorities for environmental services such as carbon storage as well as for wood.

Table 8.4. How Responses Can Differ in Various Contexts (Mayers et al. 2002)

Prevailing Governance	Potentially Effective Response Options: Key Entry Points for Governance Change
Command and control	role, powers, and accountability of authorities legislation development extension and enforcement
Privatization to corporate or civil society interests	deregulation standards and certification market reforms, royalties, and rents ombudsmen monitoring
Nationalization of enterprises and services	major institutional and legal changes user rights compensation mechanisms
Devolution of power to local authorities and/or civil society groups	empowerment costs/transition problems of divestment capacity development
Other approaches to decentralization	empowerment rights assurance capacity development negotiation
Cross-sectoral consensus and partnerships	participation/representation mechanisms and resources availability of information capacities of civil society groups

Better *information* is also needed both about the dynamics of wood supply and demand, and about the costs and benefits of the different response options and their distributional effects. Some of the more recent responses appear to have caused significant changes, but there have been relatively few independent assessments of what they have achieved. Furthermore, there is inadequate information about how forests and other wood-producing ecosystems behave under multi-purpose production regimes, especially in terms of the best possible balance between wood and other benefits. Casting responses in stone will rarely, therefore, be a good idea. Whatever its form, sustainable forest management will be information-intensive and all response options may need to invest more in integral information and review functions. Table 8.5 summarizes the assessment of response options.

Table 8.5. Summary Assessment of Responses: Wood, Fuelwood, and Non-wood Forest Products

Response	Pre-conditions	Degree of Uptake	Constraints	Links and Trade-offs	Quality of Evidence	Assessment of Effectiveness	Key Policy Challenges
Title of activity	*Key contextual factors required before response can be effected*	*Indication of spread and degree of adoption of response*	*Key obstacles preventing up-take*	*Relationship with other responses*	*Strength and credibility of information on the response and its impacts*	*Impact of response in improving ecosystem health and human well-being*	*Governance actions required to support response*
International forest policy processes and development assistance	High level of public concern. Willingness of stakeholders to engage, particularly governments. Moderate national political commitment.	Thirteen years of inter-governmental policy dialogue. Several major NGO-led and private sector-led initiatives. Forestry-specific development assistance declining.	Excessively dominated by governments. Engagement with extra-sectoral frameworks still weak. No consensus on a legally binding international instrument in forestry.	Strong links with national forest governance initiatives/programs and certification; moderate with natural forest and plantation management; rather weak with trade liberalization and others.	Strong, in international convention and U.N. Forum on Forests secretariats and among monitoring and "watchdog" NGOs.	Weak direct impact. Moderate effect in setting overarching framework and catalyzing other response actions. Moderate effect in establishing common language on sustainable forest management.	Integration of agreed forest management principles and practices in financial institutions, trade rules, global environment programs, and global security decision-making. Implementation at regional to national levels.
Trade liberalization	Few factors required—since trade liberalization is pushed by most international agencies.	Widespread adoption of liberalization prescriptions. When non-tariff measures and effect of subsidies are taken into account, the net trend internationally is probably toward increased protection rather than liberalization.	Agricultural trade policies and regional development policies have greater impacts than forest trade policies.	Weak links to international forest policy dialogue. Moderate links and some trade-offs with national forest governance initiatives, natural forest management, and plantation management.	Weak. Trade flows information is strong—but impacts information is weak. Beginning to improve in international agencies and NGOs.	Impact contingent on governance. Magnifies the effect of governance—making already good forest governance better, making bad forest governance worse. Tends to concentrate control over forest management. More positive impacts when linked to improved, impartially administered property rights.	Improve engagement of "underpowered" groups in trade policy decision-making. Ensure institutional strengthening occurs before trade liberalization. Require cost internalization as well as liberalization, and consider the case for protection to achieve the social component of sustainability.
National forest governance initiatives and national forest programs	High level political commitment and stakeholder willingness to engage required.	Major forestry process, acknowledged by many countries; at best, main overarching response into which others fit; at worst, irrelevant to more focused responses.	Lack of political engagement of forest planners; weak institutional, human, and financial capacity; weak stakeholder negotiation processes.	Strong to international policy processes; weak to local-level implementation.	Strong, on the formal national steps taken (FAO database); weak on local assessment of impact.	Moderate in many contexts and promising to be strong in some, but as yet uncertain. Previous related planning processes have had limited impact.	Foster genuine stakeholder negotiation and buy-in; keeping objectives strategic, politically high-profile and focused; implementing agreed actions. Ensure NFPs drive progress to good forest governance.

(continues)

Table 8.5. Continued

Response *Title of activity*	Pre-conditions *Key contextual factors required before response can be effected*	Degree of Uptake *Indication of spread and degree of adoption of response*	Constraints *Key obstacles preventing up-take*	Links and Trade-offs *Relationship with other responses*	Quality of Evidence *Strength and credibility of information on the response and its impacts*	Assessment of Effectiveness *Impact of response in improving ecosystem health and human well-being*	Key Policy Challenges *Governance actions required to support response*
Direct management of forests by indigenous peoples	State exerting no control over territory, or state concedes significant sovereignty to native polities.	Limited to date; prevails in a few "failed states" and in North America Australia, and New Zealand.	Poor recognition of indigenous peoples by states. Weak uptake of methodologies that can improve management in indigenous areas.	Weak links to many other responses. Strong links to specific international conventions.	Weak on impacts—heavily dependent on coarse-filter, secondary data. Strong information on indigenous peoples in forest areas.	Uncertain since information is weak (recognition of indigenous land claims does not often lead to direct forest management by indigenous peoples). Initial evidence suggests that is as effective as strict-use protected area.	Culturally and technically sound cooperation between indigenous and non-indigenous organizations for natural resource management.
Collaborative forest management and local movements for access and use of forest products	Effective local institutions (for collaborative forest management) and concerted government devolution and support for devolved arrangements.	Widespread around the globe although not an abundant response.	Internal conflict and lack of leadership in many communities. Over-bureaucratic government—or reluctance to cede sufficient control.	Strong links to national governance initiatives and moderate links to natural forest management in the tropics and commercialization of non-wood forest products.	Strong—many rigorous case studies and situation analyses.	Strong on improved resource management and access to forest resources for participating groups, but much more uncertain in impact on poverty and human well-being.	Develop the institutional arrangements and capacities for people in forest areas to have the rights and power to bring about a fair division of control, responsibility, and benefits between government and themselves.
Small-scale private and public-private ownership and management of forests	Private long-term tenure and transfer rights over forest and plantation areas.	Country specific, dependent on tenurial system—small-scale private ownership prevalent in Europe and North America, and increasing in Latin America and Asia. Public-private ownership and conservation concessions limited in area and maturity.	Economic viability of small forest areas—opportunity cost of other land uses often too great. Public-private arrangements constrained by few private investors. Conservation concessions untested as yet.	Strong links to national forest governance initiatives and plantation management. Conservation concessions link or trade-off with natural forest management in the tropics.	Strong in Europe and North America. Experimental response only for conservation concessions.	Strong for small-scale private management where tenure is secure, information is well focused, and economies of scale are achieved through association of owners.	Support for security of tenure, effective market and technological information, and development of management capability.

Company–community forestry partnerships	Companies requiring secure forest asset base but restricted in own abilities to control it. Degree of organization of smallholders or communities.	Evident in several countries in each of: North America, South America, Europe, Africa, Asia and the Pacific. Increasing emphasis in some countries (e.g., China); shifts from tight partner contracts to looser arrangements in others (e.g., India).	Lack of policy or public pressure for companies to engage. Mistrust of companies by potential local partners. Weak third party "brokering" agencies.	Strong links to plantation management. Moderate links to national governance initiatives, public and consumer action and natural forest management in the tropics.	Moderate. Information base on the response only recently developing through case study and global review work.	Moderate. Where preconditions are met, better than solely corporate or small-scale farm forestry. Impacts often indirect (e.g., companies improving long-term survival and communities better securing tenure through partnerships).	Development of governance frameworks that require accountability of partnerships; support for the bargaining power of community-level associations; enabling the emergence of third-party support agencies.
Public and consumer action	Effective use of analysis of problems in generating appropriate messages for the public and consumers.	Widespread in the north, growing in the south.	Relies on strength of NGOs to mobilize public and consumers. Periodicity of interest from public and media, and from governments and corporations in responding.	Strong links as stimulus to national forest governance initiatives and forest certification. Moderate links to international forest policy processes, natural forest management in the tropics, and commercialization of non-wood forest products.	Moderate–extensive media reports on the effectiveness of campaigns, rather weaker cause-effect evidence.	Strong but specific impact, and sometimes perverse: has caused the emergence of certification, some trade initiatives, and improvements in practices of some companies; has stimulated some policies that undermine local tenure security and benefits to poor communities from forests.	Enable public mobilization strengths of NGOs. Government action to sustain information flows and enable improvements in practices that public actions highlight.
Third-party voluntary forest certification	Market signals demanding improved and verified forestry practice.	Rapid spread since 1993. Occurs in all continents but most certified forests are in the North, managed by larger companies and exporting to Northern retailers.	Little incentive to adopt in contexts of tropical deforestation. Proliferation of certification programs, diluting credibility. Threat that competitiveness of small and medium-sized enterprises reduced where certification only used by larger companies.	Strong links to public and consumer action and plantation management. Moderate links to international forest policy processes and national governance initiatives. Weak links to natural forest management in the tropics.	Moderate. Certification has a short history; experience to date is much-analyzed, but few independent assessments.	Moderate. Has improved already good practice rather than tackling bad practice and has done so for larger rather than smaller operations. Key knock-on benefits in improving forest policy debates and provisions.	Development of step-wise approaches to certification, and approaches more appropriate for smaller and community operations. Integrate landscape-wide priorities better. Explore mutual recognition of schemes and support more education of consumers about sustainable forestry.

(continues)

Table 8.5. Continued

Response	Pre-conditions	Degree of Uptake	Constraints	Links and Trade-offs	Quality of Evidence	Assessment of Effectiveness	Key Policy Challenges
Title of activity	***Key contextual factors required before response can be effected***	***Indication of spread and degree of adoption of response***	***Key obstacles preventing up-take***	***Relationship with other responses***	***Strength and credibility of information on the response and its impacts***	***Impact of response in improving ecosystem health and human well-being***	***Governance actions required to support response***
Wood technology	Changing market demand. Research and development.	Technology developments have stimulated a wide range of new plantations and processing methods—the trend being toward locations in the tropics.	Dependent on high levels of capital outlay for research and development, and hence not available to smaller enterprises.	Strong link to plantation management and dendro power (see fuelwood management). Moderate link to national forest governance initiatives and carbon management.	Strong information on utility of technology in production, weaker on impacts of this on ecosystems and human well-being.	Moderate. Technology development leads to changes in locations of production (trend to fiber production in the tropics) and structure of the wood industry with winners and losers in terms of jobs and environments.	Enable private sector investment in technology that internalizes environmental and social costs and is accessible to a greater range of scales of enterprise.
Commercialization of non-wood forest products	High demand for NWFPs over time leads to overexploitation of the naturally regenerating resource base first, and then specialized production if market access, secure tenure over the resource base, sufficient labor and capital to invest, and entrepreneurial skills are available.	Significant growth in some NWFP markets with extension of market system to more remote areas; growing interest in natural products such as herbal medicines, wild foods, handcrafted utensils, and decorative items; and development projects focused on production, processing, and trade of NWFPs.	Few NWFPs have large and reliable markets, and those tend to be supplied by specialized producers using more intensive production systems. Many other NWFPs are vital to the livelihoods of the poor but have little scope for commercialization.	Strong links to national forest governance initiatives and collaborative forest management. Moderate on impacts of commercialization, weak on basic biophysical research for many valuable NWFPs, thus constraining efforts to improve management.	Moderate. Impacts for livelihoods through combinations of technical and capacity-building interventions to improve raw material production, processing, trade, and marketing, and through development of cooperatives and improved policy and institutional frameworks. Problems with stronger groups gaining control at the expense of weaker groups and overexploitation of resources. Increased value does not automatically translate into effective incentives for conservation.	Improved understanding of the role of natural resources in rural livelihoods in poverty reduction strategies and related frameworks. Policy that more effectively gives incentives for sustainable management of NWFPs, including exploration of joint production of timber and non-wood forest products.	

Natural forest management in the tropics	Needs to be focused on a range of forest goods and services, not just timber, to be economic.	Large investments since the early 1990s to promote improved management of natural forests, and an increasing area now under some form of management.	Low timber yields, complexity of system, high labor inputs. Uncertain financial viability of reduced impact logging.	Strong links to national governance initiatives and collaborative forest management. Moderate links to public and consumer action, company–community partnerships, and commercialization of NWFPs. Moderate trade-offs with plantation management.	Moderate to weak. Information often not readily available. Information being acquired on reduced impact logging	Moderate. Natural forest management implemented by some of the best transnational corporations and by some local enterprises has been successful in maintaining ecosystem health when it has also provided direct benefits to local communities. Poor management characterizes many of the operations in the spectrum between these two scales of enterprise.	Identification of long-term funding sources. Support for diverse, locally tailored solutions to securing both wood supplies and forest environmental services—giving greater weight to locally determined approaches that provide solutions to the trade-offs associated with the management of natural forests.
Tree plantation management	Surplus or marginal agricultural and degraded lands. Supportive policy and investment frameworks	Widespread—some 187 million hectares in 2000, constituting 5% of global forest cover but supplying about 35% of global roundwood.	Conversion of natural forest. Undermining customary ownership. Without good planning and management, forest plantations may be grown in the wrong sites, with the wrong species or provenances, by the wrong growers, for the wrong reasons.	Strong links to company–community partnerships, carbon management wood technology. Moderate links with trade liberalization and national governance initiatives.	Strong data on areas, species, and long-term trends. Weaker on social impacts.	Strong impacts as response to growing wood demand and as available natural forest areas decline through deforestation, designation for protection and economic inaccessibility. Most positive impacts where developed on degraded sites where rural investment most needed. Moderate increasing local returns from smallholder plantation management.	Development of policy and incentive structure to ensure environmental and social costs are internalized; development responds to both growing consumption and declining harvest from natural forest; and bio-safety is secured.

(continues)

Table 8.5. Continued

Response	Pre-conditions	Degree of Uptake	Constraints	Links and Trade-offs	Quality of Evidence	Assessment of Effectiveness	Key Policy Challenges
Title of activity	***Key contextual factors required before response can be effected***	***Indication of spread and degree of adoption of response***	***Key obstacles preventing up-take***	***Relationship with other responses***	***Strength and credibility of information on the response and its impacts***	***Impact of response in improving ecosystem health and human well-being***	***Governance actions required to support response***
Fuelwood management	Market incentive and supportive policy.	Fuelwood is one of the larger outputs of the forest sector. Its consumption appears to have reached a global peak. It is rising in Africa and declining in developing Asia. Charcoal use continues to rise. Attempts to manage fuelwood are as widespread as its use. Power from wood-based materials (dendro power) spreading—still on a pilot scale.	Often faces competition from subsidized woodfuel supplies from government; taxes to generate government revenue from fuelwood trade; restrictions imposed in the name of conservation; and regulations governing private sale of and trading in woodfuels. Improved woodfuel stoves not valued for fuel economy by rural people. Dendro power requires huge quantities of wood.	Strong links to national forest governance initiatives and commercialization of NWFPs. Moderate links to collaborative forest management. Dendro power in the North strongly linked to plantation management.	Moderate. Information on impacts of attempts to manage fuelwood recently greatly improved. Strong information on dendro power—but unproven impact to date.	Weak: a general failure of control measures; initiatives to raise prices closer to replacement values unsuccessful; transaction costs of trying to differentiate in the marketplace between fuelwood from natural and planted sources are too high; uncertainty as to the extent and nature of the impacts of improved stoves. Dendro power likely to spread, mostly in the North.	Support studies that help map the location, nature, and causes of woodfuel problems, and interactions between woodfuel use, energy policies, and forestry and livelihood interventions. Development of policies that enable users to evolve new "tenurial niches" that give them some access to woodfuel resources in the new landscapes; remove distortions introduced by regulations on trade and markets; and balance dendro power with other energy sources.
Carbon management	Afforestation and reforestation activities allowed into the Clean Development Mechanism under the Kyoto Protocol for the first commitment period.	Pilot activities only—distribution of activities to date shows a developing country bias, with a particular focus in Central and South America.	Major uncertainties: technical (validity of land use activities as a carbon sink and the quantification of net greenhouse gas reduction benefits); policy (eligibility of forestry activities in mitigating climate change); and market (reflecting the ongoing technical and policy uncertainties).	Strong link to international policy processes and plantation forest management.	Moderate. Information base being rapidly developed.	Weak to date. The likely impact of Joint Implementation and the CDM is largely dependent on rules still being developed and response of carbon market to increased supply of forestry based carbon offsets. Project development has high transaction costs that act as a barrier to many projects, specifically small or development oriented projects.	Promote policy processes that install livelihood priorities as well as environmental safeguards in carbon management; integrate carbon benefits with other objectives at the landscape level; and foster collective approaches for small-scale projects to achieve viable scale.

References

Amazon Financial Information Services, 2001–2002: Peru: Camisea gas field and pipeline project, updates. Accessed 12 July 2004. Available at http://www.redlisted.com/peru_camisea.html.

Anderson, K. and R. Blackhurst (eds.), 1992: *The Greening of World Trade Issues,* University of Michigan Press, Ann Arbor, MI.

Angelsen, A. and S. Wunder, 2003: Exploring the forest-poverty link: Key concepts, issues and research implications, CIFOR occasional paper no. 40, CIFOR, Bogor, Indonesia.

ANON, 2003: *The Role of Planted Forests in Sustainable Forest Management,* Report of the UNFF Intersessional Experts Meeting, 25–27 March, Wellington, New Zealand.

Balad, J. and J. Platteau, 1996: *Halting Degradation of Natural Resources: Is There a Role for Rural Communities,* Clareden Press, Oxford, UK.

Bass, S., K. Thornber, M. Markopoulos, S. Roberts, and M. Grieg-Gran, 2001: *Certification's Impacts on Forests, Stakeholders and Supply Chains,* IIED, London, UK.

Bass, S.M.J., 2003: International commitments, implementation and cooperation, Proceedings of the XII World Forestry Congress, Quebec City, September 2003, FAO, Rome, Italy.

de Beer, J.H. and M. McDermott, 1989: *The Economic Value of Non-timber Forest Products in Southeast Asia with Emphasis on Indonesia, Malaysia and Thailand,* Manuscript by Netherlands Committee for the IUCN, Amsterdam, The Netherlands.

Belcher, B.M., M. Ruiz-Perez, and R. Achdiawan, 2003: Global patterns and trends in NTFP development, Paper presented to the international conference rural livelihoods, forests, and biodiversity, 19–23 May 2003, Bonn, Germany.

Belcher, B.M., 2003: What isn't an NTFP? *International Forestry Review,* **5(2),** pp. 161–168.

Bhattacharya, S., 2001: Commercialization options for biomass energy technologies in ESCAP countries, Background paper, Regional seminar on commercialization of biomass technology, 4–7 June 2001, Guangzhou, China.

Bird, N., 2002: National forest programmes, Key sheets for sustainable livelihoods no. 17, Overseas Development Institute, London, UK.

Bourke, I.J., 2003: *Trade Restrictions and Trade Agreements Affecting International Trade in Forest Products,* Report for IIED, April, IIED, London, UK.

Britt, C., 1998: Community forestry comes of age: Forest-user networking and federation-building experiences from Nepal, Paper presented at the 7th Common Property Conference of the International Association for the Study of Common Property, 10–14 June, Vancouver, BC, Canada.

Brown, C., P.D. Durst, and T. Enters, 2002: *Forest Out of Bounds: Impacts and Effectiveness of Logging Bans in Natural Forests in Asia-Pacific,* RAP Publications, October 2001, FAO, Bangkok, Thailand.

Brown, D., K. Schreckenberg, G. Shepherd, and A. Wells, 2002: Forestry as an entry point for governance reform, ODI forestry briefing no. 1, Overseas Development Institute, London, UK.

Brown, D. and A. Williams, 2003: The case for bushmeat as a component of development policy: issues and challenges, *International Forestry Review,* **5(2),** pp. 148–155.

Bruenig, E.F., 1996: *Conservation and Management of Tropical Rainforests: An Integrated Approach to Sustainability,* CAB International, Wallingford, UK, 339 pp.

Cavendish, W., 1989: Rural livelihoods and non-timber forest products, Paper presented at CIFOR workshop on the contribution of non-timber forest products to socioeconomic development, October, Zimbabwe.

CIFOR (Center for International Forestry Research), 2003: *Fast-Wood Forestry: Myths and Realities,* C. Cossalter and C Pye-Smith (eds.), CIFOR, Indonesia, Earthscan, London, UK.

Colchester, M., M. Boscolo, A. Contreras-Hermosilla, F. Del Gatto, J. Kill, et al., 2004: *Justice In The Forest: Rural Livelihoods and Forest Law Enforcement,* CIFOR, Bogor, Indonesia.

Colchester, M., T. Apte, M. Laforge, A. Mandondo, and N. Pathak, 2003: Bridging the gap: Communities, forests and international networks, CIFOR occasional paper no. 41, CIFOR, Bogor, Indonesia.

Confederation of European Paper Industries, 2002: *Comparative Matrix of Forest Certification Schemes,* CEPI, Brussels, Belgium.

Connelly, M.W., D. Cha, and S. McGovern, 2004: *The Holy Grail,* Credit Suisse/First Boston Equity Research, Beijing, China.

Covington, W.W. and M.M. Moore, 1994: Postsettlement changes in natural fire regimes and forest structure: Ecological restoration of old-growth ponderosa pine forests, *Journal of Sustainable Forestry,* **2(1–2),** pp. 153–182.

Cromack, K., Jr., J.D. Landsberg, et al., 2000: Assessing the impacts of severe fire on forest ecosystem recovery, *Journal of Sustainable Forestry,* **11(1–2),** 177–228.

Dawkins, H.C., 1957: The management of natural tropical high-forest with special reference to Uganda, Institute paper no. 34, Imperial Forestry Institute, Oxford, UK, 155 pp.

Dawkins, H.C. and M.S. Philip, 1997: *Tropical Moist Forest Silviculture and Management: A History of Success and Failure,* CAB International, Wallingford, UK, 359 pp.

Dewees, P.A., 1989: The woodfuel crisis reconsidered: Observations on the dynamics of abundance and scarcity, *World Development,* **17(7),** pp. 1159–1172.

Dove, M., 1993: A revisionist view of tropical deforestation and development, *Environmental Conservation,* **20(1),** pp. 17–24.

Dykstra, D., 2002: Reduced impact logging: Concepts and issues. In: *Applying Reduced Impact Logging to Advance Sustainable Forest Management,* T. Enters, P.D. Durst, G. Applegate, P.C.S. Kho, and G. Man (eds.), RAP Publications, October 14, FAO, Bangkok, Thailand, pp. 9–17.

Dykstra, D., 2003: *RILSIM User's Guide: Software for Financial Analysis of Reduced-impact Logging Systems,* Blue Ox Forestry, Portland, OR.

Eba'a, A.R. and M. *Simula, 2002: Forest Certification: Pending Challenges for Tropical Timber,* ITTO technical series no. 19, ITTO, Yokohama, Japan.

Edmunds, D. and E. Wollenberg (eds.), 2003: *Local Forest Management: The Impacts of Devolution Policies,* Earthscan Publications, London, UK.

Elliott, C., 2000: *Forest Certification: A Policy Perspective,* CIFOR, Bogor, Indonesia.

Evans, J., 1999: *Sustainability of Forest Plantations: The Evidence,* Issues paper, Department for International Development, London, UK.

Evans, J. and J. Turnbull, 2004: *Plantation Forestry in the Tropics: The Role, Silviculture and Use of Planted Forests for Industrial, Social and Environmental, and Agroforestry Purposes,* Oxford University Press, Oxford, UK.

Falconer, J., 1990: The major significance of minor forest products: Examples from West Africa, *Appropriate Technology,* **17(3),** pp. 13–16.

FAO (Food and Agriculture Organization of the United Nations), 1971: *Forest Resources Assessment Project 1970,* FAO, Rome, Italy.

FAO, 1995: *Forest Resources Assessment 1990: Tropical Forest Plantation Resources,* FAO, Rome Italy.

FAO, 1996: *Formulation, Execution and Revision of National Forest Programmes: Basic Principles and Operational Guidelines,* FAO, Rome, Italy.

FAO, 2000: *The Global Outlook for Future Wood Supply from Forest Plantations,* Working paper GFPOS/WP/03, Prepared for the 1999 Global Forest Products Outlook Study, FAO, Rome, Italy.

FAO, 2001a: Global forest resources assessment 2000: Main report, Forestry paper 140, FAO Rome, Italy.

FAO, J. Evans, 2001b: Biological sustainability of productivity in successive rotations, Forest plantations thematic working paper, D.J. Mead (ed.), FAO, Rome, Italy.

FAO, T. Waggener, 2001c: Role of forest plantations as substitutes for natural forests in wood supply: Lessons learned from the Asia-Pacific region, Forest plantations thematic paper series, Rome, Italy.

FAO, 2004: National forest programmes. Available at http://www.fao.org/forestry/foris.

FAO, Past trends and future prospects for the utilization of wood for energy, Working paper no. GFPOS/WP/05, Global Forest Products Outlook Study, FAO, Rome. In press.

Fernando, K., 2003: *An Analysis of Sri Lanka's Power Crisis,* Citizens Trust, Sri Lanka.

Fisher, R., 1999: Devolution and decentralization of forest management in Asia and the Pacific, *Unasylva,* **50(4),** pp. 3–5.

Gibson, C.C., A. M. McKean, and E. Ostrom (eds.), 2000: *People and Forests: Communities, Institutions, and Governance,* MIT Press, Cambridge, MA.

Giovannini, G., 1994: The effect of fire on soil quality. In: *Soil Erosion and Degradation as a Consequence of Forest Fires,* M. Sala, and J.L. Rubio (eds.), Papers from the International Conference on Soil Erosion and Degradation as a Consequence of Forest Fires, Barcelona, 1991, Geoforma Ediciones, Logroño, Spain, pp. 15–27.

Glück, P., A. Mendes, and I. Neven (eds.), 2003: *Making NFPs Work: Supporting Factors and Procedural Aspects,* Report on COST Action, National Forest Programmes in a European Context, Publication series of the Institute of Forest Sector Policy and Economics, Vol. 47, Vienna, Austria.

Godoy, R. and K.S. Bawa, 1993: The economic value and sustainable harvest of plants and animals from the tropical forests: Assumptions, hypotheses and methods, *Economic Botany,* **47,** pp. 215–219.

Gray, D.A., 2002: Forest concession policies and revenue systems: Country experience and policy changes for sustainable tropical forestry, World Bank technical paper no. 522, World Bank, Washington, DC.

Hair, D., R.N. Sampson, and T.E. Hamilton, 1996: Summary: forest management opportunities for increasing carbon storage. In: *Forests and Global Change, Volume 2: Forest Management Opportunities for Mitigating Carbon Emissions,* R.N. Sampson and D. Hair (eds.), American Forests, Washington, DC, pp. 237–254.

Harvey, A.E., 1994: Integrated roles for insects, diseases and decomposers in fire dominated forest of the inland Western United States: Past, present and future forest health, *Journal of Sustainable Forestry,* **2(1–2),** pp. 211–220.

Holmes, T.P., G.M. Blate, J.C. Zweede, R. Perreira, Jr., P. Barreto, et al., 2002: Financial and ecological indicators of reduced impact logging performance in the eastern Amazon, *Forest Ecology and Management,* **163,** pp. 93–110.

Homma, A.K.O., 1992: The dynamics of extraction in Amazonia: A historical perspective. In: *Non-Timber Products from Tropical Forests: Evaluation of a Conservation and Development Strategy,* D.C. Nepstad and S. Schwartzman (eds.), Advances in Economic Botany, **9,** pp. 23–32.

Humphreys, D. (ed.), 2004: *National Forest Programmes in a Pan-European Context,* Earthscan, London, UK.

Hyde, W. *The Global Economics of Forestry.* Forthcoming. Draft available from the author by emailing wfhyde@aol.com.

IEA (International Energy Agency), 2002: Energy and poverty, Chapter 13. In: *World Energy Outlook 2002,* OECD, Paris, France.

IIED (International Institute for Environment and Development), 2003: How can trade promote sustainable forest management? Unpublished report prepared by IIED for FAO, IIED, London, UK.

IPF (Intergovernmental Panel on Forests), 1996: *Underlying Causes of Deforestation and Forest Degradation,* United Nations IPF, New York, NY.

ITTO (International Tropical Timber Organization), 1998: *Criteria and Indicators for Sustainable Management of Natural Tropical Forests,* ITTO, Yokohama, Japan.

ITTO, 2002: *Annual Review and Assessment of the World Timber Situation,* ITTO, Yokohama, Japan.

Kellison, R.C., 1967: A geographic variation study of yellow-poplar *(Liriodendron tulipifera)* within North Carolina, Technical report 33, North Carolina State University, School of Forestry Research, Raleigh, NC, 41 pp.

Kellison, R.C., 2001: Forestry trends in the new millennium, Proceedings of the 26th Biennial Southern Tree Improvement Conf., Athens, GA, 5 pp.

Killmann, W., Bull, G.Q., Schwab, O. and Pulkki, R.E. 2002. Reduced impact logging: does it cost or does it pay? In: *Applying Reduced Impact Logging to Advance Sustainable Forest Management.* Enters, T., Durst, P.D., Applegate, G., Kho. P.C.S. and Man, G. (eds.). RAP Publication 2002/14. Food and Agriculture Organization of the United Nations, Bangkok. pp. 107–124.

Klingberg, T. 2004: Governance of forestry for sustainability: Private ownership and certification in an institutional perspective, Working paper no. 27, University of Gävle, Gävle, Sweden.

Koch, P., 1991: *Wood vs. Non-Wood Materials in U.S. Residential Construction: Some Energy-Related International Implications,* CINTRAFOR (Center for International Trade in Forest Products) working paper 36, University of Washington, Seattle, WA.

Knuchel, H., 1953: *Planning and Control in the Managed Forest,* trans. M.L. Anderson, Oliver & Boyd, Edinburgh, UK, 360 pp.

Leslie, A., 1977: Where contradictory theory and practice co-exist, *Unasylva,* **29(115),** pp. 2–17.

Malla, Y.B., 2000: Impact of community forestry policy on rural livelihoods and food security in Nepal, *Unasylva,* **51(202),** pp. 37–45.

Mayers, J. and S. Vermeulen, 2002: *Company–Community Forestry Partnerships: From Raw Deals to Mutual Gains?* IIED, London, UK.

Mayers, J., S. Bass, and D. MacQueen, 2002: The pyramid: A diagnostic and planning tool for good forest governance, Prepared for the World Bank and World Wildlife Fund Forest Alliance, March, IIED, London, UK.

Mayers, J. and S. Bass, 2004: *Policy That Works for Forests and People: Real Prospects for Governance and Livelihoods,* Earthscan, London, UK.

Mayers, J., J. Ngalande, P. Bird, and B. Sibale, 2001: *Forestry Tactics: Lessons from Malawi's National Forestry Programme,* Policy that works for forests and people series, no.11, IIED, London, UK.

MCPFE (Ministerial Conference on the Protection of Forests in Europe), 2002: MCPFE Approach to National Forest Programmes in Europe, MCPFE.

McSweeney, K., 2004: Forest product sale as natural insurance: The effects of household characteristics and the nature of shock in eastern Honduras, *Society and Natural Resources,* **17,** pp. 39–56.

Meil, J.K., 1994**:** Environmental measures as substitution criteria for wood and nonwood building products.

Mutch, R.W., et al., 1993: Forest health in the Blue Mountains: A management strategy for fire-adapted ecosystems. In: *Forest Health in the Blue Mountains: Science Perspectives,* T.M. Quigley (ed.), Gen.technical report PNW-GTR-310, USDA Forest Service, Pacific Northwest Research Station, Portland, OR, 14 pp.

National Board of Forestry (Skogsstyrelsen), Skogsvårdsorganisationens utvärdering av skogspolitikens effekter (SUS), 2001: [The Forest Policy Evaluation made by the Forest Authorities 2001], *Meddelande,* January 2002, Jönköping, Sweden.

NCWD, 1994: *Report of the National Commission on Wildfire Disasters,* April, R.N. Sampson, chair, American Forests, Washington, DC, 29 pp.

Neuenschwander, L.F. and R.N. Sampson, 2000: A wildfire and emissions policy model for the Boise National Forest, *Journal of Sustainable Forestry,* **11(1–2),** pp. 289–309.

Neumann, R.P. and E. Hirsch, 2000: *Commercialisation of Non-Timber Forest Products: Review and Analysis of Research,* CIFOR, Bogor, Indonesia.

Ninnin, B., 1994: *Elements d'Economie Spatiale des Energies Traditionelles: Application au Cas de Cinq Pays Sahelens: Burkina Faso, Gambie, Mali, Niger, Senegal,* Review of Policies in the Traditional Energy Sector, World Bank, Washington, DC.

OECD and UNDP (Organisation for Economic Co-operation and Development and United Nations Development Programme), 2002: *Sustainable Development Strategies: A Resource Book,* Compiled by B. Dalal-Clayton and S. Bass, Earthscan, London, UK.

Oksanen, T. and C. Mersmann, 2002: Forests in poverty reduction strategies: An assessment of PRSP processes in Sub-Saharan Africa. In: *Forests in Poverty Reduction Strategies: Capturing the Potential,* T. Oksanen, B. Pajari, and T. Tuomasjukka (eds), European Forestry Institute Proceedings 47, Joensuu, Finland.

Pearce, D., F.E. Putz, and J.K. Vanclay, 2003: Sustainable forestry in the tropics: Panacea or folly? *Forest Ecology and Management,* **172(2–3),** pp. 229–247.

Peters, C.M., A.H. Gentry, and R.O. Mendelsohn, 1989: Valuation of an Amazonian rainforest, *Nature,* **339,** pp. 655–656.

Pinard, M.A. and Putz, F.E. 1996. Retaining forest biomass by reducing logging damage. *Biotropica* **27(3),** pp. 277–295.

Plotkin, M. and L. Famolare (eds.), 1992: *Sustainable Harvest and Marketing of Rain Forest Products,* Conservation International–Island Press, Washington, DC.

Poore, D., 1989: *No Timber Without Trees,* Earthscan, London, UK.

Poore, D., 2003: *Changing Landscapes: The Development of the International Timber Organization and Its Influence on Tropical Forest Management,* Stylus Publications, London, UK.

Putz, F.E., D.P. Dykstra, and R. Heinrich, 2000: Why poor logging practices persist in the tropics, *Conservation Biology,* **14(4),** pp. 951–956.

Pyne, S.J., P. Andreas, and R. Laven, 1996: *Introduction to Wildland Fire,* 2nd ed., John Wiley & Sons, Inc., New York, NY, 557 pp.

Redford, K. and J. Mansour J. (eds.), 1997: *Traditional Peoples and Biodiversity Conservation in Large Tropical Landscapes,* Island Press, Washington, DC.

Rice, T., S. Ozinga, C. Marijnissen, and M. Gregory, 2000: *Trade Liberalisation and its Impacts on Forests: An Overview of the Most Relevant Issues,* Report, Fern, Brussels, Belgium.

Richards, M., C. Palmer, C.F. Young, and K. Obidzinski, 2003: Higher international standards or rent-seeking race to the bottom? The impacts of forest product trade liberalisation on forest governance. Available at www.fao.org/forestry/foris/data/trade/pdf/richard.pdf.

Robinson, J. and E. Bennett (eds.), 2000: *Hunting for Sustainability in Tropical Forests,* Columbia University Press, New York, NY.

Ross, M.L., 2001: *Timber Booms and Institutional Breakdown in Southeast Asia,* Cambridge University Press, Cambridge, UK.

Ruiz-Pérez, M. and J.E.M. Arnold, 1996: *Current Issues in Nontimber Forest Products Research,* CIFOR, Indonesia.

Ruiz-Pérez, M., B. Belcher, R. Achdiawan, M. Alexiades, C. Aubertin, et al., 2004: Markets drive the specialization strategies of forest peoples, *Ecology and Society,* **9(2),** p. 4.

RWEDP (Regional Wood Energy Development Programme), 1997: Regional study on wood energy today and tomorrow in Asia, Field document no. 50, RWEDP in Asia, FAO, Bangkok, Thailand.

RWEDP, 2000: *Options for Dendro Power in Asia: Report on the Expert Consultation, Manila, Philippines 1–3 April,* FD57, FAO/RWEDP, Bangkok, Thailand. [On line 2003.] Available at www.rwedp.org/dendro.html.

Rytkönen, A., 2003: *Market Access of Tropical Timber,* Report submitted to the International Tropical Timber Council pursuant to ITTC decision 6 (XXXI), Helsinki, Finland.

Salafsky, N. and E. Wollenberg, 2000: Linking livelihoods and conservation: A conceptual framework and scale for assessing the integration of human needs and biodiversity, *World Development,* **28(8),** pp. 1421–1438.

Sampson, R.N. and D.L Adams (eds.), 1994**:** Assessing forest ecosystem health in the inland west, *Journal of Sustainable Forestry,* **2(1–2),** pp. 3–12.

Sarin, M., N.M. Singh, N. Sundar, and R.K. Bhogal, 2003: Devolution as a threat to democratic decision-making in forestry? Findings from three states in India, ODI working paper 197, February, Overseas Development Institute, London, UK.

Savenije, H., 2000: *National Forest Programmes: From Political Concept to Practical Instrument in Developing Countries,* Theme studies series 3, Forests, Forestry and Biological Diversity Support Group, National Reference Centre for Nature management (EC LNV), Wageningen, Netherlands. Available at www.minlnv.nl/inm.

Sedjo, R.A. and R.D. Simpson, 1999: Tariff liberalisation, wood trade flows, and global forests, Discussion paper 00–05, Resources for the Future, Washington, DC.

SEI (Stockholm Environment Institute), 2002: *Charcoal Potential in Southern Africa, CHAPOSA: Final Report,* INCO_DEV, SEI, Stockholm, Sweden.

Seymour, F.J. and N.K. Dubash, 2000: *The Right Conditions: The World Bank, Structural Adjustment, and Forest Policy Reform,* World Resources Institute, Washington, DC.

Shackleton, S. and B.M. Campbell, 2001: Devolution in natural resource management: Institutional arrangements and power shifts: A synthesis of case studies from southern Africa, CIFOR, Bogor, Indonesia, 79 pp. Available at http://www.cifor.cgiar.org/publications/zip-file/Devolution-text.exe.

Shackleton, S., B. Campbell, E. Wollenberg, and D. Edmunds, 2002: Devolution and community-based natural resource management: Creating space for local people to participate and benefit, Natural Resource Perspective no. 76, March, Overseas Development Institute, London, UK.

Shiva, V., 1987: *Forestry Crisis and Forestry Myths: A Critical Review of Tropical Forests: A Call for Action,* World Rainforest Movement.

Singh, N., 2002: Federations of community forest management groups. In: *Orissa: Crafting New Institutions to Assert Local Rights,* Forests, Trees and People Newsletter no. 46, September 2002, pp. 35–45.

Six-country Initiative, 1999: *Practitioners' Guide to the Implementation of the IPF Proposals for Action,* Prepared by the Six-Country Initiative in support of the UN Ad-hoc Intergovernmental Forum on Forests, Eschborn, Germany.

Sizer, N., 1994: *Opportunities to Save and Sustainably Use the World's Forest Through International Cooperation,* World Resources Institute, Washington, DC.

Sizer, N., 1995: *Backs to the Wall in Suriname: Forest Policy in a Country in Crisis,* World Resources Institute, Washington, DC. (Also available in Dutch.)

Sizer, N., 1996: *Profit without Plunder: How to Reap Revenue from Guyana's Forests without Destroying Them,* World Resources Institute, Washington, DC.

Sizer, N., 2004: Regional approaches to tackle illegal logging and its associated trade in Asia, *Unasylva,* **55(40–44).**

Sizer, N., D. Downes, and D. Kaimowitz, 1999: *Tree Trade: Liberalization of International Trade in Forest Products, Risks and Opportunities,* World Resources Institute, Washington, DC.

Sizer, N. and D. Plouvier, 2000: *Increased Investment and Trade by Transnational Logging Companies in Africa, the Caribbean and the Pacific: Implications for Sustainable Management and Conservation of Tropical Forests,* European Commission, Brussels, Belgium.

Stanley, D.L., 1991: Communal forest management: The Honduran resin tappers, *Development and Change,* **22(4),** pp. 757–779.

Sundar, N., 2001: Is devolution democratisation? *World Development,* **29 (12),** pp. 2007–2023.

Tay, J., J. Healey, and C. Price, 2002: Financial assessment of reduced impact logging techniques in Sabah, Malaysia. In: *Applying Reduced Impact Logging to Advance Sustainable Forest Management,* T. Enters, P.D. Durst, G. Applegate, P.C.S. Kho, and G. Man (eds.), RAP Publication 14, FAO, Bangkok, Thailand, pp. 125–140.

Thornber, K., J. Gayfer, and B. Voysey, 2001: NFPs: Agent or product of institutional change? Unpublished draft. Available at http://www.fao.org/forestry/foris.

Tockman, J., 2001: *The IMF: Funding Deforestation,* American Lands Alliance, Washington, DC.

Troup, R.S., 1940: *Colonial Forest Administration,* Oxford University Press, Oxford, UK, 476 pp.

UK Tropical Forest Forum, 1998: Record of meeting held on 9 December 1998, Working Group on CSD Intergovernmental Forum on Forests.

UNCED (United Nations Conference on Environment and Development), 1992: Available at http://www.ciesin.org/docs/008–585/unced-home.html.

UNFF (United Nations Forum on Forests), 2002: Report of the Secretary-General on national forest programmes. Available at www.un.org/esa/sustdev/unffdocs/unff_ss2-nfps.pdf.

Upreti, B., 2001: Beyond rhetorical success: Advancing the potential for the community forestry programme in Nepal to address equity concerns. In: *Social Learning in Community Forests,* E. Wollenberg, D. Edmunds, L. Buck, J. Fox, and S. Brodt (eds.), CIFOR, Bogor, Indonesia, pp. 189–209.

USDA Forest Service (US Department of Agriculture), 2000**:** *Protecting People and Sustaining Resources in Rire-adapted Ecosystems: A Cohesive Strategy,* The Forest Service Management Response to the General Accounting Office Report GAO/RCED-99–65, 13 October , USDA Forestry Service, Washington, DC, p. 85.

USTR (US Trade Representative), 1999: *Accelerated Tariff Liberalization in the Forest Products Sector: A Study of the Economic and Environmental Effects,* Office of the US Trade Representative and White House Council on Environmental Quality, Washington, DC.

Wardle, P. and B. Michie, 2001: World trade flows of forest products. In: *World Forests, Markets and Policies,* M. Palo, J. Uusivuori, and G. Mery (eds.), World Forests Volume III, Kluwer Academic Publishers, Dordrecht, The Netherlands.

WCFSD (World Commission on Forests and Sustainable Development), 1999: Final Report of the WCFSD.

White, A. and A. Martin, 2002: *Who Owns the World's Forests? Forest Tenure and Public Forests in Transition,* Forest Trends, Washington, DC.

World Bank, 1991**:** *A World Bank Policy Paper: The Forest Sector,* September, World Bank, Washington, DC.

World Bank, 2002: *A Revised Forest Strategy for the World Bank Group,* 31 October, World Bank, Washington, DC.

Wunder, S., 2003: *Oil Wealth and the Fate of Tropical Forests: A Comparative study of Eight Tropical Countries,* Routledge, London, UK.

Young, C. and V. Prochnick, 2004: O investimento direto estrangeiro (IDE) e a estrutura industrial do setor florestal brasiliero. In: *Exportando sem crises: a industria de madeira tropical brasileira e os mercados internacionais,* D. MacQueen (ed.), Small and Medium Enterprises Series No. 1, International Institute for Environment and Development, London, UK.

Zobel, B.J. and J.P. van Buijtenen, 1989: *Wood Variation: Its Causes and Control,* Springer-Verlag, New York, 363 pp.

Zobel, B.J. and J. Talbert, 1984: *Applied Forest Tree Improvement,* John Wiley and Sons, New York, NY, 511 pp.

Chapter 9

Nutrient Management

Coordinating Lead Authors: Robert Howarth, Kilaparti Ramakrishna
Lead Authors: Euiso Choi, Ragnar Elmgren, Luiz Martinelli, Arisbe Mendoza, William Moomaw, Cheryl Palm, Rabindra Roy, Mary Scholes, Zhu Zhao-Liang
Review Editors: Jorge Etchevers, Holm Tiessen

FIGURES

TABLES

*This appears in Appendix A at the end of this volume.

Main Messages

Human activity has greatly increased the flux of nutrients through the landscape, roughly doubling the global flux of nitrogen and tripling the flux of phosphorus in the landscape over natural values. Agriculture is the major driver of change for both of these nutrient cycles, although other factors also contribute, such as creation of reactive nitrogen during fossil fuel combustion and the use of phosphorus as a surfactant in detergents, with ensuing effects on the environment and human well-being. Much of the change is recent, and half the synthetic nitrogen fertilizer ever used on Earth has been utilized since 1985. The extent of alteration of nutrient cycling is not uniform over the planet and varies greatly from region to region. In many parts of Europe, Asia, and North America, nitrogen deposition from the atmosphere and nitrogen fluxes in rivers have increased 10-fold or more. On the other hand, in some regions where population levels are low and where there has not been much agricultural activity, little change, if any, has occurred in nutrient fluxes in the landscape. Some parts of the world, including much of Africa, suffer from too little fertilizer availability (particularly phosphorus fertilizer) to support agriculture needs, a stark contrast to the nutrient surpluses that characterize the developed world and East and South Asia.

The consequences of excess nutrient flows are large and varied. The effect on human health, while poorly quantified, is also varied and potentially severe. With phosphorus, the primary concern is eutrophication (excess algal growth) in freshwater ecosystems, which can lead to degraded habitat for fish and decreased quality of water for consumption by humans and livestock. For nitrogen, the range of issues is far greater. Ecological and environmental effects include eutrophication of coastal marine ecosystems, eutrophication of freshwater lakes in the tropics, contribution to acid rain with effects on both freshwater and terrestrial ecosystems, loss of biodiversity in both aquatic and terrestrial ecosystems, creation of ground-level ozone (which leads to loss of agricultural and forest productivity), destruction of ozone in the stratosphere (which leads to depletion of the ozone layer and increased UV-B radiation on Earth), and contribution to global warming. The resulting health effects include the consequences of ozone pollution on asthma and respiratory function, increased allergies and asthma due to increased pollen production, risk of blue-baby syndrome, increased risk of cancer and other chronic diseases from nitrate in drinking water, and increased risk of a variety of pulmonary and cardiac diseases from production of fine particles in the atmosphere.

Manifestations of these problems vary regionally, from too much exposure to nitrogen in the soils, atmosphere, and waters in much of industrial Europe and North America to nutrient shortages hurting subsistence farmers of Africa. Both extremes are found in Latin America and Asia, with the largest growth in demand and use of commercial fertilizers in Asia.

Technical tools exist for reduction of nutrient pollution at reasonable cost. That many of these tools have not yet been implemented on a significant scale suggests that new policy approaches are needed. Current regulatory authority for non-point source pollution is often nonexistent or very limited. Hence, increased authority to regulate such sources may be necessary to reverse pollution of surface water by nutrients. Reversal of soil nutrient depletion and consequent reduction of crop yield in Africa and many parts of Asia and Latin America can be realized through a combination of technology options and policy and institutional reforms.

Market-based instruments hold the potential for better nutrient management, but may not be relevant in all countries and circumstances. Relatively little is known empirically about the impact of these instruments on technological change. Also, much more empirical research is needed on how the pre-existing regulatory environment affects performance, including costs. Which instrument is best in any given situation depends upon the characteristics of the environmental problem, and the social, political, and economic context in which it is being regulated.

Policy responses to nutrient pollution can be addressed through uniform national approaches, or through a watershed-based approach. The latter is likely to be the most cost-effective for some sources of nutrients (such as runoff from agricultural yields), while a uniform national approach may be better for others (such as NO_x from fossil fuel combustion or phosphorus from detergents). An important question for the choice of the correct pollution control instrument is related to the implementation costs of the instruments. The more suitable measures are associated with implementation difficulties, and policymakers might evaluate the trade-offs between cost-efficiency and ease of implementation. Policies should be developed to increase the supply of fertilizer to the regions where availability has been limited, and to encourage the fertilizer to be used efficiently and with less environmental leakage than has occurred in much of the industrial world. In industrial countries, policies need to be implemented to reduce this nutrient leakage. A major focus should be the increasingly concentrated production of animal protein in many regions, both since this can be a major driver in the increased use of synthetic nitrogen fertilizer and because the animal wastes are usually poorly treated and leak substantial amounts of both nitrogen and phosphorus to the environment.

Prospects need to be explored for developing a comprehensive understanding of a nutrient management strategy that transcends geographical, economic, and political boundaries to minimize the need for extracting phosphate from limiting reserves and for introducing more biologically available nitrogen into the biosphere and to distribute those nutrients efficiently according to local, regional, and global demands. Together with the understanding of a management strategy, the existing data on nutrient mobilization, distribution, and effects need to be assessed to insure that the science used to develop management strategies is sound and complete.

9.1 Introduction

Globally, the world has seen a tremendous increase in the use of synthetic nitrogen (N) fertilizer and inorganic phosphorus (P) fertilizer over the past half century. In conjunction with the inadvertent creation of reactive N during the combustion of fossils fuels, human activity has increased N fluxes two-fold and P fluxes three-fold (Vitousek et al. 1997; NRC 2000; Smil 2001; Howarth et al. 2002b; Galloway et al. 2004). However, these changes are far from uniform, and some regions have seen increased nutrient fluxes of ten-fold or more, while other regions have seen little or no increase (Howarth et al. 1996, 2002b; NRC 2000). Many regions of the world—most notably sub-Saharan Africa—have insufficient inputs of new nutrients to support agriculture. When crops are harvested, N and P are removed with the harvest, and insufficient return of these on nutrient-poor soils leads not only to low crop production but also to further degradation of soil quality due to increased erosion. Many other regions of the world—including most of the industrial world as well as East Asia, South Asia, and southeastern South America—now suffer from greatly increased fluxes of nutrients in aquatic ecosystems and, in the case of N, the atmosphere. (See MA *Current State and Trends,* Chapter 13.) It is important to note that while the world is divided into two parts—regions where nutrients flow in excess and regions where nutrient inputs are insufficient to provide adequate food production—this division does not simply follow the classical division between developed and developing nations. While the regions that have insufficient nutrient supplies fall within the developing world, many other developing regions have problems with surplus nutrients that match those of the industrial countries.

The consequences of excess nutrient flows are large and varied. For P, the primary concern is with eutrophication (excess algal growth) in freshwater ecosystems, which can lead to degraded habitat for fish and decreased quality of water for consumption by humans and livestock (Carpenter et al. 1998). For N, the range of issues is far greater and, in fact, a single atom of N can cascade through the environment and cause multiple problems (Galloway et al. 2003). Ecological and environmental effects include eutrophication of coastal marine ecosystems, eutrophication of freshwater lakes in the tropics, contribution to acid rain with effects on both freshwater and terrestrial ecosystems, loss of biodiversity in both aquatic and terrestrial ecosystems, creation of ground-level ozone (which leads to loss of agricultural and forest productivity), destruction of ozone in the stratosphere (which leads to holes in the ozone layer and increased UV-B radiation on Earth), and contribution to global warming (Vitousek et al. 1997; NRC 2000; Howarth et al. 2000; Tartowski and Howarth 2000; Rabalais 2002; Galloway et al. 2003).

The human health effects, while poorly quantified, are also varied and potentially severe. These include the consequences of ozone pollution on respiratory function, increased allergies and asthma due to increased pollen production, risk of blue-baby syndrome if nitrate levels in drinking water are high, increased risk of cancer and other chronic diseases from the presence of nitrate in drinking water, and increased risk of a variety of pulmonary and cardiac diseases from the production of fine particles in the atmosphere (Wolfe and Patz 2002, Townsend et al. 2003).

The problems from nutrient pollution on surface water quality are particularly well recognized, especially in industrial countries (Vitousek et al. 1997; Carpenter et al. 1998; Howarth et al. 2002a). The importance of P in promoting excessive aquatic plant production in fresh water was realized early, and was already well understood in the 1960s (Vollenweider 1968). Despite some early important studies (Ryther and Dunstan 1971), public awareness of the problems caused by excessive N inputs, however, took much longer to develop (Howarth and Marino, in press). While significant progress in reducing P pollution has been made in developed countries over the past 30 years, N pollution has actually grown worse, and is particularly evident in coastal ecosystems (NRC 1993a, 2000; Nixon 1995; Howarth et al. 2000). Today, two thirds of the coastal rivers and bays in the United States are believed to be moderately to severely degraded from excess N inputs (NRC 2000). Thus improving water quality is a major focus of nutrient management in industrial countries. .

In the developed world, P pollution has been counteracted increasingly effectively since the 1960s, through P precipitation in sewage treatment, bans on P in commercial detergents, and better controls on erosion of P-rich soils from agroecosystems. Some problem areas remain due to either excessive sewage inputs or inadequate control on soil erosion. Thus while many lakes have improved greatly, in some cases increasing agricultural P losses have resulted in continued deterioration (Foy et al. 2003). In contrast, N pollution is more of a problem with coastal marine ecosystems, a problem that many managers have only recently recognized (NRC 2000; Howarth and Marino, in press.). Thus there has been less effort and little success in stemming its rise. Nitrogen fluxes to surface waters are more difficult to control than P, due to their greater mobility through the atmosphere and groundwater (NRC 2000; Howarth et al. 2002b).

Worldwide, the most dramatic reversal of N pollution from non-point sources has been inadvertent. The end of the former Soviet Union led to the economic collapse of agriculture in Eastern Europe, and fertilizer use plummeted. As a result, nutrient loading to the Black Sea, especially of N, was greatly reduced. In only a few years, the Black Sea began to recover, including fish stocks and fisheries, from the eutrophication that had grown steadily worse from 1960 until 1990 (Boesch 2002). In developing countries, nutrient pollution has often gone unchecked, being low on the list of national priorities unless drinking water supplies are affected.

Much of the developing world did not take to scientific agriculture until the 1960s. With the expansion of education and training, India and China, in particular, have shown dramatic improvements in their agricultural practices with attendant increases in the use of commercial fertilizer. The introduction of improved varieties of wheat and rice that were responsive to fertilization (Hayami and Otsuka 1994), and the knowledge of successful application breakthroughs in the industrialized world involving fertilizers (Critchfield 1982), further contributed to the increased use of synthetic fixed N. Between 1960 and 1980, global production of wheat and feed grains grew slightly faster than the population, yielding a net increase in food supplies per person of 0.8% per year. China, in just three years in the late 1970s saw 15% growth in corn production, a 20% expansion in rice production, and a 40% gain in wheat production (Insel 1985).

Other parts of the developing world did not benefit to the same extent, with Africa providing many tragic examples of lost opportunities (Paarlberg 1996). Large areas of sub-Saharan Africa are affected by nutrient depletion (Stoorvogel and Smaling 1990). The consequences include low crop yields (Sanchez 2002) and increased erosion as a result of decreasing vegetative cover. Nutrient depletion has arisen through continual nutrient removal via crop harvests with insufficient nutrient replacement. This soil nutrient mining is in part due to the removal of fertilizer subsidies in the region in the 1990s through structural adjustment programs (Scoones and Toulmin 1999). A comparison of fertilizer use in 1999, surprisingly, shows a higher average application rate per hectare in developing countries as a whole (91 kilograms per hectare per year) than in industrial countries (87 kilograms) (Dudal 2002). However, sub-Saharan Africa is distinguished by an average rate of 9 kilograms per hectare per year, with a high of 56 for Zimbabwe and a low of 0.8 for Rwanda.

9.2 Background

Human activity has greatly accelerated the cycling of both N and P globally since the start of the industrial and agricultural revolutions. For P, the largest human influence has been increased erosion due to agriculture and the application of P fertilizer to agricultural lands, although the use of P as a surfactant in detergents also had a major influence, particularly before many governments banned them. Historical trends in increased global mobilization of P can be evaluated from oceanic sediment records. Globally, human activity has probably increased fluxes of P from land to the oceans three-fold, from a natural flux of 8 Tg P per year to the current flux of 22 Tg P per year (Howarth et al. 1995, 2002b; NRC 2000). The increase in the flux that is attributable to human activity (14 Tg P per year) is roughly equivalent to the rate at which P fertilizer is now used globally in agriculture (NRC 2000).

Human activity has also had an immense effect on the global cycling of N. This is mainly due to the use of synthetic fertilizer in agriculture. However, the release of N pollution during fossil-fuel combustion also contributes significantly (Vitousek et al. 1997; Galloway et al. 2004). Prior to the industrial revolution, most reactive N on Earth was created through the natural process of bacterial N fixation, with perhaps half occurring on land and half in the oceans (Vitousek et al. 1997; Cleveland et al. 1999; Karl et al. 2002; Galloway et al. 2004). Human activity has now roughly doubled the rate of creation of reactive N on the land surfaces of Earth. (See Figure 9.1.) The rate of change is extraordinarily rapid, and the N cycle is changing faster than that of any other element (Vitousek et al. 1997). More than half of all the synthetic N fertilizer ever used on the planet has been utilized since 1985 (NRC 2000; Howarth et al. 2002).

Human alteration of nutrient cycles is not uniform across the world. The greatest changes occur where human densities and human activities such as agriculture and fossil-fuel combustion are the greatest. Some of this variation can be clearly seen in the rates of N deposition, which are far higher in Europe, East and South Asia, eastern North America, and southeastern South America than elsewhere in the world. (See Figure 9.2 in Appendix A.) Human activity (fossil-fuel combustion and volatilization of ammonia to the atmosphere from farm-animal wastes) has increased the rate of deposition over natural rates 10-fold or more in many of these regions (Holland et al. 1999). The N flux in rivers provides one of the best-integrated measures of human influences on the N cycle (Howarth et al. 1996, 2002b; NRC 2000). Whereas human activity has resulted in essentially no change in the flux of N in the rivers of some regions, such as Labrador and Hudson's Bay in Canada, it has increased fluxes by five- to ten-fold or more in many regions of North America and Europe since the start of the industrial and agricultural revolutions. (See Table 9.1.)

Taking the United States as an industrial-country example, Figure 9.3 describes how the problem grew, stabilized, and started to grow again. Growth in population, intensification of agricultural activities, and increased emissions of oxidized N pollutants (NO_x) to the atmosphere from fossil-fuel combustion were major drivers behind the increases in fluxes of N to coastal ecosystems during the 1960s and 1970s (Howarth et al. 2002a). These drivers were relatively constant in the 1980s, particularly for NO_x emissions (EPA 2000), even though emissions of NO_x in the United States did not decline as much as other air pollutants after the passage of the Clean Air Act Amendments in 1970 (NRC 2000). About half of the NO_x emissions came from mobile sources, including automobiles, buses, trucks, and off-road vehicles, and 42% from electric power generation (EPA 2000).

In the late 1990s, the United States produced approximately one third of all the NO_x released from fossil-fuel combustion globally (Howarth et al. 2002a). Since the late 1980s, the flux of N from the rivers to the coasts has increased again in the United

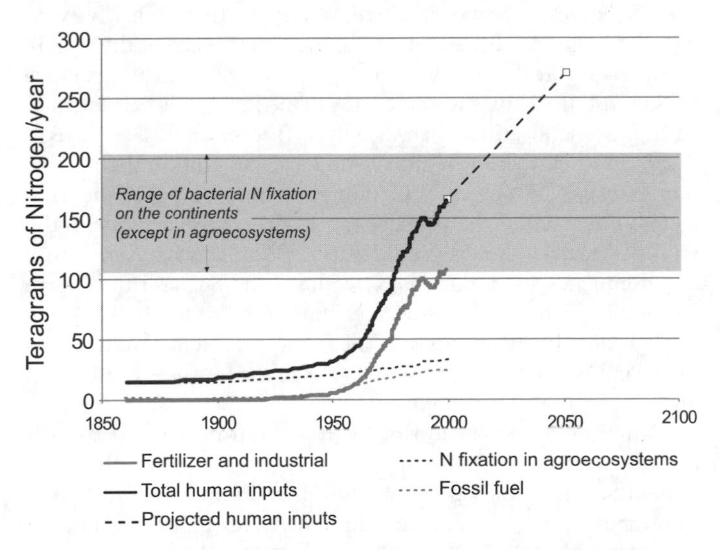

Figure 9.1. Global Trends in the Creation of Reactive Nitrogen through Human Activity, 1850–2100. The manufacturing of nitrogen by the Born Haber process for synthetic fertilizer and industrial use dominates, but the creation of reactive nitrogen as an inadvertent product of fossil-fuel combustion and the managed nitrogen fixation in agroecosystems also contribute. The natural rate of bacterial nitrogen fixation in natural terrestrial ecosystems (excluding fixation in agroecosystems) is shown for comparison. Note that human activity has roughly doubled the rate of formation of reactive nitrogen on the land surface of the planet. (Data for human creation of reactive nitrogen are from Galloway et al. 2004; data on natural rates of nitrogen fixation are from Vitousek et al. 1997 and Cleveland et al. 1999)

Table 9.1. Increase in Nitrogen Fluxes in Rivers to Coastal Oceans Due to Human Activities for Some Contrasting Regions (Howarth et al. 1996, 2002b; Bashkin 2002)

Region	Change
Labrador and Hudson Bay	no change
Southwestern Europe	3.7-fold
Great Lakes/St. Lawrence basin	4.1-fold
Baltic Sea watersheds	5.0-fold
Mississippi River basin	5.7-fold
Yellow River basin	10-fold
Northeastern United States	11-fold
North Sea watersheds	15-fold
Republic of Korea	17-fold

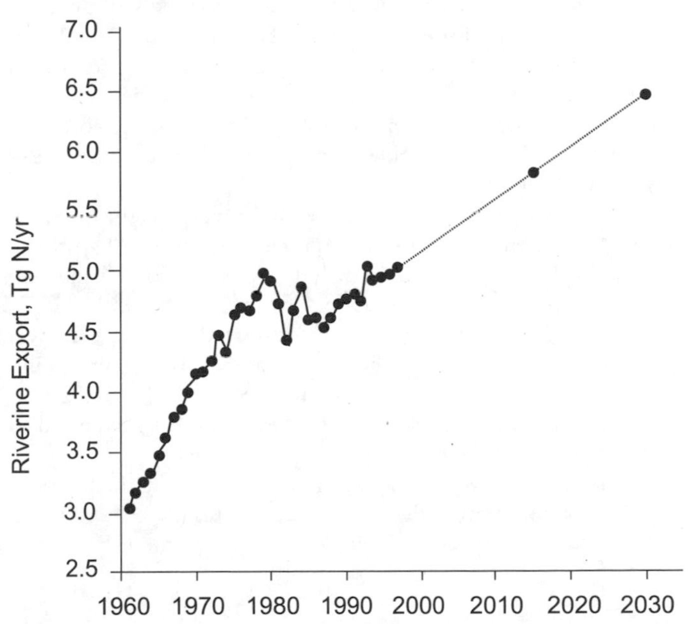

Figure 9.3. Estimated Flux of Nitrogen to Coastal Waters from the Entire United States in Rivers and from Sewage Treatment Plants, 1960–2000, with Projections to 2030. "Tg/N/yr" is equivalent to million metric tons of nitrogen per year. Future projections assume continued growth in population and export of cereal grain growth, as predicted by the U.S. Census Bureau and FAO, respectively, and no change in diet, agricultural practices, or regulation of NO_x emissions. (Howarth et al. 2002a)

States, largely due to increased synthetic fertilizer use and the increasingly industrialized production of meat protein (NRC 2000). Historical data for nitrate in major rivers such as the Mississippi River, the Susquehanna River (largest tributary of the Chesapeake Bay), and the Connecticut River (largest river input to Long Island Sound) show trends similar to those illustrated for total N use in the entire United States (Goolsby et al. 1999; Goolsby and Battaglin 2001; McIsaac et al. 2001; Jaworski et al 1997). If current trends in population growth, agricultural practices, grain exports, diet, and NO_x emissions continue, the flux of N to the coast is likely to continue to grow at the same rate as it has over the past decade (Howarth et al. 2002a). By 2030, N inputs to coastal waters in the United States could be 30% above the present level and over twice what they were in 1960.

The development of N pollution in Europe has followed a trajectory similar to that in the United States (von Egmond et al. 2002). It has been particularly troublesome for the shallow parts of the semi-enclosed seas, such as the Baltic Sea, the Black Sea, and the Adriatic Sea, whereas many coastal areas with strong tides and a good exchange of water with the open Atlantic have been much less affected. However, even the relatively open North Sea has experienced serious problems from eutrophication due to excess nutrient inputs.

Some developing countries have also experienced a rising N pollution problem as agriculture has intensified its use of synthetic fertilizers. The nitrate reaching the estuary of China's Changjiang (Yangtze) River increased four-fold from 1962 to 1990 (Shuiwang et al. 2000), and in 1995 an estimated 12 Tg of N was stored in agricultural soils in the major watersheds of China (Xing and Zhu 2002). Given current trends in N use and mobilization in China (Figure 9.4), these fluxes to the coast can be expected to continue to increase rapidly.

To date, there has been little progress in reversing the problem of coastal N pollution both in the United States (NRC 2000) and globally. N-removal technology for sewage treatment has led to water quality improvement in Tampa Bay and, to a lesser extent, in Chesapeake Bay in the United States (NRC 2000), and in Himmerfjärden Bay on Sweden's Baltic coast (Elmgren and Larsson 2001). Non-point sources of N, however, dominate inputs to most coastal waters of the United States and Europe (NRC 2000; Howarth et al. 1996, 2002b) and reducing them has proven problematic (Boesch et al. 2001a). In 1987, a target was set to reduce controllable inputs of N to Chesapeake Bay by 40% by the year 2000. This goal was not met, in part because management strategies for non-point sources were less effective than had been assumed (Boesch et al. 2001b) and because international agreements on 50% reductions for the North Sea (1987) and the Baltic Sea (1990) areas have also failed to reach their targets (OSPAR Commission 2000; Elmgren and Larsson 2001). One possible reason for these failures may be the long time needed for decreased nutrient losses from agriculture to show up as reduced riverine transport in some watersheds (Grimwall et al. 2000). Another reason may be an inadequate assessment of which management practices will work most effectively for N control (NRC 2000).

Many technical solutions for reducing N pollution exist, but few have been implemented systematically. In part, this is because nutrient management in both the United States and Europe has tended to concentrate on P pollution, which is the larger problem in freshwater ecosystems. The management community has been slow to recognize that N is often the larger problem in coastal marine ecosystems (NRC 2000; Howarth et al. 2000; Howarth and Marino, in press). Nitrogen is more mobile than P in the environment, flowing readily both through groundwater and the atmosphere, a difference in biogeochemistry that requires different management practices (NRC 2000).

Management of N pollution in coastal waters faces many policy challenges. Coastal ecosystems vary in their sensitivity to nutrient pollution, because of differences in the size of the watershed, in physical mixing regime, and in ecological structure (NRC 2000). Thus the same rate of N input will cause more harm in some locations than in others (Bricker et al. 1999; NRC 2000). Further, there is regional and local variation in the relative importance of N sources. Agriculture dominates input to the Mississippi River and is the primary cause of the hypoxic zone in the Gulf of Mexico (Goolsby et al. 1999; NRC 2000). Agriculture also dominates the inputs of N to the Baltic Sea and the North Sea, although human wastewater inputs are significant (Howarth et al. 1996). In many regions, including the Netherlands, Denmark, and Brittany in Europe and parts of the southeastern United States, animal wastes are a major part of these agricultural fluxes.

Atmospheric deposition of N from fossil-fuel combustion is the largest input to several marine ecosystems in the northeastern United States, mainly as runoff after deposition on the terrestrial landscape although, again, human wastewater inputs are important (Howarth et al. 1996; Jaworski et al. 1997; Boyer et al. 2002). And wastewater from sewage is the biggest input to the coastal waters of Korea (Bashkin et al. 2002) as well as to some major estuaries in the United States, including the Hudson River estuary in New York City and Long Island Sound (NRC 2000; Howarth et al. 2004) and rivers in the industrialized portion of Brazil (Martinelli et al. 1999; Martinelli 2003). Overall, human wastewater flows are thought to contribute 12% of the riverine N flux in the United States, 25% in Western Europe, 33% in China, and 68% in the Republic of Korea (Howarth et al. 1996, 2002a; NRC 2000; Bashkin et al. 2002; Xing and Zhu 2002). For many coastal rivers and bays, the relative importance of various inputs of N is uncertain, because the models used to estimate inputs are inexact and largely unverified (NRC 2000).

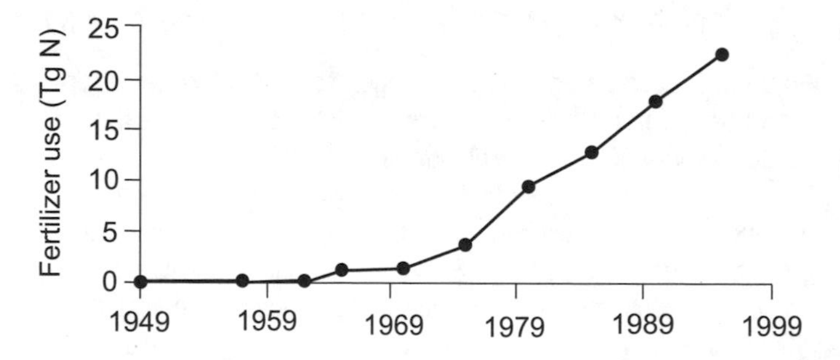

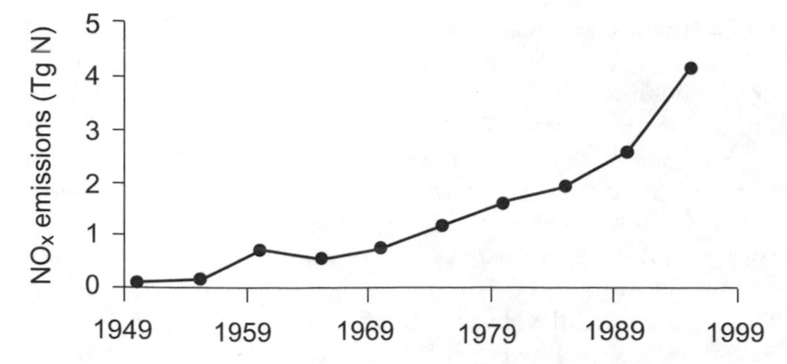

Figure 9.4. Trends in Synthetic Nitrogen Fertilizer Use and Release of NO_x to the Atmosphere from Fossil Fuel Combustion in China, 1949–99 (Xing and Zhu 2002)

9.3 Responses to Insufficient Nutrients to Support Agriculture in Some Regions

Many regions of the world, particularly Africa, are in urgent need of greater nutrient inputs to support food production. The proper use of these increased nutrients would not only increase the regional food supplies but would also improve soil characteristics and, therefore, lead to less soil loss from erosion. The challenge is how to ensure that nutrient replenishment in developing countries does not follow the pattern of excessive nutrient applications that now threatens many ecosystems. In general, there is a need for balanced fertilizer applications tailored to specific soil nutrient deficiencies and agroecosystem requirements. Whether these nutrients are applied as mineral or organic fertilizers is not as important as the management practice that assures nutrient use efficiency both at the crop and the system level. Several nutrient replenishment strategies have been or are being developed in sub-Saharan Africa. The best strategy for nutrient replenishments will depend on the soil, climate, agroecosystem, socioeconomic conditions, and policy environment. Most of these nutrient replenishment strategies entail a combination of mineral and organic inputs, with the exact mix determined in part by socioeconomic conditions as well as the realization that organic materials cannot, in general, supply sufficient P to meet crop demand (Palm et al. 1997).

Though N is the primary limiting nutrient in most soils of sub-Saharan Africa, once the N is replenished, P quickly becomes limiting to crop production (Bationo et al. 1986). P can be the primary limiting nutrient in some of the sandy soils of the semiarid area and on moderate-to-high P-fixing soils in the subhumid and humid areas (Buresh et al. 1997; Sahrawat et al. 2001). Several strategies exist for replenishing soil P: application of soluble P fertilizers, application of reactive phosphate rock (RP), or the combination of soluble P fertilizer and RP (Buresh et al. 1997). The replenishment can be achieved through a single large application with residual effects lasting several seasons or through smaller seasonal applications.

The direct application of phosphate rock is often proposed as the better alternative because of lower production costs than for soluble P fertilizers (Buresh et al. 1997). Phosphate rock deposits are found throughout Africa but they vary in their effectiveness for direct application to the soil (Mokwunye and Bationo 2002). The agronomic efficiency of the phosphate rock depends on its mineralogy, particle size, and reactivity (sedimentary forms being more reactive than igneous) as well as the soil type. Dissolution of the phosphate rock requires soils that are slightly acidic and low in calcium (Ca) and P in solution. The reactivity of the RP can also be increased through partial acidulation (Buresh et al. 1997) and also when used in combination with plants, particularly legumes, that are more efficient in accessing P (Lyasse et al. 2002).

The choice of P fertilizer depends, then, on the soil, the climate, plant species, and the comparative costs. While organic inputs do not have sufficiently high concentrations of P to replenish soil P at reasonable application rates, they can increase soil P availability above that obtained through the same application rates of mineral P. Where it is difficult and/or uneconomic to obtain P fertilizers, the combined use of organic and mineral P fertilizers has been shown to have higher P use efficiencies (Palm et al. 1997).

High rates of P application are likely to have negative environmental effects, primarily through erosion and runoff. Introduction of biological filter strips or biological terraces have proven quite effective in practically eliminating runoff and soil erosion of P; in addition, application of P increases the vegetative cover, practically eliminating runoff and loss of P by erosion. Losses of P through leaching are of concern primarily on sandy soils and medium textured soils with low soil organic matter; there is limited P movement down the soil profile on clay soils.

Biological N fixation offers an economically attractive alternative to synthetic N fertilizers (Bohlool et al. 1992; Döbereiner et al. 1995). Intercropping and rotation cropping is commonly done with N-fixing legumes. Nitrogen-fixing bacteria can be introduced in the soil to enhance N availability to both leguminous and non-leguminous plants. In Cuba, large-scale production and use of Azotobacter (free-living, N-fixing bacteria) is estimated to supply more than half of the N needed by non-legumes (Oppenheim 2001). Brazil has become the world leader in replacing chemical fertilizers with biological N fixation; mean value of N application is as low as 10 kilogram per hectare. Agriculture in Brazil is one of the main export activities, with soybeans, the largest export product of the country (Döbereiner 1997).

9.4 Responses for Management of Excess Nutrients

This section summarizes technical solutions for nutrient control and then outlines policy options for implementing them. The section relies heavily on the analysis by the U.S. National Research Council's Committee on Causes and Management of Coastal Eutrophication (NRC 2000) but uses other recent information in addition.

9.4.1 Leaching and Runoff from Agricultural Fields

Global use of synthetic N fertilizer increased steadily from about 11 teragrams N per year (million metric tons N per year) in 1960–61 to about 80 teragrams N per year in 1988–89, while fertilizer P use increased less steeply, from almost 5 to over 16 teragrams. Much of the recent growth has occurred in the developing world, particularly in China. The economic crisis after 1989 in the former Warsaw Pact countries led to an initial fall in global fertilizer use of 10–20%, but by 2000–01, global N use had recovered fully, and P use partly (IFA 2002). Global use of synthetic N fertilizer from 2001 exceeds 100 teragrams per year. Much of this fertilizer is assimilated by plants and harvested with crops, but some of it is lost to the environment. In the United States, for example, over half of the synthetic N fertilizer inputs are removed with crop harvest on average, and about 20% leaches to surface or ground waters (NRC 1993b; Howarth et al. 1996, 2002a).

The variability in leaching among fields is great, ranging from a low of 3% for grasslands with clay-loam soils to 80% for some row-crop agricultural fields on sandy soils (Howarth et al. 1996). P losses, while generally lower, are almost as variable (NRC 2000). Climate is important, with greater nutrient losses in areas of high rainfall and in wet years (Randall and Mulla 2001). These differences indicate that targeting particularly leaky types of agricultural fields can greatly reduce pollution of aquatic ecosystems.

Management practices for reducing N loss to downstream ecosystems have been reviewed by the National Research Council of the United States (NRC 1993b, 2000) and by Mitsch et al. (1999). The best management practices for reducing N pollution often differ from those for P, due to the higher mobility of nitrate-N in ground waters (NRC 2000). For instance, no-till agriculture reduces erosion and, therefore, P losses from fields, except at very high P inputs, but has little or no effect on nitrate loss (Randall and Mulla 2001). Particularly promising approaches for reducing N leaching from agricultural fields include:

- *Growing perennial crops such as alfalfa or grasses rather than annuals such as corn and soybeans.* Perennials retain N in the rooting zone and greatly reduce losses to groundwater. In Minnesota and Iowa, fields planted with perennial alfalfa lost 30- to 50-fold less nitrate than fields planted with corn and soybeans (Randall et al. 1997; Randall and Mulla 2001).
- *Planting winter cover crops, which greatly reduce the leaching of nitrate into groundwater during winter and spring, when most leaching normally occurs.* In a Maryland study, winter cover-crop plantings reduced nitrate loss three -fold (Staver and Brinsfield 1998).
- *Applying N fertilizer at the time of crop need.* In the North Temperate Zone, this is in the spring and summer, yet fertilizer, which is relatively inexpensive, is often applied in the fall, when the farmer has time, even if much of the applied fertilizer is leached to groundwater before crop growth begins in the spring. A study in Minnesota showed that fall application of fertilizer increased N leaching by 30–40% (Randall and Mulla 2001).

Much can be gained simply by eliminating excess N fertilizer. Adding more N increases crop yield only up to a point, after which the crop's need for N is saturated and further fertilization has no effect on production (NRC 2000). Land grant universities in the United States advise farmers on appropriate rates of fertilizer application for optimum economic return from crop production under local conditions. In practice, the average farmer in the upper mid-western "breadbasket" area of the United States applies significantly more synthetic N fertilizer than recommended, and 20–30% more than is required to support present crop yields. (See Figure 9.5.) The reasons include underestimation of N available from other sources, such as residues from previous crops, overly optimistic yield expectations, the relatively low cost of N fertilizer, and a tendency to apply extra fertilizer as "insurance" to guarantee maximal yield (Boesch et al. Submitted; Howarth et al. 2002a; Howarth, in press). Note that in some regions with insufficient regulation of N use, farmers overfertilize simply as a way to dispose of animal wastes (discussed below).

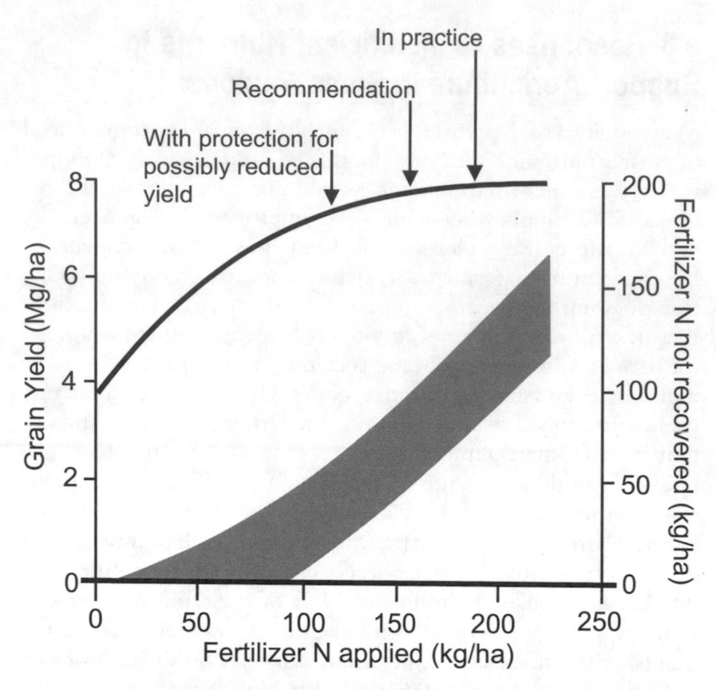

Figure 9.5. Crop Production and Leaching of Nitrogen to Surface and Ground Waters as a Function of Inputs to Agricultural Fields. "Recommendation" indicates application recommended by Land Grant universities based on optimum economic return to farmers. "In practice" indicates the actual average application by farmers. Less fertilizer than the recommended rate for maximizing economic return could be used, with great gains in reducing nitrogen leakage and only small decreases in crop yield. This figure is based on a compendium of real data for farm recommendations and crop production for the upper midwestern United States, where corn and soybean grown in rotation dominate as the cropping system, as of the mid-1990s. (Boesch et al. submitted)

Nitrogen not taken up by the crop is available for leaching to surface water and groundwater, and this increases rapidly if N inputs increase beyond the point of crop N saturation. Thus reducing fertilizer use by 20–30% would, in all likelihood, reduce the downstream N pollution by considerably more than 20–30% (Boesch et al., submitted; Howarth et al. 2002a; McIsaac et al. 2001). Such a reduction would also save farmers money, as they would get essentially the same yield but pay less for fertilizer. One promising approach to achieve this goal is the use of voluntary crop production insurance (Howarth, in press). In a trial plan run by the American Farmland Trust, farmers pay into a not-for-profit insurance fund and agree to use less N fertilizer on most of their cropland. Their payments into the fund are less than the savings from purchasing less fertilizer, so the farmers have an economic incentive to participate. Small patches of the fields are heavily fertilized, and the average yield for the entire field planted is compared with the yield in the heavily fertilized plots. If the average yield is below that of the test plots, the farmer is compensated for this lower yield.

Another promising approach for reducing N losses from agriculture fields is the use of precision agriculture, where the timing and amount of fertilization are closely matched to crop needs at relatively small spatial scales (NRC 1993b). Also, genetic engineering may hold promise for increasing the nutrient use efficiency of crops. However, as noted above, significant reduction in nutrient leakage from agroecosystems can be made with existing crops and techniques through such techniques as changing cropping systems, reducing fertilizer use, and employing cover crops.

Some agricultural lands in many regions, including the Netherlands and portions of the "breadbasket" midwest in the United States, are artificially drained with tile drains. This is necessary for the growth of most row crops, but it increases leakage of N from the fields. It may be possible to lessen this N leakage by raising the level of drains in fields, while still providing adequate drainage for crop production (Boesch et al., submitted). It is also possible to build artificial wetlands that intercept the tile drainage, provided this does not cause the tiled fields to become flooded. Such wetlands can substantially reduce the flux of nitrate to surface waters (Mitsch et al. 2001). Buffer strips, while effective for trapping P (which is largely particle bound), are not good at trapping N from drainage systems unless the drained water is fully intercepted by the buffer. The subject of artificial or reconstructed wetlands for reducing the flux of nutrients from the landscape is discussed below.

9.4.2 Animal Production and Concentrated Animal Feeding Operations

Animal wastes are a major source of nutrients in many regions, including coastal North Carolina in the United States and many areas of Western Europe, including Brittany, the Netherlands, and

Denmark. In the United States, where per capita meat consumption is among the highest in the world, over half the country's crop production is fed to animals, mostly in feedlots. Most of these crops are transported over long distances before being fed to the animals, making it expensive for the farmers to return the animal wastes to the site of the original crop production (NRC 2000; Howarth et al. 2002a). Instead it is far cheaper for farmers to purchase synthetic fertilizers to use on their fields. The production of animal protein in the United States continues to increase, in part driven by a steady increase in the per capita meat consumption of Americans (Howarth et al. 2002a). The trend for production to concentrate in fewer but larger facilities also continues. During the 1990s, production of hogs, dairy cows, poultry, and beef cattle all rose while the number of operations in each of these segments declined (NRC 2000).

Wastes from concentrated animal feeding operations are normally either spread on agricultural fields, or just held in lagoons. Some operations are beginning to compost animal wastes (NRC 2000). Animal manure can, of course, be used as fertilizer, and recycling it back to agricultural fields is desirable. In practice, however, it is difficult to apply manure at the time and rate needed by the crop, due to the uncertainty about the time of nutrient release and the difficulty of spreading it uniformly (NRC 2000). Most manure is transported only very short distances due to the expense of transporting the heavy waste and the availability of relatively inexpensive synthetic and inorganic nutrient fertilizers, making the use of manure unnecessary for crop production. This results in overfertilization of fields near animal feeding operations, and pollution of ground water and downstream aquatic ecosystems (NRC 2000).

Lagoons are a problematic approach for handling waste, due to loss during flood events, significant leakage of N to ground water, and much volatilization of N as ammonia to the atmosphere. The ammonia contributes to acid rain, the production of fine particles in the atmosphere, loss of biotic diversity in forests and grasslands (Vitousek et al. 1997), and, eventually, the flux of N to coastal waters (NRC 2000). An estimated 40% of all the N in animal wastes in the United States—whether spread on fields as manure or held in lagoons—is volatilized to the atmosphere (Howarth et al. 2002a). In Europe, animal manure is the largest source of atmospheric ammonia emissions, followed by fertilizer use. Estimates of atmospheric ammonia emissions in Europe have considerable uncertainties, but indicate a decrease by about 14% between 1990 and 1998, largely due to decreased agricultural activities in Eastern Europe after 1989. No further decrease is expected by 2010 (Erisman et al. 2003).

Animal wastes can be composted to make them easier for use as effective fertilizers. However, much ammonia is volatilized to the atmosphere during the composting, which lowers the value of the compost as fertilizer and contributes to pollution by atmospheric deposition (NRC 2000). More effective and less polluting methods for treating animal wastes are an urgent need (NRC 2000).

Some progress is being made in developing more environmentally benign approaches for animal wastes, as indicated by the proceedings of a recent symposium sponsored by the International Water Association in Seoul, Korea, on approaches for dealing with nitrogen-rich wastes, including animal wastes. The 135 papers show a wide range of creative and potentially useful approaches, including more effective agricultural re-use and production of biogas for fuel (Choi and Yun 2003). In the European Union, some dairy operations now make more money from biogas production from cow wastes than from selling milk, when the biogas subsidies are considered (Holm Tiessen, personal communication).

Van Asseldonk (1994) showed that environmental efficiency assessments of manure surpluses can be useful to: (1) demonstrate the financial benefits of improved environmental efficiency; (2) influence farmers' attitudes toward the environment; and (3) get farmers more involved in thinking about environmental targets. Leneman et al. (1993) distinguished three main categories, using the handling of pig waste as an example: the first one aims to reduce N and P excretion by changes in feeding and composition regime of the pigs; the second aims to reduce N volatilization and can be carried out in pig houses and on manure storage outside the pig house; finally, the third category aims to reduce N leaching and P leaching in the soil. Combinations of these measures can dramatically reduce N and P emissions (for example, P emissions were reduced as much as 97% in furrowing operations and 95% in finishing operations, and N leaching was reduced by 67% in furrowing operations and by 73% in finishing operations).

The redistribution of manure from areas of high livestock density to arable farming areas did take off to some extent in the Netherlands. Manure processing has turned out to be far more expensive than had been initially thought because of the high cost of developing new technology. Research on manure processing continues, however, particularly in the private sector, because, since 1998, high charges have been imposed on N and P surpluses at the farm level.

9.4.3 Fossil Fuel Sources

As Figure 9.1 shows, the emission of oxidized forms of N (NO_x) to the atmosphere from fossil-fuel combustion contributes approximately 22 teragrams of N to the global environment every year, or roughly 20% of the rate of synthetic N fertilizer use. In China, NO_x emissions, as late as the 1990s, equaled 20% of the use of synthetic N fertilizer, but the increase in NO_x emissions is even greater than that for fertilizer use (shown in Figure 9.4). In the United States, which has the higher per capita emissions of NO_x than any other nation on Earth, the emission rate is 7 teragrams N per year, or 60% of the rate of N fertilizer use in that country (Howarth et al. 2002a). Most NO_x emissions are deposited onto the landscape as rain and as dry deposition, and are the major contributors of acid rain, as well as being significant contributors of the nutrient pollution of coastal waters. A significant percentage of this deposition is exported from forests and other terrestrial systems to rivers and downstream coastal marine ecosystems (NRC 2000; Howarth et al. 2002b), particularly when total N deposition exceeds 8 to 10 kilograms N per hectare per year (Emmet et al. 1998; Aber et al. 2003). Although in most regions of the world N fluxes to the coast are dominated by agricultural sources and human wastewater, in some regions (notably the northeastern United States) atmospheric deposition is the single largest source. Atmospheric deposition of N from fossil-fuel combustion exported via watersheds is a major input to almost all coastal rivers and bays along the eastern seaboard of the United States (NRC 2000; Howarth et al. 2002b; Boyer et al. 2002).

NO_x is the only major pollutant among those in the United States that are regulated under the Clean Air Act that has not declined significantly since the Act was passed in 1970, although regulation may have stabilized the emissions (NRC 2000). Emissions rose exponentially through the 1960s and 1970s, but have been relatively constant since 1980 (EPA 2000), with about half coming from mobile sources, including automobiles, buses, trucks, and off-road vehicles and 42 per cent from electric power generation (EPA 2000). Major sources of NO_x emissions in Europe are transport, industry, and energy production. Work under the Convention on Long-range Transboundary Air Pollution resulted in a reduction of emissions in Europe (Russia excluded) by 21% between 1990 and 1998, mostly due to decreased industrial activity in Eastern Europe. Under the Convention's new Gothenburg Protocol and new EC Directives, reduction is predicted to reach 50% of 1990 emissions by 2010 (Erisman et al. 2003).

Technical controls of NO_x emissions have been much studied because they are central to the formation of ground-level ozone and also contribute to acid rain (NRC 2000; Mosier et al. 2001). The basic approaches are either to burn less fossil- fuel, through greater energy efficiency and reduced driving and/or to remove NO_x from the exhaust as with catalytic converters. NO_x emissions from fossil-fuel combustion in the United States can be almost eliminated with currently available technology (Moomaw 2002). Taking old, "grandfathered" power plants off line is a significant and inexpensive step in this direction. Stricter emission standards for sport utility vehicles, trucks, and off-road vehicles are other significant steps. Electric power generation by fuel cells rather than traditional combustion could completely eliminate NO_x emissions from that source (Moomaw 2002).

Efforts to regulate NO_x emissions have been driven largely by the ozone and air quality problems in Europe and North America. The contribution of NO_x to coastal N pollution is a reason for greater efforts, and underlines the need for year-round reductions in emissions; since ozone is a problem mostly in warm weather, current regulatory requirements often focus only on reducing emissions in the summer.

9.4.4 Urban and Suburban Sources

While non-point sources dominate inputs of N to most regions of the world, human wastes are the major source in some regions (such as the Republic of Korea) and are often the largest source to urban estuaries. Human wastes are the primary urban source of N, but atmospheric deposition of N from fossil-fuel combustion can also be substantial, since most NO_x emissions are deposited near emission sources, which are huge in urban areas (Holland et al. 1999; NRC 2000; Howarth et al. 2002b, 2004). In the older cities of both Europe and North America, sanitary wastes and storm waters are mostly combined in the same sewer system, and some of the N entering the sewage treatment plants is, therefore, derived from atmospheric deposition washed off the streets during rainstorms (NRC 2000). Currently, about half of the population of the world lives in urban areas, and the global urban population is expected to continue to grow over this century by 2% per year (UNEP 2002; Austin et al. 2003).

Most human sewage in the world enters surface water with no treatment. Thus while in North America 90% of urban wastewater is treated and in Europe 66%, in Asia only 35% of it is treated, in Latin America and the Caribbean only 14%, and in Africa it is not treated at all (Martinelli 2003). Even in North America, most sewage treatment is not aimed at nutrient reductions, but rather at the reduction of the labile organic matter that contributes to "biological oxygen demand" (NRC 2000). This "secondary treatment" is not very effective at removing N and P, and the nutrient content of the effluent from an average secondary sewage plant is substantial (NRC 1993a).

Advanced tertiary treatment of sewage for removal of nutrients will, on average, remove up to 90–95% of the P in untreated sewage. Not all tertiary treatment is effective at reducing N, but some technologies can remove up to 90% of the N in sewage. The cost for such tertiary treatment is approximately 25% above that for secondary treatment, including both capital and operating costs (NRC 1993a). For a large urban city such as New York, the additional cost is $30–60 per person per year (Howarth et al. 2004). Substantial savings could be achieved if wastewater treatment plants were designed from the start with the goal of nutrient reduction (NRC 1993a).

9.4.5 Wetlands as Nutrient Interceptors: Enhancing the Sinks

The sections above discussed approaches for reducing anthropogenic nutrient inputs to the environment. Another, complementary strategy is to enhance sinks for nutrients in the landscape. Wetlands, ponds, and riparian zones are particularly effective nutrient traps, serving both to sediment out particulate forms of N and P and to convert biologically available combined N into N_2 and N_2O gases through the process of denitrification (NRC 2000; Howarth et al. 1996; Mitsch et al. 2001). N_2 is a ubiquitous and harmless gas but, unfortunately, N_2O is a very long-lived gas in the atmosphere (about 120 years) that contributes both to climate change and to the creation of ozone holes in the stratosphere (Vitousek et al. 1997; Howarth 2002). Mitsch et al. (2001) estimate that the N load from the heavily polluted Illinois River basin to the Mississippi River could be cut in half by converting 7% of the basin back to wetlands. Restoration of water flows through the wetlands of the Mississippi River delta could also significantly reduce N fluxes onto the continental shelf, where they contribute to the Gulf's hypoxic zone (Mitsch et al. 2001; Boesch et al., submitted). Small natural streams can also be extremely effective sinks (Peterson et al. 2001).

A major unknown is whether the amount of N_2O that is produced during denitrification in wetlands and small streams is greater than or less than the amount that would be produced during denitrification in downstream coastal marine ecosystems (including on the continental shelf), which would otherwise be the fate of much of this excess N (Nixon et al. 1996). A major research goal should be to evaluate this trade-off, and to work towards designing wetlands as treatment systems that minimize the production of N_2O (Howarth et al. 2003).

9.5 Analysis and Assessment of Selected Responses

Most nations of the developed world have responded to the problem of excesses of P and N in surface waters and in the air with legislative and regulatory responses. However, these often have not proven adequate, particularly for N pollution. In this section, the various types of responses that have been proposed or tried are reviewed, and their potential usefulness is assessed. For the most part, developing-country governments have not yet managed nutrient pollution. Thus the focus of the discussion here is of necessity on the industrial world, particularly the United States but also some countries in Western Europe.

9.5.1 Watershed-based versus Nationally Uniform Approaches

The United States needs a national strategy to reduce N pollution in coastal waters (NRC 2000). Federal involvement is appropriate, because the nutrients polluting many coastal ecosystems come from large river systems that flow through many states and, for N, from large multistate air sheds. Notable examples include Long Island Sound, Chesapeake Bay, and the Mississippi River plume. Further, national agricultural policies significantly affect N pollution. However, the problem of N pollution manifests itself at the local to regional scale, so local and state governments also clearly have a role. Since coastal ecosystems vary both in sensitivity to N pollution and in their sources of N, a national goal of protecting ecosystems not yet damaged and of restoring those that have been damaged seems preferable to a goal of N reduction per se (NRC 2000; Howarth, in press). This would require a partnership of federal, state, and local authorities, cooperating with academia and industry.

This approach posits that the coastal ecosystems that should be restored first are those that are most sensitive to nutrient pollution, that have nutrient sources that can most effectively and economically be reduced, or that have the greatest ecological or societal value. The watershed (and associated air shed) is viewed as the appropriate scale for management. A benefit of this approach is that by targeting locations, limited technical and financial resources are most effectively used.

A watershed-specific approach requires estimates of maximum allowable nutrient loads to coastal rivers and bays. As a start, this could be based on the average responses of these ecosystems to increased inputs of N, which is most often the limiting nutrient (NRC 1993a, 2000; Nixon 1995). Since coastal marine ecosystems vary in their sensitivity to nutrient inputs, however, the most cost-effective protection would be obtained by setting higher allowable N load limits for ecosystems that are insensitive to N inputs and lower limits for systems that are highly sensitive (NRC 2000). This would require that coastal rivers and bays are classified according to their sensitivity to nutrient pollution and that loading limits be established by this classification (NRC 2000). An alternative but more time-consuming option is to construct site-specific models for each coastal river and bay (NRC 2000). Note that unlike many pollutants, concentrations of N are a poor predictor of the effects of N pollution in the coastal zone. A strong scientific consensus exists that N pollution should be managed on the basis of N input rates (loads) (NRC 2000). The "total maximum daily load" provision of the Clean Water Act of the United States (discussed further below) is a regulatory mechanism that uses allowable loads for reducing nutrient pollution.

A watershed-specific approach is technically challenging, since the ability to classify coastal ecosystems as to their sensitivity to nutrient pollution is lacking except in broad outline (NRC 2000). Also lacking is detailed, reliable knowledge on the sources of N to most individual coastal ecosystems, although the general patterns at the regional or national scale are clear (NRC 2000; Howarth et al. 2002b). A major strategy of the Clean Water Action Plan put forward in 1998 was to develop nutrient criteria that could aid watershed management, but nutrient criteria for coastal waters are still many years away, as discussed below. While current knowledge is sufficient for starting to restore individual coastal rivers and bays, more research on the sensitivity of ecosystems to nutrient pollution and on sources of nutrients to individual ecosystems will be needed to find cost-effective solutions (NRC 2000; Howarth et al. 2003).

The European Union decided in 2001 to adopt a mainly watershed-based approach (the "Water Framework Directive") to water quality management for its groundwater, fresh water, and coastal water (Chave 2001). The approach is a mixed one, with earlier directives, such as the technology-based "Urban Wastewater Treatment Directive" and the "Nitrate Directive" for managing agricultural N pollution, initially remaining in force, but eventually being replaced by the Water Directive. The stated objective is to achieve good, or better, water status for all covered waters by 2015.

An alternative to the watershed-based approach would be a uniform national regulation to reduce, for example, overall N fluxes to the coastal waters of the United States, by 10% by 2010 and by 25% by 2020, without regard to the effects on individual coastal ecosystems. A uniform approach requires less technical expertise and site-specific information than a watershed-specific approach. If the reductions in N flow occur to coastal systems that are insensitive to N pollution, however, they will bring little, if any, environmental benefit. Moreover, increased local loads to some sensitive areas cannot be ruled out. Consequently, this approach is likely to be less cost-effective, and may fail to protect the most sensitive coastal ecosystems (NRC 2000).

9.5.2 Voluntary Policies for Reaching Goals

Both voluntary and mandatory approaches have been used for nutrient management, and both should be considered as part of the national strategies against nutrient pollution, whether these strategies use a watershed-specific or a uniform national approach. Motivations for polluters to join a pollution abatement plan voluntarily include a commitment to environmental stewardship, a perceived payoff in the marketplace (selling a "green" product), a financial incentive or subsidy, and a fear that failure to participate will lead to stricter regulatory control (NRC 2000). The regulatory threat as a powerful motivator for voluntary compliance is an argument for hybrid approaches, which combine regulations and voluntary programs (NRC 2000).

Voluntary approaches have been used successfully to reduce N pollution in Tampa Bay. On the other hand, after trying voluntary nutrient management on farms, the State of Maryland, in 1998, moved to mandatory control (NRC 2000). The Integrated Assessment on Hypoxia in the Gulf of Mexico brought together a diverse group of scientists from governments and academia to produce consensus reports on the problem of the hypoxic zone, and on approaches for solving the problem. These reports formed the basis for negotiations between the federal government and the state governments regarding the Mississippi River basin. These resulted, as a first step, in a voluntary agreement in 2001 to reduce the size of the hypoxic zone by reducing N loading down the Mississippi River over the next few decades. Whether voluntary cooperation will prove sufficient to reach this goal remains to be seen, but the Integrated Assessment has proven that voluntary steps can be taken even at large, multistate scales.

Robinson and Napier (2002) studied the adoption of nutrient management techniques to reduce hypoxia in the Gulf of Mexico. They collected data from 1,011 landowner-operators within the three watersheds located in the north-central region of the United States to examine use of selected water protection practices; their research findings suggest that existing conservation programs are no longer useful policy instruments for motivating landowner-operators to adopt and use production systems designed to reduce agricultural pollution of waterways. They strongly suggest that policy-makers should reconsider allocation of limited funding for conservation-education and for efforts designed to increase access of farmers to training.

Financial incentives and subsidies can facilitate voluntary solutions. Incentive programs, however, risk distorting the economy in counter-productive ways. If a firm is subsidized to reduce the discharge of a pollutant, its costs and the price of its products can be kept artificially low. This can increase demand for the product, and encourage other firms to enter the industry, so that pollution actually increases (NRC 2000).

Economic incentives are a long-established part of farm policy both in the United States and the European Union, and are based on providing technical assistance and subsidies, including policies for reducing pollution from farms (NRC 2000). The Conservation Reserve Program of the United States has successfully reduced erosion and increased habitat for wildlife in the agricultural landscape through financial payments to farmers who take land out of agricultural production and create buffer strips around streams. This program is not designed to reduce N pollution, but financial incentives could clearly also be used to encourage farmers to undertake best-management practices for N reduction (NRC 2000).

A look at the use of fertilizer subsidies and their removal throughout sub-Saharan Africa in the 1990s as part of structural adjustment programs presents some useful lessons. Removal of subsidies in many cases has led to the reduction in fertilizer use but this has not been so where prices for crops increased more than that of inputs (Scoones and Toulmin 1999). Other cases, however, show that even with large subsidies for fertilizers, there was still little use; and use may be more related to availability than price (Manyong et al. 2002). In any event, throughout much of sub-Saharan Africa, current fertilizer prices are two to six times those in most other places (Sanchez 2002). Even if the prices had been at par the majority of farmers would not have been able to afford the amounts needed to raise yields sufficiently nor would it have been cost-effective even if yields had been raised substantially given relative crop prices. Although a thorough analysis has not been done, transportation and boundary-crossing costs are, most likely, two of the main reasons for these excessive costs. Efforts should be made to address these issues. Improved road infrastructure will also increase access to markets and shift relative prices and, perhaps, provide more incentive for soil fertility replenishment (Scoones and Toulmin 1999).

There has been considerable debate regarding "subsidies" (or public intervention) for soil fertility replenishment (Sanchez et al. 1997; Scoones and Toulmin 1999). Where policy distortions affect input and/or output markets and where soil nutrient depletion affects livelihoods, then public interventions may be appropriate (Scoones and Toulmin 1999). Indeed, farmers will only invest in soil fertility management if there is a perceived benefit. Even then they may not have the labor and capital to do so, and there are competing demands (education and health) for the scarce capital they do have. Scoones and Toulmin (1999) conclude that there are grounds for public intervention to intensify sustainable agriculture in sub-Saharan Africa given the importance of agriculture for providing food, incomes, and employment.

9.5.3 Mandatory Policies for Reaching Goals: Regulations

9.5.3.1 Technology-based Standards

Mandatory policies, including regulatory control and tax or fee systems, place the costs and burden of pollution control on those who generate the pollution (NRC 2000). Technology-based standards are easy to implement but tend to discourage innovation and are generally not seen as cost-effective (NRC 1993a, 2000). In the United States, regulation under the Clean Water Act has been largely technology-based since its inception (Powell 2001). Since 1977, the Clean Water Act requires point sources of pollution to meet technology-based standards, administered by the Environmental Protection Agency (Powell 2001). For publicly owned sewage treatment plants, the standard remains secondary treatment (NRC 1993a; Powell 2001), designed merely to reduce the discharge of pathogens and labile organic matter, generally referred to in terms of the biological oxygen demands (BOD) created from this organic matter, and is rather ineffective at removing nutrients (NRC 1993a). Tightening the technology standard to include effective nutrient removal could substantially reduce N pollution at modest cost (NRC 1993a, 2000). Communities that can demonstrate that a lower level of treatment results in no significant environmental deterioration could obtain waivers from the standard from EPA (NRC 1993a).

Animal feeding operations are subject to the requirements of the Clean Water Act, but compliance has been poor, with only 20% of CAFOs having the necessary permits (Powell 2001). Currently, operations with more than 1,000 "animal units" are prohibited from discharging to surface waters, except during overflow conditions expected during a 25-year storm (Powell 2001). New EPA proposed new regulations for effluents were signed in December 2002 and published in the *Federal Register* in February 2003. While the intent was to regulate land application of manure as well as lagoon systems, the new regulations contain some provisions for alternative technologies, including an option to develop an alternative technology that considers pollution losses by all media (air, surface water, ground water) and results in a net reduction (EPA 2003). Volatilization of ammonia to the atmosphere would, however, remain unregulated (Powell 2001).

9.5.3.2 Total Maximum Daily Loads

The Clean Water Act requires states to monitor for violations of ambient water quality standards and, when a standard is violated, to determine the "total maximum daily load" that could enter the water body without causing impairment. If the TMDL is exceeded, the discharges allowed from point sources are lowered in steps (Powell 2001). However, non-point sources dominate N input to most coastal waters (NRC 2000), and the Clean Water Act provides no authority for regulating non-point sources in the TMDL context (Powell 2001). New statutory authority will be required for non-point sources, if N pollution is to be reduced through regulations.

Since 1990, states participating in the Coastal Zone Management Program of the United States are required to have enforceable mechanisms for controlling non-point source pollution. In many states the coastal zone is too narrowly defined, however, for effective nutrient management. Further, many states fail to comply with this requirement, and federal agencies have no authority to force compliance (Powell 2001). Funding for the Coastal Zone Management Program is small, giving federal agencies little leverage over state actions.

Under current U.S. law, TMDLs are applied by a state based on compliance with water quality standards within that state. This is problematic when pollution from one state contributes to impairment of a water body in another state. For example, N coming down the Susquehanna River from Pennsylvania is a major source of pollution to Chesapeake Bay (Virginia and Maryland), but this has not influenced TMDLs within Pennsylvania. If the TMDL regulatory approach is to be successful in reducing coastal N pollution, not only is new authority required for non-point source pollution, but also multiple state sources must be included. Providing enforcement authority to river basin commissions or other similar watershed-based entities may help in achieving this goal.

Although long mandated by the Clean Water Act, the TMDL approach has only been applied recently, after litigation led federal courts to direct the EPA to develop TMDLs. Political opposition remains vocal, which led Congress to ask the National Research Council to assess the scientific basis for TMDLs. The appointed committee endorsed the basic usefulness of TMDLs and suggested ways in which their usefulness could be improved, such as explicitly recognizing uncertainty and relying more on biological endpoints for standards (NRC 2001).

The TMDL approach is based on water quality standards. Currently states either lack nutrient standards for coastal waters or have only loose narrative standards (NRC 2000). EPA is working to develop procedures that could be used by the states to set nutrient standards, and has published the *Nutrient Criteria Technical Guidance Manual for Estuarine and Coastal Marine Waters* (EPA 2001a). States are expected to develop nutrient standards for fresh waters by 2004 (Powell 2001), but no deadline has been set for coastal marine ecosystems. Meanwhile, TMDLs for N control are sometimes driven by other standards, for example, in Long Island Sound the current plan for reducing N pollution is designed to comply with the local dissolved oxygen standard (NRC 2000).

9.5.4 Mandatory Policies for Reaching Goals: Taxes, Fees, and Marketable Permits

Mandatory taxes and fees that could be used as regulatory instruments for inducing change include effluent charges, user or product charges, non-compliance fees, performance bonds, and legal liability for environmental damage (NRC 2000). These approaches are widely believed to be more cost-effective than command-and-control regulations and to be more likely to spur innovation. For example, gasoline taxes could be increased to reduce fuel use and, hence, NO_x emissions, and N fertilizer could be taxed to reduce its use and to encourage appropriate use of manure. It is, however, difficult to reach specific targets in pollution reduction using the tax/fee approach (NRC 2000), since regulators have difficulty predicting how polluters will react. Adjusting fees and taxes over time to achieve the desired result may create political resistance, and the rates of taxation necessary to bring about meaningful changes in behavior are likely to be seen as excessive by large segments of the public (NRC 2000).

Marketable permits for pollution avoid some of the problems of both regulation and tax/fee systems and have been used to reduce sulfur dioxide pollution from electric power plants (NRC 2000). As with the regulatory approach, marketable permits start by deciding on an allowable level of pollution, so there is an assurance that the set target will be met. By allowing trading among permit holders, however, innovation is encouraged and the most economic abatement is achieved, given a sufficient number of buyers and sellers (NRC 2000). Marketable permits are now being tried for N control to coastal waters in several locations, including Pamlico Sound (North Carolina) and Long Island Sound. To date no trading has actually occurred. A major obstacle is that the need to establish a basis for trading between point and non-point sources of pollution requires precise knowledge on the sources and extent of non-point pollution (Malik et al. 1993; NRC 2000). This is seldom available for watersheds, although the ability of models to assess sources is improving rapidly (NRC 2000).

The EPA plans to use market-based trading programs for future reduction of point and non-point source nutrient pollution in U.S. waters, with trading occurring within watersheds (EPA 2001b). The program would have caps for total nutrient pollution based on water quality standards and for impaired waters on TMDLs. The proposed trading program is thought to provide economic incentives for voluntary reductions from non-point pollution sources, and would retain permit requirements for point source pollution (EPA 2001b). The lack of statutory authority for regulating non-point sources of nutrients under the Clean Water Act (Powell 2001) will be a major problem for the plan in the majority of watersheds, where non-point sources dominate N input. The lack of authority over inter-state pollution is another challenge.

9.5.5 Hybrid Approaches for Reducing Coastal Nitrogen Pollution

Nitrogen pollution has multiple sources, and an effective national strategy may require a combination of national regulation of some sources and watershed-based management of others. Combinations of regulatory, incentive, and market-based mechanisms are possible for both national and watershed-based approaches and may be the most cost-effective and politically acceptable (NRC 2000; Howarth, in press). A brief discussion of hybrid policy options for N pollution follows.

9.5.5.1 Runoff and Leaching from Agricultural Fields

Leakage of N from farming could be substantially reduced if farmers fertilized at, or even somewhat below, the rate recommended for maximizing profit. To achieve this, national farm policy should improve the economic return of those farmers who appropriately reduce fertilizer use and reduce economic subsidies to those farmers who exceed recommended fertilization levels. The crop-insurance program of the NGO American Farmland Trust demonstrates an approach that may work when governmental leadership is lacking (Howarth, in press).

Other means of reducing N loss include winter cover crops, planting perennial rather than annual crops, discouraging fall application of fertilizer, and using wetlands to intercept farmland drainage. Any of these could be encouraged, nationally or in specific watersheds, through incentive payments. Regulations or marketable permits that charge farmers for N runoff above a set limit could also be applied. Within watersheds, either incentive systems or regulatory and market-based approaches could be tailored to the largest problems, such as farmers who grow annual row crops such as corn and soybeans on sandy soils in a wet climate.

9.5.5.2 Concentrated Animal Feeding Operations

The scale of pollution from CAFOs is nearly equal to that from municipal sewage treatment plants, and it might be sensible to apply similar national technology-based standards to CAFOs as have been applied to point source pollution under the U.S. Clean Water Act. On the other hand, a performance-based standard may encourage more innovation in treatment technologies and therefore be less expensive to the industry in the long run (NRC 1993a, 2000). Marketable permits could also be used to control pollution from CAFOs. In any case, release of nutrients to surface and ground waters and to the atmosphere should all be controlled. The fate of manure from CAFOs must also be considered. In the Netherlands and a few other countries of Europe, manure application to fields is now regulated as part of the total farm nutrient-balance programs (NRC 2000), as has also been recently mandated in Nebraska and Maryland (NRC 2000; Mosier et al. 2001). The goal is to prevent overfertilization as a disposal mechanism for the manure.

Several European countries with high concentrations of livestock have their own manure regulations. In all countries, the quality of drinking water and surface water is an important policy issue, particularly with respect to N levels. Command-and-control measures are the most common policy tools for manure regulation in the countries of the European Union. In the Netherlands, for example, economic incentives have been used since 1998 in the form of an economically significant tax on N and P surpluses at farm level. See Breembroek et al. (1996) for details on the information system to assess these surpluses.

9.5.5.3 NO_x from Fossil Fuel Combustion

Atmospheric NO_x pollution comes from multiple sources, including on- and off-road vehicles and stationary sources such as electric power plants, and can be transported over long distances, so that the airshed for pollution sources usually overlaps several countries and watersheds (NRC 2000). This suggests that NO_x emissions are, perhaps, best regulated at the national scale. Hybrid regulatory approaches are possible at the national level, for instance, relying on emission standards for vehicles and marketable permits for electric power plants. Coastal N pollution may require year-round standards rather than the summer-time standards commonly used for control of ozone and smog pollution (NRC 2000).

9.5.5.4 Urban and Suburban Sources

Municipal sewage treatment plants are currently regulated by a technology-based standard (secondary treatment) and are not related to the degree of nutrient pollution. Upgrading the national technology standard from secondary treatment to nutrient removal may be sensible, and the cost is moderate (see discussion above). Where upgrading would serve no purpose, municipalities could apply for waivers under the Clean Water Act, as they currently can for the secondary sewage treatment standard (NRC 1993a, 2000).

Another approach would be to require nutrient removal only in plants located in watersheds that contribute to known N pollution in downstream coastal rivers or bays. Currently, N-reduction technology is required on a case-by-case basis and that too only when sewage plants discharge directly into coastal ecosystems. Nitrogen discharges elsewhere in river basins have largely escaped regulation.

9.5.5.5 Wetland Creation and Preservation

Wetlands and ponds can be significant sinks of both N and P. Wetlands and ponds that directly intercept groundwater flows or tile drainage, or have effective water exchange with streams and rivers, are the greatest sinks of N, while those that intercept eroding sediments are most effective for P (Howarth, in press). Incentive programs could target the creation and preservation of wetlands designed to be good N sinks, in watersheds with N pollution problems. Federal water management programs could significantly influence the ability of wetlands and floodplains to serve as N sinks, and some authors have strongly urged that increasing N retention should be made a central concern in the planning of such projects (Mitsch et al. 1999, 2001). However, wetlands can release substantial quantities of greenhouse gases (methane, in addition to N_2O), and this should be carefully balanced against the increased nutrient sinks when planning any large-scale strategy for increased wetlands.

9.6 Lessons Learned and Synthesis

Excess nutrients cause grave pollution problems in many rivers, lakes, waterways and coastal rivers and bays in the developed countries and now, increasingly, also in the developing ones. In the United States, two thirds of all coastal systems are moderately to severely degraded by nutrient pollution.

Nutrient inputs to inland and coastal waters are increasing globally. In some developed countries, discharges of P have been effectively curtailed, but N pollution has been reduced only in some local areas. If current trends continue, N loading to the coastal waters of the United States in 2030 is projected to be 30% above what it is today and more than twice that what it was in 1960, and increases are likely to be even larger in developing countries with rapid population growth.

The strategies for replenishing the immediate N supply in the soils for crop uptake are not necessarily compatible with replenishing the longer-term storage of N in the soil organic matter (Palm et al. 2001). Fairly small amounts of high-quality organic inputs can satisfy the immediate N demand of crops but does little to build the soil organic N (Giller et al. 1997; Merckx 2002). Increasing soil organic N would be best accomplished with systems that combine large amounts of organic inputs, such as legume fallows, and minimum tillage, and even then continued applications of N to balance crop demands would be necessary (Giller et al. 1997). These systems should be designed to mitigate the potential extra losses from pooled N.

The nutrients come from a variety of sources, including agricultural fields, concentrated animal feeding operations, sewage and septic wastes, and, for N, atmospheric deposition of NO_x from fossil-fuel combustion. The relative importance of these varies among receiving water ecosystems.

Technical tools and management practices exist for reduction of nutrient pollution at reasonable cost. That many of these have not yet been implemented on a significant scale suggests that new policy approaches are needed.

The best management solution may often be a combination of some of the many voluntary and mandatory approaches available, perhaps applying different approaches to different sources of nutrient pollution.

Non-point sources dominate N inputs to most coastal waters, and are important also for P. Current regulatory authority for non-point source pollution is often nonexistent or very limited. Hence, increased authority to regulate such sources may be necessary to reverse pollution of surface waters by nutrients.

Nutrient pollution can be addressed through a uniform national approach, or a watershed-based approach, or through a combination of these (for example, by applying a uniform national approach to NO_x emissions, while also setting watershed-based N loading standards). A watershed-based approach is likely to be the most cost-effective for some sources of nutrients (such as runoff from agricultural fields), while a uniform national approach may be better for others (such as NO_x from fossil-fuel combustion or P from detergents).

While current scientific and technical knowledge is sufficient to begin making real progress toward eliminating coastal nutrient pollution, progress will be quicker and more cost-effective with increased investment in appropriate scientific research.

Since nutrient pollution in many coastal ecosystems has sources in multiple states, state and local governments may not be the most effective regulatory agencies for a watershed-based approach. River basin commissions or similar entities, if given sufficient authority, may be more appropriate. Economic research on environmental policy design has largely been concerned with the merits of emissions-based economic incentives (for example, emission charges, emission reduction subsidies, transferable discharge permits). This literature has limited relevance to nutrients from agriculture and other non-point sources since emissions in such cases are, for all practical purposes, unobservable and typically stochastic. Several features of non-point pollution problems complicate the identification of good solutions. One is the high degree of uncertainty about non-point emissions and their fate. Given current monitoring technology, non-point emissions attributable to particular decision-makers cannot be measured with reasonable accuracy at reasonable cost. The result is substantial uncertainty about the decision-makers who are responsible for non-point pollution and about the degree of each firm's or household's responsibility. This rules out the use of the kinds of emission-based instruments that economists usually advocate for cost-effective pollution control (Griffin and Bromley 1983; Shortle and Dunn 1986).

Another feature of non-point problems is the extreme spatial variation in the feasibility, effectiveness, and cost of technical options for reducing emissions. This greatly limits the applicability of the uniform technology-based regulatory approaches that are often used to control point sources (Malik et al. 1994). A variety of options are available to encourage farmers to make socially desirable choices. The menu consists of choices about who should comply, how their performance (or compliance) will be measured, and the policy tools that will be used to affect their choices. Each of these choices will affect the economic and ecological performance but problem-specific research is essential for resolving this issues.

References

Aber, J.D., C.L. Goodale, S.V. Ollinger, M.L. Smith, A.H. Magill, et al., 2003: Is nitrogen deposition altering the nitrogen status of northeastern forests? *BioScience,* **53(4),** pp. 375–389.

Austin, A., R.W. Howarth, J.S. Baron, F.S. Chapin, T.R. Christensen, et al., 2003: Human disruption of element interactions: Drivers, consequences, and trends for the 21st Century. In: *Interactions of the Major Biogeochemical Cycles: Global Change and Human Impacts, SCOPE #61,* J.M. Melillo, C.B. Field, and B. Moldan (eds.), Island Press, Washington, DC, pp. 15–45.

Bashkin, V.N., S.U. Park, M.S. Choi, and C.B. Lee, 2002: Nitrogen budgets for the Republic of Korea and the Yellow Sea region, *Biogeochemistry,* **57/58,** pp. 387–403.

Bationo, A., S.K. Mughogho, and A.U. Mokwunye, 1986: Agronomic evaluation of phosphate fertilizers in tropical Africa. In: *Management of Nitrogen and P Fertilizers in Sub-Saharan Africa,* A.U. Mokwunye and P.L.G. Vlek (ed.), Martinus Nijhoff, Dordrecht, The Netherlands, pp. 283–318.

Boesch, D.F., 2002: Reversing nutrient over-enrichment of coastal waters: Challenges and opportunities for science, *Estuaries,* **25,** pp. 744–58.

Boesch, D.F., R.H. Burroughs, J.E Baker, R.P. Mason, C.L. Rowe, et al., 2001a: Marine pollution in the United States: Significant accomplishments, future challenges, PEW Oceans Commission, Arlington, VA.

Boesch, D.F., R.B. Brinsfield, and R.E. Magnien, 2001b: Chesapeake Bay eutrophication: Scientific understanding, ecosystem restoration, and challenges for agriculture, *Journal of Environmental Quality,* **30,** pp. 303–20.

Boesch, D.F., R.B. Brinsfield, R.W. Howarth, J.L. Baker, M.B. David, et al., Submitted to *Improving Water Quality While Maintaining Agricultural Production.*

Bohlool, B.B., J.K. Ladha, D.P. Garrity, and T. George, 1992: Biological nitrogen-fixation for sustainable agriculture: A perspective, *Plant and Soil,* **141(1–2),** pp. 1–11.

Boyer, E.W., C.L. Goodale, N.A. Jaworski, and R.W. Howarth, 2002: Anthropogenic nitrogen sources and relationships to riverine nitrogen export in the northeastern USA, *Biogeochemistry.* In press.

Breembroek, J.A., B. Koole, K.J. Poppe, and G.A.A. Wossink, 1996: Environmental farm accounting: The case of the Dutch nutrient bookkeeping system, *Agricultural Systems,* **51,** pp. 29–40.

Bricker, S.B., C.G. Clement, D.E. Pirhalla, S.P. Orland, and D.G.G. Farrow, 1999: *National Estuarine Eutrophication Assessment: A Summary of Conditions, Historical Trends and Future Outlook,* National Ocean Service, National Oceanic and Atmospheric Administration, Silver Springs, MD.

Buresh, R.J., P.C. Smithson, and D.T. Hellums, 1997: Building soil P capital in Africa. In: *Replenishing Soil Fertility in Africa,* R.J. Buresh, P.A. Sanchez, and F. Calhoun, (eds.), SSSA special publication no. 51, Soil Science Society of America, Madison, Wisconsin, pp. 111–49.

Carpenter, S.R., N.F. Caraco, D.L. Correll, R.W. Howarth, A.N. Sharpley, et al., 1998: Non-point pollution of surface waters with P and nitrogen, *Issues in Ecology,* **3,** pp. 1–12.

Chave, P., 2001: *The EU Water Framework Directive: An Introduction,* IWA Publishing, London, UK, 208 pp.

Choi, E., and Z. Yun (eds.), 2003, *Proceedings of the Strong N and Agro 2003,* Two vols., IWA Specialty Symposium on Strong Nitrogenous and Agro-Wastewater, 11–13 June 2003, Seoul, Korea, 1,118 pp.

Cleveland, C.C., A.R. Townsend, D.S. Schimel, H. Fisher, R.W. Howarth, L.O. Hedin, et al., 1999: Global patterns of terrestrial biological nitrogen (N_2) fixation in natural systems, *Global Biogeochemical Cycles,* **13,** pp. 623–45.

Critchfield, R., 1982: Science and the villager: The last sleeper wakes, *Foreign Affairs,* **61(1)** (Fall).

Dobereiner, J., 1997: Biological nitrogen fixation in the tropics: Social and economic contributions, *Soil Biology and Biochemistry,* **29,** pp. 771–74.

Dobereiner, J., S. Urquigua, and R.M. Boddey, 1995: Alternatives for nitrogen nutrition of crops in tropical agriculture, *Fertilizer Research,* **42(1–3),** pp. 339–46.

Dudal, R., 2002: Forty years of soil fertility work in sub-Saharan Africa. In: *Integrated Plant Nutrient Management in Sub-Saharan Africa: From Concept to Practice,* B. Vanlauwe, J. Diels, N. Sanginga, and R. Merckx (eds.), CAB International, Wallingford, UK, pp. 7–21.

Emmett, B.A., D. Boxman, M. Bredemeir, P. Gunderson, O.J. Kjonaas, et al., 1998: Predicting the effects of atmospheric deposition in conifer stands: Evidence from the NITREX ecosystem scale experiments, *Ecosystems,* **1(4),** pp.352–360

Elmgren, R. and U. Larsson, 2001: Nitrogen and the Baltic Sea: Managing nitrogen in relation to P. In: *Optimizing Nitrogen Management in Food and Energy Production and Environmental Protection,* Proceedings of the 2nd International Nitrogen Conference on Science and Policy, The Scientific World 1 (S2), Berkshire, UK, pp. 371–77.

Erisman, J.W., P. Grennfelt, and M. Sutton, 2003: The European perspective on nitrogen emissions and deposition, *Environment International,* **29,** pp. 311–25.

Foy, R.H, S.D. Lennox, and C.E. Gibson, 2003: Changing perspectives on the importance of urban P inputs as the cause of nutrient enrichment in Lough Neagh, *Science of the Total Environment,* **310,** pp. 87–99.

Galloway, J.N., J.D. Aber, J.W. Erisman, S.P. Seitzinger, R.H. Howarth, 2003: The nitrogen cascade, *BioScience,* **53,** pp. 341–56.

Galloway, J.N., F.J. Dentener, D.G. Capone, E.W. Boyer, R.W. Howarth, et al., 2004: Nitrogen cycles: Past, present, and future, *Biogeochemistry.* In press.

Giller, K.E., G. Cacisch, C. Ehaliotis, E. Adams, W. Sakala., et al., 1997: Building soil nitrogen capital in Africa. In: *Replenishing Soil Fertility in Africa,* R.J. Buresh, P.A. Sanchez, and F. Calhoun, (eds.), SSSA special publication no. 51, Soil Science Society of America, Madison, WI, pp. 151–92.

Goolsby, D.A., W.A. Battaglin, G.B. Lawrence, R.S. Artz, B.T. Aulenbach, et al., 1999: *Flux and Sources of Nutrients in the Mississippi-Atchafalaya River Basin, Topic 3,* Report for the Integrated Assessment on Hypoxia in the Gulf of Mexico, NOAA Coastal Ocean Program Decision Analysis Series No. 17, National Oceanic Atmospheric Association, Silver Spring, MD.

Goolsby, D.A. and W.A. Battaglin, 2001: Long-term changes in concentrations and flux of nitrogen in the Mississippi River basin, USA, *Hydrological Processes,* **15,** pp. 1209–26.

Griffin, R.C. and D.W. Bromley, 1983: Agricultural runoff as a non-point externality: A theoretical development, *American Journal of Agricultural Economics,* **70,** pp. 37–49.

Grimwall, A., P. Stålnacke, and A. Tonderski, 2000: Time scales of nutrient losses from land to sea: A European perspective, *Ecological Engineering,* **14,** pp. 363–71.

Hayami, Y. and K. Otsuka, 1994: Beyond the green revolution: Agricultural development strategy into the new century. In: *Agricultural Technology: Policy Issues for the International Community,* J.R. Anderson. (ed.), CAB International, Wallingford, UK, pp. 15–42.

Holland, E.A., F.J. Dentener, B.H. Braswell, and J. M. Sulzman, 1999: Contemporary and pre-industrial global reactive nitrogen budgets, *Biogeochemistry,* **46,** pp. 7–43.

Howarth, R.W.: *The Development of Policy Approaches for Reducing Nitrogen Pollution to Coastal Waters of the USA,* Proceedings of the 3rd International Nitrogen Symposium, Nanjing, China. In press.

Howarth, R.W., 2002: The nitrogen cycle. In: *Encyclopedia of Global Environmental Change, Vol. 2, The Earth System: Biological and Ecological Dimensions of Global Environmental Change,* H.A. Mooney and J.G. Candell (eds.), Wiley, Chichester, UK, pp. 429–35.

Howarth, R.W., 2003. Human acceleration of the nitrogen cycle: Drivers, consequences, and steps towards solutions. In: E. Choi, and Z. Yun (eds.), Proceedings of the Strong N and Agro 2003 IWA Specialty Symposium, Korea University, Seoul, Korea, pp. 3–12.

Howarth, R.W., H. Jensen, R. Marino, and H. Postma, 1995: Transportation and processing of P in near-shore and oceanic waters. In: *P in the Global Environment, SCOPE #54,* H. Tiessen (ed.), Wiley & Sons, Chichester, UK, pp. 323–45.

Howarth, R.W., G. Billen, D. Swaney, A. Townsend, N. Jaworski, et al., 1996: Regional nitrogen budgets and riverine N and P fluxes for the drainages to the North Atlantic Ocean: Natural and human influences, *Biogeochemistry,* **35,** pp. 75–139.

Howarth, R.W., D. Anderson, J. Cloern, C. Elfring, C. Hopkinson, et al., 2000: Nutrient pollution of coastal rivers, bays, and seas, *Issues in Ecology,* **7,** 1–15.

Howarth, R.W., E.W. Boyer, W.J. Pabich, and J.N. Galloway, 2002a: Nitrogen use in the United States from 1961–2000 and potential future trends, *Ambio,* **31(2),** 88–96.

Howarth, R.W., D. Walker, and A. Sharpley, 2002b: Sources of nitrogen pollution to coastal waters of the United States, *Estuaries,* **25,** pp. 656–76.

Howarth, R.W., R. Marino, and D. Scavia, 2003: Priority topics for nutrient pollution in coastal waters: An integrated national research program for the United States, National Ocean Service, National Oceanic and Atmospheric Administration, Silver Spring, MD.

Howarth, R.W., R. Marino, D.P. Swaney, and E.W. Boyer, 2004: Wastewater and watershed influences on primary productivity and oxygen dynamics in the lower Hudson River estuary. In: *The Hudson River,* J. Levinton (ed.), Academic, New York, NY.

Howarth, R.W. and R. Marino, Nitrogen as the limiting nutrient for eutrophication in coastal marine ecosystems: Evolving views over 3 decades, *Limnology and Oceanography.* In press.

IFA (International Fertilizer Industry), 2002: Fertilizer Indicators, IFA Web page. Available at http://www.fertilizer.org/ifa/statistics.asp.

Insel, Barbara, 1985: A world awash in grain, *Foreign Affairs,* **63(4)** (Spring).

Jaworksi, N.A., R.W. Howarth, and L.J. Hetling, 1997: Atmospheric deposition of nitrogen oxides onto the landscape contributes to coastal eutrophication in the north-east US, *Environmental Science & Technology,* **31,** pp. 1995–2004.

Karl, D., A. Michaels, B. Bergman, E. Carpenter, R. Letelier, et al., 2002: Dinitrogen fixation in the world's oceans, *Biogeochemistry.* In press.

Leneman, H., G.W.J. Giessen, and P.B.M. Berentsen, 1993: Costs of reducing nitrogen and P emissions on pig farms in the Netherlands, *Journal of Environmental Management,* **39,** pp. 107–19.

Lyasse, O., B.K. Tosah, B. Vanlauwe, J. Diels, N. Sanginga, et al., 2002: Options for increasing P availability for low reactive phosphate rock. In: *Integrated Plant Nutrient Management in Sub-Saharan Africa: From Concept to Practice,* B. Vanlauwe, J. Diels, N. Sanginga, and R. Merckx (eds.), CAB International, Wallingford, UK, pp. 225–37.

Malik, A.S., D. Letson, and S.R. Chutchfield, 1993: Point/nonpoint sources trading of pollution abatement: Choosing the right trading ratio, *American Journal of Agriculture and Economics,* **75,** pp. 959–67.

Malik, A.S., B.A. Larson, and M.O. Ribaudo, 1994: Economic incentives for agricultural non-point source pollution control, *Water Resources Bulletin,* **30,** pp. 471–80.

Manyong, V.M., K.O. Makinde, and A.G.O. Ogungbile, 2002: Agricultural transformation and fertilizer use in the cereal-based systems of the Northern Guinea savannah, Nigeria. In: *Integrated Plant Nutrient Management in Sub-Saharan Africa,* B. Vanlauwe, J. Diels, N. Sanginga, and R. Merckx, (eds.), CAB International, Wallingford, UK, pp. 75–85.

Martinelli, L.A., 2003: Element interactions as influenced by human intervention. In: *Element Interactions: Rapid Assessment Project of SCOPE,* J.M. Melillo, C.B. Field, and B. Moldan (eds.), Island Press, Washington, DC.

Martinelli, L.A., A. Krusche, R.L. Victoria, P.B. Camargo, M.C. Bernardes, et al., 1999: Effects of sewage on the chemical composition of Piracicaba River, Brazil, *Water Air and Soil Pollution,* **110,** pp. 67–79.

McIsaac, G.F., M.B. David, G.Z. Gertner, and D.A. Goolsby, 2001: Eutrophication: Nitrate flux in the Mississippi River, *Nature,* **414,** pp. 166–67.

Merckx, R., 2002: Process research and soil fertility in Africa: Who cares? In: *Integrated Plant Nutrient Management in Sub-Saharan Africa: From Concept to Practice,* B. Vanlauwe, J. Diels, N. Sanginga, and R. Merckx (eds.), CAB International, Wallingford, UK, pp. 97–111.

Mitsch, W.J., J. Day, J.W. Gilliam, P.M. Groffman, D.L. Hey, G.W. Randall, and N. Wang, 1999: *Reducing Nutrient Loads, Especially Nitrate-Nitrogen, to Surface Water, Ground Water, and the Gulf of Mexico, Topic 5,* Report for the Integrated Assessment on Hypoxia in the Gulf of Mexico, Decision Analysis Series No. 19, NOAA Coastal Ocean Program, National Oceanic and Atmospheric Administration, Silver Spring, MD.

Mitsch, W.J., J. Day, J.W. Gilliam, P.M. Groffman, D.L. Hey, et al., 2001: Reducing nitrogen loading to the Gulf of Mexico from the Mississippi River basin: Strategies to counter a persistent ecological problem, *BioScience,* **51,** pp. 373–88.

Mokwunye, U. and A. Bationo, 2002: Meeting the P needs of the soils and crops of West Africa: The role of indigenous phosphate rocks. In: *Integrated Plant Nutrient Management in Sub-Saharan Africa: From Concept to Practice,* B. Vanlauwe, J. Diels, N. Sanginga, and R. Merckx (eds.), CAB International, Wallingford, UK, pp. 209–24.

Moomaw, W.R., 2002: Energy, industry, and nitrogen: Strategies for decreasing reactive nitrogen emissions, *Ambio,* **31(2),** pp. 184–99.

Mosier, A.R., M.A. Bleken, P. Chaiwanakupt, E.C. Ellis, J.R. Freney, et al., 2001: Policy implications of human-accelerated nitrogen cycling, *Biogeochemistry,* **52,** pp. 281–320.

Nixon, S.W., 1995: Coastal marine eutrophication: A definition, social causes, and future concerns, *Ophelia,* **41,** pp. 199–219.

Nixon, S.W., J.W. Ammerman, L.P. Atkinson, V.M. Berounsky, G. Billen, et al., 1996: The fate of nitrogen and P at the land-sea margin of the North Atlantic Ocean, *Biogeochemistry,* **35,** pp. 141–80.

NRC (National Research Council), 1993a: *Managing Wastewater in Coastal Urban Areas,* National Academy Press, Washington, DC.

NRC, 1993b: *Soil and Water Quality: An Agenda for Agriculture,* National Academy Press, Washington, DC.

NRC, 2000: *Clean Coastal Waters: Understanding and Reducing the Effects of Nutrient Pollution,* National Academy Press, Washington, DC.

NRC, 2001: *Assessing the Total Maximum Daily Load Approach to Water Quality Management,* National Academy Press, Washington, DC.

Oppenheim, S., 2001: Alternative agriculture in Cuba, *American Entomologist,* **47(4),** pp. 216–27.

OSPAR Commission, 2000: *Quality Status Report 2000,* OSPAR (Oslo-Paris) Commission, London, 108 + vii pp.

Paarlberg, R.L., 1996: Rice bowls and dust bowls: Africa, not China, faces a food crisis, *Foreign Affairs,* **75(3),** pp. 127–132.

Palm, C.A., R.J.K. Myers, and S.M. Nandwa, 1997: Combined use of organic and inorganic nutrient sources for soil fertility maintenance and replenishment. In: *Replenishing Soil Fertility in Africa,* R.J. Buresh, P.A. Sanchez, and F. Calhoun (eds.), SSSA special publication no. 51, Soil Science Society of America, Madison, WI, pp. 193–217.

Palm, C.A., K.E. Giller, P.L. Mafongoya, and M.J. Swift, 2001: Management of organic matter in the tropics: Translating theory into practice, *Nutrient Cycling in Agroecosystems,* **61,** 63–75.

Peterson, B.J., W.M. Wollheim, P.J. Mulholland, J.R. Webster, J.L. Meyer, et al., 2001: Control of nitrogen exports from watersheds by headwater streams, *Science,* **292,** 86 pp.

Powell, J., 2001: *Programs that Reduce Nutrient Pollution in Coastal Waters,* Unpublished report to the Pew Oceans Commission in Philadephia, PA, and Washington, DC.

Rabalais, N.N., 2002: Nitrogen in aquatic ecosystems, *Ambio,* **31,** pp. 102–12.

Randall, G.W. and D.J. Mulla, 2001: Nitrate nitrogen in surface waters as influence by climatic conditions and agricultural practices, *Journal of Environmental Quality,* **30,** pp. 337–44.

Randall, G.W., D.R. Huggins, M.P. Russelle, D.J. Fuchs, W.W. Nelson, et al., 1997: Nitrate losses through subsurface tile drainage in CRP, alfalfa, and row crop systems, *Journal of Environmental Quality,* **26,** pp. 1240–47.

Robinson, J.R. and T.L. Napier, 2002: Adoption of nutrient management techniques to reduce hypoxia in the Gulf of Mexico, *Agricultural Systems,* **72,** pp. 197–213.

Ryther, J.H. and W.M. Dunstan, 1971: Nitrogen, P, and eutrophication in the coastal marine environment, *Science,* **171,** pp. 1008–12.

Sahrawat, K.L., M.K. Abekoe, and S. Diatta, 2001: Application of inorganic P fertilizer. In: *Sustaining Soil Fertility in West Africa,* G. Tian, F. Ishida, and D. Keatinge, (eds.), SSSA special publication no. 58, Soil Science Society of America, Madison, Wisconsin, pp. 225–46.

Sanchez, P.A., K.D. Shepherd, M.J. Soule, F.M. Place, R.J. Buresh, et al., 1997: In: *Replenishing Soil Fertility in Africa,* R.J. Buresh, P.A. Sanchez, and F. Calhoun, (eds.), SSSA special publication no. 51, Soil Science Society of America, Madison, Wisconsin, pp. 1–46.

Sanchez, P.A., 2002: Soil fertility and hunger in Africa, *Science,* **295,** pp. 2019–20.

Scoones, I. and C. Toulmin, 1999: Policies for soil fertility management in Africa, A report prepared for the DFID (Department for International Development), IDS (Institute of Development Studies) Brighton/IIED (International Institute for Environment and Development), Edinburgh, UK, 128 pp.

Shuiwang, D., Z. Shen, and H. Hongyu, 2000: Transport of dissolved nitrogen from the major rivers to estuaries in China, *Nutrient Cycling in Agroecosystems,* **57,** pp. 13–22.

Shortle, J.S. and J.W. Dunn, 1986: The relative efficiency of agricultural source water pollution control policies, *American Journal of Agricultural Economics,* **68,** 668–77.

Smil, V., 2001: *Enriching the Earth,* MIT Press, Cambridge, MA.

Staver, K.W. and R.B. Brinsfield, 1998: Use of cereal grain winter cover crops to reduce groundwater nitrate contamination in the mid-Atlantic coastal plain, *Journal of Soil Water Conservation,* **53,** pp. 230–40.

Stoorvogel, J.J., and E.M.A. Smaling, 1990: *Assessment of Soil Nutrient Depletion in Sub-Saharan Africa: 1983–2000, Vol. I, Main report (2nd ed.), Report 28,* The Winand Staring Centre for Integrated Land, Soil, and Water Research, Wageningen, The Netherlands.

Tartowski, S. and R.W. Howarth, 2000: Nitrogen, nitrogen cycling, *Encyclopedia of Biodiversity,* **4,** pp. 377–88.

Townsend, A.R., R. Howarth, F.A. Bazzaz, M.S. Booth, C.C. Cleveland, et al., 2003: Human health effects of a changing global nitrogen cycle, *Frontiers in Ecology & Environment,* **1,** 240–46.

UNEP (United Nations Environment Programme), 2002: *Global Environment Outlook 3,* Earthscan Publications Ltd., London, UK.

USEPA (United States Environmental Protection Agency), 2000*: National Air Pollutant Emission Trends 1900–1998,* EPA-454/R-00-002, USEPA, Washington, DC.

USEPA, 2001a: *Nutrient Criteria Technical Guidance Manual for Estuarine and Coastal Marine Waters,* EPA-822-B-01-003, USEPA, Washington, DC.

USEPA, 2001b: *Market-based Approaches to Improve the Nations Waters,* Draft, 21 November 2001, USEPA, Office of Water, Washington, DC.

USEPA, 2003: Concentrated animal feeding operations (CAFO) – Final rule. Available at http://cfpub.epa.gov/npdes/afo/cafofinalrule.cfm, and http://www.epa.gov/npdes/pubs/cafo_brochure_regulated.pdf.

Van Asseldonk, M.A.P.M., 1994: Voorspellen van variatie tussen dieren met het Technisch Model Varkensvoeding, M.Sc. thesis, Department of Agricultural Economics, Wageningen Agricultural University, Wageningen, The Netherlands.

Vitousek, P.M., J.D. Aber, R.W. Howarth, G.E. Likens, P.A. Matson, et al., 1997: Human alteration of the global nitrogen cycle: Sources and consequences, *Ecological Applications,* **7,** pp. 737–50.

Vollenweider, R.A., 1968: Scientific fundamentals of the eutrophication of lakes and flowing waters, with particular reference to nitrogen and P as factors in eutrophication, Technical report DA 5/SCI/68.27, Organization for Economic Co-operation and Development, Paris, France, 250 pp.

Von Egmond, K., T. Bresser, and L. Bouwman, 2002: The European nitrogen case, *Ambio,* **31,** pp. 72–8.

Wolfe, A. H. and J.A. Patz, 2002: Reactive nitrogen and human health: Acute and long-term implications, *Ambio,* **31,** pp. 120–25.

Xing, G.X. and Z.L. Zhu, 2002: Regional nitrogen budgets for China and its major watersheds, *Biogeochemistry,* **57/58,** pp. 405–27.

Chapter 10

Waste Management, Processing, and Detoxification

Coordinating Lead Author: N.O. Adedipe
Lead Authors: M.K.C. Sridhar, Joe Baker
Contributing Author: Madhu Verma
Review Editors: Naser Faruqui, Angela Wagener

BOXES

FIGURES

TABLES

Main Messages

Human and ecosystem health can be adversely affected by all forms of waste, from its generation to its disposal. Over the years, wastes and waste management responses such as policies, legal, financial, and institutional instruments; cradle-to-cradle or cradle-to-grave technological options; and sociocultural practices have impacted on ecosystem health and human well-being. Examples are evident in all countries.

International participation and leadership in waste management and processing is essential. Waste is so diverse in its origin and forms and so pervasive in its impacts, through terrestrial, aquatic, and atmospheric ecosystems, that it has the potential to adversely affect both the inhabited and uninhabited parts of the world. These parts necessarily include the wide range of wetlands relevant to the Ramsar Convention; the species and their land and ocean habitats included in the Convention on Biological Diversity; sites important to migratory species; and grasslands, forests, and wetlands that must be protected to minimize the potential for further desertification. Without the involvement and commitment of the leaders of countries and industries, a global approach to waste management will not be achieved.

Waste management and processing involve one or more of the following processes: reduction, reuse, recovery, or disposal of waste, with practices and technologies differing according to different economic and social circumstances. The desired long-term objective of human responses should be "Avoidance of Waste." The sale of products from waste, whether by simple reuse, recycling and recovery, or by more complex technological processing, has helped to create jobs appropriate to the socioeconomic conditions of the locality or country.

Environmental awareness and educational programs have been successful in allowing consumers and resource users to make informed choices for minimizing waste in their purchasing decisions. Employers have introduced programs to encourage and recognize initiatives by the community to reduce waste. In Japan and other industrial countries, "industry clusters" have been planned, where the waste of one industry is the resource of another—an example of copying nature, or bio-mimicry.

The combined impact of these practices has been to enhance ecosystem services, improve aesthetic conditions, restore habitats for human use and for biodiversity, increase public health and well-being, create jobs, and reduce poverty. Processes for human societies to avoid waste in all its forms are not available.

Industries and governments should select indicators and standardize methods to monitor the sources, types and amounts of all wastes produced. The full costs of each type of waste produced from any proposed new product or process should also be assessed. Leaders of industry and government know that they must have precise details of waste generation, composition and characteristics, and reuse or disposal practices to manage waste, locally or internationally. Currently, the practice of transparent, participatory, and accountable decision-making for ecosystem sustainability and human well-being is lacking in many countries. Although there are gaps in the structure of waste accounting, the countries involved in State of the Environment reporting, such as Canada, The Netherlands, New Zealand, and Australia, are moving to internal standardization. The next step is to develop international standards of waste accounting to allow objective comparison of waste management.

All industries, all communities, and all countries must ensure compliance with waste management laws and regulations, and acceptance of, or changes in, such laws and regulations. Communities have shown willingness to comply with laws and regulations if there is clear understanding of the benefits of such measures, particularly if all stakeholders are involved in the formulation of such laws. Waste cannot always be confined within one locality or area of jurisdiction. Some forms of waste (particularly those associated with acid rain, greenhouse gases, and air quality in general) are transmitted in the atmosphere, which respects no political, terrestrial, or aquatic boundaries.

The rapid advances in technologies, including biotechnology, provide new opportunities for improvement in waste management. The adoption of some of these technologies may require revision of existing laws and regulations.

The dumping of waste in remote places such as deserts and oceans, and across national boundaries, is not acceptable. Moves to have the practice forbidden by international conventions should be supported by enforceable national legislation. Remote-location dumping of wastes is a classic example of the historic "out of sight, out of mind" mentality, now rejected by all internationally responsible organizations and industries. However, the challenge of safe disposal of waste often requires interim storage while new technologies are developed. Remote areas, deserts, and oceans should not be seen as convenient locations for such "interim" storage. Such international arrangements as the Basel Convention on the Control of Transboundary Movement of Hazardous Wastes and their Disposal, the Kyoto Protocol on the Protection of the Ozone Layer, and the London Convention on the Prevention of Marine Pollution by Dumping of Wastes, have yielded some positive results in creating awareness among communities and adoption of alternate technologies and compliance by industries.

The essential role of water in life processes should be valued. Wastewater is a resource in many countries and the practice of safe reuse of water should be encouraged. Water, whether freshwater or marine, is both essential to life processes and a carrier and transporter of soluble and insoluble waste, solid, liquid, and gaseous.

The challenges to removing the different types of waste from waters are diverse. Modern technologies such as bioremediation, membrane filters, trickling filters, activated sludge process, vascular aquatic vegetation, and anaerobic digestion can now be used to remove all contaminants from polluted waters. Special care should be taken in the use of different types of "gray" water and effluents for human needs, to be supported with appropriate community education programs.

10.1 Introduction

Each individual living species and each type of process or operation will have by-products in its activities, processes, or operations. In nature, diverse ecosystems (notably rainforests and coral reefs) have achieved sustainability by the coexistence of a wide range of different species, whereby the waste of one species has become the resource of another, and there is an apparent balance in the system. If for any reason one species becomes dominant, the sustainability of the system is challenged, and "nature" responds to that imbalance.

At the global scale, humans have become dominant in the ecosystem, both by their numbers and by their ability to modify systems and to extract and transform natural materials, and fabricate, use, and transport the new materials. However, humans have been slower to respond than nature can, and only in recent decades they have acknowledged the need to copy the examples of nature (bio-mimicry) to avoid accumulation of waste and address the challenge of waste management holistically.

Waste and waste management are significant components of many chapters in the Millennium Ecosystem Assessment, and specific aspects of waste generation and/or management are found in relevant chapters in this and other MA volumes. The general principles of the Responses in this chapter relate to all waste but are more specifically related to *MA Current State and Trends,* Chapter 15. However, because urban and rural wastes are issues that affect everyone, there is a further concentration of specific examples dealing with responses to urban and rural waste.

Continuously increasing quality of life and high rates of resource consumption have had an unintended and negative impact on the urban environment—by way of the generation of wastes far beyond the handling and treatment capacities of urban governments and agencies. Cities are now facing serious problems of high volumes of waste, characterized by inadequate disposal technologies/methodologies, rising costs of management, and the adverse impact of wastes on the environment. These problems, however, also have provided opportunities for cities to find solutions that involve the community and the private sector, including innovative technologies, disposal methods, behavior changes, and awareness raising. Rural areas and rural communities have been affected in many ways, including the unexpected consequences of excess use of N and P fertilizers, pesticides and herbicides, soil contaminants, and soil salinization. Chapter 9 examined the selection of responses for management of excess nutrients, mainly those from fertilizers based on N and P.

The generation of wastes and their management have attracted significant attention by local, national, sub-regional, regional, and international communities. Waste has significant impact on ecosystems, and poses threats to human health and well-being. Waste also threatens the integrity of habitats that are essential to biological diversity.

The challenge is to develop responses to waste issues that can be applied in both developing and industrial countries and that will improve the quality of human life and of the biodiversity of our lands, our seas, and our skies.

10.1.1 Waste Typology and the Main Issues

The MA volume *Current State and Trends* lists the major categories of wastes and contaminants by source and the processes involved in waste processing and detoxification (see MA *Current State and Trends,* Tables 15.1 and 15.2).

Wastes exhibit multiple impacts depending on the ecosystems, ranging from small- to medium-, and ultimately long-term scales. Responses developed must acknowledge those categories and processes and be able to deal with:

- nutrient runoff from excess fertilizers, which constitute significant pollution of terrestrial (Olson 1987) and aquatic ecosystems (Howarth et al. 1999; Ikem et al. 2002; Melillo et al. 2003);
- chemicals and their residues from agro-allied industries, which remain persistent in soils, sediments, water, and the atmosphere after their planned use (Howarth et al. 2000);
- chemical toxins and their metabolic products, and their impact on terrestrial and aquatic ecosystems (Anderson 1994; Chandler et al. 1996; WRI 1992);
- nitrogen and its oxides, which contribute to acid rain (Vitousek and Hooper 1993);
- a vast array of synthetic chemicals—widely used since the 1950s for diverse functions such as the control of pest animals and plants, fuel additives, personal hygiene, etc.—many of which remained persistent in the environment (Adesiyan 1992);
- oils and oily wastes on land and water (Burger and Peakall 1995; Chokor 1996; Colwell et al. 1978; Hay et al. 1996);
- exhaust gases from combustion engines, which have changed air quality and contributed to the enhanced greenhouse effect (UNEP 2002);
- mixed streams of N and P wastes that transform from solid to liquid and to gas, and end up in freshwater and marine ecosystems and the atmosphere (Howarth et al. 1999, 2000); and
- residual materials arising from daily living, tourism, recreation, and communal activities (UNEP 2002).

All living species generate waste. Table 10.1 summarizes the diverse sources of waste and wide range of impact and effect. The focus of this assessment, however, is predominantly on wastes generated by human activities, particularly the ubiquitous urban and rural wastes, which have been described as "a monster" (Onibokun and Kumuyi 1999) and "a nightmare" (Asomani-Boateng 2002).

Although wastes produce goods and services that affect a wide variety of ecosystems (atmospheric, aquatic, and terrestrial), and in the process, confer a negative impact on human well-being,

Table 10.1. Waste Typology, Sources, and Policy Responses

Waste Type	Sources	Policy Responses
Urban solid wastes, putrescible and non-putrescible solids, semi-solids and liquids (residential, commercial, institutional)	human activities	reduce, reuse, recycle, dispose
Inorganic nutrient runoffs	fertilizers	reduce, recycle
Demolition waste, quarry rejects	construction sites, quarries	reuse, recycle, dispose
Oils and oily wastes	industries, mining	reduce, recycle
Hazardous (including clinical) expired drugs, by-products of metabolism, chemical toxins, fecal pellets in benthos, contaminated sludge, incineration ash, leachates, ignitable, corrosive, reactive, or toxic	industries, healthcare facilities, household hazardous wastes, waste disposal facilities	reduce, dispose
Radioactive waste	spent fuel from reactors, tailings from the mining/refining of uranium, medical/ academic	reduce, reuse, dispose
E-waste	cell phones, computers, etc.	reduce, recycle
Ammonia and its oxidative products	industries, fertilizers	reduce, recycle
Mixed wastes (containing N and P) from livestock	livestock	recycle
Synthetic chemical wastes	pesticides, biocides, fuel additives, cosmetics, etc.	reduce, dispose
Waste products from combustion (greenhouse gases)	vehicle engines, sea craft, energy production	reduce, alternate fuel use

their appropriate management enhances the benefits that people obtain from ecosystems—improving the provisioning services such as food and fiber production; improving the water quality; protecting biodiversity; improving human well-being, and improving aesthetics for the promotion of recreation and tourism.

The main long-term issues in waste management relate to how humanity deals with the two extremes: "no waste" in, and the "waste-induced pollution" of, the environment. This is particularly so since waste is an inevitable product of nature (organismal existence). The "no waste" end (best-case scenario) of the spectrum is, of course, a daunting challenge to achieve, while the "pollution" end (worst-case scenario) is, equally, overwhelming in proffering solutions to the problems it poses. The practical challenge relates to minimizing the adverse impacts of waste and, if possible, transforming waste into a useful product or process. Ideally, one would characterize all the components of waste and assess the threat they pose, singly and in combination with other waste components, to humans and to the ecosystem.

Realistically, we should identify policy response options, seeking a socially acceptable, economically viable, and environmentally sustainable strategy of reducing the predominantly negative impact both on human well-being (including the factors of poverty reduction, quality of life, sound health, cordial social interactions, security of people and property, and personal liberty within the limits imposed by democratic practices and societal values) and on biological diversity (biodiversity) and its habitat.

The practical response to the challenges presented by different types of waste may differ on the basis of socioeconomic and cultural priorities of different countries. The successful high-technology solutions are, currently, generally confined to the richest of the industrial countries. Knowledge should be shared to increase the capabilities and commitment of various developing and developed countries.

The best solution to the problem would be to avoid waste in the first place, but we currently lack the knowledge, the whole-of-society commitment, or the technologies for that. The next objective is to reduce the waste and its adverse impact as far as is practicable. Methods for this reduction will vary according to whether the waste is solid, liquid, or gas, biodegradable/non-biodegradable, hazardous or non-hazardous.

Solid-waste management has traditionally been viewed as a technical (engineering) problem requiring a technical solution by the civil engineering community (Goddard 1995). But now there is a clear need to move away from *waste disposal* toward *waste processing* and *waste recycling* (aiming for eventual *waste reduction*). Some of the defining criteria for future waste minimization and waste avoidance programs will include increased community participation, improved understanding the economic benefits/recovery of waste, focus on life-cycles rather than end-of-pipe solutions, decentralized administration of waste, emphasis on minimizing adverse environmental impacts, and alignment of investment costs and long-term goals.

10.1.2 Wastes in Relation to Ecosystems and Human Well-being

The essential aspects of the generation and management of wastes are depicted in Figure 10.1, which illustrates that the goal is human well-being, which is achievable through the production of multiple *services* (such as food, fiber, water, energy) in the "niche" provided by specific *ecosystems*. These ecosystems include terrestrial (that is, cultivated/agricultural land, natural and introduced forests), aquatic (for example, freshwater and marine/coastal waters), and specialized (such as mountain, polar, and island) landscapes. There is a strong relationship between the health of the ecosystem and the health of the human system. Waste generation is moderated by drivers that can be manipulated through a wide variety of *responses* by policy actors and decision-makers to ensure the mitigation of negative impacts of wastes and the adoption/adaptation measures. The type, peculiarity, description, and characterization of wastes generated, and the goods and services provided, are detailed in the next section.

The response policies address the specific aspects of planning and implementation of chosen alternative strategies (Jacobs and Sadler 1990; Soesilo and Wilson 1995), both as they relate to actual management practices and in such ways and manners that are environment-friendly, cost-effective, and socially acceptable. Prominent among these are waste reduction at source (minimization) (Baker 1999); waste recycling (Sridhar et al. 1992; Odeyemi and Onibokun 1997); ecological impact attenuation by conversion practices, for example, wastewater treatment and composting (Asomani-Boateng 2002; Dreschel et al. 2002; Dushenkov et al. 1995; Guterstam et al. 1998; Jana 1998; Rotimi 1995; Robinson et al. 1995; Sridhar and Arinola 1991; Sridhar and Adeoye 2003); waste stream linkages (synergies and antagonisms) (Mitsch et al. 2001; Peterson et al. 2001); transportation and related technologies (Adedipe and Onibokun 1997; Onibokun et al. 2000); and the establishment of institutional partnerships (ISWA 2002) involving the public sector, the private sector, and the community through governance and social property rights linkages (Adedipe 2002; Nsirimovu 1995).

Other management considerations include legal mechanisms, financial instruments, and economic incentives (Burges et al. 1988; Chambers 2003; Costanza and Principe 1995; Reyer et al. 1990; Miranda and Aldy 1996; Onibokun et al. 2000; Panayotou 1990).

10.1.3 Key Drivers of Change in Ecosystems and Services

The major drivers of change in ecosystems services are biophysical, demographic, economic, sociopolitical, technological, social, cultural, and religious. The drivers vary by levels of socioeconomic development. Those for developing countries include:

- *demographic change*: high population size exacerbated by high growth rate of up to 3.5% coupled with rapid urbanization, as in some West African (Onibokun and Kumuyi 1999), Latin American (Onibokun 1999), and Southeast Asian (Zurbrugg 2002) countries;
- *globalization* (adverse effect as raw material producers with minimal value-added) leading to complex and disadvantaged trade relations with industrial countries;
- *frequent changes in government,* resulting in lack of continuity in commitment to waste management and to environmental pollution policy, laws, and guidelines;
- *lack of focus* on the concept of "resource recognition," that is, treating waste as an unused resource;
- *following a "hard" rather than a "soft" approach,* that is, considering waste management as a responsibility of municipal bodies than helping community-based initiatives and informal mechanisms attain recognition;
- *lack of funds and staff* to handle the ever-increasing problem of solid waste;
- *ineffective management policies and instruments*;
- *poor technology* for the collection, transfer, disposal, and processing of wastes;
- *mass illiteracy,* occasioning indifference to the environment by not adequately linking waste with deprived human well-being;

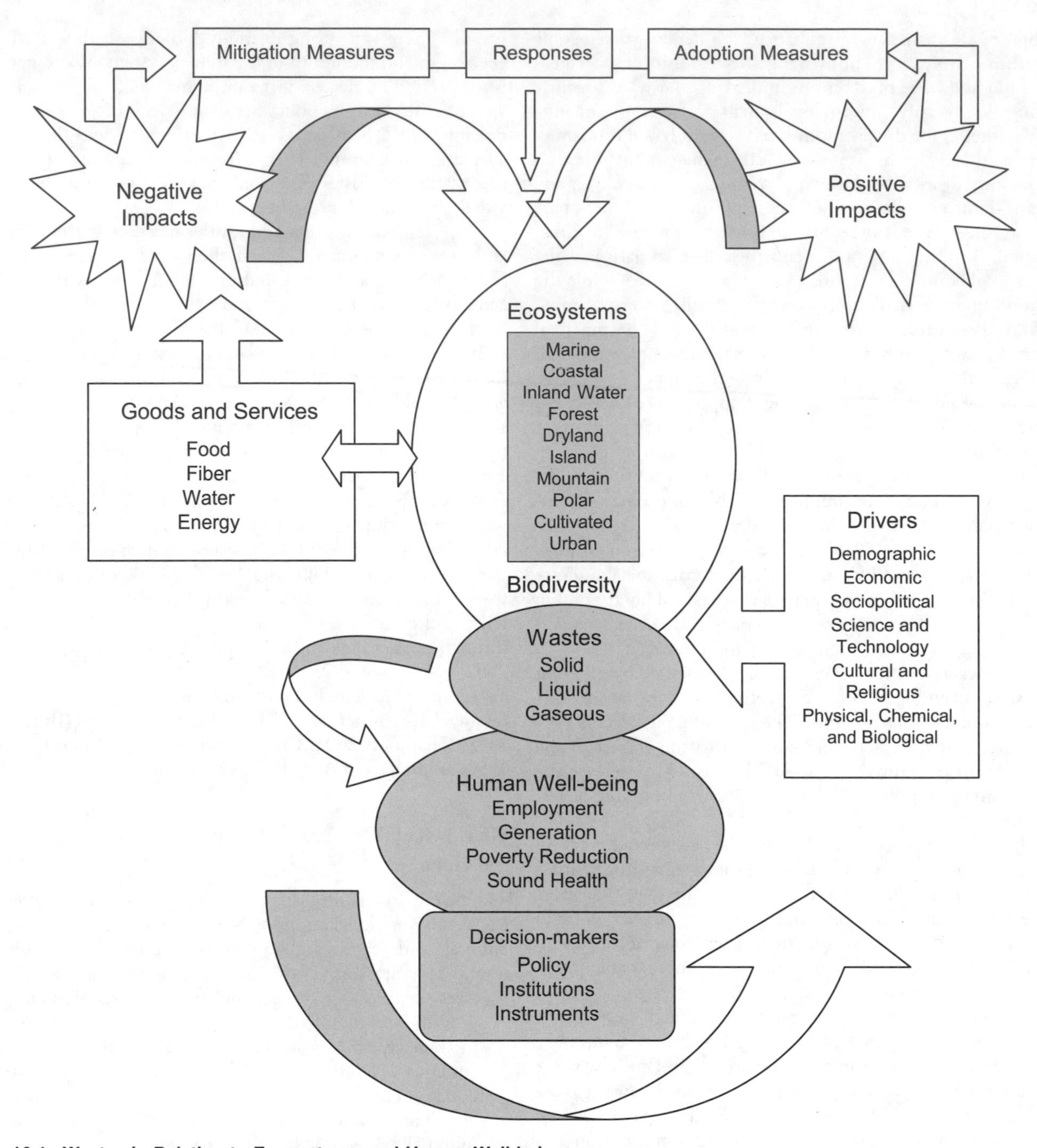

Figure 10.1. Wastes in Relation to Ecosystems and Human Well-being

- *failure to recognize, support, and integrate informal waste recyclers* into municipal solid waste systems in African cities;
- *continued use of old technologies* that continue to generate pollutants; and
- *pervasive poverty and mismanagement of public funds*.

The drivers in industrial countries include the following:

- *emphasis on wealth creation and high consumption* based on technology, industrialization, and self-created favorable international trade balance;
- *a ''use and throw away'' society* that puts little emphasis on concepts of ''reducing'' or ''reusing'';
- *regarding waste management as an ''engineering problem''*; and
- *stable political governance* characterized by transparency and accountability.

In both developing and industrial countries, national waste management policy should involve the following stages:

- waste reduction, which recognizes the costs and benefits of reducing waste;
- the optimal balance between landfill, incineration, and recycling. In developing countries, this choice will tend to be one of balancing recycling (including composting) and landfill in such a manner that recycling efforts do not use up more resources than they save;
- management of uncollected waste, an issue of some importance in developing countries but not of major significance in industrial countries; and
- the choice of regulatory measures to secure waste reduction and optimal disposal.

Around the world, developing and industrial countries have responded to the various drivers of change in ecosystems and services with a range of coping strategies, policies, and practices. These take the form of legal provisioning, improving technologies, fiscal and financial instruments, and institutionalizing waste management. "Zero waste" seems to be a difficult but not impossible task. If various options for waste management can be integrated and applied over long periods, waste minimization can be addressed effectively and sustainably.

10.2 Background and Selection of Responses

This section looks at five categories of responses: legal, technological, economic, institutional, and sociocultural. In the selection of responses, six criteria were used: historical dominance and contemporary relevance, political interest, evidence of positive impact, evidence of damaging impact, integrated design, and cutting-edge innovations.

10.2.1 Legal Responses

For the developed world, although there are copious laws dating back more than a century, there is no uniform effective enforcement (Garbutt 1995; Lieberman 1994); in the developing countries, the major problem is also that of enforcing compliance (Ajomo 1992; Adewale 1996; Onibokun et al. 2000).

The various stakeholders and actors should be involved in designing regulations and guidelines. This is because participation ensures and promotes understanding and acceptance, which helps compliance. More cooperation at the national and the international levels is also desirable to ensure that international agreements are honored. They are frequently not so honored, principally for political national interests, including the national economy, as has been the experience with the Basel Convention and the Kyoto Protocol.

A recent study of four West African countries (Gambia, Ghana, Nigeria, and Sierra Leone) identified conflicts in constitutional provisions in relation to human resource requirements and financial resource allocations, compounded by the weak revenue base of local (municipal) governments, as a major stumbling block to effective waste management and pollution abatement (Onibokun et al. 2000). The response option for developing countries lies in continuing public sector responsibility, as in the Latin American countries, or a change to privatization, particularly in mid/high-income neighborhoods. However, in low-income/high-density residential neighborhoods where waste collection and disposal is problematic, waste management has been a failure. Whatever overall policy framework is adopted, and laws are enacted, there must, in all cases, be continual processes of legal reforms to provide the underpinning for institutional and governance structures. This is to be buttressed with linkages that would promote transparent, participatory, and accountable environment-friendly decision-making for ecosystem sustainability.

10.2.2 Technological Responses

A variety of technological responses have been developed, some local and some specific to particular types of waste. In industrial countries, waste collection services often depend on prior household or workplace sorting of domestic waste into organic material, paper and cardboard, metal containers, recyclable plastics, and glass containers, placed in separate collection bins, to facilitate reprocessing and recycling, subsequent to commercial collection. The technologies extend to the design of special collection vehicles for these bins.

Whereas the concept of waste management requires a global approach, many specific waste management technologies are financially and operationally costly and thus not affordable for many developing countries. Additionally, most industrial-country technologies transferred to developing countries without expert technical and maintenance support result in failure. Examples include packaged wastewater treatment facilities for municipalities and industries, heavy-duty waste transportation equipment, and large-scale composting plants.

For the developing countries, short-term and medium-term solutions lie in the use of relevant and appropriate technologies such as biogas plants, constructed wetlands, waste stabilization ponds, appropriate solid-waste cartage and disposal machinery (Adedipe 2002 Onibokun et al. 2000), and relevant composting schemes (Zurbrugg 2004). Governments are often tempted to choose inappropriate technologies due to social pressures. One option is for developing countries to mass-produce small-scale equipment that is amenable to community access. With improvement in access over time resulting from enhanced road networks, medium and large scales of equipment characteristic of the developed economies would be justifiable. India and China are evolving ecological engineering for solid waste, sewage sludge, and wastewater. For the developed world, the issue of energy savings and operational cost-effectiveness are the response options for long-term solutions. These response options demand appropriate policy reforms toward ecosystem sustainability promotion for, and by, decision-makers. Whatever options are considered must recognize the scale and location of operations as well as the distance between the site operations and the end-users.

In situ operations are often economic and thus widely adopted (Sridhar and Adeoye 2003); see Boxes 10.1 and 10.2 for industrial-country and developing-country examples.

Certain technological or management innovations have proved beneficial in mitigating the damage to the environment. Wastewater management, with percolating filters/trickling filters, and the activated-sludge method of treatment, improved the environment many times over since 1914 (Edgar et al. 1998; Krogman et al. 2001). The advances were more pronounced during World Wars I and II, and continued during the subsequent industrial development of various European countries, the United States, Australia, and others. The cleaning up of the Thames and Rhine rivers and various lakes in the Nordic countries is a remarkable example of innovative governmental commitment.

Relatively cheap technology innovations using oxidation ponds/waste stabilization ponds, floating vascular aquatic vegetation, including gray water reuse and duckweed ponds, reed-bed or root-zone technology, and constructed wetlands have considerably helped the poorer developing countries (Aluko et al. 2003; Jana 1998; Urban Agriculture Magazine 2002). Such waste treatment systems can actually have a measurable impact on poverty. The duckweed ponds in Mirzapur, Bangladesh, or the gray water systems in Jordan, for example, reduced poverty by 10% in the homes with such systems (Faruqui and Al-Jayyousi 2002).

Development of synthetic polymers, used to make plastics such as polyethylene, polypropylenes, polyesters, and polyamides (including nylon), has revolutionized the types of containers for products, the types of material for packaging, and the materials used for carry-bags. However, most of these polymers are not biodegradable and, once used and discarded, become major waste management challenges. In cities like Cairo, Manila, Kolkata, Mumbai, Hong Kong, Beijing, Accra, Lagos, and others, the management of plastic waste has become a nightmare for city authorities. In other cities, such as in Australia, non-biodegradable plastic carry-bags are being phased out in retail stores by 2008. The introduction of biodegradable plastics is a welcome environment-friendly development. (See Box 10.3 for steps being taken in the European Union.)

BOX 10.1

Australian Strategy for Zero Solid Waste Management (www.nowaste.act.gov.au)

In December 1996, the government of the Australian Capital Territory (ACT) initiated the *No Waste By 2010—Waste Management Strategy for Canberra* for the nation's capital, with a population of 320,000. The objective is to achieve, by 2010, a society where no waste is sent to landfill.

The strategy was developed after an extensive community consultation process and reflects the views and expectations expressed by the community. It aims to build on the community's willingness to *avoid purchases* with excessive packaging, *recycle* materials, and *improve* environmental outcomes by treating all materials as potential *resources.* It is being implemented in phases. The first step involved developing an inventory of wastes, establishing recovery infrastructure, and introducing appropriate charges to provide monetary incentives. The second step was to identify current series activities. Further programs are being developed for implementation between 2003 and 2010.

ACT No Waste, a business unit in the Department of Urban Services, provides a number of services to the community and supports businesses with reuse initiatives, notably:

- curbside recycling collection,
- free drop-off services for green wastes,
- regional drop-off centers for recyclable materials,
- concrete, masonry, and steel processing facilities,
- reprocessing of old paint to an as-new product,
- community recycling of unwanted household items through "Second-hand Sunday,"
- metal and oil recovery facilities at the two landfills,
- landfill salvage operations at the landfills, where reusable goods are sold to the public, and
- a Resource Recovery Estate, which provides sites for inventive business ideas for recycling and resource recovery.

Education has been found to be a critical factor in achieving the strategy goals, given the free flow of information in the community participation process. In the first six years, significant progress was made in increasing resource recovery, which rose from 22% in 1993/94 to 64% in 2001/02. The strategy has sparked worldwide interest, with more areas and more governments setting similar goals.

BOX 10.2

Nigerian Waste-to-Wealth Scheme of Waste Conversion to Organo-mineral Fertilizer (Sridhar and Adeoye 2003)

Nigeria (population 120 million) is a federation of 36 states and the Federal Capital Territory of Abuja. Ibadan, the capital of Oyo State, is the largest city in West Africa and the third largest in Africa; it has a population of 2.5 million. Under its Environmental Planning and Management Program initiated in 1994, various stakeholders (federal, state, and local governments; the private sector; and communities) evolved a model initiative for waste management for Ibadan.

A major traditional market, *Bodija,* was selected for the collection of organic wastes. Traders in the market, local and state government representatives, professionals from the city's tertiary institutions (The University of Ibadan, The Polytechnic, Ibadan), and community leaders were constituted into the Waste Management Working Group that mapped out a strategy and action plans, all aimed at converting waste to fertilizer, creating a decent environment, promoting good health, producing goods and services, generating employment, and thereby reducing poverty.

By July 1998, an organized system of waste collection and a 25-ton per day capacity organo-mineral fertilizer facility were established. The state government provided funds, while other stakeholders contributed essential inputs. The indigenous technology was simple, employing 25 persons. The program stimulated the traders positively toward segregation of waste to feed the fertilizer plant, achieving a 90% enhancement within a year of commencement. The plant produced 10,000 kilogram and steady revenue of $20,000 per month.

The value of the project having been demonstrated, the Oyo state government has decided to increase the number of plants to ten, to serve a number of agricultural communities. The second one has been commissioned at the community level. Five other states (Akwa-Ibom, Kaduna, Kano, Lagos, and Ondo) have also shown interest in adopting the project for their capital cities. Cow dung, hitherto a nuisance, has become marketable, with a high level of patronage by small-scale peri-urban farmers, who value its good soil properties, consequently reducing the demand for chemical fertilizers in these communities.

BOX 10.3

European Union Limits Organic Waste in Landfills (Heermann 2003)

All waste sent to landfill must be treated, unless it is inert or treatment does not reduce the quantity of the waste or hazards to human health or the environment.

The Directive sets the following targets for reducing the biodegradable municipal waste added to landfills, compared to that produced in 1995:

- by 2006, reduce to 75%
- by 2009, reduce to 50%
- by 2016, reduce to 35%.

Composting has long been viewed as environmentally beneficial, particularly in tropical developing countries. Proper facility design and operation can mitigate or overcome adverse environmental impacts arising from ammonia, nitrate, phosphorus, heavy metals, and pathogen components, simultaneously reducing discharge of nutrient chemicals into the environment. However, in the industrialized countries compost is becoming unpopular because of the toxic chemicals in the waste streams (Zurbrugg et al. 2004).

10.2.3 Financial and Economic Responses

To be effective, a waste management system requires a sustainable revenue base to cover the capital (site, equipment, and facilities), maintenance, and personnel costs. The major problem in this regard involves cost-sharing among the interacting agencies either at the national and state levels (as in most developing countries of Africa and Asia) or at the unit levels of municipal (local) waste management (prevalent in industrial countries and some South American nations, notably Bolivia, Colombia, Guatemala, and Haiti). The issues include the specific roles of private-sector dominance (in the North) and public-sector prevalence (in the South) and the extent to which cost recovery (user, disposal, and refund-deposit charges) can improve waste management.

For developing countries in particular, waste management may contribute to societal well-being through poverty reduction using waste-to-wealth activities (which produce goods and services), but what are the social costs in terms of health hazards? Another important issue is the need for full-cost accounting. Where the cost estimates cover only waste management service costs and not the social and environmental costs, the selected waste management services might be harmful to the environment and human well-being in the long term. For example, in India, Nigeria, and Thailand, the open dumpsites, which produce goods and services through material recovery and recycling, make the waste-pickers vulnerable to physical injury, communicable diseases, poisoning, and zoonotic infections (Adedipe 2002). The specific economic instruments used, particularly in developing countries, must also recognize the level of poverty in relation to national income and distribution and political sensitivity, by way of infusing humanistic responses, as Panayotou (1990) has highlighted. As far as is practicable, the burden and cost of pollution must be borne by the polluter (NRC 2000).

The financial and economic responses would be more effective if we could answer the question "what is the economic nature of solid waste?" Solid wastes are consumption and production residuals. Production and consumption are driven primarily by economic variables, particularly price and incomes, although population size and concentration are also important factors (Reyer et al. 1990). If a problem is economic, it means specifically that it is characterized by scarcity and governed by choice. It follows then that solid-waste management is an important economic problem. Importantly, in our economic system, achievement of a cost-effective balance will require careful use of the market and price system to achieve waste management objectives. These objectives can be attained cost-effectively only by instituting rules that allow maximum flexibility (substitution possibilities, to the economist) in consumption and production decisions, subject to the constraint that all of the costs are paid.

The entire production and consumption cycle involves solid-waste generation and management (Goddard 1995). Governments in both developed and developing countries for the most part have relied on regulatory instruments in their efforts to mitigate the problems of solid-waste generation, collection, and disposal. But there is growing interest in some countries in the application of economic instruments in order to improve the efficiency of the waste management process.

In industrial countries, and some developing countries (for example, Nigeria) (FRN 1991), the following economic instruments have at one time or another been under consideration, or have been implemented as part of a waste management strategy: ambient standards; effluent standards; emission standards; performance standards; product standards; process standards; land use regulations; shoreline exclusion or restrictions; local ordinances; permits; protected area designations; pollution tax; product charge; administrative charge; tax incentives; subsidies; enforcement incentives; noncompliance fee; property rights; environmental assessment; need-based environmental communication and public participation programs; recycling credit (to stimulate increased recycling activity); landfill disposal levy (to reduce the amount of waste put in landfills); product charge (for example, packaging tax, to discourage overpackaging); tax concessions (to stimulate reuse/recycling or other activities); deposit refund systems (to increase the recycling of selected items such as batteries, or encourage returnable container systems); levy/tax on virgin raw materials (to influence the relative prices of primary and secondary recycled materials); and user charges (for example, household waste charges, to discourage the throwaway ethic and encourage reuse/recycling).

Financing charges (user charges) have been used to facilitate the collection, processing, and storage of waste, or the restoration of old hazardous waste sites. Incentives charges (for example, product charges) on the other hand can, among other things, be used to stimulate increased reuse/recycling.

Economic instruments also have other properties, including a revenue-raising capacity. This feature will be of particular importance in developing countries that lack basic waste treatment facilities and infrastructure. Revenue raised via waste-user charge (based on collection and/or disposal costs), for example, could be "recycled" into new or improved waste collection treatment and recycling facilities in the local area. A balance will need to be struck in terms of the level of the charge that could be levied, so that a meaningful amount of finance is raised, without at the same time stimulating extensive "illegal" dumping or corrupt practices.

Other economic instruments that appear to offer some advantages in the developing-country context are recycling credits, tax concessions, and deposit refunds. The first two instruments could involve fairly modest sums of finance but still increase recycling activities.

10.2.4 Institutional Responses

Institutional responses in their generic context embrace a set of rules that are based on the social values of the people, approved by the people through wide-ranging mechanisms, and allowed to operate freely in the society. This simple definition buttresses the concept of global governance as the sum of the many ways individuals and institutions (public and private) manage issues (Commission on Global Governance 1995). In its application to waste management, it is the process by which there is a defined structure for administering effective waste management. This demands the participation of all stakeholders concerned and creates a mechanism for transparency, accountability, and public pressure-driven commitment of government to effective and sustainable waste management operations and practices.

In industrial countries, the assumption is that there is an existing private sector to which certain public-sector functions may be transferred. But in developing countries, a viable private sector does not exist or exists in an embryonic (formative) stage. Proponents of privatization argue that it improves the efficiency and effectiveness of services previously delivered by the public sector. However, there is evidence that more often than not new inequities and ineffectiveness arise from privatization. There are therefore complexities in the transformation from public to private sector. In waste management, a shared responsibility through public-private sector partnership seems to be the answer, with the public sector providing capital equipment, promulgating laws, and conducting research, and the private sector taking responsibility for management operations, including health risk reduction and the promotion of overall human well-being (Sappington and Stiglitz 1987). Community-based organizations should be involved in this process as stakeholders, in accord with the proposal of Nsirimovu (1995) that human rights and participatory governance should be focused on the individual as the prime subject of all developments, including urban waste management.

Stakeholder participation can be encouraged by including waste prevention and resource recovery explicitly, by encouraging the analysis of interactions with other urban systems, and by promoting an integration of different habitat scales like city, neighborhood, and household (Klundert 2000).

Some of the institutional linkages that would be needed to sufficiently manage waste do not exist or are confused in developing countries. The local (municipal) governments, which are constitutionally charged with the responsibility, invariably lack the equipment base and skilled human resources. The policy response that has to be considered is the establishment of viable units of municipal governance for routine operations, while the state/provincial governments should be responsible for major cost-intensive facilities such as sanitary landfills, incinerators, and wastewater treatment plants. Consequently, state governments should work out partnerships with the national governments. Where there is need, regional/global collaboration should be encouraged. For the industrial countries, more stringent governance reforms are the response option. For example, in the United States, the *Toxic Release Inventory* has led to dramatic decreases in corporate emissions. Among developing countries, Nigeria's recent *Inventory of Hazardous Wastes* has been a tremendous response option, contributing to waste and ecosystem management (Osibanjo 2002).

At the global level, the protocols, conventions, and treaties need to be implemented with strict adherence to agreed commitments and prescribed sanctions, including diplomatic interventions. For a start, the local (municipal) and state (provisional, district, prefecture, as the case may be) governance structures should be reviewed for better compliance inculcation.

10.2.5 Sociocultural Responses

Waste management problems are closely associated with society, its beliefs, and its attitudes. The flow of waste from the place of origin to the site of disposal has human dimensions besides the application of technology, given the concept of a city or region functioning as an anthroposphere. Effective resource management must be prescribed to closely fit particular societal norms and values, since governance is in constant flux and operates in an "established milieu" even in epistemic societies, with the ultimate aim of exercising power in the management of a country's economic and social resources for development. Human rights and individual liberty, within the limits imposed by democratic principles, must also be respected (Nsirimovu 1995).

Knowledge and attitudes govern the practices. There is need to motivate change toward more environmentally sound attitudes at various levels, for example, home, school, and workplace. Education plays a key role. When developing educational programs to motivate changes in behavior, it is important for their success to include stakeholder input and to understand in-depth behavioral aspects (Okpala 1996). Obviously, the methodology options will differ with sociocultural value content of each community and nation. Generally, but particularly in developing countries, there are two sets of options for educating about waste clearance: formal and informal, these being the incorporation of waste management into the curriculum on environment and sustainable development (for the formal) and the establishment of environment and conservation clubs (for the informal) (Okpala 1996).

There is need for a shift from a single-purpose focus on ecological aspects to a more multifaceted approach that routinely considers social and risk analysis, implications, and management. The ultimate aim entails analyzing aggregate environmental degradation costs, establishing systems of national resource accounting that reflect environmental damages and losses, correcting market prices to reflect full costs (and benefits), and offering incentives such as rebates for industries adopting anti-pollution measures as well as disincentives such as taxes (Panayoutu 1990, Reyer et al. 1990).

Although sociocultural response options are theoretically sound, their practical application has not been as rewarding as globally expected; in developing countries, they have yet to be adopted as part of socioeconomic policy response regimes (Chokor 1996).

10.2.6 The Need for Integrated and Sustainable Waste Management

Responses when applied individually have been able to handle the issue of waste management only partially. Thus it becomes imperative to integrate efforts in a planned manner. Integrated waste management implies that decisions on waste handling should take into account economic (including technical in relation to its costs), environmental, social, and institutional dimensions. Economic aspects may include the costs and benefits of implementation, the available municipal budgets for waste management, and spin-off effects for other sectors in the economy in terms of investments. The environmental dimension may consist of local problems (increased risk of epidemics and groundwater pollution), regional problems (resource depletion and acid rain), and global problems (global warming and ozone depletion). Social aspects include employment effects for both the formal and the informal sectors, impact on human health, and ethical issues such as the use of child labor. Finally, the institutional dimension of integrated waste management aims at developing a system that effectively involves the main stakeholders.

"Integrated" refers to the integration of:

- different aspects of sustainability (technical, environmental/public health, financial, etc.);
- different collection and treatment options at different habitat scales, that is, household, neighborhood, and city level (operational interaction);
- different stakeholders, including governmental or nongovernmental, formal or informal, profit- or nonprofit-oriented (cooperation, linkages, alliances, economic and social interaction); and
- the waste management system and other urban systems (such as drainage, energy, urban agriculture, etc.).

10.3 Some Aspects of Responses to Waste

As identified in MA *Current State and Trends* (Chapter 15), the sources of waste and contaminants, and their impacts, are numerous and continue to grow. Society must be alert to address the challenges that new wastes present. Not every human response has been successful, and there are lessons to be learned from the past. This section considers some examples of past activities and highlights the need for holistic approaches to satisfactory responses to waste.

10.3.1 Historical Considerations

The history of waste and waste management, in general, is intricately bound with the history of solid waste, given its ubiquitous nature and visibility. It is also inevitably tied to civilization. For most of the last two million years humans generated little "garbage"; what was produced was easily disposed of through biodegradation, which was a simple ecosystem service. The rate of garbage accumulation in the ancient city of Troy was calculated to be 1.4 million tons per century (Rathje 1990).

Sanitary sewers were found in the ruins of the prehistoric cities of Crete and the ancient Assyrian cities. Storm-water sewers built by the Romans are still in service today. Although the primary function of these was drainage, the Roman practice of dumping refuse in the streets caused significant quantities of organic matter to be carried along with the rainwater runoff. Toward the end of the Middle Ages, below-ground privy vaults and, later, cesspools were developed. When these containers became full, sanitation workers removed the deposit at the owner's expense and the deposits were used as fertilizer at nearby farms or were dumped into watercourses or onto vacant land. A few centuries later, there was renewed construction of storm sewers, mostly in the form of open channels or street gutters. At first, disposing of any waste in these sewers was forbidden, but by the nineteenth century it was recognized that community health could be improved by discharging human waste into the storm sewers for rapid removal. By 1910, there were about 25,000 miles (40,225 kilometers) of sewer lines in the United States which have increased manifold today. U.S. marine waters are estimated to receive over 45 million metric tons of wastes per year (UNEP 2002).

Accumulation of waste prompted New York City to introduce incineration as a response (Breen 1990). Following this seeming success, as many as 700 cities adopted the use of incineration, further popularizing incineration as a viable societal response. Given that incineration response also created air pollution, a second phase of response necessitated the use of sanitary landfill.

While incineration was a nightmare in the early part of the last century, more advanced technologies using high temperatures have yielded wider acceptance by people and policy-makers, as modern incinerators produce less emissions and most of the toxic substances are further oxidized. This has become a very handy tool for managing industrial and infectious hazardous wastes (Tchobanoglous et al. 1993). In some cases, incineration is still not the method of choice and it has been replaced with irradiation techniques. But the latter are not affordable in the developing countries.

Legal response in the United States was later provided by the enactment of the U.S. Solid Waste Disposal Act of 1965 and the Resource Conservation and Recovery Act of 1976. This Act also established what later became the "cradle-to-grave" system of hazardous waste management (Soesilo and Wilson 1995). In developing countries, however, from the 1930s to date, their own response to the need to manage wastes has been by the use of "open dump sites," as in Africa (Onibokun et al. 1995), Asia (Hoornweg 1999; Zurbrugg 2002), and elsewhere (Lardinois and van de Klundert 1995; Klundert 2000). The point must be made, however, that the effectiveness of open dumping differs from country to country. Garbage mountains point up an important truth: "Efficient disposal is not always completely compatible with other desirable social ends—due process, human dignity, and economic modernization" (Rathje 1990, pp. 32–39).

In terms of liquid wastes, perhaps the first alarm on unsanitary conditions and their relation to health and disease is credited to Sir Edwin Chadwick of Britain, who in 1842 wrote the classic report on "Sanitary Conditions of the Labouring Population of Great Britain." The cholera epidemics of 1849 and 1853 created the need for a response when society demanded the "removal of dirt." John Simon in 1848 developed the famous British Public Health System, which was accepted by the United States and some parts of Europe (Dubos et al. 1980), and later by other countries as well.

The history of other forms of wastes is not as richly detailed as that of liquid and solid wastes. Only recently have scientific studies conclusively shown that wetlands contribute substantially to comprehensive human health and welfare (Hemond and Benoit 1988; Queen and Stanley 1995) and that such positive ecosystem functions are threatened by chemical pollution (Mitsch and Gosselink 1986). These developments were further strengthened through comprehensive and standardized analytical techniques during the 1970s (USEPA 1979), with applications to different ecosystems, notably freshwater, wetlands, forests, arid and semiarid lands, and soils (Linthurst et al. 1995).

10.3.2 Political Interest in the Responses

Since 1972, there has been reasonably sustained political interest at the national, and particularly global level, of concerns arising from the reported striking effects of DDT on wildlife (Carson 1962) and occurrence of pesticide residues in various foods, including mothers' milk (UNEP 2002). With such global sensitization, but arising from continuing degradation of the environment through a multitude of agents and causes, prominently including wastes, there was a decisive turning point with the Stockholm Conference on the Human Environment in 1972, which led to the establishment of UNEP. Despite the intensive coordinating activities of UNEP, the pollution of the environment did not abate from source, nor was it effectively controlled by the polluting economic activities of nations. Given the sad state of the continually deteriorating environment, the UN enunciated Agenda 21 in 1989, which led to the UN Conference on Environment and Development (UNCED); the first Earth Summit, in Rio de Janeiro, Brazil in 1992. For the first time, the world took the bold step to map out strategies for the integration of environment concerns with all development issues aimed at improving human well-being (Adedipe 1992). Ten years later, the whole issue was revisited at the World Summit on Sustainable Development (WSSD) in Johannesburg in 2002, given that the expected goals of Agenda 21 were yet to be substantially achieved. The WSSD came up with the Millennium Goals, which are in the continuing thrust for human well-being. (Table 10.2 summarizes the milestones in the modern history of waste management.)

10.3.3 Evidence of Positive Impacts of Responses

Waste management practices and policies over the last three decades have resulted in positive responses in terms of improvement of ecosystems. Some positive impacts of the responses identified (Hu et al. 1998; Jana 1998; Pillai et al. 1946, 1947; Smit 1996; Sommers and Smit 1996; Sridhar et al. 2002) are:

- Waste recycling activities have been found to result in improved resource conservation and reduced energy consumption as well as reduction of heavy metal contamination of water sources.
- In the Baltic Sea, the mercury levels of fish caught were reduced by 60% due to stringent pollution control measures.
- Major rivers such as the Thames have supported biodiversity, as is evident from the reappearance of salmon after rigorous pollution control measures. The ten-year "clean river" program initiated by the Singapore government in 1977 at a cost of US $200 million has brought life back to the Singapore River and the Kallang Basin, with increased dissolved oxygen levels ranging from 2 to 4 mg per liter (UNEP 1997).
- Phasing out of lead from gasoline has reduced lead emissions from vehicular sources.
- Wetlands have been widely reported to absorb significant amounts of anthropogenic pollutants.
- Ferti-irrigation practices have significantly improved the economic base of low- income communities in urban areas.

In the tropical countries in particular, controlled and judicious use of aquatic weeds such as water hyacinth (water hyacinth treatment plant for wastewater) and blue green algae (waste stabiliza-

Table 10.2. Milestones of Waste Management Responses

Year	Response
1972	• United Nations Conference on the Human Environment, Stockholm • UNEP established • UNESCO Convention Concerning the Protection of the World Cultural and Natural Heritage
1980	• International Water Supply and Sanitation Decade begins
1982	• United Nations Convention on the Law of the Sea
1989	• Basel Convention on the Control of Transboundary Movement of Hazardous Wastes and their Disposal
1990	• Eco-efficiency established as a goal for industry
1992	• United Nations Conference on Environment and Development (Earth Summit), Rio de Janeiro • Convention on Biological Diversity
1996	• United Nations Conference on Human Settlements (Habitat II), Istanbul • ISO 14000 created for environmental management systems in industry
1997	• Kyoto Protocol adopted • Rio + 5 Summit reviews implementation of Agenda 21
2000	• Cartagena Protocol on Biosafety adopted
2001	• Stockholm Convention on Persistent Organic Pollutants
2002	• World Summit on Sustainable Development, Johannesburg
2003	• World Water Forum, Kyoto, Japan

tion ponds) for treating small wastewater flows helped in improving environmental sanitation and the by-products provided protein and mineral needs of livestock.

10.3.4 Evidence of Damaging Impacts of Responses

Most of the established waste management practices such as landfills, incineration plants, and wastewater treatment plants have been shown to have negative impacts. Hazardous wastes are a particular case in point (Lieberman 1994; Soesilo and Wilson 1995). The types and quantities vary significantly from place to place and affect human health and well-being in terms of increased numbers of new cancers, abortions, mutagenic and teratogenic effects on the newborn, and increased morbidity and mortality (Adesiyan 1992).

Some waste management facilities have become difficult to maintain due to local policies and regulations, and some countries resorted to transboundary movement. In Nigeria, for example, hazardous waste from Italy was dumped in Koko (a small coastal town in southwestern Nigeria), generating a diplomatic furor between Italy (transporter) and Nigeria (recipient). The hazardous waste dumped eventually led to some deaths. Using the polluter-pays principle, the exporter of the waste was compelled to return it to Italy and to carry out remediation activities. Following the unfortunate incident, Nigeria created the Federal Environmental Protection Agency in 1989; since then, several decrees and regulations have been enunciated to improve the environment (Aina and Adedipe 1991).

The Basel Convention on the Control of Transboundary Movements of Hazardous Wastes and their Disposal aims to minimize the creation of such wastes, reduce their transboundary movement, and prohibit their shipment to countries lacking the capacity to dispose of them in an environmentally sound manner. The Convention notwithstanding, such transboundary movement keeps recurring, to the detriment of human and biodiversity well-being.

Another instance of well-documented negative impact is the practice of irrigation with wastewater. In Pakistan, for example, over-applied wastewater with insufficient drainage resulted in degradation of soil structure, visible soil salinity, and the delayed emergence of wheat and sorghum due to excessive nutrients from the ferti-irrigation practice. Further, farmers using raw wastewater for irrigation are five times more likely to be infected by hookworm than those using canal water (Scott et al. 2004). Also, irrigation with industrial wastewater has been associated with a 36% increase in enlarged livers and 100% increase in both cancer incidence and congenital malformations in China compared to areas where such practices were not adopted (Yuan 1993). Another example is in Japan, where chronic cadmium poisoning from wastewater use caused the Itai-Itai disease, a bone and kidney disorder (Osibanjo 2002; WHO 1992).

10.3.5 Integrated Design of Responses

Integrated responses—which consider the interactions and interdependencies among the regulatory, economic, technological, and institutional elements—are the beginning of a holistic approach to waste management and processing.

In a comprehensive mode of integration in the Middle East, gray water treatment and reuse in home gardens is one means of dealing with overflowing cesspits that would otherwise contaminate "wadis" and the shallow groundwater table. This response conserves fresh water by recycling it and, perhaps most importantly, helps to reduce poverty by offsetting food purchases by the poor. To be sustainable, this response must be integrated: in terms of regulations, gray water use must be made legal—it is often illegal in both developed and developing countries—and housing bylaws and standards must be changed to separate black and gray water pipes to facilitate gray water use. Second, a major appeal of the policy is economic—the poor are willing to install the gray water systems at their own cost because they know that their neighbors with installed systems are generating income as a result.

Technological aspects of the policy that must be addressed are: building the capacity of universities and research institutions to teach, train, and do research to improve gray water reuse technology, and building a cadre of plumbers who know how to install, maintain, and repair the systems. Finally, institutions such as departments of agriculture need to be reformed so that they support both traditional rural agriculture as well as peri-urban agriculture. In addition, departments of water, environment, and agriculture must develop integrated and consistent policies that facilitate gray water reuse (Faruqui and Al-Jayyousi 2002).

Ecohouses may be good models where all the wastes generated are integrated into the community for farming, livestock development and human well-being. The Australian Capital Territory's '"zero waste to landfills by 2010" strategy is another practical experiment; through a range of recycling and reuse initiatives its goal is to have very little disposable waste by 2010 (Baker 1999).

10.3.6 Cutting-edge Innovations

Certain technological or management innovations have proved beneficial in mitigating the damaging effects of waste on the environment, and in recent years, business firms and organizations have been established specifically to develop "smart ways" to deal with waste. They can include biotechnology, and may well result from smart observations of how nature itself deals with or removes waste. Bioremediation of coastal and industrial waters, where eutrophication occurs, can be affected using naturally occurring microorganisms or macroorganisms to remove the waste, and also to generate valuable and salable products.

Halogenated organic compounds (such as DDT, PCB, and the halogenated dioxins) can be degraded by a combination of bacterial, chemical, and irradiation techniques. Whereas the carbon-halogen bonds of halogenated hydrocarbons are not common in natural products from the land, they are very common natural products from the sea, and marine microbes are able to degrade many of the persistent synthetic halogenated organic substances. There is much we can learn from the sea.

Not all innovations demand high-technology input. For example, the marking of plastic containers to indicate if they can be recycled or not is a simple but essential communication idea. Also, suggestions on possible uses for containers when the original purpose is complete can stimulate the development of new implements, ornaments, or even works of art.

Innovations are simply "putting good ideas to work" and may be seen in the home, at work, or in commercial ventures. Innovations are more likely to be seen in societies that are challenged to think of better ways to overcome problems. They are often stimulated by government legislation, and can be implemented at every stage of the process from design and development to recycling and reuse. The European Union gave directions for change and different nations responded in different ways. Finland in 1998 set a target of 70% recovery of waste by 2005, with the result, there has been a large increase in recycling nation wide. The Australian Capital Territory and New Zealand targets of "zero waste to landfill" have seen the development of new industries and new uses for what was previously waste.

Sonar has been used to "remotely" determine the waste density and layering in used tanks, and remote sensing is also being investigated to detect dumping of waste in remote locations. Spectrometric fingerprinting of wastes can lead to the specific identification of the source of a particular discarded waste stream.

In the United States, the Federal Energy Technology Center performs, procures, and partners in research, development, and demonstration to advance technology for commercial waste management innovations. Other countries have similar organizations to encourage lateral thinking and application in every aspect of waste minimization.

10.4 Assessment of Selected Responses

10.4.1 Urban and Terrestrial Ecosystems

10.4.1.1 Impacts on Ecosystems and their Services

Urbanization is expanding more rapidly than any country's economy. The problem with urban waste is not only the quantity but also the change from dense and more organic to bulky and increasingly non-biodegradable. Increasing amounts of plastic, aluminum, paper, and cardboard are being discarded by households and industries (Cielap 2003), but many of these materials have been recycled in the informal sector to produce creative marketable household and novelty articles. Whereas dumpsites are typical of the developing countries, sanitary landfills and incinerators are the order in the developed world. For the developing countries, however, adoption of appropriate materials may be a long-term strategy using consensus-derived legal instruments. The primary wastes from industrial, urban, and near-urban activities, and their byproducts, the goods and services that are affected, and the nature of impacts are shown in Table 10.3.

10.4.1.2 Impacts on Human Health and Well-being

Heavy metals, mercury, cadmium, tributyl tin, and lead emanating from the waste pose serious health risks. Human activities and responses such as mining, smelting, waste dumping (including tires, used oils, electrical and electronic equipment and parts, and batteries), rubbish burning, and the addition of lead to gasoline have greatly increased the amounts of heavy metals circulating in the environment.

Mosquitoes such as *Culex quinquefasciatus, Aedes aegypti,* and others can breed in the wastewater retained in blocked drainage channels and may transmit filariasis and viral infections such as dengue and yellow fever (Cairncross and Feachem 1993). Emissions from waste burning and decomposing organic wastes may lead to gaseous emissions, thus leading to the change of pollutant status from one form to another. (See Table 10.4.)

10.4.1.3 Enabling Conditions

10.4.1.3.1 Waste reduction and recycling

With the growing unemployment, hunger, and poverty particularly in urban centers of developing countries (UNDP 2000; FAO 2003), waste may provide a short- to medium-term trade-off through reuse and recycling activities. Social conflicts in waste management are not uncommon at local, regional, and country level. To solve these problems, there is a need for economic instruments such as user charges and tax incentives for innovative practices (Heermann 2003), as discussed above. Such instruments as may be used must recognize aesthetic standards, cultural heritage, and social values.

Waste minimization is apparently the best response option for urban rejuvenation coupled with reuse and recycling of a major portion of the waste. Rising living standards and increased mass production have reduced markets for many used materials and goods in affluent countries. In most developing countries, traditional labor-intensive practices of repair, reuse, waste trading, and recycling have endured. Thus there is a large potential for waste reduction in developing countries; by contrast, the greatest potential for waste reduction currently rests with diverting biodegradable, non-biodegradable, and construction wastes. Most countries in Western Europe and North America, Australia, New Zealand, Japan, and Korea have adopted municipally sponsored source separation and collection systems. Other strategies are choice of packaging materials, packaging reduction goals in a given period of time, and mandatory separation of post-consumer materials by waste generators.

- Sensible recycling programs balance social, environmental, and commercial benefits. Paper recycling not only saves trees, it also reduces energy costs by 35 to 50% and decreases water consumption and pollution. Recycling aluminum cans brings energy savings of up to 95% and produces 95% less greenhouse gas emissions than when raw materials are used.
- Recycling also creates jobs. A "solid waste turnover sector analysis" carried out in 1994 put the British market value of solid waste at $3 billion (Jones 1995). In Britain, the recycling industry, already worth over $20 billion a year, employs 140,000 people. Getting the United Kingdom to recycle 35% of its household waste by 2010, a target under consideration by the government, will generate another 50,000 skilled and unskilled positions (Ribbans 2003).
- The motivating forces for waste recovery and recycling in the developing countries are scarcity or high cost of virgin materi-

Table 10.3. Impacts of Wastes on Urban Ecosystem Goods and Services

Urban Ecosystem	Peculiarities	Primary Wastes/Waste Byproducts Generated	Goods and Services Affected	Impacts
Urban Peri-urban Peripheral	built environment with a high human density (with human settlements of 5,000 or more)	mostly human and technology-based wastes; demolition, organic and inorganic, solid, liquid, gaseous, vehicular emissions, industrial wastes	food, biochemicals, genetic resources, climate regulation, water regulation, recreation and tourism, aesthetic, cultural heritage, spiritual and religious	population pressures rural–urban migration overuse of land environmental degradation disease burden poverty inadequate infrastructure

Table 10.4. Waste Management Responses and Health Risks

Waste Management Response	Health Risk
Collection, transportation, and disposal activities	injuries and chronic diseases due to sharps, contaminants, collapse of waste piles, cuts, infective wounds, burns, trauma, cancers, chronic respiratory infections
Mixed wastes—due to infected sharp waste, exposure to infected dust, breeding of vectors in waste-generated ponds, contaminated soil, rodents and other animals feed on waste, accidental ingestion of waste/ contaminated food, contaminated drinking water, foods grown on leachates and other waste streams	bacterial, viral, and parasitic—tetanus, staphylococcus, streptococcus, hepatitis B, AIDS, blood infections eye-related (trachoma, conjunctivitis) skin (mycosis, anthrax), respiratory infections (bacterial and viral) vector-borne (dengue, yellow fever) parasitic (malaria, filariasis, schistosomiasis) worm infestations (hookworm, ascariasis, trichuriasis) bacterial (cholera, diarrhea, typhoid) viral (dysentery) parasitic (helminthiasis, amoebiasis, giardiasis)
Toxic chemicals, leachates, and gaseous emissions from waste handling or treatment	respiratory tract infections, cancers, nuisance, particulate-induced disease conditions
Urban wastes in specific (solid wastes and their leachates, sewage, and sullage/greywater)	malaria transmitted through *Anopheles* breeding, dengue and yellow fever through *Aedes* mosquitoes, filariasis through breeding of *Culex* mosquitoes.

als, import restrictions, poverty, the availability of workers for lower wages, and the large markets for used goods and products made from recycled materials (IETC 1996).

- To solve the problems of poverty, waste recycling activities use policy and economic responses, particularly in developing countries where there is chronic unemployment. Such waste-to-wealth activities (Figure 10.2), including harnessing of nitrogen and energy from waste, need to be formalized as policy responses. However, such policies should also include presorting to protect the health of the recycling worker.

10.4.1.3.2 Urban agriculture

Urban agriculture, which was little known in the 1970s, is becoming more common (Rodrigues and Lopez-Real 1999). The number of urban farmers producing for the market is expected to double from about 200 million in the early 1990s to 400 million by 2005 (Smit 1996). Globally, more than 800 million urban dwellers are involved in this economic activity (Smit 1996; Sommers and Smit 1996; Kaspersma 2002; Redwood 2004). Many successful urban agriculture schemes are reported around the globe—from Dar es Salaam to Singapore, from Vancouver to Curitiba. Singapore is in the forefront in the development of urban aquaponics. Other examples:

- In Brazil (Saõ Paulo), urban agriculture is a planned land use activity.
- Most households in Southeast Asia and the Pacific Islands subregions practice urban agriculture.
- About 30% of the Russian Federation's food is produced on 3% of the land in suburban "dachas." In Moscow, the proportion of families engaged in agriculture grew from 20% of the city's population in 1970 to 65% in 1990.
- In Harare, when sanctions on urban agriculture were lifted temporarily in 1992, within two years the area cultivated doubled and the number of farmers more than doubled. Municipal costs for landscape maintenance and waste management were down, food prices were down, and hundreds of jobs were created.
- Urban agriculture has an important significance to global sustainability in that the production of food close to the consuming market reduces the need for transportation, thus reducing the consumption of fossil fuels and the associated emissions of CO_2. There is also a reduction in packaging, refrigeration, and the use of preserving additives (Rees 1997).
- School gardens are becoming popular and are effective learning tools to students of all ages.
- Effective wastewater use for urban agriculture cannot be implemented without creating enhanced awareness by formal and informal education, information, and awareness through a policy response (Okpala 1996; Buechler et al. 2002). In many developing countries urban agriculture is contributing to reuse of organic matter through the recycling activities. In many places, solid wastes converted to compost, and gray water emanating from open drains, are used to fertilize soils (Pillai et al. 1947; Sridhar et al. 1992; Sridhar 1995; Sridhar et al. 2002; Urban Agriculture Magazine 2002).

10.4.2 Agricultural Ecosystems

Certain agricultural products and practices are implicated in human health problems, including animal antibiotics, nitrates in

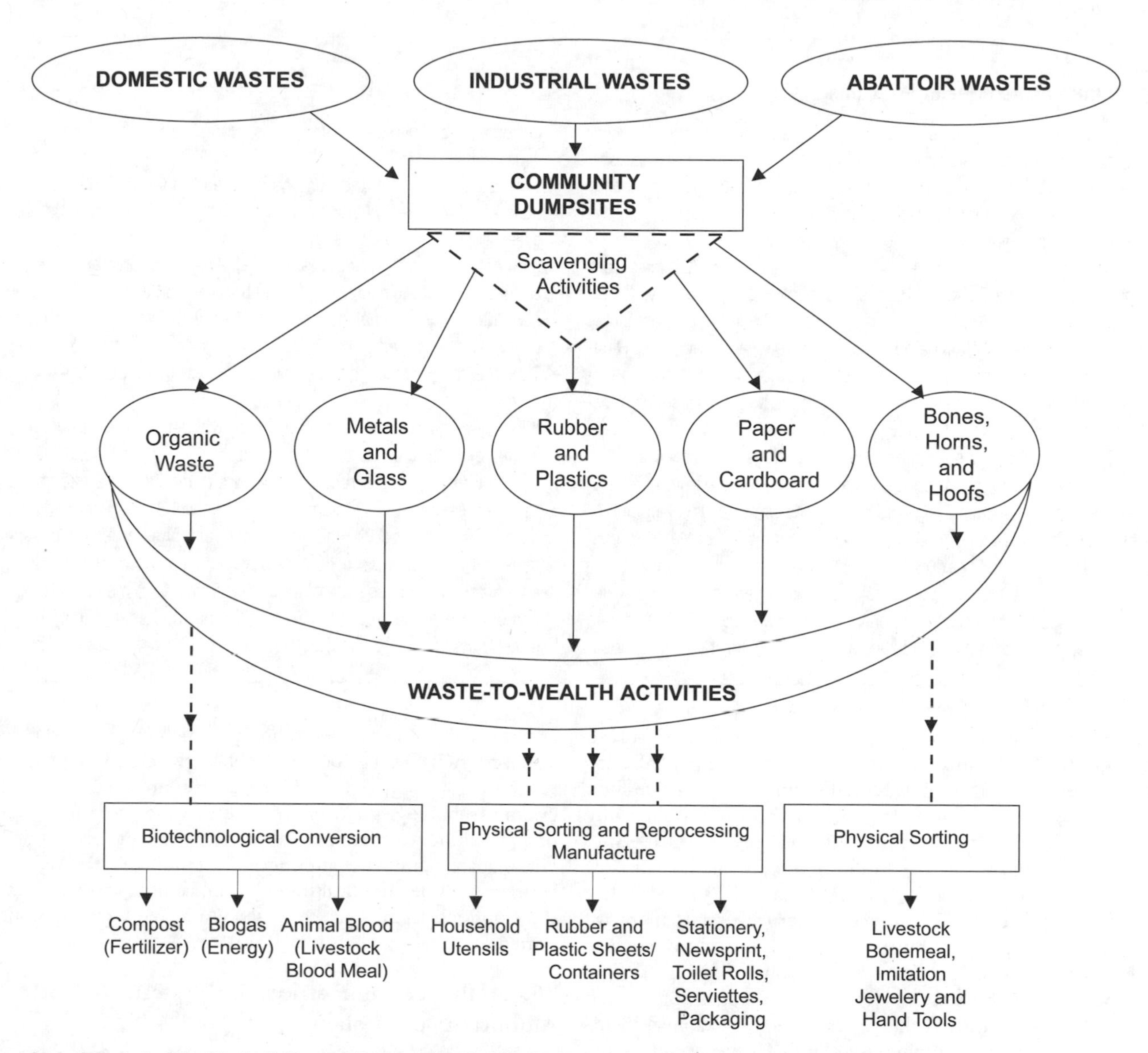

Figure 10.2. Waste-to-Wealth Activity and Processing Profiles in Developing Countries (Adedipe 2001)

groundwater, pesticide exposure in an occupational setting, pesticide residues and food additives in foods, and certain food processing techniques. (See Box 10.4.) There is some evidence to suggest that conventional soil management practices are contributing to declining nutritional value in foods (MacRae 1990).

Global Assessment of Soil Degradation estimated that about 13% or 850 million hectares of the land in the Asia and Pacific region is degraded (Olderman 1994). Improper ferti-irrigation using wastewater and poor drainage practices results in salinization and degradation of soils. Policy and technological responses should be able to offset these. Virtually every part of the globe has these problems (Pillai et al. 1946; MOAFFA 1999).

The ecosystem-positive responses to waste management lie in increased production of food and fiber, soil conservation, and/or remediation. The negative responses include soil degradation, soil erosion, weed growth, eutrophication, groundwater pollution, loss of biodiversity, and the greenhouse effects. They also include the use of agrochemicals for pest control. To this extent, as an example, the use of persistent organic pesticides is being discouraged by gradual substitution of biological pest control (Adesiyan 1992).

In Brazil, nitrogen-fixing bacteria are being used successfully to remediate nitrogen-deficient agricultural ecosystems; Doebereiner (1997) reported that Brazil has become the world leader in replacing N fertilizers by biological nitrogen fixation with large positive impacts on the production of food and biofuel crops. See Chapter 9 for a more detailed assessment of response options relative to the use of nitrogen- and phosphorus-based fertilizers in agriculture.

10.4.3 Hazardous Waste, Biosolids, and Soil Remediation

Soil contamination through waste discharges, particularly hazardous wastes, is a worldwide phenomenon. Compost is also known to remedy the contaminated soils, particularly from heavy metals and organic pollutants. Working with oil-contaminated soils, researchers showed that a combination of NPK fertilizer and straw mulch promoted natural revegetation through the restoration of soil properties, while either of them when applied singly was less effective (Odu 1972).

In industrial nations, many municipalities face the problem of disposing of increasing volumes of wastewater economically and

BOX 10.4
Agriculture-related Health Problems

- Crops irrigated with polluted water or fertilized with inadequately produced compost or animal manure may be infected with bacteria (shigella, typhoid, cholera), worms (e.g., tape- and hookworms), protozoa, enteric viruses or helminths (ascaris, trichuris); schistosomiasis is prevalent in some regions due to irrigated agriculture.
- Crops may take up heavy metals and other hazardous chemicals from soils, irrigation water, or sewage sludge polluted by industry. Crops grown close to main roads or industry, and food purchased from street vendors, may be contaminated by airborne lead and cadmium. Residues of agrochemicals may contaminate crops or drinking water (pesticides, nitrates). If waste materials are not separated at source, compost may contain heavy metals, which can be taken up by crops. Despite a 64% drop in annual global atmospheric emissions of lead since 1983, several million adults and children suffer the adverse health effects of lead poisoning, including impaired mental and physical development. In the United States alone, childhood lead poisoning is estimated to cost some $43 billion per year.
- Mosquitoes that are the vector for filariasis breed in standing water heavily polluted with organic materials. Mosquitoes that transmit dengue breed in water containers that include much solid waste, like coconut husk, rubber tires, broken bottles, water storage jars, buckets, and water butts.
- Closeness of animals and humans may lead to the occurrence of zoonotic diseases like bovine tuberculosis (cattle) and tapeworms, especially when animals are scavenging waste tips. Drinking water may get contaminated with pathogens by application of animal waste (e.g., slurries) to land. Animal products can become contaminated with pathogens due to contamination of animal feed with infected feces (salmonella, campylobacter).
- Tanneries may discharge hazardous chemicals in their wastes (tannum, chromium, aluminum).
- If fish (especially shellfish) are reared on wastewater and/or human and animal excreta, there is risk of passive transfer of pathogens (hepatitis A) and transmission of trematodes where trematodes are endemic and fish is consumed raw. Contamination of fish with human or animal fecal bacteria may occur during post-harvest operations (for example, salmonella). Poorly managed fishponds may become a breeding ground for anopheles mosquitoes. Use of antibiotics in fish feed may lead to development of antibiotic-resistant bacteria in the food chain. Fish products may be contaminated with heavy metals and agrochemicals.

in an environmentally sound manner without involving risk to human health. It is estimated that the total quantity of U.S. domestic wastewater, for example, could supply around 1.5 million tons of nitrogen alone, or about 15% of the amount currently marketed as inorganic fertilizers. But the heavy metals—zinc, lead, cadmium, and selenium—could be toxic, which is a trade-off between economic gains and health hazards (Olson 1987). Biosolids (treated residuals from wastewater treatment) are proving useful to replace wastewater residuals (formerly sewage sludge). Because of the proven fertilizer value of sewage sludge, land application of this waste material is becoming a popular and more feasible method than other alternatives (for example, incineration, landfill, ocean dumping, pyrolysis). However, sludge has the disadvantage of building up mercury levels in the soils, making it more available for plant uptake and entry into the human food chain (Cappon 1984).

A study reported from Cairo (UNCSD 1999) indicated that farmers are ready to pay for the biosolids as they are convinced of the benefits. The Greater Cairo Wastewater Project produces about 0.4 million tons of sludge or biosolids from its wastewater treatment plant and has a ready market for growing wheat, berseem clover, forage maize, and grapevines. This has reduced pressure for manures, which are scarce and more expensive to buy. India, China, and some other Asian, African and South American countries have been following this practice, though on a small scale.

Industrial waste materials are often used in fertilizers as a source of zinc and other micronutrients. Current information indicates that only a relatively small percentage of fertilizers is manufactured in United States using industrial or hazardous wastes as ingredients (National Agriculture Compliance Assistance Center 2004). The U.K. Department for Environment, Food, and Rural Affairs (DEFRA—formerly MAFF) has developed a code of practice for minimizing plant health risks through residue management. Anaerobic digestion of agricultural residues proved sound as it produces three usable products—biogas, which can be used to generate heat/electricity; fiber, which can be used as a nutrient-rich soil conditioner; and liquor, which can be used as liquid fertilizer (AGRIFOR 2004). A well-designed and cost-effective waste management system, based on resource recovery and recycling technologies as well as reuse of wastewater for irrigation or aquaculture, can produce substantial social and economic benefits that are gaining the attention of decision-makers. When total costs are considered—health, pollution, landfill, and incinerator—these options begin to make sense (World Bank 1997). Some reported successes are shown in Box 10.5.

10.4.4 Protecting Remote Ecosystems (Island, Mountain, and Polar)

There is increasing use of mountains as dumping ground for litter, hazardous and semi-hazardous wastes, and incineration ash (UNEP 2002).

The Arctic landmass is approximately 14 million square kilometers, of which the Russian Federation and Canada account for 80%. The decrease in area of glaciers in both the Arctic and the Antarctic is probably due to air pollution and lower albedo of the "dusted" glaciers' surfaces. Habitat loss, terrain modification, disturbance to wildlife, and introduction of exotic species and disease are the response impacts. There are legislative and regulatory bodies to control degradation, but implementation has been hindered by financial constraints, particularly in the Russian Federation (UNEP 2002).

10.4.5 Freshwater Ecosystems

Freshwater ecosystems are aquatic systems that contain drinkable water or water of almost no salt content. Freshwater resources include lakes and ponds, rivers and streams, reservoirs, wetlands, and groundwater. They provide drinking-water resources and water resources for agriculture, industry, sanitation, and food, including fish and shellfish. They also provide recreational opportunities and a means of transportation. Freshwater ecosystems are home to numerous organisms, for example, fish, amphibians, aquatic plants, and invertebrates. It has been estimated that 40% of

BOX 10.5

Examples of Waste Use in Agriculture (World Bank 1997)

Recycling

- Well-known examples include use of city waste in peanut-growing systems on the Kano plains of Nigeria, use of nightsoil in China, and "sewage farms" around European cities during the nineteenth century; city waste improved soils in the Netherlands.
- Urban wastewater is used in many parts of the United States, Israel, and Jordan.
- Abattoir waste is used to produce methane and compost in Senegal.
- The Kolkata sewage fisheries system is the largest single wastewater system involving aquaculture, with 4,600 hectares of sewage-fed fishponds employing 4,000 families
- In California, growers add compost to build up soil organic matter—high-quality compost is widely used in horticulture, seedling production, and general agriculture, while lower-quality material is used in city parks and to stabilize land. Sludge and compost add nutrients and organic matter to the soil, improve water retention and transmission, and improve soil structure.
- Widespread adoption of recycling technologies has been hindered by undeveloped markets, transportation costs, health and cultural issues, and inadequate regulations.
- The primary barrier is that the potential benefits are not adequately taken into account by urban planners, sanitary engineers, and farmers. Waste handling has focused on landfills and incinerators, while inorganic fertilizer and fresh water are the primary inputs used to meet the needs for soil nutrients and irrigation water.

Technology

- The current use of waste in agriculture has been mostly beneficial, but there are some potential detrimental effects, usually closely related to either the quality of the raw material, its processing, or both. The level and type of treatment varies, as does cost. Composting technologies have improved greatly in recent years and provide better quality compost with increased acceptability.

Public Health Concerns

- While disposal of municipal waste by land application is generally successful, there is evidence that irrigation with untreated municipal and industrial waste can harm crops as well as humans, either those living in the area or those who consume the products. These risks can be resolved by preventing pollutants from accumulating in the soil, taking advantage of the soil's capacity to assimilate and detoxify pollutants, and determining optimal levels of safe application. Specific attention needs to be given to the risks associated with transfer of human pathogens.

all known fish species on earth come from freshwater ecosystems (USEPA 2002).

Eutrophication leads to aquatic weed growths, prominent among them being water hyacinth (*Eichhornia crassipes*), *Salvinia molesta* and *Typha sp*. Aquatic weed growths also lead to new habitat development, particularly snails carrying schistosomiasis and mosquitoes carrying various diseases. Many tropical countries are now trying to utilize these aquatic weeds for income generation, employment, and resource utilization for energy and compost. Floating vascular aquatic weeds such as water hyacinth is a good system to treat sewage and other wastewater containing low organic content. The technology is cheap and the byproducts are easily recycled. This is a typical example of how the problem of one ecosystem can be harnessed for the benefit of other ecosystems. However, in open, uncontrolled, large bodies of water, this approach is still not practicable, given the experiences of Lake Victoria in eastern Africa, many areas of Brazil, Australia, and elsewhere.

Anthropogenic N additions to temperate ecosystems have been shown to affect a wide range of ecosystem properties and processes, especially when the inputs are large and continuous. Phosphorus is the critical limiting nutrient and the one of concern for eutrophication of freshwater systems, but evidence points to combinations of phosphorus and nitrogen. A eutrophic lake is characterized by a shift toward the dominance of phytoplankton by cyanobacteria, including noxious forms, most of which produce toxins (Anderson 1994; Chandler 1996; World Resources Institute 1992).

Increased inputs of nitrogen also manifest in relatively increased productivity but loss of biodiversity. Freshwater systems that are poorly buffered are acidified by increased deposition of nitrate and ammonia. The continuing acidification of Europe, northeastern North America, and parts of Asia is now leading to nitro-pollution rather than a sulfur pollution problem (Rabalais 2002). It is becoming increasingly apparent that the effects of eutrophication are not minor and localized, but have large-scale dimensions and are spreading rapidly (Nixon 1995). (See also Chapter 9 on this issue.)

Aquaculture, which has become a global economic venture, is also known to contribute to eutrophication. In China, polyculture of scallops, sea cucumbers, and kelp reduces eutrophication and the use of toxic antifouling compounds. Nutrients from scallop excreta are used by kelp, which used to require the addition of tons of fertilizers. Antifouling compounds and herbicides can be reduced because sea cucumbers feed on organisms which foul fishing nets and other structures. For shrimp and catfish culture, deeper ponds can be constructed to reduce weed growth to further limit herbicide use. This technique is a synergism and environment-friendly (Emerson 1999). Integration of such technologies into ecosystem restoration will bring in economic benefits.

10.4.6 Wetlands

Wetlands are among the most productive ecosystems, covering about 6% of Earth's land surface. They are characterized by marshlands, swamps, and bogs and play a vital cleansing role for the pollutants by acting as "sinks," regulating floods and providing habitat for numerous species of plants and animals. Wetlands help in water purification. Particulate matter such as suspended soil particles and associated adsorbed nutrients and pollutants settles out. Dissolved nutrients such as phosphorus and nitrogen can become adsorbed onto the particles and taken up by living organisms, or nitrogen is lost to the atmosphere. Other contaminants such as heavy metals can also be adsorbed onto sediment or organic particles. Exposure to light and atmospheric gases can break down organic pesticides or kill disease-producing organisms. The salinity of water within wetlands often increases as water levels drop, and the pollutants may become concentrated. If a lot of contaminants are flushed from the wetland by floods, they can foul the water downstream. A wetland is not a final sink for most pollutants but it may retain them for a period of days, months, or years (Australian EPA 2002).

Many wetlands have been degraded or destroyed by human activities since European settlement, through filling, draining, flooding, and clearing, and by pollution. The Ramsar Convention with 43 member signatories, signed in 1971, has protected some 200,000 square kilometers of wetlands.

10.4.7 Marine and Coastal Ecosystems

Marine ecosystems cover over 70% of Earth's surface. The habitats range from the more productive near-shore regions to the deep ocean floor, which was once thought to be uniformly barren. More recent studies and more extensive investigations by remotely controlled unmanned submarine vessels have revealed such features as deep-sea vents, which support significant marine communities. At this stage, those remote areas appear to be buffered from the adverse impacts of human activities, but we know that "dilution is not the solution" in any waste clean-up measures, and even remote and deep areas of the oceans may in time be adversely affected from some types of waste. Deep-sea dumping of waste must be banned completely. Important marine ecosystems include: oceans, sea mounts, rocky sub-tidal (kelp beds and seagrass beds), intertidal (rocky, sandy, and muddy shores), tropical communities (mangrove forests, coral reefs, and atolls), estuaries and salt marshes, and lagoons. Marine ecosystems are home to a host of different species ranging from tiny planktonic organisms that comprise the base of the marine food web (that is, phytoplankton and zooplankton) to large marine mammals like the whales, manatees, dugong, and seals. In addition, many fish species reside in marine ecosystems including flounder, scup, sea bass, monkfish, squid, mackerel, butterfish, and spiny dogfish.

Areas such as mangroves, reefs, and seagrass beds also provide protection to coastlines by reducing wave action, and helping to prevent erosion, while areas such as salt marshes, coastal lagoons, and estuaries have acted as sediment sinks, filtering runoff from the land (USEPA 2002).

Mangroves, seagrass beds, and coral reefs have all been shown to be sensitive to different types of waste, whether they are nutrients from the land or discharges of waste or oil from ships. The Australian government to the threat posed by human activities on its bio-diverse mangroves, sea-grass beds, and coral reefs and islands of the Great Barrier Reef Region was to nominate the Great Barrier Reef Region for inscription on the World Heritage List. Such listing, achieved in 1981, requires the implementation of management plans to protect these natural areas and the life they support (Baker, personal communication).

Marine and coastal degradation is caused by increasing pressure on both terrestrial and marine natural resources, and the use of the oceans for waste deposition. Some 20 billion tons of wastes, treated or untreated, end up in the sea every year through land, runoff, and other means. The total quantity of oil spilled by tankers between 1979 and 2000 is 5,322,000 tons (Encarta Encyclopedia 2003). Most tanker spills result from operations such as loading, discharging, and bunkering. However, the largest spills occur as a result of collisions and grounding. About 90% of these wastes stay in coastal waters where, particularly in coastal lagoons or marshes, they interfere with bird and marine life, mangroves, sea-grass beds and coral reefs, as well as with fisheries, tourism, and recreation.

Seventy-five percent of the pollutants in the ocean are estimated to be from the land. Increased human activities such as overfishing, coastal development, pollution, and the introduction of exotic species have also caused significant damage and pose a serious threat to marine and coastal biodiversity. The connection to land-based pollution sources can be observed visually by 58 known "dead zones"—areas where ocean life is absent, all located along coastlines (J & D Informatics 2001; Field et al. 2002). In the Gulf of Mexico, a 7,000 square mile area from the mouth of the Mississippi to the Texas border becomes so oxygen-deficient each summer that nothing can survive, creating a massive dead zone.

Efforts toward integrated coastal zone management in Africa have not made any meaningful impact, despite serious social and economic problems (Adedipe 1992, 1996; World Bank 1996). In Nigeria, the ecosystem response has been characterized by biodiversity loss and food insecurity due to overexploration and overexploitation, particularly in the Niger Delta area. A specific policy response has been a multinational project, spearheaded by UNIDO, IUCN, and UNESCO, involving six countries in the West Africa sub-region (Coté d'Ivoire, Ghana, Togo, Benin, Nigeria, and Cameroon) in the restoration of the Guinea Current Large Marine Ecosystem (UNIDO 2002). The concern is driven by population growth, urbanization, industrialization, and tourism. In 1994, an estimated 37% of the global population lived within 60 kilometers of the coast. That percentage has risen in the last decade. Fish farming in seas (for example, large salmon farms along Norwegian and Scottish coasts) is also a source of waste.

Globally, sewage has been the largest source of contamination because of the volume and the nature of organic matter it carries. Public health problems of the 1970s have been well documented (GESAMP 2001). Modern technologies can be used to remove all solid material, nutrients, and microorganisms from sewage. In industrial countries, therefore, there is no reason why untreated sewage should be discharged to the sea. Ways should be found to have these technologies made available for sustainable implementation in developing countries as well.

Coastal marine ecosystems are experiencing high rates of habitat loss and degradation. Shoreline stabilization, the development of large ports, and the existence of densely populated coastal cities—all contributed to this loss. In these zones, high human population levels coupled with increased road and pipeline densities have increased pollution as well as sedimentation and erosion rates (USEPA 2002).

Ironically, one reported dead zone between Denmark and Sweden actually increased catches temporarily, which is a positive impact (J & D Informatics 2001). Record catches of Norway lobster were being recorded when oxygen levels were very low. The lobsters normally burrow into the sea floor, but with the low oxygen levels, they swam up higher to locate oxygen, and made themselves susceptible to trawling. Thus the deteriorated conditions made the lobster more susceptible to the catch method. The yield was unsustainable.

The phenomenon of red tides appears to be becoming more frequent. Populations of marine dinoflagellates occasionally become so large that the dying organisms color the water and are referred to as tides. Such tides may lead to the poisoning of both fish and humans. Red tides occur off the western coast of Florida and in the coastal waters of New England, southern California, Texas, Peru, eastern Australia, Chile, Hong Kong, and Japan. In 1946, such a tide killed fish, turtles, oysters, and other marine organisms in the Gulf of Mexico near Fort Myers, Florida. The blue-green alga *Trichodesmus* sometimes, in dying off, imparts a reddish color to water; the Red Sea derived its name from this phenomenon.

10.5 Toward Improving Waste Management for Overall Biodiversity and Human Well-being

10.5.1 Biodiversity Conservation

A diversity of species is generally important to the natural functioning of ecosystems, and a balanced and "stable" biodiversity is therefore considered an indication of the good health of an environment. Biodiversity is also valued for aesthetic enjoyment and for natural products such as foods and drugs. Global biodiversity on the earth is well known for larger organisms such as mammals (over 4,000 species). However, total biodiversity can only be estimated, because most species of insects, deep-sea invertebrates, and microorganisms have yet to be described. Estimates of total terrestrial biodiversity range from 10 million to 100 million species, most of which are insects. Natural ecological systems generally support higher biodiversity than agricultural or urban landscapes, particularly in the tropics, where natural biodiversity is greatest. The issue of conservation of biodiversity is discussed in detail in Chapter 5.

The increasing human population threatens biodiversity (Mooney et al. 1996). Some ecologists believe that more than 50% of existing species will be lost in the next hundred years, many before they have been identified. Laws have been designed to protect threatened and endangered species, but legal and biological difficulties in defining species or other groups used for measuring biodiversity make such laws controversial. Moreover, experience has shown that species' survival depends on the preservation of their habitat. Increased nitrogen in soil and water can lead to loss of species composition of plant communities. Disappearance of salmon (an indicator fish) along with other species from River Thames in Britain in the early part of the last century is a typical example of biodiversity loss, which was restored after the introduction of stringent regulations on wastewater treatment. The international conventions such as the Basel Convention on transboundary movement of hazardous wastes and their disposal, the London Convention on dumping of wastes at sea, and the establishment of toxic waste registry have considerably reduced the threats on biodiversity. Recent data on the nature of flora at waste dumpsites revealed plant species loss (Sridhar 2004 unpublished data).

10.5.2 Poverty Reduction Strategies

As indicated earlier, poverty is "on the rampage" in developing countries. While industrial nations continue with carefree and high-consumption patterns, the developing nations are stifled by low incomes, due to massive underemployment and unemployment. Waste recycling offers ample opportunities for poverty alleviation and job creation (for example, waste-to-wealth schemes). Such opportunities already exist in Nigeria and proved particularly successful in Rio de Janeiro, where the municipality facilitated cooperatives of aluminum can collectors. Other examples include reusing newspapers, polyethylene terphthalate (PET) bottles, and other potential wastes produced with the guidance of artisans, architects, and artists. Education for adaptation should also be considered as a viable strategy for waste reduction and reuse. In addition to such income-generating activities and ventures, other approaches include promotion of cleaner environment and the reduction of negative health impact for overall human well-being.

10.5.3 Education and Enlightenment

Education and communication are essential elements of waste management, particularly waste reduction, through formal education starting with kindergarten to elementary schools (primary institutions), high schools (secondary institutions), and colleges (tertiary institutions), as well as informal and vocational training. Extension services, NGOs, the Internet, and distant learning programs contribute immensely. Also, vocational training response should be used as a tool for capacity building of the waste generators and managers. Along these lines, various professionals, particularly in the creative arts, should be involved in programs to inculcate "greener behavior" in communities. In addition, there is a need for a waste management information databank.

10.5.4 Public Health Implications

In tropical African countries, the results of improperly managed solid waste are evident in public health-related disease incidence causing ill health, sometimes leading to serious morbidity and mortality (Adedipe 2002; Onibokun et al. 2000). Poorly managed waste dumpsites, untreated sludge, and wastewater are sources of proliferation of rodents, which transmit typhoid fever, rabies, and other infectious diseases. They also lead to prolific breeding of flies, cockroaches, and mosquitoes, which transmit diarrhea, gastroenteritis, malaria, yellow fever, dengue, mosquito-borne encephalitis, filariasis and helminthiasis. Smoke emissions from burning of solid wastes are known to cause upper respiratory tract infections. In some cases, the accumulated and untreated waste penetrates the soil to contaminate groundwater or, through surface runoff, to contaminate surface water, with severe consequences for human health, for which eco-health is emerging as a viable and sustainable integrated response option involving water supply, sanitation, and waste management. A strategic policy response is needed, consisting of massive public health campaigns using all possible information, education, and communication materials.

10.6 Conclusions

This assessment covers the general challenge of waste management but concentrates on human waste and urban and rural waste. It attempts to avoid overlap with other chapters and summarizes the major issues in waste management in the context of the MA. It identifies the key drivers of ecosystem services and the responses thereto for protecting the environment and thereby improving human well-being.

The assessment shows that there are significant differences in the drivers and responses between developing and industrial countries and also within developing countries. This observation needs to be considered in any follow-up action on the MA.

The assessment views the issues of waste material reuse and recycling as a positive impact of waste management. Also, the careful and controlled use of wetlands in the management of sewage sludge and wastewater is a positive impact.

The assessment draws attention to the importance of effective governance structures, integrated responses such as harmonized institutional arrangements, relevant cost-effective civil society involvement, recognition of individual human rights, the special needs of epistemic communities and social values, private sector participation, education and public enlightenment. It stresses the overall goal of suitably modified consumption scales and patterns in the developed world and poverty reduction in the developing nations through recycling and resource recovery schemes that would reduce unemployment, in line with the millennium goal of reducing poverty by 50% in the year 2015.

References

Adedipe, N.O., 1992: Summary. In: *Environmental Consciousness for Nigerian National Development,* E.O.A Aina and N.O. Adedipe (eds.), FEPA monograph series no. 3, Federal Environmental Protection Agency, Abuja, Nigeria, pp. 241–250.

Adedipe, N.O., 1996: Summary. In: *The Petroleum Industry and the Environmental Impact in Nigeria,* E.O.A. Aina and N.O. Adedipe (eds.), FEPA monograph series no. 5, Federal Environmental Protection Agency, Abuja, Nigeria, pp. 224–232.

Adedipe, N.O., 2002: The challenge of urban solid waste management in Africa. In: *Rebirth of Science in Africa,* H. Baijnath and Y. Singh (eds.), Umdaus Press, Hatfield, South Africa, pp. 175–192.

Adedipe, N.O. and A.G. Onibokun, 1997: *Solid Waste Management and Recycling in Nigeria: Technical Capability for Small/Medium Scale Enterprises,* CASSAD technical report no. 6, Centre for African Settlement Studies and Development, Ibadan, Nigeria, 36 pp.

Adesiyan, S.O., 1992: Environmental pollution and agrochemicals. In: *Towards Industrial Pollution Abatement in Nigeria,* E.O.A. Aina and N.O. Adedipe (eds), FEPA monograph series no. 2, Federal Environmental Protection Agency, Lagos, Nigeria, pp. 115–134.

Adewale, O., 1996: Legal aspects of waste management in the petroleum industry. In: *The Petroleum Industry and the Environmental Impact in Nigeria,* E.O.A. Aina and N.O. Adedipe (eds.), FEPA monograph series no. 5, Federal Environmental Protection Agency, Abuja, Nigeria, pp. 210–220.

AGRIFOR (Agriculture, Food and Forestry), 2004: Agricultural wastes. Available at www.agriculturalwastes.html.

Aina, E. O. A. and N. O. Adedipe, 1991: *The making of the Nigerian environment policy,* Proceedings of the International Workshop on the Goals and Guidelines of the National Environmental Policy for Nigeria, Environmental Planning and Protection Division (EPPD), organized by Federal Ministry of Works and Housing, Lagos, Federal Environmental Protection Agency, pp. 1–329.

Ajomo, M.A., 1992: Legal and institutional issues of environmental management. In: *Environmental Consciousness for Nigerian National Development,* FEPA monograph series no. 3, Federal Environmental Protection Agency, Lagos, Nigeria, pp. 213–230.

Aluko, O.O., M.K.C. Sridhar, and P.A. Oluwande, 2003: Characterization of leachates from a municipal solid waste landfill site in Ibadan, Nigeria, *Journal of Environmental Health Research,* **2(1),** pp. 32–37.

Anderson, D.M., 1994: Red Tides, *Scientific American,* **271,** pp.62–68.

Asomani-Boateng, R, 2002: Urban cultivation in Accra: An examination of the nature, practices, problems, potentials and urban planning implications, *Habitat International,* **26(4),** pp. 591–607.

Australian EPA (Environmental Protection Agency), 2002: Wetlands for treating waste water. Available at www.epa.nsw.gov.au/mao.

Baker, J., 1999: *Progress Towards No Waste by 2010: A Report by the Commissioner for the Environment Act on the Implementation of the ACT Governments No Waste by 2010 Strategy,* Australian Capital Territory, Canberra, Australia, 87 pp.

Breen, B., 1990: Landfills are #1, *Garbage – The Practical Journal for the Environment,* Special issue, **September/October,** pp. 42–47.

Buechler, S., W. Hertog, and R. Van Veenhuizen, 2002: Wastewater use in urban agriculture, *Urban Agriculture Magazine,* **8,** pp. 1–4.

Burger, J. and D. Peakall, 1995: Methods to assess the effects of chemicals on aquatic and terrestrial wildlife, particularly birds and mammals. In: *Methods to Assess the Effects of Chemicals on Ecosystems,* R.A. Linthurst, P. Bourdeau, and R.G. Tardiff (eds.), SCOPE 53, John Wiley, New York, NY, pp. 292–306.

Burgess, J., M. Lim, and C.M. Harrison, 1988: Exploring environmental value through the medium of small groups: I. Theory and practices, *Environmental Planning,* **20,** pp. 109–326.

Cairncross, S. and R. Feachem, 1993: *Environmental Health Engineering in the Tropics: An Introductory Text,* 2nd ed., Wiley, London, UK, 283 pp.

Cappon, C.T., 1984: Content and chemical form of mercury and selenium in soil, sludge and fertilizer materials, *Water, Air and Soil Pollution,* **22,** pp. 95–104.

Carson, R., 1962: *Silent Spring,* Houghton Mifflin, Boston, MA, pp. 1–365.

Chambers, W.B., 2003: *Towards an Improved Understanding of Legal Effectiveness of International Environmental Treaties,* UNU/IAS International Environmental Governance working paper series, United Nations University/Institute of Advanced Studies, Tokyo, Japan.

Chandler, M., L. Kaufaman, and S. Mulsow, 1996: Human impact, biodiversity and ecosystem processes. In: *Functional Roles of Biodiversity: A Global Perspective,* H.A. Mooney, J.H. Cushman, E. Medina, O.E. Sala, and E-D. Schulze (eds.), SCOPE 55, John Wiley, New York, NY, 371–92.

Chokor, B.A., 1996: Assessment of the environmental and socioeconomic impact of the petroleum industry: A conceptual and methodological framework. In: *The Petroleum Industry and the Environmental Impact in Nigeria,* E.O.A. Aina and N.O. Adedipe (eds.), FEPA monograph series no. 5, Federal Environmental Protection Agency, Abuja, Nigeria, pp. 152–167.

Cielap, 2003: The increase in solid waste cannot be explained by urban growth alone. Available at www.cielap.org.

Colwell, R.R., A.L. Mills, J.D. Walker, P. Garcia-Tello, and P.V. Campos, 1978: Microbial ecology studies of the Metular spill in the straits of Mugellon, *Journal of Fisheries Research Board,* **35,** pp. 573–580.

Commission on Global Governance, 1995: *Our Global Neigbourhood,* The Commission on Global Governance, Oxford University Press, London, UK, pp. 246–250.

Costanza, R. and P.P. Principe, 1995: Methods for economic and sociological considerations in ecological risk assessment. In: *Methods to Assess the Effects of Chemicals or Ecosystems,* R.A. Linthurst, P. Bourdeau, and R.G. Tardiff (eds.), SCOPE 53, John Wiley, New York, NY, pp. 395–406.

Doebereiner, J., 1997. Biological nitrogen fixation in the tropics: social and economic contributions, *Soil Biology and Biochemistry,* **29,** pp. 771–774.

Dreschel, P., U. Blumenthal, and B. Keraita, 2002: Balancing health and livelihoods adjusting wastewater irrigation guidelines for resource-poor countries, *RUAF Magazine,* **8.** Available at *www.ruaf.org.*

Dubos, R., M. Pines, and Editors of Time-Life Books, 1980: *Health and Disease,* Life Science Library, Revised ed., Time-Life Books, Alexandria, VA.

Dushenkov, V., P.B.A. Nanda-Kumar, H. Motto, and I. Raskin, 1995: Rhizofiltration: The use of plants to remove heavy metals from aqueous streams, *Environment, Science and Technology,* **29,** pp. 1239–1245.

Edgar, W.W., G. De Michele, R.S. Reimers, and R.G. O'Dette, 1998: Assessment of the 1996 disinfection practice survey on biosolids, along with additional information from WEF Disinfection Committee and WEF Residuals/Biosolids Committee, Water Environment 71st Annual Conference Exposition, Water Environment Federation, Alexandria, VA.

Emerson, C., 1999: Aquaculture impacts on the environment, Cambridge Scientific Abstracts, Hot topics series. Available at www.csa.com.

Encarta Encyclopaedia, 2003: *Skill Intelligence Report,* International Tanker Owners Pollution Federation, London, UK.

FAO, 2003: *Trends in the oceanic captures and clustering of large marine ecosystems*—2 studies based on the FAO technical paper 435, pp. 1–71.

Faruqui, N.I. and O. Al-Jayyousi, 2002: Grey water reuse in Jordan, *Water International,* **27,** pp. 387–394.

Field, J.G., G. Hempel, C.P. Summerhayes, 2002. *Oceans 2020: Science, Trends and the Challenge of Sustainability,* Island Press, Washington, DC, 365 pp.

FRN (Federal Republic of Nigeria), 1991: *Guidelines and Standards for Environmental Pollution Control in Nigeria,* Federal Environmental Protection Agency, Abuja, Nigeria, 238 pp.

Garbutt, J., 1995: *Waste Management Law: A Practical Handbook,* 2nd ed., John Wiley, New York, NY, 247 pp.

GESAMP (Group of Experts on the Scientific Aspects of Marine Environmental Protection), 2001: *Protecting the Oceans from Land-based Activities: Land-based Sources and Activities Affecting the Quality and Uses of the Marine, Coastal and Associated Freshwater Environment,* GESAMP report and studies no. 71, UNEP, Nairobi, Kenya.

Goddard, H.C., 1995: The benefits and costs of alternative solid waste management policies, *Resources, Conservation and Recycling,* **13,** pp. 183–213.

Goldstein, N., 1997: State biosolids management update, *Biocycle,* **38,** pp. 5–62.

Guterstam, B., L. Forsberg, A. Bucznska, K. Frelek, R. Pilkaityte, et al., 1998: Stensund wastewater aquaculture: Studies of key factors for its optimization, *Ecological Engineering,* **11,** pp. 87–100.

Hay, B.J., P.H. Astor, S. Knauf, M. Zeff, J. Maser, et al., 1996: Natural resource damage assessment of oil spills in New York harbour. In: *The Petroleum Industry and the Environmental Impact in Nigeria,* E.O.A. Aina and N.O. Adedipe (eds.), FEPA monograph series no. 5, Federal Environmental Protection Agency, Abuja, Nigeria, 168–179.

Heermann, C., 2003: Pyrolysis and gasification: Increased potential for value recovery from ELV? International Automobile Recycling Congress, Geneva, Switzerland.

Hemond, H.F. and J. Benoit, 1988: Cumulative impacts on water quality functions of wetlands, *Environmental Management,* **12,** pp. 639–653.

Hoornweg, D.T.L., 1999: *What a Waste: Solid Waste Management in Asia,* The World Bank, Washington, DC.

Howarth, R.W., G. Billen, D. Swaney, A. Townsend, N. Jaworski, et al., 1999: Regional nitrogen budgets and riverine N and P flures for the drainages to the North Atlantic Ocean: Natural and human influences, *Biogeochemistry,* **35,** pp. 75–139.

Howarth, R.W., D. Anderson, J. Cloern, C. Elfring, C. Hopkinson, et al., 2000: Nutrient pollution of coastal rivers, bays and seas, *Issues in Ecology,* **7,** pp. 1–15.

Hu, H., M. Rabinowitz and D. Smith, 1998: Bone lead as a biological marker in epidemiologic studies of chronic toxicity: conceptual paradigms. *Environmental Health Perspectives,* **106,** pp. 1–8.

IETC(International Environmental Technology Centre), 1996: *Sustainable production and consumption,* UNEP Division of Technology, Industry and Economics, WRI Publications, Baltimore, MD, USA, UNEP-IETC Publications: *World Resources 1996–97,* www.unep.or.jp/ietc/publications/insight/spr-96/8.asp.

Ikem, A.O., O. Osibanjo, M.K.C. Sridhar, and A. Sobande, 2002: Evaluation of groundwater quality characteristics near two waste sites in Ibadan and Lagos, Nigeria, *Water, Air and Soil Pollution,* **140,** pp. 307–333.

ISWA (International Solid Waste Association), 2002: *Industry as a Partner for Sustainable Development: Waste Management,* ISWA, Copenhagen, Denmark.

J & D Informatics Inc., 2001: Report on the health of the world's oceans no. 4. Available at http://www.suite101.com/article.cfm/ecology/71821.

Jacobs, P. and B. Sadler, 1990: Sustainable development and environmental perspectives on planning for a common future, Canadian Environment Research Council, pp. 1–4.

Jana, B.B., 1998: Sewage-fed aquaculture: The Calcutta model, *Ecological Engineering,* **11,** pp. 73–85.

Jones, D.G., 1995: Environmental improvement through the management of waste, Stanley Thornes Publishers, Cheltenham, UK, 61 pp.

Kaspersma, J., 2002: Agricultural use of untreated wastewater in low income countries, *Urban Agriculture Magazine,* **8,** pp. 5–9.

Klundert, A. van de, 2000: The sustainability of alliance between stakeholders in waste management, Working paper for Urban Waste Expertise Programme/Center for Watershed Protection, May, WASTE Advisers on Urban Environment and Development, Nieuwehaven, The Netherlands.

Krogmann, U., V. Gibson, and C. Chers, 2001: Land application of sewage sludge: Perceptions of New Jersey vegetable farmers, *Waste Management Research,* **19,** pp. 115–125.

Lardinois, I. and A. vande Klundert (eds.), 1995: *Plastic Waste: Options for Small-Scale Resource Recovery,* Urban solid waste series 2, Gonda, The Netherlands, 112 pp.

Lieberman, J.L., 1994: *A Practical Guide for Hazardous Waste Management, Administration and Compliance,* CRC Press, Lewis Publishers, Boca Raton, FL, 239 pp.

Linthurst, R.A., P. Boureau, and R.G. Tardiff (eds.), 1995: *Methods to Assess the Effects of Chemicals on Ecosystems,* SCOPE 53, John Wiley, New York, NY, 416 pp.

MacRae, R., 1990: What is sustainable agriculture? Available at www.agrenv.mcgill.ca/agrecon/ecoagr.

Melillo, J.M., C.B. Field, and B. Moldan, 2003: Element interactions and the cycles of life: An overview. In: *Interactions of the Major Biogeochemical Cycles: Global Change and Human Impacts,* J.M. Melillo, C.B. Field, and B. Moldan (eds.), SCOPE 61, Island Press, Washington, DC, pp. 1–12.

Miranda, M.L. and J.E. Aldy, 1996: *Unit Pricing of Mineral Waste: Lessons from Nine Case Study Communities,* Duke University School of Environment, Durham, NC.

Mitsch, W.J. and J.G. Grosselink, 1986: *Wetlands,* Van Nostrand, Reinhold, New York, NY, 539 pp.

Mitsch, W.J., J. Day, J.W. Gilliam, P.M. Groffman, D.L. Heg, et al., 2001: Reducing nitrogen to the Gulf of Mexico from the Mississippi River Basin: Strategies to counter a persistent ecological problem, *BioScience,* **51,** pp. 373–388.

MoAFFA (Ministry of Agriculture, Fisheries and Forestry), 1999: *Serious Salinity Warning Must be Headed,* Tuckey media release, 24 June 1999, MoAFFA, Canberra, Australia.

Mooney, H. A., E. D. Schulze, E. Medina, H. Cushman and O. E. Saka (eds), 1996: *Functional roles of biodiversity, a global perspective,* John Wiley & Sons Inc., December, 493 pp.

National Agriculture Compliance Assistance Center, 2004: Fertilizers, US Environmental Protection Agency. Available at www.EPA-AgricultureTopics-Fertilizers.html.

Nixon, S.W., 1995: Coastal marine eutrophication: A definition, social causes and future concerns, *Ophelia,* **41,** pp. 199–219.

NRC (National Research Council), 2000: *Clean Coastal Waters: Understanding and Reducing the Effects of Nutrient Pollution,* NRC/National Academy Press, Washington, DC.

Nsirimovu, A., 1995: Urban development, human rights and participatory governance. In: *Governance and Urban Poverty in Anglophone West Africa,* A. Onibokun and A. Faniran (eds.), CASSAD monograph series no. 4, Centre for African Settlement Studies and Development, Ibadan, Nigeria, pp. 221–229.

Odeyemi, O. and A.G. Onibokun, 1997: *Solid Waste Management and Recycling in Nigeria: Some Current Composting and Biogas Production Practices,* CASSAD technical report no. 4, Centre for African Settlement Studies and Development, Ibadan, Nigeria, 25 pp.

Odu, C.T.I., 1972: Microbiology of soils contaminated with petroleum hydrocarbons: I. The extent of contamination and some soil and microbial properties after contamination, *Journal of Institute of Petroleum,* **58,** pp. 201–208.

Okpala, J., 1996: Enhancing environmental protection in Nigeria through environmental education. In: *The Petroleum Industry and the Environmental Impact in Nigeria,* E.O.A. Aina and N.O. Adedipe (eds.), FEPA monograph series no. 5, Federal Environmental Protection Agency, Abuja, Nigeria.

Olderman, L.R., 1994: The global extent of soil degradation. In: *Soil Resilience and Sustainable Landuse,* D.J. Greenland and T. Saboles (eds.), Commonwealth Agricultural Bureau International, Wallingford, UK.

Olson, R.A., 1987: The use of fertilizers and soil amendments. In: *Land Transformation in Agriculture,* M.G. Wolman and F.G.A. Fournier (eds.), SCOPE 32, John Wiley, New York, NY, pp. 203–226.

Onibokun, A.G. and A.J. Kumuyi, 1999: Governance and waste management in Africa. In: *Managing the Monster: Urban Waste and Governance in Africa,* A.G. Onibokun (ed.), International Development Research Center, Ottawa, ON, Canada, pp. 1–10.

Onibokun, A. G. and A. Faniran, 1995: Governance and Urban Poverty in Anglophone West Africa, Centre for African Settlement Studies And Development (CASSAD) Monograph Series 4, ISBN 878–2210–12–9, pp. 1–241.

Onibokun, A.G., N.O. Adedipe, and M.K.C. Sridhar (eds.), 2000: *Affordable Technology and Strategies for Waste Management in Africa: Lessons from Experience,* CASSAD monograph series no. 13, Centre for African Settlement Studies and Development, Ibadan, Nigeria, 133 pp.

Osibanjo, O., 2002: Perspectives on pollution and waste management for sustainable development, Proceedings of the National Forum on World Summit for Sustainable Development, Federal Ministry of Environment, Abuja, Nigeria, 29 pp.

Panayotou, T., 1990: *Incentives and Regulation: The Use of Fiscal Incentives,* Proceedings of the Conference on Environmental Management in Developing Countries, OECD, Paris, France.

Peterson, B.J., W.M. Wollheim, P.J. Mulholland, R.J. Webster, J.L. Meyer, et al., 2001: Control of nitrogen exports from watersheds by headwater streams, *Science,* **292,** pp. 86–92.

Pillai, S.C., R. Rajagopalan, and V. Subrahmanyam, 1946: Growing fodder crops on sewage *Science & Culture,* **12,** pp. 104–105.

Pillai, S.C., R. Rajagopalan, and V. Subrahmanyam, 1947: Utilization of Sewage farms for growing sugarcane and production of white sugar, *Current Science,* **16,** pp. 341–343.

Queen, W.H. and D.W. Stanley, 1995: Assessment of effect of chemicals on wetlands. In: *Methods to Assess the Effects of Chemicals on Ecosystems,* R.A. Linthurst, P. Bourdeau, and R.G. Tardiff (eds.), SCOPE 53, John Wiley, New York, NY, 416 pp.

Rabalais, N.N., 2002: Nitrogen in aquatic ecosystems, *Ambio,* **31,** pp. 102–112.

Rathje, W.L., 1990: The history of garbage, *Garbage – The Practical Journal for the Environment,* Special issue, pp. 32–39.

Redwood, M., 2004: Wastewater use in urban agriculture: Assessing current research and options for national and local governments, Cities Feeding People Series Report No. 37, International Development Research Centre, Ottawa, ON, Canada, 62 pp.

Rees, W.E., 1997: Why urban agriculture? Development Forum on Cities Feeding People: A Growth Industry, International Development Research Centre, Vancouver, BC, Canada.

Reyer, G., P. van Beukering, M. Verma, P.P. Yadav, and P. Pandey, 1990: Integrated modeling of solid waste in India, Working paper no. 26, CREED (Collaborative Research in the Economics of Environment and Development) working paper series, London, UK.

Ribbans, E., 2003: A recycling revolution, Spiked Science Debates: The Environment, Sponsored by Natural Environment Research Council, February 4, London, UK.

Robinson, H.D., C.K. Shen, R.W. Formby, and M.S. Carville, 1995: *Treatment of Leachates from Hong Kong Landfills with Full Nitrification and Denitrification,* Proceedings Sardinia '95, Fifth International Landfill Symposium, CISA, pp. 1511–1534.

Rodrigues, M.S. and J. M. Lopez-Real, 1999: Urban organic wastes, urban health and sustainable urban and peri-urban agriculture linking urban and rural by composting, *Urban Agriculture Notes,* City Farmer, Canada's Office of Urban Agriculture.

Rotimi, L., 1995: Waste to Wealth project as a means of improving the urban environment and alleviate urban poverty. In: *Governance and Urban Poverty in Anglophone West Africa,* A. Onibokun and A. Faniran (eds.), CASSAD monograph series no. 4, Centre for African Settlement Studies and Development, Ibadan, Nigeria, pp. 230–232.

Sappington, D.E. and J.E. Stiglitz, 1987: Privatization, information and incentives, *Journal of Policy Analysis and Management,* **6.**

Scott, C.A., N.I. Faruqui, and L. Raschid-Sally (eds.), 2004: *Wastewater Use in Irrigated Agriculture: Confronting the Livelihood and Environmental Realities,* International Development Research Centre–Commonwealth Agricultural Bureau, Ottawa, ON, Canada, pp. 1–188.

Smit, J., 1996: *Cities Feeding People: Report 18—Urban Agriculture: Progress and Prospect, 1975–2005,* Report 18, International Development Research Centre, Ottawa, ON, Canada.

Soesilo, J.A. and S.R. Wilson, 1995: *Hazardous Waste Planning,* Lewis Publishers, CRC Press, New York, NY, 275 pp.

Sommers, P. and J. Smit, 1996: *Cities Feeding People: Report 9—Promoting Urban Agriculture: A Strategy Framework for Planners in North America and Asia,* International Development Research Center, Ottawa, ON, Canada.

Sridhar, M.K.C., 1995: *Sullage/Waste Water in Nigeria: Problems and Scope for Utilization for Gardening,* UNICEF, Lagos, Nigeria, 63 pp.

Sridhar, M. K. C., 2004: Unpublished data from a dissertation titled "*Chemical analysis of soil and impact of municipal solid wastes on floral diversity loss at Awotan dump site*" by M. O. Olawoyin supervised by M. K. C. Sridhar and A. M. Salaam, University of Ibadan, Ibadan, Nigeria, pp. 1–111.

Sridhar, M.K.C., A.A. Ajayi, and A.M. Arinola, 1992: Collecting recyclables in Nigeria, *Biocycle,* **33,** pp. 46–47.

Sridhar, M.KC. and A.M. Arinola, 1991: Managing industrial wastes in Nigeria, *Biocycle,* **32,** p. 65.

Sridhar, M.K.C., G.O. Adeoye, and N.M. John, 2002: Organo-mineral pellets from urban wastes, *Proceedings of the 6th World Congress on Integrated Resources Management,* EMPA Swiss Federal Laboratories for Materials Testing and Research, Geneva, Switzerland, pp. 1–5.

Sridhar, M.K.C. and G.O. Adeoye, 2003: Organo-mineral fertilizers from urban wastes: Developments in Nigeria, *The Nigeria Field,* **68,** pp. 91–111.

Tchobanoglous, G., H. Theisen, and S. Vigil, 1993: *Integrated Solid Waste Management: Engineering Principles and Management Issues,* McGraw-Hill, New York, 978 pp.

UNCSD (United Nations Commission on Sustainable Development), 1999: *Cairo Sludge Disposal Study,* UNCSD, pp. 148.

UNDP (United Nations Development Programme), 2000: *Human Development Report 2000,* Oxford University Press, Oxford, UK, and New York, NY.

UNEP (United Nations Environment Programme), 1997: Global Environmental Outlook 1, Geo-1. Available at www.grida.no/geo1/ch/toc.html.

UNEP, 2002: Global Environmental Outlook 3, Geo-3, UNEP, Nairobi, Kenya, 446 pp.

UNIDO (United Nations Industrial Development Organization), 2002: *Oceans and the World Summit on Sustainable Development: The Restoration of the Guinea Current Large Marine Ecosystem,* UNIDO, Vienna, Austria, 7 pp.

Urban Agriculture Magazine, 2002: Wastewater use for urban agriculture, **8,** 46 pp. Available at www.ruaf.org.

USEPA (United States Environmental Protection Agency), 1979: *Methods of Chemical Analyses of Water and Wastes,* USEPA, Washington, DC.

USEPA, 2002: Fresh Water Ecosystems. Available at www.natureserve.org/publications. Marine Ecosystems. Available at www.epa.gov/ebtpages/wate quatimarineecosystems.html and www.epa.gov/maia/html/habitat.html.

Vitousek, P.M. and D.U. Hooper, 1993: Biological diversity and terrestrial ecosystem biogeochemistry. In: *Biodiversity and Ecosystem Function,* E-D. Shulze and H.A. Mooney (eds.), Springer, Berlin, Germany, pp. 3–14.

World Bank, 1996: *Africa: A Framework for Integrated Coastal Zone Management,* Land, Water and Natural Habitats Division, Africa Environmentally Sustainable Division, The World Bank, Washington, DC, 137 pp.

World Bank, 1997: Urban waste and rural soil management: Making the connection, *The Agriculture Notes,* **17,** (June).

WHO (World Health Organization), 1992: Environmental Health Criteria 135: Cadmium-Environmental Aspects, Geneva, Switzerland, p.156.

WRI (World Resources Institute), 1992: *World Resources 1992–93,* WRI, Washington, DC.

Yuan, Y., 1993: Etiological study of high stomach cancer incidents among residents in wastewater irrigated areas, *Environmental Protection Science,* **19,** pp. 70–73.

Zurbrugg, C. 2002: *Urban Solid Waste Management in Low-income Countries of Asia: How to Cope with the Garbage Crisis,* SCOPE workshop on urban solid waste management, University of Durban Southwest, Durban, NC.

Zurbrugg, C., S. Drescher, A. Patel, and H.C. Sharatchandra, 2004: Decentralised composting of urban waste: An overview of community and private initiatives in Indian cities, *Waste Management,* **24,** pp. 655–662.

Chapter 11

Flood and Storm Control

Coordinating Lead Authors: M. Monirul Qader Mirza, Anand Patwardhan
Lead Authors: Marlene Attz, Marcel Marchand
Contributing Authors: Motilal Ghimire, Rebecca Hanson
Review Editor: Richard Norgaard

BOXES

FIGURES

TABLES

Main Messages

Floods and storms are an integral part of ecosystem dynamics and have both positive and negative effects on human well-being. Floods interact directly with the ecosystems of a floodplain while a storm interacts with coastal, estuarine, and desert ecosystems. Floods and storm waters bring nutrients, which are beneficial to the floodplain ecosystems (wetlands, agricultural lands, and crops, fishery, etc.) and coastal ecosystems (mangroves, mudflats, reefs, fishery, etc.). They eventually contribute to human well-being by delivering a range of ecosystem services. However, flood or flood risk management options can increase the discharge of pollutants and sediments to the coastal zones. Floods and storms also cause damage to the economic and social sectors such as infrastructure, agriculture, industry, and human settlements. Prudent management approaches can reduce the extent of damage to acceptable limits.

Historically, responses to reduce the negative impacts have emphasized physical structures/measures over natural environment and social institutions. Historical responses to floods and storms have emphasized the construction of physical structures (for example, dams/reservoirs, embankments, regulators, drainage channels, and flood bypasses) over the maintenance and enhancement of environmental features and over social institutions that inform and coordinate behavior changes to reduce losses. In many cases, such efforts have been implemented without assessing their possible long-term effects on ecosystems. Such measures often create a false sense of security and encourage people to accept high risks that result from living in the floodplains and on coasts.

The preponderance of evidence indicates that, in most situations, more emphasis needs to be given to the natural environment and nonstructural measures and less to structural measures. Although physical structures (if properly designed) protect communities and infrastructure in a floodplain from flood and storm surges, they often create irreparable damage to ecosystems. Ecosystems usually lose resiliency during the long inundation-free periods after the construction of physical structures. In many cases expensive restoration efforts have failed to fully regenerate ecosystems. Overall, physical responses (in the form of human interventions) may cause more damage than benefit to ecosystems over longer time-scales, in terms of restoration and resiliency. Therefore, the focus should be shifted to use of the natural environment and nonstructural measures in mitigating flood and storm hazards. For example, nonstructural measures such as flood and storm forecasting and warning, disaster preparedness, and acquisition of lands to accommodate flood waters can reduce economic damage and loss of human life. Coastal mangroves have been found to be very effective in providing protection against storms and surges.

Sustainable approaches of flood and storm control can ensure intergenerational equity. Sustainable flood and storm control schemes could include structural and nonstructural measures. Design modifications of physical structures that allow the maintenance of natural environment to a large extent could be sustainable. This, together with the nonstructural measures (for example, water retention areas, restoration of wetlands, land use, zoning, and risk assessment, and early warning systems), can deliver benefits to humans and ecosystems over a long period of time. However, uncertainty in flood and storm forecasting can influence the decision-making procedure for design and implementation of response measures.

Drivers of change, including climate change, indicate that the geographical distribution of floods and storms and perhaps their intensity will impose new stresses, which are probably best responded to through an adaptive approach to ecosystem management and social institutions. Floods and storms are the result of extreme rainfall/snowmelt and oceanic–atmospheric disturbances. In the future, climate change may have large-scale implications for these processes. Results from climate models suggest the possibility of increased intense rainfall in many parts of the world, which may lead to increased flooding. A rise in the sea levels may cause drainage problems in many river basins as well as aggravate coastal inundation. However, the net sea level rise is dependent on a number of factors that include sediment transport to the estuaries, land accretion subsidence, and coastal protection. Although the models indicate the likelihood of increased cyclones/storms, confidence is less than for floods. There is a need for designing a comprehensive adaptive approach integrating ecosystems and social institutions. The use of advanced flood forecasting and warning, strengthening of the institutions responsible for such actions and disaster management, quick relocation of people, emergency response, coastal mangroves, afforestation in the uplands and coastal areas, conservation, restoration, and creation of wetlands can markedly reduce the threats of increased flood and storm hazards.

A more integrated approach toward managing the consequences of floods and storms is needed. This requires a range of responses that includes land use planning, financial services, information and education, use of the natural environment, and physical structures. Such an approach is likely to balance and resolve multiple objectives and goals in a better way.

11.1 Introduction

Floods and storms are intrinsic components of the natural climate system and climate variability. These are a part of the natural disturbance regime, which is an important determinant of ecosystem structure and function, particularly in the long run. Public perception and response to floods and storms are largely driven by the short-term and negative impact of these disasters. Therefore, the responses have been historically focused on interventions to modify and control natural flood regimes through structural means (for example, flood mitigation program in Bangladesh).

Floods and storms are some of the most destructive hydrometeorological phenomena in terms of their impacts on human well-being and socioeconomic activities. While floods and storms have adverse impacts on humans, infrastructure, and economic sectors and ecosystems, they also generate beneficial effects that contribute to human well-being. The impacts of floods on human well-being and the role of ecosystems in flood control are extensively discussed in MA *Current State and Trends,* Chapter 16. Some of the main impacts of floods and storms are presented here.

11.1.1 Adverse Impacts

Floods and storms may have considerable adverse impacts depending on location, intensity and duration. In 2003, floods accounted for 3,723 fatalities around the world, exceeded only by heat waves (about 22,000 due to the very extreme summer heat wave in Southern Europe) and earthquakes (about 48,000, mostly due to the Bam Iran disaster) (Munich Re 2003). The International Federation of Red Cross and Red Crescent Societies reports that weather-related disasters from a global perspective have been on the rise since 1996 and increasing from an annual average of 200 (1993–97) to 331 (1998–2002) (International Federation 2001 and 2003).

The number of disasters attributed to floods is on the rise, while on average the number of people killed due to floods remains steady (Munich Re 2003). (See Figure 11.1.) The economic costs of flood disasters have been increasing globally. Pielke et al. (2002) found that flood losses were falling as a proportion of GDP although the gross loss is on the rise. The increase in flood

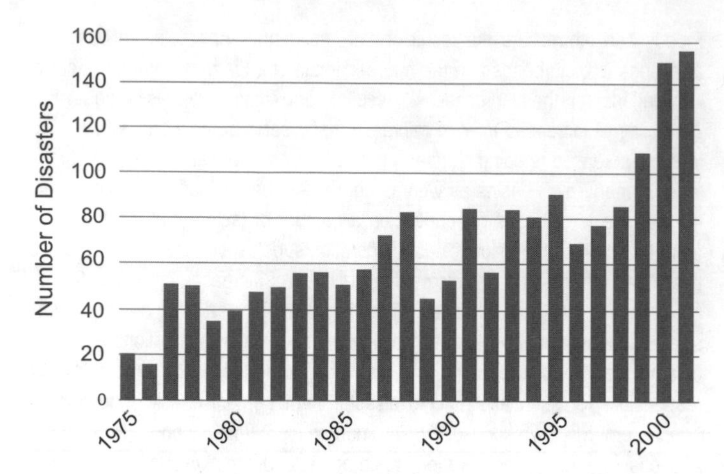

Figure 11.1. Number of Disasters Attributed to Floods, 1975–2001 (Munich Re 2003)

disasters is possibly due to more heavy rainfall events, increased economic activity, and efficiency of the use of a catchment (Green 1999; Mirza et al. 2001). Future climate change is expected to exacerbate the problem with possible increase in extreme precipitation events, perhaps in very serious ways (IPCC 2001). However, the Intergovernmental Panel on Climate Change projections put less confidence in the increase in frequency and magnitude of cyclones and storms.

Tropical cyclones are considered to be the most devastating of the natural disasters because of their capacity to cause loss of human lives and induce extensive economic losses (Gray and Landsea 1992; Diaz and Pulwarty 1997). Vulnerability to tropical cyclones is increasing due to fast population growth in the tropical coastal regions (Handerson-Sellers et al. 1998). McBride (1995) reported that each year around the globe, approximately 80–90 cyclones gain the intensity of tropical storm and about two thirds of them reach the intensity of a hurricane. Recent analysis of cyclone data for the North Atlantic and Northwest Pacific reveals an increase in windstorm activity from 1950 to 2003 (Munich Re 2003).

Floods can affect health directly, for example, by causing injuries and deaths due to drowning. These can occur during or in the aftermath of a flood disaster when the residents return to their dwellings to clean up the damage and debris. Floodwaters also can affect health indirectly, through changes in other systems (for example, waterborne infections, acute or chronic effects of exposure to chemical pollutants released into floodwaters, vector-borne diseases, food shortage, and others). Floods also can increase the risk of cholera, diarrhea, schistosomiasis, dengue, yellow fever, malaria, hantavirus, and other diseases.

After Hurricane Mitch devastated Central America, there was a widespread outbreak of communicable diseases; WHO reported 590 cases of cholera. Nicaragua had the highest number of cases, 335 (56%), followed by Guatemala, with 235 cases (40%). The remaining 4% of cases occurred in El Salvador, Honduras, and Belize (WHO 1999). Similarly, Barcellos and Sabroza (2000) identified floodwater as the cause of the outbreak of leptospirosis in 1996 in western Rio de Janeiro, Brazil. Floods in Bangladesh in the 1980s and 1990s caused an outbreak of diarrhea and other waterborne diseases that claimed the lives of thousands of people (Mirza et al. 2001). Bennet (1970) also synthesized in detail health effects of Bristol floods in the United Kingdom.

Ill health, particularly due to psychological distress, may persist for months or years following a flood. For example, the Saguenay flood in Quebec, Canada, in 1996 caused psychological distress (Auger et al. 2000). Floods and storms can also cause other kind of health problems such as carbon monoxide (CO) poisoning, as in Grand Forks, North Dakota in 1997 (Daley et al. 2001) and contamination of farmlands by pesticides as in Mississippi after the 1993 flood (Chong et al. 1998).

The actual impacts of floods and storms on human well-being are strongly dependent on the adaptive capacity of the affected groups and individuals, and their actual adaptation responses. For example, Ginexi et al. (2000) identified that the increases in symptoms as a function of flood impact were slightly greater among respondents with the lowest incomes and those living in small rural communities than among those on farms or in cities. Similarly, Kunii et al. (2002) reported widespread cases of diarrhea in Bangladesh during the 1998 floods. Some of the factors associated with developing or worsening diarrhea were: family size, poor economic condition, no distribution of water purification tablets, the type of water storage vessels, not putting a lid on vessels, no use of toilets, perceived change of drinking water, and food scarcity.

Development and urbanization patterns can exacerbate the impacts of floods. Floodplain development is increasing the number of people at risk, often because alternative (attractive) locations are not available. By contrast, in many industrial countries people put lives and property at risk by building houses and resorts on floodplains that are an attractive choice because of their aesthetic value. In many countries where land resources are scarce, expansion of human settlements and developments occur on the floodplains as there is no other choice. However, in such cases, planners should take into account the "risk factor" to reduce human and ecosystem vulnerability. For example, development and urbanization create conditions whereby runoff is greater in terms of both volume and rate of rise (speed of onset). However, efficient drainage provisions and conservation of urban wetlands can reduce vulnerability.

11.1.2 Beneficial Impacts and Well-being

Natural flooding has many beneficial effects. In Bangladesh, for example, a flood is categorized as *barsha* (beneficial flood) and *bonna* (disastrous flood). The annual flood *barsha* inundates up to 20.5% of the land area and the low-frequency, high-magnitude flood *bonna* inundates more than 35–70% of the country's area. A single flood can be both *barsha* and *bonna* (Paul 1984). Flooding has four important benefits.

First, it inundates floodplains, leaving the moisture content in the soil high at the end of the flooding season. This moisture is beneficial for agriculture depending on the crop cycle, for example, in Bangladesh. However, there are exceptions. For example, in the United Kingdom, winter flooding can make soil moisture content too high in the summer to support arable crops (Drijver and Marchand 1985). Soil moisture deficit is common in the soils of flood-free areas where irrigation is required to sustain agriculture. Second, floodwaters replenish groundwater aquifers. In many parts of the world, groundwater aquifers fully recover by natural recharge from rain or snowmelt. The replenished groundwater is used for irrigation.

Third, floodwaters contribute to increased soil fertility. However, deposition of sand carried by floodwaters on fertile agricultural land can cause serious harm (Brammer 1990). There is a notion among the farmers in Bangladesh that raw alluvium carried by floodwaters increases soil fertility. But raw alluvium is relatively infertile in the short term. It contains little organic matter and provides useable phosphorus or nitrogen. The minerals contained in river alluvium weather relatively slowly and consequently contribute to soil fertility on a long-term basis rather than in the year of their deposition (Chadwick et al. 2003). According to the World Bank (1990), the fertility associated with seasonal flooding comes mainly from the flooding itself, rather than the alluviums. Algae, including blue-green algae that are nitrogen fixing, potentially grow on the submerged soil and the stems of plants in the floodwater. The organic remnants of the algae fall on the soil surface and decompose, releasing nutrients to plants.

Fourth, natural flooding can benefit floodplain fisheries. In many countries in South and Southeast Asia, Africa, and Latin America, among others, fish is a source of animal protein. Edwards (2000) reported that fisheries supply at least 40% of all animal protein in the diet in 18 countries in Africa and Asia. Whereas the urban population has access to other sources of animal protein, many people in rural areas are highly dependent on floodplain fisheries.

Obstruction to natural flooding by the construction of high dams caused destruction of the Nile Delta (Stanley and Warne 1998). Decreased flooding also has implications for agriculture. The Aswan dam lowered the influx of nutrient rich silt to the floodplains of Egypt, where much of the food is grown.

Ecosystems play an important role in modifying and regulating hydrological and meteorological processes, and thereby affect the positive as well as negative consequences of floods and storms. The functions of ecosystems range from the regulation of surface and sub-surface flow to the modification of wave dynamics in coastal and near-shore areas. Costanza et al. (1997) listed a range of ecosystem services and functions related to floods and storms. (See Table 11.1.) Normal as well as flood flow regimes are affected by vegetation and its characteristics; hence, one important ecosystem service is to control floods and storms. This chapter aims to:

- assess the role of ecosystems in moderating or regulating storms and floods and their associated impacts, including estimates of the economic value associated with this service;
- examine the natural and anthropogenic drivers that influence this role;
- explore the range of management and policy options (for example, land use change) for ensuring the ability of ecosystems to provide these services, including the possibility of deliberate modification to enhance flood and storm protection; and
- explore the response options to reduce human vulnerability to storms and floods.

Table 11.1. Ecosystem Services and Functions (Constanza et al. 1997)

Ecosystem Service	Ecosystem Functions	Examples
Disturbance regulation	capacitance, damping, and integrity of ecosystem response to environmental fluctuations	storm protection, flood control, drought recovery and other aspects of habitat response to environmental variability mainly controlled by vegetation structure
Water regulation	regulation of hydrological flows	provision of water for agricultural (e.g., irrigation) or industrial (e.g., milling) processes or transportation
Water supply	storage or retention of water	provisioning of water by watersheds, reservoirs, and aquifers
Erosion control and sediment retention	retention of soil within an ecosystem	prevention of soil loss due to wind, runoff, or other removal processes; storage of silt in lakes and wetlands

11.1.3 Types of Events

In the context of this chapter, it is useful to distinguish between the four different types of flood events: flash, riverine, rain, and coastal floods and storms.

Flash floods occur in all climatic regions of the world. They can occur within a few minutes or hours of excessive rainfall, thunderstorms, and heavy rains from hurricanes and tropical storms; they can occur from a dam or levee failure, or from a sudden release of water held by an ice jam. Although flash flooding occurs often along mountain streams, it is also common in urban areas where much of the ground is covered by impervious surfaces and in arid areas such as North Africa.

Riverine flooding is an event of longer duration; it may last a week or more and in some cases months. (See Figure 11.2.) The riverine flooding in Bangladesh in 1998 lasted a record 68 days (Mirza 2003). In 1993, flood water stayed above the danger level for 45 days at Quad Cities, Illinois, along the Mississippi River (NOAA 1994).

Rainfall floods are a form of localized flooding due to intense rainfall occurring over a sustained period of time and the consequent drainage congestion. In 1998 (April 9–13), more 5,000 square kilometers in the regions of Midlands, Anglian, Wales, and Thames in the United Kingdom, for example, were inundated by flooding caused by heavy rainfall for three consecutive days (April 8–10) (Elahi 2000).

Coastal floods are caused by storm surge, coastal rainfall, and tidal action. Coastal flooding typically results from one or a combination of the following biophysical factors: storm surge, heavy surf, tidal piling, tidal cycles, persistence behavior of a storm that is generating flooding, topography, shoreline orientation, bathymetry, river stage or stream runoff and presence or absence of offshore reefs or other barriers. High winds can exacerbate damage.

The terms "hurricane" and "typhoon" are regionally specific names for a strong *tropical cyclone*. A tropical cyclone is the generic term for a nonfrontal synoptic scale low-pressure system over tropical or sub-tropical waters with organized convection (thunderstorm activity) and definite cyclonic surface wind circulation

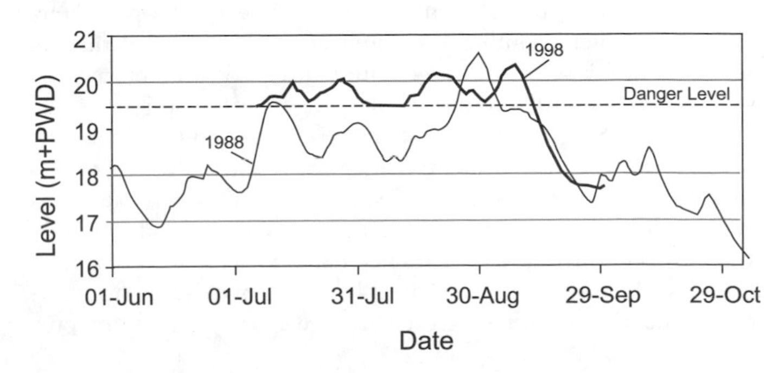

Figure 11.2. Water Levels of the Brahmaputra River at Bahadurabad in Bangladesh during the Floods of 1988 and 1998 (Mirza 2003)

(Holland 1993). Tropical cyclones with maximum sustained surface winds of less than 17 meters per second are "tropical depressions." Once the winds of the tropical cyclone reach at least 17 meters per second, they are typically called a "tropical storm" and assigned a name.

Another class of events that is particularly important in some regions is *sand and dust storms*. Dust storms are major, but understudied processes in dryland areas across the world (Goudie and Middleton 1992). They not only play an important role in desertification and land degradation, but also can cause substantial environmental impacts. Dust storms are common in the great plains of the United States, the former USSR, Morocco, the Arabian Gulf, Australia, the Sahel-Sudan Zone of Africa, China, Mongolia, and Mexico. Natural processes such as precipitation, snow cover, and wind speed are important determinants of frequency of dust events (Goudie and Middleton 1992; Qian and Zhu 2001).

Dust storms have important environmental consequences that include local climate change, nutrient additions to oceans and terrestrial ecosystems, ocean sedimentation, soil formation and loess deposition, possible rainfall suppression, health hazard, accidents, loss of agricultural outputs, and disturbance to satellite communications (Griffin et al. 2001; Goudie and Middleton 1992; USGS 2003).

11.1.4 Flood and Storm Protection Mechanisms by Ecosystems

In examining the mechanisms by which ecosystems provide flood and storm protection, it is useful to focus on two different settings: coastal regions and rivers/uplands.

11.1.4.1 Coastal Systems

In coastal regions, flood and storm protection is often provided by characteristic ecosystems that include coastal forests, mangroves, seagrass beds, coral reefs, dune systems, salt marshes, inter-tidal flats, and lagoons. Mechanisms for regulating storm and flood impacts in coastal areas include wave dissipation, absorption, reflection and resistance, barrier to flood surge, wind breaking, coastal accretion and stabilization (long-term), regulation of sediment transport, and linkage with coastal geomorphology.

11.1.4.2 Rivers and Uplands

Runoff in a catchment or flow at any given point in a channel depends on the interaction of a number of factors, the most important of which are: antecedent conditions; distribution; intensity and duration of precipitation; vegetative or other surface cover; soil type and depth; geologic structure; topography, including area, slope, and channel characteristics.

Singh (1987) assessed the role of forests in water conservation and in controlling rainfall-runoff processes. Leafy canopies intercept rain, reducing both the amount and the impact of that on the ground. Most of the interception loss develops during the initial storm period; thereafter, the rate of interception rapidly reaches zero (Singh 1987). Roots stabilize soils and form channels for rapid infiltration. Organic matter from roots and leaves improves soil structure and increases both infiltration rates and water-holding capacity, that is, the ability of the soil to retain water against gravity; water capacity can vary widely among various soils. Through transpiration, plants remove water from the soil profile, thus creating a greater storage capacity for future precipitation.

11.1.5 Drivers and Processes

Human activities and natural processes both affect ecosystem structure and function and impact services such as flood and storm protection. Following the terminology adopted in the MA, these drivers may be classified into direct and indirect drivers. The former refers to processes that directly interact with ecosystems, while the latter refers to the underlying causes. For example, habitat loss is a common direct driver, while indirect drivers might be population growth and consumption pressures.

The link between human activities and ecosystem degradation has been studied extensively and is now well established. Change in forest cover and, more generally, in land use/land cover is, perhaps, the dominant route by which human influence is expressed. In one study, Laurence and Bierregaard (1997) assessed the pattern and pace of tropical forest destruction in the Americas, Asia, and Africa and concluded that the four key drivers of forest destruction are human population pressure, weak government institutions and poor policies, increasing trade liberalization, and industrial logging. Secondary drivers include poverty and road construction. According to Tockner et al. (2002), by 2025, the increase in human population will lead to further degradation of riparian areas, intensification of the hydrological cycle, increase in the discharge of pollutants, and further proliferation of species invasions.

Urbanization has marked effects on basin runoff in terms of higher volume, higher peak discharge, and shorter time of concentration.[1] These changes are associated with the increased imperviousness and more efficient drainage that are characteristics of constructed drainage systems (Rustam et al. 2000; Singh 1987). UNESCO (1974) provides an excellent account of the hydrologic effects of urbanization. Some of the major effects are: (1) increased water demand, often exceeding the available natural resources; (2) increased wastewater, burdening rivers and lakes and endangering the ecology; (3) increased peak flow; (4) reduced infiltration; and (5) reduced groundwater recharge, increased use of groundwater, and diminishing baseflow of streams.

11.2 Response Categories and Management Approaches

Over the years, a number of management approaches and response options have been developed and followed for coping with the effects of floods and storms. These management approaches influence the extent and functioning of ecosystems, either directly through modification of ecosystems, or indirectly, by changing hydro-meteorological regimes. Five broad categories of response options may be identified, based on nature of response and familiarity of practicing managers:

- *physical structures*: river/estuary (multi-purpose storage dams/reservoirs, weirs, barriers), land protection (dikes/embankments);
- *use of natural environment*: vegetation (mangroves, wetlands, rice paddies, salt marshes, upland forests), geomorphology (natural river channels, dune systems, terrace farming);
- *information and education*: disaster preparedness, disaster management, flood and storm forecasting, early warning, evacuation;
- *financial services*: insurance, disaster relief, and aid; and
- *land use planning*: zoning, setbacks, flood-proofing (emphasis on regulation or modification of the built environment, often urban).

The actual operation and implementation of these responses and their effects on ecosystem structure and function are best examined in four distinct settings: upland/watersheds, floodplains, coastal regions, and islands. Each of these settings has distinct characteristics, biophysical as well as socioeconomic. In addition, the settings differ in terms of the institutional structures and management systems responsible for flood and storm protection as well as other services related to flood and storm protection, for example, irrigation or hydroelectricity generation.

11.3 Sustainable Flood and Storm Control: Analysis and Assessment of Responses

The concept of "sustainable development" is presently widely used and there is no common understanding of the term (Kundzewicz 1999). Table 11.2 presents alternative definitions of sustainable development, which shows the definitional diversity. Munasinghe (1993) argued that the concept of sustainable development evolved to encompass three major points of view: economic, social, and ecological. Takeuchi et al. (1998) mentioned that sustainable development entailed a blend of objectives in economic, social, and environmental areas and they had to be economically feasible, socially acceptable, and environmentally sound.

The definitions of sustainable development in Table 11.2 can be applied for floods and storms control. Although the definitions vary widely in terms of subjects, there is a general consensus on the intergenerational equity. Kundzewicz (1999) interpreted that sustainable development comprised three integral items—civilization, wealth, and environment (natural and human built) and they should be transferred to the future generations in a nondepleted condition. In terms of flood, he further argued that it was necessary for the present generation to attain freedom from the disastrous events but not at the cost of the future generations. The freedom (to a reasonable extent) can be achieved by implementing some defense schemes/response measures. The United Kingdom Environment Agency (1998, p. 9) defined a sustainable flood defense scheme as taking "account of natural processes (and the influence of human activity on them), and of other defenses and development within a river catchment . . . and which avoid as far as possible committing future generations to inappropriate options for defense." These schemes could comprise structural and nonstructural measures. (See Table 11.3.)

Takeuchi et al. (1998) criticized some flood protection infrastructure (levees, dams, etc.) in the context of sustainable development for closing options (measures that will last and generate benefits for successive generations) for future generations and introducing disturbance in the ecosystems. Kundzewicz (1999) argued that "soft" or nonstructural measures could be rated as more flexible, less committing, and more sustainable than "hard" or structural measures that might be indispensable in particular cases.

This section elaborates on the sustainable approach of flood and storm control with one or more specific examples of response and management options (structural and nonstructural). Internationally, sustainable flood and storm protection is being taken up on a priority basis. The United Nations and Economic Commission for Europe Sustainable Flood Prevention Guidelines (UN/ECE 2000) outline seven basic principles and approaches:

- Flood events are a part of nature.
- Human interference into the processes of nature has increased the threat of flooding.
- Flood prevention should cover the entire catchment area.
- Structural measures will remain important elements of flood prevention and protection, especially for protecting human health and safety, and valuable goods and property.
- Everyone who may suffer from the consequences of flood events should also take precautions on their own.
- Human uses of floodplain should be adapted to the existing hazards.
- In flood-prone areas, preventive measures should be taken to reduce the possible adverse effects on aquatic and terrestrial ecosystems.

The UN/ECE guidelines focus on recommendations for water retention areas, land use, zoning and risk assessment, structural measures and their impact, and early warning and forecast systems. Public awareness, education, and training comprise another important element of preventive strategies.

11.3.1 Physical Structures

Construction of embankments has been the most popular structural method of flood control/mitigation in many parts of the

Table 11.2. Alternative Definitions of Sustainable Development

Definition	Focus
". . . development that meets the needs of the present without compromising the ability of future generations to meet their own needs." (WCED 1987)	intergenerational equity
". . . development that secures increases in the welfare of the current generation provided that welfare in the future does not decrease." (Pearce and Warford 1993)	human welfare
". . . involves maximizing the net benefits of economic development, subject to maintaining the services and quality of natural resources over time." (Pearce and Turner 1990)	natural resource utilization
"Sustainable development [means] . . . improving the quality of human life while living within the carrying capacity of supporting ecosystems." (IUCN/UNEP/WWF 1991)	carrying capacity of ecosystems
"Sustainable development seeks to deliver the objective of achieving, now and in the future, economic development to secure higher living standards while protecting and enhancing the environment." (DOE 1997)	sustainable economic development

Table 11.3. Compliance of Components of Pre-flood Preparedness Systems with the Spirit of Sustainability (Menzel and Kundzewicz 2003)

Flood Preparedness Measures	Compliance with Sustainability
Construction of large physical infrastructure	low to medium
Zoning; development control within the floodplain	medium to high
Source control, land use planning, watershed management	medium to high
Flood forecasting and warning system	high
Flood proofing	low to high
Disaster contingency planning and maintenance of preparedness of community self-protection activity	high
Installation of insurance plan	low to high
Capacity building; improving flood awareness, understanding, and preparedness	high

world (the Netherlands, Bangladesh, China, the United States, Canada, New Zealand, etc.). They are constructed to provide protection against flooding and aim to prevent the spill of river waters. The heights of these embankments are greater than those of the annual maximum water levels along the rivers in order to minimize internal flooding through the provision of appropriate drainage structures. Such measures are provided to protect agricultural lands, rural settlements, and urban areas. (See Box 11.1.)

Physical structures can meet some elements of sustainability, depending on their design criteria. In general, the flood control drainage/flood control drainage and irrigation projects in Bangladesh (Table 11.4) created the environment for crop agriculture that has been delivering benefits to generations. However, the benefits are not equitably distributed among various groups of landowners and farm laborers. On the other hand, they have deprived people of access to animal protein as the flood control projects proved to be detrimental to floodplain fisheries (Mirza and Ericksen 1996). They have also disrupted the livelihood of fishing communities (WCD 2002). They also demonstrated other side effects (ESCAP 1997), summarized in Table 11.5.

It is vital that the sustainable engineering works will ensure minimum disruption from flooding and enhance natural habitats while providing the levels of protection demanded by the public. In order to do so, the idea of "controlled flooding" is being promoted. The U.S. Fish and Wildlife Service proposed to release a volume of water from the Missouri River reservoirs (Gavin Point) within a specified period of time to help increase population of some threatened species. However, there are critiques of "controlled flooding" who argue that it will not result in any meaningful increase in species numbers because the hydrography on the lower river has been permanently altered due to years of reservoir management (UMIMRA 2004).

11.3.2 Use of Natural Environment

11.3.2.1 Wetlands and Flood Moderation

Wetlands hold the runoff generated from heavy rainfall or snowmelt events. They reduce the possibility of flooding in downstream or moderate flooding to some extent, depending on the magnitude of runoff. Wetland vegetation slows down the flow of flood water (Ramsar 2004). Flood mitigation capacity of a wetland system is limited by a number of factors, which include water level fluctuations, plant community, habitat elements, groundwater hydrology, and downstream conditions. According to Ramsar (2004), wetlands reduce the need for expensive engineering structures. However, this is highly dependent on the hydroclimatic conditions of the area in concern, types of settlements and infrastructure, etc. It should be noted that there are various categories of wetlands, each with distinct functions; the criteria for intervention and management of these wetlands also differ. In addition to the Ramsar definition, see also Cowardin et al. (1979) and National Research Council (1995).

In the past two centuries, a substantial area of wetlands was lost due mainly due to human interventions for settlement and agricultural expansion, infrastructure development, deforestation, excessive sedimentation, etc. Reduction of wetlands in the Red River basin in Manitoba, Canada, has been audited as worsening the water level of the 1997 floods. Wetlands have been recognized to reduce the peak flows of rivers or the total flood volume, in addition to storing water. A study by the Red River Basin Task Force estimated that the basin area, constituting 12% wetland in 1870, was reduced to 3% by 1995 due to the expansion of urban settlements and conversion to agricultural lands. In the Mississippi River basin, a large number of engineering projects (thousands of levees and creation of deep navigation channels) were implemented in the past 150 years to control floods and improve and restore navigation. In this process, 6.9 million hectares of wetlands were lost (Ramsar 2004; Hey and Philippi 1995).

Table 11.4. Various Types of Flood Control Projects in Bangladesh (ESCAP 1997)

Project Type	Number	Area Covered
		(million hectares)
Flood control	29	0.210
Flood control and drainage	173	2.019
Flood control, drainage, and irrigation	42	0.711
Drainage	128	0.759

BOX 11.1

Physical Structures for Flood Control: Examples from Various Countries

- The Dutch and their ancestors have 2000 years of experience in holding back and reclaiming lands from the North Sea by building dikes (embankments) (Driessen and De Gier 1999; van Steen and Pellenberg 2004).
- Between 1960 and 2000, the Bangladesh Water Development Board constructed a total of 5,695 kilometers of embankments, including 3,433 kilometers in the coastal areas; 1,695 flood control/regulating structures; and 4,310 kilometers of drainage canals over 3.77 million hectares (Khalequzzaman 2000).
- The Yangtze River basin in China has a long series of dikes to control floods. To improve flood control capacity, about 3,600 kilometers of dykes have been repaired, heightened, and strengthened. The flood diversion and detention basins constructed for flood control in the Yangtze Valley provide an effective storage capacity of over 50 billion cubic meters (Mingguang et al. 1998).
- As a response to the disastrous Mississippi River flooding in the United States in 1927, the U.S. Army Corps of Engineers built the longest system of levees in the world and minimized flooding and improved navigability. However, during the 1993 flood, 40 of the 229 Federal levees and 1,043 of 1,347 non-Federal levees were overtopped or damaged. Damage to locks and dams was also reported (NOAA 1994).
- In Canada, when the Red River flooded in 1997, the Winnipeg floodway prevented the inundation of the city of Winnipeg. However, communities to the south of Winnipeg such as Ste. Agathe sustained serious damage; the town did not have a permanent ringed dike surrounding it. Temporary dikes were built around properties to prevent damage. Dikes surrounding the city of Winnipeg were watched closely to ensure they were not breached. The Farlinger Commission estimated the cost of the flood at CAD $500 million, CAD $39 million claimed in the city of Winnipeg itself (Haque 2000).
- In the Waikato region in New Zealand, historically, mitigation of flooding has focused on structural measures, which include, particularly, stopbanking, drainage, and pumping facilities. However, there are also extensive schemes aimed to provide protection from coastal flooding (Environment Waikato 1997).

Table 11.5. Effects of Flood Control Embankment Projects in Bangladesh

Surface Water	Groundwater	Land and Soil Resources	Fisheries and Ecosystems
Reduction in river flood within the project Increase in downstream flood risk Increase in risks from extreme flood event in schemes Reduction of dispersal contaminants inside the project Closed system needs flushing to control pollution Increased problem of agrochemical and sewage	Reduced groundwater recharge Increased chances of agro-pollution from fertilizer and pesticides	Scouring and rising bed levels Changing bank erosion Change in soil fertility status inside the project Increased occurrence of weeds	Markedly reduced floodplain fisheries Increase in cultured fisheries Changes in wetland habitat

The economic value of wetlands in flood control and moderation is not often assessed; however, it could be significant (Ramsar 2004). Costanza et al. (1997) estimated the economic value of wetlands and coastal ecosystems for disturbance regulation and water regulation along with some other services; they generalized many assumptions to derive global values, but to have a comprehensive assessment of ecosystem services at a local scale, it is best to use the primary data. In the United States, the value of wetlands in preventing serious flooding has been put at $13,500 per hectare per annum (Hails 1996).

Wetlands also deliver more direct benefits or provisioning services for human well-being. The inner Niger Delta of Mali (30,000 square kilometers) supports more than half a million people and the post-flood grasslands provide food for two million heads of livestock. In 1985, $8 million worth of cattle, sheep, and goats were exported. Floods help migration of fish, their breeding and production in the floodplains. It was estimated that in 1986 the livelihood of some 80,000 fishermen depended on fishing in the delta and in that year more than 60,000 tons of fish were landed (Dugan 1990).

Efforts have been made and are underway to restore some destroyed wetlands in Europe and the United States for flood moderation and to obtain other ecosystem benefits (Galat et al. 1998; Simons et al. 2001; Schmidt 2001). A number of steps are involved in the restoration process:

- understand the causes of the wetland deterioration or destruction (increased stormwater flow due to urbanization and flood and storm control structures);
- develop a comprehensive wetland study to identify wetland elements (hydrology, soil, and plant) requiring amendment;
- conduct hydrological modeling to analyze water level fluctuations (frequency and duration) during various flood flows;
- select the level of flood protection that does not impact the desired plant community negatively;
- design the wetland system by integrating flow reduction, desired plant community, public safety, and recreational elements;
- determine whether regulatory reviews and permits are required to ensure no net loss of wetlands; and
- develop a plan for long-term monitoring and adaptive management, which are the key elements for a successful wetland restoration project.

In 2000, the World Wide Fund for Nature launched the "Green Corridor for the Danube" project in Europe. Under the project, the governments of Romania, Bulgaria, Moldova, and Ukraine have made a pledge to create a network of at least 600,000 hectares floodplain habitats along the lower Danube River and the Prut River, and in the Danube delta. This effort will require restoring an area of 200,000 hectares (Schmidt 2001).

In Louisiana, several sites drained out for crop agriculture, fish farms, and forestry projects are on the way to returning to their original natural state, as the result of a massive $14 billion wetland restoration program called "Coast 2050." The main objective of the plan is to protect more than 10,000 square kilometers of marsh, swamp, and barrier islands (Bourne 2000). One of the components of this project is to restore and maintain Louisiana's barrier islands, which are the state's first line of defense against storm surge generated by the hurricanes. The results of computer models show that certain configurations of islands and inlets along the coast could reduce surges in the inland areas by more than a meter (Bourne 2000).

11.3.2.2 Upland Reforestation/Afforestation

Changed vegetal cover affects the hydrological behavior of a catchment. The influence of deforestation and erosion on the deterioration of flood disasters is well documented (Gade 1996; Sandstrom 1995; and Sternberg 1987). When a forested area is deforested and the forest litter removed, the interception of precipitation is virtually eliminated. Litter removal changes the infiltration capacity of soil and has a pronounced effect on raindrop impact and the resulting soil erosion. With the loss of forest mulch, the infiltration capacity is reduced and rate of erosion increased. Vegetation loss leads to less evapotranspiration. These changes directly contribute to increased direct runoff, reduced surface roughness, and decreased recharge of groundwater aquifers (Singh 1987).

Reinhart et al. (1963) investigated the effect of vegetation on storm runoff in watersheds in the Allegheny mountains of West Virginia. Due to the vegetation cover, both peak flow and total storm flow were decreased substantially. However, storm flow also depends on other factors, including the maturity of the forest and regeneration after, for example, a forest fire. Forest cover may not always affect flow volume. For example, Hirji et al. (2002) reproduced a flow time series for the Iringa catchment in Tanzania, which showed equal runoff coefficient for the forested as well as cultivated land. Runoff from the forested catchment was markedly lower in the dry season. The performance of cultivated land in reducing runoff was not that insignificant (Shaxson and Barber 2003).

Lahmer et al. (2000) investigated the impacts of environmental changes on flood generation in the 575 square kilometer floods of the Stepenitz River basin (a sub-basin of the Elbe River basin) in Germany. They simulated floods in response to an extreme precipitation event that occurred on June 12, 1993, in the basin. A 255 millimeter rainfall (39% of the mean annual sum of 650 millimeters for the period 1981–1994) was recorded in 24 hours. Scenarios were developed to analyze the event's impact under alternative conditions:

- if all arable land (about 66.4% of the total basin area) were forests when the event occurred (forest scenario), and
- if all arable land were bare land (bare land scenario).

Figure 11.3 demonstrates that the forest scenario has a considerable impact on the flood wave at the basin outlet; compared to actual land use, peak flow and discharge volume are reduced by 42.3% and 39.3%, respectively. The reductions for the bare land scenario of about 13.6% (peak flow) and 13.5% (discharge volume) are considerably smaller, but still remarkable. Even for this extreme precipitation event, the water retention of the basin is high in both cases because of the long dry period before the event, which favored percolation due to low soil saturation.

Kramer et al. (1997) examined the effects of forest cover on flooding in the Mantadia National Park, Eastern Madagascar. For this purpose, hydrologic experiments were conducted in the Perinet Experimental Watersheds, which lies in the Vohitra River basin. Land uses studied included primary and secondary forests, traditional rice agriculture with burning (swidden), and agriculture with terraces and other conservation practices. The results of an eight-year experiment suggest that flooding differs quite considerably between primary and secondary forests. Storm flow from the 30-hectare secondary forest watershed was about threefold more in water volume than from a similar-sized primary forest catchment. Reduced infiltration capacity due to soil compaction, decreased evapotranspiration, and less extensive rooting in the secondary forest could have resulted in higher stream flow. Vegetation cover delivers other benefits as well. It is important for reducing soil erosion and sedimentation downstream (for example, Loess plateau of the Yellow River in China). In non-vegetated soils, extreme rainfall can cause earth movement endangering lives and property in the downstream. Settlements on steep and unstable slopes could be more dangerous than the floodplains, depending on its location.

Vegetation cover is also important in the context of sand and dust storms. Engelstaedter et al. (2003) studied dust storm frequency data from more than 2,400 meteorological stations worldwide. Comparisons with distributions of vegetation types suggest that DSF is highest in desert/bare ground (median: 60–80 DSF per year) and shrubland (median: 20–30 DSF per year) regions, and comparatively low in grassland regions (median: 2–4 days per year). Average DSF is inversely correlated with leaf area index and net primary productivity. In non-forested regions, DSF increases as the fraction of closed topographic depressions increases, possibly due to the accumulation of fine sediments in these areas.

11.3.2.3 Mangroves, Seagrasses, and Sand Dunes in Storm Protection

Mangrove forests are diverse communities growing in the intertidal zone (between the average sea level and the high tide mark)

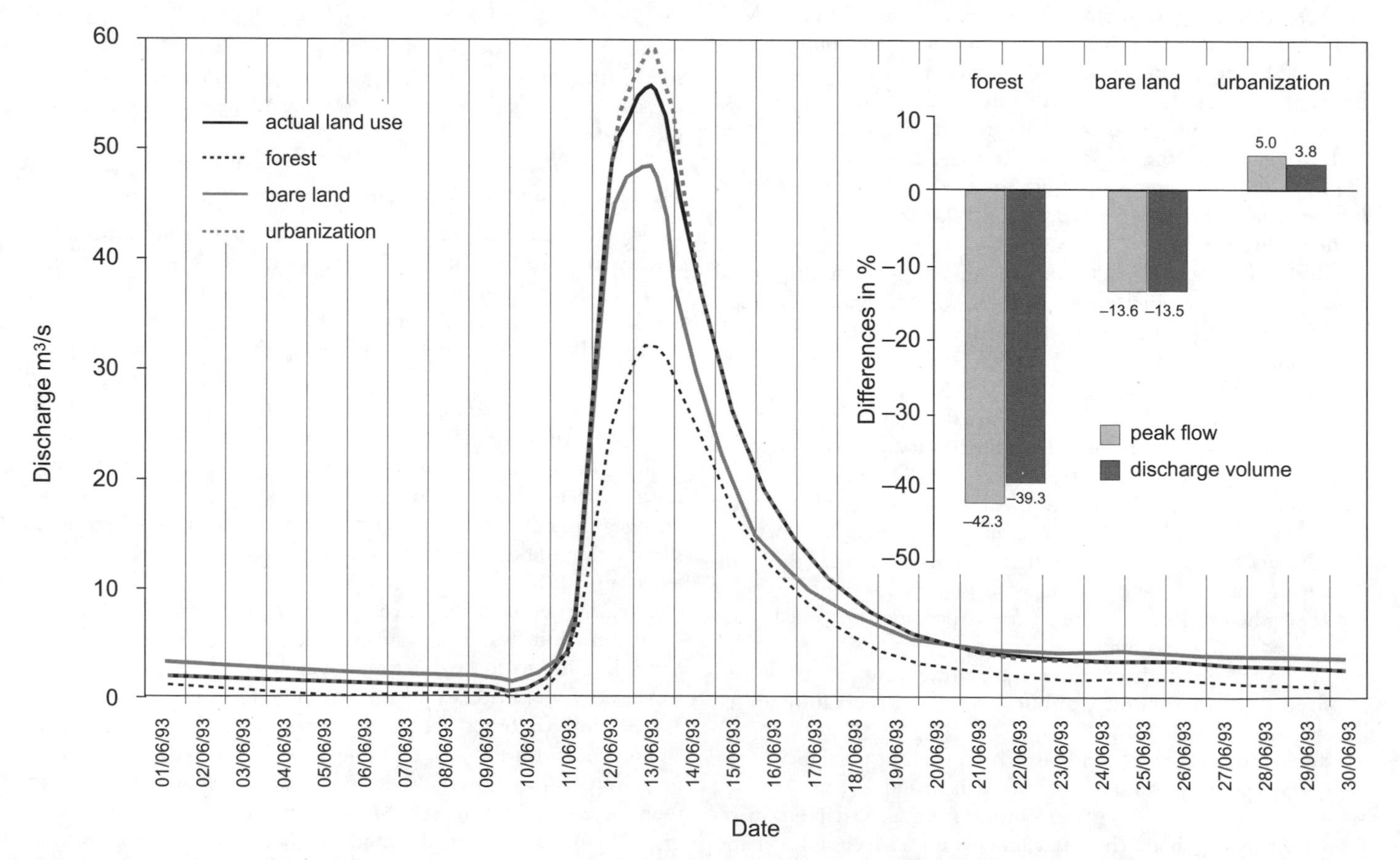

Figure 11.3. Impact of Land Use Change Measures on Stepenitz Basin Discharge following Extreme Precipitation Event on 12 June 1993. The differences for the peak flow and the discharge volume for the scenarios as compared to the actual land use are given on the right hand side of the figure. (Lahmer et al. 2000)

of tropical to sub-tropical coastal rivers, estuaries, and bays. The large amount of silt deposited by coastal rivers along the shoreline produces an environment suitable for the growth of extensive forests. Mangrove plants can also be found growing on the carbonate sediments deposited around reef-associated islands.

In 1997, the "World Mangrove Atlas" estimated approximately 18 million hectares of mangroves. South and Southeast Asia's coasts are enriched with mangroves. In 10 countries of South and Southeast Asia, the approximate area of mangroves is 4,913 thousand hectares. The highest coverage of mangroves is in Indonesia and the lowest in Sri Lanka.

Mangrove ecosystems play an important ecological role while providing a variety of services for human well-being. The benefits obtained from the mangrove ecosystems are quite broad and encompass a range of economic, environmental, and social aspects, including protection from erosion, flooding, cyclones, typhoons, and tidal waves (Primavera 2000).

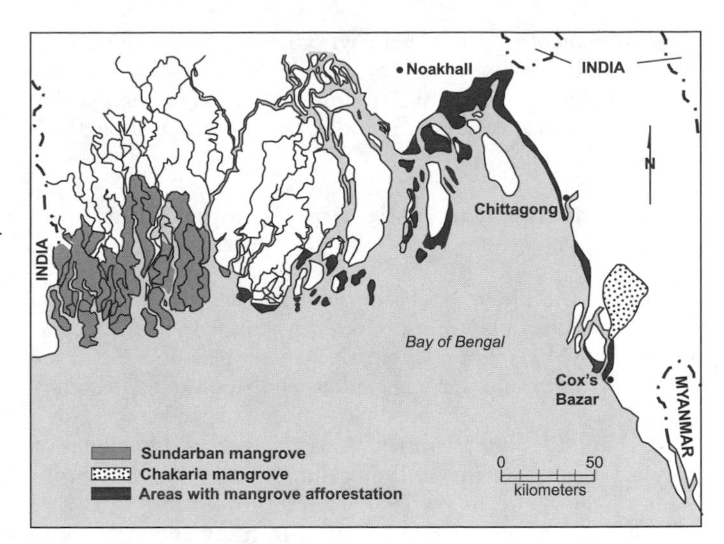

Figure 11.4. Mangroves in Bangladesh

11.3.2.3.1 Super cyclone, Orissa

On October, 29, 1999, the Indian "super cyclone" made landfall over the Indian State of Orissa (UK Met-office 1999). It was the strongest and deadliest cyclone in the region since the Bangladesh cyclone of 1991. The recorded wind speed was 356 kilometers per hour (Kriner 2000) and it generated 8–10 meter high surges. The cyclone and its aftermath led to 10,092 deaths; the demolition of millions of dwelling units; over 80% damage to standing crops, especially ready-to-harvest crops; and a loss of about 454,000 heads of cattle.

In the second half of the twentieth century, India lost more than half of its mangrove forests. In 1987, India had 674,000 hectares of mangroves. Within a period of 10 years, that amount decreased to 483,000 ha (Kumar 2000), leaving the country open to attack by the wind and waves of the cyclones that regularly hit the coast of eastern India and neighboring Bangladesh. The Orissa coastline was once covered by mangrove forests. In 1990, Orissa's coastline had a mangrove forest area of around 150 square kilometers, which had dwindled to 50 square kilometers by 1999 (Khan 1999). In the past, the mangroves would have dissipated the incoming wave energy. Mangroves trap sediment in their roots, which transforms the seabed to a shallow shape. This absorbs the energy of waves and tidal surges and thus protects the land under them. The destruction of coastal mangroves in Orissa has reduced the capacity of the coastal ecosystems to buffer storm surges and cyclonic winds (Shiva 2002). The lack of protective forest cover also made it possible for the floods to inundate large areas and cause much destruction. As forests have been lost, each consecutive cyclone has penetrated further inland (Tynkkynen 2000). However, an area near Paradeep in Orissa where forests were intact was largely saved from cyclonic damage (Tynkkynen 2000).

11.3.2.3.2 Bangladesh cyclone

Since 1822, a total of 69 extreme cyclones have landed on the Bangladesh coast, of which 10 hit the Sundarbans mangrove forest. (See Figure 11.4.) However, a cyclone that lands on the Sundarbans causes less damage than a cyclone of equal magnitude that lands on the central and eastern part of the coast. Most of the damage is caused by the surge. For example, a cyclone that landed on the Cox's Bazaar coast in 1985 generated a 4.3 meter surge and caused the deaths of 11,069 people. A similar cyclone that landed on the Sundarbans in 1988 caused half that number of fatalities.

11.3.2.3.3 Coastal flooding and Cyclone Drena, New Zealand

Coastal flooding around the Firth of Thames is reasonably frequent (annual probability of at least 3–5 %). The damage caused by flooding in July 1995 and Cyclone Drena in January 1997 was in the range of NZ$ 4–5 million or more (including damage to the settlements, agricultural lands, and roads). Dahm (1999) listed a range of natural buffer systems (beaches and wetlands, particularly mangroves) that provide effective protection from coastal flooding in the Waikato. For example, the mangroves in the southern Firth of Thames provide very effective wave and erosion protection for stop-banks protecting the Hauraki plains. It was reported that there was little or no wave action felt along the landward margins of the mangroves in Cyclone Drena despite the marked northerly waves impacting directly on the seaward margin of the mangroves (Dahm 1999).

Like mangroves, seagrasses are an important vegetation type that modifies the local hydro-meteorological regime in coastal regions. Seagrasses cover about 0.1–0.2 % of the global ocean, and contribute to developing highly productive coastal ecosystems (Duarte 2002). Seagrasses, an assemblage of marine flowering plant species, are valuable structural and functional components of coastal ecosystem and are currently experiencing worldwide decline. Widespread seagrass loss results from direct anthropogenic impacts, including dredging, fishing and anchorage, sedimentation, coastal constructions, as well as from natural causes such as cyclones (hurricanes) and floods. Over the period 1986–96, Short and Wyllie Echeverrira (1996) reported that 90,000 hectares of seagrass loss was documented; they estimate that the actual loss was greater.

Native vegetation, sand dunes, and wide beaches are the best coastal buffers for hurricanes (Theiler and Young 1991, cited in Coch 1994). The ebb surge of a storm (the flow of water returning to the sea) can be very destructive to the affected areas, causing localized flooding and damage. Ebb surge is exacerbated by coast parallel streets that trap the water, large areas without vegetation, and beach modifications such as dune overpasses and beach accesses (Coch 1994). Coastal structures can serve to worsen flooding as water gets trapped behind seawalls and is prevented from flowing back to the sea, although they can reduce the influx of the storm surge by preventing infiltration and absorbing energy from waves. Native species of vegetation are more resilient to hurricane winds and waves than the introduced species, and wide beaches help absorb the force of the storm. Sand dunes cushion wind and wave energy and prevent storm surge, or facilitate ebb surge if storm surge does occur. Natural sand dunes are more resilient than restored dunes, yet no dune is worse over all.

Management of the natural environment can reduce vulnerability to floods and storms. However, it alone may not be sufficient to reduce the losses, which can be achieved in combination with such other measures as information, institution strengthening, and education.

11.3.3 Information, Institution Strengthening, and Education

Institutions and public education programs are vital to reducing losses from floods and storms. Risk assessment is an essential component of any hazard or disaster management planning. Flood and storm forecasting and warning, and dissemination of this information play a pivotal role for saving lives, property, and crops. However, there are uncertainties in such forecasting activities. Accuracy is a vital element of forecasting to maintain public confidence. Inaccurate or partially accurate forecasting can cause more damage than reduction of losses. In many countries (especially in the developing world), flood loss occurs mainly due to the lack of institutional capability, trained manpower, and technological limitations.

In Bangladesh, flood forecasting and warning is conducted with the aid of a hydrological and hydrodynamic mathematical model (MIKE11-GIS) and the NOAA–AVHRR satellite imagery and processing system. The Flood Forecasting and Warning Cener in Dhaka is also equipped with experienced and trained personnel. FFWC is capable of issuing forecasts 30 and 72 hours in advance using real time data (Water Level and Reference Flood) from 74 stations and 44 rainfall stations. Hydrometeorological information from a limited number of Indian stations is also used. During the 1998 flood in Bangladesh, the model result was found to be very close to acceptable limits in the Brahmaputra and Ganges basin for both rising and receding time of water level. But forecasting in the flash flood areas, especially in the northeastern, southeastern, and extreme northern parts of the country suffered a setback due to the lack of hourly data.

Goswami (2000) discussed the flood forecasting and warning situation in the Brahmaputra river system in India. The Central Water Commission, the Indian Meteorological Department, and a few state government departments such as agriculture and irrigation and flood control maintain a network of hydrometeorological stations. However, the density of the network, especially in the case of automatic rain gauges, was found to be far short of the WMO suggested guideline (WAPCOS 1993).

Flood warnings are made mainly on the basis of travel time between the selected base station and the particular forecast station and the gauge-to-gauge correlations of water levels. The lead time for the forecasts varies depending on the travel time between the concerned gauges. For the main Brahmaputra river, the lead time is up to 112 hours. During the period of a high flood, flood-warning messages are broadcast through local radio and television centers and also through the printed media. Although the technology of satellite telemetry is available, it has yet to be used on an operational basis for flood forecasting purposes in the Brahmaputra basin. Mathematical models are in use for some rivers such as the Damodar, however, in the case of the Brahmaputra, there is no major effort to use mathematical modeling for the purpose of flood forecasting; disaster mitigation has been done (Goswami 2000).

Nepal established an early warning system to monitor glacier lake outbûrst floods. in the Tsho Rolpa glacier lake, located in the Rolwaling valley about 110 kilometers northeast from the capital Kathmandu. A GLOF is characterized by a sudden release of a huge amount of lake water, which eventually would rush down along the stream channel downstream in the form of devastating flood waves. The mitigation program designed to reduce GLOF vulnerability in Rolwaling valley includes: installation of test siphons, installation of the GLOF Early Warning System, nd construction of an open channel to lower the water level (Shrestha 2001). In 1996, the Meteor Burst type of early warning system was established for safe evacuation of people from the Rolwaling and Bhote/Tama Kosi valleys. The GLOF sensors are located downstream of the moraine senses and they relay a GLOF information to 19 stations distributed over 17 villages downstream of the lake.

The River Forecast Center of the Water Services Branch within the Manitoba government is responsible for preparing preliminary (outlook) and operational forecasts for all rivers in Manitoba, Canada. Preliminary forecasts are created in February and March based on model forecasts of expected flood peak data estimated using meteorological data. Operational forecasts are updated reports regarding the anticipated flood level, incorporating real time data collected from flood gauge stations, weather data, and river observers. Melt conditions, spring precipitation, air and dew point temperatures, wind data, and sunshine levels are incorporated into the operational forecasts.

Forecasts are communicated to the media through government information services, as well as to the Manitoba Emergency Measures Organization, Emergency Preparedness Canada, municipalities, aboriginal bands, and other interested stakeholders and government officials. The 47 pre-existing river monitoring sites were upgraded and supplemented with 37 more sites in response to deficiencies in forecasting the 1997 flood. Additionally, Environment Canada expanded its network of climate stations to 350, to improve the weather data collection for flood forecasting. A difficulty in the 1997 flood was the inability to accurately forecast overland flooding (Simonovic and Carson 2003). Several communities were flooded from overland flow rather than from the river.

In the United States, the National Hurricane Center in Florida is responsible for forecasting hurricáne tracks and intensity in the North Atlantic and the eastern North Pacific, east of 140(W (DeMaria 1997). The NHC obtains data from satellites, buoys, reconnaissance flights to the storms, and radar (NHC 2004a). These data are put into models to determine the likely track and intensity of the storm. Most of the models also require data from global forecast models; after the storm the various models are evaluated against the best track positions as determined by data collected during the storm and re-analyzed (DeMaria 1997).

Error is unavoidable in hurricane forecasts and increases with the length of the forecast (NHC 2004b). Error is mitigated by the inclusion of a strike probability table that provides the statistical probability that a hurricane will strike a specific location in the following three days. Error is the reason why warnings cover a wider area than the most likely strike zone. Major cities that are vulnerable to the impacts of a hurricane (for example, New Orleans) require 48 hours to evacuate; however 48 hours prior to the anticipated arrival of hurricane force winds, the strike probability of any location is only 25% (NHC 2004b).

McAdie and Lawrence discovered that error in the NHC track forecasts since 1970 have decreased by one percent a year on average (cited in Nicholls 2001). This increase in accuracy was boosted further by another improvement in accuracy in the mid-1990s, during which predictions improved twice as fast (Kerr 1999, cited in Nicholls 2001). The addition of GPS dropwindsondes in 1997 notably improved the accuracy of hurricane models 48 hours prior to landfall, some as much as 32% for track forecasts and 20% for intensity forecasts (Aberson and Franklin

1999; Kerr 1999, cited in Nicholls 2001). Despite these improvements in forecast accuracy, the length of coastline-issued hurricane warnings did not decrease (Pielke 1999, cited in Nicholls 2001).

The NHC utilizes the Sea Lake Overland Surges from Hurricane (SLOSH) model to complement hurricane forecasts and alert locations to their storm surge risk (NHC 2004c). This model evaluates pressure, size of the storm, forward speed, track, and winds in its assessment of anticipated storm surge. This information is combined with local bathymetry and shoreline configuration to determine what the storm surge will be in a location. The SLOSH model is normally accurate within 20%; however, if the track and intensity data brought in from the hurricane forecast is inaccurate, the storm surge prediction will be as well.

It is evident that effective information generation and dissemination, public education, and strengthening of institutions can markedly reduce vulnerability to floods and storms. However, other means such as financial services are needed to recover from the losses caused by these events.

11.3.4 Financial Services

Different countries have taken various approaches to insuring flood. Under the *option system,* insurers agree to extend their policy to include flood on payment of an additional premium; examples include Belgium, Australia, Germany, Italy, Canada (commercial only). In the *bundle system,* flood cover is available only if it is bundled with other perils such as storm or theft (for example, Israel, Spain, the United Kingdom). In some countries, *disaster assistance* is available from different levels of government (for example, Canada), while others rely upon the private sector (for example, Germany, Portugal, the United Kingdom). This assistance can be automatic in the event of a disaster (for example, Canada and China), or it can rely upon a government decree (for example, France and the United States) (Crichton 2002). The U.S. system is unique in having a National Flood Insurance Program that is federally based; the NFIP has been criticized, however, for encouraging development within floodplains (Larsen and Plasencia 2001) even while it provides people with the means to recover from flood disasters.

It is difficult to insure for flood because of the problem of adverse selection. Because the spatial patterns of flood risk are fairly well known since floods tend to occur in floodplains, those at risk will buy insurance, while those not at risk will not. This contradicts one of the main principles by which insurance works, that is, where risk is spread among a large population due to uncertainty related to the hazard.

Both social programs and private insurance are important coping mechanisms for flood disaster recovery. They can, however, inadvertently contribute toward community vulnerability by encouraging development within floodplains or by creating cultures of entitlement. These issues are discussed in the next section.

11.3.5 Land Use Planning

Land use planning such as floodplain zoning is a process of determining the most desirable way land should be used so that it can help to mitigate disasters and reduce risks by directing development away from hazard-prone areas. Land use planning plays a major role in regulating development and the use of land. It is normally carried out in two ways (Gunne-Jones 2003). First, it works by controlling developments through a system of issuing permits or approvals. Second, it involves planning for the future needs of a state, region, or locality through the publication and adoption of development or zoning plans.

The first step in land use planning for flood management/damage reduction is to prepare flood risk maps in which flood magnitude, water depth, flow velocity, flood duration, etc., for a specific return period are incorporated. The main purpose is to inform the public about the flood risk derived from occupying a floodplain. Flood risk maps communicate the degree of flood risk to concerned agencies and the public, enabling a dialogue on the most appropriate flood prevention and protection measures (European Environment Agency 2001). Coastal land use planning for areas vulnerable to high storms and flooding from storms includes setbacks from the shore for new developments, etc. In the United Kingdom, the findings of the Environment, Transport, and Regional Affairs Committee (2000) recommended that flood risk maps should be included in development plans and information about flood risk should become a standard part of local authority searches that are carried out by prospective property purchasers. In France, zoning is tied to insurance; in theory, integrated catchment planning has been introduced in the forms of SDAGE (general water catchment basin plans-Schémas directeur d'aménagement et de gestion des eaux) and SAGE (sub-catchment management plans-Schémas d'aménagement et de gestion des eaux). Selected land use planning mechanisms from North America, Europe, Asia, and the Caribbean are described in Box 11.2.

Land use planning can be used to serve a broad range of purposes, some of which are co-beneficial, but others of which are in conflict with each other. For example, reducing storm and flood damage by mapping flood zones and restricting development can be compatible with environmental and recreational agendas by creating natural spaces and parkland. On the other hand, zoning for commercial or residential development can enhance a community's tax base, and diversify and increase its economic base, but may ultimately increase vulnerability to extreme events.

These different agendas or needs can all be valid to a community and require a zoning process that considers multiple stakeholders to trade-off costs and benefits. Of importance is the issue of who benefits and who pays the costs. If there is no connection between these two items, as is often the case, then the decision-making process can be distorted by imbalances in power or access to power and socioeconomic status among the various stakeholders. The nature of the political system becomes critical at this point. The degree to which it is egalitarian, corrupt, or transparent will have a large bearing on the outcome, as will cultural biases toward structural versus nonstructural solutions to hazards.

Urban settlements can be particularly vulnerable to floods and storms, and need careful spatial planning. Examples of Dhaka, Bangladesh, and Toronto, Canada, stress this need.

The location of the city of Dhaka—the capital of Bangladesh—has made it particularly vulnerable to floods. It is surrounded by the Buriganga to the south, the Turag to the west, the Tongi Khal to the north, and the Balu to the east. Dhaka and its adjoining areas are composed of alluvial terraces of the southern part of the Madhupur tract and low-lying areas of *doab* of the rivers Meghna and Lakhya. The city suffered from flooding mainly due to spillage from the surrounding rivers. Local rainfall often complicates the flooding situation. Although the city was periodically flooded, adaptation and coping mechanisms are not well documented. Some initiatives, however, were taken in the wake of disastrous flooding in 1988 and 1998 (Huq and Alam 2002; Jahan 2000). During the floods of 1988 and 1998, Dhaka was severely affected. In 1998, a catastrophic flood hit the greater Dhaka area during the months of August and September. About 56% of greater Dhaka was submerged, affecting about 1.9 million

BOX 11.2

Selected Land Use Planning Mechanisms from around the World

- Construction along or near the *Florida* coastline is governed by the Standard Building Code or the National Flood Insurance Program. Compliance with these codes makes individuals and businesses within the communities eligible to purchase flood insurance. In the 1980s, Florida reinforced the stipulations contained in the building code and insurance program by establishing the Coastal Construction Control Line, which defines specific areas along the coastline that are subject to flooding and erosion etc. The CCCL was adopted throughout Florida between 1982 and 1991 and reflects storm impact zones over a 100-year period. Distinctions were made between two categories of structures based on the CCCL regulations: (1) structures located seaward of the CCCL that were built prior to enactment of the CCCL regulation were categorized as nonpermitted structures at risk of sustaining hurricane damage; and (2) structures built after the adoption of the CCCL require a special building permit to certify that the builder will adhere to a more rigid set of building standards designed to reduce the risk of structural damages that can be sustained during a hurricane.
- *New Brunswick, Canada,* completed a re-mapping of the entire coast to delineate the landward limit of coastal features. The setback for new development is defined from this limit. Some other provinces have adopted a variety of setback policies, based on estimates of future coastal retreat.
- In *Barbados,* a national statute establishes a minimum building setback along sandy coasts of 30 meters from the mean high-water mark; along coastal cliffs the setback is 10 meters from the undercut portion of the cliff. In *Aruba and Antigua,* the setback is established at 50 meters inland from the high-water mark.
- A Coastal Zone Management Plan in *Sri Lanka* identifies setback areas and no-build zones. Minimum setbacks of 60 meters from mean sea level are regarded as good planning practice.
- In the *United Kingdom,* the House of Commons in 1998 endorsed the concept of managed realignment as the preferred long-term strategy for coastal defense in some areas.
- In the United States, the states of *Maine, Massachusetts, Rhode Island,* and *South Carolina* have implemented various forms of rolling easement policies to ensure that wetlands and beaches can migrate inland as sea level rises.
- Several states in *Australia* have coastal setback and minimum elevation policies, including those to accommodate potential sea-level rise and storm surge. In South Australia, setbacks take into account the 100-year erosional trend plus the effect of a 0.3 meter sea-level rise by 2050. Building sites should be above storm surge flood level for the 100-year return interval.

people. The economic damage caused by the flood was estimated to be $10–20 million (JICA 1990).

Important nonstructural measures include flood forecasting and warning, retention ponds, natural water bodies and a drainage network, land use planning, and relief and rehabilitation. Dhaka used to have many natural water bodies, which functioned as a buffer for floodwaters. Over the years, natural water bodies dwindled markedly due to public encroachments on land. Virtually no natural water bodies are left in the old part of the city. In addition, encroachments are going on even in the new upscale residential areas—Gulshan, Banani, and Baridhara. The minimum standard for retention pond is 12% of the urban areas, but at present, the amount retained is estimated to be less than 4% (RAJUK 1995). The government has recently issued a decree banning the filling in of any wetland for urban development (Huq and Alam 2002).

Hurricane Hazel struck the city of Toronto, Canada, on October 15, 1954, with 183 millimeters of rain swelling local rivers. In the ensuing flash flood, 81 people were killed and thousands were left homeless. A comprehensive mitigation plan would provide for the construction of dams and reservoirs, structural control of river channels, improved flood forecasting, land expropriation, and changes in land use zoning in the floodplains. The mitigation plan was to cost $35 million dollars, divided between the federal, provincial, and local governments. Four of the proposed 13 dams were constructed and 844 square kilometers of floodplain were zoned to prevent future development. Emphasis of mitigation was on expropriating floodplain land from residents, thus reducing the amount of money available to build dams. Opponents of the dams argued that if floodplain land was returned to the river, it would be unnecessary to build dams, whereas, proponents argued that the construction of dams would allow more development of the floodplain. Hurricane Hazel would become the foundation of the Toronto area's flood management plan, the structural controls used would create some new habitats and destroy others, but the removal of development on the floodplain would be the most beneficial and enduring flood control initiative.

11.4 Lessons Learned and Key Research and Policy Issues

Based on the assessment of responses in the previous section, the main lessons learned and the issues to be considered in development of future responses are summarized here.

11.4.1 Substitutability

The two issues related to protection and restoration of ecosystems are: (1) the degree to which the technologies can substitute for ecosystems services, and (2) whether ecosystem restoration can re-establish not only the functions of direct use of value to humans, but also the ability of the systems to cope with future disturbances.

Moberg et al. (2003) address these two issues in their study on a number of attempts at substitution and restoration of tropical coastal "seascape" ecosystem (which generally includes a patchwork of mangroves, seagrass beds, and coral reefs); these attempts include such things as artificial reefs, aquaculture in mangroves, and artificial seawalls. They write, "Substitutions often imply the replacement of a function provided free by a solar powered, self-repairing resilient ecosystem, with a fossil fuel-powered, expensive, artificial substitute that needs maintenance. Further, restoration usually does not focus on large-scale processes such as the physical, biological, and biogeochemical interactions between mangroves, seagrass beds, and coral reefs." (p. 27) They conclude that ecosystems services cannot be readily replaced, restored, or sustained without extensive knowledge of the dynamics, multifunctionality, and interconnectedness of ecosystems. Nonetheless, they do acknowledge that restoration might be the only viable management alternative when the system is essentially locked into an undesired community state (stability domain) after a phase-shift.

Post-flood evaluations recommend that risk management plans be made more effective in reducing adverse consequences for human health and well-being, and that they take into account the multiple ways by which floods and storms can affect populations. Yet recent flooding events demonstrate that better preparation is needed, that lessons learned are not consistently applied, and that long-term solutions need to be found for effective floodplain management in the context of a changing climate.

11.4.2 Linkages among Ecosystem Services

What are the conflicts and trade-offs that emerge when ecosystems provide multiple services? For example, the ecosystems that provide flood and storm protection are also important for other services including food and fiber, fresh water, and so on. What are the challenges for management in trying to accommodate conflicts between services? At the micro level, studies have explored the potential value of benefit-cost evaluation for stormwater quality management decisions at a local level in an urban setting such as Los Angeles; a study by Kalman et al. (2000) demonstrates the economic limits of uncoordinated institutional management at the local or individual level and attests to the value of coordinated basinwide management. Macro-level studies offer valuable lessons for flood and storm management for other regions of the world; for example, Gren et al. (1995) conducted a study on economic valuation of the Danube floodplains, which are shared by several countries and provide a complex ecosystem with various habitats or biotopes. Lessons of these assessments are valuable.

11.4.3 Conflict between Short- and Long-term Objectives

Conflicts between short- and long-term objectives involve reducing immediate impacts versus maintaining long-term stability and function. Since floods and storms are part of the natural disturbance regime, they may be considered to be important for long-term ecosystem function.

11.4.4 Institutional Issues

In practice, the actual provision of flood and storm protection is often the responsibility of a number of different actors working at different levels—local, regional, national, and transnational. These institutional settings not only affect the delivery of the services, but also the manner in which they are delivered as well as the direct and indirect effects on ecosystems.

11.4.5 Climate Change

Another issue to be considered is the potential implications of climate change for the underlying hydrometeorological processes responsible for floods and storms, and the consequent implications for response and management strategies. Using a full range of 35 SRES scenarios based on a number of climate models, the IPCC has projected an increase in global mean temperature between 1990 and 2100 of s 1.4 −5.8 Celsius. Over the same period, the global mean sea level has been projected to increase by 9–88 centimeters (IPCC 2001). Table 11.6 shows IPCC's assessment of flood and storm-related changes under three different variables: (1) climate and atmospheric systems; (2) terrestrial systems; (3) economic and social systems.

The IPCC reported a statistically significant 2% change in global land precipitation in the last century but it was not uniform in spatial and temporal scales. Schönwiese and Rapp (1997) reported a significant increase in precipitation over Central and Northern Europe and western Russia. These trends were reflected in the discharge of the Rhine river. Engel (1995) reported

Table 11.6. Flood-related Variables Listed in the Third Assessment Report of the Intergovernmental Panel on Climate Change (Kundzewicz and Schellhuber 2004)

Climate and Atmospheric Systems	Terrestrial Systems	Economic and Social Systems
Total precipitation	River discharge and stage	Anthropogenic pressure
Intense precipitation events	Water storage and capacity	Adaptive capacity
Wind intensity	Runoff coefficient and infiltration capacity, portion of impervious area	Vulnerability
Seasonal distribution and climate variability (e.g., ENSO)	Impacts on ecosystems	Measures of flood losses
Sea level		Risk perception

a rising tendency of the maximum annual discharge of the Rhine at Cologne over the last 100 years. Seasonal (autumn and winter) precipitation increases were observed in the mid- and high-latitudes of the Northern Hemisphere (Kundzewicz and Schellnhuber 2004). Increases in "heavy and extreme" precipitation events were also reported from some regions where total precipitation had either decreased or remained the same (Kundzewicz and Schellnhuber 2004).

Increased precipitation could mean increased flash floods and seasonal floods, but not necessarily uniform over all regions of the world (WMO and GWP 2003). A number of studies were carried out in various regions to examine the possible effects of climate change on flood magnitude, frequency, and extent. Reynard et al. (1998) estimated changes in the magnitude of floods of different return periods in the Thames and Severn catchments in the United Kingdom. They concluded that increases in flood magnitude were due to increases in winter precipitation. Total volume of rainfall (not the peak intensity of rainfall), over several days played a major role in the flood processes in these large catchments.

Schreider et al. (1996) also reported increases in flood risk in Australia under the wettest rainfall scenarios. Using scaled precipitation change scenarios for four climate models Mirza (2002), in a study on climate change and changes in the probability of occurrence of floods in Bangladesh, concludes that climate change caused by the enhanced greenhouse effect is likely to have considerable effects on the hydrology and water resources of the Ganga, Brahmaputra, and Meghna basins and might ultimately lead to more serious floods in Bangladesh. Nicholls (1999) estimated that sea-level rise could cause the loss of up to 22% of the world's coastal wetlands.

Different kinds of responses are expected to address increased precipitation and flooding. First, modifications of design standards for future flood control/mitigation structures are required. The U.K. government has initiated flood risk management schemes in the context of climate change, and has recommended examining the effect of a probable 20% increase in flood flows. For coastal schemes, the allowance required for anticipated increases in sea level is 5 millimeters per year, which is IPCC's business-as-usual projection for the current century (DEFRA 1999, 2003). However, in many developing countries, such kinds of actions may be constrained by economic principles and available resources (WMO and GWP 2003). Second, strengthening of flood forecasting and warning system based on present vulnerability is required. Third, mapping of the vulnerable areas and associated risks should be carried out. It is likely that vulnerability may expand to new areas/settlements. Fourth, land use planning based on vulnerability maps and future socioeconomic and demographic scenarios needs to be implemented.

11.4.6 Information Failure

Turner et al. (2000), in a study on wetlands management policy, illustrate that information failure is one of the chief reasons that wetlands over the world have been lost or are threatened. The other reasons for loss of wetlands are all related to information failure; they include the public goods nature of many wetlands products and services; user externalities imposed on other stakeholders; and policy intervention failures (because of a lack of consistency among government policies in different areas). They suggest a need for integrated research combining social and natural sciences to solve (in part at least) the information failure problem. An integrated research framework combining economic valuation, integrated modeling, stakeholder analysis, and multicriteria evaluation could provide complementary insights into sustainable and welfare-optimizing ecosystem management and policy.

11.5 Conclusion

Floods and storms can cause enormous economic, social, and human losses. However, they also generate beneficial effects. In the past, structural methods of flood and storm control received priority. At present, nonstructural measures including the use of the natural environment are being emphasized to reduce vulnerability as well as economic losses. The application of integrated flood and storm management approaches can maximize social, economic, and ecosystem benefits.

Note

1. The time of concentration is the time taken by a drop of water to travel from the furthest hydrologic point in a basin to the point of discharge. Determination of the time of concentration is the summation of the individual hydraulic travel times from each section of a subdivided basin. t_c = Tt. The travel time is a function of the flow conditions, surface roughness, and the topography.

References

Aberson, S.D. and J.L. Franklin, 1999: Impact on hurricane track and intensity forecasts of GPS dropwindsonde observations from the first-season flights of the NOAA Gulfstream-IV jet aircraft, *Bulletin of the American Meteorological Society,* **80(10),** pp. 421–427.

Auger, C., S. Latour, M. Trudel, and M. Fortin, 2000: Posttraumatic stress disorder: Victims of the Saguenay flood, *Canadian Family Physician,* **46,** pp. 2420–2427.

Barcellos, C. and P.C. Sabroza, 2000: Socio-environmental determinants of the leptospirosis outbreak of 1996 in western Rio de Janeiro: A geographical approach, *International Journal of Environmental Health Research,* **10(4),** pp. 301–313.

Bennet, G., 1970: Bristol floods 1968: Controlled survey of effects on health of local community disaster, *British Medical Journal,* 21 August, pp. 454–258.

Bourne, J, 2000: Louisiana's vanishing wetlands: Going, going. . . . *Science,* **289 (5486)** (15 September), pp. 1860–1863.

Brammer, H., 1990: Floods in Bangladesh 2: Flood mitigation and environmental aspects, *Geographical Journal,* **156,** pp. 158–165.

Chadwick, M., J.G. Soussan, S.S. Alam, and D. Mallick, 2003: *From Flood to Scarcity of Water: Re-defining the Water Debate in Bangladesh,* Bangladesh Centre for Advanced Studies, Dhaka, Bangladesh.

Chong, S.K., B.P. Klubek, and J.T. Weber, 1998: Herbicide contamination by the 1993 Great Flood along the Mississippi River, *Journal of the American Water Resources Association,* **34(3),** pp. 687–693.

Coch, N.K., 1994: Hurricane hazards along the Northeastern Atlantic coast of the United States, *Journal of Coastal Research,* **12,** pp. 115–147.

Costanza, R., R. d'Arge, R. de Groot, S. Farber, M. Grasso, et al., 1997: The value of the world's ecosystem services and natural capital, *Nature,* **387,** pp. 253–260.

Cowardin, L.M., V. Carter, and E.T. LaRoe, 1979: *Classification of Wetlands and Deepwater Habitats of the United States,* Fish and Wildlife Services, US Department of Interior, Washington, DC.

Crichton, D., 2002: UK and global insurance responses to flood hazard, *Water International,* **27(1),** pp. 119–131.

Dahm, J., 1999: *Coastal Flooding Risk Mitigation Strategies,* Environment Waikato, Hamilton, New Zealand.

DEFRA (Department for Environment, Food and Rural Affairs), 1999: *Flood and Coastal Defence Project Appraisal Guidance Economic Appraisal* (FCDPAG3), DEFRA, London, UK.

DEFRA, 2003: *Supplementary Note on Climate Change Considerations for Flood and Coastal Management,* DEFRA, London, UK.

Daley, W.R., L. Shireley, and R. Gilmore, 2001: A flood-related outbreak of carbon monoxide poisoning: Grand Forks, North Dakota, *Journal of Emergency Medicine,* **21(3),** pp. 249–253.

DeMaria, M., 1997: Summary of the NHC/TPC Tropical Cyclone Track and Intensity Guidance Models: Informal reference. Available at http://www.nhc.noaa.gov/aboutmodels.shtml.

DoE (Department of Environment), 1997: *Planning Policy Guidance 1 [Revised]: General Policy and Principles,* Her Majesty's Stationary Office (HSMO), London, UK.

Diaz, H.F. and R.S. Pulwarty (eds.), 1997: *Hurricanes: Climate and Socioeconomic Impacts,* Springer-Verlag, New York, NY.

Driessen, P.P.J. and De Gier, A.A.J., 1999: Flooding, River Management and Emergency Legislation: Experiences of the Accelerated Reinforcement of Dikes in the Netherlands. *Tijdchrift voor Economische en Sociale Geografie,* **90(3),** pp. 336–342.

Drijver, A. and M. Marchand, 1985: *Taming the Floods: Environmental Aspects of Floodplain Development in Africa,* Centre for Environmental Studies, University of Leiden, Leiden, The Netherlands.

Duarte, C.M., 2002: The future of seagrass meadows, *Environmental Conservation,* **29(2),** pp. 192–206.

Dugan, P., 1990: *Wetland Conservation,* IUCN, Gland, Switzerland.

Edwards, P., 2000: Aquaculture, poverty, impacts and livelihoods, Natural Resources Perspectives 56, Overseas Development Institute, London, UK.

EEA (European Environment Agency), 2001: *Sustainable Water Use in Europe. Part 3: Extreme Hydrological Events: Floods and Droughts,* European Environment Agency, Copenhagen, Denmark.

Elahi, S., 2000: *Project on the Uninsured Elements of Natural Catastrophic Losses-Easter Floods, UK,* Centre for Environmental Strategy, University of Surrey, UK.

Engel, H., 1995: Die Hochwasser 1994 and 1995 im Rheingebiet im vieljahrigen Vergleich, *Proceedings DGFZ,* **6,** pp. 59–74.

Engelstaedter, S., K.E. Kohfeld, I. Tegen, and S.P. Harrison, 2003: Controls of dust emissions by vegetation and topographic depressions: An evaluation using dust storm frequency data, *Geophysical Research Letters,* **30(6),** Article no. 1294.

Environment Waikato, 1997: *Flood Risk Mitigation Plan (Environment Waikato Technical Report 1997/13),* Environment Waikato, Waikato, New Zealand.

ESCAP (Economic and Social Commission for Asia and the Pacific), 1997: Integrating environmental considerations into the economic decision-making process: Case study of Bangladesh, ESCAP, Bangkok, Thailand.

Gade, D.W., 1996: Deforestation and its effects in highland Madagascar, *Mountain Research and Development,* **16(2),** pp. 101–116.

Galat, D.L., L.H. Fredrickson, and D.D. Humburg, 1998: Flooding to restore connectivity of regulated, large river wetlands, *Bioscience,* **48,** pp. 721–733.

Ginexi, E.M., K. Weihs, S.J. Simmens, and D.R.Hoyt, 2000: Natural disaster and depression: A prospective investigation of reactions to the 1993 Midwest Floods, *American Journal of Community Psychology,* **28(4),** pp. 495–518.

Goswami, D.C., 2000: Flood forecasting in Brahmaputra River, India [online]. Available at www.southasianfloods.org/document/ffb/index.html.

Goudie, A.S. and N.J. Middleton, 1992: The changing frequency of dust storms through time, *Climatic Change,* **20(3),** pp. 197–225.

Gray, W.M. and C.W. Landsea, 1992: African rainfall as a precursor of hurricane-related destruction on the U.S. east coast, *Bulletin of American Meteorological Society,* **73,** pp. 1352–1364.

Green, C.H., 1999: The economics of floodplain use, *Himganga,* **1(3),** pp. 4–5.

Gren, I.M., K.H. Groth, and M. Sylven, 1995: Economic values of Danube floodplains, *Journal of Environmental Management,* **45(4),** pp. 333–345.

Griffin, D.W., V.H. Garrison, J.R. Herman, and E.A. Shinn, 2001: African desert dust in the Caribbean atmosphere: Microbiology and public health, *Aerobiologia,* **17(3),** pp. 203–213.

Gunne-Jones, A., 2003: Land use planning: How effective is it in reducing vulnerability to natural hazards [online]. Available at http://www.icdds.org/downloads/HAZARDS%20PAPER%20v.2.pdf.

Hails, A.J. (ed.), 1996: *Wetlands, Biodiversity and the Ramsar Convention,* IUCN, Gland, Switzerland.

Handerson-Sellers, A., H. Zhang, G. Berz, K. Emanuel, W. Gray, et al., 1998: Tropical cyclones and global climate change: A post-IPCC assessment, *Bulletin of the American Meteorological Society,* **79,** pp. 19–38.

Haque, C. E., 2000: Risk Assessment, Emergency Preparedness and Response to Hazards: The case of the 1997 Red River Valley Flood, Canada, *Natural Hazards,* **21 (2–3),** 225–245.

Hey, D.L. and N.S. Philippi, 1995: Flood reduction through wetland restoration: The upper Mississippi river basin as a case history, *Restoration Ecology,* **3,** pp. 4–17.

Hirji, R., P. Johnson, P. Maro, and T.M. Chiuta (eds.), 2002: *Defining and Mainstreaming Environmental Sustainability in Water Resources Management in Southern Africa,* South African Development Community/IUCN/South Africa Research Documentation Centre/World Bank, Maseru, Lesotho/Harare, Zimbabwe/Washington, DC.

Holland, G.J., 1993: Ready Reckoner. In: *Global Guide to Tropical Cyclone Forecasting, WMO/TC-No. 560, Report No. TCP-31,* World Meteorological Organization, Geneva, Switzerland.

Huq, S. and Alam, M., 2002: Flood management and vulnerability of Dhaka City, Paper presented at the conference on the Future of Disaster Risk: Building Safer Cities, organized by the Prevention Consortium, the World Bank Group, December 2002, Washington, DC.

International Federation of Red Cross and Red Crescent Societies, 2001: *World Disasters Report 2001,* International Federation of Red Cross and Red Crescent Societies, Geneva, Switzerland.

International Federation of Red Cross and Red Crescent Societies, 2003: *World Disasters Report 2003,* International Federation of Red Cross and Red Crescent Societies, Geneva, Switzerland.

IPCC (Intergovernmental Panel on Climate Change), 2001: *Climate Change 2001: Impacts, Adaptation and Vulnerability, Contribution of Working Group II to the Third Assessment Report of the Intergovernmental Panel on Climate Change,* Cambridge University Press, Cambridge, UK.

IUCN, UNEP, and WWF (International Union for Conservation of Nature and Natural Resources, United Nations Environment Programme, and World Wildlife Fund), 1991: *Caring for the Earth,* IUCN, Gland, Switzerland.

Jahan, S., 2000: Coping with flood: The experience of the people of Dhaka during the 1998 flood disaster, *Australian Journal of Emergency Management,* **15(3),** pp. 16–20.

JICA (Japan International Cooperation Agency), 1990: *Updating Study on Storm Water Drainage System Improvement Project in Dhaka City, Dhaka,* JICA, Tokyo, Japan.

Kalman, O., J. R. Lund, D. K. Lew, and D. M. Larson, 2000: Benefit-cost analysis of stormwater quality improvements [online]. Available at http://link.springer-ny.com/link/service/journals/00267/bibs/0026006/00260615.html.

Khalequzzaman, Md., 2000: Flood Control in Bangladesh through Best Management Practices, EB2000: Expatriate Bangladeshi 2000-Short Note 17 (http://www.eb2000.org/short_note_17.htm).

Khan, M.I., 1999: Orissa greens oppose coastal highway [online]. Available at www.rediff.com/news/1999/dec/17oris.htm.

Kramer, R.A., D. Richter, S. Pattanayak, and N. Sharma, 1997: Ecological and economic analysis of watershed protection in eastern Madagascar, *Journal of Environmental Management,* **49,** pp. 277–295.

Kriner, S., 2000: Three months after super cyclone, Orissa begins to rebuild: Disaster relief [online]. Available at http://www.disasterrelief.org/Disasters/000201Orissaupdate.

Kumar, R., 2000: Conservation and management of mangrove in India, with special reference to the State of Goa and the Middle Andaman Islands, Food and Agriculture Organization of the United Nations, Rome, Italy [online]. Available at http://www.fao.org/DOCREP/X8080e/x8080e07.htm.

Kundzewicz, Z.W. and H.-J. Schellnhuber, 2004: Floods in the IPCC TAR perspective, *Natural Hazards,* **31,** pp. 111–128.

Kundzewicz, Z. W., 1999: Flood protection: Sustainability issues, *Hydrological Sciences Journal,* **44(4),** pp. 559–572.

Kunii, O., S. Nakamura, R. Abdur, and S. Wakai, 2002: The impact on health and risk factors of the diarrhea epidemics in the 1998 Bangladesh floods, *Public Health,* **116(2),** pp. 68–74.

Lahmer, W., B. Pfützner, and A. Becker, 2000: Influences of environmental changes on regional flood events. In: *European Conference on Advances in Flood Research,* Bronstert, C. Bismuth, and L. Menzel (eds.), Potsdam Institute for Climate Impact Research, PIK report no. 65, A, pp. 238–254.

Larsen, L.L. and D. Plasencia, 2001: No adverse impact: A new direction in floodplain management policy, *Natural Hazards Review,* **2(4),** pp. 167–181.

Laurence, W.F. and Beirregaard, R.O. (Editors), 1997: *Tropical Forest Remanants:Ecology, Management and Conservation of Fragmented Communities,* University of Chicago Press, Chicago, USA.

McBride, J.L., 1995: Tropical cyclone formation. In: *Global Perspectives on Tropical Cyclones,* R.L. Elsberry (ed.), World Meteorological Organization, Geneva, Switzerland, pp. 63–105.

Menzel, L. and Kundzewicz, Z.W., 2003: Non-Structural Flood Protection-A Challenge. International Conference 'Towards National Flood Reduction Strategies', Warsaw, 6–13 September, 2003.

Mingguang, Z., Guowei, Y. and Hui, Z., 1998: Water Resources Development and Utilization in the Yangtze Valley. Paper presented in the Los Alamos National Laboratory, University of California, USA.

Mirza, M.M.Q. and N.J. Ericksen, 1996: Impact of water control projects on fisheries resources in Bangladesh, *Environmental Management,* **20(4),** pp. 523–539.

Mirza, M.M.Q., R.A. Warrick, N.J. Ericksen, and G.J. Kenny, 2001: Are floods getting worse in the Ganges, Brahmaputra and Meghna basins? *Environmental Hazards,* **3,** pp. 37–48.

Mirza, M.M.Q., 2003: The three recent extreme floods in Bangladesh: A hydro meteorological analysis. In: *Flood Problem and Management in South Asia,* M.M.Q. Mirza, A. Dixit, and A. Nishat (eds), Kluwer Academic Publishers, The Netherlands, pp. 34–65.

Mirza, M.M.Q., 2002: Global warming and changes in the probability of occurrence of floods in Bangladesh and implications, *Global Environmental Change-Human and Policy Dimensions,* **12(2),** pp. 127–138.

Moberg, F. and P. Rönnbäck, 2003: Ecosystem services of the tropical seascape: Interactions, substitutions and restoration, *Ocean and Coastal Management,* **46(1–2),** pp. 27–46.

Munasinghe, M., 1993: Environmental economics and sustainable development, World Bank environmental report 3, World Bank, Washington, DC.

Munich Re, 2003: Topics geo: Annual review of catastrophes, Munich Re, Germany.

National Hurricane Center, 2004a: Available at http://www.nhc.noaa.gov.

National Hurricane Center, 2004b: Forecast errors [online]. Available at http://www.nhc.noaa.gov/HAW2/english/forecast/errors.shtml.

National Hurricane Center, 2004c: SLOSH model [online]. Available at http://www.nhc.noaa.gov/HAW2/english/surge/slosh.shtml.

National Research Council, 1995: *Wetlands: Characteristics and Boundaries,* National Academy Press, Washington, DC.

Nicholls, N., 2001: Atmospheric and climate hazards: Improved monitoring and prediction for disaster mitigation, *Natural Hazards,* **23,** pp. 137–155.

Nicholls, R.J, F.M.J. Hoozemans, and M. Marchand, 1999: Increasing flood risk and wetland losses due to global sea-level rise: Regional and global analyses, *Global Environmental Change,* **9(1),** pp. S69–87.

NOAA (National Oceanic and Atmospheric Administration), 1994: *Natural Disaster Survey Report: The Great Flood of 1993,* NOAA, Rockville, MD.

Paul, B.M., 1984: Perception of and agricultural adjustments to floods in Jamuna floodplain, Bangladesh, *Human Ecology,* **12(1),** pp. 3–19.

Pearce, D.W. and R.K. Turner, 1990: *Economics of Natural Resources and Environment,* Harvester Wheatsheaf, Brighton, UK.

Pearce, D.W. and J.J. Warford, 1993: *World without End,* Oxford University Press, New York, NY.

Pielke, R.A. Jr., 1999: Nine fallacies of floods, *Climatic Change,* **42,** pp. 413–438.

Pielke, R.A., Rubiera, J., Landsea, C., Fernández, L.M. and Klein, R., 2003: Hurricane Vulnerability in Latin America and the Caribbean: Normalized Damage and Loss Potentials, *Natural Hazard Review,* **4(3),** pp. 101–114.

Primavera, J.H., 2000: Philippines mangroves: Status, threats and sustainable development, Proceedings for the UNU (United Nations University) international workshop Asia-Pacific cooperation on research for conservation of mangroves, 26–30 March, Okinawa, Japan.

Qian, W.H. and Y.F. Zhu, 2001: Climate change in China from 1880 to 1998 and its impact on the environmental condition, *Climatic Change,* **50(4),** pp. 419–444.

RAJUK (Rajdhani Unnayan Kartripakkha), 1995: *Dhaka Metropolitan Development Plan (1995–2015),* Vol. I and Vol. II, RAJUK, Dhaka, Bangladesh.

Ramsar, 2004: Wetland: Values and functions fact sheet [online]. Available at http://www.umimra.org/action.alerts/action_alerts5.htm.

Reinhart, K.G., A.R. Eschner, and G.R. Trimble, 1963: *Effect on streamflow of four forest practices in the mountains of West Virginia,* U.S. Forest Service research paper NE-1, Washington, DC.

Reynard, N.S., C. Prudhomme, and S. Crooks, 1998: The potential effects of climate change on the flood characteristics of a large catchment in the UK. In: *Proceedings of the Second International Conference on Climate and Water,* Espoo, Finland, 17–20 August 1998, Vol. 1, pp. 320–332.

Rustam, R., O.A. Karim, M.H. Ajward, and O. Jaafar, 2000: Impact of urbanisation on flood frequency in Klang river basin, ICAST 2000, Faculty of Engineering, Universiti Kebangsaan, Malaysia, pp. 1509–1520.

Sandstrom, K., 1995: The recent Lake Babati floods in semiarid Tanzania: A response to changes in land cover, *Geografiska Annaler Series A-Physical Geography,* **77A (1–2),** pp. 35–44.

Schmidt, K., 2001: Restoring the vitality of rich wetlands, *Science,* **294** (16 November), p. 1447.

Schönwiese, C.D. and J. Rapp, 1997: *Climate Trend Atlas of Europe Based on Observations 1891–1990,* Kluwer Academic Publishers, Dordrecht, The Netherlands.

Schreider, S.Y., A.J. Jakeman, A.B. Pittock, and P.H. Whetton, 1996: Estimation of possible climate change impacts on water availability, extreme flow events and soil moisture in the Goulburn and Ovens basins, Victoria, *Climatic Change,* **34,** pp. 513–546.

Shaxson, F. and R.Barber, 2003: *Optimizing Soil Moisture for Plant Production: The Significance of Soil Porosity,* FAO Soils Bulletin, Food and Agriculture Organization of the United Nations, Rome, Italy.

Shiva, B., 2002: *Water wars-Privatization, Pollution and Profit,* Between the Lines, Toronto, ON, Canada.

Short, F.T. and S. Wyllie Echeverrira, 1996: Natural and human-induced disturbance of seagrasses, *Environmental Conservation,* **23(1),** pp. 17–27.

Shrestha, A.B., 2001: Tsho Rolpa glacier lake: Is it linked to climate change? In: *Global Change and Himalayan Mountains,* K.L. Shrestha (ed.), Asia-Pacific Network for Global Change Research and Institute for Development and Innovation, Kathmandu, India, pp. 85–95.

Simons, H.E.J., C. Bakker, M.H.I. Schropp, L.H. Jans, F.R. Kok, et al., 2001: Man-made secondary channels along the Rhine River (the Netherlands): Results of post-project monitoring, *Regulated Rivers: Research and Management,* **17,** pp. 473–91.

Simonovic, S.P. and R.W., Carson, 2003: Flooding in the Red River Basin: Lessons from post flood activities, *Natural Hazards,* **28(2–3),** pp. 345–65.

Singh, V.P., 1987: *Hydrologic Systems: Watershed Modelling, Vol.II,* Prentice Hall, NJ.

Stanley, D.J. and A.G. Warne, 1998: Nile delta in its destruction phase, *Journal of Coastal Research,* **14(3),** pp. 794–825.

Sternberg, H.O., 1987: Aggravation of floods in the Amazon river as a consequence of deforestation, *Geografiska Annaler Series A-Hhysical Geography,* **69(1),** pp. 201–219.

Takeuchi, K., M. Hamlin, Z.W. Kundzewicz, D. Rosbjerg, and S.P. Simonovic, 1998: *Sustainable Reservoir Development and Management, IAHS Publication No. 251,* IAHS Press, Wallingford, UK.

Thieler, E. R., and Young, R. S., 1991: Quantitative evaluation of coastal geomorphological changes in South Carolina after Hurricane Hugo, *Journal of Coastal Research* (Special Issue) **8,** 187–200.

Tockner, K. and J.A. Stanford, 2002: Riverine flood plains: Present state and future trends, *Environmental Conservation,* **29(3),** pp. 308–330.

Turner, R.K., J.C.J. M. van den Bergh, T. Soderqvist, A. Barendregt, J. van der Straaten, et al., 2000: Ecological-economic analysis of wetlands: Scientific integration for management and policy, *Ecological Economics,* **35(1),** pp. 7–23.

Tynkkynen, O., 2000: *Orissa Cyclone: A Natural Phenomenon or a Sign of Things to Come,* Friends of the Earth, Finland.

UKEA (United Kingdom Environment Agency), 1998: *An Action Plan for Flood Defence,* UKEA, London, UK.

UMIMRA (Upper Mississippi, Illinois and Missouri River Association), 2004: UMIMRA position on controlled flooding [online]. Available at http://www.umimra.org/action.alerts/action_alerts5.htm.

UNECE (United Nations and Economic Commission for Europe), 2000: *Guidelines on Sustainable Flood Prevention, Meeting of the parties to the convention on the Protection and Land Use of Transboundary Watercourses and International Lakes,* 23–25 March 2000, Geneva, Switzerland.

UNESCO (United Nations Educational, Scientific and Cultural Organization), 1974: *Hydrological effects of urbanization: Studies and Reports in Hydrology 18,* Paris, France.

United Kingdom Meteorological Office (UK met-office), 1999: Indian cyclone fact sheet [online]. Available at http://www.met-office.gov.uk/sec2/sec2cyclone/tcbulletins/05b.html.

USGS (United States Geological Survey), 2003: African dust carries microbes across the ocean: Are they affecting human and ecosystem health, USGS penfile report 03—28, US Department of Interior, Washington, DC.

Van Steen, P.J.M. and P.H. Pellenbarg, 2004: Water management in the Netherlands: Introduction to the 2004 maps, *Tijdchrift voor Economische en Sociale Geografie,* **95(1),** pp. 127–129.

WAPCOS, 1993: *Morphological Studies of River Brahmaputra,* WAPCOS, New Delhi, India.

WHO (World Health Organization), 1999: *Controlling Diseases after Mitch Will Cost Money,* WHO news release, Geneva, Switzerland.

WMO and GWP (World Meteorological Organization and Global Water Partnership), 2003: *Integrated Flood Management: Concept Paper,* WMO and GWP, Geneva, Switzerland.

World Bank, 1990: *Flood Control in Bangladesh: A Plan for Action,* World Bank, Washington, DC.

World Commission on Dams, 2002: *Dams and Development: A New Framework for Decision-Making: The Report of the World Commission on Dams,* Earth-Scan, London, UK.

World Commission on Environment and Development, 1987: *Our Common Future,* Oxford University Press, Oxford, UK.

Chapter 12

Ecosystems and Vector-borne Disease Control

Coordinating Lead Authors: Diarmid Campbell-Lendrum, David Molyneux

Lead Authors: Felix Amerasinghe, Clive Davies, Elaine Fletcher, Christopher Schofield, Jean-Marc Hougard, Karen Polson, Steven Sinkins

Review Editors: Paul Epstein, Andrew Githeko, Jorge Rabinovich, Phil Weinstein

---

BOXES

FIGURES

TABLES

Main Messages

Actions to reduce vector-borne diseases can result in major health gains and relieve an important constraint on development in poor regions. Vector-borne diseases cause approximately 1.4 million deaths per year, mainly from malaria in Africa. These infections are both an effect of, and contribute to, poverty.

Ecosystems provide both the "disservice" of maintaining transmission cycles with cross-infection of humans and the "service" of regulating those cycles and controlling spillover into human populations. The balance between these services and disservices is influenced by the availability of suitable habitat for vectors and of reservoir hosts of infection. Transmission cycles are generally kept in a degree of equilibrium by density-dependent processes such as acquired immunity to infectious disease and by limits on the carrying capacity of the habitat to support insect vectors and reservoir hosts.

Human activities that alter natural ecosystems also affect the transmission cycles of vector-borne infectious diseases. Human settlement and deforestation patterns; the development of dam, drainage, and agricultural irrigation schemes; and climate change are all drivers influencing patterns of disease distribution and incidence. Policies that influence population size, migration, patterns of energy production/consumption, food production, and overall demand for natural resources will have significant effects on transmission of infectious agents and associated consequences in terms of disease. These changes may lead to the destabilization of natural equilibria, improved conditions for disease transmission, and disease outbreaks/elevated risks to humans, or alternatively, to the disruption of the disease transmission cycles and decreased incidence.

Present institutional structures tend to promote a narrow, sectoral approach to intervention for individual diseases. Inter-sectoral and interdisciplinary approaches can help control vector-borne diseases while maintaining ecosystem equilibrium (medium confidence). There are numerous examples of health sector institutions working together to mobilize funds and deploy appropriate interventions to significantly reduce transmission of specific infectious diseases. These include international programs that have dramatically reduced the burden of Chagas disease in South America and onchocerciasis in West Africa and created new initiatives on filariasis, schistosomiasis, African trypanosomiasis, and guinea worm. These programs often incorporate monitoring and successful management of environmental impacts. In most countries, however, the health and environment sectors are clearly divided, with little coordination of approaches to epidemiological and environmental monitoring, vector control, and associated energy, agricultural, housing, and forestry policies to further human well-being.

Actions taken to reduce the transmission of infectious diseases often have effects on other ecosystems services. Integrated vector management strategies permit a coordinated response to both health and the environment. IVM strategies use targeted interventions to remove or control vector breeding sites, disrupt vector lifecycles, and minimize vector-human contact, while minimizing effects on other ecosystem services. IVM is widely viewed as a useful approach and is increasingly promoted by international organizations (for example, the World Health Organization and the United Nations Environment Programme) and by national governments.

Environmental management and biological and chemical interventions can be highly cost-effective and entail very low environmental impacts (medium confidence). Potential interventions include use of fish and bacterial larvicides such as *Bacillus thuringiensis israelensis* (Bti) and chemical application methods that cause minimal disruption to broader ecosystems, such as insecticide-treated bed nets and indoor residual spraying. In the case of malaria, better design, management, and regulation of dams and irrigation schemes and water drainage systems can potentially reduce breeding sites, particularly in peri-urban areas and areas of less intensive transmission. In several settings, vector-human contact may potentially be reduced by location of human settlements away from major breeding sites and through the strategic management of diversionary hosts such as cattle (zooprophylaxis).

IVM will be most effective when integrated into development approaches that also improve socioeconomic status (*high confidence*). There is strong evidence that poverty and malnutrition increases vulnerability to the effects of vector-borne disease. For example, improved socioeconomic status facilitates the purchase of bed nets and other forms of personal or household protection for malaria. Better housing conditions are associated with reduced transmission of some vector-borne diseases. Disease control measures should therefore be part of integrated development strategies that improve all aspects of well-being.

Social and behavioral responses can help control vector-borne disease while also improving other ecosystem services (*medium confidence*). Public health education forms an increasingly important component of management programs and initiatives, raising awareness about individual and communal actions that may control vectors, their breeding sites, prevent disease transmission, and provide access to treatment. Aside from directly impacting disease control, health education gives individuals greater control over their lives and therefore promotes cultural services. However, it is difficult to bring about long-term behavioral changes in populations, that is, community programs of systematic and constant management of key breeding sites/containers, unless behavioral changes are reinforced by social or individual incentives/benefits.

New "cutting-edge" interventions, such as transgenic techniques to reduce or eliminate the capacity of some vector species to transmit infectious diseases, could be available within the next 5 to 10 years (*medium confidence*). However, consensus is lacking in the scientific community on the technical feasibility and public acceptability of such an approach. Transgenic techniques include the production of genetic constructs that block the expression of the pathogens (the infectious agents that can cause clinical disease) and their incorporation into gene-drive mechanisms such as transposable elements or *Wolbachia endosymbionts* that could spread them throughout mosquito populations. Because genetic replacement targets only individual mosquito species and aims to decrease vector competence rather than reduce population size, it has the potential to be an intervention with minimal environmental impact. There is no *a priori* reason to expect either that the genetic construct will affect other important characteristics of the target mosquito population (such as the ability to transmit other diseases) or that it will cross into other species. However, significant technical and logistical challenges must be overcome in order to establish an operational program. In many countries, there is also significant public opposition to transgenic products such as foods. Unless perceptions change, there is likely to be similar resistance to the release of transgenic insects.

Environmental modification not directly aimed at vector control often affects the ecosystems regulating vector-borne disease transmission (*high confidence*). The specific effects will depend on local transmission ecology. Interventions such as deforestation may cause decreases in disease incidence in one location but increases elsewhere or even increases in a different form of the same disease within the same location. Similarly, the effects of any ecosystem change will vary over time. For example, irrigation schemes in Sri Lanka initially caused an increase in malaria incidence but this later decreased due to replacement of the original *Anopheles* species with a less efficient vector, along with improved socioeconomic conditions and medical care. Assessments of local environmental, epidemiological, and socioeconomic conditions are necessary before making such modifications. This may entail studies at the cutting edge of science in determining vector species genetics and transmission potential.

The adoption of a longer-term and more holistic view of the interactions between ecosystems and infectious diseases would help to ensure sustainable benefits to human well-being (*high confidence*). The current division of responsibilities among institutions means that assessment and programmatic responses are within narrow disciplinary or sectoral frameworks. The health sector usually makes reference to scientific/logistical criteria (for example, availability of programmatic resources, immediate effectiveness in reducing disease, cost-effectiveness), and social-cultural considerations (for example, public and political feasibility). Such approaches are not ideal for a broader inter-sectoral policy that considers the effects of a particular health strategy on other ecosystem services or, conversely, the impacts of environmental and development strategies on disease transmission. An ecosystem assessment or ecohealth approach examines strategies from a trans-disciplinary perspective, considering direct and immediate effects on disease, as well as longer-term or indirect effects that may occur via alterations to ecosystems.

12.1 Introduction

12.1.1 Current Status of Vector-borne Disease

Vector-borne infections are major killers, particularly of children in developing countries. Over the past decade, more comprehensive and transparent methods of measuring health have improved understanding of the importance of these diseases. The World Health Organization reports annually on the numbers of deaths and DALYs (disability adjusted life years, a composite measure of health status combining premature death and sickness during life), by disease category in different regions of the world (for example, WHO 2004a). Despite technological advances and increasing affluence in many regions, vector-borne infectious diseases remain amongst the most important causes of global ill-health.

This burden is concentrated in the poorest regions of the world (see MA *Current State and Trends,* Table 14.1, Figure 14.1). For example, malaria alone is responsible for approximately 11% of the total disease burden in Africa, while all vector-borne diseases combined are responsible for less than 0.1% in Europe. Vector-borne disease is not only an outcome but a cause of poverty. Countries with intensive malaria have income levels averaging only 33% of those without malaria, even after accounting for the effects of tropical location, geographic isolation, and colonial history (Gallup and Sachs 2001).

12.1.2 Future Projections

The global burden of disease methodology has also been used to project the likely changes in disease impacts between 1990 and 2020 (Murray and Lopez 1996; Lopez and Murray 1998). These projections suggest that continuing economic development will be associated with an epidemiological transition period. Diet and lifestyle changes associated with growing urbanization as well as increased substance abuse, environmental degradation, population growth, and levels of regional and local conflict are expected to lead to a surge in non-communicable diseases, including those associated with cerebrovascular events, depressive illness, conflict-related conditions, road traffic accidents, and cancers.

Infectious diseases are projected to decrease in relative importance, with only malaria expected to represent a very significant proportion of the global burden of disease, and even this disease falling from a ranking of 11th to a predicted 26th. However, this may be an overly optimistic projection of progress against infections. For instance, projections made in 1996 for the burden of disease from malaria and dengue in 2000 were subsequently found to have underestimated the true burden by 32% and 11%, respectively (WHO 2001). Drug resistance, population movement, and the effects of the HIV/AIDS pandemic on immune status threaten the global effort to contain infectious diseases.

Whatever the precise nature of future trends, firm response measures to control infectious disease will be required for the foreseeable future and these measures will inevitably affect or interact with other ecosystem services. In addition, growing human populations and increased demand for ecosystem services/natural resources means that there will be continuing, and possibly increasing, human interactions with the natural processes that influence infectious disease transmission.

12.1.3 Scope and Purpose of the Assessment

All diseases, from infections to cancer and cardiovascular disease, are influenced to some degree by the environment. However, this chapter focuses on vector-borne infectious diseases for the following reasons.

- These diseases are especially ecologically sensitive, since environmental conditions, such as temperature, affect both the infectious pathogens (for example, malaria parasites or dengue viruses) that, depending on the interaction with human hosts, have the potential to cause clinical disease and the insects and other intermediate hosts that transmit them.
- Many such infections are directly linked to natural ecosystem types (such as forests and wetlands) considered elsewhere in this assessment.
- Such infections are strongly linked to poverty, and therefore other aspects of human well-being.

In addition, as noted above, vector-borne diseases are among those infectious diseases with the highest disease burden today, and may be expected to represent the highest proportionate disease burden in the future. Responses related to other ecosystem-health interactions are covered in Chapter 16 of this report.

The assessment examines potential policy responses that may reduce the burden of infectious vector-borne diseases, from an ecosystem perspective. In this perspective, ecosystem impacts from the policy responses also are considered, and the preservation of ecosystem services is viewed as a priority. As the chapter focuses on policy responses, there is only a brief introduction to the ecological concepts underpinning the relationships between environmental states and disease transmission. A more detailed description of these interactions, including for example the effects of "systemic" ecosystem disruption on complex transmission cycles involving multiple hosts, is given in MA *Current State and Trends,* Chapter 14.

It should also be noted that vector-borne disease infections are influenced by a wide range of factors not directly related to ecosystem services. These include the provision of basic public health services to prevent, detect, and treat disease, as well as "good governance" ensuring that these services are responsive to citizens needs and that resources for disease control are not lost to corruption and inefficiency. While recognizing the importance of these influences, the chapter focuses on more direct links between disease control and the environment. It includes:

- an overview of the basic ecosystem mechanisms and environmental drivers that influence and regulate vector-borne disease transmission;

- an overview of global trends (for example, trade, globalization, economic development, and urbanization) that act as indirect drivers affecting the condition of ecosystems and vector-borne disease;
- a review and assessment of integrated vector management strategies, which provide a comprehensive approach for integrating effective disease control with consideration of other ecosystem services.
- a description of social and behavioral responses to vector-borne disease management, particularly in the context of sustaining integrated policies; and
- a discussion of inter-sectoral cooperation in promoting ecosystem approaches to vector-borne disease management.

12.2 Environmental Drivers, Ecosystem States, and their Effects on Vector-borne Disease

12.2.1 The Relationship between Ecological Conditions and Vector-borne Disease

An ecosystem perspective on the risks to humans from vector-borne disease and vector management requires an appreciation of the role played by broad environmental trends and by local ecosystems in sustaining vector habitats and facilitating disease transmission.

The transmission of vector-borne diseases is governed by the same principles of population dynamics as other ecological systems. Probably the most important governing concepts are the reproductive rate (often termed r) of vectors and parasites and the carrying capacity (K) of the local habitat for each of these entities (Thomson et al. 2000). Both are influenced by broad environmental conditions. For example, in the case of malaria in an African village, seasonal temperatures affect the reproductive rate of *Anopheles* mosquitoes and of the *Plasmodium* parasites within them. Precipitation influences the availability of aquatic breeding sites.

Together, such factors influence the maximum number of adult mosquitoes that can be produced and sustained in the local environment in a given time. While these large-scale climatic factors are important, their influence in any particular setting also depends upon local characteristics. For instance, both temperature and breeding site availability are also a function of local topography and vegetation. Finally, humans are an integral part of this system. Agricultural practices will influence land use and the availability of animals that may provide blood meals for mosquitoes (along with or instead of humans). Targeted vector control interventions may reduce vector abundance and longevity vector habitats and/or vector biting rates on humans and thus prevent disease transmission, even in otherwise optimal environments. The same environmental conditions that are affected by local habitat factors also determine the *vector-carrying capacity* of the ecosystem, the point at which vector abundance is limited by density-dependent processes (for example, predation and competition for food).

From a public health perspective, ecosystems provide both "disservices" by maintaining pathogen transmission cycles, including cross-infection of humans, as well as the "service" of regulating those cycles in some degree of natural equilibrium. This state of equilibrium prevents even more explosive human disease outbreaks. (See MA *Current State and Trends,* Chapter 14.) The balance of ecosystem services and disservices depends on the nature of human interactions with the various ecosystems and the extent to which these enhance services and eliminate or manage disservices.

12.2.2 Global Trends as Indirect Drivers

When considering "responses" to manage infectious diseases, it is important to recognize that the human actions that have the most profound effects on disease transmission are often not directly aimed at vector-borne disease control. (See Figure 12.1.)

As countries develop, the transition from high-birth, high-death rate societies to low-birth, low-death rate societies, without concurrent planning programs, has led to rapid population expansion. Coupled with economic growth, this has led to new demands for energy and transport and for food and natural resource products. All of these function as underlying driving forces of environmental change (McMichael 2001). As a result of such drivers, there is consequent expansion of agricultural production, water management and irrigation schemes, hydroelectric dam construction, deforestation, urbanization, and generally greater exploitation of natural resources. All these create new pressures on ecosystems that, in turn, have profound impacts on vector habitats, the carrying capacity of the environment for vector populations, and infectious disease transmission (Molyneux 1997, 2003).

The environmental pressures generated by human activities may act through direct and/or complex mechanistic pathways, and their impacts are therefore situation-specific. The results of such processes may be trade-offs, where interventions to increase an ecosystem service are offset by increasingly frequent or severe disease outbreaks in the human population. This is illustrated by the example of dengue, transmitted by mosquitoes of the genus *Aedes* (mainly *Aedes aegypti* and to a lesser extent *Aedes albopictus*), primarily in urban areas. Rapid urbanization, the accumulation of water-retaining waste products such as plastic containers and tin cans, and increased domestic water storage, for example in large open drums (Gubler 1997), have increased the availability of *Aedes* breeding sites. Increased international travel has led to the mixing of different strains of the disease, causing more severe clinical symptoms in individuals exposed to more than one strain (Halstead 1988). Overall, the burden of dengue has increased rapidly in recent years and is a major problem even in some of the most affluent and developed settings in tropical regions, for example, Singapore (Wilder-Smith et al. 2004).

Table 12.1 describes how exploitation of some ecosystem services (mining and irrigation) has led to detrimental impacts on malaria transmission, specifically the ratio of infections with the relatively more pathogenic *Plasmodium falciparum* to the more benign *Pl. vivax*.

The long-term positive or negative effects of ecosystem changes may emerge some time later, often complicating initial assessment of the likely ecosystem impacts of development. Conversion of forest to agricultural land in Sri Lanka initially led to outbreaks of malaria, but over time these have declined considerably. (See MA *Current State and Trends,* Chapter 14.) In contrast, deforestation for wood extraction in Latin America was initially considered to disrupt the transmission cycle of cutaneous leishmaniasis, with associated reductions in transmission of the disease to humans (Sampaio 1951). However, over several decades there was a resurgence in disease transmission, particularly among women and children, as sand-fly vectors became abundant in domestic and peri-domestic habitats (Walsh et al. 1993; Campbell-Lendrum et al. 2001).

Some environmental changes triggered by human activities impact on vector habitats, hosts, or disease transmission, in ways that are even more diffuse and difficult to predict accurately. This

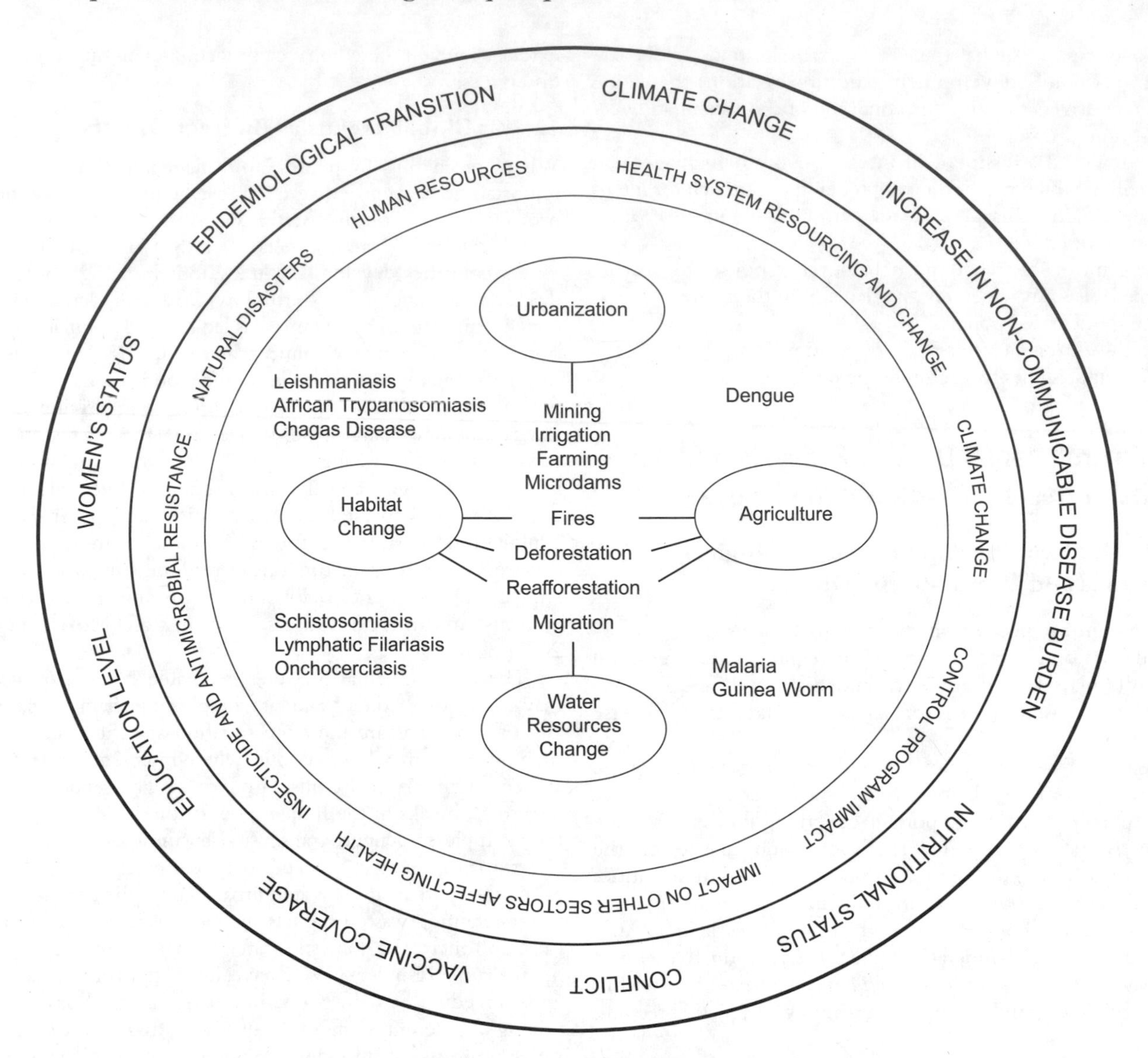

Figure 12.1. Indirect and Direct Influences on Vector-borne Disease Transmission (Molyneux 2003)

Table 12.1. Examples of Anthropogenic Change Leading to Increases in *Plasmodium falciparum: P. vivax* Ratios

Location	Anthropogenic Changes	Reference
Tajikistan	health systems disruption, conflict migration, chloroquine resistance in *Plasmodium falciparum*	Pitt et al. 1998
Afghanistan/ Pakistan	chloroquine resistance in *P. falciparum*	Rowland and Nosten 2001
Thar Desert, Rajasthan, India	irrigation on large scale; establishment of *Anopheles culcifacies*-efficient *P. falciparum* vector and dominance over *An. stephensi*, a poor vector	Tyagi and Chaudhary 1997
Amazonia, Brazil	new breeding sites for efficient vectors *An. darlingi* through mining, deforestation, road building	Marques 1987

includes, for example, the effects of disruption of ecological community structure on transmission dynamics of pathogens such as *Borrelia burgdorferi,* which is transmitted through multiple animal reservoir hosts and can cause Lyme disease in humans. As described in MA *Current State and Trends,* Chapter 14, this can in some cases lead to breakdown in ecological community structure and disruption of natural regulation, with increasing risk to humans. Similarly, it is generally accepted that the use of fossil fuels for energy production has caused changes in global climate (IPCC 2001). This is likely to have influenced the population biology of both disease vectors and pathogens throughout the world, with associated effects on clinical disease in humans (Patz et al. 2003), although in any one site, these influences may be relatively small compared to those of other ecosystem changes (Kovats et al. 2001; Reiter 2001; Hay et al. 2002).

Analytical techniques to estimate the human health risks of such widespread and pervasive ecosystem changes are improving (for example, Zavaleta, 2004) but they can only provide very approximate guidance. They also tend to exclude effects that have only a low probability of occurrence in any one location but potentially devastating consequences. This includes, for example, the possibility that human disruption of natural ecosystems may lead to emergence or re-emergence of important vector-borne diseases that can spread rapidly through the human population, as has occurred for other infectious diseases such as SARS (Severe Acute Respiratory Syndrome) and HIV/AIDS, both probably introduced to human populations through consumption of wild animals. This is a particular risk in areas that combine a high rate of development and ecosystem change with a high population density, such as China. Such low-probability large-consequence events argue in favor of responses that reflect the precautionary principle, as described in principle 15 of the 1992 Rio Declaration on Environment and Development (at the United Nations Conference on Environment and Development Rio de Janeiro, 1992) as—"Where there are threats of serious or irreversible damage, lack of full scientific uncertainty shall not be used as a reason for postponing cost-effective measures to prevent environmental degradation." In practical terms, this could include strengthening environmental protection and surveillance of ecosystems as a precautionary measure to reduce the risks of disease outbreaks.

While there are often trade-offs between disease control and protection of other ecosystem services, there are also potentially synergies where ecosystem changes for other purposes lead to disruption of the vector-borne disease transmission cycle and decreased risks to humans (Patz et al. 2000). One example is the draining of wetlands to expand agricultural production in Europe in the nineteenth and twentieth centuries. This removed large areas of potential breeding sites for *Anopheles* mosquitoes and therefore played an important role in reducing malaria transmission (Jetten and Takken 1994; Reiter 2000; Kuhn et al. 2003). More generally, increased affluence and better housing and education can reduce transmission of a wide range of diseases even when they are accompanied by ecosystem disruption. Examples include increasing wealth from irrigated agriculture leading to greater ability to purchase antimalarials in Africa (Ijumba et al. 2002) and better housing reducing exposure to malaria infections in Sri Lanka (for example, Gunawardena et al. 1998), lower rates of dengue on the more affluent U.S. side of the Mexico-U.S. border (Reiter et al. 2003), and reduced risk of Chagas disease following house improvement in Latin America (for example, De Andrade et al. 1995).

12.2.3 The Importance of Development Policies

Global, regional, and national development programs, designed to increase material well-being and alleviate poverty, have indirect and diffuse but nonetheless profound influences on infectious diseases. For instance, policies to increase trade can lead to increased deforestation and irrigation, with consequent impacts on vector-borne diseases.

In cases where trade and development policies are effective in reducing poverty, they may also protect against vector-borne disease, as higher socioeconomic status in general may increase individual and community capacity to control infectious diseases. Conversely, international policies that increase the volume of trade, energy use, urbanization, and transport, while failing to alleviate poverty or improve health may exacerbate the risk of vector-borne disease transmission.

Over the past twenty years, there have been greater efforts to integrate international development agendas with concerns over the environment, ecosystems, and health. There is increasing recognition that the link between poverty and vector-borne disease also means that the poorest populations would benefit proportionately more if development had a pro-poor focus that also addresses health and environment in an integrated manner (Gwatkin et al. 1999).

The Rio de Janeiro Earth Summit in 1992 and the Johannesburg World Summit on Sustainable Development in 2002 endorsed Agenda 21 as a comprehensive plan of action for sustainable development while protecting natural resources. Under this guiding principle, a series of internationally negotiated environmental conventions address specific themes such as the protection of biological diversity, regulation of hazardous toxic substances, and combating desertification and anthropogenic climate change. The Millennium Development goals set targets for measurable improvements in development and human and ecosystem health. Several of these goals have relatively direct links to infectious disease transmission, particularly the target of (1) halting and beginning to reverse the spread of HIV/AIDS and the incidence of malaria and other major diseases and (2) halving the proportion of people without sustainable access to safe drinking water by 2015. Other goals will have indirect effects on infectious disease. These include measures to ensure environmental sustainability and the greater use of indicators tracking forested land area and emissions of gases that contribute to climate change and ozone depletion. Although it remains to be seen whether these are feasible, the MDGs at least define specific targets that, if achieved, would represent clear improvements in human health and well-being.

12.3 Specific Responses to Vector-Borne Disease in an Ecosystems Framework

This section considers policies and strategies for vector-borne disease management. In particular, it assesses the emerging relevance of integrated vector management, which provides a conceptual approach, along with environmental management and other tools for controlling disease, within an ecosystems framework. A parallel but interrelated track relates to emerging scientific knowledge as well as behavioral and social changes that may contribute to better disease management. Case studies are used to illustrate important features of the different types of responses.

12.3.1 Integrated Vector Management

There is increasing recognition that in the developing countries that suffer the greatest burden of vector-borne disease the trend toward decentralization of decision-making, and the limitations of disease-specific control programs, including the scarcity of technical skills and resources, necessitates a flexible approach to selecting control tools.

In response, there have been rapid developments in defining integrated vector management strategies. IVM strategies have parallels with integrated pest management systems used in agriculture, where the adverse environmental and health effects of pesticides, and the development of resistance, stimulated the flexible use of all methods that have an impact on the pest problem. Such integrated approaches help to preserve ecosystem integrity and encourage the propagation of natural enemies of pest species such as pathogens and predators. An important selling point of this approach is that economic analyses have shown IPM to be more cost-effective than heavy reliance on insecticides, even ignoring the added benefits of reduced environmental contamination.

The principles of IVM can be summarized as seeking to "Improve the efficacy, cost-effectiveness, ecological soundness and sustainability of disease vector control. IVM encourages a multi-disease control approach, integration with other disease control measures and the considered and systematic application of a range of interventions, often in combination and synergistically" (WHO 2004b).

IVM works on the premise that effective control requires the collaboration of various public and private agencies and community participation rather than exclusive action by the health sector. It entails the use of a range of interventions of proven efficacy, separately or in combination. This serves to maximize cost-effectiveness and extend the useful life of insecticides and drugs by reducing the selection pressure for resistance development.

IVM harnesses precise scientific knowledge of the vector ecosystem and its interrelationship to human ecosystems to address the following issues in more environmentally sustainable ways. IVM typically uses four approaches, as appropriate, and in an interrelated manner:

- environmental management, including modification or manipulation of the environment;
- biological control methods, including bacteria and larvivorous fish;
- chemical control methods, including targeted indoor residual spraying, space spraying and larviciding (Walker 2002; WHO 2003; see also Box 12.1); as well as
- social and behavioral measures to decrease suitability for transmission.

Table 12.2 summarizes the main types of control interventions used within an IVM strategy.

Each of these interventions has different types of interactions with other ecosystem services. Table 12.3 maps the main linkages between interventions to control vector-borne disease (including vector control, prevention, and curative measures) and other ecosystem services in the MA framework. In each case, the direction of the arrow gives a rough indication of whether the intervention generally has a greater capacity to increase or decrease the provision of the various services. The following sections examine the more direct interactions in greater detail.

12.3.2 Environmental Management/Modification to Reduce Vector and Reservoir Host Abundance

The practice of using environmental management to reduce the capacity of local habitats to maintain populations of disease vectors predates the development of insecticides. Considerable success was achieved by draining swamps to remove larval breeding sites of *Anopheles* mosquitoes (Pontine Marshes in Italy) and by the use of oil to prevent larval mosquito respiration. *Glossina* (tsetse flies) have been controlled by the selective destruction of savanna and riverine forest habitats together with the destruction of host animals as well as by trapping using sticky materials on the backs of men or by attraction to visual baits since at least the beginning of the previous century.

There has been resurgent interest in environmental management techniques in recent years, stimulated partly by concerns over the sustained effectiveness and environmental consequences of insecticide use. For example, a major motivation for controlling dengue by removing, covering, or treating larval sites in and around houses has been the appreciation that outdoor application of insecticide often has poor penetration into the domestic resting sites of the vectors, has only transient effects, and is logistically demanding (Newton and Reiter 1992).

The capacity of environments to maintain vectors can be reduced by long-term physical changes, often termed environmental modification (WHO 1982; Walker 2002). This may not be the most effective approach in all epidemiological situations. For example, in many highly endemic areas for malaria, using residual insecticide spray to cause even a small increase in the mortality of adult vectors should have a disproportionately large impact on disease transmission. This is likely to outweigh the effect of the same proportional reduction in larval breeding sites.

The effectiveness of environmental management depends on how well the particular intervention is matched to the ecology of the particular disease. Large-scale modification projects tend to require significant initial investments in construction and may be effective only where the targeted area contains the overwhelming majority of breeding sites. Local modifications may also be ineffective where there are alternative breeding sites near human habitations (Mutero et al. 2004), as is the case for malaria in much of rural sub-Saharan Africa. Overall, "accidental" habitat modifica-

BOX 12.1

Indirect Policy Drivers of Vector-borne Disease: Uganda

In Uganda, cattle are typically treated with chemical pesticides, particularly synthetic pyrethroids, to prevent transmission of two types of vector-borne disease, namely, the tick-borne East Coast fever, which infects cattle, and African trypanosomiasis (sleeping sickness). The latter is transmitted by tsetse flies, with different subspecies of the *Trypanosoma* parasite causing disease in humans and cattle.

Indigenous species of East African cattle are typically more resistant to East Coast fever caused by *Thyleria parva*; they are trypanotolerant, and therefore may require fewer chemical applications. However, there has been increased development of exotic cattle breeds, which reproduce more quickly and yield more milk than native species. The acquisition and raising of livestock for milk production has become an important household-based economic activity for many poor Ugandans. Since the same synthetic pyrethroids that are used in livestock management are also used in malaria control, this has implications for human disease control efforts.

In a project sponsored by the Systemwide Initiative on Malaria in Agriculture, and funded by the Canadian-based IDRC, researchers are exploring whether the synthetic pyrethroid treatments, when strategically used on cattle, may also provide some protection against malaria infection in humans. This has been the case in some settings in South and West Asia (Hewitt and Rowland 1999; Rowland et al. 2001), and may be particularly relevant in ecosystems with zoophilic vectors, which feed upon cattle as well as humans. A second, related, question is whether the very frequent and widespread use of synthetic pyrethroids on cattle or in agriculture in general might contribute to vector resistance to the chemicals in the long term. This could potentially decrease their effectiveness in bednets, where they are also widely used. Policymakers are also concerned with how the use of chemicals on cattle for vector control may impact the broader ecosystem.

Finally, there is a national development interest in Uganda in promoting the development of a cattle industry, including native cattle breeds, which yield good quality meat and are more fly/tick resistant, thus requiring fewer pesticide/acaricide applications. Genetically mixed cattle breeds may also provide better milk production than the local species, while also providing higher levels of resistance to certain types of vector-borne disease. A joint project of the WHO/UNEP-sponsored Health and Environment Linkages Initiative (HELI) is supporting an inter-sectoral assessment of livestock industry development, chemical use, health, and environment. The project is assessing policy options for livestock industry management in light of the scientific knowledge available about ecosystems and health. The goal is to optimize the potential for creating win-win scenarios that support economic development, poverty reduction, health, and environmental protection.

Table 12.2. Components of Integrated Vector Control (based on WHO 2003, p. 5)

Type	Intervention	Targets	Products
Community education	behavioral change, application of all other interventions	all vectors	
Environmental management and sanitation	natural environment changes	mosquitoes, blackflies, snails, etc.	
	improved housing quality	vectors of Chagas disease, malaria, dengue	
	physical barriers to breeding sites	vectors of filariasis, trachoma	polystyrene beads in standing water bodies
Biological control	larvivorous fishes	mosquitoes	
	predators and competitors	snails	
	larviciding	urban mosquitoes, blackflies	microbial larvicides, organophosphates, neem extracts and other herbal insecticides
Chemical control	space spraying	urban mosquitoes	pyrethroids, organophosphates
	indoor residual spraying	vectors of malaria, lymphatic filariasis, leishmaniasis	pyrethroids, organophosphates, carbamates, DDT (malaria only)
	insecticide-treated materials	vectors of malaria, leishmaniasis, lymphatic filariasis, trypanosomiasis	pyrethroids
	household products	mosquitoes, flies, fleas	aerosols, coils, mats, repellents, natural products, etc.

tion, for example through deforestation or irrigation schemes, probably has more widespread effects on infectious disease transmission. This highlights the importance of considering vector-borne diseases within any development that causes a large-scale change to the physical environment.

There are, however, specific cases where environmental modification has been used successfully to destroy either vector or reservoir host habitat, particularly at the fringes of disease transmission. Although destructive locally, careful targeting can minimize environmental impacts. Successful examples include draining, filling, or raising brackish breeding sites of the coastal malaria vector *An. sundaicus* in Indonesia in the 1940s (Takken et al. 1991) or, more recently, the filling of peri-urban breeding sites for malaria vectors in Zambia and India (Baer et al. 1999). In each of these areas, incidence of malaria declined significantly.

Temporary changes to the environment are often termed environmental manipulation. These include flushing streams or canals, providing intermittent irrigation to agricultural fields such as rice, temporarily flooding or draining wetlands, or removing specific types of vegetation that provide larval habitats for mosquitoes. A combination of vegetation clearance, modification of river boundaries, increasing velocity of the river flow to interrupt larval development, and swamp drainage were highly successful complements to bed nets and insecticide spraying in Zambia in the early twentieth century (Utzinger et al. 2001, 2002). Intermittent irrigation has proven successful in controlling *Anopheles* in rice-growing regions in India, China, and other parts of Asia (Pal 1982; Lacey and Lacey 1990), and clearance of algae from rivers has been effective in reducing malaria transmission in Oaxaca, Mexico (IDRC 2003).

For zoonotic diseases (those with a non-human reservoir host of infection), environmental management techniques can be applied to reduce the abundance of the reservoir host as well as the vector. For example, cutaneous leishmaniasis due to *Leishmania major* in the former Soviet Union has been controlled by plowing up or flooding of colonies of the reservoir host, the great gerbil *Rhombomys opimus*. (See Box 12.2.)

Crucially, well-targeted environmental management techniques can be highly cost-effective. The analysis of malaria control in the Zambian copper belt indicated that the cost-effectiveness of sustained environmental interventions ($22–92 per healthy life saved) were comparable to those from insecticide-based methods, even excluding valuation of the reduced environmental impacts (Utzinger et al. 2001).

12.3.3 Biological Control/Natural Predators

Biological methods consist of the utilization of biological toxins and natural enemies to achieve effective vector management. Targeted biological control using larvivorous fish and copepods as well as the toxic products of bacterial agents has been successfully used to control vectors of filariasis and malaria, and notably vectors of dengue in Viet Nam (reviews by Walker 2002; Lloyd 2003). An important advantage over chemical methods is the reduction in ecosystem disturbance. Microbial larvicides can be safely added to drinking water and in environmentally sensitive areas, as they do not persist or accumulate in the environment or in body tissues and are not toxic to vertebrates (WHO 1999). Also, there is no evidence that native (as opposed to exotic) larvivorous fish pose any threat to local biodiversity or the safety of drinking water.

Biological control may be effective if breeding sites are well known and limited in number but less feasible where they are numerous. Biological control thus provides a good illustration of the importance of knowledge of local transmission ecology. Economic incentives may also be important in spurring initial interest in biological control mechanisms. In Asia, for instance, larvivorous fish have been effective where pisciculture can provide additional economic, agricultural, and nutritional benefits (Wu et al. 1991; Gupta et al. 1992; Victor et al. 1994). In China, Wu et al. (1991) found that stocking rice paddies with edible fish improved rice yield, supported significant fish production, and greatly reduced the number of malaria cases (Walker 2002). Community participation and inte-

Table 12.3. Ecosystem Services Affected by Responses to Vector-borne Diseases

Response	Disease	Ecosystem Service				
		Food	Fresh Water	Biodiversity	Cultural Services	Wood (Forestry)
Insecticidal sprays	malaria, African sleeping sickness, leishmaniasis, filariasis, Chagas, trypanosomiasis, West Nile virus, dengue	↓	↓	↓		
Larviciding	dengue, malaria, onchocerciasis	↓	↓	↓		
Insecticide-treated bednets	malaria, Leishmaniasis					
Traps/targets	African sleeping sickness, leishmaniasis					↓
Larvivorous fish	malaria, dengue	↑	↑	↓		
Bacillus sphaericus	filariasis		↓	↓		
Rodent/reservoir control	leishmaniasis			↓		
Killing hosts	dracunculiasis, schistosomiasis	↓				
Surveillance	all diseases				↑	
Chemotherapy/chem oprophylaxis	malaria, schistosomiasis, onchocerciasis, lymphatic filariasis (humans)	↓ (livestock)				
	African sleeping sickness, hydatid disease (animals)					
Vaccines	yellow fever				↑	
Personal protection	dengue, malaria, leishmaniasis				↑	
Improved housing construction	Chagas					↓
Environmental management	malaria, dengue, schistosomiasis		↑	↑	↑	
Irrigation/impoundment/ swamp drainage	schistosomiasis, malaria	↓↑	↓↑	↓		
Improvement in sanitation and hygiene practices	ectoparasitic diseases, gastrointestinal helminth parasites	↑	↑		↑	
Health education	all diseases				↑	

gration with other control methods are important, as larvivorous fish and other biological agents may need repeated restocking/ reapplication, and in some cases vegetation clearance or removal of pollution sources to maintain their habitat.

12.3.4 Chemical Control

Chemical control methods of malaria vector management can potentially be organized quickly, and can be highly cost-effective if used efficiently. The advent of insecticides in the 1940s, with the widespread use of dichlorodiphenyltrichloroethane (DDT), resulted in less emphasis on environmental and biological methods of control and the reliance, for a period of two decades, on insecticides. The WHO Malaria Eradication Campaign of the 1950s achieved eradication in several sub-tropical regions and controlled malaria transmission on much of the Indian subcontinent for several years. More recently, insecticide-based campaigns have had notable success in reducing transmission of Chagas disease in the southern cone region and elsewhere in Latin America (see Box 12.3), while the Onchocerciasis Control Program eliminated the disease from much of the program area using various insecticides in rotation. (See Box 12.4.)

Widespread application of persistent insecticides was initially undertaken with little regard for environmental consequences. Evidence has since accumulated of the wider environmental effects of insecticide spraying (for example, impacts of pyrethroid spraying against tsetse on river fauna such as Crustaceans (Molyneux et al. 1978) and poisoning of reservoirs), and greater attention is now paid to using knowledge of the ecology of vectors to develop more cost-effective and less environmentally damaging methods of application by using less toxic or persistent chemicals. Various delivery methods have thus been developed. The insecticide may be applied as a non-residual application (effective over a short time-scale, killing only insects currently exposed) or a residual (persistent) application, effective over a period of weeks or months. The latter application may kill even those insects that were in immature stages of development, and not directly exposed to the insecticide at the time of application. Synthetic pyrethroid treated nets, used in the control of malaria and other vector-borne diseases, are one example of such an approach, which has a minimal impact on broader ecosystems. Information on when and where particular vector species tend to rest and feed (indoors or outdoors) and whether they feed on humans or other animals also helps to target spraying efforts.

Certain chemical insecticides can also be safely applied to larval breeding sites. For example, temephos exhibits very low mammalian toxicity and has been used for malaria control in India

BOX 12.2

Environmental Management of Zoonotic Cutaneous Leishmaniasis in Central Asia

In the deserts of Turkmenistan, Uzbekistan, southern Kazakhstan, and southern Tajikistan in Central Asia, the distribution of zoonotic cutaneous leishmaniasis, caused by the protozoan parasite *Leishmania major*, is largely dependent on the distribution of the great gerbil *Rhombomys opimus* (Sergiev 1979; Saf'Janova 1985). This is a relatively large (200 gram) diurnal rodent that lives in family groups in vast underground burrow systems, each occupying an area of around 1,000–3,000 square meters, with around 300–400 meters of passages up to three meters in depth, with openings to the surface every one to two meters (WHO 1990). Each complex typically houses five to ten rodents during the summer months, before the young disperse; in suitable habitats, there are two to three burrow complexes per hectare. Hence, *R. opimus* has a dominant influence on the environment, with the burrows providing an enormous haven of constant temperature and humidity occupied by more than 50 species of vertebrates and hundreds of invertebrates. Among the many parasites that find hosts and are transmitted in the burrows, *L. major* is the most widely distributed, typically infecting 30–50% of *R. opimus* across its range.

The burrows provide optimal conditions for the principal sand-fly vector of *L. major, Phlebotomus papatasi*. Like all sand flies, the larvae are terrestrial and live in moist organic matter. The sand flies overwinter as diapausing fourth instar larvae, the adults emerging after the soil temperature reaches 20° Celsius at a depth of 50 centimeters. Adult sand flies rest during the day in the well-insulated burrows and feed at night. *P. papatasi* has a noted host preference for the great gerbil, being especially attracted to its fleshy ears, which when infected by *L. major* generate long-lasting non-ulcerating lesions which can be infective to sand flies for up to two years (Sergiev 1979). Parasites can therefore survive in great gerbils during the winter (that is, between transmission seasons).

Most human cases in this region are due to infections in oases located in the river valleys and foothill plains of southern Turkmenia and Uzbekistan (Sergiev 1979; Saf'Janova 1985). But humans are accidental "dead-end" hosts of *L. major*, playing no role in maintaining the transmission cycle. Hence, the highly successful Soviet Union-run control program to reduce the risk of ZCL (notably during its large programs of land reclamation in Central Asia) used to focus on rodent control. This program utilized a remarkably effective and relatively inexpensive environmental intervention that dramatically decreased the capacity of the environment to maintain the transmission cycle of *L. major*. Rodent control was achieved by three methods (Vioukov 1987; WHO 1990): complete plowing up of burrows (as in the Golodnaya Steppe in Uzbekistan); crushing of burrows with wide-track heavy caterpillar tractors (as in Turkmenia); and the use of poisoned bait (as in Karshinskaya Steppe, Uzbekistan). This latter intervention can be highly efficient. One group of three to four men can destroy all the gerbils in an area of 50,000 hectares in one season. Transmission of ZCL halted over 900 square kilometers in the Karshinskaya Steppe by poison baits after a three-year control program. For all three methods, the impact is only transitory if rodents are allowed to immigrate into the treated areas from untreated territories. Permanent environmental control is only achievable by preventing reinvasion by physical barriers, such as canals or vast agricultural fields.

Although the intervention unavoidably affects the local habitat, efforts were made to reduce wider impacts. In order to address concerns that the poison baits would affect non-target animals (birds and cattle), baits were introduced into the rodent burrows at a depth of 15–20 centimeters subsequently filling up the burrow (Saf'Janova 1985). In addition, habitats targeted for intervention are targeted by using large-scale photography to identify where burrows are likely to be and to exclude unsuitable sites—then possible sites are surveyed and burrows mapped. After treatment, follow-up surveys are conducted and treatments repeated if necessary (Vioukov 1987).

and Mauritius (Kumar et al. 1994; Gopaul 1995), in the Onchercerciasis Control Program, and is widely applied to potable water to control *Aedes aegypti*. In contrast, synthetic pyrethroids are effective but are problematic as larvicides due to their frequent high toxicity to aquatic non-target organisms (Chavasse and Yap 1997).

Insecticides therefore remain an important tool and their selective use is likely to continue within IVM. However, there are concerns over the impacts of insecticides, especially the persistent organic pollutants identified in the Stockholm Convention, on the natural environment and on exposed human populations, particularly insecticide sprayers (Chavasse and Yap 1997). There is, as well, growing evidence of insect resistance to insecticides. Such developments led to World Health Assembly resolution WHA 50.13, which called on member states to support the development and adoption of viable alternative methods of controlling vector-borne diseases and thereby reducing reliance on insecticides. IVM provides a management framework for cost-effective, rational, and environmentally sensitive use of insecticides until they can be phased out without exposing populations to increased disease risk.

12.4 Social and Behavioral Responses to Vector-borne Disease

An ecosystems perspective views the human cultural, social, and behavioral environment as integral to the sustainability of critical ecosystem services. This is particularly true in the case of vector-borne diseases, which are profoundly influenced by human patterns of settlement and behavior. Social and behavioral resources and tools are therefore increasingly recognized as relevant to the management of vector-borne disease. Improved disease control may be achieved by changing human behavior or settlement patterns in ways that reduce contact with vectors.

12.4.1 Human Settlement Patterns

Probably the most basic way in which human-vector contact may be managed is via improvements in the placement and construction of housing settlements (Rozendaal 1997). Although individual vectors are often able to disperse many kilometers, in many situations the majority of the vector population moves only a very small distance within its lifetime. Thus the abundance of insect populations can vary markedly even within a very small distance. Locating settlements on well-drained, high ground more than one to two kilometers away from major breeding sites may significantly reduce transmission of diseases such as malaria (WHO 1982). In many cases, however, housing sites are determined by other imperatives (such as the need to be close to a water source). *Aedes* vectors are particularly well adapted to breeding in urban areas, particularly in household water containers. Therefore, screening of windows, eaves, and doors to protect people from bites can help prevent the spread of the disease from infected human reservoir hosts, and improved household management of

BOX 12.3

International Initiatives to Control Trypanosomiasis

Sustainable control of vector-borne diseases is rarely achievable through small-scale intervention projects from actors outside of the local community, such as government Ministries or international agencies. This is because once an intervention project ends, or political interest wanes, the treated area remains at risk of reinfestation from untreated regions. This has been a marked problem for the control of tsetse-borne trypanosomiasis in Africa, where premature curtailment of intervention projects—even where highly successful—has typically seen steady recrudescence of transmission as tsetse returned to the project areas.

A similar problem is faced in the control of American trypanosomiasis (Chagas disease), which is by far the most socioeconomically significant parasitic disease of Latin America. Although effective methods to halt transmission by eliminating domestic vector populations have been known since the 1950s, the ease with which vectors can be passively transported from one region to another made small-scale interventions unsustainable. Even when Brazil launched a highly successful national campaign against Chagas disease transmission in 1983, it was clearly recognized that similar interventions would need to be carried out at least along the frontier regions of neighboring Argentina, Bolivia, Paraguay, and Uruguay. But even with frontier intervention agreements, the Brazilian national campaign encountered an additional problem affecting continuity. In 1986, due to changing political priorities, all 600 staff engaged in Chagas disease control in rural areas were switched to mosquito control in urban centers. The Chagas disease control program was suspended.

This experience highlighted several key points. At the technical level, it was clear that the intervention techniques and operational strategy were highly effective, while population genetic studies strongly supported the idea that a sustainable end-point could be reached through eradication of all domestic populations of the main vectors. At the political level, however, there was a clear need to find ways of maintaining continuity until that end-point could be reached, as well as a need to promote similar interventions in neighboring countries in order to eliminate all risk of reinfestation.

The enlightened response came in 1991 with a joint agreement between ministries of health of the six southernmost countries of Latin America (Argentina, Bolivia, Brazil, Chile, Paraguay, and Uruguay) to implement simultaneous interventions against Chagas disease. This program—the Southern Cone Initiative—was joined by Peru in 1996 to encompass the entire geographic distribution of the main vector, *Triatoma infestans.* All countries implement similar interventions but retain national autonomy for financing and operational approaches. Coordination is provided by the Pan American Health Organization through a system of national evaluations carried out by teams drawn from the Chagas control services of neighboring countries. By 1997, Uruguay became the first of these countries to be formally certified free of Chagas disease transmission, followed by Chile in 1999, most of the central and southern states of Brazil in 2000/1, and the first four provinces of Argentina in 2002. More provinces in Argentina and several departments in Paraguay and Bolivia are also being evaluated for similar certification.

The Southern Cone Initiative has yet to reach its final objective of eliminating all domestic populations of *Triatoma infestans* and halting Chagas disease transmission throughout the program area, some 6 million square kilometers. And it has not been without economic and political difficulties, especially the economic instability of recent years and the problems arising from decentralization of executive services. But this successful model of regional collaboration focused on a biologically defined target has paved the way for similar initiatives against Chagas disease in Central America and the Andean Pact countries (both formally launched in 1997) and in Mexico and the Amazon region. Significantly, the model has also been taken up by African countries with the launch of the Pan African Initiative against tsetse and trypanosomiasis (PATTEC) launched by the Organization of African Unity (now called the African Union) at the 2000 summit in Lomé, Togo. PATTEC represents a vision to eliminate tsetse and trypanosomiasis transmission from the whole of Africa, not by small donor-led projects but by a series of regional initiatives focused on biologically defined targets: when a vector population is eliminated over its entire range, transmission is halted and cannot be restarted by that population.

stored water (for example, putting lids on storage jars) can eliminate vector-breeding sites.

Improving housing construction is particularly important for protection against infections such as Chagas disease, where the vectors live directly in the walls of poor houses. Housing improvements shown to reduce Chagas-vector contact with humans include concrete floors, plaster and brick walls, and tiled roofs (De Andrade et al. 1995); these obviously have co-benefits in terms of raising overall living conditions. Studies in Sri Lanka have indicated that residents of poorly constructed houses were as much as 2.5 times more likely to contract malaria than neighbors in houses of good construction (Gamage-Mendis et al. 1991; Gunawardena et al. 1998). Economic analysis in Sri Lanka indicated that government investments to improve the most vulnerable houses would be compensated by savings in malaria treatment within eight years (Gunawardena et al. 1998).

Zooprophylaxis (using diversionary hosts to reduce the proportion of bloodmeals taken on humans, and thus cutting disease transmission) is also a promising tool. This technique usually makes use of domestic livestock that also provide other services such as milk and labor and causes minimal ecosystem disruption. It does, however, require knowledge of local vector ecology to select and site diversionary hosts, to ensure that increased provision of bloodmeals does not increase local vector density to an extent that outweighs the reduced proportion of bites on humans. It has been demonstrated to be effective in Indonesia (Kirnowordoyo and Supalin 1986), the Philippines (Schultz 1989), and Sri Lanka (van der Hoek et al.1998). In the most endemic regions of sub-Saharan Africa, there is some evidence for effectiveness against *Anopheles arabiensis,* but there may be fewer prospects for use against the more anthropophagic (human-biting) *Anopheles gambiae.* (See Box 12.5.)

12.4.2 Health Awareness and Behavior

Vector-borne diseases such as malaria may be more serious when the victim is malnourished, in ill health, or suffering from gastrointestinal diseases. Thus there is a strong link between overall social well-being and disease impact. In particular, improved awareness of hygiene and sanitation, particularly among women, may be important in reducing infectious disease among children. General educational levels, particularly of women, are a key determinant of awareness of health risks, and therefore health status.

In addition to general education improvements, however, messages specifically targeted at behavioral changes associated with vector protection are important tools in disease prevention,

BOX 12.4

Monitoring and Managing Environmental Effects of Insecticidal Control of Onchocerciasis in Africa

Onchocerciasis can be combated by appropriate vector control operations which consist of arresting transmission of the parasite by eliminating the vector population for the duration of the lifespan of the adult worm from the human reservoir host. This has been applied and demonstrated in parts of the original area of the Onchocerciasis Control Program in West Africa, where vector control alone has been carried out for more than fourteen consecutive years and where the disease is no longer a public health problem (Hougard et al. 2001). As black fly adults are difficult to target, the vector control operations consist of treating with appropriate insecticides the breeding sites of rivers where the reophilic larval stages develop.

The mere fact of regularly using insecticides for many years raised the concern of the potential risk such operations could have for the aquatic environment. Indeed, at the time OCP was launched, there was much evidence on biological and ecological consequences of DDT. With the awareness of the "DDT syndrome" by the international community, the participating countries as well as the donors that support the program (28 countries and foundations) had reasons to fear that repeated application of insecticides in the water courses would cause serious disturbances of the freshwater ecosystems. In 1974, just before the beginning of operational activities, OCP set up an aquatic monitoring program of rivers planned to be regularly treated with insecticides (Lévêque et al. 1979). It was implemented to satisfy three major concerns:

- to provide early warning to those carrying out treatments if toxic effects were noticed in the short term and to ensure that the insecticide release did not excessively disturb the functioning of the treated ecosystems on a long-term basis (the expected duration of OCP);
- to avoid the widespread use of chemicals that might have adverse effects on human populations near the river systems and/or might accumulate in the food chain as DDT has been known to do; and
- to prevent the irreversible loss of aquatic biodiversity in West Africa both because freshwater fish are a major source of food as well as an economic activity for West African populations, and to meet the objective of the Convention on Biodiversity that stipulates that countries are responsible for the conservation of their biodiversity.

The OCP in West Africa closed in December 2002, after 29 years of activity. There is no equivalent of a public health program benefiting from such long-term financial support from the international community. One reason for this support was that OCP always convinced the donors of the effectiveness of the control strategies used. The other reason was the OCP's ongoing to take care of the aquatic environment with the involvement of national teams and international expertise. The implementation of a long-term monitoring program to assess the potential effects of larviciding and the large-scale screening of larvicides to select that which is most efficient for *Simulium* while less drastic for the non-target fauna are unique features in large health control programs. These efforts had an economic cost in terms of insecticide consumption and operational strategies but they made it possible to preserve the quality of the water used by the riverine populations as well as the fishing resources that constitute a significant part of the foodstuff of these populations.

The goal has been achieved: onchocerciasis has been virtually eliminated from the OCP area as a disease of public health importance and as an obstacle to socioeconomic development. From the environmental point of view, the success of OCP may be jeopardized by an unsustainable use of the freed land. For example, a pilot study conducted in the Léraba area showed that 75% of the original wooded savanna was cleared for agricultural development and settlement of villages (Baldry et al. 1995). The riverine forests of many small rivers were destroyed and on some of the banks soil erosion is occurring. On the other hand, the bordering forests and the easily flooded plains of the larger rivers did not undergo any disturbance of this scale. It is therefore necessary to both take measures and sensitize the riverine populations on the need for environmental protection and biodiversity management along with the development of agricultural activities.

particularly for diseases such as dengue that lack vaccines or curative drugs. Simple poster displays in local languages can be highly effective. These may include messages such as the early identification of disease symptoms; the need to maintain attendance for drug delivery; and refraining from interfering with traps for vectors such as tsetse.

Such messages may in some cases be backed up by legislation (for example, fines for not removing potential dengue-breeding sites). In many cases, however, they will be most effective when they are developed in a participatory manner, using methods that involve the community in the identification of problems and solutions. Depending on the situation, they may best be disseminated via a range of media, such as posters, community events, or presentations, or by including the subject in popular radio or television programs such as soap operas. It should also be noted that behavioral changes may be impossible if they are not supported by government services; for example, poor urban populations will be unwilling or unable to remove dengue-breeding sites if there is no garbage collection service or reliable water supply.

The effectiveness of environmental health education in changing behavior is often measured through knowledge, attitude, and practice surveys, and ultimately through changing disease rates. KAP studies for dengue control in the Caribbean have indicated that, while knowledge has increased with respect to vector control, there is not much evidence of this knowledge being put into action (Polson et al. 2000). In many cases, people do not take action because they believe that mosquito control is the responsibility of the government (Rosenbaum et al. 1995). In addition, even community members who rid their own premises of mosquito habitats may suffer from the inaction of less vigilant neighbors, necessitating combined community efforts. While community participation and inter-sectoral collaboration is therefore perceived as a positive approach, they are challenging tasks and require concerted sustained effort by key stakeholders and professionals as well as government.

12.5 "Cutting-edge" Interventions: Genetic Modification of Vector Species to Limit Disease Transmission

In the short term, the largest strides in vector control and management may come from applied research that refines existing tools and targets them more effectively to local situations. Such applications may also be highly cost-effective. However, new tools, while requiring substantially greater investment of time and resources in order to overcome initial technical barriers, can potentially result in dramatic improvements in control. For example,

BOX 12.5

Integrated Vector Management Strategies for Malaria

IVM strategies for controlling malaria are receiving renewed attention globally among scientists, health officials, and policy-makers. This is due to the rapid spread of resistance to antimalarial drugs such as chloroquine and the apparent inability of past large-scale eradication campaigns to even hold the disease in check let alone eliminate transmission in the regions where it is most endemic, such as Africa. Indeed, over 1.2 million people, overwhelmingly African children, died from malaria in 2002, an increase in mortality in absolute terms over the previous year (WHO 2003, 2004a). There is also concern over potential vector resistance against pesticides commonly used for indoor residual spraying and insecticide-treated materials.

These limitations have spurred interest in integrated approaches to malaria vector control (Walker 2002). IVM approaches hold the potential to significantly reduce if not interrupt disease transmission as well as the burden of serious disease and death, while avoiding very severe disruptions to ecosystem services.

Environmental modification strategies, such as the drainage of wetlands, were among the first employed to deliberately control malaria vectors nearly a century ago. Of increasing interest today are strategies for less intrusive changes in wetlands environments—improving the flushing of streams or removing vegetation from water to limit the breeding areas where larvae may develop. In areas where dams and reservoirs are to be constructed, there is evidence that better engineering design of dams or irrigation schemes, to provide for alterations in level and flow of water, the flushing of reservoirs, and periodic weed removal can minimize the development of new vector habitats. Options for the best-practice design of dams and agricultural irrigation projects to control vector breeding sites, while minimizing the disruption to ecosystem services, are detailed in Tiffen (1991); Birley (1991); and Phillips et al. (1993).

In tropical Africa, where malaria is most endemic, IVM strategies may be most relevant in areas of less intense transmission and in peri-urban or urban locales, where breeding sites may be fewer, most easily identifiable, and more amenable to control (Walker 2002). Particularly in such settings, there is evidence that incremental reductions in vectorial capacity, while not eradicating the disease may play a role in reducing morbidity and mortality and improving health in specific age groups, that is, pregnant women and children under the age of two. For larvae control in such settings, greater attention is being given to the efficacy of biological controls such as larvivorous fish and biolarvicides, that is, *Bacillus thurengiensis israelensis* and *sphaericus* (Walker 2002).

However, in rural areas of intense malaria transmission, a better understanding of relationships between human habitats, farming, and livestock practices, may provide important tools for controlling malaria using IVM methods, and in a manner that sustains positive ecosystem services. For instance, alternating between cycles of irrigated and non-irrigated crops may disrupt breeding cycles (Tiffen 1991; Mutero et al. 2004). Better management and control of manmade sites where malarial mosquitoes may easily reproduce such as water bore holes may help reduce malaria breeding close to human settlements. Conversely, more strategic placement of new human settlements away from potential malaria breeding areas can also reduce transmission (Walker 2002). Recent research in Kenya on the interactions between malaria, livestock, and agriculture has highlighted the potential for livestock to act as "diversionary hosts" for certain species of malaria vectors. The study, conducted in four villages of the Mwea Division of Kenya, found malaria disease prevalence was significantly lower in the two villages with rice irrigation schemes than in the villages with no irrigated agriculture (0–9% versus 17–54%). The lower incidence occurred despite the existence of a 30–300 fold increase in the number of local malaria vectors in the irrigated locales. The likely explanation for the so-called "paddies paradox" appeared to be the tendency for the prevalent *A. arabiensis* vector to feed overwhelmingly on cattle rather than on humans (Mutero et al. 2004).

At the same time, advances in geographical information systems together with continuing research into vector entomology are permitting more precise mapping of vector species composition, spatial distribution, and transmission patterns. This is contributing to a better understanding of vector ecology, that may be used to guide control operations that maximize the use of targeted control strategies and minimize ecosystem disruptions (Shililu et al. 2003a, 2003b).

there is a need to develop new curative drugs and insecticides to both overcome problems of vector resistance and reduce disease control costs, while the development of an effective vaccine could potentially revolutionize control of malaria.

In addition to developing "new varieties of old tools," there is the potential for developing qualitatively different types of tools, with very different implications both for disease control and other ecosystem services. Genetic modification of vectors is one such method.

The practice of selecting strains of vectors that are unable to transmit disease has been applied for several decades (Collins et al. 1986; Wu and Tesh 1990). However, recent advances in molecular genetics have made it potentially much more feasible to introduce genes into vector populations that either drive the population to extinction or that reduce their capacity to maintain and transmit infections. These include the sequencing of the full genome of the malaria vector *Anopheles gambiae* (Holt et al. 2002), the stable introduction of engineered genetic constructs into the genome of a number of important mosquito species (for example, Coates et al. 1998; Catteruccia et al. 2000; Allen et al. 2001; Grossman et al. 2001), and the identification and activation of genetic constructs that block or reduce pathogen transmission of malaria parasites and dengue viruses in mosquitoes (for example, de Lara Capurro et al. 2000; Ito et al. 2002).

It is therefore theoretically possible to produce in the laboratory mosquito populations that are unable to transmit important pathogens. The next stage is to implement "gene-driving" mechanisms, which link the genes of interest to other genetic elements that have the ability to spread throughout the vector population, from initial seeding releases. Two main drive systems are under investigation. The first consists of autonomous transposable elements (Ribeiro and Kidwell 1994), genetic elements that copy themselves throughout the genome, thereby increasing the chance of inheritance into the next generation. The second is symbiotic *Wolbachia* bacteria, which are inherited through egg cells and give host females a reproductive advantage over uninfected females (Sinkins and O'Neill 2000). Genetic mechanisms that reduce or block pathogen transmission could thus be attached to such a driver in order to achieve maximum dissemination in the vector population.

Major technological challenges remain in the completion of all the stages described above in a single vector population (for example, Alphey et al. 2002). In addition, some disease control experts are concerned about the application of transgenic tools.

They highlight the potential that the transposable elements could also increase mutation rates, with unforeseen characteristics in the vector populations, and that genetically engineered constructs could cross-contaminate other species. They also stress that it may prove more difficult than assumed to spread the relevant genes through highly variable populations of parasites or vectors. Ethical issues are also involved, most importantly that it will be essential to achieve broad informed consent of the populations who will come into contact with the vector populations, as there would be no option for individuals to "opt out" of the intervention (Scott et al. 2002). Critics who raise these issues generally stress the importance of using existing tools in a wider, more sustained and targeted manner rather than developing new interventions (Curtis 2000).

Supporters of transgenic techniques point out that the kinds of inherited elements being discussed are not normally horizontally transferred between individuals, even of the same species, so that crossing species boundaries would presumably be an even more rare event (for example, O'Neill et al. 1997). If, indeed, such transfer did occur, there would be no reason to expect that an element specifically designed to interfere with pathogen-mosquito interactions would have any effects in non-target species. In addition, gene spreading mechanisms such as transposable elements and *Wolbachia* are very common in natural insect populations, and have had no detectable long-term effects on host ecology (for example, Turelli and Hoffmann 1991; Kidwell 1992). They therefore conclude that the risks of negative environmental impact associated with the kinds of replacement strategies that are being developed are low.

There is, as yet, no firm consensus among researchers on the practicality and acceptability of applying these tools in the field. As they develop further, it will become more important for the scientific community to present a rational and dispassionate view of the potential benefits and threats of this approach to decision-makers and the public so that they may be better positioned to judge whether the possible gains from this type of intervention outweigh any environmental risks.

12.6 Promoting Inter-sectoral Cooperation among Health, Environment, and Development Institutions

An ecosystems perspective on vector-borne disease control requires a reconsideration of institutional structures that manage vector control overall. Integrated vector management, in particular, operates on the premise that effective control requires the collaboration of various public and private agencies as well as community education and participation rather than exclusive action by the health sector. This is part of an increasing appreciation that the Millennium Development Goals for infectious diseases (and other goals) will only be achievable with contributions from health, environment, and development institutions at global, national, and local levels.

Yet, present institutional structures tend to promote a narrow sectoral and disease-specific approach to interventions. In most countries, health and environment sectors remain divided, with little coordination of approaches to vector control and associated energy, agricultural, housing, and forestry policies. Health sector institutions often leverage significant resources, but their actions tend to be directed towards curative treatment or interventions against specific diseases, rather than promoting integrated policies for economic development, environmental protection, and improved human health. The environment sector typically has access to fewer financial resources and political power, and focuses mainly on protection of the natural environment. Productive sectors, such as ministries of infrastructure, development, or trade, often access significant resource pools and take decisions that have profound influences (both positive and negative) on the transmission of vector-borne diseases, but these are not considered systematically. As a result, there is seldom an incentive for, for example, a malaria control program to carry out interventions that protect against other human diseases, protect the environment, or promote, say, agricultural development. Inter-sectoral collaboration is thus "blessed by everybody and funded by nobody"; it is widely acknowledged as essential, but rarely occurs because individual sectors measure success only against their own targets.

Institutionally, it is often easier to design and assess the success of "top-down" control campaigns that motivate political will and raise resources, with clear lines of institutional responsibility and measurable targets to reach a highly specific goal. In many instances, such programs have been successful, when directly judged against their stated aims (that is, eradication or specific levels of reduction).

However, such approaches have seldom led to complete eradication of a disease. This can be due to the slackening of political will once the initial visible gains of the "attack phase" have been achieved and the limits recognized of promoting a single intervention uniformly rather than a range of locally adaptable approaches. The greater use of "sector-wide approaches" to financing rather than individual project funding by international donors (Cassels 1997) and greater involvement of nongovernmental actors (such as charitable NGOs and the private sector) have also made centrally organized national or international control campaigns more difficult to execute. In a related development, the decentralization of decision-making within developing countries has resulted in the dismantling of centralized vector control mechanisms upon which control campaigns may rest, making participatory approaches far more relevant.

The advantages and disadvantages of campaign-style sectoral approaches and newer, more inter-sectoral and participatory approaches are illustrated by the Caribbean experience with dengue control. Prior to 1962, several Caribbean islands, along with most of mainland Latin America, had succeeded in eradicating the *Aedes aegypti* mosquito, vector of dengue and yellow fever, following an intensive eradication campaign. This success was attributed to centralized, vertically structured programs; use of DDT and adequate funding for insecticides; and the availability of equipment and well-trained staff. This success was, however, not sustained mainly because it was not possible to maintain the effectiveness of the limited number of control methods used (insecticide and source reduction) and achieve eradication in all countries of the region. When *Ae. aegypti* subsequently reinvaded from neighboring countries, another eradication campaign was deemed as unfeasible because of increased urbanization that had led to structural inadequacies in the provision of basic services such as garbage collection and piped water. In addition, chronic economic crises in many of the most affected countries had led to decreased capacity and political prioritization for vector control programs.

In 1992, the Pan American Health Organization in collaboration with the government of Italy and the Caribbean Cooperation in Health initiated a five-year project whose general objective was to reduce the densities of *Ae. aegypti* in fifteen English-speaking Caribbean countries to a level at which transmission of the dengue virus will not occur. The project aimed to strengthen existing national vector control programs using community participation as the focus. As a result of the project, programs in some countries, including St. Vincent and the Grenadines, shifted away from the traditional "top-down" approach to one of community involvement and partnership. Collaboration between vector control programs and environmental health programs was improved, and the scope of control was broadened to include control of pests and vectors other than *Ae. aegypti*. It should be noted, however, that there is as yet no evidence that this approach has led to a sustained decrease in the persistently high levels of *Ae. aegypti* disease in the countries involved outside the time frame of the original project.

In other settings, however, such as Viet Nam, a national approach to dengue management, based upon community participation and involvement, seems to have had greater and more sustained success. (See Box 12.6.) In small-scale projects in Ghana, an integrated community approach to malaria intervention measures has been reported to result in reduction of overall child mortality by as much as 50% (Curtis 1991).

Despite the general lack of inter-sectoral and community cooperation in vector-borne disease management, there are some promising developments. In recent decades, agencies such as the World Bank, UNICEF, and UNDP have increasingly invested in global health issues. This reflects a growing institutional recognition that healthier people are more productive, and ill health both creates and maintains poverty. The result has been several new interagency initiatives in the environment, health, and development arena. These include the WHO Commission on Macroeconomics and Health, and joint WHO and UNEP initiatives promoting linkages between environment and health in awareness raising, monitoring, and policymaking. The World Health Organization and the USAID-supported Environmental Health Project have collaborated with governments and other key international stakeholders to develop a global strategy for integrated vector management, along with practical guidelines and training tools, as an important component of the Roll Back Malaria Campaign. This initiative has already had marked success in bringing together multiple ministries in several African countries to develop "national consensus statements" on how each agency can contribute to addressing malaria. There are now also legal requirements that large-scale development projects (particularly those funded by agencies such as the World Bank or international development banks) carry out environmental impact assessments that also include consideration of human health impacts (for ex-

BOX 12.6

Advances in the Targeted Control of Mosquito Vectors of Dengue

Dengue fever and associated dengue hemorraghic fever is the world's fastest growing vector-borne disease and the most important vector-borne *viral* disease affecting humans. Dengue is found in nearly one hundred tropical countries and is responsible for the loss of over 19,000 lives and 610,000 disability adjusted life years (DALYs) annually (Lloyd 2003; WHO 2004a).

In the 1960s, *Aedes aegypti* eradication campaigns, aimed against the urban yellow fever mosquito, which also is the principal dengue-carrying vector, succeeded in eliminating the mosquito from most of Latin America. But it has resurfaced in the past three decades, and is currently found throughout much of the tropics and sub-tropics. Urbanization and population movement are responsible for much of the recent resurgence. The epidemiology and ecology of dengue is complicated by the fact that there are four primary serotypes of the disease, some or all of which may be circulating in a particular endemic region at a particular time. When a serotype is introduced to a population with low herd immunity, epidemics may occur when vector densities are relatively low. Among vulnerable populations, serial exposure to more than one dengue serotype increases, rather than reduces, the risk of severe illness (Halstead 1988).

Around human settlements, *Ae. aegypti* mosquitoes breed primarily in artificial water containers, and the mosquito's life cycle is closely associated with human activities. Larval habitats are increasing rapidly with unplanned urbanization and greater amounts of water-retaining waste products that provide a habitat for mosquitoes (Lloyd 2003). Rising global temperatures and humidity may further increase dengue disease incidence (Hales et al. 2002), as *Aedes* mosquitoes reproduce more quickly and bite more frequently at higher temperatures (Patz et al. 2003). Since there is no curative treatment for dengue, targeted environmental and ecosystem management is all the more relevant. However, in many settings, community cleanup campaigns or space-spray application of insecticides, using either pyrethroids or organophosphates, have had limited or only temporary effects on disease incidence. Recently, there has been increasing interest in identifying the most productive mosquito breeding sites and in better understanding epidemiological knowledge about the thresholds below which mosquito densities have to be reduced in order to prevent a severe outbreak.

More recently, the work of Focks et al. has led to the development of predictive computer models that may accurately determine a "threshold" limit for epidemic risk in a particular locale. This relatively inexpensive model considers local levels of population immunity to various dengue serotypes, as well as vector densities and ambient temperature, to yield a calculation of the threshold number of *Aedes* pupae (effectively equivalent to adult numbers) per person per area (Focks et al. 2000) to maintain transmission.

The same authors have also designed inexpensive household surveys that more precisely identify the most productive containers, or breeding sites, containing the highest densities of *Ae. aegypti* pupae. These are the containers from which significant numbers of adult mosquitoes are likely to emerge. Similar tools, key container and key premise indices, have been developed and tested in Viet Nam (Nam 2003). The development of such models and new indices has paralleled a growing awareness that less than 1% of the containers may produce more than 95% of the adult mosquitoes that trigger disease outbreaks. So once the most productive containers are identified, targeted environmental management of dengue becomes more affordable and feasible. Targeted vector control also minimizes disruption to other ecosystem services.

Control strategies may include more community cleanup aimed specifically at the types of containers most closely associated in the local setting with vector breeding and disease transmission, such as old tires or discarded water drums (Hayes et al. 2003). In addition, biological and targeted chemical methods of control are being used with increased precision. In Viet Nam, for instance, a small crustacean, *Mesocyclops* (Copepoda), which feeds on the newly hatched larvae of *Ae. aegypti,* has been introduced into household water tanks and water jars in three northern provinces; the intervention is reported to have resulted in eradication or near eradication of *Aedes* larvae in treated areas (Nam 2003). In Cambodia, WHO is testing, together with national and local authorities, a new long-lasting insecticide-impregnated net mesh water tank cover, using the same approach to water containers now commonly used with insecticide-treated bednets in malaria control. The cover, fitted over concrete rainwater storage tanks, is designed both to prevent mosquito breeding in these key containers and to reduce adult vector densities and longevity.

ample, consideration of the effects of dam construction on vector-borne disease).

Most importantly, a number of "bottom-up" inter-sectoral collaborations in developing countries have shown promising results against several vector-borne diseases. For example, a CGIAR System-wide Initiative on Malaria and Agriculture is exploring the links between irrigation practices, livestock practices, and malaria disease transmission and incidence in seven African countries and settings, using a transdisciplinary "ecosystems" approach. This approach emphasizes the importance of understanding socioeconomic factors, such as gender issues, substance abuse, and malnutrition, that may contribute to greater disease incidence and severity. It also stresses participatory methods that involve the local community as partners in vector and disease control, together with researchers and policymakers. This is part of a wider drive towards stakeholder-driven transdisciplinary work under the Ecosystem Approaches to Human Health initiative at Canada's International Development Research Centre (IDRC 2004).

12.7 Analysis and Assessment of the Different Responses

There is now significant experience from various parts of the world with integrated "ecosystems" approaches to vector management and control. These indicate that a more systematic use of existing scientific knowledge about vector behavior, within an ecosystems framework, may cost-effectively reduce disease burdens. At the same time, this approach may offer greater protection to other ecosystems services than has been obtained in the past from vertical "campaign-oriented" programs.

IVM strategies are promising not because they apply new and different interventions but because they provide a structure for selecting and applying the most effective existing intervention, given local epidemiological and ecological characteristics. This includes targeting control efforts in space and time, and toward specific stages of the lifecycle of vectors. There has been an additional emphasis on more precisely characterizing epidemiological zones. Geographic information systems can be useful in defining such zones and the distribution of different vector species, which may be more or less amenable to specific control interventions. IVM strategies may still rely quite heavily on the use of insecticides. These can be applied in ways that have minimal impact on the wider environment, particularly for diseases with a human-vector/human-transmission cycle (for example, use of insecticide-treated bed nets or residual spraying inside houses to control malaria), although focused application is often more difficult when the transmission cycle includes animal reservoir hosts of infection. Although these chemicals still may impact on health and environment, the relatively small amounts used, and accurate targeting, means that these effects are likely to be much smaller than those from agricultural application of pesticides. Insecticide resistance is a threat to the long-term sustainability of chemical interventions, and needs to be managed. However, there remain very few examples where it has seriously undermined control programs. While the limited use of organochloride compounds such as DDT has also been permitted under the Stockholm Convention for indoor residual spraying against malaria, a more widespread use of IVM strategies may help reduce reliance on chemical interventions.

Participatory mechanisms and behavioral tools can be important to promoting vector management at the community level, within an ecosystem framework. It is important, however, to evaluate the effectiveness of various social responses and approaches in reducing clinical disease. Behavioral changes are notoriously difficult to sustain. Some challenges are specific to particular diseases, such as the need to reduce *Aedes* populations down to specific, and generally very low, levels in order to have a significant impact on dengue transmission. Others can be extrapolated more widely. These include difficulties in engendering interest in community-wide behavioral change; what to do when noncompliance by individuals puts others at risk; and the problems in sustaining behavioral change over time, particularly when the disease transmission threat is relatively low, to prevent re-emergence of disease threats or invasion from other regions.

Along with better applied use of existing knowledge and technologies, basic research into radically different approaches may yield new tools for vector management or control over the long term. For example, research into the development of transgenic vectors is a promising new avenue that could potentially allow for the inexpensive replacement of existing vector populations with vectors that are incapable of disease transmission. However, time and financial resources are necessary to overcome various technical barriers. Secondly, there is an ongoing debate over the wider ecological and ethical implications of such introductions, as shown by the controversy over the introduction of genetically modified foods in some societies. The debate largely reflects different values rather than disputes over the scientific evidence. Proponents can demonstrate that there is no *a priori* reason to expect harmful consequences, while opponents focus on the difficulty of proving long-term safety or reversing an introduction if negative effects are detected.

Strategies to reduce infectious disease transmission should ideally be quantified in terms of overall social and economic benefits to human populations. Outcomes often remain narrowly defined and success is measured against a small set of indicators. Interventions made through the health sector often are judged primarily in terms of effectiveness in cutting incidence rates or curing disease cases—in some cases with associated measures of economic costs. While these are clearly critical considerations and are also important for making comparisons between similar kinds of health system interventions, such an assessment tends to downplay the ways a given health system intervention or policy may impact other ecosystems services. Thus important "externalities" are often ignored because they are displaced and diffused over time, geographical settings, and the affected populations, and therefore difficult to quantify.

At the same time, policy responses that have nothing to do with health overtly, often have powerful indirect impacts on ecosystem services and infectious disease rates which may also vary in time and space, and be even more difficult to assess and quantify. For example, there is clear evidence that aspects of climate regulation and biodiversity are linked to infectious disease transmission. Policies that impact climate change and biodiversity, either positively or negatively, will therefore also have an impact ultimately on disease incidence and health. It is often only possible to assess those global effects in general terms. Overall, however, policies promoting sustainable development, by increasing individual and societal wealth while decreasing inequalities and avoiding degradation of natural resources, can benefit both human and ecosystem health.

Institutional responses are clearly critical in the design of such integrated policies. Initiatives to build capacity and synergies across institutional sectors face great challenges, in design, implementation, and assessment. Inter-sectoral collaborations may yield the greatest overall benefits to society but are seldom the best way for individual sectors to demonstrate their performance. For example, ministries of health may still gain more credit for improving treatment rates for a particular disease than for collaborating with ministries of environment to ensure sustainable development and protection of natural resources, even if the latter approach yields greater overall societal benefits in the long run. Thus inter-sectoral collaboration on vector-borne disease management will be achievable only through sustained high levels of awareness and participation from intergovernmental, governmental, and donor agencies.

References

Allen, M.L., D.A. O'Brochta, P.W. Atkinson, and C.S. Levesque, 2001: Stable, germ-line transformation of Culex quinquefasciatus (Diptera: Culicidae), *Journal of Medical Entomology,* **38,** pp. 701–10.

Alphey, L., C.B. Beard, P. Billingsley, M. Coetzee, A. Crisanti, et al., 2002: Malaria control with genetically manipulated insect vectors, *Science,* **298,** pp. 119–21.

Baer, F., C. McGahey, and P. Wijeyaratne, 1999: Summary of EHP activities in Kitwe, Zambia, 1997–1999: Kitwe Urban Health Programs, US Agency for International Development, Environmental Health Project, Arlington, VA.

Baldry, D., D. Calamari, and L. Yaméogo, 1995: Environmental impact assessment of settlement and development of the Upper Léraba basin, World Bank, Washington, DC, pp. 1–38.

Birley, M., 1991: *Guidelines for Forecasting the Vector-borne Disease Implications of Water Resources Development,* WHO, Geneva, Switzerland/ Food and Agriculture Organization of the United Nations, Rome, Italy/ United Nations Environment Programme, Nairobi, Kenya/ United Nations Center for Human Settlements, Geneva, Switzerland.

Campbell-Lendrum, D., J.P. Dujardin, E. Martinez, M.D. Feliciangeli, J.E. Perez, et al., 2001: Domestic and peridomestic transmission of American cutaneous leishmaniasis: Changing epidemiological patterns present new control opportunities, *Memorias de Instituto Oswaldo Cruz,* **96,** pp. 159–62.

Cassels, A., 1997: *A Guide to Sector-wide Approaches for Health and Development: Concepts, Issues and Working Arrangements,* WHO, Danish International Development Agency, Department for International Development, European Commission, Geneva, Switzerland.

Catteruccia, F., T. Nolan, T.G. Loukeris, C. Blass, C. Savakis, F.C. Kafatos, et al., 2000: Stable germline transformation of the malaria mosquito Anopheles stephensi, *Nature,* **405,** pp. 959–62.

Chavasse, D. and H. Yap (eds.), 1997: *Chemical Methods for the Control of Vectors and Pests of Public Health Importance,* WHO, Geneva, Switzerland.

Coates, C.J., N. Jasinskiene, L. Miyashiro, and A.A. James, 1998: Mariner transposition and transformation of the yellow fever mosquito, Aedes aegypti, *Proceedings of the National Academy of Sciences, USA,* **95,** pp. 3748–51.

Collins, F.H., R.K. Sakai, K.D. Vernick, S. Paskewitz, D.C. Seeley, et al., 1986: Genetic selection of a Plasmodium-refractory strain of the malaria vector Anopheles gambiae, *Science,* **234,** pp. 607–10.

Curtis, C., 1991: *Control of Disease Vectors in the Community,* Wolfe Publishing, London, UK.

Curtis, C.F., 2000: Infectious disease: The case for deemphasizing genomics in malaria control, *Science,* **290,** p. 1508.

De Andrade, A.L., F. Zicker, R.M. De Oliveira, I.G. Da Silva, S.A. Silva, et al., 1995: Evaluation of risk factors for house infestation by *Triatoma infestans* in Brazil, *American Journal of Tropical Medicine and Hygiene,* **53,** pp. 443–7.

de Lara Capurro, M., J. Coleman, B.T. Beerntsen, K.M. Myles, K.E. Olson, et al., 2000: Virus-expressed, recombinant single-chain antibody blocks sporozoite infection of salivary glands in Plasmodium gallinaceum-infected Aedes aegypti, *American Journal of Tropical Medicine and Hygiene,* **62,** pp. 427–33.

Focks, D.A., R.J. Brenner, J. Hayes, and E. Daniels, 2000: Transmission thresholds for dengue in terms of Aedes Aegypti Pupae per person with discussion of their utility in source reduction efforts, *American Journal of Tropical Medicine and Hygiene,* **62,** pp. 11–8.

Gallup, J.L. and J.D. Sachs, 2001: The economic burden of malaria, *American Journal of Tropical Medicine and Hygiene,* **64,** pp. 85–96.

Gamage-Mendis, A.C., R. Carter, C. Mendis, A.P. De Zoysa, P.R. Herath, et al., 1991: Clustering of malaria infections within an endemic population: Risk of malaria associated with the type of housing construction, *American Journal of Tropical Medicine and Hygiene,* **45,** pp. 77–85.

Gopaul, R., 1995: [Entomological surveillance in Mauritius], *Santé,* **5,** pp. 401–5.

Grossman, G.L., C.S. Rafferty, J.R. Clayton, T.K. Stevens, O. Mukabayire, and M.Q. Benedict, 2001: Germline transformation of the malaria vector, Anopheles gambiae, with the piggy-back transposable element, *Insect Molecular Biology,* **10,** pp. 597–604.

Gubler, D.J., 1997: Human behavior and cultural context in disease control, *Tropical Medicine and International Health,* **2,** pp. A1–2.

Gunawardena, D.M., A.R. Wickremasinghe, L. Muthuwatta, S. Weerasingha, J. Rajakaruna, et al., 1998: Malaria risk factors in an endemic region of Sri Lanka, and the impact and cost implications of risk factor-based interventions, *American Journal of Tropical Medicine and Hygene,* **58,** pp. 533–42.

Gupta, D.K., R.M. Bhatt, R.C. Sharma, A.S. Gautam, and Rajnikant, 1992: Intradomestic mosquito breeding sources and their management, *Indian Journal of Malariology,* **29,** pp. 41–6.

Gwatkin, D.R., M. Guillot, and P. Heuveline, 1999: The burden of disease among the global poor, *Lancet,* **354,** pp. 586–9.

Hales, S., N. de Wet, J. Maindonald, and A. Woodward, 2002: Potential effect of population and climate changes on global distribution of dengue fever: An empirical model, *Lancet,* **360,** pp. 830–4.

Halstead, S.B., 1988. Pathogenesis of dengue: Challenges to molecular biology, *Science,* **239,** pp. 476–81.

Hay, S.I., D.J. Rogers, S.E. Randolph, D.I. Stern, J. Cox, et al., 2002: Hot topic or hot air? Climate change and malaria resurgence in East African highlands, *Trends in Parasitology,* **18,** pp. 530–4.

Hayes, J.M., E. Garcia-Rivera, R. Flores-Reyna, G. Suarez-Rangel, T. Rodriguez-Mata, et al., 2003: Risk factors for infection during a severe dengue outbreak in El Salvador in 2000, *American Journal of Tropical Medicine and Hygiene,* **69,** pp. 629–33.

Hewitt, S. and M. Rowland, 1999: Control of zoophilic malaria vectors by applying pyrethroid insecticides to cattle, *Tropical Medicine and International Health,* **4,** pp. 481–6.

Holt, R.A., G.M. Subramanian, et al., 2002: The genome sequence of the malaria mosquito Anopheles gambiae, *Science,* **298**: pp. 129–49.

Hougard, J.M., E.S. Alley, L. Yameogo, K.Y. Dadzie, and B.A. Boatin, 2001: Eliminating onchocerciasis after 14 years of vector control: A proved strategy, *Journal of Infectious Diseases,* **184,** pp. 497–503.

IDRC (International Development Research Centre), 2003: *Malaria Control in Coastal Oaxaca, Mexico,* IDRC, Ottawa, OT, Canada.

IDRC, 2004: *Ecosystems Approaches to Human Health,* IDRC, Ottawa, ON, Canada.

Ijumba, J.N., F.C. Shenton, S.E. Clarke, F.W. Mosha, and S.W. Lindsay, 2002: Irrigated crop production is associated with less malaria than traditional agricultural practices in Tanzania, *Transactions of the Royal Society of Tropical Medicine and Hygene,* **96,** pp. 476–80.

IPCC (Intergovernmental Panel on Climate Change), 2001: *Climate Change 2001: The scientific basis, Contribution of Working Group 1 to the Third Assessment Report of the IPCCC,* Cambridge University Press, Cambridge, UK.

Ito, J., A. Ghosh, L.A. Moreira, E.A. Wimmer, and M. Jacobs-Lorena, 2002: Transgenic anopheline mosquitoes impaired in transmission of a malaria parasite, *Nature,* **417,** pp. 452–5.

Jetten, T. and W. Takken, 1994: Anophelism without malaria in Europe: A review of the ecology and distribution of the genus Anopheles in Europe, Wageningen Agricultural University Papers, Wageningen, The Netherlands, pp. 1–69.

Kidwell, M.G., 1992: Horizontal transfer of P elements and other short inverted repeat transposons, *Genetica,* **86,** pp. 275–86.

Kirnowordoyo, S., and Supalin, 1986: Zooprophylaxis as a useful tool for control of A. aconitus transmitted malaria in Central Java, Indonesia, *Journal of Communicable Diseases,* **18,** pp. 90–4.

Kovats, R.S., D.H. Campbell-Lendrum, A.J. McMichael, A. Woodward, and J.S. Cox, 2001: Early effects of climate change: Do they include changes in vector-borne disease? *Philosophical Transaction of the Royal Society of London Series B - Biological Sciences,* **356,** pp. 1057–68.

Kuhn, K.G., D.H. Campbell-Lendrum, B. Armstrong, and C.R. Davies, 2003: Malaria in Britain: Past, present, and future, *Proceedings of the National Academy of Sciences,* **100,** pp. 9997–10001.

Kumar, A., V.P. Sharma, P.K. Sumodan, D. Thavaselvam, and R.H. Kamat, 1994: Malaria control utilizing Bacillus sphaericus against Anopheles stephensi in Panaji, Goa, *Journal of the American Mosquito Control Association,* **10,** pp. 534–9.

Lacey, L.A. and C.M. Lacey, 1990: The medical importance of riceland mosquitoes and their control using alternatives to chemical insecticides, *Journal of the American Mosquito Control Association Supplement,* **2,** pp. 1–93.

Lévêque, C., M. Odei, and M. Pugh Thomas, 1979: The Onchocerciasis Control Programme and the monitoring of its effects on riverine biology of the Volta River Basin. In: *Ecological Effects of Pesticides,* F. Perring and K. Mellanby (eds.), Academic Press, New York, NY, pp. 133–43.

Lloyd, L., 2003: *Best Practice for Dengue Prevention and Control in the Americas,* Environmental Health Project, USAID (Agency for International Development), Washington, DC.

Lopez, A.D. and C.C. Murray, 1998: The global burden of disease, 1990–2020, *Nature Medicine,* **4,** pp. 1241–43.

McMichael, A.J., 2001: *Human Frontiers, Environments and Disease: Past Patterns, Uncertain Futures,* Cambridge University Press, Cambridge, UK.

Molyneux, D.H., 1997: Patterns of change in vector-borne diseases, *Annals of Tropical Medicine and Parasitology,* **91,** pp. 827–39.

Molyneux, D.H., 2003: Common themes in changing vector-borne disease scenarios, *Transactions of the Royal Society of Tropical Medicine and Parasitology,* **97,** pp. 129–32.

Molyneux, D.H., D.A. Baldry, P. De Raadt, C.W. Lee, and J. Hamon, 1978: Helicopter application of insecticides for the control of riverine Glossina vectors of African human trypanosomiasis in the moist savannah zones, *Annales de la Societe Belge de Medecine Tropicale,* **58,** pp. 185–203.

Murray, C. and A. Lopez (eds.), 1996: *The Global Burden of Disease,* World Health Organization, Harvard School of Public Health, World Bank, Geneva, Switzerland.

Mutero, C., C. Kabutha, V. Kimani, L. Kabuage, G. Gitau, et al., 2004: A transdisciplinary perspective on the links between malaria and agroecosystems in Kenya, *Acta Tropica,* **89,** pp. 171–86.

Nam, V.S., 2003: Key container and key premise indices for *Ae. aegypti* surveillance and control. In: *Best Practices for Dengue Prevention and Control in the Americas,* L. S. Lloyd (ed.), Environmental Health Project, US Agency for International Development, Washington, DC, pp. 51–6.

Newton, E.A. and P. Reiter, 1992: A model of the transmission of dengue fever with an evaluation of the impact of ultra-low volume (ULV) insecticide applications on dengue epidemics, *American Journal of Tropical Medicine and Hygene,* **47,** pp. 709–20.

O'Neill, S., A. Hoffmann, and J. Werren, 1997: Influential passengers: Inherited microorganisms and arthropod reproduction, Oxford University Press, Oxford, UK.

Pal, R., 1982: Disease vector control in the People's Republic of China, *Mosquito News,* **42,** pp. 149–58.

Patz, J., A.K., M. Githeko, J.P. McCarty, S. Hussain, U. Confalonieri, et al., 2003: Climate change and infectious diseases. In: *Climate Change and Human Health: Risks and Responses,* A. McMichael, D. Campbell-Lendrum, C. Corvalan, K. Ebi, A. Githeko, et al. (eds.), WHO, Geneva, Switzerland.

Patz, J.A., T.K. Graczyk, N. Geller, and A.Y. Vittor, 2000: Effects of environmental change on emerging parasitic diseases, *International Journal of Parasitology,* **30,** pp. 1395–405.

Phillips, M., A. Mills, and C. Dye, 1993: *Guidelines for Cost-Effectiveness Analysis of Vector Control,* WHO , Geneva, Switzerland/Food and Agriculture Organization of the United Nations, Rome, Italy/ United Nations Environment Programme, Nairobi, Kenya/ United Nations Centre for Human Settlements, Geneva, Switzerland.

Pitt, S., B.E. Pearcy, R.H. Stevens, A. Sharipov, K. Satarov, et al., 1998: War in Tajikistan and re-emergence of Plasmodium falciparum, *Lancet,* **352,** p. 1279.

Polson, K., S. Rawlins, D. Chadee, and C. Brown, 2000: A case-control study of Dengue fever cases in Trinidad in 1997, *West Indian Medical Journal,* **49(Suppl. 2),** pp. 58–9.

Reiter, P., 2000: From Shakespeare to Defoe: Malaria in England in the Little Ice Age, *Emerging Infectious Diseases,* **6,** pp. 1–11.

Reiter, P., 2001: Climate change and mosquito-borne disease, *Environmental Health Perspectives,* **109(Suppl. 1),** pp. 141–61.

Reiter, P., S. Lathrop, et al., 2003: Texas lifestyle limits transmission of Dengue virus, *Emerging Infectious Diseases,* **9,** pp. 86–9.

Ribeiro, J.M. and M.G. Kidwell, 1994: Transposable elements as population drive mechanisms: Specification of critical parameter values, *Journal of Medical Entemology,* **31,** pp. 10–6.

Rosenbaum, J., M.B. Nathan, R. Ragoonanansingh, S. Rawlins, C. Gayle, et al., 1995: Community participation in dengue prevention and control: A survey of knowledge, attitudes, and practice in Trinidad and Tobago, *American Journal of Tropical Medicine and Hygene,* **53,** pp. 111–7.

Rowland, M. and F. Nosten, 2001: Malaria epidemiology and control in refugee camps and complex emergencies, *Annals of Tropical Medicine and Parasitology,* **95,** pp. 741–54.

Rowland, M., N. Durrani, M. Kenward, N. Mohammed, H. Urahman, and S. Hewitt, 2001: Control of malaria in Pakistan by applying deltamethrin insecticide to cattle: A community-randomised trial, *Lancet,* **357,** pp. 1837–41.

Rozendaal, J., 1997: *Vector Control: Methods for Use by Individuals and Communities,* WHO, Geneva, Switzerland.

Saf'Janova, V., 1985: Leishmaniasies in the USSR. In: *Leishmaniasis,* K. Chang and R. Bray (eds.), Elsevier, Amsterdam, The Netherlands.

Sampaio, L., 1951: Medicina rural: O aparecimento, a expansão e o fim da leishmaniose no estado de São Paulo, *Revista Brasileira de Medicina,* **8,** pp. 717–21.

Schultz, G.W., 1989: Animal influence on man-biting rates at a malarious site in Palawan, Philippines, *Southeast Asian Journal of Tropical Medicine and Public Health,* **20,** pp. 49–53.

Scott, T.W., W. Takken, B.G. Knols, and C. Boete, 2002: The ecology of genetically modified mosquitoes, *Science,* **298,** pp. 117–9.

Sergiev, V., 1979: Epidemiology of leishmaniasis in the USSR. In: *Biology of the Kinetoplastida,* H. Lumsden and D. Evans (eds.), Academic Press, London, UK.

Shililu, J., T. Ghebremeskel, S. Mengistu, H. Fekadu, M. Zerom, et al., 2003a: Distribution of anopheline mosquitoes in Eritrea, *American Journal of Tropical Medicine and Hygene,* **69,** pp. 295–302.

Shililu, J., T. Ghebremeskel, S. Mengistu, H. Fekadu, M. Zerom, et al., 2003b: High seasonal variation in entomologic inoculation rates in Eritrea, a semi-arid region of unstable malaria in Africa, *American Journal of Tropical Medicine and Hygene,* **69,** pp. 607–13.

Sinkins, S. and S. O'Neill, 2000: Wolbachia as a vehicle to modify insect populations, *Insect Transgenesis: Methods and Applications,* CRC Press, Boca Raton, FL.

Takken, W., W.B. Snellen, J.P. Verhave, B.G.J. Knols, and S. Atmosoedjono, 1991: *Environmental Measures for Malaria Control in Indonesia: An Historical Review on Species Sanitation,* Agricultural University Wageningen, Wageningen, The Netherlands.

Thomson, C., J. Harper, and M. Begon, 2000: *Essentials of Ecology,* Blackwell Science, Oxford, UK.

Tiffen, M., 1991: *Guidelines for the Incorporation of Health Safeguards into Irrigation Projects through Intersectoral Cooperation,* WHO , Geneva, Switzerland/ Food and Agriculture Organization of the United Nations, Rome, Italy/ UN Environmental Programme, Nairobi, Kenya/ UN Centre for Human Settlements(Habitat), Geneva, Switzerland.

Turelli, M. and A.A. Hoffmann, 1991: Rapid spread of an inherited incompatibility factor in California Drosophila, *Nature,* **353,** pp. 440–2.

Utzinger, J., Y. Tozan, and B.H. Singer, 2001: Efficacy and cost-effectiveness of environmental management for malaria control, *Tropical Medicine and International Health,* **6,** pp. 677–87.

Utzinger, J., Y. Tozan, F. Doumani, and B.H. Singer, 2002: The economic payoffs of integrated malaria control in the Zambian copperbelt between 1930 and 1950, *Tropical Medicine and International Health,* **7,** pp. 657–77.

van der Hoek, W., F. Konradsen, D.S. Dijkstra, P.H. Amerasinghe, and F.P. Amerasinghe, 1998: Risk factors for malaria: A microepidemiological study in a village in Sri Lanka, *Transactions of the Royal Society of Tropical Medicine and Hygiene,* **92,** pp. 265–9.

Victor, T.J., B. Chandrasekaran, and R. Reuben, 1994: Composite fish culture for mosquito control in rice fields in southern India, *Southeast Asian Journal of Tropical Medicine Public Health,* **25,** pp. 522–7.

Vioukov, V., 1987: Control of tranbsmission. In: *The Leishmaniases in Biology and Epidemiology,* W. Peters and R. Killick-Kendrick (eds.), Academic Press, London, UK.

Walker, K., 2002: *A Review of Control Methods for African Malaria Vectors,* Environmental Health Project, US Agency for International Development, Washington, DC.

Walsh, J.F., D.H. Molyneux, and M.H. Birley, 1993: Deforestation: Effects on vector-borne disease, *Parasitology,* **106(Suppl.),** pp. S55–75.

WHO (World Health Organization), 1982: *Manual for Environmental Management for Mosquito Control, with Special Emphasis on Malaria Vectors,* WHO, Geneva, Switzerland, 281 pp.

WHO, 1990: *Control of the Leishmaniases, Report of a WHO Expert Committee,* WHO, Geneva, Switzerland.

WHO, 1999: *Bacillus Thurinigiensis, Environmental Health Criteria,* WHO, Geneva, Switzerland, 217 pp.

WHO, 2001: *The World Health Report 2001: Mental Health: New Understanding, New Hope,* WHO, Geneva, Switzerland.

WHO, 2003: *Guidelines for Integrated Vector Management,* WHO Regional Office for Africa, Harare, Zimbabwe.

WHO, 2004a: *The World Health Report 2004: Changing History,* WHO, Geneva, Switzerland.

WHO, 2004b: *Global Strategic Framework for Integrated Vector Management,* WHO, Geneva, Switzerland.

Wilder-Smith, A., W. Foo, A. Earnest, S. Sremulanathan, and N.I. Paton, 2004: Seroepidemiology of dengue in the adult population of Singapore, *Tropical Medicine and International Health,* **9,** pp. 305–8.

Wu, N., G.H. Liao, D.F. Li, Y.L. Luo, and G.M. Zhong, 1991: The advantages of mosquito biocontrol by stocking edible fish in rice paddies, *Southeast Asian Journal of Tropical Medicine and Public Health,* **22,** pp. 436–42.

Wu, W.K. and R.B. Tesh, 1990: Selection of Phlebotomus papatasi (Diptera: Psychodidae) lines susceptible and refractory to Leishmania major infection, *American Journal of Tropical Medicine and Hygiene,* **42,** pp. 320–8.

Zavaleta, J.O. and P.A. Rossignol (2004): Community-level analysis of risk of vector-borne disease, *Transactions of the Royal Society of Tropical Medicine and Hygiene,* **98(10),** pp. 610–8.

Chapter 13

Climate Change

Coordinating Lead Authors: Ian Noble, Jyoti Parikh, Robert Watson
Lead Authors: Richard Howarth, Richard J. T. Klein
Contributing Authors: Allali Abdelkader, Tim Forsyth
Review Editors: Pavel Kabat, Shuzo Nishioka

BOXES

FIGURES

TABLES

*These appear in Appendix A at the end of this volume.

Main Messages

There is wide recognition that human-induced climate change is a serious environmental and development issue. The ultimate goal of the United Nations Framework Convention on Climate Change is "stabilization of greenhouse gas concentrations in the atmosphere at a level that would prevent dangerous anthropogenic interference with the climate system." Such a level should be achieved within a "time frame sufficient to allow ecosystems to adapt naturally to climate change, to ensure that food production is not threatened, and to enable economic development to proceed in a sustainable manner." The Kyoto Protocol, which entered into force in February 2005, contains binding limits on greenhouse gas emissions on industrial countries that agreed to reduce their emissions by an average of about 5% during 2008–2012 relative to the levels emitted in 1990.

Earth is warming, with most of the warming of the last 50 years attributable to human activities (that is, emissions of greenhouse gases); precipitation patterns are changing, and sea level is rising. Human activities have significantly increased the atmospheric concentrations of numerous greenhouse gases since the pre-industrial era, with most gases projected to increase significantly over the next 100 years (for example, carbon dioxide has increased from about 280 to 370 parts per million, and is projected to increase to between 540 and 970 parts per million by 2100). The global mean surface temperature has increased by about 0.6° Celsius over the last 100 years, and is projected to increase by a further 1.4°–5.8° Celsius by 2100. The spatial and temporal patterns of precipitation have already changed and are projected to change even more in the future, with an increasing incidence of floods and droughts. Sea levels have already risen 10–25 centimeters during the last 100 years and are projected to rise an additional 8–88 centimeters by 2100.

Observed changes in climate have already affected ecological, social, and economic systems, and the achievement of sustainable development is threatened by projected changes in climate. The timing of reproduction of animals and plants and/or migration of animals, the length of the growing season, species distributions and population sizes, and the frequency of pest and disease outbreaks have already been affected, especially by increased regional temperatures. Over the next 100 years, water availability and quality will decrease in many arid and semiarid regions, with increased risk of floods and droughts; the incidence of vector and water-borne diseases will increase in many regions; agricultural productivity will decrease in the tropics and subtropics for almost any amount of warming; and many ecological systems, their biodiversity, and their goods and services will be adversely impacted.

Adverse consequences of climate change can be reduced by adaptation measures, but cannot be completely eliminated. Even with best-practice management it is inevitable that some species will be lost, some ecosystems irreversibly modified, and some environmental goods and services adversely affected. These changes will expose human populations to risks of damage from climate change, some of which may be met with current coping systems; others may need radically new behaviors. Climate change needs to be factored into current development plans, lest hard won gains are threatened by new climatic conditions and the changes in ecosystem services that follow. Successful adaptation will require the efforts of many institutions, ranging from including adaptation in national development planning to supporting community level responses to coping better with changing conditions. Adaptation activities range from economic measures such as insurance for extreme events, capacity building for alternative crop cultivation and for managing the impact of sea level rise, infrastructure and investment for water storage, ground water recharge, storm protection, flood mitigation, shoreline stabilization, and erosion control.

Based on the current understanding of the climate system, and the response of different ecological and socioeconomic systems, *if* significant global adverse changes to ecosystems are to be avoided, the best guidance that can currently be given suggests that efforts be made to limit the increase in global mean surface temperature to less than 2° Celsius above pre-industrial levels and to limit the rate of change to less than 0.2° Celsius per decade. This will require that the atmospheric concentration of carbon dioxide be limited to about 450 parts per million and the emissions of other greenhouse gases stabilized or reduced. In suggesting these targets, it is recognized that: (1) the adverse effects of climate change on ecosystems are already apparent, and (2) the threshold from which damage to ecosystems and critical sectors such as agriculture and water resources is no longer acceptable cannot be determined precisely. Even the suggested maximum tolerable changes in global mean surface temperature will cause, on average, adverse consequences in developing countries, suggesting that adaptation will be required in these countries. All countries with significant emissions would need to reduce their projected greenhouse gas emissions. Key issues will include setting intermediate targets and an equitable allocation of emissions rights that recognizes the principle of common but differentiated responsibilities that is embodied in the UNFCCC. These long-term targets would be reviewed from time to time in light of emerging new scientific understanding.

Significant reductions in net greenhouse gas emissions are technically feasible due to an extensive array of technologies in the energy supply, energy demand, and waste management sectors, many at little or no cost to society. Reducing greenhouse gas emissions will require a portfolio of energy production technologies including fuel switching (coal/oil to gas), increased power plant efficiency, carbon dioxide capture and storage (pre- and post-combustion), and increased use of renewable energy technologies (biomass, solar, wind, run-of-the-river and large hydropower, geothermal, etc.) and nuclear power, complemented by more efficient use of energy in the transportation, buildings, and industry sectors.

Realizing the technical potential to reduce greenhouse gas emissions will involve the development and implementation of supporting institutions and policies to overcome barriers to the diffusion of these technologies into the marketplace, increased public and private sector funding for research and development, and effective technology transfer. Significant restructuring of the energy system will require several different types of responses to converge and consolidate one another. For example, legal and institutional responses give rise to economic incentives which, in turn, will push technological initiatives such as renewable energy and energy efficiency

Afforestation, reforestation, improved forest, cropland and rangeland management, and agroforestry provide a wide range of opportunities to increase carbon uptake; and slowing deforestation provides an opportunity to reduce emissions. Land use, land-use change, and forestry activities have the potential to sequester about 100 giga tons of carbon by 2050, which is equivalent to 10–20% of projected fossil emissions over the same period. However, the current financial and institutional environment, as well as competing land uses, makes it possible to deliver only a small part of this potential. The rules of the Kyoto Protocol further constrain the entry of LULUCF sequestration into the compliance system to only about 100 megatons of carbon per year.

Policies and programs are needed to facilitate the widespread deployment of climate-friendly energy production and use technologies. These include: energy pricing strategies, carbon taxes, removing subsidies that increase greenhouse gas emissions, internalizing externalities, domestic and international tradable emissions permits, voluntary programs, incentives for use of new technologies during market build-up, regulatory programs including energy-efficiency standards, education and training such as product advisories and labels, and intensified research and development. These types of policies are needed for effective penetration of renewable energy technologies and energy-efficient technologies into the market.

Market mechanisms and incentives can significantly reduce the costs of mitigation. International project-based and emissions-rights trading mechanisms allowed under the Kyoto Protocol, in combination with national and regional mechanisms, can reduce the costs of mitigation for countries belonging to the Organization for Economic Cooperation and Development. In addition, countries can reduce net costs of emissions abatement by taxing emissions (or auctioning permits) and using the revenues to cut distortion taxes on labor and capital. In the near term, project-based trading can facilitate the transfer of climate-friendly technologies to developing countries.

The long-term costs of stabilization of carbon dioxide at 450 parts per million will have a negligible effect on the growth of global gross domestic product. The cost of stabilization depends on the stabilization level, the baseline emissions scenario, and the pathway to stabilization. The reduction in projected GDP increases moderately when passing from a 750 to a 550 parts per million concentration stabilization level, with a much larger increase in passing from 550 to 450. The percentage reduction in global average GDP over the next 100 years for stabilization at 450 parts per million is about 0.02–0.1% per year, compared to projected annual average GDP growth rates of 2–3% per year.

Irrespective of the de-carbonization pathway eventually followed, some climate change is inevitable and consequently ecosystems and human societies will need to adapt to new conditions. The existing elevated greenhouse gas levels in the atmosphere are already affecting the climate and changing biological systems. There will be long time-lags as ecosystems and, in particular, oceans adjust to these new conditions.

Addressing climate change will require governments, the private sector, bilateral and multilateral agencies, the Global Environment Facility, nongovernmental organizations, and consumers to play a critical role in mitigating, and adapting to, climate change. Different actors have different roles along the research, development, demonstration, and widespread deployment value chain and pipeline for climate-friendly technologies. Innovative partnerships will be particularly important in technology transfer and financing.

13.1 Introduction

Human-induced climate change is one of the most important environmental and development issues facing society worldwide as recognized by the United Nations Framework Convention on Climate Change. The overwhelming majority of scientific experts and governments recognize that while scientific uncertainties exist, there is strong scientific evidence demonstrating that human activities are changing Earth's climate and that further human-induced climate change is inevitable. The main drivers of climate change are demographic, economic, sociopolitical, technological, and behavioral choices. These driving forces determine the future demand for energy and changes in land use which, in turn, affect emissions of greenhouse gases and aerosol precursors that, in their turn, cause changes in Earth's climate. There are several important anthropogenic greenhouse gases—including carbon dioxide, methane, nitrous oxide, ozone (changes in tropospheric and stratospheric ozone have different impacts on Earth's climate), and halogenated compounds—but the single most important one is carbon dioxide, primarily because of the large emissions resulting from energy production and use and burning associated with land use change. Changes in Earth's climate have and will continue to adversely affect ecological systems, their biodiversity, and human well-being.

This chapter follows the basic outline of the MA conceptual framework (MA 2003). It starts by briefly discussing the drivers of change, that is, the observed and projected changes in greenhouse gas emissions and concentrations and the observed and projected changes in Earth's climate. It then discusses the observed and projected impacts of these drivers on ecological systems, socioeconomic sectors, and human health. This is followed by a brief discussion of the international and national legal responses to the challenge of climate change and a discussion of the scale of response needed to limit damages to ecological systems and human well-being, suggesting limits to the magnitude and rate of change of global mean surface temperature and the implications for greenhouse gas emissions. This is followed by a discussion of adaptation and mitigation (energy and land use, land use change, and forestry) options, economic instruments and costs, and institutional responses.

This chapter has a slightly broader scope than most in the responses volume. It assesses both the overall potential responses to the threat of human-induced climate change (net emissions reductions in both energy and land management) and what responses have been used or proposed to either enhance the ecosystem's ability to provide services or to mitigate impacts that undermine its ability to provide that service (primarily through net carbon sequestration activities).This chapter also assesses the impacts on other ecosystem goods and services, for example, biodiversity.

The chapter is based extensively on the expert and government peer-reviewed comprehensive reports from the Intergovernmental Panel on Climate Change (especially 2000a, 2000b, 2000c, 2001a, 2001b, 2001c, 2001d, 2002). It is also based substantially on the CBD report on biological diversity and climate change (CBD 2003) and the World Energy Assessment (WEA 2000). Given the comprehensive nature of the work of the IPCC and the World Energy Assessment, this chapter highlights their key conclusions, summarizes recent papers that either support or refute their conclusions, and provides the reader with a guide for additional in-depth analysis.

13.1.1 Observed and Projected Greenhouse Gas Emissions and Concentrations

The atmospheric concentrations of several greenhouse gases, which tend to warm the atmosphere, have increased substantially since the pre-industrial era (around 1750) due to human activities (IPCC 2001a, Chapters 3, 4). For example, carbon dioxide has increased about 31% (from 280 to 370 parts per million) due to the combustion of fossil fuels (coal, oil, and gas); industrial processes, especially cement production; and changes in land use (predominantly deforestation in the tropics). Methane has more than doubled (from 750 to 1,750 parts per billion), mainly due to increased number of livestock, rice production, waste disposal, and leakage from natural gas pipelines. Nitrous oxide has increased by about 17% (from about 265 to about 312 parts per billion), primarily from agricultural soils, cattle feed lots, and the chemical industry.

Sulfate aerosol concentrations, which tend to cool the atmosphere, have increased regionally since the pre-industrial era, primarily due to the combustion of coal. However, since the early 1990s, emissions of sulfur dioxide are decreasing in many regions of North America and Europe because of stringent regulations.

In the year 2000, developing countries, transition economy countries, and developed countries emitted 1.6, 1.7, and 3.1 gigatons, respectively, of the fossil fuel carbon emissions (about 6.4 gigatons total). Thus per capita energy carbon emissions in developed countries are about ten times those in developing countries and about 2.8 times those in transition economy countries (0.4, 1.4, and 3.9 tons, respectively, for developing, transition, and developed countries) (Holdren 2003). In addition, another 1.6 gigatons was emitted as a result of land use changes, almost exclusively by developing countries in the tropics (IPCC 2001a, Chapters 3, 4; IPCC 2000b). Hence, industrial countries (developed countries and countries with economies in transition) having about 20% of the world's population emitted about 4.8 gigatons of carbon; in contrast, developing countries having about 80% of the world's population emitted about 3.2 gigatons. Historically, over 80% of anthropogenic emissions of greenhouse gases have emanated from industrial countries (Holdren 2003).

The quantity of emissions of carbon dioxide depends on the projected magnitude of energy services as well as the technologies used to produce and use it. The IPCC (2000a) projected emissions of greenhouse gases from 1990–2100 arising from, inter alia, energy services as well as biological resources. The demand for energy services is growing rapidly, particularly in developing countries, where cost-effective energy is critical for poverty alleviation and economic development. Indeed, nearly all of the population growth and most of the energy growth will occur in developing countries. IPCC projected world population, GDP, and energy demand under various scenarios. It projected that world population would increase from 5.3 billion people in 1990 to somewhere between 7 and 15 billion people in 2100, and world GDP would increase from $21 trillion in 1990 to between $200 and $550 trillion by 2100. IPCC projected that demand for primary energy would increase from 351 exajoule per year in 1990, to between 640 and 1,610 exajoule per year by 2050, and 515 to 2,740 exajoule per year by 2100, driven primarily by the projected increases in world GDP and changes in population. These projected increases in the demand for primary energy resulted in projected carbon dioxide emissions of 8.5–26.8 gigatons of carbon per year in 2050, and 3.3–36.8 gigatons in 2100, compared to 6.0 gigatons in 1990, assuming there are no concerted efforts internationally to protect the climate system. (See Figure 13.1A in Appendix A).

Anthropogenic methane emissions were projected to range from 359 to 671 megatons per year in 2050, and from 236 to 1,069 megatons in 2100 (compared to 310 megatons in 1990). Sulfur dioxide emissions were projected to range from 29 to 141 megatons sulfur per year in 2050, and from 11 to 93 in 2100 (compared to 70.9 megatons sulfur in 1990). These projected emissions for sulfur dioxide are significantly lower than those projected by IPCC in 1992.

The IPCC scenarios resulted in a broad range of projected greenhouse gas and aerosol concentrations (IPCC 2001a, Chapters 3, 4). For example, the atmospheric concentration of carbon dioxide was projected to increase from the current level of about 370 parts per million to between 540 and 970 parts per million by 2100 (Figure 13.1b), without taking into account the climate-induced additional releases of carbon dioxide from the biosphere in a warmer world (IPCC 2001a, Chapter 3; Cox et al. 2000; Leemans et al. 2002). The atmospheric concentration of methane was projected to change from the current level of 1,750 parts per billion to between 1,600 to 3,750 parts per billion by 2100.

13.1.2 Observed and Projected Changes in Climate

Earth's climate has warmed, on average, by about 0.6° Celsius, over the past 100 years, with the decade of the 1990s being the warmest in the instrumental record (1861 to the present). The temporal and spatial patterns of precipitation have changed, sea levels have risen 10 to 25 centimeters, most non-polar glaciers are retreating, and the extent and thickness of Arctic sea ice in summer are decreasing (IPCC 2001a, Chapters 2, 4). One contentious issue is the discrepancy between the recent ground-based and satellite-based trends in temperature. Work is continuing to resolve this issue. Fu et al. (2004) have suggested that the MSU satellite trends of tropospheric temperatures are an underestimate due to contamination of the signal by a component from the lower stratosphere, which has been cooling. In addition, Jin et al. (2004) have reported, using a different satellite technique (AVHRR), that there is no significant difference in trends between ground-based and satellite-based trends over the last 18 years. Most of the observed warming of the past 50 years can be attributed to human activities, increasing the atmospheric concentrations of greenhouse gases and aerosols, rather than changes in solar radiation or other natural factors (IPCC 2001a, Chapter 12). Changes in sea level, snow cover, ice extent, and precipitation are consistent with a warmer climate (IPCC 2001a, Chapter 11).

Projected changes in the atmospheric concentrations of greenhouse gases and aerosols are projected to result in global mean surface temperatures increases of 1.4°–5.8° Celsius between 1990 and 2100 (Figure 13.2 in Appendix A), with land areas warming more than the oceans (IPCC 2001a, Chapter 9). About half of the range of projected changes in temperature is due to uncertainties in the climate sensitivity factor and about half is due to the wide range of projected changes in greenhouse gas and aerosol precursor emissions. Globally average precipitation is projected to increase, but with increases and decreases in particular regions, accompanied by more intense precipitation events over most regions of the world; global mean sea level is projected to rise by between 8 and 88 centimeters between 1990 and 2100 (IPCC 2001a, Chapters 9, 10, 11). The climate is projected to become more El Niño–like, and the incidence of extreme weather events is projected to increase, especially hot days, floods, and droughts.

13.1.3 Impacts of Climate Change on Ecological Systems, Socioeconomic Sectors, and Human Health

Observed changes in climate, especially warmer regional temperatures, have already affected biological systems in many parts of the world (IPCC 2001b, Chapters 5, 10, 13; IPCC 2002; CBD 2003, Chapter 2). There have been changes in species distributions, population sizes, the timing of reproduction or migration events, and an increase in the frequency of pest and disease outbreaks, especially in forested systems. Many coral reefs have undergone major, although often partially reversible, bleaching episodes, when sea surface temperatures have increased by 1° Celsius during a single season (IPCC 2001b, Chapters 6, 17), with extensive mortality occurring with observed increases in temperature of 3° Celsius.

While the growing season in Europe has lengthened over the last 30 years, in some regions of Africa the combination of regional climate changes and anthropogenic stresses has led to decreased cereal crop production since 1970. Changes in fish populations have been linked to large-scale climate oscillations. For example, El Nino events have impacted fisheries off the coasts of South America and Africa, and decadal oscillations in the Pacific have impacted fisheries off the west coast of North America (IPCC 2001b, Chapters 10, 14, 15).

Climate change is projected to further adversely affect key development challenges including the provision of clean water, energy services, and food; maintaining a healthy environment; and conserving ecological systems, their biodiversity, and associated ecological goods and services—the so-called WEHAB priorities (water, energy, health, agriculture, and biodiversity) discussed at the World Summit on Sustainable Development at Johannesburg in 2002. Water availability and quality is projected to decrease in many arid and semi-arid regions, with increased risk of floods and droughts (IPCC 2001b, Chapter 4); the reliability of hydropower and biomass production is projected to decrease in many regions; the incidence of vector-borne (for example, malaria and dengue) and water-borne (for example, cholera) diseases is projected to increase in many regions and so too is heat/cold stress mortality and threats of decreased nutrition in others, along with severe weather-related traumatic injury and death (IPCC 2001b, Chapters 5, 9). Agricultural productivity is projected to decrease in the tropics and sub-tropics with almost any amount of warming (IPCC 2001b, Chapters 5, 9), and there are projected adverse effects on fisheries; and many ecological systems, their biodiversity, and their goods and services are projected to be adversely affected (IPCC 2001b, Chapters 5, 16, 17, 19).

Changes in climate projected for the twenty-first century will occur faster than they have in at least the past 10,000 years and, combined with changes in land use and the spread of exotic/alien species, are likely to limit both the capability of species to migrate and the ability of species to persist in fragmented habitats (IPCC 2001b, Chapters 5, 16, 17, 19). Climate change is projected to exacerbate the loss of biodiversity; increase the risk of extinction for many species, especially those that are already at risk due to factors such as low population numbers, restricted or patchy habitats, and limited climatic ranges; change the structure and functioning of ecosystems; and adversely impact ecosystem services essential for sustainable development.

A recent paper, using the climate envelope/species-area technique, estimated that the projected changes in climate by 2050 could lead to an eventual extinction of 15–52% of the subset of 1,103 endemic species (mammals, birds, frogs, reptiles, butterflies, and plants) analyzed (Thomas et al. 2004). As noted, other studies have shown that these changes are already occurring locally (Root et al. 2003; van Oene 2001). Some ecosystems, such as coral reefs, mangroves, high mountain ecosystems, remnant native grasslands, and ecosystems overlying permafrost, are particularly vulnerable to climate change. For a given ecosystem, functionally diverse communities are likely to be better able to adapt to climate change and climate variability than impoverished ones.

13.1.4 Approaches to Mitigation and Adaptation

Addressing the challenge of climate change will require a broad range of mitigation and adaptation activities. Mitigation involves the reduction of net emissions, while adaptation involves measures to increase the capability to cope with impacts. Greenhouse gas emissions are highly dependent upon the development pathway, and approaches to mitigate climate change will be both affected by, and have impacts on, broader socioeconomic policies and trends, those relating to development, sustainability, and equity. Lower emissions will require different patterns of energy resource development and utilization (trend toward de-carbonization) and increases in end-use efficiency.

The IPCC (2001c) concluded that significant reductions in net greenhouse gas emissions are technically feasible due to an extensive array of technologies in the energy supply, energy demand, and waste management sectors, many at little or no cost to society. However, realizing these emission reductions involves the development and implementation of supporting policies to overcome barriers to the diffusion of these technologies into the marketplace, increased public and private sector funding for research and development, and effective technology transfer (North-South and South-South). In addition, afforestation; reforestation; improved forest, cropland and rangeland management, and agroforestry provide a wide range of opportunities to increase carbon uptake, and slowing deforestation provides an opportunity to reduce emissions (IPCC 2000a; CBD 2003, Chapter 4).

Irrespective of the de-carbonization pathway eventually followed, some climate change is inevitable. The existing elevated greenhouse gas levels in the atmosphere are already affecting climate and changing biological systems. There will be long time-lags as ecosystems and, in particular, oceans adjust to these new conditions. These changes will expose human populations to risks of damage from climate change, some of which may be met with current coping systems.

Adaptation to climate change needs to be factored into current development plans and coordinated with strategies for hazard management and poverty reduction, lest hard won development gains be threatened by new climatic conditions and the changes in ecosystem services that follow.

Adaptation is becoming an increasingly important issue in the international negotiations on climate change. The burden of adaptation will fall most heavily on developing nations. Many are in regions that are most affected by additional flooding, by reduced crop yields in tropical regions, or by increased rainfall variability in marginal agricultural zones. As developing nations, they have the least resources to commit to adaptive actions and policies. The core issues are to identify the impacts of climate change, to devise responses from the community level to national planning, and to determine how the necessary responses should be funded. Clearly, adaptation measures need to be an integral part of any national program or action plan for combating climate change. Implementation of such a plan would be beneficial for all, especially the most vulnerable. Further, the International Institute for Sustainable Development (2003) suggests that coordinated strategies for disaster management, climate change, environmental management, and poverty reduction can reduce the burden to adapt.

13.2 Legal Response

13.2.1 UNFCCC and the Kyoto Protocol

The long-term challenge is to meet the goal of UNFCCC Article 2, that is, "stabilization of greenhouse gas concentrations in the atmosphere at a level that would prevent dangerous anthropogenic interference with the climate system, and in a time frame sufficient to allow ecosystems to adapt naturally to climate change, to ensure that food production is not threatened, and to enable economic development to proceed in a sustainable manner." The UNFCCC also specifies several principles to guide this process: equity, common but differentiated responsibilities, precaution, cost-effective measures, right to sustainable development, and support for an open economic system (Article 3).

The most comprehensive attempt to negotiate binding limits on greenhouse gas emissions is contained in the 1997 Kyoto Protocol, an agreement forged in a meeting of more than 160 nations, in which most developed countries (Annex 1 countries, in the Kyoto terminology) agreed to reduce their emissions by an average of about 5% between 2008 and 2012 relative to the levels emitted in 1990. In line with agreed differentiated responsibilities, the targets of the industrialized countries vary from an 8% reduction to a 10% increase. The Kyoto Protocol contains a number of core elements: these include a set of compliance rules; LULUCF activities; and flexibility mechanisms, for example, Joint Implementation for trading between developed nations and the Clean Development Mechanism for trading between developed and developing nations.

The implementation of the Kyoto Protocol provides both challenges and opportunities. A strong enabling context at the national and international level will be required to implement environmentally sound and socially equitable JI and CDM projects. Also, there is the potential for synergies between mitigation and adaptation activities in LULUCF, which would provide significant sustainable development benefits.

The Kyoto Protocol entered into force when at least 55 countries that were collectively responsible for at least 55% of Annex 1 emissions had ratified the Protocol. This occurred in February 2005 after ratification by the Russian Federation. The United States and Australia are the only major industrialized countries not to have ratified.

Provision is made within the Kyoto Protocol for parties to act jointly in achieving their emission reduction targets. The European Union formed such an agreement; often called the "EU bubble." Their target is for the EU as a whole to achieve an 8% reduction in emissions over their 1990 baseline. However, through an internal redistribution of responsibilities some member states have committed themselves to reducing their emissions by up to 28% while others will limit their increase in emissions to 27%. Europe has recently set up an emissions trading system that is legally independent of, but linked to, the Kyoto Protocol to assist sectors that are major emitters to achieve emission reductions.

13.2.2 Actions outside the Kyoto Protocol Aimed at Emission Reductions

The United States and Australia have stated that they do not intend to ratify the Kyoto Protocol. Nevertheless, they have policies and activities in place to meet obligations consistent with the UNFCCC. The United States has set a national goal to reduce its greenhouse gas intensity (measured as the ratio of greenhouse gases emitted per real GDP) by 18% by 2012 (Abraham 2004), while Australia has stated that it will achieve emission reductions equivalent to the target set in 1997 at Kyoto (Kemp 2004). Both countries have programs of policy measures including financial incentives, voluntary programs, and research priorities to reduce greenhouse gas emissions. Both are involved, along with numerous other countries, in a number of strategic research partnerships, for example, the hydrogen economy, carbon capture and storage, and global observations.

As reported in early 2004, 28 states within the United States as well as Puerto Rico have developed or are developing strategies or action plans to reduce net greenhouse gas emissions. These states have enacted legislation requiring utilities to increase their use of renewable energy sources such as wind power or biomass in generating a portion of their overall electricity or they have provided incentives for other clean technologies. Several states have set numeric goals for reducing emissions to mitigate climate change. The New England states have joined with eastern Canadian provinces in a goal of reducing greenhouse gas emissions to 1990 levels by 2010 and then another 10% lower by 2020 (Anon. 2004a). There have been similar moves at state level in Australia.

Some developing countries, such as India (Parikh 2004) and China, have also responded with policies and measures to reduce greenhouse gas emissions, often to achieve progress toward sustainable development. India has implemented measures to increase energy efficiency and conservation and to incorporate the use of renewable energy sources, especially as part of its rural electrification programs. In the late 1990s, China partly decoupled its growth in GDP from greenhouse gas emissions through, inter alia, changes in energy policies and through industrial transformation, that is, closing inefficient and polluting small- and medium-sized enterprises (Streets et al. 2001). However, more recent data indicate that China's emissions are again rising although its energy intensity continues to decline (Anon. 2004b). The challenge facing China is to achieve high effectiveness in the use of fossil fuels during a prolonged period of economic expansion. Climate policies, as is discussed below, also help contain air pollution, and lead to health benefits and efficient resource utilization, including that of financial resources.

13.2.3 Related Conventions

Strong scientific and policy interlinkages exist between the UNFCCC, the Convention on Biological Diversity, and the Convention to Combat Desertification. The three objectives of the CBD are the conservation, sustainable use, and equitable sharing of the benefits of biodiversity. Given that climate change disrupts ecosystems and their biodiversity and, in turn, changes in ecosystems can affect climate change through changes in biogeochemical cycling and surface albedo, identifying, developing, and implementing technologies and policies (activities) that have mutually positive effects is critical, while avoiding activities that positively impact on one issue but adversely affect the other. The Conference of Parties to the CBD first requested the IPCC to prepare a Technical Paper on Climate Change and Biodiversity (IPCC 2002) and then through its Subsidiary Body on Scientific, Technical, and Technological Advice established an ad hoc technical expert group on biological diversity and climate change (CBD 2003) to assess the scientific and policy interlinkages among the two issues. LULUCF activities, which play a particularly important role in both conventions, when used to sequester carbon dioxide, can have positive, neutral, or negative effects on biodiversity. LULUCF activities are directly amenable to the ecosystem approach adopted by the CBD, which is a strategy for integrated adaptive management of land, water, and living resources. It is important to note that energy technologies that have lower greenhouse gas emissions relative to fossil fuels can, in some instances, have negative effects on biodiversity.

13.3 Scale of Response Needed

Defining "dangerous" anthropogenic interference with the climate system (Article 2 of the UNFCCC) is not a simple task because the vulnerability of sectors, countries, and individuals to climate change varies significantly. As defined by IPCC, vulnerability includes the capacity of communities to adapt to climate change. One sector or group of individuals may possibly benefit from human-induced climate change, whereas another sector or group of individuals may be adversely affected. Most people and most sectors are adversely affected by climate change in the tropics, sub-tropics, and low-lying coastal areas, whereas in mid- and high-latitudes cold-related deaths may decrease and agricultural productivity may increase with small increases in temperature. The question is whether the most sensitive sectors and individuals should be protected or whether the average sector or individual should be protected. Therefore, defining dangerous anthropogenic interference with the climate system involves a value judgment determined not solely through science but invoking a sociopolitical process informed by technical and socioeconomic information.

13.3.1 Ecological Justification for Setting Targets for Limiting the Rate of Change of Climate and Absolute Climate Change

It is well recognized that: (1) scientific uncertainties exist in linking greenhouse gas emissions to regional changes in climate, and in linking changes in regional climate to sector-specific impacts; (2) there are significant variations in the responses of socioeconomic and ecological sectors to changes in climate in different parts of the world; and (3) defining what constitutes dangerous anthropogenic perturbation to the climate system as referred to in Article 2 of the UNFCCC is a value judgment determined through sociopolitical processes. However, enough is known to set a target for a "maximum tolerable" change in global mean surface temperature and the rate of change in global mean surface temperature if significant changes in ecological systems and their biodiversity and goods and services are to be avoided as mandated under the Convention and damages to socioeconomic systems and human health are to be limited in developing countries.

There are a number of cogent arguments in favor of setting a target (Pershing and Tudela 2003), including:

- providing a firm goal for current and future climate efforts,
- increasing awareness of the long-term consequences of our actions,
- calibrating short-term measures and measuring progress,
- inducing technological change,
- limiting future risks from climate change,
- mobilizing societies to understand the adverse consequences of climate change and to change their consumption patterns, and
- promoting global participation.

However, even if there is agreement that a long-term target is useful and politically feasible there is a debate as to whether the targets should be based on: utilization of specific technologies; emissions of greenhouse gases; greenhouse gas concentrations; global mean surface temperature and/or the rate of change of global mean surface temperature; or impacts on socioeconomic systems, ecological systems, or human health, or a combination of some or all of the above.

This assessment suggests that if decision-makers want to protect unique and threatened species and limit, although not avoid, the threats to development in developing countries, a "temperature derived" greenhouse gas concentration target, consistent with the approach taken in Article 2 of the Convention, that is, stabilization of the atmospheric concentration of greenhouse gases, can be established. Based on the current understanding of the climate system and how ecological and socioeconomic sectors respond to changes in regional climate, it can be argued that the maximum tolerable increase in global mean surface temperature should be about 2° Celsius above the pre-industrial level and that the rate of change should not exceed 0.2° Celsius per decade (Vellinga and Swart 1991; Smith et al. 2001). This would require that the atmospheric concentration of carbon dioxide be limited to about 450 parts per million and the atmospheric concentrations of other greenhouse gases stabilized at near current levels or lower.

This judgment is based on the conclusions of the IPCC "reasons for concern" (IPCC 2001b; Smith et al. 2001). Figure 13.2 in Appendix A shows the impact of changes in global mean surface temperature relative to 1990 (already 0.6° Celsius warmer than pre-industrial levels); it highlights that even an increase of about 2° Celsius above pre-industrial levels in global mean surface temperature would:

- pose a risk to many unique and threatened ecological systems and lead to the extinction of numerous species;
- lead to a significant increase in extreme climatic events and adversely impact agriculture in the tropics and sub-tropics, water resources in countries that are already water scarce or stressed, and human health and property;
- represent a transition between the negative effects of climate change being in only some regions of the world to being negative in most regions of the world. For example, below about 2° Celsius, agricultural productivity is projected to be adversely impacted in the tropics and sub-tropics, but beneficially impacted in most temperate and high latitude regions, whereas a warming of greater than 2° Celsius is projected to adversely impact agricultural productivity not only in the tropics and sub-tropics, but also in many temperate regions; and
- result in both positive and negative economic impacts, but with the majority of people being adversely affected, that is, predominantly negative economic effects in developing countries.

Limiting the global mean surface temperature increase to about 2°C above the pre-industrial level would result in a low probability of large-scale, high-impact events materializing, for example, the collapse of the major ice sheets or a significant change in ocean circulation.

Even changes in global mean surface temperature of 2° Celsius above the pre-industrial level and rates of change of 0.2° Celsius per decade will result in adverse consequences for many ecological systems. Hence many ecologists would argue for a more stringent target (Swart et al. 1998). Hare (2003) applied the "reason for concern" approach to local and regional vulnerabilities and emphasized stricter targets. Leemans and Eickhout (2004) analyzed the regional and global impacts of different levels of climate change on ecosystems and reported that an increase in global mean surface temperatures of between 1° and 2° Celsius would impact most species, ecosystems, and landscapes, and adaptive capacity would become limited.

The suggested target of 2° C above pre-industrial levels is consistent with the limit recommended by the German Advisory Council on Global Change, and is also consistent with the "safe corridors analysis." However, even the suggested maximum tolerable changes in global mean surface temperature will cause, on average, adverse consequences to the majority of inhabitants, especially in developing countries, with respect to food, water, human health, and livelihoods (Alcamo 1996; Toth 2003; IPCC 2001b). This suggests that adaptation assistance would be required for poor developing countries, where adverse effects would be concentrated.

Mastrandrea and Schneider (2004) reported a probabilistic assessment of what constitutes "dangerous" climate change by mapping a metric for this concept, based on the IPCC assessment of climate impacts (the five "reasons for concern," Figure 13.2), onto probability distributions of future climate change produced from uncertainty in climate sensitivity, climate damages, and discount rate. They deduced that optimal climate policy controls could reduce the probability of dangerous anthropogenic interference from approximately 45% under minimal controls to near zero.

It should be noted that a precautionary approach to protecting ecosystems and their goods and services would recognize that some impacts of anthropogenic climate change may be slow to become apparent because of inertia within the system, and some could be irreversible, if climate change is not limited in both rate and magnitude before associated thresholds, whose positions may be poorly known, are crossed (IPCC 2002; CBD 2003). For example, ecosystems dominated by long-lived species (for example, long-lived trees) will often be slow to show evidence of change. Higher rates of warming and the compounding effects of multiple stresses increase the likelihood of crossing a threshold.

13.3.2 Pathways and Stabilization Levels for Greenhouse Gas Concentrations

As noted, limiting the absolute global mean surface temperature increase to about 2° Celsius above pre-industrial levels and the rate of change to 0.2° Celsius per decade will require the atmospheric concentration of carbon dioxide to be limited to about 450 parts per million or lower (Table 13.1) and the atmospheric concentrations of other greenhouse gases stabilized at near current levels or lower, depending upon the value of the climate sensitivity factor (IPCC 2001a, d). A stabilization level of 450 parts per million of carbon dioxide corresponds to a stabilization level of about 550 parts per million carbon dioxide equivalent concentration, which includes the projected changes in the non-carbon dioxide greenhouse gases.

To stabilize carbon dioxide at 450 parts per million, a range of possible pathways could be used. Global emissions would have to peak between 2005 and 2015 and then be reduced below current emissions before 2040. In contrast, to stabilize carbon dioxide at 550 parts per million, global emissions would have to peak between 2020 and 2040 and then be reduced below current emissions between 2030 and 2100. A stabilization level of 450 parts per million of carbon dioxide would mean that global carbon dioxide emissions in 2015 and 2050, respectively, would have to be limited to about 9.5 (7–12) and 5.0 (3–7) gigatons of carbon per year, compared to emissions in the year 2000 of about 7.5 gigatons (energy and land use change).

Note, however, that different models used in IPCC (2001a) suggest that the 2015 emissions might need to be as low as 7 gigatons of carbon per year or could be as high as 12 gigatons. The large range is due to differences among the carbon models and the assumed subsequent rate of decreases in emissions—the higher the emissions are between now and 2015, the more drastic future reductions will be needed to stabilize at 450 parts per million. Similarly, the different models also suggest that in 2050 emissions might need to be as low as 3 gigatons or could be as high as 7 gigatons of carbon per year. To stabilize the atmospheric concentrations of carbon dioxide will require that emissions will eventually have to be reduced to only a small fraction of current emissions, that is, to less than 5–10% of current emissions, or less than 0.3–0.6 gigatons of carbon per year. Natural land and ocean sinks are small, that is, less than 0.2 gigatons per year (IPCC 2001a).

Even limiting the atmospheric concentration of carbon dioxide to 450 parts per million may not limit the increase in global mean surface temperature to 2° Celsius above pre-industrial or the rate of change to 0.2° Celsius per decade unless the climate sensitivity factor is toward the lower end of the range. The range of projected temperature changes shown in Table 13.1 for each stabilization level is due to the different climate sensitivity factors of the models (the climate sensitivity factor is the projected change in temperature at equilibrium when the atmospheric concentration of carbon dioxide is doubled—it ranges from 1.7° to 4.2° Celsius in Table 13.1). The uncertainty in the climate sensitivity factor is largely due to uncertainties in a quantitative understanding of the roles of water vapor, clouds, and aerosols.

Consequently, the projected changes in temperature are very sensitive to the assumed value of the climate sensitivity factor; for example, the projected change in temperature for a stabilization level of 450 parts per million and a high temperature sensitivity factor is comparable to stabilization at 1,000 parts per million with a low climate sensitivity factor. As stated earlier, if changes in the other greenhouse gases are taken into account, in 2000 it would be approximately equivalent to assuming an additional 90–100 parts per million of carbon dioxide (Prather 2004), that is, while the actual atmospheric concentration of carbon dioxide has increased from about 280 parts per million in the pre-industrial era to about 367 in 2000, the increase in the other greenhouse gases is equivalent to another 90–100 parts per million (this ignores the effects of aerosols).

IPCC commissioned a Special Report on Emission Scenarios (IPCC 2000a), which projected a range of plausible emissions of greenhouse gases and aerosol precursors up to 2100 under various assumptions of population, GDP growth, technological change, and governance structures. The lowest IPCC SRES scenario, which resulted in a projected increase in global mean surface temperature of 1.4° Celsius between 1990 and 2100, would eventually allow stabilization of carbon dioxide at about 550 parts per

Table 13.1. Pathways to Stabilize the Atmospheric Concentration of Carbon Dioxide and Implications for Changes in Global Mean Surface Temperature. These temperature changes are relative to 1990. Therefore, an additional 0.6° C would have to be added to these numbers to be relative to pre-industrial levels. These calculations include not only changes in carbon dioxide but also increases in non-carbon dioxide greenhouse gases, assuming they follow the SRES A1B scenario until 2100 and are constant thereafter.

Stabilization Level	Date for Global Emissions Peak	Date for Global Emissions to Fall below Current Levels	Temperature Change by 2100	Equilibrium Temperature Change
(parts per million)			*(degrees Celsius)*	
450	2005–2015	before 2040	1.2–2.3 (1.8)	1.5–3.9 (2.7)
550	2020–2030	2030–2100	1.7–2.8 (2.3)	2.0–5.1 (3.4)
650	2030–2045	2055–2145	1.8–3.2 (2.7)	2.4–6.1 (4.1)
750	2050–2060	2080–2180	1.9–3.4 (2.7)	2.8–7.0 (4.6)
1000	2065–2090	2135–2270	2.0–3.5 (2.8)	3.5–8.7 (5.8)

million, but none of the SRES scenarios would allow stabilization of carbon dioxide at 450 parts per million. The lowest IPCC SRES scenario could be accomplished without concerted global action to reduce greenhouse gas emissions, but only if the global population peaks near 2050 and declines thereafter; if economic growth is accompanied by the rapid introduction of less carbon-intense and more efficient technologies; and if there is an emphasis on global "sustainable and equitable solutions." This will not materialize with a business-as-usual attitude; it will require governments and the private sector worldwide to form a common vision of an equitable and sustainable world, new and innovative public-private partnerships, the development of less carbon intensive technologies, and an appropriate policy environment. Stabilization at or below 550 parts per million would require a significant change in the way energy is currently produced and consumed. IPCC concluded that known technological options could achieve a broad range of atmospheric carbon dioxide stabilization levels, such as 550 or 450 parts per million, over the next 100 years or more, but implementation would require associated socioeconomic and institutional changes.

While the four IPCC SRES scenarios were "non-climate intervention" scenarios, the lowest scenarios contained many of the features required to limit human-induced climate change, that is, a significant transition to non-fossil fuel technologies and energy efficient technologies.

13.3.3 Implications of Greenhouse Gas Emissions

Setting a stabilization target imposes some critical issues associated with equitable burden sharing and implications for economic development and human well-being.

13.3.3.1 Regional Implications of Stabilizing Greenhouse Gases

As noted earlier, even the lowest stabilization levels of carbon dioxide are projected to lead to significant changes in the magnitude and rate of change of temperature, thus threatening ecosystems, their biodiversity, and goods and services. Hence, even at the lowest stabilization levels of greenhouse gas concentrations, adaptation measures will be needed (IPCC 2001b; IPCC 2001d).

Stabilization of carbon dioxide at 450 parts per million is projected to lead to a change in global mean surface temperature of 1.2°–2.3° Celsius by 2100 and 1.5°–3.9° Celsius at equilibrium. Stabilization of carbon dioxide at 550 parts per million is projected to lead to a change in global mean surface temperature of 1.7°–2.8° Celsius by 2100 and 2.0°–5.1° Celsius at equilibrium. (The time required to reach equilibrium depends on the pathway to stabilization and on the stabilization level; for 450 parts per million, it is within a couple of hundred years, and for 550 parts per million, it would likely take an additional hundred years or so.)

One weakness is that presenting projected changes in the global mean surface temperature hides the different changes latitudinally and between land and ocean (IPCC 2001a, Chapter 10). General circulation models show that: (1) the high latitudes are projected to warm much more than the tropics and sub-tropics, and land areas are projected to warm more than the oceans; and (2) the high latitudes and tropics will tend to become wetter and most of the sub-tropics drier (IPCC 2001a, Chapter 10). Therefore, to quantitatively understand the implications of different stabilization levels of greenhouse gas concentrations on ecosystems and their biodiversity and goods and services, changes in mean temperature and precipitation, as well as changes in the variability of temperature and precipitation, and the incidence of extreme events, are needed at the regional and sub-regional scale, which requires the use of regional-scale climate models. Figure 13.2 allows for these regional differences. For marine systems, an understanding of changes in sea level is also required.

13.3.3.2 Burden Sharing/Equity Considerations

A key issue that will have to be addressed with long-term targets is the equitable allocation of emissions rights (Ashton and Wang 2003). In deciding what is equitable, a number of factors need to be considered: *Responsibility*—should those that caused the problem be responsible for mitigating the problem? *Entitlements*—should all humans enjoy equal entitlements to a global public good? *Capacity*—should those that have the greatest capacity to act bear the greatest burden? *Basic needs*—should strong nations assist poor nations to meet their basic needs? *Comparability of effort*—should the ease/difficulty of meeting a target be taken into account? *Future generations*—what is the responsibility of the current generation for future generations?

There are a series of options, each with their own political difficulties, including:

- *in proportion to current emissions,* that is, grandfathering—unlikely to be acceptable to developing countries because of their low current per capita emissions, and in many cases low total emissions;
- *in proportion to current GDP*—again, unlikely to be acceptable to developing countries given their current low GDPs;
- *current per capita emissions rights*—unlikely to be acceptable to developed countries given their current high per capita emissions; and
- *transition from grandfathering to per capita emissions.* Numerous transition schemes have been proposed, including contraction and conversion (Meyer 2000); taking into account historic emissions, for example, the Brazilian Proposal (IISD 2003); taking into account basic needs; and taking into account national circumstances, for example, ability to pay (Jacoby et al. 1999).

Negotiators could develop an allocation scheme using any one or combination of these options (submissions by Norway in 1996, by Australia in 1997, and by Iceland in 1997 to the Ad Hoc Group to the Berlin Mandate). Claussen and McNeilly (1998) proposed dividing countries into three categories based on three criteria: *responsibility* (that is, historical and current total emissions, per capita emissions, and projected emissions), *standard of living* (that is, GDP per capita), and *opportunity* (that is, related to energy intensity of the economy).

As noted, a number of different allocations schemes have been suggested. One approach that is receiving significant attention, and is endorsed by the German Advisory Council on Global Change, is some form of contraction and convergence whereby total global emissions are reduced (that is, contraction) to meet a specific agreed target, and the per capita emissions of industrialized and the developing countries converge over a suitably long time period, with the rate and magnitude of contraction and convergence being determined through the UNFCCC negotiating process. "Contraction and Convergence"* is a science-based global climate-policy framework proposed by the Global Commons Institute with the objective of realizing "safe" and stable greenhouse gas concentrations in the atmosphere; a "safe" level is defined as one that avoids dangerous anthropogenic perturbation

*For further information, see http://www.gci.org.uk; http://www.gci.org.uk/model/dl.html; http://www.feasta.org; http://www.gci.org.uk/images/CC_Demo(pc).exe; http://www.gci.org.uk/images/C&C_Bubbles.pdf.

to the climate system as defined in Article II of the UNFCCC and is to be determined through a sociopolitical process, for example, the UNFCCC. The Global Commons Institute applies principles of precaution and equity—principles identified as important in the UNFCCC but not defined—to provide the formal calculating basis of the contraction and convergence framework, which proposes:

- a full-term contraction budget for global emissions consistent with stabilizing atmospheric concentrations of greenhouse gases at a pre-agreed concentration maximum deemed to be "safe" using IPCC WG1 carbon cycle modeling;
- the international sharing of this budget as "entitlements" results from a negotiable rate of linear convergence to equal shares per person globally by an agreed date within the timeline of the full-term contraction/concentration agreement;
- negotiations for this within the UNFCCC could occur principally between regions of the world, leaving negotiations between countries primarily within their respective regions, such as the European Union, the Africa Union, the United States, etc., comparable to the current EU bubble.
- the interregional, international, and intra-national tradability of these entitlements should be encouraged to reduce costs; and
- as scientific understanding of the relationship between an emissions-free economy and concentrations develops, so contraction/conversion rates can evolve under periodic revision.

Another proposal, the "Brazilian Proposal," takes a different approach to the Kyoto Protocol. The Kyoto Protocol allocates emissions rights among parties for a particular time period. The Brazilian Proposal, which originally addressed only Annex 1 countries, could provide a framework for burden sharing among all countries, that is, Annex 1 and non-Annex 1 countries. It proposes that the criterion for the burden sharing should be measured by the country's contribution to the increase in global mean surface temperature since the emissions in a particular year do not reflect the true contribution of a country to global climate change, which is related to the cumulative emissions of greenhouse gases. The proposal by Brazil aims at sharing equally the burden of mitigation, accounting for the past contribution to global warming, that is, the cumulative historical emissions. The framework can be used to take account of all greenhouse gases, from all sources.

Deciding which allocation scheme is appropriate will have to result from negotiations involving all countries.

13.4 Adaptation to Climate Change

As described above, impacts of climate change are already being felt in some circumstances and the lag times in the global climate system mean that no mitigation effort, however rigorous, is going to prevent further climate change from happening. Thus adaptation is an essential component of our response to climate change. Adaptation in this context refers to any adjustment in natural or human systems taking place in response to actual or expected impacts of climate change, intended either to moderate harm or to exploit beneficial opportunities (IPCC 2001b).

13.4.1 Ecosystem Goods and Services

Climate change will affect the capacity of ecosystems to provide goods and services. Some changes in ecosystems in response to climate change are already being observed in that the timing of animal migratory patterns, plant phenology, and species ranges are changing in a manner consistent with observed climate change (IPCC 2001b, Chapter 5; Root et al. 2003).

Over the next few decades increasing losses of species from ecosystems are expected. Many of these losses will result from pressures that do not necessarily involve climate change, such as changes in land use and land cover or from the introduction of new species. However, some will be the result of climate-related pressures such as changes in disturbance frequency (for example, fires), or combinations of pressures, such as temperature rises and increasing nutrient and silt deposition leading to bleaching damage in coral reefs. Losses or reductions in populations of species will lead to significant reorganization within ecosystems and make them more vulnerable to invasion by species from neighboring regions or species exotic to the region. All of these processes have the potential to affect the delivery of ecosystem goods and services.

In intensively managed ecosystems such as farming lands, the options to adapt to these changes are relatively straightforward, although they may be financially and socially costly. For small changes, altering planting times or moving to new varieties of crops or livestock may be sufficient. Larger changes may require wholesale changes in farming systems such as new crops or livestock management systems or major changes in land use (for example, from farming to grazing lands). Where local populations have limited access to finance or new resources, the impacts on livelihoods are likely to be significant.

In less intensively managed ecosystems, decisions will have to be made whether to try to minimize changes in composition and functioning of the ecosystem, or to facilitate the change so as to maintain a supply of goods and services, albeit a potentially different set of goods and services from the original ecosystem. Reducing the direct drivers of change may minimize changes. These include reducing disturbances such as fires or clearing, controlling invasive species, and reducing harvesting of species most under stress. There have been few attempts to facilitate changes on ecosystems in response to climate change. The most common response is the wholesale replacement of an ecosystem after a disturbance such as a fire or prolonged drought. Damaged forests are often cleared and converted to agriculture or tree cropping. Drought-affected and overgrazed pasture lands are often abandoned or maintained in a degraded state by inappropriate grazing and other uses.

One of the great challenges will be to devise ways of managing change in these less intensively managed ecosystems. An important determinant of success will be information about the likely trends in climate as land managers already have coping strategies for a variable climate that can often be modified to help manage climate change. Another determinant will be the provision of resources and finance to make changes in management as they become necessary. There are opportunities for synergistic activities that combine adaptive actions with mitigation. For example, the restoration of degraded lands with shrubs may create a fodder reserve while it, at the same time, increases carbon storage in the woody material above and below ground. A challenge here is to mobilize financial resources to facilitate these actions (see discussion below).

Lands set aside primarily for the conservation of biodiversity pose their own special challenges. An important step in adapting to climate change is the appropriate design of the reserve system. (See Chapter 5.) Conservation areas should be designed to take into account the expected long-term shifts in the distribution of plants and animals (essentially pole-wards and upwards), although this has to be planned in detail for each conservation system. Another action is to protect reserves from disturbances or sequences of disturbances, such as drought damage followed by fire, that are likely to hasten species losses and invasions.

Ultimately the conservation priorities of each reserve need to be carefully considered. If the primary goal is to protect certain species, management may be directed at resisting change. In some cases, deliberate introduction of threatened species may be considered. If the primary goal is the maintenance of ecosystems functioning in a relatively pristine (that is, free of human interference) state, then the primary management goal is to restrict that interference especially through disturbances entering from outside and species invasions. Most conservation areas have multiple goals, including recreation and the protection of scenic values. Management plans are an essential tool for sound conservation management; a strategy for climate change should be part of such plans.

The management of biodiversity outside of formal reserve systems is likely to become increasingly important under climate change. Successful dispersal of the gene pools of local species will increase the likelihood of disturbed areas being re-colonized by local species best adapted to the new conditions rather than exotic invasive species. Successful dispersal is usually sensitive to the matrix of landscapes that surround areas managed for conservation priorities. Patches of remnant vegetation, or even appropriate exotic species, may facilitate the movement of dispersal agents such as birds.

13.4.2 Human Societies

Impacts of climate change on human societies will be superimposed on the existing vulnerabilities to climate- and non-climate-related stresses. They will vary greatly depending on people's exposure to climate change and their ability to adapt to or cope with the impacts of climate change. Many of the actions taken by individuals, communities, and institutions in response to climate change will not require external intervention. Such autonomous actions are typically triggered by changes in weather patterns that result in shifting market signals or welfare changes (such as changes in the prices of crops and in the occurrence of diseases). They take place irrespective of any broader plan or policy-based decisions. Examples of autonomous actions include changes in farming practices, the purchase of air-conditioning devices, insurance policies taken out by individuals and companies, and changes in recreational and tourist behavior.

Natural systems also undergo autonomous changes in response to changing conditions in their immediate environment. As temperatures increase and sea level rises, species migrate to higher latitudes and altitudes, and coastal wetlands re-establish on higher ground. However, in many places, human activities have reduced the potential of natural systems to adapt in this way, with settlements and other infrastructure forming barriers to the migration of species. For example, coastal protection works could block the landward migration of wetlands, causing the wetlands to be squeezed between a rising sea level and immobile infrastructure.

It is unlikely that autonomous adaptation to climate change by nature and human society will suffice to reduce the potential impacts of climate change to an acceptable level. In many parts of the world, the future impacts of climate change are projected to be significantly greater than those that have been experienced in the past as a result of natural climate variability alone. Future impacts may be more than what many natural and human systems are able to handle effectively with autonomous adaptation, particularly given additional constraints such as barriers to the migration of species, and limited information, inadequate knowledge, and insufficient access to resources for individuals, communities, and companies.

As a result, it is now widely acknowledged that there is a need for planned adaptation, aimed at preparing for the impacts of climate change and at facilitating and complementing autonomous adaptation by nature and society. Forms that such planned adaptation could take are discussed by Klein and Tol (1997), Smit et al. (2001), and Huq and Klein (2003), among others. The first classification of adaptation strategies into protection, accommodation, and planned retreat was developed for coastal zones by IPCC CZMS (1990), and it is still the basis of many coastal adaptation analyses.

Broadly, planned adaptation can serve the following objectives:

- *reduce the risk of impacts by decreasing their probability of occurrence* (examples include upgrading coastal flood protection and irrigation in the face of increasing drought);
- *reduce the risk of impacts by limiting their potential magnitude.* This can involve a broad array of actions ranging from elevating buildings in flood-prone areas, to including new crops in agricultural systems, removing barriers to the migration of plants or animals, or introducing insurance schemes. It may also involve relocating people from exposed areas such as floodplains and small islands, or changing livelihoods, such as replacing cropping by pastoralism; and
- *increase society's ability to cope with and adapt to the impacts.* This includes promoting autonomous actions by people, communities, and companies, for example, through informing the public about the risks and possible consequences of climate change, setting up early-warning systems for extreme weather events, and providing incentives for risk-reducing behavior.

IPCC (2001b) discussed adaptation for a number of sectors related to ecosystem services.

While such activities help to reduce vulnerability to climate variability and change, in many places ongoing activities increase vulnerability. The development of exposed areas and the degradation of ecosystems that protect against hazards can result in a situation where climate impacts are much more pronounced than would have been the case otherwise. Thus a good starting point for reducing vulnerability to climate variability and change would be to reverse these "maladaptive" trends.

Between and within countries, there are great differences in the level of human, technical, financial, and other resources that individuals, communities, and companies can devote to adaptation strategies. Their vulnerability to climate change is determined not only by the impacts they potentially face, but also by their ability to find the resources needed to adapt to these impacts. The ability to plan, prepare for, and implement adaptation initiatives is usually referred to as "adaptive capacity" (Smith et al. 2001; Smith et al. 2003). It is not surprising that most industrialized countries have higher adaptive capacities than developing countries. For example, Bangladesh and The Netherlands share a similar physical susceptibility to sea level rise. But Bangladesh lacks the economic resources, technology, and infrastructure that The Netherlands can call on to respond to such an event.

It thus follows that developing countries tend to be more vulnerable than industrialized countries with higher adaptive capacities. Within developing countries, the poorest people, whose livelihoods are often directly dependent on the provision of climate-sensitive ecosystem goods and services, are particularly vulnerable (Anon. 2003). Not only do they lack the benefits associated with wealth; they also often occupy locations most exposed to the impacts of climate change, such as low-lying coastal regions, marginal or arid lands, or cities with poor and overloaded infrastructure. Limited access to information and denied access to entitlements may further aggravate their situation. Moreover, poor people may lack social networks on which to draw during times of hardship.

Traditional coping mechanisms are often the starting point for developing adaptation strategies. Initially, the impacts of climate change will fall within the general range of experience of societies, although damaging events may be more frequent or intense, and stress periods may be longer. Adaptation efforts begin with the existing assets and capabilities that can be strengthened to reduce vulnerability and increase resilience to climate change. Maintaining social networks, capacity building, effective information flow, and efficient local control of limited financial assets (for example, through micro-finance and micro-insurance) appear to be important components of building and maintaining community resilience. However, the traditional coping mechanisms are already failing, so specific plans for adaptation to climate change need to be incorporated into wider regional and national development planning.

In many developing countries, the key development challenges include food security, access to clean water and sanitation, education, and health care. The integration—or mainstreaming—of policies and measures to address climate variability and change into ongoing sectoral and economic planning is needed to ensure the long-term sustainability of investments as well as to reduce the sensitivity of development activities to both today's and tomorrow's climate (Huq and Klein 2003).

Mainstreaming adaptation makes more efficient and effective use of financial and human resources by incorporating, implementing, and managing climate policy as an integral component of ongoing activities, rather than treating adaptation as a series of stand-alone actions. For example, in the Caribbean, a series of projects supported through local resources, GEF, and bilateral agencies (such as Mainstreaming Adaptation to Climate Change) aim to build national and regional capacity and facilitate governments' efforts to incorporate climate change considerations into planning and policy-making. They address a key challenge in climate policy: to build capacity and to facilitate action. Another example is the cooperation between Caribbean (CARICOM) and Pacific (SPREP) countries to integrate climate change into current environmental impact assessment procedures (Caribbean Community Secretariat 2003).

Successful implementation of adaptation options requires the presence of an enabling environment, the development of which is an important objective of national and international climate policy. Using this enabling environment, the actual implementation of options—be they technical, institutional, legal, or behavioral—is best done by sectoral planning and management agencies "on the ground" (for example, water companies, agricultural planners, coastal management agencies), as well as private companies and individuals.

Thus at the national planning level, the main challenges to effective adaptation are associated with the coordination of multiple sustainability strategies with budgetary processes, including international assistance. In most countries, developed and developing, climate change issues are the responsibility of departments and agencies that are at the periphery of government planning and, in particular, of financial decision- making. A recent view of 19 developing and developed countries (Swanson 2004) identified this disconnect to be a major impediment to sustainable development.

International negotiations over climate change are paying increasing attention to adaptation issues. It is generally accepted that the need to adapt is an additional impediment to development imposed on developing countries largely through the activities of others. Adaptation to climate change must, therefore, be firmly rooted in ongoing, broader development efforts and should build upon lessons from other activities such as disaster risk reduction, poverty reduction, and natural resource management (Hammill 2004). The immediate challenges are to explore ways to mainstream adaptation issues in the development agenda and to negotiate, both bilaterally and multilaterally, responsibilities for support through technology transfer, aid, lending arrangements, and other innovative arrangements such as disaster prevention and risk transfer.

13.5 Mitigation of Greenhouse Gas Emissions

Stabilization of the atmospheric concentrations of greenhouse gases will require emissions reductions in all regions, that is, Annex I countries cannot alone reduce their emissions enough to achieve stabilization because of the large projected increases in emissions in developing countries. Lower net emissions can be achieved through different patterns of energy resource development and utilization, increases in end-use efficiency, and land-use practices (IPCC 2001c, Chapter 3). Many of these technologies also reduce local and regional pollutants, that is, particulates, and ozone and acid deposition precursors.

Realizing greenhouse gas emissions reductions in the production and use of energy will involve overcoming technical, economic, cultural, social, behavioral, and institutional barriers (IPCC 2001c, Chapters 1, 5). National responses to climate change mitigation can be most effective when deployed as a portfolio of policy instruments to reduce greenhouse gas emissions, for example, a mix of emissions/carbon/energy taxes, tradable or non-tradable permits, provision of and/or removal of subsidies, land-use policies, deposit/refund systems, technology or performance standards, energy mix requirements, product bans, voluntary agreements, information campaigns, environmental labeling, government spending and development, and support for R&D (IPCC 2001c, Chapters 1, 5, 6). North-South and South-South technology transfer and technical assistance will be needed to facilitate the uptake of new energy technologies and alternative natural resource management practices in developing countries ((IPCC 2001c, Chapter 10; IPCC 2000c).

Pacala and Socolow (2004) argued, consistent with the IPCC, that humanity already possesses the fundamental scientific, technical, and industrial know-how to address the energy-carbon problem for the next half century. This contrasts with the more pessimistic view of Hoffert et al. (2003), who argue that current technologies are inadequate and revolutionary changes in technology are needed. Pacala and Sokolow argued that a portfolio of technologies now exists, that have already passed the laboratory bench and demonstration phases and are now being implemented in some parts of the world at full industrial scale, to meet the world's energy needs over the next 50 years and limit the atmospheric concentration of carbon dioxide to a trajectory that avoids a doubling of the pre-industrial concentration, that is, a trajectory that stabilizes at about 500 parts per million. Table 13.2 lists 15 possible strategies, each of which could, in principle, reduce carbon emissions by 2054 by 1 gigaton per year or 25 gigatons over the next 50 years (each potential intervention would steadily increase from zero today to 1 gigation by 2054). In combination, the strategies could reduce carbon emissions between 2004 and 2054 by 150–200 gigatons of carbon. Pacala and Socolow noted that fundamental research is needed now to develop the revolutionary mitigation strategies for beyond 2050 to remain on a trajectory that would eventually stabilize the atmospheric concentration of carbon dioxide at about 500 parts per million.

There is little doubt that technologies now exist that can be used to reduce current and projected levels of greenhouse gas emissions, while recognizing that an increased commitment to energy

Table 13.2. Potential Wedges: Strategies Available to Reduce the Carbon Emission Rate in 2054. The strategies aim to reduce carbon emission rates by 1 GtC per year until 2054 or by 25 GtC between 2004 and 2054. (Pacala and Socolow 2004)

	Option	Effort by 2054 for One Wedge (relative to 14 GtC/year, business-as-usual)	Comments/Issues
Energy efficiency and conservation	economy-wide carbon-intensity reduction (emissions/ $GDP)	increase reduction by additional 0.15% per year (e.g., increase U.S. goal of reduction of 1.96% per year to 2.11% per year)	can be tuned by carbon policy
	• efficient vehicles	increase fuel economy for 2 billion cars from 30 to 60 mpg	car size, power
	• reduced use of vehicles	decrease car travel for 2 billion 30-mpg cars from 10,000 to 5,000 miles per year	urban design, mass transit, telecommuting
	• efficient buildings	cut carbon emissions by one fourth in buildings and appliances projected for 2054	weak incentives
	• efficient baseload coal plants	produce twice today's coal power output at 60% instead of 40% efficiency (compared with 32% today)	advanced high-temperature materials
Fuel shift	gas baseload power for coal baseload power	replace 1,400 GW 50%-efficient coal plants with gas plants (four times the current production of gas-based power)	competing demands for natural gas
CO_2 capture and storage (CCS)	capture CO_2 at baseload power plant	introduce CCS at 800 GW coal or 1,600 GW natural gas (compared with 1,060 GW coal in 1999)	technology already in use for H_2 production
	capture CO_2 at H_2 plant	introduce CCS at plants producing 250 MtH_2/year from coal or 500 MtH_2/year from natural gas (compared with 40 MtH_2/year today from all sources)	H_2 safety, infrastructure
	capture CO_2 at coal-to-synfuels plant	introduce CCS at synfuels plants producing 30 million barrels per day from coal (200 times Sasol), if half of feedstock carbon is available for capture	increased CO_2 emissions, if synfuels are produced *without* CCS
	geological storage	create 3500 Sleipners	durable storage, successful permitting
Nuclear fission	nuclear power for coal power	add 700 GW (twice the current capacity)	nuclear proliferation, terrorism, waste
Renewable electricity and fuels	wind power for coal power	add 2 million 1-MW-peak windmills (50 times the current capacity) "occupying" $30x10^6$ ha, on land or off shore	multiple uses of land because windmills are widely spaced
	PV power for coal power	add 2,000 GW-peak PV (700 times the current capacity) on $2x10^6$ ha	PV production cost
	wind H_2 in fuel-cell car for gasoline in hybrid car	add 4 million 1-MW-peak windmills (100 times the current capacity)	H_2 safety, infrastructure
	biomass fuel for fossil fuel	add 100 times the current Brazil or U.S. ethanol production, with the use of 250×10^6 ha (1/6 of world cropland)	biodiversity, competing land use
Forests and agricultural soils	reduced deforestation, plus reforestation, afforestation, and new plantations.	decrease tropical deforestation to zero instead of 0.5 GtC/ year, and establish 300 Mha of new tree plantations (twice the current rate)	land demands of agriculture, benefits to biodiversity from reduced deforestation
	conservation tillage	apply to all cropland (10 times the current usage)	reversibility, verification

R&D is needed to develop the technologies needed to ultimately stabilize the atmospheric concentration of carbon dioxide.

13.5.1 Energy Technologies and Policies

Policy can help to further technologies in reducing greenhouse gas emissions through changes in the energy production sector and promoting increased efficiency in energy use, thus transitioning to a less carbon-intensive energy sector.

13.5.1.1 Energy Production

There are many options available to reduce greenhouse gas emissions from the energy production sector, including fuel switching (coal to oil to gas), increased power plant efficiency, improved transmission, carbon dioxide capture and storage (pre- and post combustion), increased use of renewable energy technologies (biomass, solar, wind, run-of-the-river and large hydropower, geothermal, etc.) and nuclear power (WEA 2000, Chapter 1; IPCC 2001c, Chapter 3). The solution will not come from any single technology, but rather from a portfolio of energy technologies, the mix varying in different parts of the world. This section addresses fossil fuel, renewable energy, and nuclear technologies.

13.5.1.1.1 Fossil fuel energy technologies

Energy supply and conversion will remain dominated by fossil fuels for the next several decades due to their abundance and relatively low cost. Figure 13.3 shows the projected energy use, by

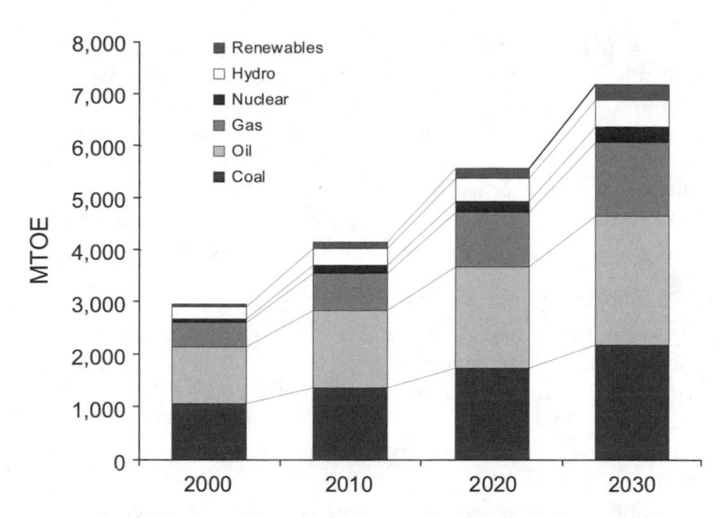

Figure 13.3. Projected Energy Use of Developing Countries in International Energy Agency Business-as-Usual Scenario

type of energy, from now to 2030 in developing countries (IEA 2003). However, there are several ways in which greenhouse gas emissions from the combustion of fossil fuels can be reduced (WEA 2000, Chapter 1; IPCC 2001c, Chapter 3). Natural gas could, where transmission is economically feasible, play a key role in reducing greenhouse gas emissions together with improved conversion efficiencies and greater use of combined cycle and/or cogeneration plants. Natural-gas-fired combined cycles offer low costs, high efficiency, and low local and regional environmental impacts. Fuel cell technologies offer significant potential for cogeneration at smaller scales, including commercial buildings. Coal gasification by partial oxidation with oxygen to produce syngas (primarily carbon monoxide and hydrogen) offers the opportunity to provide electricity through integrated gasifier combined cycle (IGCC) plants combined with carbon capture and storage, with low local air pollutant emissions. Superclean syngas-derived synthetic fuels produced in polygeneration facilities simultaneously producing multiple products may soon be economically competitive. The successful development of fuel cells, coupled with a syngas-based strategy could pave the way for the widespread use of hydrogen. Syngas-based power and hydrogen production strategies also provide an opportunity of producing energy without emissions of carbon dioxide through the separation and storage of carbon dioxide. Similarly, emissions of carbon dioxide from fossil- and/or biomass-fuel power plants could be reduced substantially through carbon capture and storage.

The viability of carbon dioxide capture and storage will depend on the cost-effectiveness and environmental sustainability of these emerging technologies. Storage in geological reservoirs (for example, depleted oil and gas wells) has enormous potential and the costs appear promising. Although physical sequestration, that is, storage in deep oceanic marine ecosystems, may offer mitigation opportunities for removing carbon dioxide from the atmosphere, the implications for biodiversity and ecosystem functioning are not understood. All proposed oceanic carbon dioxide storage schemes have the potential to cause ecosystem disturbance (Raven and Falkowski 1999), by altering the concentration of carbon dioxide and seawater pH, with potential consequences for ecosystems and marine organisms (Ametistova et al. 2002; Huesemann et al. 2002; Seibel and Walsh 2001).

13.5.1.1.2 Renewable energy technologies and nuclear power

Low- and zero-carbon sources of energy include nuclear energy and renewable energy technologies, that is, solar, wind, biomass (traditional, agricultural and forestry by-products, and dedicated plantations), hydropower (large and run-of-the-river), municipal and industrial wastes to energy, and landfill methane (WEA 2000, Chapter 1; IPCC 2001c, Chapter 3). Nuclear energy, which provides energy without emitting conventional air pollutants and greenhouse gases, currently accounts for about 6% of total energy and 16% of electricity. The future role of nuclear power will depend on its cost, solutions for and public perception of safety, radioactive waste management, and the agreed rules and effective implementation to exclude nuclear weapons proliferation.

Current renewable energy sources supply about 14% of total world energy demand, dominated by traditional biomass used for cooking and heating, with hydropower supplying about 20% of global electricity. While traditional biomass is net neutral with respect to carbon dioxide emissions, its use often places significant pressure on ecological systems, often leading to loss of biomass and biodiversity, and in some cases desertification. Improved stoves can significantly reduce this pressure on ecological systems. The potential for new hydropower is primarily in developing countries. New renewable energy sources (for example, wind and solar) contributed only about 3% of the world's energy consumption in 2000. While the potential for wind energy or solar thermal power is significant in countries along the trade wind regions or in the solar belt, even with rapid increases in installed capacity, for example, 10–30% per year, they will remain a minor supplier of total energy needs for several decades, although increasing in importance over time. What is evident is that the "learning curves" on many renewable energy technologies needed to address climate change are lowering costs and making them either competitive or justified, for example, photovoltaics in small isolated communities. (See Figure 13.4.)

13.5.1.1.3 Challenges to scaling up renewable energy technologies

Significant barriers stand in the way of an accelerated deployment of renewable energy technologies into the market, including economic risks, lack of investment, regulatory obstacles, information and technology gaps, and limited number of products (IPCC 2001c, Chapters 5, 6). Supporting policies and programs needed

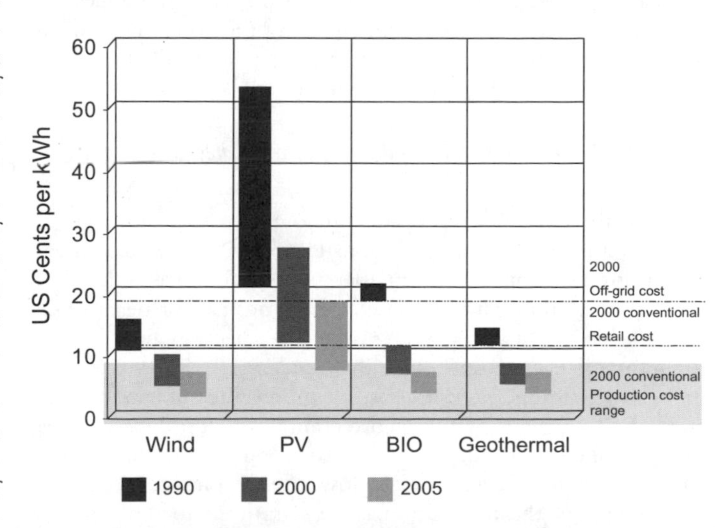

Figure 13.4. Production Cost Ranges for Fossil and Renewable Resources, 1990, 2000, 2005. Examples of renewable electricity cost competitiveness are shown in the figure. Run-of-the-River hydropower costs could range from 2–15 cents per kWh. (G8 Task Force Report, July 2001)

to overcome these barriers include: renewable portfolio standards, energy pricing strategies, carbon taxes, removing subsidies that increase greenhouse gas emissions, internalizing externalities, domestic and international tradable emissions permits, voluntary programs, incentives for use of new technologies during market build-up, and intensified R&D. These types of policies would make renewable energy technologies more competitive. Existing energy subsidies and the failure to internalize externalities are perceived as particularly problematic in several markets; they make conventional energy costs artificially low, making it harder for renewable energy to become commercially competitive.

Attracting substantial finance and investment is a prerequisite for scaling up the development of renewable energy internationally. The challenge is to introduce the right policy frameworks and financial tools to enable renewable energy to achieve its market potential. This applies both to maturing renewable energy markets in OECD countries and to emerging large-scale on-grid and small-scale off-grid markets in developing countries, where investment is put at risk from geopolitical, economic and regulatory risks, and the lack of developed financial markets and products.

Strength, clarity, and stability are decisive characteristics of the policy environment that will be needed to attract capital to renewable energy. A national policy and regulatory regime is necessary, but insufficient to tackle the issues of financing small-scale, off-grid remote renewable energy applications in less developed markets. In addition to an enhanced role for international financial institutions and regional development banks, and the development of local credit markets, public sector provision of small amounts of grant money is seen as strategically important. There is a strong argument for a blend of grant and development finance funds, particularly where renewables-based projects are also serving poverty alleviation objectives.

However, transforming the energy production sector and reducing greenhouse gas emissions will require: (1) acknowledging that uncertainties exist in estimating the costs and benefits of reducing greenhouse gas emissions; (2) addressing inter- and intragenerational equity and distributional issues; (3) overcoming the vested interests of those who benefit from, and want to protect, the status quo of reliance on current fossil fuel technologies and distortion policies; and (4) acknowledging the concerns of many governments who believe that a transition away from cheap fossil fuels will inhibit their economic growth.

13.5.1.1.4 Environmental implications of renewable energy technologies

In general, renewable energy technologies have positive effects on local and regional air pollution. However, renewable energy technologies (crop and municipal/industrial waste, solar- and wind-power and hydropower) may have positive or negative effects on biodiversity, depending upon site selection and management practices (CBD 2003, Chapters 4, 5). Substitution of fuelwood by crop waste, the use of more efficient wood stoves and solar energy, and improved techniques to produce charcoal can also take off pressure from forests, woodlots, and hedgerows. Most studies have demonstrated low rates of bird collision with windmills, but the mortality may be significant for rare species; proper site selection and a case-by-case evaluation of the implications of windmills on wildlife and ecosystem goods and services can avoid or minimize negative impacts. The potential adverse ecosystem/biodiversity impacts of specific hydropower projects vary widely and may be minimized depending on factors including type and condition of pre-dam ecosystems, type and operation of the dam (for example, water flow management), and the depth, area, and length of the reservoir. Generally, run-of-the-river hydropower and small dams have less impact on biodiversity than large dams, but the cumulative effects of many small units should be taken into account. Bio-energy plantations may have adverse impacts on biodiversity if they replace ecosystems with higher biodiversity. However, bio-energy plantations on degraded lands or abandoned agricultural sites could benefit biodiversity.

13.5.1.2 Energy Use Technologies

Opportunities to improve the efficiency of energy use exist in the conversion of useful energy to energy services, rather than in the conversion from primary energy to useful energy (WEA 2000, Chapter 1; IPCC 2001c, Chapter 3). Hundreds of opportunities exist in the residential, industrial, transportation, public, and commercial sectors to improve end-use efficiency. Over the next 20 years the amount of primary energy needed for a given level of energy services could cost-effectively be reduced by 25–40% at current energy prices, varying among industrial countries, countries with economies in transition, and developing countries, resulting in an overall improvement of 2% or more per year. This could be augmented by structural changes in the economy, that is, shifts to less energy-intensive industrial production.

The buildings sector contributed 31% of global energy-related carbon dioxide emissions in 1995, with an annual growth rate since 1971 of 1.8%. Opportunities to reduce greenhouse gas emissions in the residential and commercial building sector, many at net negative costs, include energy efficient windows, lighting, appliances, insulation, space heating, refrigeration, air conditioning, building controls, passive solar design, and integrated building design.

The transportation sector contributed 22% of global energy-related carbon dioxide emissions in 1995, and this is growing at an annual rate of about 2.5%. Technological opportunities in the transportation sector for light-duty vehicles have advanced significantly in recent years, for example, hybrid-electric cars, fuel cell vehicles, and improving fuel efficiency by advanced motor construction and lighter materials. These technological opportunities can be complemented by improved land-use planning and mass transit systems. Mass transit systems in growing urban areas of developing countries can reduce greenhouse gases, local air pollution, and congestion. Mumbai, with suburban railways and with the same population and GDP as Delhi, required only 40% of the number of vehicles and energy for transportation compared to Delhi, which did not introduce a metro system until 2003 (Parikh and Das 2003).

The industrial sector contributed 43% of global carbon dioxide emissions in 1995, with an annual growth rate of 1.5% between 1971 and 1995. However, since 1990 the rate has slowed to only 0.4% following the collapse of heavy industries in the former Soviet bloc. Improvements in energy efficiency of industrial processes offer the greatest opportunities for emissions reductions, especially in developing countries, many at net negative costs.

Current technologies are not close to reaching theoretical efficiency limits, and improvements of an order of magnitude for the whole energy system may eventually be achieved. However, the technical and economic potentials of energy efficiency have traditionally been under-realized, partly because of a number of significant barriers, primarily market imperfections. These include artificially low energy prices due to subsidies and failure to internalize environmental externalities; lack of adequate capital and financing; higher initial costs of more efficient technologies; lack of incentives for careful maintenance; differential investor/user benefits; and lack of information and training. Therefore, supporting policies and programs needed to overcome these barriers include: energy pricing strategies, energy audits, carbon taxes, internalizing externalities, regulatory programs including energy-efficiency standards, education such as product advisories and labels, staff training and energy management teams, and intensified R&D. These types of policies would stimulate the uptake of energy-efficient technologies.

13.5.1.3 Transition Rates to a Less Carbon-intensive Energy Sector

A key question is at what rate a transition to a less carbon-intensive energy sector can be accomplished and how this will compare to what has been accomplished in the past (IPCC 2001c, Chapter 2; IPCC 2001a). The historical rates of improvements in energy intensity (1–1.5% per year) are consistent with those needed for stabilization of carbon dioxide concentrations at 650 and 750 parts per million, and, in some cases, at 450 and 550 parts per million. However, the historical rates of improvements in carbon intensity (significantly less than 0.5% per year) are far slower than those needed for any stabilization level of carbon dioxide concentrations between 450 and 1,000 parts per million. Thus business-as-usual changes in technology will not achieve the desired goals of a less carbon-intensive energy system. Changes in energy intensity can arise from technological changes as well as through structural changes in the economy (for example, a move from heavy industry to a service economy), whereas changes in carbon intensity will require de-carbonizing the energy sector at a rate much faster than any historical changes.

The time taken for a transition to a less carbon-intensive energy sector is dependent upon the inertia in the energy sector, which is an inherent characteristic of socioeconomic systems. However, unlike the inertia in the climate system, inertia in the socioeconomic system is not fixed and can be changed by policies and individual choices. There is, typically, a delay of a few years to a few decades between perceiving a need and responding to it. By then, planning, researching and developing a solution, and implementing it becomes a major challenge. Technological response can be rapid, for example, the design and production of fuel-efficient cars after the oil crisis in the 1970s, but large-scale deployment of new technologies takes much longer, often dependent upon the rate of retirement of previously installed equipment. Early deployment of new technologies allows learning curve cost reductions (learning by doing), without premature retirement lock-in to existing, environmentally damaging technologies.

13.5.2 Terrestrial Sinks

Proper management of the biotic carbon cycle is essential if greenhouse gas stabilization is to be achieved. Each year roughly 60 gigatons of carbon are taken up and released by terrestrial ecosystems and 90 gigatons in the oceans. Small changes in this balance could swamp efforts to reduce current fossil fuel emissions, which are about 7 gigatons per year. Currently both the terrestrial system and the ocean systems show a net uptake of carbon. (See Table 13.3.) Some of this net uptake is a rebound from extensive clearing in many parts of the northern hemisphere over the past few centuries and from improved forest management. This is a significant factor in the estimated net uptake (IPCC 2001a, Chapter 3). Some of the net uptake is likely to be a response to the gradually rising carbon dioxide levels in the atmosphere and in the temperature as both contribute to vegetation growth and accumulation of carbon.

Table 13.3. Global Carbon Budgets. Fluxes are in GtC per year (positive is to the atmosphere) with ± standard error. (IPCC 2001a, Houghton 2003)

	IPCC 2001a		Houghton 2003	
	1980s	1990s	1980s	1990s
	(gigatons carbon per year)			
Atmospheric increase	+3.3 ± 0.1	+3.2 ± 0.1	+3.3 ± 0.1	+3.2 ± 0.2
Fossil emissions	+5.4 ± 0.3	+6.3 ± 0.4	+5.4 ± 0.3	+6.3 ± 0.4
Ocean–atmosphere flu	−1.9 ± 0.6	−1.7 ± 0.5	−1.7 ± 0.6	−2.4 ± 0.7
Net land–atmosphere flux	−0.2 ± 0.7	−1.4 ± 0.7	−0.4 ± 0.7	−0.7 ± 0.8
Land use change	+1.7 ± ?	no estimate	+2.0 ± 0.8	+2.2 ± 0.8
Residual terrestrial sink	−1.9 ± ?	no estimate	−2.4 ± 1.1	−2.9 ± 1.1

13.5.2.1 Land Use, Land Use Change, and Forestry Activities

The Kyoto Protocol recognizes that LULUCF activities can play an important role in meeting the ultimate objective of the UNFCCC. Biological mitigation of greenhouse gases through LULUCF activities can occur via three strategies: (1) conservation of existing carbon pools, for example, avoiding deforestation (2) sequestration by increasing the size of carbon pools, for example, through afforestation and reforestation or an increased wood products pool, and (3) substitution of fossil fuel energy by use of modern biomass.

The most significant sink activities include avoided deforestation, afforestation and reforestation, and forest, agricultural, and rangeland management. IPCC (IPCC 1996a, Chapter 24) estimated that LULUCF activities had the potential to sequester, or keep sequestered, about 100 gigatons of carbon by 2050, equivalent to 10–20% of projected fossil fuel emissions for the same period. However, competing land-uses, poor institutional structures and the lack of financial and legal facilities mean that only a small portion of this potential is currently being achieved.

13.5.2.1.1 Avoided deforestation

The most effective and immediate way of increasing net sequestration in terrestrial ecosystems is to reduce deforestation to only the most essential levels. Much of the 2 million gigatons of carbon released annual from forest clearance arises from the demand for agricultural and pastoral lands in developing countries (Geist and Lambin 2002). Some of this is necessary to maintain food production levels; some clearing leads to only short-term uses before the land is abandoned as degraded grasslands and often maintained that way by frequent fires. In other cases the land reverts to forest with an uptake of carbon, often to be cleared again. An immediate challenge to international institutions is to find a way to ensure that deforestation is limited to only that which leads to the long-term delivery of essential ecosystem goods and services and that the services provided by intact forests are properly valued. Perversely, the rules of the Kyoto Protocol provide incentives for landowners in developed countries to reduce deforestation but not for those in developing countries. Developed countries (Annex 1 countries) are required to account for all afforestation, reforestation, and deforestation activities and thus benefit in their accounting from any deforestation they avoid.

13.5.2.1.2 Afforestation and reforestation

The converse of avoiding deforestation is afforestation and reforestation. Globally, approximately 4.5 million hectares of previously unforested lands are reforested every year (FAO 2000). Some of this is by deliberate planting or establishment of trees, but much of it occurs through natural processes after changes in land use. Proper management of afforestation and reforestation can lead to synergies between adaptation to and mitigation of climate change. Obvious examples include rehabilitation of degraded lands with appropriate forest, woodland or shrub cover, and agroforestry systems (IPCC 2000b; IPCC 2002; CBD 2003, Chapter 4). IPCC (2000b) estimated that afforestation and reforestation could potentially store over 700 megatons of carbon per year although a much smaller amount (a few tens of megatons of carbon per year) would enter Kyoto Protocol compliance calculations.

13.5.2.1.3 Forest management

Forest degradation is another major source of carbon to the atmosphere. In many parts of the world tree densities are declining through overharvesting or overgrazing, which prevents adequate regeneration, or through shorter rotation cycles in slash and burn agricultural systems. Improved forest and woodland management could sequester an additional 170 million tons of carbon per year (Table 13.4), but little of this may enter the accounting system, as the amount for which Annex 1 countries can claim credit for activities within their borders is capped and improved forest management is excluded from the CDM.

13.5.2.1.4 Agricultural and rangeland management

Although the year-to-year storage of biomass in agricultural and pasture systems is small, changes in soil management can lead to significant carbon storage and is often accompanied by productivity benefits. Reduced tillage methods in croplands offer multiple benefits of improved soil condition and increased carbon storage. However, much of the carbon stored can be lost even due to brief periods of resumed tillage. Rangeland systems (called "grazing lands" in the Kyoto negotiations) are very extensive but have low biomass. Actions such as the management of livestock to reduce overgrazing or the exclusion of livestock to allow regeneration of trees and shrubs can lead to substantial sequestration of carbon in total. An additional 400 million tons of carbon could be sequestered per year from agricultural and rangeland systems.

13.5.2.2 Sinks and the Kyoto Protocol

The use of LULUCF activities (often called "sinks") within the Kyoto Protocol has been controversial. Some of the major concerns and counterarguments or countermeasures are summarized in Table 13.5. The Kyoto Protocol included the activities of afforestation, reforestation, and deforestation in the accounting system for Annex 1 countries, but delayed decisions on other LULUCF activities until additional information was prepared (e.g., IPCC 2000b). Annex 1 countries have the option of including a wider range of land management activities in the first commitment (accounting) period (2008 to 2012). Claims for credit from forest management are capped for each Annex 1 country. This is partly to take account of the "free-ride" issue (discussed below).

The Marrakesh Accords, agreed to in 2001, limited the eligibility of LULUCF activities under the CDM in the first commitment period to afforestation and reforestation projects, and limited the average use of carbon credits from the CDM during the first commitment period to 1% of an Annex 1 country's base year emissions. This is equivalent to about 30 million tons of carbon per year for Annex 1 countries (assuming the United States and Australia do not ratify).

This could be achieved by about 3 to 8 million hectares of new plantings in agroforestry or reforestation prior to 2008. The current rate of establishment of plantations throughout the developing world is about 4.5 million hectares per year but a high proportion of these plantings are not additional; that is, they would have occurred even without the incentives of the Kyoto Protocol and are, thus, not eligible for credit under the CDM. With strict enforcement of the additionality rule and the lack of a significant market for credits from sinks projects in the CDM, it is likely that there will be a very limited use of sinks in the CDM in the first commitment period (probably no more than 1 to 2 million tons of carbon per year, according to estimates made by the Carbon Finance Business of the World Bank).

The limitation of allowable LULUCF activities to afforestation and reforestation meant that there would be no credits through the CDM in the first commitment period for many activities that could have made significant contributions to the sustainable development goals of many developing countries. These include better forest management, reduced impact logging, forest protection (avoided deforestation), reduced tillage agriculture, or grazing management.

13.5.2.3 Environmental Implications of LULUCF Activities

LULUCF activities associated with the generation of carbon credits in either Annex 1 countries or in developing countries through the CDM can have positive, neutral, or negative impacts on the wider environment, including biodiversity (Table 13.6), depending on the specific conditions in which the activities occur (CBD 2003, Chapters 4, 5; STAP 2004).

The Kyoto Protocol and subsequent agreements include a number of clauses to prevent actions that are particularly damaging to the environment. All activities under the Protocol must be compatible with sustainable development, although this goal can be interpreted very widely. More specifically, the definition of afforestation and reforestation requires that plantings can only occur on lands that were not forested in 1990, thus existing forests cannot be cleared now to replace them with more carbon-rich forests.

Many nongovernmental organizations and some governments sought far stricter environmental regulations in the Kyoto Protocol outcomes and, in particular, in the use of CDM. These included an internationally agreed set of environmental standards and environmental assessment practices. No agreement on a specific set of standards could be reached, but the rules allow either the host-country or buyer government to reject a project if it does not meet their requirements on environmental standards. This includes issues such as the use of exotic and genetically modified organisms and activities damaging to biodiversity, such as planting forests on lands that are naturally grasslands or savannas.

LULUCF projects can have significant environmental benefits. For example, projects under consideration by the BioCarbon Fund of the World Bank for the first commitment period include the establishment of corridors to connect remnant forest patches and forest reserves, the establishment of buffer plantings to reduce intrusion into conservation areas, and several tree planting projects to rehabilitate degraded lands. Some of these projects have adaptive value, as they will also increase the resilience of the ecosystems and of the local communities to further climate change. Ironically, one of the most effective actions to reduce greenhouse gas emissions and to protect biodiversity, that is, actions to avoid

Table 13.4. A Summary of the Quantities of Sequestration and Emission Reductions through LULUCF Activities in the First Commitment Period (Watson and Noble 2004)

	Estimated Potential[a]	Indicated Use[b]	Caps	Estimated Use in First Period without the United States and Australia	Estimated Use if United States and Australia Ratify
			(million tons of carbon per year)		
Annex 1 Countries					
Afforestation, reforestation, and deforestation	40–50[c]	4[d]	—	3	4
Avoided deforestation	20[e]	15[f]	—	0	15
Forest management	100	720[g]	98[h]	70[i]	98
Crop and grazing-land management	150	18	—	10	18
Total, Annex 1		**754**	—	**83**	**127**
Non-Annex 1 Countries					
Afforestation and reforestation	Up to 700[j]	<300[k]	50[h]	<32[l]	50
Avoided deforestation	1,600	<<1,600	0	0	0
Forest management	70		0	0	0
Crop and grazing-land management	240		0	0	0
Total, Non-Annex 1			—	**<32**	**50**
Total Sinks in First Commitment Period				**c. 100**	**c. 180**
Emission reductions required below 1990 levels				**140**	**250**
Emission reductions compared with 15% business-as-usual increase over 1990 levels				**640**	**1,000**

[a] These data are from IPCC 2000b, which preceded the agreements in Marrakesh and the revised estimates prepared by parties of carbon gains and losses from forestry activities. On the whole these estimates do not include many of the factors contributing to the "free-ride."

[b] Based on national submissions prior to COP6 and FAO data in use at the time of those negotiations.

[c] This is based on an IPCC estimate of 20–30 MtC/y from uptake from A&R and 90 MtC/y emissions from deforestation; 20 MtC/y of deforestation falls under Article 3.7, as shown in the row below.

[d] Net gain from afforestation and reforestation activities and losses from deforestation under Article 3.3, as reported by Annex 1 countries at COP6.

[e] Eligible under Article 3.7, see Noble and Scholes (2000) for a detailed explanation.

[f] Based on Australia reducing land-clearing activities and the use of Article 3.7.

[g] The IPCC estimates are for increased uptake over a 1990 baseline from activities carried out since 1990 whereas the Marrakesh Accords allow forest management to be measured simply as the net uptake in managed forests. This leads to a far higher potential credit from managed forests, but the Marrakesh Accords limited total credit by applying a cap for each Annex 1 country.

[h] Including the United States cap as allocated in Marrakesh Accords.

[i] This assumes that all ratifying Annex 1 countries will use their full cap. A portion of these sinks are derived from countries that are likely to achieve compliance without the need to use sinks (e.g., Russia). Nevertheless, these sinks could enter the market either through Article 6 (JI) or Article 17 activities.

[j] Including agroforestry.

[k] Based on afforestation and reforestation activities in tropical countries, but many plantations are not eligible as they are not on land that was cleared of forest in 1990, and many others would fail a strict additionality test.

[l] Current market indications are that this cap will not be reached.

deforestation, is excluded from the CDM. However, actions that reduce deforestation in Annex 1 countries will avoid emissions and, thus, indirectly lead to credits. Similarly Annex 1 countries may also receive credit from environmentally beneficial activities such as soil carbon management in agriculture (for example, minimum till) and the restoration of degraded forests and other lands.

The use of LULUCF activities can achieve positive outcomes for mitigation, adaptation, and other environmental concerns. However, the outcomes cannot be prescribed in a set of rules and will depend on the interpretations and cooperation of national governments, including host governments, in the CDM. In the CDM, its executive board must approve all projects, LULUCF and others. Their interpretation of the sustainable development and broader environment requirements will also play a major part in establishing good practice.

13.5.2.4 Indirect Anthropogenic Effects

In looking beyond the particular rules agreed to for the first commitment period under the Kyoto Protocol, some further complications arise in the use of sinks in mitigating the increase in greenhouse gases in the atmosphere.

Human activities over the past century have led to increased atmospheric carbon dioxide and nitrogen deposition which, in turn, has increased the growth rates and carbon storage in many ecosystems. Many parties to the UNFCCC have already indicated that they wish to see the "free-ride" due to these indirect anthropogenic effects and from other effects such as changes in the age structure of forests factored out of the accounting system. If the accounting system includes only a small component of sinks-based credits, the errors and distortions from not factoring out the free-ride indirect effects would be small. This will be the case in the first commitment period.

Table 13.5. Some Major Concerns about the Inclusion of Sinks in Compliance with the Kyoto Protocol

Concern	Counter-argument or Counter-measure
Sink uptakes and emissions cannot be measured with sufficient accuracy	IPCC has produced Good Practice Guidance outlining appropriate standards for measuring and monitoring sinks (IPCC 2003).
Carbon sequestered in vegetation and soils may be re-emitted to the atmosphere via human actions (e.g., logging) or disturbances (e.g., fires)	Annex 1 countries are required to measure and account for all uptakes and losses for any sector included in their national accounting under Articles 3.3 or 3.4 of the Kyoto Protocol. Thus any losses of credited sequestered carbon will have to be replaced by new credits. In CDM projects, the continued sequestration of credited carbon must be verified every five years by an independent agent and any credits replaced if they are lost.
Planting of forests for mitigation purposes will occur on land that could be used for food production, leading to competition for land where the interests of the poor will be hard to protect	At current prices of carbon, the net value of agricultural products will usually far exceed that of the carbon. In cases where the timber value is sufficient to lead to an economic preference for forestry over agriculture, it will be rare that the value of carbon will play a significant part in the decision, and in most cases, the project will not be deemed to be additional as required by Article 12 of the Kyoto Protocol.
Sinks projects, including both afforestation/reforestation and avoided deforestation projects, lead to a loss of sovereignty over land use in the host country, particularly under the CDM	Avoided deforestation, which is the activity most often linked to this concern, is not included in the CDM. Also, the CDM now includes a temporary crediting mechanism that allows either the host or buying party to withdraw from the sequestration agreement at 5 to 20 year intervals. All sinks credits must be replaced by other credits no later than 60 years after they are created, at which time there is no penalty to the host country for releasing the sequestered carbon.
Sinks projects may lead to the planting of large areas of exotic mono-specific forest plantations, with impacts on human livelihoods and biological diversity	The increased financial value due to carbon in large-scale commercial plantations is usually too small to make any significant impact on a decision to go ahead with a planting. Thus it appears that most large-scale commercial plantations should be ruled ineligible for carbon credits as they are unlikely to be additional. There are some circumstances, for example, in projects linked to biofuels or in the rehabilitation of degraded lands where additionality conditions may be met. The Kyoto Protocol and subsequent agreements such as the Marrakesh Accords require that any such actions be compatible with sustainable development, but many sought to have stronger environmental safeguards incorporated in the agreements.
Additional forest cover will decrease the albedo of Earth (e.g., through reduced "snow cover" (Betts 2000), thus partially counteracting the effect of sequestering carbon	While this effect is real, boreal regions are likely to be the most sensitive, and these areas are not likely to be priority areas for afforestation and reforestation activities for crediting purposes due to the slow growth rates of trees in cold climates. However, current models indicate that large-scale land use changes will affect the global surface energy budget and lead to changes in local, regional, and global climates. Some of these changes will reduce the mitigation effects, such as the albedo effects in the boreal zone, while others will enhance the mitigation effect; for example, reforestation or afforestation may increase transpiration leading to cooler local climates (Marland et al. 2003).
The use of sinks credits allows further emissions from fossil fuels and thus the transfer of carbon from the long-lived and stable fossil deposits to the more dynamic atmosphere-biota-soil cycle. The equilibrium condition will thus result in more carbon in the atmosphere.	The ideal solution to the greenhouse effect is the immediate cessation of fossil fuel emissions or the quick re-sequestration of carbon in long-lived geological deposits. However, technological and institutional changes of the scale required to reduce our dependence on fossil fuels cannot be achieved other than over several decades. Sinks provide an opportunity to counteract some of the increase in greenhouse gases during the transition stage and lessen the impacts of climate change on Earth's ecosystems and human society.

IPCC was asked to assess options for factoring out the indirect anthropogenic effects. They concluded that there is considerable doubt whether the scientific knowledge to do this is available, taking into account variations across ecosystems, prevailing climate variability, and different management activities. Scientifically determined correction factors taking into account these influences will be costly to prepare, fraught with uncertainties, and controversial. An alternative approach is to use a simple discounting factor or factors and to apply these generically. This would cause inequities between some countries, which would have to be taken up in the negotiation of future targets.

13.5.2.5 Comprehensive Accounting of Biological Carbon

There has been a debate whether the accounting of sinks should be comprehensive or limited to a restricted range of activities, circumstances, and locations (referred to here as the "project approach"). A comprehensive approach would encourage complete monitoring of the biological component of the global carbon cycle and encourage actions that reduce the amount of greenhouse gas derived from these sources.

A major consequence of a comprehensive approach is that most of the free-ride and the year-by-year variation in carbon uptake would be reflected in the accounting system. Watson and Noble (2004) have estimated that the free-ride in a second commitment period could be about 0.6 gigatons of carbon per year. If a comprehensive approach were to be adopted and this additional 0.6 gigatons of carbon per year were to be factored out in the second commitment period, the amounts assigned to Annex 1 countries for the commitment period would be reduced by about 3 gigatons of carbon. This is equivalent to about another 15% of 1990 energy emissions on top of whatever reduction target is agreed to for the second commitment period. The adjustment of national targets would be an enormously risky process given the large uncertainties that remain in trying to estimate the size of the

Table 13.6. List of Possible LULUCF Activities with Potential Effects on Biodiversity or Other Aspects of Sustainable Development

Possible LULUCF Projects	Characteristics for Positive Impacts on Biodiversity	Characteristics for Negative Impacts on Biodiversity or Other Aspects of Sustainable Development
Conservation of natural forests, savannas, and woodlands	generally positive characteristics for a positive impact	
Conservation and restoration of wetlands	generally positive characteristics for a positive impact	could result in an increase in greenhouse gas emissions
Afforestation and reforestation (these are the only eligible LULUCF activities under the CDM)	on degraded lands; if fragmentation of habitats is reduced; if natural regeneration is encouraged and native species are used, reflecting the structural properties of surrounding forests; if clearing of pre-existing vegetation is minimized; if plantings are designed to create diverse landscape units; if rotation lengths are extended; if low-impact harvesting methods are used; if chemical use is minimized.	if activities occur on areas where undisturbed or non-intensively managed ecosystems are destroyed; if monocultures of exotic species are used; if there is large-scale soil disturbance; if short rotation periods are used or if harvesting operations clear complete vegetation; if sites with special significance for the in-situ conservation of agrobiodiversity are afforested; if chemicals are used abundantly.
Restoration of degraded lands and ecosystems	generally positive characteristics for a positive impact, depending upon the extent of degradation	habitats of species conditioned to extreme conditions could be destroyed; possible emission of nitrous oxide if fertilizers are used
Forest management	if natural forest regeneration occurs and "sustainable forest management" harvesting practices are applied	if monocultures of exotic species are planted and natural regeneration suppressed
Agroforestry	generally positive characteristics for a positive impact unless established on areas of natural ecosystems	negative if natural forest or other ecosystems are replaced
Cropland management	if reduced tillage is used without increased use of herbicides	if increased use of herbicides and pesticides if established on areas of natural ecosystems
Grassland and pasture management	mainly positive if no natural ecosystems are destroyed if no exotic species are used if fire management respects natural fire regeneration cycles	if established on areas that contained natural ecosystems if non-native species are introduced
Adaptation activities	generally positive characteristics for a positive impact if the activities conserve or restore natural ecosystems	

free ride even at continental scales, let alone at the individual country level. Any overestimation of the free-ride would be translated into emission reduction requirements that would not show up in the sinks accounting and would need to be made up through emission reductions through energy-based activities. Any underestimation would have the opposite effect.

Natural variation in uptake by terrestrial ecosystems adds another burden of uncertainty even when averaged over a five-year accounting period and all Annex 1 countries. The results of Bousquet et al. (2000) indicate that, averaging over all Annex 1 parties and over a five-year period, the variation in uptake will be at least plus or minus 0.25 gigatons of carbon per period (Watson and Noble 2004). Current Kyoto Protocol rules preclude banking (that is, a carryover of credits from one reporting period to the next) for credits derived from sinks. So much of this variation would show up in compliance outcomes. The variation is approximately the same size as that of the emission reduction targets in the first commitment period under the Kyoto Protocol, that is, about 5% of the 1990 Annex 1 emissions. Increasing the averaging period, allowing banking of sink credits and agreements to exchange credits between countries that are affected differently by global climate fluctuations such as El Nino, would reduce this problem. Nevertheless, variations of this size would have major impacts on trading markets and lead to significant price uncertainty.

There is still a high degree of uncertainty associated with these estimates. The calculation of the free-ride may be significantly in error and it remains possible that the size of the free-ride is not increasing. Until we understand the mechanisms and quantities involved in the full global carbon cycle and the variation year by year, anticipating the impact of comprehensive accounting on compliance targets will be fraught with difficulties and major uncertainties. If sinks are included only on a project-by-project approach, then only a small part of the global carbon cycle will be accounted for. Also, parties will tend to include projects that are likely to result in net increases in sequestration, making the monitoring of losses of carbon from the biota to the atmosphere weak.

13.5.3 Non-carbon Dioxide Greenhouse Gases

The most significant non-carbon dioxide greenhouse gases whose emissions need to be limited to address Article 2 of the UNFCCC include methane, nitrous oxide, halocarbons, and tropospheric ozone (IPCC 2000a, Chapter 5; IPCC 2001a, Chapter 4). Particulates, such as sulfate aerosols and soot, are also important in moderating Earth's climate (IPCC 2001a, Chapter 5).

The major anthropogenic sources of methane emissions include leakage from gas pipelines and coal mines, and emissions from landfills, livestock, and rice paddies. Given that the global warming potential for methane is 23 times larger than that for carbon dioxide (using a 100-year time horizon), reductions of methane emissions are particularly important (IPCC 2001a; the value used for the purposes of accounting under the Kyoto Protocol is 21). Opportunities to reduce methane include: capturing methane from landfills and coal mines and using it as an energy source (heat and electric power); flaring methane from landfills and coal mines where it is not cost-effective to capture it and use it as an energy source; and reducing leaks from gas pipelines (IPCC 2001c, Chapter 3). In the case of gas flaring and reducing leaks from gas pipelines, the costs avoided per ton of carbon are relatively low and well within the current and projected prices of carbon in the emerging carbon market. In the agricultural sector, opportunities include improved livestock and rice paddy management (for example, water management, tillage, and fertilization practices).

Emissions of fluorinated halocarbons are growing as they are being used to replace ozone-depleting substances controlled under the Montreal Protocol in a variety of sectors, including air conditioning and refrigerants. These emissions can be significantly reduced through containment, recovering and recycling refrigerants, and/or through use of alternative fluids and technologies (IPCC 2001c, Chapter 3/3, Appendix). In addition, inadvertent by-product emissions, for example, HCF-23 in the production of HCFC-22, need to be eliminated, for example, by incineration.

Emissions of nitrous oxide, which arise primarily from animal wastes and use of fertilizers in the agricultural sector, can be reduced, assuming farmers are provided with appropriate incentives to change their traditional farming methods. Nitrous oxide emissions from chemical plants can be removed catalytically or chemically.

The major precursors of tropospheric ozone that need to be reduced are non-methane hydrocarbons, carbon monoxide, and oxides of nitrogen. The major source for all three precursors is the combustion of fossil fuels, although there are other sources for the non-methane hydrocarbons, that is, industrial processes, fugitive emissions from fuel storage, and solvents. Reductions in tropospheric ozone will have significant benefits for human health and local ecosystems.

13.5.4 Geo-engineering Options

A number of geo-engineering possibilities have been suggested but a significant amount of research needs to be undertaken to evaluate their environmental efficacy and cost-effectiveness. Suggestions to date include: increasing the oceanic uptake of carbon from the atmosphere and transporting it to the deep ocean; placing reflectors in space to modify Earth's radiation balance; and adding aerosols in the lower stratosphere to reflect incoming solar radiation (IPCC 1996b, Chapter 25; IPCC 2001c, Chapter 4).

The concept of mitigating climate change through increased biological sequestration of carbon dioxide in oceanic environments (IPCC 2001a, CBD 2003, Chapter 4) has mainly focused on adding the limiting micronutrient iron to marine waters that have high nitrate and low chlorophyll levels (Boyd et al. 2000); the aim is to promote the growth of phytoplankton that, in turn, will fix significant amounts of carbon. The introduction of nitrogen into the upper ocean as a fertilizer has also been suggested (Shoji and Jones 2001). However, the effectiveness of ocean fertilization as a means of mitigating climate change may be limited (Trull et al. 2001). In addition, the consequences of larger and longer-term introductions of iron remain uncertain. There are concerns that the introduction of iron could alter food webs and biogeochemical cycles in the oceans (Chisholm et al. 2001), causing adverse effects on biodiversity. There are also possibilities of nuisance or toxic phytoplankton blooms and the risk of deep ocean anoxia from sustained fertilization (Hall and Safi 2001). A series of experimental introductions of iron into the Southern Ocean promoted a bloom of phytoplankton (Boyd et al. 2000) but also produced significant changes in community composition and the microbial food web (Hall and Safi 2001).

The concept of adding aerosols to the lower stratosphere has largely been rejected given that it would lead to an increased loss of stratospheric ozone, an associated increase in damaging ultraviolet radiation reaching Earth's surface, and a likely increase in the incidence of melanoma and non-melanoma skin cancer (IPCC 1999).

13.6 Economic Instruments

13.6.1 Kyoto Mechanisms

The Kyoto Protocol includes a series of "flexibility mechanisms" to facilitate and reduce the costs of Annex 1 countries in meeting their targets. The simplest of the mechanisms is Article 17, which allows the transfer and acquisition of emission reductions (emissions trading) between Annex 1 parties to the Protocol that are in good standing with respect to the various rules of reporting and accounting. This form of trading simply transfers credits from one national registry to another with the agreement of both parties.

Two other mechanisms are based on individual projects that achieve emissions reductions (for example, energy or LULUCF) or removals by sinks. Article 6 (often referred to as Joint Implementation or JI) deals with trading between legal entities in one Annex 1 country through acquiring credits from a legal entity in another Annex 1 country. Article 12 (the Clean Development Mechanism or CDM) allows an entity in an Annex 1 country to accrue credits from projects in a non-Annex 1 country. In each case, the transactions have to be approved by the acquiring country and the country hosting the project.

The CDM has two goals: assisting non-Annex 1 countries achieving their sustainable development goals and assisting Annex 1 countries in achieving compliance with their emission targets. The CDM also creates opportunities for technological transfer. Much of the debate over the CDM has focused on the second goal. Some are concerned that the CDM mechanism will reduce the effort made in developed countries to achieve the core goals of the Kyoto Protocol, as the first step towards the ultimate goal of the UNFCCC, that is, the stabilization of greenhouse gases in the atmosphere, largely via modifying energy use and energy supply. Any flexibility mechanism will lower the cost of achieving a compliance target and thus reduce the incentives to invest in new research and new technologies.

The focus on the compliance goal of the CDM has often been in conflict with the better achievement of the sustainable development goal. There has been a long debate about the inclusion of certain types of practices in the CDM. In the energy-related sectors, the eligibility of large hydropower and clean coal technologies has been controversial. Annex 1 countries are to refrain from using emission reductions generated from nuclear facilities to meet their commitments in both JI and CDM activities. In the LULUCF sector of the CDM, the range of activities has been limited to only afforestation and reforestation projects, which is the establishment of new forests on lands that where not forested in 1990.

In both JI and CDM trading, the emission reductions have to be "additional" to what would otherwise have occurred. This requirement is often seen as ensuring that there is additional effort aimed at reducing greenhouse gas emissions. However, the additionality clause is fundamentally more important to atmospheric accounting in the CDM than under JI. If the emission reductions in a JI transaction would have occurred without the incentive of the trading, the effect on the atmosphere remains neutral; the host country transfers some of its emission reduction credits to the acquiring country, allowing the acquiring country to emit more and leaving the host country to carry out extra efforts to meet its target. Under the CDM, the symmetry of targets does not exist. If the emission reductions from the project are not truly additional (that is, they would have occurred without the incentive of the emission trading), then the acquiring (Annex 1) country is able to raise its emissions while no extra emission reductions occur in the host country. The atmospheric greenhouse gas concentrations will increase as a consequence of the non-additional project. The identification of additionality and the estimation of the baseline over which the additional emission reductions are measured will be a major challenge for the CDM.

In 2003, the EU formally established an emissions trading system in which each country is issued allowances as to how much carbon dioxide its energy-intensive companies (for example, power plants, oil refineries, paper mills, and steel, glass, and cement factories; about 12,000 separate installations) are allowed to emit. In the EU ETS, reductions below the limits will be tradable across the European Union and in special circumstances outside the EU bubble. This trading system obviously derives from the UNFCCC negotiations but is formally independent of the Kyoto Protocol and is expected to continue, whatever the fate of the Protocol negotiations. Penalties for non-compliance are set at 40 Euros per ton of carbon dioxide in the first trading period of 2005–2007 increasing to 100 Euros in 2008–2012. In mid-2004, a "Linking Directive" allowed extra flexibility through JI and CDM trading. This additional flexibility is expected to reduce costs by about 20% (Kruger and Pizer 2004), but it also provides a formal link between the EU ETS and the Kyoto flexibility mechanisms. The Linking Directive does not allow the full range of credits into the ETS. Nuclear power and sinks are excluded and hydropower projects are to be monitored closely. The use of JI and CDM will also be monitored, as these activities are to be only supplementary to action taken at home. A limit of 6–8% for JI/CDM contributions was widely discussed before the Directive was set up.

13.6.2 Other Instruments and Options

The Kyoto Protocol negotiations have taken a particular path towards seeking to implement the broader goals of the UNFCCC. Negotiators adopted a cap-and-trade approach whereby quantitative caps (targets) are set for each Annex 1 country whose government usually passes these targets on to various national sectors as targets for particular commercial entities. These entities would be penalized if they exceeded their allowances, so they have an incentive to cut emissions. By allowing the entities to trade emission permits, those with low marginal costs of abatement will make extra cuts and sell credits to entities that find it more costly to cut emissions.

Another decision that has to be taken in designing a cap-and-trade program is whether to apply the targets "upstream," where carbon enters the economy (when fossil fuels are imported or produced domestically) or farther "downstream," closer to the point where fossil fuels are combusted and the carbon enters the atmosphere. An analysis by the Congressional Budget Office of the United States concluded that, in general, an upstream program would have several major advantages over a downstream program (CBO 2001).

An alternative is a carbon tax approach in which commodities or activities that lead to carbon emissions are taxed, thus providing an incentive to reduce the use of these commodities or activities. This is often seen as a simpler approach to achieving incentives for emission reductions. However, taxes are usually politically unpopular in most countries. Neutral carbon taxes have been suggested, where the carbon tax is introduced along with the removal or reduction of other taxes. However, these will usually lead to changes in the distribution of tax liabilities. Some have suggested that the cap-and-trade and tax approaches may be combined to overcome the main weaknesses of both schemes. In a hybrid approach, a cap-and-trade system would be set in place, but if the cost of permits rose too high, they could be purchased at a fixed price. This amounts to using a tax as a safety valve for the cap-and-trade system (Jacoby and Ellerman 2002).

Some critics object to the entire structure of the Kyoto trading system. Victor (2001) criticizes the Kyoto Protocol for setting targets without a clear idea of the costs involved in reaching those targets; he argues that huge transfers in property rights are involved both nationally and internationally and the allocation of these rights was not seriously addressed either nationally or internationally among the non-Annex 1 countries. McKibbin and Wilcoxen (1999) make a similar criticism and propose a two-tier system of emission credits—allowing an emission in a particular year that will have to be traded at a capped price and emissions endowments that have a permanent allowance to emit and that can be traded at a flexible price.

Many other variants have been suggested. Experience in the U.S. sulfur dioxide and nitrous oxide trading system and the Montreal Protocol on Ozone have often been looked to for guidance. However, neither is a good match. The EU ETS alone is ten times larger than the U.S. sulfur dioxide and nitrous oxide trading system and brings in the complexity of dealing with many countries with different pre-existing conditions. In the Montreal Protocol, each country met its targets and there was thus no need for trading; however, there may be lessons to be learned from how the targets were met within countries (for example, auctioning emission rights).

13.6.3 Technology Transfer to Lower Costs

The transfer of environmentally sound technologies is a major element in any global strategy to combat climate change. Technology transfer between countries and regions widens the choice of mitigation options and economies of scale, and learning will lower the costs of their adoption. A framework for meaningful and effective actions for technology transfer includes: assessing technology needs; establishing a technology information system; creating enabling environments for technology transfer; providing capacity-building for technology transfer; and funding to implement the various activities. There are a number of mechanisms to facilitate technology transfer, including: national innovation systems, official development assistance, the multilateral development banks, and the Global Environmental Facility and the Clean

Development Mechanism, both of which are financing instruments associated with the UNFCCC.

13.7 Economic Costs of Reducing Greenhouse Gas Emissions

There is a wide range of estimates of the costs of mitigating climate change. The breadth of this range reflects differences in both modeling methodologies and in the policies used to reduce emissions. Given the use of well-designed policies, the IPCC estimated that half of the projected increase in global emissions between now and 2020 could be reduced with direct benefits (negative costs), while the other half could be reduced at less than $100 per ton of carbon (IPCC 2001c, Chapters 1, 3, 5, 6). Reductions in emissions can be obtained at no or negative costs by exploiting no-regrets opportunities, that is, by reducing market or institutional imperfections such as subsidies; taking into account ancillary benefits (for example, local and regional air quality improvements); and using revenues from taxes or auctioned permits to reduce existing distortionary taxes through revenue recycling. For example, in countries with significant local and regional air pollution problems, the social and economic benefits associated with using more climate-friendly technologies can be considerable through improved human health.

In the absence of international carbon trading, the estimated costs of complying with the Kyoto Protocol for industrial countries range from 0.2% to 2% of GDP; with full trading among industrial countries the costs are halved to 0.1% to 1% (IPCC 2001c, Chapters 7–10). The equivalent marginal costs range from $76 to $322 in the United States without trading and from $14 to $135 with trading. (See Figure 13.5 in Appendix A.) These costs could be further reduced with use of sinks (carbon sequestration using reforestation, afforestation, decreased deforestation, and improved forest, cropland, and grassland management), project-based trading between industrialized countries and developing countries through the CDM, and reductions in the emissions of other greenhouse gases (for example, methane and halocarbons).

There is a wide range of estimates for the likely price of carbon during the first and second phases of the EU trading scheme, that is, 2005–2007 and 2008–2012, respectively (Nicholls 2004). All experts, primarily from investment banks and consultancies, recognize that the price is dependent upon a number of factors including: whether Russia ratifies the Kyoto Protocol and it enters into force; the allocation of allowances to industry under the EU trading system; the price of coal and gas; and the extent of the use of overseas credits, which is allowed under the Linking Directive. However, there is no agreement on the price, with estimates ranging from 5–15 Euros per ton of carbon dioxide during the first phase, and rising for the second phase.

Known technological options could achieve stabilization of carbon dioxide at levels of 450 to 550 parts per million over the next 100 years. The costs of stabilization are estimated to increase moderately going from 750 to 550 parts per million, but significantly going from 550 to 450 parts per million. (See Figure 13.6 in Appendix A.) However, it should be recognized that the pathway to stabilization as well as the stabilization level itself are key determinants of mitigation costs (IPCC 2001c, Chapters 2, 8, 10). The secondary economic benefits (auxiliary benefits) of mitigation activities could reduce the costs. The costs of stabilization, based on three global models, at 450 and 550 parts per million are estimated to be between $3.5 and $17.5 trillion, and $0.5 and $8 trillion, respectively, over the next century (1990 US$, present value discounted at 5% per year for the period 1990 to 2100). These estimated costs will only have a minor impact on the rate of economic growth; for example, the percentage reduction in global average GDP over the next 100 years for stabilization at 450 parts per million ranges from about 0.02% to 0.1% per year, compared to annual average GDP growth rates of 2–3% per year. The reduction in projected GDP averaged across all IPCC storylines and stabilization levels is lowest in 2020 (1%), reaches a maximum in 2050 (1.5%), and declines by 2100 to 1.3%. The annual 1990–2100 GDP growth rates over this century across all stabilization scenarios was reduced on average by only 0.003% per year, with a maximum reaching 0.06% per year.

In contrast to the costs of mitigation/stabilization are the costs of inaction, that is, damage caused by climate change. These costs are difficult to calculate because of uncertainties in the rate and magnitude of regional climate change and the resulting impacts on ecological systems, socioeconomic sectors, and human health. In addition, some ecosystem and human health damages are hard to quantify in economic terms. Many ecosystem goods and services do not trade in the marketplace, for example, climate control, flood control, pollination, and soil formation and maintenance, while religious and aesthetic issues, and placing a value on a human life are highly controversial. IPCC (1996b, Chapter 6) estimated the economic costs associated with a doubling of atmospheric carbon dioxide and a 2.5° Celsius temperature warming to range between 1.5% and 2.0% of world GDP (1–1.5% GDP in developed countries, and 2–9% in developing countries). The marginal damage was estimated to range from $5 to $125 per ton of carbon (highly dependent upon the assumed value for the discount rate). Nordhaus (1994) organized an expert group, which estimated the economic costs for a 3° Celsius temperature warming by 2090 to range between 0% and 21% of world GDP (IPCC 1996b, Chapter 6), with a mean value of 3.6% and a median answer of 1.9%. The expert panel also estimated the economic costs for a 6° Celsius temperature warming by 2090 to range between 0.8% and 62% of world GDP (IPCC 1996b, Chapter 6), with a mean value of 10.4% and a median answer of 5.5%.

Applying cost-benefit analysis to climate change is much more difficult than for many public policy issues because many of the benefits of mitigation will not be realized for decades, whereas a significant fraction of the costs will occur soon, and the estimated costs are very sensitive to the assumed value for the discount rate. Cline (2004), using a discount rate of 1.5%, estimated the relative efficiencies of: the Kyoto Protocol, a global carbon tax, and emissions reductions that mitigate damage in 95% of scenarios (comparing the mitigation costs to the worst-case damage costs). The benefit to cost ratios were positive in all three cases, being 1.77, 2.1, and 3.8, respectively, but the costs are both significant and quite uncertain, with costs borne by this generation but benefits increasing over time.

13.8 Institutional Responses

Addressing climate change and reducing greenhouse gas emissions will require the development and implementation of multilateral agreements such as the UNFCCC and its Kyoto Protocol, and a wide range of actions by local and national governments, regional economic organizations, the private sector, NGOs, bilateral and multilateral organizations and partnerships, the Global Environment Facility, media, and consumers. (See Box 13.1.)

Different actors have different roles along the research, development, demonstration, and widespread deployment value chain and pipeline for climate-friendly technologies (PCAST 1999). In-

BOX 13.1
Potential Roles of Different Actors

The potential roles of *intergovernmental processes* in addressing climate change include:

- establishing a long-term global emissions target with intermediate targets and an equitable allocation of national and/or regional emissions rights, possibly coupled with common policies and measures; and
- finalizing the rules for carbon trading and moving toward full implementation of an international carbon trading system.

National governments can take the following steps:

- All governments should consider establishing a national policy and regulatory environment, with associated institutional infrastructure, for the efficient deployment of climate-friendly energy production and use technologies, including energy sector reform, energy pricing policies, carbon taxes, elimination of fossil fuel and transportation subsidies, internalization of environmental externalities, mechanisms for market scale-up of climate-friendly technologies (for example, short-term subsidies for use of new technologies during market build-up, and quota systems that establish a minimum share of the market), energy efficiency standards, labeling systems, education, and training.
- Governments, especially from industrial countries, should consider increasing investment in energy R&D, with a greater emphasis on energy efficiency technologies, renewable energy technologies, carbon capture and storage, and hydrogen, and establish public-private partnerships for research;
- Governments with obligations, or likely to assume obligations, under the UNFCCC and its Kyoto Protocol should consider establishing domestic allocation of emissions rights and establish a national tradable emissions system (net costs of emissions abatement can be reduced by taxing emissions or auctioning permits and using the revenues to cut distortionary taxes on labor and capital);
- Industrial countries should consider assisting developing countries access climate-friendly technologies by establishing an appropriate intellectual property rights regime, coordinating relevant bilateral aid programs so as not to distort climate-friendly technology markets, and continuing to fund the GEF; and
- All governments should consider integrating climate variability and change into national economic and sector planning, especially for water resources, agriculture, forestry, health, and coastal zone management.

Local governments can take steps to establish local markets for climate-friendly technologies (for example, through purchase agreements). Subnational governments in industrial countries (for example, states, municipalities) may wish to take the lead in promoting climate-friendly policies and assuming voluntary emissions targets as a signal to national governments of willingness and ability to move on climate policy. In developing countries, such entities, particularly in higher-income cities, may also wish to consider climate policies.

The *private sector* should consider increasing investment for research, development, demonstration, and deployment of climate-friendly technologies; establish voluntary standards (for example, for energy efficiency); and ensure efficient functioning of emissions trading systems.

International financial institutions should provide financing for climate-friendly production and use technologies; promote energy sector reform (including energy pricing policies, elimination of fossil fuel and transportation subsidies, internalization of environmental externalities); promote mechanisms for market scale-up of climate-friendly technologies; promote energy efficiency standards, training, and capacity building; stimulate the flow of climate-friendly technologies to developing countries by providing carbon financing; and assist countries in reducing vulnerability to climate change by mainstreaming climate variability and change into national economic planning.

The role of the *Global Environment Facility* is to provide grant resources to developing countries to develop regulatory and policy frameworks to promote climate-friendly technologies, demonstrate effective and innovative measures to reduce greenhouse gas emissions, aggregate markets for climate-friendly technologies, build capacity for addressing climate change mitigation and adaptation, and provide financing for adaptation measures.

Academia's role involves continued research, monitoring, and data management for improved understanding of the impact of human activities on the climate system, and the consequent implications for the vulnerability of socioeconomic systems, ecological systems, and human health. It should continue to conduct energy research.

Local communities play an important role in promoting energy conservation activities and developing coping strategies to adapt to climate variability and change.

The role of *media and nongovernmental and civil society organizations* is to promote awareness, that is, to inform civil society and government officials of the seriousness of climate change and the ramifications of their actions.

Consumers shape the market by purchasing "green energy" and energy efficient technologies; theyinfluence government policies through advocacy.

novative partnerships will be particularly important in technology transfer and financing.

A critical condition for significant investment in climate-friendly technologies is the establishment by governments of an appropriate policy and regulatory framework, for example, the elimination of perverse fossil fuel subsidies, the internalization of environmental externalities, and the provision of appropriate incentives for new technologies to overcome initial market barriers.

At the R&D (laboratory/bench) and demonstration (small to medium to commercial scale pilots) stages, there are roles for both the government and the private sector, recognizing that barriers to investment in R&D and demonstration include the difficulty of capturing the economic benefits of the R&D and demonstration, long time horizons associated with capturing the benefits, high risks, and high capital costs. At the stage of widespread deployment there are clear roles for the private sector and for aid agencies, trade agencies, the GEF, and the multilateral development banks. At this stage the major barriers are high transaction costs, the fact that the prices for competing technologies rarely include externalities, and a lack of information. However, there is a critical stage in the pipeline buy-down (reducing the cost per unit), which is normally an area of neglect by all actors, where the barriers include financing the incremental costs, cost uncertainty, and technological and other risks.

A mechanism is needed to fill this gap in the innovation pipeline. Mechanisms for technology cost buy-down could be included in energy sector reform. One approach that has been used in industrial countries where energy sector restructuring has taken place, is the establishment of small guaranteed markets to assist in launching new climate-friendly technologies, where qualifying new technologies compete for shares of these markets. One example of such a program is the Renewable Non-Fossil Fuel Obligation in the United Kingdom. A clean energy technology obligation (CETO) could be a key element in energy sector reform in developing countries and countries with economies in transition to accelerate the deployment of promising new technologies using a range of competitive instruments, such as auctions. CETO competitions could be organized by guaranteeing markets sufficiently large that clean energy technologies manufacturers could expand production capacity to levels where the economies of scale can be realized, reducing unit costs by advancing along the learning curve. The incremental costs of these competitions in developing countries could be covered by bilateral donors or through an international fund, potentially managed by the GEF.

Regional and international financial institutions should play an enhanced role in financing and attracting private capital to climate-friendly technologies, for example, renewable energy and energy efficiency technologies, in emerging markets. Carbon finance, through the emerging national and international markets, can also play a vital role in promoting these technologies by increasing the internal rate of return for investments in these technologies.

Local governments can play an important role in the development of local renewable energy markets by influencing energy demand, use, and development in their jurisdictions, for example, through policy and purchasing, through their regulatory functions, and by expediting planning procedures. There is also a gap in the insurance and risk-transfer market for new renewable energy technologies, which, because of their small scale, have difficulty in passing internal business hurdles. The opportunity, therefore, arises for the public sector to work with the finance and insurance sectors to address these and other specific barriers.

Technology transfer results from actions taken by a wide range of actors, including project developers, owners, suppliers, buyers, recipients, and users of technologies; financiers and donors; governments, international institutions, and civil society organizations (IPCC 2000c). Governments of industrial and developing countries need to provide an appropriate enabling environment to enhance technology transfer, by reducing risks, through inter alia sound economic policy and regulatory frameworks, transparency, and political stability.

References

Abraham, S., 2004: The Bush administration's approach to climate change, *Science,* **305,** pp. 616–617.

Alcamo, J. and G.J.J. Kreileman, 1996: Emission scenarios and global climate protection, *Global Environmental Change: Human and Policy Dimensions,* **6,** pp. 305–334.

Ametistova, L., J. Twidell, and J. Briden, 2002: The sequestration switch: Removing industrial CO_2 by direct ocean absorption, *Science of the Total Environment,* **289,** pp. 213–223.

Anon., 2003: *Poverty and Climate Change: Reducing the Vulnerability of the Poor,* African Development Bank/ Asian Development Bank/ Department for International Development, UK/ Directorate-General for Development, European Commission/ Federal Ministry for Economic Cooperation and Development, Germany/ Ministry of Foreign Affairs – Development Cooperation, The Netherlands/ Organization for Economic Cooperation and Development/ United Nations Development Programme/ United Nations Environment Programme/ World Bank, xii + 43 pp.

Anon., 2004a: Climate change activities in the US: 2004 update,, Pew Center for Global Climate Change, Arlington, VA. Available at http://www.pewclimate.org/what_s_being_done/us_activities_2004.cfm.

Anon., 2004b: *South-North Dialogue on Equity in the Greenhouse: A Proposal for an Adequate and Equitable Global Climate Agreement,* GTZ (Gesellschaft für Technische Zusammenarbeit), Eschborn, Germany.

Ashton, J. and X. Wang, 2003: Beyond Kyoto: Advancing the international effort against climate change. In: *Equity and Climate: In Principle and Practice.* Pew Center Report, Pew Center for Global Climate Change, Arlington, VA, pp. 61–84.

Betts, R.A., 2000: Offset of the potential carbon sink from boreal forestation by decreases in surface albedo, *Nature,* **408,** pp. 187–190.

Bousquet, P., P. Peylin, P. Ciais, C. Le Que're, P. Friedlingstein, et al., 2000: Regional changes in carbon dioxide fluxes of land and oceans since 1980, *Science,* **290,** pp. 1342–1346.

Boyd, P.W., A.J. Watson, C.S. Law, E.R. Abraham, et al., 2000: A mesoscale phytoplankton bloom in the polar southern ocean stimulated by iron fertilization, *Nature,* **407,** pp. 695–702.

Caribbean Community Secretariat, 2003: *Caribbean Risk Management Guidelines for Climate Change Adaptation Decision-Making,* Caribbean Community Secretariat, Georgetown, Guyana, p. 75.

CBD (Convention on Biological Diversity), 2003: *Interlinkages Between Biological Diversity and Climate Change: Advice on the Integration of Biodiversity Considerations into the Implementation of the United Nations Framework Convention on Climate Change and its Kyoto Protocol,* Technical series no. 10, Chapters 2 and 4, CBD, Montreal, QC, Canada.

CBO (Congressional Budget Office), 2001: An evaluation of cap-and-trade programs for reducing US carbon emissions, CBO, Washington, DC. Available at http://www.cbo.gov/showdoc.cfm?index=2876&sequence=0&from=0.

Chisholm, S.W., P.G. Falkowski, and J.J. Cullen, 2001: Dis-crediting ocean fertilization, *Science,* **294,** pp. 309–310.

Claussen, E. and L. McNeilly, 1998: Equity and global climate change: The complex elements of global fairness, Pew Center on Global Climate Change, Arlington, VA. Available at http://www.pewclimate.org/report2.html.

Cline, W.R., 2004: Meeting the challenge of global warming, Paper prepared for the Copenhagen Consensus Program of the National Environmental Assessment Institute, Copenhagen, Denmark. Available at http://www.copenhagenconsensus.com/files/filer/cc/papers/sammendrag/accepted_climate_change_300404.pdf.

Cox, P.M., R.A. Betts, C.D. Jones, S.A. Spall, and I.J. Totterdell, 2000: Acceleration of global warming due to carbon-cycle feedbacks in a coupled climate model, *Nature,* **408,** pp. 180–184.

FAO (Food and Agriculture Organization of the United Nations), 2000: *Global Forest Resources Assessment 2000,* FAO, Rome, Italy.

Fu, Q., C.M. Johanson, S.G. Warren, and D.J. Seidel, 2004: Contribution of stratospheric cooling to satellite-inferred tropospheric temperature trends, *Nature,* **429,** pp. 55–58

Geist, H.J. and E.F. Lambin, 2002: Proximate causes and underlying driving forces of tropical deforestation, *BioScience,* **52,** pp. 143–150.

Hall, J.A. and K. Safi, 2001: The impact of in situ Fe fertilization on the microbial food web in the southern ocean, *Deep-Sea Research Part II,* **48,** pp. 2591–2613.

Hammill, A., 2004: *Focusing on Current Realities: It's Time for the Impacts of Climate Change To Take Centre Stage,* IISD Commentary, March 2004, IISD, Winnipeg, MB, Canada.

Hare, B., 2003: *Assessment of Knowledge on Impacts of Climate Change: Contribution to the Specification of Art. 2 of the UNFCCC,* External Expertise Report, German Advisory Council on Global Change, Berlin, Germany.

Hoffert, M.I., K. Caldeira, G. Benford, D. R. Criswell, C. Green, H. Herzog, A.K. Jain, H. S. Kheshgi, K. S. Kackner, J. S. Lewis, H. D. Lightfoot, W. Manheimer, J. C. Mankins, M. E. Mauel, L. J. Perkins, M. E. Schlesinger, T. Volk, and T.M. L. Wigley, 2003: Advanced technology paths to global climate stability: Energy for a greenhouse planet, *Science,* **298,** pp 981–987.

Holdren, J.P., 2003: Environmental change and the human condition, *American Academy Bulletin,* **Fall,** pp. 24–31.

Houghton, R.A., 2003: Why are estimates of the global carbon balance so different? *Global Change Biology,* **9,** pp. 500–509.

Huesemann, M.H., A.D. Skillman, and E.A. Crecelius, 2002: The inhibition of marine nitrification by ocean disposal of carbon dioxide, *Marine Pollution Bulletin,* **44.**

Huq, S. and R.J.T. Klein, 2003: Adaptation to climate change: Why and how, *SciDev.Net Climate Change Dossier,* Policy brief. Available at http://www.scidev.net/dossiers/index.cfm?fuseaction=printarticle&dossier=4&policy=44.

IEA (International Energy Agency), 2003: *World Energy Investment Outlook: 2003 Insights,* IEA, Paris, France, 516 pp.

IISD (International Institute for Sustainable Development), 2003: Livelihoods and climate change: Combining disaster risk reduction, natural resource management and climate change adaptation in a new approach to the reduction of vulnerability and poverty, A conceptual framework paper prepared by the Task Force on Climate Change, Vulnerable Communities, and Adaptation, IUCN,IISD, SEI, SDC and Inter-cooperation, Winnepeg, Canada.

IISD, 2003: The Brazilian proposal and its scientific and methodological aspects, IISD, Winnipeg, MB, Canada.

IPCC (Intergovernmental Panel on Climate Change), 1996a: *Climate Change 1995: Impacts, Adaptation and Mitigation,* Contribution of Working Group II to the second assessment report of the IPCC, R.T. Watson, M.C. Zinyowera, R.H. Moss, and D.J. Dokken, (eds.), Cambridge University Press, Cambridge, UK.

IPCC, 1996b: *Climate Change 1995: Economic and Social Dimensions of Climate Change,* Contribution of Working Group III to the second assessment report of the IPCC, J.P. Bruce, H. Lee, and E. Haites (eds.), Cambridge University Press, Cambridge, UK.

IPCC, 1999: Synthesis of the reports of the scientific, environmental effects, and technology and economic assessment panels of the Montreal Protocol, D.L. Albritton. S.O. Andersen, P.J. Aucamp, S. Carvalho, L. Kuijpers, et al. (eds.), United Nations Environmental Programme, Nairobi, Kenya.

IPCC, 2000a: *Special Report on Emissions Scenarios,* N. Nakicenovic and R. Swart (eds.), Cambridge University Press, Cambridge, UK.

IPCC, 2000b: *Special Report on Land Use, Land-Use Change and Forestry,* R.T. Watson, I.R. Noble, B. Bolin, N.H. Ravindranath, D.J. Verado, et al. (eds.), Cambridge University Press, Cambridge, UK.

IPCC, 2000c: *Special Report on Methodological Issues in Technology Transfer,* B. Metz, O.R. Davidson, J.W. Martens, S.N.M. van Rooijen, and L. Van Wie McGrory (eds), Cambridge University Press, Cambridge, UK.

IPCC, 2001a: *Climate Change 2001: The Scientific Basis,* Contribution of Working Group I to the third assessment report of the IPCC, J.T. Houghton, Y. Ding, D.J. Griggs, M. Noguer, P.J.van der Linden, et al. (eds.), Cambridge University Press, Cambridge, UK.

IPCC, 2001b: *Climate Change 2001: Impacts, Adaptation, and Vulnerability.* Contribution of Working Group II to the third assessment report of the IPCC, J.J. McCarthy, O.F. Canziani, N.A.Leary, D.J. Dokken, and K.S. White (eds.), Cambridge University Press, Cambridge, UK.

IPCC, 2001c: *Climate Change 2001: Mitigation,* Contribution of Working Group III to the third assessment report of the IPCC, B.Metz, O. Davidson, R. Swart, and J. Pan (eds.), Cambridge University Press, Cambridge, UK.

IPCC, 2001d: *Climate Change 2001: Synthesis Report,* R.T. Watson (ed.), Cambridge University Press, Cambridge, UK.

IPCC, 2002: *Climate Change and Biodiversity,* IPCC technical paper V, H. Gitay, A. Suarez, R.T. Watson, and D.J. Dokken (eds.), Geneva, Switzerland.

IPCC, 2003: *Good Practice Guidance for Land-Use, Land-Use Change and Forestry,* IPCC National Greenhouse Gas Inventories Programme, Institute for Global Environmental Strategies, Kanagawa, Japan.

IPCC CZMS (Coastal Zone Management Subgroup), 1990: *Strategies for Adaptation to Sea Level Rise,* Report of the CZMS, Response Strategies Working Group of the IPCC, Ministry of Transport, Public Works and Water Management, The Hague, The Netherlands, 122 pp.

Jacoby, H.D. and A.D. Ellerman, 2002: *The Safety Valve and Climate Policy,* MIT Joint Program on the Science and Policy of Global Change, Report no. 83, February [Revised July], Massachusetts Institute of Technology, Cambridge, MA.

Jacoby, H.D., R. Schlamensee, and I.S. Wing, 1999: *Towards a Useful Architecture for Climate Change Negotiations,* MIT Joint Program on Science and Policy of Global Change, Report no. 49, May, Massachusetts Institute of Technology, Cambridge, MA.

Jin, M., 2004: Analysis of temperature using AVHRR observations, *Bulletin of the American Meteorological Society,* **85,** pp. 587–600.

Kemp, D. and I. McFarlane, 2004: Australia still on Kyoto target: Greenhouse intensity down 31%, Joint media release 13 April 2004, Department of Environment and Heritage, Canabera, Australia. Available at http://www.deh.gov.au/minister/env/2004/mr13apr04.html.

Klein, R.J.T. and R.S.J. Tol, 1997: Adaptation to climate change: Options and technologies: An overview paper, Technical paper FCCC/TP/1997/3, United Nations Framework Convention on Climate Change Secretariat, Bonn, Germany, iii + 33 pp.

Kruger, J. and W.A. Pizer, 2004: *The EU Emissions Trading Directive: Opportunities and Potential Pitfalls,* Discussion paper 04–24, Resources for the Future, Washington, DC.

Leemans, R., B.J. Eickhout, B. Strengers, A.F. Bouwman, and M. Schaeffer, 2002: The consequences for the terrestrial carbon cycle of uncertainties in land use, climate and vegetation responses in the IPCC SRES scenarios, *Science in China, Series C,* **45,** pp. 126–136.

Leemans, R. and B. Eickhout, 2004: Another reason for concern: Regional and global impacts on ecosystems for different levels of climate change, *Global Environmental Change,* **14,** pp. 219–228.

Marland, G., R.A. Pielke Sr., M. Apps, R. Avissar, R. A. Betts, K.J. Davis, P.C. Frumhoff, S.T. Jackson, L.A. Joyce, P. Kauppi, J. Katzenberger, K.D. MacDicken, R.P. Neilson, J.O. Niles, D.S. Niyogi, R.J. Norby, N. Pena, N. Sampson, and Y. Xue, 2003: The climatic impacts of land surface change and carbon management, and the implications for climate-change mitigation policy, *Climate Policy,* **3,** pp. 149–157.

Mastrandrea, M.D. and S.H. Schneider, 2004: Probabilistic integrated assessment of "dangerous" climate change, *Science,* **304,** pp. 571–575.

McKibbin, W.J. and D.J. Wilcoxen, 1999: Permit trading under the Hyoto Protocol and beyond, Brookings discussion paper in international economics no. 150, The Brookings Institution, Washington, DC.

Meyer, A., 2000: *Contraction and Convergence: The Solution to Climate Change,* Schumacher briefings no. 5, Green Books, Totnes Devon, UK.

Millennium Ecosystem Assessment (MA), 2003: *Ecosystems and Human Well-being: A Framework for Assessment,* Island Press, Washington, DC, 245 pp.

Nicholls, M., 2004: Carbon trading: Making a market, *Environmental Finance,* **March,** London, UK, p. 13.

Noble, I.R. and R.J. Scholes, 2000: Sinks and the Kyoto Protocol, *Climate Policy,* **1,** pp. 1–20.

Nordhaus, W.D., 1994: Expert opinion on climate change, *American Scientist,* **82,** pp. 45–51.

Pacala, S. and R. Socolow, 2004: Stabilization wedges: Solving the climate problem for the next 50 years with current technologies, *Science,* **305,** pp. 968–72.

Parikh, J, 2004: India's efforts to minimize GHG emissions: Policies, measures and institutions. In: *India and Global Climate Change: Resources for Future,* M. Toman, U. Chakravorty, S. Gupta (eds), Oxford University Press, New Delhi, India.

Parikh, J. and A. Das, 2003: Transport scenarios in two metropolitan cities in India: Delhi and Mumbai, *Energy Conversion and Management,* **45,** pp. 2603–2625.

PCAST (President's Committee of Advisors on Science and Technology), 1999: *Powerful Partnerships: The Federal Role in International Cooperation on Energy Innovation,* Report from the Panel on International Cooperation in Energy Research, Development, Demonstration and Deployment, The President's Committee of Advisors on Science and Technology, Washington, DC.

Pershing, J. and F. Tudela, 2003: A long-term target: Framing the climate effort. In: *Beyond Kyoto: Advancing the International Effort Against Climate Change,* Pew Center report, Pew Center for Global Climate Change, Arlington, VA, pp. 11–36.

Prather, M.J., 2004: Private communication.

Raven, J.A. and P.G. Falkowski, 1999: Oceanic sinks for atmospheric CO_2, *Plant, Cell and Environment,* **22(6),** pp. 741–755.

Root, L., J.T. Price, K.R. Hall, S.H. Schneider, C. Rosenzweig, et al., 2003: Fingerprints of global warming on wild animals and plants, *Nature,* **421,** pp. 57–60.

Seibel, B.A. and P.J. Walsh, 2001: Potential impacts of CO_2 injection on deep-sea biota, *Science,* **294,** pp. 319–320.

Shoji, K. and I.S.F. Jones, 2001: The costing of carbon credits from ocean nourishment plants, *The Science of the Total Environment,* **277,** pp. 27–31.

Smit, B., O. Pilifosova, I. Burton, B. Challenger, S. Huq, et al., 2001: Adaptation to climate change in the context of sustainable development and equity. In: *Climate Change 2001: Impacts, Adaptation and Vulnerability,* J.J. McCarthy, O.F. Canziano, N. Leary, D.J. Dokken, and K.S. White (eds.), Contribution of Working Group II to the third assessment report of the IPCC, Cambridge University Press, Cambridge, UK, pp. 877–912.

Smith, J.B., H.J. Schellnhuber, M. Qader Mirza, S. Fankhauser, R. Leemans, et al., 2001: Vulnerability to climate change and reasons for concern: A synthesis. In: *Climate Change 2001: Impacts, Adaptation, and Vulnerability,* J.J. McCarthy, O.F. Canziani, N.A. Leary, D.J. Dokken, and K.S. White (eds.), Cambridge University Press, Cambridge, UK, pp. 913–967.

Smith, J.B., R.J.T. Klein, and S. Huq, 2003: *Climate Change, Adaptive Capacity and Development,* Imperial College Press, London, UK, viii + 347 pp.

STAP (Scientific and Technical Advisory Panel), 2004: *Opportunities for Global Gain: Exploiting the Interlinkages Between the Focal Areas of the GEF,* Draft report of the STAP to the Global Environment Facility, Washington, DC.

Streets, D.G., J. Kejun, X. Hu, J.E. Sinton, X.Q. Zhang, et al., 2001: Recent reductions in China's greenhouse gas emissions, *Science,* **294,** pp. 1835–1837.

Swanson, D., L. Pinter, F. Bregha, A. Volkery, and K. Jacob, 2004: *National Strategies for Sustainable Development: Challenges, Approaches and Innovations in*

Strategic and Coordinated Action, IISD, Canada and GTZ (Gesellschaft für Technische Zusammenarbeit), Eschborn, Germany.

Swart, R.J., M.M. Berk, M. Janssen, G.J.J. Kreileman, and R. Leemans, 1998: The safe landing analysis: Risks and trade-offs in climate change. In: *Global Change Scenarios of the 21st Century: Results from the IMAGE 2.1 model,* J. Alcamo, R. Leemans, and G.J.J. Kreileman (eds.), Elseviers Science, London, UK, pp.193–218.

Thomas, C.D., A. Cameron, R.E. Green, M. Bakkenes, L.J. Beaumont, et al., 2004: Extinction from climate change, Letters to Nature, *Nature,* **427,** pp. 145–148.

Toth, F.L., 2003: Integrated assessment of climate protection strategies, *Climatic Change,* **56,** pp. 1–5.

Trull, T., S.R. Rintoul, M. Hadfiled, and E.R. Abraham, 2001: Circulation and seasonal evolution of polar waters south of Australia: Implications for iron fertilization of the southern ocean, *Deep-Sea Research II,* **48,** pp. 2439–2466.

van Oene, H., W.N. Ellis, M.M.P.D. Heijmans, D. Mauquoy, W.L.M. Tamis, et al., 2001: *Long-term Effects of Climate Change on Biodiversity and Ecosystem Processes,* NRP report no. 410 200 089, Dutch National Research Programme on Global Air Pollution and Climate Change, Bilthoven, The Netherlands.

Vellinga, P. and R.J. Swart, 1991: The greenhouse marathon: A proposal for a global strategy, *Climatic Change,* **18,** pp. 7–12.

Victor, D.G., 2001: *The Collapse of the Kyoto Protocol and the Struggle to Slow Global Warming,* Princeton University Press, Princeton, NJ, 160 pp.

Watson, R.T. and I.R. Noble, 2004: *The Global Imperative and Policy for Carbon Sequestration: The Carbon Balance of Forest Biomes,* Garland Science/BIOS Scientific Publishers/Taylor & Francis, London, UK.

WEA (World Energy Assessment), 2000: *Energy and the Challenge of Sustainability,* Report of United Nations Development Programme, UN Department of Economic and Social Affairs, and WEC, New York, NY.

Chapter 14

Cultural Services

Coordinating Lead Authors: D.K. Bhattacharya, Eduardo S. Brondizio, Marja Spierenburg
Lead Authors: Abhik Ghosh, Myrle Traverse
Contributing Authors: Fabio de Castro, Carla Morsello, Andrea D. Siqueira
Review Editors: Xu Jianchu, Hebe Vessuri

BOXES

Main Messages

Despite changes in perceptions of nature—culture links (for example, the people and parks debate), many policies and economic incentives concerning management systems and conservation strategies are still based on separating people from their environments, freezing and stereotyping both cultures and ecosystems. Such policies have a limited success in addressing the linkages between ecosystem functioning, development, and human well-being. There is a range of possibilities of interactions between humans and nature. Responses concerning economic development and conservation strategies need to take into account the historical, political, economic, and cultural contexts of these interactions. It is only too common to lay the responsibility for environmental problems and conservation either in the hands of local communities or blame the private sector, while disregarding the linkages between local, national, and international policies and economic pressures. Overcoming the idealization of cultures and the dichotomic view of local communities as either "noble" or "bad" savages is equally important to promoting sustainable responses to ecosystem management and development. Recognizing various types of knowledge (scientific, local, indigenous) and their role in conservation, production systems, and management strategies may help to avoid the extremes of either dismissing local perceptions, practices, and knowledge as "unscientific" and harmful, or idealizing them.

Standard "blueprint" or "straightjacket" approaches to integrate human and economic development and ecosystem management do not seem to work. Paying attention to the larger context in which communities and governments are operating—and are linked to (including basic needs and capabilities) formal and informal knowledge systems, forms of ownership, and institutional organization—most likely increases the chance of success of conservation and development programs. Conventional "best-practices" responses which decontextualize knowledge are less successful in addressing the needs of communities or the goals of ecosystem management.

Understanding the complexities of different cultural perceptions of landscapes, management of resources, and local institutional arrangements contributes to alternative and more effective strategies to ecosystem management and socioeconomic development. Overcoming the idealization of landscapes and ecosystems as pristine, frozen in time, and dispossessed of human culture is important for the success of ecosystem management responses. Restoration and conservation initiatives pivot on the question of how far we need to go back in time. Therefore, these initiatives hinge on historical developments and the perspective of communities on their ecosystems. For example, while the "idea" of using sacred areas as a basis for conservation is not new, there is a recent growth in translating the sacred into conservation legislation or legal institutions granting land rights. However, this approach requires extensive knowledge concerning the specific ways in which the link between the sacred, nature, and society operates in a specific locale. Sacred areas may vary from a few trees to a mountain range, and their boundaries may not be fixed. Local specifics need to be studied thoroughly in a participatory way in order to develop initiatives that suit the local situation, and care needs to be taken to avoid an approach that is too instrumental.

In an international perspective, market-economic policies and technological change are interrelated as flows of resources, goods and services transcend national and regional boundaries linking local transformations of landscapes to global environmental change. This will continue to influence intensification and commoditization of resource use, land reforms, and the substitution of local technologies, which all affect local livelihoods, human well-being, and the environment. Responses such as co-management, conservation units, and integrated rural development are not only relevant to local economies, but also influence new carbon sequestration programs based on incentives for particular production systems.

A balance between a global environmental awareness and related international institutions, and respect for the sovereignty of national and local governments over their landscapes and resources, are more likely to contribute to avoiding conflicts. Lack of cooperation and backlashes lead to undesirable outcomes for ecosystem services and human well-being. Taking this balance into account also increases the chances for successful transboundary conservation initiatives.

Many governments and environmental organizations are realizing that the biggest challenge of conservation in the twenty-first century is for it to take place outside parks and enforced boundaries, thus integrated into agricultural and urban systems. Conservation and development responses entail a mosaic of strategies that include different types of production and management systems along with the valorization of rural and urban landscapes. Conservation outside parks will continuously grow in importance, opening new economic opportunities. Examples of responses such as incentives for agrotourism may help to promote conserving cultural landscapes, to value farming systems, and to address economic needs. Responses addressing the links between the rural and the urban may provide important alternatives to address the growing complexity within which human populations and ecosystems are nested.

The literature shows that conditions that favor better outcomes of environmental management tend to include: representative participation and governance, clear definition of boundaries for management, clear goals and an adaptive strategy, flexibility to adjust to new contexts and demands, and clear rules and sanctions defined by participants. In this context, any process related to ecosystem management and economic and human development is mediated by given land tenure conditions that influence the distribution of benefits derived from local and indigenous knowledge, innovations, and practices. Co-management, joint ventures, and other forms of control of resources are nested within historical conditions of land tenure control, the nature of the resources, and institutional arrangements. Hence, these forms of management are more likely to be successful if they accommodate changes and are flexible to changes in production systems and markets.

Cultural perceptions and practices affect biodiversity, including agrodiversity and management practices of ecosystems. Agrodiversity includes cultural memory and different pathways through which knowledge is transmitted, such as oral histories, rituals, sharing experiences, arts, and so forth; these are as important as inventorying species and creating germplasm collections. Furthermore, it is well known that the emphasis on substituting local technology and knowledge instead of building upon them, for instance in relation to intensification of agricultural production, often leads to different forms of land degradation and a decline in food security.

Fostering the articulation of international and national conventions and regulations linking biodiversity and local and indigenous knowledge is important, taking into account that knowledge is produced in the dynamic context of inter- and intra-group interactions, power relations, and historical settings. Responses such as compensating for the utilization of local and indigenous knowledge and resources entails taking into account relations between companies, national and regional governments, and communities as well as the power dynamics of these relations. Responses such as certification programs are more likely to be effective in addressing local economies and human well-being if they take into account the impact of particular resource extraction upon people and communities using the same resource basis, but not necessarily sharing resource ownership. Certification programs are better served if accessible to communities and small producers' cooperatives that often are not familiar with bureaucratic and costly procedures of certification. Responses such as "Fair Trade" tools are more effective if they promote the participation of local producers in processes of commercialization and price negotiations, and the transformation and retailing of their products. Such responses are not only important for rural development and the conservation and management of natural resources, but also for commercial enterprises retailing the local producers' products.

Eco-, cultural, and agrotourism can provide important opportunities to link conservation and development. However, as the literature suggests, these forms of tourism are not necessarily the same thing as community-based tourism. Community-based tourism entails institutional capacity building in marketing and negotiation, defining access to benefits, and representative participation in decision-making processes of community members, tourism operators, and government agencies. Conflicts about resource use, development of infrastructure, the conversion of ecosystems, and dispossession of communities have negative impacts on the possibilities of ecotourism contributing to human well-being and economic development. In cultural tourism, problems may emerge in the representation and ownership of cultural symbols, the reproduction of stereotypes, consent among and within communities, and the blurring of boundaries between the public and private. In both forms of tourism, economic incentives and credit programs to foster tourism activities and capacity building could benefit from representative participation of local communities. The risks and opportunities provided by tourism are related to the economic position of communities and relations of power. Economic deprivation can lead to overexploitation of resources and the acceptance of unfavorable positions in the tourism industry (low-skilled labor, sex industry, drugs). Increase in land use value for tourism real estate development purposes may lead to displacement and dispossession. This is especially a risk for communities that enjoy informal or communal land rights.

Recreation, conservation, and environmental education can go hand in hand. Cultural tourism can serve to educate people about the importance of cultural diversity, as well as the importance of the latter for the conservation of biodiversity, provided the risks are taken into account. Tourism and recreation can be linked to environmental education, fostering knowledge about the functioning of ecosystems and provoking tourists to critically examine human–nature relations. Environmental education may serve very diverse audiences, ranging from schoolchildren to university students, protected area managers, policy-makers, and representatives of the private sector. In all cases, top-down education is less effective than education that is based on sharing experiences and attempts to reach a joint understanding of the dynamics of human–nature interactions.

There is a growing demand for the maintenance and creation of green spaces in urban landscapes. Green spaces and urban parks provide opportunities to integrate spiritual, aesthetic, educational, and recreational needs; they may also generate other ecosystem services such as water purification, wildlife habitat, waste management, and carbon sequestration. These green spaces may further contribute to human well-being by reducing stress and, hence, violence.

14.1 Introduction

14.1.1 Overcoming the Dichotomy of Nature versus Culture

Much of the thinking on nature conservation and ecosystem management is still based on separating nature from culture. Cultural perceptions of landscapes reflect a gradient ranging between the extremes of complete separation to the integration of culture and nature. These are reflected in histories of colonial occupation as well as academic developments over the past century. Transformations of landscapes have been and will continue to be influenced by cultural perceptions of nature as well as by sociopolitical and economic demands and aspirations. Species and entire land covers have been introduced or removed to "domesticate" the land and/or to recreate wilderness (Crosby 1986; Crumley 1994).

Academically, the understanding of culture and nature has changed dramatically in diverse fields such as geography, ecology, economy, and anthropology, where environmental deterministic and dichotomous views of the nineteenth and early twentieth century about the influence of environment upon culture are now being dismissed (Orlove 1980; Ellen 1982; Biersack 1999; Kottak 1999; Little 1999). The very concept of ecosystems reflects the changes in thinking about human–environment interactions, rejecting the idea of fixed equilibrium, closed systems, and static nature (Moran 1990; Golley 1993). Building a vision for the new millennium on the environment requires overcoming the dichotomy of nature versus culture, the perception that natural and anthropogenic landscapes are mutually exclusive, and instead building respect for the diversity of perspectives on environmental conservation and management.

This brief history provides context for the way we understand the relations between nature and culture. When we talk about culture ". . . we locate the reality of society in historically changing, imperfectly bounded, multiple and branching social alignments, . . . the concept of a fixed, unitary, and bounded culture must give way to a sense of the fluidity and permeability of cultural sets. In . . . social interaction, groups are known to exploit the ambiguities of inherited forms, to impart new evaluations or valences to them, to borrow forms more expressive of their interests, or to create wholly new forms to answer to changed circumstances. Furthermore, if we think of such interaction not as causative in its own terms but as responsive to larger economic and political forces, the explanation of cultural forms must take account of that larger context, that wider field of force. 'A culture' is thus better seen as a series of processes that construct, reconstruct and dismantle cultural materials, in response to identifiable determinants" (Wolf 1990, p. 387).

There are a number of relevant issues today relating to nature–culture interactions. In a global perspective, market-economic policies and social systems are interrelated as the flow of resources, goods, and services increasingly transcend and subsume national and regional boundaries. Market pressure, technological changes, and government policies influence production and consumption. Influences include the intensification of local resource use, land use reforms, and the substitution of local technologies. These affect local livelihoods, institutions, and the relations between people and nature (Arizpe 1996; Granfelt 1999). Increased global awareness of these effects in turn influences policies regarding issues such as global common resources, people, and parks issues, and regulations on biodiversity. A plethora of agents and institutions participate in bringing about and are affected by the above-mentioned changes ranging from local activist groups, business and lobby groups, civil society organizations, governmental agencies, to international bodies.

Understanding the complexities of different cultural perceptions of landscapes, management of resources, and local institutional arrangements contributes to alternative strategies to ecosystem management and socioeconomic development. In a recent review of management of common pool resources revisiting thirty years of research on this subject since Hardin's seminal article "the tragedy of the commons" (1968), Dietz et al. (2003) call attention to the need to avoid "one size fits all" when considering management and conservation of natural resources. Furthermore, it is important to realize that local communities do not operate in a vacuum, they create multilevel alliances, adopt and adapt global influences to foster their own livelihoods, yet do so on the basis of their own cultural repertoires, a process referred to by some as "glocalization" (Comaroff 2000).

In the 1990s, the focus of conservation initiatives shifted to the local level, with demands for accountability and decentralization of authority over natural resources. The call for decentralization was driven partly by a combination of democratization processes and economic demands. Problems, however, occur with many governments decentralizing only responsibilities, not budgets or real decision-making powers, and often not recognizing local knowledge and authorities (Toffler 2003; Ribot 1999). In this context, there have also been some changes in the debate on "people and parks." Conception and policies regarding the creation of conservation units have moved and continue to move back and forth between exclusion of communities and local forms of resource management (frequently in disregard of local political and cultural contexts) and inclusion. Policies range from a drive to protect pristine nature while disregarding local historical uses of resources on the one hand and, on the other hand, treating local communities as indisputable stewards of nature (West and Brechin 1991; Stevens 1997; Brandon et al. 1998; Hulme and Murphree 2001).

The above mentioned shifts coincide with changing attention from local to transnational conservation efforts: the creation of corridors for migrating species and trans-frontier conservation areas. We find examples in South and Central America, central and southern Africa, and Asia. The increased focus on trans-frontier conservation is partly based on the realization that ecosystems do not stop abruptly at national boundaries, but is also a result of wider societal debates about the importance of globalization (Draper et al. 2004).

The international dimension of ecosystem conservation has increasingly become an issue of geopolitical importance, and thus, sensitive to backlashes and conflicts. A global concern over the fate of particular ecosystems, especially tropical forests, has also created a sense of entitlement and "right of voice" beyond national boundaries. Different views, authority, and sovereignty need to integrate mechanisms that, while in tune with international treaties, also include the views and aspirations of national and local forms of use. It is important to pay attention to the way in which local communities creatively respond to socioeconomic and environmental change without losing sight of global–local interlinkages.

This chapter addresses emerging issues underlying responses to human–environment problems in policy-making and institutions. The issues are organized in relation to three overarching themes: cultural perceptions, knowledge systems, and tourism and education.

14.1.2 General Background on Drivers and Types of Responses

The definition of culture cited above provides a clear link to the way the concept of "drivers" is used in this chapter. Human–ecosystem interactions are processual and dynamic and, in this sense, drivers and responses co-evolve and are difficult to separate as one becomes the other depending on one's perspective and/or level of analysis. On the other hand, recognizing drivers and responses within particular categories (typology) helps to provide comparative insights into the way society and communities solve their problems, that is, economically, legally, etc. In introducing each topic, the chapter emphasizes the need to look at historical and contemporary contexts within which driver–response interactions develop, mediating conditions between macro and micro levels, and implications for ecosystem services at different scales.

It is fundamental to acknowledge here the differences in cultural perspectives influencing the ways in which people think about and take (economic) decisions (Wilk 1996). The discussion on "rational choice" and human behavior toward the environment has endured in the social sciences, including economics, for decades (Barlett 1982; Appadurai and Breckenridge 1986, Isaac 1993, Acheson 1994). It is important to recognize models of decision-making that do not always fit formal economic models of "rational choice." Social scientists have dealt with the underlying principles of human decision-making in several ways, from questioning economic maximization models through cultural analysis (for example, formalist–substantivist debate in anthropology; see Isaac 1993; Wilk 1996) to adopting theories of bounded rationality (for example, institutional economics; see Simon 1957, 1990). The basic idea underlying these debates acknowledges that people integrate their local context, including the resource base at their disposal, into their decision-making. In this sense, the chapter emphasizes that the concepts of "drivers" and "responses" are context-dependent.

The discussion of cultural services responses is organized following three basic guidelines. First, the topic is considered in a historical perspective and relevant issues are identified, particularly regarding main "drivers" and notable problems during the past 30 years; relevant issues include policy and geopolitical issues, technological changes, and changes in production and consumption patterns. Second, responses are identified according to the typology in Chapter 2. In most cases, however, the set of responses capture the synergetic nature of responses, for instance, those that represent an intersection between legal responses (for example, a law defining conservation units) and institutional responses (for example, related changes in land ownership and rules of access to resources resulting from legal changes). Finally, in assessing each type of response, the process presented in Chapter 3 is taken into account, that is, the binding constraints, enabling conditions, trade-offs, and synergies for each type of response are discussed.

14.2 Cultural Perceptions: Human Cognitions, Spirituality, Aesthetics, and Arts

14.2.1 Cultural Perceptions of Landscapes

Land- and waterscapes not only have physical attributes, they are subjected to and influenced by cultural perceptions as well. As Simon Schama (1995, pp. 6–7) has put it: "[t]here is an elaborate frame through which our adult eyes survey the landscape . . . Before it can ever be a repose for the senses, landscape is the work of the mind. Its scenery is built up as much from strata of memory as from layers of rock." Culture and memory play an important role in creating different, sometimes contesting meanings for any one place. Multiple identities associated with landscapes—both rural and urban—can exist simultaneously at local, regional, and national levels, with one or another being forced into dominance by historical and political circumstances (Stiebel et al. 2000; Ranger 1999). Relations between landscape and religion, for instance, have to do with both moral and symbolic imaginings, but also with staking one's claim, such as to land contested by immigrants or invading states and development agencies (Dzingirai and Bourdillon, 1997; Spierenburg 2004).

Certain cultural perceptions of landscapes become dominant or imposed through economic and political forces. In the African context, Ranger (1999) illustrates how dominant colonial views commonly saw Africans, living within their environment, as not having notions of landscape, not aesthetically appreciating the land they occupied. Yet Africans have long invested their environments with moral and symbolic qualities, and with beauty. Such views of the colonized were evident in other continents as well.

Language is among the most powerful forms of cultural mapping, and cultures provide maps of meaning through which the world is made more intelligible. Places themselves are rich cultural archives, for instance, the variety of names for a single site points to openness to cultural presences and shared histories in an increasingly multicultural world (Moore 1998; Stiebel et al. 2000). Language—including the poetry of song and dance that is part of popular historiography (Luig and von Oppen 1997)—can also unlock the secrets of the landscape; examples range from aborigine's song lines and pastoralists' oral mappings to European romantic operas. These ways of placing oneself in and on the land help stake one's claim to a part of the present, writing new histories, sometimes involving social and moral imagination (Coplan 1994; Cohen and Odhiambo 1989).

The downside of this particular discursive process of staking one's claim is that many armed conflicts, especially in Eastern Europe and the developing world, have been portrayed as "ethnic" or "tribal" wars when the opposing parties refer to certain identities and histories while fighting over certain natural resources. The label of "tribal war" blurs the vision to influences from outside the region fuelling armed conflict, as is the case with the so-called "blood diamonds" (De Boeck 1998; Lunde et al. 2003).

Recent studies have illustrated the misconception concerning pristine environments. One of the most provoking examples of the pristine versus anthropogenic debate is the Leach and Fairhead (2000) hypothesis proposing an interpretation of islands of forests in parts of Africa as signs of afforestation by people, instead of deforestation. . In the Amazonian context, Balée's interpretation of the "culture of Amazonian forests" suggests that a considerable portion of the region is composed of vegetation of anthropogenic origin resulting from long-term uses by pre-Columbian populations (Balée 1989). More recently, Willis and colleagues (2004), examining archeological and paleo-ecological studies, found considerable evidence of human uses in areas of the Amazon basin, Congo basin, and the Indo-Malay region of Southeast Asia where tropical forests are considered undisturbed. Their findings suggest a different perspective on the regenerative capacity of these areas. Attention to long-term interactions between people and landscapes has contributed to a more dynamic view of ecosystems and the role of human populations. Similarly, it provides examples of alternatives to use resources, for instance, by considering forms of forest management aiming at concentrating economic resources (Posey 1998; Crumley 1994; Balée 1999; Heckenberger et al. 2003).

The history of conservation spans various centuries and many different countries. Dating as far back as 5,000 and 4,000 years ago, China and Egypt were engaged in timber management (West and Brechin 1991; Menzies 1994; Rangarajan 2001). In the United States, regulations have been created since the 1600s (for instance, to impose limits on deer hunting). By the mid-eighteenth century, the need for conservation was particularly heightened as attention was drawn to the damaging impact of human activity on the landscape, particularly on the desecration of forests, and the need for adopting better strategies for managing and conserving the landscape. One of the first international conservation treaties—the Migratory Bird Treaty—was signed in 1916 between the United States and Great Britain. Overall, greater emphasis was given to the significance of nature and its related aesthetic appeal, and this prompted the emergence of the conservation movements and "nature appreciation" in North America (Beinart and Coates 1995). However, the economic potential of protected areas to attract tourism has also been a reason to foster conservation, as the case of Costa Rica's national conservation policy shows. (See Box 14.1.)

Conservation strategies have frequently reflected orthodox idealization of both ecosystems and people. On the one side, the majority of conservation units were created by excluding resident people, often at the cost of social conflict and displacement. Conversely, since the 1980s, recognition for the role of local communities has often led to reproducing a view of locals as "noble savages," relying on the concept of "traditional" or "indigenous populations" to assign rights to land and other resource use (Conklin and Graham 1995; Draper et al. 2004).

The use of the term "traditional" is widespread and often misinterpreted within the context of conservation and ecosystem management. Although in many instances the term is applied rightfully to stress the importance of local and indigenous knowledge systems regarding resource management, it carries different meanings depending on the context. While alternative terms such as "local" and "indigenous" may also be misused, in most contexts these terms may reflect better what is intended. Using the word "traditional" to refer to local and indigenous knowledge may be static and backward-looking, and an opposition to modernity. On the other hand, the term "traditional population" has been applied as a political tool to support local communities (and by local communities themselves) to guarantee access and rights to land and resources, particularly within conservation areas and parks. However, it is also used to disqualify some communities that share similar rights, but do not fit in whatever characteristics/traits are used to define who is "traditional." The term "community" carries similar problems of interpretation, such as assuming there is homogeneity and local consensus on belonging, boundedness, and aspirations among people settled in close proximity or participating in similar economic activities. Both the terms "traditional" and "community" require careful use and empirical understanding of the local and historical conditions within which they are applied (Sinha et al. 1997; Gibson and Koontz 1998; Sylvain 2002).

BOX 14.1

The Costa Rica Experience (Evans 1999)

The Costa Rica experience illustrates the important roles played by different key players such as environmental groups, education and nongovernmental organizations, indigenous movements, ecotourism, and the work of the National Biodiversity Institute in defining conservation policies. Interestingly, the emergence of the conservation movement reflects a process that some call the "'grand contradiction,' in the sense that conservation occurred simultaneously with massive deforestation in unprotected areas." In Costa Rica, 25% of the country's land mass is considered either a national park or protected area as a response to the rapid destruction of its tropical ecosystems due to the expansion of export-related agriculture.

14.2.2 Responses Related to Cultural Perceptions: Multilevel Policies, Institutions, and Social Identities

Perceptions of natural and anthropogenic land- and waterscapes have been changing rapidly during the past decades. There is an increased awareness of global environmental problems such as climate change. Furthermore, there are changes in the circulation of information and commodities; all of this creates linkages among local, regional, and global issues (Arizpe 1996; Moran et al. 2002). The diversity of perspectives to address current environmental problems—embedded in cultural, political, and economic differences and interests—presents challenges both in terms of defining priorities, as well as of aligning local and national socioeconomic aspirations with international agreements aimed at changing economic practices affecting the global environment. The dismissal of the Kyoto Protocol by the United States—grounded in questioning the scientific and economic basis of the protocol—illustrates the complexity of these issues.

14.2.2.1 Awareness of the Global Environment and Linking Local and Global Institutions

Awareness of the globe working as a system has motivated the need to deal with ecosystems in an integrated way. This process has been characteristic of the so-called post-Stockholm way of thinking, that is, an emphasis on the human environment concept (which actually was the title of the Stockholm 1972 conference) and the discussion of environmental problems at a global scale. Global models of management and global institutions dealing with the environment have become prominent not only in environmental management, but also in international politics. Amalgamation of scientific thinking and public awareness, civil society, and business has been present not only at international government forums from Stockholm 1972 to Rio 1992 and from Kyoto 1997 to the World Summit on Sustainable Development in 2002 in South Africa, but also in science initiatives such as the Club of Rome, the Man and the Biosphere Program, the International Geosphere–Biosphere Program, and the Millennium Ecosystem Assessment, to cite a few of the most relevant.

Awareness of global environmental problems by the larger public has also led to the emergence of different views about rights to entitlement to global ecosystems and environmental resources (Ostrom 1990). Examples range from the public engagement in discussing the fate of tropical forests to pressures regarding regulations to curb the greenhouse effect (Geores 2003). International organizations now voice their opinions regarding national policies and international bank loans to development projects, while national governments complain about their lack of sovereignty and international "ecological imperialism."

Local organizations also take advantage of emerging global institutions and conventions to bring their case to wider political arenas. One among several examples is "The Samarga Declaration." The declaration by the Udegei people of Sikhote Alin mountains of the Russian Far East aims at preventing the granting of industrial logging concessions in an area that they consider theirs; it received attention and support not only from the Russian Association of Indigenous Peoples of the North but also from international human rights and environmental groups (Taiga Rescue Network 2003; Molenaar 2002).

The strengthening of national and regional institutions, particularly NGOs, mediating international and local priorities of ecosystem management and conservation is fundamental in this process. Box 14.2 illustrates the example of a Brazilian NGO—the Socio-environmental Institute—working toward these goals.

BOX 14.2
The Socio-environmental Institute in Brazil

One of the most successful civil movements integrating environmental, sociocultural, and educational activities at various levels is represented by the Brazilian NGO Instituto Socio-Ambiental (www.socioambiental.org). Building upon long-term work with indigenous communities, ISA has articulated intervention and political work with a strong emphasis on the organization of information, databases, geographic information systems, and remote sensing for monitoring boundaries, mining and logging activities, development programs in indigenous areas, conservation units, and areas of relevant social and economic interest, such as key watersheds around the megacity of São Paulo.

ISA's work also involves a pioneer education program among indigenous communities. ISA is contributing to increased political organization and wider participation around topics of social and environmental importance in collaboration with other NGOs as well as community associations and government agencies. In a 1998 collaborative effort among dozens of national and regional institutions (NGOs and government organizations), ISA contributed to an effort that brought researchers and practitioners together to define areas of conservation priorities for the Brazilian Amazon. Not surprisingly, close to 75% of priority areas fall inside areas designated as indigenous reserves where, in many cases, centuries of continuous occupation has contributed to a diverse environmental mosaic. ISA's example serves to illustrate a middle ground approach, discussing social and environmental issues by involving a wide range of social groups and participant institutions.

Another response to a growing awareness of global linkages is the trend to establish transboundary conservation areas, as has already been mentioned. Most transboundary conservation areas have been in existence for only a relatively short period, making an assessment of impacts rather difficult. However, recent studies from southern Africa suggest that transboundary areas can contribute to local human well-being through increased revenues from tourism, but can also pose a threat to human well-being through the marginalization of local communities in (international) decision-making bodies and preferential treatment of tourists over local people (Wolmer 2003). At the same time, the creation of transnational corridors as special zones for economic cooperation and development presents a growing threat to environmental conservation and local livelihoods. Examples include the Amazonian route to the Pacific Ocean and the South Africa–Mozambique economic corridor (Grant and Söderbaum 2003), among others. Ecosystems do not exist in a vacuum and are not separate from economic and political histories. Consequently, awareness and action concerning local, national, and global environmental problems often intermingle cultural views, scientific knowledge, and economic ideologies proposing different forms of resource control and entitlement.

14.2.2.2 Multiple Responses: From Restoring Landscapes to Valuing Cultural Landscapes

As noted, landscapes are subjected to and influenced by cultural perceptions as well as political economic interests. Thus ideas about what landscapes should be conserved are also influenced by such perceptions. There is a growing recognition that a wider variety of landscapes, including agricultural, urban, and industrial landscapes, need to be conserved (see Box 14.3), and that certain species may have come to depend upon human-made environments. The importance of human environments for semi-

BOX 14.3

Plans for a New York Biosphere Reserve

The Columbia University/UNESCO Joint Programme on Biosphere and Society (CUBES) has established the Urban Biosphere Group, which is leading a discussion on how UNESCO's Biosphere Reserve concept may help the city of New York and its inhabitants to manage the enormous ecological footprint of the city and promote urban sustainability. According to the Group, the city is only metaphorically a "concrete jungle" (www.earthinstitute.columbia.edu/cubes/sites/nyc.html). Diverse flora and fauna find habitats in the built environment of the city as well as in its parks, vacant lots, community gardens, and backyards. Another important habitat is the Hudson River Estuary, which is also an economic asset to the city.

CUBES plans to organize an international partnership of cities, including New York City, Madrid, Rome, São Paolo, Seoul, and Cape Town, where scientists and stakeholders will examine how the Biosphere Reserve concept may be applied to urban areas to promote socially inclusive and environmentally sustainable urban processes.

domesticated and domesticated species has long been acknowledged (Clements 1999; Harris 1989). Now, however, there is also the recognition that urban environments have become an essential part of conservation of a variety of non-crop fauna species as well as wild animal species.

Nature conservation, however, still seems to be dominated by a search for "the pristine," a concept that often leads to conservation policies that "freeze" landscapes within enforced boundaries and attributes any disturbance to human intervention that needs to be undone. Many conservation efforts entail "restoring" landscapes to their "pristine, natural state," even though it may be hard to determine what that state actually was and how far back we need to look. There is, for instance, a lively debate going on in the Netherlands, where the government as well as NGOs are buying up farms to "restore the land to its original state." The Dutch national organization for agriculture and horticulture (LTO Nederland) recalls that most "natural landscapes" in the Netherlands are the result of a centuries-long interplay between farming, herding, and attempts to control water in this country of which large parts are situated below sea-level (LTO Nederland 2003).

Restoration of landscapes may involve the removal of species considered invasive or alien—a popular issue in conservation circles worldwide—or the reintroduction of species that have disappeared over time. (See Chapter 5.) These interventions may not always be appreciated by the local populations living in or near areas that are being restored. The reintroduction of brown bears in Spain and France is a case in point. (See Box 14.4.)

The classification of species as alien or endemic is not just a biological issue. Cultural issues and interpretations also play an important role in decisions concerning the need to eradicate or reintroduce certain species. (See Box 14.5.) The concept of alien species is, on the one hand, related to ecological changes and species composition of a particular region and, on the other hand, related to ideas of desirability, both economically and aesthetically. Alien species are sometimes referred to as "invasive" species (see Chapter 5) depending on their level of dispersion and competitiveness with species characteristic of a particular ecosystem. The concept is also a function of the temporal and spatial scale used for analysis. More often than not, people tend to incorporate "exotic" species as part of their perception of a given landscape and

BOX 14.4

The Reintroduction of Brown Bears in the Pyrenees

In 1989, a technical workshop was held in the Aran Valley of the Pyrenees on the sustainability of reintroduction of brown bears in the area. Two plans were tabled: an "island" type of reintroduction involving the release of about 60 bears, and a policy of conservation of the natural environment to facilitate the natural immigration of bears. At the workshop the first plan was rejected and the second adopted. Following the workshop, the French government released three bears into France that soon crossed the French–Spanish border, despite protests by the Autonomous Government of Catalonia. All but one city council of the Aran Valley rejected the reintroduction, claiming that it was incompatible with the economically important livestock breeding in the valley. One city council welcomed the reintroduction since it was promised grants for the establishment of a nature reserve for the bears.

Despite local objections, the Worldwide Fund for Nature presented a new, international initiative in 1996 involving 15 European countries, to reintroduce large carnivores to Europe, including brown bears in the Spanish Pyrenees. The then Director of WWF's Europe and Middle East Programme claimed: "By the end of the century, we aim to have proved to farmers and local communities that many of their fears about animals such as wolves and bears are excessive."(http://lynx.uio.no/jon/lynx/wwflynx1.htm). He furthermore claimed that agricultural subsidies encourage "irresponsible farming": "One reason why hill farmers lose sheep is because they buy stock using government aid, and then turn the animals out to graze, unsupervised. You'd never get a farmer doing that with animals he'd paid for with his own money."

BOX 14.5

"Alien" Species Control in the Cape, South Africa

In the dry winter of 2000, fires raged in the mountains surrounding Cape Town. Conservation authorities concluded that alien species such as the wattle and gum trees were to blame, since these consume far more water than the local fynbos, causing the groundwater level to drop, rendering the area vulnerable to fires. This conclusion was quickly taken up and published widely by South African newspapers and journals, which for weeks on end published horror stories about the effect of the aliens, calling for the conservation of the fynbos. Following the spate of fires, the provincial government of the West Cape proposed rather farreaching legislation that would require land-owners to completely clear their land from alien species before it could be sold. Local farmers worried whether this legislation would also apply to vineyards and apple orchards so important for the local economy.

Comaroff and Comaroff (1999) argue that the seeming obsession with alien species and the loud calls to protect the fynbos—referred to locally as an "indigenous" species rather than an "endemic" species—can be linked to the increased importance of the concept of "autochthony" in a world where transboundary flows of people, plants, animals, goods, and capital are increasing in scale. They see parallels with the widespread fear in many segments of South African society of "illegal aliens," that is economic and political refugees from other African countries. Draper (forthcoming), however, argues that the issues is more complicated. Whereas the wattle and gum tree are fought hand to tooth, another alien species, "rooigras," has come to symbolize the South African "veld."

as part of their ethnobotanical repertoire, particularly when economic, agricultural, and aesthetic motivations are involved.

14.2.2.3 Recognizing "Sacred Areas"

A recent development in nature conservation tries to link sacred areas to conservation by using sacred areas as a point of departure when creating protected areas (Mountain Institute 1998). The idea in itself is not new. In the colonial period, the British had to incorporate the concept of sacred groves and the land for local priests to collect their salaries (*pahanoi* in Jharkhand, India) or lands to feed the ancestral ghosts (*bhutkheta* in Jharkhand, India). This was enshrined in the Chotanagpur Tenancy Act (Bahuguna 1992; Chandran and Subash 2000).

While the "idea" of linking sacred areas and conservation is not new, recently there has been an increase in translating "the sacred" into legislation or into legal institutions granting land rights. However, this approach requires extensive knowledge concerning the specific way in which the link between the sacred, nature, and society operates in a specific locale. Sacred areas may vary from a few trees to a mountain range, and their boundaries may not be fixed. In some cases, access may be restricted to a few religious specialists, in other cases they are open to the public to perform acts of worship that may involve the harvesting of some of the natural resources from within the sacred area. Local specifics need to be studied thoroughly in a participatory way to develop initiatives that suit the local situation, and care need to be taken to avoid an approach that is too instrumental. (Box 14.6 discusses the role of the sacred in land reform in Zimbabwe, and Box 14.7 discusses Tibetan and Buddhist ecology.)

14.3 Knowledge Systems

14.3.1 Scientific, Indigenous, and Local Knowledge

There is increasing recognition of the validity and importance of farmers' knowledge of ecosystems, species, germplasm, and soils (Brush and Stabinsky 1996). Similarly, numerous studies have demonstrated the importance of local and indigenous knowledge of aquatic systems (Dyer and McGoodwin 1994). Nevertheless, the drive for modernization and technological change is often based on the substitution of small-scale practices. Understanding of crop and forest biodiversity lies in the oral history and cultural memory of local and indigenous communities, but is frequently disregarded as backward and unneeded.

The pace of technological, agricultural, and environmental changes, and large-scale environmental modification by infrastructure development often happens at the expense of local resources and knowledge (Scott 1998). While this has an impact on local food security and economies, it is also relevant to (national and international) issues of conservation and economy. Priorities for economic development are often based on technological modernization (such as monoculture and industrial fishing) and frequently contradict policies to promote local and indigenous knowledge, conservation of germplasm, and local management strategies. Despite their productivity, local technologies are often perceived as extensive when compared to high-input production systems. While technological change may contribute to increased food production, one needs to be careful with substituting local technologies, knowledge, and forms of production. This does not automatically mean that one or the other is better; one should avoid extremes of either dismissing or idealizing both forms of knowledge, technology, and production systems (Netting 1993; Brondizio and Siqueira 1997; Posey 1998; Nazarea 1998; Pinedo-Vasquez et al. 2001; Zarin et al. 2004; Brondizio 2004a).

Wynne argues that there is a distinct difference in the way local farmers respond to uncertain environments and the way natural scientists do: "Ordinary social life, which often takes contingency and uncertainty as normal and adaptation to uncontrolled actors as a routine necessity, is in fundamental tension with the basic culture of science, which is premised on assumptions of manipulability and control. It follows that scientific sources of advice may tend generally to compare unfavorably with informal sources in terms of the flexibility and responsiveness to people's needs" (Wynne 1992, p. 120).

BOX 14.6

Sacred Areas in the Struggle against Land Reforms (Spierenburg 2004)

In 1987, a land reform program was introduced in the north of Zimbabwe, with the aim of "rationalizing" local land use practices and rendering them " 'more efficient." The program would entail the relocation of farmers from areas that were classified by project staff as "non-arable." The relocation would cause considerable deforestation. The program would furthermore render about a third of the farmers in the area landless (Derman 1993; Spierenburg 2004). Local resentment against the program was expressed by the mediums of royal ancestral spirits, Mhondoro. They rejected the program claiming that the government could not hold authority over an area that actually belonged to the ancestral spirits. The spirits were said to be particularly angry about plans to locate farmers to Tsokoto, an area they considered sacred. Some of the mediums had joined the freedom fighters in Zimbabwe's struggle for independence, which is why it was difficult for government to dismiss their statements offhand. Attempts by government to buy the mediums' support by offering them handsome rewards for their support of the struggle failed. In the end, the program was never fully implemented, as the staff members charged with the relocation feared repercussions from the spirits.

BOX 14.7

Tibetan and Buddhist Ecology (Swearer 2001)

Religion forms a significant component of the worldview and as such contributes to the way people look at the ecology of their region. A religion which induces peace and allows no harm to come to plants or animals is likely to be much more concerned with policies and programs relating to ecological conservation. This is especially true of Buddhism. In fact, "Buddhist environmentalists assert that the mindful awareness of the universality of suffering produces compassionate empathy for all forms of life, particularly for all sentient species." (p. 226) Buddhist doctrines of *karma* and rebirth link together all sentient life forms in a moral continuum. This was why an environmental policy based primarily on a utilitarian cost-benefit analysis could not possibly be sufficient. It integrates environmental ethics as general principles, collective action guides, and takes into account particular contexts. While all species have a shared dharmic nature, they have an intrinsic value to individuals. This is an ideal principle on which to base a scheme of biodiversity. This attitude toward the environment is seen in the attitudes and practices of Buddhists in their daily life. Critics claim that this attitude is against Buddhist history and the past—that the individual should not engage with the world but should attempt personal salvation and purification. Others counter that narratives of place can make a crucial contribution to environmental ethics.

Likewise, local knowledge plays a key role in lowering the degree of uncertainty in aquatic systems. Knowledge on habitat conditions, and on the behavior and life history of the fish species comprises the base to make choices about where, when, and how to fish. Acheson and Wilson (1996) argue that the focus on how to perform the fishing activity differs strikingly from the focus of scientific management systems on how much fish can be withdrawn. Box 14.8 discusses the responses of local communities to growing commercial pressure over fishing resources and the evolution of "fishing accords" in the Amazon.

Science tries to replace "haphazard" experimentation by controlled experiments that are context-independent and thus more widely applicable. This is one of the reasons why governments and development agencies have long favored "scientific" solutions, and where they have become interested in local knowledge, often try to de-contextualize it by compiling "best practices" that can be disseminated to other parts of the country or even the world. Scientists who do acknowledge the existence of local knowledge generally apply scientific methods to verify and validate the knowledge to reach a wider acceptance in policy and academic communities.

BOX 14.8

Community-based Management of Floodplain Resources

The introduction of new fishing-related technologies such as motor boats, synthetic fibers, and ice boxes, led to the intensification of fisheries in the Amazon since the 1960s. The mounting pressure of commercial fishers in the floodplain lakes caused constant confrontations between local residents and outside fishers. Since the 1970s, the populations living along the rivers have developed a local management system to restrict commercial lake fishing. The "fishing accords," as they are locally known, are formal documents that restrict access to floodplain lakes and limit the use of fishing resources (McGrath et al. 1993). These documents vary regionally in format and rules types (Lima 1999, McDaniel 1997, Castro and McGrath 2003); yet, they have in common the strategy of discussing and voting the rules collectively, writing a document that is legitimated by the residents' signatures, and carrying out a self-monitoring system.

Unlike many community-based management systems described in the literature as traditional and based on verbal understanding, the "fishing accord" emerged recently, and its structure is framed in written documents with a clear system of rules. This unique aspect of the fishing accords is due to the influence of the Catholic Church in fomenting political organization of the rural populations in the Amazon over the last decades (Castro 2002). Until recently, the illegal status of the fishing accords, which violated the open access to the water system, combined with the lack of political support by the government, limited the scope of this local management system. In the last decade, however, the local populations gained support from local and international nongovernmental organizations (NGOs). As a result, the Brazilian government has launched a research program aimed at developing legal instruments to recognize the local management systems.

The fishing accords are by no means free of problems (Castro 2002); yet, despite many barriers still to overcome, the process of recognition of this local management system by the government has created, for the first time, the ground for participation of local users in the management of resource use, leading to the integration of local and scientific knowledge, increased legitimacy, and a basis for negotiation of conflicting interests.

Scoones (1996) attributes the failing of many land reforms and other projects trying to render the landscape more "legible" to the distrust between local farmers and the "purveyors of the scientific solution." Local farmers understand the solutions offered to them; it is not ignorance that engenders their reluctance to adopt the proposed practices, but a more fundamental disquiet about the technical rationale for the suggested solution under local circumstances and a suspicion about ulterior motives. Local farmers have preferred to follow their own informal and flexible alternatives for survival because changing their methods will disrupt their production. Top-down approaches to implementation limit the possibilities of exchanging perspectives and negotiating outcomes between local farmers and external agents. The result is the emergence of forms of resistance that are actively pursued but perceived by outsiders to represent ignorance of the "correct" solution, implying that people require education and persuasion (Scoones and Cousins 1994).

The persistent perception that ignorance is at the root of resistance or noncompliance does not lead to questioning of the assumptions behind the intervention or a reexamination of its scientific premises (Scoones 1996; Posey 1998). This does not mean that resistance is the sole or main reaction to external introduced knowledge and technologies. Actually, most farmers, indigenous or not, draw upon different sources of knowledge and integrate different technologies and techniques that best fit their interest, needs, and conditions (Reij and Waters-Bayer 2001; Scoones 2001). Experimentation and diffusion of knowledge are central tenets of livelihood strategies at any level (Netting 1993).

14.3.2 Responses: Protection, Compensation, and Certification

The growing recognition of local indigenous knowledge has also led to its commercial exploitation. Market imperatives and international monetary policies have caused developing countries to gear their economies towards export. In most cases, this has led to the exploitation of their natural resources beyond long-term sustainability. At the same time, the richness and possibilities of resources, for example medicinal herbs, and their possible economic benefits, became an important argument for the conservation of nature. The prospecting for local resources has led to exploitation of local knowledge without communities being compensated. Few mechanisms are available to feed the benefits back to local communities that in many instances contributed to the production of the knowledge concerning certain species, or even the production of the species themselves.

14.3.2.1 International Agreements and Conservation of Biological and Agropastoral Diversity

Increased exploitation as well as a growing consciousness concerning the disappearance of local resources and the knowledge about these has led to concerns for the need to protect local and indigenous knowledge. The international community has recognized the close and traditional dependence of many indigenous peoples on biological resources, notably in the preamble to the Convention on Biological Diversity, which has been ratified by over 170 countries. Article 8(j) in the CBD specifically addresses local indigenous peoples and their knowledge. The CBD adopted the facilitation of indigenous peoples' participation "in developing policies for the conservation and sustainable use of resources, access to genetic resources and the sharing of benefits, and the designation and management of protected areas."

Many governments are now in the process of implementing Article 8(j) of the Convention through their national biodiversity action plans, strategies, and programs. Some have adopted specific laws, policies, and administrative arrangements for protecting indigenous knowledge, emphasizing that prior informed consent of knowledge-holders must be attained before their knowledge can be used by others (Cunningham 1996). Other international conventions have followed suit, such as the Ramsar Convention on Wetlands; the signatories have adopted Resolution VIII 19 that fosters the incorporation of cultural values in conservation efforts with the obligation of doing so with the active participation of indigenous communities. In many cases, protection of local/indigenous knowledge is a byproduct of the protection of biodiversity, while in others the main aim is to guarantee economic benefits to communities.

Apart from (inter) national policies, there are also instances of local strategies to protect as well as transmit local and indigenous knowledge. Local knowledge is, just as scientific knowledge, produced in a context of power relations. Not everybody will have access to local and indigenous knowledge; some of it may be considered the domain of specialists. Specific groups in communities may be excluded from such knowledge on the basis of their socioeconomic position and/or gender (Clark 2003). Knowledge may not be equally shared within communities; some of it may be considered sacred and/or secret, which poses problems concerning legislation. Other problems relating to legislation are the fact that a lot of the knowledge is produced by groups of people, not by individuals; is developed over time; and continues to be developed. Legislation may "freeze" knowledge as well as the rituals and practices associated with this knowledge (Laird 1994; Brush and Stabinsky 1996).

The World Intellectual Property Organization has been the voice behind intellectual property rights. But instead of supporting "local knowledge," WIPO became a western capitalist-biased organization that envisions any type of knowledge whether it is medicinal plants, songs, crafts, or any other form, as a commodity. Local communities are concerned with the extent of exploitation of local knowledge and how it is being used or removed from its culturally appropriate context and how it is being usurped as a capitalist commodity.

An issue is the western patenting system which is used to protect the IPRs of monopolists. The scale and tendency to focus on corporations is exemplified by industrially advanced Western European countries who have been strong supporters of IPR and who have imposed this system on Third World countries. Patents could translate into wealth and power for foreign transnational companies and may have negative impacts in the biodiversity of particular areas, depending on the level and structure of market demand and exploration practices. There are about four million patents in the world, which already provide a source of wealth for companies working in ecosystems across the world (Shiva 1997; Settee 2000). Local and indigenous communities, however, increasingly are organizing themselves by establishing NGOs and lobbying governments and international organizations to change legislation concerning patents; WIPO is increasingly willing to lend an ear to the protests.

Much of the local and indigenous knowledge is not written down, but transmitted through daily practices, stories, songs, dance, theatre, and visual arts. Not only knowledge but also attitudes and perceptions are transmitted that way (Dove 1999). Increasingly, programs on local/indigenous knowledge take these forms of transmission into account, and try to incorporate these into educational activities.

Language is crucial to mapping biodiversity. There is concern that when languages disappear, knowledge may disappear as well (Cox 2000). Increased recognition of this threat has resulted in a number of initiatives to protect local and indigenous languages, including UNESCO's "Safeguarding of the Endangered Languages" program (http://www.unesco.org/culture/heritage/intangible/meetings/paris_march2003.shtml). This program specifically recognizes the possibility that dispossession and loss of habitat are important risk factors leading to the disappearance of certain languages. Box 14.9 illustrates two UNESCO programs related to intangible cultural heritages.

Local and indigenous knowledge does not only pertain to species that are harvested on an extractive basis, but also to production methods, cultivars, and germplasm (Brondizio 2004b; Padoch et al. 1998; Brookfield 2001). Production systems that evolved over long periods of time benefit from cultivars and methods adapted to particular micro-environments, as well as social conditions (Altieri and Hetch 1990, Caballero 1992, Netting 1993). An important part of indigenous knowledge about fishing systems is related to secret fishing spots. In contrast to terrestrial ecosystems, water-based environments lack clear property boundaries. Yet complex tenure systems based on environmental information such as lunar-tide pattern and triangulation of beach marks, regulated by social relations, can be observed (Cordell 1989).

Diversity of production systems increases the resilience not only to factors such as climatic change, but also facilitates alternative economic options to minimize risks in household food supply (Wilken 1987; Hladik et al 1996). In *Dynamics and Diversity, Soil Fertility and Farming Livelihoods in Africa,* Scoones (2001) and colleagues show that conventional methods using demonstration plots are inadequate to extrapolate to larger areas. They argue that researchers should look at the entire farm with all its components and the ways in which farmers invest labor and inputs to maintain or increase fertility in different parts of their land. In most cases, different levels of land management intensity co-exist to attend household and market demands; and in most areas, one finds complex environmental mosaics with a high diversity of crop and wild species. Likewise, coastal communities rely on the combination of water- and land-based activities to lower the risk of each environment. Box 14.10 illustrates the importance of accounting for social indicators such as the nutritional status of children, in understanding the impact of changes in production system and environment upon the well-being of local populations.

Some initiatives at the international level are concerned with local knowledge systems and their associated landscapes and agrodiversity. One example is the Food and Agriculture Organization program "Globally Important Ingenious Agricultural Heritage Systems." GIAHS aims at enhancing, demonstrating, and promoting these systems through a number of pilot sites representing different types of agricultural and pastoral production systems and their associated landscapes. It includes examples dealing with rice and maize based systems, root crops, pastoral systems, irrigation and soil management systems, agroforestry, and extractivist systems from around the world. This program is still in its early phase, so it is not possible to assess its impact to date (http://www.fao.org/ag/agl/agll/giahs/). The UNESCO World Heritage Program deserves mentioning for its recognition of a wide range of agricultural and other cultural landscapes, including the Rice Terraces of the Philippine Cordilleras and the forest of "Cedars of God" (Horsh Arz el-Rab) in Lebanon (p://whc.unesco.org/nwhc/pages/home/pages/homepage.htm). Finally, the People, Land Management, and Ecosystem Conservation program is in a more advanced stage of development; it was built upon under-

BOX 14.9

UNESCO's Intangible Cultural Heritage Program and Local and Indigenous Knowledge Systems

UNESCO has initiated two programs related to intangible cultural heritage. Implemented by UNESCO's Division of Culture, the first program consists of three components: safeguarding endangered language, creating "living human treasures" systems, and identifying and preserving "masterpieces of oral and intangible heritage" (www.unesco.org/culture/heritage/intangible). In the context of the Division of Culture's emphasis on intangible heritage, the term "preservation" means the safeguarding of heritage in the context and environment in which it is generated. In the Masterpieces of the Oral and Intangible Heritage of Humanity program, the "masterpieces" refer to cultural spaces that are defined as places in which popular and traditional cultural activities are concentrated, or as the time usually chosen for some regularly occurring event, or forms of popular and traditional expressions (for example, as languages, oral literature, music, dance, games, mythology, rituals, costumes, craftwork, architecture), and other arts as well as traditional forms of communication and information.

Nominations for the masterpieces program should be accompanied by detailed plans to preserve the masterpiece, for which funding can be obtained. The main challenge of the program is to prevent the stifling of dynamic practices. The nominating guidelines demand that "The expression is presented as a clearly defined corpus of the orality concerned." However, "clearly defined" is open to interpretation. For instance, in the case of the Sunyanta epic (Mali), the seventh-yearly performance by griots is preceded by rehearsals during which griots argue about the interpretation of the Sunyata stories and their meaning for the interpretation of current events in society. This process is a crucial aspect of the epic (Jansen 2000).

The second UNESCO program, "Local and Indigenous Knowledge Systems in a Global Society," is an intersectoral program whose focus is not so much on the preservation of local/indigenous knowledge as on promoting the recognition of the value of this kind of knowledge as a fundamental component of sustainable development (UNESCO/Links brochure 2002). The program aims at securing an active and influential role for local communities in sustainable development and resource management processes by strengthening dialogue among traditional knowledge holders, scientists, and decision-makers, and establishing co-operative processes. Local and indigenous knowledge systems are considered to be dynamic, and special attention is paid to the way they are generated, the context in which they are generated, as well as the way in which each generation reassesses, renews, and reinvents its knowledge. The program aims at strengthening transmission of knowledge, practices, and worldviews, from elders to youth, and developing quality education that contributes to this end, so as to sustain traditional knowledge as a living and dynamic resource within local and indigenous communities.

BOX 14.10

Nutritional Status as a Social Indicator of Well-being

Despite its importance to smallholders' food security strategies, local production systems tend to be neglected by developmental agencies—usually for not being market oriented and/or for being considered backward in terms of technological use and productivity. In some cases (exemplified by several agroforestry activities) these production systems are not even considered as agricultural work, and are labeled as "extractivism," carrying with it the prejudice and the stigma of lacking a "civilized" process of plant domestication and the necessary labor specialization. Nonetheless, local crops also tend to be considered of "bad taste" and/or of poor nutritional value. The outcome of these misperceptions tend to induce changes in the agricultural work and productivity and bio-physical environment, but not necessarily induce higher and better household food consumption.

One example is the implementation of agricultural projects among rural populations of the Amazon estuarine region which tend to favor mechanized agriculture and cash crop fields. The projects intended to increase community household income and well-being. However, instead of implementing new crops side by side with food items that were part of the local diet, the project designers opted to replace them. Manioc, a crop cultivated by slash-and-burn technique and a staple food and one of the main sources of energy among Amazonian populations, was often dismissed as an important food crop and replaced by more "nutritious" crops such as beans and corn. Beans, despite being a staple food in other parts of Brazil, were not considered a desired food by many household members and were not consumed; corn was usually fed to the animals (Murrieta et al. 1999).

Nutritional status provides a way to evaluate the efficiency of food production systems as well as household and community food security (Frisancho 1990). Through the use anthropometric data—a series of standardized techniques to measure the body and its parts—and diet surveys, one can evaluate the nutritional status of individuals, households, and communities. At the community level, growth and development of individuals, mainly of children under 10 years old, have been presented as good health and nutritional indicators of a population, especially in poor areas of the world. Being in an active growth process, a child's physical development is susceptible to and directly affected by both the availability of food (quality and quantity) and the incidence of diseases, which ultimately reflect social, cultural, economic, and physical environmental conditions. Height-for-age and weight-for-age are the measurements most widely recommended by the World Health Organization to assess protein-energy deficient children. While height-for-age is used as an indicator of the past state of nutrition, weight-for-age and weight-for-height are used as indicators of current nutritional status. For adults, body mass index (a relation between height and weight) is also used to assess nutritional status. While providing a good indicator of household and community well-being, diet surveys and anthropometric techniques also permit assessing possible differential gender and age access to food and household resources, thus allowing a broader understanding of household and community structure and dynamics.

standing and translating the experience of local farmers to their colleagues, operating across rural areas representing different international realities (Brookfield 2001, PLEC 2002, Brookfield et al. 2003). Box 14.11 illustrates this example in more detail.

There are also a number of local and regional level initiatives to conserve biodiversity, including agrodiversity, by setting up seed banks (for example, at Kew Gardens). However, storing seeds is not sufficient (Nazerea 1998). People and Plants, for instance, also compiled local and indigenous knowledge associated with the species, and provided training for local ethno-botanists. Nazerea proposes the establishment of memory-banks in addition to seed banks, including oral histories, evaluation criteria, ranking, and sorting schema and cognitive drawings (1997). Another example incorporating some of these strategies is illustrated in Box 14.12 on the cultivation of medicinal plants in India.

14.3.2.2 "Best Practices"

Though very important lessons can be learned from diverse experiences in local communities, it should be noted that local and indigenous knowledge evolves in specific contexts and one needs to be very careful with de-contextualizing it. This applies to several types of responses aiming at addressing issues of "knowledge systems" and environmental management. Conventional "best practices" methods focusing on content may not be the best way to deal with local/indigenous knowledge. A content-based best-practices approach is based on the assumption that it is possible to objectively validate or disqualify local knowledge. Yet the question arises what indicators can be used to determine whether a practice is a "best practice"? Who decides what a best practice is?

There are many aspects that determine whether a practice is a best practice such as economic performance, improvement of individual rights, the range of beneficiaries, and its sustainability over time. The social and economic context is important, since it defines who benefits from opportunities opened up by particular development programs and what factors constrain local participation. Basic issues such as land tenure conflicts and rights, institutional organization, farmer's and fishermen's access to basic services, and markets for their products, need to be considered.

BOX 14.11

The People and Land Conservation Program

The People, Land Management, and Ecosystem Conservation program has further developed the concept of farmer-to-farmer learning by creating a network of farmers, communities, and sites across 13 countries. The work emerges from the collaboration between the United Nations University, the United Nations Environmental Program, and the Global Environment Facility, with national, regional, and local organizations in these countries. Following a consistent, but adaptive research design, PLEC researchers have partnered with local farmers and extension officers to learn from and systematize local experiences, knowledge, and creative solutions to increase and sustain land productivity. The work of PLEC has involved inventories of agrodiversity and production systems. Using local forms of social organization as the basis for demonstration activities, PLEC has helped to create channels to diffuse technologies and exchange experiences. Considerable efforts have been put into the valorization of farmer's knowledge and overcoming preconceptions and stereotypes not only of extension officers in relation to small-scale farmers, but also of farmers themselves in relation to other farmers.

BOX 14.12

Project in India on the Cultivation of Medicinal Plants

A company called Gram Mooligai Company Limited (GMCL) procures herbs from traditional herb gatherers of the *vallaiyar* community in Virudunagar district of Tamil Nadu, India. It makes these herb-gatherers shareholders and offers better prices. Twenty five villages of the Virudunagar, Sivagangai, Dindigul, and Theni districts of Tamil Nadu have benefited. This public limited company is a spin-off from the Medicinal Plants Conservation Network (MCPN). It is supported by the Bangalore-based Foundation for the Revitalization of Local Health Traditions (FRLHT). It began in 1993 involving state forest departments of the Karnataka, Kerala, and Tamil Nadu states of India in medicinal plants conservation. It also involves research institutes and NGOs. In the Madurai region they were helped by Covenant Center for Development. GMCL was registered in January 2000 as a public limited company. A federation of 164 self-help groups called *mahakalasam* created *sanghas* or self-help groups in villages.

GMCL now has 0.5 million subscribed shares among 44 gatherer and 12 cultivator *sanghas* in the Virudunagar district. They avoid contamination by not collecting plants along the roadside or plants that have moisture. GMCL now sources more than 300 tons of medicinal plants for pharmaceutical and herbal companies. The prices are pre-announced, weighing is transparent and the villagers can sell directly from their villages. The company encourages sustainable harvesting. Sometimes, however, the prices offered are not competitive and rivals try to undercut. (Vijayalakshmi 2003)

14.3.2.3 Compensating for Knowledge

Compensation for the use of local and indigenous knowledge by third parties is an important, yet complicated response (Moran 1999). Given the way in which such knowledge is produced, determining who owns what knowledge may not be easy. The distribution of knowledge varies according to type, use, and access to resources. Some knowledge may be shared and produced by numerous local communities (Reyes-Garcia et al. 2003). In other cases, production and diffusion of knowledge may be restricted to certain groups or individuals within communities (Moran 1999).

Local authority structures are an important, but not the only factor, that needs to be taken into account in deciding, in close cooperation with local communities, who should be responsible for distributing benefits. As remarked above, local and indigenous knowledge may concern different domains and may be produced by different individuals or groups within communities. Though possessing certain knowledge may enhance someone's position in society, the idea that powerful people within communities are related to or responsible for local and indigenous knowledge is not necessarily correct. Thus the popular idea that local and indigenous knowledge can be promoted by strengthening "traditional" authorities may not be valid in many cases. (See Box 14.13.) Such a strategy is not always "innocent"; both governments and enterprises may find it easier to deal with such authorities than with whole communities (Ribot 1999).

Furthermore, the distribution of benefits is influenced by relationship among companies that seek local and indigenous knowledge, national legislation and authority structures, and regional government bodies (Schutz 1970; Berger and Luckman 1971; Laird 1994).

14.3.2.4 Responses Changing Resource Ownership and Control

Control over resources is another crucial issue that influences the distribution of benefits derived from local/indigenous knowl-

BOX 14.13
An NGO Imposing "Traditional" Authorities in Mali
(Kassibo 2001)

The government of Mali has initiated a decentralization process that is not yet completed. Locally elected "communes" have been established that have some decision-making powers, but forest resources are still controlled by the state.

A British NGO introducing a forestry management project into the Mopti region opted to re-invent tradition as the basis for the project instead of supporting local calls to decentralizing control over the forest to the commune. The NGO based its project on the oral traditions of one of the ethnic groups in the area. According to these traditions, in the pre-colonial era groups of young people constituted associations called *ton,* which were charged with forest protection. They were accountable to the village authorities and supported by the land chiefs in resolving conflicts. Under colonization, the French authorities banned this system, and appointed civil servants to manage the forest. Access and user rights of local populations were severely restricted. From the colonial period to the present day, local populations have never ceased to try to reclaim their participation in forest management.

The NGO appropriated the traditional approach to legitimize its intervention in the area of environmental management and to more effectively lobby the central state administration for recognition. An "outside" actor by definition, the NGO created successors to the *ton* in the form of watch brigades, and established village associations that were to supervise the brigades and report back to the NGO. The NGO financed all operations of the brigades and village associations, and supervised them. Brigades and associations were accountable to the NGO, but neither they nor the NGO itself were accountable to local government structures. The NGO had no real accountability downward to the community, although members of the brigades and associations were nominated by the community. Furthermore, the brigades and village associations had no real control over the forest; they basically fulfilled the patrolling function of the Conservation Service, but without sanctioning powers.

The NGO did not take into account the fact that the local communities had become multi-ethnic when they imposed the neo-traditional structures derived from the oral traditions of one of the groups in the area. Some groups felt misrepresented by these structures, and as a result of the lack of downward accountability there was a lot of favoritism. Furthermore, the structures set-up by the NGO were used by the state to claim that there was no need to decentralize control over the forest to the existing, democratically elected local government bodies.

edge. Various tenurial options exist, including co-management, joint ventures, and the creation of conservation and sustainable management units to assign forms of land tenure rights to communities. Factors such as historical land tenure control, complexities of local institutional arrangements, and types of resource uses, all play a role. While guaranteeing tenure rights and access to resources to particular groups, legislation may constrain the level of flexibility to allow changes in production system, adoption of new technologies, accommodation to population increases, and higher pressure over particular resources. For instance, communities that are "allowed to stay" in conservation areas are often forbidden to carry on or increase their agricultural activities (West and Brechin 1991; Stevens 1997; Agrawal and Ostrom 2001; McGrath et al. 1993; National Research Council 2002; Ostrom 1990).

Uncertainty concerning tenureship not only poses problems for the distribution of benefits, but also for strategies that are becoming more and more common, for example reinforcing traditional leadership to conserve local and indigenous knowledge. The link that is not always made explicit is that communities do need control over natural resources, but whether this should be through traditional leadership remains to be seen and depends on the local context and history. Local government institutions that are democratically elected and have real authority over resources in some cases may be better options. Yet many governments seem to have a tendency to shift responsibilities back and forth between "traditional" authorities and local government bodies, without giving any of them real decision-making powers. This decreases communities' control over resources and increases central government's control, often undermining the efforts of both sides (Ribot 1999; Spierenburg 2003). Another problem concerns the control over territories that contain resources deemed of national importance, such as oil and minerals. In most cases, central governments refuse to devolve authority over such resources.

14.3.2.5 Certification Programs

Certification programs have emerged as tools to control the source and distribution of particular products and their means of extraction (Zarin et al 2004). Examples include forest products, fisheries, and agriculture. The criteria on which certification is based include biological and ecological components of production areas and ecosystems, approved management plans and environmental impact assessments, compliance with national legislation, and the participants involved, among other issues. Less attention is given to the impact of particular resource extraction upon people/communities in those areas using the same resource base. Sustainable forest certification ensures that wood products that are being sold to consumers have gone through a checklist to guarantee that they meet the standards set by the certification process (Certified Forest Products Council 2002). Although this is a very positive response, many communities do not have access to certification programs for their products or are not aware of their existence, thus limiting their participation in a growing market. In addition, the financial costs involved in establishing a certification program reduce the chances for local communities to be able to participate independently. Capacity building at the local level to prepare, implement, and monitor certification programs could be implemented alongside the regulations requiring certification in international markets.

14.3.2.6 "Fair Trade"

"Fair trade" is a long-lasting movement initiated to help disadvantaged or politically marginalized communities; its aim is to obtain better prices and providing better trading conditions, as well as raise consumers' awareness of their potential role as buyers. Yet from the early 1990s, fair trade began to overlap with initiatives focusing on the environmental performance of trade (Robins 1999).

Among the successor perspectives is the so-called "Rainforest Harvest," which focuses on the conservation of tropical forests and their dwellers. Proposed in the late 1980s by Jason Clay, the "Rainforest Harvest" centered on helping indigenous and other rainforest communities. The baseline argument was that the introduction of fair and environment-friendly markets was a powerful approach to protect people's standards of living and promote their empowerment (Clay 1992). Prompted by conservationists and indigenous advocacy groups, and made possible by consumers' interest, a series of trade initiatives began to be installed that encompassed concerns over environmental management by indigenous and rural communities, linking issues of social justice and recognition of the stewardship of these communities over natural resources. Though this is a very positive response, care should be taken of the issues already mentioned in relation to the benefits generated by local and indigenous knowledge.

Furthermore, successful fair trade initiatives depend upon local skills concerning market negotiations as well as leveling skills vis-à-vis middlemen and retailers. Commodities such as coffee, while experiencing a growing market, have seen decreasing prices due not only to increases in production (the "Viet Nam factor"), but also as a function of the centralized control of stocks by a handful of companies. Shortening the commodity chain between producers and consumers is one of the most notable contributions to rural development that may allow communities to increase their income and to value their resources and production systems. Box 14.14 illustrates fair trade examples of the Amazon and highlights key principles facilitating the success of fair trade initiatives.

14.4 Tourism: Recreation and Education

14.4.1 Valuing the Environment and Culture in Tourism

Tourism and recreation are related to cultural perceptions of land- and waterscapes and of culture itself, both embedded in economic dimensions. Tourism is a large revenue generator, representing the most important source of income in some countries. The growing awareness of cultural diversity and environmental issues has had an impact on tourists' expectations and behaviors. In some cases, the impact of large-scale tourist enterprises on the environment and local economies has undermined the very basis of tourism. Hence, a growing number of tourists are looking for new destinations and experiences; this provides both opportunities and risks for communities and landscapes.

BOX 14.14

Fair Trade in the Amazon

The Amazon is perhaps the chief platform for forms of fair trade. The pioneer and flagship initiative was "Rainforest Crunch," a candy bar containing Brazil nuts produced by the U.S. company "Community Products." An enterprise derived from "Ben and Jerry's" ice cream industry, the newly formed enterprise followed different management rules such as returning a percentage of the profits to conservation programs. A second flagship initiative was "The Body Shop" cosmetic industry of England that implemented a "trade not aid" commercialization program with communities around the world, declaring that it would pay "Third World producers, First World prices" (Clay 1992; 1997). With several trading partners, the company became particularly known for commercializing Brazil nut oil with the Kayapo indigenous group of Brazil (Corry 1994; Turner 1995).

In addition to these well-known initiatives, numerous other agreements based on fair trade deals were implemented. In the Amazon, the leading commercial sector in this effort is perhaps the cosmetic industry. Pioneered by large corporations such as the Body Shop and Aveda, cosmetic industries have adopted declarations of "beauty with social responsibility plus environmental conservation" as a common marketing strategy. However, a wide range of other sectors are involved as well, spanning from the food industry, essential oils, medicinal plants, fibers, and resins, to the automobile industry. Retail industries are also increasingly jumping into the "fair trade" business. By now, "South–South" fair trade deals are becoming increasingly common; an example is one of Brazil's largest supermarkets, which implemented a program to directly trade products with rural communities.

Fair trade projects are frequently promoted by NGOs; the Tagua initiative in Ecuador is a famous example. In an extractive community of Afro-Americans living within a biodiversity-rich but economically extremely poor area, the NGO Conservation International fostered the production of buttons made of the Tagua nut or vegetable ivory (Hidalgo 1992; Ziffer 1992). A traditional product in the process of being abandoned, the NGO actions reversed and improved the socioeconomic situation (Robins 1999).

Although they are a growing business, rainforest harvest deals are not free from criticism. It has often been said that they (1) support consumerism by "green washing"; (2) have a hidden agenda to integrate indigenous peoples into national societies; (3) resemble a renewed version of the traditional Amazon system of debt-bondage; (4) are an unviable proposition for tropical forests; and (5) produce social and cultural disruption (Corry 1994; Entine 1994; Turner 1995).

Even those who have confidence in the approach do not pretend that the task is easy. Some common problems include irregular sourcing, structural dependence on subsidies, bureaucratic processes of export that require specialized management skills, lack of storage capacity with communities, deficiencies in quality control and lack of training capacity, low efficiency of the workforce, and high transportation costs (Anderson and Clay 2002). In addition, fair trade agreements can be troublesome, particularly to indigenous groups where cultural transformations are an important issue. The reason is that commercial production commonly occurs at the expense of subsistence in labor scarce situations, and a movement away from traditional subsistence practices may imply cultural changes (Behrens 1992; Godoy et al. 1995). Although some production levels might avoid unwanted changes, they may require the payment of unsustainable premium prices for the products (Morsello 2002).

Although they have their difficulties, fair trade deals still offer substantial advantages and opportunities: increased income as a result of shortened market chains and premium prices, improvements in management skills, resources to control the territory, economic opportunities for remote communities, and, in some cases, also reductions in inequalities (Anderson and Clay 2002; Morsello 2002). Some conditions for success include (Clay 1996; Anderson and Clay 2002): undertaking resource inventories and markets research; starting with traditional products and markets; adding value locally; improving harvesting techniques; keeping the strategy simple; diversifying market niches yet concentrating on a small number of products; establishing product standards; establishing partnerships with research and government organizations; not exaggerating the profits or benefits; requiring community investments and using loans, not grants; and adopting certification or labeling.

There are nevertheless many uncertainties about conditions for success because systematic and empirical research on the subject and comparative studies are particularly scant and difficult to set up. As a consequence, it remains unclear whether these activities are successful for one of the terms of the equation—indigenous peoples societies or the forests they live in—let alone for the combination of both. Important unresolved issues include: whether forest communities are being exploited and how they are compensated; under what conditions local transformation is beneficial; whether system of paying premium prices and the possibility of dependence on specific dealers; the exclusivity demanded by some implementers based on their high investments to start up the production; the links with biological conservation; and which forms of labeling are more suitable, certification or "ethical codes."

Fifty years ago, at the time of the first successful Himalayan expedition, there were fewer than a dozen expeditions and about a hundred trekkers. Now trekkers have increased to a staggering half a million and more than a thousand expeditions have been organized. As a result there is pollution on some Himalayan trails; the lack of anti-pollution measures by some trekking agencies, lack of environmental education, and overcrowding of certain trails has negative impacts on the ecosystems involved (Kohli 2002).

Tourism can, however, provide an alternative form of land use which decreases pressure for land use conversion (for example, tropical forests to pastoral lands). Tourism can also contribute to the maintenance and revival of lifestyles and cultural practices (Hillman 2003). Opportunities arise for education and awareness-raising to understand and respect cultural diversity and biodiversity. Conservation areas are especially valued for their educational significance and for providing recreational services. Valorization of cultural landscapes and monuments can also be an important asset to the larger society. Box 14.15 highlights the Rhön Biosphere Reserve in Germany, which aims to integrate environmental conservation and cultural landscapes through traditional agricultural systems while finding economic solutions for its long-term sustainability.

Tourism and recreation pose risks as well. Consumptive tourism activities, such as sport fishing, may represent pressure on the resource. For example, marine recreational angling in the United States alone comprise more than 15 million fishers in North America, with a total harvest of 266 million pounds in 2001 (http://www.st.nmfs.gov/st1/recreational/mrf_why.html). Conflicts between recreational and professional fishers are occurring in some parts of the world where recreational fishers retain strong political power in the fishing councils.

BOX 14.15

The Rhön Biosphere Reserve in Germany (Pokorny 2001)

Biosphere reserves are recognized areas of representative environments that have been internationally designated to promote solutions to reconcile the conservation of biodiversity and its sustainable use. They are nominated by national governments through the focal points for the Man and Biosphere Programme and UNESCO in their respective countries. The main concern of the Rhön Biosphere Reserve is the maintenance of cultural landscapes through traditional agriculture systems, currently threatened by a constant decrease in the number of farms and income of farmers. This is done by promoting on-farm tourism, but also through marketing local products. The Biosphere Reserve has been looking for business partners to develop innovative and environmentally friendly products and help create or safeguard jobs in the rural area.

"The Biosphere Reserve Business Partners" project was initiated by the local government of Hessen. This partnership involves all types of businesses, for example farms, restaurants, hotels, grocery stores, crafts, tourist agencies, and riding stables. Business partners in agriculture must meet the European Union Council regulation on organic production of agricultural products, including livestock production (EU No 1804/99). Restaurants and grocery stores must offer a minimum number of products that come from other Biosphere Reserve partners. The project is now in phase three, looking for a way to introduce a Biosphere Reserve label that would be product/service related rather than simply focused on businesses; enable the marketing of a variety of regional products in regional supermarkets, and enable the integration of non-food products or services.

Furthermore, representations of nature and culture used to entice tourists often refer to pristine ecosystems and exotic cultures, reinforcing stereotypes as marketing tools of communities subjected to tourism. Commoditization of culture can generate income, but does not always benefit those portrayed. Another problem is the blurring of boundaries between private and public. Emphasis on pristine ecosystems can furthermore easily lead to eviction of people.

The risks and opportunities provided by tourism are related to the economic position of communities and relations of power. Some of the most notable risks related to tourism activities, including ecological, cultural, and agrotourism, include pollution and waste, change in consumption pattern and nutrition, change in land use and livelihood, and the spread of infectious diseases (tourist to host and host to tourist). Increase in land use value for tourism real estate development purposes may lead to displacement and dispossession. This is especially a risk for communities that enjoy informal or communal land rights. Coastal communities have long suffered from these impacts as the tourist potential of coastal landscapes continuously pushes the local residents away from their livelihood strategies. Economic deprivation can lead to overexploitation of resources and to individuals accepting unfavorable positions in the tourism industry (low-skilled labor, sex-industry, drugs). This is particularly visible in the relationship between cultural stereotypes and sex tourism in parts of Asia and South America.

Recreation and education can go hand in hand. Cultural tourism can serve to educate people about the importance of cultural diversity, as well as the importance of the latter for the conservation of biodiversity, provided the risks are taken into account. Tourism and recreation can be linked to environmental education, fostering knowledge about the functioning of ecosystems and provoking tourists to critically examine human–nature relations. Yet tourists are not the sole targets of environmental education. The Omo Biosphere Reserve in Nigeria, for example, is the site of environmental education programs for very diverse audiences, ranging from schoolchildren to university students, protected area managers and policy-makers (Ola-Adams 2001).

The Fitzgerald Biosphere Reserve in Australia conducts awareness raising activities to help local farmers address the problem of wind erosion. West (2001) stresses that awareness raising requires development, especially among communities that experience periods of economic decline. Recently more attention is being paid to the transmission of local and indigenous knowledge. Elders are invited to schools to transmit their knowledge through stories and art, for example, instructing pupils in their mother tongue. In all cases, "top-down" education is less effective than education that is based on sharing experiences and attempts to reach a joint understanding the dynamics of human–nature interactions.

14.4.2 Responses Related to Tourism and Recreation

Economic responses have been developed side-by-side with responses working at the level of social-behavioral changes, legal and institutional incentives, and cognitive behavior. A good example is the public demand and economic-political valorization for cultural, ecological, and rural and urban tourism. Infrastructure for cultural and eco-tourism has increased in response to a growing awareness of cultural and biological diversity (Ashley et al. 2001). Environmental NGOs have contributed to awareness-raising and political pressure and are providing capacity building, with various levels of success. National governments have realized the potential benefits of tourism for their economies. Many countries have developed policies and incentives to develop tourism industries. Policy-makers need to take into account environmental, cultural, and historical variability in defining priorities and strategies. The scope of policies can be constrained, however, by the economic situation of the country. Internal and regional differences may exist in possibilities of providing tourism services and generating benefits. Box 14.16 illustrates the importance of accounting for diverse views on the use of state parks and forests in the midwestern United States.

14.4.2.1 Cultural Tourism

Cultural tourism involves representations of culture at different levels—regional, national, and local. The questions that arise pertain to who represents whom and what cultural symbols are used. A recent conflict over attempts to patent an Inuit symbol used as a marketing tool by a tour operator shows how pertinent these questions are. Attention needs to be paid to a number of issues such as revenue sharing for cultural items, rights pertaining to certain cultural symbols, and the blurring of boundaries between the private and the public. Indigenous rights interplay with national imperatives of revenue earning. Furthermore, different groups will benefit differentially from tourism depending on local arrangements.

BOX 14.16

Taking into Account Diverse Views on the Use of State and National Forests in the Midwestern United States

(Welch et al. 2001)

The diverse pool of users around many parks in densely populated areas entails a variety of opinions and views on how to use public areas and which ones to conserve. Much of the controversy centers on how people view and use natural resources. A *preservationist* approach focuses on limiting the human interference. A *conservationist* approach focusing on multiple uses entails sustainable harvesting of forest resources. *Recreational users* concentrate on access to the public land and having suitable conditions for their activities in the forest. Different coalitions are formed among various interest groups.

Disagreement over the management of the national forests has led to lawsuits and an effective cessation to logging. The loss of tax dollars from timber concessions is a sore point for county-level administrations and has had a perceived negative impact on the quality of services supplied by these entities. In the midwestern United States there is a growing debate between the proponents of the use of state and national forests for recreation and conservation and those supporting the original purpose of some parks as areas for managed logging operations. However, even recreational activities are controversial such as the opening of parks to off-road vehicles or hunting.

Interest groups often use ecological arguments to support their position on the use of public land. Preservation groups focus on the importance of unbroken tracts of mature forest for neotropical migrant birds. Other groups concerned with habitat for game species lament the shrinking area of heterogeneous forest patches of variable age and the decline of tree species reliant on disturbance. One important contribution of the academic community in collaboration with the U.S. Forest Service has been on the study and survey of different groups of users. Such studies contribute to the understanding of controversial views and opinions of different user groups and local communities and may serve to inform not only park management plans, but also educate the public benefiting from different "services" provided by the conservation area.

14.4.2.2 Ecotourism

Ecotourism can provide economic alternatives to value ecosystems services. There may be potential conflicts in resource use and the aesthetics of certain ecosystems. Different ecosystems are subjected to different types and scales of impact from tourism infrastructures. Furthermore, some ecosystems are easier to market to tourists than others are. The market value of ecosystems may vary according to public perceptions of nature. Freezing of landscapes, conversion of landscapes, dispossession, and removing of human influences may result, depending on views of what ecotourism should represent. Yet when conservation receives no budgetary subsidy, tourism can provide revenues needed for conservation purposes.

Cultural and eco-tourism are not necessarily the same thing as community-based tourism (Binns and Nel 2002). Concerning access and benefit-sharing, tensions often exist between tour operators, local communities, conservation units, and other government departments. In some cases, local communities may actually run tourism enterprises themselves; in other cases, they may only have access to low-paid menial jobs. Issues that are important are matters of local decision-making powers concerning land use, infrastructure, and dealing with externalities, for example, the choice of tour operators and types of tourists, boundaries between the private and the public, goods that are imported by tourists, and waste management. Capacity-building tends to be more successful if it not only involves individual professional training, but also includes institution building, marketing, and negotiating capacities. (Box 14.17 discusses conflicts arising from tour operations without local input.)

14.4.2.3 Rural and Urban Tourism

Rural and urban tourism have been receiving increasing support in recent years, including in terms of tax incentives and credit. Changes are occurring in perceptions concerning what types of landscapes and cultural practices are of interest to tourism. Industrial monuments and certain types of resource use are now seen as having historical, social, and environmental potential for tourism. This also broadens the scope of environmental conservation in areas outside of protected areas. Small-scale farming production, while facing difficult competition in agricultural markets, can find an alternative in providing tourist services. This is just one example of how tourism can contribute to the maintenance of certain lifestyles and production methods. Other examples include farm-tourism combinations in Italy, Germany, Israel, and Russia. In Canada, ecotourism packages available to European markets offer tourists a chance to live in teepees and experience traditional First Nation peoples' lifestyle. The tours also offer chances to go on horseback and explore traditional trails. Hunting tours are offered to European and American hunters, in which First Nation peoples are used as guides because of their local knowledge of the environmental landscape.

14.5 Conclusion

Cultural perceptions and practices affect biodiversity and ecosystem management practices. Nevertheless, many natural resource management systems and conservation strategies still separate people from their environments, freezing and stereotyping both cultures and ecosystems. Such systems and strategies are less effective in addressing linkages between ecosystem functioning, develop-

BOX 14.17

Conflicts with and over Tour Operators in Zimbabwe

In the Zambezi valley in Zimbabwe, several districts have been granted appropriate authority over wildlife and participate in the Communal Areas Programme for the Management of Indigenous Resources (CAMPFIRE). These districts have the right to utilize wildlife and other resources to generate benefits that ideally should be devolved to the so-called producer communities, that is, the local communities that live with wildlife and through good stewardship assure the presence of wildlife. In many cases revenues are derived from sports hunting operations. Some CAMPFIRE projects have been very successful, with the districts actually devolving large sums of money derived from wildlife to local communities and giving them the authority to decide how the revenues are distributed (Nabane et al.1995; Spierenburg 1999; Murphree 2001). In other districts, severe conflicts have arisen over the distribution of benefits to local communities (Dzingirai 1995).

Another bone of contention has been the selection of tour operators who are awarded hunting concessions by the district authorities. In most districts, local communities have no say in this selection. One district had started its own tourist operation, employing two licensed hunters to accompany tourists. These hunters were on good terms with the neighboring community, entering into joint ventures, always impressing upon visitors that they were actually hunting in an area that (since it is part of a Communal Area) belongs to the community, and urging them to respect that. In another part of the district, however, a concession had been awarded to a tour operator who falsely considered the area as his private property. He employed guards who at night would move into the neighboring villages, overturning cooking pots to see if villagers had been poaching (Hasler 1996). The villagers have repeatedly, but in vain, demanded that the concession be awarded to another operator.

ment, and human well-being. Overcoming the nature–culture dichotomy towards a more encompassing view of ecosystems that does not isolate cultural services from other ecosystem services is necessary to face the challenges of the new millennium. There is a range of possibilities of interactions between humans and nature. Conservation and management strategies that take into account the historical, economic, and cultural contexts (including different knowledge systems, forms of ownership, and institutional organization) of these interactions are more effective in terms of contributing to both ecosystem and human well-being. Generally, standard, blueprint approaches do not seem to work.

Furthermore, global-national-local linkages are important. In a global perspective, market-economic policies are interrelated as flows of resources, goods and services transcend national and regional boundaries. Market pressures, consumption patterns, technological changes, and (inter)government policies will continue to influence intensification of resource use, land reform, and substitution of local technologies; all affect local livelihoods, capabilities, and environments. It is only too common to lay the responsibility for environmental problems and conservation in the hands of local communities, or blame the private sector, while disregarding the linkages between economic pressures and local, national, and international policies. Neither local communities nor private enterprises operate in a vacuum; all creatively seek opportunities to have their opinions heard and to promote their interests by concluding alliances at the local, national, and global level. Understanding the complexities of different cultural perceptions of landscapes, management of resources, and local institutional arrangements contributes to alternative strategies of ecosystem management and socioeconomic development.

Nature–culture relations have been the subject of intense debates, such as the people and parks debate. As a result, ideas about what land- and waterscapes should be conserved, as well as how they should be conserved, have changed. There is a growing recognition that a wider variety of landscapes, including agricultural, industrial, and aquatic landscapes, need to be conserved. Many governments and national and international conservation organizations have recognized that the biggest challenge of conservation in the twenty-first century is for it to take place outside parks and enforced boundaries, and thus be integrated into agricultural and urban systems.

Concepts and policies regarding the creation of conservation units have moved from exclusion of communities and local forms of resource management to inclusion. It is, however, difficult to assess the impacts of community participation on ecosystems and human well-being. Studies that have attempted to do so have had mixed results. Some programs do contribute to biodiversity conservation and local economic needs. In other cases the contribution to human well-being has been easier to assess than the contribution to the conservation of biodiversity. In cases where the contribution to human well-being has been limited, this is often attributed to flaws in the decentralization process. Furthermore, numerous authors have problematized the concept of "community." Perceptions of land- and waterscapes are influenced by cultural repertoires, which in turn are influenced by and influence knowledge production (scientific as well as local and indigenous).

Local and indigenous knowledge is important in conserving ecosystems and contributing to human well-being. The understanding of crop, forest, and aquatic biodiversity lies in the oral history and cultural memories of indigenous and local communities. Yet the pace of technological and agricultural change, and of large-scale environmental modification by infrastructure development, often happens at the expense of local resources and knowledge. This has an impact on local food security and economies, and is also relevant to national and international issues of conservation and economy. Local and indigenous peoples have reacted proactively to such changes by organizing themselves (for example by establishing NGOs) and concluding alliances at local, national, and global levels.

Fostering the articulation of international and national conventions and regulations linking biodiversity and local and indigenous knowledge is important, taking into account that knowledge is produced in dynamic context of inter- and intra-group interactions, power relations, and historical settings. Compensating for the utilization of local knowledge and resources entails taking into account the power dynamics of the relations among companies, national and regional governments, and communities. Certification programs are more likely to be effective in addressing human well-being if they take into account the impact of particular resource extraction upon people and communities using the same resource basis, but not necessarily sharing resource ownership. Certification programs work better if they are made more accessible to communities and small producers' groups that are not always familiar with bureaucratic procedures, and cannot afford high costs for certification. Responses such as "Fair Trade" tools are more effective if they promote the participation of local producers in the commercialization process, price negotiations, and the transformation and retailing of their products.

The economic exploitation as well as the growing consciousness concerning the disappearance of local resources and the knowledge about these has led to growing concerns of the need to protect local and indigenous knowledge. The international community has recognized the dependence of many indigenous peoples on biological resources, notably in the preamble to the CBD. Many governments are struggling to develop national policies to protect local and indigenous knowledge. Intellectual property rights, as advocated by WIPO, entail the risk of considering any type of knowledge as a commodity, removing it from its cultural context. Many legal strategies fail to take into account local and indigenous strategies to protect as well as transmit knowledge.

Eco-, cultural, and agrotourism can provide important opportunities to link conservation and development. However, in many cases, they do not directly benefit the local communities, that is, they are not necessarily community-based. The latter entails institutional capacity building in marketing and negotiation, representation among community members, tourism operators, and governments, and defining access to benefits. Conflicts about resource use, development of infrastructure, conversion of ecosystems, and dispossession of communities have negative impacts on the possibilities of ecotourism contributing to human well-being. In cultural tourism, problems may emerge in the representation and ownership of cultural symbols, reproduction of stereotypes, consent among and within communities, and blurring of boundaries between the public and private. Studies show that the risks and opportunities provided by tourism are related to the economic position of communities and relations of power. Economic deprivation can lead to overexploitation of resources and accepting unfavorable positions in the tourism industry (low-skilled labor, sex-industry, drugs). Increase in land use value for tourism real estate development purposes may lead to displacement and dispossession. This is especially a risk for communities that enjoy informal or communal land rights.

Recreation, conservation, and environmental education can go hand in hand. Cultural tourism can serve to educate people about the importance of cultural diversity, as well as the importance of the latter for the conservation of biodiversity, provided the risks mentioned are taken into account. Tourism and recreation can be linked to environmental education, fostering knowledge about the functioning of ecosystems and provoking tourists to critically examine human–nature relations. Environmental education may serve very diverse audiences, ranging from schoolchildren to university students, protected area managers, policymakers, and representatives of the private sector. In all cases, top-down education is less effective than education that is based on sharing experiences and attempts to reach a joint understanding of the dynamics of interaction between humans and nature.

References

Acheson, J. (ed.), 1994: *Anthropology and Institutional Economics,* University Press of America, Lanham, MD.

Acheson, J.M. and Wilson, J.A., 1996: Order out of chaos: The case for parametric fisheries management, *American Anthropologist,* **98(3),** pp. 579–594.

Agrawal, A. and E. Ostrom, 2001: Collective action, property rights, and decentralization in resource use in India and Nepal, *Politics & Society,* **29(4),** 485–514.

Altieri, M. and S. Hetch, 1990: *Agroecology and Small Farm Development,* CRC Press, Boca Ratón, FL.

Anderson, A. and J. Clay, 2002: Esverdeando a Amazônia, IIEB (Instituto Internacional de Educação do Brasil)/Peiropolis, São Paulo, Brasil.

Appadurai, A., and C.A. Breckenridge (eds.), 1986: *The Social Life of Things: Commodities in Cultural Perspective,* Cambridge University Press, Cambridge, UK.

Arizpe, L. (ed.), 1996: *The Cultural Dimensions of Global Change: An Anthropological Approach,* UNESCO (United Nations Educational, Scientific and Cultural Organization) Publishing, Paris, France.

Ashley, C., D. Roe and H. Goodwin, 2001: *Pro-poor Tourism Strategies: Making Tourism Work for the Poor: A Review of Experience,* Overseas Development Institute, London, UK.

Bahuguna, S., 1992: People's programme for change, *The INTACH Environmental Series,* **19,** INTACH (Indian National Trust for Art and Cultural Heritage), New Delhi, India.

Balée, W., 1989: The culture of Amazonian forests. In: *Natural Resource Management by Indigenous and Folk Societies of Amazonia,* W. Balée and D. Posey (eds.), Advances in Economic Botany Monograph Series, Vol. 7, NY Botanical Garden, New York, NY, pp. 1–21.

Balée, W., 1999: *Footprints of the Forest: Ka'apor Ethnobotany: The Historical Ecology of Plant Utilization by Amazonian People,* Columbia University Press, New York, NY.

Barlett, P.F., 1982: *Agricultural Choice and Change; Decision Making in a Costa Rican Community,* Rutgers University Press New Brunswick, N.J.

Behrens, C., 1992: Labor specialization and the formation of markets for food in a Shipibo subsistence economy, *Human Ecology,* **20,** pp. 435–462.

Beinart, W. and P.A. Coates, 1995: *Environment and History: The Taming of Nature in the USA and South Africa,* Routledge, New York, NY.

Berger, P.L. and T. Luckmann, 1971 [1966]: *The Social Construction of Reality: A Treatise in the Sociology of Knowledge,* Allen Lane, London, UK.

Biersack, A., 1999: Introduction: from the "new ecology" to the new ecologies, *American Anthropologist,* **101(1),** pp. 5–18.

Binns, T. and E. Nel, 2002: Tourism as a local development strategy in South Africa, *Geographical Journal,* **September,** pp. 235–247.

Brandon, K., K. Redford, and S. Sanderson (eds.), 1998: *Parks in Peril: People, Politics and Protected Areas,* Island Press, Washington, DC.

Brondizio, E.S. and A.D. Siqueira, 1997: From extractivists to forest farmers: Changing concepts of Caboclo agroforestry in the Amazon estuary, *Research in Economic Anthropology,* **18,** pp. 233–279.

Brondizio E.S., 2004a: Agriculture intensification, economic identity, and shared invisibility. In Amazonian peasantry: Caboclos and Colonists in comparative perspective. *Culture and Agriculture* **26(1** and **2),** pp. 1–24.

Brondízio, E.S., 2004b: From staple to fashion food: Shifting cycles, shifting opportunities in the development of the Açaí fruit (*Euterpe oleracea* mart.) economy in the Amazon estuary. In: *Working Forests in the American Tropics: Conservation through Sustainable Management*? Zarin et al. (eds.), Columbia University Press, New York, NY. In press.

Brookfield, H., 2001: *Exploring Agrodiversity,* Columbia University Press, New York, NY.

Brookfield, H., H. Parsons, and M. Brookfield (eds.), (2003): *Agrodiversity,* United Nations University Press, Tokyo, Japan.

Brush, S.B, and D. Stabinsky, 1996: *Valuing Local Knowledge: Indigenous People and Intellectual Property Rights,* Island Press, Washington, DC.

Caballero, J., 1992: Maya homegardens: Past, present, and future, *Etnoecologica* **1(1),** pp. 35–54.

Castro, F., 2002: From myths to rules: The evolution of the local management in the lower Amazonian floodplain, *Environment and History,* **8(2),** pp. 197–216.

Castro, F. and David F. McGrath, 2003: Moving toward sustainability in the local management of floodplain lake fisheries in the Brazilian Amazon, *Human Organization,* **62,** pp. 123–133.

Certified Forest Products Council, 2002: Available at http://www.certifiedwood.org.

Chandran, M., and D. Subash, 2000: Shifting cultivation, sacred groves and conflicts. In: *Colonial Forest Policy in the Western Ghats,* R.H. Grove, V. Damodaran, and S. Sangwan (eds.), Nature and the Orient: The Environmental History of South and Southeast Asia, Oxford University Press, New Delhi, India, pp. 674–707.

Clark, G. (ed.), 2003: *Gender at Work in Economic Life,* Society for Economic Anthropology monograph series no. 20, AltaMira Press, Walnut Creek, CA.

Clay, J., 1997: Business and biodiversity: Rainforest marketing and beyond: Special forest products-biodiversity meets the marketplace. In: *Sustainable Forestry-Seminar Series,* N.C. Vance and J. Thomas, USDA Forest Service, General technical report GTR-WO-63, Washington, DC, pp. 122–145.

Clay, J.W., 1996: *Generating Income and Conserving Resources: 20 lessons from the field,* World Wildlife Fund, Washington, DC.

Clay, J., 1992: Why rainforest crunch? *Cultural Survival Quarterly,* **16(2),** pp. 31–46.

Clements, C.R., 1999: 1942 and the loss of Amazonian crop genetic resources: I. The relation between domestication and human population decline, *Economic Botany* **53(2),** pp. 188–202.

Cohen, D.W., and Odhiambo, E.S. Atiendo, 1989: Siaya, *The Historical Anthropology of an African Landscape,* James Currey, Oxford, UK/Heinemann, Nairobi, Kenya/Ohio University Press, Athens, OH.

Comaroff, J., 2000: Millennial capitalism: First thoughts on a second coming, *Public Culture,* **12(2),** pp. 291–343.

Comaroff, J. 1999: Alien-nation: Zombies, immigrants and millennial capitalism, CODESRIA Bulletin, **3(4),** pp. 17–28.

Conklin, B. and Graham, L.R., 1995: The shifting middle ground: Amazonian Indians and eco-politics, *American Anthropologist,* **97(4),** pp. 695–710.

Coplan, D., 1994: *In the Time of Cannibals: The World Music of South Africa's Basotho Migrants,* Chicago University Press, Chicago. IL.

Cordell, J. (ed.), 1989: *A Sea of Small Boats,* Cambridge University Press, Cambridge, UK.

Corry, S., 1994: Harvest hype, UNEP News, Nairobi, *Our Planet,* **6(4),** pp. 35–37.

Crosby, A.W., 1986: *Ecological Imperialism: The Biological Expansion of Europe,* Cambridge University Press, Cambridge, UK, pp. 900–1900.

Cox, P.A., 2000: Will tribal knowledge survive the millennium? *Science,* **287(5450),** pp. 44–45.

Crumley, C. L. (ed.), 1994: *Historical Ecology: Cultural Knowledge and Changing Landscapes,* School of American Research Press, Santa Fe, NM.

Cunningham, A.B., 1996: Professional ethics and ethnobotanical research. In: *Selected Guidelines for Ethnobotanical Research: A Field Manual,* Alexiades, M.N. (ed.). The New York Botanical Garden, New York, NY.

De Boeck, F., 1998: Domesticating diamonds and dollars: Identity, expenditure and sharing in southwestern Zaire (1984–1997), *Development and Change,* **29(4),** pp. 777–810.

Derman, W., 1993: *Recreating Common Property Management: Government Projects and Land Use Policy in the Mid-Zambezi Valley, Zimbabwe,* Occasional paper, Centre for Applied Social Sciences, University of Zimbabwe, Harare. Zimbabwe.

Derman, W. and J. Murombedzi, 1994: Democracy, development, and human rights in Zimbabwe: A contradictory terrain, *African Rural and Urban Studies,* **1,** pp. 119–143.

Dietz, T., E. Ostrom, and P. Stern, 2003: The struggle to govern the commons, *Science,* **302(5652),** pp. 1907–1912.

Dove, M., 1999: The agronomy of memory and the memory of agronomy: Ritual conservation of archaic cultigens in contemporary farming systems. In *Situated knowledge, Located lives,* V. Nazarea Ethnoecology, Tucson: University of Arizona Press, pp. 45–70.

Draper, M., M. Spierenburg, and H. Wels, 2004: African dreams of cohesion: The mythology of community development in transfrontier conservation areas in Southern Africa, *Culture and Organization,* **10(4),** pp. 341–353.

Dyer, C.L., and J.R. McGoodwin (eds.), 1994: *Folk Management in the World's Fisheries: Lessons for Modern Fisheries Management,* University Press of Colorado, Niwot, CO.

Dzingirai, V., 1995: Take back your CAMPFIRE, NRM (Natural Resources Management) occasional papers series, Centre for Applied Social Sciences, University of Zimbabwe, Harare, Zimbabwe.

Dzingrai, V. and M.F.C. Bourdillon, 1997: Religious ritual and political control in Binga District, Zimbabwe, *African Anthropology,* **4(2),** pp. 4–26.

Ellen, R., 1982: *Environment, Subsistence, and Systems: The Ecology of Small-scale Social Formations,* Gambridge University Press, Cambridge, UK.

Entine, J., 1994: Shattered Image, *Business Ethics,* **8,** pp. 23–28.

Evans, S., 1999: *The Green Republic: A Conservation History of Costa Rica,* University of Texas Press, Austin, TX.

Frisancho, R.A., 1990: *Anthropometric Standards for the Assessment of Growth and Nutritional Status,* University of Michigan Press, Ann Arbor, MI.

Geores, M., 2002: Scale and forest resources. In: *The Commons at the Millennium,* Dolak, N. and E. Ostrom (eds.), MIT Press, Cambridge, MA.

Gibson, C.C. and T. Koontz, 1998: When community is not enough: Communities and forests in Southern Indiana, *Human Ecology,* **26,** pp. 621–647.

Godoy, R., N. Brokaw, and D. Wilkie, 1995: The effect of income on the extraction of non-timber tropical forest products: Model, hypotheses, and preliminary findings from the Sumu Indians of Nicaragua, *Human Ecology,* **23(1),** pp. 29–52.

Golley, F., 1993: *A History of the Ecosystem Concept in Ecology: More than the Sum of the Parts,* Yale University Press, New Haven, CT.

Granfelt, T. (ed.), 1999: *Managing the Globalized Environment: Local Strategies to Secure Livelihoods,* Intermediate Technology Publications, London, UK

Grant, J.A. and F. Söderbaum (eds.), 2003: *The New Regionalism in Africa,* Ashgate, London, UK.

Hardin, G., 1968: The tragedy of the commons, *Science,* **162,** pp. 1243–1248.

Harris, D.R. 1989: An evolutionary continuum of people-plant interaction. In: D.R. Harris and G.C. Hillman (eds.), *Foraging and Farming: The Evolution of Plant Exploitation,* Unwin Hyman, London, pp. 11–26.

Hasler, R., 1996: *Agriculture, Foraging and Wildlife Resource Use in Africa: Cultural and Political Dynamics in the Zambezi Valley,* Kegan Paul, London, UK/New York, NY.

Heckenberger, M.J., A. Kuikuruo, U.T. Kuikuruo, J.C. Russell, M. Schmidt, et al., 2003: Amazônia 1492: Pristine forest or cultural parkland? *Science,* **301,** pp. 1710–1713.

Hidalgo, R.C., 1992: The Tagua Initiative in Ecuador: A community approach to tropical rain forest conservation and development. In: *Sustainable Harvest and Marketing of Rainforest Products,* M. Plotkin and L. Falamore (eds.), Island Press, Washington, DC, pp. 263–273.

Hillman, B., 2003: Paradise under construction: Minorities, myths and modernity in Northwest Yunnan, *Asian Ethnicity,* **4(2),** pp. 175–188.

Hladik, C.M., A. Hladik, H. Pagezy, O.F. Linares, J.A.K. Georgius et al., 1996: L'Alimentation en forêt tropicale: Interactions bioculturelles et perspectives de développement: Les Ressources alimentaires: Production et consommation: Bases culturelles des choix alimentaires et stratégies de développement, Man and the Biosphere (MAB), vol. 13, UNESCO (United Nations Eductional, Scientific and Cultural Organization), The Parthenon Publishing Group, Paris, France.

Hulme, D. and M. Murphree, 2001: Community conservation in Africa: An introduction. In: *African Wildlife and Livelihoods: The Promise and Performance of Community Conservation,* D. Hulme and M. Murphree (eds.), James Currey, Oxford, UK, pp. 1–8.

ISA (Instituto Socioambiental), 2000: Povos Indígenas no Brasil 1996/2000, Instituto Sociambiental, São Paulo [online]. Available at http://www.socioambiental.org.

Isaac, B., 1993: Retrospective on the formalist-substantivist debate. In: *Research in economic anthropology,* B. Isaac (ed.), JAI Press, Greenwich, CT.

Jansen, J., 2000: Masking Sujnata: a Hermeneutical Critique, *History in Africa,* **27,** pp. 131–141.

Kassibo, B., 2001: Participatory management and democratic decentralization: Management of the Samori Forest in Baye Commune, Mopti Region, Mali, Working paper, World Resource Institute, Washington, DC.

Kohli, M. S., 2002: Eco-tourism and Himalayas, *Yojana,* **August,** pp. 25–28.

Kottak, C.P., 1999: The new ecological anthropology, *American Anthropologist,* **101(1),** pp. 23–35.

Kyoto Protocol. Available at http://unfccc.int/resource/convkp.html

Laird, S., 1994: Natural products and the commercialization of traditional knowledge. In: *Intellectual Property Rights for Indigenous People: A Sourcebook,* T. Greaves (ed.), Society for Applied Anthropology, Oklahoma City, OK, pp. 147–162.

Leach, M. and J. Fairhead, 2000: Fashioned forest pasts, occluded histories? International environmental analysis in West African locales, *Development and Change,* **31(1),** pp. 35–39.

Lima, D., 1999: Equity, sustainable development and biodiversity preservation: Some questions on the ecological partnership in the Brazilian Amazon. In: *Várzea: Diversity, Development, and Conservation of Amazonia's Whitewater Floodplain,* C. Padoch, J. M. Ayres, M. Pinedo-Vasquez, and A. Henderson (eds.), The New York Botanical Garden Press, New York, NY, pp. 247–263.

Little, Paul E., 1999: Environments and environmentalism in anthropological research: Facing a new millennium, *Annual Review of Anthropology,* **(28),** pp. 253–284.

LTO Nederland (Land- en Tuinbouw Organisatie) [Agriculture and Horticulture Organization of The Netherlands], 2003: De boer natuurlijk: De duurzame ontwikkeling van het platteland, [The farmer naturally: Sustainable development of rural areas], LTO, The Hague, The Netherlands.

Luig, U. and A. von Oppen, 1997: Landscape in Africa: Process and vision: An introductory essay, *Peiduma,* p. 43.

Lunde, L., M. Taylor, and A. Huser, 2003: Commerce or Crime? Regulating economies of conflict, FAFO report 424, FAFO (Forskningsstiftelsen/Fafo Institute for Applied Social Science) Programme for International Cooperation and Conflict Resolution, Oslo, Norway.

McDaniel, J., 1997: Communal fisheries management in the Peruvian Amazon, *Human Organization,* **56,** pp. 147–152.

McGrath, D., F. Castro, C. Futemma, B. Amaral, and J. Calabria, 1993: Fisheries and the evolution of resource management on the lower Amazon floodplain, *Human Ecology,* **21(1),** pp. 67–95.

Menzies, N., 1994: *Forest and Land Management in Imperial China,* Palgrave (Macmillan), London, UK.

Molenaar, B., 2002: The wild east: The impact of illegal logging on a local population, *Human Rights Internet,* **9(1)** [online]. Available at http://www.hri.ca/tribune/viewArticle.asp?ID = 2667.

Moore, D. S., 1998: Clear waters and muddied histories: Environmental history and the politics of community in Zimbabwe's eastern highlands, *Journal of Southern African Studies,* **24(2),** pp. 377–403.

Moran, E.F., E. Ostrom, and J.C. Randolph, 2002: Ecological systems and multitier human organizations. In: *UNESCO/Encyclopedia of Life Support Systems,* EOLSS (Encyclopedia of Life Support System) Publishers, Oxford, UK.

Moran, E. (ed.), 1990: *The Ecosystem Approach in Anthropology,* University of Michigan Press, Ann Arbor, MI.

Moran, K., 1999: Toward compensation: Returning benefits from ethnobotanical drug discovery to native peoples. In: *Ethnoecology: Situated Knowledge, Located Lives,* V. Nazarea (ed.), University of Arizona Press, Tucson, AZ, pp. 249–263.

Morsello, C., 2002: *Market Integration and Sustainability in Amazonian Indigenous Livelihoods: The Case of the Kayapó,* School of Environmental Sciences, University of East Anglia, Norwich, UK, pp. 301.

Mountain Institute, 1998: *Sacred Mountains and Environmental Conservation: A Practitioner's Workshop,* Unpublished report.

Murphree, M., 2001: Community, Council & Client, a Case Study in Ecotourism and Development from Mahenye, Zimbabwe. In: *African Wildlife & Livelihoods. The Promise and Performance of Community Conservation,* David Hulme & Marshall Murphree (eds.), James Currey Publishers/Heinemann, Oxford/Portsmouth, pp. 177–194.

Murrieta, R.S., D. Dufour, and A. Siqueira, 1999: Food consumption and subsistence in three Caboclo populations on Marajo Island, Amazônia, Brazil, *Human Ecology,* **27(3),** pp. 455–475.

Nabane, N., V. Dzingirai, and E. Madzudzo, 1995: Membership in common property regimes: A case study of Guruve, Binga, Tsholotsho and Bulilimamangwe CAMPFIRE Programmes, NRM (Natural Resources Management) occasional papers series, Centre for Applied Social Sciences, University of Zimbabwe, Harare, Zimbabwe.

National Research Council, 2002: *The Drama of the Commons,* E. Ostrom, T. Dietz et al (eds.), National Academy Press, Washington, DC.

Nazerea, V., 1998: *Cultural Memory and Biodiversity,* University of Arizona Press, Tucson, AZ.

Netting, R., 1993: *Smallholders, Householders: Farm Families and the Ecology of Intensive, Sustainable Agriculture,* Stanford University Press, Stanford, CA.

Ola-Adams, B.A., 2001: Education, awareness building and training in support of biosphere reserves: Lessons from Nigeria, *Parks (IUCN),* **11(1),** pp. 18–23.

Orlove, B., 1980: Ecological anthropology, *Annual Review of Anthropology,* **9,** pp. 235–273.

Ostrom, E., 1990: *Governing the Commons: The Evolution of Institutions for Collective Action.* Cambridge University Press, Cambridge, UK.

Padoch, C., E. Harwell, and A. Susanto, 1998: Swidden, sawah, and in-between: Agricultural transformation in Borneo, *Human Ecology,* **26(1),** pp. 3–20.

Pinedo-Vasquez, M., D.J. Zarin, K. Coffey, C. Padoch, and F. Rabelo, 2001: Post-boom logging in Amazonia, *Human Ecology,* **29(2),** pp. 219–239.

PLEC, 2002: (People, Land Management, and Ecosystem Conservation). Available at http://www.unu.edu/env/plec.

Pokorny, D., 2001: Biosphere Reserves for developing quality economies: Examples from the Rhön Biosphere Reserve, Germany, *Parks (IUCN),* **11(1),** pp. 16–17.

Posey, D., 1998: Can cultural rights protect traditional cultural knowledge and biodiversity? In: *Cultural Rights and Wrongs,* UNESCO (ed.), United Nations Educational, Scientific and Cultural Organization, Paris, France.

Rangarajan, M. 2001: *India's Wildlife History: An Introduction,* Ranthambhore Foundation, Delhi, India.

Ranger, T.O., 1999: *Voices from the Rocks: Nature, Culture and History in the Matopos Hills of Zimbabwe,* Baobab, Harare, Zimbabwe/Indiana University Press, Bloomington/Indianapolis, IN/James Currey, Oxford, UK.

Reij, C. and A. Waters-Bayer, 2001: Entering research and development in land husbandry through farmer innovation. In: *Farmer Innovation in Africa: A Source of Inspiration for Agricultural Development,* C. Reij and A. Waters-Bayer (eds.), Earthscan, London/Sterling, UK, pp. 3–22.

Reyes-García, V., R. Godoy, V. Vadez, L. Apaza, E. Byron, et al., 2003: Ethnobotanical knowledge shared widely among Tsimané Amerindians, Bolivia, *Science,* **299(5613),** pp. 1707.

Ribot, J., 1999: Decentralisation, participation and accountability in Sahelian forestry: Legal instruments of political-administrative control, *Africa,* **69,** pp. 23–65.

Robins, 1999: *Who benefits? A Social Assessment of Environmentally Driven Trade,* International Institute for Environment and Development, London, UK.

Schama, S., 1995: *Landscape and Memory,* Harper Collins, London, UK/New York, NY.

Schutz, A., 1970: *On Phenomenology and Social Relations: Selected Writings,* H.R. Wagner (ed.), The University of Chicago Press, Chicago, IL/London, UK.

Scoones, I. and B. Cousins, 1994: Struggle for control over wetland resources in Zimbabwe, *Society and Natural Resources,* **7,** pp. 579–594.

Scoones, I., 1996: Range management science and policy: Politics, polemics and pasture in southern Africa. In: *The Lie of the Land: Challenging Received Wisdom on the African Environment,* M. Leach and R. Mearns (eds.), James Currey, Heinemann, Oxford/London, UK, pp. 34–53.

Scoones, I., 2001: Transforming soils: The dynamics of soil-fertility management in Africa. In: *Dynamics and Diversity: Soil Fertility and Farming Livelihoods in Africa,* I. Scoones (ed.), EarthScan, London, UK, pp. 1–44.

Scott, J., 1998: *Seeing Like a State: How Certain Schemes to Improve the Human Condition Have Failed,* Yale University Press, New Haven, CT.

Settee, P., 2000: The issue of biodiversity, intellectual property rights, and indigenous rights. In: *Expressions in Canadian Native Studies,* R. Laliberte, P. Settee, J. Waldram, R. Innes, B. Macdougall, et al. (eds.), University of Saskatchewan Extension Press, Saskatoon, Canada.

Shiva, V., 1997: *Biopiracy the Plunder of Nature and Knowledge,* South End Press, Boston, MA.

Simon, H., 1957: *Models of Man,* Wiley, New York, NY.

Simon, H.A., 1990: A mechanism for social selection and successful altruism, *Science* **250(4988),** pp. 1665–1668.

Sinha, S., G. Shubhra, and B. Greenberg, 1997: The "new traditionalist" discourse of Indian environmentalism, *The Journal of Peasant Studies,* **24,** pp. 65–99.

Spierenburg, M., 2004: *Strangers, Spirits and Land Reforms: Conflicts about Land in Dande, Northern Zimbabwe,* Brill Publishers, New York, NY.

Spierenburg, 2003: Natural resource management in the communal areas: From centralisation to de-centralisation and back again. In: *Zimbabwe, Twenty Years of Independence:The Politics of Indigenisation,* S. Darnolf and L. Laakso (eds.), Palgrave (MacMillan), London, UK, pp. 78–103.

Spierenburg, M., 1999: Conflicting environmental conservation strategies in Dande, northern Zimbabwe: CAMPIRE versus the Mid-Zambezi Rural Development Project. In: *Towards Negotiated Co-management of Natural Resources in Africa,* B. Venema and H. van den Breemer (eds.), LIT-Verlag, Hamburg, Germany.

Stevens, S., (ed.), 1997: *Conservation through Cultural Survival: Indigenous Peoples and Protected Areas,* Island Press, Washington, DC.

Stiebel, L., L. Gunner, and J. Sithole, 2000: The Land in Africa: Space, Culture, History workshop, *Transformation,* **44(i–viii).**

Swearer, D.K., 2001: Principles and poetry, places and stories: The resources of Buddhist ecology in *Daedalus,* **130(4)** (Fall), pp. 225–241.

Sylvain, R., 2002: Land, water and truth: San identity and global indigenism, *American Anthropologist,* 104(4), pp. 1074–1085.

Taiga Rescue Network, 2003: Newsletter on Boreal forests, Issue 43 [online]. Available at www.taigarescue.org/_v3/files/taiganews/TN43.pdf.

Toffler, A., 2003: Tomorrow's economy, *India Today,* **March,** pp. 28–30.

Turner, T., 1995: Neoliberal ecopolitics and indigenous peoples: The Kayapó, the "rainforest harvest," and "the body shop," *Yale F & ES Bulletin* **(98),** pp. 113–127.

UNESCO/LINKS, 2002: LINKS, Local and Indigenous Knowledge Systems (Brochure), UNESCO, Paris.

UNESCO Man and biosphere (MAB) programme (United Nations Educational, Scientific and Cultural Organization) [online]. Available at http://www.unesco.org/mab.

Vijayalakshmi, E., 2003: GMCL: A green company in a grey market, *Down to Earth,* **11(22),** April 15, p. 44.

Welch, D., C. Croissant, T. Evans, and E. Ostrom, 2001: *A Social Assessment of Hoosier National Forest,* CIPEC summary report no. 4, Center for the Study of Institutions, Population, and Environmental Change, Indiana University, Bloomington, IN.

West, G., 2001: Biosphere reserves for developing quality economies: The Fitzgerald Biosphere Reserve, Australia, *Parks (IUCN),* **11(1),** pp. 10–15.

West, P.C. and R.S. Brechin (eds.), 1991: *Resident Peoples and National Parks: Social Dilemmas and Strategies in International Conservation,* University of Arizona Press, Tucson, AZ.

Wilk, R., 1996: *Economies and Cultures: Foundations of Economic Anthropology,* Westview Press, Boulder, CO.

Wilken, G.C., 1987: *Good Farmers: Traditional Agricultural Resource Management in Mexico and Central America,* University of California Press, Berkeley, CA.

Willis, K.J., L. Gillson, T.M. Brncic, 2004: How "virgin" is virgin rainforest? *Science,* **304(5669),** pp. 402–403.

WIPO (World Intellectual Property Organization): Available at http://www.wipo.org.

Wolf, E.R., 1990: *Europe and the People without History,* University of California Press, Berkeley, CA.

Wolmer, W., 2003, Transboundary conservation: The politics of ecological integrity in the Great Limpopo Transfrontier Park, *Journal of Southern African Studies,* **29(1),** pp. 261–278.

Wynne, B., 1992: Uncertainty and environmental learning: reconceiving science and policy in the preventive paradigm, *Global Environmental Change,* **2(2),** pp. 111–127.

Zarin, D., et al., (eds.), 2004: *Working Forests in the American Tropics: Conservation through Sustainable Management?* Columbia University Press, New York, NY.

Ziffer, K., 1992: *The Tagua Initiative: Building the Market for a Rain Forest Product: Sustainable Harvest and Marketing of Rainforest Products,* M. Plotkin and L. Falamore, Island Press, Washington, DC, pp. 274–279.

Part III

Synthesis and Lessons Learned

Chapter 15

Integrated Responses

Coordinating Lead Authors: Katrina Brown, Jens Mackensen, Sergio Rosendo
Lead Authors: Kuperan Viswanathan, Lina Cimarrusti, Karin Fernando, Carla Morsello, Marcia Muchagata, Ida M. Siason, Shekhar Singh, Indah Susilowati
Contributing Authors: Maria Soccoro Manguiat, Godje Bialluch, William F. Perrin
Review Editors: Navroz Dubash, Marc Hershman

BOXES

FIGURES

TABLES

Main Messages

Integrated responses intentionally and actively address ecosystem services and human well-being simultaneously. They are gaining in importance in both developing and industrial countries, albeit with mixed results. Although many integrated responses make ambitious claims about their likely benefits, in practice the results of implementation have been varied in terms of ecological, social, and economic impacts.

Integrated responses are closely allied to the concept and implementation of sustainable development. The interrelationship between ecological, economic, and social systems and the motivation to bring them together in policy and other interventions links the two.

Trade-offs and synergies are central to the development of integrated responses. Integrated responses seek to explicitly manage trade-offs and to identify positive and negative synergies between different objectives and between ecosystem services and human well-being.

Integrated responses occur at international, national, and sub-national levels. Examples at the international level include some multilateral environmental agreements, and international agreements such as the Rio Conventions. Policy integration is a growing feature of many national governments. This is evidenced through national strategies for sustainable development and many other initiatives. Integrated responses are perhaps more usually associated with sub-national and local programs, including multisectoral approaches such as integrated coastal zone management and integrated river basin management.

Many integrated responses occur simultaneously at multiple levels. Integrated responses may be "nested" within different discrete levels, for example, the embedding of Local Agenda 21 within national strategies for sustainable development, developed under the overall framework of Agenda 21. Integrated responses may also be of a multiple scale, and not related to distinct government or administrative levels, but to geographical units such as a watershed or a transboundary marine ecosystem.

Scale issues are critical in integrated responses, and cross-scale responses are necessary. Integrated responses are often deemed successful at a small-scale, or in a particular locality. However, their effectiveness is limited when constraints are encountered at higher levels, such as in legal frameworks and in government institutions. There appear to be limits to scaling up, not only because of these higher-level constraints, but also because of so-called "leakage" problems. These occur when interventions at a local level address only direct, rather than indirect, or underlying drivers of change. Examples might be where integrated conservation and development projects cause increased migration into buffer zones, or where a carbon forestry project shifts deforestation to another location. In these cases, the problems of ecosystem degradation are merely shifted from one location to another. Cross-scale responses may be better able to deal with both the higher-level constraints and leakage problems, and simultaneously tackle the regional and national, as well as, local drivers of change. Examples of successful cross-scale responses include some co-management approaches to natural resource management in fisheries and forestry, and multistakeholder policy processes.

Integration is also about getting a wider range of actors involved in policy processes and about different forms of intervention and action. Successfully integrated responses usually include the active participation of key stakeholders. Increasingly, they are associated with the application of multistakeholder processes and with decentralization, and they may include actors and institutions from the government, civil society, and the private sector.

Implementing integrated responses is resource intensive, but the potential benefits can exceed the costs. Integrated responses are inherently complex, often entailing a combination of actions in a range of domains and at different scales. This can be very costly and requires specialized skills and knowledge. For example, the costs of bringing stakeholders to the negotiation table and of employing participatory methodologies in decision-making are often high. However, if decisions command the broad support of stakeholders, they are more likely to be successfully implemented.

Politics plays an important role in integrated responses at all scales. As integrated responses require bringing together a variety of institutions and individuals with vested interests, and negotiating trade-offs between sectors and actors, collaboration and compromise play a vital role. Successful integrated responses often incorporate conflict resolution mechanisms and deliberative inclusionary processes into their decision-making and management procedures.

Integrated responses do not necessarily bring about more equitable distribution of benefits to stakeholders. It cannot be assumed that integrated responses are more or less likely to deliver their stated objectives than non-integrated responses. In most cases integrated responses meet *some* of their objectives, but not all. Many integrated responses assume that there are synergies between objectives and fail to adequately consider and evaluate trade-offs. This results in unexpected or unanticipated problems and costs, both to ecosystems and society. Generally, the distribution of benefits is not equitable, and this stems from an inadequate consideration of the social, economic, and political dynamics of society. In a number of cases, the failure to appreciate the heterogeneity of communities, property rights, and access to resources, power, and knowledge of different sectors within society are of critical importance and need to be fully understood.

Integrated responses require multiple instruments for their implementation. Integrated responses have a complex nature, because of their multiple objectives and often multiscale characteristics. Therefore, a single instrument is rarely adequate to implement them. Market-based and economic instruments are used with increasing frequency in integrated responses, for example, in river basin management and sustainable forest management, but they usually need to be accompanied by other instruments. These are likely to include redistributive measures and property rights adjustments (for example, when setting up new markets) and institutional development and capacity building. Integrated responses, therefore, require a careful coordination of multiple instruments.

Integrated responses are long-term undertakings not short-term projects. A review of the literature indicates that integrated responses cannot be treated as finite, time-bound projects, nor can they easily be added on to existing policies and interventions. They often require a longer timescale before impacts can be realized or a broad constituency of support can be established. Integrated responses, therefore, should be seen as intrinsic components of long-term changes in environmental governance.

Integrated responses require fundamental shifts in governance institutions in terms of skills, knowledge capacity, and organization. The experience of many integrated responses shows that the conventional organization of governance institutions militates against successful design and implementation of integrated responses, because the institutions are separated along sectoral lines. This is especially true for government organizations, in both industrial and developing countries, and creates barriers in the transmission of knowledge and information and collaboration across the boundaries of organizations. Within organizations, power and prestige is maintained and conferred by defending knowledge rather than sharing it, resulting in "turf defending" behavior, which needs to change in order to better support integrated responses.

Knowledge gaps are persistent and inhibit integrated responses. Knowledge gaps are prevalent in several different dimensions and constitute significant constraints to the more widespread successful implementation of integrated responses. Science itself is defined in disciplinary terms, and this undermines more holistic inclusionary approaches to understanding complex social and ecological systems. Furthermore, information needs to be shared and coordinated across disciplines and organizations.

Assessing integrated responses, assessing trade-offs and providing decision support requires multidisciplinary methods and techniques to capture the multiple impacts and assess multiple goals. Examples of good practice can be found in a number of multidisciplinary techniques such as Multicriteria Analysis. When used collaboratively within a multistakeholder process, these can help in the analysis of trade-offs, reconciliation of conflicts, and development of adaptive management strategies.

15.1 Introduction

Integrated responses are those that intentionally and actively address ecosystem services and human well-being simultaneously. Box 15.1 expands and explains this definition, although integration will be different in each specific context. In a broader sense, terms such as "mainstreaming" or "coordination of" are often used synonymously for integration.

This chapter synthesizes and further analyses the findings presented earlier in this volume by assessing the main features of integrated responses and their effectiveness, using examples from international, national, and sub-national scales.

Attempts to address the impacts of human activities on ecosystems have traditionally been based on sector-by-sector approaches, which ultimately have resulted in fragmented actions and institutions. Such approaches have not achieved optimum results as the linkages and interactions between natural and social systems have been largely ignored, compromised, or not sufficiently strengthened.

Consequently, the necessary adoption of responses that integrate ecological, economic, and social goals to achieve multiple and cross-cutting benefits for present and future generations is increasingly recognized. Integration has become an important concern in thinking about and putting into place sustainable solutions that support human development (Folke et al. 2002; Gunderson and Holling 2002).

The concept of integration is attractive, particularly, as it implies synergies or win-win solutions for complex problems. Trade-offs or adverse consequences, as well as the costs of integration are, however, less regularly considered.

The concept of integration is strongly associated with systems thinking, which has gained much currency in debates on the environment in the past few decades (Berkes and Folke 1998; Berkes 1996; Costanza and Folke 1996). This holistic understanding also underscores "sustainability science," which can be considered as an integrative approach that blends concepts from various disciplines (Kates et al. 2001; Gunderson and Holling 2002; Adger et al. 2003). Integrative responses, therefore, serve as a significant step toward the universal goal of sustainable development.

BOX 15.1

Definition: Integrated Responses

Integrated responses are those which intentionally and actively address ecosystem services and human-well-being simultaneously.

Integrated responses ideally involve all key stakeholders and span different institutional levels horizontally and vertically.

Integrated responses operate on different scales, mainly international, national, and sub-national. These scales are interrelated and determine and influence each other.

Integrated responses, even if primarily focusing on a particular ecosystem service, aim to address their specific impacts holistically, that is, in view of the related impacts on other ecosystem services, and their consequences for human well-being.

15.2 Dimensions of Integrated Responses

Integration can be understood in several ways. First, conceptually, in the way the linkages between social and ecological systems are understood and different kinds of knowledge brought together. Second, institutionally, in the way that concerns for ecosystem services and human well-being are addressed in the formulation of legal frameworks, property rights, and in the organization of the government, civil society, and the private sector. Third, in the way in which policies, decisions, and management interventions are implemented at different scales.

15.2.1 Linkages between Social and Natural Systems

Concerns about integration have grown since sustainability became a prominent concept and paradigm guiding policy. It is widely accepted that sustainable development requires the integration of social, economic, and environmental goals. However, debates surrounding sustainability have also motivated more fundamental changes in worldviews, which call for an integrated perspective on social and ecological systems or society and nature. It is no longer sensible to view environment and society as two separate entities. Scientists have begun to argue that the distinction between them is artificial and arbitrary (Berkes and Folke 1998). Human societies affect ecosystems and environmental conditions and, likewise, environmental conditions and ecosystems both impose constraints on, and provide opportunities for, societal development. Societies co-evolve with nature through dynamic and reflexive processes occurring at a variety of scales, from local to global. An emerging body of theory defines such coevolving systems as linked socioecological systems (Berkes and Folke 1998; Gunderson and Holling 2002; Olsson et al. 2003). Integration, therefore, begins with the recognition that environment and society are closely linked.

Ecosystems are complex, heterogeneous, and evolving. They often extend across administrative boundaries and are nested in larger landscapes. In contrast, the institutions that manage them tend to formulate "blueprint"-style responses that are designed to be applicable in a wider set of circumstances and that are not context specific or sensitive to local conditions. They are also resistant to change, inflexible, and are very defensive of their administrative territory. There is often a fundamental mismatch between institutions and the dynamic characteristics of ecosystems. Inadequate responses to environmental issues often result from this mismatch, which has been designated "problems of fit" between institutions and ecosystems (Young 2002; Folke et al. 1998).

Problems of fit have two sides. One is the misfit that often exists between environmental governance regimes and the ecosystems they are concerned with managing. The other concerns the misfit between the institutions themselves and the economic, social, and political contexts in which they operate (Brown and Rosendo 2000). Misfit in institutions refers to different and often conflicting goals, inability to cooperate, and failure to consider context-specific social, economic, and environmental factors. Integration can be understood as an attempt to address problems of fit.

There is an emerging consensus about the need for a fundamentally different scientific approach to meet sustainability challenges—an approach that is capable of bridging the divide between disciplines that analyze the dynamics of ecosystems and those that analyze economics and social interactions (Scheffer et al. 2002). A concern has begun to emerge within many disciplines themselves regarding the importance of synthesis and integration with other disciplines. These concerns are being reflected in new interdisciplinary and multidisciplinary research initiatives and institutions. Thus it would appear that designing strategies to achieve sustainable environment–society interactions requires integration, in this case, of scientific disciplines (Adger et al. 2003). Sustainability science, for example, has recently emerged as an integrative approach that blends concepts and understandings from across disciplines (Kates et al. 2001).

There is also a call for integration of different kinds of knowledge. Interventions to address the decline of ecosystems have drawn mostly on western scientific knowledge and worldviews. This has resulted in the exclusion of other equally valuable and valid types of knowledge. In the social dimension, integration can be about the combination of different disciplines and knowledge systems.

15.2.2 Integration of Different Actors, Stakeholders, and Institutions

A growing body of evidence suggests that addressing environmental problems or managing natural resources often requires collaboration between different actors (see, for example, the literature on co-management and decentralization). The idea is to involve all relevant stakeholders and make use of their comparative advantages. For example, many local communities, who are dependent on natural resources for livelihoods, are deeply knowledgeable of their environment and demonstrate the capacity to define common rules and sanctions, all of which contribute to making them potentially effective resource managers (Ostrom 1990; Gibson et al. 2000). However, on their own, these actors are unlikely to be able to deal with wider pressures and constraints, such as the ones brought about by globalization. Other actors have the outreach and capability to address such constraints, examples being governments, NGOs, businesses, and donors. Underpinning stakeholder involvement is the notion of participation, which has become a central ingredient to improve the effectiveness, legitimacy, and equity of environmental governance (World Bank 1996). Participation and stakeholder inclusion can be seen as a form of integration between different actors concerned with environmental management.

15.2.3 Horizontal and Vertical Integration

Integration can be referred to as either horizontal or vertical. Horizontal integration implies achieving greater coherence within and among sectors and institutions. It is about promoting linkages within the same level of social organization. A horizontally integrated response is one that links actors, stakeholders, and institutions at the same level or scale. Vertical integration implies linking discrete levels of governance, from local to international, and institutions across different levels.

Vertical integration is important in contexts where hierarchical forms of management dominate, which in the absence of collaboration and coordination tend to lead to fragmented responses unable to deal with complex problems. It is also crucial when promoting collaboration between actors at different scales. A vertically integrated response is one that addresses a given issue at multiple scales.

Responses can also integrate across different sectors and scales (Berkes 2002). A cross-scale response is one that works across different issues and multiple scales simultaneously. Integration, therefore, can also be understood as promoting cross-scale approaches.

15.2.4 Assessing Integrated Responses

Integrated responses represent a diverse set of interventions and approaches. This assessment defines integrated responses as those that intentionally and actively address ecosystem services and human well-being simultaneously; in other words, they have more than one objective—one or more related to ecosystems, and one or more targeted at aspects of human well-being or development. Integrated responses include direct interventions or actions "on the ground," but also national and international policies and programs that in turn support these actions. Table 15.1 characterizes some of these integrated responses. Though it is a selective list, it illustrates the range of integrated responses in terms of their scale, objectives, the key actors involved, and the instruments used. It highlights the cross-scale nature of many integrated responses (for example, Agenda 21) and the participation by government and civil society actors in many of them.

As Chapter 3 explains, a variety of approaches and tools can be used to make an assessment of responses. In this section, integrated responses are assessed to identify lessons learned and the important constraints and enabling factors. Where possible, meta-analyses and reviews have been utilized to inform the assessment, then case studies are employed to exemplify or illustrate key points. The key questions which guided this assessment are as follows:

What drivers of change does the response seek to address? Which are the important actors—government, nongovernmental organizations and civil society, or the private sector? What scale is the response focused on?

Although many integrated responses include actors from different sections of society, they will be initiated by a specific set of actors. For example, although many of the national-level responses (NEAPs, Local Agenda 21) are initiated by government institutions, they will also seek to include actors from different walks of life and, often, at different scales (international-national-local) as in the case of Agenda 21.

What instruments are used to implement integrated responses?

Many different instruments are used to implement or bring about integration. The instruments used depend on the objectives of the responses and which drivers it addresses, the scale at which it occurs, and the actors involved. Integrated responses include economic, legal, and institutional instruments, voluntary approaches and partnerships, projects and mechanisms specific to particular sectors and contexts.

What impacts on ecosystems and human well-being can be identified?

Evidence on the impacts of integrated responses is reviewed, and where possible general trends are identified and lessons drawn. Trade-offs and synergies are considered and explained where they are found.

Table 15.1. Assessment of Integrated Responses

Integrated Response	Drivers Addressed	Ecosystem Services Concerned	Aspects of Human Well-being	Scale	Key Actors	Instruments Used
Multilateral Environmental Agreements, including CBD, CCD, UNFCCC	habitat conversion, trade in wildlife and other species, overharvesting	biodiversity, cultural services, and others	equity, health, and others	global and international	governments, civil society, private sector, multilateral agencies	legal instruments, voluntary instruments (for example, codes of ethical conduct, guidelines, reporting), financial and market-based instruments
Agenda 21	numerous	potentially all	all	local, national, and global	international organizations, governments, civil society, and private sectors	voluntary instruments, primarily plans of action
National Environmental Action Plans	numerous, including land use change and climate change	biodiversity, water quality, food, wood and woodfuel, nutrient cycling, waste	all	national	national governments, donors	legal and voluntary instruments including capacity building
Integrated Conservation and Development Projects	overharvesting, land conversion and fragmentation	biodiversity	economic and income generation	sub-national	civil society, NGOs, private sector, donors, and governments	market-based, legal, and institutional instruments
Sustainable Forest Management	land use change, deforestation	wood and woodfuel, water, biodiversity	incomes, equity, livelihoods, vulnerability	sub-national	government and civil society groups, some private sector	voluntary and market-based instruments, including fair trade, partnerships, labeling, and certification
Integrated River Basin Management	pollution	provision, regulation, cultural	material, health, production, and livelihoods	sub-national, occasionally transnational	government, NGOs, civil society, private sector	legal, regulatory, institutional, and market-based instruments
Integrated Coastal Zone Management	land use change, pollution, overfishing	biodiversity, water quality, food, wood and woodfuel, nutrient cycling	incomes and livelihoods, health, vulnerability	sub-national, regional, transnational, or national	government, civil society, business sector	regulatory, legal, institutional (property rights) instruments

What are the enabling conditions and constraints to integrated responses, what lessons can be drawn from the assessment?

The specific and general issues affecting responses are outlined. However, difficulties exist because many integrated responses have multiple objectives, so it is often difficult to define success. Frequently, integrated responses are successful or effective in meeting some, but not all, of their objectives. They may have unplanned or unforeseen benefits or costs. This may be where explicit recognition of trade-offs between objectives is necessary. Drawing lessons across different types of integrated responses can also be difficult given their very diverse objectives, settings, and scales.

The responses are categorized according to the scale at which they are primarily focused: international, national, and sub-national. At the sub-national level, a number of quite different responses are assessed, but they each link ecosystems and human well-being (for example, integrated coastal zone management, and integrated conservation and development projects). Many integrated responses work across scales and link levels of governance in multilevel and cross-scale responses, so although the text is arranged in three sections, the divisions between the scales or levels are often blurred and many of the responses may fit together as nested initiatives. For example, Agenda 21 is an international integrated response, linked to national strategies for sustainable development and to Local Agenda 21 programs. Integrated conservation and development projects may be nested within a regional biodiversity conservation strategy, which in turn is nested within the settings of the Convention on Biological Diversity. Sustainable forest management approaches may be reflected in national forest strategies, the implementation of sub-national projects, as well as international agreements. The linkages between the scales and levels are therefore considered.

15.3 International Responses

In the 1970s, consensus began to emerge regarding the need for concerted action at the international level to protect the life-sustaining processes of Earth's biosphere. The 1972 United Nations Conference on the Human Environment held in Stockholm was groundbreaking in this respect, pulling together scientific evidence regarding the impacts of human activities on the global environment and establishing the United Nations Environment Programme. Throughout the 1980s, various reports attempted to bring global environmental issues to the attention of governments, the most influential being the 1987 *Our Common Future* report produced by the World Commission on Environment and Development, known for having popularized the concept of sustainable development (WCED 1987). The 1992 United Nations Conference on Environment and Development, held in Rio de Janeiro, also known as the Rio or Earth Summit, resulted in internationally recognized governance structures for global environmental management. Agenda 21 and the Rio Principles were regarded as significant turning points in redirecting national and international policies toward the integration of environmental dimensions into economic and developmental objectives (UNEP 2001, p. 7). It is, however, also argued that the Rio process contributed much to the existing incoherent structure for international environmental governance.

The 2002 World Summit on Sustainable Development provided an opportunity for the international community to review the progress in the implementation of UNCED outcomes. The WSSD Plan of Implementation has been designed as a framework for action to implement the commitments of the Rio Summit and Agenda 21 (IISD 2002). Much of the criticism of WSSD focused on the fact that hardly any binding agreements were endorsed and that some issues were even negotiated backwards (IUCN 2002). The perceived accomplishments of WSSD when compared with the environment-focused Rio Summit included the stronger integration of social and economic needs, and the stronger involvement of non-state actors such as the private sector and NGOs, in concrete, implementation-oriented, public-private or "Type II" partnerships (IISD 2002; Witte et al. 2003).

The following section discusses key international response processes, namely Agenda 21 and its follow-up, international environmental governance, multilateral environmental agreements, and the debates surrounding trade and environment.

15.3.1 Agenda 21

Agenda 21, as a major outcome of 1992 Earth Summit, was designed as a comprehensive strategy to address environment and development challenges. Its main goals encompass poverty alleviation, equitable economic growth, conservation, sound management of natural resources, and stakeholder inclusion. As such, it serves as a conceptual framework for integrated responses aimed at sustainable development on different scales, across different drivers and actors.

As a concept, Agenda 21 does not provide any distinct methodology towards the development and implementation of responses across drivers, scales, or actors. However, through Agenda 21 an attempt was made to address and relate to each other the direct drivers (such as land cover change, air and water pollution, and over harvesting) and indirect drivers of change. The latter are primarily economic, sociopolitical, scientific and technological, and cultural in nature.

Chapter 8 of Agenda 21 outlines a programmatic approach for integrating environment and development at the policy, planning, and management levels. The overall objective is to improve or restructure decision-making processes so that the consideration of socioeconomic and environmental issues is fully integrated and a broader range of public participation is assured. Suggested activities include, inter alia:

- conducting a national review of economic, sectoral, and environmental policies, strategies, and plans to ensure the progressive integration of environmental and developmental issues;
- strengthening institutional structures to allow the full integration of environmental and developmental issues, at all levels of decision-making;
- developing or improving mechanisms to facilitate the involvement of concerned individuals, groups, and organizations in decision-making at all levels; and
- establishing domestically determined procedures to integrate environment and development issues in decision-making. Setting priorities is left to national governments in accordance with their prevailing conditions, needs, national plans, policies, and programs.

At the international level, Agenda 21 calls for, among other things, (1) availability of funding mechanisms, including public and private funding and international development assistance, to support the transition to sustainability; (2) the adoption of ecologically sound technologies, which in turn requires (3) research for technological innovation and technology transfer from industrial to developing countries; (4) improved scientific knowledge of social and ecological systems and their linkages, (5) education, training, and capacity building; and (6) international and national legal instruments and mechanisms. Most of these requirements are constrained by either lack of funding, limited transfer of technologies and scientific knowledge, or limited enforcement of binding legal instruments (UNESC 2002; UNU 2002a).

At the national level, Agenda 21 introduced the concept of national strategies for sustainable development (discussed below) as a means for integrating economic, social, and environmental objectives into overall planning and with the participation of non-state actors. Agenda 21 resulted in a variety of follow-up initiatives on the local level by local authorities and civil groups. Many Local Agenda 21 projects were particularly successful in implementing the often abstract concept of sustainable development into tangible results for local stakeholders. (See Box 15.2.) Key elements of these projects included (1) multisectoral engagement in the planning process, (2) consultation with stakeholders, (3) participatory assessment of local social, economic, and environmental conditions and needs, (4) participatory target setting, and (5) monitoring and reporting procedures, including local indicators (ICLEI 1996).

Over the past decade, public-private partnerships have been an emerging tool of integration between different sectors and across stakeholders on given subjects of sustainable development. Many of the Local Agenda 21 projects are pioneering public-private partnerships for that reason. While such partnerships cannot be accepted as a substitute for government and intergovernmental binding regulation, they are, however, considered by many as an effective complementary approach for implementing Agenda 21 and the WSSD Plan of Implementation (Witte et al. 2003; UN 2003). Almost 300 public-private partnerships were concluded during WSSD, involving different sets of governments, international agencies, local authorities, business, and civil society organizations (UN/DESA 2002). However, the lack of specific guidelines for the coherence, accountability, and evaluation of such partnerships caused much criticism, mainly from nongovernmental organizations (Witte et al. 2003). Since many of these partnerships have just been started, it is too early to assess their integrative function.

15.3.2 International Environmental Governance

The need for international approaches to tackling environmental issues derives from the many cases of transboundary initiatives or disputes over management of resources, such as regional waters or the transport of hazardous waste. International environmental governance as a term bundles all international efforts to set coherent and achievable policies and coordinated actions in response to

BOX 15.2

Local Agenda 21: The Case of the United Kingdom (ICLEI 1996)

The United Kingdom Local Agenda 21 National Campaign was established in 1993 by the country's five local authority associations—the Association of District Councils, the Association of County Councils, the Association of Metro Authorities, the Confederation of Scottish Local Authorities, and the Association of Local Authorities in Northern Ireland. The establishment of the Campaign followed the participation of these associations in the U.K. national delegation to UNCED. Since then, the Campaign has recruited more than 60% of the United Kingdom's local authorities to commit to a Local Agenda 21 planning process. The Campaign has also served as an organizational model for the creation of Local Agenda 21 campaigns across the world.

The first step in the creation of the Campaign was the establishment of a steering committee, made up of senior local elected officials, to govern the Campaign's activities. The steering committee recruited the Local Government Management Board—a technical agency of the local authority associations—to serve as the Campaign secretariat. Recognizing the multisector and partnership-building approach to Local Agenda 21, the voluntary membership of the steering group was soon broadened to include senior representatives of environmental NGOs, the business sector, women's groups, the educational sector, academia, and trade unions.

For their first task, the steering group defined the substantive elements of Local Agenda 21 in the U.K. context, recognizing the need to implement these elements differently according to local circumstances. The first two elements focus on the internal operations of local authorities: managing and improving municipal environmental performance and integrating sustainable development into municipal policies and activities. The other four elements focus on the local community: awareness-raising and education; public consultation and participation; partnership-building; and measuring, monitoring, and reporting on progress towards sustainability.

The Campaign then developed manuals, tools, pilot projects, and seminars to assist local authorities to take action in each of these areas. A *Step-by-Step Guide to Local Agenda 21* and also a variety of guidance documents on specific aspects of Local Agenda 21 planning were published. In addition, a national database and a monitoring scheme for the implementation of Local Agenda 21 have been set up.

The U.K. Local Agenda 21 quickly became part of everyday business for the majority of U.K. local authorities. The high rate of success in such a short period of time is ultimately based on the readiness of local authorities to actively commit themselves to a leadership role in sustainable development.

global environmental change across a wide range of drivers, actors, and scales.

There is much variation in the extent to which existing international responses can be considered integrated. IEG occurs through (1) promotion of international cooperation and coordination, (2) emphasis on sustainable development, (3) involvement of stakeholders and governance structures at multiple levels, (4) the cross-cutting nature of the issues addressed, and (5) linkages among the institutional arrangements involved.

Over the past thirty years, IEG has evolved from a focus on specific issues and regions, to an emphasis on cross-cutting themes and global strategies (UNEP 1999, 2001a). This evolution denotes a greater concern with developing integrated responses at the international level. The strategies and international agreements resulting from the Earth Summit in 1992 (Agenda 21 and the Rio conventions) are explicitly concerned with integration under the broad principle of sustainable development, including economic and social development as well as environmental protection.

IEG is primarily facilitated through the U.N. system. The United Nations Environment Programme holds the principal environmental mandate within the United Nations for environmental policy coordination (see UNEP Malmö Declaration). However, in practical terms UNEP as a U.N. program (rather than a full-fledged U.N. specialized agency or organization) does not have the financial resources to lead the international environmental governance process appropriately (UNEP 2001a; WBGU 2001). UNEP also shares the responsibilities for environmental issues with many other U.N. and Bretton Woods bodies. In fact, the U.N. system appears too fragmented in design, too sectoral in its approach, and too incoherent in its decision-making to address global environmental issues effectively (WRI 2003; WBGU 2001).

Widely accepted underlying concepts in IEG include key principles such as Rio Principle 7 ("Common but differentiated responsibilities") and Rio Principle 15 ("Precautionary approach") (UNCED 1992). Applying these principles, however, has been problematic. With respect to global responsibilities North-South disagreements are common over who is responsible for global changes, who is affected by its consequences, who should act in response, what should be done, and who should pay for it. National interests also continue to undermine efforts at creating and strengthening international environmental regimes for certain thematic areas following the precautionary approach, in particular if national social and economic interests are concerned (WBGU 2001; Figueres and Ivanova 2002).

Judged by the participation of civil society groups in large international meetings, such as WSSD and UNCED, the active involvement of such groups seems strong. Civil society continues to have little to no direct power in IEG, yet the organizations have a strong voice. Few governments include NGO representatives in their national delegations, for example, for meetings on multilateral environmental agreements. Clear and transparent rules for the selection of NGOs attending such international meetings are often lacking, as in the case of the United Nations Convention to Combat Desertification. Some convention secretariats maintain regular contacts with civil society organizations, so-called multilateral consultations or multistakeholder dialogues, and promote consultations with specific groups, such as indigenous peoples or the industry sector. Civil society involvement has made the process of negotiating, revising, and implementing MEAs more inclusive (Gemill and Bamidelle-Izu 2002). However, many technical and open-ended meetings still do not provide for the participation of civil society (Dodds 2001).

Within civil society itself, there are also clear differences in the abilities of different groups to influence outcomes in IEG. International NGOs, for example, tend to be strongly involved in negotiations—perhaps due to their access to information and ability to back their arguments with scientific evidence. The lobbying capacity of the business sector is also powerful, often indirectly through lobbying of national governments. Science and the scientific community play an important role, as both influence the way problems and their solutions are framed. Relevant regional approaches have been made to improve civil society participation in environmental governance through enhancing access to information, public participation in decision-making, and access to justice. (See Box 15.3.)

The role of the judiciary in the promotion of sustainable development and adequate integration of environmental law is frequently underestimated. (See Box 15.4.) Consequently, the need for awareness raising, cross fertilization and capacity building within the judiciary, law enforcement, and prosecution is easily overlooked.

The main mechanisms of IEG include environmental treaties, so-called Multilateral Environmental Agreements (discussed in the next section) as well as "soft law," which includes nonbinding guidelines, norms, and action plans, which are developed for voluntary compliance by national governments only. Examples include Agenda 21, the UNEP-administered Global Programme of Action to address land-based sources of marine pollution, or the guidelines of the World Commission on Dams. (See Box 15.5.)

Governments increasingly express their concern that the current environmental governance structure is no longer appropriate for tackling the international agenda on environment and development. Reforms will not only have to address compliance, enforcement, and liability, but to obey the common but differentiated responsibilities of developing countries and their right to development (UNEP 2001, p. 27; WBGU 2001). In this context, any approach to reform international environmental governance will need to be responsive on the following:

- *Credibility:* reformed institutional structures must command the universal commitment of all States, based on transparency, fairness, and confidence in an independent substantive capacity to advise and adjudicate on environmental issues;
- *Authority:* reform must address the development of an institutional mandate that is not challenged. This should provide the basis for a more effective exercise of authority in coordinating environmental activities with the United Nations;
- *Financing:* adequate financial resources linked to broader development cooperation objectives must be provided;
- *Participation of all actors:* given the importance of the environmental consequences of the actions of major groups, ways must be found to incorporate their views in decision-making.

The discussion on the reform of environmental organizations and their structures generally reflects the need for a stronger environment agency (WBGU 2001). Suggestions include, inter alia, upgrading UNEP to a fully-fledged specialized agency or World Environment Organization, equipped with suitable rules and its own budget funded from assessed contributions from member States. Other options include advanced consolidation between UNDP and UNEP or restructuring the United Nations Economic and Social Council (UNEP 2001). In order to decide on the most effective manner of strengthening international environmental governance, the following questions need to be addressed:

- How can coordination and synergies on environment-related issues among various organizations be improved?
- How can the consistency of environmental standards and agreements be enhanced, particularly in the context of envi-

BOX 15.3

UNECE Convention on Access to Information, Public Participation in Decision-making, and Access to Justice in Environmental Matters: The Aarhus Convention

(http://www.unece.org/env/pp/)

The UNECE Convention on Access to Information, Public Participation in Decision-making, and Access to Justice in Environmental Matters was adopted on June 25,1998, in Aarhus, Denmark, at the Fourth Ministerial Conference in the "Environment for Europe" process.

The Aarhus Convention is a new kind of environmental agreement. It links environmental rights and human rights. It acknowledges that we owe an obligation to future generations and establishes that sustainable development can be achieved only through the involvement of all stakeholders. The agreement links government accountability and environmental protection, and focuses on interactions between the public and public authorities in a democratic context. The Aarhus process is forging a new process for public participation in the negotiation and implementation of international agreements.

The Convention is about government accountability, transparency, and responsiveness. It grants rights to the public, and imposes obligations on treaty parties and public authorities regarding access to information, public participation, and access to justice.

The Convention entered into force October 30, 2001, and progress on ratification has been relatively rapid. The first meeting of the parties took place in Lucca, Italy, October 21–23, 2002. The meeting adopted a number of decisions, thereby establishing two working groups, on genetically modified organisms and pollutant release and transfer registers, respectively. The meeting also agreed on an innovative compliance mechanism, on rules of procedure, and on a number of other issues.

BOX 15.4

Johannesburg Principles on the Role of Law and Sustainable Development

The Johannesburg Principles on the Role of Law and Sustainable Development, which were adopted during WSSD in 2002, highlight the key role of the judiciary in implementing and enforcing applicable international and national laws in the area of environment, sustainable development, and poverty alleviation (UNEP 2002a and 2002b). The following principles were adopted:

- A full commitment to contributing toward the realisation of the goals of sustainable development through the judicial mandate to implement, develop and enforce law, and to uphold the Rule of Law and the democratic process;
- Commitment to realizing the United Nations Millennium Development Goals, which depend upon the implementation of national and international legal regimes that have been established to achieve the goals of sustainable development;
- An urgent need for a concerted and sustained program of work focused on education, training and dissemination of information, including regional and sub-regional judicial colloquia in the field of environmental law; and
- Collaboration among members of the judiciary and others engaged in the judicial process within and across regions, as essential to achieving a significant improvement in compliance with, implementation, development, and enforcement of environmental law.

An 11-point program of work was adopted, which including, among other things, included a call to improve the level of public participation in environmental decision-making, access to justice for the settlement of environmental disputes, the defense and enforcement of environmental rights, and public access to relevant information (UNEP 2002b).

BOX 15.5

World Commission on Dams (World Commission on Dams 2000)

In 1997, conflicting views of the appropriateness of dams worldwide had resulted in a significant stalemate in development planning. The World Commission on Dams was established by an initiative from the World Bank and IUCN. WCD brought together participants from governments, the private sector, international financial institutions, civil society organizations, and affected people. Public consultation and access to the Commission was a key component of the process. The WCD Forum, with 68 members representing a cross-section of interests, views, and institutions, was consulted throughout the Commission's work. Funding of the WCD, likewise, involved all interest groups. A total of 53 public, private, and civil society organizations pledged funds to the WCD process.

One of the key objectives of WCD was to review the development effectiveness of large dams and assess alternatives for water resources and energy development. The other key objective aimed at developing internationally acceptable criteria, guidelines, and standards, where appropriate, for the planning, design, appraisal, construction, operation, monitoring, and decommissioning of dams.

The WCD final report, *Dams and Development: A New Framework for Decision-Making,* was released in November 2000. The report presents a holistic assessment of when, how, and why dams succeed or fail in meeting specific development objectives. It reflects a comprehensive approach to integrating social, environmental, and economic dimensions of development. It further provides the rationale for a fundamental shift in assessment, planning, and implementation of water and energy resource management, including:

- a rights-and-risks approach as a practical and principled basis for identifying all legitimate stakeholders in negotiating development choices and agreements,
- seven strategic priorities and corresponding policy principles for water and energy resources development—gaining public acceptance; comprehensive options assessment; addressing existing dams; sustaining rivers and livelihoods; recognizing entitlements and sharing benefits; ensuring compliance; and sharing rivers for peace, development, and security, and
- criteria and guidelines for good practice related to the strategic priorities, ranging from life-cycle and environmental flow assessments to impoverishment risk analysis and integrity pacts.

ronmental and trade agreement, and how will disputes be dealt with?
- What role would civil society, particularly environmental nongovernmental organizations, have in strengthened governance of the global environment?
- What role could be accorded to the private sector?
- What level of financing could be available, and with what level of predictability and stability, to ensure that mandates are realized?

15.3.3 Multilateral Environmental Agreements

Multilateral environmental agreements constitute a very substantial body of international law and the most concrete component of international environmental governance (WRI 2003). There are over 500 MEAs, of which about 300 have distinct regional focuses. The objectives and priorities of MEAs vary significantly, although the core environmental conventions and agreements are basically divided into five thematic clusters: biodiversity, atmosphere, land, chemicals and hazardous waste, and oceans and regional seas. Table 15.2 gives a selective list of major agreements and treaties under each of these thematic clusters.

As a general trend, MEAs have evolved from focusing on single issues such as wetlands, hazardous wastes, or migratory species, to cross-cutting themes such as loss of biodiversity, land degradation, and climate change. The later generation of MEAs, especially the Rio conventions, explicitly express their relevance to sustainable development across different scales (WRI 2003).

However, any international agreement is only effective if implemented by the signatories on a national scale. This requires translation into efficient strategies at national and sub-national level. It also requires encompassing governments and civil society including NGOs, academia, and the business sector (Domoto 2001). In most cases, countries need to adapt or amend national environmental legislation to meet the objectives of MEAs. Specific programs, institutions, and funds are often created to promote their implementation. Furthermore, multilateral and bilateral aid agencies are increasingly making their loans and assistance conditional upon countries adopting environmental measures, many of which are consistent with the goals of MEAs. From this perspective, MEAs are a potent driving force for national legal and policy changes in relation to the environment (UNESC 2002).

Environmental issues such as desertification, climate change, and loss of biodiversity and forests are multiply interlinked. However, these issues are dealt with separately by different conventions and policy fora, which are negotiated and implemented independently of one another, often by different departments or agencies within national governments. Progress for joint implementation on the level of MEAs (for example, on national reporting requirements) is considered limited (UNEP 2001). Attempts to link the different levels of governance and stakeholders, from local to global, have not been successful in most countries, especially as decentralization and devolution have created additional actors in environmental governance (Dodds 2001; UNEP 2001).

The international regulation of soil resource conservation is an example of thematic and administrative defragmentation that leads to a partial negligence of key environmental issues. Soil is degraded through a range of processes, including desertification and erosion of marginal land, but also by industrial contamination, soil sealing, urban sprawl, and impacts of mining or military activities. On the international level, attention rests with the issue of desertification in arid and semiarid areas as addressed directly in the UNCCD. While the features of soil degradation reach beyond drylands, no further international regulation toward a more sustainable use of soil resources exists. Indirectly, matters related to soil biodiversity or soil contamination are covered by the Convention on Biological Diversity, the Basel Convention, and Convention on Long-range Transboundary Air Pollution.

On the regional level, few legal frameworks exist for the direct protection of soil resources (one example is the Alpine Convention and its distinct Soil Protocol), while other aspects of soil degradation are partly and only indirectly dealt with in other regional frameworks on water, biodiversity, chemicals, or atmosphere (UNEP 2004). Equally, scientific advisory processes, which are essential for adequate assessments, lack a holistic approach on the

Table 15.2. Major Multilateral Environmental Agreements

Agreement	Aim	Year Adopted	Year Ratified
Atmosphere-related			
Kyoto Protocol to the United Nations Framework Convention on Climate Change	achieve quantified emission limitation of greenhouse gases and reduction commitments (http://unfccc.int/resource/conukp.html)	1997	not yet
United Nations Framework Convention on Climate Change (UNFCCC)	stabilize greenhouse gas concentrations in the atmosphere at a level that prevents dangerous anthropogenic interference with the climate system (http://unfccc.int/index.html)	1992	1994
Montreal Protocol on Substances that Deplete the Ozone Layer	phase out ozone-depleting substances (http://www.unep.org/ozone/index.asp)	1987	1989
Vienna Convention for the Protection of the Ozone Layer	protect human health and the environment against adverse effects resulting or likely to result from human activities that modify or are likely to modify the ozone layer (http://www.unep.ch/ozone/Treaties_and_Ratification/index.asp)	1985	1988
Biodiversity-related			
Cartagena Protocol on Biosafety to the Convention on Biological Diversity	ensure the safe transfer, handling, and use of living modified organisms resulting from modern biotechnology that may have adverse effects on biological diversity and human health (http://www.biodiv.org/biosafety/default.aspx)	2001	2003
International Coral Reef Initiative	stop and reverse the global destruction of coral reefs and related ecosystems such as mangroves and seagrasses (http://www.icriforum.org/)	1995	
Agreement on the Conservation of African-Eurasian Migratory Waterbirds	coordinate measures to maintain migratory waterbird species in a favorable conservation status or to restore them to such a status in the agreement area (http://www.wcmc.org.uk/cms/aew_text.htm)	1995	
Lusaka Agreement on Cooperative Enforcement Operations Directed at Illegal Trade in Wild Fauna and Flora	reduce and ultimately eliminate illegal trade in wild fauna and flora in the agreement area and establish a permanent task force for this purpose (http://www.internationalwildlifelaw.org/lusaka.pdf)	1994	
Convention on Biological Diversity	conservation of biological diversity, the sustainable use of its components, and the fair and equitable sharing of the benefits arising out of the utilization of genetic resources (http://www.biodiv.org/)	1992	1994
Agreement on the Conservation of Small Cetaceans of the Baltic and North Seas	achieve and maintain a favorable conservation status for small cetaceans in the agreement area (http://www.ascobans.org/)	1992	
Convention on the Conservation of Migratory Species of Wild Animals	avoid any migratory species becoming endangered and improve their conservation status (http://www.unep-wcmc.org/cms/)	1979	1983
Convention on International Trade in Endangered Species of Wild Fauna and Flora	ensure that international trade in specimens of wild animals and plants does not threaten their survival (http://www.cites.org)	1973	1975
Convention Concerning the Protection of the World Cultural and Natural Heritage	protect the world's cultural and natural diversity of outstanding universal value (http://whc.unesco.org/)	1972	1975
Convention on Wetlands of International Importance, especially as Waterfowl Habitat	conservation and wise use of wetlands and their resources (http://www.ramsar.org/)	1971	1975
Chemicals and Hazardous Wastes Conventions			
Stockholm Convention on Persistent Organic Pollutants	protect human health and the environment from persistent organic pollutants (http://www.pops.int/)	2001	
Basel Protocol on Liability and Compensation for Damage Resulting from Transboundary Movements of Hazardous Wastes and their Disposal	provide for a comprehensive regime for liability as well as adequate and prompt compensation for damage resulting from the transboundary movement of hazardous wastes and other wastes (http://www.basel.int/pub/protocol.html)	1999	
Rotterdam Convention on the Prior Informed Consent Procedure for Certain Hazardous Chemicals and Pesticides in International Trade	protect human health and the environment from potential harm arising from the international trade of certain hazardous chemicals and pesticides (http://www.pic.int/)	1998	
Basel Convention on the Control of Transboundary Movements of Hazardous Wastes and their Disposal	protect human health and the environment against the adverse effects that may result from the generation and management of hazardous wastes (http://www.basel.int/)	1989	1992

(continues)

Table 15.2. Continued

Agreement	Aim	Year Adopted	Year Ratified
Land Conventions			
United Nations Convention to Combat Desertification	combat desertification and mitigate the effects of drought in countries experiencing serious drought or desertification, particularly in Africa (http://www.unccd.int/)	1994	1996
Regional Seas Conventions and Related Agreements			
Global Programme of Action for the Protection of the Marine Environment from Land-based Activities	prevent the degradation of the marine environment from land-based activities by facilitating the duty of States to preserve and protect the marine environment (http://www.gpa.unep.org/)	1995	1995
Convention for the Protection of the Marine Environment and the Coastal Region of the Mediterranean	eliminate pollution of the Mediterranean Sea area and protect and enhance the marine environment in that area (http://www.unepmap.org/)	1976	
Kuwait Regional Convention for Cooperation on the Protection of the Marine Environment from Pollution	prevent, abate, and combat pollution of the marine environment from oil and other harmful or noxious materials in the region shared by Bahrain, Iran, Iraq, Kuwait, Oman, Qatar, Saudi Arabia, and the United Arab Emirates (http://www.unep.ch/seas/)	1978	

features of global soil degradation. In a way, soil has been the victim of its own unassuming character: it is difficult to see it as distinct from other milieus, and its slow, complex process of deterioration has not aroused media or public interest (El-Swaify 2000).

With the international MEAs being highly fragmented (from local to global and within governance structures), most governments face a multitude of different departments or national agencies dealing with various international, environmental, development, and trade agreements (Hisschemoller and Gupta 1999; UNEP 2001). For example, the CBD involves measures relevant to biodiversity protection, trade, and intellectual property rights. This requires different ministries, agencies, and departments to interact. Also, those agencies responsible for negotiating with particular MEAs may differ from the agency or agencies in charge of the MEA implementation (Van Toen 2001).

Most MEAs include no binding compliance and enforcement provisions, but put an emphasis on conflict solving in a nonconfrontational manner. The tendency has been to rely mostly on a "carrot" approach by offering national governments assistance to meet their obligations (Churchill and Warren 1996). The will to develop national or international indicators and parameters for management effectiveness and compliance has been largely lacking. The failure in complying with agreements is often, but not exclusively, related to lack of resources and institutional capacity, particularly in developing countries. Funds that would enable developing countries to prepare for, participate in, and implement international agreements remain scarce. Case studies of the Pacific Islands found that the burden of meeting the reporting and partnership process requirements (conferences, correspondence, and internal reporting and follow-up) for multiple MEAs was unacceptably heavy and, given extreme limitations of skilled resources, was often addressed at the cost of actually implementing the actions required by the MEAs (UNU 2002b).

Development and implementation of MEAs is clearly relevant for sustainable development (OECD 2001a, b). However, effective goals, parameters, and indicators regarding integration between social, economic, and environmental objectives are rare (Ovejero 1999). Achieving synergies across environment and development require mainstreaming environmental agreements into national planning processes, such as sustainable development strategies and poverty reduction plans. The Rio conventions have potential for this kind of integration given the cross-cutting nature of the issues they seek to address, all of which have implications for poverty. Integrating national priorities with international priorities and obligations is, however, likely only when significant benefits are identified for major stakeholders. Integration must be demand-driven and pursued when there is adequate planning and implementing capacity, as well as resources. These are critical issues in developing countries, where governments often feel overburdened with various international commitments and lack the capacity and financial resources to pursue cross-sectoral, integrated strategies.

15.3.3.1 The Ecosystem Approach as a Broad Framework for Integrated Responses

The ecosystem approach has been developed as an overall strategy for integrated environmental management promoting conservation and sustainable use in an equitable way. In essence, the ecosystem approach modifies and broadens the multiple-resource use paradigm into a holistically conceived ecosystem management. It requires one to view landscapes in a comprehensive context of living systems and their complex interdependencies. The approach has importance beyond traditional commodity and amenity considerations. With this view, management practices that optimize the production or use of one or a few natural resources can compromise the balances, values, and functional properties of the whole.

Initial concepts on the ecosystem approach by the *International Joint Commission in the Great Lakes Basin* (see Allen and Hoekstra 1992; Allen et al. 1993; Hartig 1998; Boyle et al. 2001) were taken up and further developed by the CBD, which adopted it as its main vehicle for the holistic implementation of its objectives (see CBD Decision V/6 at the fifth Conference of the Parties). The ecosystem approach focuses on managing environmental resources and human needs across landscapes and is a response to the tendency of managing ecosystems for a single good or services, trying to balance trade-offs to both human well-being and ecosystem services. (See Figure 15.1.) Currently, the ecosystem

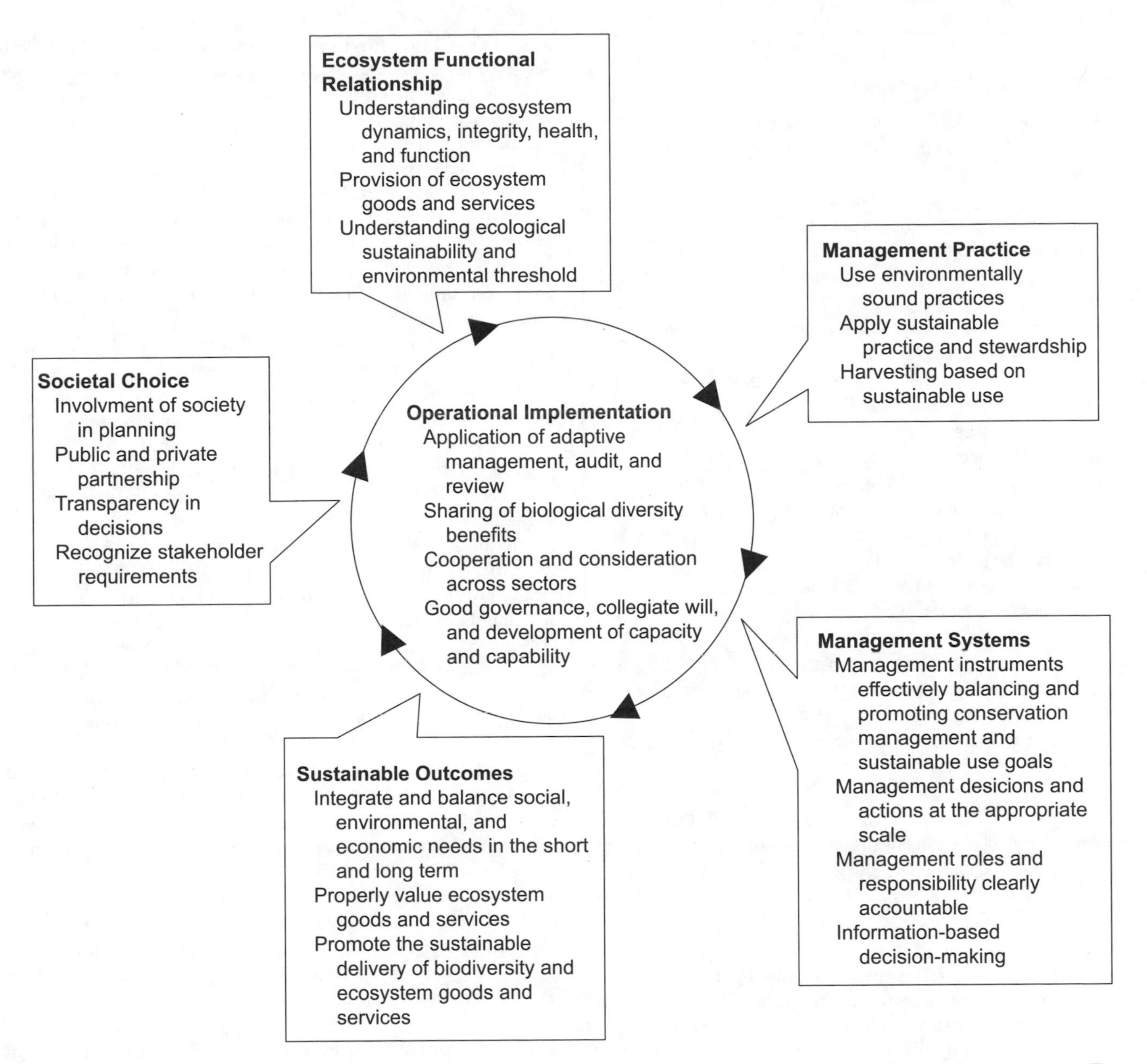

Figure 15.1. The Ecosystem Approach. The ecosystem approach contains the above elements, although it is not limited to them. The operational implementation of the ecosystem approach foresees the implementation of all principles together. Its application should be adapted to specific situations and frame conditions. (CBD Subsidiary Body on Scientific, Technical and Technological Advice 2003; Ecosystem Approach Annex 1)

approach under the CBD constitutes a set of guiding principles and strategies (see Box 15.6) rather than an applicable methodology. Criteria and indicators are, therefore, yet missing for a broader application of the ecosystem approach.

The ecosystem approach is reflected in several sectoral natural resource management concepts, such as sustainable forest management, which was mainly developed independently, but is recognized by the CBD as being largely compatible with the ecosystem approach (CBD COP 7, Decision VII/11 2004; Davey et al. 2003; Wilkie 2003). Other concepts include, for example, bioregional approaches, integrated coastal zone management, and integrated conservation and development projects.

The ecosystem approach has also been applied to health issues, recognizing the inextricable link between humans and their biophysical, social, and economic environments (Lebel 2003; IDRC 2003). Groundwater management can also be based on the ecosystem approach (Neufeld 2000). International institutions have adopted the concept in their strategies, for example, in UNESCO's Man and Biosphere Programme (UNESCO 2000), in FAO's Code of Conduct for Responsible Fisheries (FAO 2003), and UNEP's Strategy on Land Use Management and Soil Conservation (UNEP 2004). It has been pointed out that an institutional application of the ecosystem approach requires adequate organizational changes (Mullins et al. 1999).

The ecosystem approach has been criticized for being too vague and undetermined to be of practical value, while others have highlighted its flexibility (Emerton 2001; Hartje et al. 2003; Hartje 2003; Marconi et al. 2000; Smith and Maltby 2001; Smith et al. 2000a, 2000b; UNEP 2003). Negative consequences of focusing on the overall ecosystem function and processes have also been pointed out—especially the failure to consider specific areas, resources, or species that may need a more targeted approach for

BOX 15.6

Principles of the Ecosystem Approach

- Management objectives are a matter of societal choice.
- Management should be decentralized to the lowest appropriate level.
- Ecosystem managers should consider the effects (actual or potential) of their activities on adjacent and other ecosystems.
- After recognizing potential gains from management, there is a need to understand the ecosystem in an economic context. Any ecosystem management program should: (1) reduce those market distortions that adversely affect biological diversity; (2) align incentives to promote sustainable use.
- A key feature of the ecosystem approach includes conservation of ecosystem structure and functioning.
- Ecosystems must be managed within the limits to their functioning.
- The ecosystem approach should be undertaken at the appropriate scale.
- Recognizing the varying temporal scales and lag effects which characterize ecosystem processes, objectives for ecosystem management should be set for the long-term.
- Management must recognize that change is inevitable.
- The ecosystem approach should seek the appropriate balance between conservation and use of biological diversity.
- The ecosystem approach should consider all forms of relevant information, including scientific and indigenous and local knowledge, innovations, and practices.
- The ecosystem approach should involve all relevant sectors of society and scientific disciplines.

Adapted from the CBD Decision V/6 adopted at the CBD COP 5 held in Nairobi, May 2000. For the full text on the decision, including the rationale underlining each of the principles, see http://www.biodiv.org/decisions/.

their conservation. Another shortcoming highlighted is the failure to include key actors such as the private sector. Despite the emphasis on complex, dynamic ecosystems as critical natural capital assets whose functioning must be conserved, there is also much uncertainty and lack of guidance regarding how to balance conservation and sustainable use in such ecosystems. It is felt that required knowledge (for example, on the state, dynamics, and criticality of ecological and institutional aspects), is yet undeveloped or does not yet exist in many circumstances.

Further, constraints in applying ecosystem approaches include: (1) different time scales in natural ecosystem dynamics and their human utilization; (2) the requirement for broad collaboration between stakeholders, when many places are characterized by a lack of trust and poor communication between stakeholders, significant power inequalities, and divergence of interests; (3) negotiating trade-offs between stakeholders in a fair, equitable, and cost-effective way; and (4) economic under-valuation of ecosystem services. Box 15.7 summarizes a U.S. experience with recurring barriers in implementing the ecosystem approach.

15.3.3.2 Funding Mechanisms for Multilateral Environment Agreements

Funding for international environmental governance comes from bilateral development agencies, multilateral agencies such as U.N. bodies, the World Bank, as well as from domestic national budgets, private foundations, civil society groups and private investors. The Global Environment Facility has been specifically designed to facilitate integrated responses for major environmental challenges. The GEF, which is governed by its own governing council, was set up in the run up to the Rio Summit and formally established in 1994. GEF projects are implemented mainly by three agencies, the World Bank, UNDP, and UNEP. It currently concentrates on six focal areas: biodiversity, ozone, energy, international waters, and since recently, land degradation and persistent organic pollutants. In the first ten years since its inception, GEF funded some 700 projects in 150 countries, involving a budget of \$3 billion, plus an additional \$8 billion in co-financing through other sources (UNEP 2001).

In terms of integration, the most interesting GEF features include the incremental cost approach, coordination with MEA secretariats, and cross-sectoral operational programs. The incremental cost approach of the GEF is designed to support cross-scale projects with proven *global* environmental benefits. Such projects often pilot new integrated approaches. Projects that exclusively serve national development objectives are excluded from the GEF.

As the key funding mechanism for major MEAs—including the CBD, UNFCCC, and since recently, CCD—the GEF is guided by MEA governing bodies with regard to what activities are eligible for GEF projects. For example, UNFCCC member governments establish guidance for GEF spending of the UNFCCC Special Climate Change Fund or the Least Developed Country Fund. While these procedures add to the challenge of inter-institutional coordination, they also ensure activity-focused integration.

The GEF has developed operational programs (OPs), which outline specific approaches in developing and implementing projects within different focal areas. The GEF OP12, for example, on integrated ecosystem management draws on the ecosystem approach (GEF 2000). OP12 aims to create opportunities to address issues that cut across the various GEF focal areas (biodiversity, climate change, international waters, and land degradation) within a common programmatic framework. It facilitates inter-sectoral and participatory approaches to the planning and implementation of natural resource management on an ecosystem scale.

15.3.4 Integration between International Trade and Environmental Governance

The dual trends of global economic integration ("globalization") and the escalation of global environmental problems have magnified the linkage between trade and environment. Trade and environment are fundamentally related and the linkage between the two spheres is considered both complex and critical (Shahin 2002). During GATT negotiations and subsequently in the WTO, the linkage between trade and environment has been receiving growing attention.

Trade-related environmental measures include environmental taxes, environmental subsidies and procurement policies, environmental technical standards, trade bans and quarantines, and environmental labeling (UNDP 2003a, p. 323). Many developing countries perceive these measures skeptically as thinly disguised trade barriers, designed to constrain their development options. While some of these environmental policies and rules result from intergovernmental negotiations and are contained in MEAs, others are imposed unilaterally, usually by industrial countries, raising questions regarding their legitimacy and fairness. In such cases, many developing countries see such environmental impositions as green imperialism or eco-imperialism, which will endanger, in

BOX 15.7

Recurring Barriers in Implementing the Ecosystem Approach: U.S. Findings
(U.S. Interagency Ecosystem Management Task Force 1995)

A U.S. Interagency Ecosystem Management Task Force was mandated to increase understanding of the ecosystem approach and its applicability. Based on case studies, the task force identified barriers to implementing the ecosystem approach, as well as solutions that would improve the effectiveness of the approach. Among its findings:

Federal agency coordination. A coordinated and comprehensive framework is essential to implement the ecosystem approach.

Partnerships with non-federal stakeholders. The ecosystem approach requires active partnerships and collaboration with non-federal parties, particularly state, local, and tribal governments, neighboring landowners, nongovernmental organizations, and universities. Together, they must also project and articulate a desired ecosystem outcome with a shared vision for the future.

Communication between federal agencies and the public. Current outreach activities must be strengthened. Most federal employees who should be interacting with the public are not trained in the skills required to engage the broader public.

Resource allocation and management. Agency coordination in ecosystem efforts can be improved by recognizing the interdependency of agency budgets. The ability of each agency to take an ecosystem approach is affected by its ability to budget for long-term goals, organize around and fund interdisciplinary activities, and quickly modify programs in response to new information.

Knowledge base and the role of science. Existing information and knowledge bases are often inadequate for system-wide ecosystem analyses. The linkage between scientists and managers is essential in establishing or securing a shared vision of desired ecosystem conditions.

Information and data management. Managers must have coherent and complete information from all of the sources in order to make reasonable decisions on their actions that affect the ecosystem.

Flexibility for adaptive management. Adaptive management requires a willingness to undertake prudent experimentation—consistent with sound scientific and economic principles—and to accept occasional failures. This contrasts with the strongly risk-averse nature of most agencies and managers.

the long term, their growth and development (UNDP 2003a, p. 325).

Agenda 21 expressly links the economy (in particular, "an open, non-discriminatory and equitable multilateral trading system") and the environment to human well-being (UNCED 1992, Chapter 1, p. 3; Shahin 2002). The current Doha round of international trade negotiations equally underlines the mutual support of an open and non-discriminatory multilateral trading system with actions to protect the environment and the promotion of sustainable development. (See Box 15.8.)

However, the dispute on how exactly trade and environment issues should be made compatible without undermining either system is largely unsolved. The potential for conflict stems, inter alia, from the fact that environmental regimes allow for extraterritorial measures, which under WTO rules constitute flagrant violations. Furthermore, the WTO concept of nondiscrimination in trade contradicts the basic premise of global environmental regimes, where countries can, and should, discriminate against specific products and processes based on their environmental impacts (UNEP and IISD 2000, Shahin 2002: pp. 48–49). One advantage of the WTO system over MEAs is the availability of clear mechanisms for dispute resolution as compared to often insufficient enforcement and compliance regimes in environmental agreements.

Environment-concerned groups often view WTO rules, trade liberalization and globalization in general, as root causes for accelerated unsustainable consumption and production patterns, which result in resource depletion and environmental degradation (Domoto 2001). These groups are calling for an integration of environmental approaches in WTO rules. Many developing countries on the other side remain deeply suspicious that accommodation of trade-restrictive measures on environmental grounds may further limit their market access in industrial countries. Industrial countries are often thought of as neglecting the needs of developing countries following common, but differentiated responsibilities, including the right to development and basic human needs such as food, health, and education (Sampson and Chambers 2002, pp. 2–7).

The interrelationship between MEAs and the multilateral trading system, which includes the GATT trade agreements as overseen by the WTO, is one of the key issues in the trade/environment discussions (Brack 2002). Around 20 international MEAs incorporate trade measures with partly significant effects on international trade flows (UNEP and IISD 2000, p.16, Ricupero, 2001, p. 35). Three broad sets of reasons to incorporate trade measures into MEAs can be identified (Brack 2002):

- *to provide a means of monitoring and controlling trade in products where the uncontrolled trade would lead to or contribute to environmental damage.* For example, CITES requires export permits for trade in endangered species, and the Basel Convention requests prior notification and applies consent procedures for shipment of hazardous waste, which is subject to the Convention;
- *to provide a means of complying with the MEA's requirements.* The Montreal Protocol, for example, requires parties to control both consumption and production of ozone depleting substances as control measure to achieve its objectives; and
- *to provide a means of enforcing the MEA, by forbidding trade with non-parties or non-complying parties.* For example, the International Convention for the Conservation of Atlantic Tunas bans imports of certain species and products of non-parties or non-complying parties.

The effectiveness of trade measures in MEAs is difficult or "virtually impossible" to assess (Brack 2002). The necessity of MEA trade measures has not yet been challenged before the WTO (see Ricupero 2001, p. 35), although threats of such a challenge have been raised in a number of cases, for example, in the context of CITES. If judged by cases where environmentally based trade measures were imposed unilaterally, such measures were considered in the majority of cases not necessary or justifiable. However, such rulings may not necessarily be applicable in the case of multilateral agreements (Brack 2002, p. 336).

The compatibility of international trade rules and trade measures under MEAs is a long-standing, controversial issue. Major groups such as the European Union seek clarification on the interrelation between the two systems, while most developing countries reject the need for such discussions. It is argued that the perceived conflict between the multilateral trade system and MEAs may most likely be addressed only if an official WTO dispute, challenging trade measures under CITES, the Kyoto Proto-

BOX 15.8

The Doha Trade Round and the Environment

A potential start on greening global trade rules may come from the World Trade Organization's current negotiating round (called the Doha Round) launched in the Fourth WTO Ministerial Meeting in Doha, Qatar, held November 9–14, 2001 (WRI 2003). The Doha Ministerial Declaration focuses in detail on the relation between nondiscriminatory multilateral trading systems, sustainable development, and the protection of the environment:

> We strongly reaffirm our commitment to the objective of sustainable development, as stated in the Preamble to the Marrakesh Agreement. We are convinced that the aims of upholding and safeguarding an open and nondiscriminatory multilateral trading system, and acting for the protection of the environment and the promotion of sustainable development can and must be mutually supportive. We take note of the efforts by members to conduct national environmental assessments of trade policies on a voluntary basis. We recognize that under WTO rules no country should be prevented from taking measures for the protection of human, animal or plant life or health, or of the environment at the levels it considers appropriate, subject to the requirement that they are not applied in a manner which would constitute a means of arbitrary or unjustifiable discrimination between countries where the same conditions prevail, or a disguised restriction on international trade, and are otherwise in accordance with the provisions of the WTO Agreements. We welcome the WTO's continued cooperation with UNEP and other inter-governmental environmental organizations. We encourage efforts to promote cooperation between the WTO and relevant international environmental and developmental organizations, especially in the lead-up to the World Summit on Sustainable Development. (Doha Ministerial Declaration 2001, Par.6).

The Doha Declaration also established a new, if limited, mandate for negotiations on the trade-environment nexus with WTO members agreeing, "with a view to enhancing the mutual supportiveness of trade and environment" (Doha Ministerial Declaration 2001, Par. 31), to negotiate: (1) the relationship between existing WTO rules and specific trade obligations set out in MEAs; (2) procedures for regular information exchange between MEA secretariats and the relevant WTO committees, and the criteria for the granting of observer status; and (3) the reduction or, as appropriate, elimination of tariff and non-tariff barriers to environmental goods and services (Doha Ministerial Declaration 2001, Par. 31).

The outcomes of Doha have not been universally welcome. Questions have been raised whether the new trade talks it launched are really a development round that adequately reflects the needs and aspirations of developing countries. Questions have been raised about the transparency of negotiations, the pressures brought to bear on developing countries, and the potential consequences of the new trade round on local and poor communities worldwide. The failure of the Cancun meeting in September 2003 on the issue of agricultural subsidies versus regulations on investment, competition policy, government procurement, and trade facilitation (Mutume 2003; Halle 2003) may also endanger further progress in addressing crucial trade and environment issues.

col, or the Cartagena Protocol, is launched (Brack 2002, p. 350). However, the dilemma presented by the need to preserve market access opportunities for developing countries while facing the need to maintain the space to implement measures that address legitimate environmental objectives is internationally acknowledged (ICTSD 2003).

15.3.5 Enabling Conditions and Constraints at the International Level

Environmental policy integration at the international level is almost exclusively dependent on the commitment of governments to agree on binding compromises for given issues. The United Nations serves as a facilitator among sovereign states, but has limited capacity to progress beyond the expressed views of governments. The challenge of linking and effectively integrating economic, social, and environmental dimensions of development is well recognized, and appropriate international frameworks exist to enable direct national implementation. However, the international setting for addressing international environmental and development governance is fragmented, incoherent, and unbalanced. Efforts toward larger coordination or even integration are consequently limited and progress is slow.

Major challenges still ahead include a reform of the international environmental governance structure and coherence between international trade and environment mechanisms. Much of the international debate is naturally focused on feasible compromises along economic, cultural or political interests. The concept of sustainable development, while rapidly endorsed globally, still largely lacks viable criteria and indicators for its qualitative and quantitative assessable implementation, particularly on a national level. Also, more efforts are required to demonstrate benefits of a widely integrated international policy framework for concrete national development objectives.

An effective integration of international environmental policy is mainly constrained by the apparent power imbalance between international environment and economic arrangements. Independent of how much international support can be gained for upgrading UNEP to a World Environment Organization, environmental sustainability ought to be more commonly integrated into economic decision-making. Here again, standardized procedures for measuring environmental performance in relation to financial and social performance are a necessary first step, as for example provided by the Global Reporting Initiative (WRI 2003). In this regard the international business sector should be more engaged in integrating environmental aspects.

Public access and participation of all affected stakeholders is essential for fully integrated responses. In this context, public participation is not restricted to access to information and direct participation, but also includes effective representation, judicial redress, and other mechanisms that enable meaningful, democratic environmental governance. On the international level, often enough, such a degree of access and participation is not available for impoverished stakeholder groups. Clearer and more effective rules for a more meaningful access of civil society groups, for example, to United Nations–led negotiations on global environmental governance are needed. However, as civil society groups gain in influence, they will increasingly demand principles of good governance, including transparency and accountability (WBGU 2001; WRI 2003).

15.4 National Responses

Governments are increasingly adopting integrated responses, including policy-making practices, action plans, and strategies.

Many nations have initiated efforts to achieve greater coherence and integration between different policy domains. A typical example is the integration of environmental concerns into other areas of policy. This is important in order to create enabling conditions for responses linking provisions of ecosystem services and human well-being. Sometimes this is referred to as mainstreaming. Policy integration constitutes both an integrated response in itself and a central element or mechanism for other integrated responses. Some national planning initiatives also demonstrate a potential for integration. They adopt a strategic approach, linking longer-term visions to medium-term targets and short-term coordinated actions. There is a vast experience with national strategic planning, but few initiatives can be considered (or enable) integrated responses. Table 15.3 lists the main strategic planning models that have been applied in recent decades. These demonstrate different degrees of integration, from no significant integration to high integration.

Some models, like national development plans, adopt a strategic planning approach that includes fiscal targets, major infrastructure development, and economic reforms, which may contribute to improvements in human well-being. They are, however, narrowly focused on economic concerns and do not constitute integrated responses. Other types of national strategic planning are also sector-driven but demonstrate some (but still limited) potential for integration. National conservation strategies are one example; they aim to provide a comprehensive, cross-sectoral analysis of conservation and resource management issues and propose a greater integration of environmental concerns into development processes. Poverty reduction strategy papers offer another approach that seeks to address a multidimensional problem within an integrated framework. However, environmental issues are not adequately covered within PRSPs, and the main focus is on the economic dimensions of poverty (Booth 2002; Sanchez and Cash 2003).

Other national strategic planning processes demonstrate greater potential for integration. These include approaches concerned with environment that also deal with social and economic issues. National environmental action plans, for example, have been expressly designed to provide a framework for integrating environmental considerations into a nation's overall economic and social policies and programs. They still focus on environment, but give more explicit attention to links and synergies with social and economic dimensions. National strategies for sustainable development are among the few national-level initiatives that demonstrate a high degree of integration or that can be considered truly integrated responses. An NSSD is a strategic approach that aims to integrate the economic, social, and environmental objectives of society, seeking trade-offs where this is not possible, while ensuring that such trade-offs are agreed among many sectors of society.

In both industrial and developing countries, the adoption of integrated (or potentially integrated) responses has been greatly influenced by international processes discussed in the previous section on international responses. UNCED represents a major landmark in this respect by generating an international consensus regarding the need for sustainable development, which requires strategic responses capable of achieving economic, ecological, and social objectives in a balanced and integrated manner. Agenda 21 has become instrumental for the translation of sustainable development from concept to practice, calling for the adoption of actions at multiple levels (global, national, and local). It also calls upon countries to prepare national plans to implement the inter-

Table 15.3. National Strategic Planning Models (Dalal-Clayton et al. 1994)

Approach	Main Objectives	Led by
National Development Plans	to focus on fiscal targets, major infrastructural development, industrial development, etc.	national governments (often the central Ministry of Finance and/or Development Planning)
National Conservation Strategies	to provide a comprehensive, cross-sectoral analysis of conservation and resource management issues in order to integrate environmental concerns into the development process	IUCN and implemented by different sectors
National Environmental Action Plans	to provide a framework for integrating environmental considerations into a nation's overall economic and social development programs	World Bank and undertaken by host-country organizations (usually a coordinating ministry) with technical assistance from the Bank
National Tropical Forestry Action Plans	to produce informed decisions and action programs with explicit national targets on policies and practices, afforestation and forest management, forest conservation and restoration, and integration with other sectors	FAO and implemented by the country concerned
Convention-related National Plans	to define a strategy for the implementation of international conventional at the national level	conventions on climate change, biodiversity, and desertification, in collaboration with national governments
Country Energy Plans	to formulate an energy policy and coordinate energy planning at the national level	World Bank under the Energy Sector Management Assistance Program (ESMAP)
Environmental Strategies, Country Environmental Profiles, and State of the Environment Reports	to present information on conditions and trends; identify and analyze causes, linkages, and constraints; and indicate emerging issues and problems	bilateral aid donors, governments, and NGOs
Green Plans	to promote environmental improvement and resource stewardship, with government-wide objectives and commitments	produced to date by Canada and the Netherlands
Poverty Reduction Strategies	detail plans for sustained reductions in poverty	World Bank and bilateral aid donors in collaboration with national governments

national agreements reached at Rio, including those on biodiversity, climate, and forests. National implementation plans for these agreements in many instances represent integrated responses at the national level.

However, there are substantive differences in the determinants leading countries in the North and in the South to adopt integrated responses. In the South, integrated plans and strategies have often been externally conceived, motivated and promoted by multilateral development banks, development cooperation agencies, U.N.UN organizations, international NGOs, and other external organizations. Some are linked to the release of aid funds, while others have been pursued as planning mechanisms to implement international agreements (Dalal-Clayton et al. 1994). The situation in developed countries has been different, since approaches are related to international processes such as Agenda 21, but have generally been domestically driven, following national government styles and cultures, and sometimes those of businesses and networks of civil society, rather than the dictates of external agencies (Dalal-Clayton et al. 1998). This is manifest not only in the preparation of NSSDs, but also of other approaches such as Green Plans.

Also, there are differences in the particular approach taken in terms of integration. The plans and strategies of industrial countries often focus narrowly on environmental concerns, even when dealing with multidimensional challenges such as sustainable development, while in developing countries greater efforts have been made to address social and economic issues, as well as environmental concerns, in a more integrated manner. Thus there is much scope for countries from the North and South to learn from each other (OECD 2002a).

This section focuses in more detail on policy integration, national environmental action plans, and national sustainable development strategies, which are approaches that fall within the range from medium to high potential for integration. Evidently, these responses not only display different degrees of integration, but also represent substantively different approaches to integration. Therefore, they are not directly comparable, but are used as key examples of national integrative responses from which lessons can be drawn.

15.4.1 Environmental Policy Integration

Policy integration is a central element of efforts to improve the decision-making structures of government in order to reach policies that are economically viable, socially equitable and ecologically sound. In the government sphere, attempts at integration have focused primarily on enabling a more systematic consideration of the environment when decisions are made on economic, trade, fiscal, and other policies, as well as the implications of policies in these sectors for the environment. At the national level, debates on environmental policy integration predominate.

Policy integration refers both to the degree of internal coherence of policy *goals* between different domains of policy-making and the *process* of designing integrated policies (Jacob and Volkery 2003). From a process perspective, a policy is considered integrated when all the potential social, economic, and environmental consequences of that policy are recognized, aggregated into an overall evaluation that defines acceptable trade-offs, and then incorporated into the strategies of all relevant ministries and agencies. An integrated policy from a goals perspective occurs when decision-makers in a given sector recognize the complementary elements, and the repercussions of their decisions on other sectors, and adjust them appropriately so as not to undermine the policies of other sectors. In this sense, policy integration is a pre-requisite or first step toward integrated responses as defined in this chapter. Integration, therefore, can also be understood as *coordination* between policies and the ministries responsible for such policies.

Policy coordination may take different forms ranging from improved communication between departments and ministries to jointly identified policy priorities. Metcalfe (1994) has developed a scale to assess the extent to which national policies are coordinated. The scale was defined to assess EU states, but it is broadly applicable (OECD 1996). An adapted version of this policy coordination scale is shown in Box 15.9. Levels one to four are concerned with the importance of communication across government departments and ministries. Each body retains its autonomy, but joint efforts are made to avoid duplication and to achieve a level of coherence. Levels five to eight focus on deliberate attempts by ministries to work together, up to the point of developing mutually supportive policies and establishing common priorities.

Although coordination is important, integration is about more than improving communication among different bodies and minimizing contradictions between policies. Integrated policy is when there is a deliberate effort to realize mutual benefits between policies. This happens when policies generate benefits not only for the home sector, but also for other sectors. An economic policy that also enables the conservation of ecosystems, for example, would qualify as a strongly integrated policy.

In the specific case of EPI, integration has not only been promoted at the national level, but also within supra-national institutions that impact on national governments. The EU, for example, is highly committed to integrating environmental protection requirements into the definition and implementation of all EU policies and activities (Article 6 of the Treaty of the European Community). This commitment was substantiated in 1998 with the initiation of the Cardiff Process, the goal of which became to

BOX 15.9

Levels of Policy Coordination (adapted from OECD 1996)

Level 1: Independence. Each department retains autonomy within its own policy area irrespective of spillover impacts on cognate departments/areas.

Level 2: Communication. Departments inform one another of activities in their areas via accepted channels of communication.

Level 3: Consultation. Departments consult one another in the process of formulating their own policies to avoid overlaps and inconsistencies.

Level 4: Avoiding divergences in policy. Departments actively seek to ensure that their policies converge.

Level 5: Seeking consensus. Departments move beyond simply hiding differences and avoiding overlaps/spillovers to work together constructively through joint committees and teams.

Level 6: External arbitration. Central bodies are called in by, or are imposed upon, departments to settle irresolvable inter-departmental disputes.

Level 7: Limiting autonomy. Parameters are pre-defined which demarcate what departments can and cannot do in their own policy areas.

Level 8: Establishing and achieving common priorities. The core executive sets down and secures, through coordinated action, the main lines of policy.

ensure that all relevant EU bodies develop their own strategies for integrating environment and sustainable development into their respective policy areas. Table 15.4 reflects the genesis of the environmental policy integration process in the EU.

15.4.1.1 Instruments for Policy Integration

Different mechanisms and instruments are necessary depending on the kind of integration being promoted, whether it is EPI within specific ministries or within the government as a whole (meaning integration between the different ministries and other bodies that form the government). Lafferty (2002) proposes a list of key mechanisms for each of these situations. Box 15.10 gives an idea of what putting EPI in practice entails, when the aim is to promote integration within a given government ministry or sector, while Box 15.11 does the same for situations where the goal is to ensure EPI within the government as a whole.

The EU has developed specific guidelines for implementing EPI into the daily work of Community institutions. These include the introduction of detailed environmental assessments of all key policy initiatives; explicit reflection of environmental re-

Table 15.4. Environmental Policy Integration in the European Union (Cardiff European Council 1998)

Year	Event or Treaty	Description
1972	Stockholm Conference	Develops notion of "eco-development," emphasizing the interdependence between ecological and developmental goals.
1973	First Environmental Action Plan (EAP)	Establishes that effective environmental protection requires the consideration of environmental consequences in all "technical planning and decision-making processes" at national and community level.
1983	Third Environmental Action Plan	"[T]he Community should seek to integrate concern for the environment into the policy and development of certain economic activities as much as possible and thus promote the creation of an overall strategy making environmental policy part of economic and social development. This should result in a greater awareness of the environmental dimension, notably in the fields of agriculture (including forestry and fisheries), energy, industry, transport and tourism."
1986	Single European Act	New Environment Title (Article 130r) introduces the objective of integrating environment into other policies at all levels.
1987	Fourth Environmental Action Plan	Devotes a subsection to the "integration with other Community Policies" and announces that "the Commission will develop internal procedures and practices to ensure that this integration of environmental factors takes place routinely in relation to all other policy areas."
March 1992	Fifth Environmental Action Plan (CEC 1992)	Promotion of integration in five economic sectors: agriculture, energy, industry, transport, and tourism. EPI is to be achieved in a spirit of shared responsibility among all key actors and by making use of economic and communicative instrument and voluntary agreements.
1992–93	Treaty on European Union	Article 2 of the EEC treaty states: "The Community shall have as its task . . . to promote throughout the Community a harmonious and balanced development of economic activities, sustainable and non-inflationary growth respecting the environment." Article 130r(2) includes the requirement that: "Environmental protection requirements must be integrated into the definition and implementation of other Community policies."
1997–99	Amsterdam Treaty	Establishes sustainable development as one of the objectives of the EU and an overarching task of the Community. Article 6 requires that: "environmental considerations should be integrated into other policies in order to deliver sustainable development."
June 1988	Cardiff Summit	"The European Council welcomes the Commission's submission of a draft strategy [for integration of the environment into other EU policies] and commits itself to consider it rapidly in view of the implementation of the new Treaty provisions. It invites the Commission to report to future European Councils on the Community's progress" (Cardiff European Council 1988, paragraph 32). European Council invites all relevant sectoral councils to establish their own strategies for integrating the environment and sustainable development. Transport, energy, and agriculture are asked to start this progress and provided reports to the Vienna Summit.
December 1988	Vienna Summit	Transport, agriculture, and energy councils produce initial reports. Further integration plans are invited from development cooperation, internal market, and industry councils for Helsinki.
June 1999	Cologne Summit	European Council called upon the fisheries, general affairs, and ecofin (finance) councils to report on the EPI process and sustainable development in 2000. Commission submitted its report (of 26 May 1999) on mainstreaming of environmental policy (CEC 1999b).
December 1999	Helsinki Summit	European Council calls on nine Councils of Ministers (energy, transport, agriculture, development cooperation, internal market, industry, general affairs, finance, and fisheries) to complete work on environmental policy integration and to submit comprehensive strategies by June 2001. The Commission submits a report on "environment and integration indicators" (CEC 1999c), a report reviewing the integration process "from Cardiff to Helsinki and beyond" (CEC 1999d), and a "global assessment" of the results of the fifth EAP (CEC 1999a). The European Commission is invited to prepare a long-term policy on sustainable development by June 2001.

BOX 15.10

Mechanisms for Achieving Environmental Policy Integration within Ministries (adapted from Lafferty 2002, p. 17)

Sectoral report. Provides an initial mapping and specification of sectoral activity, which identifies major environmental/ecological impacts associated with key actors and processes, including the governmental unit itself.

Stakeholder forum. Establishes a system of dialogue and consultation with all relevant actors and citizens.

Sectoral strategy. Formulates a sectoral strategy for change, with basic principles, goals, targets, and timetables.

Sectoral action plan. Defines a sectoral action plan, matching prioritized goals and target-related policies with designated responsible actors.

Green budget. Incorporates the action plan into the sectoral budget and allocations.

Monitoring program. Develops a strategy-based system for monitoring impacts, implementation processes, and target results, including specified cycles for monitoring reports and revisions of the sectoral strategy and action plan.

BOX 15.11

Mechanisms for Achieving Environmental Policy Integration in Government (adapted from Lafferty 2002, p. 19)

Constitutional provisions. Sets in place the constitutional mandate for the special status given to environmental/sustainable development rights and goals.

Overarching strategy. Formulates a long-term sustainable development strategy for the domain (including timetables and targets), with a clear political mandate and the backing of the chief executive authority.

Politically responsible executive body. An option is to designate a specific governing body entrusted with the overall coordination, implementation, and supervision of the integration process, including a strategic national forum.

Information agents and programs. Aims to ensure clear communication between sectors to achieve overarching goals.

National action plan. Allocates responsibilities between sectors to achieve overall goals, including clear targets and a calendar for their achievement. Requires EPI to be implemented within sectors.

Programme for assessment, feedback, and revision. Undertakes periodic reporting of progress with respect to targets at both the central and sectoral levels.

Conflict resolution systems. These are aimed at resolving conflicts of interest between environmental and other societal objectives.

quirements in decisions and new proposals; review of existing policies and the preparation of integration strategies in key sectors; and review of current organizational arrangements to ensure policy integration (UNESC 1999). By legally requiring governments to adopt environmental considerations into their practices, EU directives can be an important driving force for EPI at the national level.

Any mechanism or combination of mechanisms for promoting EPI, however, will have limited impacts if the overall integration effort fails to adequately assess and identify the key environmental challenges for the sector, or if it fails to stipulate realistic targets, benchmarks, and measures for objective assessment of outcomes (Lafferty 2002). As with other integrated responses, an integrated policy must be viable, in both economic and political terms, and respond to real needs.

15.4.1.2 Outcomes of Policy Integration

Assessing policy integration involves looking at both process and outcome (Persson 2002). In Western Europe, for example, EPI strategies and mechanisms based on the imposition of environmental norms and criteria on policy sectors have often been unsuccessful. This is because sectors have refused to accept such norms and criteria. Sectoral strategies, which involve encouraging different sectors to develop their own programs and priorities for EPI, have a greater potential of overcoming this problem. However, this requires changing entrenched institutional norms and routines. It may be that EPI needs to be approached as an ongoing, long-term process designed to promote internal capacity and policy learning (Hertin and Berkhout 2001). Some developing countries seem to be even further away from actively tackling the challenge of policy integration. Others, however, have recognized the need to promote integration between environmental and other policies. A case study from Brazil outlined in Box 15.12 illustrates the tortuous route of policy integration. It illustrates the benefits of stakeholder involvement in the design of policies that work for both environment and development, as well as the persisting barriers to greater horizontal and vertical integration between different actors.

To date, policy integration has often consisted of adding environmental considerations to economic policy or vice-versa. Simple "add-ons" do not constitute an integrated response because economic, social, and environmental considerations are not included from the start. This is not surprising since a three dimensional approach is much more difficult to implement, requiring a careful analysis and management of trade-offs, including considerations between short-term pressures and longer-term benefits (OECD 2001a). The main stumbling block is the traditional segmentation of government that impedes integrated policy-making. Government agencies and departments are highly specialized, have accumulated knowledge to govern their particular policy field, have networks with their target groups already in place, and are often unimaginative regarding their goals and instruments (Jacob and Volkery 2003).

15.4.2 National Environmental Action Plans

A national environmental action plan is a national level planning exercise designed to integrate environmental management into the overall development objectives of a country (Lampietti and Subramanian 1995). An NEAP involves identifying the major environmental issues facing a country, defining the underlying causes of environmental degradation, setting priorities, and defining the interventions needed to address such priorities, including policies and legal and institutional reforms. NEAPs are undertaken by governments, but typically with the assistance of development institutions. The World Bank, for example, often required countries to prepare a NEAP as a pre-condition for providing development funding. Thus a NEAP typically outlines the

BOX 15.12

Integration of Agrarian and Environmental Policies in the Brazilian Amazonia

Government policies have been identified as a main driving force for deforestation and unsustainable patterns of agricultural colonization in Brazilian Amazonia (Mahar 1989; Schneider 1995; Binswanger 1991; Mahar and Schneider 1994). Policies have often been directly contradictory and conflictive, implemented by different agencies acting in isolation, and generally poorly integrated. Agrarian policies have supported agricultural development, often favoring large farms and agro-industrial enterprises, while environmental policy has focused on forest conservation. The linkages and connectedness of various economic activities or forms of land use have rarely been taken into account in policy formulation and implementation (Cavalcanti 2000).

But greater integration has been initiated in recent years as a consequence of the emergence of new actors and their reaction to various public policies. This includes rural organizations, farmers' unions, and the landless movement, among others. These actors have been able to influence the land tenure and other policies. In the Marabá region, in Eastern Amazonia, the implementation of land reform projects rose from less than 2 per year in 1987 to 17 in 2000 (INCRA 2001) due to pressure from landless framers. These organizations are very critical of development models embedded in government policies and seek to implement alternative development projects, including initiatives linking development and conservation (financed mainly by international NGOs), which they believe should become references for future public policies.

Increases in deforestation in 2000 were attributed to farmers moving into forested areas and the persistence of government incentives. Interestingly, credit for small farmers was among such incentives. Without appropriate technical support many farmers used the credit to invest in cattle, which required clearing forestland to establish pastureland. The farmers' movements recognized that a more comprehensive strategy was needed to encourage framers to switch to more sustainable patters of land use. With the support of NGOs, they successfully advocated and guaranteed a special credit program called PROAMBIENTE (pro-environment), whereby farmers would receive special credit to implement agroforestry systems, avoiding the use of fire and receiving a grant for the maintenance of environmental services.

However, integration has only happened in one direction. While agrarian measures have assimilated environmental issues, the opposite has not happened. There are still conflicts between the actors involved in the environmental and agrarian arenas. Civil society has played a pro-active role in the policy process related to agrarian policies. Only the state and environmentalist organizations have participated actively in environmental policies, such as the Forestry Code or Law on environmental crimes, which regulates natural resources use and establish penalties for mismanagement of environment goods and services. Although the environmental policy is regarded as innovative, it has been more difficult to implement and is divorced from the sub-national context and the governmental, private sector, and civil society actors at whom it is targeted.

financial and technical assistance the country requires in order to implement proposed actions, in particular the external funding needed. These initiatives were strongly promoted during the 1980s and early 1990s and share similarities with national conservation strategies, also promoted during the same period.

From the perspective of integration, NEAPs have been promoted as an important step toward integrating environmental considerations into national economic and social development strategies (Lampietti and Subramanian 1995). The World Bank, for example, expected NEAPs to evolve into an integral part of the national development policy-making framework (World Bank 2000). Some NEAPs aim explicitly to analyze and address environmental issues within a framework that considers linkages between ecosystems and human well-being. However, the majority remain largely environment oriented. Probably the most important integrative characteristics of NEAPs include their emphasis on involving key stakeholders and analyzing the causes and consequences of environmental degradation from a multidimensional and multidisciplinary angle. They also often propose actions for better compliance with and enforcement of various international agreements countries have committed to, therefore contributing to strengthening the integration between international and national policy frameworks.

15.4.2.1 Instruments Used in NEAPs

The categories of policies and instruments more commonly used in NEAPs include regulatory instruments, market-based instruments, property rights, and ways to increase stakeholder engagement in the NEAP, as well as public awareness regarding environmental issues. There is no agreed set of criteria for selecting instruments. Possible criteria include, for example, cost effectiveness, equity, institutional capacity, financial capacity, and political and social feasibility (Lampietti and Subramanian 1995).

Regarding the institutional structure for implementation, NEAPs are generally implemented by a designated environmental agency that liaises with other sectors within and outside government. Putting NEAPs into practice requires effective coordination mechanisms and a variety of structures and systems to address environmental protection at the national and local level. These structures and systems are often complex, as they require integration to span sectors and levels and varying capacities and resources. Figure 15.2 illustrates the coordination structure developed to implement Sri Lanka's NEAP.

15.4.2.2 Impacts of NEAPs

NEAPs have been successful at raising public awareness of environmental issues, strengthening national environmental management institutions, and introducing environmental policies and innovative pilot projects (OED 1996). These impacts, however, have been mixed and uneven. Public awareness of environmental issues improved particularly in those countries where the preparation process was highly participatory. Lessons from NEAPs in Africa suggest that environmental strategies had a better chance of successful implementation when a range of stakeholders participated in their implementation. The earlier generation of NEAPs, however, did not benefit from a broad participatory approach (World Bank 2001).

The impacts of NEAPs on environmental management capacity have also been mixed. NEAPs tend to rely on legislative reform to improve environmental management, particularly command-and-control instruments, to achieve environmental objectives. Institutional reform has been hampered by two important factors. On the one hand, the traditional bureaucratic institutions are often not flexible enough to accommodate the cross-sector characteristics of environmental problems. On the other hand, when restructuring involves reforming environmental pro-

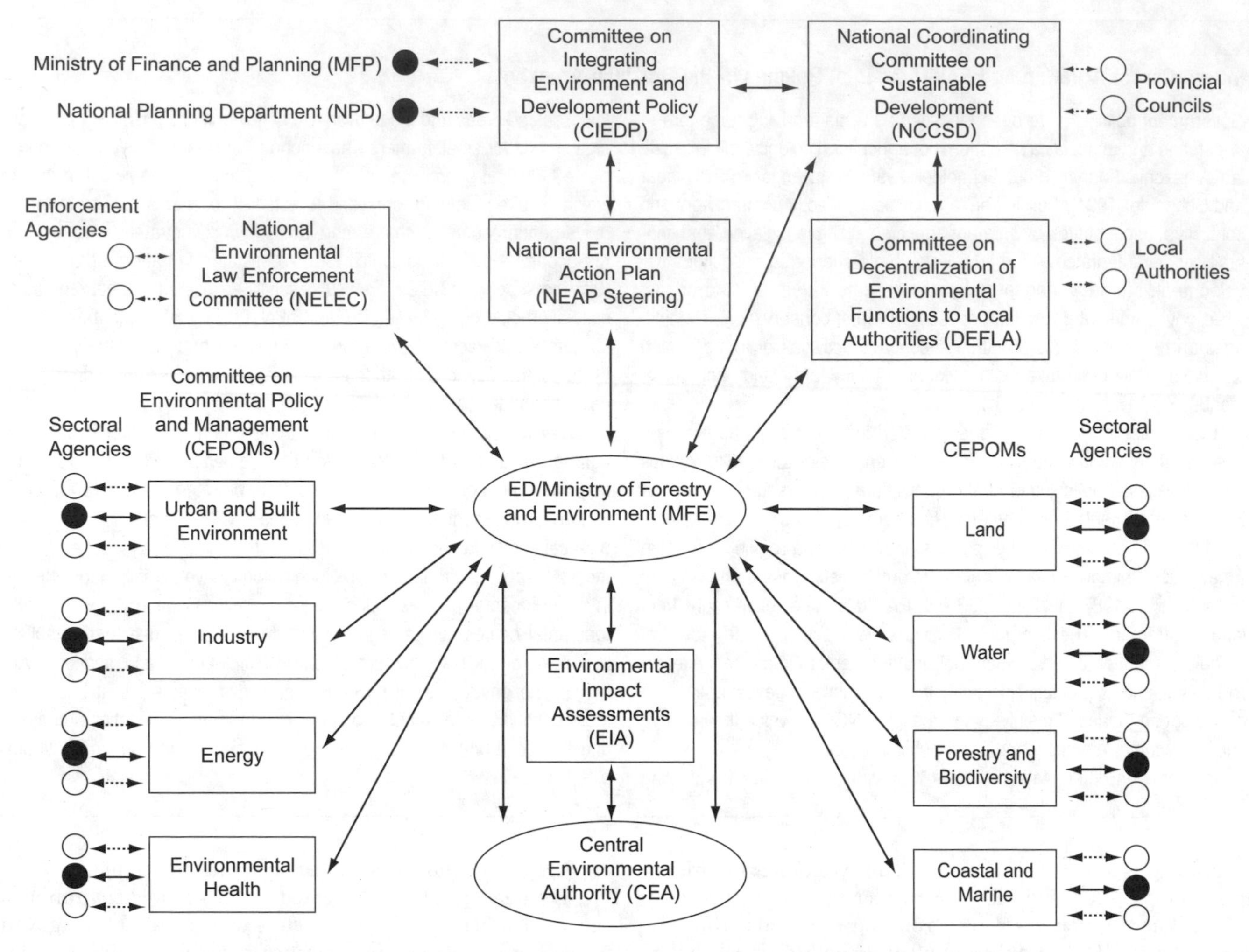

Figure 15.2. Coordination Structure Developed through the National Environmental Action Plan in Sri Lanka

tection, monitoring, and licensing, government officials may be tempted to seek new, or maintain old, rent-seeking opportunities (Lampietti and Subramanian 1995). Implementing a NEAP requires considerable environmental management and technical capacity, which many developing countries lack. Although most NEAPs include assistance to build human and institutional capacity, the NEAP processes tends to rely on international consultancy, thus failing to strengthen national capacity (OED 1996).

NEAPs have generally been supply-driven, without securing local ownership (OED 1997). Many appeared to be ''one-off efforts that ended with a document'' (OED 1996, p. 3). Few succeeded in stimulating the integration of environmental considerations into economic and social policies. More importantly, few NEAPs have resulted in an on-going, self-sustaining strategic environmental planning process at the national level (OED 1996). Many governments initiated them primarily to meet the requirements imposed by donors to provide aid loans (World Bank 2001). Ownership was often undermined by pressure to accelerate the preparation of NEAPs, while lack of systematic follow up further constrained their impact. Generally, NEAPs have not resulted in substantial long-term shifts in management style or in on-going activities.

15.4.3 National Strategies for Sustainable Development and Related Initiatives

National strategies for sustainable development aim to provide a national policy framework to tackle environment and development issues. Agenda 21 emphasized the importance of national strategies and supportive policy instruments to help translate the concept of sustainable development into practice. In the 2002 WSSD, countries reaffirmed their commitment to put into place NSSDs or similar plans, as well as coordinating bodies called National Councils for Sustainable Development bringing together representatives from government, civil society, and the economic or business sector, to facilitate implementation and monitor progress (UNEP/RRCAP 2003). The OECD Development Assistance Committee (2001b, p. 8) defines an NSSD as:

> *A coordinated set of participatory and continuously improving processes of analysis, debate, capacity-strengthening, planning and investment, which seeks to integrate the short and long-term economic, social and environmental objectives of society—through mutually supportive approaches wherever possible—and manages trade-offs where this is not possible.*

NSSDs are more comprehensive than NEAPs, which aimed to promote integration of environmental considerations into social and economic policies, but in most situations only accomplished to strengthen environmental management instruments and institutions. The main features of NSSDs are inclusion, flexibility, and integration. NSSDs aim to reflect the structures, needs, priorities, and resources of each country. It is agreed that an NSSD should comprise a set of mechanisms and processes that together offer a participatory system to develop visions, goals, and targets for sustainable development. It is also accepted that this is not a one-off initiative, but a continuing participatory process, with monitoring, learning, reviewing, and continuous innovation (OECD 2002a). Therefore, an NSSD is not supposed to be a blueprint or a master plan, but a context-specific, flexible, and on-going process.

National councils for sustainable development or similar entities are multistakeholder mechanisms or focal points for the implementation of the Earth Summit agreements (Earth Council 2004). Now active in over 90 countries, an NCSD brings together representatives from the civil society, the private sector, and governments to ensure broad-based participation in planning and policy-making, and in integrating the social, cultural, economic, environmental, and other dimensions of sustainable development into national action plans. Many NCSDs have played roles in the preparation of NSSDs (for example, in providing expertise). NCSDs often monitor the implementation of NSSDs. These institutions typically play a wide range of roles, including offering advice to government, serving as a forum for debate, acting as a vehicle for promoting awareness and information dissemination, and providing a venue for cooperative action as well as commitment to implementation (UNESCAP et al. n.d; OECD 2002a).

15.4.3.1 Instruments and Mechanisms Used in NSSDs

The OECD Development Assistance Committee identified a number of mechanisms that can be used in the preparation of effective NSSDs. Some examples include mechanisms aimed at promoting stakeholder participation; strategic assessment mechanisms to inform planning; prioritization, planning and decision-making mechanisms; mechanisms to mobilize and allocate financial resources; and monitoring and accountability mechanisms (OECD 2001b). The manner in which these mechanisms are used needs to be consistent with a set of basic strategic principles, such as those compiled by OECD (2002a):

- strategies need to be people-centered, and stakeholders need to agree on a long-term vision with a clear time-frame for implementation;
- strategies need to integrate, wherever possible, economic, social, and environmental objectives, and trade-offs need to be negotiated where integration cannot be achieved;
- strategies need to be fully integrated into the budget mechanism to ensure that plans have the financial resources to achieve their objectives;
- priorities need to be based on a comprehensive analysis of the present situation and of forecasted trends and risks, examining links between local, national and global challenges.

These principles also constitute key enabling conditions for successful sustainable development strategies.

15.4.3.2 Impacts of NSSDs

Many strategic initiatives at the national level to promote sustainable development in developing countries are externally driven and envisioned. They are often set as requirements to secure aid loans or comply with international agreements. This contributes to lack of country ownership, which in turn impacts negatively on political commitment to implementation. In some circumstances, dependency on external funds leads to competition between agencies with different agendas rather than collaboration. It also results in integration being only partly a priority, and multisectoral ideas and plans not being mainstreamed effectively. Most sustainable development strategies are little more than wish-lists with no clear objectives, achievable targets, and performance indicators. Although participation is strongly emphasized, as a general rule the participants represent only a narrow selection of all key stakeholders. These initiatives are often not supportive of existing processes, strategies, and capacities, but look to build new ones. In some cases, they are not even tailored adequately to local contexts. The necessary institutional changes to support integration have generally not been sufficient (OECD 2002a).

Although NSSDs and NCSDs have initiated national debates on sustainable development and encouraged a more intrinsic treatment of environmental concerns, in most countries economic imperatives still dominate overall development strategies. Social considerations are also underrepresented in policy-making. Therefore, stronger synergies between these three factors—with more visible impacts—need to be developed and put into practice. Political commitment is vital, and more awareness is needed of the more long-term structural change implications of integration. While institutional development can gain from one-off or project-oriented interventions, more concerted efforts are needed to establish integration as a regular feature into governance institutions.

15.4.4 Enabling Conditions and Constraints at the National Level

A number of lessons can be learned from the experience of national-level integrated responses. First, many of these responses are externally driven, either by donors as a form of "conditionality," or by the demands of compliance with international agreements. This means that they may neither necessarily or strongly reflect the priorities and interests of the country itself, nor of different sectors of its society. In turn, this may undermine the sense of ownership not only by governments, but also by civil society stakeholders. For example, OECD finds that NSSDs are generally not coordinated effectively and lack necessary national ownership (OECD 2002a). In the case of NEAPs, the links to donor assistance also mean they comply with project time scales which militates against a longer- term more strategic perspective.

Second, although many of these responses seek stakeholder participation in their formulation and implementation, a common constraint is in the mechanisms employed to widen participation, both *horizontally* across sectors within the government sector and into the private sector, and *vertically* at the sub-national and local scales. The form of participation, where any exists, is generally passive (consultative for example), not active.

Third, to be effective, the national integrated responses require political commitment at a high level. In many cases they are viewed as ineffective because they fail to be seen as important by key government departments, such as economic planning. They are not seen to contribute to the strategic development goals or economic performance of a nation.

Fourth, the compartmentalization of the government is a constraint to national integrated responses. Finally, the data and information needs, for successful national integrated responses, are demanding. Often, the capacity to collect and synthesize the type of information required does not exist, and furthermore, the process of monitoring is costly.

A review of NSSDs and related initiatives in the Asia-Pacific region illustrates these issues (UNESCAP et al. n.d.). The majority of the NSSD bodies are of an advisory nature, which does not grant them authority to enforce recommendations. Insufficient human resources and skills as well as financial resources also limit their ability to engage more effectively and productively in policy processes. Inadequate involvement of local-level actors such as local governments and NGOs has led, in some cases, to an under-representation of local concerns. Ironically, while environment has been put at the center of sustainable development debates, when it comes to priority setting, it is economic concerns that have usually been prioritized. In most countries in the Asian and Pacific region, environmental objectives are still viewed as being distinct and largely independent from economic development objectives. Most sustainable development initiatives have not managed to establish links with economic planning agencies, where power to define national development strategies is usually concentrated.

These factors are confirmed by OECD (2002b), which suggests five enabling conditions to support policy integration at the national level. These include a policy framework supportive of sustainable development; specific mechanisms to steer integration; clear commitment and leadership; effective stakeholder involvement; and effective knowledge management. The absence of one or more of these enabling conditions can frequently become a constraint. In particular, policy integration often requires modifying the government architecture so it becomes less compartmentalized. This also necessitates instituting measures for power sharing, revising roles and responsibilities, and building supportive work force dynamics. Achieving these changes may involve creating or strengthening specific mechanisms to steer and mainstream integration.

Experience shows that policy integration often does not happen spontaneously. Some have argued that, if they are to be truly effective, imperatives for integration, apart from those outlined above, should be legally binding (Klein 2001). Policy integration, however, takes place in a variety of legal and institutional settings, not all of which will accept such impositions. It is, therefore, necessary to consider, in each case, whether making integration compulsory is an enabling condition or a constraint. In a jurisdiction, subsidiarity may well be seen as mechanism of integration. Subsidiarity recognizes that action will occur at different levels of jurisdiction, depending on the nature of the issues. International policies, for example, should be adopted only when they are more effective than policy action by individual countries, or by jurisdictions within countries. Environmental policies in different jurisdictions can reflect differences in environmental conditions or development priorities, leading to variations in environmental standards within countries or among groups of countries. Harmonization of environmental standards, procedural requirements, or laws, supplemented where feasible by negotiated minimum process standards, can play an important role by ensuring that these essential differences respect a common framework.

Of particular importance are mechanisms that promote communication, collaboration, and coordination among various ministries and levels of government. Measures to assist different sectors to design their own integrated policies also enable integration, and may be more effective. Continuous, high-level political commitment and leadership is vital. Without it, policy integration becomes more formal than substantive, and environmental concerns will continue to be routinely overridden by developmental and other interests. Clear commitment and leadership at a high level also serves to influence agenda and priority-setting further down in the government hierarchy. Efficient and effective flows of information between the scientific community and decision-makers are important to design policies that integrate social, economic, and environmental considerations. Improved knowledge and public understanding of science may help to reduce dissent between the different constituencies involved, and to design policies that command the support of a wider range of stakeholders. It is also important that other types of knowledge, such as the knowledge held by traditional societies, is integrated into the policy process. Since conclusive scientific evidence related to sustainable development either is not available or is incomplete, ensuring that data gaps are filled and that information is updated has to be an ongoing part of the process. Policy that supports research and development also becomes a forward thinking strategy to aid policy-making and assessing results in the future.

Enacting policy integration has strong potential for conflicts of interest due to the demanding and multidimensional goals. At the same time, efforts at environmental policy integration can provide a crucial platform and arena for attempts to transcend such conflicts (Lafferty 2002). It certainly cannot be assumed that finding win-win solutions can always be realized and that any conflicts between different goals can be resolved to the satisfaction of all relevant interests. Trade-offs between environmental and other societal objectives may be unavoidable, and means of prioritization are essential. While other objectives may at times be deemed more important than environmental concerns, there must be means to decide policy priorities democratically.

The national integrated responses themselves may be a good starting point for cross-departmental linkages in governments. They may initiate a consultation process and the development of skills and capacity for further integrated responses. For example, the preparation of NEAPs by multidisciplinary teams comprising specialists from a range of areas is considered an important enabling condition (World Bank 2000). When essential data are lacking, the environmental action plan process may involve developing information systems and building data analysis capacity at the national level—and this should be part of the continuous process to identify gaps, strengthen, assess and then put into place the next step. In this way, these responses may be developing the capacity and know-how, which will spill over into other government activities. For example, frameworks such as Comprehensive Development Strategies and poverty reduction strategies claim to be built on the experiences of the NEAPs.

15.5 Sub-national and Multiscale Integrated Responses

Many integrated responses are implemented at the sub-national level. This is often where the impacts and outcomes of integration at different scales can be observed in terms of changes to human well-being and ecosystems. Frequently the sub-national level is where integrated responses are operationalized. As noted, many integrated responses occur at multiple scales. Some, such as integrated coastal management and watershed management work explicitly across scales and ecosystems. Others, such as Agenda 21, are at multiple or nested scales. These approaches may result from the implementation of international and national level initiatives at sub-national and local scales. This section reviews four widely adopted integrated responses: sustainable forest management, integrated conservation and development projects, integrated coastal zone management, and integrated watershed and river basin approaches.

15.5.1 Sustainable Forest Management

Sustainable forest management constitutes a set of guiding, though not legally binding forest principles, which on the international level emerged from UNCED in 1992. These principles provide a broad framework for integrated responses. They aim to ensure that forest ecosystem goods and services meet present-day needs, while securing their continued availability for, and contribution to, long-term development and human well-being (FAO 2001, 2003). There is considerable disagreement in terms of the general categories used in assessing sustainability, particularly, with regard to biodiversity, whereas agreement on silvicultural guidelines is relatively more common. SFM is considered an integrated response, as most definitions include a reference to different ecosystem services and human well-being (IIED 1996).

SFM, it is argued, allows managed forests to provide income, as well as forest resources and ecosystem services that society increasingly demands. It also intends to counteract damage to biodiversity, soil, and hydrological processes, and to mitigate global climate change through carbon sequestration (Putz et al. 2000). SFM has been particularly promoted in tropical forest regions as the standard approach to achieving biodiversity conservation outside of protected areas (Rice et al. 2001). Social forestry could be viewed as a form of SFM, in that it aims to involve local people in forestry activities. The term social forestry is used interchangeably with community forestry, farm forestry, and forestry for local development (Dankelmen and Davidson 1989; Gregersen et al. 1989).

15.5.1.1 Drivers and Scale Addressed by Sustainable Forest Management

SFM attempts to address both direct and indirect drivers of change in forest ecosystems. Direct drivers include harvesting of timber and non-wood forest products, in addition to land use changes, particularly conversion to agriculture (see Hartshorn 1998). Indirect drivers include mainly trade and market influences that lead, for instance, to timber mining usually, but by no means exclusively, in tropical regions. Other indirect drivers include the simultaneously increasing demand for ecosystem services provided by forests, including water resource protection, climate regulation, biodiversity conservation, or recreation.

SFM is essentially a sub-national scale instrument that is based on tools applied at the local level. The local level may include the project, concession, landscape, or watershed scales. SFM may be applied across scales, and it may also be a guiding principle informing and shaping national forest policies. (See Chapter 8 for examples.)

15.5.1.2 Actors Involved in Sustainable Forest Management

SFM involves a wide range of possible actors, including governments, local communities, NGOs, and private businesses. In many countries, governments are key actors, who not only control or own large areas of forest but also determine legal and economic instruments. Forest communities and forest dwellers, including small and large-scale farmers, landless families, artisans, traders, and small-scale entrepreneurs, often depend on a wide range of forests resources for their livelihoods. Other important stakeholders include forest product and service consumers in urban or peri-urban communities (MacQueen 2002). Environmental and development NGOs are usually involved in fostering SFM adoption at the local level. Finally, the private sector is involved in the commercialization of forest products and the transition to sustainable practices. SFM may be initiated by any of these actors and may involve different combinations of actors from each of these groups.

Studies on SFM implicitly agree that the integration of different stakeholders leads to processes and outcomes that are more efficient and effective (Mayers and Bass 1999; Colfer and Byron 2001; Wollenberg et al. 2001). However, until recently, few studies examined the social and economic costs of collaborative decision-making, conflict resolution, consensus building, participation, and other processes (Cooke and Kothari 2001).

15.5.1.3 Instruments and Mechanisms for SFM

SFM constitutes a large number of different instruments and mechanisms aiming at the scientific, technical, legal and administrative, economic, and social components of sustainable management of forests. (See also Chapter 8.)

Criteria and indicators for evaluating progress in implementing SFM have been developed for all key forest ecosystems in different regions. Nine such initiatives currently exist globally (Wijewardana 1998; FAO 2001). Criteria and their indicators, as well as other technical guidelines, generally address all aspects of harvesting, as well as the various pre and post-harvesting stages. They also address socioeconomic benefits and needs, as well as legal, policy, and institutional frameworks. They are designed to increase the growth of marketable timber or non-wood forest products, with efforts aimed at lowering the damage to the forest stand and critical ecosystem processes; to a somewhat lesser degree, they also include social development aspects (Putz et al. 2000; Rice et al. 2001). The degree of implementation of criteria and indicators at the national level varies considerably. In many cases, action is limited by lack of trained personnel or institutional capacity (FAO 2001).

Certification is a combined economic, legal, and behavioral instrument (See also Chapter 8). Although increasing in recent years, certified timber accounts for only a very small proportion of total tropical timber trade (Rice et al. 2001). It is likely that there is an unfulfilled demand for certified tropical timber in large consumer countries (for example, Barbier et al. 1994; Sobral et al. 2002). Key certification schemes have been noted for their expressive incorporation of community relations, labor rights, health and safety concerns, and social benefits of forest operations (for example, Forest Stewardship Council, Principle 4, Nardelli 2001). Diversifying markets for lesser-known species has been suggested as a mean to enhance the productivity per unit of forest (Buschbacher 1990). However, simply creating markets for a larger number of species may lead to more species being overexploited, and evidence from Latin America shows that multispecies exploitation cannot in itself guarantee the adoption of SFM practices (Rice et al. 2001).

Trade controls such as log export bans have been promoted as an incentive for improved forest management through larger investments in local processing (Bomsel et al. 1996). Export bans have been implemented in various tropical forest countries (Barbier et al. 1994; Rice et al. 2001). Although, these export restrictions have stimulated growth and employment in the processing timber industries, they have also led to excessive processing capacities and consequently increased logging (Rice et al. 2001, p. 20). However, temporary banning or controlled trade (for example, for endangered species) may prove an important tool (Barbier et al. 1994).

Forest concessions, as the dominant means of allocating harvesting rights in many countries, have usually failed in terms of protection and enhancement of other ecosystem services and human well-being. Security for land and resource tenure is thought to provide incentives for investment in long-term management (Buschbacher 1990). While secure tenure may be necessary to promote investments in long-term management, it is unlikely that this is sufficient for SFM (Boscolo and Vincent 2000; Rice et al. 2001). However, there is also evidence that frequent renewal of short-term concessions based on demonstrated forest management performance may provide stronger incentives for SFM (Gray 2002). Alternative approaches such as competitive allocation proved to be more effective in reducing corruption and in promoting productive and efficient management and revenue generation (Gray 2002; Landell-Mills and Ford 1999).

Improving logging and milling efficiency is seen as a way of increasing profits, while enhancing incentives for long-term management and reduced logging damage (Johnson and Cabarle

1993; Gerwing et al. 1996; Holmes et al. 2000). It is, however, also pointed out that greater efficiency may be associated with more rather than less forest destruction (Rice et al. 2001).

15.5.1.4 Impacts of Sustainable Forest Management

SFM has the potential to positively impact on a variety of ecosystem services, simultaneously. Uncertainties remain, however, in relation to the effects of exploitation on biodiversity and other key services (Bawa and Seidler 1998). Adequate understanding of ecologically sustainable and economically and socially viable forest management in complex tropical forest systems is often lacking (Boot and Gullison 1995), although much progress has been made in given cases, including, for example, the Deramakot Project in Sabah (GTZ 1994) and the Precious Woods Holding in the Amazon and Costa Rica (Freris and Laschefski 2001). While most countries have already set up SFM procedures, they usually focus on the management process rather than its impact in terms of sustainability (Gray 2002). Although governments and development agencies have devoted years of effort and hundreds of millions of dollars in promoting SFM, this form of management has failed in its attempt to curb deforestation (Rice et al. 2001). In contrast, the promotion of SFM has been perceived by many countries as a form of promoting forest exploitation instead of halting deforestation (Winterbottom 1990).

Human well-being and poverty reduction are referred to in different ways and at different scales in many SFM responses (Poschen 2000). (See also Chapter 8.) However, there are still significant uncertainties about the role of SFM on increasing the well-being of local forest dwellers and other key actors. Angelsen and Wunder (2003) argue that the role of sustainable forest management in poverty alleviation is largely overestimated. Mechanisms that explicitly refer to aspects of human well-being, such as certification, remain expensive and are rarely adopted by communities because of market risks and costs (Mallet 2000). Again, although SFM *in theory* enhances human well-being, in practice the evidence from its implementation is not conclusive.

15.5.1.5 Trade-offs between Ecosystem Services and Human Well-being in SFM

The literature on forest environmental services often assumes that services are complementary and, therefore, increased investment in one service will have positive spin-offs for others. In practice, however, relationships between ecosystem services are not well enough understood. Relationships are often dynamic and, therefore, switch between positive, negative, and neutral impacts at different levels of service supply vary, and are usually site specific. For instance, fast-growing plantations may have a detrimental impact on water supplies, while being valuable in terms of carbon sequestration. Biodiversity conservation may reduce income generation from timber and non-wood product exploitation. The most diverse forests are not necessarily optimal for landscape beauty or watershed protection (Landen-Mills and Porras 2001).

Trade-offs among ecosystem services cause problems, for instance, when markets for ecosystem services are developed. Although the commercialization of ecosystem services maximizes the returns to forest investment, merging and marketing forest services requires a clear understanding of their internal relationships. Also, allocative efficiency gains are restricted, since individual services do not have their own prices to send out signals about their relative value. Emerging trade-offs through bundling of services are likely to vary across forest types and actors. The lack of knowledge of technical relationships between services currently constrains the development of efficient markets for bundled forest services (Landen-Mills and Porras 2001).

Trade-off studies on forest services are scarce. Brown and Corbera (2003) report on research investigating the sustainable development dimensions of agroforestry carbon sequestration projects in Mexico. Although these projects have been widely promoted as a means of fostering local development and securing global benefits through carbon sequestration, a number of trade-offs were found to exist. (See Table 15.5.) The actual realization of "win-win" outcomes will depend on a careful crafting of project measures and on institutional acting as based on equity and efficiency at multiple scales ranging from local resource management and decision-making to national government frameworks.

15.5.2 Integrated Conservation and Development Projects

Integrated conservation and development projects, also termed community-based conservation projects, aim to intentionally and actively link biodiversity conservation and development of local communities (Wells et al. 1992; Hughes and Flintan 2001). ICDPs have become very popular over the last decade and absorb a major proportion of international funds available for biodiversity conservation (Wells et al. 1999; Alpert 1996).

15.5.2.1 Drivers and Scale Addressed by ICDPs

ICDPs emerged in response to the growing recognition that conventional protection approaches, which tend to ignore local needs and calls for equity, are largely inefficient or even counter-

Table 15.5. Trade-offs in Forest Carbon Sequestration Projects in Mexico (adapted from Brown and Corbera 2003)

Nature of the Trade-off	Manifestation in Mexico Forest Carbon Projects
Trade-offs between ecosystem services	maximizing carbon sequestration may jeopardize biodiversity and other ecosystem services
Trade-offs between ecosystem services and human well-being	enhancing forest carbon sequestration may make livelihoods more risky because of dependency on external finance and policy and actors
	enhancing carbon sequestration will have opportunity costs (for example, from harvesting timber and non-wood products)
Trade-offs between stakeholders	different stakeholders have different priorities in projects, include risk taking versus risk minimization; time preferences for income and investment streams
	critical differences in access to markets and decision-making are between rich and poor and between men and women
	richer farmers—who have more land and more secure property rights—are more likely than poorer farmers to capture benefits
	potentially a move from communal resource management to private property regimes, with results for equity inequities may be exacerbated by carbon projects
Trade-offs between different aspects of human well-being	maximizing income from carbon sequestration may not be compatible with diverse livelihood strategies

productive as pressure from growing rural populations is threatening the viability and integrity of protected areas (Newmark and Hough 2000; Worah 2000).

Direct drivers targeted by ICDPs include mainly unsustainable natural resource harvesting and land use changes. The scale of operation is sub-national or local and focuses primarily, though not exclusively, on areas adjacent to protected areas, so-called buffer zones (Wells et al. 1992). In some cases, ICDPs may cross national boundaries.

15.5.2.2 Actors Involved in ICDPs

ICDPs integrate local communities in buffer zones and protected areas, alongside NGOs and government organizations, and, in some instances, private enterprises. NGOs are the most common promoters and implementers of ICDPs. Buffer zones and protected areas are in many cases regulated by specific laws and administered by government agencies. Therefore, governments have at least a partial role in the implementation or regulation of ICDP activities. Private enterprises are commonly involved in commercial transactions concerned with natural resource products or services.

15.5.2.3 Instruments and Mechanisms Employed in ICDPs

ICDPs employ a set of diverse instruments ranging from the economic and institutional to the behavioral. Economic instruments are particularly important as ICDPs seek to integrate conservation and development through the provision of income-orientated incentives for local populations. Common activities fostered by ICDP projects to provide income include handicrafts, beekeeping, agroforestry, ecotourism, harvesting, and marketing of nonwood forest products, as well as sharing revenues of park entrance fees and employing local people as park rangers or wildlife guides (Wells 1995).

ICDP activities can also involve changing particular behavior, environmental education, or improving community infrastructure. The provision of schools, health services, and sanitation has been widely used as an incentive for people to cooperate with project conservation objectives. Furthermore, a number of initiatives promote alternative sources of food, fuel, and building materials in order to reduce pressure on natural resources. Finally, ICDPs are implemented or supported through institutional mechanisms. Projects often apply a combination of economic and non-economic incentives to secure the cooperation of local communities and to provide alternatives to unsustainable activities.

15.5.2.4 Impacts of ICDPs

Most ICDPs are established to alleviate pressure from local communities on protected areas and, therefore, to increase the provision of ecosystem services, among which biodiversity is the main concern. Despite the popularity of ICDPs, a number of assessments conducted over the years have concluded that most ICDPs have not achieved their objectives. Early assessments suggested that since projects were still not fully implemented, success was limited (Hannah 1992; Wells et al. 1992; Kiss 1990). Nevertheless, more recent assessments in several regions of the world continue to provide a largely negative view of the success of ICDPs (Wells et al. 1999; Newmark and Hough 2000). It appears that ICDPs have rapidly advanced from an untested idea to "best practice" in conservation, without their effectiveness ever being demonstrated and substantiated by practical results (Wells et al. 1999).

The efficiency of ICDPs to enhance ecosystem services is limited by a number of factors. While local people often pose a number of threats to biodiversity, large-scale government and business investments are generally much more serious in their impacts. There is, therefore, a fundamental mismatch between the causes of biodiversity loss and the focus of ICDPs (Kiss 1999). In Indonesia, for instance, the impact of local communities on ecosystem degradation ranks well behind road construction, mining, logging concessions, and sponsored immigration, when measured by their threats to protected areas (Wells et al. 1999). ICDPs focusing on local communities, therefore, fail to succeed in conserving biodiversity because they are often aimed at the wrong target.

Addressing the problem of biodiversity loss involves slowing, halting, and reversing land use change. This requires, consequently, a change in the behavior of a large number of people dispersed over large and ecologically significant areas over a long period of time. By contrast, ICDPs are characterized by being very limited in time and by the number of beneficiaries. Also, in terms of scale, ecosystems conservation must be realized most usually at a landscape scale, while development initiatives are often context specific and small scale (Ferraro 2001).

Another factor is the largely unproven assumption that development in areas adjacent to protected areas will necessarily lead to conservation within the protected areas. The balance of evidence suggests that efforts to establish alternative sources of income from ecosystem services can only work in combination with the adoption of strict and effective measures of resource protection (Kiss 1999). Otherwise, communities are tempted to add rather than substitute income sources and resource extraction. Often, establishing ICDPs is likely to result in "unconstructive dynamics and incentive structures" (Kiss, 1999, p. 3). (See Box 15.13.)

Many ICDPs fail because the economic incentives presented to communities are insufficient to foster behavioral changes (Gibson and Marks 1995). Sometimes the incentives offered also fail because they overlook the social and cultural importance of certain activities, such as hunting, that cannot be easily substituted.

Leakage can be a problem as ICDPs may export over-use of resources to other areas. In the Mamirauá Ecological Reserve project in Brazil, for example, negotiated fishing rules and regimes between local users and external communities led to the overexploitation of fishing resources in previously unaffected areas (Hughes and Botelho 2000). Also commercialization of nonwood forest products has been reported to induce overharvesting or overcultivation (Ferraro and Simpson 2001). On the other hand, ICDPs may also attract more pressures on resources. In countries where poverty is widespread, even modest benefits provided by ICDPs may induce higher pressures on natural resources through migration into the project area (Wells et al. 1999; Noss 1997; Barrett and Arcese 1995; Wells et al. 1992). Also, tourism may damage protected areas and buffer zones.

Overall, the bottom line is that conservation is not likely to result from *all* ICDPs (Salafsky et al. 2001). The linkages between conservation objectives and development activities, which are central to the rationale of ICDPs, are generally poorly understood or enforced. Many projects only provide nominal opportunities for community-wide participation and often do not succeed in linking development benefits directly to community conservation obligations. Simplistic ideas of making limited short-term investments in local development, then hoping this will somehow translate into sustainable resource use and less pressure on protected areas, need to be abandoned (Wells et al. 1999, p. 6).

ICDPs were developed among other things due to the concern over the impacts that protected areas implementation has on local communities. In particular, conflicts over rights to land and severe restrictions on harvesting resources called attention to the unfair distribution of the costs of biodiversity conservation (Newmark and Hough 2000).

BOX 15.13

Unconstructive Dynamics and Incentive Structures of ICDPs (adapted from Kiss 1999)

The project approach to integrating conservation and development often leads to unconstructive dynamics and incentive structures. The main problems that affect integrated conservation and development projects include:

A poor "donor/recipient" dynamic: this occurs when the main objective of the beneficiaries quickly becomes obtaining as many benefits as possible from the project and getting the project to address their most urgently felt needs (which rarely relate directly to biodiversity conservation). This raises false levels of expectation and builds dependency instead of self-sufficiency or empowerment. Instead of good relations and cooperation, it leads to worsened relations and hostility when benefits fail to meet expectations, or when they are phased out.

A "get on with it" mentality: once a project has been identified and expectations are raised (among donors and recipients) everyone involved becomes focused on getting project activities moving and getting the project going. This explains why so often we launch into implementing projects even though we realize we do not have: (1) an adequate understanding of either ecological or socioeconomic/political conditions, or (2) real consensus between project supporters and communities on the objectives and on respective roles/responsibilities. We all know that we should take the time to do this but buckle under the pressure of trying to get the project off the ground.

Misplaced "ownership": having designed and mobilized the funding for a project, the project supporters are often more intent on making it succeed than on the beneficiaries. This leads them to continue at all costs and make compromises that they should not make, just to keep it alive.

A focus on activities, rather than impacts: projects consist of specific activities, and project supporters become preoccupied with implementing them, and (inevitably?) come to measure progress and achievement in terms of implementation. The result is the general absence of effective ecological socioeconomic monitoring and evaluation in these projects. Even if an attempt is made to identify impact indicators, in practice, monitoring and evaluation usually focuses on the implementation of activities. In addition, the focus on activities usually amounts to excessive focus and time spent on community development activities, with project supporters forced into the role of "social control" (ensuring equitable use and proper accounting of development funds) for which they have no mandate and little capacity.

"Magnet effect": even very modest projects become (or become perceived as) islands of relative prosperity in the midst of poverty, attracting immigrants to the area. The result is a dissipation of project benefits, and increased demands and stress on the natural environment.

If ICDPs have in some cases been able to raise overall incomes, they have in other cases failed to evenly distribute benefits among different community groups. Failure usually derives from erroneous assumptions. Often, it is assumed that "communities" are homogenous, easily defined and recognizable and that social cohesion allows project activities to be aimed at the community as a whole. In reality, however, there may be a great deal of social differentiation within communities, which affects who benefits from project activities and who looses from restrictions on resource use. For instance, ICDPs are often biased against women in their activities and in benefit-sharing (Flintan 2000).

Another problematic aspect is that ICDPs may unintentionally promote dependency rather than reciprocity, in particular, when local communities are treated as recipients of aid rather than as partners in development (Newmark and Hough 2000).

15.5.3 Integrated Coastal Zone Management

Coastal ecosystems are critically important for human well-being. Almost half of the world's population lives in coastal areas and depends directly on coastal resources for livelihoods. The services provided by coastal ecosystems, such as protection against climate change–induced sea level rise, storm protection, and nutrient regulation are all vitally important. Many coastal zones are experiencing more rapid economic growth than inland areas. Population growth, expanding development activities, pollution, and overexploitation of natural resources are leading to the degradation of many coastal ecosystems.

Integrated coastal zone management has become a widely accepted response to sustain coastal zones (Clark 1996; Cicin-Sain and Knecht 1998; Kay and Alder 1999; Beatley et al. 2002; Harvey and Caton 2003). There are two main reasons why integrated management of coastal zones is necessary. The first concerns the impacts that coastal and ocean uses, as well as activities further inland, can have on coastal and ocean environments. The second is related to the effects that coastal and ocean users can have on one another (Cicin-Sain and Knecht 1998). ICZM is a conscious management process that acknowledges these interrelationships.

15.5.3.1 Drivers and Scale

Coastal regions are highly dynamic environments and changes in the landscape brought about by natural processes are to be expected. However, pressures exerted by humans are changing these areas dramatically and to a point that they are not able to recuperate from disturbances. This leads to the degradation of coastal ecosystems. ICZM seeks to address multiple drivers of ecosystem change (both direct and indirect) originating from processes occurring at different spatial and organizational scales.

Economic development can significantly affect the ecology of the coastal zone, ecosystem processes, and natural resources availability. For example, the removal of mangroves for aquaculture interferes with the functions that these habitats perform as buffers for storms and fish nurseries (Cicin-Sain and Knecht 1998). The types of economic activities established in a given region will often depend on subsidized credit and other incentives, as well as global market trends. Such factors are important drivers of coastal change.

The growing population of the word's coastal zones presents a major challenge to their sustainability. Coastal zones in many countries are among the most attractive places to live, both economically and aesthetically (Beatley et al. 2002). Some are seasonal tourist destinations, the population of which increases manifold during a few months of the year. More people mean expanding infrastructure, greater need for potable water and food, and larger amounts of waste. All these developments put ever-increasing pressure on natural resources, and result in competition for resources and space by different users.

Inappropriate institutional frameworks often exacerbate the problems facing coastal zones. Conventional management processes have tended to segment concerns and deal with problems on an isolated basis. Regulatory and political structures can also encourage behaviors that endanger the fragile natural resources of the coastal area (Beatley et al. 2002). Inefficient planning regulations, for example, enable the disorderly occupation of coastal areas, leading to the destruction of ecosystems and often placing people at risk from natural hazards. Another important problem that ICZM seeks to address is the inability of different agencies and levels of government involved in coastal zone management to work together.

ICZM deals with the drivers or factors that lead to the degradation of coastal resources in at least three major ways. It does so, first, by addressing conflicts between different uses and users of coastal resources. Second, by improving coastal planning and management processes in order to regulate increasing demands on resources. The third way drivers are addressed is by promoting institutional change, particularly the processes through which decisions are made about coastal zones and their resources. The latter implies fomenting more inclusive decision-making, building capacity, and promoting inter-sectoral and inter-agency coordination.

ICZM can be implemented at multiple scales. Managing complex areas such as coastal zones means focusing on geographically defined areas sharing common or interrelated resource management, pollution control, economic development, and other social, political, and environmental concerns. ICZM has often been implemented on a regional scale, sometimes using the watershed as the unit of management (Beatley et al. 2002). It is also commonly implemented on a local scale where it can respond to more locally specific interests, needs, and concerns. International ICZM plans are less common, but are urgently needed since problems often extend over national borders. Even though the focus of ICZM may be on a given scale, cross-scale interactions are considered in terms of the biophysical processes involved and the institutions responsible for management decisions and their implementation. ICZM, therefore, is best described as a cross-scale integrated response.

ICZM addresses several dimensions of integration. It involves balancing, at a number of scales, different, and very often, competing values, interests, and goals (Kenchington and Crawford 1993). ICZM promotes both horizontal and vertical integration. On the horizontal axis, integration occurs among different sectors concerned with coastal issues (for example, fisheries, tourism, environmental conservation, infrastructure development, oil exploitation) and integration between coastal sectors and land-based sectors that affect the coastal/marine environment (Cicin-Sain and Knecht 1998; OECD 1993). Integration among nations is often also a feature of ICZM, particularly when different nations share enclosed or semi-enclosed seas, or negotiated solutions must be found for fishing, transbounday pollution, and other issues.

Vertical integration implies, primarily, the integration between different levels of government (local, regional or provincial, and national). The different levels of government play different roles, address different needs, and have different perspectives. These differences may pose problems that ICZM seeks to address by promoting harmonized policy development, planning, and implementation between different national and sub-national levels (Cicin-Sain 1998; see also Sorensen 1997). The integration between different institutions and stakeholders is a central feature of coastal zone management, particularly stakeholder inclusion in decision-making (Treby and Clark 2004).

Integration between different disciplines (natural sciences, social sciences, engineering) is also a fundamental aspect of integrated coastal management. The bringing together of the natural and social sciences within an integrated framework, coupled with a learning-based management system, may enable gains to be made in the science of coastal zone management. As the knowledge base improves, so the management strategies used need to evolve (Olsen et al. 1998). Given the unsuitability of "blueprint" solutions for ICZM, a learning approach that draws on an improved knowledge base is particularly important. Essentially, ICZM is a form of adaptive management (Olsen and Christie 2000).

15.5.3.2 Instruments and Mechanisms for ICZM

A wide range of mechanisms and instruments are commonly used in coastal planning and management—ICZM draws on both well-established and more experimental ones. Kay and Alder (1999) identify a range of techniques that can be combined to assist in the integrated management of coastal zones. These techniques can be administrative, such as policies, legislation, and guidelines; social, including the use of traditional knowledge, co-management, and capacity building; and technical, such as environmental impact assessment, landscape visualization, and economic analysis (see also Clark 1996; Thia-Eng 1993).

The European Union has compiled a set of enabling mechanisms for ICZM, which range from legal and regulatory instruments to voluntary agreements and international conventions (EC 1999). Guidelines and good practices are particularly useful aides for those interested in adopting an integrated approach to coastal management (see Post and Lundin 1996; Pernetta and Elder 1993; GEF et al. 1996; UNEP 1995, 1996). Efforts have also been developed to produce guidelines for the incorporation of specific issues into coastal management such as wetlands (Ramsar 2002) and climate change (Cicin-Sain et al. 1997).

Brown et al. (2001) outline an innovative set of techniques to address dilemmas between conservation and development in managing coastal resources in developing countries. This approach has been labeled "trade-off analysis" and focuses on including the values and interests of all those concerned with coastal resources into decision-making processes (see also Brown et al. 2002). Using a framework of multicriteria analysis, the approach engages stakeholders in the research process in order to evaluate the trade-offs between users and uses of coastal resources, and to negotiate and design effective, efficient, legitimate, and equitable governance structures (Adger et al. 2003).

Trade-off analysis is particularly focused on the problems and dilemmas of those parts of the developing world where the natural resources of the coast form significant and necessary resources for livelihood resilience. Here the dilemmas and trade-offs for sustaining the coast are especially acute and immediate, given the high biodiversity and ecosystems values, and the livelihoods and dependence on coastal and marine resources. In addition to trade off-analysis, other methodologies have been proposed to facilitate the development of integrated coastal management. These are based on consensus, flexible institutional arrangements based on issues and not sectors, and more equal power in decision making (see Kay et al. 2003).

15.5.3.3 Impacts of ICZM

Assessing ICZM is challenging, not only because many (but by no means all) of these initiatives are relatively recent, but also because there is very little evaluative evidence on its effectiveness in improving the management of coastal zones and their resources (Cicin-Sain and Knecht 1998). Evaluations of ICZM have tended to focus on the quality of program implementation and the degree to which project objectives have been achieved. Some focus on management capacity to determine the adequacy of management structures and management processes as these relate to generally accepted standards and experience. For example, donor-funded ICZM initiatives usually emphasize performance evaluation, which reveal little about how interventions have impacted on coastal resources and society (Lowry et al. 1999). Evaluation difficulties are compounded by the lack of baseline studies. Without a baseline, it is difficult to analyze the impacts of management efforts, as highlighted in a recent report of the U.S. Commission on Ocean Policy (2004).

A growing number of publications on ICZM have emerged during the last decade (Clark 1996; Cicin-Sain and Knecht 1998; Kay and Alder 1999). Most cover both the theory and practice of ICZM and provide valuable planning, implementation, and operational guidelines. They provide a guide on how to develop ICZM but rarely go into examining outcomes, partly because of the problems highlighted above. Their focus, therefore, is on processes not outcomes. Thus what this assessment can say about ICZM is mainly related to the process of ICZM development. Integrated coastal zone programs appear to have positive effects on two key areas, namely the improvement of coordination between the different sectors, actors, and levels of government involved in coastal management, and greater stakeholder involvement in decision-making (Klinger 2004).

ICZM initiatives have introduced the practice of preparing strategic plans for coastal zones looking at the bigger picture and the long-term linkage between maintaining the integrity of the natural system and the provision of economic and social development options. In the United Kingdom, for example, coastal defense is currently undertaken within the framework of shoreline management plans. This is a first step toward more holistic, broader encompassing coastal management, based on coastal process cells and sub-cells rather than the administrative boundaries of coastal operating authorities. Shoreline management plans also provide an enabling framework to link the work of the many stakeholders who need to be involved, while taking into account their individual roles and responsibilities (Atkins 2004).

ICZM initiatives have created conditions for stakeholder involvement in coastal management decisions and for partnerships that have helped to break sectoral barriers. However, studies have also highlighted that a number of interests are still excluded (Atkins 2004). In some cases, the ICZM process has failed to identify all relevant stakeholders and create conditions for their effective participation. In others, the stakeholders themselves have opted out because the benefits of involvement were not clear. This is partly because of uncertainty regarding the role of consultative processes in coastal zone decision-making by government authorities. Partnerships have been important but have not always demonstrated transparent and democratic methods for selecting participating organizations and individuals that truly reflect the range of stakeholders in the area.

One of the key lessons that are emerging from ICZM experiences is that more integration per se does not guarantee better outcomes. The challenge is not only formulating improved strategies, but also implementing such strategies (Olsen 1993). Many developing nations lack the capacity to implement complex programs aiming to address many different problems simultaneously. Focusing on a few issues initially and then gradually addressing additional ones as capacity increases is often more feasible and effective. This means adopting an incremental approach to ICZM (Olsen and Christie 2000). Ultimately, ICZM is a challenge of governance that requires "modifying entrenched patterns of behavior and societal norms" (Olsen 1993, p. 203). However, the required changes can be difficult to achieve, especially when they require shifts in the distribution of power and authority. For example, competition over management control between different agencies often undermines the political feasibility of approaches requiring a high degree of intergovernmental cooperation and coordination. (See Box 15.14.)

15.5.4 Integrated River Basin Management

The need for the comprehensive management of water resources, using the river basin as the focus of analysis, has been stressed by international conferences such as the 1992 Dublin Conference on Water and the Environment and UNCED. Agenda 21 (Chapter 18) places much emphasis on the need for the integrated management of water resources, emphasizing the importance of water for both ecosystems and human development. IRBM shares many similarities with the concept of integrated water resources management, but although both concepts are interrelated, they are not necessarily identical. Integrated water resources management—which emphasizes the need for legal, institutional, and policy frameworks—is seen as the wider context within which IRBM as a concrete management approach takes place (GWP 2000; GWP 2004).

River basins are critical spatial and ecological units that sustain many important economic activities, as well as livelihoods. As freshwater and other resources provided by river basin ecosystems become scarcer, competition for their use increases. Resource degradation narrows options for future development, but these impacts are not evenly distributed. The poor, in particular, rely disproportionately on ecosystem goods and services provided by river basins, and feel the greatest effects when these are degraded (McNally and Tognetti 2002).

Conflicts among stakeholders regarding tradeoffs among different resource uses are common in river basins. Such conflicts may be exacerbated in international river basins where socioeconomic inequities between the different countries are often considerable, as are differences in power. Of the world's 263 international river basins, 158 are believed to be potential flashpoints for future disputes since cooperation between the countries covering these basins is inconsistent or absent (UNEP 2002c).

River basin approaches are not necessarily new. For example, the Murray-Darling Commission, a well-known example of river basin management institutions, has a long development history. (See Box 15.15.) However, earlier approaches tended to focus on only a few aspects of water management, such as quantity and quality, whereas the scope of IRBM is broader. IRBM is explicitly concerned with promoting integration, for example, between land and water management, of upstream and downstream water related interests, of freshwater management and coastal zone management, and across all major water use sectors. IRBM should, therefore, be linked to ICZM efforts to form a process of broad-scale, integrated ecosystem management (Ramsar 2002; Ramsar Convention Secretariat 2004).

15.5.4.1 Main Drivers Addressed by IRBM

River basins often include many different ecosystems such as river systems, riparian forests, lakes, wetlands, and deltas. IRBM seeks to address, directly or indirectly, all major drivers causing the degradation of river basins. These drivers differ from region to region, but usually include changes in land and water uses that affect ecosystems and hydrological functions (for example deforestation), population growth, pollution, and overuse of natural resources. IRBM, therefore, seeks to address a complex set of drivers that undermine the ability of river basins to provide multiple ecosystem services as well as the capacity of people to benefit from such services.

IRBM also addresses the drivers that lead to the fragmented and uncoordinated management of water and associated resources

BOX 15.14

Barriers to Inclusive, Integrated Coastal Zone Management: Evidence from the Caribbean

Inclusive, integrated coastal zone management can be undermined by a number of constraints. In Buccoo Reef, Trinidad and Tobago, the main constraints to stakeholder participation in coastal management include the high financial cost of participation; the high time costs of participation; poor skill development of leaders; poor communication within and between groups; and existence of widespread personalized conflict in communities. The prevailing policy, legal, and regulatory setting as well as governance structures and institutions also militate against the implementation of integrated coastal management.

The national legislative setting in Trinidad and Tobago is not conducive to stakeholder inclusion in coastal zone planning and management. There are no legal provisions that make stakeholder participation in environmental decision-making and policy-development a requirement. The legislation also demonstrates other important gaps. For example, property rights to inter-tidal and other areas are not clear and the roles and responsibilities of the different agencies concerned with coastal resources are poorly defined. This gives rises to disputes, duplication of work, and institutional paralysis. Resource managers also feel that the enforcement of existing laws is inadequate. Combined, these factors make implementing ICZM more than a matter of drafting addition regulations. More profound changes are needed.

Structurally, the Trinidad and Tobago government lacks specialist staff with appropriate skills to implement integrated and inclusive approaches. Officials and field staff often do not know how to engage and work with different stakeholders. Lack of training, insufficient staff numbers, and inadequate financial resources can be major impediments when undertaking participatory planning and management. Consulting with stakeholders and enabling their participation in the definition and implementation of coastal management plans is often a time consuming and resource-intensive process that requires not only appropriate skills and resources but also will. Government officials often consider the inclusion of stakeholders as impractical, making coastal management more complex.

Formal natural resource management institutions in Trinidad and Tobago impose constraints on all arms of the government. The lack of space for networks to develop is identified by many within the system as the bottleneck to the development of more integrated and inclusive approaches. But in some cases, the constraints on the expansion of networks and innovation are self-imposed by the government agencies themselves. New government agencies, especially, may need to develop public credibility to achieve "success." Consequently, they may avoid untested methods or approaches. The problem of perceived power loss by the government may diminish as more government departments start to see the potential benefits by engaging communities in making decisions about, and managing, natural resources.

In other cases, there are few possibilities to expand institutional networks because of the peculiarities of institutional structures and operational arrangements. For example, many government agencies use a project cycle approach to allocate financial resources. Project timetables are fixed in the project proposal, and funding agencies generally require project managers to deliver outputs according to strict timetables, and funding is dependent on the successful completion of intermediate targets. The time allocated to inclusionary processes might be perceived as a stumbling block to the achievement of project deadlines. Effective inclusion can be time-consuming and unpredictable in terms of length of time required to complete the project. If a project leader is determined to meet inflexible project deadlines, it might not be possible to fully engage stakeholders. The project cycle, therefore, does not support social learning or adaptive approaches to coastal management.

Low levels of social capital, as well as limited access to spaces of engagement and lack of networks linking groups with shared interests constitute significant constraints to participation in decision-making. Constraints to participation can also arise from high costs of involvement, in terms of time or money. This has implications for who participates and who is excluded. The stakeholders involved are often asked to commit a substantial amount of their time, and sometimes their finances to supporting aspects of resource management. This can create the potential problem of non-representation, through a self-selection process, whereby those who have the time or resources to attend meetings and offer input may not reflect the opinions and attitudes of others in the community or group.

Poor interpersonal communication, aggressive behavior or strained intra-community relations can all act as constraints to participation. Poor communication within and between groups as well as, between these and the government is often an important constraint. Equally important is the issue of skill development, especially of leadership qualities and relationship management at the community level. In those groups that are poorly organized, the inability to develop a coherent message and deliver it to the appropriate agency is akin to exclusion. Often, communication and winning community trust and involvement rests on the commitment, sensitivity, and leadership of one person or a very small group of people. Some groups may not work well together because of historical factors such as mistrust of public authorities. If such cases exist, ways must be found to build trust.

in river basins (Chew and Parish 2003). As a recent UNDP (2004, p. 2) publication discussing the role of water for poverty reduction argues, the water crisis that humankind is facing today has resulted mainly "not from the natural limitations of the water supply," but rather from "profound failures in water governance, that is, the ways in which individuals and societies have assigned value to, made decisions about, and managed the water resources available to them." One of the greatest challenges of IRBM, therefore, is to address institutional problems and bottlenecks. This involves changing practices and attitudes, resolving conflicts and power imbalances, and including a wider range of stakeholders in decision-making.

15.5.4.2 The Scale of IRBM

IRBM may take place at different scales: at the local to national scale it ranges from small catchments to major national basins; at the national and federal scales it focuses on intra-country transboundary issues; at the international scale it deals with transboundary river basins (examples include the Nile, Danube, and Rhine).

IRBM is inclusive of management of watersheds. Focusing on watersheds is a way to address problems that are difficult to solve at larger scales, such as relationships between land use and water flow for purposes of stabilizing stream flows, controlling erosion and sedimentation, and improving groundwater recharge (Barrow 1998). For clarification, a watershed (often called a catchment) is considered to be a topographically delineated area drained by a stream system; that is, the total land area above some (sometimes arbitrary) point on a stream system. A river basin is similarly defined, but is delineated on a larger scale and includes all the lands that drain through the tributaries into the basin.

Box 15.15

The Murray-Darling Basin Initiative

The Murray Darling Basin covers 1,019,469 square kilometers of southeastern Australia and contains the country's three longest rivers—the Darling, the Murray, and the Murrumgidgee. Key biophysical features are the presence of over 30,000 wetlands, 35 endangered birds, and 16 endangered mammals. It is also of economic significance to Australia as it generates 40% of the national income from agriculture and grazing and contains about half of the national cropland and three quarters of irrigated land, though it only drains 14% of the country's land area. It is administered by five different states that have different climatic conditions, water availability, water use requirements, and management approaches, as well as over 200 local governments. Over half of the basin is in the state of New South Wales, close to a quarter is in Queensland, and the rest in Victoria, the Australian Capital Territory, and South Australia (MDBC 2002).

A River Murray Commission was formed in 1915 to develop the Murray and Darling rivers for navigation and irrigation. Following a review and mounting environmental degradation, the Commission was reconstituted in 1988 with the Murray-Darling Basin Agreement between the four basin states and Australian Capital Territory establishing the Murray-Darling Basin Initiative and the Commission to run it. With the reinstitution of the Commission, a new program of work focused on sustainable natural resource management commenced.

The basin faced two major resource problems. First, over 70% of useable water was being diverted, mainly for irrigation. Consequently, native fisheries and water bird populations were collapsing and major wetlands had contracted by half. Pastoralists relying on beneficial flooding for livestock production on floodplains saw their productivity decline sharply. Second, the lack of dilution of pollutants and the side effects of the dams and levees were compounding the environmental impacts. Water quality was also in rapid decline with increased concentrations of nutrients, farm chemicals and salinity. Of greatest concern was an upward trend in salinity due to poor irrigation practices and deforestation in the catchment headwaters.

While these problems have not been and may never be "solved," quality of decision-making improved in the Murray Darling Basin. Among the measures that contributed to the improvements:

- including both nature conservation and resources and agriculture ministers in the MDB Ministerial Council;
- appointing an independent, authoritative Chair of the Commission, which could informally mediate in disputes;
- establishing the heads of the relevant government agencies as the MDB Commission;
- employing technical staff to advise these bodies in the Office of the Murray Darling Basin Commission;
- contracting independent national scientific authorities to report on the most controversial issues and for auditing;
- establishing a Community Advisory Committee, with representatives of key stakeholders and the chairs of the eighteen (sub-) Catchment Management Committees, with the CAC chair representing the community on the Commission and Ministerial Council;
- creating opportunities for representatives of the key stakeholder groups to meet and work together. Thus, generating a better understanding of each other's concerns and facilitating decisions where agreements could be reached.

The result of such measures is that difficult decisions can be thoroughly assessed by experts working concurrently for all the governments, and this knowledge is widely shared. Consequently, a recalcitrant government is under a lot of pressure from the community, experts, and other governments to join in difficult collective decisions.

15.5.4.3 Forms of Integration and Instruments for IRBM

IRBM is about managing interactions and integration within and between natural and social systems. The natural system is of critical importance for resource availability and quality. The social (or human) system determines resource use and allocation, waste production, and pollution of resources, and must also set development priorities (GWP 2000). From the perspective of natural systems, integrated management at the river basin scale is appropriate because it recognizes the linkages within the ecological system, such as those that exist between the various habitats and ecosystems and between different biophysical processes. However, integrated management requires more than taking into account ecosystem dynamics to include sociopolitical dynamics and how these affect resource use and decisions (Bos and Bergkamp 2001).

A range of instruments and mechanisms exist for supporting the shift toward IRBM. Useful "toolboxes" have been compiled to assist decision-makers and practitioners to put together policy packages for sustainable water management and development. The Global Water Partnership (2004) has produced one of the most comprehensive toolboxes for integrated water resources management, which is relevant for IRBM. It organizes the tools into three main types. The first set of tools comprise those aimed at creating the enabling environment for integrated management, including laws, investments, and policies. They also provide the framework for the application of other tools. The second set of tools is concerned with building appropriate institutions and strengthening their capacity. The third type includes specific management tools such as conflict resolution and consensus building mechanisms. The toolbox does not aim to be prescriptive and recognizes that the types of tools that can be used, and the way in which they can be combined, will vary from place to place.

IRBM is a particularly important approach within the Ramsar Convention on Wetlands. Ramsar has proposed a set of guidelines aimed at assisting interested parties in developing a "holistic" or integrated approach to the management of wetlands and river basins (Ramsar Convention Secretariat 2004; Chew and Parish 2003). The application of an ecosystem approach to IRBM is emphasized by the Ramsar Convention as well as other international institutions involved in promoting the sustainable use and management of water resources. Examples include the River Basin Initiative (2004), a joint work program between the Ramsar Convention on Wetlands and the Convention on Biological Diversity and IUCN's Water and Nature Initiative (2004). One of the most successful manifestations of the ecosystem approach, in fact, developed in the context of river basin management in the Great Lakes Basin (Allen et al. 2003; Kay et al. 1999; Kay and Schneider 1994; Allen et al. 1993; Hartig 1998).

15.5.4.4 Impacts of IRBM

It may take time for ongoing IRBM initiatives to reach a stage where tangible, on-the-ground benefits can be seen and comprehensively assessed at the basin-wide level. For example, integrated management efforts in the Rhine Basin began in the 1950s but evolved from addressing a relatively narrow set of concerns to a basin-wide transboundary approach that integrates water management with land-use planning and coastal protection. The more ambitious targets of the Rhine program are long-term. (See Box 15.16.)

General lessons have been drawn from IRBM experiences that can serve as a useful checklist and planning tool for both ongoing and future initiatives (see, for example, WWF 2003). River basins and associated ecosystems are extremely productive and play a vital role in sustaining livelihoods. By improving land and water resources management in river basins, IRBM can have a positive effect on human well-being and ecosystem health. IRBM within an integrated approach to water resources management has been recognized as contributing to the objectives of poverty alleviation (UNDP 2004). So far, however, few efforts at implementing IRBM have actually succeeded in achieving social, economic, and environmental objectives simultaneously (McNally and Tognetti 2002).

15.5.5 Enabling Conditions and Constraints

Integrated responses need to be implemented effectively to guarantee better outcomes. A major constraint that sub-national and multiscale responses experience is lack of implementation capacity. Lack of expertise and resources is a particularly persisting problem in developing countries. For example, developing-country forestry departments often lack trained personnel to implement sustainable forest management and are under-funded. Limited capacity, however, should not be a deterrent to the initiation of integrated responses. An incremental approach to integration can be used, initially focusing on a relatively narrow set of issues that cut across different sectors and then gradually expanding the scope of the program as experience accumulates and capacity develops. Capacity is not only important in government but also in civil society. For example, in both ICDPs and SFM, communities need skills to manage enterprises that will bring them economic benefits.

Active public participation in decision-making appears to improve the outcomes of integrated responses. It is necessary to help different stakeholders to understand each other's perspectives, work together, and make common decisions (McNally and Tognetti 2002). In order to determine the best action for society, it is necessary to balance multiple objectives and views. Methodologies such as multicriteria analysis, whereby stakeholders are engaged to consider the merits of different management strategies and explicitly determine management priorities, can yield positive results (Brown et al. 2001). Feedback mechanisms to ensure that the outcomes of participatory processes are incorporated in decision-making are essential. However, despite its merits, participation on its own is not a panacea; it has to be used in conjunction with other mechanisms. Many successful integrated approaches combine bottom-up with top-down approaches.

Policies at the national and international level can support the implementation of integrated responses. Examples include legislation enabling public participation in decision-making and com-

BOX 15.16

The Rhine Basin

The river Rhine flows for 1,320 kilometers from the Swiss Alps, through Germany and the Netherlands, to the North Sea, in a catchment area of 170,000 square kilometers with a population of over 50 million people. Other countries partly in the Rhine catchment area are Austria, Luxembourg, Italy, Liechtenstein, and Belgium. It has been heavily developed as a shipping channel and for industry and is also used to generate energy, for recreation, as a source of drinking water for 20 million people, and to dispose of municipal and industrial waste. It receives pollutants from agricultural and diffuse sources and has supported large fisheries, though most aquatic life had disappeared by the 1960s (ICPR 2001). Over the last two centuries, 90% of the functional floodplains have been lost to river regulation projects, leading to higher and more rapid flood peaks. Dam structures also prevented salmon and other migratory fish from reaching their spawning grounds. Concerns about pollution first became prominent in the aftermath of World War II.

Prior to 1950, inter-country agreements pertaining to the Rhine addressed freedom of transport and protection of fisheries. The International Commission for the Protection of the Rhine from pollution was established in 1950, but little action was taken. Measures to restore and protect the Rhine were only initiated in reaction to catastrophic events, which raised awareness of the need for basin-wide environmental impact assessments. A Rhine Action Plan to address pollutant concerns was developed in the aftermath of the 1986 Sandoz plant chemical fire in Basel, Switzerland, during which 30–40 tons of toxic substances washed into the Rhine. The action plan established ambitious goals and went beyond water quality issues to also include ecosystem goals, initiating a more integrated approach to river management. These included:

- a 50% reduction in the discharge of nitrogen, phosphorus, and dangerous substances between 1985 and 1995;
- higher safety standards in industrial facilities;
- guaranteed use of the Rhine for drinking water by 2000; and
- restoration of the ecosystem in such a way that migratory species could return and become indigenous by 2000—native migratory species that had disappeared include salmon, sea-trout, allice shad, sea-lamprey, and sturgeon; and
- reduced sediment contamination so as to restore the North Sea.

These goals were in part met. Reduction of phosphorus inputs by 66% exceeded the target. Nitrogen pollution only dropped by 26%, but much of it is stored in groundwater from which it is transported to the river very slowly. The 50% targets were reached or exceeded for point sources of most of the toxic substances. Substances that remain problematic are primarily those from diffuse sources and from contaminated sediments. There are also some signs of the return of salmon and sea trout. Current action plan targets are to reduce damages 10% by 2005, and 25% by 2020; reduce extreme flood levels by at least 30 centimeters by 2005; and 70 centimeters by 2020. New kinds of targets added to the 2020 plan are protection of groundwater quality, balancing abstraction with recharge, and restoration of habitat connectivity.

In summary, the Rhine program evolved from addressing point sources of pollution and reactive, event-driven policies, to a more proactive, basin-wide, and transboundary approach that integrates water management with land use and spatial planning to make "space for the river," and also with protection of the marine environment. It can also be considered adaptive, in that the Biannual Ministerial conferences provide an opportunity to continuously reassess and evaluate existing activities. Other developments are that the ICPR is cooperating more with NGOs as a way to promote the exchange of information and common understanding.

munal management and co-management of natural resources. A culture of transparency in decision-making is essential for the success of integrated responses. A process that is seen as fully open, based on reliable information, and accessible to all stakeholders stands a better chance of success. International policy developments have played an important role in the adoption of integrated responses at the national and sub-national level. Agenda 21, for example, has strongly advocated the adoption of ICZM and IRBM. The Forest Principles adopted at UNCED, although not legally binding, have promoted the SFM concept, while GEF has been an important provider of funds for conservation projects with development objectives.

Integrated responses must invest in building the necessary knowledge base to inform planning and implementing field operations. Key stakeholder groups must be identified; land tenure systems, drivers influencing resource management decisions, and existing institutional structures relevant to the response must be understood. One of the most important challenges in the management of natural resources is the science-policy interface. Improvements in natural resource management depend on improvements in understanding the processes involved, both ecological and social (Cicin-Sain and Knecht 1998). Increasing the success of integrated responses depends on effective monitoring and evaluation and dissemination of lessons learned for their incorporation in planning and management processes. Lack of evaluative data on integrated responses is a major constraint to successful integrated responses. Criteria and indicators to assess impacts are better developed in SFM (see CIFOR 1999; Prabhu et al. 1999). Efforts to develop frameworks and indicators to assess progress in ICZM initiatives are also being made (Olsen 2003; Olsen et al. 1997).

Working at multiple scales and using scale-dependent comparative advantages enables the success of integrated sub-national responses. The complex problems that integrated responses seek to address require action at multiple scales, including local, regional, national, and sometimes international levels. However, potential conflicts between interests and actions at different levels must be recognized. Each institution working at a particular level brings unique expertise and perspectives to the planning and management process. Local governments and communities, for example, can contribute the most detailed understanding of problems, constraints, and limitations that will affect the choice of solutions. The national government, in turn, can contribute capacity to coordinate policies and harmonize sectoral activities, funding assistance, and ties to relevant international responses.

Integrated responses cannot be fully accomplished within the scope of a typical three or five year project. They require long-term financial and technical investment. It also takes time to build sufficient trust and levels of understanding among stakeholders to enable effective planning and implementation of integrated responses. A long-term planning framework enables the success of integrated responses.

15.6 Effective Integrated Responses

15.6.1 The Limits to Integration

Many chapters in this volume recommend the implementation of integrated responses, and the assessment in this chapter has provided examples at different scales and from different contexts. However, every response cannot be integrated in all instances. In their discussion of integrated natural resource management, Sayer and Campbell (2004, p. 21) ask the pertinent question, *How integrated do we need to be?* They observe that if integrated approaches are seen to be all embracing and need to integrate *everything,* then successful examples will be very difficult, if not impossible, to find. They argue that "early attempts at integrated natural resource management sought to understand the total behavior of the system to develop the ability to predict the outcome of any management intervention. . . . In reality the skill or professionalism of integrated natural resource management lies in making judgments on what to integrate. It only makes sense to integrate those additional components, stakeholders or scales that are essential to solving the problem at hand." This assessment confirms this observation. Successful integrated responses are integrated according to the problem or issue at hand. For every context, the degree, extent, and type of integration will be specific. There are general lessons to be learned from the experiences of implementing integration, but there is no "cookie-cutter" model of how an integrated response should be.

So what are the key factors that determine when and where integrated responses are most appropriate and most likely to be successful? The assessment suggests that integrated responses are most appropriate:

- when the full costs are taken into account,
- where capacity exists in government and civil society institutions,
- where a feasible time scale to achieve objectives is possible,
- where there is compatibility and not obvious conflict between objectives,
- where the legal and institutional frameworks supporting the response are already in place, and
- when relevant and timely information is at hand and extensive new data and research are not necessary.

Furthermore, at all scales, it is apparent from the assessment that integrated responses cannot be super-imposed by external agencies, and will be more likely to be successful when key stakeholders—in government, the private sector, and civil society—possess a sense of ownership. In other words, integrated responses cannot be driven from outside—objectives must reflect stakeholder priorities.

15.6.2 Understanding Trade-offs in Integrated Responses

The assessment has shown that trade-offs may be particularly significant in integrated responses, perhaps more so than in non-integrated or single objective responses. Trade-offs between different objectives often constitute severe constraints to effective integrated responses, and trade-offs exist between different scales, and between different actors or sections of society.

In order to manage trade-offs, information and methods are necessary to assess and compare direct and indirect impacts on social, economic, and ecological aspects of different response options. This information has to be such that the decision-makers can understand it and it has to help rather than complicate their task. Single indicators of change are generally not sufficient in the case of integrated responses. Measures need to reflect multiple sectors and actors, and enable an evaluation of the impacts of responses on them. This raises the question of whether integrated responses need to be supported by integrated research. To an extent this is the case, particularly where there is a need to understand the linkages, and therefore to assess the potential trade-offs, conflicts, and synergies between different objectives, especially between different ecosystem services and aspects of human well-being.

This is increasingly recognized in the scientific and social science literature. There is an emerging consensus about the need for a fundamentally different scientific approach to meet the challenges of sustainability, one which is capable of bridging the divide between disciplines that analyze the dynamics of ecosystems and those that analyze economics and social interactions (Scheffer et al. 2002). These concerns are reflected in new interdisciplinary and multidisciplinary research initiatives and institutions (Adger et al. 2003). Sayer and Campbell (2004) maintain that integrated approaches demand a new role for science; although the components may not be new themselves, the way they are put together and conceptualized is novel. Cutting edge research is still needed, but it has to be set in local contexts and be applied in ways that recognize the special circumstances of the poor—particularly as regards risk, dependency, and long-term depletion of productive potential and environmental externalities. Sayer and Campbell identify seven critical changes necessary to affect a paradigm shift in research. The seven conditions are more widely applicable to research necessary to support integrated responses:

- acknowledge and analyze the complexity of natural resource systems,
- use action research—become actors in the system,
- consider effects at higher and lower scales,
- use models to build shared understandings and as negotiating tools,
- be realistic about the potential for dissemination and uptake,
- use performance indicators for learning and adaptation, and
- break down the barriers between science and resource users.

Ultimately trade-offs need to be not only assessed, but also evaluated and managed. This will happen through a political process, requiring transparency and legitimacy but supported by a range of decision and evaluation tools.

15.6.3 Making Better Decisions for Integrated Responses

Developing and operationalizing integrated responses makes demands on decision-makers and planners. The multiple objectives mean that information across a range of subjects is necessary and analyzing the trade-offs and costs and benefits of different options may make the adoption of new analytical tools necessary in order to support decision-making. Sayer and Campbell (2004) argue that integrated natural resource management approaches should put greater emphasis on better decision-making, maintaining options and resilience, establishing appropriate institutional arrangements for resource management, and reconciling conflicting management objectives, rather than on producing technological packages. Many conceptual models of integrated approaches, therefore, focus on the decision-making processes.

There is an increasing emphasis on deliberative and inclusionary processes and greater participation by a range of stakeholders in all aspects of designing, planning, and implementing integrated responses. Opening up the decision-making process to a larger number of actors is both advantageous and disadvantageous for integrated responses. However, most of the case studies reviewed stress the importance of including all the relevant stakeholders in the early stages of the development of responses; using established techniques to identify the appropriate stakeholders; and adopting fair, transparent processes for their inclusion. Inclusion has to go beyond tokenism and this often requires empowerment of stakeholders to take control of certain aspects of the response process, an inherently political action.

However, these actors require information to base their decisions on. Integrated responses may require different tools that enable the different impacts of responses to be weighed. Table 15.6 summarizes some of the decision-support tools commonly used and their advantages and disadvantages for integrated responses.

As with most responses, developing and implementing integrated responses requires careful consideration of issues of power, access to resources and decision-making processes, and control of information. These considerations, as the assessment has shown, are applicable at all levels of government and across all scales, including international, national, and sub-national.

15.7 Conclusion

This chapter has reviewed integrated responses. Integrated responses are those responses, which intentionally and actively address ecosystem services and human well-being simultaneously. Many different types of responses may be integrated and they will employ a range of different instruments in their implementation. Often the coordination of different responses and instruments is central to the integration approach. The promotion of integrated responses and approaches to address problems of environmental degradation and development is strongly related to the paradigm of sustainable development. In the last two decades particularly, the recognition and analysis of linkages between environmental degradation and entrenched poverty and deprivation have resulted in increased calls for the integration of responses at all scales, from the global to the local. Hand in hand with this acknowledged need for integrated responses, there is increasing emphasis and growing interest in the science of complexity for increased understanding of ecological and social dynamics and of the need to include stakeholders as full participants and partners in responses.

However, it is still in the early days, and integrated responses are relatively novel in many, although not all, areas of policy. In some contexts they constitute "policy experiments." Assessment of these experiences is difficult and requires new methods, the use of multiple or adapted techniques, and interdisciplinary and multidisciplinary approaches. This in turn might demand new institutions and new ways of working together.

The lessons from these policy experiments are encouraging, but also cautionary. Integrated responses, whether on the international, national, or sub-national level, are often expensive (in terms of resources, personnel, and time) and in many countries and sectors there is a lack of capacity to effectively implement them. Disappointing results in the short-term, often related to inadequate appreciation and assessment of trade-offs, may mean that support for integrated responses is lost, or their promotion is seriously challenged. This has been the case, for example, in integrated conservation and development projects. Nonetheless, major advances on the sub-national level are discernable in watershed and river basin management, in sustainable forest management, and in integrated coastal management. Successful implementation of integrated responses on the sub-national level require, inter alia, a strong and effective national and international framework upon which such technical approaches can be based. The lessons learned from successful integrated responses do not provide easy to duplicate "blueprints" but they make clear that operationalizing sustainable development necessitates integration.

Table 15.6. Decision Support Tools and Techniques (adapted from Pearce and Markandya 1989)

Conceptual Basis/ Method	Description	Advantages	Disadvantages
Cost-benefit analysis	evaluates options by quantifying net benefits (benefits minus costs)	considers the benefits and costs of management options translates all outcomes into commensurate monetary terms reveals the most efficient option	no direct consideration of the equity distribution of the costs and benefits ignores non-quantifiable costs and benefits assumes that all stakeholders have equal income and well-being
Cost effectiveness analysis	the least cost option that meets the goals of the decision-maker is preferred	no need to estimate the benefits of different management options cost information is often readily available	relative importance of outputs is not considered no consideration given to the social costs resulting from side effects of different options
Multicriteria analysis	uses mathematical programming techniques to select options based on objectives functions with explicit weights of stakeholders applied	allows quantification of implicit costs permits the prioritization of options model can reflect multiple goals or objectives for the resources	an unrealistic characterization of decision making theoretical difficulties associated with aggregating preferences for use as weights in the model large information needs
Risk-benefit analysis	valuates benefits associated with a policy in comparison with its risks	framework is flexible permits considerations of all risks (benefits and costs) no automatic decision rule	framework is too vague factors considered to be commensurate are not
Decision analysis	a step-by-step analysis of the consequences of choices under uncertainty	model can reflect multiple goals or objectives for the resources choices to be made are explicit explicit recognition of uncertainty	objectives are not always clear no clear mechanism for assigning weights
Environmental impact assessment	provides a detailed economic, social, and environmental statement of the impacts of management options	requires explicit consideration of environmental effects benefits and costs do not have to be monetized	difficult to integrate descriptive and qualitative analyses with monetary costs and benefits no clear criteria for using information in the decision-making process

References

Adger, N., K. Brown, J. Fairbrass, A. Jordan, J. Paavola, et al., 2003: Governance for sustainability: Towards a "thick" understanding of environmental decision making. *Environment and Planning A,* **35,** pp. 1095–1110.

Allen, T.F.H. and T.W. Hoekstra, 1992: *Toward a Unified Ecology.* Columbia University Press, New York, NY.

Allen, T.F., J. Tainter, and T. Hoekstra, 2003: *Supply-Side Sustainability.* Columbia University Press, New York, NY.

Allen, T.H.F., B. Bandursky, and A. King, 1993: *The Ecosystem Approach: Theory and Ecosystem Integrity.* Report to the Great Lakes Advisory Board, International Joint Commission, Washington, DC.

Alpert, P., 1996: Integrated conservation and development projects. *BioScience,* **46,** pp. 845–855.

Angelsen, A. and S. Wunder, 2003: Exploring the forest-poverty link: Key concepts, issues and research implications. CIFOR Occasional Paper No. 40, Bogor, Indonesia.

Atkins, 2004: ICZM in the UK: A stocktake. Queen's Printer and Controller of HMSO, London, UK. Available at http://www.defra.gov.uk/environment/marine/iczm/pdf/iczmstocktake-final.pdf.

Barbier, E.B., J.C. Burgess, J.T. Bishop, and B.A. Aylward, 1994: *The Economics of the Tropical Timber Trade.* Earthscan Publications, London, UK, p. 224.

Barrett, C.B. and P. Arcese, 1995: Are integrated conservation-development projects (ICDPs) sustainable? On the conservation of large mammals in Sub-Saharan Africa. *World Development,* **23(7),** pp. 1073–1084.

Barrow, C., 1998: River basin development planning and management: A critical review. *World Development,* **26(1).**

Bawa, K.S. and R. Seidler, 1998: Natural forest management and conservation of biodiversity in tropical forests. *Conservation Biology,* **12(1),** pp. 46–55.

Beatley, T., D.J. Brower, et al., 2002: *An Introduction to Coastal Zone Management.* Island Press, Washington, DC.

Berkes, F. and C. Folke, 1998: Linking social and ecological systems for resilience and sustainability. In: *Linking Social and Ecological Systems: Management Practices and Social Mechanisms for Building Resilience,* F. Berkes and C. Folke (eds.), Cambridge University Press, Cambridge, UK, pp. 1–25.

Berkes, F., 1996: Social systems, ecological systems and property rights. In: *Rights to Nature: Ecological, Economic, Cultural and Political Principles of Institutions for the Environment,* S. Hanna, C. Folke, and K.G. Maler (eds.), Island Press, Washington, DC, pp. 87–109.

Berkes, F., 2002: Cross-scale institutional linkages: Perspectives from the bottom-up. In: *The Drama of the Commons,* E. Ostrom, T. Dietz, N. Dolsak, P.C. Stern, S. Stonich, et al. (eds.), National Academy Press, Washington, DC, pp. 293–322.

Binswanger, H.P., 1991: Brazilian policies that encourage deforestation in the Amazon. *World Development,* **19(7),** pp. 821–829.

Bomsel, O., J.C. Carret, and S. Lazarus, 1996: La construction d'usines de transformation du bois à Lashourville: Un début d'industrialisation au Gabon. In: *Cahiers de Recherche 96-b-1,* Centre d'Economie Industrielle, École des Mines de Paris (CERNA), Paris, France.

Boot, R.G.A. and R.E. Gullison, 1995: Approaches to developing sustainable extraction systems for tropical forest products. *Ecological Applications,* **5(4),** pp. 896–903.

Booth, D. 2002: Poverty reduction and the national poverty process. In: *Handbook on Development Policy and Management,* C. Kirkpatrick, R. Clarke and C. Polidano (eds.), Edward Elgar, Cheltenham, UK.

Bos, E. and G. Bergkamp, 2001: Water and the environment. In: *Overcoming Water Scarcity and Quality Constraints,* R.S. Meinzen-Dick and M.W. Rosengrant (eds.), Focus 9, Brief 6, International Food Policy Research Institute, Washington, DC. Available online at http://www.ifpri.org/pubs/catalog.htm #focus.

Boscolo, M. and J.R. Vincent, 2000: Promoting better logging practices in tropical forests: A simulation analysis of alternative regulations. *Land Economics,* **76(1),** pp. 1–14.

Boyle, M., J.J. Kay, and B. Pond, 2001: Monitoring in support of policy: An adaptive ecosystem approach. In: *Encyclopedia of Global Environmental Change,* Vol. 4, R. Munn (ed. in chief), Wiley, London, pp. 116–137.

Brack, D., 2002: Environmental treaties and trade: Multilateral environmental agreements and the multilateral trading system. In: *Trade, Environment and the Millennium,* G.P. Sampson and W.B. Chambers (eds.), 2nd ed., United Nations University Press, Tokyo, Japan.

Brown, K. and Corbera, E. 2003: Exploring equity and development in the new carbon economy. *Climate Policy,* 3 (Supplement 1), pp. 41–56.

Brown, K. and S. Rosendo, 2000: The institutional architecture of extractive reserves in Rondonia, Brazil. *Geographical Journal,* **166(1),** pp. 35–48.

Brown, K., E. Tompkins, and W.N. Adger, 2001: *Trade-off Analysis for Participatory Coastal Zone Decision-Making.* Overseas Development Group, University of East Anglia, Norwich, UK.

Brown, K., E. Tompkins, and W.N. Adger, 2002: *Making Waves: Integrating Coastal Conservation and Development.* Earthscan Publications, London, UK.

Buschbacher, R.J., 1990: Natural forest management in the humid tropics: Ecological, social and economic considerations. *Ambio,* **19(5),** pp. 253–258.

Cardiff European Council, 1998: Presidency conclusions. Available online at http://ue.eu.int/ueDocs/cms_Data/docs/pressData/en/ec/54315.pdf

Cavalcanti, C., 2000: *The Environment Sustainable Development Public Policies: Building Sustainability in Brazil.* Edward Elgar Publishing, London, UK.

CBD COP 7 (Convention on Biological Diversity Conference of the Parties), 2004: Decisions adopted by the COP to the CBD at its seventh meeting. Available at http://www.biodiv.org/doc/decisions/COP-07-dec-en.pdf.

CEC (Commission of the European Communities), 1992: *Towards Sustainability: A European Community Programme on Policy and Action in Relation to the Environment and Sustainable Development,* COM (92) 23/fin. CEC, Brussels, Belgium.

CEC, 1999a: Europe's environment: What directions for the future? The global assessment of the European Community programme of policy and action in relation to the environment and sustainable development, Towards sustainability, Communication from the Commission, COM(1999) 543 final. CEC, Brussels.

CEC, 1999b: The Cologne report on environmental integration: Mainstreaming of environmental policy, Commission working paper addressed to the European Council, SEC(1999) 777final. CEC, Brussels, Belgium.

CEC, 1999c: Report on the environment and integration indicators to Helsinki Summit, Commission working document, SEC(1999) 1942final. CEC, Brussels, Belgium.

CEC, 1999d: From Cardiff to Helsinki and beyond, Report to the European Council on integrating environmental concerns and sustainable development into community policies, Commission working document, SEC(1999) 1941final. CEC, Brussels, Belgium.

Chew, O.M. and F. Parish, 2003: Guidelines on integrating wetland conservation and wise use into river basin management, Adapted for South East Asia region, Global Environment Centre. Kuala Lumpur, Malaysia. Available at http://www.riverbasin.org.

Churchill, R. and L. Warren, 1996: *Effectiveness of Legal Agreements to Protect Global Commons,* GEC Programme briefings no. 9, November. Global Environment Centre, Osaka, Japan.

Cicin-Sain, B. and W. Knecht, 1998: *Integrated Coastal and Ocean Management: Concepts and Practices.* Island Press, Washington, DC.

Cicin-Sain, B., C.N. Ehler, R. Knecht, R. South, and R. Weiher, 1997: *Guidelines of Integrating Coastal Management Programs and National Climate Change Action Plans,* Developed at the international workshop, Planning for Climate Change Through Integrated Coastal Management, 24–28 February 1997, Chinese Taipei. Available at http://www.globaloceans.org/guidelines/PDF_Files/Taipei.pdf.

CIFOR (Center for International Forestry Research), 1999: The CIFOR criteria and indicators generic template. CIFOR, Jakarta, Indonesia. Available at http://www.cifor.cgiar.org/acm/methods/toolbox2.html.

Clark, J.R., 1996: *Coastal Management Handbook.* CRC Press, Boca Raton, FL.

Colfer, C.J.P. and Y. Byron, 2001: *People Managing Forests: The Links Between Human Well-being and Sustainability.* Resources for the Future and CIFOR, Washington, DC.

Cooke, B. and U. Kothari (eds.), 2001: *Participation: The New Tyranny?* Zed Books, London, UK/New York, NY.

Costanza, R. and C. Folke, 1996: The structure and function of ecological systems in relation to property rights regimes. In: *Rights to Nature: Ecological, Economic, Cultural and Political Principles of Institutions for the Environment,* S. Hanna, C. Folke, and K.G. Maler (eds.), Island Press, Washington, DC, pp. 13–34.

Dalal-Clayton, B., S. Bass, B. Sadler, K. Thomsom, R. Sandbrook, et al., 1994: National sustainable development strategies: Experiences and dilemmas, IIED Environmental Planning Issues No. 6. International Institute for Environment and Development, London, UK.

Dalal-Clayton, B., S. Bass, N. Robins, and K. Swiderska, 1998: Rethinking sustainable development strategies: Promoting strategic analysis, debate and action, IIED Environmental Planning Issues, No. 24. International Institute for Environment and Development, London, UK.

Dankelmen, I. and J. Davidson, 1988: *Women and Environment in the Third World: Alliance for the Future.* Earthscan Publications Ltd., London, UK.

Davey, S.M., J.R.L. Hoare, and K.E. Rumba, 2003. Sustainable forest management and the ecosystem approach: An Australian perspective. *Unasylva,* **54 (214/215),** pp. 3–12.

Dodds, F., 2001: Inter-linkages among multilateral environmental agreements, Paper prepared for International Eminent Persons Meeting on Inter–linkages: Strategies for bridging problems and solutions to work towards sustainable development, 3–4 September 2001. United Nations University Centre, Tokyo, Japan.

Domoto, A., 2001: International environmental governance: Its impact on social and human development, Paper prepared for International Eminent Persons Meeting on Inter-linkages: Strategies for bridging problems and solutions to work towards sustainable development, 3–4 September 2001. United Nations University Centre, Tokyo, Japan. Available at http://www.unu.edu/inter-linkages/eminent/papers/WG1/Domoto.pdf.

Earth Council, 2004: National councils for sustainable development [online]. Accessed 20 October 2004. Available at http://www.ecouncil.ac.cr/ecncsd.htm.

EC (European Communities), 1999: *Towards a European Integrated Costal Zone Management (ICZM) Strategy: General Principles and Policy Options,* European Communities: Directorates-General Environment, Nuclear Safety and Civil Protection Fisheries Regional Policies and Cohesion, Luxembourg, GD of Luxembourg.

El-Swaify, S.A., 2000: The role of soil resources in global environmental conventions: A technical perspective, Expert report commissioned by GTZ, Eschborn, Germany.

Emerton, L., 2001: Using an ecosystem approach: The integration of environmental concerns into decision making. In: *Proceedings of the Workshop: Promoting Environmentally Sustainable Development in Asia,* 22–23 November 2001, Singapore, China, OECD (ed.), OECD, Singapore, pp. 26–31. Available at http://www.oecd.org/dataoecd/6/13/2483661.pdf.

FAO (Food and Agriculture Organization of the United Nations), 2001: Global forest resource assessment 2000: Main report, FAO Forestry Paper 140. FAO, Rome, Italy.

FAO, 2003: Fisheries management 2: The ecosystem approach to fisheries, FAO technical guidelines for responsible fisheries. FAO, Rome, Italy. Available at http://www.fao.org.

FAO, 2003: Sustainable forest management. FAO, Rome Italy. Available at http://www.fao.org/forestry.

Ferraro, P.J. and R.D. Simpson, 2001: Cost-effective conservation: A review of what works to preserve biodiversity. *Resources for the Future Newsletter,* **1443** (Spring), pp. 17–20.

Ferraro, P.J., 2001: Global habitat protection: Limitations of development interventions and a role for conservation performance payments. *Conservation Biology,* **15(4),** pp. 990–1000.

Figueres, C. and M.H. Ivanova, 2002: Climate change: National interests or a global regime? In: *Global Environmental Governance: Options & Opportunities,* D.C. Esty and M.H. Ivanova (eds.), Yale School of Forestry & Environmental Studies, Yale, New Haven, CT, pp. 205–224. Available at http://www.yale.edu/environment/publications.

Flintan, F., 2000: Investigating the imbalances of gender and ICDPs: Past and present research, Paper presented at the CARE-Denmark International Seminar on Integrated Conservation and Development: Contradiction of Terms? 4–5 May, Copenhagen, Denmark.

Folke, C., S. Carpenter, T. Elmqvist, L. Gunderson, C. Holling, et al., 2002: Resilience and sustainable development: Building adaptive capacity in a world of transformations, Scientific background paper on resilience for the process of the world summit on sustainable development on behalf of the Environmental Advisory Council to the Swedish Government, 16 April 2002. Available at http://www.sou.gov.se/mvb/pdf/resiliens.pdf.

Folke, K.L., J. Colding, et al, 1998: *The Problem of Fit Between Ecosystems and Institutions.* International Human Dimensions Programme on Global Environmental Change, Bonn. Germany.

Freris, N. and K. Laschefski, 2001: Seeing the wood from the trees? *The Ecologist,* **31(6),** pp. 40–43.

GEF (Global Environmental Facility), 2000: Operational program #12: Integrated ecosystem management, GEF. Available at http://www.undp.org/gef.

GEF, UNDP, and IMO (United Nations Development Programme and International Maritime Organization), 1996: *Enhancing the Success of Integrated Coastal Management: Good Practices in the Formulation, Design, and Implementation of Integrated Coastal Management Initiatives,* GEF/UNDP/IMO Regional Programme for the Prevention and Management of Marine Pollution in the East Asian Seas and the Coastal Management Center, Manila, Philippines.

Gemill, B. and A. Bamidele-Izu, 2002: The role of NGOs and civil society in global environmental governance. In: *Global Environmental Governance: Options & Opportunities,* D.C. Esty and M.H. Ivanova (eds.), Yale School of Forestry & Environmental Studies, Yale, New Haven, CT, pp. 77–100. Available at http://www.yale.edu/environment/publications.

Gerwing, J.J., Johns, J. S., and Vidal, E., 1996: Reducing waste during logging and log processing: Forest conservation in eastern Amazonia. *Unasylva,* **187(47),** pp. 17–25.

Gibson, C. and S. Marks, 1995: Transforming rural hunters into conservationists: An assessment of community-based wildlife management programs in Africa. *World Development,* **23.**

Gibson, C.C., M.A. McKean, and E. Ostrom (eds.), 2000: *People and Forests: Communities, Institutions and Governance.* MIT Press, Cambridge, MA/London, UK.

Gray, J.A., 2002: Forest concessions experience and lessons from countries around the world. In: *Manejo Integrado de Florestas Umidas Neotropicais por industria e comunidades,* C. Sabogal and N. Silva (eds.), Center for International Forestry Research, Jakarta, Indonesia/Empresa Brasileira de Pesquisa Agropecuária, Belém, Brazil, pp. 361–378.

Gregersen, H.S. and D. Draper (eds.), 1989: *People and Trees: The Role of Social Forestry in Sustainable Development.* World Bank, Washington, DC.

GTZ (Deutsche Gesellschaft fuer Technische Zusammenarbeit), 1994: Sustainability criteria for forest management in Sabah, Report 198. Malaysian-German Sustainable Forest Management Project, GTZ, Eschborn, Germany. Available at http://www.gtz.de/malaysia/projects/sabah_project_reports.htm.

Gunderson, L.H. and C.S. Holling (eds.), 2002: *Panarchy: Understanding Transformations in Human and Natural Systems.* Island Press, Washington, DC.

GWP (Global Water Partnership), 2000: Integrated water resources management, GWP Technical Committee background paper no. 4. GWP, Stockholm, Sweden.

GWP, 2004: Integrated water resources management (IWRM) and water efficiency plans by 2005: Why, what and how? GWP Technical Committee background paper no. 10. GWP, Stockholm, Sweden.

Halle, M., 2003: Cancun wrap-up, IISD. Accessed on 23 May 2004. Available at http://www.iisd.ca/whats_new/IISD_Cancun_Wrap.doc.

Hannah, L., 1992: *African people, African Parks: An Evaluation of Development Initiatives as a Means of Improving Protected Area Conservation in Africa.* Biodiversity Support Programme, Washington, DC.

Hartig, J.H., 1998: Implementing ecosystem-based management: Lessons from the Great Lakes. *Journal of Environmental Planning and Management*, **14,** pp. 45–75.

Hartje, V., 2003: The international debate on the ecosystem approach: Diffusion of a codification effort. In: *Report for the International Workshop on the Further Development of the Ecosystem Approach at the International Academy for Nature Conservation, Isle of Vilm, Germany, October 9–11, 2002,* H. Korn, R. Schliep, and J. Stadler (eds.), Federal Agency for Nature Conservation, Bonn, Germany, pp. 30–35.

Hartje, V., A. Klaphake, and R. Schliep, 2003: *The International Debate on the Ecosystem Approach: Critical Review, International Actors, Obstacles and Challenges.* Federal Agency for Nature Conservation, Bonn, Germany.

Hartshorn, G.S., 1998: Letter in response to "Logging and Tropical Forest Conservation." *Science,* **281,** pp. 1453–1458.

Harvey, N. and B. Caton, 2003: *Coastal Management in Australia.* Oxford University Press, Oxford, UK.

Hertin, J. and F. Berkhout, 2001: Ecological modernisation and EU environmental policy integration, SPRU Electronic Working Paper Series No. 72. University of Sussex, Brighton, UK. Available at http://www.sussex.ac.uk/spru/library/spru0301.html.

Hisschemoller, M. and J. Gupta, 1999: Problem solving through international environmental agreements: The issues for regime effectiveness. *International Political Science Review,* **20(2),** pp. 151–173.

Holmes, T.P., G.M. Blate, J.C. Zweede, R.J. Pereira, P. Barreto, et al., 2000: *Financial Costs and Benefits of Reduced Impact Logging Relative to Conventional Logging in the Eastern Amazon.* Tropical Forest Foundation, Washington, DC.

Hughes, R. and H. Botelho, 2000: *Environmental Impacts of Support for Phase 2 of the Mamiraua Sustainable Development Reserve: A Scoping Study,* Report to the Sociedade Civil Mamiraua, Belém, Brazil, and the Department for International Development, London, UK.

Hughes, R. and F. Flintan, 2001: *Integrating Conservation and Development Experience: A Review and Bibliography of ICDP Literature,* Biodiversity and Livelihoods Issues No. 3. IIED, London, UK.

ICLEI (International Council for Local Environment Initiatives), 1996: Local authorities and the environment. In: *The Role of Local Authorities in the Urban Environment,* R. Gilbert, Earthscan Publications, London, UK.

ICPR (International Commission for the Protection of the Rhine), 2001: Rhine 2020: Program on the sustainable development of the Rhine, Conference of Rhine Ministers, ICPR, Strasbourg, France. Available at http://www.iksr.org.

ICTSD (International Centre for Trade and Sustainable Development), 2003: Chairman's statement from high-level round table on trade and environment, ICTSD, Geneva, Switzerland. Available at http://www.ictsd.org/ministerial/cancun/docs/chair_statement_cozumel.pdf.

IDRC (International Development Research Centre), 2003. Ecosystem approaches to human health program initiatives, IDRC, Ottawa, ON, Canada. Available at http://network.idrc.ca.

IIED (International Institute for Environment and Development), 1996: *Sustainable Forest Management: An Analysis of Principles, Criteria, and Standards.* IIED, London, UK.

IISD, 2002: Summary of the World Summit on Sustainable Development, 26 August–4 September. *Earth Negotiations Bulletin,* **22(51).**

INCRA (Instituto Nacional de Colonização e Reforma Agrária), 2001: *Dados Sobre os Assentamentos da Superintendência 27.* INCRA, Marabá, Brazil.

IUCN (International Union for Conservation of Nature and Natural Resources), 2002: Mixed results: Between hope and concern, IUCN@WSSD Newsflash, Issue 4, October 2002, IUCN, Gland, Switzerland. Available at http://www.iucn.org/wssd/docs/media%20brief/newsflash4.pdf.

IUCN, 2004: Water & nature initiative. Available at http://www/iucn.org/themes/wani.

Jacob, K. and A. Volkery, 2003: Environmental policy integration (EPI): Potentials and limits for policy change through learning, Presented at the 2003 Berlin Conference on the Human Dimension of Global Environmental Change. Available at http://www.fu-berlin.de/ffu/akumwelt/bc2002/files/jacob_volkery.pdf.

Johnson, N. and B. Cabarle, 1993: *Surviving the Cut: Natural Forest Management in the Humid Tropics.* World Resources Institute, Washington, DC.

Kates, R.W., W.C. Clark, R. Corell, J.M. Hall, C.C. Jaeger, et al., 2001: Sustainability science. *Science,* **292,** pp. 641–2.

Kay, J. and E. D. Schneider, 1994: Embracing complexity: The challenge of the ecosystem approach. *Alternatives,* **20(3),** pp. 32–38.

Kay, J., H. Regier, M. Boyle, and G. Francis, 1999: An ecosystem approach for sustainability: Addressing the challenge of complexity. *Futures,* **31,** pp. 721–742.

Kay, R. and J. Alder, 1999: *Coastal Planning and Management.* E & FN Spon, London, UK/New York, NY.

Kay, R., J. Alder, D. Brown, and P. Houghton, 2003: Management cybernetics: A new institutional framework for coastal management. *Coastal Management,* **31,** pp. 213–227.

Kenchington, R. and D. Crawford, 1993: On the meaning of integration in coastal zone management. *Ocean & Coastal Management,* **21.**

Kiss, A., 1990: Living with wildlife: Wildlife resource management with local participation in Africa, World Bank Technical Paper No. 130. World Bank, Washington, DC.

Kiss, A., 1999: Making community-based conservation work, Paper presented at the Society for Conservation Biology Annual Meeting, College Park, MD.

Klein, U., 2001: Integrated resource management in New Zealand: A juridical analysis of policy, plan and rule making under the RMA. *New Zealand Journal of Environmental Law,* **5,** pp. 1–54.

Klinger, T., 2004: International ICZM: In search of successful outcomes. *Ocean & Coastal Management,* **47,** pp. 195–196.

Lafferty, W.M., 2002: *Adapting Government Practices to the Goals of Sustainable Development.* OECD, Paris, France.

Lampietti, J.A. and U. Subramanian, 1995: Taking stock of national environmental strategies. World Bank, Land Water and Natural Habitats Division, Washington DC. Available at www.eldis.org/static/DOC1887.htm.

Landell-Mills, N. and I.T. Porras, 2001: *Silver Bullet or Fools' gold? A Global Review of Markets for Forest Environmental Services and Their Impact on the Poor.* IIED, London, UK.

Landell-Mills, N. and J. Ford, 1999: *Privatising Sustainable Forestry: A Global Review of Rrends and Challenges.* IIED, London. UK.

Lebel, J., 2003: *In Focus: Health, an Ecosystem Approach.* IDRC Publishing, Ottawa, ON, Canada.

Lowry, K., S. Olsen, and J. Tobey, 1999: Donor evaluations of ICM initiatives: What can be learned from them? *Ocean & Coastal Management,* **42,** pp. 767–789.

MacQueen, D., 2002: Priority areas for integration between forest-based industries and communities: Insights from a sustainable livelihoods approach. In: *Manejo Integrado de Florestas Umidas Neotropicais por industria e comunidades,* C. Sabogal and N. Silva (eds.), Center for International Forestry Research, Jakarta, Indonesia/ Empresa Brasileira de Pesquisa Agropecuária, Belém, Brazil, pp. 441–453.

Mahar, D. and R.R. Schneider, 1994: Incentives for tropical deforestation: Some examples from Latin America. In: *The Causes of Tropical Deforestation,* K. Brown and D. Pearce (eds.), UCL (University College London) Press, London, UK, pp. 159–171.

Mahar, D.J., 1989: *Government Policies and Deforestation in Brazil's Amazon Region.* World Bank, Washington, DC.

Mallet, P., 2000: Non-timber forest products certification: Challenges and opportunities. *Forests, Trees and People,* **42,** pp. 63–66.

Marconi, M.R., R.D. Smith, and E. Maltby (eds.), 2000: *The Ecosystem Approach under the CBD: From concept to Action,* Report from South American Regional Pathfinder workshop, 18–20th September 2001, Villa de Leyva, Colombia. Available at http://www1.rhbnc.ac.uk/rhier/iucn.htm.

Mayers, J. and S. Bass, 1999: *Policy That works for Forests and People: Overview Report.* IIED, London, UK.

McNally, R. and S. Tognetti, 2002: Tackling poverty and promoting sustainable development: Key lessons for integrated river basin management. World Wildlife Fund, Godalming, Surrey, UK. Available at http://www.wwf.org.uk/filelibrary/pdf/irbm_report.pdf.

MDBC (Murray Darling Basin Commission), 2002: Murray-Darling Basin Initiative, MDBC. Available at http://www.mdbc.gov.au.

Metcalfe, L., 1994: International policy coordination and public management reform. *Review of Administrative Sciences,* **60,** pp. 271–290.

Mullins, G.W, K.J. Danter, D.L. Griest, E. Norland, 1999: Organizational change as a component of ecosystem management. *Society & Natural Resources,* **13,** pp. 537–547

Mutume, G., 2003: Hope seen in the ashes of Cancun: WTO trade talk collapse, as Africa and allies stand firm. *African Recovery,* **17(3).**

Nardelli, A., 2001: *Sistemas de Certificação e Visão de Sustentabilidade no Setor Florestal Brasileiro,* Ph.D. thesis, Federal University of Viçosa, Viçosa, Brazil, p. 136.

Neufeld, D.A., 2000: An ecosystem approach to planning groundwater: The case of Waterloo Region. *Hydrogeology Journal,* **8,** pp. 239–250.

Newmark, W.D. and J.L. Hough, 2000: Conserving wildlife in Africa: Integrated conservation and development projects and beyond. *BioScience,* **50(7),** pp. 585–592.

Noss, A.J., 1997: Challenges to nature conservation and development in Central African forests. *Oryx,* **(31),** pp. 180–188.

OECD (Organisation for Economic Co-operation and Development), 1993: *Coastal Zone Management: Integrated Policies.* OECD, Paris, France.

OECD, 1996: Globalisation: What challenges and what opportunities for government? Paper OCDE/GD (96) 64, OECD, Paris, France. Available at http://www1.oecd.org/puma/strat/pubs/glo96/toc.htm.

OECD, 2001a: Sustainable development: Critical issues. OECD, Paris, France. Available at http://www.oecd.org.

OECD, 2001b: Strategies for sustainable development: Practical guidance for development co-operation. OECD, Paris, France. Available at http://www.oecd.org.

OECD, 2002a: *Sustainable Development Strategies: A Resource Book.* Earthscan Publications, London, UK, and Sterling, VA. Available at http://www.nssd.net/working/resource/indexa.htm.

OECD, 2002b: Improving policy coherence and integration for sustainable development: a checklist. OECD Observer, Policy Brief, October 2002. Available at http://www.oecd.org/dataoecd/61/19/2763153.pdf.

OECD-CAD, 2001: *The Development Advisory Committee Guidelines: Strategies for Sustainable Development,* OECD, Paris, France. Available at *www.oecd.org/dataoecd/34/10/2669958.pdf*

OED (Operations Evaluation Department), 1996: *Environmental Assessments and National Action Plans.* OED, World Bank, Washington, DC.

OED (Operations Evaluation Department), 1997: *Review of Adjustment Lending in Sub-Saharan Africa.* OED, World Bank, Washington, D.C.

Olsen, S., J. Tobey and M. Kerr, 1997: A common framework for learning from ICM experience. *Ocean and Coastal Management,* **37(2),** pp. 155–174.

Olsen, S, J. Tobey, and L.Z. Hale, 1998: A learning-based approach to coastal management. *Ambio,* **27,** pp. 611–619.

Olsen, S. and P. Christie, 2000: What are we learning from tropical coastal management experiences? *Coastal Management,* **28,** pp. 5–18.

Olsen, S.B., 1993: Will integrated coastal management programs be sustainable: The constituency problem. *Ocean & Coastal Management,* **21,** pp. 201–225.

Olsen, S., 2003: Frameworks and indicators for assessing progress in integrated coastal management initiatives. *Ocean & Coastal Management,* **46,** pp. 347–361.

Olsson, P., C. Folke, and F. Berkes, 2003: Adaptive co-management for building resilience in social-ecological systems, Manuscript submitted to *Journal of Environmental Management,* 20 April 2003. Available at http://www.beijer.kva.se.

Ostrom, E., 1990: *Governing the Commons: The Evolution of Institutions for Collective Action.* Cambridge University Press, Cambridge, UK.

Ovejero, J., 1999: The contribution of biodiversity-related multilateral environmental agreements to sustainable development: A discussion of some of the issues, Prepared for Inter-Linkages: International Conference on Synergies and Coordination between Multilateral Environmental Agreements, 14, 15 and 16 July 1999, United Nations University, Tokyo, Japan. Available at http://www.geic.or.jp/interlinkages/docs/Ovejero.pdf.

Pearce, D.W. and A. Markandya, 1989: *Environmental Policy Benefits: Monetary Valuation.* OECD, Paris, France.

Pernetta, J. and D. Elder, 1993: *Cross-sectoral, Integrated Coastal Area Planning (CICAP): Guidelines and Principles for Coastal Area Development.* IUCN, Gland, Switzerland.

Persson, A., 2002: Environmental policy integration: An introduction, Discussion paper of the Policy Integration for Sustainability (PINTS) Project of the Stockholm Environment Institute, Stockholm, Sweden. Available at http://www.sei.se/policy/PINTS.

Poschen, P., 2000: Social criteria and indicators for sustainable forest management: A guide to ILO texts, Forest certification working paper no. 3, International Labour Office/GTZ—Programme Office for Social and Ecological Standards, Eschborn, Germany, 85 pp..

Post, J.C. and C.G. Lundin, 1996: *Guidelines for Integrated Coastal Zone Management.* World Bank, Washington, DC.

Prabhu, R., C.J.P. Colfer, and R. Dudley, 1999: Guidelines for developing, testing and selecting criteria and indicators for sustainable forest management. CIFOR, Jakarta, Indonesia. Available at http://www.cifor.cgiar.org/acm/methods/toolbox1.html.

Putz, F.E., D.P. Dykstra, and R. Heinrich, 2000: Why poor logging practices persist in the tropics. *Conservation Biology,* **14(4),** pp. 951–956.

Ramsar (The Ramsar Convention on Wetlands), 2002: Resolution VIII.4 on integrated coastal zone management (ICZM). Available at http://www.ramsar.org/key_res_viii_04_e.htm.

Ramsar Convention Secretariat, 2004: River basin management handbook. Ramsar Convention Sectretariat, Gland, Switzerland. Available at http://www.ramsar.org/wurc_handbook_index.htm.

Rice, R.E., C.A. Sugal, S.M. Ratay, and G.A. Fonseca, 2001: Sustainable forest management: A review of conventional wisdom. *Advances in Applied Biodiversity Science,* **3,** pp. 1–29.

Ricupero, R., 2001: Trade and environment: Strengthening complementarities and reducing conflicts. In: *Trade, Environment and the Millennium,* G.P. Sampson and W.B. Chambers (eds.), 2nd ed.. United Nations University Press, Tokyo, Japan.

River Basin Initiative, 2004: River Basin Initiative. Available at http://www.riverbasin.org.

Salafsky, N., H. Cauley, G. Balachander, B. Cordes, J. Parks, et al., 2001: A systematic test of an enterprise strategy for community-based biodiversity conservation. *Conservation Biology,* **15(6),** pp. 1585–1595.

Sampson, G.P. and W.B. Chambers (eds.), 2002: *Trade, Environment and the Millennium,* 2nd ed.. United Nations University Press, Tokyo, Japan.

Sanchez, D. and K. Cash, 2003: Reducing poverty or repeating mistakes? A civil society critique of Poverty Reduction Strategy Papers, Church of Sweden Aid, Diakonia, Save the Children Sweden and The Swedish Jubilee Network, December 2003.

Sayer, J. and Campbell, B. 2004: *The Science of Sustainable Development: Local Livelihoods and the Global Environment.* Cambridge University Press, Cambridge, UK.

Scheffer, M., F. Westley, W.A. Brock, and M. Holmgren, 2002: Dynamic interaction of societies and ecosystems: Linking theories from ecology, economy and sociology. In: *Panarchy: Understanding Transformations in Human and Natural Systems,* L.H. Gunderson and C.S. Holling (eds.), Island Press, Washington, DC, pp. 195–240.

Schneider, R.R., 1995: *Government and the Economy of Amazon Frontier.* World Bank, Washington, DC.

Shahin, M., 2002: Trade and environment: How real is the debate? In: *Trade, Environment and the Millennium,* G.P. Sampson and W.B Chambers, (eds.), 2nd ed.. United Nations University Press, Tokyo, Japan.

Smith, R.D. and E. Maltby, 2001: Using the ecosystems approach to implement the CBD: A global synthesis report drawing lessons from the three regional pathfinder workshops. Royal Holloway Institute for Environmental Research (RHIER), University of London, UK. Available at http://www1.rhbnc.ac.uk/rhier/iucn.htm.

Smith, R.D., E. Maltby, A.H.B.H. Shah, A.F. Mohamed, M.b. Osman, Dato, and Z.A. Hamid (eds.), 2000b: *The Ecosystem Approach under the CBD: From Concept to Action,* Report from Southeast Asia Regional Pathfinder workshop, Kuala Lumpur, Malaysia, 30 October–1 November. Available at http://www1.rhbnc.ac.uk/rhier/iucn.htm.

Smith, R.D., E. Maltby, M. Kokwe, H. Masundire, and E. Hachileke (eds.), 2000a: *The Ecosystem Approach under the CBD: From Concept to Action,* Report from Southern Africa Regional Pathfinder workshop, Victoria Falls, Zimbabwe, 17–19 July . Available at http://www1.rhbnc.ac.uk/rhier/iucn.htm.

Sobral, L., A. Veríssimo, E. LIma, T. Azevedo, and R. Smeraldi, 2002: *Acertando o Alvo 2: Consumo de madeira amazônica e certificação no Estado de São Paulo,* Imazon/Imaflora/Amigos da Terra, Belém, Brazil, p.71.

Sorensen, J., 1997: National and international efforts at integrated coastal management: Definitions, achievements and lessons. *Coastal Management,* **25,** pp. 3–41.

Thia-Eng, C., 1993: Essential elements of integrated coastal management. *Ocean & Coastal Management,* **21(1–3),** pp. 81–108.

Treby, E.J. and M.J. Clark, 2004: Refining a practical approach to participatory decision-making: An example from coastal zone management. *Coastal Management,* **32,** pp. 353–372.

UN (United Nations), 2003: UN General Assembly Resolution 58/129.

UNCED (UN Conference on Environment and Development), 1992: Report of the UNCED, 3–14 June 1992, Rio de Janeiro, Brazil.

UNDESA (UN Department of Social and Economic Affairs), 2002: Key outcomes of the summit, September, UNDESA. Accessible at http://www.johannesburgsummit.org / html / documents / summit_docs / 2009_keyoutcomes_commitments.doc.

UNDP (UN Development Programme), 2003a: *Making Global Trade Work for People.* UNDP and Earthscan Publications Ltd., New York, NY.

UNDP, 2004: *Water Governance for Poverty Reduction: Key Issues and the UNDP Response to Millennium Development Goals.* UNDP, New York, NY. Available at http://www.undp.org/water.

UNEP (UN Environment Programme), 1995: *Guidelines for Integrated Management of Coastal and Marine Areas: With Special Reference to the Mediterranean Basin.* UNEP—PAP/RAC, Split, Croatia.

UNEP, 1996: *Guidelines for Integrated Planning and Management of Coastal and Marine Areas in the Wider Caribbean Region.* UNEP Caribbean, Kingston, Jamaica, p.141.

UNEP, 2001: International environmental governance: Report to the executive director, Background paper for Open-Ended Intergovernmental Group of Ministers or their Representatives on International Environmental Governance First Meeting; New York, 18 April 2001. Available at http://www.unep.org/IEG/WorkingDocuments.asp.

UNEP, 2002a: *The Johannesburg Principles on the Role of Law and Sustainable Development,* adopted at the Global Judges Symposium held on 18–20 August 2002, Johannesburg, South Africa.

UNEP, 2002b: Report of the Global Judges Symposium on Sustainable Development and the Role of Law, UNEP/GC.22/INF/24, 12 November 2002.

UNEP, 2002c: Atlas of International Freshwater Agreements. UNEP, Nairobi, Kenya. Available at http://grid2.cr.usgs.gov/reports.php3#Treaties.

UNEP, 2003: Lessons learned from case-studies: Note by the executive director, Background document presented at the Expert Meeting on the Ecosystem Approach, 7–11 July 2003. Montreal, QC, Canada.

UNEP, 2004: UNEP's strategy on land use management and soil conservation, Policy series 4, Division of Policy development and Law. Available at http://www.unep.org/land.

UNEP and IISD (International Institute for Sustainable Development), 2000: *Trade and Environment: A Handbook.* IISD and UNEP. Available at http://www.iisd.org/trade/handbook/default.htm.

UNEP and RRCAP (Resource Centre for Asia and the Pacific), 2003: *Enviromental Assessment and Early Warning Strategy for 2003/2004: Appendix 2 and 3.* UNEP/RRCAP, Pathumthani, Thailand.

UNESC (UN Economic and Social Council), 1999: Integrating environmental considerations into sectoral policies: note by the secretariat, UNESC, Economic Commission for Europe, Committee on Environmental Policy. Available at http://www.unece.org/env/documents/1999/cep/cep.1999.3.e.pdf.

UNESC, 2002: *Implementing Agenda 21: Report to the Secretary-General,* UNESC, Document prepared for Commission on Sustainable Development acting as the preparatory committee for the World Summit on Sustainable Development Second Session 28 January–8 February 2002. Available at http://www.johannesburgsummit.org.

UNESCAP (United National Economic and Social Council for Asia Pacific), n/d: Integrating environmental concerns into economic decision making, UMC/AIT and Cardiff University, UNESCAP. Available at http://t062.cpla.cf.ac.uk/wbimages/iecedm/resources.html.

UNESCO (UN Educational, Scientific and Cultural Organization), 2000: Solving the puzzle: The ecosystem approach and biosphere reserves. UNESCO, Paris, France. Available at http://unesdoc.unesco.org/images/0011/001197/119790eb.pdf.

UNU (UN University), 2002a: Making integrated solutions work for sustainable development: UNU report to the World Summit on Sustainable Development. UNU, Tokyo, Japan. Available at http://update.unu.edu/downloads/WSSDfinal.pdf.

UNU, 2002b: The Pacific Islands countries case studies: Inter-linkages, synergies and coordination among multilateral environmental agreements. UNU, Tokyo, Japan. Available at http://www.unu.edu/inter-linkages/docs/Policy/04_PIC.pdf.

US Commission on Ocean Policy, 2004: Preliminary report of the US Commission on Ocean Policy: Governors' draft, April 2004. US Commission on Ocean Policy, Washington, DC. Available at http://oceancommission.gov/documents/prelimreport/00_complete_prelim_report.pdf.

US Interagency Ecosystem Management Task Force, 1995: *The Ecosystem Approach: Healthy Ecosystems and Sustainable Economies,* Report of the Interagency Ecosystem Management Task Force, Volume I Overview PB95–265583. National Technical Information Service, U.S. Department of Commerce, Springfield, VA..

Van Toen, C., 2001: Delegates perceptions on synergies and the implementation of MEAs: Views from the ESCAP region, Background paper prepared for the Informal Regional Consultation, Workshop on Inter-Linkages: Synergies and Coordination among Multilateral Environmental Agreements.

WBGU (Wissenschaftliche Beirat der Bundesregierung Globale Umweltveränderungen – German Advisory Council on Global Change), 2001: *The World in Transition: New Structures for Global Environmental Policy.* Earthscan, London, UK.

WCED (World Commission on Environment and Development), 1987: *Our Common Future.* Oxford University Press, Oxford, UK.

Wells, M., 1995: Biodiversity conservation and local development aspirations: New priorities for the 1990s. In: *Biodiversity Conservation,* C. A. Perrings, K.G. Mäler, C. Folke, C. S. Holling, and B.O. Jansson (eds), Kluwer Academic Publishers, Dordrecht, The Netherlands.

Wells, M., K. Brandon, and L. Hannah, 1992: *People and Parks: Linking Protected Areas with Local Communities.* The World Bank, WWF and US Agency for International Development, World Bank, Washington, DC.

Wells, M., S. Guggenheim, A. Khan, W. Wardojo, and P. Jepson, 1999: *Investing in Biodiversity: A Review of Indonesia's Integrated Conservation and Development Projects.* World Bank, East Asia Region, Washington, DC.

Wijewardana, D., 1998: *Criteria and Indicators for Sustainable Forest Management,* International Tropical Timber Organization. *Tropical Forest Update,* **8(3),** pp. 4–6.

Wilkie, M.L., 2003: Forest management and the battle of paradigms. *Unasylva* **54(214/215),** pp. 6–7.

Winterbottom, R., 1990: *Taking stock: The Tropical Forestry Action Plan after Five Years.* World Resources Institute, Washington, DC.

Witte, J.M., C. Streck, and T. Brenner (eds.), 2003: *Progress or Peril: Partnerships and Networks in Global Environmental Governance.* The Post-Johannesburg Agenda, Global Public Policy Institute, Washington, DC/Berlin, Germany.

Wollenberg, E., D. Edmunds, L. Buck, J. Fox, and S. Bodt, 2001: Social learning in community forests. Center for International Forestry Research and East-West Center, Bogor, Indonesia. Available at http://www.cifor.cgiar.org/publications.

Worah, S., 2000: International history of ICDPs. In: *Proceedings of Integrated Conservation and Development Projects Lessons Learned Workshop, June 12–13, 2000, Hanoi, Vietnam,* UNDP (ed.), UNDP/World Bank/WWF, Hanoi, Vietnam.

World Bank, 1996: *World Bank Participation Sourcebook.* World Bank, Washington, DC.

World Bank, 2000: *Environmental Action Plans: Operational Manual—OP 4.02.* World Bank, Washington, DC. Available at http://www.worldbank.org.

World Bank, 2001: Making sustainable commitments: An environment strategy for the World Bank. World Bank, Washington, DC. Available at http://lnweb18.worldbank.org/ESSD/envext.nsf/41ParentDoc/EnvironmentStrategyEnvironmentStrategyDocument?Opendocument.

World Commission on Dams, 2000: *Dams and Development: A New Framework for Decision-Making: The Report of the World Commission on Dams.* Earthscan Publications, London, UK. Available at http://www.dams.org/report/contents.htm.

WRI (World Resources Institute), 2003: *World Resources 2002–2004: Decisions for the Earth: Balance, Voice, and Power.* WRI, Washington, DC. Available at http://www.wri.org.

WTO (World Trade Organization), 2001: Doha ministerial declaration. WTO. Available at http://www.wto.org.

WWF (World Wildlife Fund), 2003: *Managing Rivers Wisely: Lessons from WWF's Work for Integrated River Basin Management.* WWF International, Washington, DC. Available at http://www.panda.org/news_facts/publications/freshwater/index.cfm.

Young, O., 2002: *The Institutional Dimensions of Environmental Change: Fit, Interplay, and Scale.* MIT Press, Cambridge, MA.

Chapter 16

Consequences and Options for Human Health

Coordinating Lead Authors: Carlos Corvalan, Simon Hales, Alistair Woodward
Lead Authors: Diarmid Campbell-Lendrum, Kristie Ebi, Fernando De Avila Pires, Colin L. Soskolne
Contributing Authors: Colin Butler, Andrew Githeko, Elisabet Lindgren, Margot Parkes
Review Editors: Paul Epstein, Jorge Rabinovich, Philip Weinstein

---

BOXES

FIGURES

TABLES

Main Messages

Human health is both a product and a determinant of well-being. Measures to ensure ecological sustainability would safeguard ecosystem services and therefore benefit health in the long term (*high confidence*). In this chapter, health is the central concern, while noting the reciprocal relationships with other determinants of well-being. Negative health effects of ecosystem disruption are already evident in many parts of the world. In the long term, some ecosystem change is inevitable. To limit the damage to human health caused by these changes, mitigation strategies that reduce the driving forces of consumption, population increase, and inappropriate technology use are needed.

Ecosystem disruption damages health through complex pathways. Local conditions exert a very strong influence on the nature, extent, and timing of the effects on health. Social adaptations may minimize, displace, or postpone effects of ecosystem disruption on human health, but there are limits to what can be achieved. Human societies have developed methods (such as agricultural systems or water supplies) that allow natural resources to be appropriated for social benefit. Piped water supplies and other man-made resource appropriation systems provide human populations with a buffer in times of environmental change. These social adaptations are usually designed to minimize local impacts. Many effects of ecosystem disruption on health are displaced, either geographically (such as the costs of rich countries' overconsumption) or into the future (for example, long-term consequences of climate change or desertification).

To understand the potential negative health impacts of ecosystem change, two aspects need to be considered: the current vulnerability of the population affected and their future adaptive capacity. These two considerations are closely related, since vulnerable populations are less able to plan and implement adaptive responses. Vulnerability and adaptive capacity are also tied to other aspects of well-being (material minimum, freedom and choice, social relations and security).

Decisions about health and ecosystems must consider how one is related to the other. Choices that are made about the management of ecosystems can have important consequences for health, and vice versa. Consideration of ecosystem change enlarges the scope of health responses by highlighting "upstream" causes of disease, injury, and premature death. The health sector can make an important contribution to reducing the damage caused by environmental disruptions, but the greatest gains will be made by interventions that are partly or wholly placed in other sectors. The health sector bears responsibility for revealing the links and indicating which interventions are needed. Decision-makers need to consider the connections between health and other sectors. Where there are trade-offs, it is important for politicians, regulators, and the public to understand the consequences of taking one path in preference to another.

Where a population is weighed down by disease related to poverty and lack of entitlement to essential resources such as shelter, nutritious food, or clean water, the provision of these resources should be the priority for healthy public policy (*high confidence*). The links between ecosystems and human health are seen most clearly among deprived communities, which lack the "buffers" that the rich can afford. Within poor communities, poverty-related diseases are more prevalent among women and children, often due to culturally related resource distribution. Poor communities are the most directly dependent upon productive ecosystems. This means that the poorest and most disadvantaged communities can be among the first to benefit from ecosystem protection. There are economic considerations also: a healthy community is more capable of sustaining local ecosystems than an unhealthy one.

Where ill-health is caused, directly or indirectly, by excessive consumption of ecosystem services, substantial reductions in overconsumption would have major health benefits while simultaneously reducing pressure on life-support systems (*high confidence*). Both human health and the environment would benefit from a reduction in overconsumption. This would improve health in the short term as well as contribute to short-term ecological sustainability. Implementing better transportation practices and systems could lead to decreased injuries, increased physical activity in sedentary populations, as well as reduction in local air pollution and greenhouse gas emissions. Integrating national agriculture and food security policies with the economic, social, and environmental goals of sustainable development could be achieved, in part, through ensuring that the environmental and social costs of production and consumption are more fully reflected in the price of food and water. Reduced consumption of animal products in rich countries would have benefits for human health and for ecosystems.

Society needs to balance technological and institutional development. To achieve this balance, governments must incorporate environmental, social, and health costs (both gains and losses) into measurements of progress.

Approaches that perpetuate or worsen social inequalities may protect the health of privileged populations, but are likely to result in the worst global health outcome overall. Globalization processes increasingly link the health and well-being of privileged with poor populations. A selective approach whereby the health and well-being of a small fraction of the global population is promoted at the expense of the majority entails a high risk for both populations.

16.1 Introduction

There are well-defined relationships between health and the other components of well-being as defined in the MA framework. Material lack, for example, is a strong determinant of health (and indeed of other aspects of well-being). Both at the country level and within countries, poorer communities have a worse health profile than richer ones. Among poor communities, women and children often bear the largest burdens of disease (WHO 2002). At the global level, poorer countries are still battling traditional hazards such as lack of clean water and sanitation, which contribute considerably to the burden of disease in these countries. For example, the African region with 11% of the world's population has over 50% of the world's burden of disease resulting from infectious and parasitic diseases; in contrast, in the European region, with 14% of the world's population, the burden of disease in this category is less than 2% of the world's total (based on WHO regions, measured in "disability-adjusted life-years" (WHO 2002). (See Box 16.1.)

On the other hand, the lack of good health is a major determinant of poverty. For example, it is estimated that Africa's GDP could have been $100 billion larger if malaria had been eliminated some 35 years ago (WHO 2000). GDP declines by about 1% when more than 20% of the population is infected with HIV (WHO 2002); see also reports of the Commission on Macroeconomics and Health (WHO 2001).

Several studies have shown an important connection between social relations and health. People with good social networks live longer and are generally healthier (Skrabski et al. 2003). Similarly, people who fall ill recover faster when good social networks are in place. Lack of security (or vulnerability) is also associated with morbidity or mortality, although the relationship is often confounded with material lack (that is, vulnerable communities are often poor). Communities and individuals can be vulnerable for other than economic reasons. For instance, all inhabitants of low-

BOX 16.1

Disease Burden and Summary Measures of Population Health

The disease burden encompasses the total amount of disease or premature death within the population. Comparing burden fractions attributable to several different risk factors requires, first, knowledge of the severity/disability and duration of the health deficit, and second, the use of standard units of health deficit. The widely used Disability-Adjusted Life Year (DALY) is the sum of:

- years of life lost from premature death (YLL)
- years of life lived with disability (YLD).

YLL takes into account the age at death. YLD takes into account disease duration, age of onset, and a disability weight that reflects the severity of the disease.

To compare the attributable burdens for disparate risk factors we need to know (1) the baseline burden of disease, in the absence of the particular risk factor, (2) the estimated increase in the risk of disease/death per unit increase in risk factor exposure (the "relative risk"), and (3) the current or estimated future population distribution of exposure. The avoidable burden is estimated by comparing projected burdens under alternative exposure scenarios.

An example of an application of this method in the field of ecosystem change and health is the assessment of the burden of disease attributable to climate change.

lying islands are vulnerable to the effects of sea level rise, although their individual responses will vary depending on economic and social conditions (Nurse and Sem 2001). Lack of control is another important cause of vulnerability. Many indigenous populations, for example, face ecosystem changes introduced by forces outside their control (such as economic interests), and these threats can have an impact on their overall well-being, including their mental health.

For most diseases, the burden of disease is not borne evenly by all members of a community. For example, children and pregnant women are at much greater risk for morbidity and mortality from malaria, particularly if malnourished, whereas morbidity and mortality due to heat waves is highest among the elderly (Kilbourne 1997; Greenwood and Mutabingwa 2002). In general, the vulnerability of a population to a health risk depends on the level of material resources, effectiveness of governance and civil institutions, quality of public health infrastructure, access to relevant information, and existing burden of disease (Woodward et al. 1998). These factors are not uniform across a region or nation; rather, there are geographic, demographic, and socioeconomic differences. Failure to understand the reasons why particular population subgroups are vulnerable to a health outcome can reduce the effectiveness of response options.

Exposure to hazards is often a function of resources, with the resource-poor most likely to be in harm's way. For example, in impoverished communities, economically tied to large urban centers, but unable to afford safe housing, mudslides cause hundreds of deaths each year (IFRCRCS 2002). Even where there is no economic or social differentiation in the exposure to hazards, the impacts on health may vary considerably from place to place or group to group.

Figure 16.1 describes the links between ecosystem services and well-being with a focus on human health. Note that ecosystem integrity and human health do not always go hand in hand, in the short term at least. Populations may flourish in degraded environments, but only where it is possible to import resources and services from elsewhere.

16.1.1 Overview of Health in the Context of Ecosystems

Definitions of health vary across cultures. Some cultures focus on physical evidence of bodily structure and function; others have a much broader conception. For instance, for the indigenous people of New Zealand, the Maori, dimensions of health include access to heritage and a sense of communion with the environment (Durie 2001). In its constitution, the World Health Organization defines health as "a state of complete physical, mental and social well-being and not merely the absence of disease or infirmity." However it is framed, health serves both as a cause of well-being and as a consequence. In the absence of good health, it is difficult to claim a state of well-being, but many of the components of well-being (such as shelter, sustenance, and social relations) are themselves important determinants of health status.

Each of the categories of services provided by ecosystems is relevant to health. For instance, provisioning services include the production of food and fiber. In their capacity as regulators of the environment, ecosystems influence the quality and flow of water and the local climate. If these services are impaired, the impacts on health tend to be direct, and relatively acute. More difficult to demonstrate are the effects of cultural services provided by ecosystems—these are nonmaterial benefits such as spiritual, recreational, and aesthetic outcomes. But even if these aspects of human experience are not encompassed in the definition of health, they do have an influence on physical and mental functioning. The history of colonization, for example, shows clear links between loss of spiritual and cultural identity and rates of disease and premature mortality (Kunitz 1994).

The links between ecosystem change and cultural services are described in detail in the next chapter. The effects of ecosystem disturbance on human health may be relatively direct, or occur at the end of long causal chains, dependent on many intermediate events, and subject at many points to modifying influences. There are also relatively direct links between human health and the health of animal populations that share a common environment (Epstein et al. 2003). (See also MA *Current State and Trends,* Chapter 14.)

As an example of long causal chains, environmental changes affecting river flows might lead to disputes over water rights, social unrest, forced migrations of large populations, conflict, and, indirectly, increased rates of disease and injury (WCD 2000). The connections between ecosystem functioning and human health may be bi-directional. Thus where there is environmental disruption leading to poor health, there may be compounding effects and "vicious cycles" established (Woodward et al. 2000). For example, land degradation and soil loss leads to crop failure, hunger, and health problems. These health problems will more likely be experienced by women and children, and, as such, can affect not only current but future health through poor growth, increased disease burdens in later life, less productivity, etc. But there are effects in the other direction also: populations with high levels of chronic health problems can put less energy and time into growing crops, preventing erosion, and managing agricultural resources.

The difference between direct and indirect effects applies to the temporal and spatial scales over which these effects occur. Owing to the many intermediate factors that may be involved,

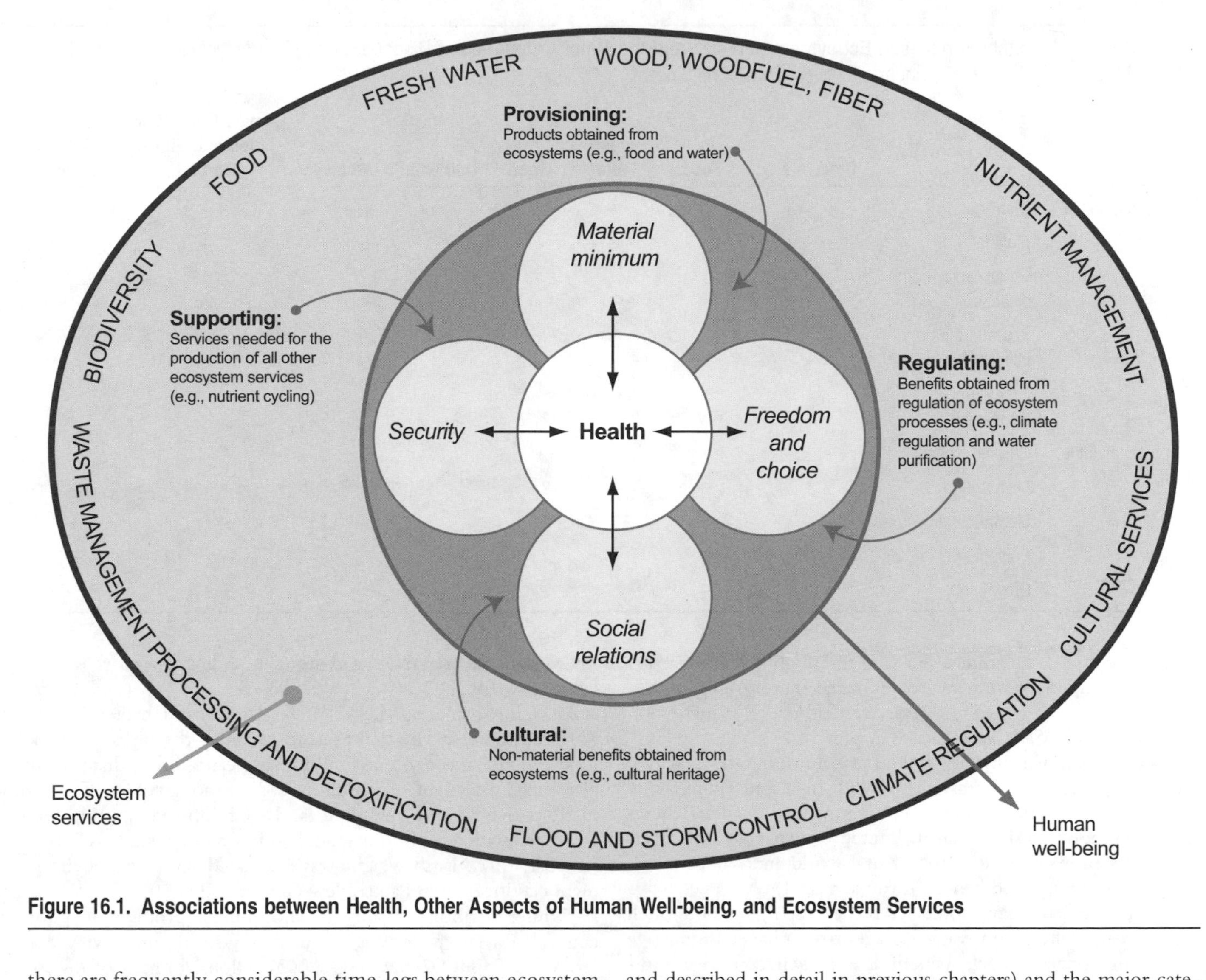

Figure 16.1. Associations between Health, Other Aspects of Human Well-being, and Ecosystem Services

there are frequently considerable time-lags between ecosystem change and health outcomes. For example, loss of biodiversity may lead to higher mortality and morbidity via diminishing supplies of bio-pharmaceuticals, but this would be apparent only after some years. In terms of spatial scales, we are most familiar with local effects (such as flooding and mudslides on steep denuded hillsides). More difficult to identify, but perhaps even more important for human health in the long term, are regional and global changes such as acid rain, stratospheric ozone depletion, and the accumulation of greenhouse gases.

Environmental and health policies are often determined without regard for one another, but there are important instances in which decisions have been swayed by health considerations. Removing lead from vehicle fuels is one case—this resulted from the accumulating evidence of risks to child health and has had far reaching consequences for ecosystems worldwide (Reuer and Weiss 2002).

16.1.2 Impacts of Ecosystem Goods and Services on Health

There are established links between the state of ecosystems and the condition of populations that depend on these ecosystems. Some of these links are shown in Table 16.1, which attempts to summarize the complex relationships between ecosystem goods and services (as defined in the Millennium Ecosystem Assessment, and described in detail in previous chapters) and the major categories of disease. The table indicates the likely strength of the associations, based on knowledge of "downstream" causal pathways leading to disease. There is a high level of uncertainty about these judgments, because few studies provide quantitative evidence of associations between ecosystem change and disease.

16.1.2.1 Biodiversity

Biodiversity underpins the resilience of the ecosystems on which humanity depends. Loss of biodiversity is occurring at an unprecedented rate, driven by overexploitation of productive ecosystems, other land use changes, climate change, pollution events such as oil spills, the transboundary migration of pollutants and hazardous substances, introduced species, and biotechnology. (See MA *Current State and Trends,* Chapter 4.) This depletion of biodiversity threatens vital ecosystem services, including food, fuel and fiber, fresh water, nutrient cycling, waste processing, flood and storm protection, and climate stability. One obvious direct impact of the loss of biodiversity is a reduction in sources of potential therapeutic chemicals. In general, the links between biodiversity loss and human health are difficult to demonstrate scientifically, due to the many factors that may confound such an association, difficulties in modeling nonlinear relationships, and lack of suitable data at appropriate scales (Sieswerda et al. 2001; Huynen et al. 2004). The clearest evidence of an association

Table 16.1. Relationships between Ecosystem Services and the Major Categories of Disease. Strength of evidence: "+ + +" High, "+ +" Medium, "+" Low, "?" Uncertain, "−" None or not known.

		Biodiversity	Food	Fresh Water	Wood	Nutrients	Waste	Climate Regulation, Flood and Storm Control	Cultural Services
Infectious parasitic diseases	Diarrhea	+ + +	+ + +	+ + +	−	−	+ + +	+ + +	−
	Malaria	+ + +	−	+ + +	−	−	+	+ +	−
	Other vector-borne disease	+ +	−	+ +	−	−	+	+ +	−
	Acute respiratory infection	+ +	+	−	+ +	−	?	?	−
	Other infectious diseases	+ +	+	+ +	?	?	−	?	−
Noncommunicable diseases	Chronic diseases	+ +	+	+	+ +	?	+ +	+	−
	Malnutrition	+ + +	+ + +	+	?	+ +	+	+ +	−
	Mental conditions	+ +	?	+	?	?	+	+ +	+ +
Injuries	Poisonings	−	+	−	?	+	+	−	−
	Drowning	−	−	−	−	−	−	+ +	−

probably comes from studies showing that high species diversity can be an important influence on reduced transmission of zoonotic diseases such as Lyme disease (Ostfeld and Keesing 2000). (See MA *Current State and Trends,* Chapter 14.)

Diversity and health are linked also in agriculture, where mono-cropping has been associated with increased vulnerability to acute food shortages and longer-term nutrient deficiencies (Waltner-Toews 2001). There is limited evidence of an association between experience of the natural world and reduced sickness rates and improved healing (Frumkin 2001).

Human societies have flourished by developing methods (such as settled agriculture and water storage) that enhance productive ecosystem services for social benefit. Especially in countries dominated by market economies, these adaptations are often designed to minimize short-term, local ecological disturbances, while maximizing profits. There is a mismatch of scale between social and ecological systems (Berkes and Folke 1998).

One result of this is that effects of ecosystem disruption on health are often displaced geographically (such as the costs of rich countries' overconsumption—climate change being a good example, in which many of the adverse health effects are likely to appear first in low carbon-emitting countries) or postponed into the future (for example, long-term consequences of climate change or desertification). But in general, the links between ecosystem change and human health are seen most clearly among impoverished communities, who lack the "buffers" that the rich can afford.

16.1.2.2 Food

The health of human populations is entirely dependent upon the services of productive ecosystems for food. This is most obvious in poor countries—especially in rural areas—where food is derived almost exclusively from local sources. Human dependence on ecosystems for nourishment is less apparent, but ultimately no less fundamental, in richer urban communities. Historically, loss of productive ecosystem services has led to the collapse of whole civilizations. For example, it has been suggested that the Mayan empire was lost near the end of the first millennium as a result of soil erosion, silting of rivers, and drought, leading to agro-ecosystem failure (UNEP 2002; Haug et al. 2003). See also MA *Current State and Trends,* Chapter 5; MA *Multiscale Assessment,* Chapter 2.

Few studies have attempted to quantify the links between food-producing ecosystems and human health. From first principles, such links might be seen most readily among vulnerable populations that live on marginal lands. Childhood stunting was associated with local land degradation in one such study (GRID/Arendal 1997). Birth weight was associated with land environment classification in Papua New Guinea (Allen 2002).

Undernutrition remains a major health problem in poor countries, where poverty is a consistently strong underlying determinant (WHO 2002; FAO 2003). Global burden of disease estimates indicate that in the year 2000, among the poorest countries, about a quarter of the burden of disease was attributable to childhood and maternal undernutrition. Among the rich countries, diet-related risks (mainly overnutrition) in combination with physical inactivity accounted for a third of the burden of disease. Worldwide, undernutrition accounted for nearly 10% of global DALYs (WHO 2002).

Aggregate food production is currently sufficient to meet the needs of all, yet of the present world population of just over 6 billion, about 800 million are underfed (FAO 2003), while hundreds of millions are overfed (WHO 2003a).

This imbalance has been driven primarily by social factors, though ecological factors may play an increasingly important role in the future. In poor countries, the number of people per hectare of arable land increased from three in 1961–63 to five in 1997–99 (WEHAB 2002a). Poverty and hunger have tended to force people onto marginal drought-prone lands with poor soil fertility. Where the conditions of poor communities are overshadowed by the need to earn foreign exchange for debt repayments, this can lead to the displacement of subsistence farming by cash crops grown for global corporations (Graber et al. 1995; McMichael 2001).

Agricultural production tripled in the last four decades, mainly through growth in yield. However, food production has not kept pace with population increase in many countries and improvements in yield appear to have slowed (UNEP 2002; WEHAB 2002a). It has been estimated that today, nearly a quarter of usable land has reduced productivity and about a billion people are affected by land degradation either through soil erosion, water logging, or salinity of irrigated land (DFID/EC/UNDP/World Bank 2002; UNEP 2002).

Providing sufficient food for an expected human population of 8–9 billion people will require major investments in poverty alleviation (Mellor 2002). There are also important trade-offs that have to be made between various possible uses of productive land. Including population health considerations in this weighing of choices could have important policy implications. The issue of overconsumption of food is relevant here, for several reasons. First, from economic first principles, overconsumption of food is encouraged by economic and trade practices, which prioritize short-term profit while externalizing longer-term environmental and social costs. Second, reductions in animal-based food consumption in rich countries could have important ecological benefits (WHO 2003a). Intensive meat production, in particular, has major adverse impacts on ecosystems (Leitzmann 2003, Reijnders and Soret 2003). (See also MA *Current State and Trends,* Chapter 8.)

16.1.2.3 Fresh Water

Fresh water is a key resource for human health; it is used for growing food, drinking, washing, cooking, and for the recycling of wastes. Of all available water globally, only 2.5% is fresh, and less than 1% is readily available in lakes, rivers, and underground. Worldwide, almost 4% of the global burden of disease is currently attributable to unsafe water, inadequate sanitation, and poor hygiene. In the next century, water resources will be strongly affected by trends in population, land use, and the management of freshwater ecosystems. Increasing demand for food, in particular, will worsen water scarcity. It is estimated that by 2025 nearly half the world population will live in river basins where water is scarce and 70% of readily available water supplies will be used (WEHAB 2002d). Water scarcity can lead to use of poorer quality sources of fresh water, which are more likely to be contaminated, tending to cause increases in water-related diseases.

At present, 1.1 billion people lack access to safe water supplies, while 2.6 billion people lack adequate sanitation (WHO/UNICEF 2004; UNESCO 2003a). Lack of improved water and sanitation is strongly associated with poverty, although this relationship varies between regions (WHO 2002). Along with sanitation, water availability and quality are well recognized as important risk factors for infectious diarrhea and other major diseases (Esrey 1996; Pruss et al. 2002; Strina et al. 2003; Thompson et al. 2003).

The associated effects on human health are severe. Poor countries, with inadequate provision of water and sanitation, will be most vulnerable to these effects that impact most severely on children. (See Table 16.2.) In addition to direct effects, there can be indirect health effects. For example, during a water shortage, women may have to walk further and spend additional time to supply households with water. This additional time and energy expenditure may affect a woman's health and her ability to earn an income and to care for household members.

The effects of climate change on water resources are difficult to forecast because of the many factors that influence rainfall, runoff, and evaporation. Nevertheless the best estimates are that climate change may increase the number of people affected by water stress by about 0.5 billion in 2025 (Arnell 1999). Increases in temperature would worsen water quality by increasing the growth of microorganisms and decreasing dissolved oxygen. Water-related disasters—droughts and floods—also have important health impacts. The frequency of heavy rainfall events is likely to increase, leading to an increase in flood magnitude and frequency and a reduction in low river flows (IPCC 2001). Heavy rainfall would tend to adversely affect water quality by increasing chemical and biological pollutants flushed into rivers and by overloading sewers and waste storage facilities. In some parts of the world, climate change also may increase requirements for irrigation water because of increased evaporation. (See also MA *Current State and Trends,* Chapter 7.)

16.1.2.4 Wood Fuel

Most of the world's population has no access, or limited access, to electricity supplies, and about two billion people must rely on wood, dung, and agricultural residues for heating and cooking, while rich countries typically consume 25 times as much energy per capita as do poor countries (WEHAB 2002b).

Lack of clean, safe power causes a range of health impacts. About half of the world's population still uses solid fuels for cooking and 0.5% of DALYs worldwide have been attributed to indoor air pollution from this source, particularly among women and children. Urban air pollution, resulting from the combustion of fossil fuels for transport, power generation, and industry, accounted for a further 0.5% of DALYs (WHO 2002). Outdoor air pollution aggravates heart and lung disease (Kunzli et al. 2000). Indoor air pollution causes a major burden of respiratory diseases among both adults and children (Ezzati et al. 2002; Smith and Mehta 2003).

Energy supplies are a fundamental factor in sustainable development and are also needed to provide and maintain modern health services. The need to spend considerable time collecting fuel can preclude proper education, especially of women, with indirect adverse effects on health through illiteracy, lost work opportunities, family health, and large family size. More indirectly still, energy use is linked to health effects via desertification, acidification, ambient air pollution, and climate change.

16.1.2.5 Nutrient Management

Application of agricultural fertilizers and organic wastes (including sewage) can improve agricultural yields but may also lead to increased concentrations of nitrogen and phosphorus in surface waters and coastal sea areas (Smil 2000). This can cause certain cancers (Wolfe and Patz 2002) and eutrophication in both marine and freshwater ecosystems, with overgrowth of bacteria, phytoplankton, macrophytes, and microalgae.

In turn, these problems can lead to increases in water-borne diseases and poisoning from harmful algal blooms (UNESCO 2003b). There are likely to be other ecological mechanisms by which increased nutrients can lead to human diseases, but further research is required to clarify these (NRC 1999; Townsend et al. 2003).

16.1.2.6 Waste Management, Processing, and Detoxification

Well-functioning ecosystems absorb and remove contaminants. For example, wetlands can remove excess nutrients from runoff, preventing damage to downstream ecosystems (Jordan et al. 2003). Inadequate management of solid waste increases human exposure to infectious disease agents (for example, via contamination of water with feces, or via disease vectors). This leads to a range of communicable diseases, especially diarrheal illness (WHO/UNICEF 2004; UNESCO 2003a). Of the 2.6 billion

Table 16.2. Water and Sanitation-related Diseases (WHO 2003c)

Disease	Disease Burden, in Disability-Adjusted Life-Years (thousands)	Mortality (deaths per year)	Relationship of Disease to Water Supply and Sanitation
Diarrheal diseases	61,966	1,797,970	strongly related to unsanitary excreta disposal, poor personal and domestic hygiene, unsafe drinking water; 90% of deaths in children under 5
Infection with intestinal helminths (ascariasis, trichuriasis, hookworm disease)	2,882	9,360	strongly related to unsanitary excreta disposal, poor personal and domestic hygiene; 133 million people suffer from high-intensity intestinal helminth infections
Schistosomiasis	1,702	15,370	strongly related to unsanitary excreta disposal and absence of nearby sources of safe water; 160 million people infected
Trachoma	2,329	150	strongly related to lack of face washing, often due to absence of nearby sources of safe water; 500 million people at risk; 6 million visually impaired
Malaria	46,486	1,272,390	related to poor water management, water storage, operation of water points, and drainage; 90% of deaths in children under 5
Onchocerciasis	484	(<5)	related to poor water management in large-scale projects
Dengue fever	616	18,560	related to drainage water organically polluted, open sewers, eutrophied ponds
Lymphatic filariasis	5,777	417	related to poor water management, water storage, operation of water points, and drainage

people who lack adequate sanitation, the majority live in Asia (Cairncross 2003).

When recycled appropriately, human waste can be a useful resource that promotes soil fertility (Esrey 2002). However, where waste contains persistent chemicals such as organochlorines or heavy metals, recycling onto land can lead to the accumulation of these pollutants and increased human exposure through food and water; this may contribute to a wide range of chronic diseases.

16.1.2.7 Climate Regulation

Climate regulation is an important property of Earth's natural systems. Each of the ecological services referred to above is sensitive to climate, and will be affected by climate change. Although climate change will have some beneficial effects on human health, most effects are expected to be negative (IPCC 2001).

Direct effects such as increased mortality from heat waves are readily predicted but indirect effects are likely to predominate (IPCC 2001; WHO/WMO/UNEP 2003). Human health is likely to be affected indirectly by changes in productive ecosystems and the availability of food, water, and energy supplies. These changes will in turn affect the distribution of infectious diseases, nutritional status, and patterns of human settlement. Changes in the geographic distribution, abundance, and behavior of plants and animals affect, and are affected by, biodiversity, nutrient cycling, and waste processing.

Attempts have been made to estimate the global burden of disease attributable to climate change (WHO 2002). But so far only a small fraction of the health outcomes associated with climate change have been included in the global burden of disease calculations, selected on the basis of sensitivity to climate variation, predicted future importance, and availability/feasibility of quantitative global models. (See Box 16.2.)

16.1.2.8 Flood and Storm Control

Climate extremes, including floods, storms, and droughts, have local and sometimes regional effects, both directly through deaths and injuries, and indirectly through economic disruption and population displacement. Extreme climate events are expected to increase as a result of climate change (WHO/WMO/UNEP 2003).

Health effects of climate extremes include physical injuries, increases in communicable diseases due to crowding, lack of clean water and lack of shelter, poor nutritional status, and adverse effects on mental health (Hajat et al. 2003).

One example was the floods along the Yangtze River in 1998. For years, loggers had been cutting forests along the river's watershed, and farmers and urban developers had gradually moved to occupy the river's flood plains by draining lakes and wetlands. Record rains fell in the Yangtze basin in the summer of 1998, and these degrading practices amplified the flooding, leaving 3,600 people dead, 14 million homeless, and $36 billion in economic losses. Restoring the ecosystem's flood control services would now take decades and billion of dollars (UNEP 2002).

Globally, the number of people killed, injured, or made homeless by natural disasters is increasing (WHO/WMO/UNEP 2003). An important reason for this is increasing settlement on coasts and floodplains that are exposed to extreme events. A number of case studies at the local scale have shown that human interactions with ecosystems have also contributed to increasing human vulnerability. Healthy ecosystems provide a buffer against the damaging effects of climate extremes. For example, forests absorb rainfall and reduce rapid increases in runoff, reducing flooding and soil erosion. Coral reefs and mangroves stabilize coastlines, limiting the damaging effect of storm surges. (See MA *Current State and Trends,* Chapters 9 and 16.)

In many areas the only land available to poor communities is that with few natural defenses against weather extremes. In recent decades, there has been a large migration to cities and more than half the world's population now lives in high-density urban areas. Such migration and increasing vulnerability means that even without increasing numbers of extreme events, losses attributable to each event will tend to increase (WHO/WMO/UNEP 2003).

BOX 16.2

Global Burden of Disease Attributable to Climate Change (WHO/WMO/UNEP 2003)

- Climate change will affect the pattern of deaths from exposure to high or low temperatures. However, the effect on actual disease burden cannot be quantified, as we do not know to what extent deaths during thermal extremes are in sick/frail persons who would have died soon. In 2030, the estimated risk of diarrhea will be up to 10% higher in some regions than if no climate change occurred. Since few studies have characterized this particular exposure-response relationship, these estimates are uncertain.
- Estimated effects on malnutrition vary markedly among regions. By 2030, the relative risks for unmitigated emissions, relative to no climate change, vary from a significant increase in the Southeast Asia region to a small decrease in the Western Pacific. Overall, although the estimates of changes in risk are somewhat unstable because of regional variation in rainfall, they refer to a major existing disease burden entailing large numbers of people.
- The estimated proportional changes in the numbers of people killed or injured in coastal floods are large, although they refer to low absolute burdens. Impacts of inland floods are predicted to increase by a similar proportion, and would generally cause a greater acute rise in disease burden. While these proportional increases are similar in industrial and developing regions, the baseline rates are much higher in developing countries.
- Changes in various vector-borne infectious diseases are predicted. This is particularly so for malaria in regions bordering current endemic zones. Smaller changes would occur in areas where the disease is currently endemic. Most temperate regions would remain unsuitable for transmission, because either they remain climatically unsuitable (as in most of Europe) or socioeconomic conditions are likely to remain unsuitable for reinvasion (for example, in the southern United States). Important causes of uncertainty in these forecasts include extrapolation between regions and the factors that translate potential into actual transmission.
- If our understanding of broad relationships between climate and disease is accurate, then climate change may already be affecting human health. The total current estimated burden is small relative to other major risk factors measured under the same framework. However, in contrast to many other risk factors, climate change and its associated risks are increasing rather than decreasing over time.

16.1.2.9 Cultural, Spiritual, and Recreational Services

Cultural services may be less tangible than material services but are nonetheless highly valued by people in all societies. People obtain diverse nonmaterial benefits from ecosystems. These benefits include recreational facilities and tourism, aesthetic appreciation, inspiration, a sense of place, and educational value. There are traditional practices linked to ecosystem services that have an important role in developing social capital and enhancing social well-being.

There is a hypothesis that stimulating contact with the rich and varied environment of ecosystems, including that of gardens, may benefit physical and mental health. There is limited evidence that this may help in the prevention and treatment of depression, drug addiction, behavioral disturbances, as well as convalescence from illness or surgery. Regular contact with pets seems to prolong and enhance the quality of life, especially in old age. Such contact with nature need not be physical; for example, some benefit may be obtained purely from visual or even visualized (imaginary) contact. On the other hand, it would follow that knowledge of the loss of valued ecosystems, even if such knowledge is indirect, may cause a profound sense of loss, and even harm health.

16.1.3 Health in Scenarios

The effects of ecosystem change on well-being in future decades are explored in this assessment using four scenarios. (See MA *Scenarios,* Chapter 11.) Each scenario describes a plausible future for the linked global socioecological system. Health is an "integrating" outcome of the distribution and interaction of ecosystem and human services. Institutions—the main legal, cultural, and attitudinal currents that flow through society—were found to be a crucial determinant of the protection of ecosystem services, human services, and human health.

In three of the four scenarios, global health was found—very broadly—to improve, while in one scenario the health of low-income populations, which currently constitute the majority of the global population, remained unchanged or worse. Importantly, caveats were identified in the three more optimistic scenarios, whereby each could have significant adverse health effects.

Society needs to balance technological and institutional development. Belief in the ability of technology alone to solve the human predicament is unwarranted. On the other hand, the size and environmental impact of the still-growing global population requires the extension and deepening of many forms of technology. The incorporation of economic, social, and health costs into measurements of progress is an important institutional change to facilitate this balance, by providing measurable feedback.

Approaches that perpetuate or worsen social inequalities may protect the health of privileged populations but are likely to result in the worst global health outcome overall. Globalization processes increasingly link the health and well-being of privileged with poor populations. A selective approach whereby the health and well-being of a small fraction of the global population is promoted at the expense of the majority entails a high risk for both populations. Humans possess the cognitive and organizational capacity to maintain or even improve global health in the next decades, but this will require substantial goodwill, cooperation, and work.

16.1.4 Typologies of Response Options and How They Apply to Health

International, national, and community responses to global ecosystem changes include policies aimed at stopping and reversing the extent and rate of change (mitigation) and response options designed to effectively reduce the current and future impacts of those changes (adaptation). It is recognized that an unusual degree of anticipatory thinking is required to develop proactive response options for reducing potential future ecosystem impacts. Such options should complement, not replace, mitigation policies to slow or avert the process of change itself.

The impacts of ecosystem changes will be site-specific and path-dependent; that is, they will depend on local circumstances (Yohe and Ebi 2005). For example, malaria epidemics occur following rainy seasons in some regions, while epidemics occur during droughts in others. Further, these impacts will not be experienced evenly across a population; there will be particularly vulnerable subgroups. Therefore, public health response options (interventions) need to be designed at spatial and temporal scales appropriate to the health outcome of concern, taking into consideration the social, economic, and demographic driving forces, and also whom the interventions should target. Interventions can focus on local, national, regional, and international scales; and within these, vulnerable subgroups.

As discussed in Chapter 3, the nature of the response options can be legal, economic and financial, institutional, social and behavioral, technological, and cognitive. As discussed in Chapter 19, within each of these, there may be gender issues that could affect not only the efficiency and effectiveness of interventions, but also future development. Effects on health may be complex, and follow a variety of causal pathways. For example, developments in agriculture that have dramatically lowered the cost of food in many countries have removed the threat of undernutrition, but have provided conditions for the emergence of new disease-causing agents (such as antibiotic-resistant *Salmonella*) (Waltner-Toews 2001).

The vulnerability of a particular population to the potential health impacts of ecosystem change will depend on the degree to which individuals and systems are susceptible to, or unable to cope with, these changes. Vulnerability depends upon the level of exposure, the sensitivity (or exposure-response relationship); and the response options in place to reduce the burden of a particular adverse health outcome (Ebi et al. 2005).

Populations, subgroups, and systems that cannot or will not adapt are more vulnerable, as are those who are more susceptible to ecosystem change. Population subgroups may not have the resilience to adapt because of a lack of material resources, lack of relevant information, lack of effective governance and civil institutions, and lack of public health infrastructure (Woodward et al. 2000). The effective targeting of interventions requires understanding which demographic or geographic subpopulations may be most at risk, the factors that contribute to their vulnerability, and which of these factors can be modified within the context of a particular time and location. Thus individual, community, and geographical factors determine vulnerability.

Response options can aim to reduce current and/or future vulnerability. Adaptive capacity describes the general ability of institutions, systems, and individuals to adjust to potential damages, to take advantage of opportunities, and minimize the long-term consequences (Smit et al. 2001). Specific options arise from the adaptive capacity of a population. Adaptive capacity encompasses coping capacity (what could be implemented now to minimize potential damage from ecosystem change) and the response options that have the potential to expand future coping capacity. Specific options arise from the coping capacity of a community, nation, or region. The primary goal of building adaptive capacity is to reduce future premature death, avoidable disease, and disease-related discomfort and disability in a population arising from ecosystem change. Examples illustrating these various concepts are shown in Table 16.3.

Response options encompass both spontaneous responses to ecosystem change by affected individuals and planned interventions by governments or other institutions. Examples of the latter include watershed protection policies or effective public warning systems for drinking water quality. In many cases, continuing and strengthening established interventions may be the best approach to reducing vulnerability and increasing adaptive capacity, while in other cases, new response options will need to be developed (Ebi et al. 2005). Increasing the adaptive capacity of a population shares similar goals with sustainable development—to increase the ability of nations, communities, and individuals to effectively and efficiently cope with the changes and challenges of ecosystem change. (See Chapter 19.) Public health scientists describe response options in terms of primary, secondary, and tertiary prevention. Primary prevention aims to prevent exposure to risk of disease in an otherwise unaffected population (for example, the supply of bednets to all members of a population at risk of exposure to malaria). Secondary prevention entails preventive actions in response to early evidence of health impacts (for example, strengthening disease surveillance and responding adequately to disease outbreaks such as the West Nile virus outbreak in North America). Tertiary prevention consists of measures to reduce long-term impairments and disabilities and to minimize suffering caused by existing disease. In general, secondary and tertiary prevention is less effective, and more expensive, than primary prevention.

The attributes of different risks affect the choice of response options, including spatial extent (the extent of land cover change or of an epidemic); speed of onset (how rapidly the event occurs, either building slowly like a drought or coming quickly like a flash flood; the slow spread of malaria or the rapid speed of an outbreak of influenza); the number of potentially affected individuals (the response to an isolated case of plague versus an epidemic of dengue fever); the onset-to-peak interval (how long it takes from the first detection to the maximum level of the hazard, such as the first impacts of a flood to its peak magnitude, or the first detected cases of a disease to its maximum prevalence); and the expected frequency or return period (frequency of drought or floods, periodicity of disease epidemics).

Other factors affecting choice of responses include knowledge and understanding of the underlying processes or causes; capacity to predict, forecast, and warn; capacity to respond (institutional and otherwise); how the risk might change over time and with ecosystem change; and ethical appropriateness.

Many of the possible response options to ecosystem change lie primarily outside the direct control of the health sector. They are rooted in areas such as sanitation and water supply, education, agriculture, trade, tourism, transport, development, and housing. Inter-sectoral and cross-sectoral integrated options are needed to reduce the potential health impacts of ecosystem change. These integrated interventions should address the social, economic, and demographic driving forces of and responses to ecosystem change.

Figure 16.2 follows an epidemiological, causal pathway approach (Corvalan et al. 2000). This highlights the main driving forces that are linked to health determinants (existing infrastructure, social values, and general social, economic, and demographic conditions); the specific exposures at different levels (either as distant, often indirect, or proximate, often direct, as well as ranging from global to local scales); the health impacts (or the positive health consequences if seen from the point of view of ecosystem protection); how these links are modified by population vulnerability; and how society (or individuals) respond, in the form of interventions at all levels (improving on the basic conditions under driving forces, reducing exposures or providing health-specific interventions).

16.2 Response Options and Actions outside the Health Sector

Factors that need to be considered when evaluating evidence that the protection of ecosystems avoided adverse health impacts include: the strength of the evidence; the plausibility of the association (that is, a probable or demonstrated etiologic chain); the presence of supporting or contradictory evidence from non-human systems; the extent that contextual factors and competing influences could explain the adverse health impact; the policies and interventions in place that could affect the exposure-response

Table 16.3. Examples of Current and Future Vulnerability and Adaptation (Kovats et al. 2003)

Definition	Current example	Future example
Vulnerability: degree to which individuals and systems are susceptible to or unable to cope with the adverse effects of climate change	populations living in areas on the fringe of the current distribution of malaria are at risk for epidemics if the range of the *Anopheles* vector changes	whether these populations might be vulnerable in the future depends, in part, on the implementation of effective prevention activities
Adaptation Baseline: the adaptation measures and actions in place in a region or community to reduce the burden of a particular health outcome	the exposure-response relationship is influenced by the current prevention measures aimed at reducing the burden of a disease; for example, the number of elderly adversely affected by a heat wave will depend on the numbers that have access to and use air conditioning during a heat wave	increasing access to and use of air conditioning will decrease the percentage of the elderly population that could be adversely affected by future heat waves; for example, the consequences of the 1995 heat wave in the midwestern United States were greater than those for a similar heat wave in 1999, in part because of programs established in the interim
Coping Capacity: the adaptation strategies, policies and measures that could be implemented now; specific adaptation plans arise from a region or community's coping capacity	a number of cities in mid-latitude countries have the level of material resources, effective institutions, and quality of public health infrastructure to establish and maintain early warning systems for heat waves; until implemented, these systems are within a city's coping capacity	over time, strategies, policies, and measures can move from being possible to being implemented (that is, being part of the adaptation baseline); for example, providing universal access to adequate quantities of clean water is not yet possible, although significant progress has been made
Adaptive Capacity: the general ability of institutions, systems, and individuals to adjust to potential damage, to take advantage of opportunities, or to cope with the consequences	adaptive capacity is the theoretical ability of a region or community to respond to the threats and opportunities presented by climate change. It is affected by a number of factors and encompasses coping capacity and the strategies, policies, and measures that have the potential to expand future coping capacity; for example, education of women provides a range of benefits to a population that results in increased ability to deal with challenges and changes	over time, it is hoped that regions and communities will increase their adaptive ability and their resilience to what future climates will bring

relationship; and the timing, scale, and location of the assessment (Scheraga et al. 2003). Assessments made at one point in time or at one location may provide different answers when the evaluations are repeated over time or over larger geographic areas.

16.2.1 Case Study: Climate Change, Land Use Changes, and Tick-borne Diseases—Illustrative Example from Sweden

Diseases transmitted by blood-sucking ticks are especially sensitive to changes in the local environment, particularly alterations caused by land use or by land cover changes and changed climatic conditions. (See MA *Current State and Trends,* Chapter 14.) The climate sets the limit for both the altitude and latitude distribution of ticks and is important for tick population density. Biodiversity and species composition may affect the transmission of pathogens in nature and, hence, the risk of disease in an area (LoGiudice et al. 2003).

Ixodid ticks, which live for up to three years, may transmit several diseases, of which the most important are Lyme disease and the severe form of tick-borne encephalitis. The latter is endemic in Europe and in most western parts of Eurasia, whereas Lyme disease is prevalent throughout the temperate zones of the Northern Hemisphere. About 85,000 cases of Lyme disease are reported in Europe annually compared to 15–20,000 cases in the United States (Steere 2001).

Over the last two decades, the incidence of Lyme disease and tick-borne encephalitis has increased in endemic regions. This is partly because of increased reporting through greater awareness among health personnel and the general public. However, case studies from Sweden have shown that a real increase in both tick population density and in disease incidence has occurred since the early 1980s, and that ticks have expanded their distribution range northward (Talleklint and Jaenson 1998). These changes have been associated with milder and shorter winters (Lindgren et al. 2000; Lindgren and Gustafson 2001).

Research findings have enabled preliminary predictions to be made every year in early spring; that is, prediction of whether the coming year is a potentially high-risk year for tick bites. Swedish newspapers, radio, and television news now address the risk of tick-borne encephalitis and Lyme disease repeatedly each year when the tick-activity season starts. High-risk areas are shown, new risk areas pointed out, and effective preventive measures are mentioned, such as removal of thick undergrowth vegetation in parks and gardens, and daily body inspection for rapid detection and removal of ticks. The latter decrease the risk for Lyme disease but do not protect against the transmission of tick-borne encephalitis. Before the high-risk season begins, tick-borne encephalitis vaccination is made easily accessible for people living or working in or visiting endemic areas.

16.2.2 Case Study: Responding to the Risk of Water-borne Campylobacteriosis

From hunter-gatherer societies through agricultural societies to industrial societies, human settlements have always centered on a

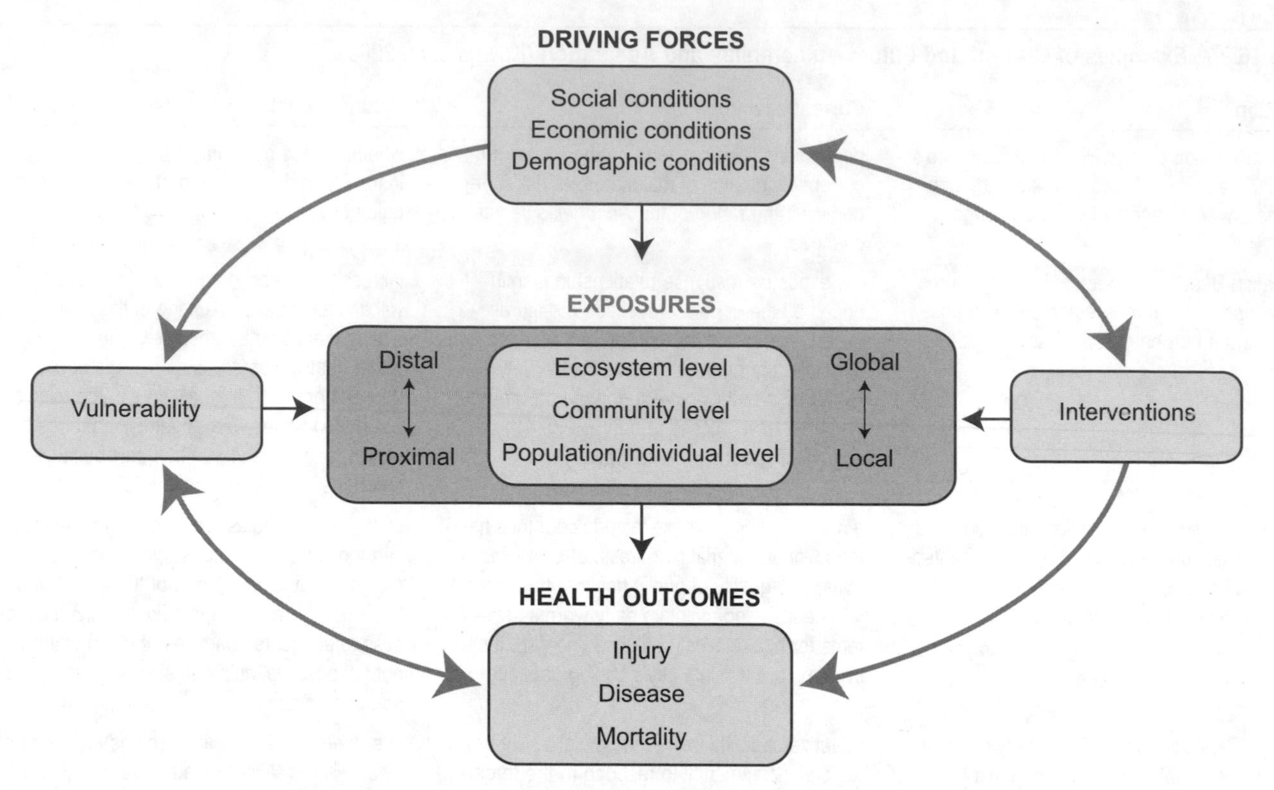

Figure 16.2. Causal Pathway from Driving Forces, through Exposures to Health Outcomes, in the Context of Ecosystem Change. The impacts are modified by the population's vulnerability and the interventions implemented.

reliable supply of good quality fresh water. When supplies have been disrupted, the effects of thirst upon health are immediate and can be rapidly fatal. When water quality has been compromised, we have seen some of the largest disease outbreaks the world has known. Human settlements have, therefore, always been dependent on healthy freshwater ecosystems to supply potable water, and water catchment protection is so ingrained in public health culture that it is often taken for granted. In modern times, water treatment plants have fulfilled a "magic bullet" role and have arguably taken the edge off the perceived importance of catchment protection—that is, until outbreaks of waterborne illness in rich countries started to seriously shake public confidence in public water supplies.

Campylobacterosis is a gastrointestinal disease that may be spread by food or by water and was first recognized as an "emerging" human disease in the late 1970s. Campyobacteriosis is now the most commonly reported infectious disease in rich countries. The disease is prevalent among domesticated animals such as poultry, sheep, and cattle, and transmission to humans depends on "survival trajectories" followed by the pathogen between excretion from the reservoir and ingestion by the case (Skelly and Weinstein 2003). The life-cycle of this organism can be complex and its survival in the environment is subject to the influence of a variety of abiotic factors. Pastoral farming has a major impact on both water flow and quality. As vegetation is lost from hillsides and riverbanks, the volume and speed of runoff increases. The natural purification of water percolating through soil and vegetation is also reduced. This exposes both livestock and humans downstream to a variety of zoonotic pathogens, including *Campylobacter, Cryptosporidium,* and *Giardia.*

Current preventive measures for controlling transmission and infection with *Campylobacter* include food and farm hygiene, thorough cooking (or irradiation) of food, use of pasteurized milk and chlorinated water supplies, and control of the disease in domestic and domesticated animals (Chin 2000). Although compliance with these measures is difficult to formally assess, there is little question that they contribute significantly to a reduction of the disease burden, and should be maintained and encouraged on that basis. However, they have failed to arrest the rapid rise of campylobacteriosis. It is appropriate, therefore, to also consider public health interventions based on restoring the health of freshwater ecosystems.

Slowing runoff is important because of the limited survival of fecal pathogens, whose half-lives are more likely to be exceeded before human exposure occurs. Waters from catchments with native vegetation are least likely to contain viable pathogens; revegetation could therefore be advocated as a public health intervention. Importantly, it is not only the direct transmission of *Campylobacter* in drinking or recreational water exposure that will be affected. If livestock infections are also decreased as a result of regrowth of native plants in water catchments, the number of human infections acquired occupationally (farm, abattoir) and by the food-borne route (animal products) will also be reduced.

The lesson from this case study is that, in many cases, scientifically based public health interventions can be devised only with an understanding of the ecology of the disease.

16.2.3 Case Study: Linking Ecosystems and Social Systems for Health and Sustainability—River Catchments

The management of river catchments poses an emerging "upstream" public health issue—spanning concerns regarding the safety and sustainability of freshwater ecosystems, socioeconomic development, and multistakeholder governance processes. As such, river catchment management has implications for both the environmental and socioeconomic determinants of health and exemplifies the importance of response options and actions *outside* the health sector.

During the 1990s, water governance priorities shifted from their developmental focus on infrastructure provision (domestic water supply, sanitation, and irrigation) to recognize the critical need for an ecosystems approach that manages freshwater resources as an integral part of natural cycles (UNCSD 1998; World Water Forum 2000; Helming and Kuylenstierna 2001). Priorities for water resource management at the turn of the twenty-first century include recognition and maintenance of: (1) catchments as critical to the management of freshwater ecosystems—enabling fresh waters to be viewed within a landscape or systems context; (2) the socioeconomic, ecological, and human health values of freshwater ecosystems, their services, and functions; (3) processes that support freshwater ecosystem integrity, structure, function, and adaptive capacity, including quality, quantity, and timing of flow (Baron et al. 2002); and (4) protecting the determinants of health through catchment management.

The place-based links between environmental and socioeconomic determinants of health were examined in a case study of catchment (ecosystems) and community (social systems) in New Zealand's Taieri River catchment. In the Taieri Catchment & Community Health Project, public health issues of concern ranged from the direct health impacts associated with the ecological determinants of water-related disease to the indirect health impacts of catchment management, freshwater ecosystem change, and rural sustainability—mediated through socioeconomic determinants of health (Duncanson et al. 2000; Hales et al. 2003; Skelly and Weinstein 2003).

The Taieri catchment case study combined knowledge generation with actions to address the social and ecological dimensions of catchment and community health issues. The multi-method study examined the links between ecosystem change and the determinants of health through socioecological analysis of knowledge strengths and deficits in the catchment; community-oriented participatory action research with diverse catchment stakeholders; and selected collaborative research initiatives—including a whole catchment questionnaire survey and specific biophysical studies. All phases of the research were based on building collaborative relationships with community reference groups (including residents living throughout the 5,650 square kilometer rural catchment) and co-researchers (included agencies, researchers, and indigenous organizations involved with science and decision-making regarding environment, health, development, and conservation issues in the catchment).

The catchment case study drew attention to the linked role of ecosystems and social systems as a mutually reinforcing basis for health, experienced as healthy living systems, livelihoods, and lifestyles. There was a transition from a research-initiated project through a "Community-University Partnership" to the "Taieri Alliance for Information Exchange and River Improvement" (the TAIERI Trust). This trust represents a shift from separate university and community interests to an integrated organization combining the interests of community, academic, and agency stakeholders to foster the health and sustainability of the river and local communities. This collaborative approach to knowledge, participation, and action demonstrates the application of successful decision-making processes into the research setting.

This case study strengthens the argument that place-based actions outside the health sector can respond to environmental and socioeconomic concerns—building resilient ecosystems and social systems that provide a double dividend for health and sustainability. Research and experience in the Taieri catchment case study led to the recommendation for ECO-PAR (Ecosystem-based Community-oriented Participatory Action Research) as a generic approach to integrated, collaborative health, and sustainability research. ECO-PAR is founded on interaction between knowledge, participation, and action and facilitates a unified approach to ecosystems, social systems, health, and sustainability (Parkes et al. 2003).

16.2.4 Case Study: Ciguatera (Fish Poisoning) and Ecological Change

Certain marine algae produce potent toxins that cause illness when consumed via contaminated fish or shellfish. The number and geographic distribution of harmful algal blooms appears to have increased in recent decades, in parallel with other changes in marine ecosystems, nutrient contamination of waterways, and climatic change (van Dolah 2000). There are several clinical syndromes associated with these events. The most common is ciguatera (fish poisoning) caused by consuming reef fish contaminated with algal toxins.

Traditional environmental health practice has focused on direct effects of pollutants on human health. Ciguatera is an example of a different kind of environmental problem. Morris (1999) writes, "Harmful algal blooms are an example of an alternative paradigm, in which human-induced stress on complex living systems leads to the emergence of new, potentially harmful microorganisms (or the reemergence of 'old' pathogens from previously restricted environmental niches), which, in turn, cause human disease."

Figure 16.3 illustrates some of the potential social and ecological drivers of ciguatera and the pathways to health impacts. Indirect drivers of change (population increase and resource consumption) affect direct drivers (global climate change and land use change). Rise in sea surface temperature, contaminated run-off, and other anthropogenic factors lead to disturbance of the marine ecosystem (including coral bleaching), increased growth of toxic algae, and contamination of reef fish. This, in turn, causes

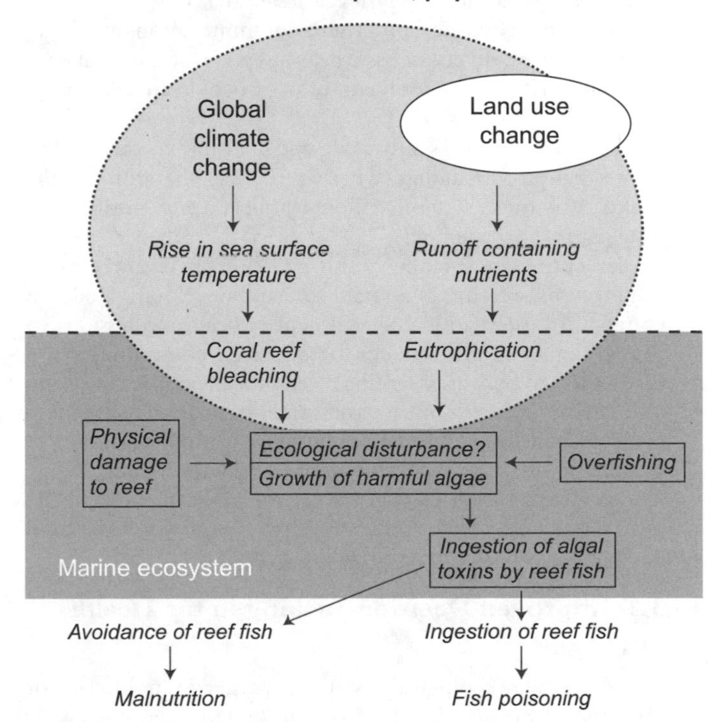

Figure 16.3. Potential Ecological Pathways in Fish Poisoning (Ciguatera)

ciguatera in people consuming the reef fish, or alternatively, it causes people in island communities to avoid this important protein source, potentially leading to malnutrition.

16.3 Response Options and Actions by the Health Sector

In order to respond effectively to threats from ecosystem change, the health sector must be able to carry out effective monitoring and surveillance of disease and risk factors for disease; interpret data provided by surveillance systems; use surveillance data in conjunction with environmental and other data to develop models to predict disease occurrence; link changes in disease rates to specific environmental factors; and intervene to remove the causes of disease or to lessen the damage they cause (Wilson and Anker 2005).

Tracking death registrations through periods of extreme weather is an example of the first condition for effective response (Hajat and Kovats 2002). An example of the second is the capacity to relate changing patterns of communicable disease to climate variability (Hales et al. 1999a; Hales et al. 1999b). The 2003/2004 epidemic of severe acute respiratory syndrome showed how quickly new pathogens can spread around the world. The source of SARS is not known but organisms of the kind that caused SARS frequently emerge from human disruption of biota-rich ecosystems. In this instance, public health systems in a large number of countries responded effectively to the threat of a global epidemic and provide an example of the third category of response options (WHO 2003b).

Pressures on the health sector as a result of ecosystem disturbance are likely to be most acute in developing countries. Ways in which these pressures could be reduced include:

- strengthening environmental health services;
- providing technical and financial assistance to implement the Health for All strategy, including health information systems and integrated databases on development hazards;
- strengthening advocacy and health communications at all levels; reviewing delivery of basic health services at the local level to ensure that priority problems of poor people are adequately addressed;
- making essential drugs affordable and available to the world's poorer nations, including (where necessary) alterations in the multilateral trade system, national policies, and institutional drug supply management;
- implementing long-range health and human resource planning to train, recruit, and retain staff and developing codes of conduct for international recruitment of health professionals;
- strengthening health services for displaced communities and those affected by war or famine or environmental degradation;
- implementing health impact assessment of major development projects, policies, and programs and monitoring indicators for health and sustainable development (WEHAB 2002c).

The following sections examine in more detail some of the actions that can be taken by the health sector to lessen harmful effects of ecosystem damage on human populations.

16.3.1 Improved Decision-making in the Health Sector

Decisions affecting ecological systems, whether by politicians or private organizations and individuals, are determined by a wide range of inputs. These include empirical evidence, value systems, and financial constraints. Despite this complexity, the health community has an important role to play in presenting evidence of likely public health consequences from any environmental change. Important policy decisions such as legislation on environmental lead, asbestos, and secondary tobacco smoke are largely dependent on health scientists measuring the links between these exposures and health outcomes, reaching a reasonably broad consensus, and presenting these findings to policy-makers. In these cases, the demonstration of a clear and significant health risk has taken precedence over other competing influences. Although most of the success stories are for environmental factors acting at a local level, examples such as the Montreal Protocol on CFC emissions show that health considerations can also be important in influencing decisions on global environmental issues.

16.3.2 Methods for Measuring and Prioritizing Environmental Influences on Health

In recent years, there have been important methodological developments in the linkages between environment and disease databases and in quantitative analytical techniques demonstrating relationships between them. (See Box 16.3.)

These linkage methods could potentially be applied to wide-area ecological measures other than climate. One such study correlated World Resources Institute measures of "ecological disintegrity" against data on life expectancy, infant mortality, and percent low-birth-weight babies for 203 countries (Sieswerda et al. 2001). There was a "modest relationship" between the ecological and health measures, but Sieswerda et al. pointed out that these relationships are inconsistent, the data are of uneven quality, and that other factors (such as GDP) appear to have a stronger influence. Another linkage study found no evidence of a negative relationship between loss of biodiversity and human health at the global scale (Huynen et al. 2004).

In the last decade, the World Health Organization promoted the use of "burden of disease" assessments. These measures express the total health effect (including both mortality and morbidity) of any disease or risk factor. The most widely used units of disease burden are DALYs, the sum of years of life lost from premature death (taking into account the age of death compared to natural life expectancy) and the number of years of life lived with a disability (taking into account the duration of the disease and weighted by a measure of the severity of the disease) (Murray et al. 1994); One advantage of these measures in the context of environmental change is that they allow impacts of different causal pathways to be combined, such as the combined effects of climate change on infectious diseases, malnutrition, and the impacts of natural disasters (WHO/WMO/UNEP 2003). This potentially allows direct comparisons of the effects of different ecological changes and can therefore help set priorities.

Burden-of-disease assessments depend on access to sufficient quantitative data to relate changes in the risk factor to the incidence of specific diseases. In the environmental health field, they have therefore been most successfully applied to discrete and relatively localized environmental factors with well-characterized health effects, such as air pollution and environmental lead. It is more difficult to apply these assessments to ecosystem changes acting through more diffuse causal pathways. For example, it is plausible, or even probable, that the reduced availability of fresh water would adversely affect health by increasing a range of water-borne diseases and through effects on agriculture, therefore negatively impacting food availability. It is, however, impossible to make accurate quantitative measurements of their contribution, in the context of the multitude of other causal factors, such as human behavior and economic influences on agricultural production.

BOX 16.3

Developments in Linking Disease to Environmental Factors

Advances in computing power and software have facilitated linkages between environmental and disease databases, and have therefore made epidemiological analyses of large-scale ecological change considerably more feasible. Exposure and disease data can be linked either in time or space. Time series methods are particularly well developed for studying the effects of air pollution. In essence, the process involves linking observations of temporal (typically daily) variations in exposure with a disease outcome measure at the same point, or with an appropriate time-lag afterwards. The quantitative relationship between the pollutant and disease outcome of concern can be defined by regression techniques (after specifying an appropriate error structure, and controlling for the effects of confounders such as seasonal variations) (Corvalan et al. 1997).

Geographic information system software can be used to link predictor and disease outcome data in space as well as over time. In addition, a wide range of satellite sensors provide detailed information on ecological characteristics such as vegetation, altitude, and climate, with complete global coverage at low or no cost. As for time series studies, regression techniques (again taking account of potential confounders and spatial auto-correlation between data points) can be used to quantify the relationship between ecological characteristics and disease outcomes.

The majority of studies of this type have been applied to specific diseases on a sub-national scale. Many of them are designed to generate predictive maps for disease control, but are equally applicable to measuring the effect of specific ecological characteristics (such as the proportion of land area covered by forest) and therefore allow estimation of the disease effects of alterations in these ecological conditions.

On a global scale, most attention has been focused on investigating the link between climate (and therefore climate change) and vector-borne disease. For example, maps of climate variables have been linked to maps of the distribution of both malaria (Rogers and Randolph 2000) and dengue (Hales et al. 2002) in order to define the climatic conditions under which each disease is most likely to occur. These statistical models can then be applied to scenarios of future climate change, to project plausible climate-driven changes in disease distribution into the future.

Linkage methods have been tried recently with broad area ecological measures, other than climate (Sieswerda et al. 2001). There was a "modest" but inconsistent relationship between the ecological and health measures, the data were of uneven quality, and other factors (such as GDP) appear to have a stronger influence. Another linkage study found no evidence of a negative relationship between loss of biodiversity and human health at the global scale (Huynen et al. 2004). Soskolne and Broemling (2002), in recognizing the importance that the health sector contributes, emphasize that methods are needed for developing sensitive measures capable of linking ecological degradation with health outcomes. Herein lies a challenge for eco-epidemiologists.

Considerations of time scale are important: the burden of disease attributable to climate change is modest compared to other risk factors over the short time scales for which most political decisions are taken (a five-year horizon, at most), but is more significant when impacts are considered over several decades (WHO/WMO/UNEP 2003). The discount rate chosen for DALY calculations has a very large effect on the rankings of long-term problems like climate change. The rate at which future gains and losses are discounted can be modified for the DALY formula, but the burden-of-disease framework fails to take into account that some environmental changes, such as biodiversity loss, are irreversible. There is no means of weighting effects from which there is no recovery. Finally, such frameworks do not account for the different valuation that people give to health risks over which they have direct individual control, compared to those controlled by the community as a whole or by other agencies. For example, there is greater concern over deaths among passive smokers rather than active smokers. Ecological changes usually fall into the category of externally imposed change.

Burden-of-disease assessments are therefore appropriate for aggregating health impacts through a range of mechanisms and can potentially aid in priority setting and decision-making in the context of ecosystem change. However, they must be considered as only one component of evidence, as they do not take full account of features such as complex causal pathways, long time-scales, potential irreversibility, and individual versus community responsibility (WHO/WMO/UNEP 2003). These important properties need to be included in the final considerations about any response to ecological change.

16.3.3 Methods for Selecting Interventions to Protect Health

Chapter 3 of this volume reviewed ways in which the effects of the environment on health and well-being may be measured. This chapter covers economic costing, environmental health indicators developed by WHO and subsequently applied in a variety of settings, and health impact assessment (Corvalan et al. 1999; Confalonieri 2001). When policy-makers contemplate decisions that impinge on human health they must make choices, and HIA is a means of laying out these choices so that significant consequences are not overlooked. These might include, for example, the effects on the health of communities and individuals of large-scale transport planning (Freeman and Scott-Samuel 2000).

HIA is a cousin of environmental impact assessment; both are related to integrated impact assessment (Hubel and Hedin 2000, Milner 2004). None require major changes to be applied to assessments of ecosystem change. For instance, Mutero (2002) adapted this approach to examine the effect of irrigation projects along the Tana River in Kenya on rates of schistosomiasis. HIA is not a "black box" for generating policy—it does not avoid the need for assumptions, approximations, improvisations, and value judgments; but it offers a systematic approach to collecting and appraising information, and for this reason, has the potential to improve the quality of decisions that affect the state of ecosystems and human health.

Cost-effectiveness analysis is increasingly used to select among different interventions to improve public health. Costs of interventions (usually measured in monetary terms) are considered alongside their resulting health gains (usually measured as deaths, or DALYs, averted). Outcomes from these analyses are quoted as cost-effectiveness ratios (for example, DALYs per dollar) as a measure of the value for money of the intervention, often along with aggregate costs and benefits, to represent the overall impact of the intervention. When applied in a rigorous and standardized manner, cost-effectiveness analysis can provide an objective ranking of the efficiency of different interventions. This allows policy-makers to select those that provide the greatest health gains for any specified level of resources.

Cost-effectiveness analysis requires quantitative data on all significant costs and benefits, which in turn requires an understanding of all the important links between the intervention and eventual health outcomes. Cost-effectiveness analysis has been employed where the intervention is clearly and directly linked to a health outcome, with relatively complete quantitative data on the relationships, such as selecting different options to improve water supplies to reduce diarrhea. Conceptually, it could equally be applied to decisions that act higher up the causal chain, such as the effect of land use policies on child health. This is seldom done, however, because the links are more diverse and complex, introducing greater uncertainty into the analysis. There are ways to determine the monetary value of nonmarket systems but these are not widely agreed upon.

16.3.4 Addressing Risk Perception and Communication

In order for any research on the health effects of ecological change to affect either official policy or individual behavior, it is necessary to take into account how risk is perceived. A deliberate and well-informed approach to community risk will maximize the chance of effective changes through policy interventions that enjoy popular support (Slovic 1999).

Any assessment of ecological change and health should be influenced by the risk perceptions of those communities that are most likely to be affected. That is, ecological assessments should involve open and frequent stakeholder participation from the beginning of the process rather than as an afterthought (Parkes et al. 2003). This approach of community engagement in the process serves the purpose of accessing local knowledge about the effects of ecological factors, ensuring that the assessment addresses issues of greatest concern to those affected and maximizing the probability that any recommended change in policy or behavior will be adopted. If a source of information is not widely trusted, it is unlikely that recommended changes will be accepted. Community surveys have shown that some groups tend to be regarded as highly trustworthy, while others (such as government agencies) are treated with caution (Maeda and Miyahara 2003). Healthcare providers tend to be one of the "high trust" groups, underlining again the important role they can play in explaining the significance of healthy ecosystems.

Any such consultation should make the best use of the expertise of both stakeholders and researchers. Stakeholders may have expert local knowledge but may have inaccurate ideas of the true nature of risks associated with different factors; researchers should have more exact knowledge of disease processes and relative risks but may inappropriately estimate the applicability of general concepts to local situations.

Accurate and accessible reporting of assessment results can remedy inaccurate risk perceptions and can enhance the public's ability to evaluate science/policy issues; the individual's ability to make rational personal choices is enhanced. In the past, poor reporting misled and disempowered a public that is increasingly affected by applications of science and technology (Myers and Raffensperger 1998). Stakeholder engagement will make it more likely that the research is credible and is translated into practice.

Technically intensive, externally driven interventions may produce rapid results but at the risk of marginalizing local communities. Interventions that engage local communities and transfer expertise are more likely to result in ecologically sustainable improvements.

16.4 Cross-sectoral Response Options and Actions

16.4.1 Health, Social Development, and Environmental Protection

Trends in inequality, resource consumption and depletion, environmental degradation, population growth, and ill health are closely interrelated (McMichael 1995). This means that better health, in the long term, will depend on cross-sectoral policies that promote ecologically sustainable development and address underlying driving forces. Agenda 21 and the Rio Declaration on Environment and Development describe a comprehensive approach to ecologically sustainable development incorporating cross-sectoral policies (McMichael 2000). The broader topic of sustainable development is discussed further in the next chapter. Examples of specific relevance to health are the following strategies, developed for the World Summit on Sustainable Development (WEHAB 2002c):

- mitigation strategies that reduce drivers of ecosystem change while simultaneously improving human health;
- adaptation strategies to reduce the effect of ecosystem disruption on health (addressing direct, mediated, and long-term health impacts);
- integrated action for health, such as health impact assessment of major development projects, policies, and programs, and indicators for health and sustainable development;
- inclusion of health in sustainable development planning efforts such as Agenda 21, in multilateral trade and environmental agreements and in poverty reduction strategies;
- improvement of inter-sectoral collaboration between different tiers of government, government departments and NGOs;
- international capacity-building initiatives, that assess health and environment linkages and use the knowledge gained to create more effective national and regional policy responses to environmental threats; and
- dissemination of knowledge and good practice on health gains from inter-sectoral policy.

The conventional indicators of population health, such as life expectancy, suggest that we have made considerable progress over the last hundred years in many parts of the world. Economic development and environmental protection are responsible for much of this improvement. An important lesson from history is that economic growth is a double-edged sword. On the one hand, it is the engine that generates wealth and opportunity; on the other hand, economic growth has tended to be socially disruptive and environmentally damaging. The experience of countries that industrialized early is that, initially, harmful effects predominated (Wohl 1983). What was needed to turn economic growth into social benefit was the development of robust, inclusive political processes and strong public institutions such as public health and local government (Szreter 1997).

What present-day summary indicators of health status fail to reveal is the gross inequalities within and among nations, between rural and urban areas, and among population subgroups. In some regions (such as southern Africa), life expectancy remains low and in, some instances, is falling further. Where gains have been made, they may be relatively fragile, as shown by the rapid deterioration of health statistics in Eastern Europe after the break-up of the Soviet Union. Underlying social and political factors include the change from politico-military colonialism to economic dependence, and migration from rural areas to urban centers resulting in unemployment, poverty, and social disruption (Avila-Pires 2003).

The accelerating rates of change brought about by high technology demand urgent solutions. On the positive side, the association of basic research with technological development proved to be a key factor in progress. But we need to find creative ways of extending its benefits to all. Technological progress implies social change and we must stimulate a corresponding effort from sociologists and philosophers to help us understand and cope with the swift pace of change.

16.4.2 Linking Health and Ecosystem Responses

For each category of ecosystem services, we have extracted from earlier sections of the report a sample of recommended responses. (See Table 16.4.) In each instance, we have listed some of the possible effects that these responses could have on human health. For simplicity these are illustrative lists, not intended to be exhaustive.

Table 16.4 makes the case that in almost every category of ecosystem response the consequences for health may be either positive or negative. The balance will depend on how the policy or regulation is framed and what account is taken of contingencies and local circumstances. Using trade and economic levers to widen food markets, for instance, has been successful in some instances and, of course, increased food supply can lead to better health (FAO 2003). However, in other settings, "globalizing" policies have led to deepening poverty, diminished food security, and deteriorating standards of public health. This illustrates the fact that national strategies to protect ecosystem services and human health can be successful only if the global policy context is supportive.

Policies addressing human health needs in relation to food and nutrition, water and sanitation, and energy services have been developed as part of the "water-energy-health-agriculture-biodiversity" process and are summarized in Box 16.4. Implementation of these policies will depend on national and local circumstances. For example, in industrialized countries, integrating national agriculture and food security policies with the economic, social, and environmental goals of sustainable development could be achieved, in part, through taxes on food products to ensure that the environmental and social costs of production and consumption are fully reflected in the price. Taxes should be one element in a package of policies designed to protect the environment without jeopardizing food security for the most vulnerable groups in society. With that proviso, a full-cost approach to food pricing may bring major benefits to health and ecosystems, for instance through reduced consumption of animal products (WHO 2003a). Improvements to traditional fuels and cooking devices could lead to the prevention or at least reduced emissions of local air pollutants, while implementing better transportation practices and systems could lead to increased physical activity in sedentary populations as well as reductions in greenhouse gas emissions (Von Schirnding and Yach 2002; WEHAB 2002b).

16.5 Conclusion

Ecosystem disruption damages health in a variety of ways and through complex pathways. The links between ecosystem change and human health are seen most clearly among impoverished communities (who lack the "buffers" that the rich can afford).

Table 16.4. Examples of Potential Health Implications of Sectoral Responses

Ecosystem services under threat	Possible Responses	Possible Consequences for Health
Floods and storm control	waste-water management	▲ improved water quality (fewer enteric infections)
	vegetation of water catchments	▼ disease vector proliferation (e.g., urban wetlands)
Food production	economic and trade policies to increase reach of global markets	▲ more food choices—improved nutrition
		▲ decreased poverty, consequent improvements in health
		▼ reduced food security—especially for the most vulnerable groups (deepening poverty and reduction in health status)
Climate regulation	reduce greenhouse gas emissions (e.g., vehicle emission standards)	▲ improved air quality
		▲ improved water quality
	carbon sequestration (e.g., reforestation)	▼ decreased access to health services for the poor
		▼ increased fire risk
		▼ displaced populations
		▼ reduced food production
Wood, woodfuel and fiber	economic incentives for re-forestation	▲ reduced flood risk
		▼ increased fire risk
Freshwaterwater	charges to reduce wasteful consumption	▲ improved access to sectors in the population
	infrastructure (e.g., dams and dikes)	▼ decreased access for low income groups—water-related diseases
		▼ new habitat for disease vectors
Wastes	increase recycling	▲ decreased toxic emissions (e.g. from incinerated waste)
	reduce amounts of waste	▼ vector-breeding sites—more mosquito-borne disease

Key: ▲ Improved health ▼ Impaired health

BOX 16.4

Examples of Responses to Improve Human Health (WEHAB 2002a, 2002b, 2002d)

Food and nutrition responses that can improve human health include:

- integrate national agriculture and food security policies with the economic, social, and environmental goals of sustainable development;
- ensure equitable access to agriculture-related services and products, with particular focus on food security and sustainable livelihood needs of the poor;
- orient market forces toward environmentally optimal solutions through appropriate policies and regulations;
- exploit and expand locally available resources for improved food security and promoting diversification for more effective risk management;
- focus on needs of rural areas through decentralized cooperative initiatives and improvements in rural infrastructure; and
- strengthen regional and international cooperation for food security and market stability.

Water and sanitation responses to improve human health:

- assign the role of water-related public awareness to the agency responsible for integrated water resource management at the country level;
- institute gender-sensitive systems and policies;
- raise awareness and understanding of the linkages among water, sanitation, and hygiene and poverty alleviation and sustainable development;
- develop in partnership with all relevant actors community-level advocacy and training programs that contribute to improved household hygiene practices for the poor;
- identify best practices and lessons learned based on existing projects and programs related to provision of safe water and sanitation services focused on children;
- create multistakeholder partnership opportunities and alliances at all levels that directly focus on the reduction of child mortality through diseases associated with unsafe water, inadequate sanitation, and poor hygiene;
- develop national, regional, and global programs related to the provision of safe water and improved sanitation services for urban slums in general, and to meet the needs of children in particular; and
- identify water pollution prevention strategies adapted to local needs to reduce health hazards related to maternal and child mortality.

Energy and fuel responses to improve human health:

- reduce poverty by providing access to modern energy services in rural and peri-urban areas;
- minimize the environmental impacts of traditional fuels and cooking devices;
- improve air quality and public health through the introduction of cleaner vehicular fuels; and
- implement better transportation practices and systems in megacities.

This extends to subpopulations within wealthier communities who have relatively less access to ecosystem resources.

Poor communities are the most directly dependent upon productive ecosystems for their health. Measures to promote ecological sustainability will (by definition) safeguard ecosystem services and therefore benefit health in the long term. This means that the poorest and most disadvantaged individuals and communities can be among the first to benefit from ecosystem protection, leading to improvements in health equity.

A healthy community is more capable of sustainable development than an unhealthy one. Therefore, where a population is weighed down by diseases related to poverty and lack of entitlement to essential resources such as shelter, nutritious food, or clean water, the provision of these resources should be the first priority for healthy public policy.

Where disease is caused by unhealthy levels of consumption (especially of food or energy), substantial reductions in this overconsumption would have major health benefits as well as reducing pressure on ecosystems. Both human health and the environment are likely to benefit from a redistribution of resources if this leads to basic entitlements being distributed more equitably and a reduction in overconsumption. Such changes could improve health in the short term as well as contribute to long-term ecological sustainability. Win-win outcomes of this kind depend on how these changes in resource use and management are achieved.

Local conditions are critical in shaping the health manifestations of ecosystem disruption. Empirical evidence supporting the link between ecosystems and health is difficult to find. Our knowledge is increasing but there are still many gaps. One reason for this is the many confounding factors (associated with environmental change and also determinants of health) that are hard to measure and to separate from the effect of interest.

The effects of ecosystem disruption on health are frequently displaced, either transferred geographically (such as the costs of rich countries' food overconsumption) or postponed (as in the case of long-term consequences of climate change or desertification). Decisions about health and ecosystems must consider how one is related to the other. Choices that are made about the management of ecosystems may have important consequences for health, and vice versa. Healthy ecosystems protect human health; healthy people protect their ecosystems.

Decision-makers need to consider the connections between health and other sectors. Where there are "win-win" options, these will be attractive to policy-makers; where there are trade-offs, it is important for politicians, regulators, and the public to understand the consequences of taking one path in preference to another. The health sector bears responsibility for revealing the links between ecological services and health and indicating which interventions are needed: this is despite the fact that responses and interventions to protect human health are often carried out in other sectors.

Consideration of ecosystem change enlarges the scope of health responses by highlighting "upstream" causes of disease and injury. This implies that health considerations should weigh heavily in decisions on ecosystem responses. History shows that health is one of the most highly valued social outcomes.

References

Allen, B., 2002: Birth weight and environment at Tari, *Papua New Guinea Medical Journal,* **45,** pp. 88–98.

Arnell, N.W., 1999: Climate change and global water resources, *Global Environmental Change,* **9(S1-S2),** pp. S31–51.

Avila-Pires, F., 2003: Health, biodiversity, and sustainable development: Making globalization sustainable, M. Pallemaerts (ed.), Brussels University Press, Brussels, Belgium, pp. 111–27.

Baron, J., N. Poff, and P. Angermeier, 2002: Meeting ecological and societal needs for freshwater, *Ecological Applications,* **12,** pp. 1247–60.

Berkes, F. and C. Folke, 1998: *Linking Social and Ecological Systems: Management Practices and Social Mechanisms for Building Resilience,* Cambridge University Press, Cambridge, UK.

Cairncross, S., 2003: Sanitation in the developing world: Current status and future solutions, *International Journal of Environmental Health Research,* **13,** pp. S123–31.

Chin, J., 2000: *Control of Communicable Diseases Manual,* American Public Health Association, Washington, DC.

Confalonieri, U., 2001: Environmental change and health in Brazil: Review of the present situation and proposal for indicators for monitoring these effects. In: *Human Dimensions of Global Environmental Change, Brazilian Perspectives,* D.J. Hogan and M.T. Tolmasquin (eds.), Brasileira De Ciencias, Rio de Janeiro, Brazil, pp.43–77.

Corvalan, C., F. Barten, and G. Zielhuis, 2000: Requirements for successful environmental health decision-making, *Decision-making in Environmental Health: From Evidence to Action,* C. Corvalan, D. Briggs, and G. Zielhuis (eds.), WHO, London, UK.

Corvalan, C., T. Kjellstrom, and K. Smith, 1999: Health, environment and sustainable development: Identifying links and indicators to promote action, *Epidemiology,* **10(5),** pp. 656–60.

Corvalan, C., M. Nurminen, and H. Pastides (eds.), 1997: *Linkage Methods for Environment and Health Analysis,* UNEP, Nairobi, Kenya/ US EPA, Washington, DC/ WHO, Geneva, Switzerland.

DFID/EC/UNDP/World Bank (Department for International Development/ European Community/United Nations Development Programme), 2002: *Linking Poverty Reduction and Environmental Management: Policy Challenges and Opportunities,* DFID, London, UK/ Directorate General for Development, EC, Brussels, Belgium/ United Nations Development Programme, Nairobi, Kenya/ The World Bank, Washington, DC.

Duncanson, M., N. Russell, and P. Weinstein, 2000: Rates of notifiable cryptosporidiosis and quality of drinking water in Aoteroa New Zealand, *Water Research,* **34,** pp. 26–34.

Durie, M., 2001: *Mauri Ora. The Dynamics of Maori Health,* Oxford University Press, Auckland, New Zealand.

Ebi, K., J. Smith, I. Burton, and S. Hitz, 2004: *Adaptation to Climate Variability and Change from a Public Health Perspective: Integration of Public Health with Adaptation to Climate Change: Lessons Learned and New Directions,* K.L. Ebi, J.B. Smith, and I. Burton (eds.), Taylor & Francis Group PLC, London, UK, pp. 1–7.

Epstein, P., E. Chivian, and K. Frith, 2003: Emerging diseases threaten conservation, *Environmental Health Perspectives,* **111,** pp. A506–7.

Esrey, S., 1996: Water, waste, and well-being: A multicountry study, *American Journal of Epidemiology,* **143,** pp. 608–23.

Esrey, S., 2002: Philosophical, ecological and technical challenges for expanding ecological sanitation into urban areas, *Water Science Technology,* **45,** pp. 225–58.

Ezzati, M., A.D. Lopez, A. Rodgers, S. Vander Hoorn, and C.J. Murray, 2002: Selected major risk factors and global and regional burden of disease, *Lancet,* **360(9343),** pp. 1347–60.

FAO (Food and Agriculture Organization of the United Nations), 2003: *The State of Food Insecurity in the World,* FAO, Rome, Italy.

Freeman, N. and A. Scott-Samuel, 2000: A prospective health impact assessment of the Merseyside Integrated Transport Strategy, *Journal of Public Health Medicine,* **22,** pp. 268–74.

Frumkin, H., 2001: Beyond toxicity: Human health and the natural environment, *American Journal of Preventive Medicine,* **20,** pp. 234–40.

Graber, D., W. Jones, and J. Johnson, 1995: Human and ecosystem health: The environment-agriculture connection in developing countries, *Journal of Agromedicine,* **2,** pp. 47–64.

Greenwood, B. and T. Mutabingwa, 2002: Malaria in 2002, *Nature,* **415,** pp. 670–2.

UNEP/GRID-Arendal, 1997: *Mapping Indicators of Poverty in West Africa,* Technical Advisory Committee working document, Consultative Group on International Agricultural Research, Washington, DC/ Food and Agriculture Organization, Rome, Italy.

Hajat, S., K. Ebi, S. Edwards, A. Haines, S. Kovats, and B.M., 2003: Review of the human health consequences of flooding in Europe and other industrialized civilizations, *Applied Environmental Science and Public Health,* **1,** pp. 13–21.

Hajat, S. and R. Kovats, 2002: Impact of hot temperatures on death in London: A time series approach, *Journal of Epidemiology and Community Health,* **56,** 367–72.

Hales, S., W. Black, and C. Skelly, 2003: Social deprivation and the quality of community water supplies in New Zealand, *Journal of Epidemiology and Community Health.* In press.

Hales, S., N. de Wet, J. Maindonald, and A. Woodward, 2002: Potential effect of population and climate changes on global distribution of dengue fever: An empirical model, *Lancet,* **360(9336),** pp. 830–4.

Hales, S., P. Weinstein, Y. Souares, and A. Woodward, 1999a: El Nino and the dynamics of vector-borne disease transmission, *Environmental Health Perspectives,* **107,** 99–102.

Hales, S., P. Weinstein, and A. Woodward, 1999b: Ciguatera (fish poisoning), El Niño, and Pacific sea surface temperatures, *Ecosystem Health,* **5,** pp. 20–5.

Haug, G.H., D. Gunther, L.C. Peterson, D.M. Sigman, K.A. Hughen, and B. Aeschlimann, 2003: Climate and the collapse of Maya civilization, *Science,* **299(5613),** pp. 1731–5.

Helming, S. and J. Kuylenstierna, 2001: *Water: A Key to Sustainable Development,* International Conference on Freshwater, Bonn, Germany.

Hubel, M. and A. Hedin, 2000: Developing health impact assessment in the European Union, *Bulletin of the World Health Organization,* **81,** pp. 463–64.

Huynen, M., P. Martens, and R. De Groot, 2004: Linkages between biodiversity loss and human health: A global indicator analysis, *International Journal of Environmental Health Research,* **14,** pp. 13–30.

IFRCRCS (International Federation of Red Cross and Red Crescent Societies), 2002: *World Disasters Report,* Eurospan, London, and IFRCRCS, Geneva, Switzerland.

IPCC (Intergovernmental Panel on Climate Change), 2001: *Climate Change 2001: Impacts, Adaptation and Vulnerability,* Contribution of Working Group II to the third assessment report of the IPCC, Cambridge University Press, Cambridge, UK.

Jordan, T., D. Whigham, K. Hofmockel, and M. Pittek, 2003: Nutrient and sediment removal by a restored wetland receiving agricultural runoff, *Journal of Environmental Quality,* **32,** pp. 1534–47.

Kilbourne, E., 1997: *Heat Waves and Hot Environments: The Public Health Consequences of Disasters,* E. Noji (ed.), Oxford University Press, New York, NY.

Kovats, R., K. Ebi, and B. Menne, 2003: *Methods for Assessing Human Health Vulnerability and Public Health Adaptation to Climate Change,* WHO, Geneva, Switzerland/WMO, Geneva, Switzerland/UNEP, Copenhagen, Denmark.

Kunitz, S.J., 1994: *Disease and Social Diversity: The European Impact on the Health of Non-Europeans,* Oxford University Press, New York, NY.

Kunzli, N., S. Medina, M. Studnicka, O. Chanel, P. Filliger, et al., 2000: Public-health impact of outdoor and traffic-related air pollution: A European assessment, *Lancet,* **356,** pp. 795–801.

Leitzmann, C., 2003: Nutrition ecology: The contribution of vegetarian diets, *American Journal of Clinical Nutrition,* **78(3 Suppl),** pp. 657S–9S.

Lindgren, E. and R. Gustafson, 2001: Tick-borne encephalitis in Sweden and climate change, *Lancet,* **358,** pp. 16–18.

Lindgren, E., L. Tälleklint, and T. Polfeldt, 2000: Impact of climatic change on the northern latitude limit and population density of the disease-transmitting European tick, *Ixodes ricinus, Environmental Health Perspectives,* **108,** pp. 119–23.

LoGiudice, K., R. Ostfeld, K. Schmidt, and F. Keesing, 2003: The ecology of infectious disease: Effects of host diversity and community composition on Lyme disease risk, *Proceedings of the National Academy of Science USA,* **100,** pp. 567–71.

Maeda, Y. and M. Miyahara, 2003: Determinants of trust in industry, government, and citizen's groups in Japan, *Risk Analysis,* **23,** pp. 303–11.

McMichael, A., 1995: Nexus between population, demographic change, poverty and environmentally sustainable development, Paper for the Third Annual World Bank Conference on environmentally sustainable development, London, UK.

McMichael, A., 2000: *Human Health: Methodological and Technological Issues in Technology Transfer,* B. Metz (ed.), IPCC, Cambridge, UK.

McMichael, P., 2001: The impact of globalization, free trade and technology on food and nutrition in the new millennium, *Proceedings of the Nutrition Society,* **60(2),** pp. 215–20.

Mellor, J., 2002: *Poverty Reduction and Biodiversity Conservation: The Complex Role for Intensifying Agriculture,* World Wide Fund for Nature, Washington, DC.

Milner, S., 2004: Using health impact assessment in local government, *Health Impact Assessment,* S. Palmer (ed.), Oxford University Press, Oxford, UK.

Morris, J., 1999: Harmful algal blooms, *Annual Review of Energy and the Environment,* **24,** pp. 367–90.

Murray, C.J.L., A.D. Lopez, and D.T. Jamison, 1994: The global burden of disease in 1990: Summary results, sensitivity analysis and future directions, *Bulletin of the World Health Organization,* 72(3), pp. 495–509.

Mutero, C., 2002: Health impact assessment of increased irrigation in the Tana River Basin, Kenya. In: *The Changing Face of Irrigation in Kenya: Opportunities for Anticipating Change in Eastern and Southern Africa,* International Water Management Institute, Colombo, Sri Lanka.

Myers, N.J. and C. Raffensperger, 1998: When science counts: A guide to reporting on environmental issues, *The Networker: Media and Environment,* **3(2).**

National Research Council, 1999: *From Monsoons to Microbes: Understanding the Ocean's Role in Human Health,* National Academies Press, Washington, DC.

Nurse, L. and G. Sem, 2001: Small island states, *Climate Change 2001: Impacts, Adaptation and Vulnerability,* Contribution of Working Group II to the third assessment report, J.J. McCarthy, O.F. Canziani, N.A. Leary, D.J. Dokken, and K.S. White (eds), Cambridge University Press, Cambridge, UK, p. 1032.

Ostfeld, R. and F. Keesing, 2000: The function of biodiversity in the ecology of vector-borne zoonotic diseases, *Canadian Journal of Zoology,* **78,** pp. 2061–78.

Parkes, M., R. Panelli, and P. Weinstein, 2003: Converging paradigms for environmental health theory and practice, *Environmental Health Perspectives,* **111,** pp. 669–75.

Pruss, A., D. Kay, L. Fewtrell, and J. Bartram, 2002: Estimating the burden of disease from water, sanitation, and hygiene at a global level, *Environmental Health Perspectives,* **110(5),** pp. 537–42.

Reijnders, L. and S. Soret, 2003: Quantification of the environmental impact of different dietary protein choices, *American Journal of Clinical Nutrition,* **78(3 Suppl.),** pp. 664S–8S

Reuer, M. and D. Weiss, 2002: Anthropogenic lead dynamics in the terrestrial and marine environment, *Philosophical Transactions of the Royal Society London, Series A (Mathematical, Physical & Engineering Sciences),* **360,** pp. 2889–904.

Rogers, D.J. and S.E. Randolph, 2000: The global spread of malaria in a future, warmer world, *Science,* **289,** pp. 1763–5.

Scheraga, J., K. Ebi, J. Furlow, and A. Moreno, 2003: From science to policy: Developing responses to climate change, *Climate Change and Human Health: Risks and Responses,* A. McMichael (ed.), WHO, Geneva, Switzerland.

Sieswerda, L., C. Soskolne, S. Newman, D. Schopflocher, and K. Smoyer, 2001: Towards measuring the impact of ecological disintegrity on human health, *Epidemiology,* **12,** pp. 28–32.

Skelly, C. and P. Weinstein, 2003: Pathogen survival trajectories: An eco-environmental approach to the modeling of human campylobacteriosis ecology, *Environmental Health Perspectives,* **111,** pp. 19–28.

Skrabski, A, M. Kopp, and I. Kawachi, 2003: Social capital in a changing society: Cross sectional associations middle-aged female and male mortality rates, *Journal of Epidemiology and Community Health,* **57,** pp. 114–19.

Slovic, P., 1999: Trust, emotion, sex, politics, and science: Surveying the risk-assessment battlefield, *Risk Analysis,* **19,** pp. 689–701.

Smil, V., 2000: *Feeding the World: A Challenge for the 21st Century,* The MIT Press, Cambridge, MA.

Smit, B, O. Pilifosova, I. Burton, B. Challenger, S. Huq, R. Klein, et al., 2001: Adaptation to climate change in the context of sustainable development and equity, *Climate Change 2001: Impacts, Adaptation and Vulnerability,* Contribution of Working Group II to the third assessment report, J.J. McCarthy, O.F. Canziani, N.A. Leary, D.J. Dokken, and K.S. White (eds), Cambridge University Press, New York, NY.

Smith, K. and S. Mehta, 2003: The burden of disease from indoor air pollution in developing countries: Comparison of estimates, *International Journal of Hygiene and Environmental Health,* **206,** pp. 279–89.

Soskolne, C. and N. Broemling, 2002: Eco-epidemiology: On the need to measure health effects from global change, *Global Change and Human Health,* **3,** pp. 58–66.

Steere, A., 2001: Lyme disease, *New England Journal of Medicine,* **345,** pp. 115–25.

Strina, A., S. Cairncross, M.L. Barreto, C. Larrea, and M.S. Prado, 2003: Childhood diarrhea and observed hygiene behavior in Salvador, Brazil, *American Journal of Epidemiology,* **157(11),** pp. 1032–8.

Szreter, S., 1997: Economic growth, disruption, deprivation, disease and death: On the importance of the politics of public health for development, *Population and Development Review,* **23,** pp. 693–728.

Talleklint, L. and T.G. Jaenson, 1998: Increasing geographical distribution and density of Ixodes ricinus (Acari: Ixodidae) in central and northern Sweden, *Journal of Medical Entomology,* **35,** pp. 521–6.

Thompson, T., M. Sobsey, and J. Bartram, 2003: Providing clean water, keeping water clean: An integrated approach, *International Journal of Environmental Health Research,* **13,** pp. S89–94.

Townsend, A., R. Howarth, F. Bazzaz, M. Booth, and C. Cleveland, 2003: Human health effects of a changing nitrogen cycle, *Frontiers in Ecology and Environment,* **1,** pp. 240–6.

UNCSD (United Nations Commission on Sustainable Development), 1998: *Report of the Export Group of Strategic Approaches to Freshwater Management,* UNCSD–6th Session of the CSD, New York, 20 April–1 May 1998.

UNEP (United Nations Environment Programme), 2002: *Global Environmental Outlook,* UNEP, Nairobi, Kenya.

UNESCO (United Nations Educational, Scientific and Cultural Organization), 2003a: *Water for People, Water for Life: UN World Water Development Report,* UNESCO,Division of Water Sciences, Paris, France.

UNESCO, 2003b: *Manual on Harmful Marine Macroalgae,* UNESCO, Paris, France.

van Dolah, F., 2000: Marine algal toxins: Origins, health effects, and their increased occurrence, *Environmental Health Perspective,* **108(Suppl. 1),** pp. 133–41.

von Schirnding, Y. and D. Yach, 2002: Unhealthy consumption threatens sustainable development, *Revista de Saúde Pública,* **36,** pp. 379–82.

Waltner-Toews, D., 2001: An ecosystem approach to health and its applications to tropical and emerging diseases, *Cadernos de Saude Publica,* **17(Suppl.),** pp. 7–36.

WCD (World Commission on Dams), 2000: *Dams and Development: A New Framework for Decision,* Earthscan, WCD, London, UK.

WEHAB (Water, Energy, Health, Agriculture, and Biodiversity), 2002a: *A Framework for Action on Agriculture,* World Summit on Sustainable Development, Johannesburg, South Africa.

WEHAB, 2002b: *A Framework for Action on Energy,* World Summit on Sustainable Development, Johannesburg, South Africa.

WEHAB, 2002c: *A Framework for Action on Health and Environment,* World Summit on Sustainable Development, Johannesburg, South Africa.

WEHAB, 2002d: *A Framework for Action on Water,* World Summit on Sustainable Development, Johannesburg, South Africa.

WHO (World Health Organization), 2000: *The African Summit on Roll Back Malaria,* WHO, Geneva, Switzerland.

WHO, 2001: *Macroeconomics and Health: Investing in Health for Economic Development,* Report of the Commission on Macroeconomics and Health, WHO, Geneva, Switzerland.

WHO, 2002: *The World Health Report 2002,* WHO, Geneva, Switzerland.

WHO, 2003a: *Diet, Nutrition and the Prevention of Chronic Diseases,* WHO, Geneva, Switzerland.

WHO, 2003b: A multicentre collaboration to investigate the cause of severe acute respiratory syndrome, *Lancet,* **361,** pp. 1730–3.

WHO, 2003c: *The World Health Report 2003,* WHO, Geneva, Switzerland.

WHO/UNICEF (UN Children's Fund), 2004: *Joint Monitoring Programme for Water Supply and Sanitation: Meeting the MDG Drinking Water and Sanitation Target: A Mid-term Assessment of Progress,* WHO and UNICEF, Geneva, Switzerland.

WHO/WMO/UNEP (World Meteorological Organization), 2003: *Climate Change and Human Health: Risks and Responses,* WHO, Geneva, Switzerland/ WMO, Geneva, Switzerland/ UNEP, Nairobi, Kenya.

Wilson, M. and M. Anker, 2004: Disease surveillance in the context of climate stressors: Needs and opportunities, *Integration of Public Health with Adaptation to Climate Change: Lessons Learned and New Directions,* K.L. Ebi, J.G. Smith, and I. Burton (eds.), Taylor & Francis Group PLC, London, UK, pp. 191–214.

Wohl, A., 1983: *Endangered Lives: Public Health in Victorian Britain,* Harvard University Press, Cambridge, MA.

Wolfe, A. and J. Patz, 2002: Reactive nitrogen and human health: Acute and long term implications, *Ambio,* **31,** pp. 120–5.

Woodward, A., S. Hales, N. Litidamu, D. Phillips, and J. Martin, 2000: Protecting human health in a changing world: The role of social and economic development, *Bulletin of the World Health Organization,* **78(9),** pp. 1148–55.

Woodward, A., S. Hales, and P. Weinstein, 1998: Climate change and health in the Asia Pacific region: Who will be most vulnerable? *Climate Research,* **11,** pp. 31–8.

World Water Forum, 2000: *Ministerial Declaration on Water Security in the 21st Century,* Second World Water Forum, The Hague, The Netherlands.

Yohe, G. and K. Ebi, 2004: Approaching adaptation: Parallels and contrasts between the climate and health communities, *Integration of Public Health with Adaptation to Climate Change: Lessons Learned and New Directions,* K.L. Ebi, J.B. Smith, and I. Burton (eds.), Taylor & Francis Group PLC, London, UK, pp. 18–43.

Chapter 17

Consequences of Responses on Human Well-being and Poverty Reduction

Coordinating Lead Authors: Anantha Kumar Duraiappah, Flavio Comim
Lead Authors: Thierry De Oliveira, Joyeeta Gupta, Pushpam Kumar, Marjorie Pyoos, Marja Spierenburg
Contributing Authors: David Barkin, Eduardo Brondizio, Rie Tsutsumi
Review Editors: Julia Carabias, Robert Kates, Kai Lee

BOXES

FIGURES

TABLES

Main Messages

Many of the economic, legal and technological responses used today to manage ecosystems emphasize economic growth—aiming at aggregate welfare as measured by income—through the efficient allocation of the provisioning services of ecosystems. There is little uncertainty that economic, legal, and technological responses have improved the material wealth and livelihoods of many people. However, there is also high certainty that the regulating, supporting, and cultural services have deteriorated in many parts of the world and this has had dire consequences on the health, security, good social relations, and the freedoms and choices of many individuals and local communities, especially in developing countries. Equity has also been deemphasized in the design of many of these responses. There have been losers and winners, with the vulnerable and the poor experiencing losses most of the time.

Responses aimed at development have had a mixed record of improving well-being and have in most instances failed to accommodate the conservation and sustainable use of all ecosystem services. Poverty reduction strategies in many developing countries have failed to integrate ecosystem related issues. However, there is an increasing trend of attempts to mainstream "environmental" in poverty reduction strategies and the accompanying budgetary frameworks. This is a promising trend with the potential for improving well-being and reducing ecosystem deterioration. The scenario in industrial countries is slightly different. One of the main drivers in industrial countries are subsidies that have dire impacts on ecosystem services within their countries as well as in the developing countries. There is increasing pressure on industrial countries to revise their subsidy programs, especially those in the agricultural sector.

Responses relying on market mechanisms require well-functioning institutional frameworks, including legal structures, to work effectively. For example, there is high certainty that market-based approaches have the potential to unlock significant supply- and demand-side efficiencies while providing cost-effective reallocation between old (largely irrigation) and new (largely municipal and in-stream) uses of water resources. But it is also well established that it is important to keep in mind that, while there is a role for the use of markets to develop efficient water markets, there is also a regulatory role for governments in providing stable and appropriate institutions for these markets to operate.

The effectiveness of international and national-local agreements depends upon decentralization of intervention strategies to the lowest level that allows successful participation, implementation, and enforcement. Evidence of international legal agreements shows that their success depends on how they are designed in both substantive and procedural terms, how they compete with other international agreements, and the context in which they are to be implemented. At the national and local level, the success of regulatory instruments depends on how well they target specific incentives and disincentives, and how well the implementation is monitored and enforced.

Technological interventions have had a mix of positive and negative impacts on ecosystems, human well-being, and poverty reduction. Unintended consequences, positive and negative, of technological interventions have been more the norm than an exception. For example, the impacts on ecosystems from attempts to increase food production by technological innovations, have been realized mostly as second round effects, and as such they represent negative externalities. Moreover, the benefits of technological interventions have varied across and within communities with some gaining more than others. On the other hand, technologies have been found to be successful in mitigating ecosystem degradation when they are designed to work within the dynamics of the ecosystems, and not "fight" it. Successful technological interventions revolve around interventions that take into account the temporal and spatial dynamics specific to local ecosystems and that are driven by a participatory, democratic, and cooperative model between local communities who have a rich database of traditional knowledge and outside technical expertise that together are able to produce incremental driven technological interventions.

Responses have been found to be more effective in reducing poverty when they respect the different degrees, and types of use, of ecosystem services by different communities. Response strategies based on capture of benefits by local people from one or more components of biodiversity (for example, products from single species or from ecotourism) have been most successful when they have at the same time created incentives for the local communities to make management decisions consistent with (overall) biodiversity conservation. For example, at a local level, protected areas have more often had a detrimental effect on poverty reduction because rural communities are excluded from a resource that has traditionally provided income-generating activities. Consequently, if conservation is to be useful, then the recommendation is for targeted incentives and involving local stakeholders in the design, implementation, and monitoring of responses.

Poverty reduction can only work if the links between ecosystems and well-being are explicitly mainstreamed into national poverty reduction strategies like Poverty Reduction Strategy Papers. Very few macroeconomic responses to poverty reduction have taken into account the importance of ecosystems to poverty reduction and improving well-being. Assets-based responses, such as the Livelihood Approach to Rural Development, have quite often been limited to narrow perceptions of natural capital as an input for economic production. Yet, there is insufficient information for policy-makers about the broader links between ecosystem services, human well-being, and poverty reduction. If policy-makers are to make informed decisions concerning ecosystem management, then a serious concerted effort to collect and collate data on ecosystem services and human well-being will need to be initiated immediately.

Rather than the "win-win" scenarios promoted (or assumed) by many practitioners, conflict is more often the norm, and tradeoffs between conservation and development need to be acknowledged. Over the last generation, responses motivated by conservation of wetlands and biodiversity have been adopted more widely. Some of these interventions have displaced resident human populations and decreased their well-being, and even in some instances pushed them into poverty. Other interventions have sought to combine conservation and development in the belief of win-win outcomes. These interventions have had a mixed record to date, and more attention may be paid to analyze how fair and equitable these responses are across all stakeholders, and to observe whether tradeoffs need to be negotiated before such responses are implemented.

In light of the urgent desire to improve well-being among the world's poor, and in light of the fragility of many ecosystems needed by people, especially those in subsistence economies, it is vital to learn from the mistakes of the past and the present. A systematic way to do this is adaptive management. Adaptive management necessitates multiple instruments, including a mix of law, economic instruments, voluntary agreements, information provision, technological solutions, research, and education—an integrated response strategy. The key word in adaptive management is flexibility. Equally important is the potential for change within institutions, property rights, and/or communities. Flexibility and change in turn is highly dependent on the availability of a number of key instrumental freedoms—transparency guarantees, participative freedom, protective security, economic facilities, social opportunities, ecological security—to all individuals.

17.1 Introduction

Societies have developed and used a wide range of innovative legal, economic, behavioral, and technological responses to manage their interactions with ecosystems. These responses in turn have had impacts on ecosystems, some good and some bad. However, the impacts of these responses have not been confined to just ecosystems but have also had positive and negative impacts on human well-being, defined to include health, security, social relations, material wealth, and freedom of choice and actions (MA 2003).

It is also increasingly becoming clear that development and poverty-reduction focused responses can have positive and negative impacts on ecosystems and human well-being. Experience shows that some responses—either ecosystem focused or development and poverty reduction—have worked well in conserving ecosystem services, improving well-being, and reducing poverty, while others have actually caused deterioration in ecosystem services, a reduction in human well-being, and increased poverty.

But the situation is much more complex in reality and it is difficult to make simplistic linear causality relationships between responses and the consequences on ecosystems and human well-being. To begin, the consequences of responses are usually not shared equally among different stakeholders. For example, some groups may have benefited from a response while others suffered a drop in one or more of their constituents and determinants of well-being. Identifying who loses and who gains is a critical component in evaluating responses. Responses directed at some ecosystem services can cause unintended impacts on other ecosystem services and subsequently on the constituents and determinants of well-being across a range of stakeholders or individuals. It should also be considered that responses implemented in one place could have impacts on ecosystem services and human well-being located further away. It should also be acknowledged that responses implemented today might have significant impacts on ecosystem services and human well-being in the future. The assessment of responses and the learning process to improve the design and implementation of responses will need to consider these complexities if the sustainability of human development, the reduction of poverty, and the integrity of ecosystems to improve human well-being is to be ensured.

Unraveling these complexities can be considered as a scientific quest by the scientific community. Decision-makers, on the other hand, want some answers to some pertinent questions to reduce or reverse the present trends of deteriorating ecosystems, reduce poverty, and improve well-being. In the event that answers are not forthcoming, then decision-makers would like some guidance from the assessment of the risks involved, and what needs to be done in the face of this uncertainty. And they need these answers now. A unique and strong feature of the MA is the guidance, in the form of key questions provided by decision-makers from both the human development and ecosystem fields. (See Box 17.1.)

The chapter is structured in the following manner. The next section provides a synthesis of the chapters in Part II of this report; the responses discussed are primarily responses directed toward the management and conservation of ecosystems and ecosystem services. There is no one systematic format in the way the various responses are presented. The economic and technological responses are structured according to the various ecosystem services, while legal responses are presented according to administrative scales. The social responses are presented along thematic lines. The following section, on the other hand, presents the main findings of a review of existing development and poverty reduction strategies. Both sections examine: (1) how effective responses have been in achieving their objectives; (2) the consequences of these responses on ecosystem services, the constituents, and determinants of human well-being and poverty; (3) the preconditions necessary for the responses to be effective; and (4) the potential for using these responses in the future.

After the two sections examining responses, another section identifies a number of context specificities and enabling conditions that play critical roles in determining the magnitude of the consequences of responses and the relative success of responses in achieving their objectives. A final section revisits the questions posed at the beginning of the chapter and presents some plausible answers.

17.2 Ecosystem Management Responses: Their Effectiveness and Consequences for Human Well-being and Ecosystem Services

Chapter 2 identified four major categories of response interventions. This section looks at the issues and sectors identified in Part II of this volume in the context of the four categories: legal responses; economic and financial responses; social, behavioral, and cognitive responses; and technological responses.

17.2.1 Legal Responses

This section synthesizes material from chapters in Part II of this report in relation to key legal responses ranging from the global through to local levels. Regulatory instruments are important in: (1) addressing the management of ecosystem services, especially the regulating, supporting, and cultural services; (2) helping in the monitoring and regulating of performance; and (3) regulating the use of proper and adapted technology. Such instruments help to provide the framework within which social actors are permitted to act.

There is implicit consensus in the literature that all instruments are embedded within a legal context and that the legal context needs to be compatible with the sociopolitical climate for an instrument to be successful. This means that separating legal from other instruments is a difficult and often artificial endeavor.

17.2.1.1 International Level

The assessment reveals that at the international level responses generally have tended to focus on specific issues rather to than take a comprehensive approach. For example, this has led to a proliferation of agreements and governance frameworks on:

- biodiversity (for example, CITES, Ramsar, CCD, CBD);
- food (for example, food aid regimes, trade regimes, technology and scientific cooperation regimes, food summits, intellectual property regime, commodity agreements, the Common Agricultural Policy of the European Union, bilateral fishery agreements, and ITPGR);
- water (more than 400 agreements in the last 200 years including the Watercourses Convention of 1997);
- wood and wood products (for example, IFAP, ITTO, ITTA, UNFF);
- waste management (for example, Basel Convention); and
- climate change (for example, UNFCC, Kyoto Protocol, bilateral agreements).

Most of these agreements focus on specific environmental and resource-related problems. Only recently is there a trend to in-

BOX 17.1
Key User Questions

Decision-makers from the human development and ecosystem fields identified key questions that MA users need to have addressed:

1. How effective are international agreements for addressing concerns related to ecosystems and human well-being, and what options exist to strengthen their effectiveness?
2. What is the scope for correcting market failures related to ecosystem services (internalizing environmental externalities)? How much of a difference would this make for ecosystems and human well-being, and what are the necessary conditions for these approaches to be successful?
3. What is the potential impact on ecosystem services and human well-being of removing "perverse subsidies" that promote excessive use of specific services? What trade-offs may exist between ecosystem and human well-being (especially poverty) goals?
4. What more can be done to strengthen national-local legal frameworks for more effective ecosystem management to reduce poverty and increase human well-being? And are there scale effects, which make legal responses at one level more effective in addressing ecosystem services?
5. What are the strengths and weaknesses of an increased focus on technological advances for addressing concerns related to ecosystems, human well-being, and poverty alleviation? What steps can be taken to increase the likely benefits for ecosystems and human well-being of technological change, and decrease the risks and costs?
6. How important is governance/institutional reform to the achievement of effective ecosystem management, and what are the characteristics of that reform that are most relevant to the pursuit of goals related to ecosystems and human well-being?
7. What are the strengths and weaknesses of greater stakeholder involvement in decision-making in the management of ecosystem services? How can decision-makers find the right balance?
8. What tools and mechanisms can promote effective cross-sectoral (water, agriculture, environment, transportation, etc.) coordination of policy and decision-making?
9. What design characteristics of ecosystem-related responses are helpful in ensuring that they provide benefits for poverty reduction?
10. What have been the consequences of macroeconomic responses like poverty reduction strategies for ecosystems and their services?
11. What has been the net impact of trade liberalization, globalization, and privatization on ecosystems, their services, and human well-being?

crease subsidiary goals of poverty alleviation (for example, CBD, MDGs, WSSD, CCD), but the operative mechanisms to achieve the latter are rare. However, in the last few years, there is increasing recognition among these legal frameworks that environmentally sensitive poverty reduction strategies can enhance the resilience of local people to environmental impacts, while also, under many circumstances, reducing the burden on the environment. (See Chapter 5.)

The assessment tends to show that the success of international legal instruments in achieving their specific goal depends on four interrelated conditions: (1) the design of the agreement; (2) the way the agreement has been negotiated; (3) policy coherence at the international level; and (4) the domestic context in which the agreement is to be implemented.

17.2.1.1.1 Instrument design

The assessment of Part II chapters, supplemented by additional literature to complete the picture (Oberthur and Ott 1999; Miles et al. 2002; IPCC 1990; Vingradov et al. 2003; Epiney 2003), shows that effectively designed international agreements:

Are legally binding. If the agreement is legally binding, like the CBD is, it is generally likely to have a much stronger effect than a soft-law agreement such as Agenda 21 and the Declaration and Action Plan of the World Summit on Sustainable Development. This is both the result of the assumption of international law that agreements made will be implemented (*pacta sunt servanda*), as well as verifiable through observation (Henkin 1979).

Have clear objectives. If the agreement has a measurable goal, it is easier to analyze whether it is meeting its goal (Bodansky 1993; Gupta 1997). The UNFCCC did not include measurable targets, though the Kyoto Protocol does. In fact, one of the key factors which contributed to the success of the Montreal Protocol on Ozone Depleting Substances and its amendments was the inclusion of targets. In similar fashion, regional water law regimes with clear objectives have been effective in meeting their goals (Kaya 2003). On the other hand, the CCD and the CBD, which have qualitative goals, allow more scope for interpretation in implementation and their effectiveness has been more difficult to measure. (See Chapter 5.)

Have clear definitions. An agreement with clear definitions and unambiguous language, like the Montreal Protocol, is easier to implement (Franck 1995). On the other hand, the weaknesses of the definitions in the Basel Convention of 1989 undermined its effectiveness. In similar manner, the complexity of the definitions in the climate change regime makes it more open to multiple interpretations. (See Chapter 13.)

Have unambiguous principles. If there are clear principles, this can help guide the parties to reach future agreement. The principles guide the implementation of an agreement. But if there is lack of clarity on the relationship between the principles, then there is no clear incentive for taking action. The interpretation of the precautionary approach in the CBD, Rio Declaration, and UNFCCC lends itself to considerable speculation (Hey and Freestone 1996). There is also conflict between the precautionary principle and the cost-effectiveness principle in the UNFCCC, and conflict between the equity principles and the no-harm principle in the Watercourses Convention. (See Chapter 7.)

Include an elaboration of obligations and appropriate rights. If there is a clear elaboration of obligations and rights, the agreement is more likely to be implemented. (See Chapters 5, 8.) The International Tropical Timber Agreement and the World Heritage Convention have relatively vague obligations, and therefore are difficult to implement. Meanwhile, even though CITES has clear objectives, the large number of species that have to be controlled makes it difficult for customs officials to monitor and implement effectively. (See Chapter 5.) In instances where language is indeterminate ("to explore" or "to encourage"), it has been difficult to control implementation.

Have financial resources. If there are financial resources in the regime, this in general increases the opportunities for implementation especially for developing countries (Brack 1996). The financial mechanism of the Montreal Protocol is a case in point. The GEF and the other funds available for international water, climate change, desertification, and biodiversity have increased the potential for implementation. However, CITES and the

Ramsar Convention, which did not have financial mechanisms, have more limited effectiveness. But sometimes, funding levels are so low that the mere inclusion of a financial mechanism is in itself not enough. A case in point is the resources allocated for desertification. (See Chapter 5.) Funding should be increased where the rich countries value the resource very highly, and where local communities have high opportunity costs of preservation.

Develop mechanisms for implementation. Where financial resources are not forthcoming, the design of market mechanisms in some instances has been found to increase the potential for implementation. For example, joint implementation, emission trading, and the clean development mechanism in the climate change agreements and Type II Agreements at the WSSD have enhanced the potential for implementation by providing the private sector an important role and incentives in the implementation of these agreements. (See Chapter 13.)

Include the establishment of bodies for the facilitation, monitoring, and implementation of the agreement. The establishment of subsidiary bodies with authority and resources to undertake specific activities to enhance the implementation of the agreements is vital to ensure continuity and preparation and follow-up to complex issues. (See Chapter 5.)

Establish good links with scientific bodies. As ecological issues become more and more complex, it becomes increasingly important to ensure that there are good institutional links established to link the legal process with the scientific community. This has been well arranged in the Long Range Transboundary Air Pollutants regime, the Montreal Protocol, and the climate change regime (Haas 1989; Gupta 2001). In the CBD regime, this is being developed. However, even the present limited links with the scientific community are limited to the mainstream natural sciences and economics. There are very few and, in many cases, no links with social science focused on development, gender, and human development studies.

Establish reporting facilities. The existence of requirements to report usually puts pressure on countries to undertake measures and exposes them in the event they have not been able to implement their goals. The inclusion of such features increases the potential for implementation. (See Chapter 5.) However, while such reporting requirements assert continuous pressure on countries, if the ultimate objective and the obligations are not clear, these documents may be less effective. But even so, if there is a reporting format, this makes it easier for implementing officials to implement the documents; and if infractions are reported, this makes it easier to keep the pressure on countries.

Have noncompliance procedures and sanctions. For an instrument to be effective, it should include sanctions for violations and/or noncompliance procedures to help countries come into compliance. (See Chapter 5.) The lack of such sanctions and/or noncompliance procedures in many environmental regimes is a major problem (for example, CBD, Watercourses Convention, CCD). On the other hand, the Montreal Protocol does have successful noncompliance mechanisms and these have played a role in the success of the Protocol. Noncompliance mechanisms are now being developed within the climate change regime.

Have dispute resolution mechanisms. The lack of compulsory jurisdiction for dispute resolution is a major weakness in international law. However, increasingly, many agreements (CBD, Watercourses Convention, SADC Protocol) are explicitly referring to specific options for dispute resolution and these can increase the potential for implementation. The Aarhus Convention, the Watercourses Convention, and the SADC Protocol allow foreigners nondiscriminatory access to the national judicial systems. It is still too early to say anything specific, but it is expected that this will enhance the effectiveness of agreements. (See Chapter 7.)

Establish coordination with other relevant agreements. The literature indicates that it is vital that these agreements are closely linked with other agreements and that solutions designed for one regime do not necessarily lead to problems in other regimes. Many regimes do indeed have articles that link up to other regimes, including CITES, CMS, the ozone regime, the Basel Convention, the UNFCCC, and the CBD.

Establish technology transfer mechanisms. Technology transfer can address environmental and social issues in developing countries. However, such technologies may not necessarily be appropriate or contribute to reducing poverty. (See Chapters 6, 7.) In the context of climate change, a literature survey of IPCC concluded that technology transfer is likely to work best in combination with national systems of innovation; social infrastructure and participatory approaches; human and institutional capacities; macroeconomic policy frameworks; sustainable markets; national legal institutions; codes, standards, and certifications; equity considerations; rights to productive resources; and research and technology development (Hedger et al. 2000; Mansley et al. 2000).

Include distributive elements. The inclusion of distributive elements may increase the effectiveness and sustainability of agreements. Access and benefit sharing with local people, as encouraged by the CBD Article 8(j) through participation of local people in the planning process and in discussion of customary and other regulatory frameworks, is one such idea. A similar provision is found in ITPGR. Benefits shared can be monetary or non-monetary. The ecosystem approach also offers a basis for successful implementation. Transboundary conservation areas may be another way to promote conservation and human well-being. (See Chapter 5.)

17.2.1.1.2 Agreement negotiation

The assessment reveals that, in addition to the design aspects of the various existing agreements, the ways in which agreements are negotiated are also critical for the effectiveness of agreements. While treaty negotiations are in theory based on the assumption of *pacta sunt servanda* (that is, the agreement will be implemented) and call for rules of procedure to be put in place, these rules are sometimes suspended for practical and/or political reasons. The assessment shows that the compliance pull (that is, the likelihood that the agreements will be voluntarily implemented) of agreements can be increased if:

There is leadership shown by specific actors. There is increasing consensus in the literature that leadership shown by the major actors, be it structural, instrumental or directional, will have an influence on the design of a regime. The more leadership in a regime, the more likely it is that the regime will take on good design features (Young 1991; Oberthur and Ott 1999). For example, in the climate change regime, the leadership from the European Union has been critical in steering the way the instrument was designed to effectively implement the goal.

Non-state actors are actively involved in the process. There is increasing consensus in the assessment of the need to include non-state actors into the negotiating process in general. This is seen as a way to increase the legitimacy and compliance pull of an agreement by ensuring that a variety of interests and information are represented at the negotiations, and this can help in the actual implementation process. (See Chapter 5.) On the other hand, the inclusion of non-state actors may actually lead to capture by vested interests (Gupta 2003). Besides, stakeholders themselves may not be willing to accept environmental goals that go against their own short-term and/or long-term local interests. (See Chapter 5.)

The instrument is negotiated in accordance with the established rules of procedure. While all international agreements must be negotiated in accordance with the established rules of procedure, the complex nature of global decision-making processes requires concurrent meetings often only in English. This has placed a big strain on developing countries, which have limited financial resources to attend and participate effectively. The negotiations on the Kyoto Protocol did not provide the developing countries adequate opportunity to negotiate effectively on the instrument as a whole, and specifically on industrial-country targets, emissions trading, and the issue of sinks (Yamin 1998; Werksman 1998). Most developing countries do not have the staying power to negotiate effectively at all these sessions (Gupta 2002).

Negotiators are given the time and expertise to prepare for the negotiations and have consulted domestic stakeholders accordingly. The Law of Treaties assumes that countries send only well informed and prepared negotiators to the negotiating table. Interviews reveal that as the complexity of the negotiations increase, most negotiators from the developing countries may be well informed about one or two aspects, but few if any have a complete grasp of all the issues that are being negotiated (Gupta 1997, 2001; Rutinwa 1997 on the Basel Convention; Nair 1997 on the Montreal Protocol). The assessment of a number of international treaties show that very often the national negotiators have not been able to adequately represent the interests and the actual problems faced by their countries. As a result. international agreements are often cast in first-world language with first-world solutions that may be ill-suited for implementation in developing countries (Agarwal et al. 1999).

The instrument includes poverty reduction and human development measures. While developing countries have tried in almost all environmental negotiations to make links to other global regimes (trade and finance) and the global order (the "unfair international economic order"), and have tried to focus attention on poverty alleviation, well-being, and development, these items either have been seen as irrelevant, and hence excluded from the agenda for discussion, or have been included to pacify the developing-country negotiators. However, even when they are included, no clear targets and timetables or mechanisms to actually achieve these goals are articulated (for example, CBD, UNFCCC, and the WSSD).

17.2.1.1.3 *Policy coherence at the international level*

The third element for influencing the effectiveness of legal agreements at the international level is policy coherence. There has been a trend toward synergetic agreements among environmental agreements and governance initiatives, for example, in the CBD, CCD, UNFCCC, WSSD, MDGs. (See Chapter 5.) However, there are contradictions with the trade regime (whose primary focus is trade rather than environmental protection); agricultural policies (where the policies of countries to subsidize their agricultural sector has given perverse incentives to farmers to overproduce in the industrial countries, while reducing markets for many developing countries; see Chapter 6); the intellectual property rights regime (which may lead to reduced access to food for people; see Chapter 7; the investment and private international law regime (where public private cooperation may actually transfer resources to the private sector, and be covered by the more restrictive rules of confidentiality and liability that operate in private international law; see Chapter 2); bilateral fisheries agreements which, combined with heavy subsidies to the fishing sector, are leading to overexploitation of the resource and reduced access for the local people; and the climate change regime (where if the progress made is slow, many ecosystems will be seriously affected and environmental refugees are likely to increase; see Chapter 5).

17.2.1.1.4 *The domestic context*

Lastly, the assessment shows that the success of an agreement depends on the contextual features of the country in which implementation is to take place. (See Chapters 2, 3.) The agreements are more likely to be successful where:

There is a high level of awareness and resources. The higher the awareness and resources (human and financial) within a country, the greater the likelihood of implementation. For example, studies by Jacobsen and Weiss (1995) and Sand (1996) found that developing countries had more difficulty in implementing international environmental agreements because of the lack of financial capacity and the lack of awareness. However, even in an environment of limited resources, the mere existence of international agreements created internal efforts within many developing countries to try to push and advance the agreements.

There is a strong institutional and legal framework. If the institutional and legal frameworks within a country are well developed, there is a higher likelihood of that an international agreement will be effectively implemented. At the local level, the question is whether non-compliance should be dealt with. Although Chapter 5 argues that, from an economic perspective, it is not optimal to prevent all non-compliance with laws and regulations because the marginal costs may exceed the marginal benefits of full enforcement, from a legal perspective, non-enforcement of rules can lead to an environment of disrespect of national laws and is in the long-term disastrous for creating an effective legal environment.

There is political will. A critical factor remains the availability of political will to actually undertake measures.

17.2.1.2 Instruments at Local and National Level

The assessment shows that one needs to be careful about the way one generalizes about the national level. Countries have legal and cultural experiences that go back more than 2000 years, at least in the case of water and land management. These experiences have shaped the way legal rights and responsibilities have developed. Each country has over the years been subjected to a vast range of influences, conquests, religious influences, colonialism, the spread of epistemic communities, the spread of codification of laws, and now globalization. While in some countries, such as the Netherlands, the influences have been integrated into one comprehensive body of law and policy, and new laws in effect rewrite the older ones, the bulk of developing countries still have multiple legal practices within their context. As such, when new laws are designed and negotiated, these may have no effect at the ground level, if they are not seen as relevant or legitimate by the local people. In other words, most developing countries have multiple levels of law-making. Furthermore, institutions are embedded in each other. Thus administrative rules are embedded in ecosystem rules that are embedded in a host of other relevant rules which are in turn embedded in the constitutional system within a country.

Legal instruments at the national level include the elaboration of rights and responsibilities (for example, who has a right to land use and water), principles (for example, the precautionary principle), obligations, standards, monitoring, the establishment of implementing bodies or the delegation of power to existing bodies, and rules in relation to violation of the agreement.

The assessment of Part II chapters shows that the following are likely to be effective:

Designing new and direct incentives within the regulatory framework. In order to protect ecosystem services, it is vital to recognize the opportunity costs for the local people and to provide with them incentives. (See Chapters 5, 8). Examples of such incentives include:

- In South Africa, direct payments are given to local people who catch and sell wild animals to game ranches instead of being killed because of the danger they pose or because of the value of their skin or hide.
- Land tenure reform. Numerous studies have shown that private land tenure rights have promoted the conservation of ecosystem services while at the same time producing direct benefits to individuals (Tiffen 1993; Kates and Haarmann 1991; Duraiappah 1998). However, it should be noted that private land tenure may not be conducive to ecosystem conservation in all ecosystems. For example, studies have shown that communal land tenure was far more conducive for ecosystem conservation than for the well-being of the local communities (Rutten 1992).

Integrating different levels of conservation of ecosystem services through regional planning. One element of regional planning is the concept of protected areas. The assessment shows that although protected areas are considered an important part of biodiversity conservation programs, notably for ensuring the survival of certain components of biodiversity, it points with a high level of certainty to the fact that protected areas by themselves are not sufficient to protect biodiversity and related ecosystem services (See Chapter 5.)

We note that protected areas were never intended as instruments for poverty reduction but rather as a way of preserving biodiversity. Nonetheless, with regard to poverty reduction, assessment indicates that they may actually exacerbate poverty by reducing access of rural people to the resources that have traditionally supported their livelihoods, thereby affecting their welfare (Bruner et al. 2001). In addition, the fact that protected areas are partly a response to degradation and habitat change due to human activities precludes the use of some ecosystem services that contribute to human well-being.

The use of protected areas as a standalone measure is not sufficient for the approach to be effective. In order to have a more efficient functioning of the area, the design should take into account the ecological, social, historical and political context, and better weigh the multiple economic values accruing to people at the local, national, and global levels, by minimizing conflicts with the needs of society and use of compensation schemes. Measures on protected areas need adequate resources, greater integration with the wider region between protected areas, and greater stakeholder engagement with the objective of giving the local communities greater rights and ownership, and local people direct payments for conservation and management

Last but not least, protected areas should be given special attention due to their prominent position within the CBD as a hybrid instrument of regulatory (legal) and economic incentives to conserve biodiversity. However, they will need to be designed in a manner complementary to the local social, economic, and political settings.

In general, we find regulatory instruments at the national, regional, and local level having the following characteristics:

- Legal instruments focusing on direct management of invasive species are more effective than an integrated approach. This tends to work best when the species in question has low dispersal ability, when sufficient economic resources are devoted to the problem and when social actors are all engaged in addressing the problem.
- Reintroducing species works best when it is based on adequate science. Availability of suitable habitats, a number of healthy and genetically diverse species members, and a long-term program for monitoring are important factors. The reintroduction of the Mexican grey wolf in the Colorado plateau is a successful case in point.
- New regulatory instruments may include regulations on ecosystems for production as a new tool for protecting ecosystem services.
- Where the private sector is engaged constructively in discussions within a strong regulatory framework, there may be potential for protecting the environment and human well-being. The case of Bioamazonia and Novartis shows that the correct modalities of involving the private sector and allowing public participation still have to be worked out.
- Regulatory options that promote community forest management can work to both protect ecosystem services and human well-being. In India, it has worked primarily to improve human well-being.
- Regulatory approaches that promote public participation in policy-making have a higher chance of being effective.
- Fair and strong law enforcement is often a powerful disincentive for exploitative behavior. Law enforcement procedures in Indonesia to stop blast fishing have increased local biodiversity.

17.2.1.3 Distributional Issues

Environmental policy and law inevitably has a distributive effect in terms of responsibilities, impacts, and benefits. Most global environmental problems are essentially "wicked" problems in that the distribution of costs and benefits are not aligned, and there is inadequate scientific certainty and a vast difference in the value systems of countries. For example, the desire to protect ecosystems and species is not always based on the same altruistic motives. The pharmaceutical industry has high financial stakes at one end of the spectrum, while the local people may face high opportunity costs that are low in absolute terms. (See Chapter 5.) For example, in the climate change regime, the costs of emission reduction, if taken seriously, are high for the developed countries, while some perceive that it is the developing countries that will benefit.

The way responsibilities are distributed between the parties also can have a distributive effect. On the one hand, the funds established under many of the agreements have a re-distributive effect, because they transfer resources to other countries to enable them to participate more effectively in the negotiations and to fulfill their obligations under the international agreements. On the other hand, the move away from the "polluter pays" principle at the international level has led to an inequitable distributive effect internationally. Thus the small island countries and the least developed countries, most of whom have negligible greenhouse gas emissions, will suffer disproportionately in comparison to the other countries, and for no fault of their own. Many may even lose large quantities of land and land-based resources and infrastructure as the impacts of climate change are experienced. In the meanwhile, the world's largest polluter, the United States, does not take on any quantitative commitment, arguing that from a domestic perspective the costs of taking policy measures far outweighs the benefits for Americans.

In the climate change agreements, the distributive effect is highly questionable because of the focus on cost effectiveness. In the other agreements, the distributive effects are similarly not always quite so clear. For instance, if African countries wish to sell ivory as national resources from a huge elephant population, to what extent may the outside world be allowed to prevent such sales under CITES? If forest-rich countries want to convert forestlands into agricultural lands, to what extent do other countries have the right to prevent such conversion, and who bears the final costs of this? There are no clear-cut answers to these issues, but the distributive impacts surrounding these concerns need to be further explored in a more rigorous and fair manner than presently done in the international arena.

17.2.1.4 Some Inferences

The key international tool for addressing global ecosystem services is treaty negotiation on an issue-by-issue basis. The proliferation of agreements on natural resources over the last two hundred years in the area of water and hundred years in the case of environmental issues has given us a vast amount of empirical evidence on the success of agreements. This evidence shows that the success of agreements depends on how they are designed in both substantive and procedural terms, how they compete with other international agreements, and the context in which they are to be implemented. At the national and local level, the success of the regulatory instruments depends on how well they target specific incentives and disincentives, and how well the implementation is monitored and enforced.

17.2.2 Economic and Financial Responses

It is common practice in economic analysis to compare and contrast regulatory or what economists call "command and control" approaches vis-à-vis economic and financial interventions. This section, however, does not apply this commonly used analytical approach. Instead it focuses on the use of economic instruments and financial interventions, as described by the Part II chapters of this volume. It does so based on the following two criteria. The first criteria will be the effect of the instrument on the continued flow of ecosystem services. The second criteria relates to the distributive issues and the impacts on human well-being and poverty reduction. Regulatory instruments are evaluated in the legal section.

Market-based instruments have been implemented to address environmental concerns ranging from solid waste management, biodiversity conservation, sustainable land use, and reducing air pollution. MBIs are said to have great potential for generating win-win situations (Rietbergen-McCracken and Abaza 2002). The win-win situations may arise in the case of the use of charges and environmental taxes, for instance, which have been singled out as providing revenue to the treasury in a cost-efficient manner and at the same time benefit the environment "by encouraging polluters or users of environmental resources to change their behavior to become less polluting and wasteful" (Rietbergen-McCracken and Abaza 2002).

Economic instruments are basically structured to achieve a mixture of three main objectives (Rietbergen-McCracken and Abaza 2002, p. 100):

- establish and enforce prices for resources consumed and environmental damages associated with production;
- address issues dealing with property rights which directly contribute to pollution or poor stewardship of resources; and
- subsidize the transition to preferred behaviors.

Against this backdrop, the use and appropriateness of MBIs is assessed, in addressing issues relating to ecosystem services, their effectiveness in different regions, the preconditions necessary for the effective use of the economic/financial instruments, and last, but not the least, how effective these instruments have been in achieving the goals related to human well-being and poverty reduction.

17.2.2.1 Biodiversity

In the case of biodiversity, two broad sets of responses are examined: indirect and direct incentives. Incentives are, broadly speaking, mechanisms to change or affect the behaviors of individuals. Incentives could be negative such as fines or positive such as tax credits.

Indirect incentives such as development interventions, integrated development projects or community-based natural resources management projects have had mixed success in terms of achieving conservation and development goals. (See Chapter 5.) On one hand, indirect incentives have had success in redirecting labor and capital away from activities such as intensive agriculture that degrade the ecosystems, encourage commercial activities such as ecotourism that supply ecosystem services, and can help in raising income to reduce dependence on resource extraction. On the other hand, redirecting labor does not necessarily mean success in reducing the level of degrading activities, as the response itself may create incentive to hire people to take advantage of the opportunities provided by the response (Muller and Albers 2003). Similarly, commercial activities that maintain ecosystem services, such as ecotourism, have limited success due to the fact that demand for such activity is not high enough to support a large fraction of the population.

Furthermore, an assessment of ecotourism in the Khao Yai National Park in Thailand points toward the fact that most income from tourism did not accrue to villagers but to tour companies instead. The information in this case seems to indicate that essential elements of poverty reduction and human well-being, that is, access to resources and income, have not been achieved. Studies find institutional settings and property rights as critical elements for making ecotourism a potential tool for poverty reduction. (See Chapter 5.)

Direct incentives or payment for conservation, which consists of cash or in-kind payment to individuals or groups, were found preferable to indirect payment because they were found to be more efficient, effective, and equitable (Simpson and Sedjo 1996; Ferraro 2001, 2002, 2003).

However, the empirical evidence of direct incentives such as transferable development rights and tax credits in successfully encouraging and achieving conservation of biodiversity in situ is unclear at best and has been the subject of criticism, notably for being relatively complex. In the case of TDRs, a new market that requires learning will need to be established. This implies high transaction costs and the establishment of new supporting institutions. It is also important to stress that TDRs need to be supported by well-defined property rights over the land and resource.

One major drawback of tax credits and TDRs, in terms of achieving goals related to ecosystems and the related services, is that both instruments are unable to target specific habitat types. There is therefore a high degree of uncertainty about whether they can protect specific kinds of biodiversity. However, and in spite of some of the criticisms, there have been successful cases in the use of property rights to conserve biodiversity. The charcoal market in St. Lucia is one example whereby use rights in the form of charcoal rights were successful in preserving biodiversity and reducing poverty. (See Box 17.2.) The main reasons why the charcoal right market was a success was because of its simplicity, using existing institutions instead of depending on new institutions, and last but not the least, it was developed through a participative process involving the local community.

BOX 17.2

Case Study of the Mankote Mangrove in St. Lucia (UNEP 2003)

The Mankote Mangrove constitutes 20% of the total mangrove area in St. Lucia. Uncontrolled charcoal harvesting through excessive tree loggings created a severe environmental decline of the mangroves, which posed a serious threat to many of the ecosystem services that the mangrove provided, including water quality, coastal stability, bird habitat, and fish breeding. Local communities, consisting primarily of poor people, undertook the practice of harvesting charcoal. These communities had no legal right to use the publicly owned mangrove resources. With no possibility for substitution, the loss of access to the mangroves by these poor populations due to resource depletion or degradation would have created permanent loss of their only source of income.

To address this problem, the following solution was implemented:

- The local communities were organized into informal cooperatives and given communal legal and exclusive rights to harvest the charcoal.
- They were involved in monitoring the program, to get accurate information on the overall health of the mangrove.
- Measures to increase the supply of wood outside the mangrove reserve were put in place, as were alternative job options for charcoal harvesters. Tourism played an important role in offering this alternative.

The effort yielded the following results:

- The decline in the Mankote Mangrove was halted and reversed.
- The density and size of trees increased.
- Charcoal harvests were maintained.
- The range of employment options for the poor population somewhat increased.

This is a clear case where a property and legal rights approach made sense, because the subsistence harvesters were the primary source of the problem due to uncontrolled harvesting of charcoal. The use of formal rights to the resource gave the poor an incentive for long-term management of the mangrove as an asset over which they had control. The introduction of a monitoring program further improved the level of and access to information they had about the general condition of the mangrove.

Moreover, the direct payment approach has been criticized because it entails on-going financial commitment by governments, multilateral donors, and firms to maintain the link between investment and conservation objectives, and can create conflicts between communities as property rights are transferred to local participants. Direct payment may also turn biodiversity into a mere commodity (Swart 2003). In other words, paying for biodiversity conservation may give local people the impression that it is something which is important because there is a "price tag" on it and this may not help in the long run to capture the real/intrinsic value of biodiversity as it plays a vital role not only in terms of provisioning services, but also in maintaining vital ecosystem equilibrium (through regulating services) and has also an impact on the spiritual lives and culture of certain communities.

In spite of the many criticisms of direct incentives, there is a medium to high certainty that direct interventions have had more success than indirect interventions. However, it will be necessary to iron out the issues of how to achieve a sustained flow of money for payments, the need to more clearly define the issue of rights, and how to make management decisions consistent with the overall biodiversity conservation and not turn biodiversity into a mere commodity.

17.2.2.2 Water

Chapter 7 of this volume finds that, market-based approaches to water have "the potential to unlock significant supply- and demand-side efficiencies while providing cost-effective reallocation between old (largely irrigation) and new (largely municipal and instream) uses." This argument is based on the principal that allocations of water have seldom taken into account its scarce nature and value. Furthermore, the argument puts forth the idea that payment for water conservation can increase water availability and that properly functioning water markets can provide price signals for reallocation not only between different uses, but also as signals to guide conservation activities.

The three main reasons identified by Bjorlund and McKay (2002) to consider the use of markets for the provision of fresh water are: (1) tradable water rights create a value for water that is distinct from land and able therefore to be conserved in its own right, (2) full recovery pricing incorporates externalities associated with the inefficient use and encourages inefficient users to leave the market, and (3) the use of market forces rather than government intervention to facilitate reallocation reduces transaction costs and delays.

Whether through the use of the water banks for direct payment of water, water markets and water exchanges where water rights and permits are traded, or the use of economic instruments for watershed services, the underlying premises behind all these types of interventions remain first and foremost the efficient allocation and use of resources. Experiments conducted in several countries (western United States, southern Australia, Mexico, and Chile), indicate that payments and incentives for water conservation can increase water availability, just as water pricing at its full marginal cost can reduce demand. (See Chapter 7.)

Functioning water markets such as those in the Limari market in Chile, with its largely laissez-faire water code, can not only provide price signals for better reallocation between different uses, but also help guide conservation through flexible management by farmers of some of the risks caused by uncertainties in water supply and agricultural markets. However, experience from Chile suggests that the imposition of an unfettered free market in water into a developing context with significant existing socioeconomic inequality may lead to further inequities. (See Chapter 7.)

At the same time, experience gathered from the California Water Bank, whose purpose was to transfer water from irrigation to municipalities, was found to have both positive and negative environmental impacts, with negative effects including damage to bird and wildlife forage and habitat, as a result of removing grain crops, reduction in groundwater recharge and groundwater quality, and also negative impact on fisheries. The positive impacts are improved surface water temperature, quantity and quality, and reduced fish entrapment (MacDonnell 1994).

Similarly, the use of economic instruments for watershed services—such as transfer payments made to landowners as compensation for the cost of adhering to specific management practices, marketable permit systems, generally in the form of cap-and-trade or credit programs, in order to allocate permitted levels of pollution and resource use—have shown their limitation in terms of achieving by themselves the goals related to ecosystems and their services, including regular flow of water, protection of water quality, and control of sedimentation (Landell-Mills and Porras 2002). The Chapter 7 assessment points out that a mix of market-based and regulatory and policy incentives are the most likely to achieve these goals, notably when the "threats are beyond the response capacity of individual communities" (Rose 2002). Among the main conclusions:

- Markets arrangements do not automatically protect ecosystems. A mix of economic and regulatory instruments are needed, especially when the scale of the threats and responses are beyond the capacity of individual communities.
- The issue of poverty reduction and human well-being need to be made explicit. Poverty reduction objectives in the form of equitable rights are not made explicit in the design of economic interventions in the water sector. Furthermore, the achievement of equity in terms of benefits and costs needs as a precondition the presence of enabling institutions (for example, good governance) in order to enable the poor to have access to market.

17.2.2.3 Food

The primary focus of economic responses in the provisioning of food service has been on increasing food production. The impact on ecosystems from attempts to increase food production has been realized largely as secondary effects, and they represent negative spin-offs or externalities of agricultural production. Furthermore, it has been found that these externalities are often ignored by individual farmers, small-scale agents, and by governments in the food and agricultural policies.

The analysis further recommends that the "need to mitigate impacts on ecosystem and sustain their capacity for future generations makes necessary the introduction of appropriate regulatory frameworks at all levels from local to global, that will control for the externalities affecting the capacity of ecosystems to sustain their food provisioning services." (See Chapter 6.) However, it is also noted that the cost associated with regulation, and representing the cost of using ecosystem services for producing food is largely unpaid due to missing markets and lack of well-defined property rights. Instruments identified as important in maintaining and expanding food production capacity include agricultural subsidies, fisheries interventions, and livestock interventions.

17.2.2.3.1 Subsidies in agriculture

Water is a critical input in the various technologies developed to increase food production from ecosystems. One of the criticisms of the present water pricing structure is the low water fees levied on farmers in order to increase food production. The low water prices contribute to inefficient use of water, water shortages, and depletion of water resources in the long run and degradation of the ecosystem (Koundouri et al .2003; Pashardes et al. 2002; Chakravarty and Swanson 2002). Moreover, there is increasing evidence that rich framers benefit more than the poor farmers from these low water fees.

There is not sufficient information from the assessment in Chapter 6 on whether or not economic-based instruments for groundwater such as pumping taxes, transferable property rights, and water markets have created greater incentive for farmers to save water and reduce the ecosystem deterioration witnessed under subsidized water regimes. It is also too early to say if these economic responses create a more equitable distribution of the use of water or if they have marginalized the poor even further.

17.2.2.3.2 Fisheries

Instruments such as total quotas or total allowable catches are used in fishery management to prevent stock depletion. Individual quotas and individual transferable quotas, describing individual annual nontransferable quota and transferable quotas between fishermen, respectively, are also commonly used.

The total allowable catch, which sets a total quota for a time period, is known to create a "race for fish," which raises the cost of fishing activity and is a problematic response to overharvesting. Individual quotas, on the other hand, have an information and compliance problem; they only work when fishing vessels have information about each other's cost structures, and thus require a large amount of information. Individual transferable quotas solve the information problems by letting fishermen trade with quotas, but continue to have compliance problem. ITQs seem to be the preferred option, but need to be combined with a strict control/regulatory and enforcement policy.

There are limited references/assessment on how effective this instrument is in achieving poverty reduction and ensuring equitable distribution across different social groups. Some evidence, notably from developing countries, tends to point toward difficulties in terms of enforcement and monitoring of these instruments. Box 17.3 provides an assessment and illustration of the use of fishery ITQs in Chile.

17.2.2.3.3 Livestock

Livestock grazing and rangeland burning strongly affect the state of vegetation in rangeland systems. Land tenure and economic policy in Mongolia, for example, has changed pastoral burning and practices, reducing forage quality and possibly diversity.

There is strong indication that conversion of rangelands into cropland by farmers leads to vegetation and wildlife losses (Serneels and Lambin 2001). The assessment points toward the fact that in Africa and semiarid India, the poorest groups compete in rangelands for land, with the farmers having the upper hand be-

BOX 17.3

Case Study of the Fisheries Sector in Chile

In the fisheries sector in Chile, fish stocks started depleting greatly after the industry was privatized in 1973. Particularly affected were artisanal fishermen who, under the individual transferable quota system, cannot compete with industrial fisheries in the market and lose their livelihoods. To address this issue, individual transferable quotes were implemented for separate subclasses of fisheries and limited to industrial/commercial fishers.

The success of the program is unclear. As structured, the ITQ policy has protected industrial-country fishing interests, but reduced the potential benefits of the market-based quotas. The issue of artisanal fishermen has not been properly addressed, and regular updating of information about fishery health remains a problem. The small percentage of total catch currently covered suggests that ITQs are not yet addressing the higher goal of protecting Chilean fisheries.

The rationale for using these measures was two-fold: to apply regulatory efforts more consistently and to control access rights.

cause land tenure regimes and property rights often favor crop cultivation over livestock grazing (Blench 2000).

This together with evidence from the tribal grazing land policy in Botswana (UNEP 2000) leads us to with a medium to high degree of certainty to conclude that the use of private land tenure and grazing land policy, as economic instruments have not adequately addressed the goals related to ecosystems conservation in arid and semi-arid ecosystem types. Moreover, these instruments have created greater inequality between the wealthy cattle owners, pastoralists, and subsistence farmers.

17.2.2.4 *Wood and Fuelwood*

Forest certification was developed initially to mitigate tropical deforestation. However, the assessment in Chapter 8 finds most certified forests in the North, managed by large companies and exporting to Northern retailers. (Chapter 8.) This seems to be a clear indication that this instrument has been more effective in the North than in the South. The assessment further points out that where certification becomes the preserve of large companies, the role and competitiveness of small and medium-sized enterprises, which may bring with them sustainability and livelihoods, may be jeopardized.

Certification standards have provided incentive only to producers who were already practicing good practices, rather than improving bad practice. Some critics suggest that small growers and community groups should not be put under the same certification as large corporations. The heart of the problem is whether small communities, who occasionally harvest for timber, should be subjected to the same level of accountability as large corporations.

Commercialization of non-wood forest products has achieved modest successes for local livelihoods—a component of human well-being—but has not always created incentives for conservation but has contributed sometimes to overexploitation of resources. It has also led to the inequitable distribution of benefits with stronger groups gaining control at the expense of socially and economically weaker groups.

The improved formal access to forests has "helped in most cases to protect a vital role of forests as safety nets for rural people to meet their basic subsistence needs." (See Chapter 8.) At the same time, there has been criticism about the need to have benefits that go beyond subsistence level and that property rights would need to extend to more secure rights over valuable resources, for the poor to really benefit.

Another criticism deals with the inability of certification to interpret social standards in complex social contexts. Norms and methods of traditional societies are not captured by predictions made by document-based certifications that rely on "scientific means of planning and monitoring." (See Chapter 8.) Some environmental and social services are produced at levels other than the forest management unit, which is outside the control of the certified company.

Fuelwood is another large output of the forest sector. Growing demand for fuelwood and charcoal remains one of the main drivers in ecosystem deterioration in regions such as the dryland forest of Africa. However, formal and informal privatization of wood and land resources have led to woodfuel resources being unavailable to fuelwood gatherers. Subsidized fuelwood supplies from government forests; taxes and other charges to generate government revenue; or restriction imposed in the name of conservation have imposed constraints on who can participate in production. (See Chapter 8.)

Low fuelwood prices combined with subsidized alternative fuels have the characteristic of generating very little surplus or incentives/disincentives for the poor engaged in selling fuelwood for their living, and discouraging sustainable management of the resource. At the same time, the low price for woodfuels does not encourage valuing the trees as "priced assets" that need to be sustained and invested in with a long-term perspective in mind.

Some of the general lessons learned is that market-based responses are redistributing rights to stakeholders (for example, allocation of used rights to public lands, voluntary certification), making these stakeholders more effective in securing wood supplies and other ecosystem services as well as helping to change the wood industries. (See Chapter 8.) However, it is usually the good practice industries that benefit the most, which does not help create incentive for the bulk of wood producers to improve forest management practice. Economic intervention will not be adequate to answer the issue of illegal traders and asset-strippers; this will require a legal action and enforcement.

There is a need to combine governance and institutional responses with development questions. Better information is needed about the dynamics of wood supply and demand, and the benefits, costs, and distributional impact of the different economic interventions.

17.2.2.5 *Nutrient Management*

Mandatory taxes and fees were found to have the potential as regulatory instruments for inducing change. These taxes and fees include effluent charges, user or product charges, noncompliance fees, performance bonds, and legal liability for environmental damage. The findings suggest that it is difficult to reach specific targets in pollution reduction using the tax/fee approach since regulators have difficulty predicting how polluters will react. There are no known examples of use of taxes and fees as a nutrient management approach for water quality, but there is a growing interest in the United States for tradable permits in nitrogen control to coastal waters. (See Chapter 9.)

Trading programs were proposed to provide economic incentives for voluntary reductions from non-point pollution sources, and permit requirements for point-source pollution. The lack of statutory authority for regulating non-point sources of nutrients under the Clean Water Act emerged as a major problem for the plan in the majority of watersheds, where non-point sources dominate nitrogen input. The lack of authority over cross-boundary pollution was another challenge.

17.2.2.6 *Responses to Climate Change*

Use of economic instruments was proposed as an effective response to climate change. (See Chapter 13.) The economic instruments basically involve a series of "flexibility mechanisms" to facilitate and reduce the costs of attaining an agreed level of emissions targets. Under the Clean Development Mechanism and Joint Implementation, there is provision of transfers and acquisition of emission reductions (emissions trading) between various parties to the Protocol that are in good standing with respect to the various rules of reporting and accounting.

With regard to the twin goals of the CDM, namely sustainable development and achieving compliance in terms of setting emission targets, there seems to be an imbalance as more focus is put on the latter. Stabilizing greenhouse gases in the atmosphere largely via modifying energy use and energy supply is a central position under this scheme. In this particular case, the assessment found out that any flexibility mechanism would lower the cost of achieving a compliance target and induce the incentives to invest in new research and new technologies.

The assessment states that in the case of a JI transaction, if the emission reductions would have occurred without the incentive of the carbon trading rights, there would have been no significant impact on the atmosphere; the host country transfers some of its emission reduction credits to the acquiring country allowing the acquiring country to emit more, leaving the host country to carry out extra efforts to meet its target.

Another decision in designing a cap-and-trade program is whether to apply the targets "upstream," where carbon enters the economy (when fossil fuels are imported or produced domestically) or farther "downstream," closer to the point where fossil fuels are combusted and the carbon enters the atmosphere. An analysis by the U.S. Congressional Budget Office concluded that, in general, an upstream program would have several major advantages over a downstream program (also see Chapter 13 for some examples). An alternative is a carbon tax approach in which commodities or activities that lead to carbon emissions are taxed, thus providing an incentive to reduce the use of these commodities or activities. This is often seen as a simpler approach to achieving incentives for emission reductions. However, taxes are usually politically unpopular in most countries. Some have suggested that cap-and-trade and taxes may be combined to overcome the main weaknesses of both schemes, namely the presence of a strong enforcement agency for the cap-and-trade scheme and political unfeasibility for the taxes.

17.2.2.7 Valuation of Ecosystem Services: A Tool for Effective Economic Responses

Many of the economic responses mentioned in the previous sections rely on finding the correct value for specific ecosystem services. For example, in order to preserve a watershed, it is necessary to find the value of not just the provisioning services but also the regulating, supporting, and cultural services it may provide. Economic valuation of ecosystem services is being increasingly used to find these appropriate values. (See Chapters 3, 4.) Market based valuation methodologies have been discussed as appropriate for provisioning services while stated preference method like contingent valuation methods have been suggested to capture regulating, cultural, and supporting services of different ecosystems.

Although both classes of techniques are valuable, it should be acknowledged that these valuation techniques view the individual purely as a self-maximizing agent in a market environment. However, in reality, individuals are known to act as moral agents making judgments and assigning values from a social perspective; this will include therefore not only there own well-being but the well-being of others and in some cases even to a loss of well-being. Moreover, the philosophy of contingent valuation methods treats an environmental good as a normal private commodity that can be purchased and consumed. However, many ecosystem services cannot be treated as private goods and have the characteristics of public goods by which it is not just an individual's value of that service that counts but the collective value society places on it that matters and reflects the true value of the ecological "good" (Sen 1995). New techniques based on deliberative community participation are being increasingly used whereby the values are based on social choices vis-à-vis market-driven individual values.

17.2.3 Social, Cultural, and Cognitive Responses

17.2.3.1 Consequences of Responses Related to Changing Perceptions of Ecosystems

Land and waterscapes do not only have physical attributes but are subjected to and influenced by cultural perceptions as well. Culture and memory play an important role in creating contesting meanings for any one place (Schama 1995; Stiebel et al. 2000). Transformations of landscapes have been and will be influenced by cultural perceptions of nature as well as by sociopolitical and economic demands and aspirations. These transformations in turn have a (differential) impact on human well-being.

Over time, ideas about the relationship between humans and nature have changed and continue to change. (See Chapters 5, 14.) The directions these changes take vary from place to place, and may lead to considerable debate within communities, whether these are communities in the sense of groups of people living together, or so-called communities of practice, that is, people who share similar professional interests. It is therefore difficult to assess the consequences of these changes. Nevertheless, some common trends can be observed in the thinking about the relationship between ecosystems and human well-being.

17.2.3.1.1 Nature–culture dichotomy

Certain cultural perceptions of landscapes become dominant or imposed through economic and political forces, often to the detriment of local praxis. For a long time a dichotomous approach to nature–culture relations was dominant in nature conservation initiatives. This dominance manifested itself in many national policies concerning the establishment of nature reserves, but also through the work of international environmental NGOs, and the selection criteria for international recognition of protected sites such as the World Heritage Sites. The establishment of conservation units where human exploitation of natural resources was reduced to a minimum in many cases positively contributed to the protection of certain species. (See Chapter 5.) This strategy has also contributed positively to the human well-being of certain sections of the population through safeguarding provisioning and cultural services. However, for local communities, the consequences have generally been less positive. If concern for environmental conservation does not take into account local dependence on the use of certain natural resources, then the human well-being of local populations is threatened through dispossession, as the case of the Maasai living in the vicinity of the World Heritage Site, the Ngorongoro crater, shows (McCabe 2003).

A dichotomous approach to human–nature relations has also been linked to a tendency in nature conservation to search for "the pristine." This concept may easily lead to conservation policies that "freeze" landscapes within enforced boundaries and attribute any disturbance to human intervention that needs to be undone. Many conservation efforts entail restoring landscapes to their "pristine, natural state," even though it may be hard to determine what that state actually was and how far back we need to look. (See Chapter 14.) In some cases, land and waterscape restoration is based on the introduction of physical elements to imitate natural habitats, such as artificial shelters to protect against natural predators, a common strategy in aquatic systems (Gore 1985); in other cases they may involve the restoration of entire wetlands. (See Chapter 7.)

Restoration may also involve the removal of invasive alien species—popular issue in conservation circles worldwide—or the reintroduction of species that have disappeared over time. Again, the impacts are difficult to assess. In some cases, restoration has contributed to human well-being by increasing species populations that can be harvested (Castro and McGrath 2003) or through increased tourism revenues (Lehouerou 1993). The restoration of riverflows has in a number of cases contributed to human well-being by reducing the risks of uncontrolled floods and improving access to (ground) water. In other cases, human well-being has decreased because access to certain natural resources was denied or because the reintroduced species posed a threat to local economic strategies. (See Chapters 5, 14.)

17.2.3.1.2 People and parks

In the past two decades, there have been notable changes in perceptions of human–nature relationships. As a result, ideas about what land- and waterscapes should be conserved, as well as how they should be conserved ("people and parks" issues) have changed. There is a growing recognition that a wider variety of landscapes, including agricultural, industrial and aquatic landscapes, need to be conserved, and that certain species may have adapted to or may even depend upon human-made environments (Daily et al. 2003). Many governments and national and international conservation organizations have recognized that the biggest challenge for conservation in the twenty-first century is for it to take place outside parks and enforced boundaries, thus integrating into agricultural and urban systems. (See Chapters 5, 14.)

Conception and policies regarding the creation of conservation units have moved from exclusion of communities and local forms of resource management to inclusion (Brandon et al. 1998). It is, however, difficult to assess the impacts of community participation on ecosystems and human well-being. Studies that have attempted to do so show mixed results. A community-managed forest in India, which contributed to a higher biodiversity as well as to local biomass needs, provides an example. (See Chapter 5, Box 5.6.) In a number of cases the contribution to human well-being has been easier to assess than the contribution to the conservation of biodiversity (Kangwana 2001; Wells et al. 1992). In cases where the contribution to human well-being has been limited, this is often attributed to flaws in the decentralization process (Barrow and Murphree 2001). Furthermore, numerous authors have problematized the concept of "community" (for example, Barrow and Murphree 2001).

Communities are notoriously difficult to define, with boundaries between them often blurred and overlapping. Within communities, differentiation exists based on socioeconomic positions, gender, and positions with local government structures. Ecosystem management may have different impacts on the well-being of different groups within communities. Local elites may have better opportunities to capture benefits, while the burden of conservation may be put on other groups in the communities. This entails that the tradeoffs and synergies between different ecosystem services may not be similar for every person with a community. Lastly, community-based ecosystem management systems may take a long time to evolve, longer than the project cycles of environmental and development organizations. In many cases, it is therefore too early to assess the impacts of such management systems on ecosystem and human well-being (Barrow and Murphree 2001). Nevertheless, literature shows that conditions that favor better outcomes of ecosystem management in terms of both biodiversity conservation and human well-being tend to include representative participation and governance; clear definition of boundaries for management; clear goals and an adaptive strategy; flexibility to adjust to new contexts and demands; and clear rules and sanctions defined by participants (Barrow and Murphree 2001; Corbridge and Jewitt 1997)

17.2.3.1.3 Links to the sacred

Recent attempts to link up with local communities' ideas about the protection of certain landscapes involve using sacred areas as a point of departure when creating protected areas (Mountain Institute 1998). While the idea is not new, what is new is a recent growth in translating the sacred into legislation or into legal institutions granting land rights (Bahuguna 1992). However, this approach requires extensive knowledge concerning the specific way in which the link between the sacred, nature, and society operates in a specific locale. Sacred areas may vary from a few trees to a mountain range, and their boundaries may not be fixed. In some cases, access may be restricted to a few religious specialists, while in other cases they are open to the public to perform acts of worship, which may involve harvesting some of the natural resources from within the sacred area.

Relations between landscape and religion have to do with moral and symbolic imaginings, but also with staking one's claim, such as to land contested by immigrants or from invading states and development agencies (Spierenburg 2004). Literature shows that the study of local specific contexts and functions of sacred areas in a participatory and democratic way increases the likelihood that initiatives to use sacred areas as a basis for nature conservation suit the local situation and contribute to human well-being.

17.2.3.1.4 Local identities and linkage to the national and global

If attempts are made to come to an understanding of the complexities of different cultural perceptions of landscapes, and the different local institutional arrangements related to natural resource management, this can (with medium certainty) contribute to alternative strategies to ecosystem management and socioeconomic development.

It is, however, only too common to either put the responsibility for environmental problems and conservation in the hands of local communities or blame the private sector, while disregarding the linkages between local, national, and international policies. Local communities do not operate in a vacuum; they create multilevel alliances, and adopt and adapt global influences to foster their own human well-being. Of course, they do so on the basis of their own cultural repertoires, a process referred by some as "glocalization." At the same time, local identities seem to acquire an increased importance in the face of globalization (Geschiere and Nyamjoh 2001), especially in relation to struggles for control over resources.

These very local identities are increasingly used to mobilize international support for the conservation of local natural resource bases, as the growing recognition of the rights of indigenous peoples shows (Sylvain 2002). International NGOs like Survival International but also the United Nations are assisting indigenous peoples in protecting their rights of access to certain territories. There are, however, risks involved; the use of the term "indigenous" (as well as the use of the term "traditional") may serve to exclude certain groups who are also dependent on certain natural resource bases, but do not fit whatever characteristics/traits are used to define who is "indigenous" or "traditional" (Sinha et al. 1997; Gibson and Koontz 1998).

As these developments suggest, there have been changes in perceptions of global-local linkages. There is a growing awareness of global environmental problems by the larger public. This has led to the emergence of different views about rights and entitlements to global ecosystems and environmental resources. Examples range from public engagement in discussing the fate of tropical forests to pressure regarding regulations to curb the greenhouse effect. Chapter 5 provides an example of joint protests from local and international NGOs against the signing of a bioprospecting contract between the government of Brazil and a Swiss-based pharmaceutical company. International organizations now voice their opinions regarding national policies and international bank loans for development projects, while national governments complain about their lack of sovereignty and international "ecological imperialism" (Geores 2002).

Increased awareness of the globe working as a system has motivated the need to deal with ecosystems in an integrated way. This process has been characteristic of the so-called post-Stockholm way of thinking, that is, an emphasis on the human environment concept (which actually was the title of the Stockholm 1972 conference) and the discussion of environmental problems at a global scale. The very concept of ecosystems reflects changes in thinking about the nature-culture relationship, dismissing the idea of fixed equilibrium, closed systems, and static nature (Moran 1990). Global institutions dealing with the environment have become prominent players not only in environmental management, but also in international politics. Amalgamation of scientific thinking, public awareness, civil society, and business has been present not only at international government forums from Stockholm 1972 to Rio 1992 and Kyoto 1997 to the World Summit on Sustainable Development in 2002, but also in science initiatives such as the Club of Rome, the Man and the Biosphere Program, the International Geosphere-Biosphere Program, and the Millennium Ecosystem Assessment, to cite a few of the most relevant.

The impacts of responses such as multilateral environmental agreements are difficult to assess. The CBD is one of the few agreements that specifically address poverty alleviation. (See Chapter 5.) The Ramsar Convention does not do so, but some of its COP recommendations do so. In a number of cases national governments have elected to let (short-term) economic interests prevail over environmental concerns (see, for instance, problems pertaining to the Kyoto Protocol), which in the short-term may increase human well-being in the countries concerned through economic growth, but whether this strategy in the long run contributes to human well-being is doubtful.

There is a changing focus of attention from local to transnational conservation efforts: the creation of corridors for migrating species and trans-frontier conservation areas. We find examples in South and Central America, central and southern Africa and Asia. The increased focus on trans-frontier conservation is partly based on the realization that ecosystems do not stop abruptly at national boundaries, but also results from wider societal debates about the importance of globalization (Draper et al. 2004). Most transboundary conservation areas have been in existence since a relatively short period only, therefore the assessment of impacts is rather difficult. However, recent studies from southern Africa suggest that transboundary conservation can contribute to local human well-being through increased revenue from tourism, but can also pose a threat to human well-being through the marginalization of local communities in (international) decision-making bodies, and preferential treatment of tourists versus local people. Again, legislation that allows for representative participation by local communities is crucial (Wolmer and Ashley 2003; see also Chapter 14).

17.2.3.2 Knowledge Systems, Ecosystem Management, and Human Well-being

17.2.3.2.1 Knowledge Systems and Management of Natural Resources

Perceptions of land- and waterscapes are influenced by cultural repertoires, which in turn both are influenced by and influence knowledge production (scientific as well as local and indigenous). Most chapters in Part II address the issue of local and indigenous knowledge, arguing that such knowledge is important in conserving ecosystems and contributing to human well-being. Nevertheless, the drive for modernization and technological change is often based on the substitution of small-scale practices. The understanding of crop, forest, and aquatic biodiversity lies in the oral history and cultural memories of indigenous and local communities, but these are often disregarded as backward and unneeded. Given the pace of technological, agricultural, and environmental change, large-scale environmental modification through infrastructure development often happens at the expense of local resources and knowledge. While this impacts local food security and economies, it is also relevant to national and international issues of conservation and economy (Brondizio and Siqueira 1997; Posey 1998; Pinedo-Vasquez et al. 2001).

Formal scientific knowledge has contributed to better ecosystem management and increased human well-being. (See Chapter 6.) However, many natural scientists have contributed to a dichotomous view on the culture–nature relation, and have proposed interventions that threaten the well-being of local communities (see Nazarea 1998; Leach and Mearns 1996). Wynne (1992, p. 120) argues that there is a distinct difference in the way local farmers respond to uncertain environments and the way natural scientists do: "Ordinary social life, which often takes contingency and uncertainty as normal and adaptation to uncontrolled actors as a routine necessity, is in fundamental tension with the basic culture of science, which is premised on assumptions of manipulability and control. It follows that scientific sources of advice may tend generally to compare unfavorably with informal sources in terms of the flexibility and responsiveness to people's needs." Many scientifically based land reform programs designed to combat soil degradation by "rationalizing" local land use patterns have had the opposite results and undermined local production (Scott 1998).

Local/indigenous knowledge pertains both to species that are harvested on an extractive basis, as well as to production methods, cultivars, and germplasm (Brondizio and Siqueira 1997; see also Chapter 14). Production and harvesting systems that evolved over long periods of time benefit from cultivars and methods adapted to particular micro-environments, as well as social conditions (Netting 1993; Altieri and Hetch 1990). Diversity of production systems most likely increases the resilience not only to factors such as climatic change, but also facilitates alternative economic options to minimize risks in household food supplies (Wilken 1987; Hladik et al. 1996).

17.2.3.2.2 Compensation for and protection of local and indigenous knowledge

Recently, there is a wider recognition of the validity and importance of farmers' indigenous knowledge (Brush and Stabinsky 1996). This growing recognition has also led to its commercial exploitation. Market imperatives and international monetary policies have pushed countries in the South to gear their economies toward export. In most cases, this has led to the exploitation of their natural resources beyond long-term sustainability, as shown in the case of fishing in the Amazon and timber cutting in Indonesia. The prospecting for local resources has led to exploitation of local knowledge without communities being compensated. The richness and possibilities of resources, for example, medicinal herbs and their possible economic benefits, became an important argument for conserving them. Few mechanisms were available, however, to feed the benefits back to local communities that in many instances contributed to the production of the knowledge concerning certain species, or even the production of the species themselves. Chapter 5, for instance, shows how the Brazilian government is struggling with legislation in this regard.

The exploitation as well as the growing consciousness concerning the disappearance of local resources and the knowledge about these has led to growing concern about the need to protect local indigenous knowledge. The international community has recognized the dependence of many indigenous peoples on biological resources, notably in the preamble to the Convention on Biological Diversity, which has been ratified by 178 countries. Article 8(j) in the CBD specifically addresses indigenous peoples and their knowledge. The CBD adopted the facilitation of indigenous peoples' participation "in developing policies for the conservation and sustainable use of resources, access to genetic resources and the sharing of benefits, and the designation and management of protected areas."

Many governments are now in the process of implementing Article 8(j) of the Convention through their national biodiversity action plans, strategies, and programs. A number of governments have adopted specific laws, policies, and administrative arrangements for protecting indigenous knowledge, emphasizing that prior informed consent of knowledge-holders must be attained before their knowledge can be used by others. In many cases, protection of local/indigenous knowledge is a by-product of protection of biodiversity, while in others the main aim was to guarantee economic benefits to communities.

Apart from national policies, there are also instances of local strategies to protect as well as transmit local and indigenous knowledge. One such strategy is restricting the transmission of certain types of knowledge to specialists; for example, in some societies knowledge of medicinal plants is considered secret and can only be transmitted from one healer to another. Such restrictions can work effectively within communities, but can also complicate the development of international legislation. (See Chapter 14.)

The World Intellectual Property Organization has been the voice behind intellectual property rights. Yet, instead of supporting "local knowledge," IPR has become a Western capitalist response that considers any type of knowledge, whether it is medicinal plants, song, crafts, or any other, as a commodity. Local communities are concerned at the extent of exploitation of local knowledge, how it is being used or removed from its culturally appropriate context, and how it is being usurped as a capitalist commodity. At issue is the Western patenting system, which is used to protect the intellectual property rights of monopolists.

The scale and tendency to focus on corporations is exemplified by industrially advanced Western European countries, which have been strong supporters of IPR and have imposed this system on developing countries. Patents translate into wealth and power for foreign transnational companies, and are likely to bring negative impacts in the biodiversity of particular areas, depending on the level and structure of market demand and exploration practices (Shiva 1997; Settee 2000). Examples are the patenting of local and indigenous knowledge concerning medicinal plants by pharmaceutical companies (see Chapter 5) and the patenting of locally developed crops (see Chapter 6).

Responses such as certification programs are more likely to be effective in addressing local economies and human well-being if they include the impact of particular resource extraction upon people and communities using the same resource basis, but not necessarily sharing resource ownership. Certification programs are better served if they are accessible by communities and small producers' co-ops, which are usually not familiar with bureaucratic and costly procedures of certification. Responses such as "fair trade" tools that promote participation of local producers in the commercialization and price negotiations, transformation, and retailing of their products is a necessary component not only of rural development and conservation and management of natural resources, but also for commercial enterprises retailing these products. (See Box 14.14 for an extensive discussion of fair trade initiatives in the Amazon. Box 14.15 describes attempts in a German Biosphere Reserve to establish a certification program.)

17.2.3.2.3 Production of knowledge and integration of knowledge systems

Studies show that local resource users often draw on a variety of knowledge sources, combining formal scientific knowledge with local and indigenous knowledge. (See Chapter 14.) This mix, often combined with a mosaic of different livelihood strategies, is important, especially for the poorer sections of the population, for human well-being. It is important to note, however, that all forms of knowledge are produced and disseminated in a context of power relations. For instance, some knowledge may be considered sacred and secret, some knowledge may be related to specific gender roles. Both community-based natural resource management programs seeking to integrate local and indigenous knowledge as well as programs aiming to protect local and indigenous knowledge are more likely to contribute to human and ecosystem well-being if they take into account the ways in which participants use a variety of knowledge systems, and pay attention to the fact that the production of and access to such knowledge systems is not always equally distributed within communities. Legislation may "freeze" knowledge as well as the rituals and practices associated with this knowledge (Laird 1994; Brush and Stabinsky 1996).

Much of the local/indigenous knowledge is not written down, but transmitted through daily practices, stories, songs, dance, theater, and visual arts. Not only knowledge, but also attitudes and perceptions are transmitted that way. Programs on local/indigenous knowledge that take these forms of transmission into account, and try to incorporate these into educational activities, are more likely to contribute to both ecosystem and human well-being (Dove 1999).

Fostering the articulation of international and national conventions, and regulations linking biodiversity with local and indigenous knowledge, are important, taking into account that knowledge is produced in the context of inter- and intra-group interactions, power relations, historical settings, and their dynamics. Responses such as compensating for the utilization of local knowledge and resources are more likely to contribute to human well-being if they take into account relations between companies, national and regional governments, and communities as well as the power dynamics of these relations. Conventional "best-practices" methods that focus on content (rather than the process of articulation knowledge from different knowledge systems) and attempt to decontextualize local knowledge are less likely to be successful.

17.2.3.3 Impact of Responses Related to Tourism, Recreation, and Education

Tourism is an economic response, but depends on cultural perceptions of land- and waterscapes. It can provide alternative land use, which decreases pressure for land use conversion (for example, tropical forests to pastoral lands). Tourism can also contribute to the maintenance and revival of lifestyles and cultural practices, including natural resource management practices. Opportunities arise for education and awareness-raising on the need to understand and respect cultural diversity and biodiversity. Conservation areas are especially valued for their educational significance and providing recreational services. (See Chapter 15.) Valorization of cultural landscapes and monuments are important assets to the larger society.

17.2.3.3.1 Tourism and recreation

Ecotourism can provide economic alternatives to value ecosystem services, but results are mixed. One of the constraining factors is potential conflict in resource use and the aesthetics of certain ecosystems. Different ecosystems are subjected to different types and scales of impact from tourism infrastructures. Furthermore, some ecosystems are easier to market to tourists than others. The market value of ecosystems may vary according to public perceptions of nature. Freezing of landscapes, conversion of landscapes, dispossession and removing of human influences may result, depending on views of what ecotourism should represent. Yet when conservation receives only limited budgetary subsidy, tourism can provide revenues that can meet its needs.

Rural and urban tourism have been receiving increasing support in recent years, including in terms of tax incentives and credit. Changes are occurring in perceptions concerning what types of landscapes and cultural practices are of interest to tourism. Industrial monuments and certain types of resource use are now seen as having historical, social, and environmental potential for tourism. In parallel, it broadens the scope of environmental conservation in areas outside of protected areas. In the North, small-scale farming production, while facing difficult competition in agricultural markets, can find an alternative in providing tourist services. The marketing and branding of agricultural products in a German Biosphere Reserve is just one example of how tourism can contribute to the maintenance of certain lifestyles and production methods and increase human well-being.

The impact of tourism on human well-being though varies as a result of several risks involved. Consumptive tourism activities, such as sport fishing, may represent pressure on the resource. For example, marine recreational angling in the United States alone comprises more than 15 million fishers, with a total harvest of 266 million pounds in 2001. (See Chapter 14.) Conflicts between recreational and professional fishers are commonplace in many parts of the world as recreational fishers detain strong political power in the fishing councils.

Non-consumptive tourism can also involve risks as representations of nature and culture used to entice tourists often refer to pristine ecosystems and exotic cultures, reinforcing stereotypes as marketing tools of communities subjected to tourism. Commoditization of culture does not always benefit those portrayed. Another serious problem is the blurring of boundaries between private and public. Emphasis on pristine ecosystems can furthermore easily lead to evictions of people. Many protected areas result from evicting local populations, but are accessible to tourists. (See Chapter 5.)

The risks and opportunities provided by tourism are related to the economic position of communities and relations of power. Studies show that economic deprivation increases the likelihood of overexploitation of resources and accepting unfavorable positions in the tourism industry (low-skilled labor, sex worker, drugs). Increase in land use value for tourism real estate development purposes are likely to lead to displacement and dispossession. This risk is much higher for communities that enjoy informal or communal land rights. Coastal communities have long been suffering from those impacts as the tourist potential of coastal landscapes continuously pushes local residents away from their livelihood strategies. (See Chapter 14.)

Cultural and ecotourism are not necessarily the same thing as community-based tourism. Concerning access and benefit sharing, tensions often exist between tour operators, local communities, conservation units, and (other) government departments. In some cases, local communities may actually run tourism enterprises themselves, in other cases they may only have access to lowly-paid menial jobs. Issues that are important are matters of local decision-making powers concerning land use, infrastructure, and dealing with externalities, for example, the choice of tour operators and types of tourists, boundaries between the private and the public, goods that are imported by tourists, and waste-management. Capacity building can successfully contribute to human well-being if it not only involves individual professional training to fulfill positions in the tourism industry, but also includes institution building, marketing, and negotiating capacities.

17.2.3.3.2 Environmental education

Recreation and education can go hand in hand. Cultural tourism can serve to educate people about the importance of cultural diversity, as well as the importance of the latter for the conservation of biodiversity, provided the risks mentioned above are taken into account. Tourism and recreation can be linked to environmental education, fostering knowledge about the functioning of ecosystems and provoking tourists to critically examine human–nature relations. Ola-Adams (2001) describes how the Omo Biosphere Reserve in Nigeria is the site of environmental education programs for very diverse audiences ranging from schoolchildren to university students, protected area managers, and policy-makers. West (2001) describes awareness raising activities among local farmers in the Fitzgerald Biosphere Reserve, helping them to address the problem of wind erosion. He stresses, however, that awareness raising requires development, especially among communities that experience periods of economic decline (West 2001, p. 15). "Top-down" education is less effective than education that is based on sharing experiences and attempts to reach a joint understanding of the dynamics of human–nature interactions.

17.2.4 Technological Interventions[1]

Technological interventions directed at ecosystems have primarily been focused on increasing the economic productivity and efficiency of ecosystem services, reducing vulnerability of human societies from extreme environmental events, and, more recently, protecting ecosystems by decreasing pollution and the intensity of material use in production activities. For example, in the second half of the twentieth century, the Green Revolution enabled world cereal production to increase threefold on about the same land acreage. Without this success, world farmers would have had to increase cropland use from 600 million hectares to some 1,800 million hectares. The ecological implications of such an expansion could have been catastrophic. We shall cover some of these impacts later in this section.

On the other hand, a weight of evidence has been building up, arguing for decision-makers to design reference frames with a longer-term and more systemic view of the wider impacts of technology interventions rather than seeking short-term solutions. The assessment illustrates that increases in the production capacity of ecosystems through technological interventions often introduce stresses in the form of "second-round ecological feedbacks." In the long run, this could impair the functioning and utility of the ecosystem. In other words, a technological response can become a driver of ecosystem deterioration if not carefully designed and there is growing acceptance that the critical issue involving the introduction of technological interventions is the management of social, political, and economical tensions that arise as technology drives higher "service" yields from ecosystems.

Summarized below are some of the key ecological and social impacts arising from the use of technological interventions in a number of critical ecosystem services. We have only covered the ecosystem services in which we found technological responses being used extensively, and in which they have significant impacts on the ecosystem services themselves as well as on human well-being and poverty.

17.2.4.1 Food

The Green Revolution is the best example of such tensions associated with a technological intervention. An assessment of the Green Revolution as a response establishes beyond doubt that in terms of impact, the combined technologies used in the Green Revolution increased food production significantly. However, from the perspective of human well-being and sustainable development, increased production does not automatically translate to food security. Food security as defined by the United Nations is, "a situation that exists when all people, at all time, have physical, social and economic access to sufficient, safe and nutritious food which meets their dietary needs and food preferences for an active and healthy life" (FAO 1996). The critical component in food security revolves around the access to food and what Amartya Sen defines as entitlements in his seminal book, *Poverty and Famines* (Sen 1981). But this is not to discredit the importance of food supply. It is critical, and it is a necessary condition for food security but, as illustrated in the definition, not a sufficient condition.

It is well established that the Green Revolution did increase global production of cereals and has improved societal welfare at the aggregate. It is also acknowledged that the Green Revolution did provide some relief from the conversion of more ecosystems into agrosystems. However, there are competing explanations of the distributive impacts of the Revolution, with many experts saying that the technology benefited those with access to financial resources, who were then able to exploit the intensely profit-driven commodities trading environment. There is a temporal element as well in these competing explanations. More recent evaluations suggest that the benefits of the Green Revolution in South Asia were more widely diffused than originally thought (Hazell and Haddad 2001).

In terms of ecosystem impact and projections for the future, many studies have found that the introduction of the new high-yielding seeds together with higher dependence on irrigation and fertilizer regimes have also caused many ecosystems to deteriorate. Increasing soil salinity has been one major side-effect of the Green Revolution. Other effects on ecosystems from the Green Revolution include contamination of soil and water due to use of fertilizers and pesticides, human health alteration because of agrochemicals, and the loss of germplasm diversity and seed banks. The distribution of the effects of these ecological deteriorations has been asymmetrical across stakeholders, with the poor farmers being most acutely affected. (See Chapter 6.)

17.2.4.1.1 *Alternative farming-food production systems*

On the positive side, in many developing and industrial countries, especially EU countries, urban agriculture—in an "organic revolution"—is contributing to improving the environment through the recycling of organic matter. Solid wastes converted into compost and gray water emanating from the open drain are used to fertilize soils (Sridhar et al. 1985; Coker 2003). Also, strong agricultural production output has been recorded in a significant number of cases in tribal areas in India where natural resource management practices have been introduced. But we must keep in mind that organic farming by itself will be insufficient to meet the growing demands for food, and the land area needed will also be substantial. The ecological implications of converting more ecosystems into agroecosystems may have a significant impact on the regulation and support of ecosystems as well as the cultural services related to them. The impact on biodiversity will also be disastrous.

17.2.4.1.2 *Genetically modified organisms and seed banks*

The growing controversy over genetically modified organisms warrants special attention in this section. Experiences to date point to difficulties in assessing the advantages or risks associated with GMOs in food production in general. Rather they must be addressed case by case for specific agroecological and, equally important, socioeconomic conditions. GMOs have also come under increasing criticism for reducing the diversity of the seed bank and for increasing the dependency of many farmers, especially in the developing countries, on large multinational companies who control the patent rights for the GMOs. (See Chapter 6.) Developing countries do stand to gain tremendously in terms of food security and broader human well-being, particularly in the area of health with improved nutrition, if the level of uncertainty associated with these technologies can be reduced significantly. What is recognized as critical from the very limited information, and bad publicity, is that a great deal of work needs to be done in terms of managing the application of these technologies.

A technological intervention to reverse the loss in biodiversity created by GMOs is the development of seed banks. (See Chapter 6.) However, for many developing countries, the maintenance of gene banks is a major problem as electricity supplies are unreliable and fuel costs expensive. A recent technological response to this problem is the "ultra dry seed storage technology," which allows the storage of seed germplasm at room temperature, thereby obviating the need for refrigeration. Other research conducted on drying techniques, such as sun and shade drying (Hay and Probert 2000), offer promising alternatives to improve the capabilities of resource-poor countries to conserve their seeds.

17.2.4.1.3 *Increase in drought, salt, and pest tolerance in food production systems*

Some traditional farming systems using low inputs have improved yields while safeguarding the resource base by upgrading the subsistent food crops and adopting integrated pest management. For example, Indonesian rice farmers who adopted IPM, which reduces the need for pesticides, achieved higher yields than those who relied solely on pesticides. Biotechnology responses can introduce genes that counter soil toxicity, resist insect pests, and increase nutrient content. Still, the questions of biosafety and the ethics of manipulating genetic material need to be resolved before the potential of biotechnology and genetic engineering can be realized.

17.2.4.2 Flood and Storm Protection

Historically, responses to reduce negative impacts of natural disasters like floods and storms have emphasized the implementation of physical structures/measures (for example, dams/reservoirs, embankments, regulators, drainage channels, and flood bypasses). However, there are competing explanations, where physical responses may cause net harm to an ecosystem in the longer timescale, which may reduce anticipated (or expected) benefits of the responses. A typical case in point is construction of dams to regulate floods, which has been known with high certainty to cause wider ecosystem degradation and a deepening of poverty among local communities. (See Chapters 7, 11.)

There is increasing recognition that natural environment measures can reduce the negative impacts of natural disasters without causing the longer-term ecological and socioeconomic deteriorations. It is well established that ecosystems such as wetlands act as buffers for floodwaters. (See Chapter 11.). Coastal mangroves have been found very effective in providing protection against storms and surges especially in Bangladesh, India and Southeast Asia. The direct benefits of these natural protective measures accrue to the poor coastal communities in the form of reduced vulnerability. There are also indirect benefits in the form of food, non-wood products, and the water cleansing properties that are equally important. Two issues related to protection and restoration of ecosystems vis-à-vis technological interventions are the degree to which technologies can substitute for ecosystems services, and whether ecosystem restoration can reestablish not only the functions of direct use value to humans, but also the ability of the system to cope with future disturbances.

It should be acknowledged that with increasing frequency and magnitude of extreme events ecosystems themselves would prove to be insufficient to address and mitigate the impacts of these events. In all probability, a system where technical responses work complementarily with natural systems would provide the best protection against extreme events. In terms of the human well-being implications, developing countries need to tap advanced technology that has taken many decades to get established together with indigenous engineering and technology capacity, particularly in water harvesting and management, and develop their own capacity for such technology interventions to work.

In practice, the actual provision of flood and storm protection is often the responsibility of a number of different actors working at different levels—local, regional, national, and transnational. The institutional settings not only affect the delivery of the services, but also the manner in which they are delivered as well as the direct and indirect effects on ecosystems and human well-being.

17.2.4.3 Waste Management

There is a high level of certainty that increasing population and economic activity will result in an increase in the production of solid waste, higher levels of air and water pollution, and that the natural systems will not be able to accommodate these increases, and that technological interventions will therefore increasingly become necessary. (See Chapter 10.) From the responses it is quite evident that developed nations have attained reasonable excellence but at high monetary cost. For example, nitrogen-removal technology for sewage treatment has led to improvement in water quality in Tampa Bay and, to a lesser extent, in Chesapeake Bay. (See Chapter 10.)

For the developing countries, short-term and medium-term solutions lie in the use of ecosystems and relatively cheap technology innovations like oxidation ponds/waste stabilization ponds, and floating vascular aquatic vegetation, which have considerably helped the poorer developing countries. Another example is the utilization of wetland ecosystems as sinks for the absorption of selected nutrients (Hu et al. 1998). Assessments show that this type of response is particularly relevant to deal with solid waste management problems in regions which lack appropriate and networked road infrastructure, sophisticated equipment, operational facilities, and inadequate human resource development.

Governments are often tempted to choose inappropriate technologies due to lack of funds and pressure from private interest groups. A viable response option is for developing countries to mass-produce small-scale equipment that is amenable to community access. Private/community-sector-driven operational schemes coupled with government provisioning of basic facilities in landfill, incinerator, and sewage treatment plants appears to be a promising option. The critical point here is that technologies that have been endogenously developed taking into consideration local ecological, economic, and cultural conditions have a higher certainty of achieving their objectives.

17.2.4.4 Water

A large number of dams have been constructed all over the world. "In North America, Europe and the former Soviet Union, three-quarters of the 139 largest river systems are strongly or moderately affected by water regulation resulting from dams, inter-basin transfers, or irrigation withdrawals" In addition, hundreds of thousands of kilometers of dikes and levees have been constructed with the purpose of river training and flood protection. While these structures have clearly provided increased supply of freshwater for many uses, as well as flood control, unfortunately they have all too often had debilitating effects on surrounding ecosystems, affecting their naturally occurring services and their biodiversity.

The direct social impact of large dams is striking; they have led to the displacement of 40–80 million people worldwide and terminated access by local people to the natural resources and cultural heritage in the valley submerged by the dam. (See Chapter 7.) The perennial freshwater systems established by large dams also contribute to health problems. For example, epidemics of Rift Valley fever and Bilharzia coincided with the construction of the Diama and Manantali dams on the Senegal River. Aside from the direct impacts of large dams, the benefits of their construction have rarely been shared equitably, that is, the poor, vulnerable, and future generations are often not the same groups that receive the water and electricity services, and the social and economic benefits from dams (WCD 2000).

Another striking example of where technology may be a more expensive option than the use of ecosystems lies in the provisioning of quality water. The Catskill/Delaware watershed provides up to 90% of New York City's water demands. In 1989, New York City was faced with an order by the U.S. Environmental Protection Agency to build a filtration plant because of declining water quality. The city estimated that a filtration plant would cost approximately $6–8 billion, while maintaining the traditional water cleansing system—the Catskill watershed—was estimated to cost about $1.5 billion. This shows that it may actually be more cost effective for developing countries to choose natural options for providing clean water, especially taking into consideration their lack of financial, technological, and human resources.

17.2.4.5 Wood and Fuelwood

Technology is changing the way wood is produced, processed, and used. Biotechnology is given increasing emphasis in commercial plantations with many now based on cloned trees to standardize production, quality, and growth rates. Much experimentation using genetic modification is also occurring to develop new generations of "super-trees." Though not as controversial as genetically modified food crops, these modified trees are being criticized by pressure groups concerned about possible environmental impacts. Wood engineering is allowing the use of many more species and smaller pieces of wood in processing. Also on the positive side for human well-being, the possibility has been strongly raised that technology allowing for a concentration of fiber for fuel production on small areas of land releases other areas for the provisioning of regulating services and livelihood activities. (See Chapter 8.)

Though wood is the principal source of energy for cooking and heating for so many of the poor, it is the least efficient—a low-density form of energy used in thermally inefficient devices. Unless the poor have access to technology to convert wood and charcoal into modern forms of energy, real costs of energy from fuelwood can be high for the poor. Lack of access to more efficient energy sources can also be an important constraint to livelihood and broader economic improvement.

The technology to use wood as a fuel is improving and it could become a major contributor to sustainable energy sources globally. Wood fiber gasifying energy generators are being developed and could one day produce large amounts of renewable electricity using trees harvested from fast growing plantations. Dendro power involves the use of wood-based materials for power generation. One useful feature of dendro power is the potential to use sustainably grown fuelwood sources. Dendro power is gathering momentum due to its multiple benefits of renewable power, reforestation, and income generation (especially in rural areas). However, we are uncertain of the impact these "fast growing" plantations may have on other ecosystem services and the subsequent impact on human well-being.

If energy policies are to become more efficient and equitable, then emphasis should be on helping the poor to move from woodfuels to more efficient fuels, or at a minimum to more efficient forms of woodfuel. This will release some of the pressures on ecosystems for providing woodfuel and "free" up some of the other ecosystem services. Lack of access to more efficient energy sources can also be an important constraint to livelihood and broader social improvement.

Information technology has been instrumental in bringing together the common interest of forest owners and society as a whole in promoting efficient and equitable wood production and at the same time securing a sustainable and diversified forest ecosystem. The relatively rapid modification of Swedish forestry in the 1970s and 1990s to more environmentally adapted practices clearly illustrates without any uncertainty the positive evidence of the impact of information technology on sustainable forest management and the distribution of the benefits of the management regime among a host of involved stakeholders.

BOX 17.4

The Citizen's Eye-view of Technology

(Winner 1986, pp. 55–56)

"The important task becomes . . . not that of studying the 'effects' and 'impacts' of technical change, but one of evaluating the material and social infrastructures specific technologies create for our life's activity. We should try to imagine and seek to build technical regimes compatible with freedom, social justice, and other key political ends . . . Faced with any proposal for a new technological system, citizens or their representatives would examine the social contract implied by building the system in a particular form. They would ask, how well do the proposed conditions match our best sense of who we are and what we want this society to be? Who gains and who loses power in the proposed change? Are the conditions produced by the change compatible with equality social justice, and the common good? To nurture this process would require building institutions in the claims of technical expertise and those of a democratic citizenry would regularly meet face to face?"

17.2.4.6 Technology and its Consequences on Poverty

Technology choices have been to a large extent dictated by efficiency issues with very little concern shown for equity and distributive issues. Technology has caused a widening of the equity gap between developed and developing countries, and within the developing countries, between the rich and the poor. This has become increasingly serious over the last few decades, as copyright right laws have been developed with the interests of the developed countries—the main developers of modern technology—taking primary precedence.

Technology development and adoption in developing countries have not addressed the "needs" of the poor but have been primarily driven by the need for quick profits. Technology is not value neutral. Technological systems in fact favor the interest of some groups over that of other groups. It is therefore imperative that societies, in their quest to address equity and effectiveness, understand the implications new technologies carry before introducing them (See Box 17.4; also Winner 1986; Smith 1998). It may in essence be the nature of the beast as most technology development in developing countries is spearheaded by the private sector. For example, although the Green Revolution significantly increased rice production, it also required substantial inputs which were beyond the reaches of the poor.

At the global and national levels, donors may be able reconsider their funding for research or technological development which may respond to the needs of poor local communities but still meet global goals. At the local level, the challenge is to provide the institutional mechanisms that will allow local communities to adopt their "own" choice for the local context. This will in turn require legitimization and utilization of local knowledge and technology, decentralization of decision-making power, and prioritization of local needs.

It must be considered whether or not the new changes will bring positive impact to the local ecosystems and human-well being (and including their capability). Modern technologies that embrace and complement the use of traditional knowledge are known to do better in reducing poverty and maintaining local ecological integrity. Ideally, technology should be flexible enough to evolve under local conditions. It is also important to prevent oneself from falling into the trap of letting technology become an end in itself and not just a means to the end—goals of choice, human agency, poverty reduction, and improving human well-being through the sustainable management of ecosystem services.

It can also be said with high certainty that if technological interventions are to achieve their objectives, then certain catalysts are needed for the development and use of technology at the local level to reduce poverty through the sustainable management of ecosystem services. These enabling conditions include institutions, micro-credit, good governance, and efficient and non-corruptive bureaucracy.

Last but not least, if technology is developed for and/or introduced to communities, then the following basic questions must be addressed and answered to guide the development and implementation of technological interventions:

- Was the decision to adopt the technology driven by a participatory process involving all relevant stakeholders, especially the poor?
- Who are the beneficiaries of the technology and does the technology address the needs of the poor?
- What are the ecological implications of the technology and are there any groups of people who will be adversely affected by these ecological changes?

- Who will fund the research and development of the technology?
- Who will regulate the technology and how?

17.2.4.7 Technology and Some Lessons Learned

It is well established that unconditional transfer of technology from developed to developing countries is not always the solution. The emphasis should be on technologies adapted for local ecosystems, economic conditions, and finally the social and cultural environment. Generally, the successes and failures of technological interventions are often determined by the degree of planning, consultation, and resource commitment on the choice and application of technologies.

A summary of the findings is given in Table 17.1. The assessment examines the maturity of the technologies with a score ranging from 1 to 3. Technologies widely practiced and not just considered routine but also as a minimum requirement are scored at 3. These technologies can be considered to be mature technologies. Patents on these technologies are not an issue. Technologies with a score of 2 have had their effectiveness demonstrated and proven, but factors such as cost or the high level of human capacity required for application have prevented them from being widely used. Local application of these technologies still offers scope for incremental innovations to the technology. Technologies with a score of 1 often carry a great deal of uncertainty in relation to their impact on ecosystems. Being at the cutting edge, the level of scientific expertise and the number of cases of successful application with no second round impacts on the ecosystem required to address this uncertainty suggests that there is still a long way to go before these technologies are widely taken up. It should be noted that a score of 1 is very much associated with technologies of advanced, wealthier economies of the developed world.

Consideration is also given to the barriers to adoption of the technologies. The range is again 1 to 3. A score of 1 indicates a low barrier and 3 a very high barrier. With respect to infrastructure, a score of 1 is indicative of low costs as well as a low level of complexity involved in establishment, operation, and maintenance.

Technical skill requirements for a technology is based on a score of 1 if a low level of expertise (perhaps even accommodating illiteracy) is required. The higher the scientific expertise needed for a technology, the closer the score gets to 3. The same scoring structure applies to barriers of access. Barriers include cost and ease of access to intellectual property. The higher the barrier, the higher is the rating.

17.2.5 Consequences of Response Options on Human Well-being

This concludes the assessment of ecosystem management responses. The complexity of the links between ecosystem services and the various constituents of well-being together with the added convolution of drivers and responses mediating the links makes such a synthesis difficult. Nevertheless, some critical findings on the consequences of the four main classes of instruments assessed in the report on the various constituents and determinants of well-being do emerge and these are reflected in Table 17.2.

Positive implies an improvement in the constituent while *negative* constitutes a drop. *Conditional* means that the results are dependent on a number of context specificities which are elaborated in the concluding section. *Mixed* results indicate that both positive and negative impacts were experienced with some groups of individuals benefiting from the response while other groups fared badly.

Before discussing Table 17.2 findings in detail, it is important to distinguish the various types of poverty acknowledged in the poverty literature. Box 17.5 provides definitions of the various types of poverty that can exist. However, a systematic assessment of the impacts across these criteria is not tried here due to the lack of data. Instead, we attempt to get a grasp of key trends of the consequences on the various constituents and determinants of human well-being as used in the MA.

17.2.5.1 Security

Security as defined by the MA includes secure access to natural and other resources, safety of person and possessions, and living in a predictable and controllable environment with security from natural and human-made disasters.

The privatization of ecosystem services through the use of economic instruments have been found to perform relatively well in defining secure rights and access to provisioning services of ecosystems like access to timber, ownership of land, and access to water resources. However, in many cases it was found that access to these services was not equally distributed across stakeholders dependent on these services. For example, there have been instances whereby introduction of private land tenure—a combination of an economic and legal response—has reduced the ability of many communities to access common ecosystem services during times of economic and ecological stress (Rutten 1992). There is no doubt that economic responses have improved the security of some groups but at the expense of increased vulnerability and insecurity to other groups.

Technological and economic responses were found to be successful responses in increasing agricultural productivity. However, it is difficult to say with a high degree of confidence that a larger food supply had translated to a higher level of food security. Many other factors like good governance, increased transparency, and well-functioning markets were found to be critical in making sure that food security and a reduction in absolute poverty is achieved. Technological responses with a high degree of mechanization are conducive to food security in a very broad sense. However, these technological innovations may not guarantee, for example, through the creation of jobs, the necessary entitlements of the poor. Thus food can be available, but people might not have the means to access it if responses are not labor inclusive and, therefore, ineffective as a poverty reduction strategy.

Social responses have indirectly contributed toward enhancing the security of many local communities. Sacred groves—a form of a social response—have traditionally served as social nets during times of ecological as well as economic shocks. Although perceived as sacred and religious, traditional rules governing these groves allow individuals to have access to the ecosystem services during times of hardships. However, privatization has eroded many of these ecological safety nets in many developing countries leading to increasing levels of insecurity and poverty. If economic responses are to be adopted, then some form of social nets need to be institutionalized in order to replace sacred groves and other forms of social responses which provide buffers to increase security during times of stress.

17.2.5.2 Health

Economic and technological responses were found to be the main drivers causing land use changes as well as changes in biodiversity composition and levels. Changes in biodiversity have been to

Table 17.1. Popular Technologies for Sustainable Use of Biological Diversity and Ecosystem Conservation and Management. *Maturity*: 1 = application is still limited and impact on ecosystems uncertain; 2 = effectiveness demonstrated, but cost of high skill level needed may inhibit widespread use. 3 = mature technology that is widely practiced and considered a minimum requirement worldwide. *Barriers to adoption:* 1 = low; 2 = medium; 3 = high.

Technologies	Purposes of Some Applications	Maturity Level	Problems and Conditions Associated with Success or Failure of the Technology	Barriers to Adoption		
				Technical Infrastructure Required	Technical Skills Required	Cost of Access to Technology
Production Technologies						
Precision agriculture: sensor technologies for differential application of inputs to cropping systems: "controlled application as machines move across field"; GPS monitoring systems; software tools for analysis; and automated tools for application of chemicals, fertilisers, etc.	intensification of production (e.g., increasing yields)	1	generally believed to have the potential for supportive effects on sustainability; high level of site-specific agroeconomic conditions; high costs associated with infrastructure; technical facilities and particularly high costs associated with "customization" needs; high level of expertise required in agricultural sciences as well as for technology intense operation systems; developing countries lack research and extension services.	3	3	3
Biotechnology: cellular and molecular biology	shortening time and cut costs of introducing innovative food varieties; breeding without limitations by species barriers; enabling improved varieties and breeds with introduction of traits for specific socioecological situations	1	costs associated with technical facilities and licensing of technologies; high level of scientific expertise.	3	3	3
Biotechnology: genetically modified organisms	more precise modification than conventional biotechnology; development of crops, animals, or bacteria with traits that cannot be introduced through classical breeding lines (e.g. herbicide tolerance, insect resistance, altered nutrition content, etc.)	1	biodiversity loss associated with monocropping; high level of uncertainty although no conclusive evidence exists at this stage about the environmental advantages and risks in the use of GM crops; considerable work has still to be done to address high levels of uncertainty related to controlling the spread of transgenes (with specific resistance traits) to non-targeted crops and weeds; genetically engineered fish (e.g., Atlantic salmon) and the difficulties of "containing" these populations has given rise to concerns that animals that can easily escape, that are highly mobile, and that form feral populations could compete more successfully for food and mates than their natural counterparts.	3	3	3

Integrated methods: organic farming	prohibition of synthetic inputs and mandatory "soil building" crop rotation system designed to maintain the functional integrity of the ecosystem	2	results on higher yields not conclusive; therefore serious consideration is needed on possibility of trade-off between lower yields and enhanced sustainability; probably stronger argument for system to be critical component of larger national mix of production systems; results show total energy use is less than for conventional farming. Less conclusive about energy efficiency (i.e., energy use per kg of output).	2	2	1
Fishing gear technology and fishing vessel technology			impact on environment and non-target species			
Aquaculture: fish stocking technologies	increasing intensity of fishing; raising yield kg/ha and fish/ha		conflicts for use of coastal resources and environmental pollution; high input demand for land, water, seed, capital investments in technologies (including research), and feeds and chemicals (antibacterials, parasiticides, fungicides, and anaesthetics); impact on environment of multiple users of land and water resources: need better integration of land, water management.			
Industrial production systems (feedlots, broilers)	intensifying production of meat	2	very high growth in these production systems—problems related to impact on ecosystems; although technologies exist, very weak commitment to internalizing costs associated with nutrient loading impacts on environment (absorptive capacity), particularly water resources.	2	2	1
	food processing (includes issues of access and nutrition); preservation and packaging (includes issues of access, health, and safety)	3	costs associated with development and enforcement of standards for food quality, balanced nutritional content, and labeling; unhealthy levels of salt, sugar, and fat contents.	2	2	1
Process Technologies						
Structural water supply and flood control systems; desalination plants; water reclamation plants; water harvesting techniques	integrated watershed planning and stabilization relevant to conservation of forest, dryland, mountain ecosystems, and inland waters.	3	costs of infrastructure; need for engineering and technician-based maintenance services	2	2	1
Artificial but specifically designed "wetlands" as a means of sewage treatment	wastewater management	3		1	3	1
Drip irrigation; sprinkler irrigation; underground and sub-underground irrigation; traditional methods such as Indian "pick-ups" or weirs	irrigation management for agriculture	2	traditional methods quite low cost; at commercial scale agriculture as well as for watershed planning purposes, costs of infrastructure and control mechanisms.	2	2	1
	erosion control, soil improvement technologies, and conservation farming	2		1	2	1

(continues)

Table 17.1. Continued.

Technologies	Purposes of Some Applications	Maturity Level	Problems and Conditions Associated with Success or Failure of the Technology	Barriers to Adoption: Technical Infrastructure Required	Barriers to Adoption: Technical Skills Required	Barriers to Adoption: Cost of Access to Technology
Protection and Conservation Technologies						
Use of host-specific weed feeders and pathogens as well as mechanical approaches; Use of micro-organisms to control insects and pathogens on crops; Biotechnological approaches.	control of weeds, invasive alien species and situational pests in agro-ecosystems; integrated pest management; genetic engineering to introduce resistance genes.	1	site-containment of application; scientific expertise; costs of licensing in technologies.	3	3	3
Dry room and cooling technologies (cryogeny, lyophilization); Aspirators to separate the mature seeds from surrounding dead material; X-ray machines to assess the proportion of empty or damaged seed in a collection; Genetic monitoring, in vitro insemination, gene pool sampling; gene storage, DNA banks.	preservation and storage technologies for ex situ storage of seed and gene collections; conservation technologies at sub-cellular level; breeding in captivity.	2	cost of equipment; for developing countries—infrastructure, overhead costs, and security of supply of electricity.	3	2	1
Identification, Observation, Monitoring, and Assessment Technologies						
Ecosystem identification and classification technologies (software)	characterization of heterogeneous landscapes in order to adapt management strategies that are sensitive to the spatial patterns and interactions	2	costs of licensing technologies; science-specific and high level of analytical skills.	2	3	1

Remote sensing (aerial surveys, satellite imagery, infrared photography)	landscape biological approach for biodiversity characterization; mapping of watercourses; assessment of susceptibility of agricultural land to soil erosion, vegetation changes, changes in the form and extent of inland waters and glaciers.	2	high cost of science and technology infrastructure; science-specific and high level of analytical skills.	3	3	1
Geographic Information Systems	simultaneous viewing of several geographic themes (e.g., water catchments and vegetation) for inference purposes; assessment of the spread of an invasive species.	2	high costs of technology platforms; long-term nature of research; science-specific and high level of analytical skills.	3	3	1
Geo-referencing, palm pilots	species distribution and range maps	2		3	2	1
Bioinformatics	graphics oriented tools for molecular sequence analysis, techniques in phylogenetic tree estimation, and testing; assigning functions to unidentified protein sequences	1	IPR; high costs of technology platforms; science-specific and high level of analytical skills.	3	3	3
Habitat, vegetation, and gene variation mapping, regional mapping	knowing spatial variation in vegetation characteristics is important to understanding and modeling ecosystem functions and their responses to climate change.	1		1	3	1
Telemetrics	range mapping (e.g., radiocollars, sensors placed in body parts of animals)	1		2	3	1
Monitoring techniques for fish stocks	fisheries and ecosystem modeling, fishery management, influence of fisheries on the ecosystem, fish population genetics	1		3	3	1

Table 17.2. Consequences of Responses on Constituents of Well-being

Response Category	Security	Health	Social Relations	Material Wealth and Livelihoods	Freedoms and Choices
Legal	mixed distributive impacts	limited information	negative; conflict with informal laws	positive, but conditional	mixed distributive impacts across different ecosystems
Economic	mixed distributive impacts	mixed distributive impacts	negative	mixed distributive impacts	mixed distributive impacts
Technological	mixed distributive impacts	mixed distributive impacts	negative	mixed distributive impacts	conditional
Social	positive	limited information	limited information	positive, but takes a long time	positive

BOX 17.5

Concepts and Measures of Poverty

Absolute poverty is the state of deprivation based on absolute universal standards. It is usually associated with hunger and undernutrition (and other standards of simple physical needs). It can be established across different cultures and societies. It is based on objective measures of poverty.

Relative poverty is the state of deprivation defined by social standards. It is fixed by a contrast between those others in the society who are not considered poor. Poverty is then seen as lack of equal opportunities. It is based on subjective measures of poverty.

Incidence of poverty is a measure of poverty based on the number of individuals who earn below a certain threshold. This measure is widely used but suffers from the limitation that it does not give any indication about the distribution of the poverty among the poor.

Depth of poverty is a measure of the average income gap of the poor in relation to a certain threshold. It defines how poor are the poor. It gives the amount of resources needed to bring all poor to the poverty-line level.

Temporary poverty is poverty characterized by a short-term deprivation, usually seasonal. It could be due to a shortage of water or food that is available only occasionally.

Chronic poverty is poverty defined by a permanent status of deprivation. It has both intra-generational and intergenerational features. It is much harder reduce due to its structural characteristics.

Monetary poverty is poverty expressed as an insufficiency of income or monetary resources. Most indicators like $1 per day or national poverty lines are defined in those terms.

Multidimensional poverty is poverty conceived as a group of irreducible deprivations that cannot be adequately expressed as income insufficiency. It combines basic constituents of well-being in a composite measure, such as the HPI (Human Poverty Index).

Other characteristics of poverty commonly used in the literature include rural and urban poverty, extreme poverty (or destitution), female poverty (to indicate gender discrimination), food-ratio poverty lines (with calorie-income elasticities), et cetera. Other indices that combine two dimensions—incidence and depth—of poverty are also widely used; examples include the FGT (Foster, Greer, and Thorbecke) and the Sen Index (Ray 1998; Fields 2001).

known to increase the occurrences of dome diseases. The most well known examples cited in Chapter 16 are Lyme borreliosis and the severe form of tick-borne encephalitis. Rural communities, and in particular individuals who have problems accessing medical facilities, are more vulnerable to these emergent diseases than individuals who have easy and cheap access to medical facilities.

Modern agriculture driven by technology and motivated by economic incentives has a major impact on both water supply and quality. The loss of natural vegetation from hillsides and riverbanks has increased the volume and speed of runoff. Increased occurrences of floods increase the prospect of diseases like cholera. Moreover, loss of natural vegetation decreases the natural purification of water percolating through soil and vegetation. This exposes both stock and humans downstream to a variety of zoonotic pathogens, including *Campylobacter, Cryptosporidium,* and *Giardia.* (See Chapter 16.) However, a cost-benefit analysis of the benefits accrued in the form of increased material well-being generated by technological and economic responses found that the costs of reduced health are few, and it is difficult to conclude with any certainty that the net benefit has been either positive or negative to local communities.

There is little and scattered information on the impacts of legal and social interventions on health status to make any definitive conclusions. For example, existing studies shed very little light on the health outcomes of communities living near protected areas. The focus has been on increasing the material and income welfare of the local communities with very little information on health status. It is an area worth studying in the future. An important lesson we can learn, however, is the need to address the multiple services ecosystems offer, and lack in one constituent may imply an increase in another constituent of well-being.

17.2.5.3 Material Wealth and Livelihoods

The MA defines material wealth and livelihoods to include the necessary material for a good life, secure and adequate livelihoods, income and assets, enough food at all times, shelter, furniture, clothing, and access to goods.

Responses designed to improve access to natural resources are central to increasing material wealth, providing livelihoods to the rural poor, and are central to the reduction of rural poverty. It is known with a high degree of certainty that at a local level, protected areas have often had a detrimental effect on poverty reduction because rural people have been excluded from a resource that has traditionally provided income generating and livelihood activities. This is not to say that protected areas themselves have caused poverty, but the way in which protected areas are presently designed have contributed to poverty or inhibited poverty reduction efforts. The effects of local equity and rural development are mixed, and financial returns have often proven insufficient to lift community partners out of poverty. Responses need to be developed within efficient and accountable systems of governance, raising the bargaining power of communities in order to reduce poverty.

In forest management, a combination of legal, social, and economic responses has demonstrated the potential for poverty reduction. Small private ownership of forests can deliver more local economic benefits and better forest management than ownership by larger corporate bodies. This is true only when a good knowledge base, capacity to manage, effective market information, good organization between smallholders to ensure economies of scale, and long-term tenure and transfer rights are available. Private ownership (or "family forestry") is very common in Western Europe and in the southern United States, and is increasingly common in Latin America and Asia.

Responses allowing the use of wood as a fuel contribute to secure livelihoods and reduce poverty by guaranteeing the subsistence and domestic energy of poor rural households. However, wood is the least efficient source of energy for the poor. Technologies helping them to convert wood and charcoal into modern forms of energy can improve their livelihoods. Interventions based on market demand for fuelwood can provide an important source of income for the poor, but reliance on it can also impede progress out of poverty, especially with large and rapid structural changes in urban market demand for fuelwood.

Waste management policies are not only important in developing countries. Responses involving recycling and resource recovery schemes can provide jobs and help reduce unemployment and poverty in industrial countries. For developing countries, it is often argued that a fruitful response option would be private sector-driven operational schemes coupled with government provision of basic facilities in landfill, incinerator, and sewage treatment plants, among others.

The legal responses of the CCD (in terms of reversing desertification processes, such as soil erosion) are central to the poor's livelihoods, and their struggle to conserve and use dryland ecosystems resources in a sustainable fashion. Those policies are central to reducing chronic poverty. However, strategies based on integrated responses, developed through participatory mechanisms, have been found to be more effective in reducing chronic rural poverty. For instance, integration into mainstream national planning and/or poverty reduction processes, such as sustainable development strategies or the poverty reduction strategy papers have been found to provide a coherent and effective framework for poverty reduction efforts. In this manner, issues relating to the distribution of benefits in an equitable and fair manner are taken explicitly into account in the planning framework.

Social responses, such as consumer campaigns and media exposes, addressing the underlying causes of forest loss and degradation, can help with the provision of solutions compatible with forest community management, especially in developing countries. However, it should be acknowledged that these responses could take a long time to effectively change individuals' behavior.

17.2.5.4 Social Relations

The MA defines good social relations to include social cohesion, mutual respect, good gender and family relations, and the ability to help others and provide for children.

The limited articulation of international and national conventions and regulations linking biodiversity and local and indigenous knowledge has contributed to a breakdown of local rules over the use of ecosystem services, leading to conflicts within local communities as well as with outside communities. The previous section on social responses concludes that these conflicts have deteriorated the social fabric of many communities. However, recent initiatives like Article 8(j) of the CBD specifically address indigenous peoples and their knowledge. The formal recognition as well as encouraging the participation of local communities "in developing policies for the conservation and sustainable use of resources, access to genetic resources and the sharing of benefits, and the designation and management of protected areas" is hoped to reduce the conflicts and improve the social fabric that used to hold communities together. Of course, it is too early to comment on how successful these initiatives are.

The forces of globalization and privatization have improved the material welfare and livelihoods of many communities in developing countries but have also initiated a collapse of many traditional institutions in these countries. These in turn have tended to stratify many societies and cause a breakdown in social relations and an increase in communal conflicts as witnessed in the Narok district in Kenya (Amman and Duraiappah 2001; Rutten 1992).

The Green Revolution no doubt increased the livelihoods and food security for many people in developing countries. However, the extent to which the technology addressed the needs of the rural agrarian poor has increasingly come into question, as evidence points out that the increased need for irrigation and fertilizers excluded many of the poor from using this technology due to limited financial resources. The dichotomy between technological responses and traditional knowledge has been another major cause for the breakdown of social relations within communities, especially rural communities.

There is limited information on the impact social responses have had on social relations. The expectations of a strong positive correlation between the two did not emerge but this is more due to lack of research on the topic rather than the lack of a relationship.

17.2.5.5 Freedoms and Choices

The MA defines freedom and choice to include having control over what happens and being able to achieve what a person values doing or being.

There is no doubt that markets increase choices. But the central question we need to answer is do markets discriminate across individuals? The answers are mixed. Some individuals have definitely benefited from the liberalization of markets and the use of economic responses to allocate ecosystem resources. However, there is also increasing evidence of large segments of populations in developing countries being excluded from this process.

Multilateral environmental agreements—a legal response—play crucial roles in the conservation and protection of the environment, and they are inextricably linked to the alleviation of poverty in developing countries. However, few MEAs address poverty reduction as a priority in these countries. As emphasized in the discussion of legal responses, many MEAs (for example, the CBD) recognize "that economic and social development and poverty eradication are the first and overriding priorities of developing countries" (CBD, Article 20.4). However, those responses are based on the role of economic growth in reducing poverty. There is by now an extensive body of literature on "pro-poor growth" (Fields 2001; Ravallion 2004) that highlights the limiting effects of those responses that are capital-intensive and based on highly qualified jobs. In other words, MEAs based on approaches to conservation that ignore the need to create jobs in the short run will have a limited impact on reducing poverty. To date, protected areas have reduced the options and choices available to local rural communities for securing security and livelihood opportunities.

At the national level, land tenure programs have contributed to better ecosystem management and are instrumental in poverty reduction in some ecosystems. However, their success has been found to be dependent on the type of ecosystem. For example, private land tenure systems in drylands have been less successful in ensuring that pastoralists have access to various ecosystem services like provisioning of water and pasture. But even in the case of communal land tenure systems, studies point toward greater transparency and fairness in the allocation of resources within the communal system across all stakeholders.

Genetically modified organisms have also come under increasing criticism for reducing the diversity of the seed bank and for increasing the dependency of many farmers, especially in the developing countries, on large multinational companies who control the patent rights for the GMOs. In this regard, the vulnerability of developing countries is exacerbated by resource and skills constraints in terms of studying and developing national and local impact assessment methodologies and regulation frameworks. The freedom and choices for farmers in developing countries have been reduced. However, developing countries do stand to gain tremendously in terms of food security and broader human well-being, particularly in the area of health with improved nutrition, if the level of uncertainty over ownership associated with these technologies can be reduced significantly.

Social responses which advocate participation and the use of democratic structures to change have contributed to the freedom and choices of individuals. For example, setting up community-based forest management initiatives allows the flexibility of local communities to harvest the forest for income opportunities but also allows the communities to use non-wood forest products for personal use.

17.3 Development and Poverty Reduction Responses: Effectiveness and Consequences for Human Well-being and Ecosystem Services

Development and poverty reduction strategies have assumed a variety of forms and shapes. It is important to note the impact of these responses on ecosystem services. Quite often poverty reduction policies have been based on different conceptions of development, and on different measures of poverty. It is common to find poverty reduction responses that identify poverty not as a deprivation of constituents of human well-being but as a deprivation of income. In these cases, the optimal strategy consists in increasing the economic entitlements of the poor. This can be achieved by improving governance, enhancing the asset bases of the poor, improving the quality of growth, and reforming international and industrial-country policies. These recommendations, as put forward by the World Bank (2002) report, *Linking Poverty Reduction and Environmental Management,* are centered in the economic dimension of well-being. Emphasis is given to the notion of "livelihoods" in analyzing the impact of poverty reduction strategies on ecosystem services and human well-being.

It has been widely acknowledged by policy-makers and other stakeholders that poor people are heavily dependent on the state of ecosystems and on ecosystem services, and to a large extent the overall responses address this interdependence. When the ecosystem is degraded, the well-being of these poor populations is severely affected not only in economic terms, but also in cultural and spiritual terms. This seems to be the core message behind those responses that articulate actions toward poverty reduction.

One important development intervention that encompasses all different dimensions of human well-being is the Millennium Development Goals. The MDGs consist of a set of goals and targets to be achieved by most countries by 2015. The goals are not restricted to a livelihood analysis, instead encompassing a broader, multidimensional view of development. The goals that are closer to the management of ecosystem services are: halving extreme poverty and hunger, achieving universal primary education, promoting gender equality, reducing child mortality, improving maternal health, combating major diseases, and achieving environmental sustainability. It is important to note the impact of those policies on the provision of ecosystem services.

The world has seen a proliferation of policies addressing the problem of increasing environmental degradation and increasing poverty in these last years. As Perrings (1996) has argued, "One of the most striking coincidences of the last decade has been . . . deepening poverty and accelerating environmental degradation in the arid, semiarid and subhumid tropical regions—the drylands—of sub-Saharan Africa." There are several elements of the MDGs that are closely linked with ecosystem services. For instance, it is known that there are 1.3 billion people living in fragile areas (World Bank 2003), with a majority living with less than one dollar a day. For these people, improvement in their food security and health status depends on the state of the ecosystems in which they live and the services these ecosystems are able to provide. Of course, access to other assets such as human, physical, financial, social, "state provided" infrastructure, in particular education and health services, political and institutional, also play an important part in defining the well-being of the poor.

Similarly, some of the central effects of poverty reduction strategies are on people's nutrition and on women's degree of independence within the household. Poor women are also often subjected to indoor pollution and heavy workload that increases their exposure to other environmental risks.

Warburton (1998) has shown how the use of successful responses to ecosystems depends on community values and community participation. According to her, a participatory approach to sustainable development, based on community-driven strategies, would promote capacity building at the local level, and improve mechanisms of sharing experience and knowledge between community groups at national and international levels. The same reasoning is pursued by *Agenda 21.*

Sectoral responses would then be defined according to the particular characteristics of the natural resource sectors. For instance, a development strategy involving secure land tenure would be conducive to sustainable activities preventing land degradation. A more comprehensive understanding of the development and poverty reduction strategies would then allows us to conclude "without a doubt that the poor do not initially or indirectly degrade the environment" (Duraiappah 1998). This is an important conclusion that suggests that responses should focus on (1) correcting institutional and market failures, and (2) providing proper incentives to agents to use the resources sustainably.

The impact of regional variations in the poverty-environment links—adding the gender dimension—has been explored by Agarwal (1997). She analyzes the particular processes of nationalization and privatization developed in India that have undermined traditional institutional arrangements of resource use and management. Responses to development problems in India have led to a continuing commercial exploration of forests and appropriation of village commons. Her work corroborates Jodha's (1986) study on the village commons. The Green Revolution in India has led to high environmental costs over time. Some of these responses emerged from the organization of civil society groups, such as women's and environmental groups, indigenous movements and peasants' participation. Agarwal calls attention to the fact that, "the processes of environmental degradation and appropriation of natural resources by the State, and by small numbers of individuals, have specific class-gender implications, manifest in the erosion of both livelihood systems and knowledge systems, but it is women and female children of poor rural households who are affected most adversely." (Agarwal 1997)

This gender bias not fully addressed by some responses is also confirmed by Ezzati and Kammen (2001), who have shown that women are twice as likely as men to have an acute respiratory infection due to their exposure to high emissions from domestic activities. Some development policies dismiss the distributive impacts of environmental degradation, ignoring that the poor are most dependent on environmental income in relative terms and the non-poor who use more the natural resources in absolute terms (Cavendish 1999). If these impacts are assessed in the space of the constituents of well-being, there are great inequalities found among the rich and the poor in terms of environmental health problems, such as illnesses due to unsafe water, indoor air pollution, or exposure to disease vectors.

Development practices will have an intrinsic role in the promotion of human well-being and an instrumental role in the fostering of ecosystems and ecosystem services. The promotion of adaptive participatory management has important consequences on stakeholder behavior and capacity building strategies that affect directly the well-being of societies (Gadgil 1999).

Distributive issues are at the core of the analysis of the impact of development policies on ecosystems and human well-being. The use of watershed projects, as a development strategy, has been widely pursued by governmental agencies. On the one hand, there are potential benefits for the poor that could arise from a watershed or a catchment: it could raise agricultural production and conserve common natural resources that are so important to the poor. On the other hand, it could benefit landholders while harming landless people. Kerr finds the projects most successful in achieving conservation and productivity benefits also have the strongest evidence of skewed distribution of benefits toward larger landholders (Kerr 2002). The analysis is limited and the extent of trade-offs is not known, but it is clear that satisfaction with watershed projects is positively correlated to land holding size, and many landless people strongly resent their loss of access to common lands."

This might happen for a variety of reasons. It is important to remark, however, that most watershed projects have not benefited the poor directly. Moreover, they have not been very friendly toward women, since they have only employed male staff members who are less sensitive to gender concerns.

In the context of tropical forests, emphasis should be given to what Wunder (2001: 1818) named "the dynamic potential of natural forests," that is, the sustainable production of benefits to the poor. This could be challenged by the existence of "vicious circles" where the poor need to use environmental resources, causing degradation, which aggravates poverty, and so on. In terms of development policies and responses, Wunder argues that "at the global level, the shift toward a people-centered conservation paradigm has not succeeded in slowing down the loss of tropical forests and their biological diversity," with perverse consequences to the poor. With regard to deforestation, evidence from Brazil shows a pro-cyclical pattern of deforestation.

Much work has been done on the construction of indexes for assisting responses to the poverty-environment nexus. (See Box 17.6.) Good examples comprise Lindenberg's (2002) index of household livelihood security at the family and community level, Sullivan's (2002) index of water poverty, and Levett's (1998) sustainability indicators. Although these examples are far from being comprehensive, they properly illustrate the proliferation of general attempts to integrate well-being and ecosystem services in the formulation of development policies. It must, however, be noted that their impact has been systematically ignored by mainstream economists.

A recent study that synthesized data from 54 household income studies from 17 countries, mostly in East and Southern Africa and South Asia, found that 22% of household income comes from "forest environmental resources" such as harvesting wild foods (bushmeat, insects, and wild fruits, and vegetables), fuelwood, fodder, medicinal plants, and timber (Vedeld et al. 2004). The income is about evenly split between cash and products consumed directly. Wealthier families harvest more forest products. However, these activities generate a much higher proportion of poorer families' total income. Villages farther away from markets and with lower educational levels get more of their income from forests. Forest environmental income was of particular significance as an additional income source in periods of both predictable and unpredictable shortfalls in other livelihood sources. This study concluded that the omission of forest environmental income in national statistics and in poverty assessments leads to an underestimation of rural incomes, and a lack of appreciation of the value of the environment in the context of poverty reduction strategies.

Poverty alleviation policies based on structural adjustment policies are very controversial because they have not been able to solve "the problem of the counterfactual" (Killick 1995), that is, "Do adjustment programs result in a better poverty situation than would have obtained without them?" Whereas it is correct to point out the adverse effects that accompanied adjustment policies, it is technically difficult to isolate the effect of the causes of the initial crisis. The subsistence farmer, the urban very poor, the smallholders in isolated rural areas, all seem to suffer with the introduction of structural adjustment programs. The impact on ecosystems has rarely been discussed by analyses of such programs. The emphasis has been given to short-term macroeconomic policy measures with fundamental fiscal reforms in the public sector.

New poverty reduction strategies have been developed in these last years, covering a wide range of policies, scales, and

BOX 17.6

Weighting Human Well-being in Development Policies

One important issue in assessing the impact of ecosystem services on human well-being is how the well-being of different individuals is weighted in a social well-being function. When handling multidimensional aspects of well-being, another aspect to be considered is how to weight those different elements. Should we give the same weight to each and every aspect, or should we examine alternatives weights? John Rawls (1971) has put forward the view that a lexicographical ranking called *maxmin* should be adopted, according to which, social well-being increases as the well-being of the worst off in society increases.

More recently, participatory approaches have suggested that emphasis should be given to the *mechanism* of setting priorities rather than on priorities itself. This is a fundamental issue for assessing the impact of the interlinkages between well-being and ecosystem services.

actors. More often than not, as mentioned above, the impact of those policies on ecosystems and ecosystem services is largely ignored. Their focus seems to be on institutional and macroeconomic stability, generation of sectoral growth, and reduction of the number of people living with less than $1 per day in poor countries. The most articulated of those strategies is the poverty reduction strategy paper. PRSPs aim to target poverty reduction by articulating macroeconomic, structural, and social policies in those countries interested in following a partnership with the World Bank and the IMF. They are comprehensive and results oriented, and promote a sense of ownership among those governments that are committed to them. Countries are supposed to prepare an interim PRSP (I-PRSP) and then update their PRSP every three years.

PRSPs are quite important elements in poverty reduction strategies for more than 40 countries all over the world. Many of these countries depend on the approval of their PRSPs or I-PRSPs as a condition to the IMF's Poverty Reduction Growth Facility program, the World Bank's concessional lending, and the Highly Indebted Poor Countries Initiatives. PRSPs' analysis of the impact of poverty on ecosystems is assessed through poverty-environment links. As mentioned in the *PRSP Sourcebook,* "In the context of a PRSP, environment and poverty are linked in two major ways. One is that poverty alleviation should not damage the environment of the poor, which would only undercut gains in one area with losses in another. The other main link is that improving environmental conditions can help to reduce poverty." (World Bank 2004: 376)

PRSPs link environmental impacts on health and security of poor people, proposing a "systematic mainstreaming of environment in PRSPs." PRSPs include issues like water supply and wastewater disposal, solid waste removal, indoor and urban air pollution, land degradation, deforestation, loss of coastal ecosystems, and fisheries. However, they reduce proper environmental management to the provision of 'sustainable livelihoods'. In policy terms, they focus on sectoral policies and programs on health, infrastructure, public works, agriculture, et cetera, giving a very broad meaning to environmental policies. They explore poverty dimensions, such as those of economic opportunity, natural resource utilization, empowerment and public actions. The potential consequences for the poor are potentially quite striking. Yet successful responses depend on the implementation of democratic participatory processes at a local level.

Participatory poverty assessments are a promising tool of poverty reduction strategies. PPAs use participatory techniques to evaluate poverty assessments. They are able to reveal many different aspects of poverty, such as vulnerability, powerlessness, isolation, insecurity, ill health, physical violence, and fragile environments that condemn the poor to lives of multiple deprivations (Robb 1999). They are usually good to reveal the depth and severity of poverty, which are normally hidden by monetary poverty lines. RRA (rapid rural appraisal) techniques and PRA (participatory rural appraisal) have developed the core elements behind current uses of PPAs. Here again, not much has been said on the impact of those strategies in terms of ecosystems and ecosystem services. This remains a challenge to be met.

To conclude, it could be said that the impact of poverty reduction strategies on ecosystems and ecosystem services has not been fully explored in the international policy-making processes. More information is needed to assess the impact of those responses on poverty reduction according to specific analytical categories. The most important poverty reduction strategies tend to emphasize institutional and macroeconomic aspects at the expense of poverty-ecosystem links. When these strategies refer to ecosystems or the environment, they tend to frame the consequences of poverty alleviation strategies in terms of the livelihoods approach, narrowing down the poverty issue to its economic dimension. Successful responses should therefore include broader notions of poverty and try to mainstream the role of ecosystem services in the main poverty reduction programs.

17.4 Critical Choices for Response Success

The assessment identified a number of elements that play significant roles in the selection of responses and the subsequent consequences of these responses on human well-being and poverty. The five more critical elements identified from the assessment include: (1) heterogeneity of stakeholders, (2) spatial, temporal, and administrative scales, (3) type of ecosystems and the underlying dynamics, (4) policy coherence and interdependency, and (5) enabling conditions or instrumental freedoms.

17.4.1 Heterogeneity of Stakeholders

Most responses, particularly economic and market-based responses, focused on improving the aggregate welfare of communities or societies. However, numerous authors have complicated the concept of "community." Communities are notoriously difficult to define, with boundaries between them often blurred and overlapping (See Chapter 14; Cavendish 1999; Barrow and Murphree 2001). Within communities, differentiation exists based on socioeconomic positions, gender, and positions with local government structures. Ecosystem management may have different impacts on the well-being of different groups within communities. Local elites may have better opportunities to capture benefits, while the burden of conservation may be put on other groups in the communities. This entails that the tradeoffs and synergies between different ecosystem services may not be similar for every person, especially women (see Box 17.7), within a community. Decision-makers will need to incorporate these differences when designing responses and evaluating their consequences.

The assessment of Part II chapters shows very clearly that a variety of stakeholders live in ecosystems and use ecosystem services for different purposes and in varying degrees and intensity. For example, in the case of forests ecosystems and the provision-

BOX 17.7

The Gender Dimension

Each of the response options has a gender dimension that can be more or less relevant to poverty reduction according to the nature of ecosystem services. The phenomenon known as the "feminization of poverty" should be acknowledged since there is great inequality in the distribution of resources and capabilities along gender lines. Considering gender issues means not only awareness of women's disadvantages in terms of a wide range of variables, but also incorporating the issue of discrimination against women in the use of resources. This might be revealed in the disproportionate time that many women in dryland areas spend searching for clean water. Interestingly, responses that take into account an active role for women provide a very positive balance for soil conservation, promoting biodiversity with the introduction of new technologies (UNFPA, 2001). Responses involving women in environmental and health decisions have proved to be important to target poverty reduction and to conserve ecosystems successfully.

ing service of fuelwood, the poorest tend to be disadvantaged by shifts to bring remaining common pool resources under sustainable management through economic (prices) and legal (tenural) responses. (See Chapter 8.) Fuelwood harvesting tends to be restricted in this process, and women's needs for fuelwood commonly have lower priority than those of men, who use forest products for sale. Weak tenure rights over the resource can also mean that poor (low-income) rural producers and traders are progressively excluded from access to the resource and markets as trade grows. (See Chapter 8.)

Another example is the establishment of protected areas in upper watersheds. Protected areas, a form of regulatory control over land use, has had mixed results because many of them fail to recognize the rights of local populations or communities who have depended on such areas to support their livelihoods, and exclude them from access to benefits. (See Chapter 5.) Age, gender, ethnic group, and wealth status play a role in who maintains and decides what diversity to maintain and where.

In similar manner, culture has been found to play a significant role in the success or failure of technological interventions, which change landscapes and disrupt the cultural and religious relationships communities have with ecosystems. (See Chapter 14.) Moreover, it was shown that technological responses usually cater for quick profits and require certain amounts of investment that usually exclude the low income.

Table 17.3 tells us which characteristics of stakeholders should be considered when designing and implementing a response. The table is read as follows. If water markets—an economic response—are being considered to improve the efficient allocation of water, then high considerations must be made for individuals with low income and more specifically women who may actually experience a drop in well-being with market determined prices. Table 17.3 also tells us, for example, that low considerations can be given to the cultural and religion background of individuals when water markets are created. There was limited information available to consider if formal education and the rural/urban divide were important attributes to consider when designing economic responses.

17.4.2 Temporal, Spatial, and Administrative Scales

Temporal, spatial, and administrative scale effects were also found to play critical roles in determining the success or failure of responses. The role these scales have on the consequences on human well-being and poverty are indirect in the sense that failure of the response to mitigate or reverse ecosystem degradation inadvertently causes a drop in human well-being and an increase in poverty.

17.4.2.1 Temporal Scales

Changes in ecosystems and their corresponding services can occur over a range of time horizons. Losses in provisioning services occur in a relatively shorter time span than losses in the regulating and supporting services. For example, the extinction of the Peruvian anchovy stock within a span of 15 years is a telling story of how fast a renewable resource can come so close to extinction. On the other hand, climate change caused by the emission of anthropogenic greenhouse gases occurs over a time scale of a century or more. It is also known that the longer the time frames, the higher the degree of uncertainty over the loss in ecosystem services and the consequences on human well-being.

Do temporal properties have an influence on the type and success of intervention strategy to be adopted? The answers are mixed. The main factor that underlies temporal dimension is the level of uncertainty. (See Figure 17.1.) If the loss in ecosystem service and/or the impact on human well-being is highly uncertain or unclear but is expected to occur within a short time span, then the precautionary principle is recommended to be adopted and a policy regime consisting of a hybrid of regulatory, economic and legal interventions can work relatively well. An example of the precautionary principle in practice is the Cartagena Protocol. This protocol requires advanced informed agreement procedures and careful assessment of risks before movements of living modified organisms are permitted. Such risk assessment should be based on the precautionary principle.

However, if there is uncertainty and the impacts are expected to occur only in the distant future, then economic interventions in the form of permits, taxes, and incentives for technological innovation, together with social interventions in the form of changing values and attitudes, work well.

In the event that there is very little uncertainty whether an impact will occur, and it is expected to happen in the near future, then regulations supported by legal interventions have proved to

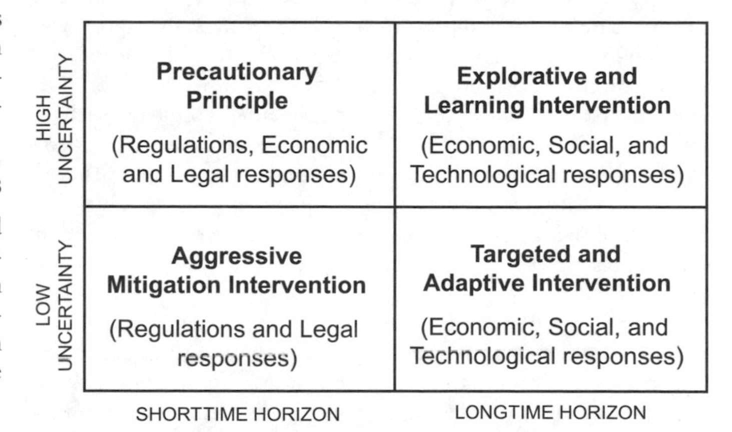

Figure 17.1. Relationship between Uncertainty, Time, and Interventions

Table 17.3. Stakeholder Characteristics That Can Influence the Consequences of Responses on Human Well-being and Poverty

	Economic and Financial Responses	Legal Responses	Technological Responses	Cultural Responses
Gender	high consideration	high consideration	medium consideration	high consideration
Culture and religion	low consideration	medium consideration	high consideration	high consideration
Formal education	imited information	high consideration	high consideration	high consideration
Income level	high consideration	low consideration	high consideration	limited information
Urban/rural divide	limited information	low consideration	limited information	high consideration

be the best intervention strategy. The Montreal Protocol to protect the ozone layer is an example of a legal response. However, if the impact is known to occur but only in the distant future, then a combination of social, economic, and technological intervention strategies will be the best response strategy. The adaptive strategy to mitigate the impacts of climate change in many developing countries is an example of response strategy falling into this class. It is therefore critical in the design of responses that the degree of uncertainty is explicitly recognized and incorporated in the design stage of the response.

17.4.2.2 Spatial and Administrative Scales

Spatial scale relates to the geographical boundary over which ecosystem services are provided while *administrative scale* denotes the type and level of authority overseeing the management of the ecosystem or ecosystem service. Chapter 5 on biodiversity stresses the importance of the principles of decentralization or subsidiarity as guiding principles for the management of ecosystems for reasons of efficiency, democratic legitimacy, and ethics. Institutional diversity and nested institutional arrangements may be the ways forward, as all levels of authority have their strengths and weaknesses.

The wood and fuelwood chapter shows that national government policy and regulation used to create the major signals influencing wood producers have often had negative environmental and social impacts. (See Chapter 8.) Now, the set of response options has widened to include voluntary, market-based, and multistakeholder actions at decentralized and international levels. The chapter concludes by saying that it is important to have a set of responses, so that each targets a specific problem, but also works well with the others. Such coordination is particularly needed at the national policy level (and can be done through national forest plans and processes) and at the forest manager level. Decentralization of authority and responsibility to local government, communities, and the private sector is also a common shift in many parts of the world, especially larger countries such as Indonesia, China, and Brazil. This shifts power closer to the people most affected by local resource use and can improve management where local institutions are adequate and accountable. But it is critical that there is a close relationship between the different scales of responsibility in order to minimize the potential for conflicts.

In the case of water resources, the objective of integrated water resource management is a general response to conflicts between various interests in water resources. These may include conflicts among individual users, among different sectoral interests, among political jurisdictions, between livelihood interests found at local scales, and larger scale interests driven by geopolitical considerations, and more generally between environmental and economic objectives associated with regulatory and provisioning services of freshwater ecosystems. Therefore, IWRM is primarily a problem of developing more appropriate forms of governance and institutional arrangements for management of shared waters, which cuts across spatial and administrative scales.

Lack of authority is particularly prevalent in transboundary basins, given the absence of a single legal authority and the need to rely on voluntary agreements among countries that occupy the basin. An example of this is the International Commission for the Protection of the Rhine. Given the generally heterogeneous nature of environmental as well as socioeconomic conditions, effective management of freshwater resources to support multiple and often conflicting uses often requires numerous site-specific responses that are beyond the capacity of centralized authorities. Although a basin-wide approach is necessary for some aspects of management of freshwater resources such as overall water allocation, flood forecasting, and emission of persistent pollutants, others, such as problems associated with land and water relationships, and operations and maintenance of irrigation canals, may best be resolved locally because it allows for more direct engagement of stakeholders and more appropriate responses to site-specific circumstances.

In general, land-water relationships are more effectively addressed at smaller scales at which land use impacts can be more readily detected, benefits are more tangible, and at which agreements can be tailored to local conditions. Interventions at these smaller scales are also unlikely to be effective beyond the level at which their consequences are at least measurable. For example, soil conservation management practices aimed at reducing erosion may be beneficial on-site but will not be of significance in reducing downstream sedimentation in basins where erosion rates are naturally high or where there is insufficient cooperation by upstream stakeholders so as to implement practices over a wide enough area.

The implied solution to spatial and administrative scale conflicts is the need both for decentralized approaches for many aspects of resource management and decision-making, and for mutually supportive governance at more centralized levels, so as to insure that human well-being and poverty reduction measures and other localized interests are adequately represented and considered in basin-wide decision-making processes. Therefore, the issue at hand is not the type of response, which should be used for different spatial or administrative scales but more of a concern of getting into place the appropriate governance and institutional frameworks to support the implantation of responses in an efficient, democratic, and ethical manner so that human well-being is increased and poverty reduced.

17.4.3 Ecosystem Dynamics and Interaction

The biodiversity chapter explicitly states that the success or failure of any response to conserve biodiversity will depend on the ecological setting in which it is applied. One of the main reasons for the limited success of protected areas is the failure to account for the distribution of biodiversity within different ecosystems. Therefore, the fixed land percentage criteria used for determining protected areas fails to capture the fact that different ecosystems vary in terms of internal heterogeneity or diversity with more diverse types arguably deserving a higher target. Moreover, different ecosystems vary in terms of likelihood of persistence in the absence of conservation action (for example, because of differences in geographic extent); percent targets can run counter to types with greatest need for protection, because models of probabilities of persistence suggest that geographically extensive habitat types may have a reasonable probability of persistence of their components even in the absence of action, and so require a *smaller* not larger percentage area protection. (See Chapter 5.)

Ecosystems are not homogenous. For example, dryland ecosystems cannot be perceived as being in equilibrium, but in continual transition—in climatic conditions, production systems, demography, and social institutions (Mortimore 1998). Attempts to use economic responses like private land ownership may work well in specific areas within ecosystems, which are characterized by stable rainfall and predictable climatic variations. However, if the larger ecosystem is characterized by more unstable precipitation patterns and introducing private land tenure in these systems may prove fatal (Amman and Duraiappah, 2004).

It should also be recognized that there is a high degree of synergy and interdependency among the different services, for example, the well-established strong inter-linkages among biodiversity and hydrological and nutrient cycles (see MA *Framework for Assessment*). Overuse of provisioning services, such as excessive biomass harvesting, can impair the productivity of the regulating and supporting services, such as water and nutrient cycling, which in turn have a negative impact on the ability of the ecosystem to produce goods, such as biomass. This feature of interdependency among ecosystem services is normally not taken into account in management and policy decisions, especially at watershed and national scales.

Integrated responses that address degradation of ecosystem services across a number of ecosystems simultaneously and also explicitly include objectives to enhance human well-being are increasingly being used. Integrated responses occur at different scales and across scales, and use a range of instruments for implementation. Increasingly they are associated with the application of multistakeholder processes and with decentralization; they may include actors and institutions from government, civil society, and the private sector. However, the limited information available to date on integrated responses makes it difficult to come to any decisive conclusion at this moment.

17.4.4 Policy Coherence and Interdependency

The assessment of the well-being–poverty–ecosystem nexus reveals that it is governed by a complex system of institutions, organizations, and instruments (Duraiappah 2004). These range from the international scale right down to the local community. Moreover, within each level, there is again a multitude of instruments, organizations, and institutions at work. Figure 17.2 shows that the links between scales is not linear, but there are many instances where there is a direct link between the international level and the local level policies. But the critical point to keep in mind is the need for coherence at all these scales and levels if the goals of poverty reduction through the sustainable management of ecosystem services are to be even remotely achieved.

Chapter 8 on wood and fuelwood finds that policies, institutions, and interventions that work toward ecosystem well-being are uncoordinated with those that work toward human well-being. This same finding is similar to the one that many PRSPs make no explicit mention of ecosystems.

Frameworks for natural resource management that are developed more locally and then linked to national objectives have shown to be more flexible and responsive to local interests. Relevant local stakeholders can develop these frameworks, with special support given to the disadvantaged poor to negotiate for their interests.

17.4.4.1 Instruments and Institutional Coherence

There are 13 global multilateral environmental agreements and/or conventions and approximately 500 international treaties or other agreements related to the environment. Couple this with an equally large number of poverty reduction plans and development strategies, and we get a complex policy arena with the potential for many conflicting objectives and goals. There has been a concerted action by the secretariats of the CBD, CCD, and Ramsar to coordinate their action plans, but our findings suggest that more is needed. A valuable input to the process will be to consolidate the various action plans under each of the conventions into an integrated response strategy and this strategy to be integrated within the larger and broader development and/or poverty reduction frameworks used in the respective countries.

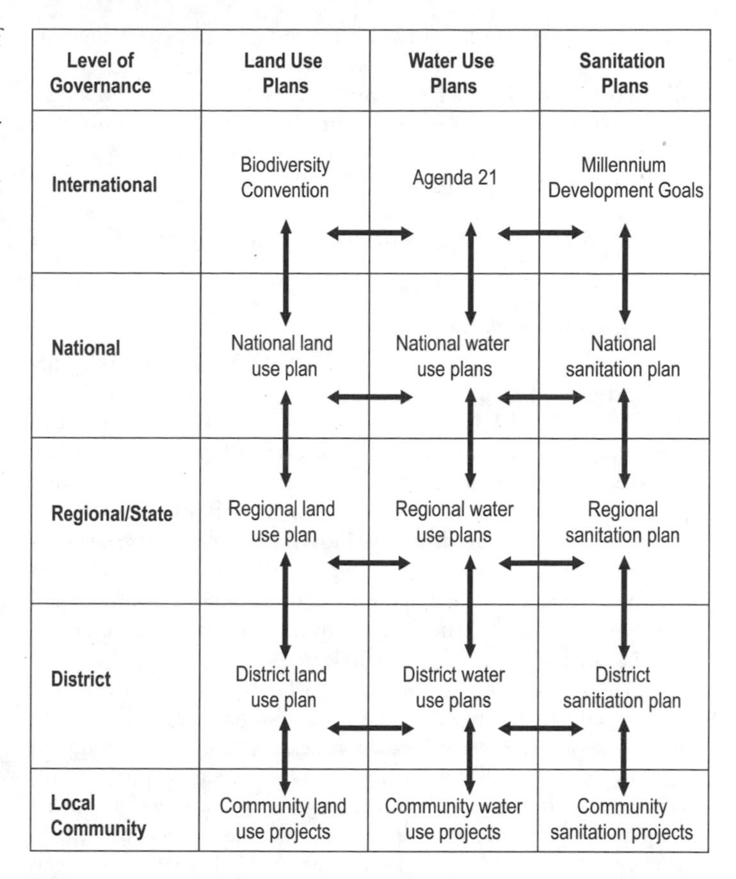

Level of Governance
Land Use Plans
Water Use Plans
Sanitation Plans
International
Biodiversity Convention
Agenda 21
Millennium Development Goals
National
National land use plan
National water use plans
National sanitation plan
Regional/State
Regional land use plan
Regional water use plans
Regional sanitation plan
District
District land use plan
District water use plans
District sanitiation plan
Local Community
Community land use projects
Community water use projects
Community sanitation projects

Figure 17.2. A Schematic Illustration of Vertical and Horizontal Policy Coherence

The initial findings at the national levels point to a fragmented approach with a few countries recognizing the need for policy coherence but they do not have the capacity and resources to carry out such a process. If the international community wants to see an improvement in the management of ecosystems as well as reductions in poverty, then additional resources should be set aside for developing systems that encourage the integration and coherence of instruments, organizations, and institutions.

17.4.4.2 Organizational Coherence

At the international level, each MEA has its own secretariat. At the national level, the responsibilities for the environment, poverty reduction, and development strategies are spread across a variety of ministries. It is imperative for these conventions at the international and ministries at the national level to work together toward common goals and objectives. If policy-makers are to reduce inter-ministerial conflicts, then it is imperative that an organizational matrix should be drafted to allow policy-makers to see who (which organization) is responsible for what (institutions and instruments).

17.4.4.3 Vertical Coherence

There is a need for vertical integration of the various policies, plans, and/or strategies. International conventions must be coherent with national policies and these in turn must be coherent with local policies. The many cases reported in the chapter on biodiversity show the impact of international environmental law upon national law, demonstrating that domestic courts play a vital role in the application of international environmental agreements. (See Chapter 5.)

In the case of biodiversity conservation, the greatest benefits were realized when intentional ecosystem planning achieved coordinated adoption over large areas. However, even when adoption was limited to individual farm-level activities, significant benefits to "wild" biodiversity were recorded. Of the 36 cases, 28 principally benefited poor, small-scale farmers. Enhanced ecosystem productivity and stability reduced production-associated risks, raised food and fiber production, and improved human well-being.

17.4.4.4 Horizontal Coherence

Horizontal coherence refers to coherence at the international level, at the national level, and the local level. This requires actors at the international level to work together and make an effort to ensure that their policies complement each other (OECD 2001). International cooperation through multilateral environmental agreements requires increased commitment to implementation to effectively conserve biodiversity. However, unless these are implemented on the ground, they are not effective. The CBD, in its Preamble, recognizes "that economic and social development and poverty eradication are the first and overriding priorities of developing countries." The CBD (Article 20.4) and UNFCCC (Article 4.7) also state that "eradication of poverty" is one of the commitments by the parties. The CCD also has several provisions toward alleviating poverty and creating an enabling environment to achieve sustainability objectives. A positive signal has been recent attempts (for example, through joint work plans) to create synergies between conventions. However, the link between biodiversity conventions and other international legal institutions that have an impact (such as WTO) are weak.

The same is true for the national level. Ministries must work together to aim for a common goal. Their plans and strategies must be complementary to each other, and trade-offs among their plans must be highlighted, discussed, and agreed upon before actions are implemented. For example, the chapter on biodiversity (see Chapter 5) stresses the importance of integrating biodiversity issues in agriculture, fishery, and forestry management—in many countries the responsibility of various ministries—in order to achieve two-fold gains in encouraging sustainable harvesting and in minimizing the negative impacts on biodiversity.

Responses have in general been developed and implemented in a sectoral manner. The sectoral approach has usually led to many conflicting objectives, and even when objectives are similar, different approaches and responses inadvertently lead to a worsening of the situation rather than an improvement. The overall objective for policy coherence will be to:

- reduce fragmentation,
- reduce duplication, and
- reduce transaction costs.

17.4.5 Enabling Conditions: Toward Adaptive Management and Evolving Strategies

Adaptive management has been cited as an ideal approach to address poverty reduction and ecosystem management in many of the chapters in Part II of this volume. The chapter on water (see Chapter 7) defines adaptive management as an integrated response option that provides a way to build on a base of imperfect knowledge, while the integrated responses chapter (see Chapter 15) defines it as learning by doing. Adaptive management necessitates multiple instruments, including a mix of law, economic instruments, voluntary agreements, information provision, technological solutions, research, and education. The key word in adaptive management is flexibility. Equally important in adaptive management are the management structures that exist and the potential for change within those structures, whether they are institutions, property rights, or communities. In order to provide flexibility and the environment for change to occur, a number of "enabling conditions" are required.

Chapter 15 highlights integrated coastal zone management as an adaptive management approach. ICZM initiatives created enabling conditions for stakeholder involvement in coastal management decisions, and for partnerships that have helped to break sectoral barriers. However, the chapter also highlights where ICZM processes have failed to identify all relevant stakeholders and create conditions for their effective participation. This is because these partnerships have not always demonstrated transparent and democratic methods for selection of the participating organizations and individuals who truly reflect the needs of the coastal area.

Many of the various enabling conditions found to be critical in the successful implementation of response strategies within an adaptive management strategy were found to be closely related to what Amartya Sen calls "instrumental freedoms" (Sen 1999). Sen identified five instrumental freedoms—participative freedom and political feasibility, social opportunities, economic facilities, transparency guarantees, and protective security—critical for reducing poverty and improving human well-being. He stresses that these freedoms should not be seen as exclusive of each other or as substitutes, but as inclusive and complementary. Therefore, some or all of the other instrumental freedoms—participative freedom and transparency guarantees—are required in order to provide, for example, economic facilities (Dreze and Sen 2002).

Individuals live and operate in a world of institutions and, therefore, there is a close relationship between instrumental freedoms and institutions. For example, Sen goes on to state that our opportunities and prospects depend crucially on what institutions exist and how they function. Not only do institutions contribute to our freedoms, their roles can be sensibly evaluated in the light of their contributions to our freedom (Sen 1999).

But the relationship is more complex than appears at first glance. Although institutions are responsible for the provisioning of the freedoms, it should also be acknowledged that the type, efficiency, and effectiveness of institutions are determined by the instrumental freedoms available in the country (Sen 1999, Chopra and Duraiappah forthcoming). (See Figure 17.3.) Therefore, communities having limited instrumental freedoms would find difficulties in changing, revising or adapting existing institutional structures, which are needed to successfully implement adaptive

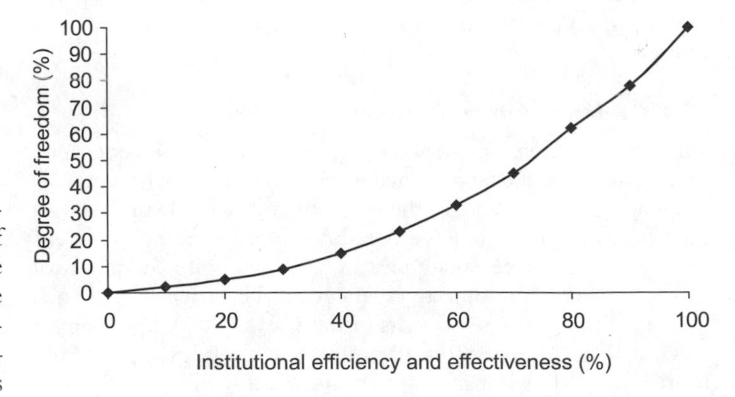

Figure 17.3. Relationship between Degree of Freedom and Institutional Efficiency and Effectiveness (Chopra and Duraiappah forthcoming)

strategies to improve well-being and better management of ecosystem services.

The assessment of the various intervention strategies in Part II of this report point to the fact that no intervention strategy operates in a vacuum. The assessment repeatedly highlights that the success or failure of many responses were largely influenced by the various institutional frameworks that were in place in the country for providing the various instrumental freedoms. For example, in the case of biodiversity, one of the critical assessment findings that evolved is the democratic and representative participation of local communities. The participation process worked relatively well in conserving biodiversity, and worked well for improving the well-being of local communities when institutions supporting democratic participation were in place. (See Chapters 5, 7.)

Another example of a successful intervention strategy that worked relatively well is the use of water permits in the United States. Local communities together with local authorities developed a water permit system that worked relatively well. The main reason for the success of this initiative was the presence of well-functioning markets—an economic facility, together with a trusted legal system, the transparency guarantee, which individuals could rely on in cases of litigation. In many developing countries, these institutional frameworks come at a very high cost in the form of high transaction costs and corruption. (See Figure 17.4.) These high costs practically prevent poor communities from adopting adaptive management strategies, which require as a prerequisite flexible institutional structures which can be changed or revised without too much cost and effort.

Moreover, the clarification of property rights or tenure security—another form of transparency guarantee—was considered as a prerequisite for the use of economic instruments. It is well established, however, that tenure security is absent in many developing countries and explains the limited success market-based instruments have had in these countries. In a majority of instances, the use of market-based intervention strategies often resulted in critical water supplies being captured by a small minority who had resources and political power. It was found that, in cases where formal tenure security is unreliable and uncertain, intervention strategy in the form of informal arrangements among the relevant stakeholders worked relatively well. However, it should be stressed that institutions ensuring equitable participation in formulating these informal institutions were present to ensure fair representation and access to ecosystem services. Experiences from the use of ICZM—an example of an adaptive management strategy—suggest that this approach can only work effectively if institutional structures can be revised, amended, or changed relatively quickly and efficiently; and this was found to occur in cases whereby a number of critical instrumental freedoms, especially participatory freedoms and transparency guarantees, are available.

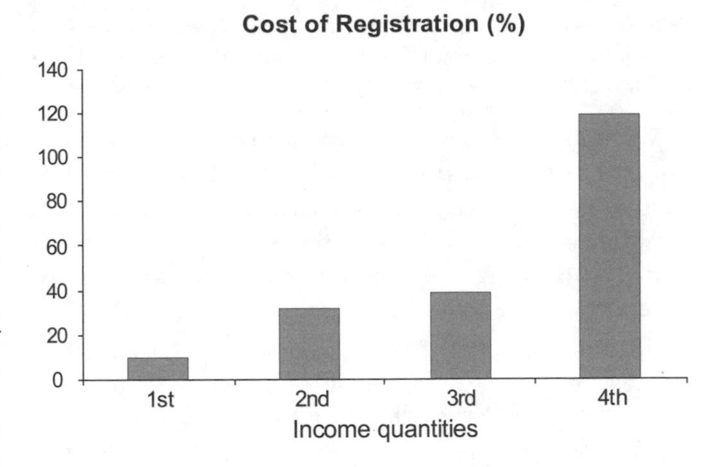

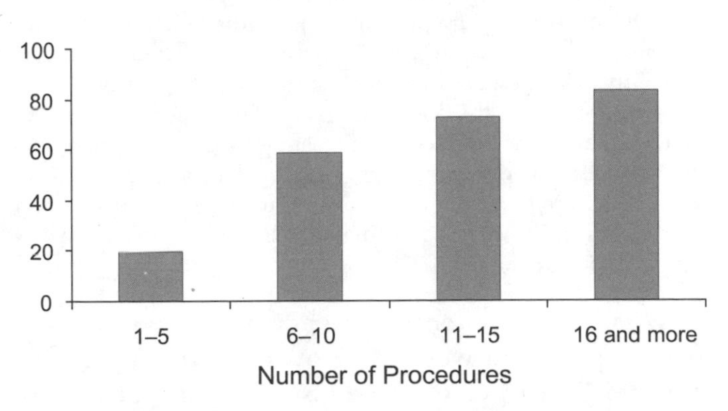

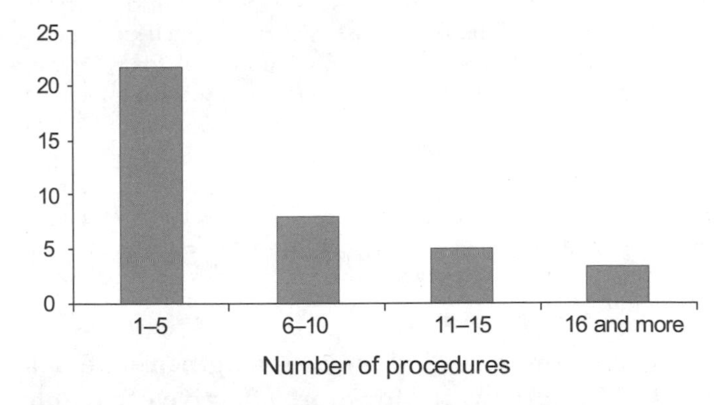

Figure 17.4. Relationship between Corruption, Transaction Cost, and Administrative Bureaucracy (World Bank 2002)

17.5 Key Findings on Consequences and Effectiveness of Responses

We shall not end the chapter with a conclusion but with some answers to the key questions (see Box 17.1) posed by decision-makers and potential users of the MA.

1. How effective are international agreements for addressing concerns related to ecosystems and human well-being, and what options exist to strengthen their effectiveness?

Bilateral and multilateral instruments have been, in general, effective in achieving their stated goals especially when reciprocity is high. It was found that in most treaties, the limited and focused nature of the goals and mechanisms developed were not able to address the broader issue of ecosystem services and human well-being. It has to be acknowledged that these agreements were never intended to address such broad issues. However, where appropriate (for example, water, forestry, biodiversity), considerations of such issues may need to be incorporated explicitly in the agreement if poverty reduction efforts related to ecosystem services are to be effective. Options to strengthen the effectiveness of international environmental agreements in this regard include the inclusion of appropriate goals in the design of the agreement (both in terms of objectives and operational mechanisms), the way the instrument was negotiated, the consistency with other governance initiatives (for example, PRSPs and trade and finance regimes), and the context in which it is to be implemented.

2. What is the scope for correcting market failures related to ecosystem services (internalizing environmental externalities)? How much of a difference would this make for ecosystems and human well-being, and what are the necessary conditions for these approaches to be successful?

There is considerable scope for correcting market failures and internalizing environmental externalities. However, a number of caveats need to be mentioned. First, markets work relatively well in mitigating polluting activities, which degrade ecosystems. Markets were also found to do a relatively good job in the efficient allocation of provisioning services (natural resources), but only when effective and efficient institution structures are in place. However, markets have difficulty in assigning right prices and allocations for regulating, supporting, and cultural services. It is also recognized that market instruments are not designed to address distributive issues especially in the constituents of security, health, social relations, and freedom and choices. It is therefore recommended that efforts be directed at creating markets for provisioning services, while implementing regulatory mechanisms/instruments for the maintenance of regulating, supporting, and cultural services as well as ensuring equitable allocations of all services across stakeholders.

3. What is the potential impact on ecosystem services and human well-being of removing "perverse subsidies" that promote excessive use of specific services? What tradeoffs may exist between ecosystem and human well-being (especially poverty) goals?

There is no doubt that many subsidies (especially in the agricultural, forestry, and fisheries sectors) have negative effects on ecosystem services. There is tremendous potential for removing perverse subsidies, but there will be a need for compensatory mechanisms for the poor who may be adversely affected by the immediate removal of subsidies. Studies show that the removal of some subsidies—especially in the fuel, food, and water sectors—will have distributive impacts and that compensation in the form of monetary and non-monetary measures will be required in the short term in order to prevent some individuals or groups from falling into poverty.

4. What more can be done to strengthen national-local legal frameworks for more effective ecosystem management to reduce poverty and increase human well-being? And are there scale effects which make legal responses at one level more effective in addressing ecosystem services?

In general, the success of legal frameworks is higher when they are designed at the lowest possible administrative level because they can then take the contextual factors into account. At the same time, in order to ensure a level playing field and access to global knowledge and resources, bilateral and multilateral agreements may also need to be made. It is important to note that what may appear to be a national issue (for example, an agreement between a government and a company) may actually be subject to international trade, investment, and environmental laws, and also to private international law. This, in turn, requires a careful design, whereby decentralized legal frameworks are closely coordinated with higher-level legal structures.

5. What are the strengths and weaknesses of an increased focus on technological advances for addressing concerns related to ecosystems, human well-being, and poverty alleviation? What steps can be taken to increase the likely benefits for ecosystems and human well-being of technological change, and decrease the risks and costs?

It is clear that technology can improve the flow of provisioning services. Technology can substitute to a limited extent supporting and regulating services but usually at high costs, and are unsustainable in the long run. It should also be recognized that technology responses can clearly have distributive impacts with losers and winners if improperly implemented. The assessment recommends that if technologies are to be introduced, and then efforts should be made to integrate local knowledge and preferences through a participatory process.

6. How important is governance/institutional reform to the achievement of effective ecosystem management, and what are the characteristics of that reform that are most relevant to the pursuit of goals related to ecosystems and human well-being?

Institutional reform is critical in both industrial and developing countries. The transition to ecosystem services requires new institutions that can accommodate scale issues that transcend traditional administrative or political boundaries, and be flexible to changing ecological conditions. Access to institutions in a relatively efficient, transparent, and costless manner was found to be quite effective in encouraging sustainable management of ecosystem services by empowering the rural poor. Institutional reform to increase instrumental freedoms provides local communities the flexibility to adopt adaptive management strategies in response to ongoing social, ecological, and economic changes.

7. What are the strengths and weaknesses of greater stakeholder involvement in decision-making in the management of ecosystem services? How can decision-makers find the right balance?

Stakeholder participation does not guarantee well-functioning ecosystems. It is a misnomer to assume that local communities, poor and rich, will conserve and manage ecosystems in a sustainable manner. There is a need for education, knowledge sharing, and information dissemination if participatory processes are to work efficiently and effectively. However, participatory processes are time consuming and costly. There is a need to clarify issues that can benefit from a participatory process, and not adopt an approach, which requires all issues and decisions to be made through a participatory process. This is a challenge, as little information is available to make any distinctive recommendations on the lessons learned.

8. What tools and mechanisms can promote effective cross-sectoral (water, agriculture, environment, transportation, etc.) coordination of policy and decision-making?

Integrated responses are gaining in importance in both developing and industrial countries, but they have had mixed results. Integrated responses address degradation of ecosystem services across a number of systems simultaneously and also explicitly include objectives to enhance human well-being. Integrated responses occur at different scales and across scales, and use a range of instruments for implementation. However, experience shows that place-based integrated assessments show more promise in addressing the ecological and human well-being issues at hand than the present sectoral or thematic approaches. Increasingly, integrated responses are associated with the application of multistakeholder processes, decentralization, and the inclusion of actors from government, civil society, and private sector. Examples include some Multilateral Environmental Agreements, environmental policy integration within national governments, and multisectoral approaches such as ICZM and IRBM. Although many integrated responses make ambitious claims about their likely benefits, in practice the results of implementation have been mixed in terms of ecological, social, and economic impacts.

9. What design characteristics of ecosystem-related responses are helpful in ensuring that they provide benefits for poverty reduction?

Responses need to:

- *Recognize complexity.* Responses must serve multiple objectives and/or sectors; they must be integrated within national poverty reduction strategies.
- *Be implemented at the appropriate scale.* The scale of a response must match the scale of the process; often, a multiscale response will be most effective.
- *Acknowledge uncertainty.* In choosing responses, we must understand the limits to our knowledge, and we must expect the unexpected.
- *Be made through an inclusive and participatory process.* It is instrumental that information is available and understandable to a wide range of affected stakeholders and that responses are designed in an open and transparent fashion.
- *Enhance adaptive capacity.* Resilience is increased if we put in place institutional frameworks that allow and promote the capacity to learn from past responses and adapt accordingly.
- *Establish supporting instrumental freedoms.* Responses do not work in a vacuum, and it is therefore critical to identify the necessary supporting instrumental freedoms needed in order for the response to work efficiently and equitably.
- *Establish legal frameworks.* If the agreement is legally binding, it is generally likely to have a much stronger effect than a soft law agreement such as the Declaration and the Action Plan of the World Summit on Sustainable Development.
- *Have mechanisms for implementation:* Where financial resources are not forthcoming, the design of market mechanisms may increase the potential for implementation.
- *Establish implementing and monitoring agencies.* The establishment of subsidiary bodies with authority and resources to undertake specific activities to enhance the implementation of the agreements is vital to ensure continuity and preparation and follow-up to complex issues.
- *Be coordinated with other responses.* The literature indicates that it is further vital that responses designed for one regime do not necessarily lead to problems in other regimes.
- *Integrate traditional and scientific knowledge.* Identify opportunities for incorporating traditional and local knowledge in designing responses.

10. What have been the consequences of macroeconomic responses like poverty reduction strategies for ecosystems and their services?

Most macroeconomic responses, especially structural adjustment programs and poverty reduction strategies, have paid relatively little or no attention to the sustainable management of ecosystem services. However, there is an increasing trend toward better integration of the environment in poverty reduction strategies, although the main emphasis has been on the efficient allocation of provisioning services. In many instances, this has led to deterioration in the regulating and supporting services. Therefore, communities and individuals who depend on these ecosystem services directly for some of their constituents of well-being face the possibility of a downward spiral of poverty when these ecosystem services deteriorate.

11. What has been the net impact of trade liberalization, globalization, and privatization on ecosystems, their services, and human well-being?

Trade liberalization has no doubt increased the rate of use of ecosystem services. The ecological footprint of many industrial countries is much larger than their carrying capacity. This has primarily been made possible through increased trade. Privatization, which goes hand in hand with globalization and liberalization, has improved the efficient use of the provisioning services of ecosystems. However, this has led to an increase in relative poverty. For example, privatization of land in countries with weak institutional frameworks has caused many of the poor to lose the lands they have been living on for generations. The fundamental premise for globalization, liberalization and privatization to be equitable is the presence of a number of key instrumental freedoms supported by strong institutions.

Note

1. This section draws from the work on UNEP/CBO/SGSTA/9/INF/13.

References

Agarwal, A., S. Narain and A. Sharma, 1999: Green politics: Global environmental negotiations, Centre for Science and Environment, New Delhi, India.

Agarwal, Bina, 1997: Gender, environment, and poverty interlinks: Regional variations and temporal shifts in rural India, 1971–1991, *World Development,* **25(1),** pp. 23–52.

Altieri, M., and S. Hetch, 1990: *Agroecology and Small Farm Development,* CRC Press, Boca Ratón, FL.

Amman, H. and A.K. Duraiappah, 2001: *Land tenure and conflict resolution: A game theoretic approach in the Narok district in Kenya.* CREED Working Paper No. 37, International Institute for Environment and Development, London, 30pp.

Amman, H and A.K Duraiappah., 2004: *Environment and Development Economics,* Vol. 9 Núm. 3, Cambridge University Press, Cambridge, U.K, pp. 383–409.

Bahuguna, S., 1992: People's programme for change, *The INTACH Environmental Series,* **19,** Indian National Trust for Art and Cultural Heritage, New Delhi, India.

Barrow, E., and M. Murphree, 2001: Community conservation: From concept to practice. In: *African Wildlife & Livelihoods: The Promise & Performance of Community Conservation,* D. Hulme and M. Murphree (eds.), James Currey, Oxford, UK/ Heinemann, Portsmouth, NH, pp. 24–37.

Blench, R. 2000: You can't go home again, extensive pastoral livestock systems: Issues and options for the future, Overseas Development Institute, London, UK/ FAO, Rome, Italy.

Bodansky, D., 1993: The United Nations Framework Convention on Climate Change: A commentary, *Yale Journal of International Law,* **18,** pp. 451–588.

Bjornlund, H. and J. McKay, 2001: *Operational Mechanism for the efficient working on water markets – Some Australian Experiences.* Paper presented to the joint conference of the International Water & Resource. Economics Consortium and the Seminar on Environmental and Resource Economics, Girona, Spain, June.

Brack, D., 1996: *International Trade and the Montreal Protocol,* Earthscan, London, UK.

Brandon, K., K. Redford, and S. Sanderson (eds.), 1998: *Parks in Peril: People, Politics and Protected Areas,* Island Press, Washington, DC.

Brondizio, E. S., and A. D. Siqueira, 1997: From extractivists to forest farmers: Changing concepts of Caboclo agroforestry in the Amazon estuary, *Research in Economic Anthropology,* **18,** pp. 233–279.

Bruner, A. G., R.E. Gullison, R.E. Rice, and G.A.B. da Fonseca, 2001: Effectiveness of parks in protecting tropical biodiversity, *Science,* **291** (5 January), pp. 125–128.

Brush, S.B., and D. Stabinsky, 1996: *Valuing Local Knowledge: Indigenous People and Intellectual Property Rights,* Island Press, Washington, DC.

Castro, F., and D. McGrath, 2003: Moving toward sustainability in the local management of floodplain lake fisheries in the Brazilian Amazon, *Human Organization,* **62,** pp. 123–133.

Chakravarty, U. and T.M. Swanson, 2002: The economics of water: Environment and development, Introduction to the special issue. In: *Environment and Development Economics,* **7 (4),** pp. 617–624.

Cavendish, W., 1999: *Empirical Regularities in the Poverty-Environment Relationship of African Rural Households,* Centre for the Study of African Economies, University of Oxford, Oxford, UK.

Chopra, K. and A. Duraiappah: Operationalising capabilities in a segmented society: The role of institutions. In: *Operationalising Capabilities,* F. Comim (ed.), Paper presented at the Conference on Justice and Poverty: Examining Sen's Capability Approach, June, Cambridge University, Cambridge, UK. In press.

Coker, A.O, M.K.C Sridhar, and E.A.Martins., 2003: *Management of Septic Sludge in Southwest Nigeria,* 24th WEDC International Conference, "Towards the Millennium Development Goals". Abija, Nigeria.

Corbridge, S., and S. Jewitt, 1997: From forest struggles to forest citizens? Joint forest management in the unquiet woods of India's Jharkhand, *Environment and Planning A,* **29,** pp. 2145–2164.

Daily, G.C., G. Ceballos, J. Pacheco, G. Suzan, and A. Sanchez-Azofeifa, 2003: Countryside biogeography of neotropical mammals: Conservation opportunities in agricultural landscapes of Costa Rica, *Conservation Biology,* **17(6),** pp. 1814–1826.

Dove, M., 1999: The agronomy of memory and the memory of agronomy: Ritual conservation of archaic cultigens in contemporary farming systems. In: *Ethnoecology: Situated knowledge, Located Lives,* V. Nazarea (ed.), University of Arizona Press, Tucson, AZ, pp. 45–70.

Draper, M., M. Spierenburg and H. Wels, 2004: African dreams of cohesion: The mythology of community development in transfrontier conservation areas in southern Africa, *Culture and Organization,* **10(4).** In press.

Dreze, J., and A. Sen, 2002: *India: Development and Participation,* Oxford University Press, Oxford, UK.

Duraiappah, A.K, 1998: Poverty and environmental degradation: A review and analysis of the nexus, *World Development,* **26(12),** pp. 2169–2179

Duraiappah, A.K, 2004: *Exploring the Links: Human Well-being, Poverty and Ecosystem Services,* United Nations Environment Programme, Nairobi, Kenya.

Epiney, A., 2000: Flexible Integration & Environmental Policy in the EU. Legal Aspects. In *Environmental Policy in a European Union of Variable Geometry? The Challenge of the next Enlargement,* Holzinger, K and Knoepfel,P (eds), Basel. Genf, Munchen, Helbing and Lichtenhahn, pp. 39–64.

Ezzati, M., and D. Kammen, 2001: Indoor air pollution from biomass combustion and acute respiratory infections in Kenya: An exposure-response study, *The Lancet,* **358,** pp. 619–624.

FAO (Food and Agriculture Organization of the United Nations), 1996: *Rome Declaration on World Food Security,* World Food Summit, FAO, Rome, Italy.

Ferraro, P.J. 2001: Global Habitat Protection: Limitations of development interventions and a role for conservation performance payments, *Conservation Biology,* **15(4),** pp. 990–1017.

Ferraro, P.J. and A. Kiss, 2003: Will direct payments help biodiversity? Response, 19, *Science,* **299** (28 March), pp. 1981–1982.

Ferraro, P.J. and R.D. Simpson, 2002: The cost-effectiveness of Conservation 21: Performance payments, *Land Economics,* **78(3),** pp. 339–353.

Ferraro, P.J. and A. Kiss, 2002: Direct payments for biodiversity conservation, *Science,* **23(298)** (29 November), pp. 1718–1719.

Fields, G. 2001: *Distribution and Development: A New Look at the Developing World,* Russell Sage Foundation, New York, NY; The MIT Press, Cambridge, MA.

Franck, T.M., 1995: Fairness to "peoples" and their right to self-determination. In: Franck, T.M, *Fairness in International Law and Institutions,* Clarendon Press, Oxford, UK, pp. 140–169.

Gadgil, M., 1999: Promoting adaptive participatory management. In: *Proceedings of the Norway/UN Conference on the Ecosystem Approach for Sustainable Use of Biological Diversity,* P. Schei, O. Sandlund, and R. Strand (eds.), pp. 51–55.

Geores, M., 2002: Scale and forest resources. In: *The Commons in the New Millennium,* N. Dolak and E. Ostrom (eds.), MIT Press, Cambridge, MA, pp. 77–99.

Geschiere, P. and F. Nyamnjoh, 2001: Capitalism and autochthony: The seesaw of mobility and belonging. In: *Millennial Capitalism and the Culture of Neoliberalism,* J. Comaroff and J.L. Comaroff (eds.), Duke University Press, Durham, NC, pp. 159–190.

Gibson, C.C. and T. Koontz, 1998: When community is not enough: Communities and forests in Southern Indiana, *Human Ecology,* **26,** pp. 621–647.

Gore, J.A. (ed.), 1985: *The Restoration of Rivers and Streams,* Butterworth, Boston, MA.

Gupta, J., 1997: *The Climate Change Convention and Developing Countries: From Conflict to Consensus?* Environment and Policy Series, Kluwer Academic Publishers, Dordrecht, The Netherlands, 256 pp.

Gupta, J., 2001: Effectiveness of air pollution treaties: The role of knowledge, power and participation. In: *Knowledge, Power and Participation,* M. Hisschemöller, J. Ravetz, R. Hoppe, and W. Dunn (eds.), Policy Studies Annual, Transaction Publishers, Somerset, NJ, pp. 145–174.

Gupta, J., 2002: The climate convention: Can a divided world unite? In: *Managing the Earth: The Eleventh Linacre Lectures,* J. Briden and E.D. Thomas, Downing (eds.), Oxford University Press, Oxford, UK, pp. 129–156.

Gupta, J., 2003: The role of non-state actors in international environmental affairs, *Heidelberg Journal of International Law,* **63(2),** pp. 459–486.

Haas, P.M., 1989: Do regimes matter? Epistemic communities and Mediterranean pollution control, *International Organization,* **43(3)** (Summer), pp. 377–403.

Hay, F., and R. Probert, 2000: Keeping seeds alive. In: *Seed Technology and Its Biological Basis,* M. Black and J.D. Bewley (eds.), Sheffield Academic Press, Sheffield, UK.

Hazell, P. and L. Haddad, 2001: Agricultural research and poverty reduction, International Food Policy Research Institute, Washington, DC, vi + 41 p.

Hedger, M., E. Martinot, T. Onchan, D. Ahuja, W. Chantanakome, et al., 2000: Enabling environments for technology transfer: Methodological and technological issues in technology transfer, IPCC, Cambridge University Press, Cambridge, UK, pp. 105–141.

Henkin, L., 1979: "Is it Law or Politics?" *How Nations Behave, 2nd ed,* Columbia University Press, New York, pp. 88–98.

Hey, E. and D. Freestone (eds), 1996: *The Precautionary Principle and International Law: The Challenge and Implementation,* Kluwer Law International, The Hague, Netherlands.

Hladik, C.M., A. Hladik, H. Pagezy, O.F. Linares, G.J.A. Koppert, et al., 1996: L'alimentation en forêt tropicale: Interactions bioculturelles et perspectives de développement, Les ressources alimentaires: Production et consommation, Bases culturelles des choix alimentaires et stratégies de développement, Man and the Biosphere (MAB), 13, UN Educational, Scientific and Cultural Organization, The Parthenon Publishing Group, Paris, France.

Hu, D., R. Wang, J. Yan, C. Xu, Y. Wang. A pilot ecological engineering project for municipal solid waste reduction, disinfection, regeneration and industrialization in Guanghan City, China, *Ecological Engineering,* **11(1)** (October 1998), pp. 129–138.

IPCC (Intergovernmental Panel on Climate Change), 1990: The IPCC response strategies: Report of Working Group III. Island Press, Washington, DC, pp. 270.

Jacobson, H.K. and E.B. Weiss, 1995: Strengthening compliance with international environmental accords: Preliminary observations from a collaborative project, *Global Governance,* **1(2)** (May–August), pp. 119–148.

Jodha, NS. 1986. Common property resources and rural poor in dry regions of India. *Econ. Polit. Wkly,* **21(27),** pp. 1169–1181.

Kangwana, K., 2001: Can community conservation strategies meet the conservation agenda? In: *African Wildlife & Livelihoods: The Promise & Performance of Community Conservation,* D. Hulme and M. Murphree (eds.), Heinemann, Portsmouth, NH/ James Currey, Oxford, UK, pp. 256–266.

Kates, R.W. and V. Haarmann, 1991: *Poor People and Threatened Environments: Global Overviews, Country Comparisons, and Local Studies,* Research report RR-91-2, Alan Shawn Feinstein World Hunger Program, Brown University, Providence, RI.

Kaya, I., 2003: *Equitable Utilization: The Law of the Non-navigational Uses of International Watercourses,* Ashgate Publishing, Hampshire, UK.

Kerr, J., 2002: Watershed development, environmental services, and poverty alleviation in India, *World Development,* **30(8),** pp. 1387–1400.

Killick, Tony, 1995: *IMF Programmes in Developing Countries: Design and Impact,* Routledge, London, UK.

Koundouri, P., T.M. Swanson, and A. Xepapadeas, 2003: *Economics of Water Management in Developing Countries: Problems, Principles and Policies, Edward-Elgar 2, Publishers,* Cheltenham, UK, p. 3.

Kyoto Protocol: Available at http://unfccc.int/resource/convkp.html.

Laird, S., 1994: Natural products and the commercialization of traditional knowledge. In: *Intellectual Property Rights for Indigenous People: A Sourcebook,* T. Greaves (ed.), Society for Applied Anthropology, Oklahoma City, Oklahoma, pp. 147–162.

Landell-Mills, N., and I. Porras, 2002: Silver bullet or fools' gold? A global review of markets for forest environmental services and their impact on the poor, International Institute for Environment and Development, Forestry and Land Use Program, London, UK.

Leach, M. and R. Mearns, 1996: Environmental change & policy: Challenging received wisdom in Africa. In: *The Lie of the Land: Challenging Received Wisdom on the African Environment,* M. Leach and R. Mearns (eds.), James Currey, Oxford, UK/ Heinemann, Portsmouth, NH, pp. 1–33.

Lehouerou, H.N., 1993: Land degradation in Mediterranean Europe: Can agroforestry be a part of the solution? *Agroforestry Systems,* **21(1),** pp. 43–61.

Levett, R., 1998: Sustainability indicators: Integrating quality of life and environmental protection, *Journal of the Royal Statistical Society,* **161(3),** pp. 291–302.

Lindenberg, M., 2002: Measuring household livelihood security at the family and community level in the developing world, *World Development,* **30(2),** pp. 301–318.

MA (Millennium Ecosystem Assessment), 2003: *Ecosystems and Human Well-being: A Framework for Assessment,* Island Press, Washington, DC.

MacDonnell, L., (1994): Using water banks to promote more flexible water use, United States Geological Survey final project report, Award no. 1434922253, Reston, VA.

Mansley, M., E. Martinot, D. Ahuja, W. Chantanakome, S. DeCanio, et al., 2000: Financing and partnerships for technology transfer. In: *Methodological and Technological Issues in Technology Transfer,* IPCC, Cambridge University Press, Cambridge, UK, pp. 143–174.

McCabe, J.T., 2003: Sustainability and livelihood diversification among the Maasai of northern Tanzania, *Human Organization,* **62(2),** pp. 100–111.

Miles, E.L., A. Underdal, S. Andresen, J. Wtterstad, J.B. Skjaerseth, et al. (eds.), 2002: *Environmental Regime Effectiveness: Confronting Theory with Evidence,* MIT Press, Cambridge, MA.

Moran, E. (ed.) 1990: *The Ecosystem Approach in Anthropology,* University of Michigan Press, Ann Arbor, MI.

Mortimore, M., 1998: *Roots in the African Dust: Sustaining the Drylands,* Cambridge University Press, Cambridge, UK.

Mountain Institute, 1998: *Sacred Mountains and Environmental Conservation: A Practitioner's Workshop,* unpublished report.

Muller, J. and H.J. Albers, 2003: Enforcement, payments, and development projects near protected areas: How the market setting determines what works where, *Resource and Energy Economics,* **26(2)** (June), pp. 185–204.

Nazarea, V. D., 1998: *Cultural Memory and Biodiversity,* The University of Arizona Press, Tucson, pp. 189.

Netting, R., 1993: *Smallholders, Householders: Farm Families and the Ecology of Intensive, Sustainable Agriculture,* Stanford University Press, Stanford, CA.

Oberthur, S and H.E.Ott, 1999: *The Kyoto Protocol,* Springer, Berlin, pp. 359.

OECD, 2001: *Poverty Reduction: The DAC Guideline,* OECD, Paris, France.

Ola-Adams, B.A., 2001: Education, awareness building and training in support of biosphere reserves: Lessons from Nigeria, *Parks (IUCN),* **11(1),** pp. 18–23.

Pashardes, P., T. Swanson, and A. Xepapadeas (eds.), 2002: *Economics of Water Resources,* Kluwers Academic Publishers, Dorecht, The Netherlands.

Perrings, C., 1996: *Sustainable Development and Poverty Alleviation in Sub-Saharan Africa: The Case of Botswana,* ILO (International Labor Organization) studies series, Macmillan Press, London, UK.

Pinedo-Vasquez, M., D.J. Zarin, K. Coffey, C. Padoch, and F. Rabelo, 2001: Post-boom logging in Amazonia, *Human Ecology,* **29(2),** pp. 219–239.

Posey, D., 1998: Can cultural rights protect traditional cultural knowledge and biodiversity? In: *Cultural Rights and Wrongs,* UNESCO (ed.), United Nations Eductional, Scientific and Cultural Organization, Paris, France.

PRSP Sourcebook (Poverty Reduction Strategy Paper), Available at http://web.worldbank.org/WBSITE/EXTERNAL/TOPICS/EXTPOVERTY/EXTPRS/0contentMDK:20175742~pagePK:210058~piPK:210062~theSitePK:384201,00.html.

Ravallion, M., 2004: Pro-poor growth: A primer, Policy Research working paper series 3242, The World Bank, Washington, DC.

Rawls, J., 1971: *A Theory of Justice,* Harvard University Press, Cambridge, MA.

Ray, D., 1998: *Development Economics,* Princeton University Press, Princeton, NJ.

Rietbergen-McCracken, J. and H. Abaza, 2002: Environmental valuation: A worldwide compendium of case studies, Earthscan Publishers, London, UK.

Robb, C., 1999: *Can the Poor Influence Policy? Participatory Poverty Assessments in the Developing World,* World Bank, Washington, DC.

Rose, C.M., 2002: Common property, regulatory property, and environmental protection: Comparing community-based management and tradable environmental allowances. In: *The Drama of the Commons,* E. Ostrom, T. Dietz, N. Dolak, P. Stern, S. Stonich, et al. (eds.), National Academy Press, Washington, DC.

Rutinwa. 1997: Liability and Compensation for Injurious Consequences of the Transboundary Movement of Hazardous Waste. *RECEIL* **6:1,** pp. 7–13.

Rutten, M., 1992: *Selling Wealth to Buy Poverty,* Verlag Breitenbach Publishers, Saarbrucken, Germany/ Fort Lauderdale, FL.

Sand, P. H., 1992: *The Effectiveness of International Environmental Agreements: A Survey of Existing Legal Instruments,* Grotius Publications Limited, Cambridge.

Schama, S., 1995: *Landscape and Memory,* Harper Collins, London, UK/New York, NY.

Scott, J., 1998: *Seeing Like a State: How Certain Schemes To Improve the Human Condition Have Failed,* Yale University Press, New Haven, CT.

Sen, A., 1981: Poverty and famines: An essay on entitlement and deprivation, Clarendon Press, Oxford, UK.

Sen, A., 1995: Environmental evaluation and social choice: Contingent valuation and the market analogy, *The Japanese Economic Review,* Blackwell Publishers, UK, **46(1),** pp. 23–37.

Sen, A., 1999: *Development as Freedom,* Oxford University Press, New Delhi, India.

Serneels, S., and E.F. Lambin, 2001: Impact of land-use changes on the wildebeest, *Biogeography,* **28,** pp. 391–407.

Settee, P., 2000: The issue of biodiversity, intellectual property rights, and indigenous rights. In: *Expressions in Canadian Native Studies,* R. Laliberte, P. Settee, J. Waldram, R. Innes, B. Macdougall, et al. (eds.), University of Saskatchewan Extension Press, Saskatoon, SK, Canada.

Shiva, V., 1997: *Biopiracy, the Plunder of Nature and Knowledge,* South End Press, Boston, MA.

Simpson, R.D. and R.A. Sedjo, 1996: Paying for the conservation of endangered ecosystems: A comparison of direct and indirect approaches, *Environment Development Economics,* **1,** pp. 241–257.

Sinha, S., G. Shubhra, and B. Greenberg, 1997: The "new traditionalist" discourse of Indian environmentalism, *The Journal of Peasant Studies,* **24,** pp. 65–99.

Smith, M.R., 1998: Technological determinism in American culture. In: *Does Technology Drive History?* M.R. Smith and L. Marx (eds.), MIT Press, Cambridge, MA, pp. 1–35.

Spierenburg, M., 2004: *Strangers, Spirits and Land Reforms: Conflicts about Land in Dande, Northern Zimbabwe,* Brill Publishers, Leiden, The Netherlands.

Sridhar, M.K.C., A.O. Bammeke, M.A. Omishakin, 1985: A Study on the Characteristics of Refuse in Ibadan, Nigeria. *Waste Management and Research,* Denmark **3,** pp. 191–201.

Stiebel, L., L.Gunner, J.Sithole, 2001: The 'Land in Africa: space, culture, history' workshop, Transformation, 2001: No. 44.

Sullivan, C., 2002: Calculating a water poverty index, *World Development,* **30(7),** pp. 1195–210.

Swart, J.A.A., P.J. Ferraro, and A. Kiss, 2003: Will direct payments help biodiversity? *Science,* **299,** pp. 1981–1982.

Sylvain, R., 2002: Land, water and truth: San identity and global indigenism, *American Anthropologist,* **104(4),** pp. 1074–1085.

Tiffen, M., 1993: Productivity and environmental conservation under rapid population growth: A case study of Machakos district, *Journal of International Development,* **5(2),** pp. 207–223.

UNEP (United Nations Environment Programme), 2003: Nairobi, Kenya. Available at http://www.washingtonpost.com/wp-dyn/articles/A37099-2005Jan26.html.

UNEP, 2000: *Economic instruments for environmental management : A worldwide compendium of case studies.* Earthscan Publications, London.

UNFPA, 2001: *Population, Environment and Poverty Linkages: Operational Challenges.* UNFPA, New York, pp. 79.

Vedeld, P., A. Angelsen, E. Sjaastad, and G. Kobugabe Berg, 2004: Counting on the environment: Forest incomes and the rural poor, Environment Economics Series Paper 98, World Bank, Washington, DC, pp. 95.

Warburton, D., 1998: *Community & Sustainable Development,* Earthscan, London, UK.

Wells, M., and K. Brandon, with L. Hannah, 1992: *People and Parks: Linking Protected Area Management with Local Communities,* World Bank, World Wildlife Fund, US Agency for International Development, Washington, DC.

Werksmann,J., 1998: 'The Clean Development Mechanism: Unwrapping the "Kyoto Surprise"', *RECIEL* **7:2,** 1998, pp. 147–158.

West, G., 2001: Biosphere reserves for developing quality economies: The Fitzgerald Biosphere Reserve, Australia, *Parks (IUCN),* **11(1),** pp. 10–15.

Wilken, G.C., 1987: *Good Farmers: Traditional Agricultural Resource Management in Mexico and Central America,* University of California Press, Berkeley, CA.

Winner, L., 1986: *The Whale and the Reactor: A Search for the Limits in an Age of High Technology,* University of Chicago Press, Chicago, IL.

Wolmer, W. and C. Ashley, 2003: Part III: Resources and policies 3: Wild resources management in southern Africa: Participation, partnerships, ecoregions and redistribution, *IDS Bulletin,* **34(3),** pp. 31–40.

World Bank, 2003: *World Development Report 2002/2003: Sustainable Development in a Changing World,* World Bank, Washington, DC.

World Bank, 2004: A sourcebook for poverty reduction strategies, World Bank, Washington, DC, **1(11).** Available at http://web.worldbank.org/WBSITE/EXTERNAL/TOPICS/EXTPOVERTY/EXTPRS/0contentMDK:20177457.

World Bank et al., 2002: *Linking Poverty Reduction and Environmental Management,* World Bank, Washington, DC.

WCD (World Commission on Dams), 2000: *A New Framework for Decision-making: The Report of the World Commission on Dams,* Earthscan Publications, London, UK, xxxvii + 404 pp.

WIPO (World Intellectual Property Organization): Geneva, Switzerland. Available at http://www.wipo.org.

Wunder, S., 2001: Poverty alleviation and tropical forests: What scope for synergies? *World Development,* **29(11),** pp. 1817–1833

Wynne, B., 1992: Uncertainty and environmental learning: Reconceiving science and policy in the preventive paradigm, *Global Environmental Change,* **2(2),** pp. 111–127.

Yamin, F., 1998: The Kyoto Protocol: Origins, assessment and future challenges, *Review of European Community and International Environmental Law,* **7(2),** pp. 113–127.

Young, O., 1991: Political leadership and regime formation: On the development of institutions in international society, *International Organisation,* **45(3).**

Chapter 18

Choosing Responses

Coordinating Lead Authors: Bedrich Moldan, Steve Percy, Janet Riley
Lead Authors: Tomás Hák, Jorge Rivera, Ferenc L. Toth
Review Editors: Neil Leary, Victor Ramos

TABLES

Main Messages

The decision-making process itself, and the actors involved in it, influence the intervention chosen. Elements of decision-making processes related to ecosystems and their services that improve decisions reached and their outcomes for ecosystems and human well-being include: using the best available information; ensuring transparency and participation of important stakeholders; recognizing that certain important values cannot be quantified, but must be considered; striving for both efficiency and effectiveness in the decision-making process; considering stakeholder equity and vulnerabilities; ensuring accountability; providing for monitoring and evaluation; and considering cross-scale effects.

There is a cascade of uncertainties associated with legal, market, institutional, behavioral, and other responses. **Integration across response strategies can mitigate and reduce elements of uncertainty, but it is unlikely that uncertainty can be eliminated in any important context.** The choice of appropriate decision-making processes can help to address uncertainties inherent in ecosystem management and ensure more equitable and sustainable outcomes. A wide range of deliberative tools can now assist decision-making concerning ecosystems and their services. These include tools that facilitate transparency and stakeholder dialogue; information gathering tools, which are primarily focused on collecting data and opinions; and planning tools, which are typically employed for the evaluation of potential policy options.

The use of decision-making methods that adopt a pluralistic perspective is particularly pertinent, since these techniques do not privilege any particular viewpoint. These tools can be employed at a variety of scales. However, the context of decision-making about ecosystems is changing rapidly while old challenges must still be addressed. For national governments, the greatest benefits are likely to be gained from several types of actions. Economic incentives need to be aligned with the goal of sound ecosystem management. **In particular, two kinds of actions are needed: eliminating subsidies that promote excessive use of specific ecosystem services and correcting market failures.** Subsidies to agriculture, forestry, and fisheries in many countries lead to overproduction and promote overuse of inputs that may harm other services. Because many ecosystem services are not traded in markets, markets fail to provide appropriate signals that might otherwise contribute to the efficient allocation and sustainable use of the services.

The transparency and accountability of government and private sector performance need to be increased in ecosystem management through greater involvement of concerned stakeholders. **Institutions need to be developed that enable effective coordination of decision-making at multiple scales and across multiple sectors.** Problems of ecosystem management have been exacerbated both by overly centralized and overly decentralized decision-making. Many ecosystem services tend to be managed in a highly sectoral arrangement that does not provide for appropriate analysis of the cross-sectoral trade-offs inherent in decisions.

Increased emphasis is needed on both demand-side management and adaptive management. As the per capita supply of services drops and the costs associated with production increase, emphasis should shift from actions designed to increase production of the services to actions designed to reduce demand. Management interventions should always include a significant monitoring component, which allows for greater learning about the consequences of the intervention and improved management with time.

Businesses can take action which will both improve their "triple bottom line" (economic, social, and environmental gains) and reduce degradation of ecosystems. They can encourage, through business decisions and support for legislation, resource management policies that reflect the social value of the use of natural resources. They can promote technologies, which reduce demand for ecosystem services and reduce pressures on ecosystems. Human demand for ecosystem services will continue to grow. Significant increases in the efficiency of the use of ecosystem services will be needed to cope with that demand without undermining ecosystems.

To gain competitive advantage, business could take decisions that anticipate the eventual strengthening of regulations (or establishment of market mechanisms) to significantly reduce carbon emissions, reduce nitrogen and phosphorous loading, and increase water and energy use efficiency.

Businesses can provide objective information on their operations and encourage access to this information. Trust and transparency can help create a value-adding reputation. **Reporting environmental performance and meeting certain standards such as those found in eco-labeling and/or certification schemes are responses that leading corporations are pursuing with reputation and brand image in mind.** They can also pursue partnerships with civil organizations for the mutual advantage of all parties. Partnerships help accelerate learning and leverage resources.

Civil society organizations can take actions that further human well-being and the conservation and restoration of ecosystems. They can raise awareness among the public and decision-makers of "emerging issues" such as nutrient loading and invasive species or they can encourage greater access to information on the status and trends in ecosystem services and provide greater quantification of the nonmarketed benefits obtained from ecosystem services. They can facilitate the involvement of stakeholders at the highest risk and greatest vulnerability to the effects of ecosystem change. They can also help build coalitions and partnerships. The consensus-building coalitions of nongovernmental organizations and other like-minded stakeholders greatly increase the leverage of individual members. Partnership with businesses can encourage the best practices necessary to achieve environmentally benign products, support environmental innovation, and look into new "sustainable" business opportunities.

18.1 Introduction

The objective of this chapter is to identify the ingredients of good decision-making with regard to choosing responses regarding ecosystem services and human well-being. It draws from the responses identified and discussed in the chapters on ecosystem services and human well-being in Part II of this volume, and focuses on identifying the factors that enhance the quality of the processes for choosing responses and whose absence diminishes the quality of those processes. It also builds upon the chapters in Part I dealing with typology, assessment of responses, and uncertainties. In addition, this chapter builds on the MA conceptual framework (MA 2003). Chapter 8 of this report identifies the following issues:

- the desirable properties of decision-making processes such as considerations of equity, attention to vulnerability, transparency, accountability, and participation;
- primary influences in choosing among responses such as temporal and physical scale, cultural context, uncertainty, and considerations of equity;
- the key steps in the policy-making cycle, including problem identification and analysis, policy option (that is, response) identification, policy choice, policy implementation, and monitoring and evaluation in an iterative fashion;

- the range of analytical tools useful to the choice of responses and the contexts that will help determine the appropriate tool;
- the balance that must be struck between the need for policy adaptability and flexibility to hedge uncertainty and risk with the need for a predictable and stable policy regime against which to plan and invest; and
- the need for indicators to link policy and action with impacts on ecosystem services and human well-being and the role of traditional and practitioner knowledge.

The conceptual framework report (MA 2003) makes clear that some old challenges must still be addressed. Perhaps the most important of these challenges is the complex trade-off faced when making decisions about how to manage ecosystems with the goal of enhancing the flow of services while allocating benefits, costs, and risks equitably. Increasing the flow of one service from a system, such as provision of timber, may decrease the flow from others, such as biodiversity and the provision of habitat.

While some benefits of ecosystem services are hard to capture locally, some others may be easily captured by those who have access to the system. For example, it may be relatively easy for local people to capture the direct-use value of timber in a forest via market prices—they are capturing the value of provisioning services. At the same time, people around the world may benefit from the many aspects of biodiversity provided by the forest—an indirect-use value of a supporting service. Under many institutional arrangements, the people near the forest have no way to capture this other value. Because the direct-use value—revenues from logging—can easily be converted into income for local people, for local and national governments, and for local, regional, and multinational firms, there is a strong incentive to log the forest. In contrast, the value inherent in protecting biodiversity is much harder to translate into income for anyone. As a result, there will be a tendency for decisions to favor the direct use even though a full analysis of the total value of ecosystem services might favor preserving or enhancing the indirect use retained by not logging (MA 2003).

The characteristics of the ecosystem, the technologies available for using it and monitoring such use, and the institutional arrangements that distribute values across groups have consequences for decisions that are made (Ostrom et al. 1999). A great deal is understood about these problems, and the state of the science often provides guidance on the design of institutions to promote capturing the full value of an ecosystem (Costanza and Folke 1996; Stern et al. 2002; MA 2003).

The analysis of the process of choosing responses may be started by answering the basic question: Why is an intervention needed? Four possibilities arise:

Sustaining the existing ecosystem service. If the current level of a given ecosystem service is satisfactory then there may be a natural interest in safeguarding the service for any foreseeable future by managing the regular renewal of the service in concert with natural processes The sustainability of the ecosystem service may be threatened by overexploitation of the service or the degradation of the ecosystem because it provides some other service. An important consideration may be thresholds beyond which provision of the service may be severely diminished.

Enhancing existing services or developing new ones. This may arise in response to growing demand caused by increasing population, increasing wealth of the population, changing tastes of people, and/or the increasing need for human well-being. An effort to enhance the existing volume of a service may also be caused by purely economic reasons in order to exploit the comparative advantage established by a marketable product based on the given service. The same applies for the development of a new service, classical examples being the shrimp plantations in Southeast Asia or tourist development on many tropical islands. Another reason for an attempt to enhance ecosystem services may be social or political, for example, to help poor or vulnerable groups or, on the other hand, to benefit certain preferred groups of stakeholders, for example, the local elite.

Restoring degraded or damaged services. If an important or even vital ecosystem service is downgraded, an attempt may be made to restore it to its original state even if exact reconstruction is not possible. For cultural reasons (tradition, natural pride, regional trademark, etc.) the efforts may be much more costly than the value of the service recovered.

Adapting to the situation when a given ecosystem service is damaged beyond any sensible repair. Sometimes restoration is not physically possible for technical or financial reasons. The response may, therefore, be oriented toward some other measures, such as seeking substitutes, securing imports, or reducing (even to zero) demand for the given service.

When deciding what should be done, the choice of "doing nothing" should always be considered.

This chapter focuses mostly on explicit responses and interventions made by decision-makers in all sectors and at all levels. The typology of the responses, by the nature of the intervention, by the impact on drivers, and by the actors and their scale of operation, is the subject of Chapter 2, which also discusses that responses can be implicit, such as those made by consumers of products that depend on or impact upon ecosystems and their services.

When making decisions, the responses are evaluated on the basis of their costs, on the one hand, and their effects, on the other. The costs could be initial, short-term, or long-term (for example, the regular management or operational costs). They may be not only financial but also (sometimes more important) societal, political, cultural, or others. The response may bring not only benefits but also side effects with negative consequences. As we are dealing with ecosystem services where a long-time scale is the rule rather than an exception, the conflicts between short-term (political) and long-term (mostly ecological) perspectives occur quite often and may be very difficult to resolve. The effects of responses are discussed in depth in Chapter 3.

In addition, it should not be forgotten that various decision-makers (in a very broad sense) are making decisions on responses having very different needs and objectives. Many people seek benefits for themselves or the stakeholder group with which they identify (for example, a firm, an NGO, or a small community). On the other hand, some personalities do exist whose goal is to achieve benefits (or greater glory) for some abstract entity called "Society," "the Only True Belief," or "Mother Nature."

18.2 Decision-making Processes

There is a significant literature on the nature of the rational approach to decision-making in the environmental policy realm. Referred to as the "decision-making" or "policy-making" cycle, it consists of at least four stages: agenda setting, policy formulation, policy implementation, and policy evaluation (Barkenbus 1998; Dale and English 1998).

Feedback loops occur at each stage in the cycle. Further refinement of this concept suggests that the agenda-setting stage can be divided into "problem identification" and "public awareness/problem acknowledgement" (UNEP/DPCSD 1995; Moldan and Billharz 1997). In addition, the policy formulation stage suggests further sub-division into identification of alternatives, gathering and analyzing of alternatives, and application of decision-making tools (Barkenbus 1998).

A synthesis of this research suggests that rational decision-making processes are comprised broadly of the following elements:

- defining the problem, including gaining public awareness and recognition of the problem;
- determining the range of options appropriate to the problem, taking into account the scale, actors, and primary and proximate drivers in play;
- assessing the efficacy of the options based on political feasibility, capacity for governance, economic and social impacts, and other barriers and limitations to their use;
- choosing the appropriate response option through the aid of decision-making tools whether they are normative, descriptive, deliberative, or ethical/cultural (MA 2003);
- implementation of the option chosen;
- monitoring and evaluating results; and
- adjusting the problem definition, and range, assessment, choice, and implementation of responses.

18.2.1 Problem Definition: Agenda-setting and Policy-formulation

There is, again, much research on the agenda-setting process for environmental policy formulation as well as the extent to which science and expertise play a role (Downs 1998; Kingdom 1984; Dearing and Rogers 1996; Barkenbus 1998). In idealized decision-making, evidence and scientific fact with respect to impact and risk play key roles in the setting of agendas and defining of problems. However, public perception has played an important role, as has scientific evidence. Events, as portrayed by the "media" (for example, Exxon Valdez, Love Canal), have been key contributors to public perception. Politicians, by virtue of their positions and responsibilities, have reacted to public perception/opinion as well as other pressures brought to bear by powerful constituencies such as NGOs and business communities. In this context, science and expertise have often been used to add legitimacy to agendas that are already established.

A critical challenge for environmental decision-making is the integration of environmental considerations into virtually every major business, resource, or economic development decision. Because the wide range of decisions in every sector of the economy affects ecosystems, ecosystem management and environmental protection cannot be concerns of environmental policy-makers alone. Ecosystems must be the responsibility of private business as much as public agencies, and of financial investors as much as fisheries or forest managers. The "integration principle" has been known since the United Nations Conference on Environment and Development (UN 1992). However, only transparent and open decision-making regarding economic issues gives people with environmental concerns the chance to raise them (WRI 2003).

Nevertheless, the introduction of science (that is, comparative risk studies) to improve the process of setting agendas and defining problems is increasing in a number of jurisdictions (Barkenbus 1998). The success of these efforts remains unclear (Davies and Mazurek 1998). When it comes to helping the public form an opinion based on scientific knowledge, the role of clear, simple, and unambiguous information seems obvious. In addition, as certain environmental problems are not easily observed (for example, climate change), information is required to raise public consciousness. In each case, indicators that are understandable, valid and verifiable, relevant, and technically feasible/efficient can play a key role in rational decision-making (Moldan and Billharz 1997).

18.2.2 Implementation

Participation and accountability are two key concepts underpinning the principles and practice of environmental governance.

Meaningful participation brings influence. Those who participate in decision-making processes that affect ecosystems stand the best chance of having their interests represented. Public participation brings legitimacy, credibility, and effectiveness to the decision-making process. Public involvement in some form is required for any broad-based consensus behind the final decision, especially for large or controversial projects. Failure to provide for public input can bring just the opposite result: conflict and resistance. A common challenge to ensuring participation in environmental decision-making is that not all affected stakeholders are equally well positioned to express their views. For some people, there are still barriers of distance, time, language, literacy, and connectivity that might prevent full participation. To participate meaningfully, people need access to information: about the environment, about the decisions made and their environmental implications, and about the decision-making process itself. A recent survey shows that citizens, by and large, feel that governments do not provide them with as much environmental information, or opportunity to participate in environmental decision-making, as they would like (Petkova et al. 2002).

Good governance requires making decisions at the appropriate level. Generally, the appropriate level for decision-making is determined by the scale of the natural system to be managed. However, decision-making still tends to be centralized and isolated from the people and places affected. In many instances, drawing on local knowledge can result in more informed decisions that would serve local people and ecosystems better. This "subsidiarity principle" is often necessary; in many other cases it may be best to let a higher-level authority specify the outcome of the decision-making (maximum quota), while a lower-level authority specifies the procedure (how licenses are awarded).

The ability to seek redress or challenge a decision, if stakeholders consider it flawed or unfair, is of the same importance as access to information and appropriate level for decision-making. In practice, it requires public access to judicial or administrative remedies, existence of an independent arbiter, etc. (WRI 2003)

As already discussed, the effective implementation of responses is dependent on decision-making processes that have a wide input and consider all those that might be affected. In addition, the support of those affected, including the public, is required for successful implementation of responses (Moldan and Billharz 1997). Effective information flows, indicators, and reports are the key to maintaining the requisite support.

Accountability refers to the way in which the public and the private sectors are held responsible for their decisions and actions. Accountability involves the provision to sanction the responsible party. Also, effective implementation of responses and policies requires bringing scientific information and expertise to bear at the administrative level so as to add specificity and definition to regulations and standards that flow from more broadly stated policy mandates (Barkenbus 1998).

18.2.3 Monitoring and Evaluation

While monitoring and evaluation are often not given adequate attention, they are an integral part of an effective decision-making process, and a systematic approach is essential.

There are two types of evaluation—the ex ante and the ex post evaluation (EEA 2001). The ex ante evaluation of intended policies is part of the formulation/choice stage of the process with the use of scenarios as a possible tool. With regard to ex post evaluation, the legitimacy of the institution performing the evaluation is the key to facilitating any adjustments to the decision or policy that might be indicated.

Evaluation should start with the establishment of the "base line," that is, the precise description of the situation before the policy was implemented. This starting place needs to take into account any change in status due to the announcement (or unintended statement) of a possible change in policy, as this can have a powerful effect on behavior.

The achievement (or non-achievement) of targets is the next key consideration. As monitoring and evaluation are key elements of the decision-making process, earlier stages should include the establishment of transparent and well-communicated targets, which are specified as objectively as possible.

Finally, the evaluation should examine the impact of unexpected factors that may have helped or hindered the achievement of objectives and targets or had other ancillary impacts.

The most important tools for monitoring and evaluating the effects of decisions and policies are "performance indicators." It is critical that these indicators be specified at the time a response is chosen and targets are set and allow for an assessment of the "distance" between the actual state and the desired one (that is, the target). Use of this practice by nations and international bodies is increasing (Adriaanse 1993; EEA 1999). Consistent underperformance suggests a need to adjust the decision or policy.

The adjustment stage of the decision-making or policy-making process is really the link to a new decision-making process with the problem definition/agenda setting stages triggered as a result of the monitoring and evaluation stage of the previous one. This fact argues for the need for effectiveness, objectivity, and legitimacy when monitoring and evaluating decisions.

18.2.4 Some Practical Considerations

For most people, it is not obvious who is "in charge" of the environment, and how decisions are made about developing, using, or managing ecosystems. Decisions that shape environmental and natural resource policy are not made by a small group of enlightened government officials (WRI 2003). Many actors at different levels, in and outside the government make and/or influence the array of choices that form constantly evolving environmental and natural resource policies. Officials in different departments of the government, business representatives, environmentalists, politicians, scientists, and local communities are traditionally involved in the environmental policy process. Often they bring to the table conflicting interests, ideologies, knowledge, and levels of influence. Hence, the selection of response options to manage ecosystems is an inherently political process in which actors intensively compete to advance their economic, ideological, social, and cultural goals (Ascher 1999; WRI 2003; Rivera 2002).

The political nature of the selection of response options to manage ecosystems is also enhanced by the intrinsic uncertainty and complexity of the environmental and natural resource policy-making process. Contrary to the normative assumptions of economic and bureaucratic rationality, policy issues do not follow a linear pattern that is divided into specific well-defined steps (Lasswell 1947; Simon 1976 and 1985; Lindblom and Woodhouse 1993). Decision-makers, confronted with this highly uncertain and complex political reality, display behaviors that significantly diverge from the rational choice ideal (Simon 1985; North 1990; Rivera 2004). Beliefs, motivations, personalities, and ideologies (a person's cognitive base) that emerge from family upbringing, life experiences, education, religion, and economic interests are used by policy-makers to simplify reality (Cyert and March 1963; Hambrick and Mason 1984). This simplification process helps decision-makers to avoid lengthy assessments of all existing information and alternatives. Yet it can also generate biases and blind spots that significantly affect their preferences and decisions (Starbuck and Milliken 1988; Walsh 1988). Thus environmental decision-making is characterized by the following:

- environmental problems that are very difficult to identify and define as they are not easily separated from other issues and are seldom confronted in isolation;
- values and objectives of natural resource and environmental policy in conflict with other valid objectives such as funding economic development projects or protecting local producers from the pressures of global trade (Ascher 1999);
- a limited number of policy options;
- policy options and consequences not clearly defined or evaluated;
- policy selection favoring the influential, powerful, and well-connected (Lindblom and Woodhouse 1993; Lasswell 1947; Ascher 1999);
- little attention to implementation during the selection of policy alternatives (although intense political struggle during implementation can significantly change the impact of sound policy); and
- limited evaluation of environmental decisions and political influence with respect to that which is undertaken.

These characteristics of actual environmental decision and natural resource policy-making are exacerbated by the unique contextual conditions prevailing in developing countries. In some countries, democratic processes may not be fully in place. Environmental protection must compete with the need for economic development. Environmental groups are fewer in number, less powerful, and have limited channels to participate in the environmental policy process. Consequently, environmental policy decisions can be designed to pursue other goals and priorities as well.

Similarly, in developing countries, environmental agencies frequently lack clear mandates and their capacity to enforce and monitor natural resource and environmental regulations is very weak (Ascher 1999; Rivera 2002). Some communities and local groups view violence as a legitimate resource to oppose implementation of government regulations. Corruption is a more widespread problem than in industrial countries. On the technical side, developing countries, in addition, may have fewer highly trained people to deal with environmental problems. The quality and availability of scientific knowledge and information can also be poor. This can lead to the symbolic adoption of "canned" policy instruments designed by foreign experts that fail to consider the political, economic, administrative, and technical limitations intrinsic to environmental and natural resource agencies operating in the developing world.

18.3 Key Ingredients to Good Decision-making

As experience in decision-making for ecosystem management and in the analytical work to support the related decision-making processes accumulated over the past decades, increasing attention was devoted to questions concerning the key criteria for success (The Social Learning Group 2001; Clark et al. 2001). The bulk of ecosystems-related decision-making is deeply permeated with complexity, uncertainty, and the incompleteness of science. Accordingly, any assessment process intended to serve decision-making needs to take these facts of life fully into account. This section draws on recent critical appraisals as organized into a synoptic framework by Toth (2004), partly inspired by Chapter 8 of the conceptual framework report (MA 2003) and also drawing on Dietz (2002). We also draw on results of recent research on decision analysis, decision-making, and environmental governance. Conceptual work, analytical efforts, and case studies presented in earlier chapters of this report are also important sources. The objective is to specify a set of ingredients that have characterized successful decision-making in the past and are likely to lead to environmentally effective, socially fair, economically efficient, and politically feasible decisions in the future.

Table 18.1 lists the fundamental criteria and their implications for the two large domains: the decision-making process per se and the decision analysis/support activities. These criteria and the implied guidelines may appear to be far too general at the first glance. Without doubt, the relative importance, feasibility, and practicality of the individual points differ from case to case. Yet the guidelines draw on a large body of critical appraisal of environmental management (NRC 1996; Ostrom et al. 2002; Dietz et al. 2003) so that they have general validity in the human management of environmental systems. In particular, these principles and criteria are valid for decision-making processes (and decision analyses conducted to support them) for all public policy-makers and private stakeholders.

The relative importance of many criteria differs depending on which social actor or group has the primary right or mandate to make the decision. Public policy-makers (local and national governments) are mandated to pursue the interests of the community as a whole and to give special attention to vulnerable or poor social groups, but they are also required to use public funds efficiently. In the mirror case: private stakeholders tend to pursue their own interests and focus on economic efficiency but many of them pay increasing attention to the social and environmental implications of their decisions in the spirit of emerging corporate responsibility and because of the increasing importance of their company's public image. Neither public nor private decision-makers who are concerned with ecosystems services can ignore the social context in which they want to implement their decisions. As subsequent sections in this chapter show, the actual fulfillment of these broad criteria varies immensely not only across societies and development levels but also across the types of decision-making entities.

A number of ingredients that are key to good decision-making with respect to the protection and enhancement of ecosystem services and human well-being are broadly and strongly supported by the chapters in Part II. The following discussion focuses on various ingredients of successful decision-making and analysis. It

Table 18.1. Ingredients for Good Ecosystem Management: Analysis and Decision-making (based on Toth 2004)

Criteria	Implications for Decision-making	Implications for Decision Analysis
Use the best available information about social context	design decision-making process consistent with prevailing social, economic, political, technological, and institutional situation	choose decision-making framework according to prevailing social, economic, political, technological, and institutional situation
Use the best available ecosystem/biophysical information	devise decision-making process so as to allow using the best available information	choose the decision-making tool to allow the incorporation of best available information
Consider efficiency concerns and implications	devise an efficient decision-making process to save time and costs (procedural efficiency); respect the prevailing economic principles (outcome efficiency)	select the analytical tools and the decision criteria according to the relative importance of efficiency concerns in the decision-making context
Strive for effectiveness	devise an effective decision process with clear and flexible procedures that foster finding compromises	present complete results in understandable form
Consider equity concerns and implications	devise a fair decision-making process to allow stakeholder participation (procedural equity) and understanding of the outcome (transparency); respect the prevailing equity principles (consequential equity)	select the analytical tools and the decision criteria according to the relative importance of fairness concerns in the decision-making context; consider participatory assessment techniques
Use the best available information about values	recognize values, beliefs, aspirations of affected stakeholders in the decision-making process	choose the decision analysis tools and decision criteria according to the existing values, beliefs, and aspirations of stakeholders
Pursue accountability	establish clear responsibility assignments during and after the decision process	set up quality control and good practice regimes for assessments
Consider vulnerability concerns and implications	beware of the interests of vulnerable groups/communities.	assess the implications of different options for vulnerable groups/communities
Consider uncertainties	conduct a flexible decision-making process to accommodate new information about the ecosystem and possible changes in values or positions of stakeholders	choose the analytical framework so that it allows an adequate representation of uncertainties; define decision options that allow policy corrections as new information becomes available
Consider cross-scale effects	expand the decision-making process to initiate/comply with relevant policies at lower/higher levels	choose the analytical tools to incorporate constraints from higher decision-making levels and to explore decision needs at lower decision-making levels

is important to note the critical importance of judgment and scientific ignorance in these processes. NRC (1996) presents the concept of an analytic-deliberative process and argues that in order to understand policy choices involving risks to environmental quality and human health, it is necessary to employ a process in which scientists, decision-makers, and the interested and affected parties to the decision deliberate about the nature of the questions that require analysis, the forms of analysis that would be relevant and useful for the decision, the assumptions that should be incorporated in the analysis when the correct assumptions are unknown or disputed, the appropriate interpretation of the results of the analysis, etc. In other words, a process of public participation is required for decision analysis (not just decision-making) to ensure that decisions are well informed. These issues are further addressed in the context of ecosystems in NRC (1999b) and Dietz and Stern (1998). The main conclusions of NRC (1999a) are also integrated into other studies on environmental decision analysis (NRC 1999b; Stern and Easterling 1999). Other countries (CSA 1997; RCEP 1998) and international organizations (OECD 2002) also provide interesting sources.

The proposition to use the best available information in the analytical work to support decision-making and in the decision process itself sounds rather obvious. Yet decisions concerning ecosystems services often suffer from information deficits ranging from insufficient effort to obtain relevant information to inadvertently ignoring or purposefully withholding information. Four main information domains are important to draw on for successful decisions: *biophysical information* about the ecosystem status and processes; *impact assessment information* about economic, social, and political consequences of both the ecosystems changes and of different policy options; *socioeconomic information* about the sociopolitical context in which and for which the decision will be made; and, as an important subset of the latter, *information about the values, norms, and interests of key stakeholders* shaping decisions and affected by them.

For most ecosystems and environmental risks, there is a large body of information available in natural sciences that should be identified and used. Similarly, social science can offer not only information about which policies would be acceptable and feasible, but also information about how ecological changes (whether or not policy-driven) affect such important human outcomes as economic growth, distribution of jobs, availability and price of food, organizational viability, cultural change, and the potential for social conflict. At the same time, however, it is important to recognize how much natural and social sciences do not know about ecological processes and their effects on the ecosystems goods and services that humans value. Therefore, when we argue below for using the best knowledge, it inherently implies making the best use of ignorance as well, that is, the knowledge of what is not known (Ravetz 1986). This underlines the importance of analytical methods (for example, decision analysis or value-of-information calculations) that can inform decision-makers about the implications of the different types of looming uncertainties, of the resolution of uncertainties in the future as knowledge improves, and of the potential course corrections that might be required in the light of new knowledge.

The metastrategy presented in NRC (1996) involves a process which entails the best decision-relevant information from the various perspectives of those involved or affected and which considers this information from a variety of relevant perspectives. This NRC report emphasizes the need to get the science right, but also the need to get the right science. The former requires that the "underlying analysis meets high scientific standards in terms of measurement, analytic methods, data bases used, plausibility of assumptions, and respectfulness of both the magnitude and the character of uncertainty" (pp. 7–8), whereas the latter implies that the analysis needs to address "the significant risk-related concerns of public officials and the spectrum of interested and affected parties, such as risks to health, economic well-being, and ecological and social values" (p. 8). For complex ecosystems-management problems that are plagued with profound uncertainties, interested and affected parties should, early in the process, be involved in defining the questions to be subjected to analysis.

The relative importance of the criteria in Table 18.1 differs depending on the temporal and spatial scale of the ecosystem or resource management problem, on the number and relative power of the stakeholders involved, on the institutional capacity to implement and enforce the emerging decisions, and many other factors. Yet, at least a modest amount of all these ingredients can be recognized in the assessment and decision-making processes that led to successful decisions. Similarly, it is easy to identify *a posteriori* which ingredients had been missing from analytical and decision-making processes that failed or were outright disasters.

In summary: this section argues on the basis of recent literature on environmental decision-making that the process of choosing a strategic intervention or a broader policy in response to potential or emerging environmental problems needs to be informed by the best available information that is responsive to the concerns of those who may be interested in or affected by the ultimate decision. Accordingly, the analytical work to support choosing responses should also incorporate the perspectives, values, and interests of those affected by the final outcome.

These ingredients are expanded and elaborated upon below in the context of the responses in the chapters of Part II. The order of discussion (as above) is a "rough signal" of the importance placed on these ingredients by the service chapter responses.

18.3.1 Using the Best Available Information on the Sociopolitical Context

The decision-making process must be realistic in the sense that it observes and accommodates prevailing social customs and practices, economic realism (power, interests), political situations (authority, control), technological conditions (availability, feasibility), and institutional status (implementation, enforceability). The same features influence the choice of analytical framework because its underlying principles must be congruent with the social situation. Moreover, these features also determine the range of options that can be meaningfully assessed to help decision-making because only strategies and measures viable in the given social and political context will be considered.

This is especially true when economic incentive and substitute economic opportunities are being considered. Regional plans, environmental impact assessments, and education and communications programs could consider which social context would be most effective. Obviously, decisions about responses that pertain to sustainable production practices must be taken with a range of stakeholders in mind beyond just producers and users.

The *social context* is defined as a large group of people who live together in an organized way, making decisions about how to do things and sharing the work that needs to be done. The *political context* is defined as the relationships within a group or organization, which allow particular people to have power over others.

So the *sociopolitical context* is defined as relationships and decisions between people sharing the work that needs to be done within the group with some people having power over others. The relationships will be different for different stakeholders and, in some cases, there will be more vulnerable people and, in others, more powerful people. The relationships will not only be different at each scale level, both temporal and spatial, but will change as they are scaled up and will also change with time at the different scale levels.

Taking account of the sociopolitical context is important to decision-making in many areas. Beekeeping in southern Africa provides one example. Nel and Illgner (2004) show that social, economic, cultural, and natural landscapes can be combined and related to the complex and diverse relationships among rural people. Beekeeping is a commercial activity that supplements what rural people derive from subsistence farming and fishing but it is often overlooked (Quong 1993).

Responses with respect to wood also support the sociopolitical context as important to good decision-making (Chapter 8). National forest programs were devised in the context of favoring national action over an international approach. However, within this, institutional capacity to implement these national programs continues to be a key constraint. International forest policy processes and development assistance may be less effective given their focus at the national versus local level.

The sociopolitical context is a key consideration in deciding to allow direct management of forests by indigenous people or decentralizing authority for forest management to local communities. National forest governance initiatives and national forest programs make significant policy changes for participatory forestry. The greatest positive effects were felt in countries of low forest cover, such as Nepal and Tanzania where the capacity of the local people to manage forests was given greater policy support and the condition of the resource also improved (Brown 2002) Again, capacities to manage are the key issue here. Capacity is also the key issue when deciding whether forest-planning techniques will be effective in tropical areas. Public and consumer action is premised on sociopolitical calculations. Balancing the needs of the poor with respect to harvesting and using fuelwood versus forest protection must be undertaken in a political and social context.

Social and behavioral responses can play an important role in controlling infectious diseases while optimizing other ecosystem services (that is, with respect to sanitation).

In yet another example, the sociopolitical context is also important when it comes to the protection of local knowledge and devising landscape conservation and restoration schemes, especially when it involves the removal or the reintroduction of species, implementing eco-tourism enterprises, instituting certification programs, and establishing "fair trade" standards. For example, recreation and education are complementary. Cultural tourism can educate people about cultural diversity. Ola-Adams (2001) describes how the Omo Biosphere Reserve in Nigeria is creating programs for diverse audiences, ranging from school children to university students, from protected area managers to policy-makers.

18.3.2 Using the Best Available Ecosystem Biophysical Information

The decision-making process needs to open communication channels to the diverse sources of relevant information about the biophysical status and processes of the ecosystem concerned. In addition to state-of-the-art modern science, traditional knowledge should also be used where it is relevant and available. The mirror implication on the analytical side is the need to choose analytical frameworks that are capable of incorporating and handling the diverse sets of information from different sources required for the assessment of a useful range of decision options. Closely related to the criterion above, it is essential to collect and evaluate information about the socioeconomic implications of ecosystems changes as well as about the economic and social impacts of the feasible policies and measures to manage them. This requires integration of knowledge of widely diverging uncertainties from different scientific disciplines and sociopolitical perspectives and its consolidation in a form that is acceptable to all stakeholders. Complex decision problems can be usefully supported by analytical frameworks that are specifically developed to incorporate diverse sets of data, tools, and perspectives, like integrated environmental assessments (Rotmans and Vellinga 1998).

A solid body of reliable information on ecosystems and their function in the broadest sense is, clearly, the first and foremost prerequisite for any successful response regarding ecosystem services. This involves more than a mere collection of data and information on biological, chemical, and geological properties of the ecosystems, as it includes the transformation of such information into useable knowledge. It is knowledge that addresses the particular concerns of a user, deals with different spatial scales, time frames, and organizational levels. The principal findings are seldom easily transferable from one scale or level to another. One example is an evaluation of the regional or local impact of global climate change or other global phenomena. The recently stressed notion of a "place-based" science for sustainability—which should be relevant for local policy-making—points in this direction (ICSU 2002a). It is equally important—and difficult—to translate long-term impacts that may affect only future generations into terms that are relevant to day-to-day decision-making.

The majority of "usable knowledge" is in the form of numerical or other quantitative information (ICSU 2002b). Among various forms of such information, indicators play an important role. For example, an environmental indicator is air quality measured by ozone levels in parts per billion compared to its threshold value of 50 parts per billion. A component of the necessary knowledge system is the theoretical and institutional framework capable of handling the diverse set of information from different sources. Data gathering; their transfer, validation, and translation into useful information; and, finally, the presentation of the information are all part of such a framework.

Gaps may exist between the sources of usable knowledge and the potential users. Organizations that synthesize and translate scientific research and explore its policy implications are able to bridge this gap. They are sometimes called "boundary organizations" because they facilitate the transfer of usable knowledge between science and policy and they give both policy-makers and scientists the opportunity to cross the boundary between their domains.

Choosing responses should be based on both formal scientific information and traditional or local knowledge. To be credible and useful to decision-makers, all sources of information, whether scientific or traditional, must be critically assessed and validated as part of the assessment process through procedures relevant to the forms of knowledge. When speaking about the "best" biophysical information it should be made clear that in no case can such information be absolutely certain. Starting from the not fully assessed quality of basic data, the level of uncertainty increases up to the peak of the "information pyramid" (ICSU 2002b). The degree of uncertainty is mostly not known. Chapter 4 speaks on the "cascade of uncertainty."

With respect to responses regarding *biodiversity,* the use of concrete biophysical information on ecosystems is critical. Having the appropriate biophysical information is most important when it comes to responses that include the management of wild animals, in situ conservation (including the need to improve storage technologies), habitat restoration, and sustainable production. However, it is also important when it comes to regional planning, environmental impact assessments (including business biodiversity action plans), and devising habitat and area protection schemes (Chapter 5).

In relation to *food production,* all responses—if they are to be successful—must be based on precise and long-term biophysical information. Detailed information on local conditions, which are mostly incorporated in traditional knowledge, is critical. In particular, introduction of the new technology responses (biotechnology, genetically modified organisms, precision agriculture, integrated pest management) must be based on reliable and detailed biophysical information if failures are to be prevented (Chapter 6).

Valid biophysical information is critical for most of the responses regarding *water services* (see Chapter 7). In particular, the determination of environmental flows is based on such information. Decision-makers responsible for water allocation often seek the minimum flow that must remain in a river to maintain environmental quality. However, such thresholds of flow are very illusive and may not exist in reality. In any case, the desired condition should be decided prior to the application of an environmental flow methodology, preferably with the involvement of a broad array of stakeholders. Market-based incentives must also rely on very robust biophysical information to be successful.

A number of the responses regarding *wood* take biophysical information as a key ingredient to good decision-making. In some cases, responses such as national forest programs have not been successful because of a lack of sound information and there is a need for research into traditional knowledge and improved forest information systems. With respect to the understanding of the effectiveness of direct management of forests by indigenous people there is little information on outcomes upon which to make assessments. A key gap in our biophysical information is in measuring biodiversity. With respect to responses such as small-scale private ownership and private-public partnerships in forest management, dissemination of existing information to the practitioners in the field is an important issue. Obviously, biophysical information is the life-blood for improving wood technology and biotechnology responses. It is critical in improving forest plantation development and management especially with regard to the impact of monocultures, as well as determining how traditional forest planning techniques might be applied to tropical forests. Finally, understanding the promise that forests hold for carbon capture depends greatly on biophysical information (Chapter 8).

With regard to *nutrient cycling,* an example of the response that is very much dependent on biophysical information is the management practices aiming to minimize leaching and run-off of nitrogen and phosphorus fertilizers from agriculture fields (Chapter 9).

All responses to *floods and storms,* in particular, to the use of natural environment and non-structural measures in order to reduce negative impacts, depend on detailed information on biophysical conditions. Elements of the natural environment such as wetlands act as buffers against floodwaters. Coastal mangroves have been found to be very effective in providing protection against storms and surges in Bangladesh, India, and Southeast Asia. These measures include land-use planning through zoning, setbacks, and flood-proofing with emphasis on regulation or modification of the built environment, often urban. Insurance, as a response option, is as critically dependent on this type of information as any other response option (Chapter 11).

Since the elucidation of the life cycles of parasites and the recognition that insects transmit infectious agents, the vectors (insects, ticks, and snails) have been the targets through which the control of diseases has been attempted. Initial attempts at *vector control,* before insecticides became available and application techniques were developed, depended on environmental management to reduce vector population. Considerable success was achieved by draining swamps, by the use of oil to prevent larval mosquito respiration, and by the selective destruction of savannah and riverine forest habitats of these vectors. The advent of insecticides in the 1940s resulted in less emphasis on environmental and biological methods of control and the reliance, for a period of two decades, on insecticides (Chapter 12).

Having appropriate biophysical information is also important when it comes to all the identified responses with respect to *cultural services.* Protection of local knowledge has the protection of biophysical information at its core. Landscape conservation and restoration schemes including the introduction and removal of species, ecotourism, sustainable production practices, and locally based management schemes rely on biophysical information. Multilateral science initiatives, local data gathering and integration programs, and knowledge diffusion efforts all have biophysical information as an objective (Chapter 14).

Biophysical information is inevitably an important element within all *integrated responses,* but it is a critical factor for responses at the local level and cross-scale issues as well as for multilateral environmental agreements. In particular, the new generation of MEAs is critically dependent on precise biophysical information as they deal with difficult cross-cutting issues like climate change or loss of biodiversity.

Sustainable forest management is an example of an integrated response at the local level. In forestry, a range of examples can be found that address more than one ecosystem service at the same time. Sustainable forest management is an approach that seeks to integrate several ecosystem services and different stakeholders through innovative institutional arrangements, methods, and tools. Another example is integrated coastal zone management. Coastal zones involve a diverse set of ecosystems and habitats, which provide rich services and functions to society, and are associated with multiple uses and users (Chapter 15).

18.3.3 Pursuing Efficiency and Effectiveness

The basic principle of devising efficient decision-making is to conduct fast and thrifty decision processes (procedural efficiency). This implies designing the decision process so as to allow for fast and clear exchange of information and views, to allow flexibility for shifting positions, while progressing towards compromise solutions. The assessment activities can enhance and support the efficiency of decision-making by presenting the multitude of feasible decision options with all relevant implications, uncertainty features (qualitative characterization and quantitative ranges), and preconditions for and possible pitfalls of implementation and enforcement.

However, there is often a trade-off between the principles of procedural equity and efficiency. There are conflicting claims about stakeholder participation and the efficiency of the decision-making process. Some maintain that stakeholder participation is cumbersome and slows down the process while others claim that such involvement is controllable and may even turn out to be faster if the consensus-based outcome is implemented as soon as the decision is made as opposed to the long delays resulting from several rounds of rebuffs and revisions instigated by excluded stakeholder groups. Moreover, the emerging policy or regulation needs to be compatible with prevailing economic values and principles (outcome efficiency). This is especially important in cases when (re)distribution of public funds is involved. In order to help fulfill these objectives, the assessment framework and the decision criteria should be chosen so that they can properly handle the relative importance of economic and financial concerns in the given decision-making context. Typical efficiency criteria include balancing costs and benefits or identifying least-cost solutions under a given set of constraints. An important but often neglected factor in cost and efficiency calculations are the transaction costs required for implementation, enforcement, etc.

Effective decisions result in policies and measures that can be, and will be, realistically implemented to achieve the intended outcomes. The effectiveness of the decisions is, therefore, dependent on the extent to which the decision-making process is able to fulfill all the criteria above, ranging from the acquisition and use of the best available information to accommodating the appropriate mix of concerns (efficiency, equity, etc.). Decisions based on appealing ideals but void of pragmatic aspects are bound to fail and are, therefore, ineffective. The assessment process can foster the effectiveness of the decision by performing "reality checks" of the policy options by adopting analytical tools from disciplines like political science or game theory.

Responses discussed in Part II of this report illustrate that the efficiency and effectiveness of decision-making is important for achieving desirable outcomes. A number of responses to the issue of *biodiversity* provide good examples. Habitat protection schemes through indirect incentives such as integrated conservation and development projects are designed to integrate, optimize trade-offs, and create synergy. The same can be said of regional planning approaches to habitat protection. Eco-agriculture techniques including organic farming, integrated crop management, and conservation farming are also designed and pursued with integration, trade-off, and synergy as objectives (Chapter 5).

In a related way, the *food* responses in Chapter 6 recognize fundamental trade-offs that can arise from the demands placed on agricultural systems to produce food efficiently while sustaining ecosystem health and sustainability. For example, natural resource constraints include shortages of arable land, water, fisheries, and biodiversity. However, current trends in a significant number of agricultural indicators suggest threats to long-term economic, social, and environmental sustainability of the food system. Further, the overall effect of agricultural trade liberalization on the environment is ambiguous, and trade-offs must be weighed in order to find ways to limit the adverse effects of trade while enabling the collection of its benefits. There are similar concerns about the development and use of genetically modified organisms. Given the complexities that abound in this domain, efficient and effective decision frameworks need to allow both for the interaction between a large number of sectors and actors at multiple scales and for uncertainties at these different scales, both spatial and temporal.

The chapter on *water services* (Chapter 7) highlights responses that, increasingly, have the efficiency and the effectiveness of the decision-making process as objectives. Basin-wide river management schemes that are integrative are becoming prevalent. Also, market-based incentives for the provision of freshwater services are increasing in popularity, partly because of the fact that market forces are inclined to reduce transaction costs and delays as against those found with government intervention and regulation.

The chapter on *wood* (Chapter 8) points to missed opportunities to pursue responses that are more efficient, integrative, and synergy seeking. It also points out numerous opportunities to improve outcomes through responses that have these attributes. In general, the realm of multilateral agreements and initiatives has lacked the facility to cross cut, integrate, create synergy, and gain economies of scale. However, many of the responses that operate at a more localized level can be highly integrative. For example, company/community partnerships are premised on "win-win" ideas with each party taking away benefits suited to its particular needs. Responses for improving the technology of growing or using wood work in much the same way by trying to accomplish economic as well as environmental objectives. Plantation forests can be highly integrative and synergistic by satisfying wood demand, addressing lost habitat, combating desertification, and providing carbon sinks.

The success of the small-scale private owner as a forest management response may depend to a large extent on whether lack of economies of scale can be overcome. One solution to this situation is to band the small operators together into cooperative arrangements. This, of course, will depend on the receptivity of the small operators to such an organizational form. Certification programs are seen as expensive for small producers and local communities. In addition, there has been a proliferation of certification programs, which adds to cost and to confusion in the minds of consumers. Rationalization approaches may have large benefits for all stakeholders.

The positive *vector-borne disease control* outcomes discussed in Chapter 12 rely on integrative social, behavioral, and environmental responses and, therefore, decision-making processes reflect this need for integration, trade-off, and synergy. For example, the extent and prevalence of gastrointestinal infections throughout the world pose a massive problem. Only an integrated approach via mass treatment, safe disposal of waste, and provision of latrines effectively address this problem. Health education has been a vital element in successful outcomes. Environmental hygiene through protection of food from cockroaches and flies can also play a significant role. Effective decision-making processes consider this wide portfolio of responses.

Finally, the integrative and synergy-building aspects of good decision-making are found in several of the *cultural service* responses in Chapter 15. The consistent message is that decision-making and outcomes with respect to ecosystem services would be enhanced if local cultures were given a larger say in the process. This is particularly true when considering landscape restoration and eco-tourism schemes, and certification programs. Good examples of highly integrative and effective decisions and responses found in the chapter include the cultivation of medicinal plants in India and the Rhön Biosphere Reserve in Germany. In these examples, ecosystem conservation has been well integrated with local culture and economies.

18.3.4 Using the Best Available Information on Values

A crucial field of the social context for ecosystems decisions is information about the norms, beliefs, values, and aspirations of the affected communities. Even the best intended and, from a different perspective, perfectly rational decisions or measures will inevitably fail if they run counter to the norms and rules, which the affected stakeholders follow. These aspects need to be recognized in decision-making. Accordingly, prevailing norms and values influence the choice of the decision analytical tool and the decision criteria adopted in the assessment.

The responses that have been reviewed make it clear that this ingredient is very important to good decision-making. In these responses the concept of "values" goes beyond quantifiable costs and economic benefits and includes a broad range of determinants of human well-being, that is, all that humans value or need. Therefore, "values" range from personal security, sustenance, and health to material and economic goods to beliefs, traditions, rituals, and aesthetics. However, "values" also go beyond the individual to the collective to include organizations, corporations, communities, and even nations, and bring in the concepts of influence, power, tenure, and reputation. In this way, the use of the best available information on values is strongly related to the use of the best available information on the sociopolitical context discussed above.

Different stakeholders (or sets of stakeholders) bring different mixes of values (value systems) to particular circumstances, problems, or decisions. Also, different stakeholders have differing stakes in any particular issue. The responses discussed in the chapters in Part II provide evidence that participation and transparency in decision-making is the most effective way to develop the best information on values and respective stakes.

Chapter 5 demonstrates that *biodiversity* responses that seek to change the nature of use of habitat (that is, area protection) or species or provide alternatives to the use of targeted habitats or species are most effective when built with human values as a central theme. Biodiversity responses that seek to change productive behaviors to those that have greater sustainability (that is, certification and labeling schemes) take into account human values. A particularly good example of a mix of human values in choosing a response deals with reclamation where costs are incurred to restore aesthetic values. Local knowledge of biodiversity is another type of human value that can be considered in making decisions about ecosystems.

According to Chapter 7, values are an ingredient that is important for all responses in connection with *water services*. Property rights, and the human value systems they imply, are seen as an important response for sharing the benefits of freshwater services. These rights determine whether those who pay the costs of management have access to any of the benefits and, therefore, have an incentive for cooperation in the conservation activities needed to provide them. In the case of watershed services, which play a critical role in the provision of freshwater, rights over both land and water have been considered. In addition, there are compelling reasons to consider the use of markets for incorporating values into choosing responses for the provision of fresh water. Tradable water rights create a "visible or discoverable" value for water, and the concept of full cost recovery pricing incorporates externalities and, thus, a broad range of values (despite the inherent difficulties in quantifying all costs). Water exchanges, banks, and leasing and trading programs have developed to address water quantity and quality issues. Although recognition of the right of access to water for basic human needs may be undecided as a matter of international law, a number of nation states have directly or indirectly given formal recognition to the right of access to water as a fundamental human right. South Africa is one example. Therefore, this human value is being "wired" into its water responses.

Values including economic values are highlighted as key to good decision-making with respect to *wood and fuelwood* (Chapter 8). Efforts such as national forest programs and international forest policy processes have been either more or less effective depending on the extent to which local values and human well-being have or have not been incorporated into their development. Small-scale private ownership and forest management schemes are based on the premise that property rights lead to greater stewardship. The delegation of public forest management rights through conservation concessions is based on determining the full opportunity cost of lost public use and enjoyment. Organized public and consumer action as a response is based on both the value that politicians/companies place on reputation and the value that consumers place on environmentally sensitive production practice. Dendro power and fuelwood activities are grounded in the need for poverty reduction and economic development. Finally, calculations and decisions about forest protection are being aided by the "internalization" of the value of sequestering carbon by way of a developing carbon trading market under the Kyoto regime.

18.3.5 Considering Equity Concerns and Implications

The most direct way of using "the best available information about values" is to devise a fair decision-making process and to involve stakeholders directly in it. Different disciplines and different schools in ethics define what is fair in many different ways (Rayner and Malone 1998; Toth 1999). In the present context, "fair" is simply what those who are involved in or affected by the decision-making find to be fair. This entails giving a fair chance to all affected groups to participate, to present their values and concerns, and to protect their interests (procedural equity). Participation has become a buzzword in recent years and evidence is accumulating that it increases the overall quality of decisions concerning environmental assets and natural resources (World Bank 1996). In addition to the possibility of mobilizing local knowledge that is not otherwise accessible, and of increasing the acceptance of the decision, broad participatory approaches also facilitate dealing with the diversity of values, interests, conflicting interpretations of biophysical and social science analyses, and perspectives on how to cope with uncertainty.

Even if the participation of all stakeholders is impractical or impossible, the decision-making process needs to be open so that all affected parties can understand how a decision came about, its rationale, and how it affects different social or stakeholder groups (transparency). Irrespective of whether direct participation is possible and/or meaningful, the decision outcome needs to obey prevailing fairness principles in the society (consequential equity). The corresponding axiom in the analysis domain is the requirement to choose the assessment framework and the decision criteria according to the relative importance of fairness concerns in the decision-making context. Exploring outcomes under different criteria provides valuable insights into the trade-offs among them while multicriteria frameworks can help progress towards compromise solutions. In recent years, a variety of participatory assessment techniques have been proposed and are being increasingly used (Toth and Hizsnyik 1998) in which stakeholders jointly investigate the problem and the range of available options in preparation for the decision-making process. Participatory techniques are particularly worth considering in complex and controversial decision-making situations.

A review of the responses discussed in the chapters in Part II supports the importance of including a concern for equity, participation, and transparency in the decision-making process.

For example, with respect to *biodiversity,* Chapter 5 points out that habitat and area protection responses that rely on indirect incentives such as alternative economic development opportunities, integrated conservation and development projects, or ecotourism are designed with equity concerns as key considerations. Habitat and area protection responses that rely on direct incentives, such as the purchase of easements, tax incentives, tradable development rights, or direct land acquisition, also incorporate equity considerations because the focus of these responses is that of sharing the benefits of global biodiversity values locally with those whose well-being is tied to some exploitation of the targeted habitat or area. Participation of local people and communities in the design of such responses is also important for successful outcomes. Equity is a key aspect of the Convention on Biological Diversity's explicit protection of local knowledge. Also, equity is a key consideration when the reintroduction of fauna is a response option (for example, between cattle ranchers and wolves).

Failure to consider questions of equity and participation with regard to *water services* (Chapter 7) can cause a major problem for several important responses such as sharing water in a transboundary context, command-and-control regimes, assigning property rights in freshwater services, making changes to infrastructure, or using market-based incentives. For example, a specific response to the challenge of transboundary water management is the strengthening of provisions for public involvement, which includes access to information, public participation, and access to justice or legal recourse. An important tool for public involvement is the development of a process for transboundary environmental impact assessment. Given the general heterogeneity of environmental and socioeconomic conditions, effective management of freshwater resources to support multiple uses often requires numerous site-specific responses that are beyond the capacity of centralized authorities. Although a basin-wide approach is necessary for some aspects of management of freshwater resources (that is, flood forecasting), many aspects may best be resolved locally because it allows for more direct engagement of stakeholders. Although the use of water for basic human needs has not been recognized as a fundamental human right in international law, there should be no debate about the fact that human beings cannot survive without access to potable drinking water. This is an important issue of equity with respect to water responses.

Many of the responses on *wood,* found in Chapter 8, discussed equity as the key consideration while making decisions. The need for full participation of the affected parties and for multistakeholder processes was cited numerous times. The discussion of multilateral processes that have led to national programs delineates between success and failure based upon the degree of participation afforded by local people. Public-private partnerships likewise have been more or less effective to the extent that the economic rights of local people are considered and protected. The same can be said of traditional forest planning approaches applied in the tropics as well as plantation developments. Insuring that proceeds from royalty concessions find their way to those most affected by the change in the use of a public forest was highlighted. Attempts at collaborative forest management and decentralization have equity considerations at their core. It is also acknowledged that when company/community partnerships are pursued getting the right balance of benefits between the parties is often difficult because the nature of the respective benefits may be very different. Forest certification schemes have been criticized for failure to include the views of local people or to consider small producers who do not have the economies of scale to be able to afford to participate in what have been very expensive programs. In fact, the "paper-based" approach to certification is often a barrier to indigenous peoples who do not have the resources or skills necessary to comply with the detailed reporting requirements. Finally, some see the competing uses of wood (as between products and fuelwood) as an issue of gender equity with men relying on the former use and women on the latter.

With respect to *flood and storm services* (Chapter 11), equity may play an increasing role in deciding whether and how disaster relief and aid will be provided as the concern about extreme hydro-meteorological events tied to climate change grows.

Considerations of equity are an important influence with respect to finding responses for *vector-borne disease control* (Chapter 12). The role of the community and health education is a vital element of success with respect to sanitation. Dissemination of information to all people plays an increasingly important role generally.

Cultural service responses found in Chapter 14 are tied closely to considerations of equity in ways very similar to both the biodiversity and wood responses. The focus of these responses is the sharing of the benefits of global value locally with those whose well-being is tied to the exploitation of a local resource that might be restricted or those whose culture is being "marketed." These responses allow local people to share the fruits of tourism and eco-tourism schemes in a substantial way. They also include proper participation in decisions that might affect the continued habitation of a particular area or decision about production standards incorporated in certification schemes that might have an impact on local practices. In addition, these responses suggest participation by local people in decisions that would lead to the exploitation of a group's culture (that is, for tourism purposes).

One outcome of such participation might be revenue sharing with respect to the use of a group's cultural symbols. Another important aspect of the responses is the building of capacity to allow for the aforementioned participation in a meaningful way. Equity is a key motivator of the CBD's explicit protection of property rights regarding local knowledge. It is also a key consideration in the reintroduction of fauna or the elimination of alien species when it comes to landscape restoration. (For example, with respect to this latter point, equity might dictate the compensation of livestock farmers when wild animals such as bears and wolves are reintroduced to an area, or to fruit farmers when a non-indigenous tree is eliminated from a restored landscape).

18.3.6 Assigning Clear Accountability and Providing for Monitoring and Evaluation

Responsibility for ecosystems decisions and their implications is an elusive issue if one takes into account the multitude and magnitude of uncertainties about the biophysical process, social behavior, and the poor controllability of the underlying processes in both domains. Yet a reasonable level of accountability for at least the manageable aspects of the decisions would encourage decision-makers to use the best available information, involve relevant stakeholders, and keep the decision process transparent. In relatively simple regulatory or resource allocation cases, the responsibility rests with the decision-maker who has the ultimate authority to put policies and measures in place. In more complex situations involving several organizations, each should be accountable for the formulation and implementation of the decision component in its own domain or mandate. Similar principles of accountability would motivate analysts to use the most suitable tools and the best available data and to expose their results to extensive reviews.

Decisions with respect to responses are made within a complicated web of different levels of governance in different sectors and at different scales. Some decision-makers have both official and genuine power while others are mere representatives for others with power who stay in the background. Moreover, the consequences of decisions may be so remote, indirect, and time-lagged that it is very difficult to clearly assign accountability for their outcome. Aligning accountability with decision-making will improve this process, and attaining transparency in the decision-making process is a way of achieving this alignment.

A prerequisite for effective accountability is a full evaluation of policies based on reliable monitoring. However, to monitor properly and objectively the outcome of policies is difficult to accomplish. First, there is always a lack of money and other resources that are mostly assigned to other, more visible purposes. Second, some people may not welcome monitoring as it may reveal irregularities (including corruption). Evaluation of the effects of a measure or action requires establishment of a causal link between the action and its impact (for example, introduction of catalytic converters in cars—reduction of carbon monoxide emissions). It is certainly not so simple to distinguish between causality and simple association. A careful analysis to discount the effects of confounding factors is necessary.

The achievement (or non-achievement) of targets is one of the key considerations. A crucial prerequisite is that the targets are transparently and clearly stated, preferably in quantitative terms. If this is done, the evaluation is a relatively easy task provided that proper indicators both for the targets and for the actual state are available. The evaluation should also examine the impact of unexpected factors that may have helped or hindered the achievement of objectives and targets or had other ancillary impacts.

To cover properly the overall impact of a given policy is probably the most difficult part of the evaluation. First, the different scales (spatial and temporal) must be taken into account. In particular, the evaluation process must take in account long time scales. Second, there could be direct but also indirect effects. Third, most policies have impacts in all the environmental, economic, and social realms. Fourth, many stakeholders may exist as a result of which the impact may be very different (distributive effects). Even if it is sometimes stressed that a certain action will result in a "win-win" situation there always are some losers. In this context, these questions—among others—should also be answered: Are the achieved objectives justified in terms of financial and other costs? Are the impacts enhancing human well-being and/or bringing economic benefits besides improving the environment? For example, has reduction of emissions had any effect in decreasing health problems? (Clancy et al. 2002; WRI 2003)

Insurance and other financial markets play an increasing role in the area of environmental accountability and performance monitoring (UNEP 2004). With respect to accountability, insurers and providers of financial capital are beginning to charge premiums in accordance with expected environmental liability. In addition, trends in overall premiums and claims will provide an explicit signal with regard to the success of the responses.

Central to the responses on *biodiversity* is monitoring and evaluation of policies, especially in the habitat and species protection schemes based on direct incentives. For example, property rights that have been created or regulations that have been promulgated need enforcing. Sustainable production practices as embodied in certification programs require that standards be maintained. An important problem in many cases is the lack of clear baseline indicators and quantitative targets (Chapter 5).

An example of an important response regarding *water services* is the basin-wide river basin management. Management of river basins is mostly performed by different river basin organizations, of which several examples are given in the Chapter 7. A pattern that is often observed is the tendency of basin-level management to be dominated by more tangible and economically dominant interests. However, recently, the integrated approach is more prevalent. Also important in this respect is the so-called shared water in a transboundary context. An important tool for public involvement is the development of a process for transboundary environmental impact assessment. The issue of water resources management is presently high on both the international environmental and development agendas. In part, this is due to necessity—261 major river basins are shared by two or more sovereign states worldwide. The accountability issue is essential when assigning rights to fresh water services and applying market-based incentives.

In the area of *forest management and protection,* accountability, monitoring, and enforcement are important aspects of response design and selection. Multilateral agreements and initiatives recognize the necessity of accountability including the codes of conduct for the private sector. The International Tropical Timber Organization has created indicators and has tried its hand at enforcement of sound forestry practice. Neither traditional forest planning (when applied to the tropics) nor reduced impact logging approaches can be effective without adequate enforcement regimes and resources. It is often the failure to provide adequate resources for monitoring where enforcement breaks down. Certification responses are based on standards and monitoring. Conservation concessions tie royalty payments to the maintenance of certain parameters of protection (Chapter 8).

Financial services regarding *flood and storm* responses include insurance, disaster relief, and aid. Insurance, in particular in connection with floods, is an increasingly important response. Its significance has grown in recent years, with more frequent threats of extreme hydro-meteorological events in connection with global climate change. Disaster relief and aid is getting more international recognition. Connected with these responses are large sums of money. Therefore, the requirement of accountability is very important (Chapter 11).

Because large financial resources are attached to international programs, accountability issues play an important role with respect to *vector-borne diseases* (Chapter 12).

18.3.7 Considering Vulnerabilities and Risks

A crucial aspect of equity issues is related to vulnerable groups and communities. Vulnerable here refers to people who are sensitive to changes in ecosystems services and lack the ability to cope with those changes, that is, recognize preliminary signals in time, consider response options, and adapt to emerging changes or counteract them. The interests of the vulnerable communities are much better respected when defended by a credible, legitimate advocate, coming ideally from the concerned community or communities. Yet vulnerable groups are often unable to engage even in open and receptive decision-making processes because they lack the basic knowledge, or the necessary information and communication tools. Special representatives or legitimate assigned advocates are, therefore, required to speak for their interests in order to prevent top-down decisions being imposed on them. In the assessment work, extended analyses framed from the perspectives of vulnerable groups are required to estimate the implications of the different options for them.

Vulnerability and risk pertain to human populations as well as ecosystems and their services. Vulnerability is defined as the capacity to be wounded by socioeconomic and ecological change. It has three main elements: exposure, sensitivity, and resilience. Resilience is particularly important—if resilience is not maintained within the system or the person then they will become more vulnerable. Vulnerability is, therefore, a property of coupled social-ecological systems. An example from South Africa illustrates the point.

In the South African Development Community (regional level) during 2002–03, the complex system of outside pressures contributed to the complexities associated with climate stress and food insecurity. Many donors have provided early warning systems and are managing food insecurity and risks, thus reducing vulnerability in the region. However, the contributions of adverse synergies including droughts and politics that have precipitated famines have become more prevalent and endemic in sub-Saharan Africa. In the Vhembe District in Limpopo Province (district-local scale) research results have shown that there are gaps and weaknesses with regard to improved resilience to climatic risk.

Identifying the reasons for the lack of action is the key to understanding the drought effects that occurred at the national and regional levels specified above. First, it is clear that forecast alone is not enough. There needs to be more activity in broadcasting the forecast by different media, for example, the radio, newspapers, videos, to district institutions, and to the community level. Second, farmers may be constrained by lack of resources from responding to information about climate stress. The resource constraints include lack of access to credit, land, and markets as well as lack of decision-making power. There are, however, encouraging signs in the Vhembe District and at the national level for building adaptive capacity under conditions of climatic and environmental stress. There are signs that research on ways to improve adaptive capacity in South Africa will produce generalized recommendations that will improve policy (Vogel and Smith 2002).

Vulnerability and risk considerations may go directly to the heart of several of the *biodiversity* responses (Chapter 5). Habitat protection schemes that are based on direct incentives such as easement and land acquisition target vulnerable places, ecosystems, and species. The participation of the local people and communities in the design of any responses to biodiversity is also very important for successful outcomes. An example is the Misiones Region in northeastern Argentina where the forest has been replaced with agriculture although the soils are fragile (Rosenfeld 1998). Two major types of peasants are distinguished and they have designed very different farming systems and control strategies that interact in the wider context in which they operate.

With respect to *food,* agricultural research could be prioritized with the participation of farmers and begin with an integrated evaluation of their socioeconomic needs and their natural resource endowments in order to provide an equitable and effective process (Chapter 6). For example, poverty and vulnerability among smallholder farmers is high because the soils are of poor quality and are drought-prone. Low productivity affects hunger and poverty and leads to low economic growth. This leads to poor health, which, in turn, leads to low productivity. A response might be to apply integrated pest management to reduce the need for pesticides, but this may be subject to uncertainty as IPM has not been very successful in the past. In Kenya and Tanzania, indigenous plants were a source of raw material to allow people to cope when the harvest failed. This provided a crucial safety net; for example, indigenous fruits provide important nutrients for children when meals are reduced at home. The sale of livestock and poultry and engaging casual labor are often indirectly dependant on ecosystem services. Data in MA *Current State and Trends* show the increasing percentage of households who depended on indigenous plants in Kenya and Tanzania (Eriksen 2000).

With respect to *wood,* the vulnerability of human well-being arises in the context of multilateral agreements and initiatives and any objectives for poverty reduction (Chapter 8). Environmental vulnerability is raised explicitly in the context of the protection of habitat as an objective of certain public-private partnerships and conservation concessions. The drive to better manage forest resources is implicitly directed at the vulnerability of those resources and vulnerable people and the risk that we may pass some tipping point with respect to ecosystem services.

Vulnerability of ecosystems and human populations are prominent when it comes to *flood and storm control* (Chapter 11). Flood plain and coastal zone development increases the number of people at risk. Human beings are increasingly occupying regions and localities that are exposed to extreme events, and are likely to become more poverty-stricken as a result. Many of the datasets on extreme events show that impacts are increasing around the world, and studies show that human vulnerability is the primary factor explaining trends in impacts. Case studies at the local scale have shown that human interactions with ecosystems have increased the vulnerability of humans and impacts on human well-being, and that appropriate management of ecosystems can reduce vulnerability and contribute to increased human well-being.

With regard to ecosystems and *vector-borne disease control* (Chapter 12), responses that affect the state of ecosystems are also likely to affect the health of people, thereby putting them at risk. On the other hand, responses aimed at promoting human well-being through the eradication of vectors can have profound effects on vulnerable ecosystems such as wetlands. All responses should be measured in terms of their effectiveness on human well-being in its broadest sense, including the provision of ecosystem services. The International Red Cross Federation (2002) has shown that the death toll from infectious diseases such as HIV/AIDS, malaria, diarrhea, and respiratory diseases was 160 times the number of people killed due to natural disasters in 1999.

With regard to *climate* regulation, there is a close interrelationship between climate and ecosystems (Chapter 13). When climate variation increases outside its "normal" bounds, vulnerability increases. Adaptation serves to reduce vulnerability to climate change by minimizing exposure or maximizing adaptive capacity. The poor will have less capacity to adapt and mitigate the impacts of climate-induced changes to ecosystems. Desertification is an example of a coupled socioecological system that threatens livelihoods. It is a good example of issues in understanding vulnerability (Downing and Ludeke 2002). Diversifying and strengthening local livelihoods will contribute to climate change policy by providing greater adaptive capacity and reduced vulnerability to change.

The *cultural service* chapter (Chapter 14) views vulnerability and risk in the context of fragile cultures or those without capacities or sufficient power to be meaningful participants in decision-making or negotiations. Therefore, the chapter stresses that responses such as those dealing with certification of sustainable production practice or fair trade should address this issue in decision-making. It does so, likewise, with respect to responses that involve relocation of local cultures in the light of landscape restoration or that which involve eco-tourism that capitalizes upon elements of local culture. Responses that seek to protect local knowledge and language directly address the vulnerability of traditional cultures and the risk to society in general from their atrophy or absorption. Conversely, UN programs designed to diffuse knowledge and best practice are intended to protect the viability of indigenous and local human well-being and, thereby, indirectly, their cultures.

With regard to *integrated responses,* the well-being of people around the world is strongly related to the environment in terms of livelihoods, health, and vulnerability (Chapter 15). The poor are highly vulnerable to droughts and floods, the frequency and severity of which may be expected to increase with climate change. The chapter stresses the need for more fresh water, the absence of which can lead to illness, malnutrition, famine, and greater incidences of floods and droughts. UNDP (2003) formulates a disaster risk index to assess global patterns of natural disasters and the relationship to development. The disaster risk index calculates the relative vulnerability of a country to a given hazard as the number of people killed by the hazard divided by the number of people exposed to it.

The best available data on a global scale confirm that during the past four decades the number of great disasters has increased four times, while economic losses have increased by ten. (Swisse Re 2003; Munich Re 2003; CRED 2002) Although comprehensive global databases are not available for smaller-scale hazards the significance of these more common events to the social vulnerability of exposed human populations is a significant concern among vulnerability analysts (ISDR 2002; Wisner et al. 2004).

18.3.8 Dealing with Uncertainties

Decision-making about ecological management and the use of ecosystems services is plagued by inherent uncertainties. Even if the functioning of an ecosystem is relatively well understood under the prevailing conditions, the ecosystem behavior might shift as a result of changes in some external driving forces or conditions (Walker and Steffen 1996). Moreover, the values and valuation of ecosystems and their services by the relevant communities might change or stakeholders may revise their positions. The implication of all these uncertainties for decision-making is that both the process and its outcome must be flexible so that they can respond to newly available information about the biophysical system (ecological or scientific uncertainties), about the social system (value- and behavior-related uncertainties), and about the effectiveness of the decision itself (regulatory uncertainties) (NRC 1996).

The sources, nature, and magnitude of uncertainties involved in a given decision-making problem also have implications for choosing the analytical framework (Morgan and Henrion 1990). In order to provide useful insights, the assessment tool needs to be suitable for accommodating decision-making under uncertainty and hedging, and multiple decision criteria reflecting differing values of the different stakeholder groups. Ideally, a single assessment framework should be chosen that is sufficiently flexible to accommodate and help consolidate a diversity of relevant perspectives on ecosystem change. If this is not possible, multiple frameworks are needed but this raises the important problem of how to consolidate their results. The range of decision options explored by the analytical tool should also take adaptation possibilities into account, including the feasibility and costs of mid-course corrections in the light of new information and give special consideration to irreversibilities, uncertain thresholds, etc. Dealing with risk and uncertainty is considered a very important part of the overall framework for the whole millennium ecosystem assessment. (See MA 2003, Chapters 4, 8.)

First of all, uncertainties arise regarding information both on the biophysical systems and the social and economic contexts including changing values and behavior. Second, the effectiveness of the decisions themselves and their implementation introduce uncertainty to outcomes. Decisions at all levels and scales should, therefore, allow for the policies to be flexible and adaptive, to allow learning, to incorporate results of evaluation, and to make necessary adjustments to accommodate new situations and/or new information. On the other hand, there is always a trade-off between flexibility and responsiveness of the policy and its stability without which it loses credibility and, therefore, all effectiveness. There is a difference between rigid policies that insist on nonessential requirements and policies that are reliable such that rules do not change in the middle of the game.

In contrast to human perspectives, ecosystem services issues are long-term and, therefore, the uncertainties caused, starting with the limited knowledge of the evolutionary processes and external influences, are inherently large. In this context, one of the solutions is the use of the precautionary approach defined by Principle 15 of the 1992 Rio Declaration (UN 1992).

With respect to *biodiversity* responses, better information on levels of uncertainty about biodiversity and its values could greatly assist decision-making. For example, we do not know which species are most likely to go extinct but we may say with *high certainty* that the rapid loss of biodiversity threatens the functioning of natural systems and human welfare. Many sources of uncertainty affect decisions in this case (missing data, random sampling errors, unknown functional relationships within ecosystems and between ecosystems and humans, unknown future consumption patterns, etc.) and due to complexity of the issue, our ability to choose the right options will be always imperfect. Thus it is necessary to avoid irreversible actions until uncertainty is resolved. Also integration across response options can mitigate and reduce uncertainty. Regional planning approaches to habitat protection and environmental impact assessments will be much more effective if uncertainty and adaptability are the key elements. Reclamation and rehabilitation in and of themselves demonstrate the quality of adaptability. However, the most telling example of the importance of considering unintended consequences and uncertainty comes when the introduction of a non-native species is chosen as a response to eradicate another invasive and damaging species (Chapter 5).

The aspect of uncertainty, flexibility, and adaptation is not strongly expressed within the responses regarding *food*. However, new approaches such as novel technology like introduction of genetically modified organisms and biological control methods should be guided by a principle of precaution as the level of uncertainty is high. Flexibility and adaptation is very much needed in successful development of effective and environmentally sound methods of aquaculture (Chapter 6).

Responses regarding *water services* are inherently chosen under the condition of high uncertainty due to rapidly changing and highly unpredictable hydro-meteorological conditions. High levels of flexibility and adaptability can be achieved by decentralization of management and decision-making (that is, democratic decentralization, deconcentration, and privatization). Given the general heterogeneity of environmental as well as socioeconomic conditions, effective management of freshwater resources to support multiple and often conflicting uses often requires numerous site-specific responses that are beyond the capacity of centralized authorities.

Although a basin-wide approach is necessary for some aspects of management of freshwater resources such as overall water allocation, flood forecasting, and emission of persistent pollutants, others, such as problems associated with land and water relationships, and operations and maintenance of irrigation canals, may best be resolved locally because it allows for more direct engagement of stakeholders and more appropriate responses to site-specific circumstances. For example, in North America, the United States and Canada are developing and advancing a number of large-scale watershed ecosystem approaches along their extensive inland border, which are moving toward carrying out holistic approaches that include addressing the interrelated challenges, goals, and problems of water flows, quantities and levels, water quality, and protection of aquatic wildlife and their habitats. This long-term Canada-U.S. cooperation on shared watershed ecosystems helps respond to many calls from national, international, regional, and global levels to develop and implement sustainable development approaches between and among countries (Chapter 7).

With respect to *wood,* considerations of uncertainty are important in the context of forest management strategies, in particular, of both indigenous people and small-scale owners/managers. These types of managers tend to diversify the products and benefits they take from the forest resources they manage and, therefore, they are able to adapt to unforeseen circumstances and cope with uncertainty. Unintended consequences arise in the context of public delegation of forest management rights through conservation concessions in that governments/communities have ceased other conservation efforts in the face of granting these concessions (Chapter 8).

In response to the highly uncertain nature of *flood and storm* only such non-structural measures such as forecasting and warning have the characteristic of flexibility. Most of the other responses are relatively rigid and non-flexible. The modern types of physical structures like dikes, weirs, and barriers are striving to achieve some flexibility through modern technology. Even if the physical structures, in general, are not adaptive as such we will continue to rely on them through the twenty-first century. The important point is to place them within an integrated system including warning and other measures (Chapter 11).

Unintended consequences are prominently illustrated by the failure of insecticide use when it comes to controlling *vector-borne diseases* (Chapter 12).

18.3.9 Considering Cross-scale Effects

The overwhelming majority of new decisions about ecosystems management have to be incorporated in the hierarchy of existing policies and regulations. Accordingly, the decision-making process needs to be open to comply with relevant policies already in place or to initiate appropriate changes in them. Similarly, the decision-making process has to be extended to initiate relevant decisions at lower levels that might be required for effective implementation. On the analytical side, the selected tools must be capable of incorporating the hierarchical conditions of the decision-making problem at hand. They must be able to accommodate constraints provided by higher-level regulations and to explore decision needs and options at lower levels required to achieve the goals of the decision problem explored.

With respect to choosing responses to protect, conserve, and enhance habitat and species, the more important scale dimension is, in fact, that of time. Obviously, multilateral environmental agreements such as the Convention on Biological Diversity work across jurisdictional and geographic scale, primarily global to national. Regional planning approaches to habitat protection are, in fact, intended to integrate scale (that is, regional to local to site-specific responses such as certification programs that target sustainable practice in forestry and fisheries work from global to regional to local scales). See Berkes (2004) relating to cross-scale interactions and certification programs. With regard to *biodiversity* the incorporation of biodiversity policies into integrated regional planning will promote cross-scale effects and make sound trade-offs between all the different scales. Local biodiversity may be useful but global biodiversity ignores the local biodiversity values. Vermeulen and Koziell (2002) see the focus on global values as a consequence of the fact that the global consensus relates to wealthy countries that recommend biodiversity in terms of services derived from it and not as an end in itself (Chapter 5).

Scale is a very important concept with respect to *water.* Management of river basins that stretch across jurisdictional bounds is mostly performed by different river basin organizations. A specific response to the challenge of transboundary water management is the strengthening of provisions for various aspects of public involvement, which includes access to information, public participation, and access to justice or legal recourse. Because 261 major river basins are shared by two or more sovereign states worldwide (Wolf et al. 1999), the development of a process for transboundary environmental impact assessments is an important tool for public involvement. In Africa, where 57 shared international river basins cover 60% of the continent (Gleick 1993), management of transboundary water is not a new challenge and the local people are encouraged to cooperate and manage in a transboundary context. The issue of water resources management is presently high on both the international environmental and development agendas (Chapter 7).

With respect to *wood,* the time scale comes prominently into play when considering the stewardship motivations (perpetuating the family asset) of the small-scale private owners and managers of forests. However, the size of forests is also the key to their management strategies, and influences their abilities to satisfy requirements associated with certification programs. The duration of the term is a key issue when discussing the delegation of public forest management through conservation concessions. Finally, the time scale is important to proper forest plantation development and management (Chapter 8).

With regard to *nutrient cycling,* the problem of nitrogen pollution manifests itself at the local to regional scale, so local and regional governments clearly have a role to play. For example, the technologies for nitrogen removal for sewage treatment in the Tampa Bay have led to water quality improvement but to a lesser extent in Chesapeake Bay (NRC 2000). These U.S. examples are at a local scale (Chapter 9).

Physical responses to *flood and storm control* such as dams and levees may cause net harm to ecosystems in the longer time-scale in terms of restoration and resiliency. In turn, this may reduce the anticipated (or expected) benefits of the responses (Chapter 11).

In terms of *cultural services,* multilateral environment agreements such as the CBD work across jurisdictional and geographic scales, primarily global to national. Responses such as certification and fair trade programs work from global to regional to local scales (and vice versa). Local organizations can take advantage of emerging global institutions and conventions to bring their case to wider political arenas (Chapter 14). An example is "The Samarga Declaration" to prevent the granting of industrial logging in an area they consider theirs (Taiga Rescue Network 2003; Molenaar 2002).

Scale issues are critical in *integrated responses* and cross-scale responses may be necessary. Integrated responses are long-term in nature, and require fundamental shifts in governance institutions with regard to skills, knowledge, capacity, and organization. Integrated responses also occur at different geographic and jurisdictional scales and across scales and use a range of instruments for implementation. However many attempts at integration are sector-based and do not address multiple ecosystem services and human well-being simultaneously. Implementing integrated responses may be resource-intensive but the benefits can outweigh the costs. Thus it requires the bringing together of many different stakeholders at different levels and the need to provide decision-making and management procedures at all levels. Integrated responses do not necessarily bring about equitable distribution of benefits to stakeholders (Chapter 15, especially Table 15.1).

18.4 Considering Business Motivations

Business is positioned to be a positive force in the resolution of key trade-offs. It can play a role through the development and deployment of new technology, pursuit of new business models, reduction of operational footprints, provision of leadership, setting of examples, and coalescing of partnerships. For example, as environmental pressures build up, the developed world and its consumers may begin to demand more cyclic models of activity and begin to define quality of life in less material ways such as leisure, experiences, knowledge acquisition, and relationships. Changes such as these could create business opportunities in service, "reverse flows," education, and travel. In addition, supporting public policy that raises industry environmental performance standards could advantage leaders and first movers while raising the standing of the industry as a whole with its important constituencies. Business leadership with respect to reducing poverty, improving human well-being, and protecting the environment can be in business' self interest. For example, this leadership could help secure stable and safe societies, preserve open and free markets, insure access to critical resources, provide new product and business opportunities, optimize social and environmental transitions, and, for the most astute and agile, carve out competitive advantage.

18.4.1 Reputation and Brand Risk, Partnerships, and Investor Confidence

In a fast changing business and market environment, a firm's reputation can be the certainty that it can provide customers, investors, employees, suppliers, and communities. In this way, reputation, as signaled through its brand, acts as a magnet. A good reputation can help differentiate a firm in crowded markets, both product and capital. A very tangible indicator of the value of reputation can be found in market shares, price premiums for otherwise similar products, or higher price/earning multiples for companies in the same sector. The right reputation can attract the best employees and partners and, therefore, provide access to the best ideas. In this way, reputation might be considered a key corporate asset to defend and enhance (Ottman 1998).

"Value adding" and strategic partnerships can be important to successfully achieving corporate objectives. Partnerships help accelerate learning and leverage resources. Important relationships must be designed for the mutual advantage of all partners, and with the idea that a "bigger pie" may be more important than a "bigger slice." Finding good partners can be a source of competitive advantage (Rondinelli and London 2003).

Investors of capital do not like uncertainty or surprises and, therefore, steer investment away from sectors or from firms within those sectors whose risks and potential contingent liabilities are not well understood. In order to attract capital, these sectors and firms must pay higher rates. The uncertainties introduced by questions of sustainability, potential costs and liabilities for the use of common environmental resources which are currently not taken into formal accounting statements, potential regulatory constraints on products and operations, and the prospect of restricted access to natural resources or sites are playing a larger role in the investors' calculus. Corporations are increasingly aware of the impact that reputation for business practices that address these risks and uncertainties can have on their cost of capital (Reed 2001).

Trust and transparency can help create a value-adding reputation, and environmental performance reporting (that is, Global Reporting Initiative) and meeting certain standards such as those found in eco-labeling schemes are responses that leading corporations are pursuing with reputation and brand image in mind (Chapter 8).

18.4.2 Access to Raw Materials and Operational Impacts

The availability and access to clean water is likely to change the way private enterprises in the developing world and the industrial countries conduct business in the twenty-first century. For industries as different as food and agriculture and high technology (for example, semiconductor plants require enormous amounts of water for chip production), water will increasingly be a factor in determining where, how, and with whom private enterprises conduct their business (MA *Scenarios,* Summary).

While ecological degradation is often portrayed as a conflict between "public environmental interests" and "private business goals," different types of "business conflicts" are likely to emerge in the future. For example, with tourism becoming the world's largest employer and an important economic factor in many developing countries, native forestland and other natural resources will be increasingly perceived as "vital business assets" of many private companies (MA *Scenarios,* Summary).

Non-point source pollution associated with agriculture is under greater scrutiny (Chapter 6).

Development of farm wood-lots and large-scale plantations is an increasingly widespread response to the growing demand for wood, and the decline of available natural forest areas. Not all afforestation projects have positive economic, environmental, social, or cultural impacts. Without adequate planning and management, the wrong growers, for the wrong reasons, may grow forest plantations in the wrong sites, with the wrong species. In areas where land degradation has occurred, afforestation may play an important role in delivering economic, environmental, and social benefits to communities reducing poverty and enhancing food security. In these instances, forests and trees must be planted in ways that will support livelihoods, agriculture, landscape restoration, and local development aspirations. There is increasing recognition that semi-natural, mixed-species, and mixed age plantings can provide a larger range of products, "insurance" against unfavorable market conditions or insect and disease attacks, diversity of flora and fauna, protection against the spread of wildfires, and provision of greater variety and aesthetic value in the landscape (Chapter 8).

18.4.3 Opportunities and Incentives

18.4.3.1 Technology

Technology has helped to increase food production from cultivated ecosystems and is expected to continue to do so in the future. The experiences of the last Green Revolution, combined with the best of new agricultural sciences, could support a future agricultural revolution to meet worldwide food needs in the twenty-first century. Increased pressures on the resource base (land, water, fisheries, biodiversity) and the potentially serious effects from climate change add to the importance of the role technology can play (MA *Current State and Trends,* Summary).

Technology has made possible a rapid rate of "development" of water resources with a view towards maximizing freshwater provisioning services (for example, water supply, irrigation, hydropower, and transport) to meet rising populations and human needs. However, it is the re-examination and alterations of existing infrastructure that offers the most opportunity in the short and medium term (Chapter 7).

An extensive array of technologies is now available in the energy supply, energy demand, and waste management sectors, many at little cost to society. Significant reductions in net greenhouse gas emissions are technically feasible given a portfolio of energy production technologies including fuel switching (coal/oil to gas), increased power plant efficiency, carbon dioxide capture and storage, pre- and post-combustion, and increased use of renewable energy technologies (biomass, solar, wind, run-of-the-river and large hydropower, geothermal, etc.) and nuclear power, complemented by more efficient use of energy in the transportation, buildings, and industry sectors (Chapter 13).

Similarly, technical tools exist for reduction of nutrient pollution at reasonable cost. That many of these tools have not yet been implemented on a significant scale suggests that new policy approaches are needed, but also that business opportunities may exist (Chapter 9).

18.4.3.2 Market and Other Economic Incentives

Market-based approaches have the potential to unlock significant supply- and demand-side efficiencies while providing cost-effective allocation of scarce resources. Supporting legal and economic institutions need to be in place. Also, market driven instruments do not automatically address poverty and equity issues related to the use of provisioning ecosystem services.

Functioning water markets can provide price signals for reallocation not only between different uses, but also signals to guide conservation activities. Water exchanges, water banks, and water leasing have emerged as arrangements for promoting market activity (Chapter 7).

Market mechanisms and economic incentives can significantly reduce the costs of mitigation in the context of climate change (Chapter 13) and market-based instruments hold the potential for better nutrient management (Chapter 9).

Consumer preferences operating through the market have resulted in some important forest and trade policy initiatives and improved practices in some large forest corporations. Forest certification has become widespread in many countries and forest conditions (Chapter 8).

Reforestation, improved forest, cropland, and rangeland management, and agroforestry provide a wide range of opportunities to increase carbon uptake, and slowing deforestation provides an opportunity to reduce emissions. Land use and its change and forestry activities have the potential to sequester about 100 gigatons of carbon by 2050, which is equivalent to about 10–20% of projected fossil emissions over the same period. Evolving markets for carbon reduction credits raises the prospect of market opportunities (Chapter 13).

Biological resources supply all of our food, much of our raw materials, and a wide range of goods and services including genetic materials for agriculture, medicine, and industry. Potential future uses convey option values. In the light of current and future uses of biological resources, it is important to understand the implications of the loss (at an accelerated pace) of species. The private sector is showing greater willingness to contribute to biodiversity conservation, due to the influence of shareholders, customers, and government regulation. Many companies are now preparing their own biodiversity action plans for biodiversity conservation, supporting certification schemes that promote more sustainable use, and accepting their responsibility for addressing biodiversity issues in their operations (Chapter 5).

18.4.4 Examples of New Business Opportunities

Organic farming can contribute to enhancing sustainability of production systems and agricultural biodiversity. In several industrial countries, organic agriculture contributes a growing portion of the food system. Agroforestry, which is a low-input farming system with greater sustainability than "slash-and-burn" or high-input monocultures, is an alternative technology for increased food production, using nitrogen-fixing trees to increase soil fertility and nutrient cycling. New crops developed from indigenous trees producing traditionally important foods and other marketable products enhance food and nutritional security, and also allow farmers the opportunity to increase the productivity of their staple food crops. Aquaculture is an example of a novel food production system that has evolved into a well-known production system, but the present situation is accompanied by serious impacts on ecosystems, including loss of vegetation, deterioration of water and soil quality, and loss of biodiversity (Chapter 6).

Environmental awareness and educational programs have been successful in allowing consumers and resource users to make well-informed choices for minimizing waste in their purchasing decisions. Employers have introduced programs to encourage and recognize initiatives by the community to reduce waste. In Japan and other industrial countries, "industry clusters/technology platforms" have been planned where the waste of one industry is the resource of another. The sale of products from waste, whether by simple re-use, recycling, and recovery, or by more complex technological processing, has helped to create jobs appropriate to the socioeconomic conditions of various localities or countries (Chapter 10).

There has been a significant growth in some non-wood forest product markets with the extension of the market system to more remote areas; a growing interest in natural products such as herbal medicines, wild foods, handcrafted utensils, and decorative items; and development projects focused on production, processing, and trade of non-timber forest products (Chapter 8).

If technology continues to develop, industrial-scale fuel derived from forest products could become a major contributor to sustainable energy sources. Consumption of fuelwood has recently been shown to be growing less rapidly than had been estimated earlier. Increasing urbanization and rising income have contributed to a slowing in the rate of increase in the use of fuelwood as users switch to more efficient and convenient sources of energy. In some regions, including much of developing Asia, total consumption is now declining. Efforts to encourage adoption of improved wood burning stoves have had some impact in the urban areas of some countries, but there has been little success in rural areas due to cultural and economic obstacles. Recent attention to improved stoves has shifted from increasing efficiency of fuelwood use to reducing damage to health from airborne particulate matter and noxious fumes associated with the burning of wood and charcoal. In industrial-country contexts, as renewable options gather more momentum, and the technology becomes more fine-tuned, it can be expected that "dendro power" options will become more competitive and investor-friendly (Chapter 8).

The biggest challenge for conservation in the twenty-first century is for it to take place outside parks and other protected areas and, thus, become integrated into agricultural and urban systems. Conservation outside parks could become important in opening new economic opportunities. Ecotourism could provide important opportunities to link conservation and development. An example is agrotourism, which could help conserve cultural landscapes, add value to farming systems, and address economic needs (Chapter 14).

Recreation, conservation, and environmental education can go hand in hand. Cultural tourism can serve to educate people about the importance of cultural diversity, as well as the importance of the latter for the conservation of biodiversity, provided the risks mentioned above are taken into account. Tourism and recreation can be linked to environmental education, fostering knowledge about the functioning of ecosystems and provoking tourists to critically examine human–nature relations. Environmental education may serve very diverse audiences, ranging from schoolchildren to university students, protected area managers, policy-makers, and representatives of the private sector. In all cases, top-down education is less effective than education that is based on sharing experiences and attempts to reach a joint understanding of the dynamics of human–nature interactions (Chapter 14).

18.4.5 Considering Business Impacts in Public Policy

Despite the potential to positively engage business in providing solutions to questions about pressures on ecosystem services and human well-being, the financial impact that different response options have on corporations has received relatively little attention by the MA and by the public policy literature in general (Andrews 1998; Khanna 2001; Rivera 2002). This oversight is specifically highlighted in MA *Multiscale Assessments,* and arises, perhaps, because estimating business benefits is seen as more important for corporations than for decision-makers interested in ecosystem management. For instance, Chapter 3 of this volume does not explicitly address how to evaluate the cost and benefits of the response options for corporations.

Yet taking into account private sector benefits and costs is critical for the selection and implementation of response options. Response options that are too costly for firms exacerbate the traditional resistance from the business community to ecosystem protection measures making their enactment and implementation very difficult (Andrews 1998; Henriques and Sadorsky 1996; Highley and Leveque 2001; Rivera 2002). Conversely, win-win alternatives that promote ecosystem protection and provide direct incentives for businesses are more likely to have successful implementation (Chapters 2, 5, 8, and 15). For example, guaranteeing the sustainability of supplies may present one of the most persuasive cases for businesses to proactively protect biodiversity (Chapter 5).

The reaction of corporations to different response options is also affected by the combination of regulatory enforcement and consumer preferences. South Africa and Costa Rica are examples that exhibit the synergetic potential between ecotourism demand, increasingly stringent protection of national parks, and proactive environmental protection by tourism-related business (Chapter 5). On the other hand, firms operating in countries or regions with weak oversight from government, environmental groups, and/or other stakeholders show little interest in adopting ecosystem management practices even when they may have a positive effect on their bottom line (Cashore and Vertinisky 2000; Khanna et al. 1998; Henriques and Sadorsky 1996; Rivera 2001; Rivera and deLeon 2004).

Empirical findings from studies implemented in different parts of the world consistently suggest that besides offering financial incentives to corporations, traditional mandatory pressures are key ingredients for encouraging the proactive protection of ecosystems by the business sector (Chapter 15) (Wheeler 1999; Cashore and Vertinisky 2000; Khanna et al. 1998; Henriques and Sadorsky 1996; Rivera 2004; Rivera and deLeon 2004). Consumer preferences can also reinforce the pressures from regulators and stakeholders to promote proactive ecosystem management by the private sector. Markets with sizeable segments of environmentally aware (or "green") consumers significantly increase the incentives for proactive protection of ecosystems by corporations (Reinhardt 1998; Rivera 2002). For example, certification programs have taken advantage of increased demand for environmentally friendly wood products to promote sustainable forestry management practices in different parts of the world (Chapters 8 and 15).

Finally, empirical research also highlights the importance of training and technical assistance to promote proactive ecosystem protection practices among businesses. Virtually all chapters of the MA highlight that the lack of ecosystem management expertise is a fundamental barrier to improving protection of ecosystems. Higher education and environmental expertise appear to increase CEOs' recognition of the intrinsic value of nature and their perceived sense of ethical duty to protect it (Ewert and Baker 2001; Rivera and deLeon 2005; Cottrell 2003; Wiersema and Bantel 1992; Hambrick and Mason 1984). CEOs with higher education and natural resources management expertise can also be expected to be more aware of innovative technologies that lead to cost savings in the form of reduced waste, energy savings, and use of recycled materials (Hart 1995; Rivera and deLeon 2005). These CEOs may also have a better understanding of how an enhanced "green" reputation, generated by proactive ecosystem management, would create differentiation advantages in the form of price premiums and higher sales for their companies (Reinhardt 1998; Rivera 2002).

18.5 Summary Conclusions for Governments and Civil Society Organizations

Decisions or responses regarding ecosystem services are made at different levels by decision-makers identified in Chapter 2 by their scale of operation. This section briefly summarizes the main messages for decision-makers in governments (including, in principle, not only national but also international and sub-national levels) and civil society.

Government decision-makers should consider the factors that can facilitate effective responses. The most important ones include:

1. *Developing institutions that enable effective coordination of decision-making across multiple sectors.* Many ecosystems are managed in a sectorally arranged structure, (for example, by various ministries such as agriculture, environment, or industry) which is not conducive to effective horizontal coordination. In this way, the cross-sectoral trade-offs are difficult to resolve.
2. *Strengthening of institutions at a lower level of governance.* Regional and local governments often lack both sufficient capacity and empowerment to work properly. The decision-making at the sub-national and community level is better suited to holistic approaches. On the other hand, overly decentralized decision-making could also lead to poor ecosystem service management.
3. *Extending participation procedures focusing on the earliest phases of the decision-making cycle.* This includes increasing transparency and accountability of government decision-making, encouraging and supporting independent monitoring and assessment of government performance, and securing access to information and justice for all stakeholders.
4. *Promoting "win-win" solutions by creating an economic framework that supports proper management of ecosystem services.* This includes correcting market failures and internalizing negative environmental externalities. Because many ecosystem services are not traded, markets often fail to provide appropriate signals for optimal allocation of services. This unfavorable situation is exacerbated by harmful subsidies that promote the excessive use of some ecosystem services.

Agriculture subsidies promoting overproduction and/or overuse of fertilizers and pesticides are an example.

5. *Increasing emphasis on demand-side management and on the reduction of negative trade-offs.* As the per capita supply of services drops and the costs associated with production increase, greater gains can often be achieved through actions designed to reduce demand for harmful trade-offs rather than actions aimed at further increases in production. For example, in agriculture, the net economic gains from steps taken to reduce post-harvest losses, to reduce water pollution associated with fertilizer use, or to increase water use efficiency may often exceed the net gains from further investment in increased productivity.
6. *Building human and institutional capacity to assess the consequences of ecosystem change for human well-being and to properly manage ecosystems.* Current human and institutional capacity is extremely limited in all countries. To improve the situation, more and better-trained natural and social scientists and appropriate institutions are needed, as are effective mechanisms for incorporating local and traditional knowledge, dissemination of information, and dialogue with involved stakeholders.
7. *Requiring companies to publicly report on their environmental performance.* Asking companies to report on emissions in key areas and disclosing environmental liabilities (such as hazardous materials use) increase incentives for improved ecosystems management.
8. *Increasing emphasis on adaptive management.* Management interventions should always include a significant monitoring component, which would allow greater learning about the consequences of the interventions and improved management with time.

Civil society organizations should consider the following (based on WRI 2003):

1. *Stimulating demand for access to information, participation, and justice.* There may be gaps in national practices of access and so the corrective actions have to be encouraged. It is necessary to build the capacity of the community to engage in the public participation system.
2. *Providing objective information.* As many opinion polls show, the public considers the information provided by NGOs to be the most reliable. Undertaking independent assessment and regular monitoring of the activities of both the governmental and private sectors regarding the management of ecosystem services and their statutes is one of the main tasks for the civil society organizations. An important prerequisite for such an activity is sufficient capacity (knowledge, interests, the right and the ability to participate, etc.).
3. *Raising awareness among the public and the decision-makers of "emerging issues" such as nutrient loading.* Civil society organizations play a unique role in bringing new issues to the attention of the public and the decision-makers through public education and lobbying. The implications of many of the changes underway in ecosystems are simply not known by the public or by decision-makers. Without greater public support it will often be difficult for government officials to take actions that they know are important. Moreover, civil society organizations can help to hold decision-makers accountable for the actions that they do take.
4. *Encouraging greater access to information on the status and trends in ecosystem services, greater monitoring of those services, and greater quantification of the non-marketed benefits obtained from ecosystem services.* Civil society organizations can also help to ensure that appropriate consideration is given to non-utilitarian values in decision-making.
5. *Embracing the same policies of accountability and transparency about its own operations as are advocated for governments and corporations.* The policy of full openness about the funding, purposes, goals, activities, and accomplishments should be a cornerstone of any civil society group. First of all, it shall be accountable to the community it lives in.
6. *Building coalitions.* The consensus-building coalitions of NGOs and other like-minded stakeholders greatly increase the leverage of individual members. Priority attention should be given to enhancing alliances with NGOs from developing countries. The involvement of stakeholders who are at the highest risk and most vulnerable to the effects of ecosystems change is essential. The coalitions can also provide assistance to such groups including detailed information on ecosystems and their services.
7. *Partnering with corporations.* NGOs are often effective public watchdogs by compiling, analyzing, and publicizing corporate environmental performance data. In addition, they may partner with industry to encourage the best practices necessary to achieve environmentally benign products, support environmental innovation, and even encourage various forms of environmental philanthropy.
8. *Initiating and implementing certification schemes.* NGOs are the most trusted institutions regarding certification of sustainably manufactured, harvested, or extracted products. In the case of forest products, the NGOs' actions are highly successful (Chapter 8).

References

Adriaanse, A., 1993: *Environmental Policy Performance Indicators: A Study on the Development of Indicators for Environmental Policy in the Netherlands,* SDU, Uitgevereji, The Hague, The Netherlands.

Andrews, R., 1998: Environmental regulation and business self-regulation, *Policy Sciences,* **31,** pp. 177–197.

Ascher, W., 1999: *Why Governments Waste Natural Resources,* The John Hopkins University Press, Baltimore, MD.

Barkenbus, J., 1998: Expertise and the policy cycle, Technical report NCEDR/99-04, University of Tennessee, Knoxville, TN. Available at http://www.ncedr.org/publications/papersreports.htm.

Berkes, F., 2004: Rethinking community-based conservation, *Conservation Biology,* **18,** pp. 621–630.

Brown, K., 2002: Innovations for conservation and development, *The Geographical Journal,* **168(1)**; pp. 6–17.

Cashore, B. and I. Vertinisky, 2000: Policy networks and firm behaviors: Governance systems and firm responses to external demands for sustainable forest management, *Policy Sciences,* **33,** pp. 1–30.

Clancy L., P. Goodman, H. Sinclair and D.W. Dockery, 2002: Effects of air pollution control on death rates in Dublin, Ireland: An intervention study, *Lancet,* **360(9341)** , pp. 1210–1214.

Clark, W., N. Eckley, A. Farrell, J. Jäger, and D. Stanners, 2001: *Designing Effective Assessments: The Role of Participation, Science and Governance, and Focus,* Report for the Global Environmental Assessment Project and the European Environment Agency, Kennedy School of Government, Harvard University Press, Cambridge, MA.

Costanza, R. and C. Folke, 1996: The structure and function of ecological systems in relation to property rights regimes. In: *Rights to Nature,* S. Hanna, C. Folke, and K.G. Maler (eds.), Island Press, Washington, DC.

Cottrell, S., 2003: Influence of sociodemographics and environmental attitudes on general responsible environmental behavior among recreational boaters, *Environment and Behavior,* **35(3),** pp. 347–375.

CRED (Center for Research on the Epidemiology of Disasters), 2002: EM-DAT: The Office of Foreign Disaster Assistance/CRED International Disaster Database, CRED. Available at http://www.em-dat.net/.

CSA (Canadian Standards Association), 1997: *Risk Management: Guideline for Decision-Makers,* CAN/CSA-Q850–97, Canadian Standards Association, Etobicoke, Toronto, ON, Canada.

Cyert, R.M. and J.G. March, 1963: *A Behavioral Theory of the Firm,* Prentice-Hall, Englewood Cliffs, NJ.

Dale, V.H. and M.R. English (eds.), 1998: *Tools to Aid Environmental Decision-Making,* Springer-Verlag, New York, NY.

Davies, J. C. and J. Mazurek, 1998: *Pollution Control in the United States: Evaluating the System,* Resources for the Future, Washington, DC.

Dearing, J.W. and E.M. Rogers, 1996: *Agenda-Setting,* Sage Publications, Thousand Oaks, CA.

Dietz, T., 2002: *What are Good Decisions? Environment, Democracy and Science,* Department of Environmental Science and Policy, George Mason University, Fairfax, VA.

Dietz, T. and P.C. Stern, 1998: Science, values, and biodiversity, *BioScience,* **48,** pp. 441–444.

Dietz, T., E. Ostrom, and P.C. Stern, 2003: The struggle to govern the commons, *Science,* **302,** pp. 1907–1912.

Downing, T.E. and M. Ludeke, 2002: International desertification: Social geographies of vulnerability and adaptation. In: *Global Desertification: Do Humans Cause Deserts?* J.F. Reynolds and D.M. Stafford Smith (eds.), Dahlem University Press, Berlin, Germany.

Downs, A., 1998: Political Theory and Public Choice. Edward Elgar, Northampton, MA, pp. 192.

EEA (European Environmental Agency), 1999: Environmental indicators: Typology and overview, Technical report no 25, EEA, Copenhagen, Denmark, 19 pp.

EEA, 2001: *Reporting on Environmental Measures: Are We Being Effective?* Environmental issue report no 25, EEA, Copenhagen, Denmark, 35 pp.

Eriksen, S., 2000: *Responding to Global Change: Vulnerability and Management of Agro-ecosystems in Kenya and Tanzania,* Ph.D. thesis, Climate Research Unit, School of Environmental Sciences, University of East Anglia, Norwich, UK.

Ewert, A. and D. Baker, 2001: Standing for where you sit: An exploratory analysis of the relationship between academic major and environment beliefs, *Environment and Behavior,* **33(5),** pp. 687–707.

Gleick, P.H. (ed), 1993: *Water in Crisis: A Guide to the World's Fresh Water Resources.* Oxford University Press, New York.

Hambrick, D. and P. Mason, 1984: Upper echelons: The organization as a reflection of its top managers, *Academy of Management Review,* **9(2),** pp. 193–206.

Hart, S., 1995: A natural resource based view of the firm, *Academy of Management Review,* **20,** pp. 986–1014.

Henriques, I. and P. Sadorsky, 1996: The determinants of an environmental responsive firm: An empirical approach, *Journal of Environmental Economics and Management,* **30,** pp. 381–395.

Highley, C.J. and F. Leveque (eds.), 2001: *Environmental voluntary approaches: Research Insights for Policy-Makers,* Fundazione Eni Enrico Mattei, Trieste, Italy.

ICSU (International Council for Science), 2002a: *Science, Traditional Knowledge and Sustainable Development,* ICSU series on science for sustainable development no. 4, ICSU, Paris, France, 24 pp.

ICSU, 2002b: *Making Science for Sustainable Development More Policy Relevant: New Tools for Analysis,* ICSU series on science for sustainable development no. 8, ICSU, Paris, France, 28 pp.

International Red Cross Federation, 2002: *World Disasters Report 2002: Focus on Reducing Risks,* International Red Cross Federation and Red Crescent Societies, Geneva, Switzerland, 239 pp.

International Strategy for Disaster Reduction Secretariat, 2002: *Living with Risk: A Global Review of Disaster Reduction Initiatives,* ISDR, Geneva, Switzerland.

Khanna, M., 2001: Non-mandatory approaches to environmental protection, *Journal of Economic Surveys,* **15(3),** pp. 291–324.

Khanna, M., W.H. Quimio, and D. Bojilova, 1998: Toxics release information: A policy tool for environmental protection, *Journal of Environmental Economics and Management,* **36,** pp. 243–266.

Kingdom, J.W., 1984: *Agendas, Alternatives, and Public Choices,* Little Brown, Boston, MA.

Lasswell, H.D., 1947: *The Analysis of Political Behavior,* Kegan Paul Co. Ltd., London, UK.

Lindblom, C. and E. Woodhouse, 1993: *The Policy-making Process,* 3rd ed., Prentice Hall, Upper Saddle River, NJ.

MA (Millennium Ecosystem Assessment), 2003: *Ecosystems and Human Well-being: A Framework for Assessment,* Island Press, Washington, DC, 245 pp.

Moldan, B. and S. Billharz (eds.), 1997: *Sustainability Indicators: A Report on the Project on Indicators of Sustainable Development, SCOPE 58,* John Wiley & Sons, Chichester, UK, 415 pp.

Molenaar, B., 2002: The wild east: The impact of illegal logging on a local population, *Human Rights Tribune, Reports from the Field,* **9(1)** (Spring). Available at http://www.hri.ca/tribune/viewArticle.asp?ID=2667.

Morgan, M.G. and M. Henrion, 1990: *Uncertainty: A Guide to Dealing with Uncertainty in Quantitative Risk and Policy Analysis,* Cambridge University Press, New York, 344 pp.

Munich Re, 2003: *Topics: Annual Review of Natural Catastrophes 2002,* Munich, Germany. Available at http://www.munichre.com.

Nel, E. and P. Illgner, 2004: The contribution of bees to livelihoods in southern Africa. In: *Rights, Resources & Rural Development,* C. Fabricius and E. Koch with H. Magome and S. Turner (eds.), Earthscan Publications, London, UK/ Sterling, Virginia, pp. 127–134.

North, D.C., 1990: *Institutions, Institutional Change, and Economic Performance,* Cambridge University Press, Cambridge, UK.

NRC (National Research Council), 1996: *Understanding Risk: Informing Decisions in a Democratic Society: Committee on Risk Characterization,* National Research Council, National Academy Press, Washington, DC, 264 pp.

NRC, 1999a: *Toward Environmental Justice: Research, Education, and Health Policy Needs,* NRC, National Academy Press, Washington, DC, 137 pp.

NRC, 1999b: *Perspectives on Biodiversity: Valuing Its Role in an Ever-changing World,* NRC, National Academy Press, Washington, DC, 168 pp.

NRC, 2000: *Clean Coastal Waters: Understanding and Reducing the Effects of Nutrient Pollution,* NRC, National Academy Press, Washington, DC.

OECD (Organisation for Economic Co-operation and Development), 2002: *OECD Guidance Document on Risk Communication for Chemical Risk Management,* ENV/JM/MONO(2002)18, OECD, Paris, France.

Ola-Adams, B.A., 2001: Education, awareness building and training in support of biosphere reserves: Lessons from Nigeria's parks, *IUCN,* **11(1),** pp. 18–23.

Ostrom, E., J. Burger, C.B. Field, R.B. Norgaard, and D. Policansky, 1999: Revisiting the commons: Local lessons, global challenges, *Science,* **284,** pp. 278–282.

Ostrom, T., N. Dietz, P.C. Dolsak, S. Stern, S. Stonich, et al. (eds.), 2002: *The Drama of the Commons: Committee on the Human Dimensions of Global Change,* National Academy Press, Washington, DC, 534 pp.

Ottman, J., 1998: *Green Marketing: Opportunity for Innovation.* McGraw Hill, New York, NY.

Petkova, E., C. Maurer, N. Henninger, F. Irwin, J. Coyle, et al., 2002: *Closing the Gap: Information, Participation, and Justice in Decision-making for the Environment,* WRI, Washington, DC, 157 pp.

Quong, A., 1993: *The Implications for Traditional Beekeeping in Tabora Region, Tanzania for Miombo Woodland Conservation,* School of International Training, Dar es Salaam, Tanzania.

Ravetz, J.R., 1986: Usable knowledge, usable ignorance: Incomplete science with policy implications. In: *Sustainable Development of the Biosphere,* W.C. Clark and T.E. Munn (eds.), Cambridge University Press, Cambridge, UK, pp. 415–434.

Rayner, S. and E.I. Malone (eds.), 1998: *Human Choice and Climate Change, Vols 1–4,* Battelle Press, Columbus, OH.

RCEP (Royal Commission on Environmental Pollution), 1998: *Setting Environmental Standards,* 21st Report of RCEP, Her Majesty's Stationery Office, Norwich, UK.

Reed, D.J., 2001: *Stalking the Elusive Business Case for Corporate Sustainability: Sustainable Enterprise Perspectives Note,* WRI, Washington, DC, 24 pp.

Reinhardt, F.L., 1998: Environmental product differentiation: Implications for corporate strategy, *California Management Review,* **40(4),** pp. 43–73.

Rivera, J., 2001: *Does It Pay to Green in the Developing World? Participation in Voluntary Environmental Programs and Its Impact on Firm Competitive Advantage,* Academy of Management Best Paper Proceedings, Washington, DC.

Rivera, J., 2002: Assessing a voluntary environmental initiative in the developing world: The Costa Rican certification for sustainable tourism, *Policy Sciences,* **35,** pp. 333–360.

Rivera, J., 2004: Institutional pressures and voluntary environmental behavior in developing countries: Evidence from Costa Rica, *Society and Natural Resources,* **17,** pp. 779–797.

Rivera, J. and P. de Leon, 2005: CEOs and environmental performance in Costa Rica's certification for sustainable tourism, *Policy Sciences.* Forthcoming.

Rivera, J. and P. deLeon, 2004: Is greener whiter? The sustainable slopes program and the voluntary environmental performance of western ski areas, *Policy Studies Journal,* **32(3),** pp. 417–437.

Rondinelli, D. and T. London, 2003: How corporations and environmental groups cooperate: Assessing cross-sector alliances and collaborations, *Academy of Management Executive,* **17(1),** pp. 61–76.

Rosenfeld, A., 1998: Evaluacion de Sostenibilidad Agroecologia de Pequenos Productores (Misiones- Argentina) Tesis de Maestria, Univisidad Internacional de Andalucia, Espana.

Rotmans, J. and P. Vellinga (eds.), 1998: Special issue: Challenges and opportunities for integrated environmental assessment, *Environmental Model Assessment,* **3(3),** pp. 135–207.

Simon, H.A., 1976: *Administrative Behavior,* Free Press, New York, NY.

Simon, H.A., 1985: Human nature in politics: The dialogue of psychology with political science, *American Political Science Review,* **79,** pp. 293–304.

Starbuck, W.H. and F.J. Milliken, 1988: Challenger: Fine tuning the odds until something breaks, *Journal of Management Studies,* **25,** pp. 319–340.

Stern, P.C., T. Dietz, N. Dolsak, E. Ostrom, and S. Stonich, 2002: Knowledge and questions after fifteen years of research. In: *Drama of the Commons,* E. Ostrom, T. Dietz, N. Dolsak, P.C. Stern, S. Stonich, et al. (eds.), National Academy Press, Washington, DC, pp. 443–490.

Stern, P.C. and W.E. Easterling (eds.), 1999: *Making Climate Forecasts Matter,* NRC, National Academy Press, Washington, DC, 192 pp.

Social Learning Group, 2001: *Learning to Manage Global Environmental Risks, Volume 1: A Comparative History of Social Responses to Climate Change, Ozone Depletion, and Acid Rain, Volume 2: A Functional Analysis of Social Responses to Climate Change, Ozone Depletion, and Acid Rain,* The MIT Press, Cambridge, MA, 376 + 226 pp.

Swiss Re, 2003: *Natural Catastrophes and Man-made Disasters in 2002,* Sigma no.2/2003. Available at http://www.swissre.com.

Taiga Rescue Network, 2003: Who owns the taiga? Who has rights to use the boreal and forests and how are those rights regulated? Articles in this issue focus on land tenure in the boreal forests, *Taiga News,* **43** (Summer), TRN, Jokkmokk, Sweden. Available at http://www.taigarescue.org/_v3/files/taiganews/TN43.pdf.

Toth, F.L. (ed.), 1999: *Fair Weather? Equity Concerns in Climate Change,* Earthscan, London, UK.

Toth, F.L. and E. Hizsnyik, 1998: Integrated environmental assessment methods: Evolution and applications, *Environmental Modeling and Assessment,* **3,** pp. 193–207.

Toth, F.L., 2004: *Environmental Management: From Assessment to Decision-making,* Interim report IR-04-076, International Institute for Applied Systems Analysis, Laxenburg, Austria.

UN (United Nations), 1992: *Rio Declaration on Environment and Development,* The UN Conference on Environment and Development, 3–14 June, Rio de Janeiro, Brazil.

UNDP (UN Development Programme), 2003: *Making Global Trade Work for People,* UNDP and Earthscan Publications, New York, NY.

UNEP (UN Environment Programme), 2004: *Statement of Environmental Commitment by the Insurance Industry,* Insurance Industry Initiative, UNEP, Nairobi, Kenya. Available at http://unepfi.net/iii/statemen.htm.

UNEP/DPCSD (Department for Policy Coordination and Sustainable Development), 1995: The role of indicators in decision-making. In: *Indicators of Sustainable Development for Decision-making,* N.Gouzee, B.Mazijn, and S.Billharz (eds.), Report of the workshop at Ghent, Belgium, 9–11 January, Bureau Fédéral du Plan, Bruxelles, Belgique.

Vermeulen, S. and I. Koziell, 2002: *Integrating Global and Local Biodiversity Values: A Review of Biodiversity Assessment,* International Institute for Environment and Development, London, UK, 104 pp.

Vogel, C. and J. Smith, 2002:The politics of scarcity: Conceptualizing the current food security crisis in Southern Africa, *South African Journal of Science,* **98(7/8),** pp. 315–317.

Walker, B.H. and W. Steffen (eds.), 1996: *Global Change and Terrestrial Ecosystems,* Cambridge University Press, Cambridge, UK.

Walsh, J.P., 1988: Selectivity and selective perception: An investigation of managers' belief structures and information processing, *Academy of Management Journal,* **31,** pp. 873–896.

Wheeler, D., 1999: *Greening Industry: New Roles for Communities, Markets, and Governments,* Oxford University Press, New York for World Bank, Washington, DC, 113 pp.

Wiersema, M. and K. Bantel, 1992: Top management team demography and corporate strategic change, *Academy of Management Journal,* **35(1),** pp. 91–121.

Wisner, B., P. Blaikie, T. Cannon, and I. Davis, 2004: *At Risk: Natural Hazards, Peoples' Vulnerability, and Disasters,* Routledge, London, UK, 464 pp.

Wolf, A., J. Natharius, and A. Al, 1999: International river basins of the world, *International Journal of Water Resources Development,* **15(4),** pp. 387–428.

World Bank, 1996: *Participation Sourcebook,* World Bank, Washington, DC, 259 pp.

WRI (World Resources Institute), 2003: *World Resources 2002–2004: Decisions for the Earth: Balance, Voice, and Power,* WRI, Washington, DC, 328 pp.

Chapter 19

Implications for Achieving the Millennium Development Goals

Coordinating Lead Authors: Diana Wall, Rudy Rabbinge

Lead Authors: Gilberto Gallopin, Kishan Khoday, Nancy Lewis, Jane Lubchenco, Jerry Melillo, Guido Schmidt-Traub, Mercedita Sombilla

Contributing Author: Lina Cimarrusti

Review Editors: Tony La Viña, Mohan Munasinghe, Wang Rusong

BOXES

FIGURES

TABLES

Main Messages

Progress toward achieving the 2015 targets of the Millennium Development Goals will need to be accelerated dramatically *(high certainty)*. In particular, sub-Saharan Africa, Central Asia, parts of South and Southeast Asia, and some regions in Latin America, are currently off track with respect to meeting the goals.

Knowledge and information about ecosystems and ecosystem services are vital for developing ways to achieve target 9 on environmental sustainability *(high certainty).* The MA provides information about the ways in which ecosystems and the services they provide to local, regional, and global communities affect human well-being. The evidence synthesized by the MA underlines that ecosystem services can only be sustained in the long term if the integrity and completeness of ecosystems are maintained or restored. This information and the tools for improved management of ecosystems need to be integrated more systematically into development strategies such as poverty and hunger reduction strategies.

The MDGs and their 15 targets form a set of highly interdependent objectives that can only be met through integrated strategies instead of isolated interventions or "silver bullets." Greater collective gains are possible through simultaneous rather than sequential interventions *(high certainty).* This integrated and synchronous approach requires a focus on improved management of ecosystems and their services. This is a particularly important prerequisite for achieving the targets relating to poverty, hunger, gender equality, water, and sanitation and health. Countries that are not on track to achieving the 2015 targets are experiencing rapid environmental degradation and loss of ecosystem services that can be slowed or reversed through improved ecosystem management. In many places, the sustainability and continuity of particularly agroecosystems is threatened by structural shortage of measures to maintain their services and productivity. By restoring those functions, there is more room for other less productive systems, but that requires clear choices at the local, regional, national, and international level.

Particular emphasis needs to be placed on the sustainable intensification of existing cultivated ecosystems to satisfy growing demand for food and other ecosystem services *(high certainty).* Maintaining the present cultivated land in lieu of expanding into new areas will be possible when intensification and modernization of present agroecosystems is promoted with higher productivity per hectare, per person hour, and per kilogram input As a result of this intensification counter intuitively total use of inputs such as fertilizers or pesticides is decreased in absolute terms and per unit of product. Thus environmental side effects decrease. Moreover the higher productivity per ha spares space for nature and untouched ecosystems, thus safeguarding as much as possible biodiversity. Thus the protection of fragile and vulnerable ecosystems such as wetlands, mangroves, and upland areas that provide many ecosystem services is facilitated.

Intensification of production systems such as agroecosystems (plant and animal) and aquaculture needs to be carried out by the most socially and economically appropriate ecological techniques such that ecosystem degradation is prevented. Various ecosystems are threatened by mismanagement, overuse, or insufficient care for continuity. That holds for many aquaecosystems (overfishing, etc.), rain fed agroecosystems by inadequate insufficient maintenance of soil fertility, and most importantly expansion of cultivated area, logging of tropical rainforest, and cultivating former coastal ecosystems such as mangrove ecosystems. The use of ecotechnological advanced production techniques requires a good understanding of the basic processes (chemical, physical, physiological, ecological) that determine agroecosystem behavior.

Modified ecosystem management as part of a strategy to achieve the 2015 targets has to consider that several drivers affect environmental change. Therefore, policies, institutions and reorientation acting at local, regional, and global scales need to address several drivers at the same time *(medium certainty).* To achieve the 2015 targets, particular attention needs to be placed on improving ecosystem management and the capacity for policy-making at the national and local level as well as addressing global challenges including long-term climate change and the depletion of international fisheries.

The complexity of human–nature interactions makes it difficult, though not impossible, to formulate quantitative targets using the best available science synthesized by the Millennium Ecosystem Assessment *(medium certainty).* In the face of this uncertainty, communities, countries, regions, and the international system need to agree on and set well defined local, national, and regional quantified goals for preserving, managing, and utilizing ecosystems. The findings of the MA may provide guidance for developing targets and for designing strategies to achieve them.

Available monitoring systems for ecosystems and the services they provide are inadequate in many parts of the developing world *(high certainty).* Monitoring and documenting progress toward achieving the 2015 targets may require strengthening monitoring systems for soil fertility, hydrological flows, biodiversity, climate, and so forth. Documenting progress will provide invaluable information for analysis, research, and the development of technology and mitigation strategies.

19.1 Introduction

This chapter explains how the MA contributes to the achievement of the 2015 targets of the Millennium Development Goals. Based on the MA conceptual framework (MA 2003) and key findings regarding current conditions of ecosystem services and alternative scenarios, this chapter explains the implications for achieving the 2015 targets. The chapter integrates the MA findings and focuses them toward the politically and widely accepted MDGs.

With 1990 as the baseline, the MDGs aim to improve human well-being by reducing poverty, hunger, child and maternal mortality; ensuring education for all, controlling and managing diseases, tackling gender disparity, ensuring sustainable development and pursuing global partnerships by 2015. (See Table 19.1.) Like the MA conceptual framework (see Chapter 2, Box 2.1), the MDGs support the multidimensional concept of sustainable development and the potential benefits of identifying the linkages between ecosystem services and HWB. As this chapter suggests, the 2015 targets are more likely to be achieved if the goals are addressed simultaneously.

While the MDGs are focused on human goals, the MA is primarily concerned with ecosystem services and how to best manage them for the benefit of HWB. Ecosystem services are categorized by supporting, provisioning, regulating, and cultural services. (See Table 19.2.) HWB, as defined by the MA, involves multiple constituents, including basic material for a good life, freedoms and choices, health, good social relations, and security (MA 2003).

In adapting the MA conceptual framework to the MDGs, this chapter identifies which goals are directly or indirectly dependent on supporting, provisioning, regulating, and cultural services. While most targets (such as poverty, hunger, gender, child mortality, disease water, and sustainable development) are dependent on ecosystem services as discussed in the goal-by-goal analysis, some

Table 19.1. Millennium Development Goals

Goals and Targets	Indicators
Goal 1: Eradicate extreme poverty and hunger	
Target 1: Halve, between 1990 and 2015, the proportion of people whose income is less than one dollar a day	1. Proportion of population below $1 per day (PPP values) 2. Poverty gap ratio [incidence x depth of poverty] 3. Share of poorest quintile in national consumption
Target 2: Halve, between 1990 and 2015, the proportion of people who suffer from hunger	4. Prevalence of underweight children (under five years of age) 5. Proportion of population below minimum level of dietary energy consumption
Goal 2: Achieve universal primary education	
Target 3: Ensure that, by 2015, children everywhere, boys and girls alike, will be able to complete a full course of primary schooling	6. Net enrolment ratio in primary education 7. Proportion of pupils starting grade 1 who reach grade 5 8. Literacy rate of 15–24 year olds
Goal 3: Promote gender equality and empower women	
Target 4: Eliminate gender disparity in primary and secondary education preferably by 2005 and at all levels of education no later than 2015	9. Ratio of girls to boys in primary, secondary, and tertiary education 10. Ratio of literate females to males of 15–24 year olds 11. Share of women in wage employment in the nonagricultural sector 12. Proportion of seats held by women in national parliament
Goal 4: Reduce child mortality	
Target 5: Reduce by two thirds, between 1990 and 2015, the under-five mortality rate	13. Under-five mortality rate 14. Infant mortality rate 15. Proportion of 1-year-old children immunized against measles
Goal 5: Improve maternal health	
Target 6: Reduce by three quarters, between 1990 and 2015, the maternal mortality ratio	16. Maternal mortality ratio 17. Proportion of births attended by skilled health personnel
Goal 6: Combat HIV/AIDS, malaria, and other disease	
Target 7: Have halted by 2015, and begun to reverse, the spread of HIV/AIDS	18. HIV prevalence among 15–24-year-old pregnant women 19. Contraceptive prevalence rate 20. Number of children orphaned by HIV/AIDS
Target 8: Have halted by 2015, and begun to reverse, the incidence of malaria another major diseases	21. Prevalence and death rates associated with malaria 22. Proportion of population in malaria risk areas using effective malaria prevention and treatment measures 23. Prevalence and death rates associated with tuberculosis 24. Proportion of TB cases detected and cured under DOTS (Directly Observed Treatment Short Course).
Goal 7: Ensure environmental sustainability	
Target 9: Integrate the principles of sustainable development into country policies and programs and reverse the loss of environmental resources	25. Proportion of land area covered by forest 26. Land area protected to maintain biological diversity 27. GDP per unit of energy use (as proxy for energy efficiency) 28. Carbon dioxide emissions (per capita) (Plus two figures of global atmospheric pollution: ozone depletion and accumulation of global warming gases)
Target 10: Halve, by 2015, the proportion of people without sustainable access to safe drinking water	29. Proportion of population with sustainable access to an improved water source
Target 11: By 2020, to have achieved a significant improvement in the lives of at least 100 million slum dwellers	30. Proportion of people with access to improved sanitation 31. Proportion of people with access to secure tenure (urban/rural disaggregation of several of the above indicators may be relevant for monitoring improvement in the lives of slum dwellers)

Goal 8: Develop a global partnership for development	
Target 12: Develop further an open, rule-based, predictable, non-discriminatory trading and financial system (includes a commitment to good governance, development, and poverty reduction, both nationally and internationally)	*Official Development Assistance* 32. Net ODA as percentage of DAC donors' GNP (targets of 0.7% in total and 0.15% for LDCs) 33. Proportion of ODA to basic social services (basic education, primary health care, nutrition, safe water, and sanitation) 34. Proportion of ODA that is untied 35. Proportion of ODA for environment in small island developing states 36. Proportion of ODA for transport sector in land-locked countries
Target 13: Address the special needs of the least developed countries (includes tariff and quota free access for LDC exports; enhanced program of debt relief for heavily indebted poor countries and cancellation of official bilateral debt; and more generous ODA for countries committed to poverty reduction)	*Market Access* 37. Proportion of exports (by value and excluding arms) admitted free of duties and quotas 38. Average tariffs and quota on agricultural products and textiles and clothing 39. Domestic and export agriculture subsidies in OECD countries 40. Proportion of ODA provided to help build trade capacity
Target 14: Address the special needs of landlocked countries and small island developing states (through Barbados Programme and 22nd General Assembly provisions)	
Target 15: Deal comprehensively with the debt problems of developing countries through national and international measures to make debt sustainable in the long term	*Debt Sustainability* 41. Proportion of official bilateral HIPC debt cancelled 42. Debt service as a percentage of exports of goods and services 43. Proportion of ODA provided as debt relief 44. Number of countries reaching HIPC decision and completion points

Table 19.2. Ecosystem Services Clustered by Major Categories

Supporting Services *(Services necessary for the production of other ecosystem services)*	Provisioning Services *(Products obtained from ecosystems)*	Regulating Services *(Benefits obtained from regulation of ecosystem processes)*	Cultural Services *(Nonmaterial benefits obtained from ecosystems)*
Soil formation	Food	Climate regulation	Spiritual and religious
Nutrient cycling	Freshwater	Disease regulation	Recreation and ecotourism
Primary production	Fuelwood	Water regulation	Aesthetic
Pollination	Fiber	Water purification	Inspirational
Habitat maintenance	Biochemicals	Landscape stabilization against erosion	Educational
Seed dispersal	Genetic resources	Binding of toxic compounds	Sense of place
		Air purification	Cultural heritage
		Pest and pathogen control	
		Carbon sequestration	

2015 targets are less dependent or indirectly linked to ecosystem services and these targets include education, maternal mortality, and global partnerships.

Emanating from the United Nations Millennium Declaration, the eight MDGs bind countries to do more and join forces in the fight against poverty, illiteracy, hunger, lack of education, gender inequality, child and maternal mortality, disease, and environmental degradation. The eighth goal, reaffirmed in Monterrey and Johannesburg, calls on industrial countries to relieve debt, increase aid, and give poor countries fair access to their markets and their technology. The MDGs are a test of political will to build stronger partnerships. Developing countries have the responsibility to undertake policy reforms and strengthen governance to liberate the creative energies of their people. But they cannot reach the goals on their own without new aid commitments, equitable trading rules, and debt relief. The goals offer the world a means to accelerate the pace of development and to measure results. While there may be alternative views on what the MDG goals and targets should be, we are confining our analysis to the official ones as established by the Millennium Declaration.

The MDGs' first focus is on priority countries that face the greatest hurdles in achieving the goals. The international community has been closely monitoring the achievement of the 2015 targets in particular regions of the world, including sub-Saharan Africa, Central Asia, parts of South and Southeast Asia, and Latin America. Recent trends published by the World Bank and the United Nations Development Programme indicate that these regions have a long way to go to reach the 2015 targets. Sub-Saharan Africa, for example, has seen an increase in maternal deaths and in poverty (those living on less than $1 per day) and it is forecasted that the number of people living in poverty will rise from 315 million in 1999 to 404 million by 2015 (World Bank 2003).

The Millennium Task Forces that were established by the U.N. Millennium Project is a major initiative addressing each MDG. They have identified specific interventions for each goal. Whereas the Task Force's interventions are distinct and consider immediate action, the interventions identified in this chapter range from short to long term and are broader in their perspective. The chapter authors have defined the interventions in consideration of those of the U.N. Millennium Project's Task Forces.

The MA has much to share on the ecological challenges and opportunities for poverty alleviation in various regions with strong focus on sub-Saharan Africa as well as other priority regions such as South Asia and Latin America where ecological change is expected to have major consequences for HWB.

The MA scenarios can serve to support analysis of the status of ecosystems by 2015, albeit only under certain MA scenario themes where 2015 has expressly been used as a target date for analysis. Though the MA is looking to 2050, there are lessons from it for 2015, and particularly for gauging what different paths to 2015 and beyond may hold for sustaining HWB in the longer-term. Having said this, it is noted that the MA scenarios were not produced with the MDGs in mind and, therefore, they cannot be expected to be directly applicable to the goals.

Achieving the 2015 targets and sustaining these targets beyond 2015 will require that we preserve the productive capacity of natural ecosystems for future generations. The challenge has two dimensions: addressing natural resource scarcity for the world's poor people and reversing environmental damage resulting from high consumption by rich people. For instance, soil degradation affects nearly two billion hectares, damages livelihoods of up to one billion people living on drylands. Around 70% of commercial fisheries are either fully or overexploited and 1.7 billion people—a third of the developing world population—live in countries facing water stress.

Also of utmost importance to achieving the 2015 targets is the fact that many ecological changes of global importance to all countries are suffering from pressures not only within poor countries but from the world's wealthy. Climate change is an example, where high emissions by industrial countries may well impact the sustainability of development activities in sub-Saharan Africa and elsewhere. However, poverty as such has a detrimental effect on many ecosystems in sub-Saharan Africa and Asia.

This chapter on the MDGs has been constructed through a goal-by-goal analysis presenting a qualitative assessment of how ecosystem services will respond to the proposed interventions. The chapter highlights those interventions that are directly or indirectly related to ecosystem services and the positive or negative effects on them.

The MDG chapter was formed near the final work of the MA, which has limited the depth of detail in the MDG chapter, and our ability to consider numerous case studies. Throughout, we note where our findings are robust and where there are gaps in knowledge. Based on this individual goal-by-goal analysis, we then determine where multiple MDGs might be dependent on the same service. We note that our suggestions are not prescriptive and may, in many cases, be geographically bound.

19.2 Goal-by-goal and Target Analysis

The goal-by-goal analysis examines each of the 2015 targets by addressing dependencies, interventions, synergies, and trade-offs. The examination of dependencies identifies the direct link the target in question has on ecosystem services. Interventions describe potential actions policy-makers can take to achieve the target. Subsequently, synergies and trade-offs identify the potential positive and negative impacts of implementing the suggested interventions.

19.2.1. MDG 1: Eradicate Extreme Poverty and Hunger, Target 1

Target 1: Halve, between 1990 and 2015, the proportion of people whose income is less than one dollar a day.

19.2.1.1 Introduction

In the Millennium Declaration, global poverty is identified as the most daunting of all problems facing the world in the new century. Despite the higher living standards that globalization has delivered in large parts of the world, hundreds of millions of people still experience economic reversals rather than advances (World Bank 2003). Millions of people fight for daily survival as they live on less than $1 per day. Many suffer from low human development, which is defined by UNDP (2003) by longevity (measured by life expectancy at birth), knowledge (measured by adult literacy rate and the net enrollment ratios in primary, secondary and tertiary levels), and standard of living (measured by GDP per capita). Low human development is often associated with poor health and lack of any sort of education as well as of freedom and political voice to choose what is best to improve their well-being (Sen 1999; Duraiappah 1998; Chapter 17). About 70% of the poor live in the rural areas. Most of them are in South Asia and sub-Saharan Africa where coincidence between deepening poverty and acceleration of ecosystem degradation has been most striking in the past decade (Chapter 17).

Helping these people escape the poverty trap requires clear understanding of the complex and dynamic relationship between poverty and environment and the identification and adoption of appropriate intervention measures that will help the poor improve their well-being. Table 19.3 summarizes the dependencies, interventions, synergies, and trade-offs for MDG Target 1.

19.2.1.2 Dependencies

The environment matters greatly to people living in poverty as they often depend directly on a wide range of natural resources and ecosystem services for their livelihoods, health, and sense of empowerment and ability to control their lives (*high certainty*; Chapter 17). Forest ecosystems, agroecosystems, grasslands, and freshwater and coastal ecosystems provide a wide range of services that are essential in increasing agricultural productivity (for example, natural predators, wild pollination, etc); providing primary source of energy (for example, wood fuel); protecting watershed and hydrological stability, including recharging of water tables and buffering of extreme hydrological conditions that might otherwise precipitate drought or flood conditions; maintaining of soil fertility through storage and recycling of essential nutrients; and breaking down waste and pollutants. Healthy ecosystems and their provision of abundant and diversified ecosystem services increase livelihood options, promote livelihood diversification, and improve food security, especially for those living in more marginal environments where access to external technology and other inputs are limited.

Agricultural productivity improvement, the cornerstone for poverty reduction strategies in many countries, is now seriously threatened by land and water degradation, nutrient mining, extensive ecosystem conversion, growing susceptibility of diseases and build up of pest resistance, and erosion of genetic diversity. Current estimates are that up to one billion people are affected by soil erosion and land degradation due to deforestation, overgrazing, and agriculture. Water scarcity is now a major issue in more than 20 developing countries. Water management decisions without regard to the ecosystems' reaction has led to shifts in the distribution of economic benefits from large numbers of local beneficiaries who obtain their livelihood at a very modest level from fisheries to a few who could afford to invest in international fish exports. Shortage of wood fuel and water imposes time and financial costs on billions of poor households, especially women and girls for their increased time, physical burden, and personal

Table 19.3. MDG 1, Target 1: Dependencies, Interventions, Synergies, and Trade-offs for Poverty

Target	Dependencies on Ecosystems for Achieving Target	Interventions	Links to Ecosystems and Ecosystem Services: Synergies (benefits to ecosystems/ ecosystem services)	Links to Ecosystems and Ecosystem Services: Trade-offs (threats to other types of ecosystems/ecosystem services)
Poverty	opportunities to expand livelihood strategies for the poor: • agricultural production expansion • rural-based and non-rural employment opportunities	economic interventions • develop infrastructure facilities to help make markets function correctly • promote agro-based industries financial interventions • establish micro-credit schemes • establish risk management institute technical interventions • develop appropriate technologies based on traditional practices social interventions • more effective extension and training services to educate the poor and enhance their capacity to manage ecosystem other interventions • effective property rights policy, for example, rational land reform system	less pressure on the use of natural resources for production and reduced exploitation of crop/animal/fish/marine species and their wild relatives more efficient ecosystem management demographic transition toward smaller families	change in land use brought about by migration and shifts in demand preferences for food and other needs more rapid industrialization development of tourism and recreational centers, etc. development of infrastructure facilities greater competition for water, energy, etc. increased pollution of air, fresh waters, estuaries, and coastal marine ecosystems

risk in having to travel greater distances to collect fuel, fodder, and clean water.

Where employment and output needs are intense, the result is exploitation of the ecosystem and the services they render. The poor allocate a low weight to the future since poverty requires prioritizing immediate needs and thus encourages the overutilization of resources. To them, their decisions are rational in the context of the constraints they face in the short term. But resource degradation and reduction of ecosystem services cannot be totally blamed on them. People living in the industrial world who keep high living standards likewise continue to consume and exploit resources, even from developing countries.

The causal relationship between poverty and environmental degradation runs in both directions (Dasgupta et al. 1994; Southgate 1998). Poverty leads to natural resources degradation/depletion through a very high rate of discount. Conversely, persistent environmental degradation (caused by industrial countries as well) can contribute to poverty, particularly among subsistence farmers, fishermen, livestock keepers, nomads, and others. Suitable adaptive intervention measures need to embrace, understand, and respect the complexity of this linkage not only in the developing countries but also in industrial countries.

19.2.1.3 Interventions

Numerous interventions have been identified to curb the unsustainable relationship between poverty and ecosystem use and to break the unsustainability spiral through the use of external inputs. In identifying the appropriate intervention measures, it is necessary to consider the different degrees and types of use of ecosystem services. This ensures that no stakeholders are marginalized in the process and avoids the ever increasing conflicts caused by appropriation of ecosystem services by some groups at the expense of other disenfranchised groups.

The key intervention measures are focused on increasing the entitlements of the poor through greater choices and enhanced sources of livelihood opportunities for farmers and rural folks (Sen 1991). Suitable measures that would enhance the capability of the poor and strengthen their resource rights to increase their economic and social entitlements have been found to play a fundamental role in the poverty–environment nexus. Individual or collective rights to some resources provide additional income opportunities and access to various services such as credit, training, and extension. Strengthening land rights, rationalizing land reform programs, providing water permits, allowing direct management of forests by indigenous people, and promoting company–community or private–public partnership of productive public resources including the protection of local scientific knowledge could facilitate more sustainable resource-management investment among the poor *(high certainty)*.

Expanding access to environmentally sound and locally appropriate technologies such as crop production technologies that conserve soil and water and minimize the use of pesticides have been the mainstay in finding substitutes for ecosystem services. But many of these technologies have been developed from outside of the areas where they would be used. This has been the primary reason why many technologies introduced in developing countries have not worked. Technology development should consider and incorporate the indigenous practices and traditional knowledge of farmers *(high certainty)*.

Economic and financial interventions that are aimed at increasing opportunities to enhance incomes will enable poor farmers to participate in the market and take advantage of the benefits of globalization *(medium certainty)*. Promotion of agro-based industries will provide alternative employment opportunities that would make farmers more financially secure in terms of the capital needs for production activities. Low prices and farmers' inaccessibility to markets have put them at the mercy of unscrupulous traders and thereby deterred them from taking advantage of reasonable prices for their produce. This can be reduced by providing efficient functioning labor and credit markets, reliable market information, and appropriate infrastructure to allow easy flow of both products and inputs. Easy access to credit, particularly for the production of traditional foods and handicrafts that tend to be more in balance with ecosystem goods and services, will reduce or eliminate dependence on middlemen that charge huge interest rates that would be paid forever and lead to more reliable and efficient intermediate structures. In addition, the availability of risk management institutions through schemes that are convivial to small producers is extremely essential to keep them from totally being wiped out in case of calamities that lead to investment losses *(medium certainty)*.

Social opportunities in the form of training, capacity building, and education enable greater understanding of the important role of ecosystem services. These opportunities would increase their knowledge and capability to manage the resources properly and safeguard them from harm and degradation. In many agroecosystems, this will require the judicious use of external inputs to maintain soil fertility and to manage water and ecosystems in an appropriate manner.

19.2.1.4 Synergies and Trade-offs

There are both positive and negative impacts on ecosystems and the services they provide of poverty reduction.

19.2.1.4.1 Positive impacts

Provision of more sustainable and more diverse livelihood opportunities that alleviate poverty would help reduce reliance of the poor on subsistence and resource-based production activities. This will lessen pressure on ecosystems and enable them to recover or be restored. This includes the preservation of various species and their wild relatives. Richer soil fertility, cleaner and more abundant water, denser forests, preservation of various species and their wild relatives, all of these will enable the poor to produce more food and meet other basic needs.

Poverty reduction can also result in better human development such as in health and education. This results in the demographic transition, as people would aim for smaller family sizes in order to attain relatively higher quality life standards (MA *Scenarios,* Chapter 7). Studies have shown, however, that poor and food-insecure households tend to be larger and to have higher numbers of dependents (Reardon 1991). This is due to cultural contexts where having many children is perceived as positive and not as a negative impact of poverty. Accompanying poverty reduction are changes in cultural and moral values. While poverty pressed the poor to exploit these resources, the poor would now be more protected through the use of better and more adaptive participatory management practices.

19.2.1.4.2 Negative impacts

Alleviating poverty can result in some negative consequences on the environment if not carefully managed. Because poverty alleviation can only occur in tandem with economic development, another dimension will be the increased demand for high value products as illustrated in the MA Global Orchestration Scenario. Increased preferences for meat and other livestock products as well as fish and other marine products take place with income growth. While such a trend is clearly favorable due to improved nutrition for low-income populations that do not currently have a balanced diet with respect to the essential nutrients, the ancillary effects on ecosystem services could be damaging, depending on the manner in which livestock production and/or marine and aquaculture cultivation take place. An example is the shift in land use toward uncontrolled land clearing or deforestation to give way for animal grazing and development of infrastructure for animal production. This could lead to both soil and water degradation that is harmful to increasing agricultural production. The most valuable polices are those that safeguard unsustainable land clearing and deforestation and improve food distribution channels to increase availability and accessibility.

Changes in land use would also be triggered by the need to expand commercial, industrial, and social services as demand for them increases with continued improvement in the well-being and economic status of the poor. The risk is, however, of diverting the rural poor into economic traps that could expose them to market uncertainty. The expanded services involve the development and improvement of infrastructure facilities like roads, better water supply system, better drainage system, better waste disposal system, etc as well as the greater provision of health centers and recreation and tourism services. Without effective policies and enforcement measures to promote proper governance and orchestration of these development efforts, the effect would not be beneficial, as these could lead, among other things, to increased competition for resources, particularly of land, water and energy, and to greater incidence of water and air pollution that could be detrimental to promoting a sustained and healthy environment.

19.2.2 MDG 1: Eradicate Extreme Poverty and Hunger, Target 2

Target 2: Halve, between 1990 and 2015, the proportion of people who suffer from hunger.

19.2.2.1 Introduction

Society is facing a daunting task as it works to reduce hunger across the globe. The Food and Agriculture Organization of the United Nations estimates that about 840 million people go to bed hungry each night. One of the major challenges of the new millennium is to meet the food needs of these people while maintaining the environment's capacity to deliver services such as the cleansing of the air and water, the recycling of nutrients, the stabilization of landscapes against erosion and so much more.

At the same time the rapidly increasing consumption of medicaments including for example, minerals and vitamins ends up through the human digestive system into waterways resulting in serious consequences for marine species, for example, tadpole growth retardation. Hence contamination of ecosystems needs to be given consideration in this context. Table 19.4 summarizes the dependencies, interventions, synergies, and trade-offs for MDG target 2.

19.2.2.2 Dependencies

Food production on land and in fresh and marine waters depends on a large number of ecosystem services that we often take for granted. On land, the maintenance of soil structure, the recycling of nutrients essential to plants, and the pollination of plants are essential for growing crops. In aquatic and marine ecosystems, fish and other animals harvested for food depend on the recycling of nutrients, detoxification of water, and the provisioning of nursery habitats.

19.2.2.3 Agricultural Interventions

Interventions to reduce hunger have been generally divided into three categories: increasing yield (and also nutrient fortification),

Table 19.4. MDG 1, Target 2: Dependencies, Interventions, Synergies, and Trade-offs for Hunger

Target	Dependencies on Ecosystems for Achieving Target	Interventions	Links to Ecosystems and Ecosystem Services: Synergies (benefits to ecosystems/ ecosystem services)	Links to Ecosystems and Ecosystem Services: Trade-offs (threats to other types of ecosystems/ecosystem services)
Hunger	soil structure, the recycling of nutrients essential to plants, and the pollination of plants	increasing inputs to increase crop yield: • fertilizer • herbicides/pesticides • water precision agriculture, nutrition, and water saving	reduced demand for new agricultural land, leading to habitat/biodiversity preservation	increased water pollution increased water demand reduced agro-biodiversity and resilience
		introducing genetically modified organisms to increase crop yield crops of relevance to the poor nutrient fortification, for example, golden rice	reduced demand for new agricultural land, leading to habitat/biodiversity preservation reduced need for herbicides/ pesticides, leading to decreased water pollution	potential threat to biodiversity through hybridization
		increasing the area of agricultural land	none apparent	potential threat to biodiversity through habitat loss increased erosion and degradation of streams draining agricultural areas increased incidence of floods local to regional climate modifications
		reducing post-harvest crop losses	reduced demand for new agricultural land, leading to habitat/biodiversity preservation	none apparent

increasing the area in agriculture, and reducing post harvest losses (through measures such as crop protection and appropriate storage of harvests). The MA has identified ways in which some of the major interventions will affect ecosystem services. Some of these effects will be positive and others negative. Here we briefly review the potential consequences of some of the major interventions

Increasing crop yield through agricultural intensification is a first major intervention. Many of the crop-yield increases attained over the last several decades of the twentieth century resulted from increasing inputs to agriculture in the form of fertilizer, pesticides, and water. Introducing new crop varieties bred from extant genetic stock has also increased crop yields. In the future, an additional intervention for increasing yield may be the introduction of genetically modified organisms.

Increased use of fertilizers and pesticides in agriculture has been documented by the FAO for several decades. In 2002, FAO estimated that just over 138 million tons of nutrients (nitrogen, phosphorus, and potash) were applied to the world's agricultural lands. Developing countries accounted for about 60% of this total. The quantity of pesticides used in global agriculture is not as well known. Many countries do not report pesticide usage information and measurement methods vary among those that do report. Pesticide use is known to be high in some tropical countries that specialize in growing cash crops for the international market. For example, Costa Rica reported to the FAO that pesticide use was 51.2 kilograms per hectare of commercial cropland (primarily fruit plantations).

High rates of crop nutrients and pesticide applications to croplands can result in water pollution and major changes in ecosystems and the degradation of ecosystem services. For example, nitrogen leaching from agricultural soils into estuaries and coastal waters is a factor of the increase in oxygen-starved bottom-waters in these places. Increased flows of nitrogen and other nutrients from agricultural runoff stimulate blooms of algae in brackish and salt waters. When the algae die, they sink to the bottom and are decomposed by microorganisms. These microorganisms also use most of the oxygen in the system, resulting in very low oxygen levels in the water, creating an inhospitable habitat for fish, shellfish, and other living things. In recent decades, large regions of coastal waters with severely depleted oxygen levels have been observed and clearly linked with agricultural runoff.

Chapter 6 recognizes both the importance of fertilizer and pesticide inputs to achieving higher crop yield, and their environmental costs. To minimize these costs in the future, it urges the use of a range of agricultural methods. Some of these methods (such as the "Leaf Color Chart" of the International Rice Research Institute) enable farmers to better manage their nitrogen application at relatively low-cost, are technologically simple, and are especially well suited for small farms. Other methods, including precision farming, cost much more, are technologically advanced, and are well suited for large industrial agriculture. Precision farming, also called site-specific management, refers to the "as-needed" application of inputs to management units (fields) that accounts for fine-scale heterogeneity, often measured in square meters, in site characteristics. Satellite-based monitoring technologies linked to farm machinery are being commercially developed to make precision farming a reality. Precision farming is an important element of production ecology where all biological mechanisms are fully used and fine-tuned to specific needs.

Another approach offered in the MA to minimize or eliminate the environmental impacts of fertilizer and pesticide inputs is organic farming. In this approach, farmers are not allowed to use synthetic pesticides or fertilizers. Rather, they rely on developing biological diversity in the agricultural fields to disrupt the habitat for pest organisms, and work toward the purposeful maintenance of soil fertility through practices such as the cultivation of nitrogen-fixing plants. With respect to fertilizer impacts, all forms of farming, including organic farming, must incorporate good nutrient management techniques that tightly couple plant nutrient demand with nutrient supply. While organic farming will not replace other forms of farming as the primary means of meeting the growing demand for food, it certainly has a place in twenty-first century agriculture, for a specific group of consumers that are willing to pay the price for a farming method that is ultimately less environmental friendly and less productive than the advanced eco-technological approach.

Crop irrigation is another important management tool used in many areas to enhance agricultural yield. Of the world's 1.6 billion hectares of arable cropland (the developing-country share is 60%), about 10% is irrigated (total irrigated land is 271 million hectares, with developing countries accounting for 75% of this). According to FAO statistics, countries in Asia, the Middle East, and North Africa have some of the highest percentages of croplands under irrigation. Virtually all cropland in some countries in these regions, including Turkmenistan and Egypt, is irrigated

Irrigation schemes often set up a competition for water such that instream water functions such as wildlife habitat are compromised. More efficient use of water through establishing realistic water markets and pricing is one of the approaches offered in Chapter 6 to reduce this competition. Many of the experiences with water markets in the developing and developed world show that it has great potential for alleviating pressures on instream water functions.

As the world works to reduce hunger and meet the food needs of a growing population with changing food preferences, the intensification of agriculture in an environmentally friendly way can help reduce the need to convert natural ecosystems (such as forests and grasslands) to agricultural land.

Increasing crop yield through genetic manipulations is a second major intervention. (See Chapter 6.) Crop yields have increased by developing new crop varieties bred from extant genetic stock, and more recently by developing GMO crops. The introduction during the 1960s of high-yielding varieties of wheat and rice to Asian and Latin American countries gave these nations the chance to provide their people with adequate supplies of food. High-yielding varieties of other crops such as potatoes, barley, and corn were also developed from extant genetic stock. These high-yielding varieties required intensive cultivation methods, including the use of fertilizers, pesticides and machines, to realize their potential. As noted above, these interventions have both costs and benefits for the environment.

Scientists have developed GMO crops in an attempt to increase yield while reducing the need for costly inputs. Globally, cultivation of GMO crops has increased to 59 million hectares in 2002 (James 2002). GMO crops offer economic and environmental benefits (Trewavas and Leaver 2001), but also raise environmental, social, and political concerns. Dramatic increases in yields have been shown for GMO crops grown widely in several countries, including Argentina and China in the developing world. Environmental benefits include reductions in the use of pesticides. One of the major environmental concerns is that hybridization with wild relatives could cause unwanted environmental changes including negative effects on ecosystem services (Raybould and Gray 1993; Dale et al. 2002).

In Chapter 6, there is a balanced discussion of the environmental and socioeconomic benefits and risks associated with the use of GMOs. The MA stresses that our present experiences suggest that it may not be possible to evaluate the benefits and risks of GMOs in general; rather they must be addressed case by case for specific agroecological and socioeconomic conditions. That does imply that a moratorium or ban on GMO crops is counterproductive in environmental and in food production terms. In the final analysis, it may be desirable to have improved regulations that allow regionally differentiated uses of specific GMOs that add nuance to a globalized trading system.

Chapter 6 also points out that GMOs are only one of the new molecular methods for improving sustainability in agriculture. Other important tools include marker-assisted methods of breeding and molecular methods for the preservation of germplasm diversity.

Increasing the area in agriculture is a third major intervention. Currently about 11% of the terrestrial biosphere has been converted to cropland (some 36% of the world's land base has crop production potential to some degree), and another 30% is used as grazing land. Some of the world's food needs may be met by expanding agricultural areas, but this comes at the expense of the services provided by natural landscapes such as forests.

Deforestation is one of the world's gravest environmental threats. Since the early 1970s, about 5,000,000 square kilometers of Amazon forest have been cleared for crops and pasture. Between the mid-1960s and the mid-1990s, all but three of the Central American countries cleared more forest than they left standing. Currently, Mexico is losing more than 10,000 square kilometers of forest a year to small farmers, and sub-Saharan Africa is losing more than 49,000. (FAO identifies deforestation during the 10-year period 1990–2000 at the following annual rate: Africa 8%, South America 4.5%, Asia 0.8%, Oceania 2%, North and Central America 1%, world 2.3%.) Forest destruction results in decreased soil fertility and increased erosion. Uncontrolled soil erosion can affect the production of hydroelectric power as silt builds up behind dams. Increased sedimentation of waterways can harm downstream fisheries, and in coastal regions can result in the death of coral reefs. Deforestation also leads to greater incidence of floods and droughts in affected regions. Clearing of forests for agricultural land contributes to loss of species, with tropical species especially vulnerable to habitat modification and destruction. The expansion of agricultural land is the most severe threat to maintaining biodiversity, and therefore sustained sophisticated intensification of the present and best lands is in fact the only realistic alternative solution *(high certainty)*.

19.2.3 MDG 2: Achieve Universal Primary Education, Target 3

Target 3: Ensure that, by 2015, children everywhere, boys and girls alike, will be able to complete a full course of primary schooling.

19.2.3.1 Introduction

The quantitative gap between the current situation and the target for 2015 is less dramatic than is the case for other targets such as hunger and poverty. In most regions of the world, the net enrolment ratio in primary education has been improving since 1990, and the values in 2000/01 were above 80% in most regions, with the exceptions of sub-Saharan Africa (57.7%) and South-central Asia (79.2%) (UNESCOa 2003; World Bank 2003). Net enrolment ratio is the number of students enrolled in a level of education who are of official school age for that level, as a percentage of the population of official school age for that level. (See Table 19.5.)

However, when the drop-out rate (as measured by the percentage of pupils starting grade 1 who reach grade 5) is factored in, the situation looks more problematic. Because of limited availability of data at the country level, regional totals cannot be calculated; country figures of pupils reaching grade 5 range from 43% to 100% (UNESCOb 2003). Table 19.6 summarizes the dependencies, interventions, synergies, and trade-offs for MDG target 3.

19.2.3.2 Dependencies

There are no obvious and direct dependencies on ecosystem services for achieving the goal of education, but the required buildup of infrastructure (including building schools, roads) in the areas lacking these facilities will use land, building materials, and energy; potentially open new land to colonization; and may have varied environmental impacts.

19.2.3.3 Interventions

Different specific interventions will be required to reach this target by 2015. The interventions needed to reach the target include:

- building schools and accessing infrastructure. This includes building the schools, installing furniture and, in many cases in remote areas, also building and maintaining roads of access;
- providing transportation facilities, particularly in rural areas. This includes provision of transportation services, acquisition and maintenance of vehicles, building stations, and providing operational elements (fuel, tires, etc.);
- training and recruiting teachers;
- preparing and distributing teaching materials, such as textbooks, stationary, uniforms, etc.;
- providing school meals and basic health services;
- providing family subsidies, when necessary, to compensate for the child's absence from home.

19.2.3.4 Synergies and Trade-offs

The positive impacts of expansion of primary education on ecosystems and their services include improved ecosystems use and management in the long term, but also short-term gains if schooling addresses the needs and means for sustainable management of ecosystems and the lessons are applied in the students' homes. Environmental care and awareness should be included in school curricula to positively influence consumption behavior and natural resources management, with important long-term benefits

On the other hand, the associated infrastructure works, if not carefully done, may have negative impacts on ecosystems and their services, including in some places the fostering of habitat destruction and biodiversity loss due to the opening of areas to land colonization through the new access provided by the roads built to connect villages to the new schools. Also, building the large number of schools that will be required is likely to have some local negative environmental impacts due to disturbances associated with the extraction and processing of building materials (for example, cement factories, which also have global effects through the release of carbon dioxide to the atmosphere).

Table 19.5. Net Enrolment Ratio in Primary Education, 1990/91 and 2000/01

	1990/01	2000/01
(primary-level enrollees per 100 children of enrollment)		
World	81.9	83.6
Industrial regions	94.9	95.6
Countries in transition	88.2	90.6
Developing regions	79.8	82.1
North Africa	82.6	91.4
Sub-Saharan Africa	54.5	57.7
Latin America and the Caribbean	86.9	96.6
East Asia	97.7	93.5
South-Central Asia	73.1	79.2
Southeast Asia	92.6	91.5
West Asia	81.8	85.1

Table 19.6. MDG 2, Target 3: Dependencies, Interventions, Synergies, and Trade-offs for Primary Education

			Links to Ecosystems and Ecosystem Services	
Target	Dependencies on Ecosystems for Achieving Target	Interventions	Synergies (benefits to ecosystems/ ecosystem services)	Trade-offs (threats to other types of ecosystems/ecosystem services)
Primary education	land, building materials, and energy	build schools and access infrastructure provide transportation facilities, particularly in rural areas train and recruit teachers prepare and distribute teaching materials provide school meals, and basic health services provide family subsidies	adequate education can improve ecosystems use and management, in the long and short term	fostering habitat destruction through opening of remote areas to land colonization local impacts of extraction and processing of building materials

19.2.4 MDG 3: Promote Gender Equality and Empower Women, Target 4

Target 4: Eliminate gender disparity in primary and secondary education preferably by 2005 and at all levels of education no later than 2015.

19.2.4.1 Introduction

MDG 3 is directed at achieving gender equity and empowering women. It has a single target: target 4, which calls for the elimination of gender disparity in primary and secondary education, preferably by 2005, and in all levels of education no later than 2015. While it is acknowledged that this target does not adequately represent the many components of gender equity and the many interventions required in achieving it, it is a target that has direct ecosystem service links. The ecosystem services involved would be primary production, food, fresh water, fuelwood, spiritual and religious benefits, education, and cultural heritage. Women (as well as men), especially the poor, are dependent on the natural environment and the ecosystem services it provides for their health and well-being (for example, fresh water, fuel, productive agricultural land, and natural biodiversity) and on the cultural, spiritual, and religious aspects of their environments.

Achieving gender equity is an essential goal in itself and it is absolutely critical for meeting all the 2015 targets. The date set for achieving gender parity in primary and secondary education is 2005, ten years before the 2015 targets (UNICEF 2004). The issue is much more than a question of equity; the challenges facing the globe demand the utilization of all human potential. There have been notable strides but we are far from achieving that goal. Of the 876 million persons over 15 who cannot read or write, nearly two thirds are women. UNICEF estimates that there are 121 million primary-school-age children out of school, 65 million girls and 56 million boys (UNICEF 2004). For the Asia-Pacific region, in the 46 (of 58) economies for which data are available, in almost all there were moderate to severe gender disparities in primary education although the situation was somewhat better in secondary school. In China, by 2000, there were actually more girls enrolled in primary education than boys. The pattern does not hold true in secondary education, however. The situation in primary education in China is offset by disparities in South Asia (UN 2003). Gender inequity reverberates through society in complex ways, impacting not only women but also their children, their families, and their societies. The gender gap in earnings persists, and employed women have less social protection and fewer employment rights than men. Violence against women remains a serious violation of women's rights across the globe, and the majority of new cases of HIV/AIDS in the 15–24 year old age group are women.

A more comprehensive view of gender equity than that employed in the MDGs conceptualizes three main domains of gender equity, all of which are amenable to policy intervention. The first, which includes education, is enhancing women's capabilities, including health, education, and nutrition. MDG 5, improving maternal health, is related. The second domain is enhancing women's opportunities—access to assets, income, and employment. The third domain is enhancing women's agency—the ability to make choices that can alter outcomes including participating in and leading the political process. Because of the daunting nature of the problem, the prevention of violence is viewed as a separate category (Grown et al. 2003). Table 19.7 summarizes the dependencies, interventions, synergies, and trade-offs for MDG target 4.

19.2.4.2 Dependencies

An ecologically balanced natural environmental base, including productive agroecosystems that support economic development will help to alleviate poverty and allow investment in girls' education. This is more than a strictly financial issue, although school fees, uniforms, books, and supplies are important concerns. In many parts of the world, one of the primary reasons that girls are not in school is because of domestic responsibilities, for example, the collection of water and fuelwood and agricultural tasks.

19.2.4.3 Interventions

The specific interventions include: expanding access to safe, affordable fuel for domestic use, expanding access to improved water supply, an improved transportation infrastructure, an improved education infrastructure, the provision of information and communications, and access to technology for distance education in rural areas.

19.2.4.4 Synergies and Trade-offs

The provision of additional fuel sources, especially modern liquid fuels such as liquid petroleum gas or renewable sources such as solar power and hydroelectricity would reduce stress on local forests and other fuelwood sources. Few, if any, negative impacts would be expected. In terms of HWB, if attention is paid to the fuel source and improved cooking stoves, this should also decrease women and children's exposure to indoor air pollution, an important problem in many parts of the world. The primary consequence of expanding access to a safe, potable water supply would be an increased demand for surface and groundwater, potentially affecting surface water quality and quantity. This could impact other components of the local ecosystem. If engineering solutions are applied (for example, dams or other waterworks), there could be considerable ecosystem service impacts. The corollary of improved access to safe water is improvements in sanitation (for example, lack of adequate toilet facilities is one of the reasons given for not sending girls to school). With respect to both increased access to potable water and improved sanitation, wastewater treatment needs to be addressed.

Wastewater treatment has definite benefits in terms of HWB; these include decreased exposure to gastroenteric pathogens in the water supply as well as decreased risk of exposure to pathogens while collecting water, for example, those causing schistosomiasis and leptospirosis. There would also be decreased risk of diseases related to lack of hygiene or the "water-washed" diseases. Safe water supplies may also lead to a decrease in the breeding sites for disease carrying vectors, for example, *Aedes* mosquitoes that transmit dengue fever. Improved health should have positive consequences for educational, agricultural, employment, and other endeavors for entire communities.

Both access to affordable fuel and improved water supply should free women and girls from the burdensome duties of collecting these basic domestic necessities.

An improved transportation infrastructure could have either positive or negative impacts on ecosystems, depending on planning and mitigation efforts. The construction of roads has been presented as one of the classic examples of "development" strategies that have had negative ecological impacts. Roads can threaten biodiversity, lead to the introduction of alien species, cause forest fragmentation and altered habitat as well as unplanned development. However, transportation infrastructure is critical for economic equity and development, for example, providing rural dwellers better access to health and education services as well as employment opportunities and access to markets. Improved trans-

Table 19.7. MDG 3, Target 4: Dependencies, Interventions, Synergies, and Trade-offs for Gender Disparity

			Links to Ecosystems and Ecosystem Services	
Target	Dependencies on Ecosystems for Achieving Target	Interventions	Synergies (benefits to ecosystems/ ecosystem services)	Trade-offs (threats to other types of ecosystems/ecosystem services)
Gender disparity in education	ecologically balanced natural environmental base, including productive agroecosystems	expand access to safe, affordable fuel for domestic use expand access to improved water supply improved transportation infrastructure (in rural areas) improved educational infrastructure—more and more well distributed schools provision of information and communication technology in rural areas/distance education	less stress on local forests and other fuelwood sources	increase demand for surface and groundwater potential impacts of water supply projects, dams, etc. threatens biodiversity introduction of alien species altered habitat forest fragmentation unplanned development local environmental impact at sites potential negative impact of new information technologies on traditional indigenous knowledge systems including ecosystem knowledge; conversely information and communication technology could enhance the preservation of this knowledge

portation infrastructure can also provide access to remote areas for environmental assessment and management. Roads provide various economic development options, for example, ecotourism that may have second order environmental impacts. In different contexts, roads may either slow or increase rural out-migration, with demographic and ecological consequences.

An improved education infrastructure including the building of new schools might have a very localized negative impact in terms of ecosystem services. Positive impacts would be second order—a more educated civil society (girls and boys, men and women) that, at least in theory, could be better stewards of the land. The infrastructure provided by the school (and schoolteachers) could be employed for education efforts aimed at ecosystem services.

There are important issues concerning the quality of the education, completion rates, grade completed, and actual learning. As noted, a major consideration, even if education is available to girls, is the financial and other costs. If girls are going to have access to education, policies will have to be put in place to assure subsidies, transfers, tuition, book fees, etc. In terms of ecosystem services, improved gender equality and education will allow access to land and other household resources increasing the likelihood of wise natural resource management decisions. If women and girls have ready access to water and fuel (some innovative schemes, site collection points near to schools) it will free their time for other activities that contribute to personal and family well-being, including education.

If women are literate, agricultural/natural resource education/outreach can be more easily targeted at women to enhance sustainable agricultural practices. Education provides opportunities for capacity building and women can be effectively engaged in local level environmental management. Education may also increase women's access to (micro) credits. While the uses to which these credits are put could have negative as well as positive impacts on ecosystem services, there are numerous examples of environmentally sensitive entrepreneurial activities undertaken by women.

Education is not only a building block for reaching the other MDGs, it is also a foundation for achieving other aspects of empowerment. Two of the indicators used for measuring progress toward MDG 3 are the share of women in wage employment in the non-agricultural sector and the proportion of women holding seats in the national parliament. Educated women benefit all of society. Hill and King (1995) found that of the countries in which the ratio of female to male enrollment in primary and secondary schools is less than 0.75, you can expect levels of GDP that, all other things being equal, are lower than where there is less discrimination in education.

The topic of women's relationship to and stewardship of the environment has been the subject of broad debate in the literature on ecofeminism. (Jackson 1993, Chapter 2, and elsewhere.) Without arguing that women are naturally attuned to making wise environmental choices, educated and empowered women in developing countries are much less likely to find themselves in circumstances that lead to unwise ecological choices. For example, the women's association of traditional healers called WAINIMATE (The Women's Association for Natural Medicinal Therapy) has ensured women in Fiji and in other parts of the Pacific traditional medicines that are most effective and safe for treating diseases (Tabunakawai 1997, Chapter 2). As well, women who do achieve local, regional, and national office can contribute to wise policy decisions with respect to ecosystem services.

In conclusion, it is worth noting that if all aspects of empowering women were adequately treated in this discussion, it would be a much more complex and nuanced discussion with regard to ecosystem services. Furthermore, a review of the chapters of MA *Current State and Trends* indicated that there were very few references to gender or women (often none). Chapter 3 on drivers is the exception, but most references to women were with regard to fertility. Chapter 5 on HWB does point out that women can be either winners or losers with respect to trade-offs. Even in Chapter 14 on health, women were mentioned only once and that was in the context of their vulnerability to malaria. Lastly, one might ask how well other "differences," for example, socioeconomic status or indigenous peoples are dealt with in the MA.

19.2.5. MDG 4: Reduce Child Mortality, Target 5

Target 5: Reduce by two thirds, between 1990 and 2015, the under-five mortality rate.

19.2.5.1 Introduction

For more than 25 years, the prevention of child mortality has been one of the leading priorities for the international health community. Despite progress made in reducing child mortality in children less than five years old, more than 10 million of them still die each year, almost all in poor countries (Black et al. 2003). In many countries, infant and childhood mortality rates are falling more slowly than before and, in some, they have stagnated or are on the rise (UNICEF 2004). The infant mortality rate has been highest in sub-Saharan Africa, and the region has shown the slowest decline (Ahmad et al. 2000). Despite a 50% drop in South Asia's mortality rate, almost one in ten children in the region still dies before his/her fifth birthday (Ahmad et al 2000). Pneumonia, diarrhea, and neonatal disorders are predominant causes of infant deaths worldwide. In sub-Saharan Africa, malaria additionally plays an important part in child mortality in many countries of the region.

Undernutrition is the underlying cause of a substantial proportion of all child deaths that arise from factors—including socioeconomic factors such as income, social status, and education—that work through an intermediate level of environmental and behavioral risk factors (Mosley et al. 1984). Better information and determinants of undernutrition remains a daunting challenge and so is the development of effective and affordable health intervention measures that could facilitate the reduction of the under-five mortality rate by two thirds between 1990 and 2015. Ecosystems information provided by the MA will have indirect impact on the under-five mortality rate reduction through maternal nutrition improvement as well as healthier prenatal and postnatal care. Table 19.8 summarizes the dependencies, interventions, synergies, and trade-offs for MDG target 5.

19.2.5.2 Dependency on Ecosystems and their Services

Unhygienic and unsafe environments place children at high risk (*high certainty*). Ingestion of unsafe water, inadequate water for hygiene, and lack of access to sanitation contribute to millions of deaths among children, a significant portion of which are from diarrhea (WHO 2002), as are the malaria and dengue outbreaks, particularly in developing countries, that have afflicted young children. Underweight status and micronutrient deficiencies among children that decrease their immune and non-immune host defenses have made many of them more vulnerable to infectious diseases. The causal link of the latter to ecosystems and their services are complex and they are seen most clearly among impoverished communities that lack the buffers that the rich can afford. (See Chapter 16.) For example, climate change that affects food production has an impact on malnutrition that can aggravate many other health problems, particularly among the poor; small children are primarily jeopardized (*medium certainty*). The alterations of land cover have produced insects and pests that transmit diseases detrimental to children's health. Degraded ecosystems, especially those that allow considerable amount of standing water, are excellent sources of waterborne diseases (like malaria, dengue, diarrheal diseases) to which children are susceptible.

Where ill health is caused, in large part, by excessive consumption of ecosystem services, then substantial reduction in overconsumption would have major health benefits, among both adults and children (*high certainty*). Conversely, healthy children are in a better position to positively influence other aspects of well-being, especially their parents' and their closest kin in terms of reducing pressure on financial resources and on other life-support systems. (See Chapter 16.)

19.2.5.3 Intervention Measures

Some intervention measures to improve human health and reduce the under-five mortality rate come from the health sector and others from outside it. The case of reduced nutrition in a community, as a result of food scarcity resulting in turn from a changed ecosystem, is one example of the latter. Malnourished children are more prone to life-threatening diseases including diarrhea, acute respiratory infections, malaria, and measles. Intervention measures that seek to sustain food production increases and to improve accessibility to food by the poor will help facilitate the under-five mortality rate reduction (*high certainty*). Provision of livelihood activities, especially to poor households, will help improve access to food but also to other amenities in life that could elevate nutritional status (*high certainty*). Another intervention relates to the need to increase awareness, monitoring and reporting not only among health personnel but also other authorities as well as the general public on the incidence or possible incidence of a disease (*medium certainty*). Increasing mother's knowledge of diseases and their prevention through rigid extension campaign and training could result to considerable disease prevention and to more healthy children (*high certainty*).

Strong community participation and support, supplemented with appropriate regulation measures, are essential to the attainment of safe and good quality water (for example, clean river or watershed catchment), clean air (for example, reduced used of wood fuel for cooking), and healthy environment (for example, regular maintenance of canals, areas where water stagnates, etc.) (*medium certainty*). The latter will necessarily involve the provision of appropriate infrastructure facilities, especially those that would facilitate the achievement of a cleaner environment.

Interventions from within the health sector are curative in nature to save children dying from diseases. As such, health care systems should be technically and financially equipped to deliver necessary and effective services to all, especially the poorer segments of the society (*medium certainty*). Infrastructure facilities are also needed to facilitate the access of such health facilities and services by the all constituents (*medium certainty*).

It should be noted that health intervention measures should have spatial and temporal dimensions and should take into consideration other factors such as the social, economic, and demographic characteristics of their target group beneficiaries, as knowledge of these variables could facilitate their spread and adoption. Prior to all these, however, is the need for a realistic

Table 19.8. MDG 4, Target 5: Dependencies, Interventions, Synergies, and Trade-offs for Child Mortality

			Links to Ecosystems and Ecosystem Services	
Target	Dependencies on Ecosystems for Achieving Target	Interventions	Synergies (benefits to ecosystems/ ecosystem services)	Trade-offs (threats to other types of ecosystems/ecosystem services)
Child mortality	healthy ecosystems and ecosystem services reduces the incidence of endemic and emerging diseases that affect the lives of children	better understanding of epidemiological profile and the capabilities of the health system in order to be able to identify: • more efficient and equitable delivery of health care services • infrastructure facilities needed to facilitate access to health services • improved delivery/access to clean water • improved sanitation including proper waste disposal practices	reduced pressure on ecosystems reduced use of wood fuel less pressure on the regulatory services of ecosystem promotes better ecosystem management	increased population that could put pressure on the use of ecosystem services, particularly on its provisioning services may involve land use change (also demographic changes) that could have negative impact on ecosystems
	healthy ecosystem that provides options to improve means of livelihood will promote well-being of children.	increased awareness and more efficient monitoring and reporting of health personnel on the incidence or possible incidence of disease. increased livelihood opportunities, especially for the poor		land use change from increased as well as shifts in demand due to economic growth competition for resources; increased demand for surface/groundwater

picture of the country's epidemiological profile and the capabilities of their health systems to be able to design and implement the appropriate public health interventions.

19.2.5.4 Synergies and Trade-offs

All these interventions have positive and negative impacts on the ecosystem. Positive impacts, as already mentioned, relate to the fact that healthy people, especially children, save families from spending money on medicines, hospitalization, etc.; these savings can then be invested in productive activities to improve economic well-being. Better-off farm families have the tendency to put fewer burdens for their children to be involved in household chores, especially in collecting woodfuel and fetching clean water. Mothers, too, can have the luxury of more modern appliances or delegating household chores to others and thus gaining more time to take care of their children. The negative impacts, on the other hand, relate to the externalities that result from an increase in population as mortality rates are reduced (but which can be neutralized by lower fertility rates as households start to appreciate smaller family sizes) as well as from shifts in demand due to economic growth and as poverty is reduced that may involve environments to be disturbed. Examples include the change in land use to cater to more diversified production activities, diversified demand for food and other commodities, and increased demand for services and other amenities.

19.2.6 MDG 5: Improve Maternal Health, Target 6

Target 6: Reduce by three quarters, between 1990 and 2015, the maternal mortality rate.

19.2.6.1 Introduction

A report of the International Conference on Population and Development, endorsed by 179 countries, states that women, including their health and well-being, are valuable as people and citizens (ICPD 1994). Currently, some 350 million women have no access to safe and affordable contraception, contributing to undesirable fertility for women, and half a million deaths occur per year during pregnancy and childbirth, particularly in sub-Saharan Africa, where maternal deaths account for 53% of all deaths. A woman's risk of death during pregnancy and childbirth varies greatly between industrial and developing countries. The risk of dying from pregnancy in Africa is 1 in 16; in Asia, 1 in 65; and in North America, 1 in 3,700. The overall risk for developing countries is 1 in 48, whereas in industrial countries the risk is 1 in 1,800 (ICPD 1994). Providing conditions that enable women to have a healthy pregnancy and childbirth is, as agreed by the ICPD, a fundamental right. Their involvement and decision on when to have children, and how many to have, is dependent on information access to reproductive health and society's commitment to gender equity.

Many health problems incurred by women during pregnancy are a result of complications arising from but not limited to HIV/AIDS, sexually transmitted disease, respiratory infections, and malaria. These problems can be lessened through investment in economic and social infrastructure such as access to health aids and facilities, and also through the development and conservation of ecosystem services. Decreasing maternal mortality, however, requires a focus on preventative care since there are no direct links with ecosystem services and maternal deaths (MA *Scenarios,* Chapter 7).

To decrease maternal mortality, improvements to women's health during pregnancy and throughout the post-natal years are critical, as these periods in a woman's life are essential to her well-being. The Millennium Project Task Force on Child Health and Maternal Health identifies where there are low maternal mortality rates in particular countries it is primarily due to the availability and access to skilled attendants, emergency obstetric care, and referral systems, so that women can reach emergency obstetric care for life saving situations (Freedman et al. 2003). As short-term solutions, these services are extremely important in reducing maternal mortality. By ensuring access to and providing safe water, soils, and food, policy-makers can ensure long-term benefits. Pregnant mothers will have better chances at having healthy children and economic spending can be put toward improving health services and strengthening the entire health system. Table 19.9 summarizes the dependencies, interventions, synergies, and trade-offs for MDG target 6.

19.2.6.2 Interventions

Improving infrastructure, that is, provision of high-quality and accessible medical care and information, in clinics and hospitals located for easy access to women can result in habitat modification and water use locally. Transportation and road development is also necessary to ensure that medical emergency services can reach their destinations efficiently and women will have the opportunity to access transport to clinics or hospitals. Providing emergency care services will dramatically reduce the maternal mortality rate (Freedman et al. 2003).

Improving access to nutrition through food quality and quantity places demand on land and aquatic habitats for agricultural land of higher soil fertility, for chemicals to increase plant productivity. Alternatives to chemical usage such as natural pest control can be considered.

Increasing health and gender equity through provision of knowledge about reproductive alternatives, nutrition, and disease can lead to healthy pregnancies and choices about reproductive health. Gender equity can broaden appreciation of educational and cultural services, indirectly leading to wise use of ecosystems and services. As already noted, empowering women through training and education will allow them to manage resources wisely for themselves and their families).

19.2.6.3 Trade-offs

Where there is a need for economic development such as building and improving infrastructure to improve access to emergency obstetric care, achieving the maternal health goal may require converting natural ecosystems resulting from an increased demand on land, water and biodiversity. This is the single most relevant trade-off concerning maternal mortality. There is a need for more medical facilities that offer emergency care, roads, and transport for women to access them.

19.2.7 MDG 6: Combat HIV/AIDS, Malaria, and Other Diseases, Targets 7 and 8

Target 7: Have halted by 2015, and begun to reverse, the spread of HIV/AIDS.

Target 8: Have halted by 2015, and begun to reverse, the incidence of malaria and other major diseases.

19.2.7.1 Introduction

Health, defined differently in different cultures, is fundamental to HWB. It is also both a cause of well-being and a consequence. In Chapter 16, the links between health and ecosystem services are listed as biodiversity, food, fresh water, wood, wood fuel, fiber, nutrient management, waste management and detoxification, flood and storm regulation, and cultural services. To these could be added climate regulation. This long list is indicative of the complex relationship between human health and well-being and ecosystem services. Related MDGs are goal 4, Reducing Child Mortality, and goal 5, Improving Maternal Health. Even taking these three goals together captures only a portion of the complex relationship between health and ecosystem services. The United Nations and the World Health Organization have identified HIV/AIDS, malaria, and tuberculosis as major threats to health across the globe. MDG 6 explicitly addresses HIV/AIDS and malaria and reduction in the prevalence of and death rates associated with tuberculosis, which are used as a measure of the success in meeting this goal.

For HIV/AIDS (primarily a sexually transmitted disease, although also transmitted perinatally and by intravenous drug use and blood transfusions) and TB (a communicable disease), most of the interventions are in the realm of public health, medicine, education, and politics and are not directly related to ecosystems. Malaria is the most serious and common vector-borne disease in the world, responsible for 2 million deaths a year, 90% of which occur in sub-Saharan Africa, the majority in children under five. Transmitted by the *Anopheles sp.* mosquito, malaria provides a classic example of the relationship between vector-borne disease and ecosystem services. Malaria epidemics are often the result of changes in the eco-epidemiological system, for example, abnormal meteorological conditions (WHO 2000) or deforestation. Other globally important vector-borne diseases include dengue, filariasis, schistosomiasis, and yellow fever.

Another important group of ecosystem sensitive diseases are those that are related to water supply and sanitation, for example, those such as cholera that are waterborne or those related to poor personal hygiene, that is, the "water-washed diseases." MA *Current State and Trends,* Chapter 14, addresses the role of altered ecosystem and emerging and re-emerging infectious disease. Food production, an ecosystem service, and availability are directly related to human health and there is a synergism between malnutri-

Table 19.9. MDG 5, Target 6: Dependencies, Interventions, Synergies, and Trade-offs for Maternal Health

			Links to Ecosystems and Ecosystem Services	
Target	**Dependencies on Ecosystems for Achieving Target**	**Interventions**	**Synergies (benefits to ecosystems/ ecosystem services)**	**Trade-offs (threats to other types of ecosystems/ecosystem services)**
Maternal health	There are not direct links to ecosystem services. Maternal mortality is caused by complications at birth and the lack of access to skilled birth attendants and other health services.	build medical facilities and emergency care facilities provide transportation facilities, particularly in rural areas for women train and recruit medical professionals for preventative care	no direct benefits to ecosystem services; access to medical care will improve maternal health	alternating natural habitats for infrastructure development local impacts of extraction and processing of building materials

tion and infection, especially in children. Overnutrition, as well as some other ecosystem-relevant factors, for example, exposure to carcinogens, are implicated in the increase in noncommunicable diseases, for example, cardiovascular disease, cancer, hypertension, and diabetes. These are becoming increasingly prevalent not only in the developed world but also in the developing world, resulting in what some refer to as developing countries' "double burden of disease."

19.2.7.2 Dependencies

For target 7, halting and beginning to reverse the spread of HIV/AIDS by 2015, the direct links between intervention strategies that are largely aimed at behavior and ecosystem services are not evident. In a very general sense, to the degree that countries will need to rely on their own national fiscal resources to fight the spread of HIV/AIDS, ecosystem services that support sustainable development will allow resources to be targeted at the HIV/AIDS epidemic. On a community level, as fewer and fewer individuals are living in poverty, individuals will not have to resort to livelihood strategies (for example, prostitution and male migration) that increase risk of HIV/AIDS. In those countries where the HIV/AIDS epidemic is acute, as in sub-Saharan Africa, HIV/AIDS has a direct impact on ecosystem services. With a considerable segment of the population too ill to work, subsistence production and agricultural activities decline as too few hands are available for weeding and other agronomic activities. In terms of the natural environment, this could have positive or negative effects, but in terms of human health and well-being the implications will be disastrous. Table 19.10 summarizes the dependencies for HIV/AIDS, as well as interventions, synergies, and trade-offs.

An estimated 29.4 million (of a global total of 42 million) persons are living with HIV/AIDS in sub-Saharan Africa, and 3.5 million new infections occurred in 2002. Prevalence rates vary across the continent but reach 20% of the adult population in South Africa and Zambia.(UN AIDS 2003). HIV/AIDS is a threat to reaching the MDGs on the African continent. However, while more than half of the HIV/AIDS cases occur in sub-Saharan continent, there is no reason for complacency elsewhere. Over seven million people are living with HIV/AIDS in Asia and the Pacific and while rates may be low (0.1% in China and 0.6% in India,) this represents 8,500,000 and 4,000,000 persons, respectively (UNDP 2003). If appropriate policies are not put into place, there is reason to be very concerned about the epidemic in Asia, although countries like Thailand and Cambodia have made great strides in stemming the epidemic (Brown 2003).

Malaria is endemic in over 100 countries and two and a half billion people are at risk. Malaria is responsible for 11% of the disease burden in Africa, and countries with high rates of malaria have income levels averaging 33% of those without malaria. (See Chapter 12.) Malaria is caused by four species of the plasmodium parasite; *Plasmodium vivax* is most widely distributed and *Plasmodium falciparum* is the most clinically dangerous. Both of these are becoming increasingly drug resistant and the anopholine species that transmit malaria are becoming increasingly resistant to pesticides as well. There are a complex of socioeconomic, geographic, and ecological conditions that are conducive to the transmission of malaria. As with HIV/AIDS, high rates of malaria endemicity have serious social and economic impacts.

The vectors are sensitive to temperature, precipitation, and extreme weather events. Relative humidity becomes important at higher temperatures. The extrinsic incubation cycle of the plasmodium is also sensitive to temperature and this can serve as a limiting factor in disease transmission. Temperature is important with respect to the survival of both the vector and the parasite and precipitation directly influences both the abundance of breeding sites and vector densities (McMichael et al. 1996). Deforestation and irrigation schemes can lead to increased malaria transmission. The presence or absence of alternate hosts can also be important. There is concern that with global warming some vector species will increase their range. In Papua New Guinea, for instance, if temperatures in the relatively densely populated highlands were to increase allowing vector species to populate higher altitudes, a large population with no immunity would be at risk. Typically populations on the margins of endemic areas are at the greatest risk.

19.2.7.3 Negative Impacts

The impacts of HIV/AIDS on ecosystem services are mediated through the devastating impact on the human population, on families, on the social fabric, and on the economy. In many countries, AIDS is erasing decades of progress. Families are loosing income earners. Many of those who die leave behind HIV-infected partners, and when they die, orphans. Children of HIV-positive mothers may well be infected themselves. Children of

Table 19.10. MDG 6, Targets 7 and 8: Dependencies, Interventions, Synergies, and Trade-offs for HIV/AIDS, Malaria, and Other Diseases

			Links to Ecosystems and Ecosystem Services	
Target	**Dependencies on Ecosystems for Achieving Target**	**Interventions**	**Synergies (benefits to ecosystems/ ecosystem services)**	**Trade-offs (threats to other types of ecosystems/ecosystem services)**
HIV/AIDS	no direct links evident; intervention strategies are aimed at behavior	public health interventions aimed at high risk populations/behavior modification public policy interventions in support of affected families and children	in parts of southern Africa where HIV/AIDS rates are extremely high, reducing rates may allow better ecosystem management	
Malaria and other major diseases		larval and adult spraying drainage intermittent irrigation impregnated bednets		alterations in natural ecosystem with larval spraying drainage could cause alternation or destruction of wetlands

the ill take on more and more responsibility and fewer families have the resources to send their children to school. The vast majority of those with HIV/AIDS are between the ages of 15 and 49, in the prime of their working lives. HIV/AIDS has a direct impact on the economy by lowering production and earnings and hence taxes, and the resources demanded to deal with the epidemic are a further drain on the economy. Direct impacts on ecosystem services are probably most prominent in the agricultural sector and could be either positive or negative with respect to ecosystem services, as noted above, but are definitely negative in terms of food production and the fight against hunger. It is estimated in Burkina Faso, that 20% of rural families have reduced their agricultural work or even abandoned their farms because of HIV/AIDS. In Ethiopia, AIDS-affected households were found to spend 11–16 hours per week performing agricultural work, compared with an average 33 hours for non-AIDS-affected households (AVERT 2003).

HIV/AIDS-affected families have fewer resources to devote to their agricultural pursuits. This could, on the one hand, mean that, for example, fewer pesticides are used or on the other that ecologically unsound practices are used to try to increase production. Abandoned land might revert to second growth or, especially if soil erosion is a problem, become further degraded. With high mortality rates, some indigenous traditional, ecological, and agricultural knowledge may also be lost. It is worth noting that other important endemic diseases, including malaria, that affect a notable proportion of the population can have similar, if not as extreme, effects.

19.2.7.4 Interventions

The interventions for HIV/AIDS and TB are not primarily environmental. The main interventions for malaria include the prevention of mosquito bites including the provision and distribution of insecticide-treated bednets and other materials, indoor residual spraying only in well-defined high-risk or special situations, the prevention of infection by chemoprophylaxis particularly in pregnant women, the treatment of clinical episodes, and surveillance and rapid response (WHO 2000). Some of these, as the interventions for HIV/AIDS and TB, are in the realm of medicine, public health, and behavior. However, there are important environmental interventions for malaria including selective environmental management, drainage, and intermittent irrigation that are important in helping to control malaria. (See Chapter 12 for a discussion of the importance of integrated vector management.)

There have been many examples of the misuse of dichloro diphenyl trichloroethane (DDT) and other pesticides, particularly in the agricultural sector. While DDT is being phased out in many locations in favor of pyrethroids, WHO still recommends its use in indoor residual spraying in special situations (WHO 2000). The recent resurgence in malaria has been blamed on the bans on the use of DDT in many regions of the world. Inappropriate use of pesticides can have severe impacts on many ecosystem services, although the application of pesticides plays an important role in disease control in many parts of the world. Drug resistance in parasites and pesticide resistance in vectors is becoming an increasing problem. The chemicals may have negative impacts on other organisms, on groundwater, etc. In addition, it is worth noting that specific interventions for one vector may differ radically from those used for other vectors, for example, different genera of mosquitoes. The impact of drainage and intermittent irrigation strategies would be site specific. In a degraded ecosystem, with no endangered species, it could be positive, whereas in a natural wetland, it would be negative.

19.2.7.5 Conclusion

This discussion of HIV/AIDS and malaria only begins to explore the complex relationship between human health and ecosystem services. An important link that is now receiving considerable scientific attention is the relationship between human and ecosystem health and climate variability and climate change. Emerging and re-emerging infectious diseases may have a direct link to alterations in ecosystem services, for example, Lyme disease or West Nile virus. Wise management of the environment, for example, river catchments, positively affects environmental and socioeconomic determinants of health. The complexity of the relationships between human health and ecosystem services demands interdisciplinary and cross-sectoral collaboration in addressing them. In trying to assess the impact of any given set of conditions on human health, we need to understand the current vulnerability of the population affected and their future adaptive capacity. The relationship is often inverted. As in the case of sub-Saharan Africa, the links between ecosystem change and human health are seen most clearly among impoverished communities who lack the "buffers" the rich can purchase.

A recent multiyear study, under the auspices of the WHO Global Burden of Disease project, developed projections of mortality and disability for each five-year period from 1990 to 2020 (Harvard 2003). HIV/AIDS, malaria, and TB are major global problems, and there is the possibility of new communicable diseases as we were reminded by the Severe Acute Respiratory Syndrome outbreak in 2002/3. In the GBD projections, deaths from communicable, maternal, and prenatal conditions and nutritional deficiencies are expected to fall from 17.3 million in 1990 to 10.3 million in 2020. Deaths from noncommunicable diseases are expected to climb from 28.1 million deaths in 1990 to 49.7 million in 2020, an increase of 77%. Much of this increase will occur in the developing world. The important two driving forces behind these changes are population aging and the use of tobacco. As we think about ecosystem services and human health, it would be a mistake to ignore the links to noncommunicable disease. Certainly not the only, but a very important, link is through food supply and nutrition, very closely linked to ecosystem services. It is also important to remember that modern or Western conceptions of health that are largely based on physical evidence of bodily structure and function may differ greatly from indigenous conceptions based on common heritage and union with the environment.

19.2.8 MDG 7: Ensure Environmental Sustainability, Target 9

Target 9: Integrate the principles of sustainable development into country policies and programs and reverse the loss of environmental resources.

19.2.8.1 Introduction

MDG 7 contains three targets, of which target 9 refers specifically to ecosystem services. This target can clearly be subdivided into two components: (1) integration of the principles and (2) reversal of environmental loss, requiring different types of interventions and having differentiated implications for ecosystems and their services.

Target 9a: Integrate the principles of sustainable development into country policies and programs

The principles of sustainable development referred to in target 9 have been officially agreed at the Earth Summit of 1992, and defined in the Rio Declaration on Environment and Development (UN 1992). The twenty-seven principles internationally agreed are reproduced in Box 19.1.

Lack of integration in planning and policy-making has long been recognized as one major obstacle to progress toward sustainable development (UNCSD 1995). While sectoral plans and approaches have the advantages of being unambiguous, with clear objectives and a good correspondence to the specialized institutions responsible for implementation, they have the disadvantage that the assortment of plans and regulations across sectors may have inconsistent and incompatible objectives, and often the issues of integration and coordination fall between the gaps. (See Chapter 14.). Table 19.11 summarizes the dependencies, interventions, synergies, and trade-offs for MDG target 9a.

19.2.8.2 Dependencies

The fulfillment of target 9a does not exhibit any obvious direct dependencies on ecosystems for achieving it, as it implies mainly institutional changes.

19.2.8.3 Interventions

Among the interventions required, the most important are:

- reinforce environmental policies;
- introduce environmental principles in economic and social policies;
- reinforce and enforce environmental legislation at various scales;
- train and build capacity on sustainable development in the public and private sectors;
- establish economic and social incentives for the contribution of the private sector to the integration and implementation of the principles;
- support research and development on methodologies and techniques for integrated planning and policy-making and develop national information strategies for sustainable development;
- establish effective and integrated legal and normative frameworks for economic planning, market instruments, and regulatory instruments; and
- define and implement and ensure compatibility/alliance with country strategies such as country sustainable development strategies.

19.2.8.4 Synergies and Trade-offs

Most of these will have a positive impact on ecosystems and their services, as they are specifically focused on improving environmental sustainability. The increased integration of the principles of sustainable development into policies and programs will considerably reduce the negative ecological impacts of economic and social policies and actions. No direct negative impacts on the ecosystems are likely to result from those actions.

Target 9b: Reverse the loss of environmental resources

The loss of environmental resources is a serious problem abundantly documented by many studies, and full details can be found in various chapters of this and other MA volumes. As demonstrated several times, this process amounts to erosion of the ecological basis for development and directly threatens HWB. For example, an important ecosystem service provided by intact vegetation is the stabilization of landscapes against wind, soil, and water erosion. The clearing of forests and other types of vegetation in hilly and mountainous regions of the world has diminished the capacity of the land as a habitat for people and in some cases led to the death of thousands of people in mudslides associated with extreme rainfall. Intense rains falling on steep slopes cleared of forests in Caribbean and throughout Central and South America have resulted in thousands of people dying in massive mudslides in recent decades, including those resulting from Hurricane Mitch in 1998. Large-scale mudslides became the signature of Hurricane Mitch, a storm that grew to become the Atlantic basin's fourth strongest hurricane ever with sustained winds of 180 mph for more than 24 hours. Hurricane Mitch stalled off the coast of Honduras from late on October, 27, 1998, until the evening of October 29, dropping up to 25 inches of rain in one six-hour period in some places. The heavy rain led to widespread flooding and mudslides that killed at least 10,000 people.

The specific indicators identified for target 9b so far (proportion of land area covered by forest, land area protected to maintain

Table 19.11. MDG 7, Target 9a: Dependencies, Interventions, Synergies, and Trade-offs for Integrating Principles

			Links to Ecosystems and Ecosystem Services	
Target	Dependencies on Ecosystems for Achieving Target	Interventions	Synergies (benefits to ecosystems/ ecosystem services)	Trade-offs (threats to other types of ecosystems/ecosystem services)
Integrate principles into policies and programs	none	environmental policies introduce environmental principles in economic and social policies environmental legislation train and build capacity incentives for private-sector contribution support R&D on methods for integration and develop national information strategies establish integrated legal and normative frameworks implement country sustainable development strategies	reduced negative impacts of ecosystem services	none

biological diversity, energy used per unit of GDP, per capita carbon dioxide emissions and consumption of ozone-depleting chlorofluorocarbons, and population using solid fuels) are, from the viewpoint of ecosystem services, far from complete for this component of the target. Table 19.12 summarizes the dependencies, interventions, synergies, and trade-offs for MDG target 9b.

19.2.8.5 Dependencies

The fulfillment of target 9b does not show obvious dependencies on ecosystem services, except that, in some cases, the damage to ecosystem services may already be irreversible or too costly to revert, and the reversal of the loss of environmental resources may no longer be possible.

19.2.8.6 Interventions

A number of different types of will be necessary:

- increase expenditure on environmental sustainability. In many countries, the current level of public and private expenditures on the sustainability of the environment is insufficient to markedly reverse the losses;
- increase resources allocated to R&D on ecosystems and their services, and new eco-technologies;
- implement programs of ecosystem restoration and rehabilitation and ecological engineering (for watersheds, agricultural land, deforested areas, polluted aquatic ecosystems, etc).
- develop economic incentives and regulatory instruments directed to the sustainability of ecosystems services (for example, reduce the energy intensity of the economy). A number of policy instruments are available, ranging from voluntary compliance to pollution taxes and environmental protection laws. The appropriate mix will vary with the country institutional system;
- introduce the principles of sustainable development in education curricula and implement programs for raising public awareness. This is an intervention that will take some time to show its effects, but that is likely to have a long-term positive impact on ecosystem services.
- introduce environmental sustainability considerations into international trade and cooperation agreements at various scales; and
- protect areas and ecological corridors or networks essential to maintain biological diversity.

19.2.8.7 Synergies and Trade-offs

Meeting the target will have a number of positive ecosystem impacts. First, the economic impacts of economic and social policies and actions will be reduced. This is probably the positive impact that will be most important in quantitative terms, as usually the negative ecological impact of economic actions such as investment in public works, industry allocation, resettlement of population and land colonization, etc., far exceeds the positive impacts of specifically environmental policies. Second, improved management of ecosystems will contribute to enhance ecosystem services and thus help to fulfill the other goals. Third, restoration and rehabilitation of degraded ecosystems will increase the ecological basis for development. Fourth, increased understanding of ecosystem functions will provide novel management technologies and identify new productive and sustainable uses for ecosystem services.

There are no obvious negative ecosystem impacts of the interventions. However, it is known that ecosystems managed for maximum productivity or yield, without maximizing ecological (self) regulation mechanisms, may lose their resilience in the long term and they may even collapse (Holling and Meffe 1996).

19.2.9 MDG 7: Ensure Environmental Sustainability, Target 10

Target 10: Halve, by 2015, the proportion of people without sustainable access to safe drinking water and basic sanitation.

19.2.9.1 Introduction

The baseline year for estimating the targets is 1990. Unfortunately, the terminology chosen for defining the targets does not conform with the classification adopted by the WHO/UNICEF Joint Monitoring Program, which collects official data on "improved" access to water supply and sanitation. The distinction between "improved" and the more stringent definition of "adequate" access is very important. For example, the availability of a shared toilet in an urban settlement is termed as "improved" ac-

Table 19.12. MDG 7, Target 9b: Dependencies, Interventions, Synergies, and Trade-offs for Reversing the Loss of Environmental Resources

			Links to Ecosystems and Ecosystem Services	
Target	Dependencies on Ecosystems for Achieving Target	Interventions	Synergies (benefits to ecosystems/ ecosystem services)	Trade-offs (threats to other types of ecosystems/ecosystem services)
reverse loss of environmental resources	no obvious dependencies	increase expenditure on sustainability increase resources implement ecosystem restoration, rehabilitation, and ecological engineering develop economic incentives and regulations introduce the principles in education curricula and raise awareness introduce environmental sustainability into trade and cooperation protect areas	enhanced ecosystem services increased ecological basis for development novel management technologies and new productive uses	ecosystems managed for maximum productivity may lose their resilience in the long term

BOX 19.1

The Principles of Sustainable Development from the Rio Declaration (UN 1992)

Principle 1. Human beings are at the center of concerns for sustainable development. They are entitled to a healthy and productive life in harmony with nature.

Principle 2. States have, in accordance with the Charter of the United Nations and the principles of international law, the sovereign right to exploit their own resources pursuant to their own environmental and developmental policies, and the responsibility to ensure that activities within their jurisdiction or control do not cause damage to the environment of other States or of areas beyond the limits of national jurisdiction.

Principle 3. The right to development must be fulfilled so as to equitably meet developmental and environmental needs of present and future generations.

Principle 4. In order to achieve sustainable development, environmental protection shall constitute an integral part of the development process and cannot be considered in isolation from it.

Principle 5. All States and all people shall cooperate in the essential task of eradicating poverty as an indispensable requirement for sustainable development, in order to decrease the disparities in standards of living and better meet the needs of the majority of the people of the world.

Principle 6. The special situation and needs of developing countries, particularly the least developed, and those most environmentally vulnerable, shall be given special priority. International actions in the field of environment and development should also address the interests and needs of all countries.

Principle 7. States shall cooperate in a spirit of global partnership to conserve, protect and restore the health and integrity of the Earth's ecosystem. In view of the different contributions to global environmental degradation, States have common but differentiated responsibilities. The developed countries acknowledge the responsibility that they bear in the international pursuit of sustainable development in view of the pressures their societies place on the global environment and of the technologies and financial resources they command.

Principle 8. To achieve sustainable development and a higher quality of life for all people, States should reduce and eliminate unsustainable patterns of production and consumption and promote appropriate demographic policies.

Principle 9. States should cooperate to strengthen endogenous capacity-building for sustainable development by improving scientific understanding through exchanges of scientific and technological knowledge, and by enhancing the development, adaptation, diffusion and transfer of technologies, including new and innovative technologies.

Principle 10. Environmental issues are best handled with the participation of all concerned citizens, at the relevant level. At the national level, each individual shall have appropriate access to information concerning the environment that is held by public authorities, including information on hazardous materials and activities in their communities, and the opportunity to participate in decision-making processes. States shall facilitate and encourage public awareness and participation by making information widely available. Effective access to judicial and administrative proceedings, including redress and remedy, shall be provided.

Principle 11. States shall enact effective environmental legislation. Environmental standards, management objectives and priorities should reflect the environmental and developmental context to which they apply. Standards applied by some countries may be inappropriate and of unwarranted economic and social cost to other countries, in particular developing countries.

Principle 12. States should cooperate to promote a supportive and open international economic system that would lead to economic growth and sustainable development in all countries, to better address the problems of environmental degradation. Trade policy measures for environmental purposes should not constitute a means of arbitrary or unjustifiable discrimination or a disguised restriction on international trade. Unilateral actions to deal with environmental challenges outside the jurisdiction of the importing country should be avoided. Environmental measures addressing transboundary or global environmental problems should, as far as possible, be based on an international consensus.

Principle 13. States shall develop national law regarding liability and compensation for the victims of pollution and other environmental damage. States shall also cooperate in an expeditious and more determined manner to develop further international law regarding liability and compensation for adverse effects of environmental damage caused by activities within their jurisdiction or control to areas beyond their jurisdiction.

Principle 14. States should effectively cooperate to discourage or prevent the relocation and transfer to other States of any activities and substances that cause severe environmental degradation or are found to be harmful to human health.

Principle 15. In order to protect the environment, the precautionary approach shall be widely applied by States according to their capabilities. Where there are threats of serious or irreversible damage, lack of full scientific certainty shall not be used as a reason for postponing cost-effective measures to prevent environmental degradation.

Principle 16. National authorities should endeavor to promote the internalization of environmental costs and the use of economic instruments, taking into account the approach that the polluter should, in principle, bear the cost of pollution, with due regard to the public interest and without distorting international trade and investment.

Principle 17. Environmental impact assessment, as a national instrument, shall be undertaken for proposed activities that are likely to have a significant adverse impact on the environment and are subject to a decision of a competent national authority.

Principle 18. States shall immediately notify other States of any natural disasters or other emergencies that are likely to produce sudden harmful effects on the environment of those States. Every effort shall be made by the international community to help States so afflicted.

Principle 19. States shall provide prior and timely notification and relevant information to potentially affected States on activities that may have a significant adverse transboundary environmental effect and shall consult with those States at an early stage and in good faith.

Principle 20. Women have a vital role in environmental management and development. Their full participation is therefore essential to achieve sustainable development.

Principle 21. The creativity, ideals and courage of the youth of the world should be mobilized to forge a global partnership in order to achieve sustainable development and ensure a better future for all.

BOX 19.1 Continued

Principle 22. Indigenous people and their communities, and other local communities, have a vital role in environmental management and development because of their knowledge and traditional practices. States should recognize and duly support their identity, culture and interests and enable their effective participation in the achievement of sustainable development.

Principle 23. The environment and natural resources of people under oppression, domination and occupation shall be protected.

Principle 24. Warfare is inherently destructive of sustainable development. States shall therefore respect international law providing protection for the environment in times of armed conflict and cooperate in its further development, as necessary.

Principle 25. Peace, development and environmental protection are interdependent and indivisible.

Principle 26. States shall resolve all their environmental disputes peacefully and by appropriate means in accordance with the Charter of the United Nations.

Principle 27. States and people shall cooperate in good faith and in a spirit of partnership in the fulfillment of the principles embodied in this Declaration and in the further development of international law in the field of sustainable development.

cess to sanitation without taking into account the number of people who may need to share the facility. Likewise the presence of user fees and their affordability to the poor are not taken into consideration. As a result, "improved" access to water supply and sanitation does not necessarily imply effective or "adequate" access, since users may be prevented from using the facilities by high fees or overcrowding. These problems are particularly acute in urban areas (Bartlett 2003). In fact it has been shown that "improved" sanitation does not "greatly reduce the risk of oral-fecal diseases" (Pruess and Fewtrell 2002).

In the absence of data on "adequate" or "basic" access, we use the data on "improved" access provided by the JMP. The JMP definition is in line with the U.N. Secretary-General's Report on the MDGs, but will underestimate the number of people who do not have effective access to clean water and sanitation—particularly in urban areas. Based on the JMP data, a considerable expansion of infrastructure services is required in developing countries to meet the needs of 1.2 billion people in developing countries for improved access to drinking water and 1.8 billion people who need to be connected to sanitation by 2015. Each year until 2015, it will be necessary to provide access to water and sanitation to 100 million and 140 million people, respectively. In comparison, an average of 85 million additional people per year received access to water and sanitation facilities during the 1990s. Hence considerable additional investments will be required to increase the pace of providing water and particularly sanitation facilities in developing countries. (See Table 19.13.)

19.2.9.2 Dependencies

Achieving the water and sanitation targets will depend on a number of ecosystem services. Table 19.14 summarizes the impact of meeting the targets on ecosystems.

Table 19.14. Ecosystem Services Contributing to the Water and Sanitation Targets

Ecosystem Services Contributing to Meeting the Water and Sanitation Targets	Ecosystems Directly Influenced by Progress toward the Water and Sanitation Targets
provisioning services • provision of fresh water	freshwater ecosystems and wetlands (through increased water withdrawal and pollution)
regulating services • water flow regulation (to ensure sufficient and regular supply) and aquifer recharge • water filtration and purification	coastal and marine ecosystems (through increased water withdrawal and water pollution)
supporting services • water cycling	agroecosystems (through increased demand for water)
	urban ecosystems through increased or reduced fecal pollution (depending on sanitation and water treatment technologies)
	drylands, forests, and other ecosystems that compete with humans over water

From an ecosystem perspective, meeting MDG target 10 and the WSSD sanitation target poses two related challenges. One necessary, but not sufficient input into achieving the target is the provision of sufficient quantities of clean water through ecosystem services. In addition, rising water pollution through domestic

Table 19.13. Access to Improved Water Supply and Sanitation in Developing Countries. This table illustrates improved access to water supply and sanitation between the years 1990 and 2002. Despite the increases, there still remains an additional 6% and 21% of the growing population that will require additional investments in improved water supply and sanitation, respectively, in developing countries. (UNICEF/WHO 2000)

	Population	Access to Improved Water	Access to Improved Water	Access to Improved Sanitation	Access to Improved Sanitation
	(billion)		*(billion)*		*(billion)*
1990	4.1	71%	2.9	34%	1.4
2000	4.9	79%	3.9	49%	2.4
MDG 2015 Targets	6.0	85%	5.1	70%	4.2
Difference, 2015–2000	1.1	6%	1.2	21%	1.8

and industrial effluents must be managed to protect ecosystem services and to improve human health. These two challenges are discussed in detail here. Table 19.15 summarizes the dependencies, as well as interventions, synergies, and trade-offs for MDG target 10.

19.2.9.2.1 Water quantity

As described in MA *Current State and Trends,* Chapter 7, fresh water that is readily accessible for human use is limited to approximately 9,000 cubic kilometers a year with an additional 3,500 cubic kilometers stored by dams and reservoirs. Currently, humans consume about 50% of the combined 12,500 cubic kilometers of fresh water that are readily available each year.

To ensure basic needs for drinking, food preparation, and personal hygiene, a minimum of 20 to 50 liters of water free from harmful contaminants is required per person every day. Assuming an incremental consumption of 50 liters per day, the 1.4 billion people who need to be provided with access to improved water supply to meet target 10 would require a total of 0.07 cubic kilometers of water, or less than 0.001% of readily available fresh water. Since many of the 1.4 billion people currently without access to improved water supply already consume some fresh water, this overestimates the additional demand for water to meet basic human needs. Hence at the global level, overall fresh water availability does not appear to represent a binding constraint on providing access to the minimum amount of drinking water that is required to achieve the water and sanitation targets.

To underline this point, Figure 19.1 and Figure 19.2 provide simple plots of available water resources per capita and access to improved water supply and sanitation, respectively, showing no discernible correlation between access to water or sanitation and overall per capita cater availability. This finding is too robust to control for per capita income and regional effects.

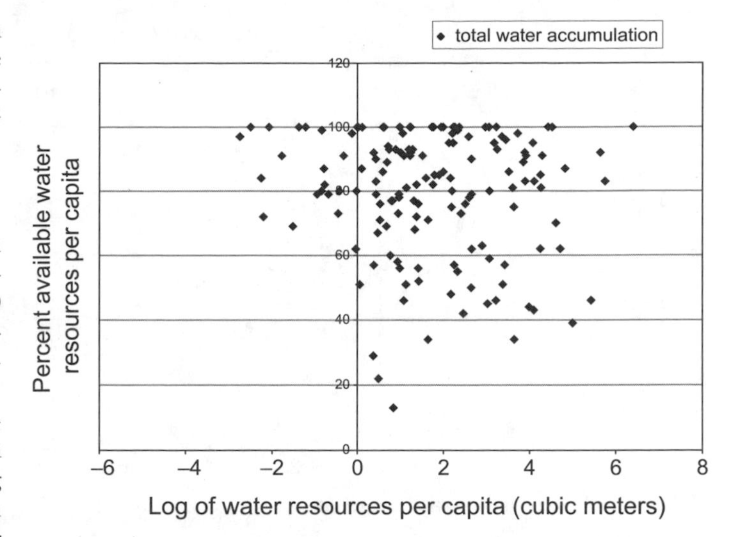

Figure 19.1. Percent Access to Water Supply against Log of Per Capita Water Resources in 2000 (FAO Aquastat; UNICEF/WHO 2004)

However, domestic per capita consumption of fresh water tends to rise considerably beyond the minimum needs of 50 liters per day once access to water supply has been established. Coupled with increased agricultural and industrial demand for water, water withdrawals increase much faster than populations. For example, between 1900 and 1995, water withdrawals increased by over six

Table 19.15. MDG 7, Target 10: Dependencies, Interventions, Synergies, and Trade-offs for Access to Water

Target	Dependencies on Ecosystems for Achieving Target	Interventions	Links to Ecosystems and Ecosystem Services: Synergies (benefits to ecosystems/ ecosystem services)	Links to Ecosystems and Ecosystem Services: Trade-offs (threats to other types of ecosystems/ecosystem services)
Access to water and to sanitation	freshwater ecosystems and wetlands coastal and marine ecosystems agroecosystems urban ecosystems drylands, forests, and other ecosystems that compete with humans over water	expand access to improved water supply expand access to improved sanitation greywater treatment and disposal industrial wastewater treatment additional water storage for provision of drinking water	improved sanitation systems can reduce local microbial pollution and total nutrient load by using improved technologies (for example, lined pits or improved sewer systems) reduces microbial water pollution as well as nutrient loads. reduces chemical water pollution	will increase demand for surface and groundwater can lead to a dramatic increase in demand for water, particularly if traditional sewer technology is used unless accompanied by wastewater treatment, increased access to (urban) sewers can lead to an increase in microbial, nutrient and other pollution of freshwater ecosystems in some areas, improving access to water supply may require additional water storage capacity, which can have negative impacts on ecosystems and their services

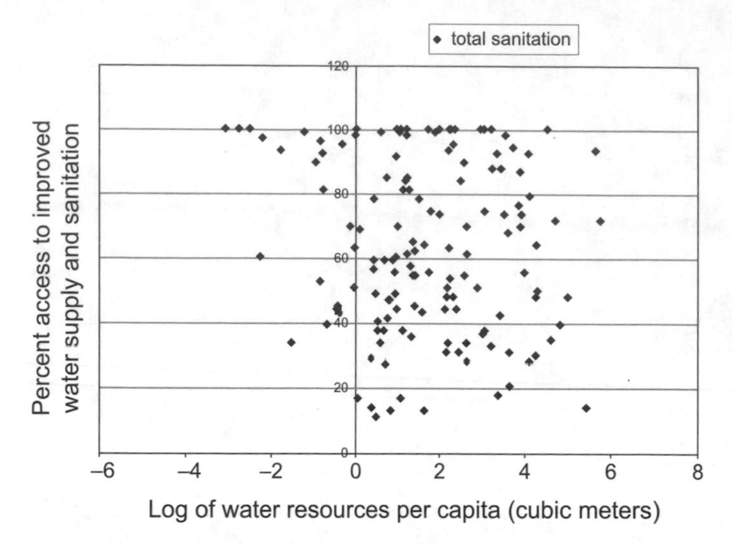

Figure 19.2. Percent Access to Sanitation against Log of Per Capita Water Resources in 2000 (FAO Aquastat; UNICEF/WHO 2004)

times, more than double the rate of population growth (MA *Current State and Trends,* Chapter 7). If this trend continues unabated, human consumption of fresh water at the global level may soon be constrained by the availability of fresh water.

Importantly, global averages hide important regional variations in terms of demand for water as well as fresh water availability. As described in MA *Current State and Trends,* Chapter 7, prior assessment (Revenga et al. 2000) shows that, as of 1995, some 41% of the world's population, or 2.3 billion people, were living in river basins under water stress, with some 1.7 billion of these people residing in highly stressed river basins. Another 29 basins will descend further into scarcity by 2025. This problem is exacerbated by the fact that a large part of the expected increase in human population will occur in regions where water is already scarce and/or erratically available. Any changes to ecosystem services resulting from increased demand for water will have a disproportionate impact on the poor who depend most on ecosystem services for well-being.

In summary, the availability of fresh water at the global level is unlikely to prevent low- and middle-income countries from achieving the targets on water and sanitation, since the minimum per capita needs addressed by these targets are very low compared with current total consumption, which is dominated by agricultural and industrial water demand. However, water scarcity at the catchment level may occur in some areas where demand for water can exceed the readily available freshwater resources. This is particularly true for some of the rapidly growing cities in low-income countries, which may experience water shortage or need to import water from other catchment areas. A particular area of concern in some of the fastest growing cities is the growing mining of groundwater resources beyond sustainable levels.

In most cases, though, water shortages are the result of excessive water consumption for agricultural and sometimes industrial use rather than domestic consumption. In such cases, water consumption for non-domestic use may need to be balanced with need through ecosystem-wide integrated water resources management to meet the water and sanitation target. Consequently, the Johannesburg Plan of Implementation urges all countries to adopt IWRM plans by 2005. That also implies a substantial increase of water use efficiency in agroecosystems.

19.2.9.2.2 Water quality

Increased access to water and sanitation may have notable impacts on surface water quality and affect groundwater in areas where the water table is high. The extent and direction of this effect can depend on the types of technologies used. In general, properly constructed and maintained sanitation systems can reduce water pollution through human waste. For example, improved double-pit latrines or septic tanks can process domestic wastewater to avoid any adverse impact on surface and groundwater. The same applies to water-borne sewers, provided they are equipped with appropriate water treatment systems that can remove microbial pollutants and reduce nutrient load.

On the other hand, sewer systems that lack appropriate trunk infrastructure and water treatment facilities can lead to high levels of microbial pollution with potentially adverse impacts on human health through the spread of infectious diseases. In some ecosystems such as lakes, the increased inflow of nutrients can lead to eutrophication with important consequences for ecosystem services. Since septic tanks and bore sewers are often not properly operated and maintained, they cannot remove solid particles or microbial pollutants and thereby contribute to the pollution of water systems.

These problems are particularly acute in developing countries where an estimated 90% of wastewater is discharged directly to rivers and streams without any waste processing treatment (WMO 1997, p.11). Threats of water quality degradation are most severe in areas where water is scarce because the dilution effect is inversely related to the amount of water in circulation. (MA *Current State and Trends,* Chapter 8). In these areas, achieving the MDGs on sanitation without parallel investments in water treatment can threaten fresh water and coastal ecosystems and the services they provide.

In many parts of the world, agricultural inputs such as fertilizers and pesticides, as well as accelerated soil erosion from faulty land management practices can degrade surface and groundwater quality. A broad range of site-specific interventions is required to minimize the use of agricultural inputs and to improve soil management. The World Overview of Conservation Approaches and Technologies has an extensive database of such interventions.

19.2.9.3 Interventions

Critical interventions for achieving the water and sanitation targets include the expansion of appropriate water and sanitation infrastructure services accompanied by improved hygiene education. In particular, the design standards for sanitation infrastructure used to achieve the 2015 target can have a substantial impact on the extent to which ecosystems and their services are affected. Conventional waterborne sewers can dramatically increase per capita demand for water and thereby exacerbate water stress in some areas. If untreated, sewer systems can further lead to increased water pollution.

In many regions where access to water supply is currently very low, there may be a need for additional water storage capacity to ensure a steady and perennial supply of fresh water. Dams can have severe adverse impacts on fresh water and other ecosystems (MA *Current State and Trends,* Chapter 7; World Commission on Dams 2000). The WCD has laid out detailed guidelines that can help prevent unnecessary degradation of ecosystems. Adequate water demand management can further reduce the need for additional storage capacity.

19.2.9.4 Conclusions

On balance, achieving the water and sanitation targets will require an ambitious increase in the number of people provided with access to water supply and sanitation each year. The provision of clean water depends on a number of provisioning and regulating ecosystem services that will need to be protected in order to ensure a regular supply of clean water.

While water scarcity is likely to worsen in some areas, global freshwater ability appears to be sufficient for achieving the water and sanitation targets. The overall impact that achieving the 2015 targets may have on water demand and quality depends critically on appropriate water demand management policies such as pricing that can help contain demand for fresh water beyond the minimum basic needs of 20–50 liter per capita. In addition, integrated governance and institutional frameworks such as basin management authorities, IWRM, and other multisectoral approaches to managing water needs, can reduce pressure on ecosystems and their services. Finally, the nature of sanitation technologies and the extent to which they are properly operated can help mitigate adverse impacts on water quality.

19.2.10 MDG 7: Ensure Environmental Sustainability, Target 11

Target 11 has two components. The first focuses on slum upgrading as stated in the text of the MDG Target, "By 2020, to have achieved significant improvement in the lives of at least 100 million slum dwellers." The second component concentrates on prevention by stopping the formation of slums by 2006 as stated by the Cities Alliance for Cities without Slums (UN 2000; Annan 2000).

19.2.10.1 Introduction

Today 3 billion people live in urban areas (United Nations 2002), which is close to 50% of the world's population, compared with only 15% in 1900 (Graumann 1977). Urbanization is fundamentally transforming the world: most of the world's largest cities are now in Asia, not in Europe or North America. U.N. projections suggest that urban populations are growing so much faster than rural populations that 85% of the growth in the world's population between 2000 and 2010 will be in urban areas and virtually all of this growth will be in Africa, Asia, and Latin America. Moreover, most of these new urban dwellers are likely to be poor, resulting in the increasing urbanization of poverty. Slums are a physical and spatial manifestation of increasing urban poverty and intra-city inequality.

Slums, like poverty, are multidimensional in nature. Some of the characteristics of slums, such as access to physical services or density, can be clearly defined, while others such as social capital cannot. UN-Habitat proposes that a person whose living conditions suffer from at least one of the following five characteristics is be considered a "slum dweller":

- inadequate access to safe water,
- inadequate access to sanitation and other infrastructure,
- poor structural quality of housing,
- overcrowding, and
- insecure residential status.

Based on these criteria, it is estimated that currently over 900 million people—roughly one third of the world's urban population—live in slums. More than 70% of the urban population in least industrial countries and sub-Saharan Africa (UN-HABITAT 2003) lives in slum-like conditions. This number is set to increase to roughly 2 billion by 2020 unless current trends change substantially. The challenge is particularly acute in sub-Saharan Africa, where urbanization proceeds at a very high pace.

Of course, these projections are fraught with uncertainty, but they suggest that in addition to improving the lives of 100 million slum dwellers an incremental urban population of up to 1.1 billion who may otherwise end up living in slum-like conditions will need to gain access to decent housing, adequate infrastructure, and basic services. While achieving this target can have important benefits for HWB in urban areas, it is not clear to what extent it will have an impact on the rate of urbanization. On balance, it seems most likely that cities will continue to grow rapidly, so that by 2020, approximately 4.2 million people corresponding to 56% of the world's population will live in urban areas (UN 2001; UN 2002).

19.2.10.2 Dependencies

An analysis of the linkages between slum formation or urbanization more broadly and ecosystem services is best disaggregated according to spatial dimensions. MA *Current State and Trends,* Chapter 24, distinguishes interactions between urban systems and (1) ecosystems contained within the urban areas; (2) adjoining non-urban ecosystems; and (3) distant ecosystems. Urban systems including slums depend on a number of ecosystem services, which they consume. In turn, urbanization and the achievement of the slum dwellers target will impact on ecosystems, as summarized in Table 19.16.

As shown in MA *Current State and Trends,* Chapter 24, urbanization is not inherently bad for ecosystems. For example, urban centers can facilitate human access to ecosystem services through, for example, the scale economies of piped water systems. While urban demographic and economic growth increases pressures on ecosystems globally, the same demographic and economic growth would probably be even more stressful if the same people, with similar consumption and production patterns, were dispersed over the rural landscape. Urbanization usually reduces the demand for land relative to population. Indeed, the world's current urban population, corresponding to half the total population, would fit into an area of 2,000,000 square kilometers—roughly the size of Senegal or Oman—at densities similar to high-class residential areas in European cities (Hardoy et al. 2001).

The relatively lower per capita demand for ecosystem services of people with a given consumption and production pattern tends to have the highest effect on distant ecosystems. Meanwhile, adverse environmental impacts of urban systems are mostly confined

Table 19.16. Ecosystem Services Contributing to Improving the Lives of Slum Dwellers

Ecosystem Services in and around Urban Areas Contributing to Meeting the Slum Dwellers Target	Ecosystems Directly Influenced by Improving the Lives of Slum Dwellers
provisioning services • provision of food, clean water, construction materials, and other ecosystem services	ecosystems within urban areas (for example, grasslands and freshwater ecosystems) freshwater, coastal, and marine ecosystems (through increased water withdrawal and water pollution)
regulating services • alter filtration and purification • provision of clean air • provision of a healthy urban environment	ecosystems adjoining urban areas (through urban expansion, water and air pollution)
cultural services • recreational • cultural heritage	agroecosystems and other ecosystems further away from urban systems (through demand for food and other ecosystem services)

to ecosystems within cities and towns as well as the immediate hinterland. In general, the most critical environmental burdens of slums and cities tend to be local, such as inadequate and unsafe piped water supply, lack of proper sewerage and storm water drainage, insufficient garbage collection and disposal, indoor air pollution that results from burning biomass, poor health care services, etc. (Bartone et al. 1994; McGranahan and Songsore 1994). As a result, adjoining ecosystems and those contained within urban systems experience increased pressure and stress with rising urbanization and slum formation. For example, increased demand for water and rising levels of water pollution caused by cities and slums in particular have adverse effects on freshwater and coastal ecosystems surrounding urban centers.

The most extreme forms of environmental degradation tend to be found in slums. Chronic pollution of water sources, high disease prevalence and deterioration of public health conditions are common features in many of these illegal urban settlements. In addition, many slum dwellers reside in protected and/or fragile areas such as protected watersheds, wetlands located next to rivers and lagoons, steep hillsides vulnerable to landslides and soil erosion, and mangroves and valleys subject to flooding or tidal inundation. Table 19.17 summarizes the dependencies, as well as the interventions, synergies, and trade-offs for MDG target 11.

19.2.10.3. Interventions

Meeting target 11 will require a combination of slum upgrading and prevention of the formation of new slums. Slum upgrading refers to the provision of basic services such as solid waste disposal, improved infrastructure including water supply and sanitation, and improvements in housing. In many cases, successful slum upgrading requires considerable improvements in the security of tenure extended to and perceived by the slum dwellers. In addition to producing dramatic improvements in HWB, slum upgrading can lessen pressure on freshwater ecosystems, grassland, forests, and other ecosystems found within urban areas. As a result, critical ecosystem services such as water purification and disease control can be maintained and will contribute to improved human health.

In some instances, improving the lives of slum dwellers may require the negotiated relocation of urban populations away from fragile land such as steep slopes or floodplains and other vulnerable or protected ecosystems within urban areas. Such carefully planned interventions can generate important improvements in HWB within cities provided that the communities concerned are closely involved in the preparation, negotiation, and execution of resettlements.

Preventing the formation of new slums will require a broad range of interventions targeted at making land available at minimal cost to low-income households, improving urban planning and design, and providing adequate access to basic services as well as infrastructure for transport, energy, water supply, and sanitation. Managing the spatial layout and design of urban agglomerations can help reduce pressure on vulnerable ecosystems such as adjoining wetlands, riverbanks, or steep slopes.

19.2.10.4. Conclusions

On balance, meeting target 11 will have profound and positive effects on ecosystems and their services found within and adjoining to urban areas. Further preventing the formation of new slums is likely to have an even greater impact on ecosystems and their services through improving the urban environment and reducing pollution of water and air. The overall impact of target 11 on rates of urbanization and total urban demand for ecosystem services is likely to be limited since the targeted 100 million slum dwellers

Table 19.17. MDG 7, Target 11: Dependencies, Interventions, Synergies, and Trade-offs for Improving the Lives of Slum Dwellers

			Links to Ecosystems and Ecosystem Services	
Target	**Dependencies on Ecosystems for Achieving Target**	**Interventions**	**Synergies (benefits to ecosystems/ ecosystem services)**	**Trade-offs (threats to other types of ecosystems/ecosystem services)**
Improving the lives of slum dwellers	no dependencies exist	provision of secure tenure extension of urban infrastructure provision of basic services (for example, solid waste disposal, transportation, protection services) improved and strengthened institutions for urban management and planning	slum dwellers will have stronger incentives to invest in the improved management of the urban ecosystem (for example, through improved sanitation) improved sanitation combined with water treatment can help control water pollution improved waste disposal can reduce pressure on urban and related water ecosystems improved institutions for urban management will strengthen slum dwellers' voice in local decision-making; as a result local deterioration in ecosystem services will figure more prominently in decision-making	impact of urban infrastructure (for example, transport and energy) depends critically on design and implementation; negative impacts can occur if environmental needs are not taken into account at the outset reduced availability and higher prices for the local poor

account for only a relatively small share of the total urban population.

19.2.11 MDG 8: A Global Partnership for Development, Targets 12–15

19.2.11.1 Introduction: The Link between MDG 8 and Ecosystems

The MDGs are meant to address the central challenge we face today—"*to ensure that globalization becomes a positive force for all the world's people.*" Globalization is a multifaceted collection of processes, a central part of which is the expansion of world trade and technology. A more open and fair trading system can expand developing countries' growth rates, providing revenues needed to finance MDGs 1 through 7. But least industrial countries, including many landlocked, small island, and highly indebted nations, lack the basic health, education, and infrastructure capacities needed to adequately access expanded markets and make the most of more open trade regimes. Table 19.18 summarizes the dependencies, interventions, synergies, and trade-offs for targets 12 through 15.

19.2.11.2 Target 12: Develop Further an Open Trading and Financial System: Including a Commitment to Good Governance, Development, and Poverty Reduction, both Nationally and Internationally

What do ecosystems have to do with this? How are ecosystem services essential to reaching MDG 8, and how may achieving MDG 8 affect ecosystems? Of the 1.2 billion people living in extreme poverty on less than $1 per day, approximately 900 million live in rural areas and are thus highly dependent on primary sector economic activities such as agriculture, forestry, and fisheries. Access to markets and expanded trade can lead to increased incomes for these communities, and greater revenues for developing countries to achieve MDGs 1 through 7. Meanwhile, high agriculture subsidies in the industrialized world continue to distort global market dynamics and promote unsustainable forms of agriculture in many developing countries, instigating a cycle of land degradation and loss of ecosystem functions.

But the critical link between MDG 8 and ecosystems is by no means limited only to those in extreme poverty. There is no question that further increases in growth and trade expansion is required to raise the incomes needed for poverty reduction, food security, etc., particularly in least developed, landlocked, and small island nations. However, the aggregate consumption levels of an expanded global consumer society could soon push critical aspects of the planet's life support system over an "ecological cliff."

Equal attention should be placed on countries that may be progressing well toward the MDGs' social and economic development goals, but in the process are having a relatively larger impact on global ecosystem functioning. Areas of high biodiversity and critical global ecosystem functions are often located in or severely affected by such countries. The industrial and developing countries within WTO that are involved in expanding global trade patterns, including Brazil, China, the European Union, India, Indonesia, Japan, Russia, the United States, and various other countries increasingly drive global ecological change through their trade policies.

19.2.11.2.1 Policy interventions to achieve target 12 and maintain ecosystem services

The main factor for understanding the link between trade and ecosystem services relates to the problem of scale. Unchecked, today's global economy may be five times bigger only one generation or so hence. Our challenge today is to expand international trade in a way that does not cause irreparable harm to global ecosystems. Notable improvements in the efficiency of production and consumption patterns will be necessary, in both industrial and developing countries.

The focus of target 12 is not only on more open trade but also on the rules required to manage such trade with equity and sustainability, through a process of good governance. What communities need is the space to shape their own path to HWB. The drive toward expanded growth and trade brings marked changes to local culture, often relegating traditional local knowledge of HWB and ecosystems to the background. The ability to engage excluded communities in the process of globalization will largely depend on means by which traditional knowledge systems can be integrated into local growth and human development strategies. It will also involve greater engagement of the private sector in this process of creating an inclusive form of globalization, as market-based development becomes an increasing reality.

19.2.11.2.2 Trade-offs between human development and ecosystem services

The extent to which ecosystems services are disrupted is an outcome of policy options, clear trade-offs, and choices. The expansion of trade is often based on the assumption that environmental impacts will be corrected for as affluence and investments in sustainability increase. But direct links between rising incomes and improved environmental protection is only relevant for certain types of pollution, and does not apply to many critical ecosystem services such as biodiversity and climate regulation, both of which appear to be proportional to rising consumption levels. In this regard, proactive regulation is needed to avoid undesirable ecological feedbacks, recognizing the value of such services for underpinning human welfare. Recent estimates place the value of the world's ecosystems to be more than the total value of the world's economy. There is a need to use ecological valuation methods to account for the economic values of ecological services ignored by markets.

Where trade-off mechanisms fall short, the challenge is to ensure that a minimum level of "ecological security" is maintained, based on scientific knowledge of HWB/ecosystem service linkages, and preserved without trade off. This would be critical to ensure a long-term, minimum level of ecological stock to meet basic human development. Such an ecological security system could be integrated into trade and finance policies.

19.2.11.3 Targets 13–15: Address the Special Needs of Least Developed Countries, Landlocked and Small Island Developing States, and Heavily Indebted Poor Countries

Most least developed countries rely heavily on agriculture for their socioeconomic well-being, making them directly dependent on ecosystem services, particularly land and water sustainability. Many least developed countries suffer from a lack of options beyond improving agriculture due to various barriers—inaccessible markets, lack of infrastructure, lack of private investment, and underdeveloped value-added industries, to name a few.

Achievement of targets 12 through 15 would reduce tariffs and subsidies in agricultural products, but even with more open markets, many communities may well remain excluded from the losses and benefits. Least developed countries, particularly landlocked and small island states, are geographically isolated from large markets, and lack the infrastructure needed for expansion. Furthermore, many least developed countries now suffer from se-

Table 19.18. MDG 8, Targets 12–15: Dependencies, Interventions, Synergies, and Trade-offs for Global Partnerships

Target	Dependencies on Ecosystems for Achieving Target	Interventions	Links to Ecosystems and Ecosystem Services: Synergies (benefits to ecosystems/ ecosystem services)	Links to Ecosystems and Ecosystem Services: Trade-offs (threats to other types of ecosystems/ecosystem services)
Rule-based trading and financial system	global trade includes significant level of products and services derived directly or indirectly from ecosystem goods and services global growth depends heavily on carrying capacity of planet and ability of trading system to adapt to temporary or permanent disruptions in ecological goods and services negative feedbacks can damage sustainability of such systems, thereby jeopardizing long-term global economic systemd global, regional, national, and local trade flows are dependent on natural flows of ecosystem goods and services, both of which are changing rapidly many developing countries rely increasingly on foreign direct investments for growth in trade and finance multinational corporations have an increasing dependence on ecosystem services that may affect the sustainability of the goods or services they produce or trade with increasing role in global and national governance processes, multinational corporations also play a significant role in making trade-offs between commercial and ecological values	global and national decision-making processes whereby more just and effective trade-offs can be made between socioeconomic and ecological values, including expanded role for civil society in decision-making processes include use of ecological valuation methods for making trade-offs in decision-making processes integrate an ecological security system, whereby minimum required level of ecosystem services can be protected without trade-off, to ensure basic human well-being in long-term include approaches to map trade flow–ecosystem service flow links and areas where high trade growth rates collide with critical ecosystem services build partnerships among business, government, and civil society to integrate ecosystem approach into investment planning at macro (industrial trade policy) and micro (foreign direct investment project) levels. establish local learning forums among businesses and partners to develop and use benchmarks and indicators for monitoring ecosystem service issues and exploring lessons for achievement of human well-being	improved well-being in terms of non-income aspects of poverty related to exclusion provides decision-makers with quantitative ability to compare choices with resulting increases chance of preserving critical ecosystem services in the long-term trade-offs and choices may well lead to degradation of ecosystem services; however, core ecosystem services can be preserved addresses fact that ecosystems know no political and trade boundaries so that decisions may be taken with reference to local and global scales of ecosystem service and impact reduces risk of conflict among business and civil society if ecosystem services valuable to their well-being are sustained ability to gauge drop in ecosystem services related to business decisions and take corrective actions	not applicable
Special needs of least developed countries Landlocked countries, small island countries Debt sustainability	LDCs, including landlocked and small island states, have high dependence on agriculture and related ecosystem services these countries lack infrastructure (irrigation, energy, roads, etc.) needed to take advantage of expanded markets from a drop in tariffs and subsidies in industrial countries these countries are burdened by land and water stress caused by unsustainable agriculture from decades of low market access	synergies in official development assistance and debt relief programs to ensure ecosystem approach for infrastructure and agricultural expansion initiatives, including strategic ecological assessment methods for infrastructure, and sustainable land management measures to ensure sustained productivity to stay out of debt traps. integrate ecosystem approach into other macro (structural adjustments, poverty/human development strategies) and micro-level (local ODA projects) interventions	prevention of costly or irreparable change to least developed country ecosystems more sustainable land and water use has benefits related to other ecosystem services as well.	not applicable

rious land and water stresses, and these may well increase should the pace of global change increase. And open markets are not enough—increased official development assistance and real debt relief are required to help finance new infrastructure, and address land and water issues. Considerable opportunity now exists to synergize new ODA, debt relief, and innovative "debt for MDG" swaps to ensure that these issues are addressed side by side.

MDG 8 serves to re-focus attention on rural agriculture as a vital part of achieving human development. This includes increased coordination of ODA to nationally designed and executed MDG programs. Such programs should use strategic ecological assessment methods to improve the likelihood of long-term benefits. This would be relevant at macro- (national structural adjustments and poverty reduction/human development strategies) and micro-levels (local ODA-financed projects).

ODA should focus on agricultural sustainability, as well as infrastructure needed in least developed countries to take advantage of increased market access. New infrastructure will be required, but should be developed in a way that integrates ecosystem approaches into grant and lending processes through strategic ecological assessment methods. Sustaining agricultural outputs will depend on sustainable land and water use, in turn depending on an array of governance issues—secure access to land, water, and local markets; access to market information; rural enterprise development and credit; and safety-net policies to adapt to market reform.

19.3 Multi-goal Analysis

The Millennium Development Goals strive to improve human well-being. Ecosystems provide many of the services needed to meet these goals, and many ecosystem services and MDGs are interlinked and interdependent. The previous section considered a goal-by-goal analysis. Here we consider a multi-goal analysis. Table 19.19 illustrates all MDGs and targets, and the ecosystem services needed to achieve those targets. Where there exists a positive or negative impact, each cell is highlighted in black. General trade-offs and synergies associated with achieving the 2015 target are analyzed.

19.3.1 Crosscutting Analysis

How does meeting the target of one or more of the MDGs affect the ability to meet all 2015 targets?

Achieving target 9 is essential to simultaneously achieving targets 1, 2, 5, 7, 8, and 12–15, as depicted in Table 19.19. Target 9's emphasis on integrating principles of sustainable development corresponds with direct investment in supporting, provisioning, regulating, and nonmaterial services. Ultimately, simultaneous contributions will occur with reduced hunger, child deaths, and disease, and an increase in income levels, trade, and reduced debt.

Can all the targets be met simultaneously?

Meeting all the targets simultaneously may be possible if ecosystem services are maintained and/or protected (in particular provisioning, supporting, and regulating services) *and* if institutional capacity building and good governance are initiated. Capacity building and good governance are essential in achieving the 2015 targets and, in particular, essential to those targets not directly related to ecosystem services. These targets include gender, education, maternal mortality, youth employment, and access to medicines and technology. By incorporating principles of sustainable development into national polices and by integrating MDG programs into country strategies such as poverty reduction strategy papers, institutions can be created to address the above targets.

How dependent are the 2015 targets on ecosystems and ecosystem services?

Targets 1, 2, 5, 7–9, and 12–15 are directly dependent on ecosystem services. It is well-established that these targets cannot be met without essential ecosystem services such as primary production and nutrient cycling, food, water purification and regulation, and erosion control. These benefits and more are the very basics of survival.

By investing primarily in provisioning, supporting, and regulating services, the targets most likely to be achieved simultaneous by 2015 are targets 1, 2, 5, 7–9, 12–15. The synergistic effects of investing in these three ecosystem services are noted among reducing poverty, hunger, child mortality, disease, opening trading systems, least developed countries, land locked/small island states and LDC debt. In terms of MDG 1 (eradicating extreme hunger and poverty), expanded agricultural production, more inputs, and new technologies such as precision farming will be required to increase crop yield and promote rural and non-rural employment. Achieving target 9 requires incorporating sustainable agricultural policies in areas of international trade and restoration and rehabilitation, and training programs to yield greater productivity. With investment in supporting, provisioning, and regulating services, this will ensure ecosystems are maintained to provide resilience and production while increasing quality of life.

Targets 4, 6, 7, and 12–15 require financial and institutional investment in supporting, provisioning, and regulating services. The synergistic effects of establishing such investment involves increasing life expectancy, establishing new markets, reducing debt and increasing financial instability and contributing to capacity building for decision-making.

The trade-offs incurred by investing in supporting, provisioning, and regulating benefits include fewer initiatives in polices and programs that would be created for nonmaterial benefits. Nonmaterial benefits provide important meaning to many individuals in society. In addition, employment is generated through such services as recreation and ecotourism. In achieving the 2015 targets, however, priority in provisioning, supporting, and regulating services may be substantial in achieving the 2015 targets. These services are essential to the poor for ensuring entitlements.

A major trade-off is also incurred in infrastructure development. Most developing societies will require infrastructure to ensure public facilities, distribution channels including roads and ports, and other building structures. Polices targeted at infrastructure development may alter natural ecosystems and decision-makers will have the difficult task of ensuring sustainable principles and practices are captured.

How can knowledge of the dependency on ecosystems to provide the services help to achieve the 2015 targets?

Decision-makers have a challenging task of ensuring that the products and services that the ecosystem provides are constantly available and maintained while ensuring good governance and institutional capacity building. Understanding the benefits and the roles of regulating, provisioning, and supporting services provides decision-makers with the opportunity to:

- maintain natural ecosystems through sustainable practices to ensure food and water are readily available for distribution;
- invite new institutions and organizations to aid in the maintenance of ecosystem services and institutional capacity building;

Table 19.19. Cross-cutting Goals and Targets: Ecosystem Services Affected in Implementing the 2015 Targets. Shaded cells indicate a positive or negative impact on the ecosystem service in question.

Ecosystem Services	MDG1		MDG 2	MDG3	MDG4	MDG5	MDG 6		MDG 7			MDG 8						
	T1: Poverty	T2: Hunger	T3: Primary Education	T4: Gender	T5: Child Mortality	T6: Maternal Health	T7: HIV/ AIDS	T8: Malaria/ Other Diseases	T9: Sustainable Development	T10: Safe Drinking Water	T11: Slum Dweller Life Improved	T12: Open Trading/ Financial	T13: Least Developed Countries	T14: Landlocked/ Small Island States	T15: LDC Debt Sustainability	T16: Youth Employment	T17: Access to Medicines	T18: New Techn. ICT
Supporting Services																		
Soil formation																		
Nutrient cycling																		
Primary production																		
Pollination																		
Habitat maintenance																		
Seed dispersal																		
Provisioning Services																		
Food																		
Fresh water																		
Fuelwood																		
Fiber																		
Biochemicals																		
Genetic resources																		

Regulating Services																		
Climate regulation																		
Disease regulation																		
Water regulation																		
Water purification																		
Landscape stabilization against erosion																		
Binding of toxic compounds																		
Air purification																		
Pest and pathogen control																		
Carbon sequestration																		
Cultural Services																		
Spiritual and religious																		
Recreation and ecotourism																		
Aesthetic																		
Inspirational																		
Educational																		
Sense of place																		
Cultural heritage																		

- create new markets for local and foreign investors and create employment for rural and non-rural people;
- ensure the protection of ecosystems such as watershed protection to contribute to disease regulation and fresh water so that children in particular will live much longer to contribute to their developing societies;
- invest in technologies that are efficient, nonpolluting, and cost-effective, such as wastewater treatment technology and solar and wind technology;
- initiate policies and programs that provide fuel such as biogas that will benefit women through less physical labor and provide more access to communication systems;
- initiate policies and programs that reduce deforestation through agricultural intensification and invest in nursery planting and restoration programs to achieve a multitude of benefits including climate regulation and medicinal products.

What are the respective roles and responsibilities of industrial and developing countries in achieving the 2015 targets?

Principle 7 of the Rio Declaration identifies that the industrial world must acknowledge its considerable consumption and exploitation of natural resources. The industrial world, therefore, can initiate policies and plans that acknowledge its role in minimizing ecosystem services and identify strategies to achieving the 2015 targets. This will include contribution to monitoring, research, and analysis in both industrial and developing countries, which can be conducted through existing institutions and governance structures.

Developing countries will benefit in achieving the 2015 targets by ensuring plans and policies geared at achieving the 2015 targets are integrated vertically and horizontally into national polices such as PRSPs and within governance *(high certainty)*.

19.3.2 Trade-offs and Synergies

Meeting the MDGs by 2015 may cause time trade-offs. The trade-offs listed below are more concerned with achieving the 2015 deadline than with ensuring sustainable methods of achieving the MDGs. The trade-offs include:

- As identified above, meeting the food target for MDG 1 by 2015 through unsustainable changes in production such as the increase in use of pesticides and rapid agriculture intensification may meet the target; however, there will be more stress on the environment and people's livelihoods. Therefore, "leapfrogging" toward sound ecological sound production techniques and approaches is needed.
- Plans to achieve all the MDGs by 2015 could be "rushed," requiring decision-makers to look for policies that are "one size fits all." Policies will differ according to the regions in question and therefore a "one size fits all" approach could be detrimental to specific ecosystems.
- Plans to achieve MDG 1 through increasing fish aquaculture by 2015 is one way of increasing fish stocks and reducing hunger; however, it may also lead to the depletion of various fish species since aquaculture breeding requires fish as a food resource.

Correspondingly, general synergistic effects include:

- Investing in watershed protection means benefits to drinking water and less water-borne disease such as malaria, etc.
- Intensification instead of extensification of agriculture means reducing expansion of agriculture into forests and the benefits of doing so include maintenance of resources such as medicine and other services such as climate regulation.
- Providing fuel such as biogas for households will benefit women through less physical labor, as there will be less indoor and outdoor pollution.
- Increase in access to safe water means healthy and more hygienic environments and people.

19.3.3 Direct and Indirect Impacts

The fulfillment of the 2015 targets (more specifically, of the interventions made to reach the specified targets for each goal) will have varied impacts (positive and negative) upon ecosystems and their resources. Conversely, the current state of the ecosystems and the availability in quantity and quality of their services may put additional constraints to, as well as provide opportunities for, the achievement of the 2015 targets.

The impacts of achieving the 2015 targets upon ecosystems and their services well include *direct effects* of the interventions made, for example, to build schools or apply agrochemicals) but also a plethora of *indirect effects* associated with more diffuse processes (such as capital investment and redistribution, changes in trade, transportation, economic growth, changes in expectations), potentially affecting the fulfillment of all goals and the state and dynamics of many ecosystems and resources. These processes are also likely to imply a higher throughput of materials through the world economy, as well as increased energy consumption, resulting in higher demands for ecosystem services.

While the specification of the indirect impacts is not feasible within the time and resource constraints of the MEA (and perhaps also in principle), those impacts will be very real and must be included as part of the ecosystem implications of fulfilling the target. Target 8, in particular, will have mostly indirect impacts upon ecosystem services.

Regarding the direct impacts, it is important to note that in general they are not defined by the goals themselves, not even by the targets, but by the specific menu of interventions chosen to fulfill them. In other words, for many goals, there are degrees of freedom in the selection of interventions and their relative weight; therefore, different strategic choices will have different balances of negative to positive impacts. In the same vein, different regional and local contexts will require different combinations of interventions to fulfill the goals. In general, it is possible to organize in all ecosystems interventions such that win–win situations are attained. Sustainable intensification may explicitly address MDG 1, targets 1 and 2 has neutral effects on MDGs 2 through 6, and will have positive effects on MDG 7. It is possible to look explicitly for such situations and reduce or limit negative effects on ecosystems.

19.4 Discussion

Clearly, achieving each 2015 target requires a multitude of inputs, while in turn each input can contribute to more than one goal. In their current formulation, the MDGs do not set explicit targets for several broad categories of inputs that are critical for their achievement. Examples include good governance, secondary education, and energy and transport services. Since the latter two exhibit strong dependencies and trade-offs with ecosystem services, they are discussed briefly here.

19.4.1 Energy Services

Improved energy services are a necessary input for achieving most MDGs. In many cases a shift toward modern sources of energy such as clean fuels for cooking and space heating, as well as electricity, is required for meeting the following goals: 1 (eradicating extreme poverty and hunger), 3 (promoting gender equality and empowering women), 4 (reducing child mortality), 5(improving maternal health), and 7 (ensuring environmental sustainability).

At the household level, clean energy can improve health and general well-being, particularly of women and children, by lowering indoor-air pollution levels associated with the use of biomass and solid fuels, which currently account for 1.6 million deaths each year—a death toll that exceeds the global health burden caused by malaria (WHO 2002). Improved energy services also reduce the burden on women and young girls who often spend several hours each day gathering biomass for cooking and thus free up time for their participation in education and income-generation activities. It follows that energy services are central to the achievements of the 2015 targets even though no individual goal addresses them explicitly.

Energy supply in low-income countries depends critically on the provisioning services such as primary production of firewood and other biomass, provided by a broad range of ecosystems including drylands, forests, and cultivated ecosystems. Worldwide, 2.4 billion people burn wood, charcoal, dung, or crop residues for cooking and heating (International Energy Agency 2002). In many poor countries, biomass accounts for 90% of household energy consumption. Hence, ecosystem services not only sustain energy supply in low-income countries, but they are also critically affected by the predominant choice of energy carrier and aggregate consumption levels.

A lot of controversy has surrounded the link between fuelwood use and deforestation (MA *Current State and Trends,* Chapter 9). While the relationship between both is extremely complex and data continue to be sparse, recent evidence suggests that the extent of deforestation caused by wood-based energy services is less severe than thought in the past (Arnold et al. 2003). Global demand for fuelwood appears to have peaked, while demand for charcoal continues to grow in many areas. In Africa and South America, the combined per capita consumption of wood and charcoal is stagnating while it is declining in most parts of Asia (Arnold et al. 2003).

These aggregate figures, however, mask important local variation. Particularly in the vicinity of urban areas, where total consumption of charcoal and wood is high and continues to increase in line with rapid population growth, the impact on ecosystems and the services they provide can be very severe. In light of rapid growth of urban populations in Africa and most other parts of the developing world, this problem is likely to be exacerbated over the coming years.

To meet the 2015 targets, access to improved cookstoves and adequate ventilation of homes and kitchens will need to be improved. In the medium term, the transition from biomass and other solid fuels toward improved fuels or solar efficient technology will need to be accelerated through improving access to liquid petroleum gas and other fuels that result in lower indoor air pollution. This shift will be necessary to meet several 2015 targets, including the ones on health, gender equality, and environmental sustainability, while at the same time providing a sufficient energy supply for expanding per capita consumption as incomes and rates of urbanization continue to grow on low-income countries. Critically, moving up the energy ladder will lessen the growing pressure on ecosystems that provide fuel for rapidly growing cities.

Perhaps the most important global impact of energy consumption is long-term climate change caused by unsustainable emission levels of greenhouse gases. While most of these emissions are caused by rich countries, emission levels in many developing countries are growing rapidly. However, per capita energy consumption in the "top priority" and "high priority" countries identified in the *Human Development Report 2003* continues to be only a fraction of rich countries' consumption levels (UNDP 2003). Shifting toward fossil fuels in these countries, though not desirable in the long term, will not have any substantial effect on global emissions of greenhouse gases. In cases where renewable non-biomass energy sources such as solar or hydro are not available and commercially viable, the shift away from biomass toward cleaner fossil fuels for cooking and heating in low-income countries can help reduce pressure on ecosystems.

19.4.2 Transport Services

Improvements in transport services such as road and rail transport, shipping services, as well as air transport, play an important role in meeting the 2015 targets. These services are required to provide effective access to social services such as emergency obstetric care; lower the household transport burden and reduce time poverty, especially of women and young girls; reduce the cost of agricultural inputs and raise farmgate prices for produce; facilitate the creation of export-based manufacturing and service industries including tourism; and increase the size of the captive market for the local private sector by lowering transport costs. In summary, improved transport services are necessary for achieving the 2015 targets including target 1, reducing income poverty.

It is no coincidence that many of the regions that are furthest away from meeting the 2015 targets also have extremely low levels of transport infrastructure. For example, sub-Saharan Africa has a mere 0.23 kilometers of paved roads per 1,000 people, while the average road density for South Asia is 1.08 kilometers per 1,000 people. Combined with the small size of African vehicle fleets, this results in transport costs that are much higher than in typical Asian countries and constitutes a major obstacle to reducing poverty.

Improving transport services to meet the 2015 targets will require substantial public investments in transport infrastructure, including roads, ports, and rail networks. Such investments will need to be complemented by policies and institutions that promote the promulgation of motorized vehicles and intermediate means of transport such as bicycles, animal carts, etc. For example, effective access to microcredit can play a critical role in improving access to means of transport.

However, improved transport infrastructure can have adverse impacts on ecosystems and the services they provide through ecosystem fragmentation, the opening of ecosystems to human settlement and exploitation, or increases in transport-related emissions of greenhouse gases. In particular, the construction of new road networks into areas of high biodiversity with very low human population densities (for example, tropical forests in the Amazon, Congo basin, Borneo, etc.) is likely to lead to an accelerated degradation of these ecosystems and their services. In comparison, the upgrading of roads in relatively densely populated areas consisting predominantly of cultivated ecosystems is likely to have limited adverse environmental impacts.

In all likelihood, this trade-off between HWB and the protection of ecosystems cannot be easily overcome and therefore needs to be managed appropriately to balance between the competing needs of humans and the environment. One way of achieving this is through environmental impact assessment in which the trade-offs of construction and upgrading of transport infrastructure is examined. This can be achieved through subjecting the construction and upgrading of transport infrastructure to environmental impact assessments.

19.4.3 Strengthening Capacity for Policy-making

Response options in this chapter have been identified as interventions under each goal. Essentially they are interventions that encourage education and knowledge sharing and direct investment or expenditure. In conjunction with those interventions identified by the United Nations Millennium Project, each will depend on the capacity of local and national governance and institutions to execute them (Chapter 4).

An important finding of the MA *Scenarios* is that while ecosystem goods and services will decrease globally, HWB will generally increase in all the scenarios (*MA Scenarios,* Chapter 13). A major reason for a global increase in HWB is due to strengthened institutions at various scales (MA *Scenarios,* Chapter 11; MA *Current State and Trends,* Chapter 28). In particular, the strengthening of local governance can lead to avoiding loss of ecosystem services and biological diversity, through the implementation of effective monitoring systems (MA *Scenarios,* Chapter 14).

Strengthening institutions and governance is also positively correlated with improvement of performance within the private sector, service delivery sectors, and macroeconomic management (World Bank and IMF 2004). Achieving the 2015 targets, therefore, depends on the capacity of institutions and governance to effectively implement interventions that will most likely aid in achieving the 2015 targets *(high certainty)*.

Addressing the goals collectively may require an integrative approach in which policy linkages are drawn horizontally and vertically, within ministries and agencies and across stakeholders from civil society, government, and private sector. (See Chapter 14.) It would also mean that the trade-offs and interventions associated with each goal are addressed by all stakeholders and endorsed into policy- and decision-making.

It is also well-established that achieving the 2015 targets will require existing institutions and governance structures in the developed world to address and execute immediate concerns affecting ecosystem services such as overconsumption.

19.5 Conclusion

The MDG chapter established the importance of ecosystem services to HWB in terms of specific goals, primarily poverty, hunger, gender, child mortality, disease, and water. Goals such as education, maternal mortality, sustainable development, and global partnerships have less direct links to ecosystem services.

The goal-by-goal analysis provided a qualitative assessment of how ecosystem services will respond to interventions. The list of trade-offs and synergies identified in the multi-goal analysis highlights the positive and negative impacts on ecosystems, services, and people of achieving the goals.

19.5.1 Specific Conclusions

- The achievement of targets 1 and 2, of MDG 1 (eradicate extreme poverty and hunger), is probably the best condition for maintaining ecosystems. Appropriate use and management of ecosystems requires the breaking of unsustainability spirals due to poverty.
- MDGs 2 through 6 are only indirectly affected by ecosystem services and ecosystem management. These indirect effects are, however, very important for the well-being of humans and that has considerable impact on the functioning of ecosystems and their continuity.
- MDG 7 is directly related to management of ecosystems. The chapter shows how various human activities such as energy use, transport, mobility, food production, etc., affect ecosystems. A drastic increase of efficiency in the use of land, energy, and natural resources is required to change the mega trend of enormous increase of the use of scarce resources. Many examples in the chapter illustrate the way ecosystems may be utilized to fulfill human objectives, but careful management and clear political choices at all integration levels are then required.

19.5.2 Key Messages

- The goals are inherently linked (whether directly or indirectly) to ecosystem services. The management of ecosystem services and the drawing of continuous links to HWB will therefore provide synergistic effects that will aid countries and regions of the world to get back on track to meeting the goals *(high certainty)*.
- The interventions identified within the goal-by-goal analysis to reduce natural resource degradation and protect ecosystem services are key to achieving the MDGs. The final impact of fulfilling the goals will depend upon the specific menu of operational options (interventions) selected, and upon the ecological and societal local context. Some interventions, however, are more sustainable than others.
- The achievement of the MDGs will be dependent on the interventions used, the social and ecological context, and the consideration of spatial and temporal issues. For example, labor-intensive employment may lead to a reduction of poverty and extreme hunger in specific regions of the world such as Africa. The achievement of various MDGs is working in the same direction. To produce more with less is doable, but needs political will and support of the political leaders.
- Some interventions will provide considerable intensification. At present, there is no action plan to reach these MDGs. The various task forces that report next year may help, but for MDGs 1 and 7, the MDG panel is willing to take the lead.

Knowledge gaps are evident within the findings. Strengthening local and national institutions and governance is one way to ensure the implementation of monitoring systems that provide data on specific indicators of ecosystem services.

19.5.3 A Research Agenda

As a result of our findings, a research agenda is proposed with the following priorities:

- The relationship between the MDGs and the MA should be explicitly addressed in studies and research programs; there should be a clear responsibility at the highest level of the United Nations to achieve this.
- The industrial world should contribute more to investigations, analysis, and research programs in developing countries; they can do this directly through research organizations and also by building new alliances (InterAcademy Council Report 1 and 2 2004)
- The following specific research issues should be addressed: (1) analyses and quantitative methodology should be applied to further define the relationship between ecosystems, ecosystem services, and HWB; (2) the relationship of energy, transportation, and ecosystem services toward meeting the MDG targets should be further explored; and (3) research should be undertaken on the resilience of ecosystems under pressure from overpopulation and/or increasing human drives for material goods.
- Studies aiming at systems innovation, including all aspects of ecosystems behavior, should be initiated for the most fragile and most threatened ecosystems. Specifically, research is needed in international agricultural research institutions on futher use and intensification of agroecosystems with minimization of environmental side effects.

- Many examples of applied technology introduced in the 1900s should be reintroduced. It is now time for a considerable increase in the introduction of appropriate technology.

References

Ahmad, O.B., A.D. Lopez, and M. Inoue, 2000: The decline in child mortality: A reappraisal, *Bulletin of the World Health Organization,* **78,** pp. 1175–91.

Annan, Kofi, 2000: *Opening Speech of the Millennium Assembly: We the Peoples,* United Nations, New York, NY.

Arnold, M., G. Köhlin, R. Persson, and G. Shepherd, 2003: Fuelwood revisited: What has changed in the last decade? CIFOR occasional paper no. 39, Center for International Forestry Research, Jakarta, Indonesia, pp. 35.

AVERT, 2003: *HIV/AIDS in Africa,* [online]. Accessed 28 February 2005. Available at http://www.avert.org/aidsimpact.htm.

Bartlett, S., 2003: Water, sanitation and urban children: The need to go beyond "improved provision," *Environment and Urbanization,* **15(2),** pp. 57–70.

Bartone, C., J. Bernstein, J. Leitmann, and J. Eigen, 1994: Toward environmental strategies for cities: Policy consideration for urban environmental management in developing countries, UNDP/UNCHS/World Bank, Urban Management Programme 18, World Bank, Washington, DC, 132 pp.

Black, R.E., S.S. Morris, and J. Bryce, 2003: Where and why are 10 million children dying every year? *Lancet* **361,** pp. 226–34.

Brown, T., 2003: HIV/AIDS in Asia, *Asia Pacific Issues,* No. 68, East-West Center, Honolulu, HI, 8 pp.

Dale, P.J., B. Clarke, E.M.G. Fontes, 2002: Potential for the environmental impacts of transgenic crops, *Nature Biotechnology,* **20,** pp. 567–74

Dasgupta, P., C. Folke, and K.G. Maler, 1994: *The Environmental Resource Base and Human Welfare,* Beijer reprint series no. 35, Oxford University Press, Oxford, UK, pp. 25–50.

Duraiappah, A., 1998: Poverty and environmental degradation: A review and analysis of the nexus, *World Development,* **26(12),** pp. 2169–79.

Freedman, L., M. Wirth, R. Waldman, M. Chowdury, and A. Rosenfield, 2003: Background paper on the task force on child health and maternal health [online]. The Millennium Project, UNDP, New York, NY. Available at http://www.unmillenniumproject.org/documents/tf04apr18.pdf.

Graumann, J. V., 1977: Orders of magnitude of the world's urban and rural population in history, UN Populations Bulletin 8, UN, New York, NY.

Grown, C., G. Rao Gupta, and Z. Khan, 2003: Background paper of the millennium project task force on gender equity, UNDP, New York, NY, pp. 52.

Hardoy, J.E., D. Mitlin, D. Satterthwaite, 2001: *Environmental Problems in an Urbanizing World,* EarthScan, London, UK, 464 pp.

Harvard University, 2003: *WHO Global burden of disease publication series* [online], Center for Population and Development Studies at Harvard University School of Public Health, Boston, MA. Available at http://www.hsph/harvard.edu/organization/bdu/GBD.series.htm.

Hill, A. and E. King, 1995: Women's education and economic well-being, *Feminist Economics,* **1(2),** pp. 1–26.

Holling, C.S. and G.K. Meffe, 1996: Command and control and the pathology of natural resource management, *Conservation Biology,* **10,** pp. 328–37.

ICPD (International Conference on Population and Development), 1994: Report of the international conference on population and development [online], A/CONF.171/13, 1994: UN, New York, NY. Accessed 28 February 2005. Available at http://www.un.org/popin/icpd/conference/offeng/poa.html.

InterAcademy Council, 2004: Inventing a better future: A strategy for building worldwide capacities in science and technology, January 2004, and Realizing the promise and potential of African agriculture: Science and technology strategies for improving agricultural productivity and food security in Africa, June 2004 [online], The InterAcademy Council, Amsterdam, The Netherlands. Available at http://www.interacademycouncil.net/reports.asp.

International Energy Agency, 2002: *World Energy Outlook,* International Energy Agency, Paris, France, 391 pp.

Jackson, C., 1993: Doing what comes naturally: Women and environment in development, *World Development,* **21(12),** pp. 1947–63.

James, C. 2002: Global status of commercialized transgenic crops: 2002, ISAAA briefs no. 27 [online], International Service for the Acquisition of Agribiotech Applications, Ithaca, NY. Accessed 28 February 2005. Available at http://www.icsu.org/1_icsuinscience/GMO/html/isaaa.htm.

MA (Millennium Ecosystem Assessment), 2003: *Ecosystems and Human Well-being: A Framework for Assessment,* Island Press, Washington, DC, 245 pp.

McGranahan, G. and J. Songsore, 1994: Wealth, health, and the urban household: Weighing environmental burden in Jakarta, Accra, Sao Paulo, *Environment,* **36(6),** pp. 4–11, 40–5.

McMichael, A.J, Haines, A., Sloof, R and Kovats, S., 1996: Climate change and human health: An assessment prepared by a task group on behalf of the World Health Organization, the World Meteorological Organization and the United Nations Environment Programme, WHO, Geneva, Switzerland.

Mosley, W.H. and L.C. Chen, 1984: An analytical framework for the study of child survival in developing countries, *Population and Development Review,* **10** (Suppl.), pp. 25–45.

Pruess, A., D. Kay, L. Fewtrell, and J. Bertram, 2002: Estimating the burden of disease from water, sanitation, and hygiene at a global level, *Environmental Health Perspectives,* **110(5),** pp. 537–42; cited in Editors' Introduction, *Environment and Urbanization,* 2003, **15(2),** p. 5.

Raybould, A.F. and A.J. Gray, 1993: Genetically modified crops and hybridization with wild relatives, *Journal of Applied Ecology,* **30,** p.199.

Reardon, T, 1991: Income sources of the malnourished rural poor in a drought year in Burkina Faso. In: *Income Sources of Malnourished People in Rural Areas: Microlevel Information and Policy Implications,* J. von Barun and R. Pandya-Lorch (eds.), Working papers on commercialization of agriculture and nutrition no. 5, International Food Policy Research Institute, Washington DC, pp 95–104.

Revenga, C., J. Brunner, , N. Henniger, K. Kassem, and R. Payner, 2000: *Pilot Analysis of Global Ecosystems: Freshwater Systems,* World Resources Institute, Washington, DC.

Sen, A., 1999: *Development as Freedom,* Oxford University Press, Oxford, UK, 384 pp.

Sen, A., 1991: *Inequality Reexamined,* Oxford University Press, UK, 207 pp.

Southgate, D., 1998: The economics of land degradation in the Third World, Environment Working Paper, No 2, The World Bank, Washington, DC, 17 pp.

Tabunakawai, K., 1997: The importance of preserving traditional medical knowledge (WAINIMATE), Proceedings Women, Science and Development Symposium: From Indigenous Knowledge to New Information Technologies, VIII Pacific Science Inter-Congress, 17–18 July, Suva, Fiji, *Pacific Science Association Information Bulletin,* **3(4),** pp. 39–42.

Trewavas, A. and C. Leaver, 2001: Is opposition to GM crops science or politics? An investigation into arguments that GM crops pose a particular threat to the environment, *EMBO Reports,* **2,** pp. 455–9.

UN (United Nations), 1992: *Rio Declaration on Environment and Development* Report of the UN Conference on Environment and Development; A/CONF.151/26 (Vol. I), UN, New York, NY.

UN, 2000: *Cities without Slums Alliance,* General Assembly Resolution 55/, 18 September 2000, UN, New York, NY.

UN, 2001: World population prospects: The 2001 revision [online], Population Division, Department of Economic and Social Affairs, UN, New York, NY. Accessed 28 February 2005. Available at http://www.un.org/esa/population/publications/wup2001/WUP2001report.htm.

UN, 2002a: World population prospects: The 2002 revision [online], Population Division, Department of Economic and Social Affairs, UN, New York, NY. Available at http://www.un.org/esa/population/publications/wpp2002/WPP2002-HIGHLIGHTSrev1.P DF. Accessed February 28, 2005.

UN, 2002b: *World Urbanization Prospects: The 2001 Revision: Data Tables and Highlights,* ESA/P/WP/173, UN Secretariat, Department of Economic and Social Affairs, Population Division, New York, NY.

UN, 2003: *Promoting the Millennium Development Goals in Asia and the Pacific: Meeting the Challenges of Poverty Reduction,* UN, New York, NY, 111 pp.

UN AIDS, 2003: *Aids Epidemic Update,* December, 2003, Geneva, Switzerland, 39 pp.

UNCSD (UN Commission on Sustainable Development), 1995: Chapter 8: Integrating environment and development in decision-making [online], Background paper, UNCSD, New York, NY. Available at http://www.un.org/documents/ecosoc/cn17/1995/background/ecn171995-bpch8.htm.

UNDP (UN Development Programme), 2003: The human development report 2003: Millennium development goals a compact among nations to end human poverty [online], UNDP, New York, NY. Available at http://www.undp.org/hdr2003/.

UNESCO (UN Educational, Scientific and Cultural Organization), 2003a: World and regional trends: Data for years around 1990 and 2000 [online], UNESCO Statistics Division, New York, NY. Available at http://millenniumindicators.un.org/unsd/mi/mi_worldregn.asp.

UNESCO, 2003b: Series: Education, percentage of pupils starting grade 1 reaching grade 5, both sexes [online], UNESCO Statistics Division, New York, NY. Available at http://millenniumindicators.un.org/unsd/mi/mi_series_results.asp?rowID=591.

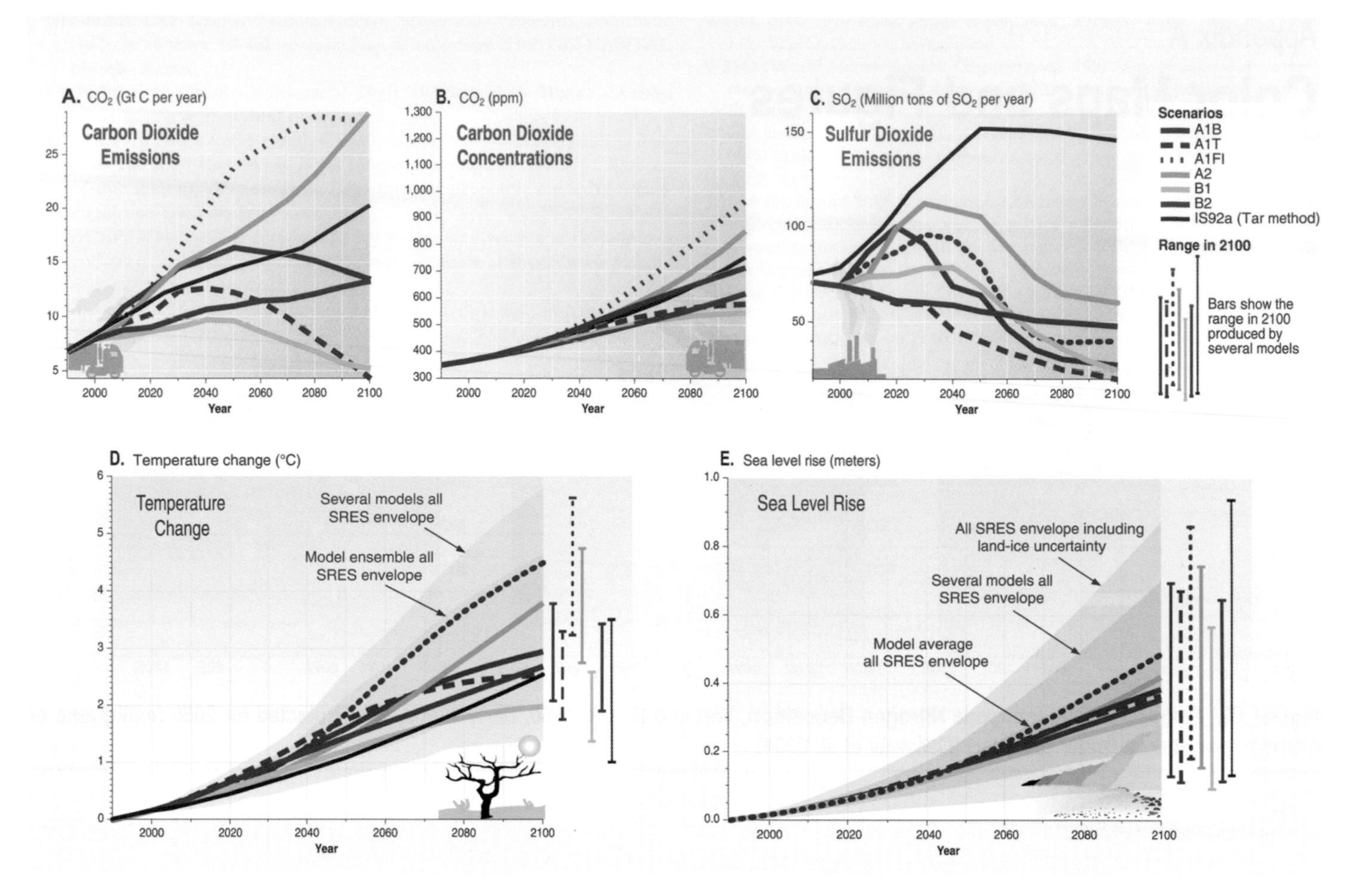

Figure 13.1. The Global Climate Models for the Twenty-first Century. The global climate in this century will depend on natural changes and the response of the climate system to human activities. Climate models project the response of many variables—such as increases in global surface temperature and sea level—to various scenarios of greenhouse gases and other human-related emissions. Graph A shows the carbon dioxide emissions of the six illustrative SRES scenarios; B shows the projected carbon dioxide concentrations; C shows anthropogenic sulfur dioxide emissions. Emissions of other greenhouse gases and aerosols were included in the model but are not shown in the figures. D and E show the temperature and sea level responses, respectively.

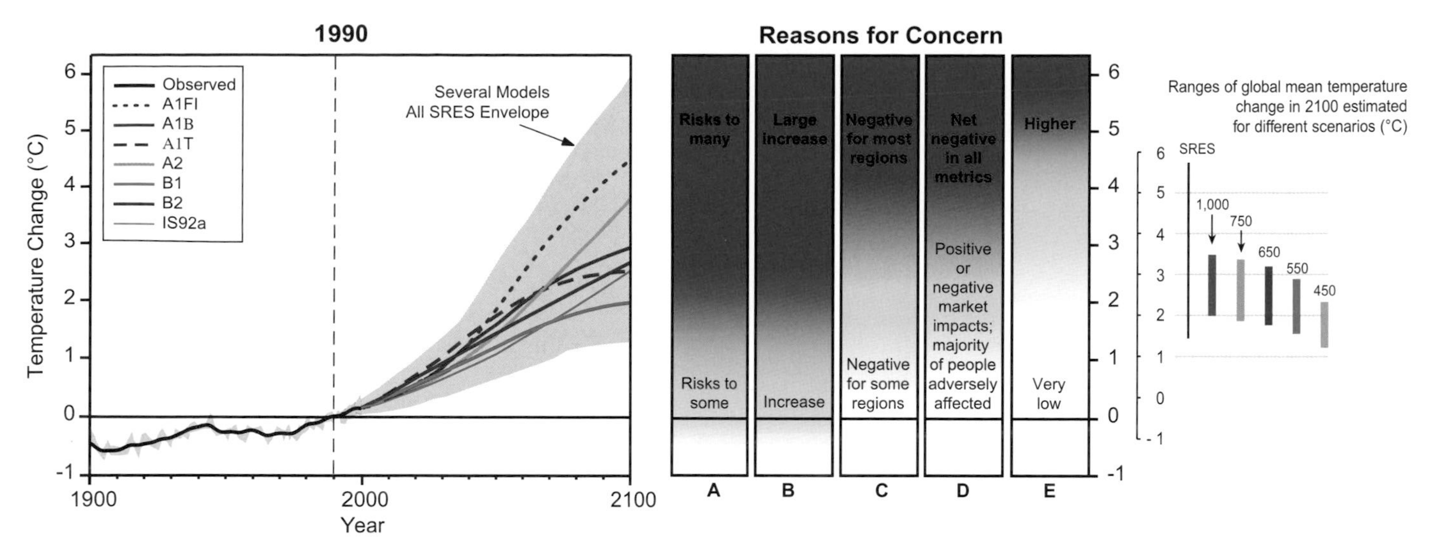

Figure 13.2. Reasons for Concern about Projected Climate Change Impacts. The risks of adverse impacts from climate change increase with the magnitude of climate change. The left part of the figure displays the observed temperature increase relative to 1990 and the range of projected temperature increase after 1990 as estimated by Working Group I of the IPCC for scenarios from the *Special Report on Emissions Scenarios.* The middle panel displays conceptualizations of five reasons for concern regarding climate change risks evolving through 2100. White indicates neutral or small negative or positive impacts or risks, yellow indicates negative impacts for some systems or low risks, and red means negative impacts or risks that are more widespread and/or greater in magnitude. The assessment of impacts or risks takes into account only the magnitude of change and not the rate of change. Global mean annual temperature change is used in the figure as a proxy for the magnitude of climate change, but projected impacts will be the function of, among other factors, the magnitude and rate of global and regional changes in mean climate, climate variability and extreme climate phenomena, social land economic conditions, and adaptations. The right panel shows estimates of global mean temperature change by 2100 relative to 1990 for scenarios that would lead to stabilization of the atmospheric concentration of carbon dioxide, as well as the full set of SRES projections, which are shown in the left panel. As shown in Table 13.1, the equilibrium changes in temperature associated with each of these stabilization levels is significantly higher than the projected increase by 2100, for example, stabilization at 750 ppm is projected to result in an increase of 2.8–7.0C, compared to an increase of 1.9–3.4C by 2100. Reasons for Concern: A. Risks to Unique and Threatened Systems: Extinction of species, loss of unique habitats and coastal wetlands, and bleaching and death of coral; B. Risks from Extreme Climate Events: Health, property, and environmental impacts from increased frequency and intensity of some climate extremes; C. Distribution of Impacts: Cereal crop yield changes that vary from increases to decreases across regions but that are estimated to decrease in most tropical and sub-tropical regions; decrease in water availability in some water-stressed countries, increase in others; greater risks to health in developing countries than in industrial countries; net market sector losses estimated for many developing countries; mixed effects estimated for industrial countries up to a few degrees warming and negative effects for greater warming; D. Aggregate Impacts: Estimates of globally aggregated net market sector impacts are positive and negative up to few degrees warming and negative for greater warming. More people adversely affected than beneficially affected even for warming of less than a few degrees; E. Risks from Future Large-Scale Discontinuities: Significant slowing of thermohaline circulation possible by 2100; melting and collapse of ice sheets adding substantially to sea level rise (very low probability before 2100; likelihood higher on multi-century time scale).

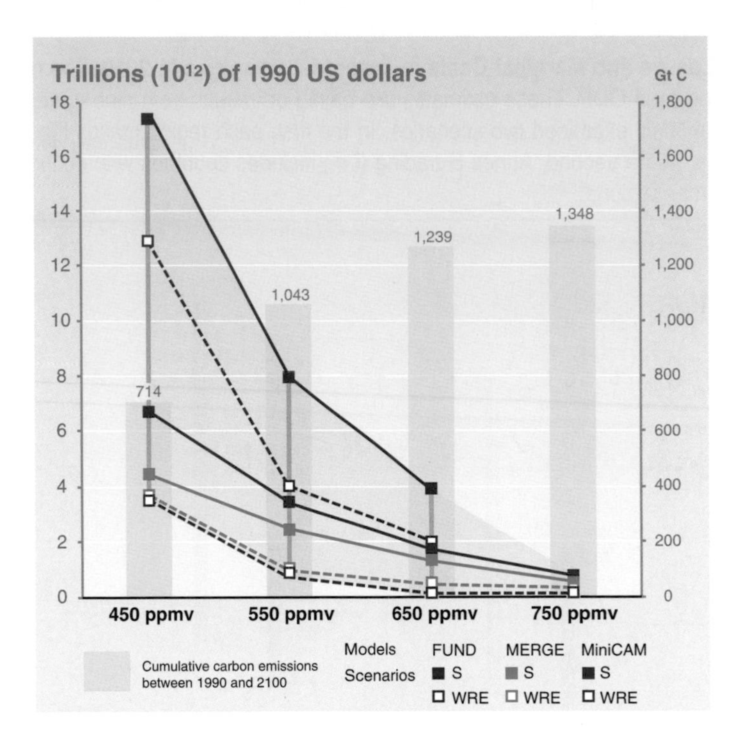

Figure 13.6. Projected Costs of Stabilizing Carbon Dioxide Concentrations. The mitigation costs (1990 US dollars, present value discounted at 5% per year for the period 1990–2100) of stabilizing carbon dioxide concentrations at 450 to 750 ppmv are calculated using three global models, based on different model-dependent baselines. Avoided impacts of climate change are not included. In each instance, costs were calculated based on two emissions pathways for achieving the prescribed target (S and WRE). The bars show cumulative carbon emissions between the years 1990 and 2100. Cumulative future emissions until the carbon budget ceiling is reached are reported above the bars in Gt C.

Appendix B
Authors

Australia
Joseph Baker, Department of Primary Industries and Fisheries
Colin Butler, Australia National University
Angela Cassar, Phillips Fox
Daniel P. Faith, Australian Museum
Matthew Kendall, Murray-Darling Basin Commission
Mark Anthony Siebentritt, Murray-Darling Basin Commission

Austria
Gunther Fischer, International Institute for Applied Systems Analysis
Alexander Haslberger, University of Vienna
Mahendra Shah, International Institute for Applied Systems Analysis
Ferenc Toth, International Atomic Energy Agency (IAEA)
Harrij van Velthuizen, International Institute for Applied Systems Analysis
David Wiberg, International Institute for Applied Systems Analysis

Bolivia
William Powers, Fundacion Amigos de la Naturaleza (FAN)

Brazil
David Cleary, The Nature Conservancy
Fernando De Avila Pires, Universidade Federal de Santa Catarina
Luiz Martinelli, Universidade de São Paulo
Carla Morsello, Universidade de São Paulo
Marcia Muchagata, Consultora PCT MDA/FAO

Canada
Lina Cimmarusti, Millennium Ecosystem Assessment
Debra Davidson, University of Alberta
Hadi Dowlatabadi, University of British Columbia
Anantha Kumar Duraiappah, International Institute for Sustainable Development
Ted Gullison, Hardner Gullison and Associates
Rebecca Hanson, University of Toronto
Monirul Qader Mirza, University of Toronto
Margot Parkes, University of British Columbia
John Robinson, University of British Columbia
Colin Soskoln, University of Alberta
Myrle Traverse, IIFB

Chile
Gilberto Gallopin, Economic Commission for Latin America and the Caribbean
Benjamin Kiersch, Food and Agriculture Organization of the United Nations
Sergio Pena-Neira, Universidad del Mar

China
Ian Hunter, International Network for Bamboo and Rattan (INBAR)
Kishan Khoday, United Nations Development Programme
Zhu Zhao-liang, Institute of Soil Science

Czech Republic
Tomas Hak, Charles University
Bedrich Moldan, Charles University

Denmark
Frank Jensen, University of Southern Denmark
Tipparat Pongthanapanich, University of Southern Denmark

Ethiopia
Ernesto Gonzalez-Estrada, International Livestock Research Institute
Don Peden, International Livestock Research Institute
Thomas Randolph, International Livestock Research Institute
D. Romney, International Livestock Research Institute

France
Salvatore Arico, United Nations Educational, Scientific and Cultural Organization
Peter Börkey, Organisation for Economic Co-operation and Development
Florent Engelmann, Centre for International Cooperation in Agronomic Research for Development
Jean-Marc Hougard, LIN-IRD
Lilian Saade, Organisation for Economic Co-operation and Development

Georgia
Merab Machavariani, Georgia Forests Development Project

Germany
Pierre Ibisch, University of Applied Sciences Eberswalde
Richard Klein, Potsdam Institute for Climate Impact Research
Dagmar Lohan, United Nations University
Jens Mackensen, KfW Entwicklungsbank (German Development Bank)
Maria Socorro Manguiat, IUCN—The World Conservation Union
Arisbe Mendoza, University of Bonn
Klaus Riede, University of Bonn

Ghana
Rose Enma Mamaa Entsua-Mensah, Council for Scientific and Industrial Research

Greece
Eftichios Sartzetakis, University of Macedonia
Anastasios Xepapadeas, University of Crete

India
Jayanta Bandyopadhyay, Indian Institute of Management
D.K.Bhattacharya, University of Delhi
Abhik Ghosh, Panjab University

Pushpam Kumar, Institute of Economic Growth
Vinod Mathur, Wildlife Institute of India
K.S Murali, Indian Institute of Science
Jyoti Parikh, Integrated Research and Action for Development (IRADe)
Anand Patwardhan, Indian Institute of Technology-Bombay
Asha Rajvanshi, Wildlife Institute of India
Shekhar Singh, National Campaign for People's Right to Information
Madhu Verma, Indian Institute of Forest Management

Indonesia
Brian Murray Belcher, Center for International Forestry Research
Bruce Campbell, Center for International Forestry Research
Patricia Shanley, Center for International Forestry Research
Nigel Sizer, The Nature Conservancy
Indah Susilowati, Diponegoro University (UNDIP)
Brent Swallow, World Agroforestry Centre
Eva Wollenberg, Center for International Forestry Research

Italy
Jim Carle, Food and Agriculture Organization of the United Nations
Mohammad Ehsan Dulloo, International Plant Genetic Resources Institute
Jan Engels, International Plant Genetic Resources Institute
Pablo Eyzaguirre, International Plant Genetic Resources Institute
Angela Guimaraes-Pereira, European Commission
Toby Hodgkin, International Plant Genetic Resources Institute
Devra Jarvis, International Plant Genetic Resources Institute
Wulf Killmann, Food and Agriculture Organization of the United Nations
Rabindra Roy, Food and Agriculture Organization of the United Nations

Japan
W. Bradnee Chambers, United Nations University
Sofia Hirakuri, United Nations University
Hiroji Isozaki, Meijigakuin University
Alphonse Kambu, Ishikawa International Cooperation Research Centre
Yumiko Kura

Kenya
Thierry de Oliveira, United Nations Environment Programme
Andrew Githeko, Kenya Medical Research Institute
John McDermott, International Livestock Research Institute
Robin Reid, International Livestock Research Institute
Tom Tomich, World Agroforestry Centre (ICRAF)

Malaysia
Helen Leitch, WorldFish Center
V. Ramanatha Rao, International Plant Genetic Research Institute
Blake Ratner, WorldFish Center
K.Kuperan Viswanathan, WorldFish Center

Mexico
David Barkin, Universidad Autónoma Metropolitana
Eduardo Mestre Rodriguez, Red Latinoamericana de Organizaciones de Cuenca
Cecilia Tortajada, Third World Centre for Water Management

Morocco
Abdelkader Allali, Ministry of Agriculture and Rural Development and Fishing

Nepal
Motilal Ghimire, Tribhuvan University

The Netherlands
Fabio de Castro, Pilot Program Tropical Forests
Joyeeta Gupta, UNESCO—IHE Policy and Law in Water Resources and the Environment
Marcel Marchand, WL/Delft Hydraulics
Rudy Rabbinge, Wageningen University
Henk Simons, RIVM
Marja Spierenburg, Free University of Amsterdam

New Zealand
Simon Hales, Wellington School of Medicine and Health
Alistair Woodward, University of Auckland

Nigeria
Nimbe Adedipe, National Universities Commission
MKC Sridhar, Medical Statistics and Environmental Health College of Medicine

Norway
Indra de Soysa, Norwegian University of Science and Technology

Philippines
Ben Malayang, University of the Philippines Los Banos
Ida Siason, University of the Philippines in the Visayas
Mercedita A. Sombilla, Southeast Asian Regional Center for Graduate Study and Research in Agriculture (SEARCA)

Singapore
Prisna Nuengsigkapian, International Institute for Sustainable Development

South Africa
Marjorie Pyoos, Science and Technology for Economic Impacts
Belinda Reyers, Council for Scientific and Industrial Research
Mary Scholes, University of the Witwatersrand
Sheona Shackleton, Rhodes University
Robyn Stein, Edward Nathan and Friedland
Gina Ziervogel, University of Cape Town

South Korea
Euiso Choi, Korea University

Spain
Diana Elizabeth Marco, Estacion Experimental del Zaidin—CSIC

Sri Lanka
Felix Amerasinghe, International Water Management Institute
Karin Fernando, South Asian Co-operative Environment USAID

Sweden
Ragnar Elmgren, University of Stockholm
Tage Klingberg, University of Gävle
Elisabet Lindgren, Stockholm University

Switzerland
Diarmid Campbell-Lendrum, World Health Organization
Carlos Corvalan, World Health Organization
Elaine Fletcher, World Health Organization
Wendy Goldstein, IUCN—The World Conservation Union
Pedro M. Rosabal Gonzales, IUCN—The World Conservation Union

Susan Mainka, IUCN—The World Conservation Union
Jeffrey McNeely, IUCN—The World Conservation Union
Thomas McShane, WWF International
Frederik Schutyser, IUCN—The World Conservation Union
Rie Tsutsumi, United Nations Development Programme

Thailand
Patrick Durst, Food and Agriculture Organization of the United Nations
Thomas Enters, Food and Agriculture Organization of the United Nations

Trinidad and Tobago
Marlene Attzs, University of the West Indies
Karen Polson, Caribbean Epidemiology Centre

United Kingdom
Neil Adger, University of East Anglia
Tony Allan, King's College
Lousie Auckland, Ecosecurities Ltd
Robert Barrington, F&C Asset Management
Stephen Bass, International Institute for Environment and Development
Carl Bauer, Resources for the Future
Neil Maclean Bird, Overseas Development Institute
Katrina Brown, University of East Anglia
Flavio Comim, University of Cambridge
Simon Counsell, The Rainforest Foundation UK
Clive Davies, London School of Hygiene and Tropical Medicine
Tim Forsyth, London School of Economics and Political Science
Brian Groombridge, UNEP—World Conservation Monitoring Centre
John Hudson, Department for International Development
Saleemul Huq, International Institute for Environment and Development
Joy Hyvarinen, Royal Society for the Protection of Birds
Valerie Kapos, UNEP—World Conservation Monitoring Centre
Izabella Koziell, Department for International Development
Julian Laird, Earthwatch Institute (Europe)
Clay Landry, WestWater Research LLC
James Mayers, International Institute for Environment and Development
David Molyneux, Liverpool School of Tropical Medicine
Dominic Moran, Scottish Agricultural College
Adrian Phillips, Cardiff School of City and Regional Planning
Ina T. Porras, International Institute for Environment and Development
Janet Riley, Rothamsted Research
Elizabeth Robinson, University of Oxford
Sergio Rosendo, University of East Anglia
Dale Rothman, The Macaulay Institute
Christopher Schofield, London School of Hygiene and Tropical Medicine
David Simpson, University College London
Steven Sinkins, University of Oxford
Kerry Ten Kate, Insight Investment
Sonja Vermeulen, International Institute for Environment and Development
Bhaskar Vira, University of Cambridge

United States
Robin Abell, World Wildlife Fund
Heidi Albers, Oregon State University
Bruce Aylward, Deschutes Resources Conservancy
Juan Carlos Belausteguigoitia, The World Bank
Eduardo Brondizio, Indiana University
Carl Bruch, United Nations Environment Programme—RONA
Mang Cheuk, University of Oklahoma
Carmen Cheung, Wesleyan University
Christopher Delgado, International Food Policy Research Institute
Dennis Dykstra, USDA Forest Service
Kristie Ebi, Exponent Health Group
David Edmunds, University of California
Paul Ferraro, Georgia State University
Dana Fisher, Columbia University
Jessica F. Green, United Nations University—Institute of Advanced Studies
Thomas P. Holmes, USDA Forest Service
Richard Howarth, Dartmouth College
Robert Howarth, Cornell University
Robert Clay Kellison, Institute of Forest Biotechnology
Sarah Laird
Nancy Lewis, East-West Center
Jane Lubchenco, Oregon State University
Laura Meadors, The Deschutes Resources Conservancy
Ruth Meinzen-Dick, International Food Policy Research Institute
Jerry Melillo, The Ecosystems Center
Suzanne Moellendorf, The Deschutes Resources Conservancy
William Moomaw, Tufts University
Ian Noble, The World Bank
Jennifer Olson, Michigan State University
Carlton Owen, The Environmental Edge
Stefano Pagiola, The World Bank
Cheryl Palm, The Earth Institute at Columbia University
Steve Percy, CEO, BP America (formerly)
William F. Perrin, Southwest Fisheries Science Center
Steve Polasky, University of Minnesota
Robert Pontius Jr., Clark University
Francis Putz, University of Florida
Kilaparti Ramakrishna, Woods Hole Research Center
Kent Redford, Wildlife Conservation Society
Carmen Revenga, The Nature Conservancy
Claire Rhodes, Ecoagriculture Partners
Jorge Rivera, George Washington University
Andrew Shea, Poseidon Resources Corporation
David Simpson, Environmental Protection Agency
Neil Sampson, The Sampson Group, Inc.
Guido Schmidt-Traub, United Nations Development Programme
Andrea Siqueira, Indiana University
Kenneth Strzepek, University of Colorado
Sylvia Tognetti, Environmental Science and Policy
Nikolay Voutchkov, Poseidon Resources Corporation
Diana Harrison Wall, Colorado State University
Robert Watson, The World Bank
Gary Yohe, Wesleyan University

Appendix C

Abbreviations and Acronyms

AI aridity index

AKRSP Aga Khan Rural Support Programme

AMF arbuscular mycorrhizal fungi

ASB alternatives to slash-and-burn

ASOMPH Asian Symposium on Medicinal Plants, Spices and Other Natural Products

AVHRR advanced very high resolution radiometer

BCA benefit-cost analysis

BGP Biogeochemical Province

BII Biodiversity Intactness Index

BMI body mass index

BNF biological nitrogen fixation

BOOT build-own-operate-transfer

BRT Bus Rapid Transit (Brazil)

BSE bovine spongiform encephalopathy

Bt *Bacillus thuringiensis*

C&I criteria and indicators

CAFO concentrated animal feeding operations

CAP Common Agricultural Policy (of the European Union)

CAREC Central Asia Regional Environment Centre

CBA cost-benefit analysis

CBD Convention on Biological Diversity

CBO community-based organization

CCAMLR Commission for the Conservation of Antarctic Marine Living Resources

CCN cloud condensation nuclei

CCS CO_2 capture and storage

CDM Clean Development Mechanism

CEA cost-effectiveness analysis

CENICAFE Centro Nacional de Investigaciones de Café (Colombia)

CFCs chlorofluorocarbons

CGIAR Consultative Group on International Agricultural Research

CIFOR Center for International Forestry Research

CITES Convention on International Trade in Endangered Species of Wild Fauna and Flora

CMS Convention on the Conservation of Migratory Species of Wild Animals (Bonn Convention)

CONICET Consejo de Investigaciones Científicas y Técnicas (Argentina)

COP Conference of the Parties (of treaties)

CPF Collaborative Partnership on Forests

CSIR Council for Scientific and Industrial Research (South Africa)

CV contingent valuation

CVM contingent valuation method

DAF decision analytical framework

DALY disability-adjusted life year

DDT dichloro diphenyl trichloroethane

DES dietary energy supply

DHF dengue hemorrhagic fever

DHS demographic and health surveys

DMS dimethyl sulfide

DPSEEA driving forces-pressure-state-exposure-effect-action

DPSIR driver-pressure-state-impact-response

DSF dust storm frequency

DU Dobson Units

EEA European Environment Agency

EEZ exclusive economic zone

EGS ecosystem global scenario

EHI environmental health indicator

EIA environmental impact assessment

EID emerging infectious disease

EKC Environmental Kuznets Curve

EMF ectomycorrhizal fungi

E/MSY extinctions per million species per year

ENSO El Niño/Southern Oscillation

EPA Environmental Protection Agency (United States)

EPI environmental policy integration

EU European Union

EU ETS European Union Emissions Trading System

FAO Food and Agriculture Organization (United Nations)

FAPRI Food and Agriculture Policy Research Institute

FLEGT Forest Law Enforcement, Governance, and Trade

FRA Forest Resources Assessment

FSC Forest Stewardship Council

GATS General Agreement on Trade and Services

GATT General Agreement on Tariffs and Trade

GCM general circulation model

GDI Gender-related Development Index

GDP gross domestic product

GEF Global Environment Facility

GEO *Global Environment Outlook*

GHG greenhouse gases

GIS geographic information system

GIWA Global International Waters Assessment

GLASOD Global Assessment of Soil Degradation

GLC Global Land Cover

GLOF Glacier Lake Outburst Flood

GM genetic modification

GMO genetically modified organism

GNI gross national income

GNP gross national product

GPS Global Positioning System

GRoWI *Global Review of Wetland Resources and Priorities for Wetland Inventory*

GSG Global Scenarios Group

GSPC Global Strategy for Plant Conservation

GtC-eq gigatons of carbon equivalent

GWP global warming potential

HDI Human Development Index

HIA health impact assessment

HIPC heavily indebted poor countries

HPI Human Poverty Index

HPS hantavirus pulmonary syndrome

HWB human well-being

IAA integrated agriculture-aquaculture

IAM integrated assessment model

IBI Index of Biotic Integrity

ICBG International Cooperative Biodiversity Groups

ICDP integrated conservation and development project

ICJ International Court of Justice

ICRAF International Center for Research in Agroforestry

ICRW International Convention for the Regulation of Whaling

ICSU International Council for Science

ICZM integrated coastal zone management

IDRC International Development Research Centre (Canada)

IEA International Energy Agency

IEG international environmental governance

IEK indigenous ecological knowledge

IFPRI International Food Policy Research Institute

IGBP International Geosphere-Biosphere Program

IIASA International Institute for Applied Systems Analysis

IK indigenous knowledge

ILO International Labour Organization

IMF International Monetary Fund

IMPACT International Model for Policy Analysis of Agricultural Commodities and Trade

IMR infant mortality rate

INESI International Network of Sustainability Initiatives (hypothetical, in *Scenarios*)

INTA Instituto Nacional de Tecnología Agropecuaria (Argentina)

IPAT impact of population, affluence, technology

IPCC Intergovernmental Panel on Climate Change

IPM integrated pest management

IPR intellectual property rights

IRBM integrated river basin management

ISEH International Society for Ecosystem Health

ISO International Organization for Standardization

ITPGR International Treaty on Plant Genetic Resources for Food and Agriculture

ITQs individual transferable quotas

ITTO International Tropical Timber Organization

IUCN World Conservation Union

IUU illegal, unregulated, and unreported (fishing)

IVM integrated vector management

IWMI	International Water Management Institute
IWRM	integrated water resources management
JDSD	Johannesburg Declaration on Sustainable Development
JI	joint implementation
JMP	Joint Monitoring Program
LAC	Latin America and the Caribbean
LAI	leaf area index
LARD	livelihood approaches to rural development
LDC	least developed country
LEK	local ecological knowledge
LME	large marine ecosystems
LPI	Living Planet Index
LSMS	Living Standards Measurement Study
LULUCF	land use, land use change, and forestry
MA	Millennium Ecosystem Assessment
MAI	mean annual increments
MBI	market-based instruments
MCA	multicriteria analysis
MDG	Millennium Development Goal
MEA	multilateral environmental agreement
MENA	Middle East and North Africa
MER	market exchange rate
MHC	major histocompatibility complex
MICS	multiple indicator cluster surveys
MIT	Massachusetts Institute of Technology
MPA	marine protected area
MSVPA	multispecies virtual population analysis
NAP	National Action Program (of desertification convention)
NBP	net biome productivity
NCD	noncommunicable disease
NCS	National Conservation Strategy
NCSD	national council for sustainable development
NDVI	normalized difference vegetation index
NE	effective size of a population
NEAP	national environmental action plan
NEP	new ecological paradigm; also net ecosystem productivity
NEPAD	New Partnership for Africa's Development
NFAP	National Forestry Action Plan
NFP	national forest programs
NGO	nongovernmental organization
NIH	National Institutes of Health (United States)
NMHC	non-methane hydrocarbons
NOAA	National Oceanographic and Atmospheric Administration (United States)
NPP	net primary productivity
NSSD	national strategies for sustainable development
NUE	nitrogen use efficiency
NWFP	non-wood forest product
ODA	official development assistance
OECD	Organisation for Economic Co-operation and Development
OSB	oriented strand board
OWL	other wooded land
PA	protected area
PAH	polycyclic aromatic hydrocarbons
PCBs	polychlorinated biphenyls
PEM	protein energy malnutrition
PES	payment for environmental (or ecosystem) services
PFT	plant functional type
PNG	Papua New Guinea
POPs	persistent organic pollutants
PPA	participatory poverty assessment
ppb	parts per billion
PPI	potential Pareto improvement
ppm	parts per million
ppmv	parts per million by volume
PPP	purchasing power parity; also public-private partnership
ppt	parts per thousand
PQLI	Physical Quality of Life Index
PRA	participatory rural appraisal
PRSP	Poverty Reduction Strategy Paper
PSE	producer support estimate
PVA	population viability analysis
RANWA	Research and Action in Natural Wealth Administration
RBO	river basin organization
RIDES	Recursos e Investigación para el Desarrollo Sustentable (Chile)
RIL	reduced impact logging
RLI	Red List Index
RO	reverse osmosis

RRA rapid rural appraisal

RUE rain use efficiency

SADC Southern African Development Community

SADCC Southern African Development Coordination Conference

SAfMA Southern African Millennium Ecosystem Assessment

SAP structural adjustment program

SAR species-area relationship

SARS severe acute respiratory syndrome

SBSTTA Subsidiary Body on Scientific, Technical and Technological Advice (of CBD)

SEA strategic environmental assessment

SEME simple empirical models for eutrophication

SES social-ecological system

SFM sustainable forest management

SIDS small island developing states

SMS safe minimum standard

SOM soil organic matter

SRES Special Report on Emissions Scenarios (of the IPCC)

SSC Species Survival Commission (of IUCN)

SWAP sector-wide approach

TAC total allowable catch

TBT tributyltin

TC travel cost

TCM travel cost method

TDR tradable development rights

TDS total dissolved solids

TEIA transboundary environmental impact assessment

TEK traditional ecological knowledge

TEM terrestrial ecosystem model

TESEO Treaty Enforcement Services Using Earth Observation

TEV total economic value

TFAP Tropical Forests Action Plan

TFP total factor productivity

TFR total fertility rate

Tg teragram (10^{12} grams)

TK traditional knowledge

TMDL total maximum daily load

TOF trees outside of forests

TRIPS Trade-Related Aspects of Intellectual Property Rights

TSU Technical Support Unit

TW terawatt

UMD University of Maryland

UNCCD United Nations Convention to Combat Desertification

UNCED United Nations Conference on Environment and Development

UNCLOS United Nations Convention on the Law of the Sea

UNDP United Nations Development Programme

UNECE United Nations Economic Commission for Europe

UNEP United Nations Environment Programme

UNESCO United Nations Educational, Scientific and Cultural Organization

UNFCCC United Nations Framework Convention on Climate Change

UNIDO United Nations Industrial Development Organization

UNRO United Nations Regional Organization (hypothetical body, in *Scenarios*)

UNSO UNDP's Office to Combat Desertification and Drought

USAID U.S. Agency for International Development

USDA U.S. Department of Agriculture

VOC volatile organic compound

VW virtual water

WBCSD World Business Council for Sustainable Development

WCD World Commission on Dams

WCED World Commission on Environment and Development

WCMC World Conservation Monitoring Centre (of UNEP)

WFP World Food Programme

WHO World Health Organization

WIPO World Intellectual Property Organization

WISP weighted index of social progress

WMO World Meteorological Organization

WPI Water Poverty Index

WRF white rot fungi

WSSD World Summit on Sustainable Development

wta withdrawals-to-availability ratio (of water)

WTA willingness to accept compensation

WTO World Trade Organization

WTP willingness to pay

WWAP World Water Assessment Programme

WWF World Wide Fund for Nature

WWV World Water Vision

Appendix D

Glossary

Abatement cost: See *Marginal abatement cost.*

Abundance: The total number of individuals of a taxon or taxa in an area, population, or community. Relative abundance refers to the total number of individuals of one taxon compared with the total number of individuals of all other taxa in an area, volume, or community.

Active adaptive management: See *Adaptive management.*

Adaptation: Adjustment in natural or human systems to a new or changing environment. Various types of adaptation can be distinguished, including anticipatory and reactive adaptation, private and public adaptation, and autonomous and planned adaptation.

Adaptive capacity: The general ability of institutions, systems, and individuals to adjust to potential damage, to take advantage of opportunities, or to cope with the consequences.

Adaptive management: A systematic process for continually improving management policies and practices by learning from the outcomes of previously employed policies and practices. In active adaptive management, management is treated as a deliberate experiment for purposes of learning.

Afforestation: Planting of forests on land that has historically not contained forests. (Compare *Reforestation.*)

Agrobiodiversity: The diversity of plants, insects, and soil biota found in cultivated systems.

Agroforestry systems: Mixed systems of crops and trees providing wood, non-wood forest products, food, fuel, fodder, and shelter.

Albedo: A measure of the degree to which a surface or object reflects solar radiation.

Alien species: Species introduced outside its normal distribution.

Alien invasive species: See *Invasive alien species.*

Aquaculture: Breeding and rearing of fish, shellfish, or plants in ponds, enclosures, or other forms of confinement in fresh or marine waters for the direct harvest of the product.

Benefits transfer approach: Economic valuation approach in which estimates obtained (by whatever method) in one context are used to estimate values in a different context.

Binding constraints: Political, social, economic, institutional, or ecological factors that rule out a particular response.

Biodiversity (a contraction of biological diversity): The variability among living organisms from all sources, including terrestrial, marine, and other aquatic ecosystems and the ecological complexes of which they are part. Biodiversity includes diversity within species, between species, and between ecosystems.

Biodiversity regulation: The regulation of ecosystem processes and services by the different components of biodiversity.

Biogeographic realm: A large spatial region, within which ecosystems share a broadly similar biota. Eight terrestrial biogeographic realms are typically recognized, corresponding roughly to continents (e.g., Afrotropical realm).

Biological diversity: See *Biodiversity.*

Biomass: The mass of tissues in living organisms in a population, ecosystem, or spatial unit.

Biome: The largest unit of ecological classification that is convenient to recognize below the entire globe. Terrestrial biomes are typically based on dominant vegetation structure (e.g., forest, grassland). Ecosystems within a biome function in a broadly similar way, although they may have very different species composition. For example, all forests share certain properties regarding nutrient cycling, disturbance, and biomass that are different from the properties of grasslands. Marine biomes are typically based on biogeochemical properties. The WWF biome classification is used in the MA.

Bioprospecting: The exploration of biodiversity for genetic and biochemical resources of social or commercial value.

Biotechnology: Any technological application that uses biological systems, living organisms, or derivatives thereof to make or modify products or processes for specific use.

Biotic homogenization: Process by which the differences between biotic communities in different areas are on average reduced.

Blueprint approaches: Approaches that are designed to be applicable in a wider set of circumstances and that are not context-specific or sensitive to local conditions.

Boundary organizations: Public or private organizations that synthesize and translate scientific research and explore its policy implications to help bridge the gap between science and decision-making.

Bridging organizations: Organizations that facilitate, and offer an arena for, stakeholder collaboration, trust-building, and conflict resolution.

Capability: The combinations of doings and beings from which people can choose to lead the kind of life they value. Basic capability is the capability to meet a basic need.

Capacity building: A process of strengthening or developing human resources, institutions, organizations, or networks. Also referred to as capacity development or capacity enhancement.

Capital value (of an ecosystem): The present value of the stream of ecosystem services that an ecosystem will generate under a particular management or institutional regime.

Capture fisheries: See *Fishery.*

Carbon sequestration: The process of increasing the carbon content of a reservoir other than the atmosphere.

Cascading interaction: See *Trophic cascade.*

Catch: The number or weight of all fish caught by fishing operations, whether the fish are landed or not.

Coastal system: Systems containing terrestrial areas dominated by ocean influences of tides and marine aerosols, plus nearshore marine areas. The inland extent of coastal ecosystems is the line where land-based influences dominate, up to a maximum of 100 kilometers from the coastline or 100-meter elevation (whichever is closer to the sea), and the outward extent is the 50-meter-depth contour. See also *System.*

Collaborative (or joint) forest management: Community-based management of forests, where resource tenure by local communities is secured.

Common pool resource: A valued natural or human-made resource or facility in which one person's use subtracts from another's use and where it is often necessary but difficult to exclude potential users from the resource. (Compare *Common property resource.*)

Common property management system: The institutions (i.e., sets of rules) that define and regulate the use rights for common pool resources. Not the same as an open access system.

Common property resource: A good or service shared by a well-defined community. (Compare *Common pool resource.*)

Community (ecological): An assemblage of species occurring in the same space or time, often linked by biotic interactions such as competition or predation.

Community (human, local): A collection of human beings who have something in common. A local community is a fairly small group of people who share a common place of residence and a set of institutions based on this fact, but the word 'community' is also used to refer to larger collections of people who have something else in common (e.g., national community, donor community).

Condition of an ecosystem: The capacity of an ecosystem to yield services, relative to its potential capacity.

Condition of an ecosystem service: The capacity of an ecosystem service to yield benefits to people, relative to its potential capacity.

Constituents of well-being: The experiential aspects of well-being, such as health, happiness, and freedom to be and do, and, more broadly, basic liberties.

Consumptive use: The reduction in the quantity or quality of a good available for other users due to consumption.

Contingent valuation: Economic valuation technique based on a survey of how much respondents would be willing to pay for specified benefits.

Core dataset: Data sets designated to have wide potential application throughout the Millennium Ecosystem Assessment process. They include land use, land cover, climate, and population data sets.

Cost-benefit analysis: A technique designed to determine the feasibility of a project or plan by quantifying its costs and benefits.

Cost-effectiveness analysis: Analysis to identify the least cost option that meets a particular goal.

Critically endangered species: Species that face an extremely high risk of extinction in the wild. See also *Threatened species*.

Cross-scale feedback: A process in which effects of some action are transmitted from a smaller spatial extent to a larger one, or vice versa. For example, a global policy may constrain the flexibility of a local region to use certain response options to environmental change, or a local agricultural pest outbreak may affect regional food supply.

Cultivar (a contraction of cultivated variety): A variety of a plant developed from a natural species and maintained under cultivation.

Cultivated system: Areas of landscape or seascape actively managed for the production of food, feed, fiber, or biofuels.

Cultural landscape: See *Landscape*.

Cultural services: The nonmaterial benefits people obtain from ecosystems through spiritual enrichment, cognitive development, reflection, recreation, and aesthetic experience, including, e.g., knowledge systems, social relations, and aesthetic values.

Decision analytical framework: A coherent set of concepts and procedures aimed at synthesizing available information to help policymakers assess consequences of various decision options. DAFs organize the relevant information in a suitable framework, apply decision criteria (both based on some paradigms or theories), and thus identify options that are better than others under the assumptions characterizing the analytical framework and the application at hand.

Decision-maker: A person whose decisions, and the actions that follow from them, can influence a condition, process, or issue under consideration.

Decomposition: The ecological process carried out primarily by microbes that leads to a transformation of dead organic matter into inorganic mater.

Deforestation: Conversion of forest to non-forest.

Degradation of an ecosystem service: For *provisioning services,* decreased production of the service through changes in area over which the services is provided, or decreased production per unit area. For *regulating* and *supporting services,* a reduction in the benefits obtained from the service, either through a change in the service or through human pressures on the service exceeding its limits. For *cultural services,* a change in the ecosystem features that decreases the cultural benefits provided by the ecosystem.

Degradation of ecosystems: A persistent reduction in the capacity to provide ecosystem services.

Desertification: land degradation in drylands resulting from various factors, including climatic variations and human activities.

Determinants of well-being: Inputs into the production of well-being, such as food, clothing, potable water, and access to knowledge and information.

Direct use value (of ecosystems): The benefits derived from the services provided by an ecosystem that are used directly by an economic agent. These include consumptive uses (e.g., harvesting goods) and nonconsumptive uses (e.g., enjoyment of scenic beauty). Agents are often physically present in an ecosystem to receive direct use value. (Compare *Indirect use value.*)

Disability-adjusted life years: The sum of years of life lost due to premature death and illness, taking into account the age of death compared with natural life expectancy and the number of years of life lived with a disability. The measure of number of years lived with the disability considers the duration of the disease, weighted by a measure of the severity of the disease.

Diversity: The variety and relative abundance of different entities in a sample.

Driver: Any natural or human-induced factor that directly or indirectly causes a change in an ecosystem.

Driver, direct: A driver that unequivocally influences ecosystem processes and can therefore be identified and measured to differing degrees of accuracy. (Compare *Driver, indirect.*)

Driver, endogenous: A driver whose magnitude can be influenced by the decision-maker. Whether a driver is exogenous or endogenous depends on the organizational scale. Some drivers (e.g., prices) are exogenous to a decision-maker at one level (a farmer) but endogenous at other levels (the nation-state). (Compare *Driver, exogenous.*)

Driver, exogenous: A driver that cannot be altered by the decision-maker. (Compare *Driver, endogenous.*)

Driver, indirect: A driver that operates by altering the level or rate of change of one or more direct drivers. (Compare *Driver, direct.*)

Drylands: See *Dryland system.*

Dryland system: Areas characterized by lack of water, which constrains the two major interlinked services of the system: primary production and nutrient cycling. Four dryland subtypes are widely recognized: dry sub-humid, semiarid, arid, and hyperarid, showing an increasing level of aridity or moisture deficit. See also *System.*

Ecological character: See *Ecosystem properties.*

Ecological degradation: See *Degradation of ecosystems.*

Ecological footprint: An index of the area of productive land and aquatic ecosystems required to produce the resources used and to assimilate the wastes produced by a defined population at a specified material standard of living, wherever on Earth that land may be located.

Ecological security: A condition of ecological safety that ensures access to a sustainable flow of provisioning, regulating, and cultural services needed by local communities to meet their basic capabilities.

Ecological surprises: unexpected—and often disproportionately large—consequence of changes in the abiotic (e.g., climate, disturbance) or biotic (e.g., invasions, pathogens) environment.

Ecosystem: A dynamic complex of plant, animal, and microorganism communities and their non-living environment interacting as a functional unit.

Ecosystem approach: A strategy for the integrated management of land, water, and living resources that promotes conservation and sustainable use. An ecosystem approach is based on the application of appropriate scientific methods focused on levels of biological organization, which encompass the essential structure, processes, functions, and interactions among organisms and their environment. It recognizes that humans, with their cultural diversity, are an integral component of many ecosystems.

Ecosystem assessment: A social process through which the findings of science concerning the causes of ecosystem change, their consequences for human well-being, and management and policy options are brought to bear on the needs of decision-makers.

Ecosystem boundary: The spatial delimitation of an ecosystem, typically based on discontinuities in the distribution of organisms, the biophysical environment (soil types, drainage basins, depth in a water body), and spatial interactions (home ranges, migration patterns, fluxes of matter).

Ecosystem change: Any variation in the state, outputs, or structure of an ecosystem.

Ecosystem function: See *Ecosystem process*.

Ecosystem interactions: Exchanges of materials, energy, and information within and among ecosystems.

Ecosystem management: An approach to maintaining or restoring the composition, structure, function, and delivery of services of natural and modified ecosystems for the goal of achieving sustainability. It is based on an adaptive, collaboratively developed vision of desired future conditions that integrates ecological, socioeconomic, and institutional perspectives, applied within a geographic framework, and defined primarily by natural ecological boundaries.

Ecosystem process: An intrinsic ecosystem characteristic whereby an ecosystem maintains its integrity. Ecosystem processes include decomposition, production, nutrient cycling, and fluxes of nutrients and energy.

Ecosystem properties: The size, biodiversity, stability, degree of organization, internal exchanges of materials, energy, and information among different pools, and other properties that characterize an ecosystem. Includes ecosystem functions and processes.

Ecosystem resilience: See *Resilience*.

Ecosystem resistance: See *Resistance*.

Ecosystem robustness: See *Ecosystem stability*.

Ecosystem services: The benefits people obtain from ecosystems. These include *provisioning services* such as food and water; *regulating services* such as flood and disease control; *cultural services* such as spiritual, recreational, and cultural benefits; and *supporting services* such as nutrient cycling that maintain the conditions for life on Earth. The concept "ecosystem goods and services" is synonymous with ecosystem services.

Ecosystem stability (or ecosystem robustness): A description of the dynamic properties of an ecosystem. An ecosystem is considered stable or robust if it returns to its original state after a perturbation, exhibits low temporal variability, or does not change dramatically in the face of a perturbation.

Elasticity: A measure of responsiveness of one variable to a change in another, usually defined in terms of percentage change. For example, own-price elasticity of demand is the percentage change in the quantity demanded of a good for a 1% change in the price of that good. Other common elasticity measures include supply and income elasticity.

Emergent disease: Diseases that have recently increased in incidence, impact, or geographic range; that are caused by pathogens that have recently evolved; that are newly discovered; or that have recently changed their clinical presentation.

Emergent property: A phenomenon that is not evident in the constituent parts of a system but that appears when they interact in the system as a whole.

Enabling conditions: Critical preconditions for success of responses, including political, institutional, social, economic, and ecological factors.

Endangered species: Species that face a very high risk of extinction in the wild. See also *Threatened species*.

Endemic (in ecology): A species or higher taxonomic unit found only within a specific area.

Endemic (in health): The constant presence of a disease or infectious agent within a given geographic area or population group; may also refer to the usual prevalence of a given disease within such area or group.

Endemism: The fraction of species that is endemic relative to the total number of species found in a specific area.

Epistemology: The theory of knowledge, or a "way of knowing."

Equity: Fairness of rights, distribution, and access. Depending on context, this can refer to resources, services, or power.

Eutrophication: The increase in additions of nutrients to freshwater or marine systems, which leads to increases in plant growth and often to undesirable changes in ecosystem structure and function.

Evapotranspiration: See *Transpiration*.

Existence value: The value that individuals place on knowing that a resource exists, even if they never use that resource (also sometimes known as conservation value or passive use value).

Exotic species: See *Alien species*.

Externality: A consequence of an action that affects someone other than the agent undertaking that action and for which the agent is neither compensated nor penalized through the markets. Externalities can be positive or negative.

Feedback: See *Negative feedback, Positive feedback,* and *Cross-scale feedback*.

Fishery: A particular kind of fishing activity, e.g., a trawl fishery, or a particular species targeted, e.g., a cod fishery or salmon fishery.

Fish stock: See *Stock*.

Fixed nitrogen: See *Reactive nitrogen*.

Flyway: Areas of the world used by migratory birds in moving between breeding and wintering grounds.

Forest systems: Systems in which trees are the predominant life forms. Statistics reported in this assessment are based on areas that are dominated by trees (perennial woody plants taller than five meters at maturity), where the tree crown cover exceeds 10%, and where the area is more than 0.5 hectares. "Open forests" have a canopy cover between 10% and 40%, and "closed forests" a canopy cover of more than 40%. "Fragmented forests" refer to mosaics of forest patches and non-forest land. See also *System*.

Freedom: The range of options a person has in deciding the kind of life to lead.

Functional diversity: The value, range, and relative abundance of traits present in the organisms in an ecological community.

Functional redundancy (= functional compensation): A characteristic of ecosystems in which more than one species in the system can carry out a particular process. Redundancy may be total or partial—that is, a species may not be able to completely replace the other species or it may compensate only some of the processes in which the other species are involved.

Functional types (= functional groups = guilds): Groups of organisms that respond to the environment or affect ecosystem processes in a similar way. Examples of plant functional types include nitrogen-fixer versus non-fixer, stress-tolerant versus ruderal versus competitor, resprouter versus seeder, deciduous versus evergreen. Examples of animal functional types include granivorous versus fleshy-fruit eater, nocturnal versus diurnal predator, browser versus grazer.

Geographic information system: A computerized system organizing data sets through a geographical referencing of all data included in its collections.

Globalization: The increasing integration of economies and societies around the world, particularly through trade and financial flows, and the transfer of culture and technology.

Global scale: The geographical realm encompassing all of Earth.

Governance: The process of regulating human behavior in accordance with shared objectives. The term includes both governmental and nongovernmental mechanisms.

Health, human: A state of complete physical, mental, and social well-being and not merely the absence of disease or infirmity. The health of a whole community or population is reflected in measurements of disease incidence and prevalence, age-specific death rates, and life expectancy.

High seas: The area outside of national jurisdiction, i.e., beyond each nation's Exclusive Economic Zone or other territorial waters.

Human well-being: See *Well-being*.

Income poverty: See *Poverty*.

Indicator: Information based on measured data used to represent a particular attribute, characteristic, or property of a system.

Indigenous knowledge (or local knowledge): The knowledge that is unique to a given culture or society.

Indirect interaction: Those interactions among species in which a species, through direct interaction with another species or modification of resources, alters the abundance of a third species with which it is not directly interacting. Indirect interactions can be trophic or nontrophic in nature.

Indirect use value: The benefits derived from the goods and services provided by an ecosystem that are used indirectly by an economic agent. For example, an agent at some distance from an ecosystem may derive benefits from drinking water that has been purified as it passed through the ecosystem. (Compare *Direct use value.*)

Infant mortality rate: Number of deaths of infants aged 0–12 months divided by the number of live births.

Inland water systems: Permanent water bodies other than salt-water systems on the coast, seas and oceans. Includes rivers, lakes, reservoirs wetlands and inland saline lakes and marshes. See also *System.*

Institutions: The rules that guide how people within societies live, work, and interact with each other. Formal institutions are written or codified rules. Examples of formal institutions would be the constitution, the judiciary laws, the organized market, and property rights. Informal institutions are rules governed by social and behavioral norms of the society, family, or community. Also referred to as organizations.

Integrated coastal zone management: Approaches that integrate economic, social, and ecological perspectives for the management of coastal resources and areas.

Integrated conservation and development projects: Initiatives that aim to link biodiversity conservation and development.

Integrated pest management: Any practices that attempt to capitalize on natural processes that reduce pest abundance. Sometimes used to refer to monitoring programs where farmers apply pesticides to improve economic efficiency (reducing application rates and improving profitability).

Integrated responses: Responses that address degradation of ecosystem services across a number of systems simultaneously or that also explicitly include objectives to enhance human well-being.

Integrated river basin management: Integration of water planning and management with environmental, social, and economic development concerns, with an explicit objective of improving human welfare.

Interventions: See *Responses.*

Intrinsic value: The value of someone or something in and for itself, irrespective of its utility for people.

Invasibility: Intrinsic susceptibility of an ecosystem to be invaded by an alien species.

Invasive alien species: An alien species whose establishment and spread modifies ecosystems, habitats, or species.

Irreversibility: The quality of being impossible or difficult to return to, or to restore to, a former condition. See also *Option value, Precautionary principle, Resilience,* and *Threshold.*

Island systems: Lands isolated by surrounding water, with a high proportion of coast to hinterland. The degree of isolation from the mainland in both natural and social aspects is accounted by the *isola effect.* See also *System.*

Isola effect: Environmental issues that are unique to island systems. This uniqueness takes into account the physical seclusion of islands as isolated pieces of land exposed to marine or climatic disturbances with a more limited access to space, products, and services when compared with most continental areas, but also includes subjective issues such as the perceptions and attitudes of islanders themselves.

Keystone species: A species whose impact on the community is disproportionately large relative to its abundance. Effects can be produced by consumption (trophic interactions), competition, mutualism, dispersal, pollination, disease, or habitat modification (nontrophic interactions).

Land cover: The physical coverage of land, usually expressed in terms of vegetation cover or lack of it. Related to, but not synonymous with, *land use.*

Landscape: An area of land that contains a mosaic of ecosystems, including human-dominated ecosystems. The term cultural landscape is often used when referring to landscapes containing significant human populations or in which there has been significant human influence on the land.

Landscape unit: A portion of relatively homogenous land cover within the local-to-regional landscape.

Land use: The human use of a piece of land for a certain purpose (such as irrigated agriculture or recreation). Influenced by, but not synonymous with, *land cover.*

Length of growing period: The total number of days in a year during which rainfall exceeds one half of potential evapotranspiration. For boreal and temperate zone, growing season is usually defined as a number of days with the average daily temperature that exceeds a definite threshold, such as 10° Celsius.

Local knowledge: See *Indigenous knowledge.*

Mainstreaming: Incorporating a specific concern, e.g. sustainable use of ecosystems, into policies and actions.

Malnutrition: A state of bad nourishment. Malnutrition refers both to undernutrition and overnutrition, as well as to conditions arising from dietary imbalances leading to diet-related noncommunicable diseases.

Marginal abatement cost: The cost of abating an incremental unit of, for instance, a pollutant.

Marine system: Marine waters from the low-water mark to the high seas that support marine capture fisheries, as well as deepwater (>50 meters) habitats. Four sub-divisions (marine biomes) are recognized: the coastal boundary zone; trade-winds; westerlies; and polar.

Market-based instruments: Mechanisms that create a market for ecosystem services in order to improving the efficiency in the way the service is used. The term is used for mechanisms that create new markets, but also for responses such as taxes, subsidies, or regulations that affect existing markets.

Market failure: The inability of a market to capture the correct values of ecosystem services.

Mitigation: An anthropogenic intervention to reduce negative or unsustainable uses of ecosystems or to enhance sustainable practices.

Mountain system: High-altitude (greater than 2,500 meters) areas and steep mid-altitude (1,000 meters at the equator, decreasing to sea level where alpine life zones meet polar life zones at high latitudes) areas, excluding large plateaus.

Negative feedback: Feedback that has a net effect of dampening perturbation.

Net primary productivity: See *Production, biological.*

Non-linearity: A relationship or process in which a small change in the value of a driver (i.e., an independent variable) produces an disproportionate change in the outcome (i.e., the dependent variable). Relationships where there is a sudden discontinuity or change in rate are sometimes referred to as abrupt and often form the basis of thresholds. In loose terms, they may lead to unexpected outcomes or "surprises."

Nutrient cycling: The processes by which elements are extracted from their mineral, aquatic, or atmospheric sources or recycled from their organic forms, converting them to the ionic form in which biotic uptake occurs and ultimately returning them to the atmosphere, water, or soil.

Nutrients: The approximately 20 chemical elements known to be essential for the growth of living organisms, including nitrogen, sulfur, phosphorus, and carbon.

Open access resource: A good or service over which no property rights are recognized.

Opportunity cost: The benefits forgone by undertaking one activity instead of another.

Option value: The value of preserving the option to use services in the future either by oneself (option value) or by others or heirs (bequest value). Quasi-option value represents the value of avoiding irreversible decisions until new information reveals whether certain ecosystem services have values society is not currently aware of.

Organic farming: Crop and livestock production systems that do not make use of synthetic fertilizers, pesticides, or herbicides. May also include restrictions on the use of transgenic crops (genetically modified organisms).

Pastoralism, pastoral system: The use of domestic animals as a primary means for obtaining resources from habitats.

Perturbation: An imposed movement of a system away from its current state.

Polar system: Treeless lands at high latitudes. Includes Arctic and Antarctic areas, where the polar system merges with the northern boreal forest and the Southern Ocean respectively. See also *System*.

Policy failure: A situation in which government policies create inefficiencies in the use of goods and services.

Policy-maker: A person with power to influence or determine policies and practices at an international, national, regional, or local level.

Pollination: A process in the sexual phase of reproduction in some plants caused by the transportation of pollen. In the context of ecosystem services, pollination generally refers to animal-assisted pollination, such as that done by bees, rather than wind pollination.

Population, biological: A group of individuals of the same species, occupying a defined area, and usually isolated to some degree from other similar groups. Populations can be relatively reproductively isolated and adapted to local environments.

Population, human: A collection of living people in a given area. (Compare *Community (human, local)*.)

Positive feedback: Feedback that has a net effect of amplifying perturbation.

Poverty: The pronounced deprivation of well-being. Income poverty refers to a particular formulation expressed solely in terms of per capita or household income.

Precautionary principle: The management concept stating that in cases "where there are threats of serious or irreversible damage, lack of full scientific certainty shall not be used as a reason for postponing cost-effective measures to prevent environmental degradation," as defined in the Rio Declaration.

Prediction (or forecast): The result of an attempt to produce a most likely description or estimate of the actual evolution of a variable or system in the future. See also *Projection* and *Scenario*.

Primary production: See *Production, biological*.

Private costs and benefits: Costs and benefits directly felt by individual economic agents or groups as seen from their perspective. (Externalities imposed on others are ignored.) Costs and benefits are valued at the prices actually paid or received by the group, even if these prices are highly distorted. Sometimes termed "financial" costs and benefits. (Compare *Social costs and benefits*.)

Probability distribution: A distribution that shows all the values that a random variable can take and the likelihood that each will occur.

Production, biological: Rate of biomass produced by an ecosystem, generally expressed as biomass produced per unit of time per unit of surface or volume. Net primary productivity is defined as the energy fixed by plants minus their respiration.

Production, economic: Output of a system.

Productivity, biological: See *Production, biological*.

Productivity, economic: Capacity of a system to produce high levels of output or responsiveness of the output of a system to inputs.

Projection: A potential future evolution of a quantity or set of quantities, often computed with the aid of a model. Projections are distinguished from "predictions" in order to emphasize that projections involve assumptions concerning, for example, future socioeconomic and technological developments that may or may not be realized; they are therefore subject to substantial uncertainty.

Property rights: The right to specific uses, perhaps including exchange in a market, of ecosystems and their services.

Provisioning services: The products obtained from ecosystems, including, for example, genetic resources, food and fiber, and fresh water.

Public good: A good or service in which the benefit received by any one party does not diminish the availability of the benefits to others, and where access to the good cannot be restricted.

Reactive nitrogen (or fixed nitrogen): The forms of nitrogen that are generally available to organisms, such as ammonia, nitrate, and organic nitrogen. Nitrogen gas (or dinitrogen), which is the major component of the atmosphere, is inert to most organisms.

Realm: Used to describe the three major types of ecosystems on earth: terrestrial, freshwater, and marine. Differs fundamentally from *biogeographic realm*.

Reforestation: Planting of forests on lands that have previously contained forest but have since been converted to some other use. (Compare *Afforestation*.)

Regime shift: A rapid reorganization of an ecosystem from one relatively stable state to another.

Regulating services: The benefits obtained from the regulation of ecosystem processes, including, for example, the regulation of climate, water, and some human diseases.

Relative abundance: See *Abundance*.

Reporting unit: The spatial or temporal unit at which assessment or analysis findings are reported. In an assessment, these units are chosen to maximize policy relevance or relevance to the public and thus may differ from those upon which the analyses were conducted (e.g., analyses conducted on mapped ecosystems can be reported on administrative units). See also *System*.

Resilience: The level of disturbance that an ecosystem can undergo without crossing a threshold to a situation with different structure or outputs. Resilience depends on ecological dynamics as well as the organizational and institutional capacity to understand, manage, and respond to these dynamics.

Resistance: The capacity of an ecosystem to withstand the impacts of drivers without displacement from its present state.

Responses: Human actions, including policies, strategies, and interventions, to address specific issues, needs, opportunities, or problems. In the context of ecosystem management, responses may be of legal, technical, institutional, economic, and behavioral nature and may operate at various spatial and time scales.

Riparian: Something related to, living on, or located at the banks of a watercourse, usually a river or stream.

Safe minimum standard: A decision analytical framework in which the benefits of ecosystem services are assumed to be incalculable and should be preserved unless the costs of doing so rise to an intolerable level, thus shifting the burden of proof to those who would convert them.

Salinization: The buildup of salts in soils.

Scale: The measurable dimensions of phenomena or observations. Expressed in physical units, such as meters, years, population size, or quantities moved or exchanged. In observation, scale determines the relative fineness and coarseness of different detail and the selectivity among patterns these data may form.

Scenario: A plausible and often simplified description of how the future may develop, based on a coherent and internally consistent set of assumptions about key driving forces (e.g., rate of technology change, prices) and relationships. Scenarios are neither predictions nor projections and sometimes may be based on a "narrative storyline." Scenarios may include projections but are often based on additional information from other sources.

Security: Access to resources, safety, and the ability to live in a predictable and controllable environment.

Service: See *Ecosystem services*.

Social costs and benefits: Costs and benefits as seen from the perspective of society as a whole. These differ from private costs and benefits in being more inclusive (all costs and benefits borne by some member of society are taken into account) and in being valued at social opportunity cost rather than market prices, where these differ. Sometimes termed "economic" costs and benefits. (Compare *Private costs and benefits*.)

Social incentives: Measures that lower transaction costs by facilitating trust-building and learning as well as rewarding collaboration and conflict resolution. Social incentives are often provided by bridging organizations.

Socioecological system: An ecosystem, the management of this ecosystem by actors and organizations, and the rules, social norms, and conventions underlying this management. (Compare *System*.)

Soft law: Non-legally binding instruments, such as guidelines, standards, criteria, codes of practice, resolutions, and principles or declarations, that states establish to implement national laws.

Soil fertility: The potential of the soil to supply nutrient elements in the quantity, form, and proportion required to support optimum plant growth. See also *Nutrients*.

Speciation: The formation of new species.

Species: An interbreeding group of organisms that is reproductively isolated from all other organisms, although there are many partial exceptions to this rule in particular taxa. Operationally, the term *species* is a generally agreed fundamental taxonomic unit, based on morphological or genetic similarity, that once described and accepted is associated with a unique scientific name.

Species diversity: Biodiversity at the species level, often combining aspects of species richness, their relative abundance, and their dissimilarity.

Species richness: The number of species within a given sample, community, or area.

Statistical variation: Variability in data due to error in measurement, error in sampling, or variation in the measured quantity itself.

Stock (in fisheries): The population or biomass of a fishery resource. Such stocks are usually identified by their location. They can be, but are not always, genetically discrete from other stocks.

Stoichiometry, ecological: The relatively constant proportions of the different nutrients in plant or animal biomass that set constraints on production. Nutrients only available in lower proportions are likely to limit growth.

Storyline: A narrative description of a scenario, which highlights its main features and the relationships between the scenario's driving forces and its main features.

Strategies: See *Responses.*

Streamflow: The quantity of water flowing in a watercourse.

Subsidiarity, principle of: The notion of devolving decision-making authority to the lowest appropriate level.

Subsidy: Transfer of resources to an entity, which either reduces the operating costs or increases the revenues of such entity for the purpose of achieving some objective.

Subsistence: An activity in which the output is mostly for the use of the individual person doing it, or their family, and which is a significant component of their livelihood.

Subspecies: A population that is distinct from, and partially reproductively isolated from, other populations of a species but that has not yet diverged sufficiently that interbreeding is impossible.

Supporting services: Ecosystem services that are necessary for the production of all other ecosystem services. Some examples include biomass production, production of atmospheric oxygen, soil formation and retention, nutrient cycling, water cycling, and provisioning of habitat.

Sustainability: A characteristic or state whereby the needs of the present and local population can be met without compromising the ability of future generations or populations in other locations to meet their needs.

Sustainable use (of an ecosystem): Human use of an ecosystem so that it may yield a continuous benefit to present generations while maintaining its potential to meet the needs and aspirations of future generations.

Symbiosis: Close and usually obligatory relationship between two organisms of different species, not necessarily to their mutual benefit.

Synergy: When the combined effect of several forces operating is greater than the sum of the separate effects of the forces.

System: In the Millennium Ecosystem Assessment, reporting units that are ecosystem-based but at a level of aggregation far higher than that usually applied to ecosystems. Thus the system includes many component ecosystems, some of which may not strongly interact with each other, that may be spatially separate, or that may be of a different type to the ecosystems that constitute the majority, or matrix, of the system overall. The system includes the social and economic systems that have an impact on and are affected by the ecosystems included within it. For example, the Condition and Trend Working Group refers to "forest systems," "cultivated systems," "mountain systems," and so on. Systems thus defined are not mutually exclusive, and are permitted to overlap spatially or conceptually. For instance, the "cultivated system" may include areas of "dryland system" and vice versa.

Taxon (pl. taxa): The named classification unit to which individuals or sets of species are assigned. Higher taxa are those above the species level. For example, the common mouse, *Mus musculus,* belongs to the Genus *Mus,* the Family Muridae, and the Class Mammalia.

Taxonomy: A system of nested categories (*taxa*) reflecting evolutionary relationships or morphological similarity.

Tenure: See *Property rights,* although also sometimes used more specifically in reference to the temporal dimensions and security of property rights.

Threatened species: Species that face a high (*vulnerable species*), very high (*endangered species*), or extremely high (*critically endangered species*) risk of extinction in the wild.

Threshold: A point or level at which new properties emerge in an ecological, economic, or other system, invalidating predictions based on mathematical relationships that apply at lower levels. For example, species diversity of a landscape may decline steadily with increasing habitat degradation to a certain point, then fall sharply after a critical threshold of degradation is reached. Human behavior, especially at group levels, sometimes exhibits threshold effects. Thresholds at which irreversible changes occur are especially of concern to decision-makers. (Compare *Non-linearity.*)

Time series data: A set of data that expresses a particular variable measured over time.

Total economic value framework: A widely used framework to disaggregate the components of utilitarian value, including *direct use value, indirect use value, option value,* quasi-option value, and *existence value.*

Total factor productivity: A measure of the aggregate increase in efficiency of use of inputs. TFP is the ratio of the quantity of output divided by an index of the amount of inputs used. A common input index uses as weights the share of the input in the total cost of production.

Total fertility rate: The number of children a woman would give birth to if through her lifetime she experienced the set of age-specific fertility rates currently observed. Since age-specific rates generally change over time, TFR does not in general give the actual number of births a woman alive today can be expected to have. Rather, it is a synthetic index meant to measure age-specific birth rates in a given year.

Trade-off: Management choices that intentionally or otherwise change the type, magnitude, and relative mix of services provided by ecosystems.

Traditional ecological knowledge: The cumulative body of knowledge, practices, and beliefs evolved by adaptive processes and handed down through generations. TEK may or may not be indigenous or local, but it is distinguished by the way in which it is acquired and used, through the social process of learning and sharing knowledge. (Compare *Indigenous knowledge.*)

Traditional knowledge: See *Traditional ecological knowledge.*

Traditional use: Exploitation of natural resources by indigenous users or by nonindigenous residents using traditional methods. Local use refers to exploitation by local residents.

Transpiration: The process by which water is drawn through plants and returned to the air as water vapor. Evapotranspiration is combined loss of water to the atmosphere via the processes of evaporation and transpiration.

Travel cost methods: Economic valuation techniques that use observed costs to travel to a destination to derive demand functions for that destination.

Trend: A pattern of change over time, over and above short-term fluctuations.

Trophic cascade: A chain reaction of top-down interactions across multiple tropic levels. These occur when changes in the presence or absence (or shifts in abundance) of a top predator alter the production at several lower trophic levels. Such positive indirect effects of top predators on lower tropic levels are mediated by the consumption of mid-level consumers (generally herbivores).

Trophic level: The average level of an organism within a food web, with plants having a trophic level of 1, herbivores 2, first-order carnivores 3, and so on.

Umbrella species: Species that have either large habitat needs or other requirements whose conservation results in many other species being conserved at the ecosystem or landscape level.

Uncertainty: An expression of the degree to which a future condition (e.g., of an ecosystem) is unknown. Uncertainty can result from lack of information or from disagreement about what is known or even knowable. It may have many types of sources, from quantifiable errors in the data to ambiguously defined terminology or uncertain projections of human behavior. Uncertainty can therefore be represented by quantitative measures (e.g., a range of values calculated by various models) or by qualitative statements (e.g., reflecting the judgment of a team of experts).

Urbanization: An increase in the proportion of the population living in urban areas.

Urban systems: Built environments with a high human population density. Operationally defined as human settlements with a minimum population density commonly in the range of 400 to 1,000 persons per square kilometer, minimum size of typically between 1,000 and 5,000 people, and maximum agricultural employment usually in the vicinity of 50–75%. See also *System*.

Utility: In economics, the measure of the degree of satisfaction or happiness of a person.

Valuation: The process of expressing a value for a particular good or service in a certain context (e.g., of decision-making) usually in terms of something that can be counted, often money, but also through methods and measures from other disciplines (sociology, ecology, and so on). See also *Value*.

Value: The contribution of an action or object to user-specified goals, objectives, or conditions. (Compare *Valuation*.)

Value systems: Norms and precepts that guide human judgment and action.

Voluntary measures: Measures that are adopted by firms or other actors in the absence of government mandates.

Vulnerability: Exposure to contingencies and stress, and the difficulty in coping with them. Three major dimensions of vulnerability are involved: exposure to stresses, perturbations, and shocks; the sensitivity of people, places, ecosystems, and species to the stress or perturbation, including their capacity to anticipate and cope with the stress; and the resilience of the exposed people, places, ecosystems, and species in terms of their capacity to absorb shocks and perturbations while maintaining function.

Vulnerable species: Species that face a high risk of extinction in the wild. See also *Threatened species*.

Water scarcity: A water supply that limits food production, human health, and economic development. Severe scarcity is taken to be equivalent to 1,000 cubic meters per year per person or greater than 40% use relative to supply.

Watershed (also catchment basin): The land area that drains into a particular watercourse or body of water. Sometimes used to describe the dividing line of high ground between two catchment basins.

Water stress: See *Water scarcity*.

Well-being: A context- and situation-dependent state, comprising basic material for a good life, freedom and choice, health and bodily well-being, good social relations, security, peace of mind, and spiritual experience.

Wetlands: Areas of marsh, fen, peatland, or water, whether natural or artificial, permanent or temporary, with water that is static or flowing, fresh, brackish or salt, including areas of marine water the depth of which at low tide does not exceed six meters. May incorporate riparian and coastal zones adjacent to the wetlands and islands or bodies of marine water deeper than six meters at low tide laying within the wetlands.

Wise use (of an ecosystem): Sustainable utilization for the benefit of humankind in a way compatible with the maintenance of the natural properties of the ecosystem

Index

Italic page numbers refer to figures, tables, and boxes. Bold page numbers refer to the Summary.

F

T

U

SOUTH ORANGE PUBLIC LIBRARY
3 9507 00131268 4